PRINCIPLES OF
PHYSIOLOGY

SECOND EDITION

PRINCIPLES OF
PHYSIOLOGY

Edited by

ROBERT M. BERNE, M.D., D.Sc. (Hon.)

Professor of Physiology, Emeritus
Department of Physiology,
University of Virginia School of Medicine,
Charlottesville, Virginia

MATTHEW N. LEVY, M.D.

Chief of Investigative Medicine,
Mount Sinai Medical Center;
Professor of Physiology and Biophysics and of
Biomedical Engineering, Emeritus
Case Western Reserve University,
Cleveland, Ohio

With 620 illustrations

 Mosby

St. Louis Baltimore Boston Carlsbad Chicago Naples New York Philadelphia Portland
London Madrid Mexico City Singapore Sydney Tokyo Toronto Wiesbaden

Mosby
Dedicated to Publishing Excellence

A Times Mirror
Company

Publisher: Anne S. Patterson
Editor: Emma D. Underdown
Developmental Editor: Christy Wells
Editorial Assistant: Alicia E. Moten
Project Manager: Mark Spann
Production Editor: Stephen C. Hetager
Designer: David Zielinski
Manufacturing Supervisor: Betty Richmond
Cover Design: Bill Schraeder

SECOND EDITION

Printed in the United States of America
Composition by The Clarinda Company
Printing/binding by Von Hoffmann Press

Mosby-Year Book, Inc.
11830 Westline Industrial Drive
St. Louis, Missouri 63146

ISBN 0-8151-0523-1

97 98 99 00 / 9 8 7 6 5 4 3 2

CONTRIBUTORS

SAUL M. GENUTH, M.D.
Professor, Department of Medicine,
School of Medicine,
Case Western Reserve University,
Cleveland, Ohio

BRUCE M. KOEPPEN, M.D., Ph.D.
Dean, Academic Affairs and Education,
Department of Medicine and Physiology,
University of Connecticut Health Center,
Farmington, Connecticut

HOWARD C. KUTCHAI, Ph.D.
Professor,
Department of Physiology,
University of Virginia School of Medicine,
Charlottesville, Virginia

RICHARD A. MURPHY, Ph.D.
Professor,
Department of Physiology,
University of Virginia School of Medicine,
Charlottesville, Virginia

BRUCE A. STANTON, Ph.D.
Dean, Academic Affairs and Education,
Associate Professor,
Department of Physiology,
Dartmouth College Medical School,
Hanover, New Hampshire

NORMAN C. STAUB, M.D.
Professor of Physiology, Emeritus,
University of California School of Medicine,
San Francisco, California

WILLIAM D. WILLIS, JR., M.D., Ph.D.
Professor and Chairman,
Department of Anatomy and Neurosciences,
The University of Texas Medical Branch at Galveston,
Galveston, Texas

PREFACE

Principles of Physiology has been carefully designed to present the important features of mammalian physiology clearly and concisely. General principles and underlying mechanisms are emphasized, and nonvital details are minimized. Considerable attention is directed to cell physiology, which serves as the basis for body functions. The first section of the text is devoted to this topic, and a new chapter on the mechanisms of cell signaling has been added to this section. Furthermore, the relevant cell physiology has been included in each of the succeeding sections. We have tried to show that the processes that take place in living cells in general are usually also applicable to the specific cell types in the various organ systems.

The major emphasis in *Principles of Physiology* is on regulation. The mechanisms that regulate the functions of the individual organ systems are thoroughly described. These mechanisms are then applied to the complex interactions among the systems as they maintain the internal environment constant, a process that is so important for the optimal function of the constituent cells.

In this edition, Dr. Norman Staub has rewritten the section on respiratory physiology. At the end of the section on cardiovascular physiology, a chapter has been added on the responses to physical exercise and to hemorrhage. The purpose of this chapter is mainly to illustrate the ways in which the various components of the circulatory system are coordinated to allow the body to adapt to certain substantial stresses.

To contribute to our goal of clarity, multicolored illustrations are used to portray concepts as simply as possible. When sequential mechanisms are involved, multipaneled diagrams have been designed to illustrate each step clearly. Block diagrams are used to depict the interrelationships among the various factors that may affect a specific function. Finally, figures are included to illustrate some of the concepts that appear in the text and to inform the reader about important investigative techniques.

Because the intent of this text is to offer, clearly and concisely, all information needed to master a complete course in physiology, the use of mathematics has been minimized, and succinct, lucid descriptions have been substituted wherever feasible. Controversial issues have been omitted to allow ample room for the explanation of important, generally accepted physiological mechanisms. We have refrained from citing the sources of the statements or assertions that appear in the text. Throughout the book, we have used italics to emphasize important concepts, and we have used boldface to denote new terms and definitions. We have also emphasized many of the important concepts by citing clinical conditions in which such concepts are involved. These clinical illustrations are highlighted in screened areas throughout the text.

Summaries are provided at the end of each chapter to emphasize the key points in the chapter, and brief bibliographies are included to direct the student to more detailed information. The references listed in these bibliographies are mainly review articles or recent, relevant scientific papers. At the end of the book, we have included a number of multiple-choice review questions and answers. These questions can serve as a guide for the readers to evaluate their comprehension of the material covered in the text.

Robert M. Berne

Matthew N. Levy

CONTENTS

CELL PHYSIOLOGY

HOWARD C. KUTCHAI

PART I

Cellular Membranes and Transmembrane Transport of Solutes and Water

CELLULAR MEMBRANES

Each cell is surrounded by a plasma membrane that separates it from the extracellular milieu. The **plasma membrane** serves as a permeability barrier that allows the cell to maintain a cytoplasmic composition far different from the composition of the extracellular fluid. The plasma membrane contains enzymes, receptors, and antigens that play central roles in the interaction of the cell with other cells and with hormones and other regulatory agents in the extracellular fluid.

The membranes that enclose the various organelles divide the cell into discrete compartments and allow the localization of particular biochemical processes in specific organelles. Many vital cellular processes take place in or on the membranes of the organelles. Striking examples are the processes of electron transport and oxidative phosphorylation, which occur on, within, and across the mitochondrial inner membrane.

Most biological membranes have certain features in common. However, in keeping with the diversity of membrane functions, the composition and structure of the membranes differ from one cell to another and among the membranes of a single cell.

Membrane Structure

Proteins and phospholipids are the most abundant constituents of cellular membranes. A **phospholipid** molecule has a polar head group and two very nonpolar, hydrophobic fatty acyl chains (Figure 1-1, *A*). In an aqueous environment phospholipids tend to form structures that allow the fatty acyl chains to be kept away from contact with water. One such structure is the **lipid bilayer** (Figure 1-1, *B*). Many phospholipids, when dispersed in water, spontaneously form lipid bilayers. Most

of the phospholipid molecules in biological membranes have a lipid bilayer structure.

The phospholipid bilayer is responsible for certain passive permeability properties of biological membranes. Substances that are highly soluble in water typically permeate cellular membranes very slowly, while nonpolar compounds that are more soluble in nonpolar organic solvents cross cell membranes more rapidly. High concentrations of barium salts are administered by mouth or by enema in order to make the interior of the gastrointestinal tract opaque to x-rays and improve the contrast of diagnostic x-ray films of the gastrointestinal tract. Barium ions in this concentration would be highly toxic, but because barium is highly water-soluble, it is barely absorbed at all from the gastrointestinal tract. Hence, the concentration of barium in the blood rises very little after administration of barium salts.

Figure 1-2 depicts the **fluid mosaic model** of membrane structure. This model is consistent with many of the properties of biological membranes. Note the bilayer structure of most of the membrane phospholipids. The membrane proteins are of two major classes: (1) **integral** or **intrinsic membrane proteins** that are embedded in the phospholipid bilayer and (2) **peripheral** or **extrinsic membrane proteins** that are associated with the surface of the membrane. The peripheral membrane proteins interact with the membrane predominantly by charge interactions with integral membrane proteins. Thus peripheral proteins may often be removed from the membrane by altering the ionic composition of the medium. Integral membrane proteins have important hydrophobic interactions with the interior of the membrane.

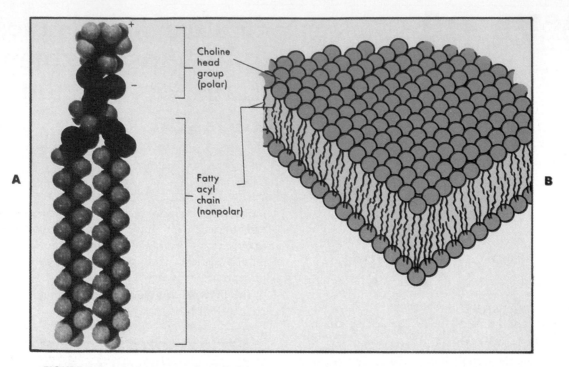

FIGURE 1-1 **A,** Structure of a membrane phospholipid molecule, in this case a phosphatidylcholine. **B,** Structure of a phospholipid bilayer. The blue circles represent the polar head groups of the phospholipid molecules. The wavy lines represent the fatty acyl chains of the phospholipids.

These hydrophobic interactions can be disrupted only by detergents that make the integral proteins soluble by interacting hydrophobically with nonpolar amino acid side chains.

Cellular membranes are fluid structures in which many of the constituent molecules are free to diffuse in the plane of the membrane. Most lipids and proteins can move freely in the bilayer plane, but they "flip-flop" from one phospholipid monolayer to the other at much slower rates. A large hydrophilic moiety is unlikely to flip-flop if it must be dragged through the nonpolar interior of the lipid bilayer.

In some cases membrane components are not free to diffuse in the plane of the membrane. Examples of this motional constraint are the sequestration of acetylcholine receptors (integral membrane proteins) at the motor endplate of skeletal muscle and the presence of different membrane proteins in the apical and basolateral plasma membranes of epithelial cells. The cytoskeleton appears to tether certain membrane proteins. The **anion exchanger,** a major protein of the human erythrocyte membrane, is bound to the spectrin network that undergirds the membrane via a protein called **ankyrin.**

If the motor nerve that innervates a skeletal muscle is accidentally severed, the acetylcholine receptors are no longer sequestered at the motor endplate, but instead they spread out over the entire plasma membrane of the muscle cells. Then the entire surface of the cell becomes excitable by acetylcholine, a phenomenon known as **denervation supersensitivity.**

MEMBRANE COMPOSITION

Lipid Composition

Major Phospholipids In animal cell membranes the most abundant phospholipids are often the choline-containing phospholipids: the lecithins (phosphatidylcholines) and the sphingomyelins. Next in abundance are frequently the amino phospholipids: phosphatidylserine and phosphatidylethanolamine. *The phospholipid bilayer is primarily responsible for the passive permeability properties of the membrane.* Other important phos-

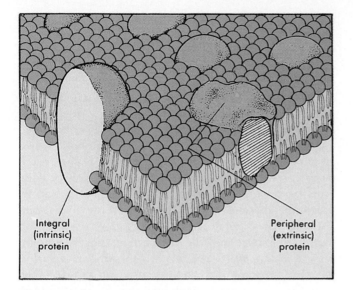

FIGURE 1-2 Schematic representation of the fluid mosaic model of membrane structure. The integral proteins are embedded in the lipid bilayer matrix of the membrane, and the peripheral proteins are associated with the external surfaces of integral membrane proteins.

pholipids present in smaller amounts are phosphatidylglycerol, phosphatidylinositol, and cardiolipin.

Certain phospholipids present in tiny proportions in the plasma membrane play a vital role in cellular signal transduction processes. **Phosphatidylinositol bisphosphate,** when cleaved by a receptor-activated phospholipase C, releases **inositol trisphosphate (IP$_3$)** and **diacyl glycerol.** IP$_3$ is released into the cytosol, where it acts on receptors in the endoplasmic reticulum to cause release of stored Ca^{++}, which affects a wide variety of cellular processes. Diacylglycerol remains in the plasma membrane, where it participates, along with Ca^{++}, in activating **protein kinase C,** an important signal transduction protein.

Cholesterol Cholesterol is a major constituent of plasma membranes, and its steroid nucleus lies parallel to the fatty acyl chains of membrane phospholipids. Cholesterol functions as a "fluidity buffer" in the plasma membrane in that its presence tends to keep the fluidity of the acyl chain region of the phospholipid bilayer in an intermediate range in the presence of agents, such as alcohols and general anesthetics, that tend to fluidize biological membranes.

Glycolipids Glycolipids are not abundant, but they have important functions. Glycolipids are found mostly in plasma membranes, where their carbohydrate moieties protrude from the external surface of the membrane. The carbohydrate parts of glycolipids frequently function as receptors or antigens.

The receptor for cholera toxin (Chapter 34) is the carbohydrate moiety of a particular glycolipid, ganglioside (G$_{M1}$). The A and B blood group antigens (Chapter 15) are the carbohydrate moieties of other gangliosides on the human erythrocyte membrane.

Asymmetry of Lipid Distribution In many membranes the lipid components are not distributed uniformly across the bilayer. The glycolipids of the plasma membrane are located almost exclusively in the outer monolayer. Phospholipids are also distributed asymmetrically between the inner and outer monolayers of membranes. In the red blood cell membrane, for example, the outer monolayer contains most of the choline-containing phospholipids, whereas the inner monolayer contains most of the amino phospholipids.

Membrane Proteins The protein composition of membranes may be simple or complex. The functionally specialized membranes of the sarcoplasmic reticulum of skeletal muscle and the disks of the rod outer segment of the retina contain only a few different proteins. By contrast, plasma membranes, which perform many functions, may have more than 100 different protein constituents. Membrane proteins include enzymes, transport proteins, and receptors for hormones and neurotransmitters.

Glycoproteins Some membrane proteins are glycoproteins with covalently bound carbohydrate side chains. As with glycolipids, the carbohydrate chains of glycoproteins are located almost exclusively on the external surfaces of plasma membranes. The carbohydrate moieties of membrane glycoproteins and glycolipids have important functions. The negative surface charge of cells is caused by the negatively charged sialic acid of glycolipids and glycoproteins.

Fibronectin is a large fibrous glycoprotein that helps cells attach, via cell surface glycoproteins called **integrins,** to proteins of the extracellular matrix. This linkage mediates communication between the extracellular matrix and the cell's cytoskeleton during embryonic development.

The major membrane proteins of enveloped viruses are glycoproteins. Their carbohydrate moieties stud the outer surface of the virus with "spikes" that are required for the virus to bind to a host cell.

Asymmetry of Membrane Proteins The Na^+-K^+-ATPase of the plasma membrane and the Ca^{++} pump protein (Ca^{++}-ATPase) of the sarcoplasmic reticulum membrane are examples of the asymmetric disposition of membrane proteins. In both cases ATP is split on the cytoplasmic face of the membrane, and some of the energy liberated is used to pump ions in specific directions across the membrane. In the case of the Na$^+$-K$^+$-ATPase, K$^+$ is pumped into the cell and Na$^+$ is pumped out, whereas the Ca^{++}-ATPase actively pumps Ca^{++} into the sarcoplasmic reticulum.

MEMBRANES AS PERMEABILITY BARRIERS

Biological membranes serve as **permeability barriers.** Most of the molecules present in living systems are highly soluble in water and poorly soluble in nonpolar solvents. Such molecules are poorly soluble in the nonpolar environment in the interior of the lipid bilayer of biological membranes. As a consequence, biological membranes pose a formidable barrier to most water-soluble molecules. *The plasma membrane is a permeability barrier between the cytoplasm and the extracellular fluid.* This barrier allows the maintenance of large concentration differences for many substances between the cytoplasm and the extracellular fluid.

The localization of various cellular processes in certain organelles depends on the barrier properties of cellular membranes. For example, the inner mitochondrial membrane is impermeable to the enzymes and substrates of the tricarboxylic acid cycle, and thus it allows the localization of the tricarboxylic cycle in the mitochondrial matrix. The spatial organization of chemical and physical processes in the cell depends on the barrier functions of cellular membranes, much as the walls of a house separate rooms with different functions.

The passage of important molecules across membranes at controlled rates is central to the life of the cell. Examples are the uptake of nutrient molecules, the discharge of waste products, and the release of secreted molecules. As discussed in the next section, molecules may move from one side of a membrane to another without actually moving through the membrane itself. In other cases molecules cross a particular membrane by passing through or between the molecules that make up the membrane.

TRANSPORT ACROSS, BUT NOT THROUGH, MEMBRANES

Endocytosis

Endocytosis is the process that allows material to enter the cell without passing through the membrane (Figure 1-3); it includes **phagocytosis** and **pinocytosis.** The uptake of particulate material is termed phagocytosis (Figure 1-3, *A*). The uptake of soluble molecules is called pinocytosis (Figure 1-3, *B*). Sometimes special regions of the plasma membrane are involved in endocytosis. In these regions the cytoplasmic surface of the plasma membrane is covered with bristles made primarily of a protein called **clathrin.** These clathrin-covered regions are called **coated pits,** and their endocytosis gives rise to coated vesicles (Figure 1-3, *C*). The coated pits are involved in **receptor-mediated endocytosis.** Proteins to be taken up are recognized and bound by specific membrane receptor proteins in the coated pits. The binding often leads to aggregation of receptor-ligand complexes, and the aggregation triggers endocytosis. *Endocytosis is an active process that requires metabolic energy.* Endocytosis also can occur in regions of the plasma membrane that do not contain coated pits.

Most cells cannot synthesize cholesterol, which is needed for synthesis of new membranes (see also Chapter 41). Cholesterol is carried in the blood, predominantly in low-density lipoproteins (LDL). Many cells have LDL receptors in their plasma membranes. When LDL binds to these receptors, the receptor-LDL complexes migrate to coated pits, where they aggregate and are taken into the cell by receptor-mediated endocytosis. Individuals who lack LDL receptors have high levels of cholesterol-laden LDL in their blood. Consequently such individuals tend to develop arterial disease **(atherosclerosis)** at an early age, which makes them more likely to experience heart attacks prematurely.

Exocytosis

Molecules can be ejected from cells by **exocytosis,** a process that resembles endocytosis in reverse. The release of neurotransmitters, which is considered in more detail in Chapter 4, takes place by exocytosis. Exocytosis is responsible for the release of secretory proteins by many cells; the release of pancreatic enzymes from the acinar cells of the pancreas is a well-studied example. The pancreatic enzymes play vital roles in digestion of protein, carbohydrates, and lipids (see Chapter 33). In such cases the proteins to be secreted are stored in secretory vesicles in the cytoplasm. *A stimulus to secrete causes the secretory vesicles to fuse with the plasma membrane and to release the vesicle contents by exocytosis.*

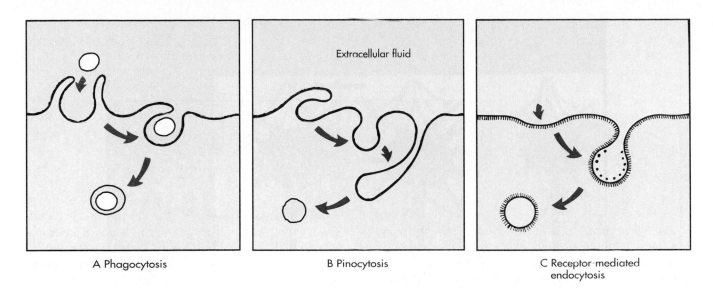

FIGURE 1-3 Schematic depiction of endocytotic processes. **A,** Phagocytosis of a solid particle. **B,** Pinocytosis of extracellular fluid. **C,** Receptor-mediated endocytosis by coated pits.

Fusion of Membrane Vesicles

The contents of one type of organelle can be transferred to another organelle by fusion of the membranes of the organelles. In some cells, secretory products are transferred from the endoplasmic reticulum to the Golgi apparatus by fusion of endoplasmic reticulum vesicles with membranous sacs of the Golgi apparatus. Fusion of phagocytic vesicles with lysosomes allows the phagocytosed material to be digested by proteolytic enzymes in the lysosomes. The turnover of many normal cellular constituents involves their destruction in lysosomes, followed by their resynthesis.

Influenza viruses have membrane proteins that undergo a dramatic conformational change to insert a "fusion peptide" into the host cell. The fusion peptide promotes the fusion of the viral membrane with the plasma membrane of the host cell, allowing entry of the viral genome into the host cell.

TRANSPORT OF MOLECULES THROUGH MEMBRANES

The traffic of molecules through biological membranes is vital for most cellular processes. Some molecules move through biological membranes simply by diffusing among the molecules that make up the membrane, whereas the passage of other molecules involves the mediation of specific transport proteins in the membrane.

Oxygen, for example, is a small molecule that is fairly soluble in nonpolar solvents. It crosses biological membranes by diffusing among membrane lipid molecules. Glucose, on the other hand, is a much larger molecule that is not very soluble in the membrane lipids. Glucose enters cells via specific glucose transport proteins in the plasma membrane.

Diffusion

Diffusion is the process whereby atoms or molecules intermingle because of their random thermal motion, also called **Brownian motion.** Imagine a container divided into two compartments by a removable partition (Figure 1-4). A much larger number of molecules of a compound is placed on side A than on side B, and then the partition is removed. Every molecule is in random thermal motion. It is equally probable that a molecule that begins on side A will move to side B in a given time and that a molecule beginning on side B will end up on side A. Because many more molecules are present on side A, the total number of molecules moving from side A to side B will be greater than the number moving from side B to side A. In this way the number of molecules on side A will decrease, whereas the number of molecules on side B will increase. This process of net diffusion of molecules will continue until the concentration of molecules on side A equals that on side B. Thereafter the rate of diffusion of molecules from A to B will equal that from B to A, and

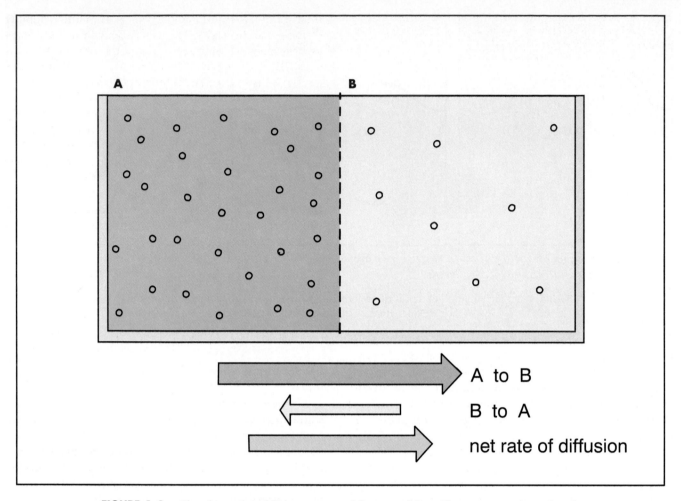

FIGURE 1-4   Chambers A and B are separated by a partition. The concentration of molecules in chamber A is much greater than that in chamber B. For that reason, when the partition is removed, the rate of diffusion of molecules from A to B is much greater than that from B to A. There is thus a net flux of molecules from A to B.

no further net movement will occur; a dynamic equilibrium exists.

Range of Diffusion   Diffusion is a rapid process when the distance over which it takes place is small. A rule of thumb is that a typical molecule takes 1 msec to diffuse 1 μm. However, the time required for diffusion increases with the square of the distance over which diffusion occurs. Thus *a tenfold increase in the diffusion distance means that the diffusion process will require about 100 times longer to reach a given degree of completion.*

Table 1-1 shows the results of calculations for a typical, small, water-soluble solute. It can be seen that diffusion is extremely rapid on a microscopic scale of distance. For macroscopic distances, diffusion is rather slow. A cell that is 100 μm away from the nearest capillary can receive nutrients from the blood by diffusion with a time lag of only 5 seconds or so. This is sufficiently fast to satisfy the metabolic demands of many cells. However, a skeletal muscle cell that is 1 cm long cannot rely on

diffusion for the intracellular transport of vital metabolites, because the 14 hours required for diffusion over the 1 cm distance is too long on the time scale of cellular metabolism. Some nerve fibers are longer than 1 m. Therefore, it is no wonder that intracellular axonal transport systems are involved in transporting important molecules along nerve fibers. Because of the slowness of diffusion over macroscopic distances, it is not surprising that even small multicellular organisms have evolved circulatory systems to bring the individual cells of the organisms within a reasonable diffusion range of nutrients.

Diffusion Coefficient   The **diffusion coefficient (D)** is proportional to the speed with which the diffusing molecule can move in the surrounding medium. The larger the molecule and the more viscous the medium, the smaller is D. For small molecules, D is inversely proportional to $MW^{1/2}$ (MW refers to molecular weight.) For macromolecules, D is inversely proportional to $MW^{1/3}$. Thus *a protein that has one-eighth the mass of*

Table 1-1 Time Required for Diffusion to Occur over Various Diffusion Distances*

Diffusion Distance (μm)	Time Required for Diffusion
1	0.5 msec
10	50 msec
100	5 seconds
1000 (1mm)	8.3 minutes
10,000 (1cm)	14 hours

*The time required for the "average" molecule (with diffusion coefficient taken to be 1×10^{-5} cm/sec) to diffuse the required distance was computed.

another molecule will have a diffusion coefficient only two times larger than the bigger molecule.

Diffusion Across a Membrane Diffusion leads to a state in which the concentration of the diffusing species is constant in space and time. Diffusion across cellular membranes tends to equalize the concentrations on the two sides of the membrane (see Figure 1-4). The diffusion rate across a membrane is proportional to the area of the membrane and to the difference in concentration of the diffusing substance on the two sides of the membrane. **Fick's First Law of Diffusion** states that

$$J = -DA\frac{\Delta c}{\Delta x} \qquad (1)$$

where

J = Net rate of diffusion in moles or grams per unit time
D = Diffusion coefficient of the diffusing solute in the membrane
A = Area of the membrane
Δc = Concentration difference across the membrane
Δx = Thickness of the membrane

Diffusive Permeability of Cellular Membranes

Permeability to Lipid-Soluble Molecules The plasma membrane serves as a diffusion barrier that enables the cell to maintain cytoplasmic concentrations of many substances that differ greatly from their extracellular concentrations. As early as the turn of the century, the relative impermeability of the plasma membrane to most water-soluble substances was attributed to its "lipoid nature."

The hypothesis that the plasma membrane has a lipoid character is supported by experiments showing that compounds that are soluble in nonpolar solvents (e.g., benzene or olive oil) enter cells more readily than do water-soluble substances of similar molecular weight. Figure 1-5 shows the relationship between membrane permeability and solubility in a nonpolar solvent for a

number of different solutes. The ratio of the solubility of the solute in olive oil to its solubility in water is used as a measure of solubility in nonpolar solvents. This ratio is called the olive oil/water partition coefficient. The permeability of the plasma membrane to a particular substance increases with the "lipid solubility" of the substance. For compounds with the same olive oil/water partition coefficient, permeability decreases with increasing molecular weight. As described previously, the fluid mosaic model of membrane structure envisions the plasma membrane as a lipid bilayer with proteins embedded in it (see Figure 1-2). The data of Figure 1-5 support the idea that the lipid bilayer is the principal barrier to substances that permeate the membrane by simple diffusion.

Fat-soluble vitamins are absorbed by the epithelial cells of the small intestine by simply diffusing across their luminal plasma membranes. Water-soluble vitamins, by contrast, do not readily diffuse across biological membranes, and thus special membrane transport proteins are required for absorption of water-soluble vitamins (see Chapter 34).

Permeability to Water-Soluble Molecules Very small, uncharged, water-soluble molecules pass through cell membranes much more rapidly than predicted by their lipid solubility. For example, water permeates cell membranes about 100 times more rapidly than is predicted from its molecular radius and its olive oil/water partition coefficient. There are two reasons for the unusually high permeability to water. Water and certain very small water-soluble molecules can pass between adjacent phospholipid molecules without actually dissolving in the region occupied by the fatty acid side chains. Moreover, the plasma membranes of most cells contain membrane proteins that form channels that permit a high rate of water flow.

The permeability of membranes to uncharged, water-soluble molecules decreases as the size of the molecules increases. *Most plasma membranes are essentially impermeable to water-soluble molecules whose molecular weights are greater than about 200.*

Because of their charge, ions are relatively insoluble in lipid solvents, and thus membranes are not very permeable to most ions. Ionic diffusion across membranes occurs mainly through protein **ion channels** that span the membrane. Some ion channels are highly specific with respect to the ions allowed to pass, whereas others allow all ions below a certain size to pass. *Some ion channels are controlled by the voltage difference across the membrane, and others are controlled by neuro-*

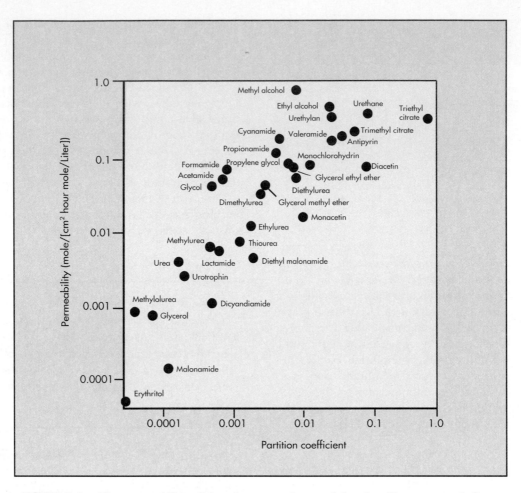

FIGURE 1-5 The permeability of the plasma membrane of the alga *Chara ceratophylla* to various nonelectrolytes as a function of the lipid solubility of the solutes. Lipid solubility is represented on the abscissa by the olive oil/water partition coefficient. (Redrawn from Christensen HN: *Biological transport,* ed 2, Menlo Park, Calif, 1975, WA Benjamin. Data from Collander R: *Trans Faraday Soc* 33:985, 1937.)

transmitters or certain other regulatory molecules (Chapters 3 and 4).

Although certain water-soluble molecules such as sugars and amino acids are essential for cellular survival, they do not cross plasma membranes appreciably by simple diffusion. Plasma membranes have specific proteins that allow the transfer of vital metabolites into or out of the cell. The characteristics of membrane **protein-mediated transport** are discussed later.

OSMOSIS

Osmosis is defined as the flow of water across a semipermeable membrane from a compartment in which the solute concentration is lower to one in which the solute concentration is greater. A **semipermeable membrane** is defined as a membrane permeable to water but

impermeable to solutes. *Osmosis takes place because the presence of solute decreases the chemical potential of water.* Water tends to flow from where its chemical potential is higher to where its chemical potential is lower. Other effects caused by the decrease in the chemical potential of water (because of the presence of solute) include reduced vapor pressure, lower freezing point, and higher boiling point of the solution as compared with pure water. Because these properties, and osmotic pressure as well, depend primarily on the concentration of the solute present rather than on its chemical properties, they are called **colligative properties.**

Osmotic Pressure

In Figure 1-6 a semipermeable membrane separates a solution from pure water. Water flows from side B to side A by osmosis because the presence of solute on side A

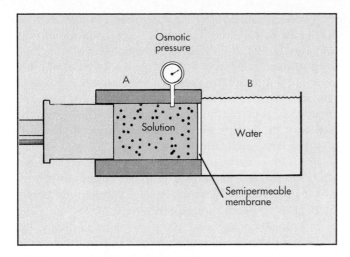

FIGURE 1-6   Schematic representation of the definition of osmotic pressure. When the hydrostatic pressure applied to the solution in chamber A is equal to the osmotic pressure of that solution, there will be no net water flow across the membrane.

Table 1-2 Osmotic Coefficients (ϕ) of Certain Solutes of Physiological Interest

Substance	i	Molecular Weight	ϕ
NaCl	2	58.5	0.93
KCl	2	74.6	0.92
HCl	2	36.6	0.95
NH_4Cl	2	53.5	0.92
$NaHCO_3$	2	84.0	0.96
$NaNO_3$	2	85.0	0.90
KSCN	2	97.2	0.91
KH_2PO_4	2	136.0	0.87
$CaCl_2$	3	111.0	0.86
$MgCl_2$	3	95.2	0.89
Na_2SO_4	3	142.0	0.74
K_2SO_4	3	174.0	0.74
$MgSO_4$	2	120.0	0.58
Glucose	1	180.0	1.01
Sucrose	1	342.0	1.02
Maltose	1	342.0	1.01
Lactose	1	342.0	1.01

Reproduced with permission from Lifson N and Visscher MB: Osmosis in living systems. In Glasser O, ed: *Medical physics*, vol 1, Chicago, 1944, Year Book Medical Publishers, Inc.

reduces the chemical potential of water in the solution. Pushing on the piston will increase the chemical potential of the water in the solution of side A and slow the net rate of osmotic water flow. If the force on the piston is increased gradually, a pressure is eventually reached at which net water flow stops. Application of still more pressure will cause water to flow in the opposite direction. The pressure on side A that is just sufficient to keep pure water from entering is called the **osmotic pressure** of the solution on side A.

The osmotic pressure of a solution depends on the number of particles in solution. Thus the degree of ionization of the solute must be taken into account. A 1 M solution of glucose, a 0.5 M solution of NaCl, and a 0.333 M solution of $CaCl_2$ have approximately the same osmotic pressure. (Actually their osmotic pressures will differ somewhat because of the deviations of real solutions from ideal behavior.) One form of **van't Hoff's law** for calculation of osmotic pressure is

$$\pi = RT(\phi ic) \qquad (2)$$

where

π = Osmotic pressure
R = Ideal gas constant
T = Absolute temperature
ϕ = Osmotic coefficient
 i = Number of ions formed by dissociation of a solute molecule
 c = Molar concentration of solute (moles of solute per liter of solution).

The osmotic coefficient (ϕ) accounts for the deviation of the solution from ideality. ϕ depends on the particular compound, its concentration, and the temperature. Values of ϕ may be greater or less than one. The value is less than one for electrolytes of physiological importance, and for all solutes ϕ approaches one as the solution becomes more and more dilute. *The term ϕic can be regarded as the osmotically effective concentration, and ϕic is called the* **osmolarity** *of the solution, expressed in osmoles per liter.* Sometimes a less precise estimate of osmotic pressure is computed assuming that ϕ is equal to 1.

Values of ϕ can be obtained from handbooks that list values of ϕ for different substances as functions of concentration. Solutions of proteins deviate greatly from ideal behavior, and different proteins may deviate to different extents. Values of the osmotic coefficient depend on the concentration of the solute and on its chemical properties. Table 1-2 lists osmotic coefficients for several solutes. These values apply fairly well at the concentrations of these solutes in the extracellular fluids of mammals.

Measurement of Osmotic Pressure

The osmotic pressure of a solution can be obtained by determining the pressure required to prevent water from entering the solution across a semipermeable membrane (Figure 1-6). More often, however, the osmotic pressure is estimated from another colligative property, such as depression of the freezing point. The relation that describes the osmolarity (ϕic) of a solution in terms of the depression of the freezing point of water by the solute is

$$\phi ic = \Delta T_f / 1.86 \qquad (3)$$

where ΔT_f is the freezing point depression in degrees centigrade. When the freezing point depression of a multicomponent solution is determined, the effective osmolarity (in osmoles per liter) of the solution as a whole can be obtained.

If the total osmotic pressures of two solutions (as measured by freezing point depression or by the osmotic pressure developed across a semipermeable membrane) are equal, the solutions are said to be **isoosmotic** (or **isosmotic**). If solution A has greater osmotic pressure than solution B, A is said to be **hyperosmotic** with respect to B. If solution A has less total osmotic pressure than solution B, A is said to be **hypoosmotic** to B.

Osmotic Swelling and Shrinking of Cells

The plasma membranes of most of the body's cells are relatively impermeable to many of the solutes of the extracellular fluid but are highly permeable to water. Therefore, when the osmotic pressure of the extracellular fluid is increased, water leaves the cells by osmosis and the cells shrink. Thus the cellular solutes become more concentrated until the effective osmotic pressure of the cytoplasm is again equal to that of the extracellular fluid. Conversely, if the osmotic pressure of the extracellular fluid is decreased, water enters the cells. The cells will continue to swell until the intracellular and extracellular osmotic pressures are equal.

Red blood cells are often used to illustrate the osmotic properties of cells, because they are readily obtained and are easily studied. Within a certain range of external solute concentrations, the red cell behaves as an osmometer, because its volume is inversely related to the solute concentration in the extracellular medium. In Figure 1-7 the red cell volume, as a fraction of its normal volume in plasma, is shown as a function of the concentration of NaCl solution in which the red cells are suspended. At an NaCl concentration of 154 mM (308 mM osmotically active particles), the volume of the cells is the same as their volume in plasma; this concentration of NaCl is said to be **isotonic** to the red cell.

Isotonic NaCl solution (also known as **isotonic saline**) is used for intravenous rehydration or for administration of drugs to patients. Almost every patient undergoing surgery will have an intravenous drip of isotonic saline.

A concentration of NaCl greater than 154 mM is called **hypertonic** (greater strength, causes cells to shrink), and a solution less concentrated than 154 mM is termed **hypotonic** (cells swell). When red cells have swollen to about 1.4 times their original volume, some cells lyse (burst). At this volume the properties of the red cell membrane abruptly change; hemoglobin leaks out of the cell, and the membrane becomes transiently permeable to other large molecules as well.

The intracellular substances of the erythrocyte that produce an osmotic pressure that just balances the osmotic pressure of the extracellular fluid include hemoglobin, K^+, organic phosphates (e.g., ATP and 2,3-diphosphoglycerate), and glycolytic intermediates. Regardless of the chemical nature of its contents, the red cell behaves as though it were filled with a solution of impermeant molecules with an osmotically effective concentration of 286 milliosmolar, which is the same as the osmolarity of isotonic saline:

$$\phi_{NaCl} i_{NaCl} C_{NaCl} = 0.932 \times 2 \times 0.154 \text{ M}$$
$$= 0.286 \text{ osmolar} = 286 \text{ milliosmolar} \qquad (4)$$

Osmotic Effects of Permeant Solutes Permeating solutes eventually equilibrate across the plasma membrane. For this reason permeating solutes exert only a transient effect on cell volume.

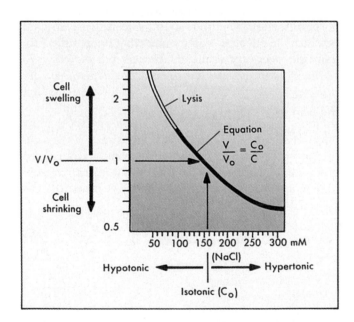

FIGURE 1-7 The osmotic behavior of human red blood cells in NaCl solutions. At 154 mM NaCl (isotonic), the red cell has its normal volume. It shrinks in more concentrated (hypertonic) solutions and swells in more dilute (hypotonic) solutions. V_o and C_o are the red cell volume and intracellular solute concentration, respectively, for the red cell in the blood or in an isotonic solution. V and C are, respectively, the cell volume and intracellular solute concentration in a solution that is not isotonic.

Consider a red blood cell placed in a large volume of 0.154 M NaCl that contains 0.050 M glycerol. Initially, because of the extracellular NaCl and glycerol, the osmotic pressure of the extracellular fluid will exceed that of the cell interior, and the cell will shrink. With time, however, glycerol will equilibrate across the plasma membrane of the red cell, and the cell will swell back toward its original volume. *The steady-state volume of the cell will be determined only by the impermeant solutes in the extracellular fluid.* In this case the impermeant solutes (NaCl) have a total concentration that is isotonic, so the final volume of the cell will be equal to the normal red cell volume. Because the red cell ultimately returns to its normal volume, the solution (0.050 M glycerol in 0.154 M NaCl) is isotonic. Because the red cell initially shrinks when put in this solution, the solution is hyperosmotic with respect to the normal red cell. The transient changes in cell volume depend on equilibration of glycerol across the membrane. Had we used urea (a more rapidly permeating substance), the cell would have reached steady-state volume sooner.

The following rules help predict the volume changes a cell will undergo when suspended in solutions of permeant and impermeant solutes:

1. *The steady-state volume of the cell is determined only by the concentration of impermeant solutes in the extracellular fluid.*
2. *Permeant solutes cause only transient changes in cell volume.*
3. The greater the permeability of the membrane to the permeant solute, the more rapid the transient changes.

Magnitudes of Osmotic Flows Caused by Permeating Solutes

In the preceding example it was explained that permeants, such as glycerol, exert only a transient osmotic effect. It is sometimes important to determine the rate of the osmotic flow caused by a particular permeant.

When a difference in hydrostatic pressure (ΔP) causes water flow across a membrane, the rate of water flow ($\dot{V}_w$) is

$$\dot{V}_w = L\Delta P \qquad (5)$$

where L is a constant of proportionality, called the **hydraulic conductivity.**

Osmotic flow of water across a membrane is directly proportional to the osmotic pressure difference ($\Delta\pi$) between the solutions on the two sides of the membrane; thus

$$\dot{V}_w = L\Delta\pi \qquad (6)$$

Equation 6 holds only for osmosis caused by impermeant solutes. *Permeant solutes cause less osmotic flow. The greater the permeability of a solute, the smaller the osmotic flow it causes.* Table 1-3 shows the osmotic water flows induced across a porous membrane by solutes of different molecular sizes. The solutions have identical freezing points, so the total osmotic pressures are the same. The larger the solute molecule, the more impermeable the membrane is to the solute, and the greater the osmotic water flow it causes.

Reflection Coefficients Equation 6 can be rewritten to take solute permeability into account by including σ, the **reflection coefficient.**

$$\dot{V}_w = \sigma L\Delta\pi \qquad (7)$$

σ is a dimensionless number that ranges from 1 for completely impermeant solutes to 0 for extremely permeant solutes. σ is a property of a particular solute and a particular membrane and represents the osmotic flow induced by the solute as a fraction of the theoretical maximum osmotic flow (Table 1-3).

Table 1-3 Osmotic Water Flow Across a Porous Dialysis Membrane Caused by Various Solutes*

Gradient Producing the Water Flow	Net Volume Flow (μl/min)*	Solute Radius ($\overset{\circ}{A}$)	Reflection Coefficient (σ)
D_2O	0.06	1.9	0.0024
Urea	0.6	2.7	0.024
Glucose	5.1	4.4	0.205
Sucrose	9.2	5.3	0.368
Raffinose	11	6.1	0.440
Inulin	19	12	0.760
Bovine serum albumin	25.5	37	1.02
Hydrostatic pressure	25		

Data from Durbin RP: *J Gen Physiol* 44:315, 1960. Reproduced from *The Journal of General Physiology* by copyright permission of The Rockefeller University Press.
*Flow is expressed as microliters per minute caused by a 1 M concentration difference of solute across the membrane. The flows are compared with the flow caused by a theoretically equivalent hydrostatic pressure.

The mechanism by which the kidney produces urine that is more concentrated than the extracellular fluid (see Chapter 36) involves various parts of the nephron that have different reflection coefficients for important solutes, such as NaCl and urea. The osmotic water flows induced by NaCl and urea in a particular segment of the nephron depend on the values of σ for the epithelium in that segment to these solutes.

PROTEIN-MEDIATED MEMBRANE TRANSPORT

Certain substances enter or leave cells by way of specific carriers or channels that are intrinsic proteins of the plasma membrane. Transport via such protein carriers or channels is called **protein-mediated transport** or simply **mediated transport.** Specific ions or molecules may cross the membranes of mitochondria, endoplasmic reticulum, and other organelles by mediated transport. Mediated transport systems include **active transport** and **facilitated transport** processes, which have several properties in common. The principal distinction between these two processes is that *active transport is capable of "pumping" a substance against a gradient of concentration (or electrochemical potential), whereas facilitated transport tends to equilibrate the substance across the membrane.*

Properties of Mediated Transport

1. A substance that is transported by mediated transport is transported much more rapidly than molecules that have a similar molecular weight and lipid solubility but that cross the membrane by simple diffusion.
2. The transport rate shows **saturation kinetics:** as the concentration of the transported compound is increased, the rate of transport at first increases, but eventually a concentration is reached after which the transport rate increases no further (Figure 1-8). At this point the transport system is said to be saturated with the transported compound.
3. The mediating protein has **chemical specificity:** only molecules with the requisite chemical structure are transported. The specificity of most transport systems is not absolute, and in general it is broader than the specificity of most enzymes. The lock-and-key relationship between an enzyme and its substrate applies to transport proteins as well.
4. Structurally related molecules may compete for transport. Typically, one transport substrate will decrease

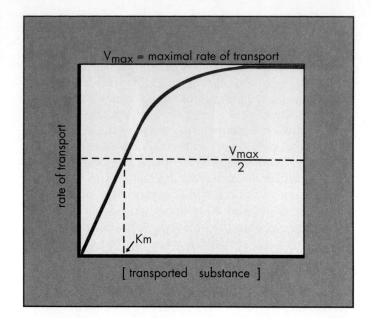

FIGURE 1-8 Transport via a transport protein shows saturation kinetics. As the concentration of the transported substance increases, the rate of its transport approaches a maximal value, the V_{max} for the transporter. The concentration of the transported substance required for the transport rate to be half-maximal is termed the K_m of the transporter.

the transport rate of a second substrate by competing for the transport protein. The competition is analogous to competitive inhibition of an enzyme.
5. Transport may be inhibited by compounds that are not structurally related to transport substrates. An inhibitor may bind to the transport protein in a way that decreases the affinity of the protein for the normal transport substrate. The compound phloretin does not resemble a sugar molecule, yet it strongly inhibits red cell sugar transport. Active transport systems, which require some link to metabolism, may be inhibited by metabolic inhibitors. The rate of Na^+ transport out of cells by the Na^+-K^+-ATPase is decreased by substances that interfere with ATP generation.

Facilitated Transport

Sometimes called **facilitated diffusion,** facilitated transport occurs via a *transport protein that is not linked to metabolic energy.* Facilitated transport has the properties discussed previously, except that it is not generally depressed by metabolic inhibitors. Because facilitated transport processes are not linked to energy metabolism, they cannot move substances against concentration gradients. Facilitated transport systems act to equalize concentrations of the transported substances on the two sides of the membrane.

Monosaccharides enter muscle cells by facilitated transport. Glucose, galactose, arabinose, and 3-O-methylglucose compete for the same carrier. The rate of transport shows saturation kinetics. The nonphysiological stereoisomer L-glucose enters the cells very slowly, and nontransported sugars, such as mannitol or sorbose, enter muscle cells very slowly, if at all. Phloretin inhibits sugar uptake, and insulin stimulates it.

A major action of the hormone **insulin** is to stimulate the transport of glucose across the plasma membranes of muscle and fat cells. People with **type 1 diabetes** secrete insulin at markedly subnormal rates (see Chapter 41). In this disease the rate of glucose uptake by muscle and adipose cells is so slow that the ability of these tissues to use glucose as a metabolic fuel is markedly impaired. Some of the pathological consequences of type 1 diabetes are caused by the inability to metabolize glucose at normal rates.

Current evidence suggests that most transport proteins span the membrane and are multimeric. Figure 1-9 depicts a hypothetical model that has been proposed for the monosaccharide transport protein of the membrane of the human red blood cell. Conformational changes of the protein, induced by monosaccharide binding, may allow a sugar molecule to enter and leave the central cavity of the transport protein.

Active Transport

Active transport processes have most of the properties of facilitated transport. In addition, active transport systems can concentrate their substrates against concentration or electrochemical potential gradients. This requires energy; hence *active transport processes must be linked to energy metabolism in some way.* Active transport systems may use ATP directly, or they may be linked more indirectly to metabolism. Because of their dependence on metabolism, active transport processes may be inhibited by any substance that interferes with energy metabolism.

Primary Active Transport An active transport process that is linked rather directly to cellular metabolism—for example by using ATP to power the transport—is called **primary active transport.**

The Na$^+$-K$^+$ pump (also called the **Na$^+$-K$^+$-ATPase**) is a primary active transporter, *because it uses ATP di-*

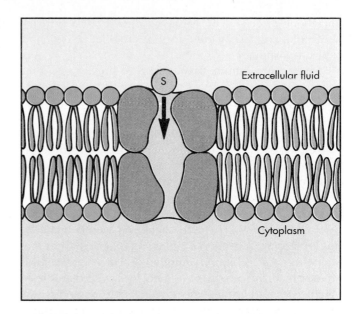

FIGURE 1-9 Hypothetical model of a transport protein. This is a model of the mechanism of monosaccharide transport by the sugar transport protein of the human red blood cell. The protein is postulated to be a tetramer. Binding of sugars is proposed to cause conformational changes that allow sugar molecules to enter and leave the central cavity of the transport protein.

rectly to power the transport of Na$^+$ and K$^+$. In the cytoplasm of most animal cells the concentration of Na$^+$ is much less and the concentration of K$^+$ is much greater than their extracellular concentrations. These concentration gradients are brought about by the action of an Na$^+$-K$^+$ pump in the plasma membrane; Na$^+$ is pumped out of the cell, and K$^+$ is pumped into the cell. The Na$^+$-K$^+$ pump activity is the result of an integral membrane protein called the Na$^+$-K$^+$-ATPase. The Na$^+$-K$^+$-ATPase transports three sodium ions out of the cell and transports two potassium ions into the cell for each molecule of ATP hydrolyzed. The cyclic phosphorylation and dephosphorylation of the protein causes it to alternate between two conformations, E1 and E2. In the E1 conformation the ion-binding sites of the protein have a high affinity for Na$^+$ and a low affinity for K$^+$, and the binding sites face the cytoplasm. In the E2 conformation the ion-binding sites face the extracellular fluid, and their affinities favor the binding of K$^+$ and the dissociation of Na$^+$. In this way the Na$^+$-K$^+$-ATPase alternates between the E1 and the E2 conformations and transports K$^+$ into the cell and Na$^+$ out of the cell by a process resembling "molecular peristalsis."

Because the Na$^+$-K$^+$-ATPase uses the energy in the terminal phosphate bond of ATP to power the transport cycle, it is said to be a primary active transport system. A transport process powered by some other high-energy metabolic intermediate or linked directly to a primary metabolic reaction would also be classified as primary active transport.

Secondary Active Transport The previous section emphasized that energy is required to create a concentration gradient for a transported substance. Once created, *a concentration gradient represents a store of chemical potential energy that can be harnessed to do work* (Chapter 2). In many cell types the concentration gradient of Na^+ created by the Na^+-K^+-ATPase is used to actively transport certain other solutes into the cell. Many cells take up neutral, hydrophilic amino acids by membrane transport proteins that link the inward transport of Na^+ down its electrochemical potential gradient to the inward transport of amino acids against their gradients of concentration (Figure 1-10). The energy for the transport of the amino acid is not provided directly by ATP or some other high-energy metabolite, but indirectly, from the gradient of Na^+ that is itself actively transported. Hence the amino acid is said to be transported by **secondary active transport.** In the secondary active transport of amino acids, both the rate of amino acid transport and the extent to which the amino acid is accumulated depend on the electrochemical potential gradient of Na^+.

Other Membrane Transport Processes

Calcium Transport Under most circumstances the concentration of Ca^{++} in the cytosol of cells is maintained at low levels, below 10^{-7} M, whereas the concentration of Ca^{++} in extracellular fluids is of the order of 10^{-3} M. Plasma membranes contain a Ca^{++}-ATPase that helps to maintain the large gradient of Ca^{++} across the plasma membrane. The plasma membrane Ca^{++}-ATPase is a close relative of the Ca^{++}-ATPase that is responsible for sequestering Ca^{++} in the sarcoplasmic reticulum of muscle (see Chapter 13). The plasma membrane Ca^{++}-ATPase shares several important properties with the Ca^{++}-ATPase of the sarcoplasmic reticulum and with the Na^+-K^+-ATPase of plasma membranes. These proteins carry out the primary active transport of ions across membranes, and they use the energy of the terminal phosphate bond of ATP to accomplish this task.

In addition, most cells store Ca^{++} in endoplasmic reticulum or other intracellular storage vesicles. Ca^{++} is concentrated in these vesicles by a Ca^{++}-ATPase that is closely related to the Ca^{++}-ATPase of the sarcoplasmic reticulum of muscle cells. Because Ca^{++} is a key second messenger, many hormones or agonists elevate the intracellular level of Ca^{++} by opening Ca^{++} channels in the plasma membrane and/or in the membranes of Ca^{++}-storage vesicles.

Certain electrically excitable cells, such as those of the

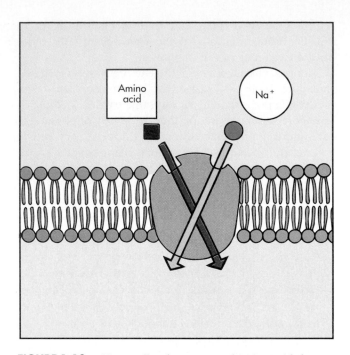

FIGURE 1-10 Many cells take up neutral amino acids by secondary active transport. The transport protein binds both Na^+ and the amino acid. Na^+ is transported down its electrochemical gradient, and the transport protein uses the energy released by Na^+ flux to transport the amino acid against a concentration gradient.

heart, have an additional mechanism for controlling the level of intracellular Ca^{++}. A sodium/calcium exchange protein in the plasma membrane uses the energy in the Na^+ gradient to extrude Ca^{++} from the cell. In heart cells the rapid, transient changes in intracellular Ca^{++} appear to be mediated by the sodium/calcium exchange protein, whereas the resting level of intracellular Ca^{++} is set mainly by the Ca^{++}-ATPases of plasma membrane and sarcoplasmic reticulum (see Chapters 14 and 18).

In most cells the rate at which Ca^{++} leaks into the cell down its electrochemical potential gradient is slow, so that the energy cost of maintaining a low intracellular level of Ca^{++} is not great. This contrasts with the cost of pumping Na^+ and K^+; *running the Na^+-K^+ pump is a major item in the energy budget of many cells.* Kidneys have an extremely high metabolic rate; the largest fraction of the energy expended by the kidney is consumed by the Na^+-K^+-ATPase.

Sugar Transport Glucose is a primary fuel for most of the cells of the body, but glucose diffuses across plasma membranes very slowly. The plasma membranes of many cell types contain sugar transport proteins that mediate the facilitated transport of glucose and related monosaccharides. Red blood cells, hepatocytes, adipocytes, and muscle cells (skeletal, cardiac, smooth) all possess glucose transporters. The uptake of glucose in these

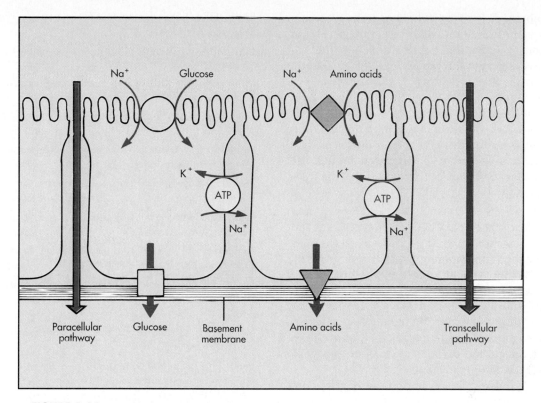

FIGURE 1-11   Epithelial transport processes that occur in the small intestine and renal tubules. Epithelia are polarized such that the transport processes on one side of the cell differ from those on the other side. Glucose and neutral amino acids enter the epithelial cell at the brush border via Na^+-powered secondary active transport, but they leave the cell across the basolateral membrane by facilitated transport.

cell types depends neither on the electrochemical potential difference of Na^+ across the plasma membrane nor in any direct way on cellular metabolism. In adipocytes and muscle cells, transport of glucose across the plasma membrane is increased by insulin. *Insulin increases glucose transport by causing more glucose transport proteins to be inserted into the plasma membrane.* The source of the newly inserted protein is a preformed pool of transporters in the membranes of the endoplasmic reticulum within the cell.

Amino Acid Transport   Most of the cells in the body synthesize proteins and therefore require amino acids. The synthesis of proteins is required for the turnover of cells and tissues and in processes such as wound healing. Several different amino acid transport proteins are present in plasma membranes. The amino acid transport systems include three distinct classes of transporters: for neutral, for basic, and for acidic amino acids (Chapter 34). Amino acid transport proteins overlap significantly in specificities, and the distribution of the different transport proteins varies from one cell type to another. Some of these transport proteins are secondary active trans-

porters powered by the concentration gradient of Na^+. Others are facilitated transport proteins.

Transport Across Epithelia

Epithelial cells are polarized with respect to their transport properties; that is, the transport properties of the plasma membrane facing one side of the epithelial cell layer are different from those of the membrane facing the other side.

The epithelial cells of the small intestine (Chapter 33) and the proximal tubule of the kidney (Chapter 36) are good examples of this polarity. The complement of membrane transport proteins in the brush border that faces the lumen of the small bowel or the renal tubule differs from the transport protein composition of the basolateral plasma membrane of the cell. The tight junctions that join the epithelial cells side to side prevent mixing of the transport proteins of the luminal and basolateral plasma membranes. The brush border plasma membranes of these epithelia contain very few Na^+-K^+-ATPase molecules, which reside mainly in the basolateral plasma membrane. Glucose (and galactose) and neutral amino acids enter these epithelial cells at the brush border by secondary active transporters driven by the Na^+ gradient. However, these substances leave the cells at the basolateral membrane primarily by facilitated transporters (Figure 1-11).

The tight junctions that join the cells are leaky to water and small water-soluble molecules and ions. There are thus two types of pathways for transport across the epithelia: (1) **transcellular pathways,** through the cells, and (2) **paracellular pathways,** in between the cells (Figure 1-11).

SUMMARY

1. Biological membranes are phospholipid bilayers with integral membrane proteins imbedded in the bilayer and peripheral membrane proteins adherent to the surfaces of the membrane. Membranes serve as permeability barriers that separate the cell from the extracellular milieu and divide the cell into biochemically specialized compartments.

2. Endocytosis and exocytosis permit material to enter or leave the cell without passing through the membrane.

3. Diffusion is an effective biological transport process on the microscopic scale of distances. Only very small water-soluble molecules and lipid-soluble molecules can diffuse across biological membranes at appreciable rates.

4. Gradients of solutes across membranes power the flow of water by osmosis. The steady-state volumes of cells are determined by impermeant solutes, while permeant solutes have only transient effects. The osmotic water flow caused by a particular solute depends on the permeability of the membrane to that species: the greater the permeability, the smaller the osmotic water flow.

5. Biological membranes contain transport proteins (transporters) to promote the permeation of various classes of molecules. Facilitated transporters allow the transported substance to equilibrate across the membrane. Active transporters can pump the transported species against a concentration or energy gradient. This requires a linkage to metabolism. Primary active transport proteins have a direct link to metabolism, frequently by consuming ATP. Secondary active transport proteins use the gradient of another substance, frequently Na^+, to power the transport of substances such as sugars and amino acids.

BIBLIOGRAPHY
Journal Articles

Carruthers A: Facilitated diffusion of glucose, *Physiol Rev* 70:1135, 1990.

Christensen HN: Role of amino acid transport and countertransport in nutrition and metabolism, *Physiol Rev* 70:43, 1990.

Fambrough DM: The sodium pump becomes a family, *Trends Neurol Sci* 11:325, 1988.

Finkelstein A: Water movement through membrane channels, *Curr Top Membr Transp* 21:295, 1984.

Griffith JK: Membrane transport proteins: implications of sequence comparisons, *Curr Opin Cell Biol* 4:684, 1992.

Handler JS: Overview of epithelial polarity, *Annu Rev Physiol* 51:729, 1989.

Henderson PJF: The 12-transmembrane helix transporters, *Curr Opin Cell Biol* 5:708, 1993.

Lodish HF: Anion-exchange and glucose transport proteins: structure, function, and distribution, *Harvey Lect* 82:19, 1988.

Mercer RW: Structure of the Na,K-ATPase, *Int Rev Cytol* 137C:139, 1993.

Pedersen PL, Carafoli E: Ion motive ATPases. I. Ubiquity, properties, and significance to cell function, *Trends Biochem Sci* 12:146,1987.

Sachs G, Munson K: Mammalian phosphorylating ion-motive ATPases, *Curr Opin Cell Biol* 3:685, 1991.

Walter A, Gutknecht J: Permeability of small nonelectrolytes through lipid bilayer membranes, *J Membr Biol* 90:207, 1986.

Wright EM, Hager KM, Turk E: Sodium co-transport proteins, *Curr Opin Cell Biol* 4:696, 1992.

Books and Monographs

Andreoli TE et al, eds: *Physiology of membrane disorders,* ed 2, New York, 1986, Plenum Press.

Finean JB, Michell RH, eds: *Membrane structure,* New York, 1981, Elsevier/North-Holland Biomedical Press.

Finkelstein A: *Water movement through lipid bilayers, pores, and plasma membranes: theory and reality,* New York, 1987, John Wiley.

Kaplan JH, De Weer P, eds: *The sodium pump: structure, mechanism, and regulation— 44th Symposium of the Society of General Physiologists,* New York, 1990, Rockefeller Press.

Kotyk A, Janacek K, Koryta J: *Biophysical chemistry of membrane functions,* New York, 1988, Wiley Interscience.

Läuger P: *Electrogenic ion pumps,* Sunderland, Mass, 1991, Sinauer Associates.

Martonosi AN, ed: *The enzymes of biological membranes,* ed 2, New York, 1985, Plenum Press.

Stein WH: *Channels, carriers, and pumps: an introduction to membrane transport,* San Diego, 1990, Academic Press.

CHAPTER 2

Ionic Equilibria and Resting Membrane Potentials

Most animal cells have an electrical potential difference (voltage) across their plasma membranes. *The cytoplasm is usually electrically negative relative to the extracellular fluid.* The electrical potential difference across the plasma membrane in a resting cell is called the **resting membrane potential.** The resting membrane potential plays a central role in the excitability of nerve and muscle cells and in certain other cellular responses.

IONIC EQUILIBRIA

Electrochemical Potentials of Ions

A membrane separates aqueous solutions in two chambers (A and B). The ion X^+ is at a higher concentration on side A than on side B (Figure 2-1). If no electrical potential difference exists between side A and side B, X^+ will tend to diffuse from side A to side B, just as if it were an uncharged molecule. If, however, side A is electrically negative with respect to side B, the situation is more complex. The tendency of X^+ to diffuse from side A to side B because of the concentration difference remains, but now X^+ also tends to move in the opposite direction (from B to A) because of the electrical potential difference across the membrane. The direction of net X^+ movement depends on whether the effect of the concentration difference or the effect of the electrical potential difference is larger. By comparing the two tendencies—concentration and electrical—one can predict the direction of net X^+ movement.

The quantity that allows us to compare the relative contributions of ionic concentration and electrical potential is called the **electrochemical potential (μ)** of an ion. The electrochemical potential difference of X^+ across the membrane is defined as

$$\Delta\mu(X) = \mu_A(X) - \mu_B(X) = RT\ln\frac{[X]_A}{[X]_B} + zF(E_A - E_B) \quad (1)$$

where

$\Delta\mu$ = Electrochemical potential difference of the ion between sides A and B of the membrane

R = Ideal gas constant

T = Absolute temperature

$\ln\dfrac{[X]_A}{[X]_B}$ = Natural logarithm of concentration ratio of X^+ on the two sides of the membrane

z = Charge number of the ion ($+2$ for Ca^{++}, -1 for Cl^-, etc.)

F = Faraday's number

$E_A - E_B$ = Electrical potential difference across the membrane

The first term, $RT\ln\dfrac{[X]_A}{[X]_B}$, on the right-hand side of equation 1 is the tendency for X^+ ions to move from A to B *because of the concentration difference,* and the second term, $zF(E_A - E_B)$, is the tendency for the ions to move from A to B *because of the electrical potential difference.* The first term represents the potential energy difference between a mole of X^+ ions on side A and a mole of X^+ ions on side B as a result of the concentration difference. The second term represents the potential energy difference between a mole of X^+ ions on side A and a mole of X^+ ions on side B caused by the electrical potential difference between A and B. Thus $\Delta\mu(X)$ describes the difference that exists in potential energy between a mole of X^+ ions on side A and a mole of X^+ ions on side B and that results from both concentration and electrical potential differences; hence the name **electrochemical potential difference.** The unit of electrochemical potential, and of both terms on the right hand side of equation 1, is energy/mole.

The X^+ ions will tend to move spontaneously from higher to lower electrochemical potential. We defined $\Delta\mu$ as the electrochemical potential of the ion on side A minus that on side B. If $\Delta\mu$ is positive, the ions will

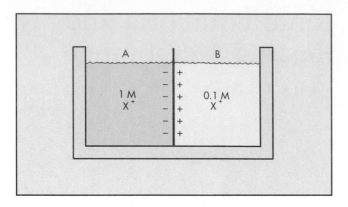

FIGURE 2-1 X^+ is present at 1 M in chamber A and at 0.1 M in chamber B. A concentration force for X^+ tends to cause X^+ to flow from A to B. However, chamber A is electrically negative with respect to chamber B, so an electrical force tends to cause X^+ to flow from B to A.

tend to move from A to B; if $\Delta\mu$ is zero, there is no net tendency for the ions to move at all; and if $\Delta\mu$ is negative, the ions will tend to move from side B to side A.

If μ_A is greater than μ_B, ions will tend to flow spontaneously from side A to side B. To cause ions to flow from B to A, work must be done. Specifically, $\mu_A - \mu_B$ is the minimal amount of work that must be done to cause 1 mole of ions to flow from B to A. When ions flow from A to B, on the other hand, energy is dissipated. In fact, this energy can be harnessed to perform work. The maximal amount of work that can be done by 1 mole of ions flowing from A to B is $\mu_A - \mu_B$. *An electrochemical potential difference of an ion across a membrane thus represents potential energy that can be used to perform work.*

What sort of work can be done by the electrochemical potential energy stored in an ion gradient? In Chapter 1 we mentioned that the electrochemical potential of the Na^+ gradient is used to power the secondary active transport of sugars and amino acids. In mitochondria, the action of the electron transport enzymes creates an electrochemical potential gradient of H^+ across the mitochondrial inner membrane. The H^+ ions flow back into the mitochondrial matrix via the ATP synthase enzyme complex in the mitochondrial inner membrane. The ATP synthase uses the energy released by the H^+ ions to drive the synthesis of ATP. Drugs, such as the poison dinitrophenol, that increase the permeability of the mitochondrial inner membrane to H^+ collapse the H^+ gradient and prevent the synthesis of ATP.

Electrochemical Equilibrium and the Nernst Equation

In equation 1, $\Delta\mu$ may be thought of as the net force on the ion, whereas $RT\ln\dfrac{[X]_A}{[X]_B}$ is the force caused by

the concentration difference, and $zF(E_A - E_B)$ is the force caused by the electrical potential difference. When the two forces are equal and opposite, $\Delta\mu = 0$, and there is no net force on the ion. When there is no net force on the ion, no net movement of the ion will occur, and the ion is said to be in **electrochemical equilibrium** across the membrane. At equilibrium $\Delta\mu = 0$. From equation 1, therefore, at equilibrium:

$$RT\ln\frac{[X]_A}{[X]_B} + zF(E_A - E_B) = 0 \qquad \textbf{(2)}$$

Solving for E_A-E_B, we obtain

$$E_A - E_B = -\frac{RT}{zF}\ln\frac{[X]_A}{[X]_B} = \frac{RT}{zF}\ln\frac{[X]_B}{[X]_A} \qquad \textbf{(3)}$$

Equation 3 is called the **Nernst equation.** The condition of equilibrium was assumed in its derivation, and *the Nernst equation will be satisfied only for ions at equilibrium.* It allows one to compute the electrical potential difference, $E_A - E_B$, required to produce an electrical force, $zF(E_A - E_B)$, that is equal and opposite to the concentration force, $\dfrac{RT}{zF}\ln\dfrac{[X]_A}{[X]_B}$.

Use of Nernst Equation It is often convenient to convert the Nernst equation to a form involving logarithm to the base 10 (log) rather than natural logarithms (ln). The formula for this conversion is $\ln x = 2.303 \log x$. Because biological potentials are usually expressed in millivolts (mV), the units of R may be selected so that RT/F comes out in millivolts. At 29.2° C the quantity 2.303 RT/F is equal to 60 mV. Because this quantity is proportional to the absolute temperature, it changes by approximately 1/273 (0.36%) for each centigrade degree. Thus the value of 60 mV for 2.303 RT/F holds approximately for most experimental conditions in biology, and a useful form of the Nernst equation is

$$E_A - E_B = \frac{-60mV}{z}\log\frac{[X]_A}{[X]_B} = \frac{60mV}{z}\log\frac{[X]_B}{[X]_A} \qquad \textbf{(4)}$$

Examples of Uses of Nernst Equation
Example 1 In Figure 2-2, K^+ is 10 times more concentrated in chamber A than in chamber B. Following is a calculation of the electrical potential difference that must exist between the chambers for K^+ to be in equilibrium across the membrane. Because we have specified that K^+ should be in equilibrium, the Nernst equation will hold.

$$E_A - E_B = \frac{-60mV}{+1}\log\frac{[K^+]_A}{[K^+]_B} = -(60mV)\log\frac{0.1}{0.01} \qquad \textbf{(5)}$$

$$= -60mV \log(10) = -60\ mV$$

The Nernst equation tells us that at equilibrium, side A must be 60 mV negative relative to side B. We can see

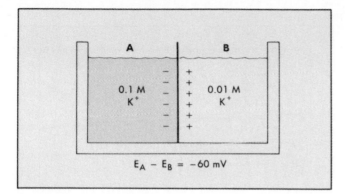

FIGURE 2-2 A membrane separates chambers containing different K^+ concentrations. At an electrical potential difference $(E_A - E_B)$ of -60 mV, K^+ is in electrochemical equilibrium across the membrane.

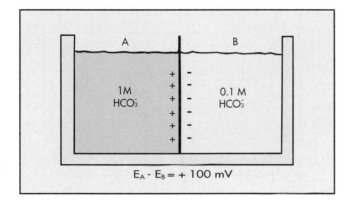

FIGURE 2-3 A membrane separates chambers that contain different HCO_3^- concentrations. $E_A - E_B = +100$ mV. HCO_3^- is not in electrochemical equilibrium. If $E_A - E_B$ were $+60$ mV, HCO_3^- would be in equilibrium. $E_A - E_B$ ($+100$ mV) is stronger than it needs to be ($+60$ mV) to just balance the tendency for HCO_3^- to move from A to B because of its concentration difference. Thus net movement of HCO_3^- from B to A will occur.

that this polarity is correct because K^+ will tend to move from B to A driven by this electrical force, which will counteract the tendency for it to move from A to B because of the concentration difference.

This example shows that *an electrical potential difference of about 60 mV is required to balance a tenfold concentration difference of a univalent ion.* This is a useful rule of thumb.

Example 2 In Figure 2-3 the Nernst equation can help decide whether HCO_3^- is in equilibrium. If HCO_3^- is not in equilibrium, the Nernst equation allows us to predict the direction of net flow of HCO_3^-.

The Nernst equation tells us the electrical potential difference, $E_A - E_B$, that will just balance the concentration difference of HCO_3^- across the membrane.

$$E_A - E_B = \frac{-60\text{mV}}{-1} \log\frac{[HCO_3^-]_A}{[HCO_3^-]_B} = +(60\text{mV})\log\frac{1}{0.1} \quad \textbf{(6)}$$

$$= +60\text{mV} \log(10) = +60\text{mV}$$

Thus a potential difference of $+60$ mV between A and B would just balance the tendency of HCO_3^- to move from A to B because of its concentration difference. However, $E_A - E_B$ is actually $+100$ mV. Therefore the electrical force is in the right direction to balance the concentration force, but it is 40 mV larger than it needs to be to just balance the concentration force. Because the electrical force on HCO_3^- is larger than the concentration force, the electrical force will determine the direction of net HCO_3^- movement. Net HCO_3^- flow will occur from B to A.

In summary, the Nernst equation can be used to predict the direction that ions will tend to flow:

1. If the potential difference measured across a membrane is *equal* to the potential difference calculated from the Nernst equation for a particular ion, then that ion is in *electrochemical equilibrium* across the membrane, and no net flow of that ion will occur across the membrane.

2. If the measured electrical potential is of the same sign as that calculated from the Nernst equation for a particular ion but is *larger* in magnitude than the calculated value, then the electrical force is larger than the concentration force, and net movement of that particular ion will tend to occur in the direction *determined by the electrical force.*

3. When the electrical potential difference is of the same sign but is *numerically* less than that calculated from the Nernst equation for a particular ion, then the concentration force is larger than the electrical force, and net movement of that ion tends to occur in the direction *determined by the concentration difference.*

4. If the electrical potential difference measured across the membrane is of the *opposite sign* to that predicted by the Nernst equation for a particular ion, then the electrical and concentration forces are in the same direction. Thus that ion *cannot be in equilibrium,* and it will tend to flow in the direction determined by both electrical and concentration forces.

Gibbs-Donnan Equilibrium

Cytoplasm typically contains proteins, organic polyphosphates, nucleic acids, and other ionized substances that cannot permeate the plasma membrane. The majority of these impermeant intracellular ions are negatively charged at physiological pH. The steady-state properties of this mixture of permeant and impermeant ions are described by the **Gibbs-Donnan equilibrium.**

As a model of a cell with impermeant anions, consider a membrane separating a solution of KCl from a solution of KY, where Y^- is an anion to which the membrane is completely impermeable (Figure 2-4, *top*). The membrane is permeable to water, K^+, and Cl^-. Suppose that initially chamber A contains a 0.1 M solution of KY and that chamber B contains an equal volume of 0.1 M KCl. Because $[Cl^-]_B$ exceeds $[Cl^-]_A$, there will be a net flow of Cl^- from chamber B to chamber A. Negatively charged Cl^- ions flowing from side B to side A will create an electrical potential difference (side A negative) that will cause K^+ also to flow from side B to side A. Given enough time, K^+ and Cl^- will come to equilibrium. At equilibrium both $\Delta\mu_{K^+}$ and $\Delta\mu_{Cl^-}$ must equal zero. When both K^+ and Cl^- are at equilibrium,

$$[K^+]_A[Cl^-]_A = [K^+]_B[Cl^-]_B \qquad (7)$$

Equation 7 is called the **Donnan relation** or the **Gibbs-Donnan equation,** and it holds for any pair of univalent cation and anion in equilibrium between the two chambers. If other univalent ions that could attain an equilibrium distribution were present, the same reasoning and an equation similar to equation 7 would apply to each cation-anion pair of them as well.

For the model situation we are considering, application of the Gibbs-Donnan equation will result in the final concentrations shown in Figure 2-4, *bottom.*

In this Gibbs-Donnan equilibrium both K^+ and Cl^- (but not Y^-) are in electrochemical equilibrium. This means that both K^+ and Cl^- must satisfy the Nernst equation, so that the equilibrium transmembrane potential difference can be computed from the Nernst equation for either K^+ or Cl^-. Applying the Nernst equation to either K^+ or Cl^- will result in

$$E_A - E_B = -60mV \log(2) = -18 \ mV$$

The presence of the impermeant Y^- anions results in a negative electrical potential in the chamber that contains them. In this way the impermeant anions in the cytoplasm contribute on the order of -10 mV to the resting membrane potential of the cytoplasm relative to the extracellular fluid.

Note that only the permeant ions (K^+ and Cl^- in this example) attain equilibrium. The impermeant anion, Y^-, cannot reach an equilibrium distribution. It may not be evident that water also will not achieve equilibrium, unless provision is made for that to occur. The sum of the concentrations of K^+ and Cl^- ions on side A in the preceding example exceeds that on side B. This is a general property of Gibbs-Donnan equilibria. Taking the impermeant Y^- into account as well, the total concentration of osmotically active ions is considerably greater on

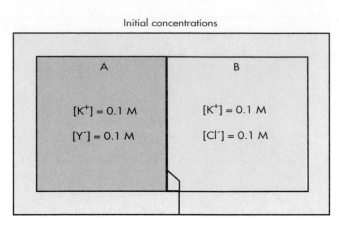

Initial concentrations

Membrane permeable to H_2O, K^+, and Cl^- but impermeable to Y^-

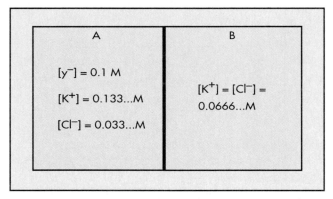

Equilibrium concentrations

FIGURE 2-4   *Top,* Before a Gibbs-Donnan equilibrium is established, a membrane separates two aqueous compartments. The membrane is permeable to water, K^+, and Cl^- but impermeable to Y^-. *Bottom,* Ion concentrations after Gibbs-Donnan equilibrium has been attained.

side A than on side B. Water will tend to flow by osmosis from side B to side A until the total osmotic pressure of the two solutions is equal. However, ions will then flow to set up a new Gibbs-Donnan equilibrium, and that requires there be more osmotically active ions on the side with Y. All the water from side B will end up on side A unless water is restrained from moving.

This can be done by enclosing the solution on side A in a rigid container (Figure 2-5). Then, as fluid flows from side B to side A, pressure will build up in chamber A and that pressure will oppose further osmotic water flow. The pressure in chamber A at equilibrium will be equal to the difference between the total osmotic pressures of the solutions in chambers A and B. The rigid cell wall of plant cells allows turgor pressure to build up in the cell and to partly compensate for the osmotic effects of the Gibbs-Donnan equilibrium. Left to its own devices, the Gibbs-Donnan equilibrium will result in an

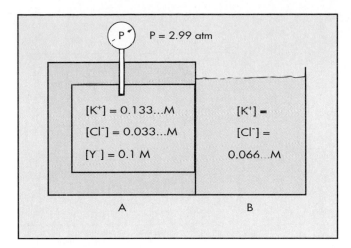

FIGURE 2-5 A hydrostatic pressure of 2.99 atmospheres is required to prevent water from flowing from chamber B to chamber A in the Gibbs-Donnan equilibrium in Figure 2-4. This 2.99 atmosphere is equal to the osmotic pressure in chamber A minus that in chamber B.

The plasma membranes of red blood cells from patients with **hereditary spherocytosis** (HS) are about three times more permeable to Na^+ than red cells from normal individuals. The level of Na^+-K^+-ATPase in the erythrocyte membranes of hereditary spherocytosis patients is also substantially elevated. When HS red cells have sufficient glucose to maintain normal ATP levels, they extrude Na^+ as rapidly as it diffuses into the cell cytosol, and the red cell volume is maintained. However, when HS erythrocytes are delayed in the venous sinuses of the spleen, where glucose and ATP are present at low levels, the intracellular ATP concentrations fall, and Na^+ cannot be pumped out by the Na^+-K^+-ATPase as rapidly as it enters, and the red cells swell. The swollen erythrocytes are prone to being destroyed by the spleen and, as a consequence, HS patients become anemic.

osmotic pressure in the cytoplasm that is in excess of that in the extracellular fluid. This poses a threat to the maintenance of the normal cellular volume. Animal cells do not have cell walls and have thus evolved other ways that involve ion transport processes to deal with the osmotic consequences of the Gibbs-Donnan equilibrium.

Regulation of Cell Volume

Both K^+ and Cl^- are nearly in equilibrium across many plasma membranes, and their distribution is influenced by the predominantly negatively charged impermeant ions, such as proteins and nucleotides, in the cytoplasm. This being the case, why does the osmotic imbalance discussed previously not cause the cells to swell and finally burst? One reason is that *cells actively pump Na^+ out of the cytoplasm to the extracellular fluid. The extrusion of Na^+ decreases the osmotic pressure of the cytoplasm and increases that of the extracellular fluid.* Much of the pumping of Na^+ is done by the Na^+ pump (i.e., the Na^+-K^+-ATPase) in the plasma membrane. The Na^+-K^+-ATPase splits an ATP and uses some of the energy released to extrude 3 Na^+ from the cytoplasm and to pump 2 K^+ into the cell. Whereas K^+ is only slightly removed from an equilibrium distribution, Na^+ is pumped out against a large electrochemical potential difference.

When the ATP production of a cell is compromised (in the presence of metabolic inhibitors or low O_2 levels), or when the Na^+-K^+-ATPase is specifically inhibited, Na^+ enters the cell more rapidly than it can be pumped out. As a result, the cell swells.

RESTING MEMBRANE POTENTIALS

Communication between nerve cells depends on an electrical disturbance, called an **action potential,** that is propagated in the plasma membrane of the nerve cell. In striated muscle an action potential propagates rapidly over the entire cell surface and allows the cell to contract synchronously. The action potential in nerve and muscle cells and the ionic mechanisms that account for its properties are discussed in Chapter 3. All cells that can produce action potentials have sizable resting membrane potentials (cytoplasm negative) across their plasma membranes. Inexcitable cells also have negative resting membrane potentials.

The resting membrane potential of a skeletal muscle cell is about -90 mV. By convention we express membrane potential difference as the voltage in the cytoplasm minus that in the extracellular fluid. A negative value denotes that the cytoplasm is electrically negative relative to the extracellular fluid. The resting membrane potential is necessary for the cell to fire an action potential.

Ions that are actively transported are not in electrochemical equilibrium across the plasma membrane. It is shown later that the flow of ions across the plasma membrane, down their electrochemical potential gradients, is directly responsible for generating much of the resting membrane potential. To understand how the electrochemical potential gradient of an ion can give rise to a transmembrane difference in electrical potential, let us first consider a model system known as a **concentration cell.**

Concentration Cells

In Figure 2-6 the membrane that separates chambers A and B is permeable to cations but not to anions. Initially no electrical potential difference exists across the membrane. K^+ will flow from A to B because of the concentration force acting on it. Cl^- has the same force on it, but it cannot flow because the membrane is impermeable to anions. The flow of K^+ from A to B will transfer net positive charge to side B and leave a very slight excess of negative charges behind on side A. Side A will thus become electrically negative to side B (Figure 2-6). This electrical force is oppositely directed to the concentration force on K^+. The more K^+ that flows, the larger the opposing electrical force. Net K^+ flow will stop when the electrical force just balances the concentration force—that is, when the electrical potential difference is equal to the equilibrium (Nernst) potential for K^+. That is, when

$$E_A - E_B = \frac{-60mV}{+1} \log \frac{[K^+]_A}{[K^+]_B} = -(60mV)\log\frac{0.1}{0.01} = -60mV$$

Only a very small amount of K^+ flows from A to B before equilibrium is reached. This is because the separation of positive and negative charges requires a large amount of work. The potential difference that builds up to oppose further K^+ movement is a manifestation of that work.

The K^+ concentration difference in this example acts like a battery. The natural tendency for any ion that can flow is to seek equilibrium; thus K^+ tends to flow until its equilibrium potential difference is established. As explained later, when more than one type of ion can permeate a membrane, each ion "strives" to make the transmembrane potential difference equal to its equilibrium potential. *The more permeant the ion, the greater its ability to force the electrical potential difference toward its equilibrium potential.*

Distribution of Ions Across Plasma Membranes

In most tissues a number of ions are not in equilibrium between the extracellular fluid and the cytoplasm. Table 2-1 gives the concentrations of Na^+, K^+, and Cl^- in the extracellular fluid and in the cytoplasmic water of frog skeletal muscle. Ion concentrations for mammalian muscle are similar to those for frog muscle.

Cl^- is nearly in equilibrium across the plasma membrane of frog muscle. This is known because chloride's equilibrium potential, as calculated from the Nernst equation, is about equal to the measured transmembrane potential difference. K^+ has a concentration force that tends to make it flow out of the cell. The electrical force

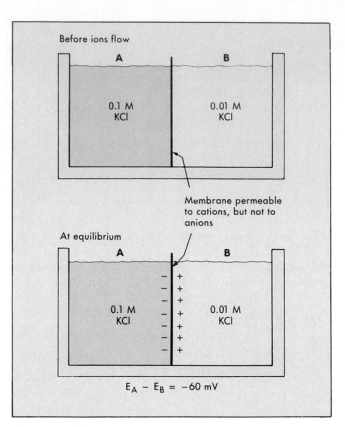

FIGURE 2-6 *Top,* A concentration cell. A membrane, which is permeable to cations but not to anions, separates KCl solutions of different concentrations. *Bottom,* The concentration cell after electrochemical equilibrium has been established. The flow of an infinitesimal amount of K^+ generated an electrical potential difference across the membrane that is equal to the equilibrium potential for K^+.

on K^+ is oppositely directed to the concentration force. If the $E_{in} - E_{out}$ in frog muscle were -105 mV, electrical and concentration forces on K^+ would exactly balance. Because $E_{in} - E_{out}$ is only -90 mV, the concentration force on K^+ is greater than the electrical force. Therefore K^+ has a net tendency to flow out of the cell. Both the concentration and the electrical forces on Na^+ tend to cause it to flow into the cell. Na^+ is the ion farthest from an equilibrium distribution. The larger the difference between the measured membrane potential and the equilibrium potential for an ion, the larger the net force tending to make that ion flow.

Active Ion Pumping and Resting Potential

The Na^+-K^+-ATPase, located in the plasma membrane, uses the energy of the terminal phosphate ester bond of ATP to extrude Na^+ actively from the cell and to take K^+ actively into the cell. The Na^+-K^+ pump is responsible for the high intracellular K^+ concentration and the low intracellular Na^+ concentration. Because the pump moves a larger number of Na^+ ions out than K^+ ions in (3 Na^+ to 2 K^+), it causes a net transfer of positive charge out of the cell and thus contributes to the resting membrane potential. Because it brings about net movement of charge across the membrane, the pump is termed **electrogenic.**

The size of the pump's electrogenic contribution to the resting potential can be estimated by completely inhibiting the pump with a cardiac glycoside, such as **ouabain.** Such studies show that in some cells the electrogenic Na^+-K^+ pump is responsible for a large fraction of the resting potential. In most vertebrate nerve and skeletal muscle cells, however, the direct contribution of the pump to the resting potential is usually small—less than 5 mV. The resting membrane potential in nerve and skeletal muscle results mainly from the diffusion of ions down their electrochemical potential gradients. The ionic gradients are maintained by active ion pumping. In other types of excitable cells, electrogenic pumping of ions may contribute more to the resting membrane potential. In certain smooth muscle cells, for example, the electrogenic effect of the Na^+-K^+ pump is responsible for 20 mV or more of the resting membrane potential.

Cardiac glycosides, such as **digitalis** and related drugs, are able to increase the strength of contraction of the heart (Chapter 18). These compounds inhibit the Na^+-K^+ pump. As a consequence the intracellular level of Na^+ in cardiac cells is elevated. Each contraction of the heart is initiated by an increase in the cytosolic concentration of Ca^{++} (Chapter 17). In order for cardiac muscle to relax, Ca^{++} must be removed from the cytosol. The removal of Ca^{++} from the cytosol of the cardiac cells is accomplished by its being pumped into the sarcoplasmic reticulum (SR) by a Ca^{++}-ATPase in the SR membrane and out across the plasma membrane by a plasma membrane Ca^{++}-ATPase and by Na^+/Ca^{++} exchangers in the plasma membrane. (These ion transporters were described in Chapter 1). In the presence of cardiac glycosides, because of the elevated cytosolic Na^+ concentration, the Na^+/Ca^{++} exchanger is not as effective in extruding Ca^{++} from the cell. Consequently, the Ca^{++}-ATPase can accumulate more Ca^{++} in the SR, so that more Ca^{++} is released from the SR to power the next cardiac contraction, which is stronger than normal because of the greater peak level of Ca^{++} in the cytosol.

Generation of Resting Membrane Potential by Ion Gradients

The earlier discussion of concentration cells shows how an ion gradient can act as a battery. When a number of ions are distributed across a membrane, all being removed from electrochemical equilibrium, each ion will tend to force the transmembrane potential toward its own equilibrium potential, as calculated from the Nernst equation. The more permeable the membrane to a particular ion, the greater strength that ion will have in forcing the membrane potential toward its equilibrium potential. In frog muscle (Table 2-1) the Na^+ concentration difference can be regarded as a battery that tries to make $E_{in} - E_{out}$ equal to +67 mV. The K^+ concentration difference resembles a battery that attempts to make $E_{in} - E_{out}$ equal to −105 mV. The Cl^- concentration difference resembles a battery trying to make $E_{in} - E_{out}$ equal to −90 mV.

Chord Conductance Equation The way in which the interplay of ion gradients creates the resting membrane potential (E_m) is illustrated by a simple mathematical model. If we consider the distribution of K^+, Na^+, and Cl^- across the plasma membrane of a cell, then the following equation predicts the transmembrane potential difference across the membrane:

$$E_m = \frac{g_K}{\Sigma g}E_K + \frac{g_{Na}}{\Sigma g}E_{Na} + \frac{g_{Cl}}{\Sigma g}E_{Cl} \qquad \textbf{(8)}$$

where $\Sigma g = (g_K + g_{Na} + g_{Cl})$ and the g's represent the conductances of the membrane to the ions indicated by the subscripts and the E's represent the equilibrium potentials of the ions denoted by their subscripts. Conductance is the reciprocal of resistance ($g = 1/R$). The more permeable the membrane to a particular ion, the greater the conductance of the membrane to that ion.

Equation 8 is called the **chord conductance equation.** It states that the membrane potential is a weighted average of the equilibrium potentials of all the ions to which the membrane is permeable, in this case K^+, Na^+, and Cl^-. The weighting factor for each ion is the fraction of the total ionic conductance of the membrane (the sum of the individual ionic conductances) that results from the conductance of the ion in question. Note that the sum of the weighting factors for the ions must equal 1, so that if one weighting factor grows larger, the others must become smaller. The chord conductance equation shows that *the greater the conductance of the membrane to a particular ion, the greater the ability of that ion to bring the membrane potential toward the equilibrium potential of that ion.*

For the frog muscle fiber discussed earlier, $E_{in} - E_{out} = -90$ mV. The membrane potential is much closer to E_K (−105 mV) than to E_{Na} (+67 mV), because

Table 2-1 Distribution of Na^+, K^+ and Cl^- Across the Plasma Membranes of Frog Skeletal Muscle				
	Extracellular Fluid (mM)	**Cytoplasm (mM)**	**Approximate Equilibrium Potential (mV)**	**Actual Resting Potential (mV)**
FROG MUSCLE				
$[Na^+]$	120	9.2	+67	
$[K^+]$	2.5	140	−105	
$[Cl^-]$	120	3 to 4	−89 to −96	−90

Data from Katz B: *Nerve, muscle, and synapse,* New York, 1966, McGraw-Hill Book Co.

in the resting cell g_K is larger than g_{Na}. The chord conductance equation predicts that in resting muscle g_K is about 10 times larger than g_{Na}. This has been confirmed by ion flux measurements with radioactive tracers. In other types of excitable cells the relationship between g_K and g_{Na} may be somewhat different. Other ions also may play a role in generating the resting membrane potential. Resting membrane potentials vary from about −10 mV or so in human erythrocytes to around −40 mV in some types of smooth muscle and up to −90 mV or more in vertebrate skeletal muscle and cardiac ventricular cells.

We have seen that K^+ has the largest resting conductance and thus has the largest influence on the resting membrane potential. For this reason changes that occur in the concentration of K^+ in a patient's extracellular fluid will affect the resting membrane potentials of all cells. An increase in extracellular K^+ will partially depolarize cells (decrease the magnitude of the resting membrane potential), whereas a decrease in the level of extracellular K^+ will hyperpolarize cells (increase the magnitude) of the resting membrane potential. Either a depolarization or a hyperpolarization of cardiac cells (see Chapter 17) may lead to cardiac arrhythmias, some of which are life-threatening. **Hypokalemia** (low serum K^+) may be a result of long-term diuretic use. **Hyperkalemia** (elevated serum K^+) occurs in acute renal failure and in a disorder called **hyperkalemic periodic paralysis,** which is characterized by episodes of muscle weakness and flaccid paralysis.

Roles of Na^+-K^+-ATPase in Establishing Resting Membrane Potential: Direct versus Indirect

The Na^+-K^+ pump establishes gradients of Na^+ and K^+ across the plasma membranes of cells. Because the amount of Na^+ pumped out is larger than the amount of K^+ pumped in, the pump transfers net charge across the membrane and in this way contributes *directly* to the resting membrane potential. In vertebrate skeletal and cardiac muscle and in nerve, this electrogenic activity of the pump is directly responsible for only a small fraction of the resting membrane potential. The major portion of the resting membrane potential in these tissues is a result of the diffusion of Na^+ and K^+ down their electrochemical potential gradients, with each ion tending to bring the transmembrane potential toward its own equilibrium potential. This contribution to the resting membrane potential is *indirectly caused by the Na^+-K^+-ATPase.* The relative magnitudes of the direct and indirect contributions of the Na^+-K^+-ATPase to the resting membrane potential vary from one cell type to another.

SUMMARY

1. An ion will tend to flow across a membrane if there is a concentration difference of that ion or an electrical potential difference across the membrane. The electrochemical potential difference ($\Delta\mu$) of an ion across a membrane includes the contributions of both the concentration difference and the electrical potential difference to the tendency of the ion to flow across the membrane.

2. An electrochemical potential difference of an ion across a membrane represents a difference of chemical potential energy. This potential energy difference can be harnessed to do work.

3. An ion that is distributed in equilibrium across a membrane will satisfy the Nernst equation. We can use the Nernst equation to tell whether an ion is in equilibrium or to compute what the electrical potential difference across the membrane would have to be in order for a particular ion to be in equilibrium.

4. Cytoplasm contains an excess of negative ions that are impermeant to the plasma membrane. A permeant univalent ion pair, X^+, Z^-, that can attain

equilibrium across the membrane will satisfy the Gibbs-Donnan equilibrium, which is represented by the relationship: $[X]_{in}[Z]_{in} = [X]_{out}[Z]_{out}$, where in and out refer to cytoplasm and extracellular fluid, respectively.

5. All cells have a negative resting membrane potential; that is, the cytoplasm is electrically negative relative to the extracellular fluid. The diffusion of ions across the plasma membrane and down their electrochemical potential gradients contributes to the resting membrane potential.

6. The flow of each ion across the plasma membrane tends to bring the resting membrane potential toward the equilibrium potential for that ion. The more conductive the membrane to a particular ion, the greater the ability of that ion to bring the membrane potential toward its equilibrium potential. This is described by the chord conductance equation.

7. Three processes contribute to generating the resting membrane potential: (a) ionic diffusion as just described (major), (b) the electrogenic effect of the Na^+-K^+-ATPase (variable in importance), and (c) the Gibbs-Donnan equilibrium (minor in excitable cells).

BIBLIOGRAPHY
Books and Monographs

Aidley DJ: *The physiology of excitable cells,* ed 3, Cambridge, 1990, Cambridge University Press.

Hille B: *Ion channels of excitable membranes,* ed 2, Sunderland, Mass, 1992, Sinauer Associates.

Junge D: *Nerve and muscle excitation,* ed 2, Sunderland, Mass, 1981, Sinauer Associates.

Kandel ER, Schwartz JH, Jessell, TM: *Principles of neural science,* ed 3, New York, 1991, Elsevier Science Publishing Co, Inc.

Katz B: *Nerve, muscle, and synapse,* New York, 1966, McGraw-Hill Book Co.

Keynes RD, Aidley DJ: *Nerve and muscle,* ed 2, New York, 1991, Cambridge University Press.

Läuger P: *Electrogenic ion pumps,* Sunderland, Mass, 1991, Sinauer Associates.

Nicholls JG, Martin AR, Wallace BG: *From neuron to brain,* ed 3, Sunderland, Mass, 1992, Sinauer Associates.

Shepherd GM: *Neurobiology,* ed 2, New York, 1988, Oxford University Press.

Generation and Conduction of Action Potentials

ACTION POTENTIALS IN DIFFERENT TISSUES

An **action potential** is a rapid change in the membrane potential followed by a return to the resting membrane potential (Figure 3-1). The size and shape of action potentials differ considerably from one excitable tissue to another. An action potential is propagated with the same shape and size along the whole length of a nerve or muscle cell. The action potential is the basis of the signal-carrying ability of nerve cells. An action potential allows all parts of a long muscle cell to contract almost simultaneously. Voltage-dependent ion channel proteins in the plasma membrane are responsible for action potentials. Different action potentials in the cell types shown in Figure 3-1 occur because these cells have different populations of voltage-dependent ion channels.

MEMBRANE POTENTIALS

Observations of Membrane Potentials

Our knowledge of the ionic mechanisms of action potentials was first obtained from experiments on the squid giant axon. The large diameter (up to 0.5 mm) of the squid giant axon makes it a convenient model for electrophysiological research with intracellular electrodes. The frog sartorius muscle is another useful preparation.

If the plasma membrane of a single muscle cell of a frog sartorius muscle is penetrated by a microelectrode (tip diameter less than 0.5 μm), a potential difference is observed between the microelectrode whose tip is inside the cell and an extracellular electrode. The internal electrode is about 90 mV negative with respect to the external electrode. This 90 mV potential difference is the **resting membrane potential** of the muscle fiber. In the absence of perturbing influences, the resting membrane potential remains at 90 mV.

Subthreshold Responses: the Local Response

Figure 3-2 illustrates the results of an experiment in which the membrane potential of an axon of a shore crab is perturbed by passing rectangular pulses of current across the plasma membrane. Current pulses are depolarizing or hyperpolarizing, depending on the direction of current flow. The terms **depolarizing** and **hyperpolarizing** may be confusing. A change in the membrane potential from −90 mV to −70 mV is a depolarization because it is a *decrease* in the potential difference, or polarization, across the cell membrane. If the membrane potential changes from −90 mV to −100 mV, the polarization of the membrane has *increased;* this is hyperpolarization.

The larger the current passed, the larger the perturbation of the membrane potential. As shown in Figure 3-2, in response to depolarizing current pulses above a certain **threshold** strength, the cell fires an action potential.

When subthreshold current pulses are passed, the size of the potential change observed depends on the distance of the recording electrode from the point of current passage (Figure 3-3, *A*). *The closer the recording electrode to the site of current passage, the larger is the potential change observed.* The size of the potential change is found to decrease exponentially with distance from the site of current passage (Figure 3-3, *B*). The response is said to be *conducted with decrement.* The distance over which the potential change decreases to 1/e (37%) of its maximal value is called the **length constant** (or space constant). (e is the base of natural logarithms and is equal to 2.7182.) *A length constant of 1 to 3 mm is typical for mammalian nerve or muscle cells.* Because these potential changes are observed primarily near the

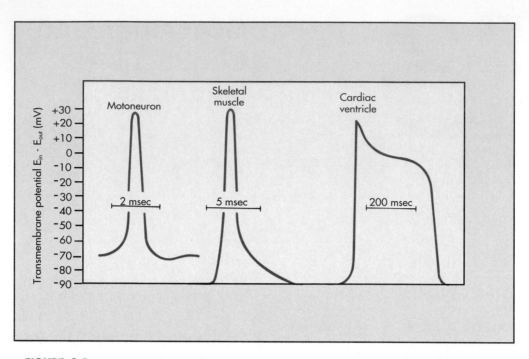

FIGURE 3-1 Action potentials from three vertebrate cell types. Note the different time scales. (Redrawn from Flickinger CJ et al: *Medical cell biology,* Philadelphia, 1979, WB Saunders Co.)

site of current passage and the changes are not propagated along the length of the cell (as are action potentials), they are called **local responses.**

Action Potentials

If progressively larger depolarizing current pulses are applied, a condition is reached at which a different sort of response, the action potential, occurs (Figures 3-2 and 3-4). *An action potential is triggered when the depolarization is sufficient for the membrane potential to reach a threshold value,* which is near -55 mV for squid giant axon. The action potential differs from the local response in two important ways: (1) it is a much larger response, with the polarity of the membrane potential actually reversing (the cell interior becoming positive with respect to the exterior); and (2) the action potential is propagated without decrement down the entire length of the nerve or muscle fiber. *The size and shape of an action potential remain the same as it travels along the fiber;* it does not decrease in size with distance, unlike the local response. When a stimulus larger than the threshold stimulus is applied, the size and shape of the action potential do not change; the size of the action potential does not increase with increased stimulus strength. A stimulus either fails to elicit an action potential (a subthreshold stimulus), or it produces a full-sized action potential.

For this reason the action potential is an **all-or-none response.**

IONIC MECHANISMS OF ACTION POTENTIALS

Action Potentials in Squid Giant Axon

The form of an action potential of a squid giant axon is shown in Figure 3-4. Once the membrane is depolarized to the threshold, an explosive depolarization occurs, which completely depolarizes the membrane and even overshoots, so that the membrane becomes polarized in the reverse direction. The peak of the action potential reaches about $+50$ mV. The membrane potential then returns toward the resting membrane potential almost as rapidly as it was depolarized. After repolarization, a transient hyperpolarization occurs that is known as the **hyperpolarizing afterpotential.** It persists for about 4 msec. The following section discusses the ionic currents that cause the phases of the action potential.

Ionic Mechanism of Action Potential in Squid Giant Axon In Chapter 2 the resting membrane potential was seen to be a weighted sum of the equilibrium potentials for Na^+, K^+, Cl^-, etc. The weighting factor for each ion is the fraction that its conductance contributes to the total ionic conductance of the membrane (the

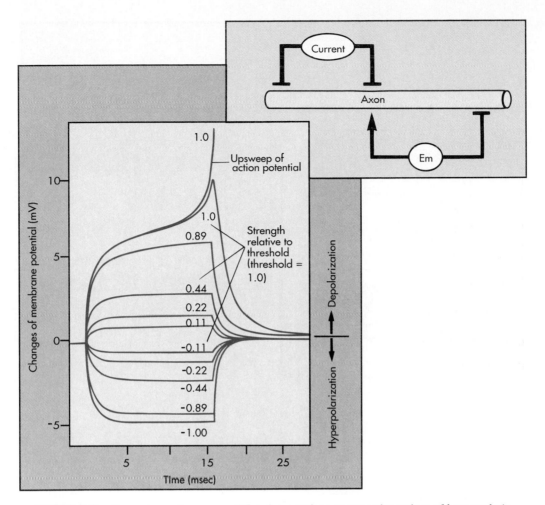

FIGURE 3-2   Responses of an axon of the shore crab to rectangular pulses of hyperpolarizing or depolarizing current. The change in membrane potential as recorded by an extracellular electrode is shown as a function of time. The numbers on the curves give the strength of the current relative to threshold. Note that when stimulated to threshold, the axon sometimes fires an action potential, but sometimes does not. This is because the threshold represents a metastable state where the membrane potential is delicately poised. Hence when the cell is stimulated just to the threshold, the cell may remain at the threshold level of depolarization for a variable time period, after which it may return to the resting potential or may fire an action potential. (Redrawn from Hodgkin AL, Rushton WAH: *Proc R Soc* B133:97, 1946.)

chord conductance equation, equation 8 in Chapter 2). In squid giant axon the resting membrane potential (E_m) is about -70 mV. E_K is about -100 mV in squid axon, so an increase in g_K would hyperpolarize the membrane, and a decrease in g_K would tend to depolarize the membrane. E_{Cl} is about -70 mV, so an increase in g_{Cl} would stabilize E_m at -70 mV. An increase in g_{Na} of sufficient magnitude would cause depolarization and reversal of the membrane polarity, because E_{Na} is about $+65$ in squid giant axon.

In the 1950s Hodgkin and Huxley showed that the action potential of squid giant axon is caused by successive increases in conductance to sodium and potassium ions. They found that the conductance to Na^+, g_{Na}, increases very rapidly during the early part of the action potential (Figure 3-5). The sodium conductance reaches a peak about the same time as the peak of the action potential; then it decreases rapidly. The potassium conductance, g_K, increases more slowly, reaches a peak at about the middle of the repolarization phase, and then returns more slowly to resting levels.

As described in Chapter 2, the chord conductance equation shows that the membrane potential is a result of the opposing tendencies of the K^+ gradient to bring E_m toward the equilibrium potential for K^+ and the Na^+ gradient to bring E_m toward the equilibrium potential for

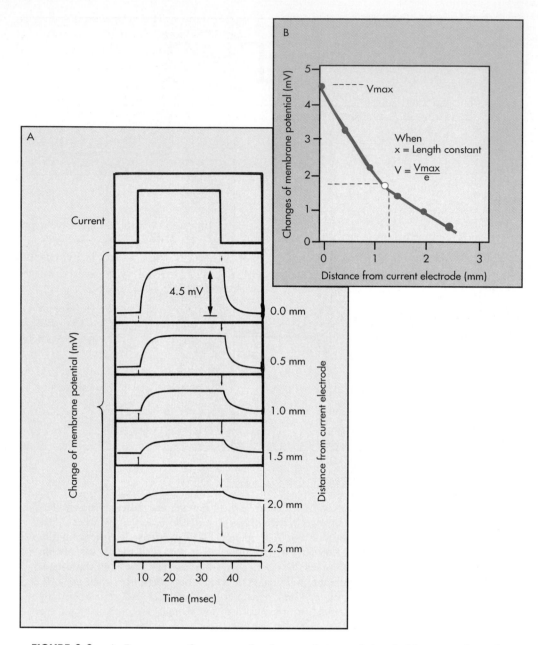

FIGURE 3-3 **A,** Responses of an axon of a shore crab to a subthreshold rectangular pulse of current recorded extracellularly by an electrode located different distances from the current-passing electrode. As the recording electrode is moved farther from the point of stimulation, the response of the membrane potential is slower and smaller. **B,** The maximal change in membrane potential from **A** is plotted versus distance from the point of current passage. The distance over which the response falls to 1/e (37%) of the maximal response is called the length constant. (**A** redrawn from Hodgkin AL, Rushton WAH: *Proc R Soc* B133:97, 1946.)

Na$^+$. Increasing the conductance of either ion will increase its ability to pull E_m toward its equilibrium potential. The rapid increase in g_{Na} during the early part of the action potential causes the membrane potential to move toward the equilibrium potential for Na$^+$(+65 mV). The peak of the action potential reaches only about +50 mV because g_{Na} quickly decreases toward resting

levels and because g_K increases to provide an opposing tendency to the depolarization.

The rapid return of the membrane potential toward the resting potential is caused by the rapid decrease in g_{Na} and the continued increase in g_K. These conductance changes decrease the size of the Na$^+$ term in the chord conductance equation and increase the size of the

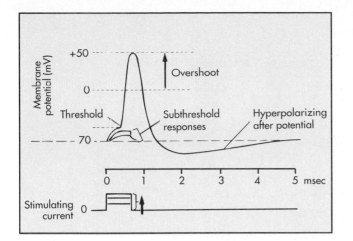

FIGURE 3-4   Responses of the membrane potential of a squid giant axon to increasing pulses of depolarizing current. When the cell is depolarized to threshold, it fires an action potential.

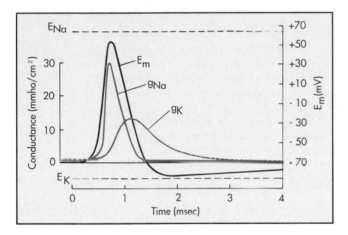

FIGURE 3-5   The action potential (E_m) of a squid giant axon is shown on the same time scale with the associated changes in the conductance of the axon membrane to sodium and potassium ions. (Redrawn from Hodgkin AL, Huxley AF: *J Physiol* 117:500, 1952.)

K^+ term. During the hyperpolarizing afterpotential, when the membrane potential is actually more negative than the resting potential (more polarized), g_{Na} has returned to baseline levels, but g_K remains elevated above resting levels. Thus E_m is pulled closer to the K^+ equilibrium potential (-100 mV) as long as g_K remains elevated.

Ion Channels and Gates

Hodgkin and Huxley proposed that the ion currents pass through separate Na^+ and K^+ channels, each with distinct characteristics, in the plasma membrane. Subsequent research has supported this interpretation and has determined some of the properties of proteins that form

the channels. The amino acid sequences of several K^+ and Na^+ channels have been determined, and our knowledge of the structure of ion channels is rapidly expanding (Figure 3-6). Although the three-dimensional structure of the Na^+ channel remains to be determined, its intramembrane domain is known to consist of a number of α-helices that span the membrane and probably surround the ion channel. The Na^+ channel has both an **activation gate** and an **inactivation gate** that account for the changes in g_{Na} during an action potential (Figure 3-6). Groups of charged amino acid residues that form the activation and inactivation gates have been tentatively identified.

To enter the channel's narrowest part, known as the **selectivity filter,** it is believed that K^+ and Na^+ must shed most of their waters of hydration. In order to strip a K^+ or Na^+ ion of its associated water molecules, negative amino acid residues that line the pore of the channel must have a particular geometry, the precise geometry being different for K^+ than for Na^+. This requirement is believed to confer the specificity of ion channels.

Tetrodotoxin (TTX), one of the most potent poisons known, is a specific blocker of the Na^+ channel. TTX binds to the extracellular side of the sodium channel. **Tetraethylammonium** (TEA^+) blocks the K^+ channel. TEA^+ enters the K^+ channel from the cytoplasmic side, and TEA^+ blocks the channel because it cannot pass through the channel.

The ovaries of certain species of puffer fish, also known as blowfish, contain tetrodotoxin. Raw puffer fish is highly prized in Japan. Connoisseurs of puffer fish enjoy the tingling numbness of the lips that is caused by minuscule quantities of tetrodotoxin present in the flesh. Sushi chefs who are trained to remove the ovaries safely are licensed by the government to prepare puffer fish. Each year several people die from eating improperly prepared puffer fish.

Saxitoxin is another blocker of Na^+ channels. Saxitoxin is produced by reddish-colored dinoflagellates that are responsible for the so-called red tide. Shellfish eat the dinoflagellates and concentrate saxitoxin in their tissues. A person who eats these shellfish may experience life-threatening paralysis 30 min after the meal.

Behavior of Individual Ion Channels

It is possible to study the behavior of individual ion channels. One way to do this is to incorporate either purified ion channel proteins or bits of membrane into pla-

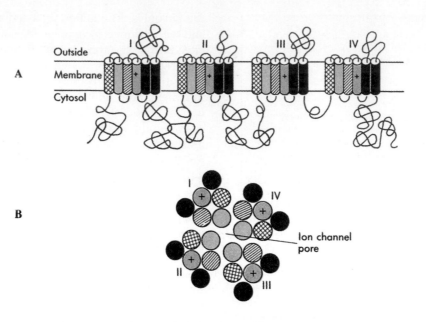

FIGURE 3-6   A model of the voltage-dependent Na$^+$ channel protein. **A,** A two-dimensional model of the Na$^+$ channel. The cylinders represent transmembrane α-helices. There are 4 repeats of 6-cylinder domains of homologous α-helices. The S4 helices, marked with + signs, function as voltage-sensors, and movements of these helices are responsible for activation (opening) of the channel. The intracellular loop connecting domains III and IV functions as the inactivation gate: following depolarization, with a slight delay, this loop apparently swings up into the mouth of the channel to block ion conduction. **B,** Top view of this model of the Na$^+$ channel, showing how the central ion channel pore is believed to be lined by one of the helices from each of the 4 domains. (Modified from Noda M et al: *Nature* 312:121, 1984, and from Levitan IB, Kaczmarek LK: *The neuron: cell and molecular biology,* New York, 1991, Oxford University Press.)

nar lipid bilayers that separate two aqueous compartments. Then electrodes placed in the aqueous compartments can be used to monitor or impose currents and voltages across the membrane. Under some conditions only one, or a few, ion channels of a particular type may be present in the planar membrane. The ion channels spontaneously oscillate between two conductance states, an open state and a closed state (Figure 3-7).

Another way to study individual ion channels involves the use of so-called patch electrodes. A fire-polished microelectrode is placed against the surface of a cell, and suction is applied to the electrode. A high-resistance seal is formed around the tip of the electrode. The sealed patch electrode can then be used to monitor the activity of whatever channels happen to be trapped inside the seal. Sometimes the patch trapped inside the electrode contains more than one functional ion channel of a particular type (Figure 3-8).

During an action potential in a skeletal muscle cell, there is a rapid influx of Na$^+$ ions that persists for only about a millisecond. The time course of this inward Na$^+$ current resembles the time course of the change in the Na$^+$ conductance shown in Figure 3-5. The overall Na$^+$ current that flows into the muscle cell is caused by the opening of thousands of Na$^+$ channels in response to the depolarization. By contrast, the behavior of each individual Na$^+$ channel is random, like the behavior of the channels shown in Figures 3-7 and 3-8. *The probability of each Na$^+$ channel being in the open state is increased when the membrane is depolarized to threshold.* In response to a step depolarization of a muscle cell plasma membrane (Figure 3-9), some of the Na$^+$ channels show an opening event, some do not open at all, and some open more than once. When the currents of a large number of channels are averaged (Figure 3-9, tracing B), the resulting behavior resembles the macroscopic behavior of Na$^+$ channels discussed earlier. That is, the "average channel" opens (activates) promptly in response to depolarization; then after a short time delay, the channel closes (inactivates), even though the applied depolarization is maintained.

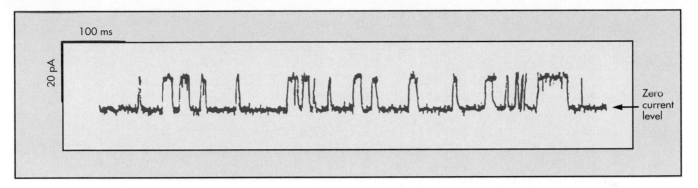

FIGURE 3-7   Ionic current through a single ion channel from rat muscle incorporated into a planar lipid bilayer membrane. The channel opens and closes spontaneously. The fraction of time this channel spends in the open state is a function of calcium ion concentration and membrane potential. (Reproduced from Moczydlowski E, Latorre R: *J Gen Physiol* 82:511-542,1983.)

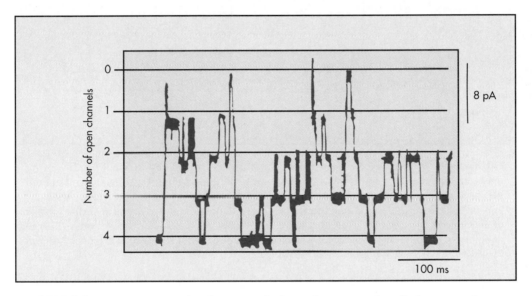

FIGURE 3-8   A current recording from a patch electrode on a muscle cell plasma membrane. The four different current levels show that this particular patch of membrane contains four different ion channels, each opening and closing independently of the others. (Redrawn from Hammill OP et al: *Pflugers Arch* 391:85, 1981.)

Action Potentials in Cardiac and Smooth Muscle

Cardiac Muscle An action potential in a cardiac ventricular cell is schematically shown in Figure 3-1. The initial rapid depolarization and overshoot are caused by the rapid entry of Na^+ through channels that are very similar to the Na^+ channels of nerve and skeletal muscle. After the initial depolarization and overshoot, the cardiac ventricular action potential has a plateau phase. The plateau is caused by another set of channels that are distinct from the fast Na^+ channels. These channels open and close much more slowly than the fast Na^+ channels

and are sometimes called slow channels. The slow channels belong to a particular class of Ca^{++} channels called **L-type Ca^{++} channels** (for long lasting). The Ca^{++} that enters the cell via the L-type Ca^{++} channels during the plateau phase helps to initiate contraction of the ventricular cell and stimulates release of more Ca^{++} from the sarcoplasmic reticulum of the heart cell. The repolarization of the ventricular cell is brought about by the closing of the L-type Ca^{++} channels and by a much delayed opening of K^+ channels. The ionic mechanisms of cardiac action potentials are discussed in more detail in Chapter 17.

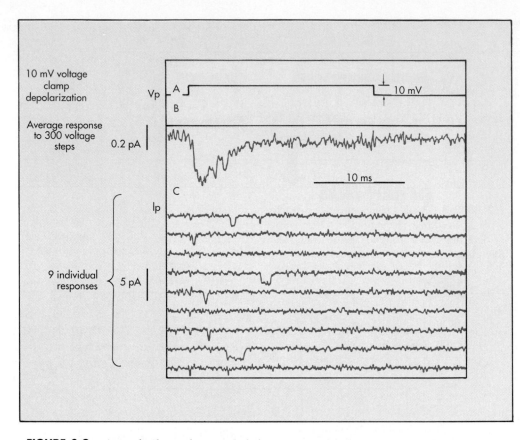

FIGURE 3-9 A patch electrode recorded the currents that flowed in small patches of rat muscle membrane in response to a 10 mV depolarization (tracing *A*). Tetraethylammonium was used to block potassium channels that might have been present in the patch. The tracings in *C* show responses to nine individual 10 mV depolarizations. Tracing *B* is the average of 300 individual responses. Note that this average response resembles the response of large numbers of sodium channels, as seen in measurement of whole-cell Na^+ currents. (Redrawn from Sigworth FJ, Neher E: *Nature* 287:447-449, 1980.)

Smooth Muscle Action potentials vary considerably among different types of smooth muscle (see Chapter 14). Characteristically, action potentials in smooth muscle have slower rates of depolarization and repolarization and less overshoot than skeletal muscle action potentials. Most smooth muscle cells lack Na^+ channels. The depolarizing phase of smooth muscle action potentials is caused primarily by Ca^{++} channels, like those that contribute to the plateau phase in cardiac cells. These channels open and close slowly. The Ca^{++} that enters via these channels is often vital for excitation-contraction coupling in smooth muscle, because some smooth muscle cells have little sarcoplasmic reticulum. Repolarization is caused by the closing of the slow Ca^{++} channels and by a simultaneous delayed opening of K^+ channels.

PROPERTIES OF ACTION POTENTIALS

Voltage Inactivation

If a neuron or skeletal muscle cell is partially depolarized—for example, by increasing the concentration of K^+ in the extracellular fluid—its action potential has a slower rate of rise and a smaller overshoot than does the action potential of the normally polarized cell. This is a result of two factors: (1) a smaller electrical force driving Na^+ into the depolarized cell and (2) voltage inactivation of some of the Na^+ channels. The increase in g_{Na} in response to a depolarization is self-inactivating; that is, the inactivation gates close soon after the activation gates open. Once the Na^+ channels are inactivated, the membrane must be repolarized toward the normal resting membrane potential before the channels can be reopened. As the membrane potential is restored toward

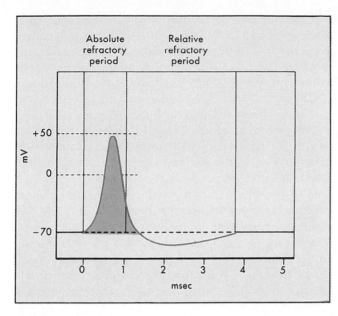

FIGURE 3-10   The action potential of a nerve, illustrating the associated absolute and relative refractory periods.

normal resting levels, more and more of the Na^+ channels again become capable of being activated.

The explosive depolarizing phase of the action potential may be compared to a chemical explosion. A chemical explosion requires a critical mass of material; the spike of the action potential can be generated only if a critical number of Na^+ channels are recruited. When a cell is partly depolarized, the pool of activatable Na^+ channels is reduced; consequently a stimulus may not be able to recruit a sufficient number of Na^+ channels to generate an action potential. This is called **voltage inactivation** of the action potential, which results from voltage inactivation of the Na^+ channels. *Voltage inactivation of Na^+ channels partially accounts for important properties of excitable cells, such as refractory periods and accommodation.*

Refractory Periods

During much of the action potential the membrane is completely refractory to further stimulation. This means that no matter how strongly the cell is stimulated, it is unable to fire a second action potential. This unresponsive state is called the **absolute refractory period** (Figure 3-10). The cell is refractory because a large fraction of its Na^+ channels is voltage inactivated and cannot be reopened until the membrane is repolarized.

During the latter part of the action potential the cell is able to fire a second action potential, but a stronger than normal stimulus is required. This is the **relative refractory period.** Early in the relative refractory period,

before the membrane potential has returned to the resting potential level, some Na^+ channels are voltage inactivated, so a stronger than normal stimulus is required to open the critical number of Na^+ channels needed to trigger an action potential. Throughout the relative refractory period the conductance to K^+ is elevated, which opposes depolarization of the membrane. This also contributes to the refractoriness.

Accommodation

When a nerve or muscle cell is depolarized slowly, the normal threshold may be passed without an action potential being fired; this is called **accommodation.** Na^+ and K^+ channels are both involved in accommodation. During slow depolarization some of the Na^+ channels that are opened by depolarization have enough time to become voltage inactivated before the threshold potential is attained. If depolarization is slow enough, the critical number of open Na^+ channels required to trigger the action potential may never be attained. In addition, K^+ channels open in response to the depolarization. The increased g_K tends to repolarize the membrane, making it still more refractory to depolarization.

In an inherited disorder called **primary hyperkalemic paralysis,** patients suffer episodes of painful spontaneous contractures of muscles followed by periods of paralysis of the affected muscles. These symptoms are accompanied by elevated levels of K^+ in the plasma and extracellular fluid. The elevation of extracellular K^+ causes depolarization of skeletal muscle cells. Initially the depolarization brings muscle cells closer to threshold, so that spontaneous action potentials and contractions are more likely. As depolarization of the cells becomes more marked, the cells accommodate because of the voltage-inactivated Na^+ channels. Thus they become unable to fire action potentials and are unable to contract in response to action potentials in their motor axons.

CONDUCTION OF ACTION POTENTIALS

A principal function of neurons is to transfer information by the conduction of action potentials. The axons of the motor neurons of the ventral horn of the spinal cord conduct action potentials from the cell body of the neuron to a number of skeletal muscle fibers. The distance from the motor neuron to one of the muscle fibers it innervates may be greater than 1 m.

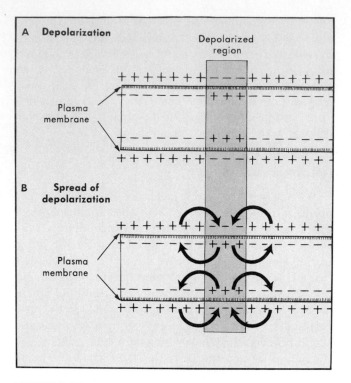

FIGURE 3-11 Mechanism of electrotonic spread of depolarization. **A,** The reversal of membrane polarity that occurs with local depolarization. **B,** The local currents that flow to depolarize adjacent areas of the membrane and allow conduction of the depolarization.

Action potentials are conducted along a nerve or muscle fiber by local current flow, just as occurs in electrotonic conduction of subthreshold potential changes. Thus the same factors that govern the velocity of electrotonic conduction also determine the speed of action potential propagation.

The Local Response: Conduction with Decrement

Figure 3-11, *A,* shows the membrane of an axon or muscle fiber that has been depolarized in a small region. In this region the external surface of the membrane is negative relative to the adjacent membrane, and the internal face of the depolarized membrane is positively charged relative to neighboring internal areas. The potential differences cause local currents to flow (Figure 3-11, *B*), which depolarize the membrane adjacent to the initial site of depolarization. These newly depolarized areas then cause current flows that depolarize other segments of the membrane still farther removed from the initial site of depolarization. This spread of depolarization is called the **local response.** This mechanism of conduction is known as **electrotonic conduction.**

A subthreshold depolarization will be conducted electrotonically, and it will diminish in strength as it moves along the cell. Thus it is conducted with decrement. As shown in Figure 3-3, *B,* an electrotonically conducted signal dies away to 37% of its maximal strength over a distance of one length constant (about 1 to 2 mm) and decreases to almost nothing over about 5 mm.

Determinants of the Length Constant A nerve or muscle fiber has some of the properties of an electrical cable. In a perfect cable the insulation surrounding the core conductor prevents all loss of current to the surrounding medium so that a signal is transmitted along the cable with undiminished strength (Figure 3-12). The plasma membrane of an unmyelinated nerve or muscle fiber serves as the insulation, the cytoplasm being the core conductor. The membrane has a resistance much higher than the resistance of the cytoplasm, but (partly because of its thinness) the plasma membrane is not a perfect insulator. The higher the ratio of R_m to R_{in}, the less current is lost across the plasma membrane, the better the cell can function as a cable, and the longer the distance that a signal can be transmitted electrotonically without significant decrement. R_m/R_{in} determines the length constant of a cell: the length constant is equal to

$$\sqrt{\frac{R_m}{R_{in}}}.$$

Action Potential as Self-Reinforcing Signal

Many nerve and muscle fibers are much longer than their length constants (1 to 2 mm). Skeletal muscle cells can be as long as 1 to 2 cm. Nerve axons can be of the order of 1 m in length. Conduction with decrement will not do for such long cells! The action potential serves to conduct an electrical impulse with undiminished strength along the full length of these cells. To do this, the action potential reinforces itself as it is conducted along the fiber. The action potential may be said to be **propagated,** as well as conducted. The conduction of the action potential occurs via local circuit currents by the electrotonic mechanism depicted in Figure 3-11. When the areas on either side of the depolarized region reach threshold, these areas also fire action potentials, which locally reverses the polarity of the membrane potential. By local current flow, the areas of the fiber adjacent to these areas are next brought to threshold, and then these areas in turn fire action potentials. A cycle of depolarization occurs by local current flow followed by generation of an action potential in a restricted region that then is conducted along the length of the fiber, with "new" action potentials being generated as they spread. In this way the action potentials are regenerated as they spread, and the action potential propagates over long distances, keeping the same size and shape.

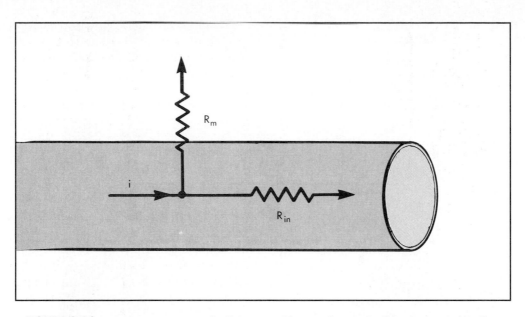

FIGURE 3-12   An axon or a muscle fiber resembles an electrical cable. Currents that flow across the membrane resistance (R_m) are lost from the cable. Currents that flow through the longitudinal resistance (R_{in}) carry the electrical signal along the cable. The larger the ratio R_m/R_{in}, the more efficient is signal transmission along the fiber.

Because the shape and size of the action potential are usually invariant, *only variations in the frequency of the action potentials can be used in the code for information transmission along axons.* The maximum frequency is limited by the duration of the absolute refractory period (about 1 msec) to about 1000 impulses/second in large mammalian nerves.

Conduction Velocity

The speed of electrotonic conduction along a nerve or muscle fiber is determined by the electrical properties of the cytoplasm and of the plasma membrane that surrounds the fiber. The same electrical properties determine the velocity of propagation of an action potential. The following discussion focuses on the mechanism of electrotonic conduction, but it applies equally well to the mechanism of propagation of the action potential.

Fibers that are larger in diameter have a greater conduction velocity. This is principally caused by the decrease in resistance to conduction in the cytoplasm along the length of the fiber as the radius (and hence the cross-sectional area) of the fiber increases.

Effect of Myelination on Conduction In vertebrates certain nerve fibers are coated with **myelin;** such fibers are said to be **myelinated.** Myelin is formed from multiple wrappings of the plasma membranes of **Schwann cells** that wind themselves around the nerve fiber (Figure 3-13). The myelin sheath consists of several to more than 100 layers of plasma membrane. Gaps that occur in the sheath every 1 to 2 mm are known as **nodes of Ran-**

vier. Nodes of Ranvier are about 1 μm wide and are the lateral spaces between adjacent Schwann cells along the axon. *Myelin alters the electrical properties of the nerve fiber and results in a great increase in the conduction velocity of the fiber.*

A squid giant axon with a 500 μm diameter has a conduction velocity of 25 m/sec and is unmyelinated. If conduction velocity were directly proportional to fiber radius, a human nerve fiber with a 10 μm diameter would conduct at 0.5 m/sec. With this conduction velocity a reflex withdrawal of the foot from a hot coal would take about 4 seconds. Even though our nerve fibers are much smaller in diameter than squid giant axons, our reflexes are much faster than this. The myelin sheath that surrounds certain vertebrate nerve fibers results in a much greater conduction velocity than that of unmyelinated fibers of similar diameters. A 10 μm myelinated fiber has a conduction velocity of about 50 m/sec, which is twice that of the 500 μm squid giant axon. The high conduction velocity permits reflexes that are fast enough to allow us to avoid dangerous stimuli. A myelinated axon has a greater conduction velocity than an unmyelinated fiber that is 100 times larger in diameter (Figure 3-14). As discussed next, the myelin sheath increases the velocity of action potential conduction by (1) increasing the length constant of the axon, (2) decreasing the capacitance of the axon, and (3) restricting the generation of action potentials to the nodes of Ranvier.

Myelination greatly alters the electrical properties of the axon. The many wrappings of membrane around the axon increase the effective membrane resistance, so that

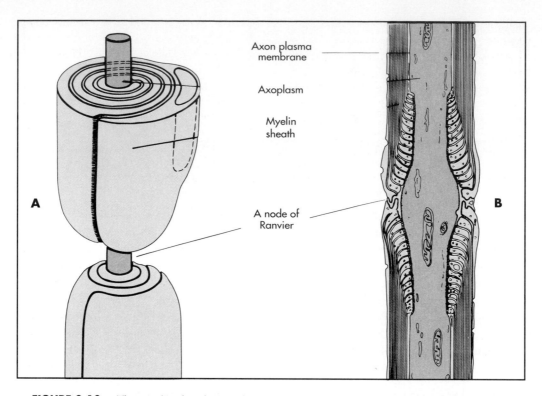

FIGURE 3-13   The myelin sheath. **A,** Schematic drawing of Schwann cells wrapping around an axon to form a myelin sheath. **B,** Drawing of a cross section through a myelinated axon near a node of Ranvier.

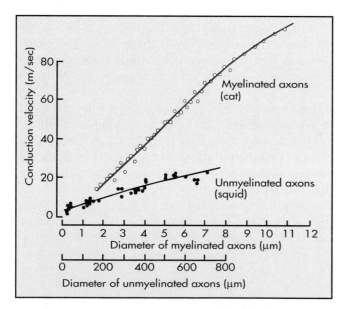

FIGURE 3-14   Conduction velocities of myelinated and unmyelinated axons as functions of axon diameter. Myelinated axons are from cat saphenous nerve at 38° C. Unmyelinated axons are from squid and are at 20° C to 22° C. Note that myelinated axons have greater conduction velocities than unmyelinated axons 100 times greater in diameter. (Data for myelinated axons from Gasser HS, Grundfest H: *Am J Physiol* 127:393, 1939. Data for unmyelinated axons from Pumphrey RJ, Young JZ: *J Exp Biol* 15:453, 1938.)

R_m/R_{in} and thus the length constant are much greater. Less of the conducted signal is lost through the electrical insulation of the myelin sheath, so that the amplitude of a conducted signal declines less with distance along the axon. The myelin-wrapped membrane has a much smaller electrical capacitance than the naked axonal membrane, so that the local currents can more rapidly depolarize the membrane as a signal is conducted. For this reason *the conduction velocity is greatly increased by myelination.* Because of the increase in length constant and in conduction velocity, an action potential is conducted with little decrement and at great speed from one node of Ranvier to the next.

The resistance to the flow of ions across the many layers of Schwann cell membrane that make up the myelin sheath is so high that the ionic currents are effectively localized to the short stretches of naked plasma membrane that occur at the nodes of Ranvier. For this reason the action potential is regenerated only at the nodes of Ranvier (1 to 2 mm apart), rather than being regenerated at each place along the fiber, as is the case in an unmyelinated fiber. The action potential is rapidly conducted from one node to the next (in about 20 μsec) and "pauses" to be regenerated at each node. The action potential appears to "jump" from one node of Ranvier to the next, a process called **saltatory conduction.**

Myelinated axons are also more efficient metabolically than unmyelinated axons. The sodium-potassium pump extrudes the sodium that enters and reaccumulates the potassium that leaves the cell during action potentials. In myelinated axons, ionic currents are restricted to the small fraction of the membrane surface at the nodes of Ranvier. For this reason far fewer Na^+ and K^+ ions traverse a unit area of fiber membrane, and much less ion pumping is required to maintain Na^+ and K^+ gradients.

In some diseases, known as **demyelinating disorders,** the myelin sheath deteriorates. In **multiple sclerosis** scattered progressive demyelination of axons in the central nervous system results in loss of motor control. The neuropathy common in severe cases of **diabetes mellitus** is due to demyelination of peripheral axons. When myelin is lost, the length constant, which is dramatically increased by myelination, becomes much shorter. Hence when the action potential is electrotonically conducted from one node of Ranvier to the next, it loses amplitude. If demyelination is sufficiently severe, the action potential may arrive at the next node of Ranvier with insufficient strength to fire an action potential. The axon will then fail to propagate action potentials.

SUMMARY

1. Different cell types have differently shaped action potentials because their populations of voltage-dependent ion channels differ.
2. The action potential in a squid giant axon is generated by the rapid opening and subsequent voltage-inactivation of voltage-dependent Na^+ channels and the delayed opening and closing of voltage-dependent K^+ channels.
3. Ion channels are integral membrane proteins that have ion-selective pores. Charged polypeptide regions of an ion channel protein act as gates that are responsible for the activation and inactivation of the channel.
4. An ion channel typically has two states: high conductance (open) and low conductance (closed). The channel oscillates randomly between the open and closed states. For a voltage-dependent channel the fraction of time the channel spends in the open state is a function of the transmembrane potential difference.
5. Cardiac and smooth muscle cells have L-type Ca^{++} channels that open and close slowly and are responsible for the long duration of the action potential in these cell types.

6. The voltage inactivation of Na^+ channels is an important factor in the absolute and relative refractory periods and in the accommodation of an excitable cell to a slowly rising stimulus.
7. Local circuit currents are the mechanism of electrotonic conduction. This is the mechanism by which both subthreshold signals and action potentials are conducted along the length of a cell.
8. A subthreshold signal is conducted with decrement. It will die away to 37% of its maximum strength over a distance of 1 length constant. The length constant is equal to $\sqrt{\dfrac{R_m}{R_{in}}}$. A typical value for the length constant is 1 to 2 mm.
10. The action potential is propagated, rather than merely conducted: it is regenerated as it moves along the cell. In this way an action potential remains the same size and shape as it is conducted.
11. The velocity of conduction is determined by the electrical properties of the cell. A large-diameter cell has a faster conduction velocity.
12. Myelination dramatically increases the conduction velocity of a nerve axon. Because of myelination an action potential is conducted very rapidly, and with little decrement, from one node of Ranvier to the next. Action potentials are regenerated only at the nodes of Ranvier; the internodal membrane cannot fire an action potential. Because it takes much longer to generate an action potential at each node than it does for the action potential to be conducted between nodes, the action potential appears to jump from node to node; this is saltatory conduction.

BIBLIOGRAPHY
Journal Articles

Barchi RL: Probing the molecular structure of the voltage-dependent sodium channel, *Annu Rev Neurosci* 11:455, 1988.

Bean BP: Classes of calcium channels in vertebrate cells, *Annu Rev Physiol* 51:367, 1989.

Catterall WA: Structure and function of voltage-sensitive ion channels, *Science* 242:50, 1988.

Catterall WA: Cellular and molecular biology of voltage-gated sodium channels, *Physiol Rev* 72:S15, 1992.

Hoffman F, Biel M, Flockerzi V: Molecular basis for Ca^{++} channel diversity, *Annu Rev Neurosci* 17:399, 1994.

Jan LY, Jan YN: Structural elements involved in specific K^+ channel functions, *Annu Rev Physiol* 54:537, 1992.

Neher E, Sakmann B: The patch clamp technique, *Sci Am* 266(3):28, 1992.

Perney TM, Kaczmarek LK: The molecular biology of K^+ channels, *Curr Opin Cell Biol* 3:663, 1991.

Stuhmer W: Structure-function studies of voltage-gated ion channels, *Annu Rev Biophys Biophys Chem* 20:65, 1991.

Books and Monographs

Aidley DJ: *The physiology of excitable cells,* ed 3, Cambridge, 1990, Cambridge University Press.

Hille B: *Ionic channels of excitable membranes,* ed 2, Sunderland, Mass, 1992, Sinauer Associates.

Hodgkin AL: *The conduction of the nervous impulse,* Springfield, Ill, 1964, Charles C Thomas, Publisher.

Kandel ER, Schwartz JH: *Principles of neural science,* ed 3, New York, 1991, Elsevier.

Katz B: *Nerve, muscle, and synapse,* New York, 1966, McGraw-Hill.

Levitan IB, Kaczmarek LK: *The neuron: cell and molecular biology,* New York, 1991, Oxford University Press.

Nicholls JG, Martin AR, Wallace BG: *From neuron to brain,* ed 3, Sunderland, Mass, 1992, Sinauer Associates.

Stevens CF: *Neurophysiology: a primer,* New York, 1966, John Wiley & Sons.

CHAPTER 4

Synaptic Transmission

A **synapse** is a site at which an electrical signal is transmitted from one cell to another. There are two types: electrical and chemical synapses. At an **electrical synapse** two excitable cells communicate by the direct passage of electrical current between them. This is called **electrotonic transmission. Gap junctions** link electrotonically coupled cells and provide low-resistance pathways for current flow directly between the cells.

Electrical signals are also transferred between excitable cells by means of chemical synapses. At a chemical synapse an action potential causes a **transmitter substance** to be released from the presynaptic neuron. The transmitter diffuses across the extracellular **synaptic cleft** and binds to receptors on the membrane. Chemical synapses have **synaptic delay**—the time required for these events to occur.

NEUROMUSCULAR JUNCTIONS

The synapses between the axons of motor neurons and skeletal muscle fibers are called **neuromuscular junctions, myoneural junctions,** or **motor endplates.** The neuromuscular junction, the first vertebrate synapse to be well characterized, serves as a model chemical synapse that provides a basis for understanding more complex synaptic interactions among neurons in the central nervous system.

Structure of the Neuromuscular Junction

Near the neuromuscular junction the motor nerve loses its myelin sheath and divides into fine terminal branches (Figure 4-1). The terminal branches of the motor axons lie in **synaptic troughs** on the surfaces of the muscle cells. The plasma membrane of the muscle cell lining the trough is thrown into numerous **junctional folds.** The axon terminals contain many 40 nm smooth-surfaced **synaptic vesicles** that contain acetylcholine, the neurotransmitter employed at this synapse. The axon terminal and the muscle cell are separated by the **junctional cleft,** which contains a carbohydrate-rich amorphous material.

Acetylcholine receptor molecules, integral proteins of the postjunctional muscle cell plasma membrane, are concentrated near the mouths of the junctional folds. Acetylcholinesterase, the enzyme that cleaves acetylcholine into acetate and choline, is distributed on the external surface of the postjunctional membrane. The synaptic vesicles in the nerve terminals and specialized release sites (called **active zones**) on the prejunctional membrane are concentrated opposite the mouths of the junctional folds.

Overview of Neuromuscular Transmission

The action potential is conducted down the motor axon to the presynaptic axon terminal. Depolarization of the plasma membrane of the axon terminal transiently opens voltage-gated calcium channels. Ca^{++} from the interstitial fluid flows down its electrochemical potential gradient into the axon terminal. The increased concentration of Ca^{++} in the nerve terminal causes synaptic vesicles to fuse with the plasma membrane and to empty their acetylcholine into the synaptic cleft by exocytosis. Acetylcholine diffuses across the synaptic cleft and combines with a specific acetylcholine receptor protein on the external surface of the muscle plasma membrane of the motor endplate. The combination of acetylcholine with the receptor protein transiently increases the conductance of the postjunctional membrane to Na^+ and K^+. Ionic currents (Na^+ and K^+) result in a transient depolarization of the endplate region. The transient depolarization is called the **endplate potential,** or **EPP** (Figure 4-2). The EPP is transient because the action of acetylcholine is ended by the hydrolysis of acetylcholine to

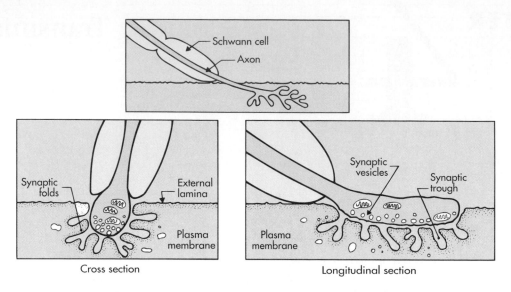

FIGURE 4-1   The structure of the neuromuscular junction in skeletal muscle.

form choline and acetate. The hydrolysis of acetylcholine is catalyzed by the enzyme *acetylcholinesterase,* which is present in high concentration on the postjunctional membrane.

> The importance of the influx of Ca^{++} into the nerve terminal to initiate the release of transmitter is illustrated by a disease, known as **Lambert-Eaton syndrome,** in which there are circulating antibodies against the type of voltage-gated Ca^{++} channels present in nerve terminals. Patients with this disorder experience muscular weakness and diminished stretch reflexes.

The postjunctional plasma membrane of the neuromuscular junction is not electrically excitable and does not fire action potentials. After it is depolarized, adjacent regions of the muscle cell membrane are depolarized by electrotonic conduction (Figure 4-2, *B*). When those regions reach threshold, action potentials are generated. Action potentials are propagated along the muscle fiber at high velocity and initiate the chain of events that leads to muscle contraction (Chapter 12). The steps involved in neuromuscular transmission are listed in the box on p. 46.

Synthesis of Acetylcholine

Motor neurons and their axons synthesize acetylcholine. Most other cells are not able to make acetylcholine. The enzyme **choline-O-acetyltransferase** in the motor neuron catalyzes the condensation of acetyl coenzyme A (acetyl CoA) and choline. Acetyl CoA is produced by

the neuron, as it is by most cells. However, choline cannot be synthesized by the motor neuron, but it is obtained by active uptake from the extracellular fluid. The plasma membrane of the motor nerve terminal has a transport system that can accumulate choline against a large electrochemical potential gradient.

Quantal Release of Acetylcholine

The amount of acetylcholine released by the prejunctional nerve ending does not vary continuously; rather the amount varies in steps, with each step corresponding to the release of one synaptic vesicle. The amount of acetylcholine contained in one vesicle corresponds to a **quantum** of acetylcholine.

Even if the motor neuron is not stimulated, small depolarizations of the postjunctional muscle cell occur spontaneously. These small spontaneous depolarizations are known as **miniature endplate potentials,** or MEPPs (Figure 4-3). They occur at random times with a frequency that averages about 1 per second. Each MEPP depolarizes the postjunctional membrane by only about 0.4 mV on average, not nearly enough to trigger an action potential in the adjacent muscle plasma membrane. The MEPP has the same time course as an EPP that is evoked by an action potential in the nerve terminal. The MEPP is similar to the EPP in its responses to most drugs. The EPP and the MEPP are both prolonged by drugs that inhibit acetylcholinesterase, and both are similarly depressed by compounds that compete with acetylcholine for binding to the receptor protein. The frequencies of MEPPs vary, but their amplitudes are within a relatively narrow range (Figure 4-3). An MEPP is caused by the spontaneous release of one quantum of acetylcholine into the junctional cleft.

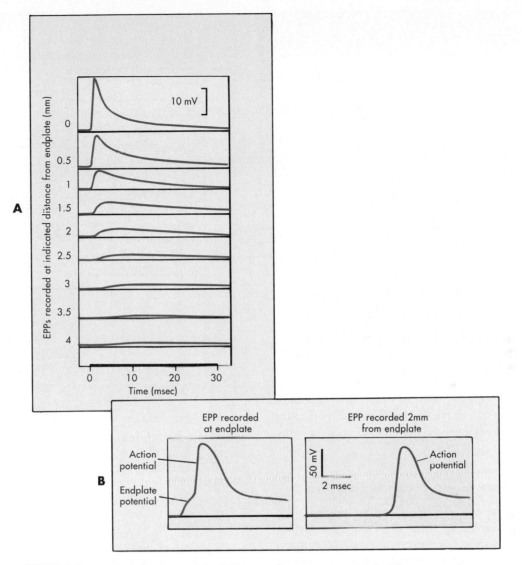

FIGURE 4-2 **A,** Endplate potentials (EPPs) in a frog sartorius muscle. The preparation was treated with curare to bring the EPP just below threshold for eliciting an action potential. The EPP, recorded at increasing distances from the neuromuscular junction, decreases in amplitude and rate of rise. **B,** Intracellular recordings made at the motor endplate *(left panel)* and 2 mm away *(right panel)* in a muscle fiber of frog extensor digitorum longus. When the motor nerve was stimulated, an EPP occurred, which triggered an action potential. Both the EPP and the resulting action potential can be recorded at the endplate, but 2 mm away from the endplate only the action potential can be seen because the EPP is conducted with decrement and has substantially decayed before reaching this point on the muscle fiber. (**A** redrawn from Fatt P, Katz B: *J Physiol* 115:320, 1951; **B** redrawn from Fatt P, Katz B: *J Physiol* 117:109, 1952.)

Action of Cholinesterase and Reuptake of Choline

Acetylcholinesterase is concentrated on the external surface of the postjunctional membrane and in the external lamina. Drugs that inhibit the enzyme are called **anticholinesterases.** In the presence of an anticholinesterase, the EPP is larger and dramatically prolonged.

The motor neuron cannot synthesize choline, so the reuptake from the synaptic cleft provides choline needed for the resynthesis of acetylcholine. **Hemicho-**

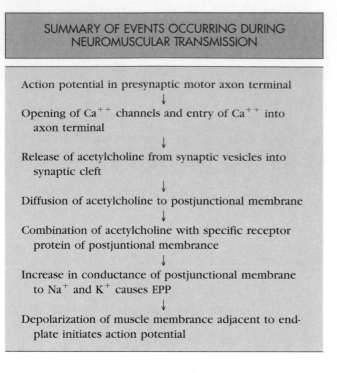

SUMMARY OF EVENTS OCCURRING DURING
NEUROMUSCULAR TRANSMISSION

Action potential in presynaptic motor axon terminal
↓
Opening of Ca^{++} channels and entry of Ca^{++} into axon terminal
↓
Release of acetylcholine from synaptic vesicles into synaptic cleft
↓
Diffusion of acetylcholine to postjunctional membrane
↓
Combination of acetylcholine with specific receptor protein of postjuntional membrane
↓
Increase in conductance of postjunctional membrane to Na^+ and K^+ causes EPP
↓
Depolarization of muscle membrane adjacent to endplate initiates action potential

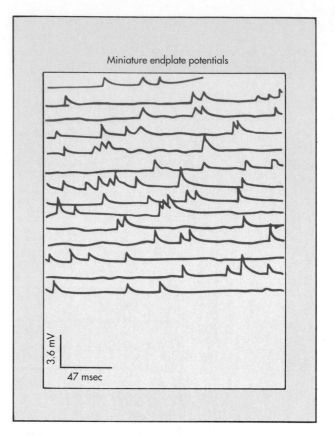

FIGURE 4-3 Spontaneous miniature endplate potentials (MEPPs) recorded at a neuromuscular junction in a fiber of frog extensor digitorum longus. (Redrawn from Fatt P, Katz B: *Nature* 166:597, 1950.)

liniums are drugs that block the choline transport system and inhibit choline uptake. Prolonged treatment with hemicholiniums depletes the store of transmitter and ultimately decreases the acetylcholine content of the quanta.

Ionic Mechanism of the Endplate Potential

The cation channels that acetylcholine opens in the postjunctional membrane differ from the cation channels of nerve and muscle, in that they are independent of the membrane potential. *The postjunctional channels are gated by the action of acetylcholine rather than by the transmembrane potential.* Acetylcholine receptors belong to the superfamily of **ligand-gated ion channels.**

Acetylcholine increases the permeability of the postsynaptic membrane to both Na^+ and K^+. At the cell's resting potential, the driving force for Na^+ to enter the cell is much larger than the net force that causes K^+ to leave the cell. Thus a net inward ionic current will flow through the open acetylcholine receptor protein channels, which will depolarize the postjunctional membrane.

Acetylcholine Receptor Protein

The acetylcholine receptor protein has been studied intensively. Development of methods for isolating and purifying hydrophobic membrane proteins and the availability of snake venom neurotoxins that bind very tightly to the acetylcholine receptor have been essential in these studies.

So-called α-**toxins** in cobra venoms are responsible for paralyzing the snakes' prey. α-toxins bind to the acetylcholine binding site on the acetylcholine receptor protein and prevent acetylcholine from acting. Poison arrows whose tips are dipped in **curare,** an α-toxin extracted from certain plants, are used by some South American Indians to paralyze their prey.

There are 10^7 to 10^8 acetylcholine receptor proteins per motor endplate; they are highly concentrated near the mouths of the postjunctional folds. The acetylcholine receptor protein is an integral membrane protein and is deeply embedded in the hydrophobic lipid matrix of the postjunctional membrane. Cholinesterase, on the other hand, is loosely associated with the surface of the postjunctional membrane by hydrophilic interactions.

The acetylcholine receptor consists of five subunits (Figure 4-4), two of which are identical, so that there are four different polypeptide chains.

> Patients with a disorder called **myasthenia gravis** are unable to maintain prolonged contraction of skeletal muscle. These individuals have circulating antibodies against the acetylcholine receptor protein.

SYNAPSES BETWEEN NEURONS

Chemical transmission between neurons has many of the same properties that characterize the neuromuscular junction. Electrical synapses are also present in the central nervous systems of animals, from invertebrates to mammals.

Electrical Synapses

At an electrical synapse, a change in the membrane potential of one cell is transmitted to the other cell by the direct flow of current. Because current flows directly between two cells that make an electrical synapse, there is essentially no synaptic delay. In general, electrical synapses allow conduction in both directions. In this respect they differ from chemical synapses, which must be unidirectional. Certain electrical synapses conduct more readily in one direction than in another; this property is called **rectification.**

Cells that form electrical synapses typically are joined by **gap junctions.** Gap junctions are plaque-like structures in which the plasma membranes of the coupled cells are very close (less than 3 nm). Freeze-fracture electron micrographs of gap junctions display regular arrays of intramembrane protein particles. The intramembrane particles consist of six subunits surrounding a central channel that is accessible to water. The hexagonal array is called a **connexon.** Each of the six subunits is a single protein (one polypeptide chain) called **connexin** (molecular weight of about 25,000). At the gap junction the connexons of the coupled cells are aligned to form **connexon channels** (Figure 4-5, *A*). The channels allow the passage of water-soluble molecules up to molecular weights of 1200 to 1500 from one cell to the other. These channels are the pathways for electrical current flow between the cells.

Cells that are electrically coupled may become uncoupled by closing of the connexon channels. The chan-

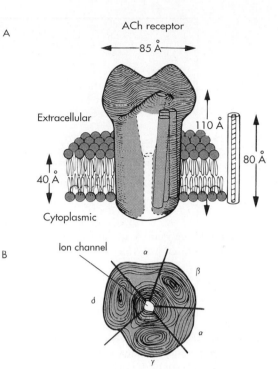

FIGURE 4-4   A model of the structure of the acetylcholine receptor protein. **A,** Viewed from the side and, **B,** viewed looking down on the acetylcholine receptor from the extracellular surface. The closed curves are electron density profiles. The five subunits surround a central ion channel. (Redrawn from Kistler J et al: *Biophys J* 37:371, 1982).

nels may close in response to increased intracellular Ca^{++} or H^+ in one of the cells or in response to depolarization of one or both of the cells. A model for the mechanism of closing the channels is shown in Figure 4-5, *B*.

Electrical synapses are widespread in the peripheral and central nervous systems of invertebrates and vertebrates. Electrical synapses are particularly useful in reflex pathways in which rapid transmission between cells (little synaptic delay) is necessary or when the synchronous response of a number of neurons is required. Among the many nonneuronal cells that are coupled by gap junctions are hepatocytes, myocardial cells, intestinal smooth muscle cells, and the epithelial cells of the lens.

Chemical Synapses

When one neuron makes a chemical synapse with another, the presynaptic nerve terminal characteristically broadens to form a **terminal bouton.** At the synapse itself the presynaptic and postsynaptic membranes are closely apposed and lie parallel to one another. Substan-

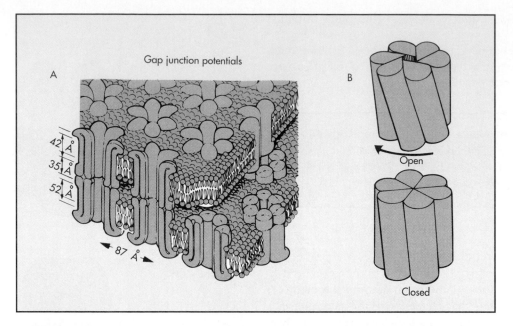

FIGURE 4-5   **A,** A model for the structure of the gap junction channels. Each plasma membrane contains connexons, each of which consists of a hexagonal array of six connexin polypeptides. The connexons of the two membranes are aligned at the gap junction to form channels between the cytosolic compartments of the two cells. **B,** A model of the opening and closing of the gap junction channel. The individual connexin subunits of the connexon are thought to twist relative to one another to open and close the central channel. (**A** redrawn from Makowski L et al: *J Cell Biol* 74:629, 1977; **B** redrawn from Unwin PNT, Zampighi G: *Nature* 283:45, 1980.)

tial structures stabilize the synapse, so that when nervous tissue is disrupted, the relationship of the presynaptic and postsynaptic membranes at the synapse is often preserved.

Because of the structure and organization of chemical synapses, conduction is necessarily one way. One-way conduction of chemical synapses contributes to the organization of central nervous systems of vertebrates. The synaptic delay at chemical synapses is about 0.5 msec. Synaptic delay is mainly caused by the time required for the release of transmitter. In polysynaptic pathways, synaptic delay accounts for a significant fraction of the total conduction time.

At chemical synapses the transmitter released by the presynaptic neurons alters the conductance of the postsynaptic plasma membrane to one or more ions. A change in the conductance of the postsynaptic membrane to an ion that is not in equilibrium across the membrane alters the current carried by that ion, and the change in ionic current alters the membrane potential of the postsynaptic cell. *In most cases transmitters produce their effects by increasing the conductance of the postsynaptic membrane to one or more ions.* However, some transmitters may act by decreasing the postsynaptic conductance to specific ions.

The part of the membrane of the postsynaptic neuron that forms the synapse is specialized for chemical sensitivity rather than electrical sensitivity. Action potentials are not produced at the synapse. The change in membrane potential, whether depolarization or hyperpolarization, that occurs at the synapse is conducted electrotonically over the membrane of the postsynaptic neuron. The part of the neuron where its axon originates is called the **axon hillock;** the part of the axon very near to the neuronal cell body is called the **initial segment.** In many neurons the axon hillock–initial segment region of the cell has a lower threshold than the rest of the plasma membrane of the postsynaptic cell. An action potential will be generated at that site if the sum of all the inputs to the cell exceeds threshold. Once the action potential has been generated, it is conducted back over the surface of the soma of the postsynaptic cell and is propagated along its axon.

Input-Output Relations

The neuromuscular junction is representative of a very simple type of synapse, in which one action potential in the presynaptic cell (the input) elicits a single action potential in the postsynaptic cell (the output). In

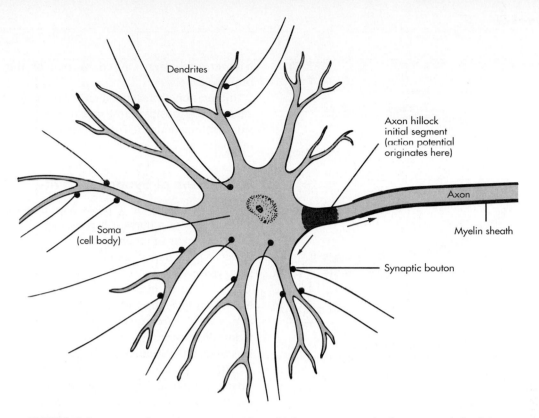

Dendrites

Axon hillock
initial segment
(action potential
originates here)

Axon

Myelin sheath

Soma
(cell body)

Synaptic bouton

FIGURE 4-6   A spinal motor neuron with multiple synapses on both soma and dendrites. The axon hillock–initial segment region has the lowest threshold, and as a result, action potentials tend to originate here.

other types of synapses the output may differ from the input. Synapses can be classified as one-to-one, one-to-many, or many-to-one, based on the relationship between input and output.

In a **one-to-one synapse,** such as the neuromuscular junction, the input and the output are the same. A single action potential in the presynaptic cell evokes a single action potential in the postsynaptic cell. Because the output is the same as the input, no integration occurs at this type of synapse.

In a **one-to-many synapse** a single action potential in the presynaptic cell elicits many action potentials in the postsynaptic cell. Axon collaterals of motor neurons make one-to-many synapses on **Renshaw cells** in the spinal cord. One action potential in the motor neuron induces the Renshaw cell to fire a burst of action potentials. The Renshaw cell synapses with the same motor neuron; the burst of action potentials in the Renshaw cell inhibits the motor neuron and prevents the motor neuron from being fired too frequently.

In a **many-to-one** synaptic arrangement, one action potential in the presynaptic cell is not enough to make the postsynaptic cell fire an action potential. The nearly simultaneous arrival of presynaptic action potentials in several input neurons that synapse on the postsynaptic cell is necessary to depolarize the postsynaptic cell to

threshold. The spinal motor neuron has this type of synaptic organization. One hundred or more presynaptic axons synapse on each spinal motor neuron (Figure 4-6). Some of these are excitatory inputs that depolarize the postsynaptic cell and bring it closer to its threshold. Other inputs are inhibitory and hyperpolarize the motor neuron; they take it farther away from threshold.

Excitatory and Inhibitory Postsynaptic Potentials

The changes in postsynaptic potential of a spinal motor neuron caused by an action potential in a single presynaptic input are about 1 to 2 mV. Thus no one excitatory input can bring the motor neuron to threshold. A transient depolarization of the postsynaptic neuron evoked by an action potential in a presynaptic axon is called an **excitatory postsynaptic potential** (EPSP) (Figure 4-7). The transient hyperpolarization elicited by an action potential in an inhibitory input is called by an **inhibitory postsynaptic potential** (IPSP) (Figure 4-7). At any instant the postsynaptic cell integrates the various inputs. If the momentary sum of the inputs depolarizes the postsynaptic cell to its threshold, it will fire an action potential. This is integration at the level of a single postsynaptic neuron.

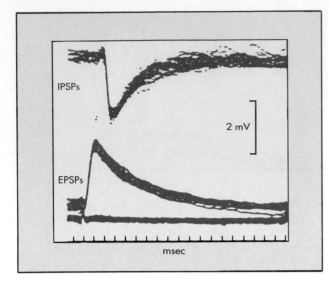

FIGURE 4-7 Inhibitory postsynaptic potentials (IPSPs) and excitatory postsynaptic potentials (EPSPs) recorded with a microelectrode in a cat spinal motor neuron in response to stimulation of appropriate peripheral afferent fibers. Forty tracings are superimposed. (Redrawn from Curtis DR, Eccles JC: *J Physiol* 145:529, 1959.)

Summation of Synaptic Inputs

The summation (or integration) of inputs can occur by either spatial summation or temporal summation (Figure 4-8). **Spatial summation** occurs when two separate inputs arrive almost simultaneously. The two postsynaptic potentials are added, so that two simultaneous excitatory inputs will depolarize the postsynaptic cell about twice as much as either input alone. However, if one EPSP and one IPSP occur simultaneously, they tend to cancel one another. Even postsynaptic potentials from synapses at opposite ends of the postsynaptic cell body act in this way. The postsynaptic potentials (EPSPs and IPSPs) are conducted rapidly over the entire cell membrane of the postsynaptic cell body with almost no decrement. This is because cellular dimensions (less than 100 μm) are much smaller than the length constant (about 1 to 2 mm) for electrotonic conduction. Synaptic potentials that originate in fine dendritic branches decrease in magnitude as they are conducted to the cell body; the finer the dendrite, the greater the decrement.

Temporal summation occurs when two or more action potentials in a single presynaptic neuron occur in rapid succession, so that the resulting postsynaptic potentials overlap in time. A train of impulses in a single presynaptic neuron can cause the potential of the postsynaptic cell to change in a stepwise

manner, each step caused by one of the presynaptic impulses.

Integration at the spinal motor neuron takes place because many positive and negative inputs impinge on a single motor neuron. This permits fine control of the firing pattern of the spinal motor neuron.

Modulation of Synaptic Activity

The responses of a postsynaptic neuron to single stimulations of a particular presynaptic neuron are relatively constant in magnitude and time course. However, when a presynaptic axon is stimulated repeatedly, the postsynaptic response may grow with each stimulation. This phenomenon is called **facilitation** (Figure 4-9, *A*). As shown in Figure 4-9, *B*, the extent of facilitation depends on the frequency of presynaptic impulses. Facilitation dies away rapidly, within tens to hundreds of milliseconds after stimulation stops.

When a presynaptic neuron is stimulated **tetanically** (many stimuli at high frequency) for several seconds, **posttetanic potentiation,** a longer-lived enhancement of postsynaptic response, occurs (Figure 4-9, *C*). Posttetanic potentiation persists much longer than facilitation; it lasts tens of seconds to several minutes after cessation of tetanic stimulation.

Facilitation and posttetanic potentiation are the result of the effects of repeated stimulation on the presynaptic neuron. These phenomena do not involve a change in the sensitivity of the postsynaptic cell to transmitter. With repeated stimulation, an increased number of quanta of transmitter are released. Increased levels of intracellular calcium enhance transmitter release during repetitive stimulation.

Repetitive stimulation of certain synapses in the brain increases the efficacy of transmission at those synapses. This phenomenon, called **long-term potentiation,** can persist for days to weeks. Long-term potentiation is believed to be involved in memory. The increased synaptic efficacy that occurs in long-term potentiation seems to involve both presynaptic (greater transmitter release) and postsynaptic (greater response to transmitter) changes.

When a synapse is repetitively stimulated for a long time, a point is reached at which each successive presynaptic stimulation elicits smaller postsynaptic responses. This phenomenon is called **synaptic fatigue** (neuromuscular depression at the motor endplate). The postsynaptic cell at a fatigued synapse responds normally to transmitter applied from a micropipette; thus the defect is presynaptic. In some cases a decrease in quantal content (the amount of transmitter per synaptic vesicle) contributes to synaptic fatigue. A fatigued synapse typically recovers in a few seconds.

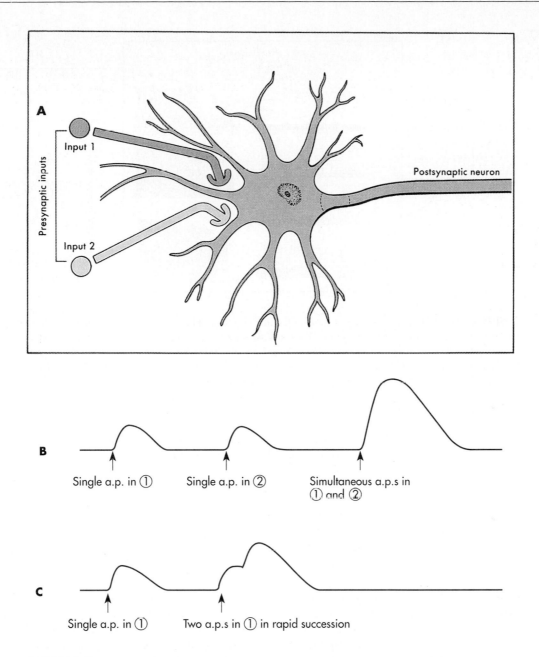

FIGURE 4-8 **A,** Spatial and temporal summation at a postsynaptic neuron with two synaptic inputs *(1* and *2).* **B,** Spatial summation. The postsynaptic potential in response to single action potentials in inputs 1 and 2 occurring separately and simultaneously. **C,** Temporal summation. The postsynaptic response to two impulses in rapid succession in the same input.

Ionic Mechanisms of Postsynaptic Potentials in Spinal Motor Neurons

Much of our knowledge of synaptic mechanisms in the mammalian central nervous system is derived from studies of cat spinal motor neurons.

Excitatory Postsynaptic Potentials (EPSPs) The EPSP (see Figure 4-7) of the cat spinal motor neuron is caused by a transient increase in the conductance of the postsynaptic membrane to both Na^+ and K^+ in response to the neurotransmitter. At the cell's resting potential, the driving force for Na^+ to enter the cell is much greater than the force for K^+ to leave. Hence, in response to the neurotransmitter, a net inward flow of Na^+ ions occurs and this depolarizes the postsynaptic cell.

Inhibitory Postsynaptic Potentials The IPSP (see Figure 4-7) of cat spinal motor neurons is caused by an increased Cl^- conductance of the postjunctional

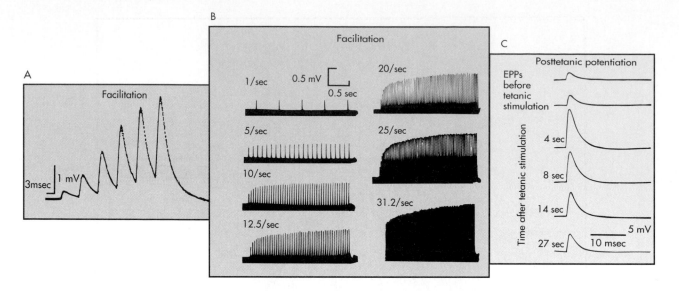

FIGURE 4-9 **A,** Facilitation at a neuromuscular junction. EPPs at a neuromuscular junction in toad sartorius muscle were elicited by successive action potentials in the motor axon. Neuromuscular transmission is depressed by 5 mM Mg^{++} and 2.1 μM curare, so that action potentials do not occur. **B,** EPPs at a frog neuromuscular junction elicited by repetitively stimulating the motor axon at different frequencies. Note that facilitation fails to occur at the lowest frequency of stimulation (1/sec) and that the degree of facilitation increases with increasing frequency of stimulation in the range of frequency employed. Neuromuscular transmission was inhibited by bathing the preparation in 12 to 20 mM Mg^{++}. **C,** Posttetanic potentiation at a frog neuromuscular junction. The top two tracings indicate control EPPs in response to single action potentials in the motor axon. Subsequent tracings indicate EPPs in response to single action potentials following tetanic stimulation (50 impulses/sec for 20 seconds) of the motor neuron. The time interval between the end of tetanic stimulation and the single action potential is shown on each tracing. The muscle was treated with tetrodotoxin to prevent generation of action potentials. (**A** redrawn from Belnave RJ, Gage PW: *J Physiol* 266:435, 1977; **B** redrawn from Magelby KL: *J Physiol* 234:327, 1973; **C** redrawn from Weinrich D: *J Physiol* 212:431, 1971.)

membrane. At rest the net tendency is for Cl^- to enter the cell. The increase in Cl^- conductance, as the result of transmitter release at the inhibitory synapse, allows Cl^- to enter the postsynaptic cell and hyperpolarize it.

Presynaptic Inhibition Inhibitory interactions are vital in stabilizing the central nervous system. Another type of inhibition is called **presynaptic inhibition.** If an inhibitory input to a spinal motor neuron is stimulated tetanically and then an excitatory input is stimulated once, the EPSP elicited by stimulating the excitatory input may be reduced in magnitude after the inhibitory volley. This is believed to occur by a mechanism in which axon collaterals of the inhibitory axons synapse on the excitatory nerve terminals (Figure 4-10). Action potentials in the inhibitory nerve depolarize the excitatory nerve terminal for a long time. This brings the excitatory nerve terminal closer to threshold and the partly depolarized excitatory terminal will release less transmit-

ter in response to an action potential. The smaller release of transmitter diminishes the EPSP. The phenomenon of decreased transmitter release from a partially depolarized nerve terminal commonly occurs at the neuromuscular junction.

NEUROTRANSMITTERS AND NEUROMODULATORS

Identification of Transmitter Substances

Compounds that may function as neurotransmitters are called **candidate** or **putative neurotransmitters.** Candidate neurotransmitters are usually concentrated in specific neurons or in specific neuronal pathways. Microapplication of putative transmitters to particular areas of the central nervous system (CNS) may evoke specific responses. Correlation of information about the

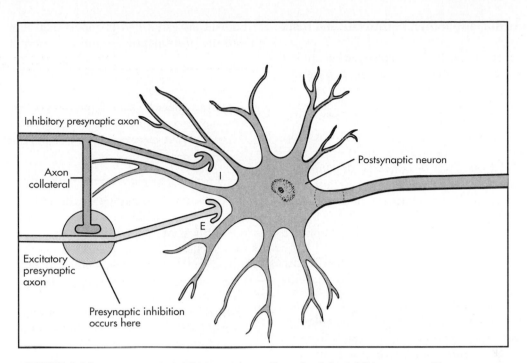

FIGURE 4-10   Presynaptic inhibition. Axon collaterals of the inhibitory axon *(I)* synapse on the excitatory axon terminal *(E)*. An action potential in the inhibitory axon depolarizes the excitatory axon terminal. The depolarized excitatory axon terminal will release less transmitter in response to an action potential in the excitatory neuron.

localization of the candidate transmitter with knowledge of the location of neurons that respond to this substance, as well as the ways in which they respond, provides clues about the functions of a putative neurotransmitter.

It is often difficult to prove that a substance is the transmitter at a particular synapse. A putative transmitter (X) must satisfy the following criteria before it is accepted as a proven transmitter at a particular synapse:

1. The presynaptic neurons must contain X and must be able to synthesize it.
2. X must be released by the presynaptic neurons on appropriate stimulation.
3. Microapplication of X to the postsynaptic membrane must mimic the effects of stimulation of the presynaptic neuron.
4. The effects of presynaptic stimulation and of microapplication of X should be altered in the same way by drugs.

Our knowledge of neurotransmitters has increased greatly in recent years. Some transmitters have rapid and transient effects on the postsynaptic cell. Other transmitters have effects that are much slower in onset and may last for minutes or even hours. Most known neurotransmitters fall into three major chemical classes: amines, amino acids, and oligopeptides.

Neurotransmitters

Acetylcholine As discussed previously, acetylcholine is the transmitter used by motor axons that arise from the spinal cord. Acetylcholine also plays a central role in the autonomic nervous system; it is the transmitter for all autonomic preganglionic neurons and also for postganglionic parasympathetic fibers. The Betz cells of the motor cortex use acetylcholine as their transmitter. The basal ganglia, which are involved in the control of movement, contain high levels of acetylcholine, and acetylcholine is believed to be a transmitter there. In addition, acetylcholine may be the transmitter in a large number of central pathways.

> Deficits in pathways involving acetylcholine (**cholinergic pathways**) in the brain have been implicated in some forms of **senile dementia** (such as **Alzheimer's disease**). Treatment with long-lasting anticholinergic drugs that penetrate the blood-brain barrier may improve cognitive function in some individuals suffering from dementia.

Biogenic Amine Transmitters Among the amines that may serve as neurotransmitters are **norepineph-**

rine, epinephrine, dopamine, serotonin, and histamine.

Dopamine, norepinephrine, and epinephrine are catecholamines, and they share a common biosynthetic pathway that starts with the amino acid tyrosine. Tyrosine is converted to L-dopa by tyrosine hydroxylase. L-Dopa is converted to dopamine by a specific decarboxylase. In dopaminergic neurons the pathway stops here. Noradrenergic neurons have another enzyme, dopamine β-hydroxylase, that converts dopamine to norepinephrine. Norepinephrine is the primary transmitter for postganglionic sympathetic neurons. Chromaffin cells in the adrenal medulla add a methyl group to norepinephrine to produce the hormone epinephrine.

Neurons that contain high levels of dopamine are prominent in the midbrain regions known as the **substantia nigra** and the **ventral tegmentum.** Some of the axons of these terminate in the **corpus striatum,** where they participate in controlling complex movements. The degeneration of dopaminergic synapses in the corpus striatum occurs in **Parkinson's disease** and may be a major cause of the muscular tremors and rigidity that characterize this disease. Treatment of some Parkinson's patients with L-dopa, a precursor of dopamine, improves motor control.

Serotonin (5-hydroxytryptamine)–containing neurons are present in high concentration in certain nuclei located in the brainstem. Serotonergic neurons may be involved in temperature regulation, sensory perception, onset of sleep, and control of mood.

Histamine is present in certain neurons in the hypothalamus. The functions of these presumably histaminergic neurons are not yet known.

Amino Acid Transmitters Glycine, the simplest amino acid, is an inhibitory neurotransmitter released by certain spinal interneurons.

γ-Aminobutyric acid (GABA) is not incorporated into proteins, nor is it present in all cells (as are the other naturally occurring amino acids). GABA is produced from glutamate by a specific decarboxylase present only in the central nervous system. Among the cells that contain GABA are some cells in the basal ganglia, the cerebellar Purkinje cells, and certain spinal interneurons. In all known cases, GABA functions as an inhibitory transmitter. It is the most common transmitter in the brain. GABA may be the neurotransmitter at as many as one third of the synapses in the brain.

The postsynaptic receptors for glycine and GABA are both ligand-gated Cl^- channels that allow Cl^- to flow into the postsynaptic neuron to hyperpolarize it.

General anesthetics prolong the open time of GABA receptor chloride channels, and thus prolong the inhibition of the postsynaptic neurons at GABA-ergic synapses. GABA receptors may be the principal targets of general anesthetics.

Glutamate and **aspartate,** dicarboxylic amino acids, strongly excite many neurons in the brain. Glutamate is the most common excitatory neurotransmitter in the brain. There are five identified classes of **excitatory amino acid receptors.** One of these, the NMDA (N-methyl-D-aspartate) receptors, consists of ligand-gated Ca^{++} channels.

Nitric Oxide Nitric oxide (NO) is a recently discovered neurotransmitter. NO is a transmitter at synapses between inhibitory motor neurons of the enteric nervous system and gastrointestinal smooth muscle cells (Chapter 32). NO may also function as a neurotransmitter in the central nervous system. NO is an unusual neurotransmitter, because it is neither packaged into synaptic vesicles nor released by exocytosis. NO is highly permeant and simply diffuses from its site of production to neighboring cells. **NO synthase** is the enzyme that catalyzes the production of NO as a product of the oxidation of arginine to citrulline. This enzyme is stimulated by an increase in cytosolic Ca^{++}.

In addition to serving as a neurotransmitter, NO can serve as a cellular signal transduction molecule in both neurons and nonneuronal cells (such as vascular smooth muscle). One way NO does this is by regulating guanylyl cyclase, the enzyme that produces cyclic GMP from GTP. NO binds to a heme group in soluble guanylyl cyclase and potently stimulates the enzyme. This leads to an elevation of cyclic GMP in the target cell. The elevated cyclic GMP can then influence multiple cellular processes. Thus NO can serve as a neuromodulator, as well as functioning as a neurotransmitter.

Neuroactive Peptides

Certain cells release peptides that act at very low concentrations to excite or inhibit neurons. To date, more than 25 of these so-called **neuropeptides,** ranging from 2 to about 40 amino acids long, have been identified. Some of these neuropeptides are listed in the box on p. 55.

Neuropeptides typically affect their target neurons at lower concentrations than the classical neurotransmitters discussed previously, and the neuropeptides usually

NEUROACTIVE PEPTIDES

GUT-BRAIN PEPTIDES

Vasoactive intestinal polypeptide (VIP)
Cholecystokinin octapeptide (CCK-8)
Substance P
Neurotensin
Methionine enkephalin
Leucine enkephalin
Motilin
Insulin
Glucagon

HYPOTHALAMIC-RELEASING HORMONES

Thyrotropin-releasing hormone (TRH)
Luteinizing hormone-releasing hormone (LHRH)
Somatostatin (growth hormone release-inhibiting hormone)

PITUITARY PEPTIDES

Adrenocorticotropin (ACTH)
β-Endorphin
α-Melanocyte-stimulating hormone (α-MSH)

OTHERS

Dynorphin
Angiotensin II
Bradykinin
Vasopressin
Oxytocin
Carnosine
Bombesin

Modified from Snyder SH: *Science* 209:976, 1980.

Table 4-1 Examples of the Coexistence of a Classic Transmitter and Neuropeptide Within the same Nerve Terminal*

Transmitter	Neuropeptide
Acetylcholine	Vasoactive intestinal peptide (VIP)
Norepinephrine	Somatostatin
	Enkephalin
	Neurotensin
Dopamine	Cholecystokinin (CCK)
	Enkephalin
Epinephrine	Enkephalin
Serotonin	Substance P
	Thyrotropin-releasing hormone (TRH)

Reprinted by permission of the publisher from *Chemical messengers: small molecules and peptides* by Schwartz JH. In Kandel ER and Schwartz JH. editors: Principles of neural science. Copyright © 1981 by Elsevier Science Publishing Co. Inc.
*Evidence for the coexistence of a classic transmitter substance with a neuroactive peptide has been reported for these combinations. With the information thus far available, it is not yet possible to determine the specificity of the pairs and their physiological significance.

act longer. A number of the neuropeptides listed in the box are more familiar as hormones. Neuropeptides may act as hormones, as neurotransmitters, or as neuromodulators. A hormone is a substance that is released into the blood and that reaches its target cells via the circulation. A neurotransmitter or neuromodulator is typically released near the surface of its target cell and diffuses to the target cell. Neurotransmitters, as discussed earlier, act to change the conductance of the target cell to one or more ions, and in that way they change the membrane potential of the target cell. A neuromodulator modulates synaptic transmission. The neuromodulator may act presynaptically to change the amount of transmitter released in response to an action potential, or it may act on the postsynaptic cell to modify its response to the neurotransmitter. A number of neuropeptides act as true transmitters at particular synapses and as neuromodulators at other synapses.

In several instances, neuropeptides coexist in the same nerve terminals with classical transmitters (Table 4-1). In some of these cases the neuropeptide is released along with the transmitter in response to nerve stimulation.

Synthesis of Neuropeptides Nonpeptide neurotransmitters are synthesized in nerve terminals by pathways that involve soluble enzymes and simple precursors. Neuropeptides are synthesized in the neuronal cell body. They are encoded in the cell's DNA and transcribed into messenger RNA, which is translated on polyribosomes bound to the endoplasmic reticulum. Secretory vesicles containing the neuropeptide are released from the mature face of the Golgi complex. The secretory vesicles are moved by fast axonal transport to the axon terminal, where they function as synaptic vesicles.

Some neuropeptides are synthesized as preprohormones (see also Chapter 40). Cleavage of the signal sequence converts the preprohormone to a prohormone. Proteolytic cleavage of the prohormone may then release one or more active peptides. In some cases one prohormone may contain several active peptide sequences. For example, the prohormone of the opioid peptide β-endorphin is a 31,000-dalton polypeptide that contains several active sequences. One cleavage of the prohormone releases adrenocorticotropic hormone (ACTH) and β-lipotropin. Cleavage of ACTH releases a hormone, melanocyte-stimulating hormone (α-MSH), and cleavage of α-lipotropin releases β-MSH and a number of active β-endorphins.

Opioid Peptides Opiates are drugs that are derived from the juice of the opium poppy.

Opiates are useful therapeutically as powerful **analgesics** (pain relievers). They exert their analgesic effect by binding to specific opiate receptors. The binding of opiates to their receptors is stereospecifically inhibited by a morphine derivative called **naloxone.**

Compounds that do not derive from the opium poppy, but that exert direct effects by binding to opiate receptors, are called **opioids.** Operationally, opioids are defined as direct-acting compounds whose effects are stereospecifically antagonized by naloxone.

The three major classes of endogenous opioid peptides in mammals are enkephalins, endorphins, and dynorphins. **Enkephalins** are the simplest opioids; they are pentapeptides. **Dynorphins** and **endorphins** are somewhat longer peptides that contain one or the other of the enkephalin sequences at their N-terminal ends.

Opioid peptides are widely distributed in neurons of the central nervous system and intrinsic neurons of the gastrointestinal tract. Opioid peptides are found in vesicles that resemble synaptic vesicles. The endorphins are discretely localized in particular structures of the CNS, whereas the enkephalins and dynorphins are more widely distributed. Opioids also inhibit structures in the brain involved in the perception of pain.

Nonopioid Neuropeptides Most of the known neuropeptides are not opioids. **Substance P,** a peptide of 11 amino acids, is present in specific neurons in the brain, in primary sensory neurons, and in plexus neurons in the wall of the gastrointestinal tract. Substance P was the first so-called gut-brain peptide to be discovered. The wall of the gastrointestinal tract is richly innervated with neurons that form networks or plexuses (see also Chapter 32). The intrinsic plexuses of the gastrointestinal tract exert primary control over its motor and secretory activities. These enteric neurons contain many of the neuropeptides, including substance P, that are found in the brain and spinal column.

Substance P is the suspected transmitter at synapses made by primary sensory neurons (their cell bodies are in the dorsal root ganglia) with spinal interneurons in the dorsal horn of the spinal column. Enkephalins act to decrease the release of substance P at these synapses and thereby inhibit the pathway for pain sensation at the first synapse in the pathway.

Vasoactive intestinal polypeptide (VIP) is a member of a family of neuropeptides related to the hormone secretin. VIP was first discovered as a gastrointestinal hormone, but it is now known to be a neuropeptide as well.

VIP is widely distributed in the CNS and in the intrinsic neurons of the gastrointestinal tract. In neurons in the brain, VIP has been localized in synaptic vesicles. VIP may function as an inhibitory transmitter to vascular and nonvascular smooth muscle and as an excitatory transmitter to glandular epithelial cells.

Secretin, glucagon, and **gastrointestinal inhibitory peptide,** molecules whose functions as hormones have been well characterized, have sequence homology with VIP. These peptides have also been found in particular neurons in the central nervous system, but their functions in neurons remain undetermined.

Cholecystokinin (CCK) is a member of a group of neuropeptides that includes gastrin and cerulein, which have similar C-terminal sequences. CCK is a well-known gastrointestinal hormone that elicits contraction of the gallbladder (see Chapter 33). One form of CCK is present in particular neurons of the central nervous system.

Neurotensin, one of the most recently discovered neuropeptides, is present in enteric neurons and in the brain. When neurotensin is injected into cerebrospinal fluid at low concentrations, it lowers body temperature. Thus neurotensin may function in temperature regulation.

Other Neuromodulators

Some important neuromodulators are not peptides. Purines and purine nucleotides (ATP) and nucleosides (adenosine) function as neuromodulators in the central, autonomic, and peripheral nervous systems. Substances that serve as neurotransmitters may also act as neuromodulators. In some cases the transmitter binds to receptors on the presynaptic neuron that released it and thereby regulates its own release.

SUMMARY

1. The neuromuscular junction is the best-characterized chemical synapse in vertebrates. Acetylcholine released by the prejunctional nerve terminal binds to acetylcholine receptors in the postjunctional membrane to open ion channels conductive to Na^+ and K^+. The resulting ion flow across the postjunctional membrane causes a depolarization, called an endplate potential.
2. The endplate potential is terminated by the hydrolysis of acetylcholine by the enzyme acetylcholinesterase. When acetylcholine is hydrolyzed, the choline liberated in the synaptic cleft is actively transported back into the nerve terminal.
3. The release of acetylcholine is quantal. A quantum

corresponds to the amount of acetylcholine in a single presynaptic vesicle.

4. Direct electrical transmission between neighboring cells is mediated by gap junctions.

5. An action potential in an excitatory input to a spinal motor neuron causes an excitatory postsynaptic potential that depolarizes the motor neuron and brings it closer to threshold. An action potential in an inhibitory input causes an inhibitory postsynaptic potential that hyperpolarizes the motor neuron.

6. The efficacy of synaptic transmission depends on the timing and frequency of action potentials in the presynaptic neuron. Facilitation, posttetanic potentiation, and long-term potentiation are examples of increased efficacy of synaptic transmission in response to previous multiple stimulations of a synapse.

7. Acetylcholine, biogenic amines, glutamate, glycine, and γ-aminobutyric acid (GABA) are important neurotransmitters in the central nervous system.

8. Glycine and γγ-aminobutyric acid are the major transmitters at inhibitory synapses in the central nervous system. Glutamate is the most common transmitter at excitatory synapses.

9. Many neuroactive peptides function as neuromodulators or neurotransmitters in the central nervous system.

BIBLIOGRAPHY

Journal Articles

Amara SG, Kuhar MJ: Neurotransmitter transporters: recent progress, *Annu Rev Neurosci* 16:73, 1993.

Barnard EA: Receptor classes and the transmitter-gated ion channels, *Trends Biochem Sci* 17:368, 1992.

Baxter DA, Byrne JH: Ionic conductance mechanisms contributing to electrophysiological properties of neurons, *Curr Opin Neurobiol* 1:105, 1991.

Bennett MK, Scheller RH: A molecular description of synaptic vesicle membrane trafficking, *Annu Rev Biochem* 63:63, 1994.

Bredt DS, Snyder SH: Nitric oxide: a physiologic messenger molecule, *Annu Rev Biochem* 63:175, 1994.

Changeux JP: The nicotinic acetylcholine receptor: an allosteric protein prototype of ligand-gated ion channels, *Trends Pharmacol Sci* 11:485, 1990.

Froehner SC: Regulation of ion channel distribution at synapses, *Annu Rev Neurosci* 16:347, 1993.

Gingrich JA, Caron MG: Recent advances in the molecular biology of dopamine receptors, *Annu Rev Neurosci* 16:299, 1993.

Hökfelt T: Neuropeptides in perspective: the last ten years, *Neuron* 7:867, 1991.

Hollman M, Heinemann S: Cloned glutamate receptors, *Annu Rev Neurosci* 17:31, 1994.

Jahn R, Sudhof TC: Synaptic vesicles and exocytosis. *Annu Rev Neurosci* 17:219, 1994.

Jessel TM, Kandel ER: Synaptic transmission: a bidirectional and self-modifiable form of cell-cell communication, *Cell* 72(Suppl 1):1, 1993.

Kennedy MB: The biochemistry of synaptic regulation in the central nervous system, *Annu Rev Biochem* 63:571, 1994.

Kupferman I: Functional studies of cotransmission, *Physiol Rev* 71:683, 1991.

Lester HA: The permeation pathway of neurotransmitter-gated ion channels, *Annu Rev Biophys Biomol Struct* 21:267, 1992.

Nakanishi S, Masu M: Molecular diversity and functions of glutamate receptors, *Annu Rev Biophys Biomolec Struct* 23:319, 1994.

Nicoll RA, Malenka RC, Kauer JA: Functional comparison of neuroreceptor subtypes in mammalian central nervous system, *Physiol Rev* 70:513, 1990.

Sakmann B: Elementary steps in synaptic transmission revealed by currents through single ion channels, *Science* 256:28, 1992.

Schuman EM, Madison DV: Nitric oxide and synaptic function, *Annu Rev Neurosci* 17:153, 1994.

Stevens CF: Quantal release of neurotransmitter and long-term term potentiation, *Cell* 72(Suppl):55, 1993.

Unwin N: Neurotransmitter action: opening of ligand-gated ion channels, *Cell* 72(Suppl):31, 1993.

Young AB, Fagg GE: Excitatory amino acid receptors in the brain: membrane binding and receptor autoradiographic approaches, *Trends Pharmacol Sci* 11:126, 1990.

Books and Monographs

Aidley DJ: *The physiology of excitable cells,* ed 3, Cambridge, 1990, Cambridge University Press.

Eccles JC: *The physiology of synapses,* Berlin, 1964, Springer Verlag.

Hall Z: *An introduction to molecular neurobiology,* Sunderland, Mass, 1991, Sinauer Associates.

Kandel ER, Schwartz JH, Jessell TM: *Principles of neural science,* ed 3, New York, 1991, McGraw-Hill Book Co.

Katz B: *Nerve, muscle, and synapse,* New York, 1966, McGraw-Hill Book Co.

Levitan IB, Kaczmarek LK: *The neuron: cell and molecular biology,* New York, 1991, Oxford University Press.

Nicholls JG, Martin AR, Wallace BG: *From neuron to brain,* ed 3, Sunderland, Mass, 1992, Sinauer Associates.

CHAPTER 5

Membrane Receptors, Second Messengers, and Signal Transduction Pathways

OVERVIEW

Basic cellular processes are regulated by a host of regulatory substances. Some regulatory substances, such as steroid hormones, enter the cell and influence the transcription of certain genes (Chapter 46). In this chapter we will discuss regulatory substances that exert their influences from outside the cell.

The first step in the action of such substances is to bind to specific protein **receptors** on the extracellular surface of the plasma membrane of the target cells. The neurotransmitters discussed in Chapter 4 and their receptors are examples. For the neurotransmitters discussed so far, the receptor is a ligand-gated ion channel, and the response of the cell is a ligand-induced ionic current. In such cases the ligand-gated ion channel is both the receptor and the effector for the action of the neurotransmitter.

For most regulatory molecules, however, a more complex series of events links the binding of the substance to its specific membrane receptor to its final effects on cellular function. Extracellular regulatory molecules exert their effects on cells via **signal transduction pathways.** Although many regulatory substances exist, there are relatively few signal transduction pathways whereby a regulatory substance binds to its plasma membrane receptor and elicits a cellular response by altering the activities of particular proteins in the cell. Our knowledge of signal transduction mechanisms is increasing at such a rapid rate that a comprehensive discussion of this subject is beyond the scope of this book. Therefore, only the most common and the best understood signal transduction pathways will be emphasized, especially those that are relevant to topics discussed in subsequent chapters.

The extracellular regulatory compounds we will consider are often classified as endocrine, neurocrine, or paracrine substances. **Endocrine** regulatory substances (hormones) are released by endocrine cells and reach their target cells, which may be far from the endocrine cells, by way of the bloodstream. **Neurocrine** regulators are released by neurons in the immediate vicinity of the target cells. Neurotransmitters are neurocrine substances, as are most of the neuromodulators discussed in Chapter 4. **Paracrine** substances are released by cells that are not immediately adjacent to the target cells, but sufficiently close that the paracrine substance can reach the target cells by diffusion. Histamine is a paracrine agonist of gastric HCl secretion (Chapter 33). Histamine is released by ECL (enterochromaffin-like) cells in the gastric mucosa, and it reaches the acid-secreting parietal cells by diffusion.

PROTEIN KINASES AND SECOND MESSENGERS IN SIGNAL TRANSDUCTION PATHWAYS

Frequently the final step in a signal transduction pathway involves the phosphorylation of particular proteins that play central roles in a biological process. When these proteins are phosphorylated, their activities may be enhanced or suppressed. **Protein kinases** in the cell are responsible for phosphorylating particular proteins, and **protein phosphatases** are responsible for cleaving phosphates from proteins. The state of phosphorylation of an effector protein depends on the balance of the activities of the kinase that phosphorylates it and the phosphatase that dephosphorylates it. Protein phosphatases will be discussed later in this chapter.

A signal transduction pathway frequently alters activity of a protein kinase in response to the binding of the regulatory molecule, often called an **agonist,** to its membrane receptor. The major classes of agonist-activated protein kinases are shown in Figure 5-1.

Among the receptor-mediated signals that regulate

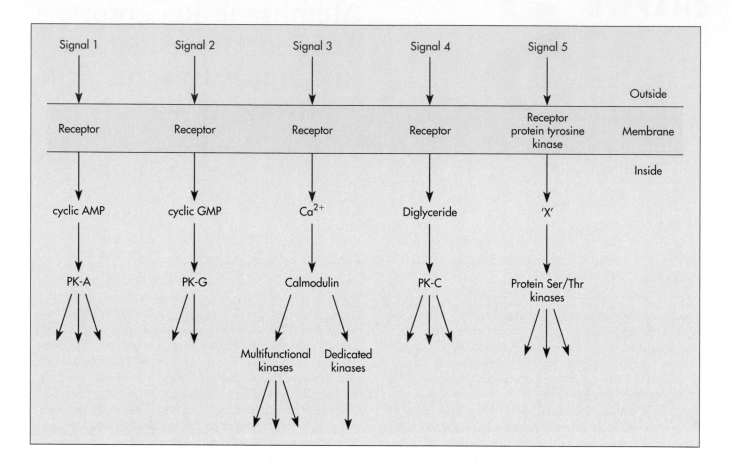

FIGURE 5-1   Frequently the final step in a signal transduction pathway is the phosphorylation of an effector protein by a protein kinase. Five major signal transduction pathways of mammalian cells that involve protein kinases are depicted in this illustration. *PK-A,* Cyclic AMP–dependent protein kinase; *PK-G,* cyclic GMP–dependent protein kinase; *PK-C,* protein kinase C; *X,* hypothetical signaling pathway. (Adapted from Cohen P: *Trends Biochem Sci* 17:408, 1992.)

the activities of protein kinases are the following **second messengers: cyclic AMP, cyclic GMP, Ca^{++}, inositol trisphosphate** (IP$_3$), and **diglycerides** (also called diacylglyercols). Cells contain protein kinases that are modulated by each of these second messengers. Binding of an agonist to its membrane receptor often changes the intracellular level of one of the second messengers, which then modulates the activity of a protein kinase.

Cells contain protein kinases whose activities are enhanced by cyclic AMP and cyclic GMP. These protein kinases are called **cyclic AMP–dependent protein kinases** and **cyclic GMP–dependent protein kinases,** respectively.

The activities of **calmodulin-dependent protein kinases** are enhanced when they bind the complex of

Ca^{++} with a protein called calmodulin. Calmodulin (MW 16,700) is present in all cells and binds four Ca^{++} ions. The complex of Ca^{++} and calmodulin then regulates a host of other intracellular proteins, many of which are not kinases.

Protein kinases of the **protein kinase C** class are activated by Ca^{++} and by diglycerides, and certain other breakdown products of membrane phospholipids.

Insulin and certain growth factors bind to membrane receptors that are themselves protein kinases that phosphorylate protein substrates on tyrosine residues (the other protein kinases we have discussed phosphorylate proteins on serine or threonine residues). Binding of an agonist growth factor activates the **protein tyrosine kinase** activity of the receptor.

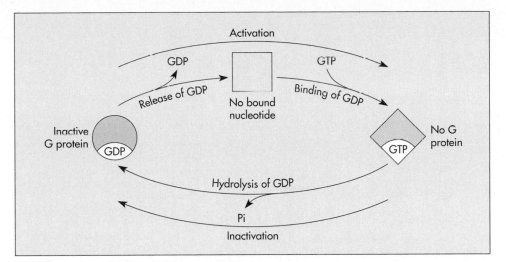

FIGURE 5-2   The activity cycle of a GTP-binding protein (G-protein). The inactive form of the G-protein *(colored circle)* binds GDP. Interaction of the G-protein with a ligand-bound membrane receptor promotes a conformational change leading to the release of GDP and the binding of GTP. The GTP-bound form of the G-protein *(colored diamond)* is the active form, which interacts with proteins such as adenylyl cyclase and ion channels to alter their activity. The G-protein has an intrinsic GTPase activity; hydrolysis of GTP converts the G-protein back to its inactive state.

G-PROTEIN–MEDIATED SIGNAL TRANSDUCTION PATHWAYS

Many hormones, neuromodulators, and other regulatory molecules that alter cellular processes do so by signal transduction pathways that involve heterotrimeric **GTP-binding proteins,** or **G-proteins.** (Another class of GTP-binding proteins, monomeric GTP-binding proteins, will be discussed later.) A G-protein is a molecular switch (Figure 5-2) that can exist in two states: in its activated ("on") state a G-protein has a higher affinity for GTP, whereas the inactivated ("off") G-protein preferentially binds GDP. When they have agonist molecules bound to them, some membrane receptors can interact with a G-protein to promote conversion of the G-protein to its activated state and its binding of GTP. The activated G-protein can then interact with many **effector proteins,** most notably enzymes or ion channels, to alter their activities. The activated G-protein has GTPase activity, so that eventually the bound GTP is hydrolyzed to GDP and the G-protein reverts to its inactive state (Figure 5-2).

Among the most important targets of activated G-proteins are molecules that change the cellular concentrations of the second messengers cyclic AMP, cyclic GMP, Ca^{++}, IP_3, and diglyceride (Figures 5-1 and 5-3). **Adenylyl cyclase** and **cyclic GMP phosphodiesterase,** the enzymes responsible for the synthesis of cyclic AMP

and for the breakdown of cyclic GMP, respectively, are powerfully modulated by G-protein–mediated mechanisms. Ca^{++} channels may be modulated directly by G-proteins or indirectly by second messenger–dependent protein kinases. Other effectors that are modulated by G-proteins include certain K^+ channels and phospholipases C, A_2, and D.

Characteristics of a general G-protein–protein kinase–mediated signal transduction pathway (see Figure 5-3):

1. A hormone or other regulatory molecule binds to its plasma membrane receptor.
2. The ligand-bearing receptor interacts with a G-protein and activates it. The activated G-protein binds GTP.
3. The activated G-protein interacts with one or more of the following to activate or inhibit them: adenylyl cyclase, cyclic GMP phosphodiesterase, Ca^{++} or K^+ channels, or phospholipases C, A_2, or D.
4. The cellular level of one or more of the following second messengers increases or decreases: cyclic AMP, cyclic GMP, Ca^{++}, IP_3, or diglyceride.
5. The increase in a second messenger changes the activity of one or more of the second messenger-dependent protein kinases: cyclic AMP–dependent protein kinase, cyclic GMP–dependent protein kinase, calmodulin-dependent protein kinase, or protein kinase C.

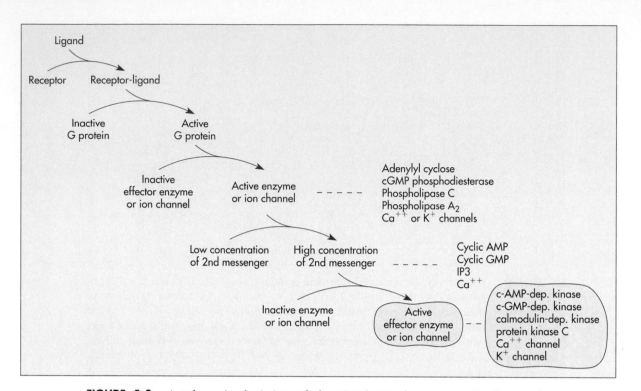

FIGURE 5-3 A schematic depiction of the signal transduction cascade by which an extracellular ligand such as a peptide hormone can bind to its receptor so as to activate a G-protein and, via the cascade, lead to the activation or inactivation of an ion channel, a protein kinase, or a phospholipase. At each level of the cascade amplification can occur: each receptor-ligand complex may activate more than one G-protein, each activated G-protein may activate more than one enzyme or ion channel, etc. For the sake of simplicity, the illustration indicates that occupation of receptor by ligand leads to an *increase* in a second messenger, which causes an *activation* of an enzyme or ion channel. In reality there are numerous cases in which receptor occupation leads to a *decreased* concentration of second messenger and cases in which increased second messenger results in *inactivation* of an enzyme or ion channel.

6. The level of phosphorylation of an enzyme or an ion channel is altered or an ion channel activity changes because of interaction with an activated G-protein, which brings about the final cellular response.

MEMBRANE PHOSPHOLIPIDS AND SIGNAL TRANSDUCTION PATHWAYS

Another class of extracellular agonists binds to receptors that activate, via a G-protein, a specific **phospholipase C** that cleaves phosphatidylinositol bisphosphate (a phospholipid present in minute quantities in the plasma membrane) into **inositol-1,4,5-trisphosphate** (IP_3) and **diglyceride** (Figure 5-4). Both IP_3 and diglyceride are second messengers. IP3 binds to specific ligand-gated Ca^{++} channels in the endoplasmic reticulum and releases Ca^{++} to increase its cytosolic level. The Ca^{++} channel of the endoplasmic reticulum has a structure similar to the Ca^{++} channel of the sarcoplasmic reticulum that is involved in excitation-contraction coupling of skeletal and cardiac muscle (Chapters 12 and 18). Diglyceride, together with Ca^{++}, activates another important class of protein kinases called **protein kinase C.** Among the substrates of protein kinase C are certain proteins involved in the control of cellular proliferation.

Phospholipase A_2 (PLA_2) and **phospholipase D** are also activated by some agonists via G-protein–dependent pathways. The products of the action of these enzymes on membrane phospholipids also activate protein kinase C (discussed later). PLA_2 cleaves the number 2 fatty acid from membrane phospholipids. Because some of the phospholipids are species with arachidonic

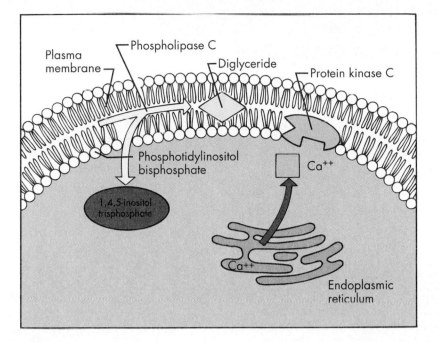

FIGURE 5-4   The signal transduction pathway activated by the hydrolysis of inositol phospholipids of the plasma membrane. A specific phospholipase C that hydrolyzes the minor membrane phospholipid phosphatidylinositol bisphosphate is stimulated by receptor-activated G-proteins. This cleavage releases IP_3 and diglyceride, both of which are second messengers. IP_3 binds to a specific Ca^{++} channel in the endoplasmic reticulum membrane, causing Ca^{++} to be released from the endoplasmic reticulum. Diglyceride, together with Ca^{++}, activates protein kinase C to phosphorylate important cellular effector proteins.

acid esterified to the number 2 carbon of the glycerol-backbone, PLA_2 releases significant amounts of **arachidonic acid.** Arachidonic acid, which is a regulatory molecule in its own right, is also the precursor for the cellular synthesis of **prostaglandins, prostacyclins,** and **leukotrienes,** important classes of potent regulatory molecules. Arachidonic acid is also a product of diglyceride breakdown.

Cleavage of membrane phosphatidylcholine by phospholipase D produces choline and phosphatidic acid. Phosphatases rapidly convert phosphatidic acid to diglyceride, which can activate protein kinase C.

MEMBRANE RECEPTORS FOR REGULATORY MOLECULES

The membrane receptors that mediate agonist-dependent activation of G-proteins are members of a protein family with more than 500 members. This family includes α- and β-adrenergic receptors, muscarinic acetylcholine receptors, serotonin receptors, adenosine receptors, receptors for peptide hormones, and rhodopsin. Members of the G-protein–coupled receptor family

(Figure 5-5) have 7 transmembrane helices of 22 to 28 amino acids that are predominantly hydrophobic.

GTP-BINDING PROTEINS (G-PROTEINS)

Heterotrimeric G-Proteins

A heterotrimeric G-protein has three subunits: an α *subunit* (40,000 to 45,000 daltons), a β *subunit* (about 37,000 daltons), and a γ *subunit* (8,000 to 10,000 daltons). Currently we know of about 20 different genes that encode α subunits, at least four genes that encode β subunits, and seven genes that encode γ subunits in mammals. The function and specificity of a G-protein are usually, but not always, determined by its α subunit. In most cases the β and γ subunits are tightly associated with one another. Heterotrimeric G-proteins function as intermediaries between the plasma membrane receptors for over 100 different extracellular regulatory substances (hormones, neuromodulators, etc.) and the intracellular processes they control. Binding of the regulatory substance to its receptor activates the G-protein, and the activated G-protein then either stimulates or inhibits an en-

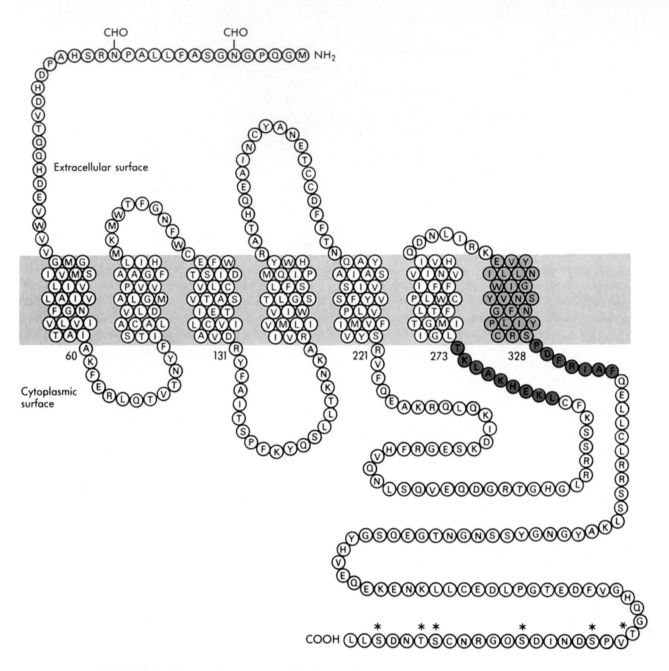

FIGURE 5-5 Proposed structure of the human β₂-adrenergic receptor. This receptor is a member of the family of G-protein–linked receptors that includes the other adrenergic receptors, the muscarinic acetylcholine receptor, the serotonin receptor, rhodopsin, and more than 500 other receptors. Among the structural features the members of this protein family have in common are an extracellular N-terminus with multiple N-linked glycosylation sites, seven transmembrane α-helices, a long intracellular loop connecting the sixth and seventh transmembrane helices, and an intracellular C-terminus with serines and threonines that are potential phosphorylation sites. The seventh transmembrane helix *(light color)* plays a role in agonist recognition. The C-terminal end of the third cytoplasmic loop and the N-terminal end of the cytoplasmic tail *(solid color)* appear to be involved in interacting with G-proteins. Serine and threonine residues near the C-terminus (indicated by *), when phosphorylated by β-adrenergic receptor kinase, promote desensitization of the receptor. The desensitized receptor has a diminished response to the agonist. (Modified from O'Dowd BF, Lefkowitz RJ, Caron MG: *Annu Rev Neurosci* 12:67, 1989.)

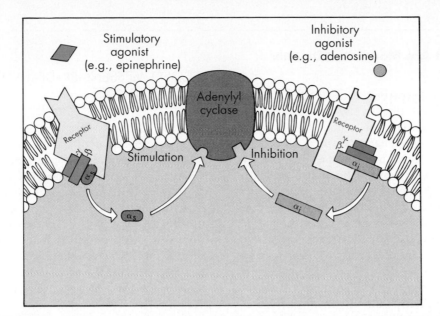

FIGURE 5-6 Adenylyl cyclase may be stimulated or inhibited via signal transduction pathways. Receptors for agonists that result in stimulation of adenylyl cyclase activate $G_s\alpha$, whose α_s subunit dissociates from $\beta\gamma$ and then interacts with adenylyl cyclase to stimulate it. Receptors for agonists that produce inhibition of adenylyl cyclase activate $G_i\alpha$, whose α_i subunit inhibits adenylyl cyclase.

zyme or an ion channel. In most instances the α subunit is the "business end" of the heterotrimeric G-protein (Figure 5-6). The inactive G-protein exists primarily as the $\alpha\beta\gamma$ hetcrotrimer, with GDP in its nucleotide binding site. The interaction of the heterotrimeric G-protein with a ligand-bearing receptor causes a conformational change in the α subunit to the active form, which has a higher affinity for GTP and a lower affinity for the $\beta\gamma$ pair. Therefore the activated α subunit releases GDP, binds GTP, and then dissociates from $\beta\gamma$. In most cases the dissociated α subunit then interacts with the next protein in the signal transduction pathway. In some cases, however, the $\beta\gamma$ dimer is responsible for all or some of the receptor-mediated response.

Regulation of Adenylyl Cyclase Cyclic AMP was the first second messenger to be discovered, and the regulation of adenylyl cyclase, the enzyme that produces cyclic AMP, is the prototype for G-protein–mediated signal transduction pathways. Adenylyl cyclase is subject to both positive and negative control by G-protein–mediated pathways (Figure 5-6). The binding of a stimulatory ligand, such as epinephrine acting through β-adrenergic receptors, results in activation of heterotrimeric G-proteins with α subunits of the type called α_s (s for stimulatory). Activation of the G_s G-protein by the ligand-bearing receptor causes its α_s subunit to bind GTP and then to dissociate from $\beta\gamma$. α_s then interacts with adenylyl cyclase to activate it.

A regulatory substance, such as adenosine, can inhibit adenylyl cyclase by binding to A_1 adenosine receptors,

thereby activating G_i-type G-proteins. G_i has α-subunits of a type, called α_i (i for inhibitory). Binding of the inhibitory ligand to its receptor activates the G_i-type G-protein and causes its α_i subunit to dissociate from the $\beta\gamma$ dimers. The activated α_i binds to and inhibits adenylyl cyclase (Figure 5-6). In addition, the $\beta\gamma$ dimers may bind to α_s and thus prevent the stimulation of adenylyl cyclase by stimulatory ligands.

Patients afflicted with **cholera** produce a watery diarrhea that can rapidly lead to dehydration and death if it is not promptly treated. The diarrhea is caused by a toxin produced by the bacterium *Vibrio cholerae*. Cholera toxin causes a massive secretion of salt and water into the lumen of the small bowel. A component of the cholera toxin enters the cells and catalyzes the covalent addition of ADP-ribose to the $G_s\alpha$ subunit and thereby permanently activates $G_s\alpha$, which results in permanent activation of adenylyl cyclase. As a consequence, cyclic AMP is permanently elevated. The brush border membrane that faces the lumen of the small intestine contains an electrogenic chloride channel that is stimulated to open by elevated cyclic AMP. In cholera the persistent activation of this Cl^- channel causes a secretion of Cl^-, Na^+, and water into the lumen of the small intestine. This secretion results in a persistent watery diarrhea of up to 20 liters per day.

Ion Channels That Are Modulated Directly by G-Proteins

In the previous chapter several ligand-gated ion channels that are modulated directly by an extracellular agonist, such as acetylcholine or γ-aminobutyric acid, were discussed. Other ion channels are regulated by second messenger–mediated mechanisms that involve G-proteins in the second step of the signal transduction cascade. Some ion channels, however, are directly modulated by G-proteins without the involvement of a second messenger. The binding of acetylcholine to M_2 muscarinic receptors in the heart and in certain neurons leads to the activation of a particular type of K^+ channels. Acetylcholine binding to the muscarinic receptor leads to activation of a G-protein of the G_i subclass. The activated α_i subunit then dissociates from the $\beta\gamma$ dimer. The $\beta\gamma$ dimer directly interacts with a particular class of K^+ channels to increase their probability of opening. The role of the activated α_i subunit in acetylcholine action on K^+ channels is controversial.

The L-type (dihydropyridine-sensitive) Ca^{++} channels in the heart (Chapter 17) and in skeletal muscle are activated in response to agonists that bind to the β-adrenergic receptor. The L-type channel may be regulated both directly by G-proteins and indirectly by a second messenger signal transduction cascade involving cyclic AMP–dependent protein kinase. For both the direct and indirect effects of β-adrenergic agonists on the L-type Ca^{++} channels, the G-protein involved is $G_s\alpha$.

Monomeric GTP-Binding Proteins

Cells contain another family of GTP-binding proteins called monomeric GTP-binding proteins; they are also known as low-molecular weight-G-proteins or small G-proteins (MW 20,000 to 35,000 daltons). Table 5-1 lists the major subfamilies of monomeric GTP-binding proteins and some of their properties. The Ras-like and Rho-like monomeric GTP-binding proteins are involved in the signal transduction pathways that link growth factor receptor tyrosine kinases to their intracellular effects. Among the processes that are regulated by pathways that involve monomeric GTP-binding proteins are polypeptide chain elongation in protein synthesis, proliferation and differentiation of cells, neoplastic transformation of cells, control of the actin cytoskeleton and linkages between the cytoskeleton and the extracellular matrix, transport of vesicles among different organelles, and exocytotic secretion.

Monomeric GTP-binding proteins, like their heterotrimeric cousins, operate via a cycle of activation and inactivation similar to the one shown in Figure 5-2.

SECOND MESSENGER–DEPENDENT PROTEIN KINASES

Cyclic AMP was first identified as a second messenger in investigations of the mechanisms involved in the hormonal control of glycogen synthesis and breakdown (Chapter 42). The phosphorylation of rate-determining enzymes in these metabolic pathways by cyclic AMP–dependent protein kinases is responsible for the hormonal regulation of glycogen metabolism.

It is now recognized that many other enzymes and effector molecules are regulated by cyclic AMP–dependent protein kinase and that some protein kinases depend on the concentrations of other second messengers (Figure 5-1).

Cyclic AMP–Dependent Protein Kinase

In the absence of cyclic AMP, cyclic AMP–dependent protein kinase is composed of 4 subunits: 2 regulatory subunits and 2 catalytic subunits. Most cell types contain the same catalytic subunit, but their regulatory subunits differ significantly. The presence of the regulatory subunits greatly inhibits the enzymatic activity of the complex. In the presence of micromolar levels of cyclic AMP each regulatory subunit binds 2 molecules of cyclic AMP. The binding of cyclic AMP causes a conformational change in the regulatory subunits and diminishes their affinity for binding the catalytic subunits. Hence, the regulatory subunits dissociate from the catalytic subunits, and in this way the catalytic subunits become activated (Figure 5-7). The active catalytic subunit phosphorylates target proteins on particular serine and threonine residues.

Comparison of the amino acid sequence of cyclic AMP–dependent protein kinase with representatives of the other classes of protein kinases shows that, in spite of vast differences in their regulatory properties, the different classes of protein kinases share a common core with high amino acid homology (Figure 5-8). The core structure includes the ATP-binding domain and the enzymes's active center, where the transfer of phosphate from ATP to the acceptor protein occurs. Regions of the kinases outside the catalytic core are involved in regulation of the kinase activities.

Calmodulin-Dependent Protein Kinases

A host of vital cellular processes, including release of neurotransmitters, secretion of hormones, and muscle contraction, are regulated by the cytosolic level of Ca^{++}. One way that Ca^{++} exerts control is by binding to calmodulin. The complex of Ca^{++} and calmodulin then can

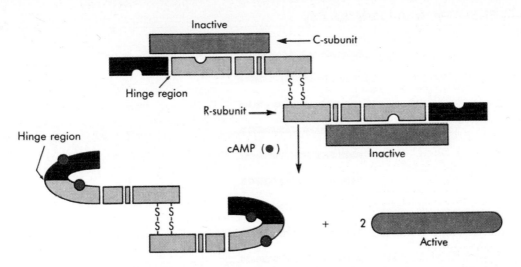

FIGURE 5-7 Activation of cyclic AMP–dependent protein kinase. The two regulatory subunits (R-subunit) of the R_2C_2 complex are held together by two disulfide bonds. Binding of two molecules of cyclic AMP to each R-subunit causes flexion in a hinge region of each R-subunit and releases the two active catalytic subunits (C-subunit). (Redrawn from Taylor S: *J Biol Chem* 264:8443, 1989.)

Table 5-1 Subfamilies of Monomeric GTP-Binding Proteins and the Intracellular Processes They Regulate

Subfamily	Cellular Effects
Ras-like proteins	Control growth and differentiation
Rho-like proteins (including Rac)	Control polymerization of actin filaments and their assembly into particular structures like focal adhesions.
Rab-like proteins	Control vesicle trafficking by helping target vesicles to particular membranes
ARF-like proteins	Regulate the assembly and disassembly of vesicle coat proteins and thereby control vesicle traffic

influence the activity of many different proteins, among them a group of protein kinases known as **calmodulin-dependent protein kinases** (Figure 5-9). Dedicated calmodulin-dependent protein kinases, such as myosin light-chain kinase, have only one cellular substrate. Multifunctional calmodulin-dependent protein kinases phosphorylate more than one substrate protein.

Myosin light-chain kinase plays a central role in the regulation of contraction of smooth muscle (Chapter 14). Elevation of the cytosolic Ca^{++} concentration in a smooth muscle cell stimulates the activity of myosin light-chain kinase; the resulting phosphorylation of the regulatory light chains of myosin allows contraction of smooth muscle cells to proceed.

Protein Kinase C

The activation of protein kinase C by Ca^{++} and diglycerides, produced as a result of the receptor-activated hydrolysis of phosphatidylinositol bisphosphate, was briefly described above. The best-known pathway of ac-

tivation of protein kinase C is as follows. In an unstimulated cell much of the protein kinase C is present in the cytosol and is inactive. When cytosolic levels of Ca^{++} rise, Ca^{++} binds to protein kinase C. This causes protein kinase C to bind to the inner surface of the plasma membrane, where it can be activated by the diglyceride that is produced by the hydrolysis of phosphatidylinositol bisphosphate. Membrane phosphatidylserine is also a potent activator of protein kinase C, once the enzyme has bound to the membrane.

Protein kinase C plays a vital role in the control of certain cellular processes. The primary action of certain lipophilic tumor-promoting substances, most notably the **phorbol esters,** is to activate protein kinase C directly. This powerfully stimulates cell division in many cell types and converts normal cells with controlled growth properties to transformed cells that resemble tumor cells in their uncontrolled growth.

At the present time 10 different isoforms of protein kinase C have been discovered. Although certain of the subtypes are present in many or most mammalian cells,

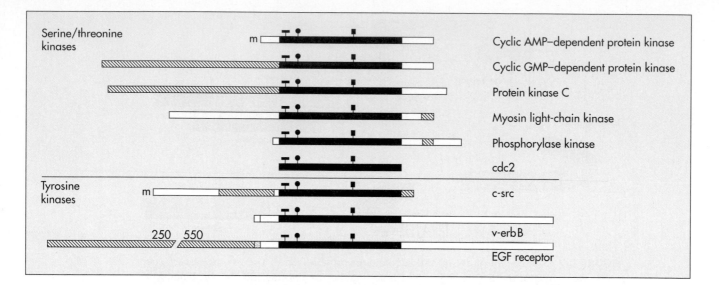

FIGURE 5-8 The protein kinase family. All known protein kinases share a common catalytic-core *(black)* that contains ATP- and peptide-binding domains and the active site where phosphoryl transfer occurs. Conserved residues are aligned with lysine 72 *(black circle)*, aspartate 184 *(black square),* and the glycine-rich loop *(small black rectangle)* of the catalytic subunit of cyclic AMP–dependent protein kinase. Regions important for regulation are cross-hatched. The membrane-spanning segment of the EGF receptor is stippled. Sites of myristoylation are indicated by *m.* A covalently attached myristic acid residue helps to anchor the protein kinase in the plasma membrane. (Adapted from Taylor S et al: *Annu Rev Cell Biol* 8:429, 1992.)

other subtypes are localized predominantly in certain tissues or cell types. Some of the subtypes are bound to the plasma membrane in unstimulated cells and thus do not require elevated Ca^{++} for activation. The subtypes also require different activators. Some of the isoforms of protein kinase C are activated by arachidonic acid, by other unsaturated fatty acids, or by lysophospholipids.

A current view of the multiple pathways of phospholipid hydrolysis that may be involved in stimulating protein kinase C is shown in Figure 5-10.

PROTEIN TYROSINE KINASES

Proteins with intrinsic **protein tyrosine kinase** activity constitute another family of membrane receptors, receptors that are not linked to G-proteins. When these receptors bind agonist, their tyrosine kinase activity is turned on and they phosphorylate specific effector proteins on particular tyrosine residues. The other protein kinases that have already been discussed phosphorylate proteins on serine and threonine residues exclusively.

The receptors for certain peptide hormones and growth factors are proteins with a glycosylated extra-cellular domain, a single transmembrane sequence, and an intracellular domain with protein tyrosine kinase activity. Members of this superfamily (Figure 5-11) of peptide receptors include the receptors for insulin and related growth factors, epidermal growth factor (EGF), platelet-derived growth factor (PDGF), colony-stimulating factor (CSF), and fibroblast growth factor (FGF). The binding of hormone or growth factor to its receptor triggers multiple cellular responses, including Ca^{++} influx, increased Na^+/H^+ exchange, stimulation of the uptake of sugars and amino acids, and stimulation of phospholipase C and hydrolysis of phosphatidylinositol bisphosphate.

The known protein tyrosine kinase receptors fall into 8 subfamilies, four of which are depicted in Figure 5-11. Binding of ligand to the receptor results in dimerization of the receptor-ligand complexes. The dimerization enhances binding affinity and activates the protein tyrosine kinase activity. In subclass II receptors, the insulin receptor family, the unliganded receptor exists as a disulfide-linked dimer, and binding of insulin results in a conformational change in both "monomers." This change enhances ligand binding and activates the tyrosine kinase activity. The tyrosine kinase activity of the receptors is required for the signal transduction function of the receptors.

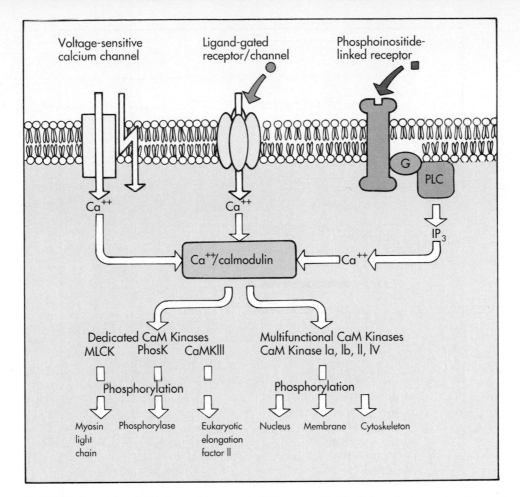

FIGURE 5-9 Calmodulin-dependent protein kinases are the final steps in many signal transduction pathways that are elicited by increases in cytosolic Ca^{++} levels. Cytosolic Ca^{++} may rise because of Ca^{++} influx via a voltage- or ligand-gated ion channel or by the release of Ca^{++} from internal stores by IP_3. The complex of Ca^{++} with calmodulin activates calmodulin-dependent protein kinases (CaM kinases). Dedicated CaM kinases phosphorylate specific effector proteins such as the regulatory light chain of myosin, phosphorylase, and elongation factor II. Multifunctional CaM kinases phosphorylate multiple proteins of the nucleus or cytoskeleton or membrane proteins. *PLC,* Phospholipase C; *MLCK,* myosin light-chain kinase; *PhosK,* phosphorylase kinase; *CaMKIII,* calmodulin-dependent kinase III. (Adapted from Schulman H: *Curr Opin Cell Biol* 5:247, 1993.)

Protein tyrosine kinases that are "out of control" play a central role in cell transformation and **cancer.** In some cell types, a mutation of the receptor renders it active in phosphorylating tyrosines, regardless of the presence or absence of a growth factor. Other tumor cells secrete a growth factor *and* overexpress its receptor. This overexpression leads to abnormally high rates of protein tyrosine kinase activity.

Mutations in Ras may produce overactive Ras forms that constitutively activate the downstream effectors that normally are active only in the presence of growth factors. In such cases cell growth may be uncontrolled. Approximately 30% of human cancers involve mutated Ras proteins.

Monomeric GTP-binding proteins of the Ras family (Table 5-1) are involved in coupling the binding of mitogenic ligands to their tyrosine protein kinase receptors to the resultant intracellular effects on cell proliferation. When Ras is inactive, cells cannot respond to the growth factors that operate via receptor tyrosine kinases.

PROTEIN PHOSPHATASES AND THEIR MODULATION

Phosphorylation of proteins is one of the most significant means of regulating their activities. The extent of phosphorylation of a regulated protein is the result of the activities of the protein kinase that phosphorylates

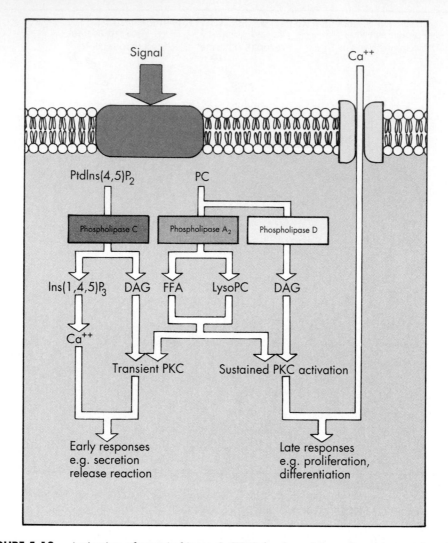

FIGURE 5-10 Activation of protein kinase C *(PKC)* by degradation of membrane phospholipids. Rapid and transient activation of PKC is effected by IP$_3$ *[Ins(1,4,5)P$_3$]* and diglyceride *(DAG)* formed by the degradation of phosphatidylinositol bisphosphate *[PtdIns(4,5)P$_2$]* by a specific receptor-activated phospholipase C. Slower and more sustained activation of PKC is caused by free fatty acids *(FFA)*, lysolecithin *(lysoPC)*, and DAG released by phospholipases A$_2$ and D acting on phosphatidylcholine *(PC)*. (Adapted from Asaoka Y et al: *Trends Biochem Sci* 17:414, 1992.)

Table 5-2 Properties of Subtypes of Serine-Threonine Protein Phosphatases

Subtype	PP-1	PP-2A	PP-2B	PP-2C
Preference for the α or β subunit of phosphorylase kinase	β subunit	α subunit	α subunit	α subunit
Inhibition by I-1 and I-2	Yes	No	No	No
Absolute requirement for divalent cations	No	No	Yes (Ca^{++})	Yes (Mg^{++})
Stimulation by calmodulin	No	No	Yes	No
Inhibition by okadaic acid (K$_1$)	Yes (20 nM)	Yes (0.2 nM)	Yes (5 μM)	No
Phosphorylase phosphatase activity	High	High	Very low	Very low

Modified from Cohen P: *Annu Rev Biochem* 58:453, 1989.

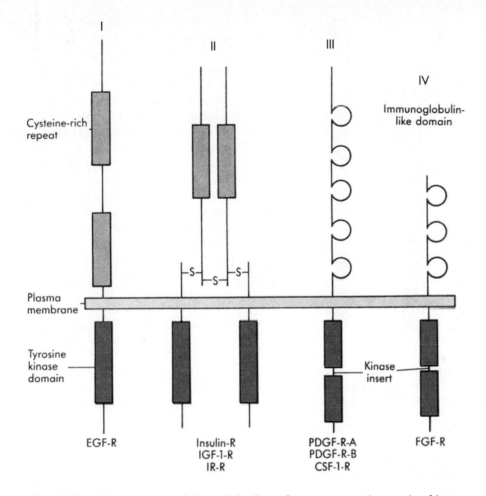

FIGURE 5-11   The structures of four subfamilies of receptor protein tyrosine kinases are depicted; eight subfamilies have been identified. Subfamilies I and II have extracellular domains with cysteine-rich repeat sequence domains *(light color)*, while the extracellular domains of subclasses III and IV have immunoglobulin-like regions *(loops)*. The protein tyrosine kinase domains *(solid color)* are the most conserved sequences. The short intracellular region just inside the membrane, the kinase insert region (quite variable in length) and the carboxyl terminal tail are sites of regulation of protein kinase activity. *EGF,* Epidermal growth factor; *IGF-1,* insulin-like growth factor-1; *IR,* insulin-related protein; *PDGF,* platelet-derived growth factor; *CSF-1,* colony-stimulating factor 1; *FGF,* fibroblast growth factor. (Adapted from Ullrich A, Schlessinger J: *Cell* 61:203, 1990.)

that protein and the **protein phosphatase** that dephosphorylates it. In addition to the different types of protein kinases that we have discussed, all cells also contain protein phosphatases whose task is to reverse the effects of protein phosphorylation. In keeping with the classification of protein kinases, protein phosphatases are classified as **serine-threonine protein phosphatases** and **tyrosine protein phosphatases.**

Serine-Threonine Protein Phosphatases

The serine-threonine protein phosphatases are a large family of structurally related molecules. At present they

are classified as type 1 **(PP-1)** or type 2 **(PP-2)**; the type is based on which subunit of phosphorylase kinase they prefer to dephosphorylate. PP-1's prefer the β-subunit, but PP-2's prefer the α-subunit of phosphorylase kinase (Table 5-2). The PP-2's are subclassified into PP-1A, PP-1B, and PP-1C; the type is based on their regulation by divalent cations (Table 5-2). PP-1 and the PP-2's can also be distinguished by their inhibition by **okadaic acid** (Table 5-2), a complex fatty acid produced by marine dinoflagellates. Okadaic acid is as potent a tumor promoter as the phorbol esters, presumably because both okadaic acid and phorbol esters enhance the phosphorylation of certain substrates of protein kinase C.

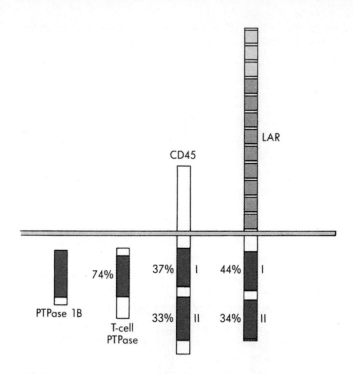

FIGURE 5-12   Protein tyrosine phosphatases (PTPases) schematically depicted. About 65 different PTPases have been identified. Shown are two small cytosolic PTPases: PTPase 1B, from human placenta, and T-cell PTPase, from human T-lymphocytes. Also shown are two transmembrane PTPases: CD45 (the leukocyte common antigen) and LAR (leukocyte common antigen related protein). The solid-colored cytosolic segments of each protein are the PTPase catalytic domains. The extracellular domains of LAR are homologous to N-CAM (neural cell adhesion molecule). The intermediate-colored domains are homologous to the IgG-like domains, and the light-colored domains are homologous to the non-IgG-like domains of N-CAM. (Adapted from Tonks NK, Charbonneau H: *Trends Biochem Sci* 14:497, 1989.)

Protein Tyrosine Phosphatases

Protein tyrosine phosphatases (PTPases) are not structurally homologous to serine-threonine protein phosphatases. Recall that, by contrast, all the protein kinases may derive from a common ancestor protein kinase. Figure 5-12 schematically depicts 4 of the 65 PTPases that are currently known. Note that, although two of the PTPases are small cytosolic proteins, two other PTPases are larger transmembrane proteins. Based on this structure, the transmembrane PTPases are probably receptors whose PTPase activity is modulated by extracellular ligands.

ATRIAL NATRIURETIC PEPTIDE RECEPTOR AND GUANYLYL CYCLASES

Earlier in this chapter we described the actions of certain extracellular agonists to bind to membrane recep-

tors that either stimulate adenylyl cyclase via G_s heterotrimeric G-proteins or inhibit adenylyl cyclase via G_i. Membrane receptors for **atrial natriuretic peptide** (ANP) are notable because the receptors themselves possess guanylyl cyclase activity that is stimulated when ANP binds to the receptor. Thus the binding of ANP elevates intracellular levels of the second messenger cyclic GMP without the intermediation of a G-protein or any other signal-transducing proteins. ANP is released by cells of the atrium of the heart in response to an elevation of blood pressure (Chapter 19). This hormone then acts to increase the excretion of NaCl and water by the kidney (Chapter 37) and to diminish the constriction of certain blood vessels.

SUMMARY

1. Many regulatory substances exert their effects on cellular processes via few signal transduction pathways.
2. Heterotrimeric GTP-binding proteins serve as intermediaries between a receptor that has been activated by binding an agonist and enzymes and ion channels whose activity is modulated in response to agonist binding.
3. A GTP-binding protein that has been activated by interacting with an agonist-bearing receptor, then changes the activity of an enzyme or an ion channel to alter the intracellular concentration of a second messenger such as cyclic AMP, cyclic GMP, Ca^{++}, IP_3, or diglyceride.
4. An increased level of one or more of the second messengers may increase the activity of a second messenger–dependent protein kinase: cyclic AMP–dependent protein kinase, cyclic GMP–dependent protein kinase, calmodulin-dependent protein kinase, or protein kinase C.
5. Myriad cellular processes are regulated via the phosphorylation of enzymes and ion channels.
6. Certain membrane receptors for hormones and growth factors are protein tyrosine kinases that are activated directly by binding of the agonist.
7. Monomeric GTP-binding proteins are intermediaries between the binding of growth factors to their protein tyrosine kinase receptors and the downstream effects on cellular proliferation. The small G-proteins also regulate the function of the actin cytoskeleton and intracellular vesicular trafficking.
8. Protein phosphatases, themselves subject to complex regulation by agonists and second messengers, reverse the effects of protein phosphorylation.

Journal articles

Berridge MJ: Inositol triphosphates and calcium signalling, *Nature* 361:315, 1993.

Birnbaumer L: Receptor-to-effector signaling: roles for beta gamma dimers as well as alpha subunits, *Cell* 71:1069, 1992.

Bourne HR, Sanders DA, McCormick F: The GTPase superfamily: conserved structure and molecular mechanism, *Nature* 349:117, 1991.

Brown AM, Birnbaumer L: Ionic channels and their regulation by G-protein subunits, *Annu Rev Physiol* 52:197, 1990.

Charbonneau H, Tonks NK: 1002 protein phosphatases? *Annu Rev Cell Biol* 8:463, 1992.

Cohen, P: Signal integration at the level of protein kinases, protein phosphatases and their substrates, *Trends Biochem Sci* 17:408, 1992.

Fantl WJ, Johnson DE, Williams LT: Signalling by receptor tyrosine kinases, *Annu Rev Biochem* 62:453, 1993.

Ferris CD, Snyder SH: Inositol 1,4,5-trisphosphate-activated calcium channels, *Annu Rev Physiol* 54:469, 1992.

Hall A: The cellular functions of small GTP-binding proteins, *Science* 249:635, 1990.

Hall A: Ras-related proteins, *Curr Opin Cell Biol* 5:265, 1993.

Harden TK: G-protein regulated phospholipase C: identification of component proteins, *Adv Sec Messenger Phosphoprot Res* 26:11, 1992.

Hepler JR, Gilman AG: G proteins, *Trends Biochem Sci* 17:383, 1992.

Hille B: G protein-coupled mechanisms and nervous signaling, *Neuron* 9:187, 1992.

Iniguez-Lluhi J, Kleuss C, Gilman AG: The importance of G-protein beta-gamma subunits, *Trends Cell Biol* 3:230, 1993.

Lamb TD, Pugh EN, Jr: G-protein cascades: gain and kinetics, *Trends Neurosci* 15.291, 1992.

Lefkowitz RJ: G-protein receptor kinases, *Cell* 74:409, 1993.

Linder ME, Gilman AG: G proteins, *Sci Am* 267:36, 1992.

Lowy DR, Willumsen BM: Function and regulation of Ras, *Annu Rev Biochem* 62:851, 1993.

Meldolesi J: Multifarious IP$_3$ receptors, *Curr Biol* 2:393, 1992.

Michell RH: Inositol lipids in cellular signalling mechanisms, *Trends Biochem Sci* 17:274, 1992.

Nishida E, Gotoh Y: The MAP kinase cascade is essential for diverse signal transduction pathways, *Trends Biochem Sci* 18:128, 1993.

Nishizuka Y: Intracellular signaling by hydrolysis of phospholipids and activation of protein kinase C, *Science* 258:607, 1992.

Nishizuka Y: Signal transduction: crosstalk, *Trends Biochem Sci* 17:367, 1992.

Posada J, Cooper JA: Molecular signal integration: interplay between serine, threonine, and tyrosine phosphorylation. *Mol Biol Cell* 3:583, 1992.

Ruderman JV: MAP kinase and the activation of quiescent cells, *Curr Opin Cell Biol* 5:207, 1993.

Schlessinger J, Ullrich A: Growth factor signaling by receptor tyrosine kinases, *Neuron* 9:383, 1992.

Schulman H: The multifunctional Ca^{2+}/calmodulin-dependent protein kinases, *Curr Opin Cell Biol* 5:247, 1993.

Sternweis PC, Smrcka AV: Regulation of phospholipase C by G proteins, *Trends Biochem Sci* 17:502, 1992.

Szabo G, Otero AS: G-protein-mediated regulation of K^+ channels in heart, *Annu Rev Physiol* 52:293, 1990.

Tang W-J, Gilman AG: Adenylyl cyclases, *Cell* 70:869, 1992.

Taylor CW, Marshall ICB: Calcium and inositol 1,4,5-trisphosphate receptors: a complex relationship, *Trends Biochem Sci* 17:403, 1992.

Taylor SS, Knighton DR, Zheng J, TenEyck LF, Sowadski J: Structural framework for the protein kinase family, *Annu Rev Cell Biol* 8:429, 1992.

Walton KM, Dixon JE: Protein tyrosine phosphatases, *Annu Rev Biochem* 62:101, 1993.

BIBLIOGRAPHY
Books and monographs

Barritt GJ: *Communication within animal cells,* Oxford, 1992, Oxford Science Publications.

Cohen P, Klee C, eds: *Calmodulin* (Molecular aspects of cellular regulation, vol. 5), New York, 1988, Elsevier.

Hardie, DG: *Biochemical messengers: hormones, neurotransmitters and growth factors,* London, 1990, Chapman and Hall.

Houslay MD, Milligan G, eds.: *G-proteins as mediators of cellular signalling processes,* 1990, New York, Wiley.

Michell RH, Drummond AH, Downes CP, eds.: *Inositol lipids in cell signaling,* New York, 1989, Academic Press.

Peroutka SJ, ed.: *G-protein-coupled receptors,* Boca Raton, Fla, 1994, CRC Press.

NERVOUS SYSTEM

WILLIAM D. WILLIS, JR.

PART II

CHAPTER 6

Cellular Organization of the Nervous System

The nervous system is a communication network that allows an organism to interact in appropriate ways with the environment. This system has *sensory components* that detect environmental events, *integrative components* that process sensory data and information stored in memory, and *motor components* that generate movements and other activity. The nervous system can be subdivided into peripheral and central parts, each with a number of further subdivisions.

The functional unit of the nervous system is the **neuron,** a cell whose processes, the **dendrites** and **axon,** make synaptic connections with other neurons to form the communications network. **Neuroglia** are supportive cells of the nervous system. They assist neurons in performing their functions; for example, certain neuroglia provide **myelin sheaths** that surround axons and that speed the conduction of nerve impulses. Neural activity is coded, and information is passed from one neuron to the next by **synaptic transmission,** which generally involves the release of a chemical substance from the axon terminals of one neuron and an action of that substance on the dendrites of the next neuron. The released chemical substance is called a **neurotransmitter.** Axons not only transmit information but also move chemical substances by a special energy-dependent transport process involving microtubules and called **axonal transport.** Damage to axons may lead to an **axonal reaction** in the cell bodies, **wallerian degeneration** of the distal part of the axon, and sometimes **transneuronal degeneration.** Axonal transport and the reactions to injury can be used experimentally to trace neural pathways. **Regeneration** of neurons is more effective in peripheral than in central neurons.

The environment of neurons in the central nervous system is highly regulated. The extracellular fluid can exchange substances readily with the **cerebrospinal fluid** (CSF), which is formed within the ventricular spaces of the brain by the choroid plexuses and which circulates over the surface of the brain and spinal cord. However, the composition of the CSF is determined by secretory processes in the choroid plexuses. Interchanges with the blood are restricted by the blood-brain barrier, which is formed by tight junctions between capillary endothelium cells and by neuroglial processes.

GENERAL FUNCTIONS OF THE NERVOUS SYSTEM

Functions of the nervous system include *sensory detection, information processing,* and *behavior.* Learning and memory permit behavior to change appropriately in response to environmental challenges that are based on past experience. Other systems, such as the endocrine and immune systems, share these functions, but the nervous system is specialized for them.

Excitability is a cellular property that enables neurons to perform their functions. Excitability is manifested by such electrical events as **nerve impulses** (or action potentials), **receptor potentials,** and **synaptic potentials** (see Chapter 4). Chemical events often accompany these electrical ones.

Sensory detection is accomplished by special nerve cells called **sensory receptors.** Various forms of energy are sensed, including light, sound and other mechanical events, chemicals, temperature gradients, and, in some animals, electrical fields.

Information processing in neural circuits depends on intercellular communication, which is accomplished by nerve cells as they respond to and generate chemical signals. The mechanisms involved require both electrical and chemical changes.

Behavior may be covert, as in cognition or memory, but it is often readily observable as a motor act, such as a movement or an autonomic response. In humans a particularly important set of behaviors are those involved in language.

ORGANIZATION OF THE NERVOUS SYSTEM

The nervous system consists of a highly complex aggregation of cells, part of which forms a communication network and another part, a supportive matrix. The communication network is formed by **neurons.** The cells involved in communication are specialized for receiving and making decisions on information and for transmitting signals to other neurons or to effector cells. The human brain contains approximately 10^{12} neurons. The supportive cells of the nervous system include the **neuroglia** (meaning nerve glue). These cells help maintain an appropriate local environment for neurons, or they ensheath axons to increase the speed of nerve impulse propagation (Chapter 4). The human brain has 10 times as many neuroglia as neurons.

The nervous system in most animals can be subdivided into a **peripheral nervous system** and a **central nervous system.** The emphasis here is on the mammalian nervous system, especially that of humans.

Peripheral Nervous System

The peripheral nervous system provides an interface between the central nervous system and the environment, including both the external world and the body apart from the nervous system. The peripheral nervous system includes a sensory component, formed by **sensory receptor organs** and **primary afferent neurons,** and motor components to command effector organs to perform muscular or glandular activity. Motor components include **somatic motor fibers, autonomic ganglia,** and **autonomic motor fibers.** Autonomic neurons can be further subdivided into **sympathetic, parasympathetic,** and **enteric neurons.** Somatic motor fibers cause contraction of the skeletal muscle fibers. Autonomic motor fibers excite or inhibit cardiac or smooth muscle and glands. The actions of the sympathetic nervous system include preparing the organism for emergency action, whereas the parasympathetic and enteric nervous systems promote more routine activities, such as digestion. Ordinarily the sympathetic and parasympathetic nervous systems work together in regulating visceral function.

Central Nervous System

The central nervous system (CNS) includes the **spinal cord** and the **brain** (Figure 6-1). The brain can be further subdivided into five regions, based on embryologic development: the **myelencephalon, meten-**

cephalon, mesencephalon, diencephalon, and te-**lencephalon.** In the adult brain the myelencephalon becomes the **medulla;** the metencephalon, the **pons** and **cerebellum;** the mesencephalon, the **midbrain;** the diencephalon, the **thalamus** and **hypothalamus;** and the telencephalon, the **basal ganglia** and the various lobes of the **cerebral cortex.** The thalamus and basal ganglia are hidden from view in Figure 6-1.

COMPOSITION OF THE NERVOUS TISSUE

Nervous tissue consists largely of neurons and neuroglia. Neurons are responsible for communications. Neu-

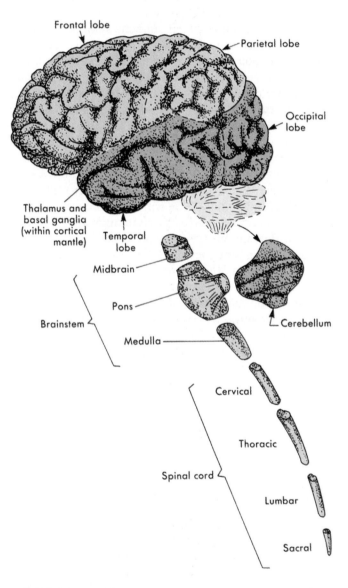

FIGURE 6-1 Exploded view showing the major components of the central nervous system. Also shown are the four major divisions of the cerebral cortex: the occipital, parietal, frontal, and temporal lobes.

rons that communicate with the periphery include **sensory receptor cells** and **somatic** and **autonomic motoneurons.** Other neurons perform integrative tasks through activity in central neural networks. Not surprisingly, given the many different roles that neurons play, neurons display a great variety of shapes and sizes (Figure 6-2).

Support for neurons is provided by **neuroglial cells** (Figure 6-3). These include **Schwann cells** and **satellite cells** in the peripheral nervous system and **astrocytes** and **oligodendroglia** (oligodendrocytes) in the CNS. **Microglia** and **ependymal cells** are also considered neuroglia.

Myelin is a spiral structure that enwraps many axons and is composed of the lipoprotein surface membranes of Schwann cells and oligodendroglia (Figure 6-3, *C*). It allows nerve impulses to be conducted rapidly. Satellite cells encapsulate dorsal root and cranial nerve ganglion cells and regulate their microenvironment. Astrocytes (Figure 6-3, *A* and *B*) play a similar role in the CNS, although astrocytes contact only a part of the surface of central neurons. However, their processes surround groups of synaptic endings and isolate them from adjacent synapses. Astrocytes have foot processes that contact capillaries and the connective tissue at the surface of the CNS (Figure 6-3,*A*). These foot processes may help limit the free diffusion of substances into the CNS.

Microglia (Figure 6-3, *D*) are phagocytes that remove the products of cellular damage from the CNS. They are probably derived from the circulation.

Ependymal cells (Figure 6-3, *E*) form an epithelium that separates the CNS from the ventricles, a series of cavities within the brain; these cavities contain **cerebrospinal fluid** (CSF). Many substances diffuse readily across the ependyma between the extracellular space of the brain and the CSF. CSF is secreted by specialized ependymal cells of the **choroid plexuses,** which are found in certain parts of the ventricular system.

Delivery of nutrients and removal of wastes are accomplished by the vascular system. Capillaries and other blood vessels are abundant in nervous tissue. Diffusion of many substances between the blood and the CNS is limited by the **blood-brain barrier.** This barrier is formed chiefly by tight junctions between capillary endothelial cells.

The external surface of the CNS is covered by several layers of connective tissue. These layers form the **pia mater, arachnoid,** and **dura mater.**

MICROSCOPIC ANATOMY OF THE NEURON

Most neurons have the following parts: a **cell body,** or soma; one or more **dendrites;** and an **axon.**

The **cell body** (Figure 6-4) contains the nucleus and

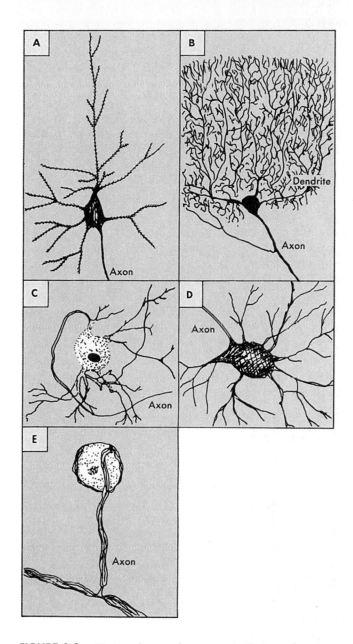

FIGURE 6-2   Various forms of neurons. **A,** Neuron characterized by a cell body that has a roughly pyramidal shape. This type of neuron, called a pyramidal cell, is typical of the cerebral cortex. Note the many spinous processes lining the surface of the dendrites. **B,** Cell type first described by the Czech neuroanatomist Purkinje and since known as the Purkinje cell. Purkinje cells are characteristic of the cerebellar cortex. The cell body is pear shaped, with a rich dendritic plexus originating from one end and the axon from the other. The fine branches of the dendrites are covered with spines (not shown). **C,** A sympathetic postganglionic motoneuron. **D,** An α-motoneuron of the spinal cord. Both **C** and **D** are multipolar neurons with radially arranged dendrites. **E,** A sensory dorsal root ganglion cell; no dendrites are present. The axon branches into a central and a peripheral process. Because the axon results from fusion of two processes during embryonic development, these cells are described as pseudounipolar neurons rather than unipolar.

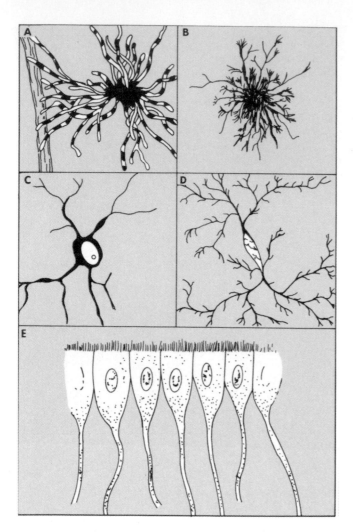

FIGURE 6-3 Different types of neuroglial cells of the central nervous system. **A,** Fibrous astrocyte; note the glial foot processes in association with a capillary. **B,** Protoplasmic astrocyte. **C,** An oligodendrocyte. Each of the processes is responsible for the production of one or more myelin sheath internodes about central axons. **D,** Microglial cell. **E,** Ependymal cells.

nucleolus of the neuron. It possesses a well-developed biosynthetic apparatus for the manufacture of membrane constituents, synthetic enzymes, and other chemical substances needed for the specialized functions of the nerve cell. The neuronal biosynthetic apparatus includes **Nissl bodies,** which are stacks of **rough endoplasmic reticulum,** the organelle that is responsible for protein synthesis. The soma also contains a prominent **Golgi apparatus,** which packages materials into vesicles for transport to other parts of the cell, and numerous mitochondria and cytoskeletal elements, including neurofilaments and microtubules. **Neurofilaments** are thin, rod-like structures, whereas **microtubules** are larger cylinder-like structures. **Lipofuscin** is a pigment formed from incompletely degraded membrane components, and it accumulates in some neurons. A few groups

of neurons in the brainstem contain melanin pigment.

The **dendrites** (Figure 6-4) are extensions of the cell body. In some neurons dendrites may be about 1 mm in length, and they account for more than 90% of the surface area of many neurons. The proximal dendrites (near the cell body) contain Nissl bodies and parts of the Golgi apparatus. However, the main cytoplasmic organelles in dendrites are microtubules and neurofilaments.

The **axon** (Figure 6-4) arises from the soma (or sometimes from a dendrite) in a specialized region called the **axon hillock.** The axon hillock and axon differ from the soma and proximal dendrites because they lack rough endoplasmic reticulum and **free ribosomes,** as well as the Golgi apparatus. The axon contains smooth endoplasmic reticulum and a prominent cytoskeleton. Axons may be short and, as with dendrites, terminate near the soma (Golgi type 1 neurons), or they may be long (Golgi type 2 neurons) and extend for a meter or more.

Axons may be ensheathed or bare. In the peripheral nervous system, axons are always ensheathed by **Schwann cells.** Many axons are surrounded by a spiral, multilayered wrapping of Schwann cell membrane called a **myelin sheath.** In the CNS, myelinated axons are ensheathed by oligodendroglia (Figure 6-5); other axons are unmyelinated. In the peripheral nervous system, unmyelinated axons are embedded in Schwann cells but are not wrapped in myelin (Figure 6-6). A group of such axons and the accompanying Schwann cells are called a **bundle of Remak.** In the CNS, unmyelinated axons are bare.

In some diseases of the nervous system, myelin may be lost over one or more internodes of many axons without interruption of the axons. In such cases, conduction of nerve impulses may be slowed or blocked, and the function of the affected axons is therefore abnormal. Such demyelination occurs in the peripheral nervous system in the **Guillain-Barré syndrome** and in **diphtheria.** An important demyelinating disease of the central nervous system is **multiple sclerosis.**

TRANSMISSION OF INFORMATION

A major role of axons is to transmit information from the region of the cell body and dendrites of a neuron to synapses on other neurons or effector cells. The information is generally transmitted as a series of **nerve impulses.**

The speed of transmission of information depends partly on the conduction velocity of the axon. **Conduction velocity** in turn depends on the diameter of the axon and whether it is unmyelinated or myelinated. Un-

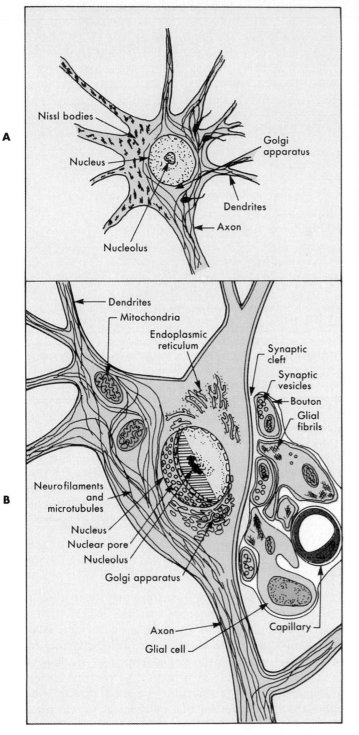

A

B

FIGURE 6-4   Organelles of the neuron. In **A** the organelles typical of a neuron, as seen with the light microscope, are shown. The portion of the illustration to the left of the soma represents structures seen with a Nissl stain. These include the nucleus and nucleolus, Nissl bodies in the cytoplasm of the cell body, and proximal dendrites, and as a negative image, the Golgi apparatus. The absence of Nissl bodies in the axon hillock and axon is also shown. To the right of the soma are structures seen with a heavymetal stain; these include neurofibrils. The appropriate heavymetal stain may demonstrate the Golgi apparatus (not shown). On the surface of the neuron several synaptic endings are indicated, as stained by the heavy metal. In **B** are shown structures visible at the electron microscopic level are shown. The nucleus, nucleolus, chromatin, and nuclear pores are represented. Mitochondria, rough endoplasmic reticulum, Golgi apparatus, neurofilaments, and microtubules are in the cytoplasm. Along the surface membrane are such associated structures as synaptic endings and astrocytic processes.

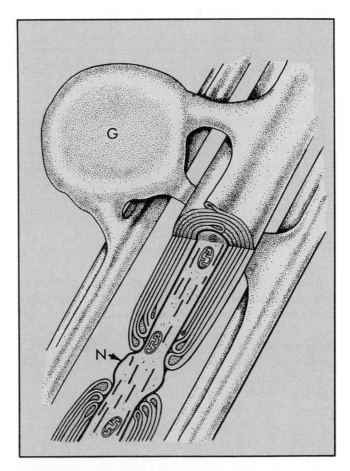

FIGURE 6-5   The myelin sheath in the CNS. Each oligodendroglial process forms the internode for one axon. *G*, Oligodendroglial cell; *N*, node of Ranvier.

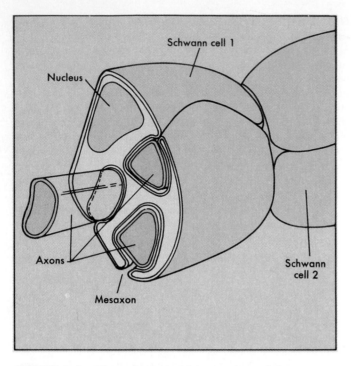

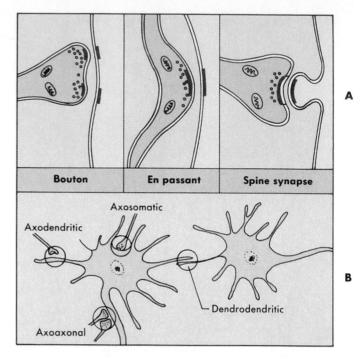

FIGURE 6-6 Three-dimensional impression of the appearance of Remak's bundle. The cut face of the bundle is seen to the left. One of the three unmyelinated axons is represented as protruding from the bundle. A mesaxon is indicated, as is the nucleus of the Schwann cell. To the right, the junction of adjacent Schwann cells is depicted.

FIGURE 6-7 **A,** Several types of synaptic configurations. *Left,* The synaptic bouton, which consists of the distended terminal of an axon ending on a postsynaptic structure. *Center,* An en passant ending. The presynaptic element does not terminate; instead synaptic specializations exist along the course of an axon. *Right,* An example of a synapse with a postsynaptic specialization, a dendritic spine. The spine synapse involves either a synaptic bouton or an en passant synapse presynaptically and a protrusion of the postsynaptic element. The spine may contain a special organelle, a cisternal structure called a spine apparatus (not illustrated). **B,** Types of synapses according to the parts of neurons contributing to the presynaptic and postsynaptic elements.

myelinated axons are generally less than 1 μm in diameter and conduct at speeds less than 2.5 m/sec. About 1 second would be required for a signal in an unmyelinated axon that supplies a sensory receptor in a person's foot to reach the spinal cord if the axonal conduction velocity was 1 m/sec. Myelinated axons have diameters of 1 to 20 μm and conduct at speeds of 3 to 120 m/sec. A spinal motoneuron with an axon that conducts at 100 m/sec would be able to trigger the contraction of a toe muscle in about 10 msec.

In the CNS certain neurons (**amacrine cells**) that lack axons signal information by electrical current flow rather than by generating action potentials. This current flow produces a **local potential,** which decays over a short distance (millimeters to hundreds of micrometers, depending on the length constant of the neuron involved). Local potentials differ from action potentials in that they are nonpropagating and cannot spread over long distances. In contrast, action potentials can propagate over long distances along axons (see also Chapters 3 and 4).

Signaling by local potentials is also characteristic of sensory receptors, which produce **receptor potentials,** and of communications between nerve cells, **synaptic potentials.**

Coding

Information conveyed by axons may be coded in several ways. Sets of neurons may be dedicated to a general function, such as a particular sensory modality. For example, the visual pathway includes the retina, the optic nerve and tract, the lateral geniculate nucleus of the thalamus, and the visual part of the cerebral cortex. The normal means of activating the visual system is by light striking the retina, but mechanical or electrical stimulation of the visual system will also produce a visual response, although a distorted one. Thus neurons of the visual system can be regarded as a **labeled line,** which, when activated, causes a visual sensation. *A labeled line consists of a set of neurons, including sensory receptors and central nervous system processing neurons, that together are responsible for signaling a particular type of sensation.* The usual way in which a labeled line is activated is by sensory stimula-

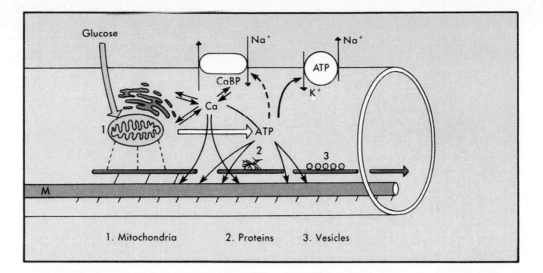

FIGURE 6-8 Axonal transport has been proposed to depend on the movement of transport filaments. Energy is required and is supplied by glucose. Mitochondria control the level of cations in the axoplasm by supplying adenosine triphosphate *(ATP)* to the ion pumps. An important cation for axonal transport is calcium. The concentration of free calcium is limited by the presence of calcium binding proteins *(CaBP)* and by pumps that extrude calcium from the cell or sequester it in organelles. Transport filaments *(blue bars at bottom of drawing)* move along the microtubules *(brown tube)* by means of crossbridges. Transported components attach to the transport filaments. These transported components include (1) mitochondria, (2) proteins, and (3) vesicles. Some vesicles become synaptic vesicles, which accumulate at synaptic specializations and release neurotransmitter during synaptic transmission.

tion. However, a labeled line may also be activated artificially (e.g., you can cause a person to see flashes of light by pressing on the eye or by stimulating the visual processing areas of the cerebral cortex with electric shocks). Other sensory systems provide further examples of labeled lines.

A second way in which information is coded by the nervous system is through **spatial maps.** The body surface may be mapped by an array of neurons in a sensory or a motor system. This is termed a **somatotopic map** (or, for humans, a **homunculus;** see Chapter 7). In the visual system **retinotopic maps** exist. In the auditory system, frequency of sound is represented in **tonotopic maps.**

A third method for coding information is by **patterns of nerve impulses.** Axons transmit a sequence of nerve impulses that results in synaptic transmission of information to a new set of neurons. The information that is communicated is coded in terms of the structure of the nerve impulse trains. Several different types of nerve impulse codes have been proposed. A common code is likely to depend on the mean discharge frequency. Other candidate codes depend on the time of firing, the temporal pattern, and the duration of bursts.

Synaptic Transmission

Neurons communicate with each other at specialized junctions called **synapses** (see Chapter 4). Typically, synapses are formed between the terminals of the axon of one neuron and the dendrites of another (Figure 6-7); these are called **axodendritic synapses.** However, many other types of synapses occur, including **axosomatic, axoaxonal,** and **dendrodendritic** (Figure 6-7, *B*). The synapse between a motoneuron and a skeletal muscle fiber is called an **endplate** or **neuromuscular junction.**

Axonal Transport

Many axons are too long to allow the movement of substances from the soma to the synaptic endings simply by diffusion. Certain membrane and cytoplasmic components that originate in the biosynthetic apparatus of the soma and proximal dendrites must be distributed along the axon and especially to the presynaptic elements of synapses to replenish secreted or inactivated materials. A special transport mechanism, called **axonal transport,** accomplishes this distribution.

Several types of axonal transport exist. Certain membrane-bound organelles and mitochondria are trans-

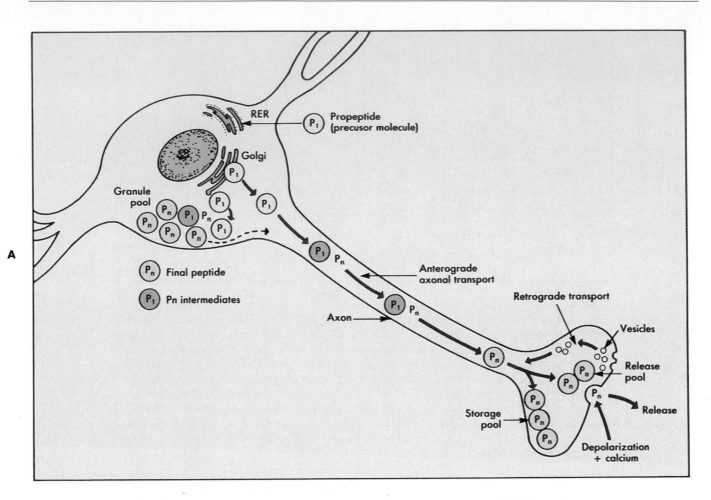

FIGURE 6-9 **A,** Axonal transport and its relation to the synthesis of peptides in the cell body and their release from terminals. *RER,* Rough endoplasmic reticulum.

ported relatively rapidly by **fast axonal transport.** Substances (e.g., proteins) that are dissolved in cytoplasm are moved by **slow axonal transport.** In mammals fast axonal transport proceeds as rapidly as 400 mm/day, whereas slow axonal transport occurs at about 1 mm/day. This means that synaptic vesicles can travel from a motoneuron in the spinal cord to a neuromuscular junction in a foot in about 2½ days, whereas the transport of many soluble proteins over the same distance would take nearly 3 years.

Axonal transport requires metabolic energy and involves calcium ions. The cytoskeleton, particularly microtubules, provides a system of guide-wires along which membrane-bound organelles move (Figure 6-8). These organelles may attach to microtubules through a linkage similar to that between the thick and thin filaments of skeletal muscle fibers; calcium triggers the movement of the organelles along the microtubules.

Axonal transport occurs in both directions. Transport from the soma toward the axonal terminals is called **anterograde axonal transport** (Figure 6-9, *A*). This pro-

cess allows the replenishment of synaptic vesicles and enzymes responsible for neurotransmitter synthesis in synaptic terminals. Transport in the opposite direction is **retrograde axonal transport** (Figure 6-9, *B*). This process returns synaptic vesicle membrane to the soma for lysosomal degradation. Marker substances, such as the enzyme horseradish peroxidase, can be transported anterogradely or retrogradely and can be used experimentally to trace neural pathways.

Axonal transport is important in pathology. Primary afferent neurons and motoneurons link the central nervous system with the periphery and thus form a protoplasmic bridge that crosses the blood-brain barrier (see below). Certain viruses, such as the **rabies** and **polio viruses,** and toxins, such as **tetanus toxin,** can enter the central nervous system from the periphery if they are taken up and transported in the axons of these neurons.

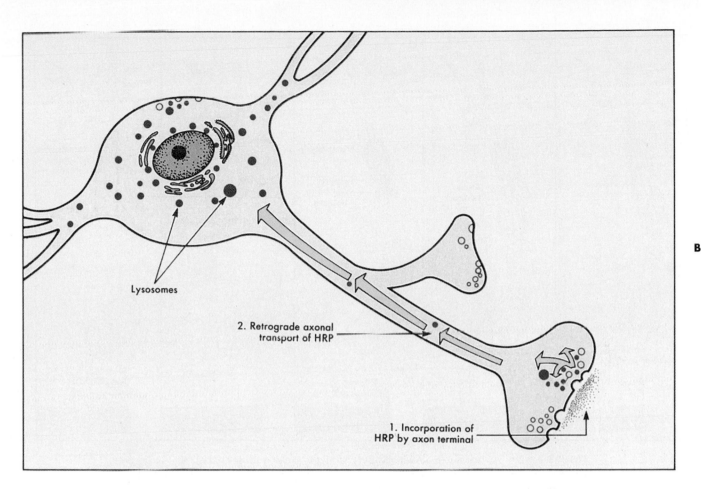

B

FIGURE 6-9, cont'd   B, Schematic summary of incorporation, retrograde axonal transport, and lysosomal accumulation of horseradish peroxidase *(HRP)* in neurons. Anterograde axonal transport of HRP from the soma is not illustrated.

REACTIONS TO INJURY

Injury to nervous tissue elicits responses by neurons and neuroglia. Severe injury causes cell death. Once a neuron is lost, it cannot be replaced because neurons are postmitotic cells; that is, neurons are fully differentiated and no longer undergo cell division. Most neurons complete their differentiation before birth, although neuroglial cells continue to divide even in adulthood. Thus, most tumors of the central nervous system originate from neuroglial precursor cells rather than from neurons.

Axonal Reaction

When an axon is transected, the soma of a neuron may show the **axonal reaction.** Normally, Nissl bodies stain well with basic aniline dyes, which attach to the ribonucleic acid of the ribosomes (Figure 6-10, *A*). During the axonal reaction the cisterns of the rough endoplasmic reticulum become distended with the products of protein synthesis. The ribosomes become disorganized, and thus the Nissl bodies are stained weakly by basic aniline dyes. This alteration in staining is termed **chromatolysis** (Figure 6-10, *C*). The soma may also become swollen and rounded, and the nucleus may assume an eccentric position. These morphologic changes reflect the cytologic processes that accompany protein synthesis. The damaged neuron is repairing itself.

Wallerian Degeneration

If a nerve fiber is cut, the axon distal to the transection dies (Figure 6-10, *B*). Within a few days the axon and all the synaptic endings formed by the axon disintegrate. If the axon was myelinated, the myelin sheath becomes fragmented and it is eventually phagocytized and removed. However, the neuroglial cells that formed the myelin sheath remain viable. This sequence of events was originally described by Waller and is called **wallerian degeneration.**

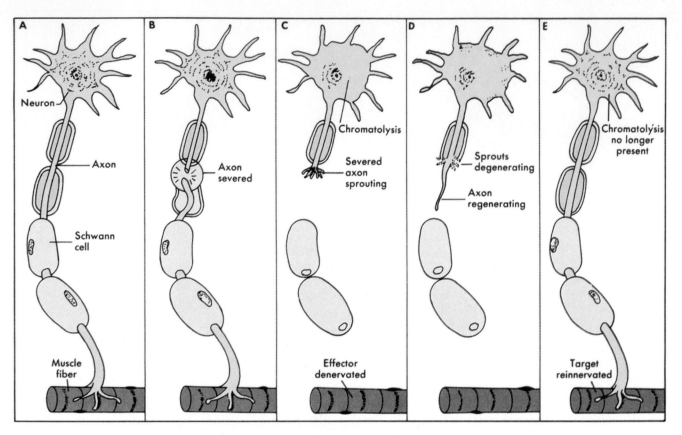

FIGURE 6-10   **A,** Normal motoneuron innervating a skeletal muscle fiber. **B,** Motor axon has been severed, and the motoneuron is undergoing chromatolysis. **C,** This is associated in time with sprouting and, in **D,** with regeneration of the axon. The excess sprouts degenerate. **E,** When the target cell is reinnervated, chromatolysis is no longer present.

If the axons that provide the sole or predominant synaptic input to a neuron or an effector cell are interrupted, the postsynaptic cell may undergo degeneration and even death. The best known example of this is the atrophy of skeletal muscle fibers after their innervation by motoneurons is interrupted.

These pathological changes have been useful in neuroanatomic investigations to trace neural pathways. For example, retrograde chromatolysis has been used to reveal groups of neurons whose axons have been deliberately interrupted. The projection target of axons can be determined by following the course of interrupted axons undergoing wallerian degeneration. Synaptic targets can also be mapped if neurons undergo transneuronal degeneration after an axonal bundle is transected.

Regeneration

Many neurons can regenerate a new axon if the axon is lost through injury. The proximal stump of the damaged axon develops sprouts (Figure 6-10, *C*). In the pe-

ripheral nervous system these sprouts elongate and grow along the path of the original nerve if this route is available. The Schwann cells in the distal stump of the nerve not only survive the wallerian degeneration but also proliferate and form rows along the course previously taken by the axons. Growth cones of the sprouting axons find their way along the rows of Schwann cells and may eventually reinnervate the original peripheral target structures (Figure 6-10, *D* and *E*). The Schwann cells then remyelinate the axons. The rate of regeneration is limited by the rate of slow axonal transport to about 1 mm/day.

Nerve growth factor plays an important role in the growth of axons during development, in their maintenance, and in their regeneration after injury.

Transection of axons in the CNS also results in sprouting. Proper guidance for the sprouts is lacking, however, because the oligodendroglia do not form a path along which the sprouts can grow, because a single oligodendroglial cell myelinates many central axons, whereas a given Schwann cell provides myelin for only a single axon in the periphery. Alternately, different chemical sig-

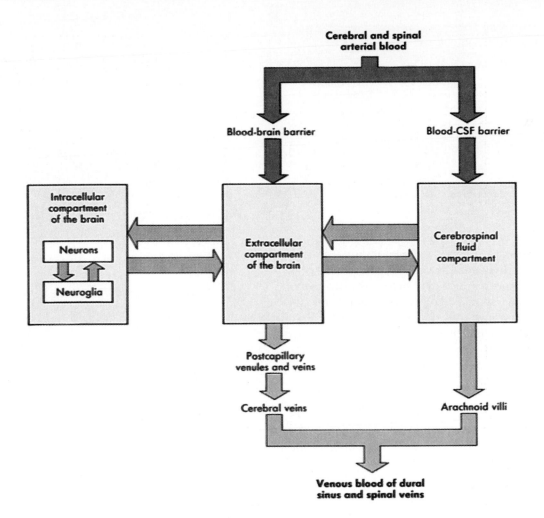

FIGURE 6-11   The structural and functional relationships involved in the blood-brain and blood-CSF barriers. Substances entering the neurons and glial cells (i.e., intracellular compartment) must pass through the cell membrane. Arrows indicate direction of fluid flow under normal conditions.

nals, such as different growth factors or growth inhibiting factors, may affect the peripheral and central attempts at regeneration differently. Another obstacle is the formation of glial scars by astrocytes.

ENVIRONMENT OF THE NEURON

The local environment of most neurons is controlled so that neurons are normally protected from extreme variations in the composition of the extracellular fluid that bathes them. This control is provided by regulating the circulation of the CNS (see Chapter 25), the presence of a blood-brain barrier, the buffering function of astrocytes, and the exchange of substances with the CSF. Substances exchange freely between the extracellular fluid of the CNS and the CSF. However, the entry of sub-

stances from the blood into the CNS is controlled by secretory processes in the choroid plexuses, which form the CSF, and by the blood-brain barrier (Figure 6-11).

Fluid Compartments of the Cranium

The cranial cavity contains the brain, blood, and CSF. The human brain weighs about 1350 g, of which approximately 15%, or 200 ml, is extracellular fluid. The intracranial blood volume is about 100 ml, and the cranial CSF volume another 100 ml. Thus the extracellular fluid space in the cranial cavity totals approximately 400 ml.

Blood-Brain Barrier

The movement of large molecules and highly charged ions from the blood into the brain and spinal cord is se-

Table 6-1 Constituents of Cerebrospinal Fluid and Blood

Constituent	Lumbar CSF	Blood
Na^+ (mEq/L)	148	136-145
K^+ (mEq/L)	2.9	3.5-5
Cl^- (mEq/L)	120-130	100-106
Glucose (mg/dl)	50-75	70-100
Protein (mg/dl)	15-45	$6\text{-}8 \times 10^3$
pH	7.3	7.4

verely restricted by a **blood-brain barrier** (Figure 6-11). The restriction is at least partly caused by the presence of tight junctions between the capillary endothelial cells of the CNS. Astrocytes may also help limit the movements of certain substances. For example, astrocytes can take up potassium ions and thus regulate the K^+ concentration in the extracellular space. Some substances are removed from the CNS by transport mechanisms.

Cerebrospinal Fluid

The extracellular fluid within the CNS communicates directly with the cerebrospinal fluid (CSF). Thus the composition of the CSF indicates the composition of the extracellular environment of neurons in the brain and spinal cord. The main constituents of CSF in the lumbar cistern are shown in Table 6-1. For comparison, the concentrations of the same constituents in the blood are also given. The CSF has lower concentrations of K^+, glucose, and protein but greater concentrations of Na^+ and Cl^- than does blood. Furthermore, CSF contains practically no blood cells. The increased concentrations of Na^+ and Cl^- allow the CSF to be isotonic to blood, despite the much lower concentration of protein in the former.

The CSF is formed largely by the **choroid plexuses,** which are capillary loops covered by specialized ependymal cells and located in the ventricular system of the brain. The **ventricular system** includes the two **lateral ventricles** in the telencephalon, the **third ventricle** of the diencephalon, and the **fourth ventricle** of the metencephalon and myelencephalon. The lateral ventricles connect to the third ventricle by way of the **interventricular foramina,** and the third ventricle connects with the fourth through the **cerebral aqueduct.** Choroid plexuses are found in the lateral ventricles, third ventricle, and fourth ventricle (Figure 6-12).

The CSF escapes from the fourth ventricle through openings in its connective tissue roof (Figure 6-12). These openings are the unpaired **median aperture** and the paired **lateral apertures.** After leaving the ventricu-

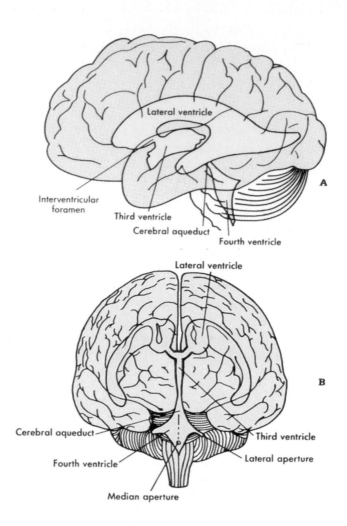

FIGURE 6-12 The ventricular system in situ as seen from the side (**A**) and from the front (**B**).

lar system, the CSF circulates through the subarachnoid spaces surrounding the brain and spinal cord. Part of the CSF is removed by bulk flow through the valvular **arachnoid villi** into the dural venous sinuses.

The volume of the CSF within the cerebral ventricles is approximately 35 ml, and that in the subarachnoid spaces of the brain and spinal cord is about 100 ml. The rate at which CSF is produced is about 0.35 ml/min. This allows the CSF to be turned over approximately four times daily.

The pressure in the CSF column is about 120 to 180 mm H_2O when a person is recumbent. The rate at which CSF is formed is relatively independent of the pressure in the ventricles and subarachnoid space and of the systemic blood pressure. However, the absorption rate of CSF is a direct function of CSF pressure.

Obstruction of the circulation of CSF leads to increased CSF pressure and hydrocephalus. In **hydrocephalus** the ventricles become distended. In young children, the intracranial volume may be increased because the sutures are not closed and the head can enlarge. However, if the increase continues, brain substance may be lost. In adults, an increase in the ventricular size compromises the flow of blood and causes the loss of brain tissue. When the obstruction is within the ventricular system or in the roof of the fourth ventricle, the condition is called **noncommunicating hydrocephalus.** If the obstruction is in the subarachnoid space or arachnoid villi, it is known as **communicating hydrocephalus.**

SUMMARY

1. The general functions of the nervous system are sensory detection, information processing, and behavior.
2. The functional unit of the nervous system is the neuron.
3. The nervous system is subdivided into the peripheral nervous system (PNS; sensory, motor, and autonomic neurons) and the central nervous system (CNS; spinal cord and brain).
4. Nervous tissue contains not only neurons, but also supporting cells or neuroglia (Schwann and satellite cells in the PNS; astrocytes, oligodendroglia, microglia, and ependymal cells in the CNS) and blood vessels.
5. The parts of neurons include the cell body, dendrites, and the axon. Axons may be myelinated or unmyelinated.
6. Information is conducted along axons by nerve impulses, whose conduction velocity depends on the presence or absence of myelin and the diameter of the axon.
7. Several coding mechanisms are used by neurons to convey information in neural networks. These include labeled lines, spatial maps, and patterns of nerve impulses.
8. Information is transmitted from neuron to neuron or from neuron to effector organ by means of synaptic transmission.
9. Materials are moved along axons by axonal transport. The movements can be rapid or slow and away from or toward the cell body.

10. Neurons respond to damage either by cell death or by less drastic changes, such as the axonal reaction (chromatolysis) and wallerian degeneration. Viable axons may subsequently regenerate.
11. The extracellular environment of neurons in the CNS is highly regulated. The cerebrospinal fluid is secreted by the choroid plexuses. The blood-brain barrier restricts the entry of substances into the brain.

BIBLIOGRAPHY
Journal Articles

Bray GM et al: Interactions between axons and their sheath cells, *Annu Rev Neurosci* 4:127, 1981.

Fawcett JW, Keynes RJ: Peripheral nerve regeneration, *Annu Rev Neurosci* 13:43, 1990.

Partridge WM: Brain metabolism: a perspective from the blood-brain barrier, *Physiol Rev* 63:1481, 1983.

Udin SB, Fawcett JW: Formation of topographic maps, *Annu Rev Neurosci* 11:289, 1988.

Vallee RB, Bloom GS: Mechanisms of fast and slow axonal transport, *Annu Rev Neurosci* 14:59, 1991.

Books and Monographs

Bullock TH et al: *Introduction to nervous systems,* San Francisco, 1977, WH Freeman and Co.

Cajal SR: *Degeneration and regeneration of the nervous system,* New York, 1959, Hafner Publishing Co.

Cajal SR: *The neuron and the glial cell,* Springfield, Ill, 1984, Charles C Thomas, Publisher.

Heimer L, Robards MJ, eds: *Neuroanatomical tract-tracing methods,* New York, 1981, Plenum Press.

Kandel ER, Schwartz JH, Jessell TM: *Principles of neural science,* ed 3, New York, 1991, Elsevier Science Publishing Co, Inc.

Millen JW, Woollam DHM: *The anatomy of the cerebrospinal fluid,* New York, 1962, Oxford University Press.

Paxinos G: *The human nervous system,* San Diego, 1990, Academic Press.

Shephard GM: *The synaptic organization of the brain,* ed 3, New York, 1990, Oxford University Press.

Whitfield IC: *Neurocommunications: an introduction,* New York, 1984, John Wiley & Sons, Inc.

Willis WD, Grossman RG: *Medical neurobiology,* ed 3, St Louis, 1981, Mosby.

CHAPTER 7

The General Sensory System

The nervous system can be regarded as a complex of several different subsystems with different functional roles. These subsystems interact, however, and their activity leads to unified behavior. Some of the neural subsystems are concerned with sensory functions. The general sensory system, or **somatosensory system,** analyzes sensory events relating to mechanical, thermal, or chemical stimulation of the body and face. If both visceral and somatic organs are included, the term **somatovisceral sensory system** is perhaps more appropriate. The special sensory systems include the **visual system,** which analyzes patterns of light detected by the eye; the **auditory system,** which interprets sounds that impinge on the ear; the **vestibular system,** which responds to the position of the head in space; and the **chemical sensory systems,** which are special sensory apparatuses for taste and olfaction.

Sensory systems are activated by **sensory receptors,** which respond to various stimuli by a process of sensory transduction. The sensory receptors then provide sensory information about these stimuli to the central nervous system. Sensory receptors have **receptive fields,** which are the areas that when stimulated activate the receptors. Sensory information is encoded in various ways, and this encoded information is then transmitted through sensory pathways within the central nervous system.

The sensory receptors in the somatovisceral sensory system include **mechanoreceptors, thermoreceptors,** and **nociceptors** that supply the skin, muscles, joints, and viscera (see pp. 97-98). The peripheral endings of primary afferent fibers that innervate the skin are distributed in a segmental pattern derived from the embryonic **dermatomes.** Each dermatome receives its main supply from the **dorsal root ganglion** of a particular spinal cord segment, although a dermatome is also innervated by collateral branches of the nerves to adjacent dermatomes. The trigeminal nerve innervates regions of the skin of the face that are equivalent to dermatomes.

The central branches of neurons in a dorsal root ganglion enter the spinal cord through a **dorsal root.** They terminate in the gray matter of the spinal cord and also give rise to collaterals that ascend and descend in the white matter of the spinal cord. Many of the branches that ascend in the dorsal white matter synapse in nuclei of the caudal **medulla.** Neurons of these nuclei then project to the contralateral **thalamus,** which in turn projects to the somatosensory areas of the cerebral cortex. This ascending pathway is known as the **dorsal column–medial lemniscus path.** It is largely responsible for the sensations of *flutter-vibration* and *touch-pressure,* as well as for *proprioception.*

Fine primary afferent fibers, many of which are **nociceptors** or **thermoreceptors,** project into the spinal cord through a dorsal root and synapse in the dorsal horn of the spinal gray matter. Some of the neurons activated by these afferents send their axons to the contralateral side of the spinal cord, where they ascend in the lateral and ventral white matter to the thalamus. This pathway is the **spinothalamic tract,** and it is responsible for *pain* and *temperature sensations.*

Neural pathways descend from the brainstem to control transmission in the ascending somatosensory pathways. The **endogenous analgesia system** involves pathways of this kind. The endogenous analgesia system includes projections from midline regions of the brainstem, and it depends in part on the release of opioid neurotransmitters and such amines as serotonin and norepinephrine.

PRINCIPLES OF SENSORY PHYSIOLOGY

Transduction

Sensory systems are designed to respond to the environment. Useful features of the environment are detected by **sensory receptors,** which then provide that

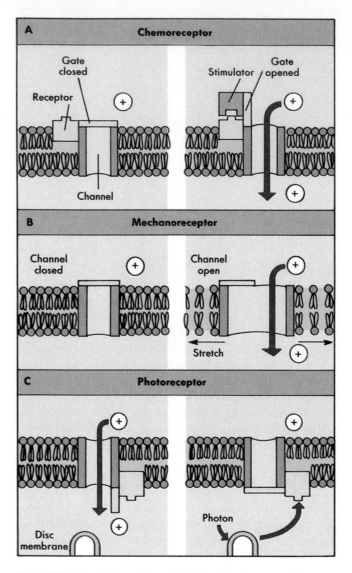

FIGURE 7-1   Conceptual models of transducer mechanisms in three types of receptors. **A,** Chemoreceptor. **B,** Mechanoreceptor. **C,** Vertebrate photoreceptor. (See text.)

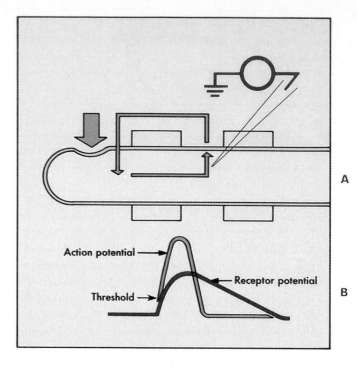

FIGURE 7-2   **A,** The terminal region of a myelinated mechanoreceptor afferent fiber is shown as if cut in longitudinal section. The surface membrane is indicated by a light blue line, the axoplasm by the light blue area, and the myelin internodes by the light pink rectangles. The current flow produced by stimulation of a mechanoreceptor at the site indicated by the large blue arrow is shown by the smaller blue arrows. The tip of an intracellular microelectrode placed next to a node of Ranvier is indicated by the thin red lines, and the recording system by the thick red line. **B,** The receptor potential produced by the current *(red line)* and an action potential that may be superimposed on the receptor potential if the latter exceeds threshold *(blue line).*

information to the central nervous system (CNS). The interaction of environmental energy with a sensory receptor is called a **stimulus.** The effect of the stimulus on the sensory receptor may lead to a **response.** The process that enables a sensory receptor to respond usefully to a stimulus is called **sensory transduction.**

Environmental events that lead to sensory transduction can involve mechanical, thermal, chemical, or other forms of energy, depending on the sensory apparatus. Although humans cannot sense electrical and magnetic fields, other animals, such as fish, can respond to such stimuli. Figure 7-1 shows how different types of stimuli can alter the membrane properties of sensory receptor neurons specialized to transduce such stimuli. In Figure 7-1, *A,* a **chemoreceptor** responds when a mol-

ecule of a chemical stimulant reacts with a receptor molecule on the sensory receptor. The reaction results in the opening of an ion channel and a consequent influx of ionic current. In Figure 7-1, *B,* the ion channel of a **mechanoreceptor** opens in response to the application of a mechanical force along the membrane. In Figure 7-1, *C,* the ion channel of a **photoreceptor** is open in the dark but closes when a photon is absorbed by pigment on the disc membrane.

Sensory transduction generally induces a **receptor potential** in the peripheral terminal of a primary afferent sensory neuron. A receptor potential is usually a depolarizing event that results from inward current flow, and it brings the membrane potential of the sensory receptor toward or past the threshold needed to trigger a nerve impulse. For example, in Figure 7-2, *A,* a mechanical stimulus *(large arrow)* distorts the ending of a mechanoreceptor and causes inward current flow at the terminal, and longitudinal and outward current flow along the axon *(small arrows)*. The outward current produces a depolarization, the receptor potential *(red line in B),* which may or may not exceed threshold for an action potential *(blue line in B)*. In this case the action potential is generated at a trigger zone in the first node of Ranvier of the afferent fiber. However, in photoreceptors the cessation of inward current flow during transduction leads to a hyperpolarization of the receptor.

In some sensory receptor organs the peripheral terminal of a primary afferent fiber contacts a separate, peripherally located sensory cell. For example, in the cochlea, primary afferent fibers contact hair cells. Sensory transduction in such sense organs is made more complex by this arrangement. In the cochlea a receptor potential is produced in the hair cells in response to sound. The receptor potential is a depolarization of the hair cell's membrane, and the depolarization liberates an excitatory neurotransmitter onto the primary afferent terminal. This produces a **generator potential,** which in turn depolarizes the primary afferent fiber. This depolarization brings the membrane potential of the primary afferent fiber toward or beyond the threshold for firing nerve impulses.

Sensory receptors have the property of **adaptation** to maintained stimuli. A long-lasting stimulus may produce either a prolonged repetitive discharge or a short-lived response (one or a few discharges), depending on whether the sensory receptor is slowly or rapidly adapting. The adaptation rates differ because a prolonged stimulus may produce either a maintained or a transient receptor potential in the sensory receptor. The functional implication of the adaptation rate is that different temporal features of a stimulus can be analyzed by receptors with different adaptation rates. For example, during an indentation of the skin, a slowly adapting receptor may respond repetitively at a rate proportional to the amount of indentation (Figure 7-3,*A*). On the other hand, rapidly adapting receptors in the skin respond best to transient mechanical stimuli. The information signaled may reflect stimulus velocity (Figure 7-3, *B*) or acceleration (Figure 7-3, *C*), rather than the amount of skin indentation.

Receptive Fields

The relationship between the location of a stimulus and the activation of particular sensory neurons is a major theme in sensory physiology. The **receptive field** of a sensory neuron is the region that, when stimulated, affects the discharge of the neuron. For example, a sensory receptor might be activated by indentation of only a small area of skin. That area is the excitatory receptive field of the sensory receptor. A neuron in the CNS might be excited by stimulation of a receptive field several

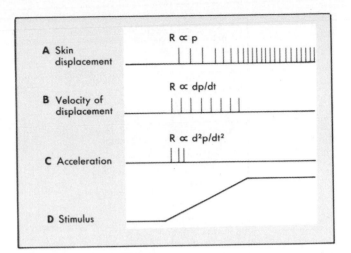

FIGURE 7-3   Responses of slowly and rapidly adapting mechanoreceptors to displacement of the skin. The discharges of the primary afferent fibers supplying the receptors in response to the ramp and hold stimulus (shown at the bottom) are termed the responses *(R)*. **A,** *R* is proportional to skin position *(p)*. The receptor is slowly adapting and signals skin displacement. **B,** *R* is a function of the velocity of displacement *(dp/dt)*. **C,** *R* is a function of the acceleration *(d²p/dt²)*. The receptors in **B** and **C** are rapidly adapting, but they signal different, dynamic features of the stimulus.

times as large. The receptive fields of sensory neurons of the CNS are typically larger than those of sensory receptors, because the central neurons receive information from many sensory receptors, each with a slightly different receptive field. The location of the receptive field is determined by the location of the sensory transduction apparatus responsible for signaling information about the stimulus to the sensory neuron.

Generally the receptive fields of sensory receptors are excitatory. However, a central sensory neuron can have either an **excitatory** or an **inhibitory receptive field** (Figure 7-4). Inhibition results from data processing in sensory neural circuits and is mediated by inhibitory interneurons.

Sensory Coding

Sensory neurons encode stimuli. In the process of sensory transduction, one or more aspects of the stimulus must be encoded in a way that can be interpreted by the CNS. The **encoded information** is an abstraction based on the responses of sensory receptors to the stimulus and on information processing within the sensory pathway. Some of the aspects of stimuli that are encoded include *modality, spatial location, threshold, intensity, frequency,* and *duration*. Other aspects encoded are presented in reference to particular sensory systems.

A **sensory modality** is a readily identified class of

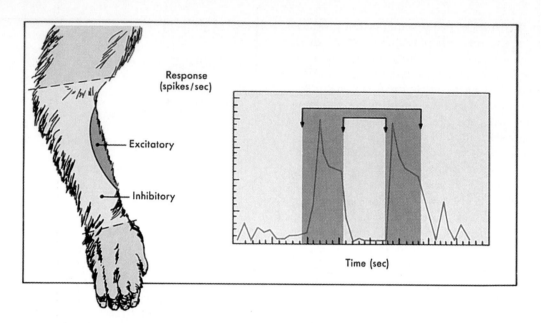

FIGURE 7-4   Excitatory and inhibitory receptive fields of a central somatosensory neuron located in the primary somatosensory cerebral cortex (SI). The excitatory receptive field is on the forearm and is surrounded by an inhibitory receptive field. The graph shows the response to an excitatory stimulus and the inhibition of that response by a stimulus applied in the inhibitory field.

sensation. For example, maintained mechanical stimuli applied to the skin result in a sensation of *touch-pressure,* and transient mechanical stimuli may evoke a sensation of *flutter-vibration.* Other cutaneous modalities include *cold, warm,* and *pain.* Vision, audition, position sense, taste, and smell are examples of noncutaneous modalities. Coding for modality is signaled by labeled-line sensory channels in most sensory systems (see Chapter 6). A **labeled-line sensory channel** consists of a set of neurons devoted to a particular sensory modality.

The **location of a stimulus** is often signaled by the activation of the particular population of sensory neurons whose receptive fields are affected by the stimulus (Figure 7-5, *A*). In some cases an inhibitory receptive field or a contrasting border between an excitatory and an inhibitory receptive field can have localizing value. Resolution of two different adjacent stimuli may depend both on excitation of partially separate populations of neurons and on inhibitory interactions (Figure 7-5, *B*).

A **threshold stimulus** is the weakest that can be detected. For detection, a stimulus must produce receptor potentials large enough to activate one or more primary afferent fibers. Weaker intensities of stimulation can produce subthreshold receptor potentials; however, such stimuli would not excite central sensory neurons. Furthermore, the number of primary afferent fibers that need to be excited for sensory detection depends on the requirements for spatial and temporal summation in the

sensory pathway (see Chapter 4). Thus a stimulus at threshold for detection may be much greater than threshold for activation of the most responsive primary afferent fibers. Conversely, a stimulus that excites some primary afferent fibers may not lead to perception of that stimulus.

Stimulus intensity may be encoded by the mean frequency of discharge of sensory neurons. The relationship between stimulus intensity and response can be plotted as a **stimulus-response function.** For many sensory neurons, the stimulus-response function approximates an exponential curve (Figure 7-6). The general equation for such a curve is

$$\text{Response} = \text{Stimulus}^n \times \text{Constant}$$

The exponent, n, can be less than, equal to, or greater than 1. Stimulus-response functions with fractional exponents characterize many mechanoreceptors (Figure 7-6). Thermoreceptors have linear stimulus-response curves. Nociceptors may have linear or positively accelerating stimulus-response functions; that is, the exponent for these curves is 1 or more.

Another way in which stimulus intensity is encoded is by the number of sensory receptors activated. A stimulus that is threshold for perception may activate just one or a few primary afferent fibers, whereas a strong stimulus may excite many similar receptors. Central neurons that receive input from a particular class of sensory receptor would be more powerfully activated as more pri-

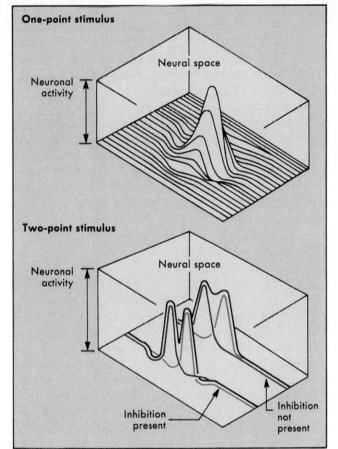

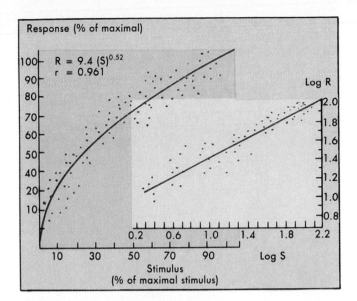

FIGURE 7-6   Stimulus-response function for slowly adapting cutaneous mechanoreceptors. The rate of discharge is plotted against stimulus strength (normalized to maximal). The plots are on linear and on log-log scales. The stimulus-response function is $R = 9.4(S)^{0.52}$.

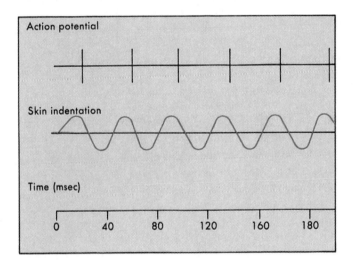

FIGURE 7-7   Coding for the frequency of stimulation. Discharge of a rapidly adapting cutaneous mechanoreceptor in phase with a sinusoidal stimulus. The action potentials are shown at the top, and the stimulus is shown in the tracing in the middle.

FIGURE 7-5   **A,** Representation of the activity of a large population of neurons distributed three-dimensionally in neural space. The activity is in response to stimulation of a point on the skin. Note that an excitatory peak is surrounded by an inhibitory trough; these are determined by the excitatory and inhibitory fields of sensory neurons in the central pathways. **B,** The activity in response to stimulation of two adjacent points on the skin. Note that the sum of the activity *(solid line)* is separated better into two peaks when inhibition is present than when it is not.

mary afferents are caused to discharge, and a greater activity in central sensory neurons results in the perception of a stronger stimulus.

Stimuli of different intensities may activate different sets of receptors. For example, a weak mechanical stimulus applied to the skin might activate only mechanoreceptors, whereas a strong mechanical stimulus might activate both mechanoreceptors and nociceptors. In this case the sensation evoked by the stronger stimulus would be more intense, and the quality would be different.

Stimulus frequency can be encoded by the intervals between the discharges of sensory neurons. Sometimes the interspike intervals correspond exactly to the intervals between stimuli (Figure 7-7), but in other cases a given neuron may discharge at intervals that are multiples of the interstimulus interval.

Stimulus duration may be encoded in slowly adapting sensory neurons by the duration of enhanced firing. In rapidly adapting neurons the beginning and end of a stimulus may be signaled by transient discharges.

Sensory Pathways

A **sensory pathway** can be viewed as a set of neurons arranged in series (Figure 7-8). First-, second-, third-,

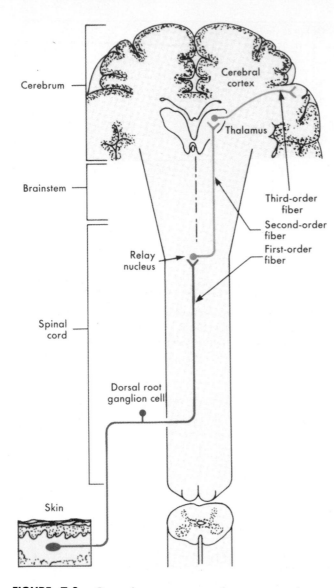

FIGURE 7-8 General arrangement of sensory pathways. First-, second-, and third-order neurons are shown. Note that the axon of the second-order neuron crosses the midline, so that sensory information from one side of the body is transmitted to the opposite side of the brain.

thalamus. The information may be transformed at the level of the second-order neuron by local neural processing circuits. The ascending axon of the second-order neuron typically crosses the midline, and thus sensory information that originates on one side of the body reaches the contralateral thalamus.

The **third-order neuron** is located in one of the sensory nuclei of the thalamus. Again, local circuits may transform information from second-order neurons before the signals are transmitted to the cerebral cortex.

Fourth-order neurons in the appropriate sensory receiving areas of the cerebral cortex and **higher-order neurons** in the same and other cerebral cortical areas process the information further. At some undetermined site the sensory information results in **perception,** which is a conscious awareness of the stimulus.

SOMATOVISCERAL SENSORY SYSTEM

The somatovisceral sensory system includes sensory units that have sensory receptor organs in the skin, muscle, joints, and viscera. Information arising from these sensory receptors reaches the CNS by way of first-order neurons, which are the primary afferent neurons. The cell bodies of the primary afferent neurons are generally located in dorsal root ganglia or cranial nerve ganglia. Each ganglion cell gives off a neurite that bifurcates into a peripheral process and a central process. The peripheral process has the structure of an axon and terminates peripherally as a sensory receptor. The central process is also an axon and enters the spinal cord through a dorsal root or the brainstem through a cranial nerve. The central process typically gives rise to numerous collateral branches that end synaptically on several second-order neurons.

The processing of somatovisceral sensory information involves a number of CNS structures, including the spinal cord, brainstem, thalamus, and cerebral cortex. The ascending pathways arise from second-order neurons that are located in the spinal cord and brainstem and that project to the contralateral thalamus. The most important ascending somatosensory pathways that carry somatovisceral information from the body are the **dorsal column–medial lemniscus path** and the **spinothalamic tract.** The main somatosensory projection that represents the face is the **trigeminothalamic tract.** The organization of these pathways will be described in the following sections. Ancillary somatovisceral pathways include the spinocervicothalamic path, the postsynaptic dorsal column path, the dorsal spinocerebellar tract, the spinoreticular tract, and the spinomesencephalic tract.

The somatovisceral sensory system can be regarded as a general sensory system. The sensory modalities me-

and higher-order neurons serve as sequential elements in a given sensory pathway. Furthermore, several parallel sensory pathways often transmit similar sensory information. The **first-order neuron** in a sensory pathway is the primary afferent neuron. The peripheral endings of this neuron form a sensory receptor (or receive input from an accessory sensory cell, such as a hair cell), and thus the neuron responds to a stimulus and transmits encoded information to the CNS. The primary afferent neuron often has its soma in a dorsal root ganglion or a cranial nerve ganglion.

The **second-order neuron** is likely to be located in the spinal cord or brainstem. It receives information from first-order neurons and transmits information to the

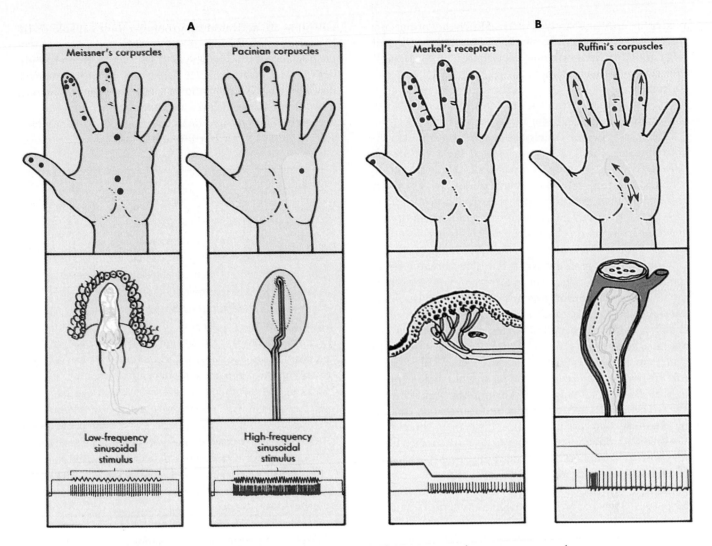

A   B

| Meissner's corpuscles | Pacinian corpuscles | Merkel's receptors | Ruffini's corpuscles |

Low-frequency sinusoidal stimulus

High-frequency sinusoidal stimulus

FIGURE 7-9 The receptive fields of several types of cutaneous mechanoreceptors are shown in the top row of drawings. Only a few representative receptive fields are shown. Many more receptors are present than are indicated. Their density is greater on the more distal parts of the fingers. The red dots show the centers of the receptive fields and the yellow areas the extent of the receptive fields. **A,** Rapidly adapting mechanoreceptors: Meissner's corpuscles and Pacinian corpuscles. **B,** Slowly adapting mechanoreceptors: Merkel's receptors and Ruffini's corpuscles. Note that the receptive fields of Meissner's corpuscles and Merkel's receptors are small, whereas those of pacinian corpuscles and Ruffini's corpuscles are large. The red arrows on the receptive fields of Ruffini's corpuscles show the directions of skin stretch that activated the afferent fibers. The second row of drawings shows the morphology of the receptors; the third row, the responses to sinusoidal stimuli **(A)** or to step indentations of the skin **(B)**.

diated by the somatovisceral sensory system include touch-pressure, flutter-vibration, position sense, joint movement, thermal sense, pain, and visceral distension.

Receptors

The somatovisceral sensory system includes various types of sensory receptor organs in the skin, muscle, joints, and viscera.

Cutaneous receptors can be subdivided according to the type of stimulus to which they respond. The major types of cutaneous receptors include mechanoreceptors, thermoreceptors, and nociceptors. **Mechanoreceptors** respond to such mechanical stimuli as stroking or indenting the skin and can be rapidly adapting or slowly adapting. Rapidly adapting cutaneous mechanoreceptors include **hair follicle receptors** in the hairy skin, **Meissner's corpuscles** in the nonhairy (glabrous) skin, and **Pacinian corpuscles** in subcutaneous tissue (Figure 7-9, *A*). Hair follicle receptors and Meissner's corpuscles respond best to stimuli repeated at rates of about 30 to 40 Hz, whereas pacinian corpuscles prefer stimuli repeated at approximately 250 Hz. Slowly adapting cutaneous mechanoreceptors include **Merkel's cell endings** and **Ruffini's corpuscles** (Figure 7-9, *B*). Merkel's cell receptors have punctate receptive fields, whereas Ruffini's corpuscles can be activated by stretching the skin some distance from the receptor terminals. The axons of all of these receptor types are myelinated.

The two types of **thermoreceptors** in the skin are *cold receptors* and *warm receptors*. Both classes are slowly adapting, although they also discharge phasically when skin temperature is changed rapidly. These are among the few receptor types that discharge spontaneously under normal circumstances. Cold receptors are supplied by small myelinated axons and warm receptors by unmyelinated axons.

Nociceptors respond to stimuli that threaten or actually produce damage. There are two major classes of cutaneous nociceptors: the **A-δ mechanical nociceptors** and the **C polymodal nociceptors.** A-δ mechanical nociceptors are supplied by finely myelinated (or A-δ) axons, whereas C polymodal nociceptors are supplied by unmyelinated (or C) fibers. The A-δ mechanical nociceptors respond to strong mechanical stimuli, such as pricking the skin with a needle or crushing the skin with forceps. They typically do not respond to noxious thermal or chemical stimuli unless they have been previously sensitized. C polymodal nociceptors, on the other hand, respond to several types of noxious stimuli, including mechanical, thermal, and chemical.

Skeletal muscle also contains several types of sensory receptors. These are chiefly mechanoreceptors and nociceptors, although some muscle receptors may possess thermosensitivity or chemosensitivity. The best-studied muscle receptors are the stretch receptors, which include **muscle spindles** and **Golgi tendon organs.** Although these play an important role in proprioception, they may be more important in motor control. Therefore their structure and function are discussed in Chapter 9.

Other sensory receptors in muscle include nociceptors. These respond to pressure applied to the muscle and to release of metabolites, especially during **ischemia** (inadequate blood flow). Muscle nociceptors are supplied by medium-sized and small myelinated (group II and III) axons or by unmyelinated (group IV) afferent fibers.

Joints are associated with several types of sensory receptors, including rapidly and slowly adapting mechanoreceptors and nociceptors. The rapidly adapting mechanoreceptors are pacinian corpuscles, which respond to mechanical transients, including vibration. The slowly adapting joint receptors are Ruffini's endings, which respond best to movements of a joint to extremes of flexion or extension; these endings signal pressure or torque applied to the joint. Joint mechanoreceptors are innervated by medium-sized (group II) afferent fibers. Joint no-

ciceptors are activated by probing a joint capsule or by hyperextension or hyperflexion, although many articular nociceptors fail to respond to joint movements under normal conditions. If sensitized by inflammation, however, they can respond to innocuous stimuli, such as movements or weak pressure. Joint nociceptors are innervated by finely myelinated (group III) or unmyelinated (group IV) primary afferent fibers.

Arthritis is a common painful condition caused by inflammation of one or more joints. The nociceptors become sensitized by the release of a number of chemical substances from nerve endings, mast cells, and blood elements. These substances include the neuropeptides, substance P and calcitonin gene-related peptide, histamine, bradykinin, serotonin, and prostaglandins. The sensitized nerve endings cause the joint to develop **hyperalgesia,** a condition in which the threshold for pain is lowered and the amount of pain produced by a given stimulus is increased. The joint also becomes swollen. This is caused by **neurogenic edema,** which is the collection of edema fluid that follows an increase in capillary permeability caused by the release of neuropeptides from joint nociceptors. These peptides also cause **vasodilation,** which increases the temperature of the joint. Arthritic pain is often treated successfully with substances, such as acetylsalicylic acid, that block the synthesis of prostaglandins.

Viscera are supplied with sensory receptors. These are usually involved in reflexes and have little to do with sensory experience. However, some visceral mechanoreceptors are responsible for the sensation of distension, and visceral nociceptors produce visceral pain. Pacinian corpuscles are present in the mesentery and in the capsules of visceral organs such as the pancreas; these presumably signal mechanical transients. Whether some forms of visceral pain result from overactivity of the mechanoreceptor afferents is still controversial. Some viscera, however, clearly have specific nociceptors.

Dermatomes, Myotomes, and Sclerotomes

Primary afferent fibers in the adult are distributed systematically, as determined during embryologic development. The mammalian embryo becomes segmented, and each body segment is called a **somite.** A somite is innervated by an adjacent segment of the spinal cord or, in the case of a somite of the head, by a cranial nerve. The portion of a somite destined to form skin is called a **dermatome.** Similarly, the part of a somite that will form muscle is a **myotome,** and the part that will form bone,

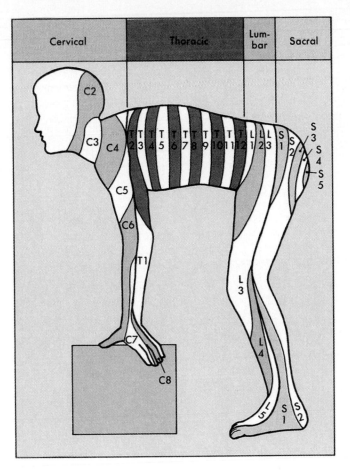

FIGURE 7-10 Dermatomes represented on a drawing of a person assuming a quadrupedal position.

a **sclerotome.** Viscera are also supplied by particular segments of the spinal cord or particular cranial nerves.

Many dermatomes become distorted during development, chiefly because of the way the upper and lower extremities are formed and because humans maintain an upright posture. However, the sequence of dermatomes can be understood if pictured on the body in a quadrupedal position (Figure 7-10).

Although a dermatome receives its densest innervation from the corresponding spinal cord segment, the dermatome is also supplied by several adjacent spinal segments. Thus transection of a dorsal root produces little sensory loss in the corresponding dermatome. Anesthesia of any given dermatome requires interruption of several successive dorsal roots.

Spinal roots and spinal cord

Axons of the peripheral nervous system enter or leave the CNS through the **spinal roots** (or through cranial nerves). The dorsal root on one side of a given spinal segment is composed entirely of the central processes of **dorsal root ganglion** cells. The **ventral root** con-

sists chiefly of motor axons, including α-motor axons, γ-motor axons, and, at certain segmental levels, autonomic preganglionic axons. Ventral roots also contain many primary afferent fibers, whose role is still unclear.

The spinal cord can be subdivided into gray matter and white matter (Figure 7-11). The **gray matter** includes the cell bodies and dendrites of the intrinsic neurons of the spinal cord. It is here that synaptic connections are made by primary afferent fibers and by pathways descending from the brain. The gray matter is subdivided into the **dorsal horn,** the **intermediate region,** and the **ventral horn.** The neurons of the gray matter form layers, or **laminae.**

The gray matter of the spinal cord is surrounded by **white matter** (Figure 7-11). The white matter located between the dorsal midline of the spinal cord and the entry line of the dorsal roots is called the **dorsal funiculus.** The white matter between the dorsal root entry line and the ventral root exit line is the **lateral funiculus.** The white matter found between the ventral root exit line and the ventral midline is the **ventral funiculus.** The **dorsolateral fasciculus** is a zone of fine nerve fibers that caps the dorsal horn. The white matter contains axons that belong to primary afferent fibers, spinal cord interneurons, and long ascending and descending pathways that connect the spinal cord and brain.

Trigeminal Nerve

The arrangement for primary afferent fibers that supply the face is comparable to that for fibers supplying the body. Peripheral processes of neurons in the **trigeminal ganglion** pass through the ophthalmic, maxillary, and mandibular divisions of the trigeminal nerve to innervate dermatome-like regions of the face (Figure 7-12, *A*). The trigeminal nerve also innervates the oral and nasal cavities and the dura mater.

The large myelinated fibers that supply mechanoreceptors of the skin and structures of the oral and nasal cavities synapse in the **principal sensory nucleus** of the trigeminal nerve (Figure 7-12, *B*). Small myelinated and unmyelinated primary afferent fibers of the trigeminal nerve terminate in the nerve's **spinal nucleus** (Figure 7-12, *C*). Primary afferent fibers from stretch receptors have their cell bodies in the **mesencephalic nucleus** of the trigeminal nerve (Fig. 7-12, *D*). This arrangement is exceptional, because all other primary afferent cell bodies in the somatovisceral sensory system are in peripheral ganglia. The central processes synapse in the **motor nucleus** of the trigeminal nerve.

Dorsal Column–Medial Lemniscus Pathway

The ascending branches of many large myelinated nerve fibers travel rostrally in the dorsal column to the

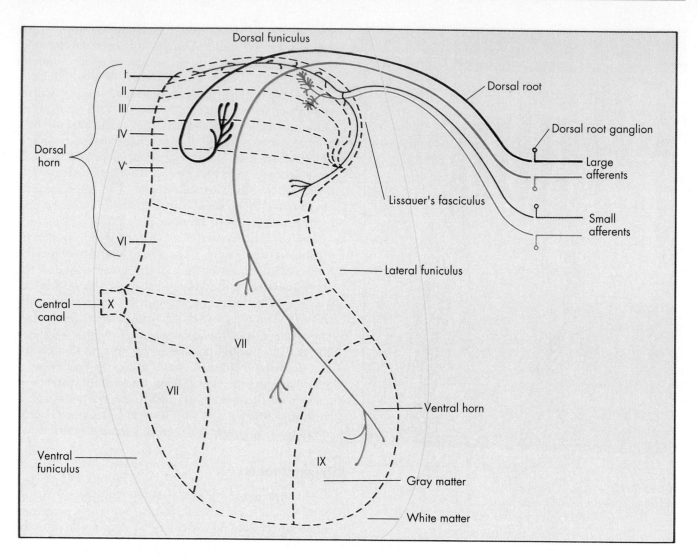

FIGURE 7-11 Distribution of large and small primary afferent fibers in the spinal cord. The dorsal, lateral, and ventral funiculi and the dorsolateral fasciculus of the white matter are indicated, as are the dorsal root ganglion and dorsal root. The dashed line shows an outline of the gray matter on one side of the spinal cord and the boundaries between the laminae. The laminae are indicated by the Roman numerals. Laminae I to VI form the dorsal horn; parts of laminae VI and VII make up the intermediate region, and part of lamina VII, plus laminae VIII and IX, form the ventral horn; lamina X is the gray matter surrounding the central canal. Terminals of two large primary afferent fibers are shown. One, from a hair follicle receptor, synapses in laminae III to V. The other, from a muscle spindle, synapses in laminae VI, VII, and IX. Terminals from two fine afferent fibers are also shown. The small myelinated fiber (A-δ), from a cutaneous mechanical nociceptor, ends in laminae I and V, whereas the unmyelinated (C) afferent fiber synapses in laminae I and II.

medulla (see a standard textbook of neuroanatomy). The dorsal column can be subdivided into two smaller components, the **fasciculus gracilis** and the **fasciculus cuneatus.** Axons that innervate sensory receptors of the lower extremity and the lower trunk (T7 segment and caudally) ascend in the gracile fasciculus, whereas fibers from receptors of the upper extremity and upper trunk ascend in the cuneate fasciculus (T6 segment and ros-

trally). These axons are the first-order neurons of the dorsal column–medial lemniscus path. The second-order neurons are in the **nucleus gracilis** and **nucleus cuneatus,** which are collections of neurons in the caudal medulla. These nuclei are often called collectively the dorsal column nuclei. Axons of the fasciculus gracilis synapse in the nucleus gracilis, and axons of the fasciculus cuneatus synapse in the nucleus cuneatus.

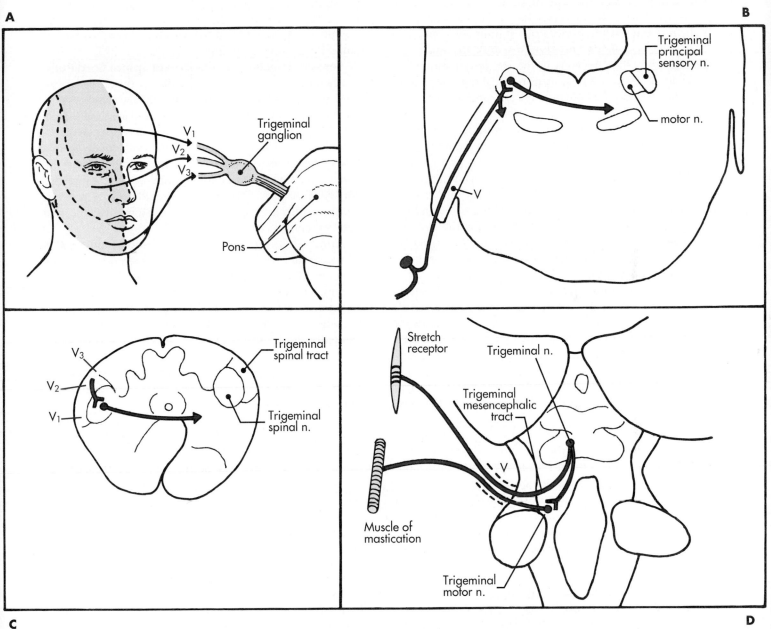

FIGURE 7-12   In **A** are seen the dermatome-like areas of distribution of the ophthalmic (V_1), maxillary (V_2), and mandibular (V_3) divisions of the trigeminal nerve. **B** shows the synaptic terminations of large myelinated primary afferent fibers of the trigeminal nerve in the main sensory nucleus. **C** shows the descending branches of fine myelinated and unmyelinated primary afferent axons of the trigeminal nerve synapsing in the spinal nucleus. **D** shows the location of the cell bodies of proprioceptive neurons of the trigeminal nerve in the mesencephalic nucleus. Synaptic connections of collaterals of these neurons are made with motoneurons in the trigeminal motor nucleus.

Neurons of the dorsal column nuclei respond in much the same way as the primary afferent fibers that synapse on them. Some behave like rapidly adapting receptors, responding to hair movement or to mechanical transients applied to the glabrous skin. Others discharge at high frequencies when vibratory stimuli are applied to their receptive fields and thus resemble pacinian corpuscles. Still other neurons in the dorsal column nuclei have slowly adapting responses to cutaneous stimuli. In the cuneate nucleus many neurons are activated by stretching muscles. The main differences between the responses of dorsal column neurons and the primary afferent fibers are that (1) dorsal column neurons have larger receptive fields, because more than one primary afferent fiber synapses on a given dorsal column neuron; (2) they sometimes respond to more than one class of sensory receptor because of convergence of several different types of primary afferent fibers on the second-order neurons; and (3) they often have inhibitory receptive fields mediated through interneuronal circuits in the dorsal column nuclei.

The dorsal column nuclei project to the contralateral thalamus by way of the **medial lemniscus.** The medial lemniscus terminates in the **ventral posterolateral (VPL) nucleus** of the thalamus. Neurons in the VPL nucleus in turn project to the **primary somatosensory (SI) cortex.**

Projections of Trigeminal Nuclei As already mentioned, large myelinated primary afferent fibers that supply mechanoreceptors in the skin of the face synapse in the principal sensory nucleus of the trigeminal nerve. Second-order neurons in the principal sensory nucleus project to the contralateral thalamus by way of the **trigeminothalamic tract.** Some neurons project ipsilaterally. The projections are to the ventral **posteromedial (VPM) thalamic nucleus.** Third-order neurons of the VPM nucleus project to the somatosensory cortex. This pathway of the trigeminal system is equivalent for the face to the dorsal column–medial lemniscus path for the body.

Other Somatosensory Pathways of the Dorsal Spinal Cord Three other pathways that carry somatosensory information ascend in the dorsal part of the spinal cord on the same side as the afferent input: (1) the spinocervical tract, (2) the postsynaptic dorsal column pathway, and (3) the dorsal spinocerebellar tract.

The cells of origin of the **spinocervical tract** receive input largely from cutaneous mechanoreceptors, although some of these cells are activated by nociceptors as well. The cells of origin of the **postsynaptic dorsal column path** receive information similar to that reaching spinocervical tract neurons. The information conveyed by these cells is eventually conveyed to the VPL nucleus of the thalamus.

The **dorsal spinocerebellar tract** responds to input from muscle and joint receptors of the lower extremity. The main destination of the tract is the cerebellum, but it also provides proprioceptive information from the leg to the VPL nucleus of the thalamus after a relay in the medulla. Proprioceptive information from the arm is signaled by the dorsal column path.

Sensory Functions of the Dorsal Spinal Cord Pathways The sensory qualities mediated by dorsal spinal cord pathways include flutter-vibration, touch-pressure, joint movement and position sense, and visceral distension. Each of these qualities of sensation depends on activity in a set of sensory neurons that collectively form a labeled-line sensory channel. A sensory channel may involve several parallel ascending pathways, and it includes particular primary afferent neurons and sensory processing mechanisms at spinal cord, brainstem, thalamic, and cerebral cortical levels.

Flutter-vibration is a complex sensation. Flutter refers to recognition of events that have low-frequency components. The sensory receptors detecting flutter include hair follicles and Meissner's corpuscles. Ascending sensory tracts that convey the information needed for flutter sensation include the dorsal column–medial lemniscus pathway, the spinocervical tract, and the postsynaptic dorsal column path. High-frequency vibration is primarily detected by pacinian corpuscles. Branches of Pacinian corpuscle afferents ascend in the dorsal columns.

Touch-pressure sensation involves the recognition of skin indentation by Merkel's cell and Ruffini's receptors. The ascending pathways that convey information from these receptors include the dorsal column–medial lemniscus path and the postsynaptic dorsal column tract.

Proprioception, the senses of joint movement and joint position, is complex and depends on sensory information that arises from muscle, joint, and cutaneous receptors. For some joints, such as the knee, the most important information is derived from muscle spindles in the muscles that move the joint. In other joints, however, such as those of the digits, Ruffini's endings and joint receptors also contribute.

Higher Processing of Tactile and Proprioceptive Information

As mentioned earlier, the medial lemniscus synapses in the VPL nucleus of the thalamus. The responses of many neurons in the VPL nucleus resemble those of the first- and second-order neurons of the dorsal column–medial lemniscus pathway. The responses may be dominated by a particular type of receptor, and the receptive field may be small, although larger than that of a primary afferent fiber. Thalamic neurons often have inhibitory receptive fields. A notable difference between neurons in the VPL nucleus and neurons at lower levels of the dorsal column–medial lemniscus path is that the excitability of the thalamic neurons depends on the stage of the sleep-wake cycle and on the presence or absence of anesthesia.

The primary somatosensory receiving area (SI) of the cortex is located in the parietal lobe of the cerebral cortex. Within any particular area of the SI cortex, all the neurons along a line perpendicular to the cortical surface have similar response properties and receptive fields. The SI cortex is thus said to have a **columnar organization.** A comparable columnar organization has also been demonstrated for other primary sensory receiving areas, including the primary visual and auditory cortices.

The location of cortical columns in the SI region is related systematically to the location of the receptive fields on the body surface. This relationship is called a **somatotopic organization.** The body surface is mapped in the SI cortex. The lower extremity is represented in the human in the medial aspect and the apex of the postcentral gyrus, whereas the upper extremity is mapped on the dorsolateral aspect of the postcentral gyrus and the face is mapped just dorsal to the lateral fissure (Figure 7-13). This somatotopic map for humans is called a **homunculus,** meaning a little man. The somatotopic organization of the SI cortex is responsible for the coding of stimulus location. The somatotopic organization at the cortical level reflects the same organization at lower levels of the somatosensory system, including the dorsal column nuclei and the VPL nucleus of the thalamus.

Besides being responsible for the initial processing of somatosensory information, the SI cortex also begins higher-order processing, such as **feature extraction,** the recognition of special features of a stimulus. For example, certain neurons in area 1 respond preferentially to a stimulus moving in one direction across the receptive field but not in the opposite direction (Figure 7-14). This response pattern is the result of the organization of inhibitory circuits in the cortex. Such neurons might contribute to the perceptual ability to recognize the direction of an applied stimulus.

Spinothalamic Tract

The **spinothalamic tract** originates from spinal cord neurons that project mainly to the contralateral thalamus. The place where the axon of a spinothalamic tract cell crosses is within the same segment as the cell body, and the axon ascends in the lateral or ventral funiculus. The spinothalamic tract terminates in several nuclei of the thalamus, including the VPL nucleus and several nuclei of the medial thalamus.

The cells of origin of the spinothalamic tract are found chiefly in spinal cord laminae I and V. Effective stimuli include noxious mechanical, thermal, and chemical stimuli. Some spinothalamic neurons are excited by activity in cold or warm thermoreceptors or sensitive mechanoreceptors.

Spinothalamic tract cells often receive a convergent excitatory input from several different classes of sensory receptors. For example, a given spinothalamic tract neuron may be activated weakly by tactile stimuli but more powerfully by noxious stimuli (Figure 7-15). Such neurons are called **wide-dynamic-range cells** because they are activated by stimuli having a great range of intensities. Wide-dynamic-range neurons mainly signal noxious events, the weak response to tactile stimuli perhaps being ignored by higher centers. However, in pathological conditions these neurons may be activated sufficiently by normally innocuous stimuli to evoke a sensation of pain. This would explain some pain states in which activation of mechanoreceptors causes pain **(allodynia).** Other spinothalamic tract cells are activated only by noxious stimuli. Such neurons are often called **nociceptive-specific** or high-threshold cells (Figure 7-15).

An example of a pathological state in which **allodynia** is prominent is **central pain,** which can result from damage to the central nervous system. For example, a central pain syndrome called **thalamic pain** may be caused by a lesion involving the VPL nucleus. The pain is typically burning in quality, although it can be sharp. Pain is often evoked by very weak stimulation, such as contact of the clothing with the skin. Thalamic pain patients often prefer moistened clothing to dry clothing and sometimes wear a wet cotton glove on the side affected (the **Michael Jackson sign**).

Spinothalamic tract cells often have inhibitory receptive fields. Inhibition may result from weak mechanical stimuli, but usually the most effective inhibitory stimuli are noxious ones. The nociceptive inhibitory receptive fields may be very large and include most of the body and face (Figure 7-15).

The **gate control theory of pain** explains how innocuous stimuli may inhibit the responses of dorsal horn neurons that transmit information about painful stimuli to the brain. In this theory pain transmission is prevented by innocuous inputs mediated by large myelinated afferent fibers, whereas pain transmission is enhanced by in-

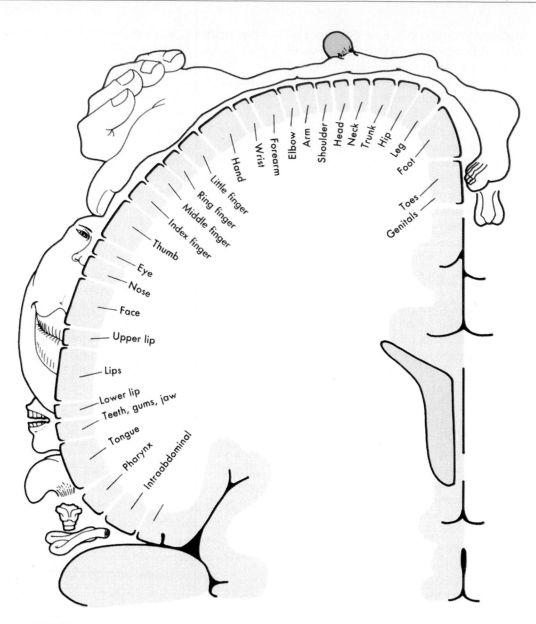

FIGURE 7-13 Sensory homunculus of the primary somatosensory cortex (SI). One cerebral hemisphere is drawn as if it were sectioned in the coronal plane. The cortex is shown by the pink area. A sensory homunculus is a representation of the body surface as it is mapped systematically on to neurons in the SI cortex. For example, neurons in cortex on the lateral surface of the SI cortex just above the lateral fissure have receptive fields on the face; neurons on the more superior part of the lateral SI cortex have receptive fields on the hand and arm; finally, neurons in the cortex of the medial surface of the hemisphere have receptive fields on the foot and leg. The sizes of different parts of the homunculus indicate the proportionate amounts of cortex dedicated to these parts of the body.

puts carried over fine afferent fibers. The inhibitory interneurons of lamina II serve as a gating mechanism. The circuit diagram originally proposed has been criticized, but the basic notion of a gating mechanism is still viable.

Many of the spinothalamic tract cells that project to medial thalamic nuclei have very large receptive fields that often include much of the surface of the body and face. The large receptive fields of these spinothalamic neurons suggest that they function to trigger **motivational-affective responses** to painful stimuli, rather than participating in **sensory discrimination.**

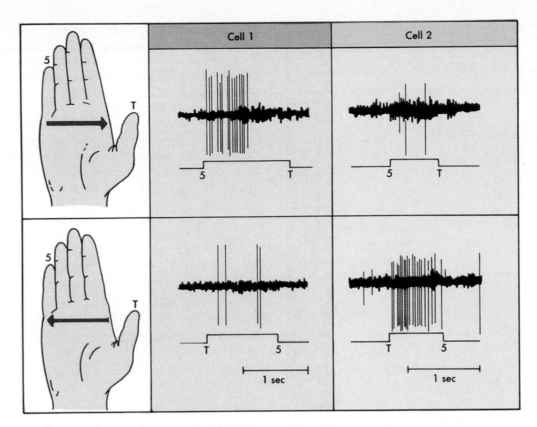

FIGURE 7-14 Feature extraction by cortical neurons. The responses of two cortical neurons are shown to stimuli moved across the palm. Cell 1 was excited strongly by movement of the stimulus from the fifth digit toward the thumb, but only weakly by movement of the stimulus in the opposite direction. Cell 2 showed the converse behavior.

Various therapeutic techniques have been developed for the treatment of pain. Some of these involve the surgical interruption of nociceptive pathways, such as the spinothalamic tract. For example, the operation of **cordotomy** involves a surgical lesion in the ventrolateral white matter of the spinal cord at the level rostral to the spinal cord segments that receive the pain signals and on the opposite side (since the spinothalamic tract crosses within the spinal cord). After such a lesion, pain and temperature sensations are lost on the opposite side of the body, below the level of the lesion. Cordotomies can be effective, at least for a few months. However, pain may recur, and so cordotomies are of limited value when survival time is extended. For chronic pain cases, other modalities of treatment are often tried, such as **transcutaneous electrical nerve stimulation (TENS)** or stimulation of the dorsal funiculus with an implanted electrode. These approaches are based on the gate theory of pain and depend upon inhibitory processing of nociceptive signals in the central nervous system.

Projection of Trigeminal Nuclei Nociceptors and thermoreceptors of the face enter the brainstem with the trigeminal nerve and synapse in the spinal nucleus of the trigeminal. Second-order neurons of the spinal nucleus project to the contralateral VPM thalamic nucleus and medial thalamus through the trigeminothalamic tract. Third-order thalamic neurons in turn project to the face area of the somatosensory cortex. This pathway for the face is equivalent to the spinothalamic tract for the body.

Other Somatosensory Pathways of the Ventral Spinal Cord Two other pathways, the spinoreticular tract and the spinomesencephalic tract, transmit somatosensory information and ascend in the ventral part of the spinal cord.

The cells of origin of the **spinoreticular tract** are often difficult to activate, but when receptive fields are found, these are generally large, sometimes bilateral, and the effective stimuli include noxious ones. The reticular formation is located within the core of the brainstem, and it is involved in attentional mechanisms and arousal. Ascending fibers from the reticular formation extend to the medial thalamus and from there to wide areas of the cerebral cortex.

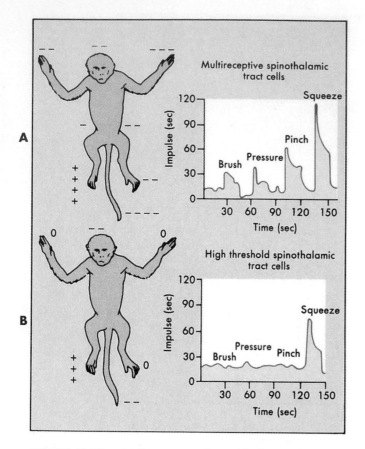

FIGURE 7-15 **A,** Responses of a wide-dynamic-range or multireceptive spinothalamic tract cell. **B,** Responses of a high-threshold spinothalamic tract cell. The figurines at the left indicate the excitatory *(plus signs)* and inhibitory *(minus signs)* receptive fields. The graphs at the right show the responses to graded intensities of mechanical stimulation. *Brush,* A camel's hair brush repeatedly stroked across the receptive field. *Pressure,* Attachment of an arterial clip to the skin; this is a marginally painful stimulus to a human. *Pinch,* Attachment of a stiff arterial clip to the skin; distinctly painful. *Squeeze,* Compressing a fold of skin with forceps; damaging to the skin.

Many cells of the **spinomesencephalic tract** respond to noxious stimuli, and the receptive fields are generally small. The terminations of the tract are in several midbrain nuclei, including the **periaqueductal gray** (the area around the cerebral aqueduct), which is an important component of the *endogenous analgesia system* (see later discussion). Information from the midbrain is also relayed to the **limbic system.** This may provide one pathway by which noxious stimuli can trigger emotional responses. Motivational-affective responses may also result from activation of the periaqueductal gray and midbrain reticular formation. The latter is an

important part of the arousal system, and stimulation in the periaqueductal gray causes vocalization and aversive behavior.

Sensory Functions of the Ventral Spinal Cord Pathways The most important sensory modalities mediated by ventral spinal cord pathways are pain and thermal sensations. *Thermal sense* depends on input from cold and warm receptors to spinothalamic tract neurons. *Pain* resulting from stimulation of nociceptors is mediated partly by spinothalamic tract cells and partly by the spinoreticular and spinomesencephalic tracts.

The motivational-affective responses to painful stimuli include attention and arousal, somatic and autonomic reflexes, endocrine responses, and emotional changes. These collectively account for the unpleasant nature of painful stimuli. The motivational-affective responses depend on several ascending pathways, including the component of the spinothalamic tract that projects to the medial thalamus, the spinoreticular tract, and the spinomesencephalic tract. As indicated previously, these pathways have access to attentional, orientational, and arousal systems, as well as to the limbic system.

Pain that originates from the skin is generally well localized, presumably because spinothalamic tract cells have relatively discrete cutaneous receptive fields. Also, the ascending system through which they signal is somatotopically organized. However, pain that originates from deep structures, including muscle and viscera, is poorly localized and is often mistakenly attributed to superficial structures **(referred pain).**

Angina pectoris is a type of visceral pain that results from ischemia (inadequate blood flow) to the heart. The pain is often referred to the inner aspect of the left arm, although other regions of referral, such as the jaw, the abdomen, or the back are not uncommon. The ischemia, generally caused by arteriosclerotic narrowing of one or more coronary arteries, releases algesic chemicals that sensitize visceral nociceptors that supply the heart. The nociceptors activate spinothalamic tract neurons in the left upper thoracic spinal cord. These neurons are also excited by sensory receptors that supply the left upper trunk and the T1 dermatome, which courses along the inner aspect of the arm.

One explanation of referred pain is that many spinothalamic neurons receive excitatory input not only from the skin but also from muscle and viscera. The spinal cord segments that innervate the dermatomes containing the cutaneous receptive field of the cell correspond well to the segments innervating the muscle or viscus. The activity in a population of spinothalamic tract cells may be interpreted as originating from the skin, based on the learning of this association during childhood. Subsequently, activation of these neurons by pathological input from visceral nociceptors is misinterpreted as resulting from stimulation of superficial parts of the body. This idea is known as the **convergence-projection theory.**

Pain sometimes occurs in the absence of nociceptor stimulation. This is most likely to happen after damage to peripheral nerves or to the parts of the CNS involved in transmitting nociceptive information. The mechanism of such pain caused by neural damage (**neuropathic pain**) is poorly understood. However, it appears that many afferent nerve fibers become sensitive to the sympathetic transmitter *norepinephrine* after nerve damage. Also dorsal root ganglion cells and damaged afferent fibers may develop abnormal spike generators that cause spontaneous discharges in unstimulated primary afferent neurons. This activity may well cause abnormal sensory experience, including pain and **paresthesias** (feelings of pins and needles).

The Trigeminal System

Sensory processing for the face, oral cavity, and dura mater is organized in a fashion similar to that for the rest of the body. Neurons in the **principal sensory nucleus** are apparently responsible for flutter-vibration and touch-pressure sensations from the face, and neurons in the **subnucleus caudalis of the spinal nucleus** mediate pain and temperature sensations that originate from the face, oral cavity, and dura mater. *Pain in the trigeminal distribution is of particular importance, because this includes both tooth pain and headaches.*

Centrifugal Control of Somatovisceral Sensation

Sensory experience is not just the passive detection of environmental events; instead, it more often depends on exploration of the environment. Tactile cues are sought by moving the hand over a surface. Visual cues result from scanning visual targets with the eyes. Thus sensory information is often received as a result of activity in the motor system. Furthermore, sensory transmission in pathways to the sensory centers of the brain is regulated by descending control systems. This allows the brain to control its input by filtering the incoming

sensory messages. Important information can be processed and unimportant information ignored.

The tactile and proprioceptive somatosensory pathways are regulated by descending pathways. However, of particular interest are the descending control systems that regulate the transmission of nociceptive information. These systems presumably serve to reduce excessive pain under certain circumstances.

It is well known that soldiers on the battlefield, athletes in competition, accident victims, and others facing stressful circumstances often feel little or no pain at the time a wound occurs or a bone is broken. Later, however, pain may be severe. Although the descending regulatory pathways that control pain are part of the more general centrifugal system that modulates all forms of sensation, the pain control system is so important medically that it is distinguished as a special **endogenous analgesia system.**

Several descending pathways contribute to the endogenous analgesia system. The activity of these descending pathways inhibits spinothalamic tract neurons. The **raphe nuclei,** which are located at the midline of the medulla, and the **periaqueductal gray,** located near the midline in the midbrain, give rise to direct and indirect projections to the medullary and spinal dorsal horns, and these pathways inhibit nociceptive neurons, including trigeminothalamic and spinothalamic tract cells (Figure 7-16, *A*).

The endogenous analgesia system can be subdivided into pathways that release one of the **endogenous opioids** and those that do not. The endogenous opioid substances are neuropeptides that activate one of the several forms of opiate receptors. Some of the endogenous opioids are **enkephalin, dynorphin,** and β-**endorphin.** Opioid analgesia can generally be antagonized with the narcotic antagonist *naloxone.* Therefore naloxone is used as a test of whether analgesia is mediated through an opioid mechanism.

Opiates typically inhibit neural activity in nociceptive pathways. Two sites of action have been proposed for opiate inhibition: presynaptic and postsynaptic (Figure 7-16, *B*). The presynaptic action of opiates on nociceptive afferent terminals is thought to prevent the release of excitatory transmitters, such as **substance P.** The postsynaptic action produces an inhibitory postsynaptic potential. How can an inhibitory neurotransmitter activate descending pathways? One hypothesis is that the descending analgesia system is under tonic inhibitory

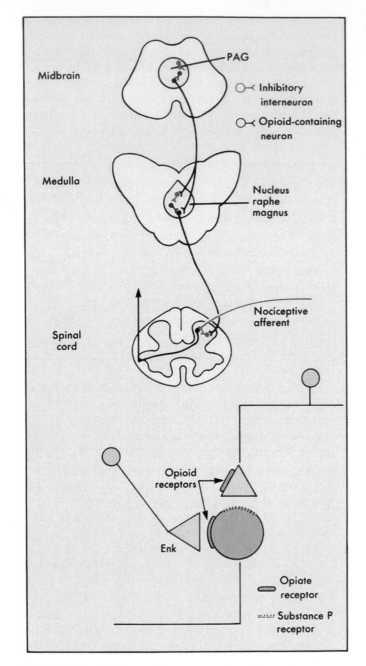

A

B

FIGURE 7-16 **A,** Some of the neurons thought to play a role in the endogenous analgesia system. Neurons in the periaqueductal gray *(PAG)* activate a raphe-spinal pathway, which in turn inhibits spinothalamic tract cells. Interneurons containing opioid substances are involved in the system at each level. **B,** Possible presynaptic and postsynaptic sites of action of enkephalin *(Enk).* The presynaptic action prevents the release of substance P (Sub P) from nociceptors, and the postsynaptic site causes inhibition of a nociceptive neuron.

control by inhibitory interneurons in both the midbrain and the medulla. The action of opiates would inhibit the inhibitory interneurons and thereby disinhibit the descending analgesia pathways.

Opiate receptors are found in the brain and also in the spinal cord. At the spinal level, opiate receptors are located not only on dorsal horn neurons but also on the terminals of nociceptors within the dorsal horn. Because of the presence of these opiate receptors, pain can sometimes be treated by application of morphine to the spinal cord. This is done by use of a **morphine pump** connected to a catheter introduced into the epidural space. The morphine can diffuse across the dura and arachnoid to enter the cerebrospinal fluid space around the spinal cord. The advantage of this route of delivery of morphine is that the morphine has a direct action on pain processing circuits without seriously affecting thought processes. However, an overdose of morphine can still depress respiration by rostral spread in the CSF.

The dose applied by the morphine pump can be controlled by the patient (patient-controlled analgesia). It turns out that the total dose of morphine that successfully minimizes pain is less when regulated by the patient than when regulated by the health care team.

Some endogenous analgesia pathways operate by neurotransmitters other than opioids and thus are unaffected by naloxone. One way of engaging a nonopioid analgesia pathway is through any of several types of stress. The analgesia so produced is called **stress-induced analgesia.**

Many neurons in the raphe nuclei release *serotonin* as a neurotransmitter. Serotonin is able to inhibit nociceptive neurons and presumably plays an important role in the endogenous analgesia system. Other brainstem neurons release catecholamines, such as *norepinephrine.* Catecholamines also inhibit nociceptive neurons, and therefore catecholaminergic neurons may contribute to the endogenous analgesia system. Undoubtedly many other substances also affect the analgesia system. Furthermore, evidence now shows that endogenous opiate antagonists exist that can prevent opiate analgesia.

SUMMARY

1. Sensory receptors respond to stimuli by various transduction mechanisms. The specific mechanism depends on the type of receptor.
2. Transduction causes a receptor potential, which is usually a depolarization that causes the receptor's primary afferent fiber to discharge.

3. In some receptor organs, the primary afferent fiber develops a generator potential in response to the release of synaptic transmitter from a receptor cell.

4. The responses of sensory receptors can be slowly or rapidly adapting, depending on the type of information to be signalled.

5. A sensory receptor has a receptive field, which is the region that when stimulated effects the discharge of the neuron. Central sensory neurons may have both excitatory and inhibitory receptive fields.

6. Sensory information may be encoded so that modality, spatial location, threshold, intensity, frequency, and duration are all recognized.

7. Sensory pathways involve not only primary afferent fibers (first-order neurons), but also neurons in the spinal cord or brainstem (second-order neurons), thalamus (third-order neurons), and cerebral cortex (fourth- and higher-order neurons).

8. The main sensory pathways of the somatovisceral (general sensory) system are the dorsal column–medial lemniscus path, the spinothalamic tract, and the trigeminothalamic tract.

9. Somatovisceral sensory receptors include mechanoreceptors, thermoreceptors and nociceptors.

10. Cutaneous sensory receptors include mechanoreceptors (hair follicle receptors, Meissner's corpuscles, pacinian corpuscles, Merkel's cell endings, Ruffini's corpuscles), thermoreceptors (cold and warm receptors), and several kinds of nociceptors.

11. Skeletal muscle contains stretch receptors, as well as nociceptors. Joints and viscera are also supplied with mechanoreceptors and nociceptors.

12. Nociceptors can be sensitized by release of chemical substances by injury or inflammation.

13. The segmental development of the body leads to a segmental pattern of innervation of the skin, the dermatomes.

14. The spinal cord contains gray and white matter. The gray matter is divided into the dorsal horn, the intermediate region, and the ventral horn.

15. Large myelinated primary afferent fibers synapse in the dorsal horn, the intermediate region, and the ventral horn. The small myelinated and unmyelinated afferents synapse in the dorsal horn.

16. The trigeminal nerve supplies dermatome-like regions of the face.

17. The dorsal column–medial lemniscus path mediates the sensations of flutter-vibration, touch-pressure, and proprioception. The path is somatotopically organized. The cortical representation of the body forms a sensory homunculus.

18. The spinothalamic tract mediates pain and temperature sensations. The motivational-affective part of the pain response involves nonsomatotopic connections with the medial thalamus. Visceral pain is often referred to somatic structures.

19. The somatovisceral sensory pathways, including the spinothalamic tract, are controlled by pathways that descend from the brainstem. The regulation of pain sensation is by the endogenous analgesia system. The endogenous analgesia system releases endogenous opioid substances, as well as amines, such as serotonin and norepinephrine, as neurotransmitters.

BIBLIOGRAPHY
Journal Articles

Akil H et al: Endogenous opioids: biology and function, *Annu Rev Neurosci* 7:223, 1984.

Amit Z, Galina ZH: Stress-induced analgesia: adaptive pain suppression, *Physiol Rev* 66:1091, 1986.

Basbaum AI, Fields HL: Endogenous pain control systems: brainstem spinal pathways and endorphin circuitry, *Annu Rev Neurosci* 7:309, 1984.

Besson JM, Chaouch A: Peripheral and spinal mechanisms of nociception, *Physiol Rev* 67:67, 1987.

Cervero F: Sensory innervation of the viscera: peripheral basis of visceral pain, *Physiol Rev* 74:95, 1994.

Johnson KO, Hsiao SS: Neural mechanisms of tactual form and texture perception, *Annu Rev Neurosci* 15:227, 1992.

Kaas JH: What, if anything, is SI? Organization of first somatosensory area of cortex, *Physiol Rev* 63:206, 1983.

McCloskey DI: Kinesthetic sensibility, *Physiol Rev* 58:763, 1978.

Snyder SH, Childers SR: Opiate receptors and opioid peptides, *Annu Rev Neurosci* 2:35, 1979.

Books and Monographs

Boivie JJG, Perl ER: Neural substrates of somatic sensation. In Hunt CC, ed: *Neurophysiology,* MTP Physiology series one, vol 3, International Review of Science, Baltimore, 1975, University Park Press.

Bonica JJ: *The management of pain,* ed 2, Philadelphia, 1990, Lea & Febiger.

DeGroot J, Chusid JG: *Correlative neuroanatomy,* East Norwalk, Conn, 1988, Appleton & Lange.

Foreman RD: Organization of the spinothalamic tract as a relay for cardiopulmonary sympathetic afferent fiber activity. In Ottoson D, ed: *Progress in sensory physiology,* ed 9, Berlin, 1989, Springer-Verlag.

Kandel ER, Schwartz JH, Jessell TM: *Principles of neural science,* ed 3, New York, 1991, Elsevier Science Publishing Co, Inc.

Kenshalo DR: *The skin senses,* Springfield, Ill, 1968, Charles C Thomas, Publisher.

Shepherd GM: *The synaptic organization of the brain,* ed 3, New York, 1990, Oxford University Press.

Uttal WR: *The psychobiology of sensory coding,* New York, 1973, Harper & Row, Publishers, Inc.

Willis WD: *Control of nociceptive transmission in the spinal cord,* Berlin, 1982, Springer-Verlag.

Willis WD: *The pain system,* Basel, 1985, Karger.

Willis WD, Coggeshall RE: *Sensory mechanisms of the spinal cord,* ed 2, New York, 1991, Plenum Press.

Willis WD, Grossman RG: *Medical neurobiology,* ed 3, St Louis, 1981, Mosby.

CHAPTER 8

Special Senses

In the evolution of the nervous system, an important trend was **encephalization,** in which special sensory organs developed in the heads of animals, along with appropriate neural systems in the brain. These special sensory systems, which included the **visual, auditory, olfactory,** and **gustatory systems,** allowed the animal to detect and analyze light, sound, and chemical signals in the environment. In addition, the **vestibular system** evolved to signal the position of the head.

VISUAL SYSTEM

The visual system detects and interprets photic stimuli. In vertebrates effective photic stimuli are electromagnetic waves between 400 and 700 nm long, or **visible light.** Light enters the eye and impinges on **photoreceptors** of a specialized sensory epithelium, the retina. The photoreceptors are the **rods** and **cones.** *Rods have low thresholds for detecting light and thus operate best under conditions of reduced lighting (scotopic vision).* However, rods neither provide well-defined visual images nor contribute to color vision. *Cones, by contrast, are not as sensitive as rods to light but operate best under daylight conditions (photopic vision). Cones are responsible for high visual acuity and color vision.* Information processing within the retina is done by the **retinal interneurons,** and the output signals are carried to the brain by axons of the **retinal ganglion cells** through the **optic nerves** and **optic tracts.** The main visual pathway is through the **lateral geniculate nucleus** of the **thalamus** to the visual receiving areas of the **cerebral cortex** (Figure 8-1).

Structure of the Eye

The wall of the eye is formed of three concentric layers (Figure 8-2). The outer layer is the fibrous coat, which includes the transparent **cornea,** with its epithelium, the conjunctiva, and the opaque **sclera.** The middle layer is the vascular coat, which includes the **iris** and the **choroid.** The iris contains both radially and circularly oriented smooth muscle fibers, which constitute the **pupillary dilator** and **sphincter;** the iris forms a diaphragm to control the size of the **pupil.** *The dilator is activated by the sympathetic nervous system and the sphincter by the parasympathetic nervous system (oculomotor nerve).* The choroid is rich in blood vessels that supply the outer layers of the retina. The inner retinal layers are nourished by tributaries of the central artery and veins of the retina; these vessels course with the optic nerve.

The inner layer of the eye is the neural coat, the **retina.** The functional part of the retina covers the entire posterior eye except for the blind spot, which is the optic nerve head. Visual acuity is highest in the central part of the retina, the **macula lutea.** The **fovea** is a pit-like depression in the middle of the macula where visual targets are focused; it is the **fixation point,** or point at which light rays are focused when the eyes are directed at a visual target of interest (Figure 8-2).

Besides the retina, the eye contains a **lens** to focus light on the retina, **pigment** to reduce light scatter, and fluids called **aqueous humor** and **vitreous humor** that help maintain the shape of the eye. Externally attached **extraocular muscles** aim the eye toward an appropriate visual target.

The lens is held in place behind the iris by the **suspensory ligaments** (or **zonule fibers**), which attach to the wall of the eye at the **ciliary body** (Figure 8-2). *When the ciliary muscles are relaxed, the tension exerted by the suspensory ligaments tends to flatten the lens. When the ciliary muscles contract, tension on the suspensory ligaments is reduced, allowing the lens to assume a more spherical shape because of its elastic properties.* The ciliary muscles are activated by the parasympathetic nervous system by way of the oculomotor (third cranial) nerve.

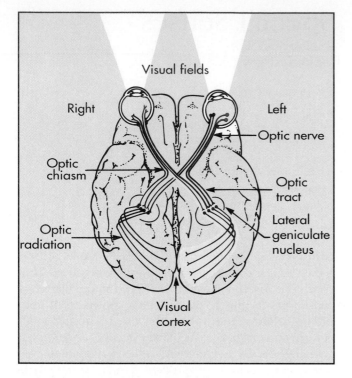

FIGURE 8-1 The main visual pathway as viewed from the base of the brain.

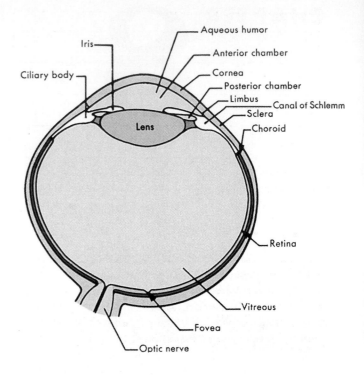

FIGURE 8-2 The right eye as viewed from above.

Light scattering within the eye is minimized by pigment. The choroid contains an abundance of pigment. In addition, the outermost layer of the retina is a pigment-containing epithelium. In addition to light absorption, the pigment cells of the **retinal pigment layer** are involved in the turnover of photoreceptor outer segments and in the regeneration of rhodopsin.

The space around the iris is filled with aqueous humor, which is a clear fluid that resembles cerebrospinal fluid. The aqueous humor is actively secreted by the **ciliary processes,** which form an epithelium that is posterior to the iris and protrudes into a space called the **posterior chamber** (Figure 8-2). The aqueous humor circulates through the posterior chamber, out the pupil, and into the **anterior chamber.** It is then reabsorbed into the **canal of Schlemm** and returned to the venous circulation.

Imbalance in the secretion and reabsorption of aqueous humor can increase the pressure in the eye, a condition that threatens the viability of the retina. This malady is known as **glaucoma.** Reabsorption can be increased surgically, and secretion can be decreased by drug therapy. Cholinergic drugs, like pilocarpine, which constrict the pupil, are helpful since they reduce the resistance to drainage of the aqueous humor.

The space behind the lens contains a gelatinous material, the **vitreous humor.** The vitreous humor turns over very slowly.

The extraocular muscles insert on the sclera from their origins on the bony orbit. Details concerning the organization and operation of the eye movement control system are described in Chapter 9.

Physiological Optics

The eye is often compared to a camera. Both are devices that capture images by using a lens system to focus light on a photosensitive surface. The quality of the image is enhanced by use of a diaphragm to reduce the effect of spherical aberrations of the lens and to increase depth of field. The diaphragm also controls the amount of entering light.

The eye, as with the camera, produces an inverted image of an object (Figure 8-3). The inversion is caused by the light rays from the object crossing at a nodal point within the lens. The image is inverted both from side to side and from above downward.

The ability of a lens to bend light is called its **refractive power.** The unit of refractive power is the **diopter.** For an image to be in focus on the retina, light coming from any point on the object and passing through the cornea and lens of the eye must be refracted just enough so that it falls on a corresponding point on the retina. *The cornea is the main refractive surface of the*

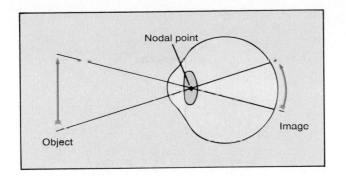

FIGURE 8-3   Image formation in the eye. The image is reversed as rays of light pass through the nodal point of the lens.

eye; it has a refractive power of 43 diopters. *However, the lens is crucial for focusing images on the retina,* because its refractive power can be varied (from 13 to 26 diopters). *The refractive power of the lens is altered by changes in the shape of the lens through relaxation or contraction of the ciliary muscles.* **Accommodation** is the process by which contraction of the ciliary muscle causes the lens to become more rounded. The result of accommodation is that images of nearby objects are brought into focus on the retina.

During aging the lens tends to lose its elasticity. This loss reduces the ability of the eye to accommodate. This visual disturbance is called **presbyopia.** Other common defects in focusing ability are myopia, hypertropia, and astigmatism. In **myopia** (nearsightedness) images are focused in front of the retina because the eye is disproportionately long for the refractive system. In **hypermetropia** (farsightedness) images are focused behind the retina because the eye is short relative to the refractive system. **Astigmatism** is the result of asymmetric focusing, usually because the cornea lacks radial symmetry.

Retina

The outermost of the 10 retinal layers is the **pigment epithelium.** The pigment cells capture stray light. They also phagocytize photoreceptor membrane shed from the outer segments of the rods and cones. Substances that move between the photoreceptors and the blood vessels within the choroid must pass through the pigment cell layer. Interactions between pigment cells and photoreceptor cells are very important in visual function.

Individual photoreceptor cells can be subdivided into three regions: the **outer segment,** the **inner segment,**

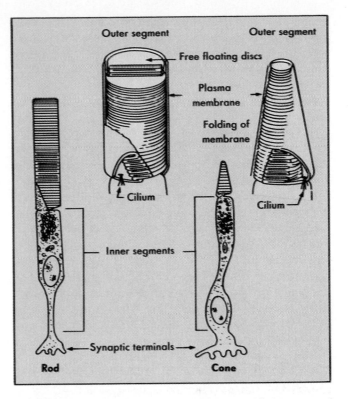

FIGURE 8-4   Structure of rods and cones. The inner and outer segments and the synaptic terminals are shown, as are details of the membranous discs in the outer segments.

and the **synaptic terminal** (Figure 8-4). The outer segment contains a stack of **membranous discs** that are rich in photopigment. The inner segment connects with the outer segment by way of a modified cilium containing nine pairs of microtubules, but lacking the two pairs of central microtubules seen in most cilia. The inner segment of the photoreceptor cell contains the nucleus, mitochondria, and other organelles. The synaptic ending contacts one or more **bipolar cells.**

The rod is so sensitive to light that it can respond to a single photon. The greater sensitivity of rods than cones is partly caused by the long outer segments of rods. Consequently rods contain more photopigment, which is arranged in a monomolecular layer on each outer segment disc. The pigment is **rhodopsin,** which is composed of a chromophore, retinal, and a protein, opsin. **Retinal** is the aldehyde form of **vitamin A.** In the dark, retinal is bound to opsin in the 11-*cis*-retinal form. Absorption of light causes a change to the all-*trans*-retinal form, which no longer binds to opsin. Before the photopigment can be regenerated, the all-*trans*-retinal must be transported to the pigment cell layer, reduced, isomerized, and esterified.

Cones also contain 11-*cis*-retinal attached to an opsin. However, three different **cone opsins** are found in three different types of cones, each sensitive to a different part of the visible light spectrum. *One cone type responds* *best to blue light (420 nm), another to green (531 nm), and the third to red (558 nm).* The presence of three types of cones gives the retina a mechanism for **trichromatic color vision.** Light causes a series of changes in the photopigment of cones; these changes resemble the sequence in rods, but the reactions and recovery are quicker.

Color vision requires at least two photopigments. A single pigment absorbs light over much of the spectrum but absorbs best at a particular **wavelength.** The amount of light absorbed depends on its wavelength and **intensity.** Light of one wavelength and a given intensity could produce the same effect on a particular photoreceptor as another light of a different wavelength and intensity. Therefore the signal is ambiguous because intensity can substitute for wavelength. However, with at least two different photoreceptors that have two different pigments, different wavelengths can be distinguished if the intensity of the light that falls on both photoreceptors is the same. Three different photoreceptor types reduce the ambiguity even more.

Color blindness is often based on a genetic defect that results in the loss of one or more of the cone mechanisms or in a change in the absorption spectra of one or more photopigments. People are normally **trichromats** because they have three cone mechanisms. **Dichromats** have two cone mechanisms and cannot distinguish between red and green. **Monochromats** generally lack all three cone mechanisms, but in rare instances they lack two. **Protanopia** is the loss of the long wavelength system; **deuteranopia,** the loss of the medium wavelength system; and **tritanopia,** the loss of the short wavelength system. Any of these causes a person to be a dichromat.

The commonest type of color blindness is the red-green form. This occurs in 8% of the male population and is a sex-linked recessive trait, since the genetic defect is on the X-chromosome. The main difficulty experienced by people with red-green color blindness occurs when there is a serious need to distinguish these colors, such as at traffic lights.

The cones are most concentrated in the fovea, where all of the photoreceptors are cones (Figure 8-5). This is the region of the retina that provides the greatest visual acuity. In the fovea the retina is thinned to just the outer four layers; thus the image here is of the highest quality. Rods are most concentrated in the parafoveal region.

No photoreceptors exist in the optic disc (Figure 8-5), which is the site where the ganglion cell axons collect

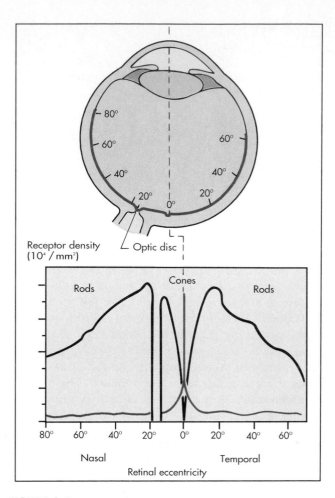

FIGURE 8-5   Density of cones and rods in different parts of the retina.

to leave the eye as the **optic nerve.** The optic disc is therefore a **blind spot.** The optic disc is in the medial retina. Therefore the part of the visual field that would be imaged on the blind spot would be on the temporal side of the field of vision of that eye. The blind spot is not noticed in binocular vision, because the region of the visual field that fails to be seen by the blind spot in one eye is seen by the opposite eye because the light falls on the temporal side of that retina.

Information Processing in the Retina The most direct route for information flow through the retina is from **photoreceptors** to **bipolar cells** and then to **ganglion cells.** The ganglion cells provide the output of the retina to the **thalamus.**

The neural pathways in the retina can be subdivided into **rod pathways** and **cone pathways.** Convergence from photoreceptors onto bipolar cells is greater in the rod than in the cone pathways. This convergence enhances the sensitivity of the rod pathways. Cone path-

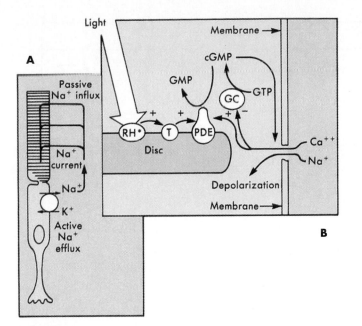

FIGURE 8-6 **A,** The dark current in a photoreceptor, caused by passive influx of sodium ions. The sodium ions are returned to the extracellular space by pumping. Light closes the sodium channels and thus reduces the dark current. **B,** The second messenger system underlying phototransduction. When light reacts with rhodopsin *(Rh)*, the G-protein transducin *(T)* is activated. This in turn activates phosphodiesterase *(PDE)*, which breaks down cyclic guanosine monophosphate *(cGMP)* into GMP. The dark current depends on cGMP, and thus a fall in cGMP concentration reduces the dark current. A decrease in dark current causes a hyperpolarization of the photoreceptor. *GTP,* Guanosine triphosphate; *GC,* guanylate cyclase.

ways display much less convergence, in keeping with their role in visual acuity.

Because the photoreceptor cells and many of the retinal interneurons have short processes, action potentials are not required for transmitting information to the next cell in the circuit. Instead, local potentials alter neurotransmitter release, which in turn provides for information transfer. In darkness, photoreceptor cells have open sodium channels, which result in a **dark current** and consequently a tonic release of neurotransmitter onto bipolar cells and **horizontal cells** (Figure 8-6). When light is absorbed, the sodium channels are closed, which leads to a hyperpolarization of the photoreceptor cells and a decrease in the release of transmitter, probably glutamate.

This information processing involves an amplification mechanism that depends on a **second messenger system.** Cyclic guanosine monophosphate (cGMP) maintains the sodium channels in an open configuration (Figure 8-6). Light activates a G-protein, called **transducin,** in the photoreceptor membrane, by the conversion of guanosine triphosphate (GTP) to guanosine diphosphate

(GDP). Transducin in turn activates a phosphodiesterase, which hydrolyzes cGMP. This causes the sodium channels to close and the membrane to hyperpolarize.

The **receptive field** of a photoreceptor is generally a small circular area that is coextensive with the area of the retina occupied by the photoreceptor. Bipolar cells are of two types: *on-center* and *off-center* (Figure 8-7). An on-center bipolar cell is depolarized when light is shined in the center of its receptive field and hyperpolarized when light is shined in an annulus around the center of the receptive field. An off-center bipolar cell behaves in the converse manner. Exactly how these responses are generated is unclear.

Ganglion cells, like bipolar cells, may have center-surround antagonistic receptive fields (Figure 8-7) or, like amacrine cells, may have large receptive fields. The type of receptive field presumably reflects the dominant input. Ganglion cells can be classified as X-cells, Y-cells, and W-cells. Both X- and Y-cells have center-surround receptive fields. **X-cells** have smaller receptive fields than Y-cells, respond more tonically to stimuli, have slower axons, sum multiple responses in a linear fashion, and distinguish between colors. They are responsible for high visual acuity and color vision. **Y-cells** respond to complex stimuli in an unpredictable fashion. They are motion detectors. **W-cells** often have large, diffuse receptive fields. They signal the intensity of ambient light.

Central Visual Projections of the Retina The **optic nerves** from the two eyes converge at the optic chiasm (see Figure 8-1). Some of the optic nerve fibers decussate in the chiasm and some continue posteriorly on the same side as the eye of origin. The fibers that cross originate from the **nasal hemiretinas** of the two eyes. The uncrossed fibers originate from the **temporal hemiretinas.** Because of this arrangement, each **optic tract** contains both uncrossed and crossed fibers.

Lateral Geniculate Nucleus

Most neurons in the **lateral geniculate nucleus (LGN)** project to the visual cortex; however, some are interneurons. A given LGN neuron receives a dominant input from one or a few retinal ganglion cells, and the responses resemble those of the ganglion cells. Thus the LGN neurons can be classified as X- or Y-cells, and they have on- or off-center receptive fields. However, LGN neurons are subject to inputs from regions other than the retina. These other regions include the visual cortex, several brainstem nuclei, and the **reticular nucleus of the thalamus.** Inhibitory actions originating from the brainstem or reticular nucleus can prevent visual signals from reaching the cortex or can reduce these signals. In effect, the LGN serves as a filter for visual information before it accesses the visual cortex.

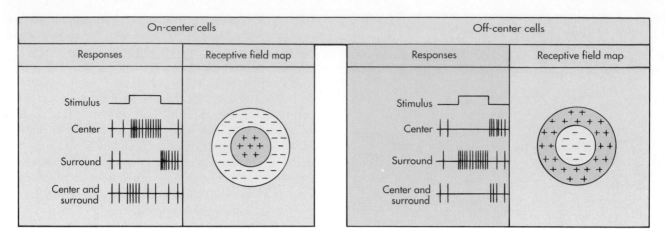

FIGURE 8-7   Center-surround receptive field organization of retinal ganglion cells. On the left are the responses to light stimuli on the center or in a surrounding annulus for on-center and off-center ganglion cells. The effect of stimulating the entire field is also shown. On the right are the receptive fields; plus signs indicate excitation and minus signs inhibition.

Visual Field Deficits

Because the central visual pathway extends from the base of the brain just above the pituitary gland, through the temporal and parietal lobes to the occipital lobe, damage to a wide area of the brain can cause loss of vision. The visual loss can be in one eye if the retina or optic nerve on one side is involved, but the visual loss can be in both eyes if the optic chiasm, optic tract, lateral geniculate nucleus, optic radiation, or visual cortex is involved. The particular pattern of visual loss depends on the exact site and extent of damage.

A visual field defect is described in terms of the part of the visual world that the patient is unable to see. Each eye has a visual field that can be subdivided into a temporal and a nasal visual field. Loss of vision in one eye is simply blindness in that eye. Interruption of an optic nerve causes blindness in that eye or a **scotoma** (partial blindness in one visual field). A lesion affecting the optic chiasm causes loss of vision in the temporal fields of both eyes, a condition called **bitemporal hemianopsia.** This can happen, for example, because of a pituitary tumor. Destruction of an optic tract causes loss of vision in the contralateral half of the visual field of each eye, or **contralateral homonymous hemianopsia.** A similar visual field deficit results from destruction of a lateral geniculate body or the entire optic radiation or the primary visual cortex. Macular vision may be spared in cortical lesions, perhaps because of the very large size of the macular representation or because of collateral circulation in the case of a vascular lesion.

Striate Cortex

The optic radiation ends chiefly in layer IV of the **primary visual cortex.** The dense axon terminals form a white stripe (the **stripe of Gennari**) that can be seen grossly and that gives rise to the name **striate cortex** for this region of cortex. Projections from magnocellular (large-celled) and parvocellular (small-celled) layers of the LGN are separate, and axons that carry information from the two eyes end in alternating patches of cortex called **ocular dominance columns** (Figure 8-8, *B*). Recordings from neurons in area 17 reveal that usually a given cell receives input from both eyes, although one eye is dominant.

The retina is mapped onto the striate cortex (**retinotopic map**). The macular region is represented at and for a distance anterior to the occipital pole. The remainder of the retina is represented still more anteriorly along the medial aspect of the occipital lobe. More neural representation is devoted to the macula than to the rest of the retina because of the requirements for visual acuity.

Most neurons in the striate cortex respond best to elongated stimuli. A rectangular visual target or an edge evokes a much more vigorous response than a small spot. The orientation of the stimulus is an important factor. Neurons in a region of striate cortex perpendicular to the cortical surface will all respond best to elongated stimuli having the same orientation (Figure 8-8, *B*). These neurons form **orientation columns.**

Higher Processing of Visual Information

The striate or primary visual cortex receives visual information from the LGN and begins the analysis of that

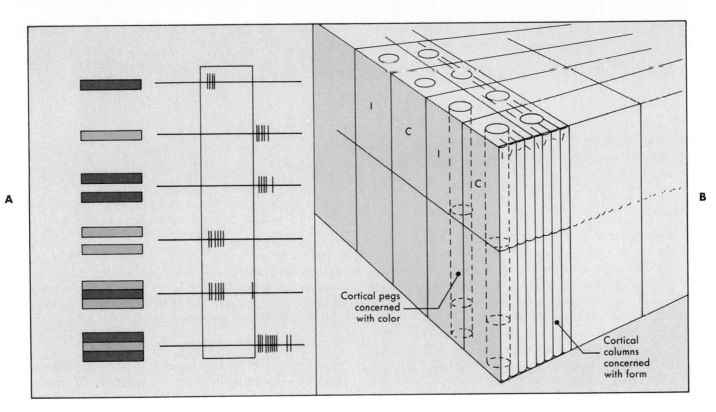

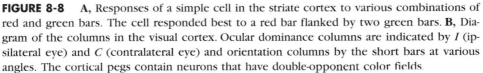

FIGURE 8-8 **A,** Responses of a simple cell in the striate cortex to various combinations of red and green bars. The cell responded best to a red bar flanked by two green bars. **B,** Diagram of the columns in the visual cortex. Ocular dominance columns are indicated by *I* (ipsilateral eye) and *C* (contralateral eye) and orientation columns by the short bars at various angles. The cortical pegs contain neurons that have double-opponent color fields.

information. The striate cortex connects with many other cortical areas, known as the **extrastriate visual cortex,** which participates in the further processing of visual information. These cortical areas are interconnected with other nuclei of the thalamus.

Stereopsis is binocular depth perception. It depends on slight differences in the images in the two eyes, such that a given cortical neuron will have its receptive field at points on the two retinas that are slightly out of correspondence. This provides the brain with a signal that can be used to judge differences in the distances of objects.

Color vision depends on discrimination of wavelengths of light. Retinal ganglion cells and LGN neurons may respond selectively to one wavelength and be inhibited by another. These cells are called **spectral opponent neurons.** An example would be a neuron that is excited by red light that is shined in the center of its receptive field and by green light in the surrounding part of the field (Figure 8-8, *A*). Spectral opponent cells belong to the X-cell category. Neurons in the cortex discriminate wavelength and also brightness, and thus they permit the perception of true color. Such neurons are concentrated in **cortical pegs,** which are sets of

neurons within the ocular dominance columns (Figure 8-8, *B*).

Superior Colliculus

The superior colliculus is a layered midbrain structure that serves as a visual center and as a coordination center for orientation reflexes that occur in response to visual, auditory, and somatic stimuli. The dorsal three layers are involved in visual processing, whereas the deeper four layers also process other sensory inputs.

Retinal ganglion cells project to the upper layers of the superior colliculus. The ganglion cells include both Y-cells and W-cells. The superior colliculus also receives a projection from the cerebral cortex. The cortical neurons involved in this projection are activated by Y-cells. Thus *the visual input to the superior colliculus is concerned with motion detection and light intensity.* The output of the upper layers of the superior colliculus influences visual processing in the cortex. Experiments in animals suggest that the superior colliculus is important in determining the location of objects in visual space, whereas the cortex determines what the objects are.

The deep layers of the superior colliculus are considered with the motor system (see Chapter 9).

AUDITORY SYSTEM

The auditory system is designed to analyze sound. Audition is important not only for recognition of environmental cues, but also for communication, especially language in humans.

Sound

Sound is produced by alternating waves of pressure in the air. Sound waves are composed of the sum of a set of sinusoidal waves of the appropriate amplitudes, frequencies, and phase. Thus sound can be regarded as a mixture of pure tones. *The human acoustic system acts as a filter that is sensitive to pure tones within a range of frequencies from approximately 20 to 15,000 Hz.* Threshold varies with frequency. Sound intensity is measured in **decibels** (dB), which are expressed in terms of a reference level of sound pressure (P_r), often 0.002 dyne/cm^2, the threshold for hearing. The formula for sound intensity is

$$\text{Sound pressure (decibels)} = 20 \log(P/P_r)$$

The ear is most sensitive to tones from 1000 to 3000 Hz. At these frequencies, threshold is by definition 0 dB. Threshold is higher at frequencies less than 1000 Hz and greater than 3000 Hz (Figure 8-9). For example, threshold at 100 Hz is approximately 40 dB. Speech has an intensity of about 65 dB. Damage to the acoustic apparatus can be produced by sounds that exceed 100 dB, and discomfort results from sound pressures that exceed 120 dB.

Structure of the Ear

The ear can be subdivided into the **external ear,** the **middle ear,** and the **inner ear.** The external ear includes the **pinna** and the **external auditory meatus,** which leads by way of the **auditory canal** to the outer surface of the tympanic membrane (Figure 8-10). The auditory canal contains glands that secrete **cerumen,** a wax that guards the ear from invasion by insects.

The middle ear is a cavity that extends deep to the tympanic membrane. It contains a chain of ossicles, the **malleus, incus,** and **stapes** (Figure 8-10, *A*), which connect the tympanic membrane with another membrane that covers the **oval window,** an opening into the inner ear (Figure 8-10, *B*). A second opening between the middle and inner ears, also covered by the **secondary tympanic membrane,** is the **round window.** The

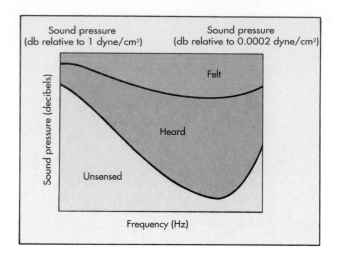

FIGURE 8-9 Sound levels for human hearing as a function of frequency. Below the range for hearing, sound is not sensed; above the hearing range, it is detected by both the auditory and the somatosensory systems.

middle ear contains two muscles, the **tensor tympani** and the **stapedius;** the former attaches to the malleus and the latter to the stapes. Contraction of the middle ear muscles dampens movements of the ossicular chain. The **eustachian tube** provides an opening from the middle ear to the nasopharynx. This permits pressure differences between the environment and the middle ear to be equalized.

The inner ear is a cavity within the temporal bone and contains the **cochlea** and the **vestibular apparatus** (Figure 8-10). The cochlea is the organ of hearing formed by elements of both the **bony labyrinth** and the **membranous labyrinth.** The space in the bony labyrinth just inside the oval window is the **vestibule.**

The cochlea is a coiled structure formed by subdivision of the bony labyrinth into two compartments. The partition between the compartments is formed by a component of the membranous labyrinth; this component is called the **cochlear duct,** or **scala media.** The portion of bony labyrinth in continuity with the vestibule is the **scala vestibuli.** This extends along the two and a half turns of the human cochlea to the end of the cochlear duct. At this point the scala vestibuli connects with the **scala tympani** by way of a space called the **helicotrema.** The scala tympani spirals back to the bony interface with the middle ear and ends at the secondary tympanic membrane that covers the round window. The base of the cochlea is near the oval and round windows and the apex is at the helicotrema (see Figure 8-12). The bony core of the cochlea is the **modiolus.**

The cochlear duct is a tube and is part of the membranous labyrinth (Figure 8-11, *A*). The **basilar membrane** forms the base of the cochlear duct and can be

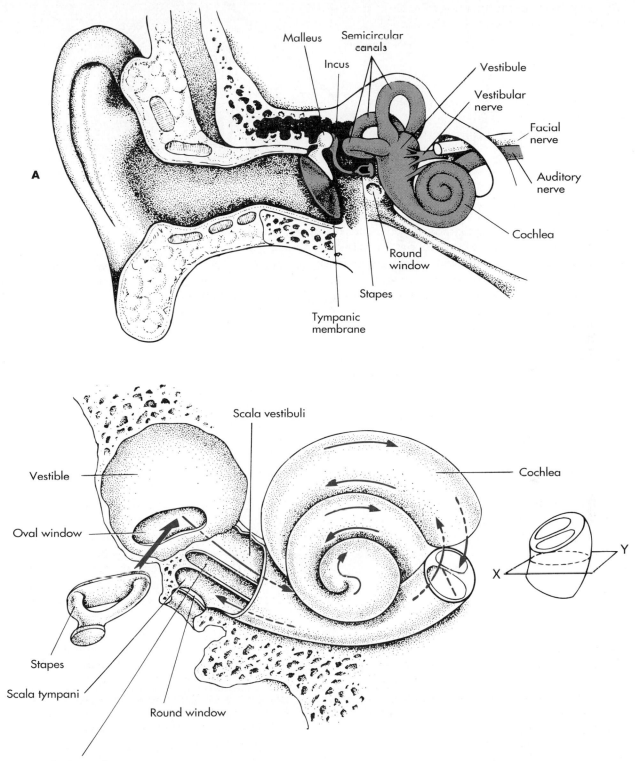

FIGURE 8-10 Structure of the cochlea. **A,** Components of the ear, including the membranous labyrinth. **B,** The cochlea in more detail. The arrows indicate the path of fluid movements that would result from movement of the stapes into the oval window.

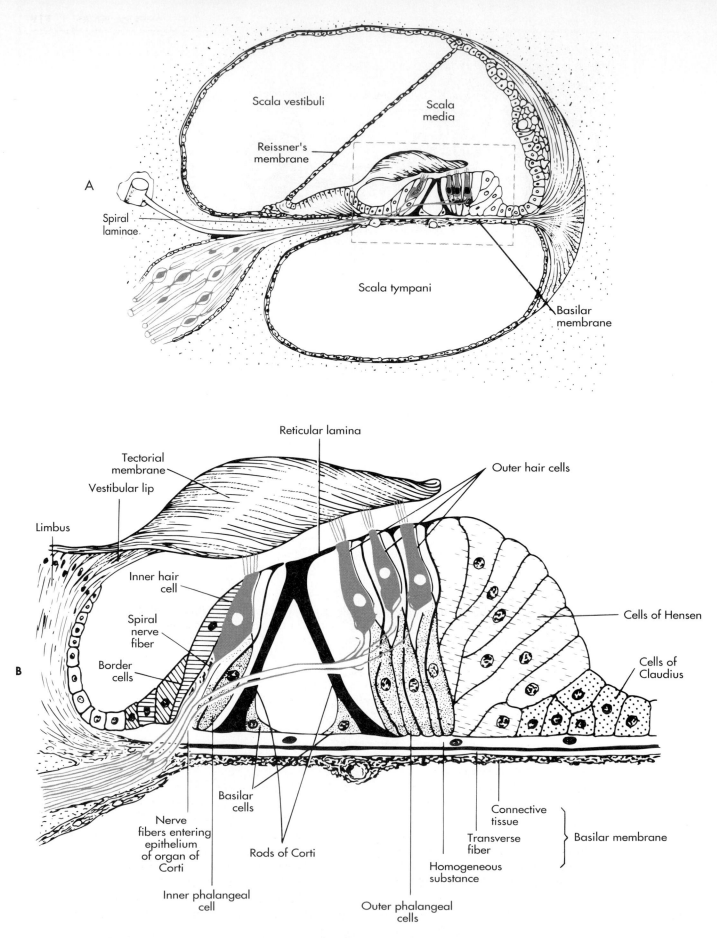

FIGURE 8-11 **A,** The organ of Corti within the cochlear duct (scala media). **B,** Enlargement of the area outlined by the dashed rectangle.

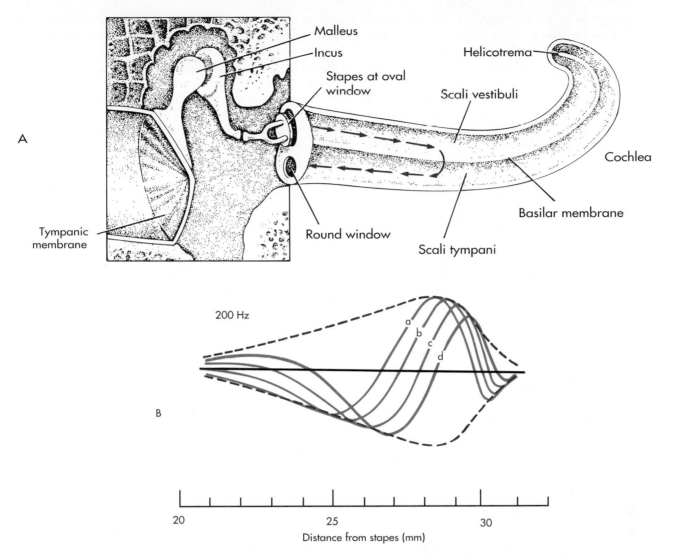

FIGURE 8-12 **A,** The impedance matching arrangement of the ear. The tympanic membrane and ossicular chain oscillate in response to sound waves in the air. The movements of the stapes in the oval window produce comparable oscillations of the fluid columns within the cochlea. The distance along the basilar membrane at which the oscillation is maximal depends on the frequency of the sound. The largest displacements of the basilar membrane are near the base of the cochlea for high frequencies and near the apex for low frequencies. **B,** The traveling wave produced by a 200 Hz sound at four different times. The dashed line is the envelope of the peaks of the successive positions of the wave, showing a maximal deflection of the basilar membrane about 29 mm from the stapes.

regarded as the main partition between the scalae vestibuli and tympani. The basilar membrane is narrowest near the base of the cochlea and widest near the helicotrema. The basilar membrane is attached internally to a ledge, the **spiral lamina,** that arises from the modiolus. Externally the basilar membrane is anchored to the wall of the cochlea by the **spiral ligament.** Contained within the spiral ligament is a vascular structure, the **stria vascularis.** The roof over the cochlear duct is formed by Reissner's membrane. The cochlear duct contains **endolymph,** a fluid with a high concentration of potassium ions; the endolymph is secreted by the stria vascularis. The bony labyrinth contains **perilymph,** which resembles cerebrospinal fluid.

The **organ of Corti** is the sense organ for hearing (Figure 8-11, *B*). It lies within the cochlear duct along the basilar membrane. The organ of Corti consists of **hair cells,** the **tectorial membrane,** a stiff framework, and several types of supportive cells. The **stereocilia** of the hair cells contact the tectorial membrane. The hair cells are innervated by primary afferent fibers and also by efferent fibers of the cochlear nerve. The cell bodies of the primary afferent fibers are in the **spiral ganglion,** which is contained in the modiolus. The spiral ganglion cells are bipolar neurons whose peripheral processes reach the hair cells through the spiral lamina. The central processes join the **cochlear nerve,** which projects into the brainstem.

Sound Transduction

The external ear acts as a filter that is tuned to frequencies between 800 and 6000 Hz. The pinna serves little function in humans, although it is important in many animals. Pressure waves that reach the tympanic membrane cause it and the ossicular chain to vibrate at the frequency of the sound. The ossicular chain in turn causes an oscillation of the oval window and of the fluids within the cochlea. The round window completes the hydraulic pathway.

The middle ear mechanism serves as an **impedance matching device** to couple airborne sound waves with those conducted through the cochlear fluids (Figure 8-12, *A*). If sound waves were to be conducted directly from air to the oval window, most of the energy would be reflected and lost. With the mechanical advantage provided by the ratio of the area of the tympanic membrane to that of the oval window, plus that provided by the lever action of the ossicular chain, only 10 to 15 dB are lost in the impedance matching process of the ear.

Within the cochlear duct, the maximal amplitude of the oscillations extends for various distances along the basilar membrane; the distance depends on the frequency of the sound (Figure 8-12, *B*). Although much of the basilar membrane oscillates in a **traveling wave** in response to a particular frequency of sound, high frequencies result in movements that are largest in the basal part of the cochlea, whereas low frequencies induce movements that are largest near the apex of the cochlea.

As the basilar membrane oscillates, the stereocilia of the hair cells in the organ of Corti are subjected to shear forces at their junctions with the tectorial membrane (Figure 8-13). When the stereocilia are bent in a direction toward the longest cilia, a hair cell will become depolarized because of an increased conductance of the apical membrane to cations. This depolarization is a **receptor potential,** and it causes the release of an excitatory transmitter that produces a **generator potential** in the primary afferent nerve fibers synapsing on the hair cell. As the oscillations of the basilar membrane move in the opposite direction, the membrane of the hair cell

is hyperpolarized and less transmitter is released. The generator potential in the primary afferent terminals is thus an oscillatory one, and if its amplitude is sufficient during the depolarizing phases, it will trigger action potentials in primary afferent nerve fibers.

The difference in potential between the endolymph and the intracellular fluid of the hair cells is unusually high. This potential difference is an important factor in the sensitivity of the auditory system. If the perilymph is considered the reference potential, the endolymph has a positive steady potential of about 85 mV. This is called the **endocochlear potential** and is the result of electrogenic pumping by the stria vascularis. The resting potential of the hair cells is approximately 85 mV with reference to the perilymph. Because of the positive potential in the endolymph, however, the transmembrane potential across the apical membrane of the hair cells can be as great as 170 mV. This increases the ionic driving forces across the transducer membrane.

An oscillatory potential, called the **cochlear microphonic potential,** can be recorded from the bony labyrinth of the cochlea. This potential results from the current flow associated with the activity of the hair cells in response to sound. The cochlear microphonic potential has the frequency of the sound stimulus, and its amplitude is graded with the sound intensity.

Cochlear nerve fibers that innervate hair cells at different points along the length of the organ of Corti are tuned to different frequencies of sound. The tuning properties of the primary afferent fibers can be demonstrated by constructing tuning curves that relate the threshold for activation of the fiber to the frequencies of sound stimuli. The frequency that activates the fiber at the lowest intensity is called the **characteristic frequency** of the fiber. Cochlear nerve fibers that innervate the organ of Corti near the base of the cochlea have high characteristic frequencies, whereas those innervating the apex have low characteristic frequencies. The organ of Corti is thus organized **tonotopically.**

For the lower part of the frequency range detected by the cochlea (less than 4000 Hz), the discharges of a given cochlear nerve fiber show **phase locking.** That is, they occur consistently at a particular phase of the sound oscillation. The discharges of a population of afferent nerve fibers could actually signal the stimulus frequency. This is **volley coding of acoustic signals.** However, cochlear afferent fibers with higher characteristic frequencies do not show phase locking. Coding in these depends on **place coding;** the afferent fibers that innervate regions near the base of the cochlea signal frequencies that depend on the site innervated. **Intensity coding** depends on the number of discharges evoked by sounds of different intensities and presumably also on the number of neurons that discharge.

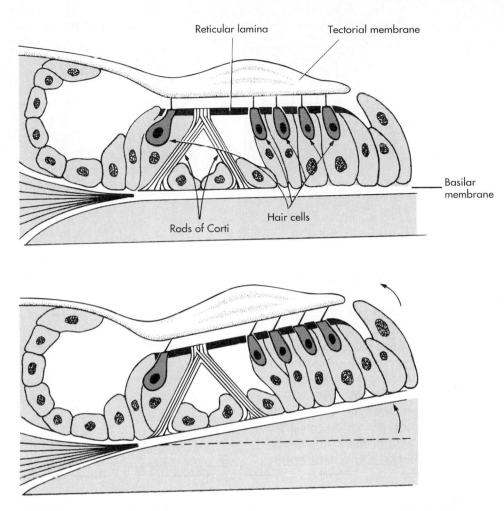

Reticular lamina

Tectorial membrane

Basilar
membrane

Rods of Corti

Hair cells

FIGURE 8-13 Transduction in the organ of Corti. An upward movement of the basilar membrane causes the development of shear forces between the stereocilia of the hair cells and the tectorial membrane, resulting in displacement of the cilia.

Central Auditory Pathway

Branches of individual primary afferent fibers of the cochlear nerve synapse in the **dorsal** and **ventral cochlear nuclei** (Figure 8-14). The cochlear nuclei project to several brainstem nuclei, including the **superior olivary complex** and the **inferior colliculus.** Many of the axons from the cochlear nuclei cross the midline in the **trapezoid body** to innervate the contralateral superior olivary complex or to ascend in the **lateral lemniscus.** The ascending auditory system relays in the inferior colliculus, the **medial geniculate nucleus,** and the **primary auditory cortex.** As in the visual system, several additional cortical areas also contribute to auditory processing.

Central Auditory Processing

The superior olivary complex is concerned with sound localization. Neurons in the medial superior olivary nuclei compare the arrival times of sound in the two ears, whereas neurons in the lateral superior olivary nuclei compare differences in the intensity of sounds that reach the two ears. A sound originating from a source located to the left will reach the left ear first, and the head will provide an acoustic shield that lowers the intensity of the sound that reaches the right ear. By means of these **binaural cues,** signals from the superior olivary nuclei allow the central auditory pathways to judge the location of the sound source.

Binaural processing occurs also in the cortex, as shown by the presence of **summation** and **suppression columns** in the auditory cortex. The responses of the neurons in these columns depend on whether sounds are introduced into the left or right ear or both. In summation columns, neurons respond better when sound reaches both ears rather than only one. Neurons of suppression columns respond better to

A

B

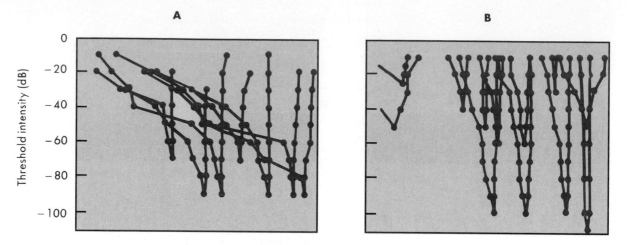

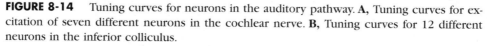

FIGURE 8-14 Tuning curves for neurons in the auditory pathway. **A,** Tuning curves for excitation of seven different neurons in the cochlear nerve. **B,** Tuning curves for 12 different neurons in the inferior colliculus.

sound in one ear than to sound simultaneously in both.

Frequency analysis within the central auditory pathways is reflected in the tonotopic maps characteristic of many auditory structures. The tonotopic map of the cochlea is also reflected in tonotopic maps in the cochlear nuclei, inferior colliculus, medial geniculate nucleus, and several regions of the auditory cortex.

> The bilateral organization of the central auditory pathways is the reason that neurologic lesions of the brainstem at levels rostral to the cochlear nuclei do not produce unilateral deafness (although large unilateral lesions of the auditory cortex do interfere with the localization of sounds in space). Unilateral deafness, for example, implies a defect in the sound conduction system (e.g., in the tympanic membrane or the ossicle chain) or in the initial stages of the auditory pathway (organ of Corti, cochlear nerve, or cochlear nuclei). These conditions are called **conduction deafness** and **nerve deafness,** respectively.

The degree of deafness and the frequencies affected can be determined by **audiometry,** the testing of the patient in each ear with pure tones of different frequencies and intensities. By comparing auditory thresholds for different sample frequencies with those expected in normal subjects, deficits can be described in terms of decibel losses for a certain range of frequencies or for the entire frequency spectrum.

> The type of deafness can be assessed by the Weber and Rinné tuning fork tests. For the **Weber test,** a vibrating tuning fork is placed on the forehead. Normally, the sound is not localized to either ear. In **conduction deafness,** the sound is localized to the deaf ear; in **nerve deafness,** the sound is localized to the normal ear. For the **Rinné test,** the base of the tuning fork is placed against the mastoid process. In normal subjects, when the sound disappears, it can be heard again if the tuning fork is moved to a position in the air near the external auditory meatus (**air conduction** is better than **bone conduction).** In conduction deafness, bone conduction is better than air conduction, and so sound is not restored when the tuning fork is moved.

VESTIBULAR SYSTEM

The sensory role of the vestibular system is a form of proprioception. *The vestibular apparatus detects head movements and the position of the head in space.* To accomplish this, it uses two sets of sensory epithelia to transduce angular and linear accelerations of the head. The vestibular apparatus is part of the membranous labyrinth of the inner ear.

Structure of the Vestibular Apparatus

The vestibular apparatus is contained within the bony labyrinth, but unlike the cochlea, its function depends mainly on the membranous labyrinth. The vestibular ap-

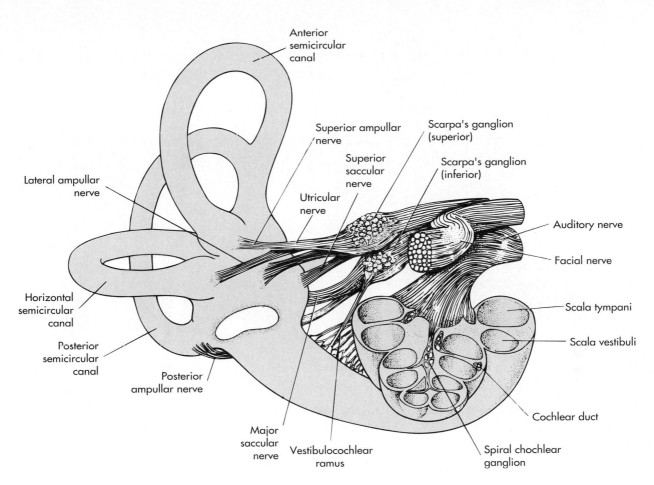

FIGURE 8-15 Labyrinth and nerves of the inner ear.

paratus is connected with the cochlear duct, contains endolymph, and is surrounded by perilymph. The vestibular apparatus includes three pairs of **semicircular canals** on each side: the **anterior, posterior,** and **horizontal canals** (Figure 8-15). The anterior and posterior canals are oriented in vertical planes that are perpendicular to each other, as well as perpendicular to the plane of the horizontal canals. Thus the canals are well positioned to sense events in the three dimensions of space. The superior canal on one side is parallel to the posterior canal on the other side; the horizontal canals are in the same plane.

Each of the semicircular canals has a dilatation, called an **ampulla.** Within the ampulla is a sensory epithelium, known as an **ampullary crest** (Figure 8-16). The apical surface of each of the **hair cells** of the sensory epithelium has both **stereocilia** and a single **kinocilium** (unlike cochlear hair cells, which lack kinocilia). The arrangement of the kinocilium with respect to the stereocilia gives a **functional polarity** to the vestibular hair cell. The cilia are all oriented in the same way relative to the axis of the semicircular duct. The cilia contact a

gelatinous mass, the **cupula,** which extends across the ampulla and occludes it completely. Pressure shifts in the endolymph produced by angular accelerations of the head distort the cupula (Figure 8-16) and bend the cilia of the ampullary crest.

The semicircular canals connect with the **utricle,** one of the otolith organs. The sensory epithelium of the utricle is the **utricular macula,** which is oriented horizontally along the floor of the utricle. The **otolithic membrane** is a gelatinous mass that contains numerous **otoliths** formed from crystals of calcium carbonate. The hair cells of the macula are oriented in relation to a groove, called the **striola,** along the length of the macula. The kinocilia in the utricle are on the striola side of the hair cells. The **saccule** is a separate part of the membranous labyrinth, and the **saccular macula** is oriented vertically. Linear accelerations of the head shift the otolithic membranes with respect to the hair cells. This shift results in bending of the cilia and sensory transduction. Angular accelerations do not affect the otolithic membrane substantially, because the otolithic membranes do not protrude into the endolymph.

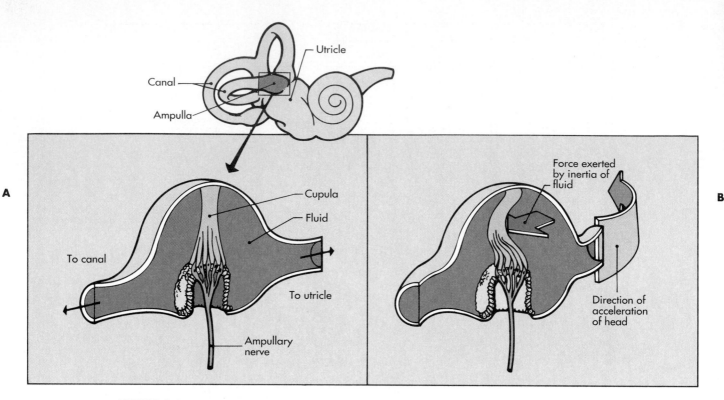

FIGURE 8-16   **A,** The relationship of the cupula to the ampulla when the head is stationary. **B,** The displacement of the cupula when the head is rotated.

Vestibular Transduction

When the stereocilia on the vestibular hair cells are bent toward the kinocilium, the hair cell is depolarized because of an increased conductance of the hair cell membrane to cations (Figure 8-17). Bending of the cilia in the opposite direction leads to hyperpolarization. When vestibular hair cells are depolarized, they release more neurotransmitter (probably an excitatory amino acid such as glutamate), and when they are hyperpolarized, they release less. The neurotransmitter excites primary afferent fibers that end on the hair cells. In the absence of overt stimuli, vestibular primary afferent fibers are spontaneously active. Thus vestibular stimuli modulate afferent activity (Figure 8-17). The activity either increases or decreases, depending on the direction in which the cilia are bent.

In the ampullary crest of the horizontal semicircular duct, the kinocilia are arranged so that they are on the utricular side of the ampulla (Figure 8-18). If the head is rotated to the left, inertial forces will shift the endolymph relatively to the right in both the horizontal canals. In the left ear this means that the stereocilia of the hair cells of the left horizontal canal will bend toward the kinocilium (toward the utricle), and the discharges of the primary afferent nerve fibers that supply the left ampullary crest will increase. Conversely, the stereocilia in the crest of the right horizontal duct will bend away from their

kinocilia (away from the utricle), and thus the discharge of the primary afferent fibers of this crest will be reduced.

The orientation of the kinocilia in the utricular macula is toward the striola. The orientation in the saccular macula is away from the striola. That is, hair cells on the two sides of the striola are functionally polarized in opposite directions. The changes in the discharges of vestibular afferents from a macula produced by linear acceleration of the head differ for different hair cells. The pattern of input to the central nervous system is analyzed and interpreted by the central vestibular pathways in terms of head position.

Central Vestibular Pathways and Vestibular Sensation

Primary afferent fibers from the vestibular apparatus reach the brainstem by way of the **vestibular nerve** (VIII). Most of the afferents terminate in the **vestibular nuclei.** The vestibular nuclei connect with the **cerebellum** and **reticular formation,** the **oculomotor nuclei,** and the **spinal cord.** These connections are very important for the vestibular control of eye and head movements and posture. A pathway to the cerebral cortex by way of the thalamus is responsible for vestibular sensation.

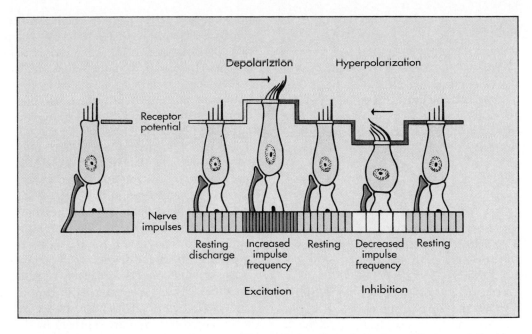

FIGURE 8-17   Directional selectivity of hair cells. Bending of the stereocilia toward the kinocilium depolarizes the hair cell and increases the firing rate in its afferent fiber. Bending of the stereocilia away from the kinocilium hyperpolarizes the hair cell and decreases the firing rate in its afferent fiber.

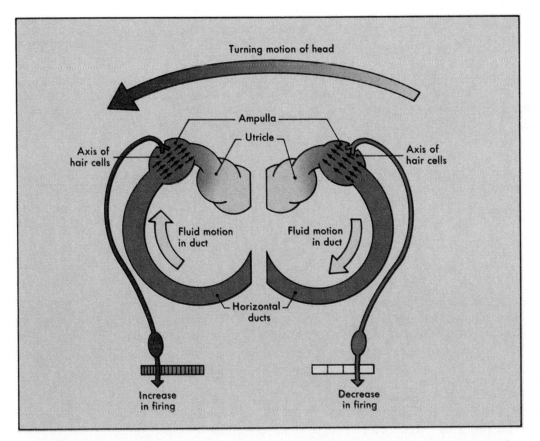

FIGURE 8-18   Effect of head movement on the hair cells in the ampullae of the horizontal semicircular ducts. The functional polarity of the hair cells is indicated by the arrows.

CHEMICAL SENSES

The chemical senses include *taste* (gustation) and *smell* (olfaction). They permit detection of chemical substances in food, water, and the atmosphere. Humans are less adept at chemical detection than many animals, but the chemical senses contribute substantially to the affective aspects of life, and their malfunction may be significant in disease.

Taste

The human gustatory system recognizes many different taste stimuli. However, these can generally be classified as one of four primary taste qualities: *sweet, salty, sour,* and *bitter.*

The sensory receptors for taste are the **taste buds.** Most taste buds are on the tongue, but some are on the palate, pharynx, larynx, and upper esophagus. Taste buds occur in groups on papillae (Figure 8-19). **Fungiform papillae** are mushroom-like structures, several hundred of which are present on the anterior two thirds of the tongue. The taste buds of the fungiform papillae respond mainly to sweet and salty substances but also to sour. The taste buds on fungiform papillae are innervated by the **chorda tympani** branch of the **facial nerve. Foliate papillae** are folded structures on the posterior edge of the tongue, and their taste buds respond best to sour stimuli. **Circumvallate papillae** are large, round structures encircled by a depression; they are on the posterior tongue and respond to bitter substances. The foliate and circumvallate papillae are innervated by the **glossopharyngeal nerve.** Taste buds in the region of the epiglottis and upper esophagus are supplied by the **superior laryngeal branch** of the **vagus nerve.**

A taste bud consists of a group of some 50 gustatory receptor cells in association with supporting cells and basal cells (Figure 8-20). The gustatory cells are continuously turned over and are replaced by differentiation of

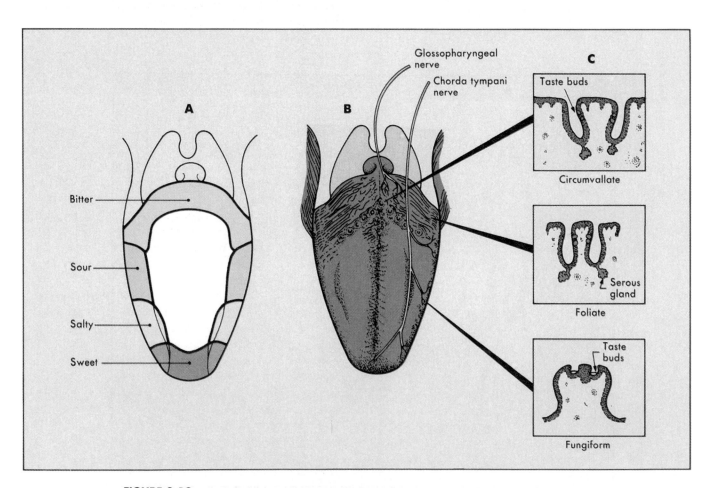

FIGURE 8-19   Peripheral sensory apparatus for gustation. **A,** Taste qualities associated with different regions of the tongue. **B,** The innervation of taste buds in the anterior two thirds and posterior one third of the tongue by the facial and glossopharyngeal nerves. **C,** Arrangement of the taste buds on the three types of papillae.

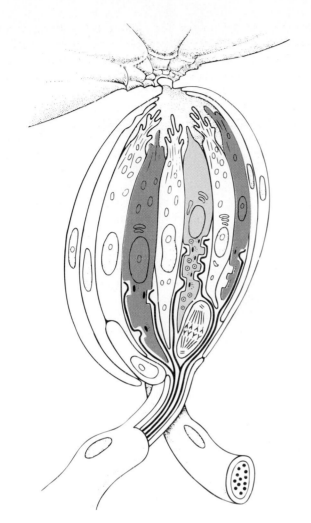

FIGURE 8-20 Taste bud. The receptor cells are darker and the supportive cells lighter.

supporting cells from basal cells. The apical membranes of the gustatory cells have microvilli that protrude into a **taste pore,** where they come into contact with saliva.

Receptor molecules on the microvilli recognize chemical substances in the saliva. The gustatory cells are in synaptic contact with primary afferent nerve terminals. Gustatory signals apparently evoke a receptor potential in the gustatory cell, which leads to transmitter release, a generator potential, and a coded pattern of nerve impulses in the primary afferent fiber. Individual gustatory cells do not appear to be completely selective for a particular primary taste. Rather, they respond best to one type of taste stimulus and less well to others. The recognition of a particular taste quality depends on the activity of a population of gustatory cells. This is a modified labeled-line system.

The primary afferent fibers from the taste buds enter the brainstem and travel caudally in the **solitary tract,** ending in the **nucleus of the solitary tract.** Ascending gustatory fibers reach a special part of the **ventral pos-** teromedial nucleus of the thalamus. This nucleus projects to the **postcentral gyrus,** ending adjacent to the area representing the tongue. An unusual feature of the gustatory projection is that it is ipsilateral rather than crossed.

Smell

The human olfactory system can recognize many odors. These are difficult to classify, but there are at least seven primary odors: *camphoraceous, musk, floral, peppermint, ethereal, pungent,* and *putrid.*

The sensory receptors for olfaction are located in the **olfactory mucosa,** a specialized area of about 2.5 cm^2 in each nasal mucosa. The **olfactory receptor cells** are themselves primary afferent neurons. They have an apical process with cilia that extend into a layer of mucus, in which are dissolved chemical substances that elicit olfactory responses. The base of the olfactory receptor cells gives rise to an axon that projects centrally to end in the **olfactory bulb.** Associated with the olfactory receptor cells are supporting and basal cells that replace olfactory receptor cells as they turn over.

Olfactory transduction depends on the binding of odorants (dissolved in the mucous layer) to receptor molecules on the cilia of olfactory receptor cells. The resulting receptor potential increases the firing rate of the primary afferent fiber. The firing rate is a function of the concentration of the odorant.

The coding mechanism for odors is a modified labeled-line system similar to that for taste. Olfactory receptors respond best to a particular type of odorant and less well to others. Olfactory receptors are grouped according to sensitivity to the class of odorant and are located in different regions of the olfactory mucosa. The central nervous system is presented with a spatially coded input that partly represents odor qualities.

The central olfactory pathway is complex. An unusual feature is that the primary afferent neurons synapse directly on neurons of the telencephalon, whereas in all other sensory systems, sensory processing occurs at several lower stages before information reaches the telencephalon. The primary afferent axons from olfactory receptor cells are unmyelinated axons that collect into filaments of the **olfactory nerve** (Figure 8-21). The olfactory nerve bundles pass through the base of the skull and synapse in the **olfactory bulb.** The main projections of the olfactory bulb form the **olfactory tract.** Terminations are made in a number of structures at the base of the brain. The orbitofrontal region of the neocortex also receives olfactory information by way of the thalamus. Presumably the limbic system projections of the olfactory system are involved in the affective responses to odors, whereas the neocortex is concerned with the discrimination of odors.

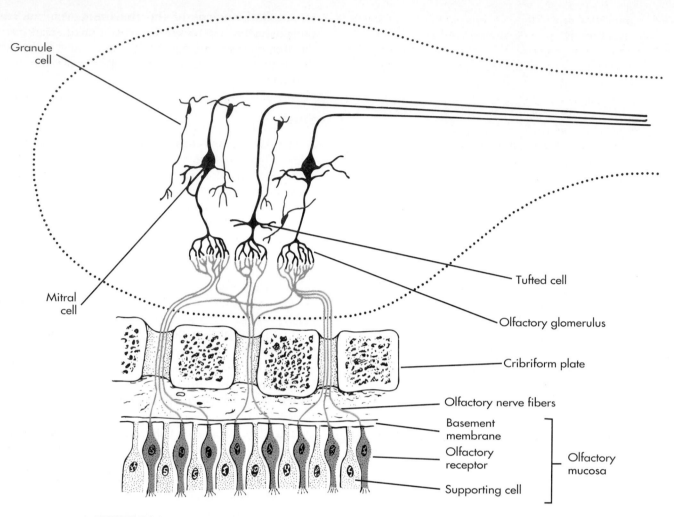

Granule cell

Mitral cell

Tufted cell

Olfactory glomerulus

Cribriform plate

Olfactory nerve fibers

Basement membrane

Olfactory receptor

Supporting cell

Olfactory mucosa

FIGURE 8-21   Initial part of the olfactory pathway, showing the olfactory receptor cells, their projections to the olfactory bulb, and their synapses in the glomeruli with tufted and mitral cells. Also shown are some of the granule cells, which serve as inhibitory interneurons.

Important disturbances in olfaction include **anosmia,** the loss of olfaction, and **uncinate fits,** seizures originating in the temporal lobe that cause **olfactory hallucinations.** Head trauma can cause anosmia on one or both sides, because olfactory nerve filaments can be torn as they enter the cranial cavity through the cribriform plate of the ethmoid bone or the cribriform plate may be fractured. Unilateral anosmia can be produced by compression of an olfactory bulb or tract by a tumor, such as an **olfactory groove meningioma.** Uncinate fits are epileptic seizures that originate in a region of the temporal lobe near the olfactory cortex. These seizures begin with a **sensory aura** in which the person undergoing the seizure has an olfactory hallucination of an unpleasant odor, such as burning rubber. Following this there may be automisms, such as movements of the lips and chewing.

SUMMARY

1. Vision depends on the detection of visible light by photoreceptors in the eye. Rods are very sensitive but do not provide high-resolution images or information about color. Cones are less sensitive but have good resolving power and allow color vision.

2. Light is focused on the retina by the refractive surfaces of the eye: the cornea and the lens. The lens has a variable refractory power so that focus can be changed depending on the distance of the object to be imaged on the retina.

3. The pigment layer of the retina not only reduces light scatter but also has the important functions of phagocytosing the cast-off photopigment discs of rods and resynthesis of rhodopsin, the rod visual pigment.

4. Cones have any of three photopigments, which give cones the ability to distinguish color.

5. There are no photoreceptors in the optic disc, the exit region of optic nerve fibers from the eye. The optic disc is thus a blind spot.

6. Retinal interneurons assist in visual information processing in the retina.

7. Photoreceptors release an excitatory neurotransmitter under conditions of darkness because of a membrane conductance for sodium ions. Light stops this dark current, and thus reduces transmitter release and hyperpolarizes the bipolar and horizontal cells. These changes depend on a second messenger system that involves a G-protein and cyclic GMP.

8. Many photoreceptors and retinal ganglion cells have a center-surround receptive field organization. Some ganglion cells (X-cells) provide signals appropriate for high visual acuity and color; others (Y-cells) have nonlinear responses that are appropriate for motion detection; still others (W-cells) may have diffuse receptive fields and signal brightness.

9. The auditory system analyzes sound. Sound frequency determines pitch. Human ears can detect tones with frequencies of 20 to 15,000 Hz, but are most sensitive to frequencies of 1000 to 3000 Hz. Loudness is measured in decibels.

10. Sound causes oscillatory movements of the tympanic membrane. These movements are transmitted to the oval window by a chain of ossicles. Oscillations of the oval window are transmitted to the fluids within the cochlea, resulting in oscillations of the basilar membrane.

11. Transduction of sound occurs when movements of the basilar membrane cause the stereocilia on the hair cells of the organ of Corti to bend. This causes a change in the release of neurotransmitter by the hair cells and an alteration in the discharges of cochlear nerve fibers that supply the hair cells.

12. Cochlear nerve fibers transmit signals to the cochlear nuclei in the brainstem. Ascending connections are made bilaterally in the superior olivary complex, inferior colliculus, thalamus, and cortex. Processing of sound localization begins in the superior olivary complex. Sound frequency is processed at each level of the auditory pathway.

13. The semicircular ducts of the vestibular apparatus detect angular accelerations of the head because of an inertial shift in the endolymph, which causes the cilia of the hair cells to bend.

14. The otolith organs (utricle and saccule) detect linear accelerations of the head because of shifts in the otolithic membrane in response to changes in gravitational forces.

15. The vestibular apparatus sends signals to the brainstem that are used to control eye movements and posture, as well as to evoke vestibular sensations.

16. The chemical senses include taste and smell. Taste-provoking substances are dissolved in saliva and activate taste receptors located on taste buds. Taste receptors are most sensitive to one of the primary tastes: sweet, salty, sour, and bitter.

17. Olfactory receptor cells in the nasal mucosa signal smell. Odorants dissolved in a mucous layer that contacts the cilia of the receptor cells cause the discharge of nerve impulses that are carried to the olfactory bulb by olfactory nerve filaments. Higher processing is done both by the limbic system (to produce affective responses to odors) and by the neocortex (to discriminate odors).

BIBLIOGRAPHY
Journal Articles

Allman J et al: Stimulus specific responses from beyond the classical receptive field, *Annu Rev Neurosci* 8:407, 1985.

Brugge JF, Geisler CD: Auditory mechanisms of the lower brainstem, *Annu Rev Neurosci* 1:363, 1978.

Getchell TV: Functional properties of vertebrate olfactory receptor neurons, *Physiol Rev* 66:772, 1986.

Gilbert CD: Microcircuitry of the visual cortex, *Annu Rev Neurosci* 6:217, 1983.

Hudspeth AJ: Mechanoelectrical transduction by hair cells in the acousticolateralis sensory system, *Annu Rev Neurosci* 6:187, 1983.

Imig TJ, Morel A: Organization of the thalamocortical auditory system in the cat, *Annu Rev Neurosci* 6:95, 1983.

Lancet D: Vertebrate olfactory reception, *Annu Rev Neurosci* 9:329, 1986.

Maunsell JHR, Newsome WT: Visual processing in monkey extrastriate cortex, *Annu Rev Neurosci* 10:363, 1987.

Merigan WH, Maunsell JHR: How parallel are the primate visual pathways? *Annu Rev Neursci* 16:369, 1993.

Moulton DG: Spatial patterning of response to odors in the peripheral olfactory system, *Physiol Rev* 56:578, 1976.

Patuzzi R, Robertson D: Tuning in the mammalian cochlea, *Physiol Rev* 68:1009, 1988.

Schwartz EA: Phototransduction in vertebrate rods, *Annu Rev Neurosci* 8:339, 1985.

Sherman SM, Spear PD: Organization of visual pathways in normal and visually deprived cats, *Physiol Rev* 62:738, 1982.

Sterling P: Microcircuitry of the cat retina, *Annu Rev Neurosci* 6:149, 1983.

Stryer L: Cyclic GMP cascade of vision, *Annu Rev Neurosci* 9:87, 1986.

Travers JB et al: Gustatory neural processing in the hindbrain, *Annu Rev Neurosci* 10:595, 1987.

Wässle H, Boycott BB: Functional architecture of the mammalian retina, *Physiol Rev* 71:447, 1991.

Books and Monographs

Dowling JE: *The retina, an approachable part of the brain,* Cambridge, Mass, 1987, Harvard University Press.

Kandel ER, Schwartz JH, Jessell TM: *Principles of neural science,* ed 3, New York, 1991, Elsevier Science Publishing Co, Inc.

Nicholls JG, Martin AR, Wallace BG: *From neuron to brain,* ed 3, Sunderland, Mass, 1984, Sinauer Associates, Inc.

Shepherd GM: *The synaptic organization of the brain,* ed 3, New York, 1990, Oxford University Press.

Wilson VJ, Jones GM: *Mammalian vestibular physiology,* New York, 1979, Plenum Press.

CHANGE 9

The Motor System

The term **motor system** refers to the neural pathways that control the sequence and pattern of contraction of skeletal muscles. Skeletal muscle contractions result in posture, reflexes, rhythmic activity (e.g., locomotion), and voluntary movements. A given motor act may involve several of these. Motor acts make up a substantial part of the readily observable behavior of an organism. Motor behaviors that are especially important in humans include speech and movements of the digits and eyes.

The basic element of motor control is the **motor unit,** which comprises an α-motoneuron and the skeletal muscle fibers it innervates. The speed and force of contraction, and the fatigability of motor units, vary in correspondence to their histochemical types. Contractile force can be increased by increasing the firing rates of α-motoneurons or by recruiting more. Recruitment is orderly, according to the **size principle:** small α-motoneurons are recruited first, then larger ones. The consequence is that contractions are initially weak but can maintain posture; with increasing activity, the contractions become more forceful but they may fatigue.

Muscle stretch receptors include **muscle spindles** and **Golgi tendon organs.** Muscle spindles are complex sense organs that signal the rate and extent of stretch. **γ-motoneurons** can cause the intrafusal muscle fibers to contract, preventing unloading of the spindle. Golgi tendon organs innervate tendons and respond to muscle stretch and contraction. They signal the tension in the tendon.

Reflex pathways involving muscle stretch receptors and a variety of other sensory receptors are organized within the spinal cord. The **stretch reflex** is activated by muscle stretch, such as can be produced by using a reflex hammer to tap a tendon briskly. The stretch reflex depends on muscle spindles and regulates muscle length. Golgi tendon organs are responsible for the **inverse myotatic reflex,** which regulates muscle tension.

The **flexion reflex** is activated by flexion reflex afferents, which include nociceptors.

The higher motor centers of the brain superimpose commands on activity intrinsic to the spinal cord. These centers include the **brainstem,** the **cerebral cortex,** the **cerebellum,** and **basal ganglia**. Voluntary movements are initiated by way of commands generated in the cerebral cortex, in concert with activity in a number of cortical control systems. Motor programs are developed by these cortical areas so that the proper sequence and organization of muscle contractions for particular movements are coordinated. The **cerebellum** helps regulate movements by controlling the activity of descending motor pathways from the brainstem and cortex. The **basal ganglia** also regulate movement, primarily by feedback to the motor areas of the cortex via the thalamus.

SPINAL CORD MOTOR ORGANIZATION

The Motor Unit

The basic element in motor control is the motor unit (see Chapter 13). This consists of an α-**motoneuron,** its **motor axon,** and all the skeletal muscle fibers that it innervates. A **muscle unit** is the set of skeletal muscle fibers in a motor unit. The discharge of an α-motoneuron will normally result in the contraction of each of the muscle fibers that it supplies, because the **endplate potential** in skeletal muscle fibers is normally suprathreshold (see Chapter 13). In mammals and other vertebrates, no inhibitory synapses exist on skeletal muscle fibers, although they do exist in many invertebrates. This means that all decisions about whether a skeletal muscle fiber will contract are normally made by the α-motoneuron. Furthermore, each time an α-motoneuron discharges, the entire muscle unit will contract. This means that the

Table 9-1 Muscle Fiber Contractile Properties

Type	Speed	Strength	Fatigability	Motor Unit
I	Slow	Weak	Fatigue resistant	S
IIB	Fast	Strong	Fatigable	FF
IIA	Fast	Intermediate	Fatigue resistant	FR

smallest gradation of force that can be generated by a muscle depends on the force of contraction of the weakest muscle units in that muscle.

A given skeletal muscle will contain a number of muscle units. The ratio between the number of α-motoneurons and the total number of skeletal muscle fibers in a muscle is the **innervation ratio.** This gives the number of muscle fibers in the average muscle unit. The number is large for muscles that are used for coarse movements (e.g., 2000 fibers for the gastrocnemius muscle) and small for muscles that produce finely graded movements (e.g., three to six for the eye muscles). The muscle fibers in a muscle unit are distributed widely in a muscle, and they are separated by fibers belonging to other motor units. All the skeletal muscle fibers in a muscle unit are of the same histochemical type. That is, they are either all type I, type IIB, or type IIA. The contractile properties of these muscle fiber types are summarized in Table 9-1. The motor units that twitch slowly and resist fatigue are classified as **S (slow)** and have type I fibers. S motor units depend on oxidative metabolism for their energy supply and have weak contractions (Figure 9-1). The motor units with fast twitches are **FF (fast, fatigable)** and **FR (fast, fatigue resistant).** FF motor units have type IIB fibers, use glycolytic metabolism, and have strong contractions, but they fatigue easily. FR motor units have type IIA fibers and rely on oxidative metabolism; their contractions are of intermediate force, and these motor units resist fatigue (Figure 9-1).

α-Motoneurons

The only way in which the central nervous system can cause skeletal muscle fibers to contract is by evoking discharges in α-motoneurons. Therefore all motor acts depend on neural circuits that eventually impinge on α-motoneurons. This is why α-motoneurons are called the final common pathway.

Motor Nucleus α-Motoneurons are large neurons found in **lamina IX** of the spinal cord ventral horn and in **cranial nerve motor nuclei** that supply skeletal muscles. Each muscle or group of synergistic muscles (those having a similar action) has its own **motor nucleus.** α-Motoneurons that supply a given muscle are

generally arranged as a longitudinal column of cells, often extending two to three segments in the spinal cord and several millimeters in the brainstem. The set of α-motoneurons that innervates a muscle is called the **motoneuron pool** of the muscle.

The motor nuclei of different muscles or muscle groups are located in different parts of the ventral horn. That is, motor nuclei have a **somatotopic organization.** Motor nuclei that supply the axial muscles of the body are in the medial part of the ventral horn in the cervical and lumbosacral enlargements (Figure 9-2) and in the most ventral part of the ventral horn in the upper cervical, thoracic, and upper lumbar segments of the spinal cord. The innervation ratio for the motor units in these muscles is large, because the role of the axial muscles includes such gross activities as maintenance of posture, support for limb movements, and respiration.

Motor nuclei that innervate the limb muscles are in the lateral part of the ventral horn, in the cervical and lumbosacral enlargements (Figure 9-2). The most distal muscles are supplied by motor nuclei located in the dorsolateral part of the ventral horn, whereas more proximal muscles are innervated by motor nuclei in the ventrolateral ventral horn. The innervation ratios of these muscles are small for the distal muscles and large for the proximal ones.

Motoneurons The individual α-motoneuron (Figure 9-3) is a cell with a large soma (up to 70 μm in diameter). Each of the 5 to 22 dendrites may be as long as 1 mm. The large myelinated axon has a diameter of 12 to 20 μm, and its conduction velocity is 72 to 120 m/sec. The axon of an α-motoneuron is often called an α-motor axon. It arises from an axon hillock on the soma or a proximal dendrite and has a short, unmyelinated **initial segment** before the myelin sheath begins. The axons of α-motoneurons collect in bundles that leave the ventral horn, pass through the ventral white matter of the spinal cord, and enter a filament of the **ventral root.** Just before leaving the ventral horn, some α-motor axons give off **recurrent collaterals.** Recurrent collaterals typically project dorsally and synapse on interneurons, called **Renshaw cells,** in the ventral part of lamina VII.

Synaptic Integration The dendrites and soma of the α-motoneuron are covered with synapses from primary afferent fibers, interneurons, and pathways that descend from the brain. Most of the synapses are from interneurons. Approximately half of the surface membrane lies beneath synaptic endings. Some of the synapses are excitatory, whereas others are inhibitory.

The part of the α-motoneuron membrane with the lowest threshold is thought to be the **initial segment,** which therefore serves as a **trigger zone** for the gen-

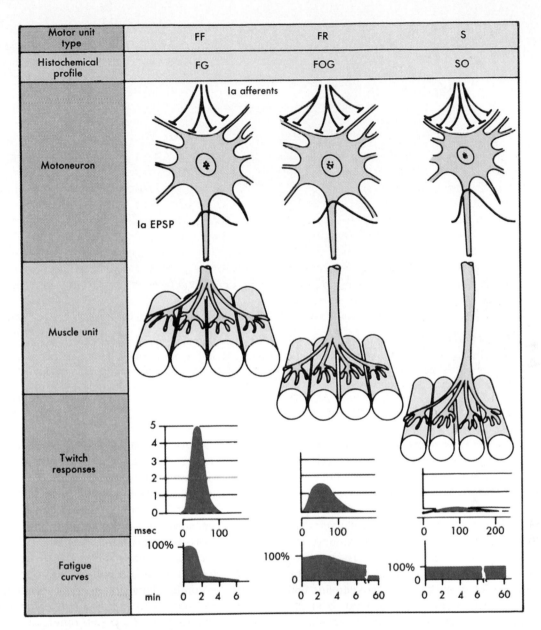

Motor unit type	FF	FR	S
Histochemical profile	FG	FOG	SO
Motoneuron			
Muscle unit			
Twitch responses			
Fatigue curves			

FIGURE 9-1 Summary of features of motor units in a mixed muscle (medial gastrocnemius of cat). Relative sizes are shown for motoneurons, muscle fibers, monosynaptic excitatory postsynaptic potentials evoked by volleys in group Ia afferent fibers, and twitch responses. *EPSP,* Excitatory postsynaptic potential; *FG,* fast glycolytic; *FOG,* fast oxidative-glycolytic; *SO,* slow oxidative.

eration of action potentials. Excitatory synaptic currents depolarize all parts of the membrane of the motoneuron, but in terms of the initiation of an action potential, the depolarization of the initial segment is crucial. The **excitatory postsynaptic potentials** (EPSPs) produced by activation of more than one excitatory pathway to a motoneuron may sum (Figure 9-4, *A*), and the summed EPSPs may exceed threshold for discharge (**spatial summation**). Alternatively, repetitive activation of an excita-

tory pathway can produce **temporal summation** (see Chapter 4). **Inhibitory synaptic potentials** (IPSPs) interfere with the excitatory ones and tend to prevent the discharge of an action potential (Figure 9-4, *B*). The interactions of excitatory and inhibitory synaptic currents in determining whether a neuron discharges is termed **synaptic integration.**

The location of a synapse on the membrane of a neuron such as an α-motoneuron may determine the effec-

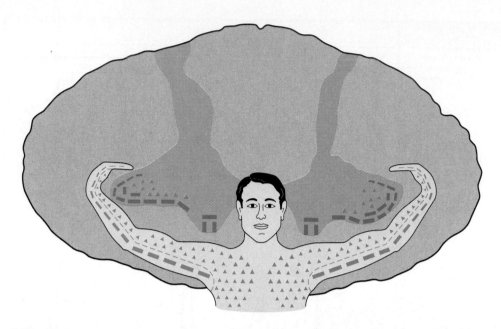

FIGURE 9-2 Somatotopic organization of spinal cord motoneurons. Motoneurons to axial muscles are in the medial ventral horn. In the lateral part of the motor nucleus, motoneurons to more proximal muscles are indicated by the larger symbols. Extensor muscles are supplied by motoneurons indicated by solid rectangles and flexor muscles by motoneurons indicated by triangles.

tiveness of that particular synapse in synaptic integration. Analysis of the **passive electrical properties** of α-motoneurons indicates that synaptic currents can reach the initial segment from even the most distant part of the dendritic tree. However, synaptic potentials produced by distal synapses are smaller and slower than those produced by proximal synapses. For example, if a synapse on a dendrite is one **length constant** (see Chapter 3) away from the initial segment, the size of the membrane potential change in the initial segment will be only about one third (1/e) of that generated in the dendrite. Furthermore, the postsynaptic potential will be slowed considerably. In many neurons, inhibitory synapses, which prevent action potential generation, tend to be located near the initial segment. Another arrangement is that excitatory and inhibitory synapses from neural pathways with antagonistic functions are located near each other on a given dendrite, but away from the initial segment. This allows different pathways to influence the motoneuron independently. In still another arrangement, excitatory synaptic endings of one pathway receive **axo-axonal synapses** from another. These axo-axonal synapses cause **presynaptic inhibition,** which can reduce the effectiveness of one pathway to the motoneuron without altering the excitability of the motoneuron, allowing it to participate in other pathways.

Action Potential Generation When a motoneuron discharges in response to synaptic excitation, the potentials follow a characteristic sequence. A recording from the soma reveals first an EPSP (Figure 9-5, *A*). Arising from this is a spike potential with two phases: an initial small spike, on which is superimposed a slightly delayed but larger spike. The first, small spike is believed to represent the action potential generated by the initial segment; it is small in a recording from the soma because of **electrotonic decrement.** The larger spike is thought to represent invasion of the soma by the action potential. Unseen in the recording (Figure 9-5, *A*) is the axonal action potential, which is generated simultaneously and conducted distally from the initial segment and toward the muscle supplied by the motoneuron. The activation of a motoneuron in this fashion by an EPSP is called **orthodromic activation,** because the sequence is in the normal, or orthograde, direction.

Under experimental conditions an action potential may be initiated in the motor axon and conducted retrogradely to the motoneuron. This is called **antidromic activation** (Figure 9-5, *B*). A recording from the soma of the motoneuron reveals that the antidromic action potential arises directly from the resting membrane potential and has the same initial small spike and slightly delayed large spike as the orthodromic action potential. The missing part is the EPSP. After the spike, a large, long-lasting **afterhyperpolarization** occurs. This is also present in orthodromic action potentials, but it is often obscured by the EPSP.

The afterhyperpolarization is a very important feature of the motoneuronal action potential because it helps de-

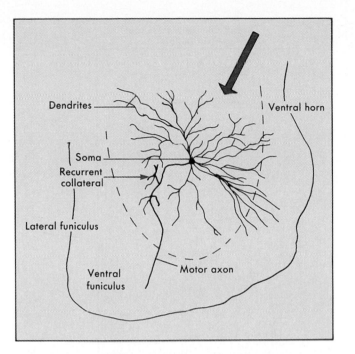

FIGURE 9-3 α-Motoneuron injected intracellularly with horseradish peroxidase. A small arrow indicates a recurrent collateral of the motor axon. The large arrow shows the direction followed by the microelectrode.

termine the characteristic firing rate of the neuron. Large α motoneurons have shorter-lasting afterhyperpolarizations (about 50 msec long) than do small α-motoneurons (about 100 msec long). Therefore, large α-motoneurons tend to discharge at rates of about 20 Hz, whereas small motoneurons discharge at approximately 10 Hz.

Muscle fibers contract with a twitch when a motoneuron discharges once. However, repetitive discharges of a motoneuron result in a **tetanic contraction** (see Chapter 13) of the muscle (see Figure 12-8). The contractile force of a tetanic contraction increases with the rate of discharge of the motoneuron up to a limit imposed by the properties of the muscle. When the tetanic contraction is submaximal, muscle force increases with each motoneuronal discharge. This condition is called an **unfused tetanus.** When the tetanic contraction is maximal, the contraction becomes a **fused tetanus.** The motoneuronal firing rate that produces a fused tetanus in the muscle unit causes the greatest contractile force possible for that motor unit. A higher firing rate produces no greater action.

The characteristic firing rates of α-motoneurons match the mechanical properties of the skeletal muscle fibers. For example, large motoneurons fire at fast rates and innervate fast-twitch muscle fibers; that is, the muscle units of large motoneurons are either the FF or the FR type (see Table 9-1). Small motoneurons fire at

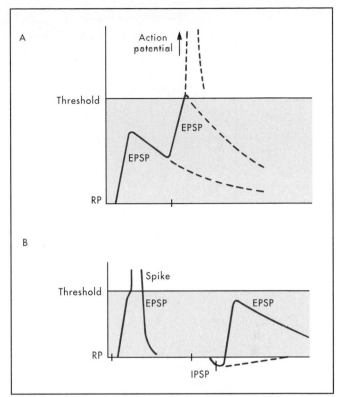

FIGURE 9-4 Synaptic integration. **A,** Summation of two excitatory postsynaptic potentials recorded from an α-motoneuron. The second EPSP exceeds threshold, and so an action potential is triggered. The two EPSPs can be the result of stimulation of separate pathways or activation of a single pathway repetitively. Thus summation can be spatial or temporal. **B,** An EPSP and the action potential it triggers at the left, as well as the results of an interaction between the EPSP and inhibitory postsynaptic potential *(IPSP)*. Note that the IPSP prevents the EPSP from triggering the spike. *RP,* Resting potential.

slow rates and innervate slow-twitch muscle fibers; the muscle units of small motoneurons are the S type.

The contraction of a muscle is regulated by the nervous system in two ways. The first is by the **firing rates** of α-motoneurons. As already indicated, the effects of changing the firing rate are limited by the firing rate at which a tetanus in a given motor unit becomes fused. The second means of regulating muscle tension is by changing the number of active α-motoneurons. The activation of additional motoneurons is called **recruitment.**

The recruitment of α-motoneurons is orderly. Small motoneurons are usually recruited more easily than large motoneurons. This may be related to differences in the membrane properties of small and large motoneurons or may reflect the synaptic organization that controls their discharges. This difference in the effective excitability of small and large α-motoneurons is called the **size principle.** Not only are the small motoneurons recruited be-

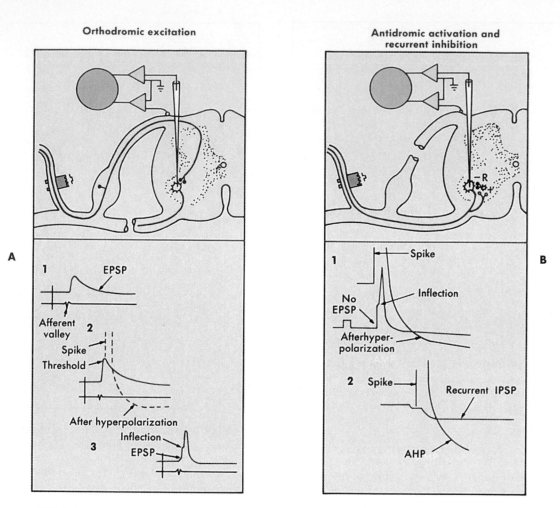

FIGURE 9-5 Orthodromic and antidromic action potentials recorded from motoneurons. **A,** Monosynaptic EPSP *(1)*, a larger EPSP that reaches threshold on some trials *(2)*, and a low-gain recording of an orthodromic action potential *(3)*. In *3* the arrows indicate the EPSP *(lower)* and the inflection between the initial segment and the soma-dendritic spikes *(upper)*. The upper drawing shows the recording arrangement and the interruption of the ventral root to prevent antidromic activation. **B,** Recordings in *1* show an antidromic action potential in a motoneuron at high gain (spike truncated) and at low gain. Note the inflection on the rising phase of the spike. Also note that the spike is succeeded by a large afterhyperpolarization. This is best seen in the high-gain record. In *2* most of the records are with the stimulus subthreshold for the motor axon, and thus the potentials recorded are IPSPs caused by the activity of Renshaw cells excited by other motor axons. The upper drawing shows the experimental arrangement. Note that the dorsal root is cut to prevent orthodromic excitation. *R,* Renshaw cell.

fore the large ones during excitation, but the activity of the small motoneurons persists longer than that of the large motoneurons during inhibition. Because the diameters of the motor axons of large motoneurons are greater than those of small motoneurons, the action potentials recorded extracellularly from the ventral root are greater for the large than for the small motoneurons (Figure 9-6). This allows an evaluation of the recruiting sequence by recordings from the ventral root.

Because of the progressive and orderly recruitment of small and then large α-motoneurons, a weak activation of a motoneuronal pool will discharge only the small α-motoneurons. This activity will produce a weak, slow contraction of slow-twitch muscle fibers. This type of muscle activity is suited to the maintenance of posture and to slow movements such as walking. The recruitment of large α-motoneurons will activate powerful, fast-twitch muscle fibers. The contractions of these add to

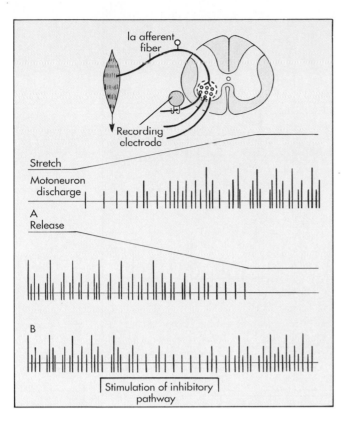

FIGURE 9-6   The size principle and motoneuron recruitment. The drawing shows the experimental arrangement. **A,** Stretching the muscle is shown to activate several motoneurons. The motor axon with the smallest action potential in the ventral root filament is activated first; then progressively larger units begin to discharge. The converse occurs when the muscle is released from the stretch: the large units stop firing first. **B,** An inhibitory input causes cessation of discharge of the larger units but not of the small unit.

the initial force evoked by the slow-twitch fibers, and the resulting movements are appropriate to vigorous activity such as running and jumping.

Several diseases of the central nervous system cause weakness by destroying α-motoneurons. One of these is **poliomyelitis.** For unknown reasons, the polio virus selectively kills α-motoneurons, paralyzing the muscles supplied by these α-motoneurons. Usually, the α-motoneurons affected are concentrated in only a few motor nuclei of the spinal cord, but more widespread loss can occur, and sometimes cranial nerve motor nuclei are involved (bulbar polio). The denervated muscle undergoes atrophy (although some muscle fibers may be reinnervated by collateral axonal sprouts of α-motoneurons that did not die).

Another disease that affects α-motoneurons is **amyotrophic lateral sclerosis** or **ALS** (often called **Lou Gehrig's disease,** after the baseball player of the New York Yankees). In ALS, α-motoneurons at all levels of the spinal cord and brainstem gradually die. As they die, they discharge erratically, causing **fasciculations** (visible contractions of muscle units). After α-motoneurons have died, the denervated muscle fibers atrophy and develop **fibrillations** (spontaneous contractions of individual muscle fibers that cannot be seen through the skin but that can be observed by electromyography). In ALS, cortical pyramidal cells that give rise to the corticospinal tract (see below in this chapter) also die, resulting in further weakness of voluntary movements and pathological reflexes.

Muscle Stretch Receptors

Skeletal muscles and their tendons contain specialized sensory receptors, called **stretch receptors,** that discharge when the muscles are stretched. These receptors include muscle spindles and Golgi tendon organs. These receptors are involved in sensory experience and contribute to proprioception (see Chapter 7). However, they are discussed here because of their importance in motor control. The most complex muscle receptor is the muscle spindle. Muscle spindles are composed of elongated bundles of narrow muscle fibers, called **intrafusal muscle fibers,** enclosed within a connective tissue capsule. The spindles are richly innervated with both sensory and motor endings. Most of the muscle spindle lies freely within the space between the regular, or **extrafusal, muscle fibers,** but its distal ends merge with connective tissue in the muscle. This parallel arrangement is important for the operation of the muscle spindle. When the whole muscle is stretched, the muscle spindle is also stretched (Figure 9-7, *B*). However, when the extrafusal fibers of the muscle contract, the muscle spindle will be unloaded, unless the intrafusal muscle fibers also contract (Figure 9-7, *A*).

The intrafusal muscle fibers are of two main types, called **nuclear bag fibers** and **nuclear chain fibers,** so named because of the arrangement of their nuclei (Figure 9-8). Nuclear bag fibers are larger than nuclear chain fibers and have a cluster of nuclei near the midpoint (resembling a "bag" of oranges). Nuclear chain fibers have a single row of nuclei near the midpoint.

The sensory endings in a muscle spindle are of two types (Figure 9-8): a **primary ending** and one or more **secondary endings.** The primary ending has annulospiral-shaped terminals on both nuclear bag and

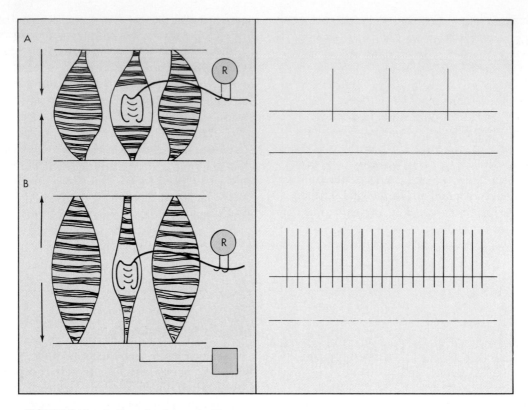

FIGURE 9-7 **A,** Extrafusal muscle fibers in a muscle contract, unloading the muscle spindle and reducing the discharge of an afferent fiber from the spindle. **B,** The muscle is stretched, activating the afferent fiber. *R,* Recording electrode.

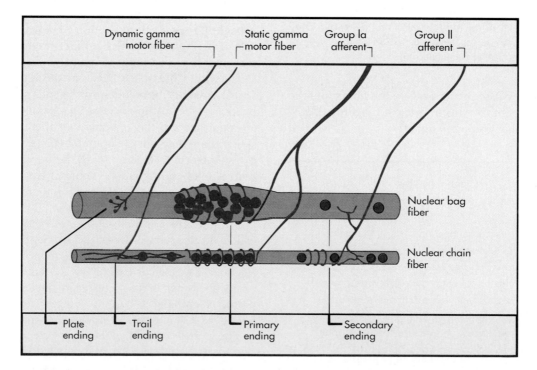

FIGURE 9-8 Innervation of nuclear bag and nuclear chain fibers of the muscle spindle.

nuclear chain intrafusal muscle fibers, and it is innervated by a large, myelinated nerve fiber, called a **group Ia fiber.** Secondary endings have spray-like terminals that are primarily on nuclear chain fibers. The secondary endings are supplied by medium-sized, **group II primary afferent nerve fibers)**

γ-Motoneurons provide the motor innervation of muscle spindles (Figure 9-8). The endings may be small endplates or elongated trail endings. There are two types of γ-motoneurons: **dynamic γ-motoneurons** innervate chiefly the nuclear bag intrafusal muscle fibers, and **static γ-motoneurons** supply nuclear chain fibers.

Primary endings respond to maintained muscle stretch with a slowly adapting discharge that has both a dynamic and a static component (Figure 9-9). The dynamic response signals the rate of stretch of muscle. Secondary endings have only a static response and thus signal muscle length. The dynamic response probably results from properties of nuclear bag fibers, which show an elastic rebound after an initial rapid elongation during muscle stretch.

γ-Motoneurons regulate the sensitivity of muscle spindles to muscle stretch (Figure 9-10). They can also prevent the unloading effect of muscle contraction by contracting the intrafusal muscle fibers during or just before contraction of the extrafusal muscle fibers. Dynamic

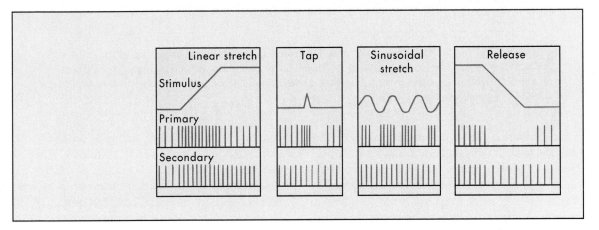

FIGURE 9-9   Responses of primary and secondary endings of a muscle spindle to linear stretch, tap, sinusoidal stretch, and release of the muscle.

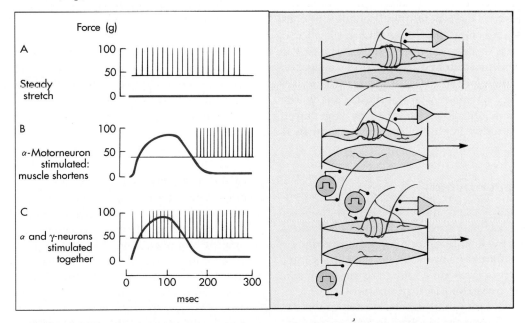

FIGURE 9-10   Effect of activation of γ-motoneurons. **A,** Stretch of the muscle activates an afferent fiber supplying a muscle spindle. **B,** The discharge stops during a muscle contraction produced by activity in α-motoneurons. **C,** The effect of unloading is avoided because α- and γ-motoneurons are coactivated.

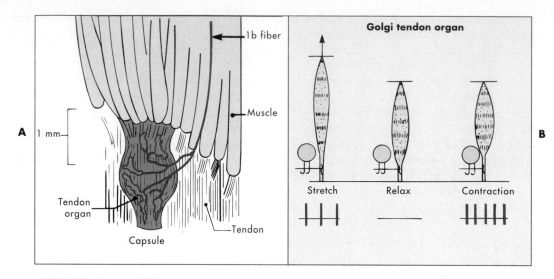

FIGURE 9-11   **A,** The structure of a Golgi tendon organ and its relationship to the tendon of a muscle. **B,** The recordings at the bottom show that a Golgi tendon organ can be activated by either muscle stretch or muscle contraction. The activity of Golgi tendon organs signals muscle tension.

γ-motoneurons enhance the dynamic responses of the primary endings, and static γ-motoneurons increase the static responses of both primary and secondary endings. *The central nervous system can thus regulate dynamic and static responses of muscle spindles independently.*

Another type of muscle stretch receptor is the **Golgi tendon organ.** These receptors are found in tendons, in the connective tissue inscriptions within skeletal muscles, and around joint capsules. The terminals of Golgi tendon organs interdigitate with bundles of collagen fibers, an arrangement that allows the application of mechanical force to the terminals when the muscle is either contracted or stretched (Figure 9-11). Golgi tendon organs are therefore arranged in series with the skeletal muscle. They signal force in the muscle and its tendon. Golgi tendon organs are supplied by large, myelinated primary afferent nerve fibers, called **group Ib fibers.**

Spinal Cord Interneurons

As mentioned earlier, most of the synapses on α-motoneurons originate from spinal cord interneurons. **Interneurons,** by definition, are neurons interposed between primary afferent neurons and motoneurons. Interneurons whose processes are confined to the spinal cord are often called **propriospinal neurons.**

Most spinal cord interneurons are located in the dorsal horn. Many of these are involved in sensory processing and contribute directly or indirectly to the transmission of sensory information to the brain. However, neurons in the dorsal horn also project to the intermediate nucleus and ventral horn and affect the discharges of mo-

toneurons. Furthermore, axons in pathways descending from the brain only rarely terminate directly on motoneurons. Axons in descending pathways usually end on interneurons and alter motor output by changing the level of activity in spinal cord circuits.

Various types of interneurons involved in motor control have been well characterized. The Renshaw cell has already been mentioned. **Renshaw cells** are inhibitory interneurons located in the part of lamina VII that protrudes ventrally between the lateral part of lamina IX and lamina VIII. Recurrent collaterals from α-motor axons synapse on Renshaw cells. When the motor axons discharge, they release acetylcholine at the synapses on Renshaw cells and excite these cells. The Renshaw cells in turn synapse on and inhibit α-motoneurons; thus, when motoneurons discharge, this causes an inhibitory feedback by way of Renshaw cells. This is called **recurrent inhibition** (see Figure 9-5, *B*).

Another well-studied interneuron is the **group Ia inhibitory interneuron.** These interneurons are located in the dorsal part of lamina VII. They are excited monosynaptically by group Ia primary afferent fibers from the primary endings of muscle spindles. The term **monosynaptic** implies a neural pathway in which only one synapse intervenes between one element of the pathway and the next. Group Ia inhibitory interneurons in turn synapse on the α-motoneurons that supply the muscle or muscle group that serves as the antagonist to the muscle giving rise to the group Ia afferent fibers. Thus a **disynaptic pathway** is formed involving group Ia fibers from one muscle group, inhibitory interneurons, and α-motoneurons to the antagonist muscle group (Figure 9-12).

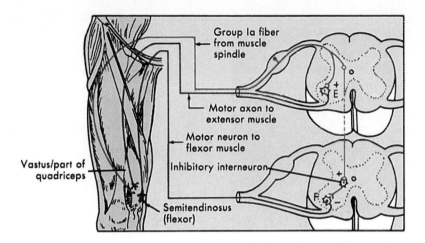

FIGURE 9-12 Reflex pathway for the stretch reflex. The illustration is for the quadriceps stretch reflex, but similar connections would be made for other muscles, including flexor muscles. A muscle spindle is shown to be supplied by a group Ia fiber that enters the spinal cord through a dorsal root and makes monosynaptic excitatory connections with an α-motoneuron to the quadriceps muscle in the same segment and a group Ia inhibitory neuron in another segment. The inhibitory interneuron synapses with an α-motoneuron to the antagonistic flexor muscle, semitendinosus. *E,* Extensor motoneuron; *F,* flexor motoneuron.

Spinal Cord Reflexes

A **reflex** is a relatively simple, stereotyped motor response to a defined sensory input. Some of the reflexes mediated by spinal cord circuits are described here. However, many other reflexes are also organized at the level of either the spinal cord or the brain (see below).

Stretch Reflex A particularly important spinal reflex is the **stretch reflex** (sometimes called the **myotatic reflex**). Stretching a muscle causes a reflex contraction of that muscle and a reflex relaxation of the antagonistic muscles. The stretch reflex has two components, the phasic and the tonic stretch reflexes. The **phasic stretch reflex** is elicited by stretching the muscle quickly. This is often done clinically by tapping the tendon of the muscle with a reflex hammer; when the patellar tendon is struck, a knee jerk reflex results. The **tonic stretch reflex** results from a slower stretch of a muscle, such as occurs during passive movement of a joint. *The tonic stretch reflex is important in the maintenance of posture.*

Hinge joints, such as the knee and the ankle, are extended or flexed by extensor and flexor muscles, which are antagonists because they produce opposite movements of the joint. A set of extensor and flexor muscles about a joint is called a **myotatic unit.** A phasic stretch reflex produced by stretching an extensor muscle results in contraction of the extensor muscle and relaxation of the flexor muscle. Conversely, a phasic stretch reflex of the flexor muscle involves a concomitant relaxation of

the extensor muscle. The reciprocal organization of the stretch reflex pathways is called **reciprocal innervation.** The neural basis of a spinal reflex is the **reflex arc,** a circuit that includes a set of primary afferent fibers, interneurons, and α-motoneurons. The reflex arc for the phasic stretch reflex of a particular muscle includes (1) group Ia afferent fibers from primary endings of muscle spindles located within that muscle, (2) a monosynaptic excitatory connection of these afferents with α-motoneurons that innervate the muscle, and (3) a disynaptic inhibitory pathway involving group Ia inhibitory interneurons that synapse with α-motoneurons that innervate the antagonistic muscles (Figure 9-12).

Group Ia afferent fibers from muscle spindles in synergistic muscles (i.e., muscles that have a similar function) also participate in the excitation of α-motoneurons to a particular muscle, but generally the excitation is not sufficiently powerful to cause the motoneurons to discharge.

The reflex arc for the tonic stretch reflex involves the same connections just described for the phasic stretch reflex. However, group II afferent fibers from secondary endings of muscle spindles also contribute and make monosynaptic excitatory connections with the α-motoneurons that supply the muscle containing the muscle spindles.

As previously mentioned, the sensitivity of the primary and secondary endings of muscle spindles is controlled by dynamic and static γ-motoneurons. Activation of γ-motoneurons can result in a sufficiently strong excitatory input in group Ia afferent fibers to discharge the α-motoneurons. The pathway linking γ- to α-motoneurons by way of group Ia primary afferent fibers is called the γ-loop. However, the group Ia fibers in humans discharge after contractions of skeletal muscle. This pattern of discharge indicates that γ-motoneurons do activate muscle spindles during voluntary movements, but at about the same time as the activation of α-motoneurons. Presumably the shortening of the muscle spindles functions to prevent unloading. Thus voluntary and other movements depend on coactivation of α- and γ-motoneurons.

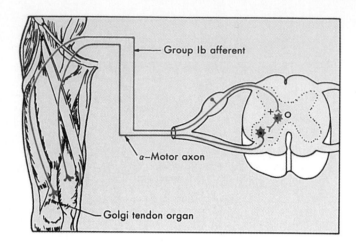

FIGURE 9-13 Pathway for the inverse myotatic reflex produced by Golgi tendon organs. A Golgi tendon organ is shown in the patellar tendon. Its group Ib afferent fiber enters the spinal cord through a dorsal root to terminate on an inhibitory interneuron that synapses on a motoneuron to the quadriceps muscle.

The stretch reflexes are used routinely in the neurological examination. A reflex hammer is commonly employed to elicit phasic stretch reflexes. The limb to be examined is placed in a position that allows relaxation of the joint operated on by the muscles tested. The tendon of each muscle tested is struck briskly with the reflex hammer, and the subsequent contraction of the muscle is observed (or felt). Responses on the two sides are compared. The fact that tendons are struck gave rise to a misleading terminology for the stretch reflex ("deep tendon reflex"); this terminology should be avoided. The sensory receptors responsible for the phasic stretch reflex are muscle spindles located within the muscle and not receptors in the tendon. Muscles that are often tested include the biceps brachii, the quadriceps, and the triceps surae. When stretch reflexes appear to be reduced, it is sometimes possible to enhance them by the **Jendrassik maneuver,** in which the subject hooks the fingers of the two hands and pulls the hands apart against resistance. Tonic stretch reflexes are examined by flexing and extending joints.

In pathological conditions, the stretch reflexes may be either diminished or hyperactive. Causes of decreased stretch reflexes include interruption of peripheral nerves or spinal roots and motoneuron disease. Increased stretch reflexes are seen in diseases that affect the descending motor pathways, such as cerebrovascular accidents (strokes) that interrupt the internal capsule.

Inverse Myotatic Reflex An important reflex whose afferent limb is group Ib afferent fibers from Golgi tendon organs is sometimes called the **inverse myotatic reflex.** Group Ib afferent fibers from extensor muscles synapse monosynaptically on inhibitory interneurons (Figure 9-13). The inhibitory interneurons in turn synapse on the α-motoneurons that supply the same and other extensor muscles in the limb. Flexor muscles are

relatively unaffected by this pathway. This pathway is activated by increases in muscle force, which can be produced either by stretch or by contraction of the muscle, rather than by stretch alone. Therefore the word "myotatic" (which refers to muscle stretch) is probably inappropriate.

The stretch reflex and the group Ib (inverse myotatic) reflex are examples of **negative feedback loops.** In a negative feedback loop the output of the system is compared with the desired output (Figure 9-14). Any difference (error) is fed back to the input so that a corrective action can be made. The variable regulated by a negative feedback system is the **controlled variable.** In the stretch reflex the controlled variable is the muscle length; in the group Ib pathway it is muscle force. An example of how these negative feedback loops might operate is a soldier at attention. The knees tend to flex because of gravitational forces. If slight flexion occurs, the knee extensor muscles are stretched, activating the stretch reflex in these muscles. The result is restoration of knee extension. In terms of muscle length, gravitational forces tend to stretch the knee extensors; the stretch reflex acts as a negative feedback mechanism to control muscle length, keeping it constant. On the other hand, if the knee extensor muscles begin to fatigue, the force that they exert on the patellar tendon decreases. This will cause the knee to flex. However, a reduction in the tension in the patellar tendon will reduce the activity of Golgi tendon organs in the tendon. Decreased activity in group Ib afferent fibers will reduce the inverse myotatic reflex (which inhibits α-motoneurons to the knee extensor muscles), allowing the knee extensor

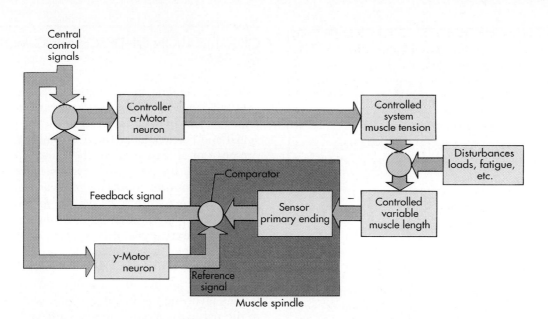

FIGURE 9-14   Diagram showing the operation of the stretch reflex as a negative feedback system that regulates muscle length. Length is determined by the level of muscle tension set by activity in α-motoneurons. Muscle length is detected by the muscle spindle, which has its sensitivity set by γ-motoneurons. If the muscle length is increased, such as by a changed load, feedback conveyed by the primary endings will increase the discharges of α-motoneurons, which reduces muscle length through further contraction of the muscle.

muscles to contract more vigorously. Thus, a reduction in tension in the patellar tendon causes a negative feedback resulting in restoration of the tension.

The concurrent regulation of muscle length and muscle force makes it possible to control muscle stiffness. Muscle has mechanical properties that resemble those of a spring. When a spring is slack, it exerts no force. When the spring is lengthened beyond a threshold point, known as the **set point** (or resting length), each increment of stretch is associated with the development of an increment of force. The relationship between length and force in an ideal spring is linear, and the slope of the curve is the stiffness of the spring. The stiffness of different springs varies, being greater if more force is produced for a given increment of stretch.

Muscles also have characteristic **length-force curves** (see Chapter 12). They are determined with the muscle relaxed or activated by nerve stimulation, and such curves can be highly nonlinear. If a muscle is passively stretched, the muscle will develop force as it is stretched beyond the set point (resting length). When the nerve to the muscle is stimulated, the length-force curve shifts, now having a lower set point and a greater slope. The increased slope indicates that contraction has increased the stiffness of the muscle. It is the interplay between length and force that regulates muscle stiffness and that sets joint position.

In joints whose position is controlled by sets of agonist and antagonist muscles, a particular joint angle

(equilibrium point) can actually be attained in several different ways. Because the myotatic unit is reciprocally innervated, neural circuitry is available to cause one muscle to be activated and the antagonist to be relaxed. The combination of these two events determines the equilibrium point of the joint. Another possibility is the **co-contraction** of the agonist and antagonist muscles. Although this mechanism requires more energy than the one using reciprocal innervation, co-contraction can provide stability in case of unanticipated changes in load because the stiffness of the joint is increased. One typically performs a new task using co-contraction until the task is learned, when co-contraction is replaced by the strategy of relaxing the antagonist.

Flexion Reflex   Other important reflexes, such as the **flexion reflex,** also operate at the spinal cord level. In the flexion reflex the **physiological flexor muscles** of one or more joints in a limb contract and the **physiological extensor muscles** relax. *The physiological flexor muscles are those that tend to withdraw the limb from a noxious stimulus.* The flexion reflex has several different uses. The **flexor withdrawal reflex** consists of a defensive removal of a limb from a threatening or damaging stimulus. For example, if you step on a nail, the foot will be withdrawn reflexly by flexion of the ankle, knee and hip. This reflex takes precedence over other reflexes. It may be accompanied by a **crossed extensor reflex,** which involves the contraction of the extensor muscles and the relaxation of the flexor muscles

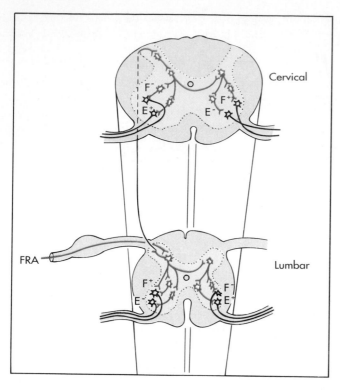

FIGURE 9-15 Flexion reflex pathway. The receptive field of flexion reflex afferents on the lower extremity is not shown. Activation of these afferents leads to input over dorsal roots of the lumbar enlargement. By way of polysynaptic connections, flexor motoneurons of the lower extremity on the side stimulated are activated and extensor motoneurons inhibited. This results in flexion of the lower extremity. Crossed connections cause activation of extensor motoneurons and inhibition of flexor motoneurons to the opposite lower extremity, the crossed extensor reflex. A reverse pattern of reflex activity may be produced in the upper extremities. *E,* Extensor motoneuron; *F,* flexor motoneuron; *FRA,* flexion reflex afferents.

of the contralateral limb. The crossed extension serves as a postural adjustment to compensate for the loss of the antigravitational support by the limb that flexes. In quadrupeds the converse pattern can occur in the other pair of limbs. The flexion reflex is also involved in locomotion and in the scratch reflex.

The flexion reflex is initiated by the **flexion reflex afferent fibers,** which supply high-threshold muscle and joint receptors, many cutaneous receptors, and also nociceptors. It seems likely that the lower-threshold receptors help modulate locomotion and that the nociceptors are crucial for evoking the flexor withdrawal reflex. The flexion reflex pathway from primary afferent fibers to the motoneurons is polysynaptic and involves both excitatory and inhibitory interneurons (Figure 9-15). There are both uncrossed and crossed components of the pathway. The motoneurons involved are those appropriate to the reflex movements already described.

ORGANIZATION OF DESCENDING MOTOR PATHWAYS

The descending motor pathways have traditionally been subdivided into pyramidal and extrapyramidal components. The **pyramidal system** includes the corticospinal and corticobulbar tracts; pyramidal refers to the presence of at least parts of these tracts in the medullary pyramid. *The pyramidal system is the main pathway for mediating voluntary movements of the distal parts of the extremities, as well as for mimetic movements of the face muscles and movements of the tongue.* The **extrapyramidal system** originally referred to motor pathways other than the corticospinal and corticobulbar tracts. At present, however, extrapyramidal is most usefully applied to motor disorders associated with lesions involving the **basal ganglia,** without reference to the particular motor pathways affected.

A helpful classification of descending motor pathways is based on their site of termination in the spinal cord. One set of pathways ends on α-motoneurons in the lateral part of lamina IX or on the interneurons that project to them (Figure 9-16). *This lateral system of descending pathways controls muscles of the distal part of the limbs. These muscles subserve fine movements used in manipulation and other precise actions, especially of the digits.* A parallel control system ends in the brainstem and regulates the part of the facial motor nucleus that supplies the muscles of the lower part of the face, as well as the hypoglossal nucleus, which innervates the tongue. The other set of pathways ends on motoneurons in the medial part of lamina IX or on interneurons that project to them (Figure 9-16). *This medial system of descending pathways controls axial and girdle muscles, as well as most cranial nerve motor nuclei. The muscles of the body regulated by the medial system contribute importantly to posture, balance, and locomotion.* Muscles in the head are involved in such activities as closure of the eyelids, chewing, swallowing, and phonation.

Lateral System

The lateral system includes two pathways from the brain to the spinal cord: the lateral corticospinal tract and the rubrospinal tract. In addition, the part of the corticobulbar tract that controls the lower face and the tongue can be considered part of the lateral system.

The motor cortex has a somatotopic organization (Figure 9-17) resembling that of the somatosensory cortex (see Chapter 7). The component of the **lateral corticospinal tract** that controls the upper extremity originates from the dorsolateral aspect of the precentral gyrus (arm

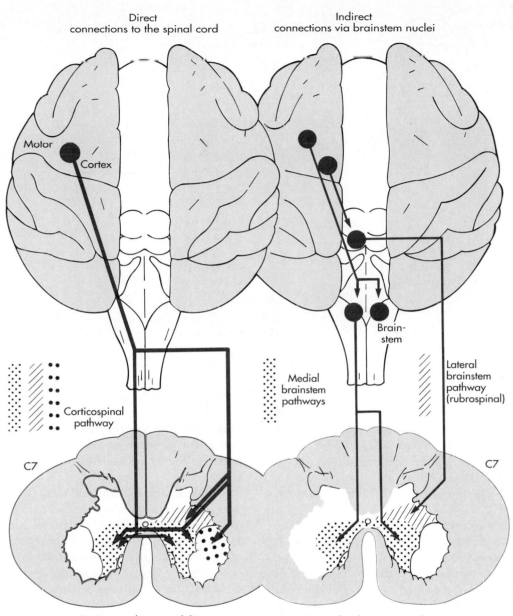

Direct
connections to the spinal cord

Indirect
connections via brainstem nuclei

Motor
Cortex

Corticospinal
pathway

Medial
brainstem
pathways

Lateral
brainstem
pathway
(rubrospinal)

Brain-
stem

C7

C7

Corticospinal (pyramidal) tract

Descending brainstem pathways

FIGURE 9-16   Lateral and medial motor control systems. Descending motor pathways in the lateral funiculus include the lateral corticospinal tract *(left)* and rubrospinal tract *(right)*. The lateral corticospinal tract projects directly to motoneurons innervating distal muscles, as well as to interneurons controlling these motoneurons. The rubrospinal tract projects onto lateral interneurons. Medial pathways include the ventral corticospinal tract *(left)* and several pathways from the medial brainstem *(right)*. These pathways end in the medial ventral horn and control motoneurons to axial and proximal muscles.

representation), and most of the fibers terminate in the cervical enlargement. The component that controls the lower extremity arises from the vertex and medial part of the precentral gyrus (leg area) and passes caudally in the spinal cord in the lateral part of the tract to end in the lumbosacral enlargement.

The lateral corticospinal tract synapses in the dorsal horn, the intermediate region, and the ventral horn. The terminations in the dorsal horn come from corticospinal neurons in the postcentral gyrus. Thus, they permit the somatosensory cortex to regulate sensory transmission in the spinal cord. Most of the terminations of axons from the motor areas of the cerebral cortex are on interneurons in the base of the dorsal horn and in the

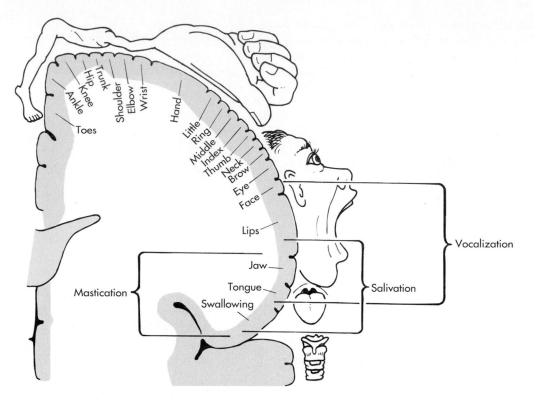

FIGURE 9-17   Somatotopic organization (homunculus) of the motor cortex.

intermediate region. However, some monosynaptic excitatory connections with α-motoneurons exist in lamina IX. These are especially prominent on α-motoneurons to the muscles of the hand. Such connections allow a direct influence by the motor cortex on fine hand movements.

The **rubrospinal tract** originates in the red nucleus. The **red nucleus** has a somatotopic organization with different components that project to the cervical and lumbosacral enlargements. The tract decussates as it leaves the ventral aspect of the red nucleus and descends through the brainstem and the dorsal part of the lateral funiculus of the spinal cord. Terminations are in the base of the dorsal horn and the intermediate region. The tract ends on interneurons and thus regulates movements indirectly by actions on reflex pathways. Neurons of the red nucleus receive a prominent corticorubral projection, and therefore the rubrospinal pathway is strongly influenced by motor commands that originate in the cerebral cortex. The rubrospinal tract appears to be much less significant in humans than in experimental animals.

Medial System

The medial system includes the ventral corticospinal tract, much of the corticobulbar tract, and several pathways descending from the brainstem, including the tectospinal tract, the lateral and medial vestibulospinal tracts, and the pontine and medullary reticulospinal tracts. The descending projections of monoaminergic nuclei in the brainstem form an additional modulatory system.

The **ventral corticospinal tract** originates from neurons of the motor areas of the cortex that are separate from those that project in the lateral corticospinal tract. The ventral corticospinal tract can influence the activity of motoneurons to both sides of the body. This is an important arrangement, because the axial muscles of both sides often function together.

Much of the **corticobulbar tract** can be considered to belong to the medial system. The corticobulbar projections of the medial system provide a bilateral innervation to the trigeminal motor nucleus, the component of the facial nucleus that supplies the frontal and orbicularis oculi muscles, the nucleus ambiguus, and the spinal accessory nucleus. The cortical projection from the frontal eye fields to the gaze centers (see later discussion) can also be regarded as a component of the corticobulbar projection, because it exerts bilateral motor control, in this case of conjugate eye movements.

The **tectospinal tract** originates from neurons of the deep layers of the superior colliculus. The tectospinal tract helps control head movements. A tectobulbar tract also helps control eye position.

The **lateral** and **medial vestibulospinal tracts** arise, respectively, from two of the vestibular nuclei, the lateral and medial vestibular nuclei. *The lateral vestibulospinal tract enhances the activity of extensor (antigravity) muscles and thus functions in postural adjustments to vestibular signals. The medial vestibulospinal tracts are influenced primarily by the semicircular ducts (see later discussion); these tracts cause postural adjustments of the neck and upper limbs in response to angular accelerations of the head.*

The **pontine** and **medullary reticulospinal tracts** arise from neurons of the pontine and medullary reticular formation. The main action of the pontine reticulospinal tract is the excitation of extensor (antigravity) motoneurons. The medullary reticulospinal tract inhibits several reflexes and also sensory transmission.

BRAINSTEM CONTROL OF POSTURE AND MOVEMENT

A hierarchic organization of the motor system can be demonstrated by the effects of lesions at different levels of the neuraxis. A lesion can result in a particular effect either (1) by abolishing functions subserved by a structure whose influence is removed by the lesion or (2) by allowing an action to appear through removal of an inhibitory influence. Such latter responses are called **release phenomena.** Lesions that are particularly instructive include spinal cord transection and decerebration.

Spinal Cord Transection

Spinal cord transection at a cervical level but below the phrenic nucleus has several important effects on the motor system, in addition to a complete loss of sensation from the body over the dermatomes at and below the level of the transection. Assuming that respiration is spared, the most important motor loss is of voluntary movement. *Immediately after transection, especially in humans, a period of spinal shock may follow in which reflex activities are absent.* This is presumably caused by the loss of the excitatory actions of descending pathways from the brain. After a time, up to months in humans, **hyperactive stretch** and **flexion reflexes** develop. Hyperactive stretch reflexes may result in **clonus,** an alternating sequence of contractions of extensor and then flexor muscles. **Mass reflexes** are associated with hyperactive flexion reflexes and are characterized by flexion of one or both limbs and evacuation of the bladder and bowel. These changes may reflect loss of descending inhibition and also rearrangements in spinal cord circuits, possibly including sprouting of primary afferent fibers and the formation of new synaptic connections within the spinal cord. Other release phenomena include the appearance of pathological reflexes,

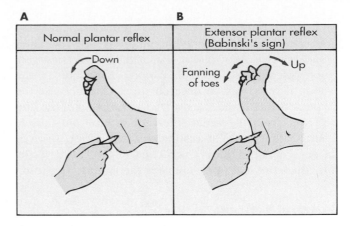

FIGURE 9-18 Babinski's sign. **A,** The normal response to stroking the plantar surface of the foot. **B,** Babinski's sign (extensor plantar reflex) in a person with an interruption of the corticospinal tract.

such as **Babinski's sign** (Figure 9-18), which results from interruption of the lateral corticospinal tracts. Although locomotion is not regained in humans, a capability for locomotion reappears in some animal species because the locomotor pattern generator is contained within the neural circuitry of the spinal cord. A **pattern generator** is a neural circuit that controls a specific type of motor behavior, often a rhythmic behavior such as locomotion or respiration. In animals with chronic spinal transections, locomotion is triggered by afferent signals rather than by the midbrain locomotor center.

Spinal cord injury is unfortunately a relatively common occurrence and generally affects young adults. Frequent causes are automobile and motorcycle accidents and gunshot wounds. Although incomplete transections are more frequent than complete ones, incomplete lesions may nevertheless be disastrous. When spinal injuries affect the upper cervical spinal cord, they are often fatal because of interruption of the respiratory control system that descends from the brainstem to the phrenic motor nucleus. A lesion below the phrenic nucleus may result in paralysis of all four extremities **(quadriplegia),** whereas a lesion of the thoracic spinal cord causes **paraplegia,** which is paralysis of the lower extremities.

Decerebrate Rigidity

Transection of the brainstem at a midbrain level results in a condition known as **decerebrate rigidity.** This develops immediately after the brainstem is transected, and in animals the "rigidity" is expressed as an exaggerated extensor (antigravity) posture caused by hyperactive stretch reflexes. The term decerebrate rigidity is unfortunate, because the condition more closely resembles spasticity than the rigidity that results from basal ganglion disease (see later discussion). Activation of the γ-loop is important in decerebrate rigidity, because the hyperactive reflexes are lost after transection of the dorsal roots (which would interrupt input from muscle spindle afferents). The vestibular system is also involved in decerebrate rigidity; destruction of the lateral vestibular nuclei reduces or eliminates the extensor posture.

Postural Reflexes

Various reflexes assist in postural adjustments that occur as the head is moved or the neck is bent. The receptors that trigger these reflexes include the **vestibular apparatus** and **stretch receptors in the neck.** The visual system also contributes to postural adjustments, but the reflexes described here are elicited in the absence of visual cues.

Angular accelerations of the head activate the sensory epithelia of the semicircular ducts and elicit the **acceleratory reflexes.** These reflexes cause eye, neck, and limb movements that tend to oppose changes in position. For example, if the head is turned to the left (Figure 9-19), the eyes will be reflexly rotated a similar degree to the right. This reflex action will help maintain stability of the visual field. The movements of the two eyes are **conjugate,** meaning that the eyes move together in the same direction and through the same angle. When the head rotation exceeds the range of eye movement, the eyes are quickly deflected to the left and another visual target is found. If the head continues to rotate to the left, there will be an alteration of slow eye movements to the right, followed by rapid eye movements to the left. These alternating slow and fast eye movements are called **nystagmus.** The reflex is called the **vestibuloocular reflex.** A similar response affects the neck muscles and is called the **vestibulocollic reflex.**

The same stimulus, rotation of the head to the left, will also tend to increase contraction of the extensor (antigravity) muscles on the left. This response will oppose the tendency to fall to the left as the head rotation continues.

The neural mechanism that underlies the acceleratory reflexes depends on stimulation of the sensory epithelia in the **semicircular canals.** In the case of rotation of the head in a plane parallel to the ground, the semicircular canals primarily involved are the horizontal ones (Figure 9-19). The inertia of the endolymph within the horizontal semicircular canals causes the endolymph to lag behind as the head rotates. This relative shift of the endolymph will deflect the cupulas of the ampullae of the horizontal canals and cause the stereocilia of the hair

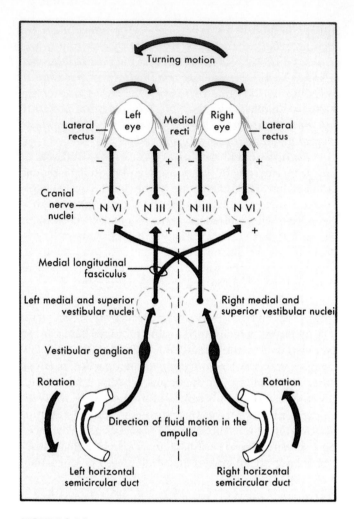

FIGURE 9-19 Neural circuit for the vestibuloocular reflex, an acceleratory reflex. The horizontal semicircular ducts, brainstem pathways, and eyes are viewed from above. Rotation of the head to the left is indicated by the arrow at the top and the arrows next to the drawings of the semicircular ducts. Fluid movement within the ducts is indicated by the arrows in the opposite direction. The neural activity produced by the vestibular system causes the eyes to move conjugately to the right. As the eyes reach the limit of their movement, they are quickly returned to the left.

cells in the ampullary crest to bend. The hair cells of one duct will depolarize and thereby increase the discharge of the vestibular afferent fibers. The opposite will occur in the other horizontal canal. The mismatch in input from the left and right canals to the brainstem will result in reflex discharges that tend to counteract the positional changes that result from the head rotation.

Several reflexes can be elicited by linear accelerations of the head and activation of the otolith organs. If an animal is dropped, the stimulation of the utricles leads to extension of the forelimbs; this is the **vestibular placing reaction.** The response prepares the animal for landing. If the head is tilted, the otolith organs cause the eyes to rotate in the opposition direction, the **ocular counter-rolling response.** Ocular counter-rolling tends to keep the visual axes aligned with the horizon. If the head and body of a quadruped are tilted forward (without bending the neck), the forelimbs are extended and the hindlimbs are flexed. As a consequence of these actions, the body is restored toward a more normal position. Conversely, if the head and body are tilted back (without bending the neck), the hindlimbs extend and the forelimbs flex.

The **tonic neck reflexes** are another class of positional reflexes. The neck muscles contain the largest concentration of muscle spindles of any muscles of the body; these are presumably responsible for the tonic neck reflexes. In the absence of vestibular reflexes (head position normal), if the neck of a quadruped is extended relative to the body, the forelimbs extend and the hindlimbs flex. The converse happens if the neck is flexed relative to the body. These changes are opposite to those expected from the vestibular reflexes. If the neck is turned to the left, the extensor muscles in the limbs on the left will contract more, and the flexor muscles in the limbs on the right will relax.

The **righting reflexes** tend to restore the position of the head and body in space to normal. The vestibular apparatus and the neck stretch receptors are involved in the righting reflexes, as are mechanoreceptors in the body wall.

Locomotion

As mentioned earlier, the pattern generator for locomotion is contained within the neural circuitry of the spinal cord. Actually, separate pattern generators exist for each limb. The activity of these is coupled so that the movements of the limbs are coordinated during locomotion.

The pattern generators for locomotion and for other rhythmic activities (e.g., respiration) are regarded as **biological oscillators.** Many biological oscillators operate on the basis of reciprocal inhibition of circuits, called **half-centers,** that control antagonistic muscles. Excitation of an extensor muscle by one half-center is accompanied by reciprocal inhibition of the half-center for the antagonistic flexor muscle. When the excitation of the extensor muscle decreases, the antagonist is less inhibited and thus it can be activated by the second half-center. This results in reciprocal inhibition of the first half-center. The details of this mechanism and the factors that cause switching between the two half-centers vary with the particular oscillator being considered.

The locomotor pattern generator is normally activated by commands that descend from the brainstem. Neurons in a circumscribed region, called the **midbrain locomotor center** (Figure 9-20), are crucial in the initiation of locomotion. The midbrain locomotor center activates neurons that are located in the pontomedullary reticular formation and that are involved in transmitting the descending commands. The locomotor pattern generator converts tonic activity in the descending pathways into rhythmic discharges of motoneurons to the muscles involved in locomotor activity.

The midbrain locomotor center can be brought into action by voluntary commands that originate in the motor regions of the cerebral cortex. It can also be engaged by afferent signals. The activity of the locomotor pattern generator in the spinal cord is also influenced by afferent signals. These signals modify the ongoing motor program so that motor performance is altered in accord with environmental demands.

Control of Eye Position

Movements of the eyes are generally **conjugate** (in the same direction), although sometimes they are **convergent** or **divergent** when they are targeted on nearby objects (as in reading) or on distant objects.

A rapid conjugate movement of the eyes is called a **saccade.** Usually a saccade causes a visual target to be imaged on the fovea. However, saccades can be made in the dark. Once the eyes have located a visual target, fixation is maintained by **smooth pursuit movements.** Actually, during fixation the eyes drift somewhat and are returned to the target by **microsaccades.** Without these small movements, the retina would adapt and lose vision of the target. Smooth pursuit movements do not take place in the dark, because they require a visual target.

Misalignment of the two visual axes can cause double vision, or **diplopia.** Such misalignment, or **strabismus** (cross-eyed), can result from weakness of the muscles of one eye, causing its visual axis to differ from that of the other eye. Over time the misaligned eye may lose visual acuity, a condition called **amblyopia.**

Several central nervous system structures influence eye position. These structures include the **gaze centers,** the **vestibular system,** the **superior colliculus,** and **frontal** and **occipital lobes of the cerebrum.** Horizontal eye movements are organized by the **pontine horizontal gaze center,** located near the abducens nucleus. There is also a **midbrain ventrical gaze center** in the pretectum.

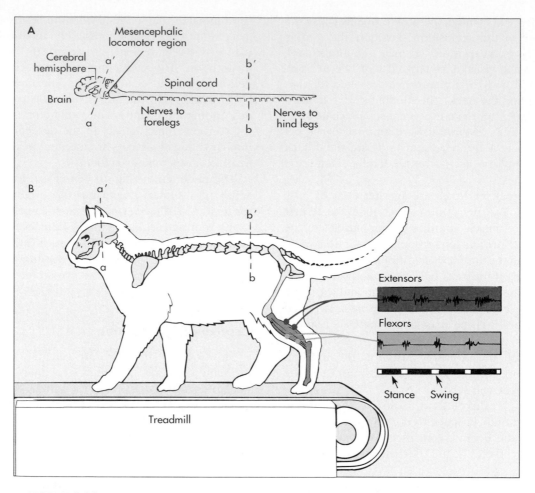

FIGURE 9-20 Control of locomotion. **A,** Location of the mesencephalic locomotor center, which can trigger locomotion even in the decerebrate state (transection at *a'-a*). **B,** Locomotion can also be elicited by afferent input produced by a treadmill after the spinal cord is transected at *b'-b*. Electromyographic recordings from extensor and flexor muscles of the hind limb show bursts of activity during the stance and swing stages of the step cycle.

Neurons with different response properties have been found in the horizontal gaze center (Figure 9-21). **Burst cells** discharge rapidly just before saccades, which burst cells are thought to initiate. **Tonic cells** discharge during slow pursuit movements and fixation. **Burst-tonic cells** show a burst discharge during saccades and tonic activity during fixation. **Pause cells** stop firing during saccades and seem to inhibit burst cells. Saccades occur when pause cells stop firing, resulting in a release of activity in burst cells and a discharge of eye muscle motoneurons. Feedback when the eye is on target inhibits the burst cells and reactivates the pause cells.

Only the vestibular nuclei and the gaze centers send direct projections to the motor nuclei that supply the eye muscles. The superior and medial vestibular nuclei project axons rostrally in the medial longitudinal fasciculus to the abducens, trochlear, and oculomotor nuclei (see Figure 9-19). The circuitry is reciprocally organized,

so that a signal causing an eye movement will excite a set of agonistic motoneurons and inhibit the antagonistic ones. Similarly, neurons of the pontine horizontal gaze center send projections directly to abducens motoneurons on the same side to excite them and cause abduction of the ipsilateral eye. These neurons also send ascending projections through the medial longitudinal fasciculi to excite motoneurons of the contralateral medial rectus muscle (Figure 9-22). Appropriate connections are made to inhibit the motoneurons of the antagonistic muscles.

Neurons located in the deep layers of the superior colliculus and activated by visual, auditory, and somatosensory stimuli project to the horizontal gaze center. They produce saccadic eye movements that are part of an orientation response to a novel or threatening stimulus.

The **frontal eye fields** in the premotor region of the frontal lobe trigger voluntary saccadic eye move-

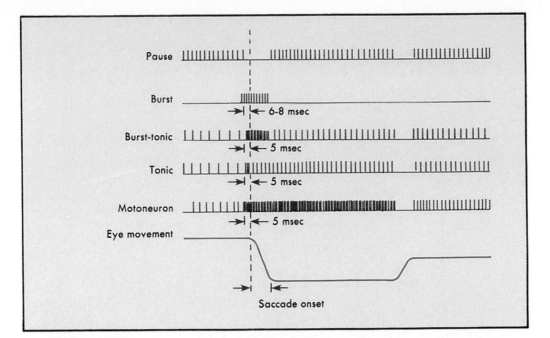

FIGURE 9-21 Types of neurons found in the pontine horizontal gaze center. Pause cells are thought normally to inhibit burst cells. A saccade begins with cessation of activity in pause cells and an explosion of activity in burst cells. Slightly later, burst-tonic and tonic cells discharge. Motoneurons to the muscles involved in the saccade are excited in the burst-tonic fashion, which causes the eye muscle to contract quickly and then to maintain its contraction. This results in the saccadic eye movement.

ments by way of a projection to the contralateral pontine horizontal gaze center. The occipital eye fields are involved in smooth pursuit movements, optokinetic nystagmus, and visual fixation. Adjustments for near vision include vergence, pupillary constriction, and rounding of the lens. The **occipital eye fields** are connected with the superior colliculus and the pretectal region and influence the vertical and horizontal gaze centers.

Damage to the frontal eye field, such as by a stroke that affects one frontal lobe, can result in tonic conjugate deviation of the eyes toward the damaged side (a stroke victim may look toward the side of the stroke). This is presumably because of the activity of the intact frontal eye field in the contralateral hemisphere. Electrical stimulation of the frontal eye field on the side causes forced conjugate deviation of the eyes toward the opposite side. A similar movement can occur during the onset of an epileptic seizure that originates in the frontal lobe. Tonic conjugate deviation of the eyes can also occur following damage to the horizontal gaze center in

the pons. The eyes in this case would deviate toward the side opposite that which is damaged. This condition can be distinguished from that caused by frontal lobe damage by the distribution of paralysis of the extremities and by cranial nerve signs.

CORTICAL CONTROL OF VOLUNTARY MOVEMENT

The **corticospinal** and **corticobulbar tracts** are the most important pathways used in the initiation of voluntary movements. *The lateral corticospinal tract and the comparable part of the corticobulbar tract control the fine movements produced by muscles of the contralateral distal extremities, face, and tongue. The ventral corticospinal tract and part of the corticobulbar tract, as well as more indirect pathways such as the corticorubrospinal and corticoreticulospinal paths, provide for postural support of voluntary movements.*

Corticospinal and corticobulbar neurons do not op-

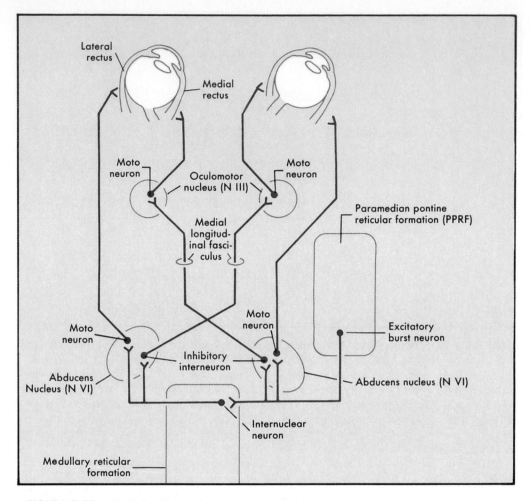

FIGURE 9-22   Organization of the pontine horizontal gaze center. Activation of burst neurons in the right gaze center results in direct excitation of abducens motoneurons on the right and oculomotor motoneurons to the medial rectus muscle on the left through an interneuronal path by way of the medial longitudinal fasciculus. Concurrently, the contralateral saccadic mechanism is inhibited by way of the reticular formation.

erate in isolation. Their discharges represent decisions based on inputs from many sources. The motor cortex receives projections from the **ventral lateral nucleus** of the thalamus, the **postcentral gyrus,** the **posterior parietal cortex,** the **supplementary motor cortex,** and the **premotor cortex.** The ventral lateral thalamic nucleus is part of the circuitry by which the cerebellum and basal ganglia regulate movements (see later discussion). The postcentral gyrus processes and then transmits somatosensory information to the motor cortex. This provides feedback about the movements and about contacts between the skin and objects being explored. The posterior parietal cortex, the supplementary motor cortex, and the premotor cortex help program movements.

Motor Programs

Voluntary movements require contractions and relaxations in the proper sequence, not only of the muscles directly involved in the movements but also of the appropriate postural muscles. Therefore a mechanism is needed for programming these complex events. The cortical areas thought to be responsible for cortical motor programs include the **posterior parietal lobe,** the **supplementary motor cortex,** and the **premotor cortex** (Figure 9-23).

The posterior parietal lobe receives somatosensory information from the postcentral gyrus and visual information from the occipital cortex. The posterior parietal cortex connects with the supplementary motor cortex and the premotor cortex (Figure 9-23) and is important for

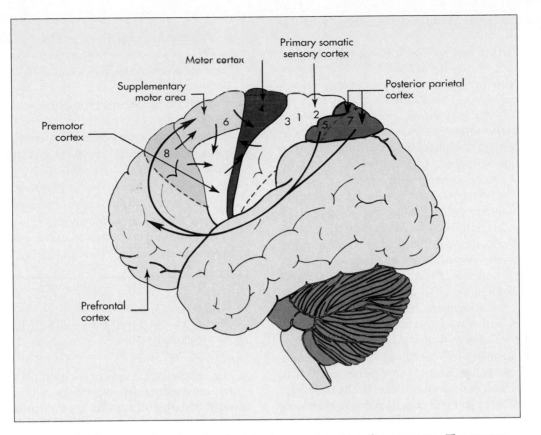

FIGURE 9-23   Cortical regions involved in the programming of movements. The arrows show some of the interconnections of these regions. The numbers refer to Brodmann's areas.

processing sensory information that leads to goal-directed movements.

> A lesion of the posterior parietal cortex will cause deficits in visually guided movements. Humans may develop a **neglect syndrome** (especially if the lesion is in the nondominant hemisphere for speech), in which a patient will be unable to recognize objects placed in the contralateral hand and unable to draw three-dimensional objects accurately. In fact, the patient may deny that the contralateral limbs even belong to him.

Electrical stimulation of the supplementary motor cortex produces complex, often bilateral, movements. Recordings from the supplementary cortex show that neurons in this region discharge in association with complex activity, such as movements of either arm. No relationship exists between the discharges and the details of the movement. This finding suggests that the activity evoked in the supplementary motor cortex causes deficits in orientation during movements and impairment of bilateral coordination.

The premotor cortex receives input from the part of the ventral lateral nucleus of the thalamus controlled by the cerebellum. It also receives input from the posterior parietal lobe and the supplementary motor cortex. The premotor cortex projects to the motor cortex and also to the spinal cord and brainstem. It is thought to be especially involved in the control of axial and proximal muscles. A lesion of the premotor cortex in humans or monkeys results in the appearance of the **grasp response,** in which touching the palm or extending the fingers elicits a grasping movement of the hand.

Motor Cortex

The motor cortex is recognizable microscopically by the presence of the **giant pyramidal,** or **Betz, cells.** However, many more projections of this area arise from small and medium-sized pyramidal cells than from the Betz cells. The corticospinal and corticobulbar tracts originate from pyramidal cells in layer 5 of the motor cortex, as well as from other cortical regions, including the premotor cortex and supplementary motor cortex and the postcentral gyrus. The somatotopic organization of the motor cortex has already been described (see Figure 9-17).

The motor cortex controls both distal and proximal muscles. However, *the corticospinal and corticobulbar projections of the lateral system are especially important for the activation of distal muscles of the contralateral upper and lower extremities and of the contralateral lower face and tongue.* A lesion that interrupts the corticospinal and corticobulbar tracts eliminates movements of distal muscles, but other pathways can still be used to activate proximal and axial muscles.

The lateral corticospinal tract makes monosynaptic excitatory connections with α-motoneurons, especially those that are located in the dorsolateral part of the ventral horn and that innervate distal muscles. The same pathway also excites γ-motoneurons, probably through an interneuronal pathway. Thus when the lateral corticospinal tract commands a voluntary movement, it co-activates α- and γ-motoneurons. In addition, interneurons in various reflex pathways are activated, so that the corticospinal tract influences spinal cord activity by regulating reflex transmission. Recordings from neurons in the motor cortex during learned movements reveal that some pyramidal tract neurons discharge just before a particular phase of the movement (e.g., flexion or extension of a joint). The activity encodes the force exerted by the muscles involved in the movement rather than encoding the position of the joint. Some neurons code for the rate at which force is developed, whereas others code for the steady-state force.

Neurons in the motor cortex receive input from somatosensory receptors in the skin, muscles, or joints. The receptive field of a given cortical neuron is related to the muscles activated from the same area of cortex. The somatosensory information presumably reaches the motor cortex by way of the somatosensory cortex, although a more direct pathway through the thalamus is also possible. The interaction between sensory feedback and motor cortical output is important for exploratory behavior. The tactile placing reaction is mediated by the motor cortex. When the dorsal surface of the paw of an animal held off the ground makes contact with a surface, the limb is flexed and then extended to establish support for the body.

When the corticospinal and corticobulbar tracts are completely interrupted, the distal muscles of the contralateral upper and lower extremities and muscles of the contralateral lower face and tongue are paralyzed (**hemiplegia**). Unless the lesion is restricted to these tracts, the deficit is a **spastic paralysis.** Spasticity usually accompanies hemiplegia produced by a lesion of the internal capsule or at other levels of the nervous system, because the corticoreticulospinal pathway is interrupted along with the pyramidal tract. Spasticity is also present in spinal cord injuries when transection at an upper cervical level causes paralysis of all four extremities (**quadriplegia**) or transection below the cervical enlargement causes paralysis of both lower extremities (**paraplegia**). Spastic paralysis is associated with an increase in muscle tone and increased phasic stretch reflexes. The latter may lead to **clonus,** such as in the ankle, in response to a brisk passive movement. Interruption of the corticospinal tract at any level causes an important release response, **Babinski's sign** (see Figure 9-18).

CEREBELLAR REGULATION OF POSTURE AND MOVEMENT

The cerebellum assists in the performance of coordinated movements by receiving sensory information about the status of movements and then adjusting the activity of the various descending motor pathways to optimize performance. These functions improve with practice, and thus the cerebellum is involved in the learning of motor skills. Destruction of the cerebellum produces no sensory deficits; therefore it has no essential role in sensation.

Organization of the Cerebellum

Afferent fibers from other parts of the central nervous system approach the cerebellar cortex through the cerebellar white matter. Two types of afferent fibers are found: **mossy fibers** and **climbing fibers** (Figure 9-24). Mossy fibers originate from various sources, but all climbing fibers are derived from the contralateral **inferior olivary nucleus.** In the cerebellar cortex the mossy fibers synapse in the granular layer on the dendrites of **granule cells.** There is considerable divergence, because a given mossy fiber branches repeatedly and synapses on many different granule cells. An individual climbing fiber synapses at many sites on the soma and dendritic tree of one or a few **Purkinje cells.** Thus the climbing fiber pathways show little divergence.

The granule cell axons form bundles of parallel fibers that synapse on the dendrites of Purkinje cells and of several classes of interneurons: Golgi cells, basket cells, and stellate cells. Granule cells are the only excitatory interneurons in the cerebellar cortex. The mossy fiber–granule cell pathway and the climbing fiber pathway are

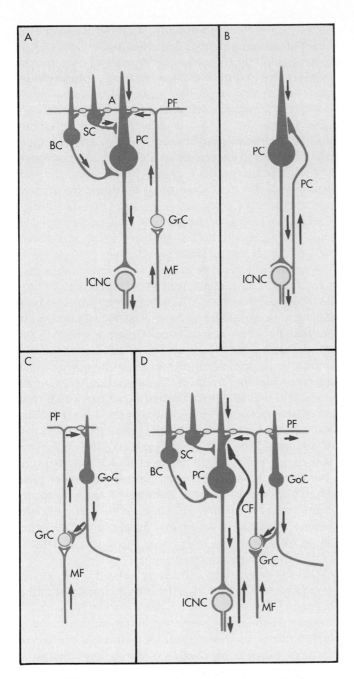

FIGURE 9-24 Excitatory and inhibitory circuits in the cerebellar cortex. Excitatory neurons are shown in outline; inhibitory neurons are dark. **A,** Connections made by mossy fibers through granule cells to Purkinje cells, stellate cells, and basket cells. Purkinje cells inhibit neurons of the deep cerebellar nuclei. **B,** The climbing fiber input to a Purkinje cell. **C,** The excitation of a Golgi cell by the mossy fiber through the granule cell path, with inhibition of granule cells by the Golgi cell. **D,** Combination of these circuits. *PC,* Purkinje cell; *SC,* stellate cell; *BC,* basket cell; *PF,* parallel fiber; *GrC,* granule cell; *MF,* mossy fiber; *ICNC,* deep cerebellar nuclear cell; *CF,* climbing fiber; *GoC,* Golgi cell.

able to excite Purkinje cells, and hence these may be regarded as the excitatory circuits of the cerebellar cortex. Mossy fiber–granule cell excitation typically elicits single action potentials in a Purkinje cell (**simple spike response**), whereas a climbing fiber evokes a high-frequency burst of action potentials in a Purkinje cell (**complex spike**).

The Golgi cells, basket cells, and stellate cells are all inhibitory interneurons of the cerebellar cortex. The Golgi cells inhibit granule cells; the basket cells inhibit Purkinje cell somata; the stellate cells inhibit Purkinje cell dendrites. All these inhibitory interneurons are activated by the mossy fiber–granule cell pathway.

A surprising observation is that although Purkinje cells are the only output neurons of the cerebellar cortex, their synaptic actions are inhibitory. Most Purkinje cells project to the deep cerebellar nuclei, but some synapse in the lateral vestibular nucleus. The cells of the deep cerebellar nuclei are tonically active; they receive excitatory input from collaterals of the mossy fibers and climbing fibers. Thus the inhibitory action of Purkinje cells modulates the discharges of the neurons of the deep cerebellar nuclei.

Functional Systems of the Cerebellum

The cerebellum can be considered on phylogenetic and functional grounds to be composed of three major components: the archicerebellum, the paleocerebellum, and the neocerebellum.

The **archicerebellum** is the earliest part of the cerebellum to evolve and is related in function primarily to the vestibular system. The archicerebellum is thus often referred to as the **vestibulocerebellum.** It corresponds in the human to the flocculonodular lobe and parts of the vermis in addition to the nodule. *The archicerebellum helps control axial muscles and thus balance. It also coordinates head and eye movements.* A lesion of the archicerebellum can result in a "drunken" stagger, called an **ataxic gait,** and also in **nystagmus.**

The **paleocerebellum** receives somatotypically organized information from the spinal cord; hence it is often called the **spinocerebellum.** The paleocerebellum regulates both movement and muscle tone. Lesions of the paleocerebellum produce deficits in coordination similar to those seen after damage to the neocerebellum.

The **neocerebellum** is the dominant component of the human cerebellum. It occupies the hemispheres of the cerebellum. The input is from wide areas of the cerebral cortex, and hence this region is sometimes called the **cerebrocerebellum.** *The neocerebellum modulates the output of the motor cortex.* Because the right side of the neocerebellum controls activity in the left cortex, and because the left cortex influences movements of the right limbs, the neocerebellum regulates motor activity of the same side of the body. The neocerebellum probably interacts with neurons of the premotor cortex in programming movements.

> Lesions of the neocerebellum affect chiefly the distal limbs. The deficits include delayed initiation of movements, **ataxia of the limbs** (incoordination), and reduced muscle tone. The limb ataxia results in **asynergy** (lack of synergy in movements), **dysmetria** (inaccurate movements), **intention tremor** (oscillations at the end of a movement), and **dysdiadochokinesia** (irregular performance of pronation and supination movements of the forearm). Reduced muscle tone leads to **pendular phasic stretch reflexes** in the lower extremity. Bilateral lesions of the neocerebellum may result in **dysarthria** (slow, slurred speech; synonymous with scanning speech). These classical neocerebellar signs are often seen in **multiple sclerosis.**

REGULATION OF POSTURE AND MOVEMENT BY THE BASAL GANGLIA

As with the neocerebellum, *the basal ganglia help regulate the activity of the motor cortex.* Unlike the archicerebellum and paleocerebellum, the basal ganglia exert only a minor influence on descending motor pathways other than the corticospinal and corticobulbar tracts. Judging from the effects of lesions, the role of the basal ganglia is often opposed to that of the neocerebellum.

Organization of the Basal Ganglia

The basal ganglia are the deep nuclei of the telencephalon. They include the caudate nucleus and putamen (neostriatum) and the globus pallidus (paleostriatum). The caudate nucleus and putamen are often collectively called the striatum because of the "striations" formed by fibers that pass between these nuclei in the human. The role of the basal ganglia in motor control has been inferred more on the basis of the effects of disorders of the basal ganglia than from experimental evidence.

Disturbances Caused by Basal Ganglia Diseases

Basal ganglia diseases can produce various motor disturbances. These can be categorized as disorders of movement and disorders of posture. Movement disturbances include **tremor** (rhythmic, "pill-rolling" oscillations at rest), **chorea** (rapid flicking movements), **atheto-**

sis (slow, writhing movements of limbs), **ballism** (violent, flailing movements), and **dystonia** (slow, twisting movements of the torso), and movements that are delayed in initiation and slow to reach completion, or **bradykinesia.** The disorders of posture produced in basal ganglion disease are forms of **rigidity.** The rigidity may be the **cogwheel** type; as a joint is moved, resistance occurs throughout the range of the movement, although there may be repeated alterations in the amount of resistance. These alterations produce a ratchetlike effect as the joint is passively moved. Alternatively, a **leadpipe** rigidity may occur, in which resistance is constantly present through the range of motion of the joint. The rigidity of basal ganglion disease should be distinguished from "decerebrate rigidity," which is more similar to spasticity.

Parkinson's disease is caused by a lesion of the substantia nigra and is characterized by tremor, rigidity, and bradykinesia. The loss of dopaminergic projections to the striatum is thought to be crucial. **Hemiparkinsonism** results when one substantia nigra is affected; the manifestations are contralateral, because these are caused by inappropriate regulation of the corticospinal tract. Destruction of part of the subthalamic nucleus on one side results in a **hemiballism,** characterized by ballistic movements on the contralateral side. **Huntington's chorea** is a genetic disorder in which there is loss of striatopallidal and striatonigral neurons containing γ-aminobutyric acid; cholinergic striatal interneurons also are lost. Loss of inhibitory input to the globus pallidus is thought to underlie the choreiform movements characteristic of the disease. The cerebral cortex also degenerates, leading to severe *mental deterioration.* In cerebral palsy, athetosis often occurs because of damage of the striatum and globus pallidus.

SUMMARY

1. The basic element of motor control is the motor unit, which comprises an α-motoneuron and all of the muscle fibers that it innervates.
2. Motor units have a variable number of muscle fibers, depending on how coarse or fine are the movements made by the particular muscle.
3. All of the muscle fibers of a motor unit are of the same histochemical type (I, IIB, or IIA). Slow-twitch motor units have slow contractions but resist fatigue; these have type I muscle fibers, which depend on oxidative metabolism. Fast-twitch motor units may be fatigable, with type IIB muscle fibers that use glycolytic metabolism, or fatigue resistant, with type IIA muscle fibers that rely on oxidative metabolism.
4. α-motoneurons must be activated when motor acts of any sort are performed. Hence, α-motoneurons form the final common pathway for movements.

5. α-motoneurons are found in motor nuclei of the spinal cord ventral horn and in cranial nerve motor nuclei. Spinal cord motor nuclei are arranged somatotopically.

6. Excitatory synapses tend to end on dendrites far away from the initial segment of the axons of α-motoneurons, whereas inhibitory synapses end near the initial segment. Excitatory postsynaptic potentials can sum spatially or temporally, and the synaptic currents may reach threshold at the initial segment. Inhibitory postsynaptic potentials may prevent threshold from being reached.

7. The action potentials of α-motoneurons are initiated at the initial segment, regardless of whether the neurons are activated orthodromically, antidromically, or directly.

8. The afterhyperpolarization in α-motoneurons is often quite large and helps to regulate the discharge rate.

9. Repetitive firing of an α-motoneuron can produce an unfused or a fused tetanus of the muscle fibers that it innervates. Contractile force can therefore be increased by increasing the discharge rates of α-motoneurons up to the point that a fused tetanus is produced. Contractile force can also be increased by recruitment of additional α-motoneurons.

10. Recruitment of α-motoneurons is orderly. Small α-motoneurons are generally activated before large ones in most motor acts. This is called the size principle. The motor units of small α-motoneurons do not generate much tension, but their fatigue resistance allows them to maintain posture. The motor units of larger α-motoneurons generate more tension, but may fatigue.

11. Muscles and tendons contain stretch receptors. These include muscle spindles, which are arranged to lie in parallel with the regular muscle fibers, and Golgi tendon organs, which are located in tendons and so are in series with the muscle.

12. The sensory endings of muscle spindles signal the rate of change of muscle length, as well as muscle length. Activation of γ-motoneurons causes intrafusal muscle fibers to shorten. This can prevent unloading of the muscle spindle when the regular muscle fibers contract, with the result that the sensory endings continue to discharge rather than becoming silent.

13. Golgi tendon organs are supplied by group Ib afferent fibers. They respond to both muscle stretch and muscle contraction and signal the tension in the tendon.

14. Spinal cord reflexes are an important foundation for motor control. They are relatively simple, stereotyped responses to a defined sensory input.

15. The stretch reflex involves a monosynaptic excitatory connection between group Ia afferent fibers from muscle spindles in a muscle and the α-motoneurons supplying the muscle, as well as synergistic muscles. There is also a disynaptic inhibitory pathway from the same group Ia afferent fibers, by way of group Ia inhibitory interneurons, to α-motoneurons that supply antagonistic muscles. A phasic stretch reflex can be elicited clinically by causing quick stretches of a muscle by striking its tendon with a reflex hammer. A tonic stretch reflex may be elicited by bending a joint.

16. The inverse myotatic reflex involves a disynaptic pathway from Golgi tendon organs through inhibitory interneurons to α-motoneurons of the same muscle. This reflex controls muscle tension.

17. Both the stretch reflex and the inverse myotatic reflex represent negative feedback loops. The controlled variable of the stretch reflex is muscle length, and that of the inverse myotatic reflex is muscle tension.

18. Another important spinal reflex is the flexion reflex. This involves flexion reflex afferent fibers, which excite flexor α-motoneurons and inhibit extensor α-motoneurons in the limb. There may be an accompanying crossed extensor reflex during which contralateral extensor α-motoneurons are excited and flexor α-motoneurons inhibited.

19. The motor pathways that descend from the brain can be grouped into lateral and medial systems. The lateral system acts on motoneurons and reflex circuits that control muscles of the distal extremities, as well as muscles of the lower face and the tongue. The medial system controls the proximal limb and axial muscles and so provides for postural support.

20. Spinal cord transection causes sensory loss and paralysis below the level of the lesion. Stretch and flexion reflexes become hyperactive, and pathological reflexes, such as the sign of Babinski, appear.

21. Transection of the upper brainstem may lead to decerebrate rigidity, in which the stretch reflexes also become exaggerated.

22. Several types of postural reflexes are organized by neural circuits in the brainstem and upper cervical spinal cord. These include vestibular reflexes, such as the vestibuloocular and vestibulocollic reflexes, and the ocular counterrolling reflex, as well as the tonic neck reflexes and the righting reflexes.

23. Locomotion is organized by a pattern generator in the spinal cord. Locomotion can be triggered by activity in the midbrain locomotor center, and it can be modified by segmental afferent input.

24. Eye position is controlled by several motor systems. Movements of the eyes are generally conjugate, but they can be vergent. Conjugate eye movements include saccades, smooth pursuit movements, and optokinetic movements.

25. The motor cortex issues commands that are transmitted by the corticospinal and corticobulbar tracts and that evoke voluntary movements.

26. The lateral corticospinal tract and the equivalent part of the corticobulbar tract originate largely in somatotopically organized parts of the precentral and postcentral gyri. The descending projections from the precentral gyrus make monosynaptic excitatory connections with α- and γ-motoneurons and with interneurons of spinal reflex circuits. These pathways control fine movements of distal muscles of the limbs and of the face and tongue. If other descending motor pathways, such as the reticulospinal tracts, are also interrupted, the paralysis becomes a spastic one, with hyperactive phasic stretch reflexes.

27. The cerebellum helps coordinate movements and is involved in the learning of motor skills. The neural circuits in the cerebellar cortex involve two types of afferents, the mossy and the climbing fibers. Mossy fibers excite granule cells, which in turn excite the Purkinje cells. Climbing fibers excite Purkinje cells directly.

28. The cerebellum can be subdivided into the archicerebellum (vestibulocerebellum), paleocerebellum (spinocerebellum), and neocerebellum (corticocerebellum). Damage to the vestibulocerebellum can produce an ataxic (staggering) gait and nystagmus. Damage to the neocerebellum causes incoordination of the limbs and reduced muscle tone.

29. The basal ganglia regulate movements largely by their influence on the motor cortex.

30. Disturbances of the basal ganglia result in disorders of movement and posture, including tremor, chorea, athetosis, ballism, dystonia, bradykinesia, and rigidity.

BIBLIOGRAPHY
Journal Articles

Alexander GE et al: Parallel organization of functionally segregated circuits linking basal ganglia and cortex, *Annu Rev Neurosci* 9:357, 1986.

Burke RE: Motor unit properties and selective involvement in movement, *Exerc Sport Sci Rev* 3:31, 1975.

Cullheim S, Kellerth JO: Combined light and electron microscopic tracing of neurons, including axons and synaptic terminals, after intracellular injection of horseradish peroxidase, *Neurosci Lett* 2:307, 1976.

Gerfen CR: The neostriatal mosaic: multiple levels of compartmental organization in the basal ganglia, *Annu Rev Neurosci* 15:285, 1992.

Hunt CC: Mammalian muscle spindle: peripheral mechanisms, *Physiol Rev* 70:643, 1990.

Ito M: Cerebellar control of the vestibulo-ocular reflex around the flocculus hypothesis, *Annu Rev Neurosci* 5:275, 1982.

Jami L: Golgi tendon organs in mammalian skeletal muscle: functional properties and central actions, *Physiol Rev* 72:623, 1992.

Keifer J, Houk JC: Motor function of the cerebellorubrospinal system, *Physiol Rev* 74:509, 1994.

Lisberger SG et al: Visual motion processing and sensorimotor integration for smooth pursuit eye movements, *Annu Rev Neurosci* 10:97, 1987.

Lüscher HR, Clamann HP: Relation between structure and function in information transfer in spinal monosynaptic reflex, *Physiol Rev* 72:71, 1992.

Moschovakis AK, Highstein SM: The anatomy and physiology of primate neurons that control rapid eye movements, *Annu Rev Neurosci* 17:465, 1994.

Penney JB, Young AB: Speculations on the functional anatomy of basal ganglia disorders, *Annu Rev Neurosci* 6:73, 1983.

Shik ML, Orlovsky GN: Neurophysiology of locomotor automatism, *Physiol Rev* 56:465, 1976.

Soechting JF, Flanders M: Moving in three-dimensional space: frames of reference, vectors, and coordinate systems, *Annu Rev Neurosci* 15:167, 1992.

Sparks DL, Mays LE: Signal transformations required for the generation of saccadic eye movements, *Annu Rev Neurosci* 13:309, 1990.

Thach WT, Goodkin HP, Keating JG: The cerebellum and the adaptive coordination of movement, *Annu Rev Neurosci* 15:403, 1992.

Wilson VJ, Peterson BW: Peripheral and central substrates of vestibulospinal reflexes, *Physiol Rev* 58:80, 1978.

Wise SP: Premotor cortex: past, present, and preparatory, *Annu Rev Neurosci* 8:1, 1985.

Wurtz RH, Albano JE: Visual-motor function of the primate superior colliculus, *Annu Rev Neurosci* 3:189, 1980.

Books and Monographs

Brooks VB: *The neural basis of motor control*, New York, 1986, Oxford University Press.

Burke RE: Motor units: anatomy, physiology, and functional organization. In *Handbook of physiology,* section 1: The nervous system, vol II: Motor control, part 1, Bethesda, Md, 1981, American Physiological Society.

Eccles JC et al: *The cerebellum as a neuronal machine,* New York, 1967, Springer Publishing Co.

Kandel ER, Schwartz JH, Jessell TM: *Principles of neural science,* ed 3, New York, 1991, Elsevier Science Publishing Co, Inc.

Matthews PBC: *Mammalian muscle receptors and their central actions,* Baltimore, 1971, Williams & Wilkins.

Phillips CG, Porter R: *Corticospinal neurons: their role in movement,* New York, 1977, Academic Press.

Roberts TDM: *Neurophysiology of postural mechanisms,* ed 2, London, 1978, Butterworth Publishers.

Willis WD, Grossman RG: *Medical neurobiology,* St Louis, ed 3, 1981, Mosby.

Wilson VJ, Jones GM: *Mammalian vestibular physiology,* New York, 1979, Plenum Press.

CHAPTER 10

The Autonomic Nervous System and Its Control

The autonomic nervous system is a motor system concerned with the regulation of smooth muscle, cardiac muscle, and glands. The autonomic nervous system is not directly accessible to voluntary control. Instead, it operates in an automatic fashion on the basis of **autonomic reflexes** and **central control.** *A major function of the autonomic nervous system is* ***homeostasis,*** *which is the maintenance of the internal environment in an optimal state.* For instance, the autonomic nervous system, in cooperation with the somatic motor system, helps keep the body temperature of homeothermic animals relatively constant. Another important role of the autonomic nervous system is to make the appropriate adjustments in smooth muscle tone, cardiac muscle activity, and glandular secretion for different behaviors. For example, the autonomic activity that can be observed during digestion is very different from that during a sprint.

The autonomic nervous system proper consists of the **sympathetic nervous system,** the **parasympathetic nervous system,** and the **enteric nervous system.** The final common pathway in the sympathetic and parasympathetic nervous systems consists of a sequence of two types of motoneurons: **preganglionic neurons,** whose cell bodies are located in the spinal cord or brainstem, and **postganglionic neurons,** whose cell bodies are in peripheral autonomic ganglia. Accompanying the autonomic motor axons in visceral nerves are **visceral afferent fibers,** which enter the central nervous system in dorsal roots or in cranial nerves. Visceral afferent fibers form the afferent link in many autonomic reflexes. Somatic afferent fibers can also trigger somatovisceral reflexes that employ an autonomic output. The enteric nervous system is a peripheral reflex network located in the wall of the gastrointestinal tract. Afferent neurons, postganglionic motoneurons, and interneurons are present in the enteric ganglia of **Meissner's submucosal** and **Auerbach's myenteric plexuses** (see Chapter 32).

The preganglionic neurons of the sympathetic nervous system are located in the **intermediolateral** (and **intermediomedial**) cell columns at spinal cord segments T1-L2. The preganglionic axons leave the spinal cord in white communicating rami and synapse on postganglionic neurons in **paravertebral** or **prevertebral ganglia.** The neurotransmitter used is acetylcholine. Postganglionic axons reenter the spinal nerve through gray communicating rami and distribute to the body wall, or they pass through splanchnic nerves to innervate viscera of the body cavities. They use norepinephrine as a neurotransmitter (except for those to sweat glands and those producing vasodilation, which use acetylcholine).

Parasympathetic preganglionic neurons are located either in cranial nerve nuclei or in the sacral parasympathetic nucleus. Their axons distribute to postganglionic neurons in cranial nerve ganglia or in ganglia near visceral organs in the pelvic cavity (and lower abdominal cavity). Postganglionic axons end in the enteric plexus or directly on cardiac or smooth muscle or glands. Both preganglionic and postganglionic neurons use acetylcholine as their neurotransmitter.

The peripheral component of the autonomic nervous system is under the control of several central nervous system structures; important ones include the **brainstem reticular formation,** the **hypothalamus,** and the **limbic system.** However, other motor system structures of the brain, including the cerebellum and the basal ganglia, also strongly influence autonomic activity.

ORGANIZATION OF THE AUTONOMIC NERVOUS SYSTEM

Sympathetic System

The sympathetic nervous system is a widely distributed motor system. It reaches not only the viscera contained in the body cavities, but also the skin and muscles of the body wall.

The cell bodies of the **preganglionic neurons** are located in the thoracic and upper lumbar spinal cord (T1 to about L2) in the **intermediolateral** and **intermedio-medial cell columns** (Figure 10-1). The motor axons of the sympathetic preganglionic neurons leave the spinal cord in the T1 to L2 ventral roots. The motor axons are small myelinated **B fibers** or in some cases unmyelinated **C fibers.** They pass from the spinal nerves into the **white communicating rami.**

When the sympathetic preganglionic axons in a given white ramus reach the **sympathetic paravertebral ganglion** of the same segment, they may (1) synapse in that ganglion, (2) turn rostrally or caudally in the sympathetic chain and synapse in a paravertebral ganglion at another segmental level, or (3) continue through a splanchnic nerve to synapse in a **prevertebral ganglion** (Figure 10-1). In this way preganglionic axons that originate from motoneurons limited to spinal cord segments T1 to L2 are able to synapse on postganglionic neurons located in the entire chain of paravertebral sympathetic ganglia (including the superior, middle, and inferior cervical sympathetic ganglia and the ganglia below L2 that do not receive white communicating rami), as well as in the prevertebral ganglia of the abdominal cavity (Figure 10-2). Preganglionic axons also directly innervate the chromaffin cells of the **adrenal medulla,** which are developmentally comparable to sympathetic ganglion cells.

Sympathetic postganglionic axons are unmyelinated (C) fibers that originate from ganglion cells of the sympathetic ganglia. Axons from ganglion cells in the paravertebral ganglia may distribute either to the body wall or to the viscera in the body cavities (Figure 10-2). If they are destined for the body wall, they pass from a ganglion into a spinal nerve by way of a **gray communicating ramus.** Gray rami are found on all ganglia of the sympathetic chain and connect with the appropriate spinal nerves. Sympathetic postganglionic axons destined for viscera in the body cavities enter **splanchnic nerves** and distribute to their targets. These axons may traverse prevertebral ganglia without synapsing en route to their targets. Postganglionic axons originating in prevertebral ganglia distribute to their targets through the sympathetic plexuses near the target organs.

The **sympathetic preganglionic neurons** that supply the head are in the upper thoracic segments. Their axons leave the spinal cord in the **white communicating rami** at T1 and T2, enter the **sympathetic chain,** and ascend to the superior cervical sympathetic ganglion, where they synapse on **postganglionic neurons.** The postganglionic axons pass into the head through a

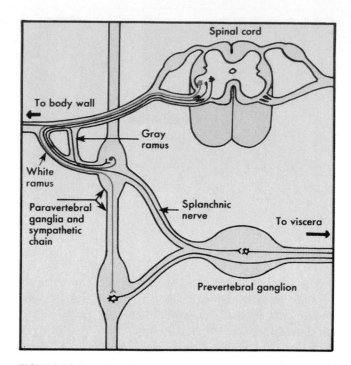

FIGURE 10-1   Distribution of sympathetic preganglionic projections to paravertebral and prevertebral ganglia.

plexus around the great vessels; they synapse on smooth muscle and glands of the face, eyes, and other structures of the head.

Interruption of the sympathetic supply to the head (or of descending pathways from the hypothalamus that control sympathetic activity) results in **Horner's syndrome,** which consists of a **partial ptosis** (drooping of the eyelid caused by paralysis of the superior tarsal muscle of the eyelid), **pupillary constriction** (because the unopposed parasympathetic supply of the iris is intact), **anhydrosis of the face** (caused by interruption of the innervation of the sweat glands of the face, and **enophthalmos** (retraction of the globe of the eye because of the interruption of the innervation of the smooth muscle of the orbit).

Parasympathetic System

The parasympathetic nervous system is less widely distributed than the sympathetic nervous system. A parasympathetic supply exists for various structures in the head and neck, but much of the distribution is to the viscera contained in the body cavities. No parasympathetic outflow reaches the skin or muscles of the body wall or extremities.

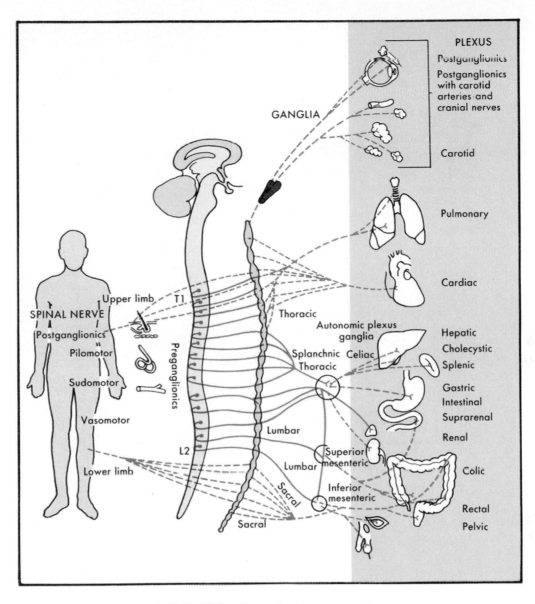

FIGURE 10-2   Sympathetic nervous system.

As with the sympathetic nervous system, the parasympathetic outflow involves a sequence of preganglionic and postganglionic neurons (Figure 10-3). The cell bodies of the parasympathetic preganglionic neurons are located either in the **brainstem** or in the **sacral spinal cord** (S2 to S4). Cranial nerve nuclei that contain preganglionic parasympathetic neurons include the **Edinger-Westphal nucleus** (cranial nerve III), **superior salivatory nucleus** (cranial nerve VII), **inferior salivatory nucleus** (cranial nerve IX), and **nucleus ambiguus** and **dorsal motor nucleus of the vagus** (cranial nerve X). **Sacral parasympathetic preganglionic neurons** are located in the **sacral parasympathetic nucleus.** No lateral horn exists in the sacral spinal cord.

The cranial parasympathetic preganglionic axons leave the brainstem in the appropriate cranial nerves and synapse on ganglion cells in the **cranial parasympathetic ganglia** (Figure 10-3; III: **ciliary ganglion**; VII: **sphenopalatine** and **submaxillary ganglia**; IX: **otic ganglion**) or on ganglia in or near the walls of target viscera in the thoracic and abdominal cavities (X). For the gastrointestinal tract the vagal preganglionic axons synapse on neurons belonging to the **enteric nervous system** (see following discussion). The sacral parasympathetic preganglionic axons distribute to the abdominal cavity and pelvis and synapse on ganglion cells located in the walls of viscera in these regions. The **splenic flexure** of the colon is the boundary between

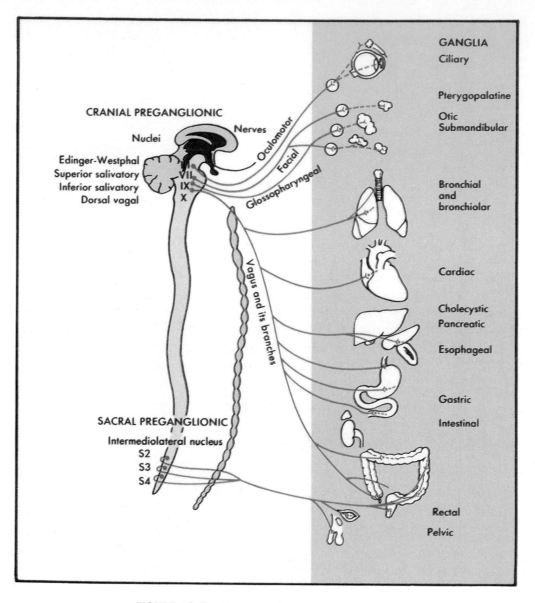

CRANIAL PREGANGLIONIC

Nuclei

Edinger-Westphal
Superior salivatory
Inferior salivatory
Dorsal vagal

VII
IX
X

Nerves

Oculomotor
Facial
Glossopharyngeal

Vagus and its branches

SACRAL PREGANGLIONIC

Intermediolateral nucleus
S2
S3
S4

GANGLIA
Ciliary

Pterygopalatine

Otic
Submandibular

Bronchial
and
bronchiolar

Cardiac

Cholecystic
Pancreatic

Esophageal

Gastric

Intestinal

Rectal

Pelvic

FIGURE 10-3 Parasympathetic nervous system.

the gastrointestinal organs supplied by the vagus nerve and those supplied by sacral parasympathetics. **Parasympathetic postganglionic neurons** directly innervate their nearby target organs.

The **parasympathetic preganglionic neurons** that control pupil size are located in the **Edinger-Westphal nucleus,** which is in the midline just ventral to the cerebral aqueduct. The preganglionic axons leave the midbrain with the **oculomotor nerve** and synapse in the **ciliary ganglion,** which is in the orbit just behind the eye. Axons of the **postganglionic neurons** of the ciliary ganglion pass into the eye through short ciliary nerves, and some terminate on the smooth muscle cells of the iris that form the **pupillary sphincter.**

When there is a large increase in intracranial pressure, such as happens during a **cerebral hemorrhage** or because of a **brain tumor,** the brain may shift, causing a herniation of the uncus through the tentorial notch. This may cause a compression of the oculomotor nerve and a sudden unilateral dilation of a pupil. A fixed dilated pupil is an indicator of impending death unless the high intracranial pressure is relieved by surgical intervention.

The **enteric nervous system** is a miniature nervous system within the wall of the gastrointestinal tract (see

Chapter 32). Reflex networks in this system organize gut movements that can occur even when the gut is removed from the body. Afferent neurons, interneurons, and motoneurons are included in the system. Parasympathetic and sympathetic connections to the enteric nervous system permit autonomic control. The component of the enteric nervous system in **Auerbach's myenteric plexus** controls the activity of the muscular layers, and that in **Meissner's submucosal plexus** controls the muscularis mucosae and intestinal glands.

AUTONOMIC FUNCTIONS

The sympathetic and parasympathetic nervous systems act continuously to adjust activity in smooth muscles, cardiac muscle, and glands, often by exerting a reciprocal control.

The sympathetic nervous system actively regulates visceral function under normal circumstances. The parasympathetic nervous system often acts contrary to the sympathetic nervous system when a given organ is innervated by both systems. However, it is more appropriate to view concurrent control of organs by activity in the sympathetic and parasympathetic nervous systems as a means of coordinating visceral activity.

The preganglionic neurons of the autonomic nervous system, as with α-motoneurons of the somatic motor system, use acetylcholine as their neurotransmitter (Figure 10-4). The **acetylcholine receptors** on postganglionic neurons, as with those of skeletal muscle fibers, are of the **nicotinic** type. Nicotinic receptors are activated by low doses of nicotine and blocked by curare. Parasympathetic and some sympathetic postganglionic neurons also use acetylcholine. However, the receptors on target organs are of the **muscarinic type.** Muscarinic receptors are activated by muscarine and blocked by atropine. The **cholinergic sympathetic fibers** include those to sweat glands and the vasodilator fibers in skin and skeletal muscle. Parasympathetic neurons also release neuropeptides, such as **vasoactive intestinal polypeptide** (VIP).

Sympathetic postganglionic neurons generally use norepinephrine as their neurotransmitter (Figure 10-4). Receptors for norepinephrine include α- **and** β-**adrenergic receptors.** α-Receptors are more powerfully activated by norepinephrine than by isoproterenol; the converse is true of β-receptors. Agents such as phenoxybenzamine block α-receptors, whereas drugs such as propranolol block β-receptors. α-Receptors can be further divided into α_1 and α_2 subtypes, and β-receptors into β_1 and β_2. Sympathetic neurons also release neuropeptides, such as **neuropeptide Y** (NPY).

The **adrenal medulla** is supplied by sympathetic

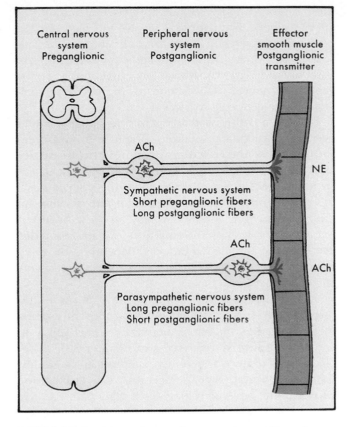

FIGURE 10-4   Transmitters of autonomic ganglia and postganglionic synapses. *Ach,* Acetylcholine; *NE,* norepinephrine.

preganglionic axons, which release acetylcholine as their neurotransmitter. The **chromaffin cells** of the adrenal medulla are developmentally similar to sympathetic postganglionic neurons, and they secrete epinephrine and norepinephrine into the circulation, where these agents act as hormones. In humans the ratio of epinephrine to norepinephrine is 4:1.

Neurons of the enteric nervous system release not only acetylcholine and norepinephrine, but also serotonin, adenosine triphosphate (ATP), and a variety of peptides as neurotransmitters and neuromodulators.

Control of Autonomic Function

The operation of the autonomic nervous system is regulated hierarchically in much the same way as the somatic motor system. The most direct neural control of many organs is by means of **autonomic reflexes.** However, these are often regulated by descending pathways from the brainstem. In addition, autonomic function is controlled by higher autonomic centers, including the **hypothalamus** and other parts of the **limbic system.**

Autonomic reflexes are mediated by neural circuits in the spinal cord and brainstem. The afferent limbs of these reflex pathways include both visceral and somatic afferent fibers. The pathways involve interneurons that receive a convergent input from visceral and somatic sensory receptors. The efferent limbs are formed by sympathetic and parasympathetic pre-ganglionic and post-ganglionic neurons. The actions of the two autonomic systems are generally reciprocal.

Brainstem pathways that regulate the activity of autonomic preganglionic neurons originate from several sites, including the **reticular formation, raphe nuclei,** and the **locus ceruleus complex.** These brainstem structures receive information about the visceral activities they regulate by way of ascending tracts. Some autonomic functions depend strongly on these brainstem pathways. For example, **micturition** (urination) and **defecation** (the elimination of feces) depend on the integrity of pathways that interconnect the sacral spinal cord and the pons.

The urinary bladder is emptied by means of the **micturition reflex,** which involves both the sympathetic and parasympathetic nervous systems and a descending control system (see also Chapter 32). As the bladder fills to near its capacity, receptors in the bladder wall are activated. Signals ascend to the **micturition center** in the pons, activating descending pathways that cause a parasympathetic contraction of the detrusor muscle of the bladder and relaxation of the internal and external sphincters. Simultaneously, the sympathetic system relaxes the neck of the bladder and no longer causes constriction of the internal sphincter. The bladder can thus empty.

> Following spinal cord injury, the micturition reflex is prevented by disruption of the long pathways connecting the sacral cord and the micturition center. The bladder now fills excessively **(atonic neurogenic bladder)** and must be drained by catherterization. Later, a spinal reflex pathway becomes operative, but it is overactive **(spastic bladder).** However, the bladder does not empty completely and so is liable to infection, and there is frequent **incontinence.**

Higher centers that regulate autonomic function include the hypothalamus and other components of the limbic system. Limbic structures are interconnected with nonlimbic parts of the nervous system, including the neocortex, cerebellum, and basal ganglia. The hypothalamus projects to the brainstem (e.g., to the reticular formation) and to the spinal cord. The limbic system controls motivation directly through neural pathways and indirectly through the endocrine system.

FUNCTIONS OF THE HYPOTHALAMUS

The hypothalamus (see Chapter 44) has several broadly defined functions, which include the regulation of homeostasis, motivation, and emotional behavior. These functions are mediated through hypothalamic control of autonomic and endocrine activity, as well as by interactions between the hypothalamus and other parts of the limbic system.

If the hypothalamus is stimulated electrically, particular regions are shown to be related to particular autonomic responses. For example, stimulation in the lateral and posterior hypothalamus produces responses mediated by the sympathetic nervous system. Stimulation in the anterior hypothalamus activates parasympathetic output. The responses include changes in heart rate and blood pressure. Other global functions controlled by the hypothalamus include food and water intake, emotional behavior, and regulation of the immune system.

The neurons in some hypothalamic nuclei release peptides, either as hormones or as neuromodulator substances. Such neurons are classified as **neuroendocrine cells.** Neuroendocrine structures include the paraventricular and supraoptic nuclei, which give rise to the **hypothalamohypophyseal tract** from the hypothalamus to the posterior pituitary gland. This tract releases the peptide hormones **oxytocin** and **vasopressin** into the circulation (see Chapter 44). The paraventricular nucleus also sends peptide-containing axons to the various sites within the central nervous system, including the solitary nucleus, the dorsal motor nucleus of the vagus, and the intermediolateral cell column of the spinal cord. Oxytocin and vasopressin apparently are used both as hormones and as neuromodulators of autonomic function.

Neuroendocrine cells in a number of hypothalamic nuclei secrete hormones into the **portal system** that supplies the **anterior pituitary gland** (see Chapter 44). These hormones release or inhibit the release of pituitary hormones into the circulation, and they are very important in endocrine regulation. As in the case of oxytocin and vasopressin, the same hypothalamic hormones can be used as neuromodulatory substances at synaptic terminals within the central nervous system.

Temperature Regulation

Homeothermic animals regulate their body temperature. *When the environmental temperature decreases, the body adjusts by reducing heat loss and by increasing heat production. Conversely, when the temperature increases, the body increases its heat loss and reduces heat production.*

Information about the external temperature is provided by **thermoreceptors** in the skin (and probably other organs, such as muscle). Internal temperature is monitored by **central thermoreceptive neurons** in the anterior hypothalamus. The central thermoreceptors monitor the temperature of the blood. The system acts as a servomechanism with a set point at the normal body temperature. Error signals, representing deviation from the set point, lead to responses that tend to restore body temperature toward the set point. These responses are mediated by the autonomic, somatic, and endocrine systems.

Cooling causes **shivering,** which consists of asynchronous muscle contractions that lead to increased heat production. There is also an increase in the activity of the thyroid gland and an increase in sympathetic activity, both of which tend to increase heat production metabolically. Heat loss is reduced by **piloerection** (effective in animals with fur, although not in humans; in the latter, the result is goose-bumps) and by **cutaneous vasoconstriction.**

Warming the body causes changes in the opposite direction. There is a reduction in the activity of the thyroid gland, leading to reduced metabolic activity and less heat production. Heat loss is increased by *sweating* and *cutaneous vasodilatation.*

The hypothalamus serves as the temperature servomechanism. The heat loss responses are organized by the **heat loss center,** which is thought to be composed of neurons in the preoptic region and anterior hypothalamus. Lesions here prevent sweating and cutaneous vasodilatation, resulting in **hyperthermia** when the individual is placed in a warm environment. Conversely, electrical stimulation here causes cutaneous vasodilatation and inhibits shivering. Neurons in the posterior hypothalamus form a **heat production and conservation center.** Lesions in the area dorsolateral to the mammillary body eliminate heat production and conservation, leading to **hypothermia** in a cold environment. Electrical stimulation in this region evokes shivering.

Thermoregulatory responses are also produced when the hypothalamus is locally warmed or cooled. This is consistent with the presence of central thermoreceptive neurons in the hypothalamus. Many of the details of the interconnections of the hypothalamus involved in the heat loss and the heat production and conservation centers are as yet unknown.

In fever, the set point for body temperature is elevated. A mechanism for this is the release of a **pyrogen** by microorganisms, such as certain bacteria. The pyrogen changes the set point, leading to increased heat production by shivering and heat conservation by cutaneous vasoconstriction.

THE LIMBIC SYSTEM

The limbic system includes the **limbic lobe** of the telencephalon, as well as the **hypothalamus** and several **midbrain nuclei.** The limbic components of the telencephalon include the **cingulate, parahippocampal,** and **subcallosal gyri,** as well as the **hippocampal formation** (hippocampus, dentate gyrus, and subiculum).

The functions of the limbic system include regulation of aggressive behavior and sexuality. More generally the limbic system appears to be concerned with motivational states, which in turn are vital for survival of both the individual and the species.

Bilateral removal of temporal lobe structures, including the amygdaloid nuclei, results in a complex set of changes in behavior called the **Klüver-Bucy syndrome.** Animals previously wild become tame; they develop a pronounced tendency to put objects into their mouths; and they become sexually hyperactive. These changes result chiefly from damage to the amygdaloid nuclei.

The **hippocampus** appears to be important for the storage of recent memory (see also Chapter 11). *Memories are stored in the following sequence: short-term memory, recent memory, and long-term memory.* Short-term memory is easily disrupted and is presumed to depend on ongoing neural events. Long-term memory seems to result from a permanent functional or structural change in the nervous system. Bilateral lesions of the hippocampus may not interfere with either short- or long-term memory but may prevent the process by which short-term memories are permanently stored. The process of recollection of memories may also be disrupted, resulting in **amnesia.**

Bilateral lesions of the temporal lobes that damage the hippocampus, and degenerative diseases, notably **Alzheimer's disease,** that destroy hippocampal circuits, can lead to deficits in the incorporation of recent memory. Such a patient may be able to remember a conversation for a short time, but minutes later may repeat the same conversation as if it had never happened. However, memories of childhood events may still be relatively clear.

SUMMARY

1. Sympathetic preganglionic neurons are located in the intermediolateral (and intermediomedial) cell columns of the T1 to L2 segments of the spinal cord. Their axons leave the spinal cord through ventral roots and enter the sympathetic chain through white communicating rami. They synapse on postganglionic neurons in paravertebral ganglia or in prevertebral ganglia. Postganglionic axons synapse in target organs.

2. Parasympathetic preganglionic neurons are located in cranial nerve nuclei, and in the sacral preganglionic nucleus. Postganglionic axons synapse in target organs.

3. The enteric nervous system is in the wall of the gastrointestinal tract in the myenteric and submucosal plexuses. It coordinates movements and glandular secretions of the gut.

4. The sympathetic and parasympathetic nervous systems regulate the activity of cardiac muscle, smooth muscle, and glands. Often, these components of the autonomic nervous system act in a reciprocal fashion.

5. Preganglionic sympathetic and parasympathetic neurons release acetylcholine as their neurotransmitter. This acts on nicotinic cholinergic receptors on postganglionic neurons. These receptors are blocked by curare.

6. Parasympathetic and some sympathetic postganglionic neurons (supplying sweat glands) also release acetylcholine. The postsynaptic receptors on target tissue in this case are muscarinic and can be blocked by atropine.

7. Most sympathetic postganglionic neurons release norepinephrine. The action is on α-adrenergic or β-adrenergic receptors.

8. The adrenal medulla receives sympathetic preganglionic input and releases epinephrine and norepinephrine into the general circulation.

9. The autonomic nervous system operates reflexly and in response to descending control systems, similarly to the somatic nervous system. A number of structures of the brain are involved in autonomic control. However, the hypothalamus and other parts of the limbic system are of particular importance.

10. The hypothalamus regulates homeostasis, motivation, and emotional behavior through control of the autonomic nervous system, the endocrine system, and the somatic nervous system. Some of the functions regulated include body temperature, cardiovascular activity, appetite, water intake, and immune responses.

11. The hypothalamus controls endocrine function both by direct release of hormones in the posterior pituitary and by release of peptides into the portal circulation of the anterior pituitary gland.

12. The limbic system includes not only the hypothalamus but also a number of forebrain structures, including the hippocampus, and several nuclei in the midbrain. Some of the functions of the limbic system include the regulation of aggressive behavior and sexuality. The hippocampus is involved in the storage of recently acquired memories.

BIBLIOGRAPHY
Journal Articles

Gershon MD: The enteric nervous system, *Annu Rev Neurosci* 4:227, 1981.

Dampney RAL: Functional organization of central pathways regulating the cardiovascular system, *Physiol Rev* 74:323, 1994.

Davis M: The role of the amygdala in fear and anxiety, *Annu Rev Neurosci* 15:353, 1992.

Elfin LG, Lindh B, Hökfelt T: The chemical neuroanatomy of sympathetic ganglia, *Annu Rev Neurosci* 16:471, 1993.

Hayward JN: Functional and morphological aspects of hypothalamic neurons, *Physiol Rev* 57:574, 1977.

Jaenig W, McLachlan E: Organization of lumbar spinal outflow to distal colon and pelvic organs, *Physiol Rev* 67:1332, 1987.

Lopes da Silva FH, Witter MP, Boeijinaga PH, Lohman AHM: Anatomic organization and physiology of the limbic cortex, *Physiol Rev* 70:453, 1990.

Simon E et al: Central and peripheral thermal control of effectors in homeothermic temperature regulation, *Physiol Rev* 66:235, 1986.

Smith OA, DeVito JL: Central neural integration for the control of autonomic responses associated with emotion, *Annu Rev Neurosci* 7:43, 1984.

Books and Monographs

Bannister R: *Autonomic failure,* Oxford, 1983, Oxford University Press.

Carpenter MB, Sutin J: *Human neuroanatomy,* ed 8, Baltimore, 1983, Williams & Wilkins.

Kandel ER, Schwartz JH, Jessell TM: *Principles of neural science,* ed 3, New York, 1991, Elsevier Science Publishing Co, Inc.

Loewy AD, Spyer KM: *Central regulation of autonomic functions,* New York, 1990, Oxford University Press.

Willis WD, Grossman RG: *Medical neurobiology,* ed 3, St Louis, 1981, Mosby.

CHAPTER 11

Higher Functions of the Nervous System

The central nervous system is responsible for the higher functions that characterize humans. These functions include consciousness, thought, perception, learning, memory, and language. States of consciousness, including the sleep-wake cycle, are generally studied with the help of neurophysiological techniques, such as the electroencephalogram and evoked potentials.

Learning and memory depend on alterations in neural functions and even in structure. The human brain is actually two brains. One hemisphere is dominant for certain functions, including handedness and speech, and the opposite hemisphere is dominant for others, such as spatial relations and music.

THE ELECTROENCEPHALOGRAM

The higher functions of the human brain are expressed on the background of continuous thalamocortical interactions. Neurons of nearly all the nuclei of the thalamus project to the cerebral cortex, and the cerebral cortex projects back to the thalamus.

Recordings from the surface of the cerebral cortex (**electrocorticogram**) or from the scalp (**electroencephalogram** or **EEG**) reveal the incessant oscillations of extracellular potentials caused by membrane potential oscillations in large numbers of cortical neurons in response to the rhythmic alterations of activity in thalamocortical circuits. On a single neuron level, activity corresponding to the EEG consists of alternating **excitatory** and **inhibitory postsynaptic potentials.** The excitatory potentials often result in discharges of cortical neurons.

The normal EEG can be described in terms of its frequency composition. Several characteristic frequency ranges can be recognized (Figure 11-1). These are called **alpha waves** (8 to 13 Hz), **beta waves** (more than 13 Hz), **theta waves** (4 to 7 Hz), and **delta waves** (less than 4 Hz). Other transient waves are also seen. The dominant frequency depends on several factors, including age, state of consciousness, recording site, the action of drugs, and the presence of disease. During the early years of life the EEG is dominated by low frequencies. In mature individuals at rest with eyes closed, the EEG recorded from the posterior region of the brain shows an alpha rhythm, whereas that recorded from the anterior part of the brain has a beta rhythm. If the individual is aroused, lower-voltage, higher-frequency beta rhythms take the place of the synchronized, lower-frequency alpha waves (Figure 11-1). Slower waves of theta and delta frequencies are associated with deeper levels of sleep. One definition of **brain death** is a *persistent isoelectric EEG* in the absence of depressant drugs.

EVOKED POTENTIALS

The EEG represents spontaneous activity that is not linked to a particular event. Similar activity can be evoked in response to a stimulus that activates a neural pathway to the thalamus or circuits within the cerebral cortex. Such stimulus-linked activity recorded from the cortex is called a **cortical evoked potential.** Evoked potentials can easily be produced in human subjects by stimulating (1) a peripheral nerve with electric shocks, (2) the retina with flashes of light, or (3) the ear with an acoustic stimulus such as a click. The waveform that is recorded is largest over the appropriate region of the brain.

STATES OF CONSCIOUSNESS

Mental processes occur in the brains of conscious subjects. Consciousness is not understood, but it is required for perception, thought, and the use of language.

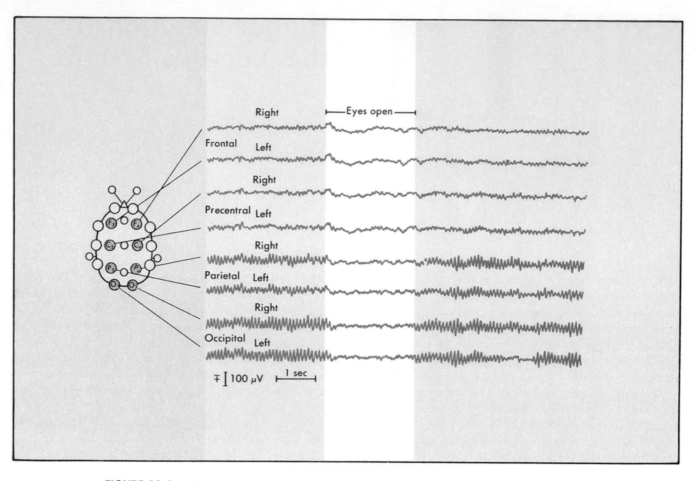

FIGURE 11-1 EEG recorded from a normal human subject. Recordings are from eight sites on the scalp. In the resting condition an alpha rhythm is prominent over the parietal and occipital lobes. When the eyes are opened, the alpha rhythm is blocked and replaced by a beta rhythm.

The conscious state appears to depend on an interaction between the **brainstem reticular formation** and **thalamocortical circuits.** When consciousness is depressed, the EEG becomes more synchronous and slowed in frequency. During behavioral arousal, as in response to a painful stimulus, the EEG changes to a low-voltage, high-frequency pattern (EEG arousal).

Sleep

Sleep is an alteration, rather than a loss, of consciousness. This is shown by the ease with which sleep is interrupted by significant environmental events, such as a baby's cry. Sleep has a circadian rhythm, as well as a more rapid oscillation. **Circadian rhythms** repeat at approximately daily intervals. The sleep rhythm, along with many other biological rhythms, is normally entrained by the **light-dark cycle.** When a person rapidly changes location to a different time zone, it takes days for the circadian rhythms to be reentrained. The sleep disturbance and other disorders that result are collectively known as **jet lag.**

The various stages of sleep are characterized by different types of motor and autonomic, EEG, and psychological activity. The major distinction is between **rapid-eye-movement (REM) sleep** and **non-REM sleep.** When an individual falls asleep, the initial type is non-REM sleep. The EEG becomes more synchronized and slows (Figure 11-2, *A*). There are four levels of non-REM sleep. In the first level **(stage 1),** the person is drowsy and the EEG shows 7 to 10 Hz rhythms. Over time the depth of non-REM sleep increases; that is, the EEG becomes progressively slower, and the person becomes difficult to arouse. In **stage 2,** or light sleep, a person is easily aroused; the dominant EEG frequencies are 3 to 7 Hz, with bursts of 12 to 14 Hz sleep spindles. During **stage 3** sleep, muscle tone and reflex activity are depressed, blood pressure falls, the heart rate slows, and the pupils constrict; the EEG shows 1 to 2 Hz, high-voltage waves. **Stage 4** is the deepest level of sleep, and it is also charac-

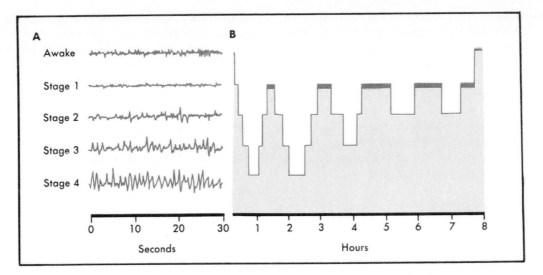

FIGURE 11-2   Stages of sleep and changes during the course of a night. **A**, EEG recordings during the waking state and during progressively deeper levels of non-REM sleep. The EEG in REM sleep would resemble that shown for the awake individual. **B**, Different sleep stages experienced during a typical night for a young adult. The black bars represent periods of REM sleep.

terized by 1 to 2 Hz EEG waves. The amount of time spent in the deepest stage of non-REM sleep decreases with age, and this stage may entirely disappear after age 60.

After about 90 minutes, sleep lightens and changes to a period of REM sleep that lasts approximately 20 minutes. During REM sleep the EEG becomes desynchronized and has a low voltage; the pattern is similar to that seen during arousal (Figure 11-2). Tone in many muscles disappears, and reflexes are inhibited. Interrupting this tonic inhibition are phasic motor events, including rapid movements of the eyes and brief contractions of other muscles. Transient waves (pontine-geniculate-occipital waves) can be recorded from the brainstem and occipital cortex at the times of the REMs. Autonomic events include irregular respiration, reduced blood pressure interrupted by episodes of hypertension, and penile erection in males. Dreams tend to occur in REM sleep, although nightmares usually occur during deep non-REM sleep. It is difficult to awaken individuals from REM sleep, but spontaneous awakening often occurs. REM sleep recurs about six times during a night. The proportion of time spent in REM sleep is greatest in the fetus and newborn, but it declines sharply during early infancy and then further with aging.

Sleep appears to be triggered by an active mechanism that involves the **reticular formation** and **monoamine neurons** in the brainstem. Some of the neurotransmitters associated with sleep include serotonin, norepinephrine, and acetylcholine. Several sleep-inducing peptides have been discovered as well.

Attention

Attention is the process by which perception is directed at particular events. It involves orientation to stimuli that are potentially significant, such as novel stimuli or stimuli that are likely to lead to a reward or a punishment. Attention to a stimulus may result in the loss of these changes (**habituation**). Application of a threatening stimulus enhances attention; this process is termed **sensitization.**

Epilepsy

Epilepsy refers to disease states characterized by behavioral and EEG seizures. The seizures may be partial or generalized. In **partial seizures** only part of the brain shows abnormal activity, and consciousness is retained. In **generalized seizures** large regions of the brain are involved, and consciousness is lost.

Partial seizures may originate in a damaged area of the motor cortex. Such seizures are characterized by contractions of muscles in the somatotopically appropriate region on the contralateral side and a focal EEG spike train (an EEG spike is a synchronous wave resulting from simultaneous activity in many neurons). The seizures often spread to adjacent areas in a *march* of convulsive activity to other contralateral parts of the body. For example, the seizure may start with contractions of the fingers, but the movements may then spread to the arm, shoulder, and face and lower extremity. Partial seizures may also originate from the somatosensory cortex and produce focal sensory experiences contralaterally. Psychomotor seizures are partial seizures that originate in the limbic lobe. These are characterized by semipurposeful movements, changes in consciousness, hallucinations, and illusions. A common hallucination is an unpleasant odor.

Generalized seizures include **grand mal** and **petit mal seizures.** Grand mal attacks may be preceded by an aura. Consciousness is soon lost, followed by tonic and clonic contractions of muscles on both sides of the body. Petit mal attacks are brief losses of consciousness, accompanied by a characteristic EEG pattern.

LEARNING AND MEMORY

Learning is a process by which behavior is modified on the basis of experience. **Memory** is the storage of information that has been learned. There are several stages of memory, including short-term memory, recent memory, and long-term memory. **Short-term memory** appears to depend on ongoing neural activity because it is easily disrupted (e.g., by head trauma). **Recent memory** refers to the process by which information in short-term memory is transformed into long-term memory. This process seems to depend on activity transmitted by the hippocampal formation, because damage to the hippocampus and related structures prevents the consolidation of short-term into long-term memory. **Long-term memory** apparently depends on permanent changes in widely distributed sets of neurons. The changes may include morphological as well as functional changes. In addition to memory stores, mechanisms must exist for accessing these stores, retrieving the information, recalling it to consciousness, comparing it with other information, and using the information for decisions.

Little is known about the mechanisms of learning and memory. Experiments on the simple nervous systems of invertebrates have shed some light on the neural basis of simple forms of learning. Habituation and sensitization are examples of **nonassociative learning,** because these do not require learning an association between two events. In habituation a response to a particular stimulus diminishes with repetition of the stimulus. Habituation is thus the process of learning that a stimulus is unimportant. Conversely, sensitization is the process by which a person learns that a stimulus is important. For example, with repetition of a painful stimulus, an individual quickly learns to respond.

In **associative learning** the relationship between two different stimuli is learned. In **classical conditioning** a conditioned stimulus is paired with an unconditioned stimulus. The latter initially produces an unconditioned response (e.g., food produces salivation in a hungry dog). After conditioning, the conditioned stimulus may now produce the same response (e.g., ringing a bell at the time food is presented ultimately causes salivation even if the food is omitted). In **operant conditioning** reinforcement of a response changes the probability of the response. Operant behaviors are not reflexes but rather spontaneous actions. An example of operant conditioning would be an animal's avoidance of a wire grid that induces an electric shock when the animal happens to step on the grid. In this case the conditioning stimulus provides **negative reinforcement.**

Habituation, sensitization, and classical conditioning have all been demonstrated in invertebrate models, and the neural mechanisms that underlie both nonassociative and associative learning are being studied. A major theme of such work is that synaptic efficacy changes during these simple forms of learning. These changes de-

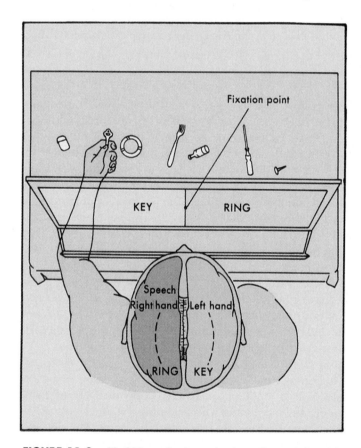

FIGURE 11-3 Technique for investigation of a patient with a disconnection syndrome caused by transection of the corpus callosum. The subject is asked to fix his or her vision on a fixation point at the center of a screen. Pictures are projected on the screen. If the picture is in the left visual field, the image is processed in the right hemisphere (key in this instance). The subject can also reach under the screen to reach objects that can be identified by tactile cues. (See also text.)

pend on the activation of second messenger systems. Long-term changes are accompanied by structural as well as functional changes. A parallel experimental approach in mammals involves the enhancement of synaptic transmission for hours to days or even longer after activation of particular pathways in the hippocampus. The mechanisms that underlie this **long-term potentiation** are also under active study.

CEREBRAL DOMINANCE

The two halves of the human brain are not equivalent. *In a real sense the human has two brains that communicate with each other by way of the cerebral commissures.* The left hemisphere is dominant in most individuals with respect to control of the preferred hand (right in most people) and in language. However, the right hemisphere can be considered dominant for other functions (music, spatial relationships). A structural correlate of cerebral dominance is the greater size of the left than of the right **planum temporale** (temporal plane, the superior surface of the temporal lobe; see Figure 11-4).

Cerebral dominance has been studied best in patients whose left and right hemispheres have been disconnected by surgical division of the corpus callosum. Visual images can be shown separately to the left and right visual fields of such individuals, and they can be asked to identify objects placed in the right or left hand (Figure 11-3). If a picture of an object, such as a ring, is presented to the left hemisphere, the subject can identify the object verbally as a ring. If a picture of a key is presented to the right hemisphere, the subject cannot identify it verbally. This is because information about the key that reaches the right hemisphere about the key does not gain access to the language centers of the left hemisphere. However, the subject can identify the picture in another way, by picking up a key with the left hand after feeling a group of objects.

Language

Language depends on activity in the left hemisphere in most people. This can be demonstrated by injecting local anesthetic into the carotid circulation on the left while an individual is speaking. The anesthetic stops speech (causes **aphasia**).

Analysis of aphasia that occurs after damage to the left hemisphere of adults has revealed that several major zones are important for language. One of these is called **Broca's area,** which is located in the inferior frontal gyrus just anterior to the face representation in the motor cortex (Figure 11-4). The other important region for the control of language is **Wernicke's area,** which is in the supramarginal and angular gyri of the temporal lobe and the posterior part of the superior temporal gyrus.

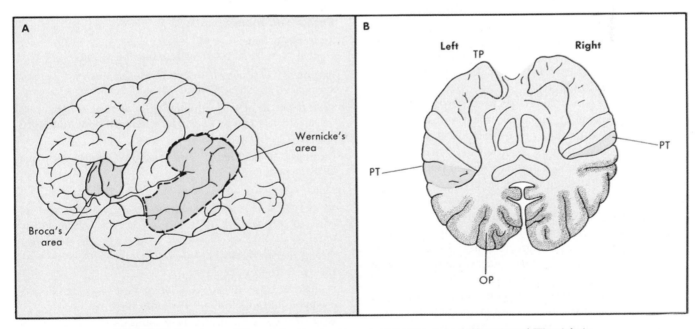

FIGURE 11-4   Areas of the cerebral cortex important for language. **A,** Broca's and Wernicke's areas. **B,** The relative size of the planum temporale *(PT)* on the two sides of the brain. *TP,* Temporal pole; *OP,* occipital pole.

Damage to Broca's area diminishes the ability of the individual to speak and write. The person understands spoken or written words, and there is nonfluent speech. Although the lesion may also result in a hemiplegia, there is not necessarily an impediment to sound production. Vocabulary is often reduced to expletives. This type of aphasia is called **expressive** or **Broca's aphasia.** Damage to Wernicke's area diminishes the comprehension of spoken or written language. However, the person has fluent speech, much of which is meaningless, with frequent paraphasias and neologisms. This type of aphasia is called **receptive** or **Wernicke's aphasia.**

SUMMARY

1. The electroencephalogram (EEG) can be recorded using scalp electrodes and is produced by ongoing activity in thalamocortical neural circuits. It represents the summed synaptic potentials of numerous cerebral cortical neurons.

2. The EEG has waves of different characteristic frequencies, which range from >13 Hz (beta waves), to 8 to 13 Hz (alpha waves), 4 to 7 Hz (theta waves) and <4 Hz (delta waves). Beta waves are observed in the aroused state and in rapid-eye-movement (REM) sleep, alpha waves in quiet wakefulness, and delta waves in non-REM sleep.

3. Events in cortical neurons can be triggered by stimulation of sensory pathways. When recorded in a way similar to that used for the EEG, these events are called evoked potentials.

4. Consciousness is a state that results from activity of the brain. It may be impaired or lost in disease states.

5. Sleep is an alteration of consciousness and occurs with a circadian rhythm. Sleep is subdivided into REM and non-REM forms. There are several stages of non-REM state. REM sleep occurs several times per night.

6. Attentional mechanisms allow the brain to orient a person to novel or threatening stimuli.

7. Learning and memory allow the modification of behavior based on experience. Memory includes short-term, recent, and long-term stages. Learning can be nonassociative or associative, depending on whether pairing of conditioned or unconditioned stimuli is involved. Long-term potentiation is an enduring synaptic change that may underlie learning and memory.

8. The two cerebral hemispheres differ in that one is dominant in some functions and the other in different functions. The left hemisphere determines handedness and also language. The right hemisphere may be dominant for music and for spatial relationships.

9. Damage to certain regions of the brain results in the loss of the ability to use language. Loss of Broca's area, in the inferior frontal gyrus, results in nonfluent or expressive aphasia, whereas loss of Wernicke's area, in the supramarginal, angular, and posterior-superior temporal gyri, results in fluent or receptive aphasia.

BIBLIOGRAPHY
Journal Articles

Birbaumer N, Elbert T, Canavan AGM, Rockstroh B: Slow potentials of the cerebral cortex and behavior, *Physiol Rev* 70:1, 1990.

Byrne JH: Cellular analysis of associative learning, *Physiol Rev* 67:329, 1987.

Damasio AR, Geschwind N: The neural basis of language, *Annu Rev Neurol* 7:127, 1984.

Posner MI, Petersen SE: The attention system of the human brain, *Annu Rev Neurosci,* 13:25, 1990.

Steriade M, Llinas RR: The functional states of the thalamus and the associated neuronal interplay, *Physiol Rev* 68:649, 1988.

Teyler TJ, DiScenna P: Long-term potentiation, *Annu Rev Neurosci* 10:131, 1987.

Thompson RF, Krupa DJ: Organization of memory traces in the mammalian brain, *Annu Rev Neurosci* 17:519, 1994.

Werker JF, Tees RC: The organization and reorganization of human speech, *Annu Rev Neurosci,* 15:547, 1993.

Zola-Morgan S, Squire LR: Neuroanatomy of memory, *Annu Rev Neurosci* 16:547, 1993.

Books and Monographs

Anderson P, Andersson SA: *Physiological basis of the alpha rhythm,* New York, 1968, Appleton-Century-Crofts.

Bergamini L, Bergamasco B: *Cortical evoked potentials in man,* Springfield, Ill, 1967, Charles C Thomas, Publisher.

Creutzfeldt OD et al: Electrophysiology of cortical nerve cells. In Purpura DP, Yahr MD, eds: *The thalamus,* New York, 1966, Columbia University Press.

Gazzaniga MS, LeDoux JE: *The integrated mind,* New York, 1978, Plenum Press.

Jones EG: *The thalamus,* New York, 1985, Plenum Press.

Kandel ER, Schwartz JH, Jessell TM: *Principles of neural science,* ed 3, New York, 1991, Elsevier Science Publishing Co., Inc.

Magound HW: *The waking brain,* ed 2, Springfield, Ill, 1963, Charles C Thomas, Publisher.

Penfield W, Jasper H: *Epilepsy and the functional anatomy of the human brain,* Boston, 1954, Little, Brown & Co.

Willis WD, Grossman RG: *Medical neurobiology,* ed 3, St Louis, 1981, Mosby.

MUSCLE

RICHARD A. MURPHY

PART III

CHAPTER 12

The Molecular Basis of Contraction

Movement, produced by the neuromuscular system, is perhaps the most striking difference between plants and animals. The basis for movement is a biologic energy transformation called **chemomechanical transduction.** In this process most of the body's metabolic production of adenosine triphosphate (ATP) is converted into force or movement by muscle cells. Evolution has led to specialization of muscle cells to minimize the ATP consumption required for specific functions. Nevertheless, the basic molecular process underlying contraction is the same in all muscle cells. The objective of this chapter is to show how muscle cells contract by the **sliding filament–crossbridge mechanism.**

THE CONTRACTILE UNIT

The basic structure involved in contraction consists of organized arrays of insoluble structural proteins. One set of proteins forms a **cytoskeleton** that serves as an anchor and force-transmitting structure for the contractile proteins organized in the **myofilaments.** The contractile unit in **striated muscle** cells is called a **sarcomere** (Figure 12-1; see electron micrograph in Figure 13-2). Enormous numbers of sarcomeres are linked together by the cytoskeleton. **Z-disks** mechanically link sarcomeres end to end. **Intermediate filaments** (polymers of the proteins desmin or vimentin) connect the Z-disks of adjacent **myofibrils** (see Figure 13-2) within a striated muscle cell. The transverse alignment of sarcomeres and their constituent myofilaments gives these cells their striated appearance.

Thin Filaments

Thin filaments are ubiquitous cell structures that always contain **actin** and **tropomyosin** (Figure 12-2). The thin filaments in vertebrate striated muscle are anchored

in the Z-disks, and a molecule of **troponin** is bound to each tropomyosin molecule. Troponin, a regulatory protein, contains Ca^{++}-binding sites that are involved in control of contraction and relaxation in vertebrate striated muscle.

Thick Filaments

Myosin is a large, complex molecule consisting of tail and head regions (Figure 12-2). The tails aggregate to form thick filaments, with the heads projecting out toward the thin filaments. Each head, termed a **crossbridge,** contains an actin-binding site and an enzymatic site that can hydrolyze adenosine triphosphate (ATP) to adenosine diphosphate (ADP) and inorganic phosphate (P_i). Both sites are involved in chemomechanical transduction. The interactions between the crossbridges and the thin filaments draw the thin filaments toward the sarcomere's center, and thereby shorten the sarcomere as the Z-disks come closer together (Figure 12-1).

THE CROSSBRIDGE CYCLE

Hydrolysis of ATP occurs when purified myosin and thin filaments are mixed in a solution that approximates the ionic content of the cytoplasm, often called the **myoplasm** in muscle cells (Figure 12-3, *A*). This cycle can be represented by four steps. ATP binds to myosin and is hydrolyzed to form the myosin-ADP-P_i complex. This complex, characterized by a high level of free energy, has a great affinity for actin and rapidly binds to the thin filament *(step 1)*. ADP and P_i are released after myosin attaches to the thin filament, and the myosin head undergoes a conformational change *(step 2)*. The resulting actin-myosin complex has a low level of free energy. In *step 3* the actin-myosin complex binds ATP. The resulting actin-myosin-ATP complex has a low binding affinity,

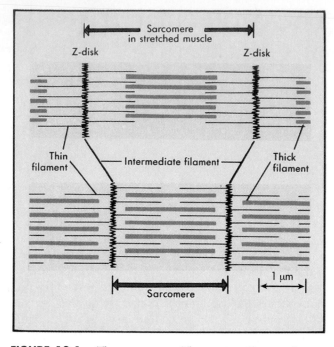

FIGURE 12-1 The sarcomere. The contractile proteins are found in interdigitating arrays of thick and thin filaments that slide past each other during contraction and relaxation. Thin filaments are attached to the Z-disks. Other cytoskeletal proteins stabilize the thick filaments.

so that the crossbridge dissociates from the thin filament. Internal hydrolysis of the bound ATP regenerates the high-energy myosin-ADP-P_i complex to complete the cycle *(step 4)*.

In this biochemical cycle the release of free energy occurring in the overall reaction, ATP to ADP plus P_i, is lost as heat. The conversion of part of this energy into mechanical work depends on the sarcomere structure of the muscle cells. The orientation of the myosin heads incorporated into a thick filament is constrained. The preferred orientation of the high-energy complexes (myosin-ADP-P_i and actin-myosin-ADP-P_i) is perpendicular (90 degrees) to the thick filament (Figure 12-3, *B*). However, the preferred (i.e., lowest level of free energy) conformation of the actin-myosin complex after release of ADP and P_i occurs when the crossbridge is oriented 45 degrees to the filaments. Thus part of the energy from ATP is translated into conformation changes in the crossbridges. This "bending" of the crossbridges generates forces that draw the thin filaments past the thick filaments and toward the center of the sarcomere. The force is transmitted by the cytoskeleton to the ends of the cell to exert a force on the skeleton. A single crossbridge cycle moves a thin filament on a molecular scale (about 10 nm or 1×10^{-8} m) and develops a minute force estimated to be between 5 and 10×10^{-12} newtons (about

0.0000000001 gram). Nevertheless, *millions of cross-bridges cycling asynchronously can generate great forces and shorten muscle cells considerably.*

Determinants of Crossbridge Cycling

The crossbridge cycle illustrated in Figure 12-3 will continue until all the ATP is consumed and the cycle is arrested (after step 2). This occurs only after death, when the ATP supplies are not replenished. This state is characterized by **rigor mortis,** or muscular rigidity, because the crossbridges are permanently attached.

Resting or relaxed muscle contains detached crossbridges in the myosin-ADP-P_i state and is freely extensible. These crossbridges are prevented from attaching to the thin filaments by Ca^{++}-dependent regulatory systems, which differ among muscle types (see Chapters 13 and 14). Such regulatory systems control the number of crossbridges interacting with the thin filaments.

Crossbridge cycling rates determine how fast a muscle shortens. The maximal shortening velocities occur when no load opposes filament sliding. A load on a muscle cell is transmitted by the cytoskeleton to the sarcomere and opposes bending of the crossbridges (Figure 12-3, *B*). Increasing the load slows crossbridge cycling. Velocities fall to zero when the load prevents the transition from the 90- to the 45-degree conformation. Different types of muscle cells vary in their maximal unloaded shortening velocities. Such functional differences are determined by the isoenzymatic variant of myosin expressed in a particular muscle cell.

Contractile protein isoform expression changes developmentally. Fetal, neonatal, and adult myosin isoforms constitute a normal progression. Pathological conditions can lead to changes in the expression of myosin genes, with altered muscle performance. For example, **poliomyelitis** is a viral disease that can cause the death of skeletal muscle nerves. The denervated muscle cells subsequently **atrophy,** or waste away, with a reversal in the developmental expression of contractile protein isoforms.

THE BIOLOGICAL RESPONSE: CHARACTERIZING CONTRACTION

Contracting muscle cells can perform several actions. They may develop a force without shortening, may shorten at various velocities, or may lengthen while opposing a larger force than the muscle can generate. The response depends on the loading. A simple mechanical

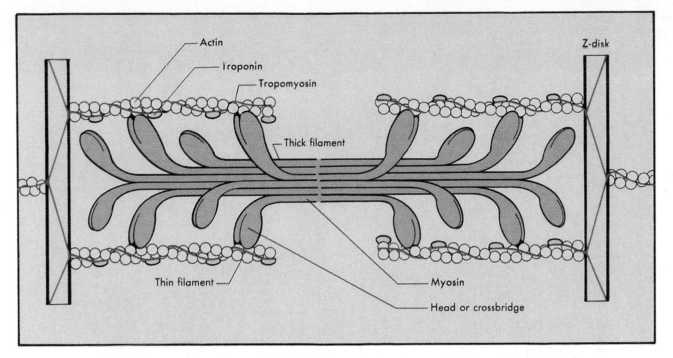

FIGURE 12-2 The core of the thin filament is a twisted, two-stranded chain of polymerized actin molecules. Each molecule of the long, rigid tropomyosin binds with six or seven actin monomers plus one troponin. Thick filaments are composed of 300 to 400 myosin molecules. The tails of the very large myosin molecules aggregate to form the filament. The heads protrude as crossbridges that can interact with the thin filament. Note the bipolar structure of the thick filament: the central zone, lacking crossbridges, divides the filament into two halves, where the crossbridges have opposite orientations.

analysis of contraction describes the output of the muscle cells and helps show how muscle functions.

Only three variables are needed to describe the output of a muscle:force, length, and time (Table 12-1). The analysis is simplified by experimentally holding one of the three variables constant and determining the relationship between the other two. This experimental constraint yields two types of contractions: **isometric** (*iso,* constant or equal; *metric,* length) and **isotonic** (at constant force or load).

Isometric Contractions: the Dependence of Force on Length

A muscle cell develops a characteristic force when maximally stimulated at a fixed length. The force-length relationship depicts this steady-state behavior (Figure 12-4). Stimulated skeletal muscle cells develop no force if they are first stretched to sarcomere lengths greater than 3.7 µm. At shorter lengths *force is proportional to the number of crossbridges that interact with the thin filament in each half-sarcomere.* Force generation is also lower at muscle lengths less than the optimal length (L_O); disturbances of the sarcomeric structure and failure of the activation processes are responsible.

The forces generated depend on the size of the muscle cell and the number of filaments. When normalized for size as force per cell cross-sectional area, vertebrate muscle cells generate about 3×10^5 newtons/m^2 (or about 3 kg/cm^2) at their L_O. These remarkable forces are attributed to high concentrations of crossbridges that individually generate very small forces. *Increases in strength associated with growth or exercise result from the synthesis of more thick and thin filaments and an increase in the cross-sectional area of muscle cells.*

Isotonic Contractions: the Dependence of Velocity on Load

A lever simplifies measurement of shortening of a muscle at a constant load (Figure 12-5, *A*). The relaxed muscle is adjusted to the length that is optimal for force development (see Figure 12-4), and different loads are attached to the lever before the muscle is stimulated. If the load is greater than the muscle can lift, the maximal force (F_O) is developed in an isometric contraction, as just discussed. If the load is somewhat smaller, force development occurs with no shortening until the force developed by the muscle is equal to the load. The muscle then begins to shorten isotonically (Figure 12-5, *B*). The

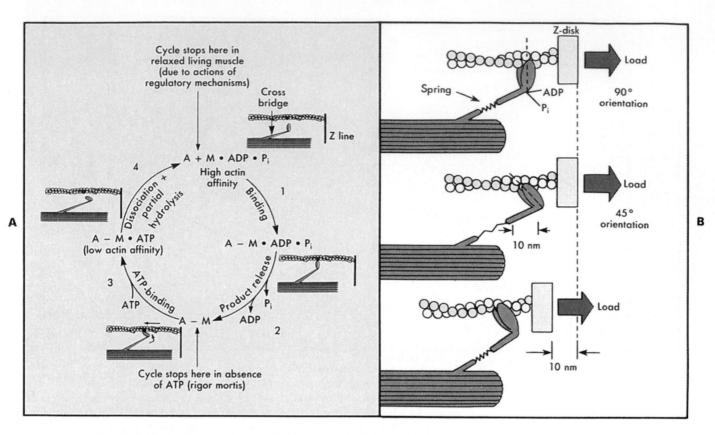

FIGURE 12-3 Steps in the crossbridge cycle. **A,** Relationship between steps *(1 to 4)* in the hydrolysis of adenosine triphosphate *(ATP)* and crossbridge conformations. **B,** Preferred or minimal free energy conformations of attached crossbridges. The transition from the 90- to the 45-degree conformation that occurs on release of adenosine diphosphate *(ADP)* plus inorganic phosphate *(P_i)* generates a force in the crossbridge represented by the stretched spring. This force can be translated into shortening by movement of the thin filament past the thick filament. *A,* Actin; *M,* myosin.

Table 12-1 Basic Mechanical Variables in Muscle Contraction		
Parameter (Symbol)	**Units**	**Definition**
Force (F)	Newton (N)	(=102 g weights)
Length (L)	Meter (m)	
Time (T)	Second (sec)	
DERIVED VARIABLES		
Velocity (V)	m/sec	Change in length/ change in time
Work (W)	N × m	Force times distance

slope of the length-versus-time record gives the shortening velocity. A third contraction with an even lighter load gives a higher shortening velocity. The complete dependence of velocity on load obtained from many contractions is shown in Figure 12-5, *C.*

The velocity-force relationship is the mechanical manifestation of the sum of all the crossbridge interactions in the cell. Shortening velocity is proportional to the average crossbridge cycling rate. Maximal velocities (V_O) occur with zero load. A load slows the average cycling rate as it opposes the transition from the 90- to the 45-degree conformation *(step 2* in Figure 12-3, *A).*

Resisting Imposed Loads

Crossbridge cycling produces force development and shortening. Nevertheless, the force of gravity acting on the body, contraction of opposing muscles, or other external forces can impose large loads that may lengthen contracting muscle cells. A contracting muscle cell can briefly resist imposed loads 60% greater than the force the cell can develop (Figure 12-5, *C*). High forces are required to break the bonds to the thin filaments. Stretched crossbridges are unable to undergo a conformational change to 45 degrees. After the link is broken, the cross-

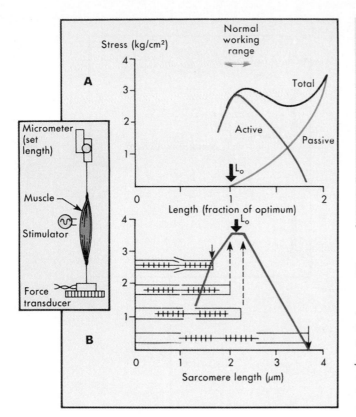

FIGURE 12-4 **A,** Relaxed muscles exhibit an elastic counterforce as the cell is lengthened, like a rubber band. The curve describing this elasticity is termed the passive force-length relationship. This is primarily caused by the presence of connective tissue fibrils of collagen and elastin surrounding the muscle and individual muscle cells, although intracellular cytoskeletal elements contribute. Contracting muscles develop greater forces; this curve is known as the total force-length relationship. The difference between the total force and the passive force curves is the active force-length relationship and represents the force-generating properties of the crossbridges. Inset shows experimental setup. **B,** A sophisticated analysis of single cells or sarcomeres reveals that the generated force depends on the overlap of thick and thin filaments. At an optimal length (L_O), the cell can develop the most force because all the crossbridges can interact with the thin filaments in each half of a sarcomere.

bridges reattach to resist further stretch. The crossbridge cycle, with release of ADP and P_i and binding of ATP, is never completed when the muscle is stretched (see Figure 12-3). The work is done on the muscle cells rather than by the muscle cells, and no ATP cost is incurred.

Injuries occur when external forces on the musculoskeletal system cause bones to break, tendons to rupture, or muscles to tear. Such injuries may occur when the muscles are relaxed if the musculoskeletal system is

forced beyond its normal range of movements. However, most injuries happen when muscles are maximally contracted, as may occur when the weight of the body is concentrated on one extended arm in a fall. The highest forces that occur in a muscle are the result of imposed external loads (Figure 12-5) that stretch contracting muscle cells and are transmitted by the tendons to joints and bones. Stretching contracting muscle cells is a normal occurrence and does not usually result in injury. Nevertheless, such situations occasionally push the musculoskeletal system beyond its limits.

Energy Cost of Contraction

Crossbridge cycling accounts for most of the body's ATP consumption. Functional specialization among muscle types minimizes this cost. One measure of energy cost is the **efficiency** of contraction. *Efficiency is defined as the ratio of the mechanical work performed to the chemical energy released by ATP hydrolysis.* The efficiency of contraction varies. Work equals force times distance shortened. Work is zero in an isometric contraction (distance shortened equals 0) and in an unloaded contraction (force equals 0). Under these conditions the efficiency is also zero. The greatest efficiency is approximately 45% conversion of chemical to mechanical energy by the crossbridges. The maximal efficiency occurs with a moderately loaded muscle.

At very low loads many crossbridges attach and cycle without exerting their maximal force on the rapidly sliding thin filaments (*record 3* in Figure 12-5). At high loads, a crossbridge is likely to exert its maximal effect in the cycle, but it has a high probability of reattaching at the same point on the very slowly moving thin filament (*record 1* in Figure 12-5). An optimal loading of about 30% of the maximum force a cell can generate gives the highest probability of capturing the maximal amount of energy from ATP hydrolysis as useful work.

The efficiency of chemomechanical transduction in the crossbridge cycle is quite high (comparable to modern gasoline engines). Nevertheless, more than half the potential free energy is dissipated as heat, even under optimal conditions. The resulting increase in temperature of muscle cells is a significant physiological burden during exercise. However, muscles can be used to generate heat by shivering to maintain normal body temperatures in cold environments. The crossbridge cycle is not the only energy-consuming reaction in muscle cells, although it is the only one that yields mechanical work. These other reactions lower the overall efficiency of muscular contraction to about 20% to 25%.

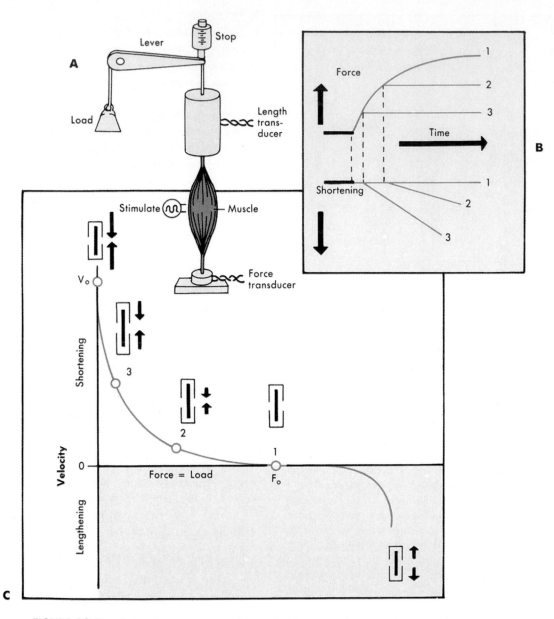

FIGURE 12-5   **A,** Device to measure shortening in a muscle at predetermined loads. Note the stop that supports the load until the muscle contracts. **B,** Transducers in the apparatus detecting force and length show the response to stimulation at very high *(1)*, moderate *(2)*, and low *(3)* loads. **C,** The velocity-force relationship for a shortening muscle is hyperbolic. The cell can shorten rapidly or can develop high forces, but not at the same time. Contracting muscles can withstand higher forces than they develop (loads > maximal force, F_O). High imposed loads are associated with a slow lengthening up to loads of approximately 1.6 F_O, where the contractile system yields and rapid lengthening occurs. The relative rates of movement at various loads are indicated by the lengths of the arrows beside the five sarcomere diagrams. V_O, Maximal shortening velocity at zero load.

No work is done in isometric contractions that must be characterized energetically in terms of their economy, defined as force maintained/ATP consumed. The more time a crossbridge spends in the attached, force-generating configuration (Figure 12-3), the greater is the economy of the contraction. High econo-

mies are associated with slow crossbridge cycling rates. Vertebrate muscle can be divided into two broad classes: One consists of **fast muscle cells,** characterized by high rates of doing work, high ATP consumptions, and high efficiencies (Chapter 13). These cells are typically attached to the skeleton and are striated. The other class contains

slow muscle cells, which are specialized to contract more or less continuously and which use little ATP (Chapter 14). These (smooth) muscle cells are typically part of the walls of hollow organs and exhibit a remarkable economy in maintaining force.

SUMMARY

1. The elementary contractile unit has an array of thick filaments containing myosin. The thick filaments interdigitate with thin filaments that consist of actin, tropomyosin, and other proteins. The filaments are attached to cytoskeletal elements. In striated muscles this contractile unit is called the sarcomere.
2. Contraction is caused by molecular motors called crossbridges that consist of the heads of myosin molecules projecting from thick filaments. The crossbridges cyclically interact with actin in the thin filaments.
3. Each crossbridge cycle converts part of the free energy associated with the hydrolysis of an ATP molecule into a conformational change in the crossbridge. This change imposes a tiny force (5 to 10×10^{-12} newtons) that pulls the thin filaments toward the center of the sarcomere in a tiny step (about 10 nm).
4. Muscles are organized to link enormous numbers of sarcomeres so that the summed contributions of individual crossbridges generate very high forces and/or large movements.
5. Maximum force generation is proportional to the cross-sectional area of the cells in a muscle.
6. Shortening velocities are determined by crossbridge cycling rates. These rates depend on the particular isoform of myosin expressed in a cell and on the load on the muscle: lower loads allow higher shortening velocities.
7. Contracting muscles sometimes lengthen when they are opposing greater external forces to decelerate the body or stabilize complex motions. Crossbridges can bear loads that are about 1.6 times greater than the maximum force the crossbridges can generate.
8. Crossbridges are the primary consumers of the total body ATP production, and ATP hydrolysis is proportional to crossbridge cycling rates. Muscle cells are specialized for working efficiently during rapid shortening or for maintaining sustained forces economically.

BIBLIOGRAPHY
Journal Articles

Bandman E: Contractile protein isoforms in muscle development, *Dev Biol* 154:273, 1992.

Block BA: Thermogenesis in muscle, *Annu Rev Physiol* 56:535, 1994.

Burton K: Myosin step size: estimates from motility assays and shortening muscle, *J Muscle Res Cell Motility* 13:590, 1992.

Eisenberg E, Hill TL: Muscle contraction and free energy transduction in biological systems, *Science* 227:999, 1985.

Finer JT et al: Single myosin molecule mechanics: piconewton forces and nanometre steps, *Nature* 368:113, 1994.

Gordon AM et al: The variation in isometric tension with sarcomere length in vertebrate muscle fibres, *J Physiol* 184:170, 1966.

Huxley AF: Muscular contraction, *Annu Rev Physiol* 50:1, 1988.

Ishijima A et al: Single-molecule analysis of the actomyosin motor using nano-manipulation, *Biochem Biophys Res Commun* 199:1057, 1994.

Josephson RK: Contraction dynamics and power output of skeletal muscle, *Annu Rev Physiol* 55:527, 1993.

Obinata T: Contractile proteins and myofibrillogenesis, *Int Rev Cytol* 143:153, 1993.

Rayment I et al: Three-dimensional structure of myosin subfragment-1: a molecular motor, *Science* 261:50, 1993.

Rayment I et al: Structure of the actin-myosin complex and its implications for muscle contraction, *Science* 261:58, 1993.

Schoenberg M: Equilibrium muscle crossbridge behavior: the interaction of myosin crossbridges with actin, *Adv Biophys* 29:55-73, 1993.

Books and Monographs

Amos LA, Amos WB: *Molecules of the cytoskeleton,* New York, 1991, Guilford Press.

Kreis T, Vale R, eds: *Guidebook to the cytoskeletal and motor proteins,* Oxford, 1993, Oxford University Press.

Paul RJ et al, eds: *Muscle energetics,* New York, 1989, Alan R. Loss, Inc.

Peachey LD et al, eds: *Handbook of physiology,* section 10: Skeletal muscle, Bethesda, Md, 1983, American Physiological Society.

Pollack GH: *Muscles and molecules,* Seattle, Wash, 1990, Ebner & Sons.

Squire JM: *Molecular mechanisms in muscular contraction,* Boca Raton, Fla, 1990, CRC Press, Inc.

Sugi H, Pollack GH: *Mechanism of myofilament sliding in muscle contraction,* New York, 1993, Plenum Press.

Woledge RC: Energy transformations in living muscle. In Wieser W, Gnaiger E. *Energy transformations in cells and organisms,* Stuttgart, 1990, Thieme.

CHAPTER 13

Muscles Acting on the Skeleton

Skeletal muscle cells are sometimes called **striated muscle** because of their appearance. They are also called **voluntary muscle,** reflecting their neural connections to the motor cortex. A muscle cell acting on a skeleton to produce movement has a specific role that dictates many of its properties. Starting with the structure of skeletal muscle, these properties are examined in this chapter. The ways in which crossbridge cycling is controlled to produce contraction and relaxation are then explored. Finally, specializations that allow functional diversity and adaptation are considered.

MUSCULOSKELETAL RELATIONSHIPS

The skeleton serves as a supporting lever system on which most muscle cells act (Figure 13-1). Exceptions include skeletal muscle in the lips and esophagus, where the muscles participate in the voluntary acts of speaking and swallowing. Characteristically, striated muscle cells in the limbs bridge two joints before they attach to the skeleton via **tendons** or other mechanical connections.

The relationship between the muscle cells and the skeleton dictates several important characteristics of skeletal muscle:

1. Because cells are not connected to each other, they act independently in response to a nerve impulse. The force of contraction can be increased by recruiting more cells.
2. Skeletal muscle cells are usually relaxed, and the skeleton bears most gravitational loads.
3. Skeletal muscle cells typically act on the short end of the skeletal lever system (Figure 13-1). Thus they must develop forces that are much greater than the load moved. However, large movements can result from limited cell shortening.
4. Most contractions produce movement and do mechanical *work* (= force × distance). The rate of

doing work (*power* = work/time) can be great. Skeletal muscles are characterized by a high *efficiency* (work done/ATP consumed).

Skeletal Muscle Cells

Muscles, such as the biceps, consist of bundles of muscle cells that are linked together and that are separated from other muscles by connective tissues. The contraction of some or all of the cells in a muscle yields complex movements. Mammals typically have more than 400 different muscles, with several attached to one bone. (An elephant's trunk is a notable exception, with thousands of muscles and no bones.) Discrete movements are the result of coordinated contractions involving many muscles. Muscles may act together as **synergists** to produce the same movement, or they may function as **antagonists** to other muscles in decelerating a motion. The summed actions of several muscles stabilize joints and produce precisely controlled movements. **Flexors** and **extensors** (antagonist muscles that act on the limb joints) are normally both involved in a movement. A surprising variation exists among individuals in the muscles they use to accomplish a specific motion. These patterns reflect differences in neuromuscular learning or training and are a factor in coordination and athletic performance.

Embryonic muscle cells fuse end to end to form the enormous multinucleated, differentiated skeletal muscle cells. Although their diameter is that of a fine thread (50 to 100 µm), these multinucleated cells may be many centimeters long. The contractile units, or sarcomeres (Chapter 12), of skeletal muscle are linked in series along a **myofibril** (Figure 13-2). The cytoskeleton links the Z-disks of the myofibrils so that the sarcomeres are aligned. The resulting alternating dark and light stripes correspond to regions that contain thick filaments separated by regions that contain only thin filaments. These

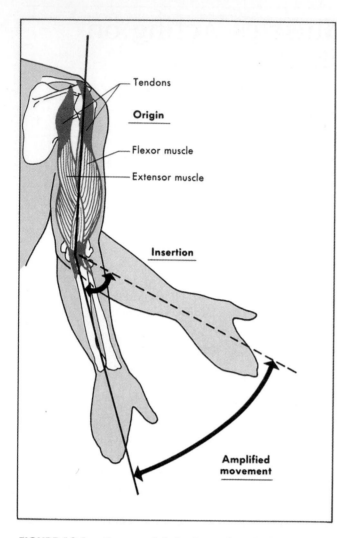

FIGURE 13-1 Groups of skeletal muscle cells form discrete muscles, which span anchor sites on the skeletal lever system. Note that the muscles act on the short end of the skeletal lever system (see text for consequences). Tendons connecting muscle cells to the skeleton contain inextensible collagen fibrils.

stripes are clearly visible when viewed through the light microscope. Looking at the changing scales in Figure 13-2, one can envision the enormous numbers of sarcomeres present in a cell. For example, a 10-cm-long cell would have more than 4500 sarcomeres.

Membranes

Four structurally and functionally distinct membranes in the cell are involved in regulation of contraction:

1. The **neuromuscular junction,** or **motor endplate,** is a specialized region of the plasma membrane (Figures 13-2 and 13-3).
2. The plasma membrane, or **sarcolemma,** along which action potentials are propagated, is the sec-

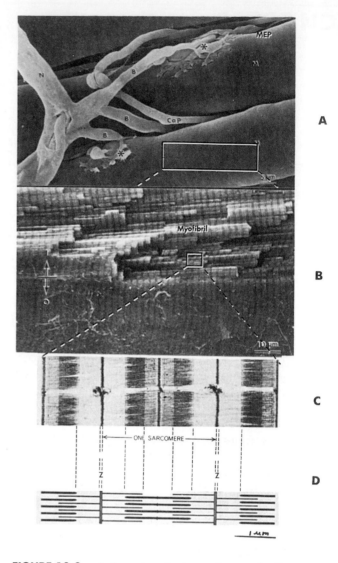

FIGURE 13-2 **A,** Scanning electron micrograph of segments of three mammalian skeletal muscle cells *(M)* showing striations. Each cell receives a branch *(B)* of a motor nerve *(N)* in a complex structure termed the motor endplate or neuromuscular junction *(asterisks)*. Note close association of the giant cells with capillaries *(cap)*. **B,** Scanning electron micrograph of skeletal muscle cell that was broken open *(I,* interior; *O,* extracellular space) to show interior packed with large numbers of striated myofibrils. **C,** Higher-magnification transmission electron micrograph of thin section through two myofibrils. Individual filaments are difficult to discern at this magnification. **D,** Shows the structure of the sarcomere arising from thin filaments attached to Z-disks interdigitating with a central lattice of thick filaments. (**A** from Desaki J, Uehara Y: *J Neurocytol* 10:107, 1981. **B** from Swada H, Ishikawa H, Yamada E: *Tissue Cell* 10:183, 1978. **C** from Huxley HE: *Sci Am* 213:18, 1965.)

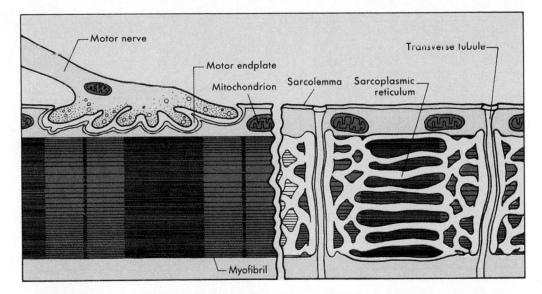

FIGURE 13-3   The membranes of skeletal muscle. The plasma membrane separating the extracellular space from the intracellular space (myoplasm) has three specialized portions. These are the motor endplate, the sarcolemma, and the transverse tubules (T-tubules) that form a network at the ends of each sarcomere in mammals and are contiguous with the extracellular space. The sarcoplasmic reticulum is a distinct membrane system enclosing a separate intracellular compartment surrounding each myofibril and is intimately associated with the T-tubular system.

ond membrane. Neuromuscular transmission (Figure 4-2) and action potential generation are described in Chapter 4.

3. Tiny openings in the sarcolemma lead into the **transverse-tubular (T-tubular) network** located at the Z-disk in mammals. This network defines an extracellular space within the cell (Figure 13-3). The extensive T-tubular network virtually encircles each myofibril. *Depolarization spreads into the cell via this system when the action potential travels down the sarcolemma. By this means, excitation spreads to the level of the myofibrils.*

4. The **sarcoplasmic reticulum,** a distinct membrane system, connects intimately with the T-tubules (Figure 13-3). The sarcoplasmic reticulum surrounds an intracellular compartment that forms a sleeve around each myofibril.

Nerves

The basic neuromuscular relationships are described in Chapter 9. During early development, a motor nerve axon originating from the spinal cord makes contact with a muscle cell, and an endplate develops. Connections with other growing axons are inhibited, and thus each muscle cell has one **neuromuscular junction.** Individual axons branch in the muscle, and therefore

each nerve controls many muscle cells. The resulting functional grouping of a nerve and its associated muscle cells is called a **motor unit.** Motor units may contain only a few muscle cells or up to many thousands of cells.

Muscle weakness or hyperactivity **(spasms)** is usually symptomatic of problems in the central or peripheral neural motor systems. The effect is to block or enhance excitatory or inhibitory neurotransmitter release at synapses and ultimately to determine acetylcholine release at the neuromuscular junction. For example, food poisoning resulting from the bacteria-produced **botulism toxin** leads to weakness and paralysis associated in part with reduced acetylcholine release. Another bacterium produces **tetanus,** an infectious disease characterized by muscular spasms. Although the tetanus toxin can experimentally block neuromuscular transmission, its main effect is to block transmitter release at inhibitory synapses in the central motor pathways. This action evokes pathological trains of action potentials in the motor axons.

An action potential is elicited in a motor nerve when the sum of the excitatory and inhibitory synaptic inputs to the cell body produces a critical depolariza-

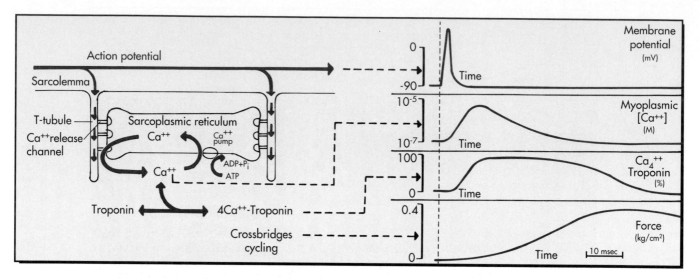

FIGURE 13-4   Excitation-contraction coupling in skeletal muscle. The action potential depolarizing the sarcolemma spreads into the interior of the cells via T-tubular membranes. The T-tubules are anatomically coupled to Ca^{++}-release channels in the sarcoplasmic reticulum that open momentarily with the fall in voltage. Ca^{++} is the intracellular messenger coupling electrical events at the cell membranes to activation of the contractile apparatus. Ca^{++} rapidly diffuses down its concentration gradient from the sarcoplasmic reticulum store to the myoplasm, where it binds to sites on troponin, a thin filament protein. Ca^{++} binding to troponin induces a conformational change in the thin filament that allows crossbridge attachment and cycling with force development and shortening. Increases in the myoplasmic Ca^{++} also activate Ca^{++} pumps in the sarcoplasmic reticulum membrane, and Ca^{++} is returned to the sarcoplasmic reticulum. The fall in myoplasmic Ca^{++} causes bound Ca^{++} to dissociate from troponin, and the thin filaments return to the "off" conformation with cessation of crossbridge attachment and relaxation.

tion (see Chapters 3 and 9). That action potential releases sufficient acetylcholine to produce an endplate potential and generate an action potential in all the muscle cells in the motor unit, and synchronous contractions result. *Motor units rather than cells are the basic functional contractile elements that can be recruited individually.*

Myasthenia literally means muscle weakness. **Myasthenia gravis** is a serious, progressive disease characterized by extreme muscle weakness. An autoimmune mechanism is typically involved. Circulating antibodies to the acetylcholine receptors in the endplate membrane markedly reduce receptor numbers, and neuromuscular transmission fails. The symptoms of weakness can be alleviated by anticholinesterase drugs, such as neostigmine. Acetylcholine released at the endplate persists after treatment, and its concentration rises to a level that restores neuromuscular transmission.

REGULATION OF CONTRACTION AND RELAXATION

The process that links the action potential to crossbridge cycling and contraction is called **excitation-contraction coupling** (Figure 13-4). The events involved are (1) signal transduction at the cell membrane and (2) generation of a second messenger that (3) acts on myofibrillar regulatory mechanisms that control crossbridge cycling. Excitation-contraction coupling in skeletal muscle is comparatively simple, in the sense that only one event is critical in each of the three steps.

Regulation of Ca^{++}

Calcium is the second messenger that couples signals at the cell membrane to crossbridge cycling in all muscles. Skeletal muscle cells are too large and contract too rapidly for Ca^{++} to diffuse through channels in the sarcolemma from the extracellular space to the myofibrils. The cellular compartment enclosed by the sarco-

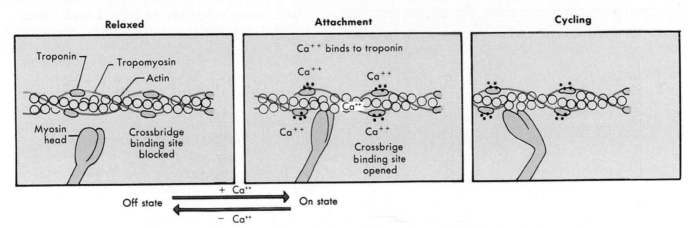

Relaxed

Troponin — Tropomyosin

— Actin

Myosin head — Crossbridge binding site blocked

Attachment

Ca^{++} binds to troponin

Ca^{++} Ca^{++}

Ca^{++}

Ca^{++} Ca^{++}

Crossbrige binding site opened

Cycling

Off state $\xrightarrow{+ \text{ Ca}^{++}}$ On state $\xleftarrow{- \text{ Ca}^{++}}$

FIGURE 13-5   Ca^{++} can switch crossbridge cycling on by reversible binding to the regulatory protein troponin. Troponin has 4 Ca^{++}-binding sites and changes its shape when they are occupied. Binding also induces a conformational change in the associated tropomyosin and actin constituting the thin filaments. By this mechanism Ca^{++} acts as a molecular switch initiating contraction.

plasmic reticulum (see Figure 13-3) contains the Ca^{++} pool involved in activation. Signal transduction in skeletal muscle cells is the process by which an action potential triggers Ca^{++} release from the sarcoplasmic reticulum. Depolarization of T-tubules opens anatomically coupled channels in the sarcoplasmic reticulum. Channel opening allows Ca^{++} ions to diffuse down their electrochemical gradient into the myoplasm (see Figure 13-4). This process is rapid (1 to 2 msec), because the Ca^{++} concentration gradient is huge (about 10^5) and the distances are short (<1 µm).

In addition to the voltage-sensitive Ca^{++} channels, the sarcoplasmic reticular membrane contains large amounts of a protein complex that pumps Ca^{++} from the myoplasm back into the sarcoplasmic reticulum. This active transport depends on the hydrolysis of adenosine triphosphate (see Figure 13-4).

Ca^{++} pumping and crossbridge cycling account for the hydrolysis of a large fraction of an estimated 40 kg of ATP that is hydrolyzed by a 68 kg man during a restful day (in vivo, a small pool of ATP is continuously resynthesized from ADP). This generates considerable heat that normally must be dissipated by sweating and other mechanisms. However, muscle plays a key role in thermoregulation. Exposure to cold causes shivering, which produces heat but no useful work. A genetic disorder, termed **malignant hyperthermia,** is characterized by a rapid and frequently lethal rise in body temperature. Malignant hyperthermia is often triggered by volatile anesthetics

during surgery in humans. The disease is associated with a mutation in the gene for the Ca^{++}-release channel in the sarcoplasmic reticulum. The heat reflects pathological increases in the ATP-dependent Ca^{++} pumping necessitated by excess Ca^{++} release caused by stress or anesthesia. The Ca^{++} pump may normally play a role in nonshivering thermogenesis.

Crossbridge Regulation

Two basic crossbridge states are possible in skeletal muscle: free and attached. A Ca^{++} switch effectively allows the transition from an **off state,** in which crossbridges cannot attach in a relaxed muscle, to an **on state,** in which attachment and cycling are possible (Figure 13-5).

Troponin, a regulatory protein bound to tropomyosin in the thin filament, has four high-affinity Ca^{++}-binding sites. These sites are filled very rapidly when Ca^{++} is released from the sarcoplasmic reticulum (see Figure 13-4). The result is that all the thin filaments are quickly turned on (see Figure 13-5). The crossbridges can then attach and cycle until the transport pump lowers the Ca^{++} concentration so that Ca^{++} dissociates from troponin (see Figure 13-4). The thin filaments return to the off conformation, and the cell relaxes. *Activation in skeletal muscle is an all-or-none process.* Action potentials give uniform Ca^{++} transients. This switches the thin filaments on for a brief period and leads to a consistent mechanical response called a **twitch** (Figure 13-6).

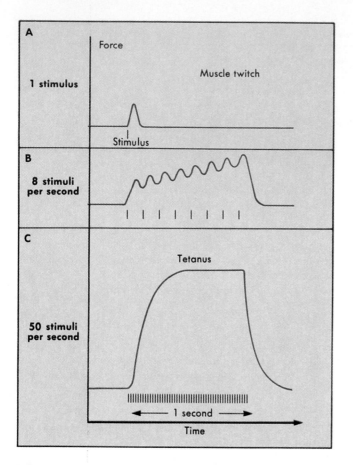

FIGURE 13-6 The force of contraction of a motor unit can be increased by more frequent recruitment so that twitches (**A**) sum in an incomplete (**B**) or complete (**C**) tetanus.

Grading Contractile Force

Skeletal muscles must generate different forces, sometimes for considerable periods. Two mechanisms control the amount of force generated by a muscle. First, because muscles contain large numbers of motor units, force can be varied over a wide range by recruitment of more motor units. The second way to increase force and prolong a contraction is to increase the frequency of firing of the motor nerves (Figure 13-6). Even though all the crossbridges are cycling during a twitch, the maximal force is not attained before the Ca^{++} levels fall and the contractile apparatus is turned off (see Figure 13-4). There is simply not enough time during the Ca^{++} transient produced by one action potential for sufficient crossbridge cycles to generate the full force. Firing the motor nerves at higher frequencies elicits further Ca^{++} transients. This allows the mechanical responses to sum and produce a greater, prolonged contraction, termed a **tetanus.** (Note that this is not the same as the disease termed tetanus.) The maximal tetanic force may be up to eight-fold greater than the twitch force.

Several factors contribute to finely graded contractions. The motor units recruited for weak contractions are the smallest, giving small increments in force. *The presence of large numbers of motor units and of tetanization in a muscle allow continuous gradation in force generation over a wide range.*

FUNCTION DIVERSITY IN SKELETAL MUSCLE

The transition from rest to contraction in skeletal muscle triggers an extraordinary jump in ATP consumption. This must be matched instantaneously by an increase in ATP resynthesis. Like all cells, muscle has three pathways to regenerate ATP (Figure 13-7).

Metabolism

Direct phosphorylation of adenosine diphosphate (ADP) from creatine phosphate by creatine phosphoryltransferase is not a net synthetic reaction. Creatine phosphate serves as a large storage pool of almost instantaneously available high-energy phosphate. Direct phosphorylation buffers the cellular ATP levels at the onset of contraction while synthetic pathways become active. The cell has only two ways to synthesize ATP, and these have very different characteristics (Figure 13-7).

Glycolysis can supply ATP at very high rates, even though the yield per mole of glucose is low. However, this pathway fails when the cellular glycogen stores are depleted (seconds to minutes in muscles).

Oxidative phosphorylation in the mitochondria uses oxygen and substrates that diffuse into the cell from the capillaries to generate ATP continuously and very efficiently. The disadvantage is that this pathway is much slower than glycolysis. Oxidative phosphorylation cannot meet the demands of very rapid crossbridge cycling rates.

Fiber Types

Skeletal muscle cells are specialized into two main types in humans and other primates. This specialization allows either high work rates (power output) or long-duration contractions. The two cell classes are differentiated on the basis of whether the gene for a slow or a fast myosin isoenzyme (i.e., having a moderate or a high ATPase activity or cycling rate) is expressed in the cell (Table 13-1). **Slow fibers,** characterized by moderate shortening velocities and power outputs, consume ATP at moderate rates. Slow fibers have a high blood supply (high capillary density), many mitochondria, and a moderate diameter. These characteristics minimize diffusion distances for oxygen and substrates. Slow fibers are

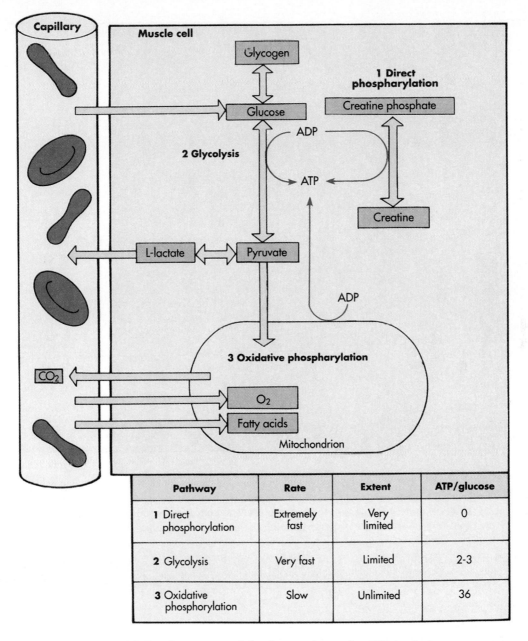

Pathway	Rate	Extent	ATP/glucose
1 Direct phosphorylation	Extremely fast	Very limited	0
2 Glycolysis	Very fast	Limited	2-3
3 Oxidative phosphorylation	Slow	Unlimited	36

FIGURE 13-7   Comparison of the three pathways for ATP production.

sometimes called **red fibers** because of the distinctive coloration provided by the iron-containing hemoglobin (blood supply), myoglobin in the myoplasm, and cytochromes in the mitochondria. If the blood supply is adequate, slow fibers provide great endurance.

The maximal ATP consumption rate of **fast fibers** can be met only by glycolysis. These large, pale **(white)** cells have a more extensive sarcoplasmic reticulum so that fast contractions are matched by rapid relaxations. These cells are pale because they contain few oxygen-binding proteins. Fast fibers fatigue rapidly as glycogen is depleted.

The motor unit rather than the cell is the functional grouping. Fast and slow motor units differ by more than their exclusive composition of either slow or fast fibers (Figure 13-8). In general, slow motor units generate low forces because of the smaller average fiber diameter and the comparatively few cells. The motor nerve determines the physiological characteristics of a motor unit (Chapter 9). Nerves with small cell bodies and narrow axons can synthesize limited amounts of acetylcholine, and they form small motor units. Small motor axons are readily excitable; relatively few excitatory postsynaptic potentials at the small cell body are needed to depolarize

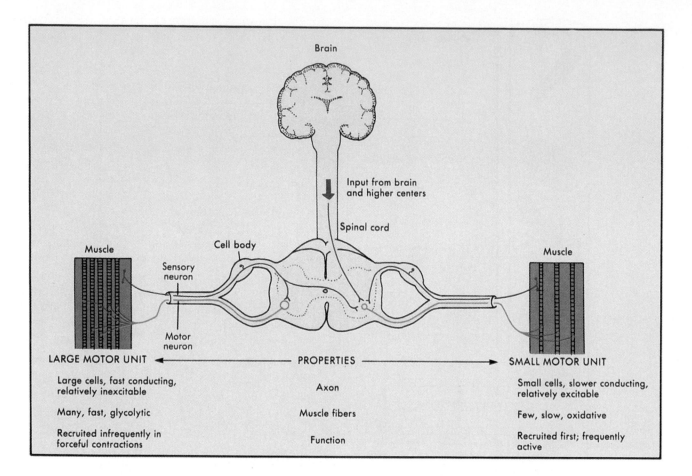

Brain

Input from brain
and higher centers

Spinal cord

Cell body

Muscle

Sensory
neuron

Motor
neuron

Muscle

LARGE MOTOR UNIT ◄—————— PROPERTIES ——————► SMALL MOTOR UNIT

Large cells, fast conducting, relatively inexcitable	Axon	Small cells, slower conducting, relatively excitable
Many, fast, glycolytic	Muscle fibers	Few, slow, oxidative
Recruited infrequently in forceful contractions	Function	Recruited first; frequently active

FIGURE 13-8   The characteristics of fast and slow motor units and some of the synaptic connections to the motor axons. Most muscles contain a mix of many slow (red) motor units with smaller numbers of fast (white) motor units. The cells of a motor unit are not anatomically discrete groups, but are dispersed among the cells of other motor units.

Table 13-1   Fiber Types in Primate Skeletal Muscle

	Slow, Oxidative (Red)	Fast, Glycolytic (White)
Myosin isoenzyme (ATPase rate)	Moderate	Fast
Sarcoplasmic reticular Ca^{++} pumping rate	Moderate	Fast
ATP consumption rate	Moderate	Extremely high
Diameter (diffusion distance)	Moderate	Large
Oxidative capacity: mitochondrial content, capillary density	High	Low
Glycolytic capacity	Moderate	High

the cell to the critical potential to fire an action potential. The large axons of fast motor units are less excitable.

Fatigue

The largest motor units may contain more than a thousand cells and develop hundreds of grams of force in humans. This means that most muscles can carry out vari-

ous types of activities. For instance, moderate workloads can be sustained with little fatigue. *The initial motor units recruited are the more excitable slow units, which are fatigue resistant. The fast motor units are also recruited for rapid, forceful contractions with a high power output. These maximal efforts cannot be sustained, and the fast units fatigue rapidly* (Figure 13-9).

Surprisingly, little is known about the causes of the

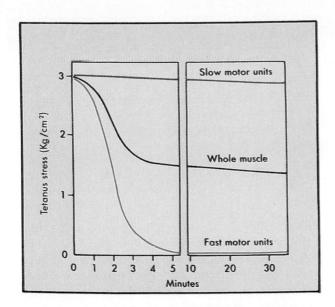

FIGURE 13-9   Fatigue in a muscle that is repeatedly tetanized for part of each second by experimentally stimulating the nerve trunk. Fatigue of the fast motor units occurs with glycogen depletion.

general physical fatigue one experiences after working vigorously. The state of disturbed homeostasis perceived as the discomfort of fatigue occurs before cells cease to contract. In **cellular fatigue,** motor nerve stimulation does not generate normal forces (Figure 13-9). However, individuals stop using their motor units before the cellular ATP concentration falls. Metabolic changes, such as an increased blood lactate concentration and a fall in pH, may contribute to the perception of fatigue, but they do not fully explain the phenomenon. Cellular creatine phosphate and glycogen contents decrease during activity, even in slow motor units. An elevated oxidative metabolism resynthesizes these stores within a brief period after cessation of exercise. Fatigue persists long after these cellular recovery processes are complete.

GROWTH AND ADAPTATION

Muscle typically constitutes 45% to 50% of body mass and can increase ATP consumption up to 100-fold on activation. Adequate matching of nutrients and metabolism to consumption dictates the properties of several major organ systems. Animals may be sprinters, with mostly fast, glycolytic motor units (the cat family). Others are adapted for comparatively slow but continuous running; they have many slow, oxidative muscle fibers and heavy-duty cardiovascular and respiratory systems (canines). Humans are intermediate and exhibit considerable genetic variability. Regular sustained exercise will enhance

oxygen transport capacity (maximum oxygen uptake and cardiac output) and will increase the vascularity of active muscles (maximum blood flow). (See also Chapter 26.) Only those muscle fibers subjected to increased recruitment will form more mitochondria. These changes are proportional to the training effort up to some limit and are rapidly reversed if training ceases. Improvements in time for an endurance event are attributable to delayed onset of fatigue, improved technique, increased tolerance to pain, and motivational factors. Cross-bridge cycling rates (myosin isoforms) are unchanged. Increases in strength attributed to muscle cell hypertrophy are modest in such training programs.

Differentiation of skeletal muscle cells depends on their pattern of contractile activity. Motor units that are frequently active express the slow myosin isoenzyme and develop a high oxidative capacity. Motor units that contract infrequently differentiate into the fast fiber phenotype. Thus differentiation depends on innervation.

> Skeletal muscle will atrophy if not used. Cell diameter and the number of myofibrils decrease. Such changes can be initiated after only 2 days of bed rest. If the motor nerve is destroyed by injury or disease, the denervated muscle cells will first atrophy and most will degenerate within a few months. This occurs in severe cases of **poliomyelitis** in which the viral infection in the central nervous system damages the motor systems. These processes are reversible if reinnervation occurs. The fiber type will change if a cell in a formerly fast motor unit is reinnervated by a small motor nerve, and vice versa.

Activity has large and diverse effects on muscle. These changes depend on the type of activity (Table 13-2). The responses are adaptive, and include neuromuscular learning, increased endurance, and increased strength. The learned ability to carry out complex movements, such as riding a bicycle, persists for years even without practice. However, regular exercise is required to maintain changes in the muscle cells. The increased physical well-being that results from moderate endurance exercise is mainly caused by enhanced capacity of the respiratory and cardiovascular systems rather than by the effects on the skeletal muscle fibers. Strength training also has broader effects, including growth in bones and tendons to bear the greater forces.

Although exercise can have profound effects on the involved motor units, weight lifting will not convert slow units into fast units or induce the formation of new muscle cells. The transformation of slow fibers to fast fibers and the reverse can occur, as shown by cross-

Table 13-2 Effects of Exercise

Type of Training	Example	Major Adaptive Response
Learning/coordination	Typing	Increased rate and accuracy of motor skills (central nervous system)
Endurance (submaximal, sustained efforts)	Marathon running	Increased oxidative capacity in all involved motor units, with limited cellular hypertrophy
Strength (brief, maximal efforts)	Weight lifting	Hypertrophy and enhanced glycolytic capacity of motor units employed

innervation experiments. However, no exercise regimen appears to change activity patterns sufficiently to alter the expression of the myosin isoenzymes.

> **Testosterone** promotes hypertrophy of skeletal muscle cells by inducing the formation of more myofibrils. This hormone contributes to the greater muscularity and strength of men than of women. **Anabolic steroids,** compounds structurally similar to testosterone, are illegally taken by many young men to increase muscle bulk and to improve performance in athletic events that require strength. However, these hormones have diverse actions (Chapter 49) and cause various deleterious side effects at the doses required to amplify the effects of strength training alone. The risks are increased by the variety and uncertain purity of illicit drugs.

SUMMARY

1. Skeletal (voluntary) muscle cells are controlled by the motor system, and most act on the skeleton to generate force and produce movement.
2. Four specialized membranes are involved in regulation of contraction:
   Motor endplate: a specialized postsynaptic membrane where acetylcholine receptors open nonspecific ion channels and induce depolarization
   Sarcolemma: the excitable plasma membrane that propagates action potentials away from the endplate region
   Transverse tubules: tiny tubules that transmit depolarization from the sarcolemma to the sarcoplasmic reticulum
   Sarcoplasmic reticulum: an intracellular Ca^{++}-storage compartment that surrounds each sarcomere of the myofibrils
3. Excitation-contraction coupling includes the fol-

lowing steps: (a) the acetylcholine-induced action potential; (b) voltage-dependent opening of Ca^{++} channels in the sarcoplasmic reticulum; (c) Ca^{++} diffusion and binding to troponin; and (d) a conformation change of the thin filament to allow crossbridge attachment and cycling.

4. In a motor unit, one nerve impulse induces a twitch in all the muscle fibers that receive branches from the nerve. Contractile force is graded by recruitment of more motor units and by tetanization with trains of nerve impulses.
5. Slow, oxidative motor units are involved in all muscular activity and depend on ATP production by oxidative phosphorylation of fatty acids delivered by the circulation. They resist fatigue because the slow myosin isoform limits crossbridge cycling and ATP consumption to rates that can be met by oxidative phosphorylation.
6. Fast, glycolytic motor units are also recruited in efforts to increase force and power output for brief periods.
7. Fatigue involves systemic changes that lead to the cessation of muscular activity.
8. The phenotype of a muscle cell depends on its activity patterns. Cells atrophy if not recruited and will degenerate after denervation. The cells develop into fast, glycolytic motor units if they are infrequently recruited, and form slow oxidative units if they are often activated.

BIBLIOGRAPHY
Journal Articles

Ashley CC, Mulligan IP, Lea TJ: Ca^{2+} and activation mechanisms in skeletal muscle, *Q Rev Biophys* 24:1, 1991.

Block BA: Thermogenesis in muscle, *Annu Rev Physiol* 56:535, 1994.

Booth FW, Thomason DB: Molecular and cellular adaptation of muscle in response to exercise: perspectives of various models, *Physiol Rev* 71:541, 1991.

Buckingham M: Making muscle in mammals, *Trends Genet* 8:144, 1992.

Fitts RH: Cellular mechanisms of muscle fatigue, *Physiol Rev* 74:49, 1994.

Florini JR, Magri KA. Effects of growth factors on myogenic differentiation, *Am J Physiol* 256 (*Cell Physiol* 25):C701, 1989.

Franzini-Armstrong C, Jorgensen AO: Structure and development of E-C coupling units in skeletal muscle, *Annu Rev Physiol* 56:509, 1994.

Gunning P, Hardeman E: Multiple mechanisms regulate muscle fiber diversity, *FASEB J* 5:3064, 1991.

Josephson RK: Contraction dynamics and power output of skeletal muscle, *Annu Rev Physiol* 55:527, 1993.

Ríos E, Pizarro G: Voltage sensor of excitation-contraction coupling in skeletal muscle, *Physiol Rev* 71:849, 1991.

Schneider MF: Control of calcium release in functioning skeletal muscle fibers, *Annu Rev Physiol* 56:463, 1994.

Westerblad H, Lee JA, Lännergren J, Allen DG: Cellular mechanisms of fatigue in skeletal muscle, *Am J Physiol* 261 (*Cell Physiol* 30):C195, 1991.

Books and Monographs

Bagshaw CR: *Muscle contraction*, London, 1993, Chapman and Hall.

Hochachka PW: *Muscles as molecular and metabolic machines*, Boca Raton, Fla, 1994, CRC Press.

Huang CL-H: *Intramembrane charge movements in striated muscle*, Oxford, 1993, Clarendon Press.

Lieber, RL: *Skeletal muscle structure and function: implications for rehabilitation and sports medicine*, Baltimore, Md, 1992, Williams & Wilkins.

Mastaglia FL, Walton JN: *Skeletal muscle pathology*, ed 2, Edinburgh, 1992, Churchill Livingstone.

McMahon TA: *Muscles, reflexes, and locomotion*, Princeton, NJ, 1984, Princeton University Press.

Needham DM: *Machina carnis: the biochemistry of muscular contraction in its historical development*, Cambridge, 1971, Cambridge University Press.

Netter FH: Musculoskeletal system, part I: Anatomy, physiology and metabolic disorders. In Dingle RV, ed: *The Ciba collection of medical illustrations*, vol 8, Summit, NJ, 1987, Ciba-Geigy Corp.

Peachey LD, ed: *Handbook of physiology*, section 10: Skeletal muscle, Bethesda, Md, 1983, American Physiological Society.

Rüegg JC: *Calcium in muscle contraction*, ed 2, Berlin, 1992, Springer-Verlag.

14

Muscle in the Walls of Hollow Organs

Muscle is a significant constituent of all organs and plays important roles in their function. The organs include the heart and vascular system, the airways, the gastrointestinal tract, and the urogenital system. These muscles also have great medical significance because of their involvement in many diseases, including asthma, hypertension, and atherosclerosis. Muscle function differs in the absence of a skeleton, and when high shortening velocities and power output are unnecessary.

MUSCLE FUNCTION IN HOLLOW ORGANS

The load on the cells in hollow organs is imposed by the pressure in the organ (compare Figure 14-1 with Figure 13-1). If the muscle cells are relaxed, the volume of the organ will increase with the volume of its contents, until the connective tissue matrix within the organ wall limits further expansion. The muscle may shorten to empty the organ by briefly increasing the pressure, in what is termed a **phasic contraction.** However, the muscle may contract isometrically for long periods to maintain the dimensions of the organ in a **tonic contraction.** In such cases the muscle serves as an "adjustable skeleton." The economy of force maintenance (force × time/adenosine triphosphate consumption) is critical in this case. In fact, *the economy of some smooth muscles is more than 300 times that of striated muscle.*

Unlike cells in skeletal muscle, the cells in hollow organs do not individually link two bones, but are connected to each other and to extracellular connective tissue. Thus they cannot be recruited individually to increase force. Each cellular link in the chain must be equally activated and develop the same force. Contraction in one part of an organ will change the pressure throughout, so muscle cell function must be coordinated.

Smooth Muscle Classification

Skeletal muscle fiber types are distinguished by the expression of the fast or slow myosin isoenzyme and the dominant metabolic pathway. Smooth muscles lack metabolic specializations, as their ATP consumption rates are low. However, smooth muscle from various tissues is characterized by an extraordinary range of distinguishing features. These differences are mainly related to the properties of the cell membranes (innervation, receptors, ion pumps and channels, and junctions) and the ways in which Ca^{++} is mobilized.

No scheme to classify smooth muscle has been satisfactory, although it is useful to make the functional distinction between phasic and tonic muscles. Most of the smooth muscles in the gastrointestinal tract and urogenital organs are phasic: normally relaxed or rhythmically active. The smooth muscle in the walls of blood vessels or airways by contrast is typically tonic: always contracted to some degree. Such sustained active force is called **tone.** However, all smooth muscles can exhibit both types of behavior. This chapter emphasizes the mechanisms that underlie this range of behavior; other chapters describe smooth muscle function in specific organ systems.

Structure-Function Relationships

The structural organization of smooth muscle in organs can be complex (Figure 14-2). The simplest case may be an **arteriole,** which consists of three main elements: endothelial cell lining, a single smooth muscle cell layer encircling the endothelial cells, and connective tissue (Figure 14-2, *C;* see Chapters 22 and 23).

In the intestinal tract the **lining of the tube** is a mucosal layer involved in the digestion and absorption of nutrients. The muscle in the walls mixes and propels the contents (Chapter 32). Two muscle layers are required: (1) an inner circular layer to determine circumference,

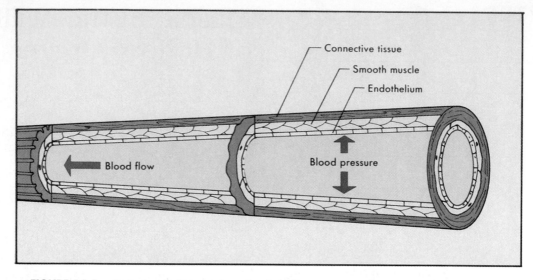

FIGURE 14-1 Functional elements in a simple hollow organ, such as many blood vessels, include (1) a connective tissue network to passively limit maximum organ distention, (2) muscle cells circumferentially arranged to alter vessel diameter and thus resistance to blood flow, and (3) a lining that interfaces with the organ contents (endothelial cells in blood vessels).

and (2) an outer longitudinal layer to control length. Coordination to mix and propel the contents depends on an elaborate neural network that originates from plexuses between the two muscle layers (Chapter 32).

Another category of organ is a sack in which the muscle is normally relaxed. The volume increases with the delivery of the contents. Contraction of the muscle empties the sack, in coordination with relaxation of a valve or **sphincter.** Examples are the urinary bladder, the uterus, and the rectum. Anatomically complex layers of muscle enable large volume changes to empty the organ.

Control and coordination of the muscle (Figures 14-2 and 14-3) depend on: (1) the intrinsic and extrinsic innervation, (2) the blood supply that provides nutrients and circulating hormones, and (3) cell junctions that allow electrical, chemical, and mechanical interactions. The nature and relative importance of these elements are highly variable.

Cell Structure

Most smooth muscle cells are 2 to 5 μm in diameter and 100 to 400 μm long at the optimal length for force generation (L_o), although there are many exceptions (Figure 14-2, *B*). All have a single central nucleus, and most taper toward the ends. The cells are comparatively featureless when viewed with a light microscope, and the general characteristics of the membranes, contractile apparatus, and cytoskeleton exhibit many differences from skeletal muscle.

Three membranes are involved in coupling extracel-

lular signals to the contractile apparatus: the sarcolemma, the caveoli, and the sarcoplasmic reticulum (Figure 14-3, *A*). The **caveoli** are tiny sac-like invaginations of the sarcolemma arranged in rows along the cell, and they communicate with the extracellular space. In smooth muscle the sarcoplasmic reticulum is a continuous irregular tubular network branching throughout the cell. The sarcoplasmic reticulum is closely associated with the sarcolemma and the caveoli, but lacks the specialized junctions that couple the T-tubules to the sarcoplasmic reticulum of skeletal muscle. Unlike skeletal muscle, there are no anatomically defined motor endplates. Neurotransmitters released from the enlarged **varicosities** spaced along autonomic nerves (see Figure 14-2) diffuse to receptors distributed in the sarcolemma.

Contractile Apparatus

The contractile apparatus lacks a myofibrillar structure with thick and thin filaments aligned to provide striations. There appear to be contractile units consisting of thin filaments attached to a cytoskeleton that overlap with much smaller numbers of myosin-containing thick filaments (Figure 14-4). The filaments are approximately aligned in the cell's long axis, along which force is generated, but shortening can lead to angular displacements. Nevertheless, the three-dimensional organization of thick and thin filaments and the cytoskeleton is uncertain.

Forces produced by crossbridges that act on the thin filaments are transmitted to the sarcolemma by the cy-

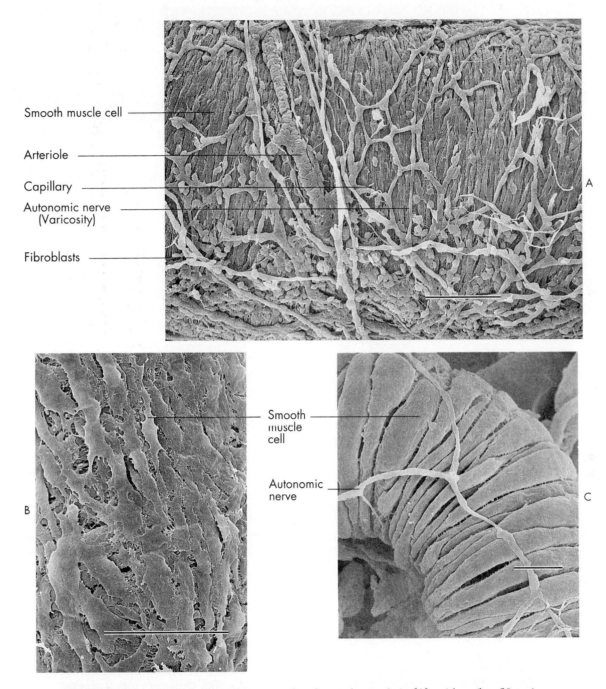

Smooth muscle cell

Arteriole

Capillary

Autonomic nerve (Varicosity)

Fibroblasts

A

B

Smooth muscle cell

Autonomic nerve

C

FIGURE 14-2   Scanning electron micrographs of smooth muscle in **(A)** oviduct (bar, 50 μm), **(B)** the outer layer of the epididymus (bar 50, μm), and **(C)** a small arteriole (bar, 5 μm). Note the irregular and varied shape of smooth muscle cells, their linkage into muscular sheets, and an extracellular matrix that is typically the product of the smooth muscle cells. Smooth muscle cells interact with other cell types, including nerves, endothelial or epithelial cells, and fibroblasts. (From Uehara Y et al. In Motta PM, ed: *Ultrastructure of smooth muscle,* Norwell, Mass, 1990, Kluwer Academic Publishers.)

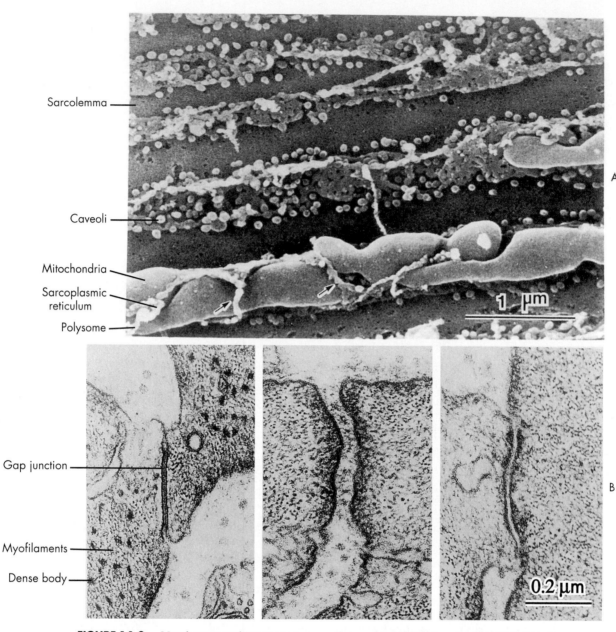

Sarcolemma

Caveoli

Mitochondria

Sarcoplasmic
reticulum

Polysome

A

Gap junction

Myofilaments

Dense body

B

1 μm

0.2 μm

FIGURE 14-3 Membranes and junctions in smooth muscle. **A,** Scanning electron micrograph of the cytoplasmic surface of the sarcolemma of an intestinal smooth muscle cell. Rows of caveoli project into the cytoplasm, but are open to the extracellular space. The intervening space is normally where junctions with other cells are formed on the outside surface and where cytoskeletal membrane dense areas anchor thin filaments. Elements of the sarcoplasmic reticulum snake around the caveoli and mitochondria. Polysomes consisting of clusters of ribosomes, sites of protein synthesis, are associated with the sarcoplasmic reticulum. **B,** Transmission electron micrographs of various types of junctions between intestinal smooth muscle cells. Left panel shows a gap junction where ions and small molecules can diffuse between cells (see Figure 4-5). Other junctions that mainly provide mechanical linkages have wider separation of the cells with darkly staining extracellular proteins in the cleft. (**A** from Inoué T and **B** from Gabella G. In Motta PM, ed: *Ultrastructure of smooth muscle,* Norwell, Mass, 1990, Kluwer Academic Publishers,)

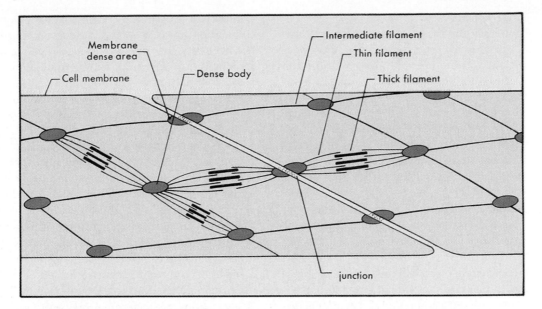

FIGURE 14-4   Possible organization of the cytoskeleton and myofilaments in smooth muscle. Note junctions linking contractile apparatus of two cells. Three-dimensional imaging techniques are needed to reconstruct the structure of cells where the filaments are not organized into parallel arrays. Smooth muscle has a high thin-filament density and few thick filaments compared with striated muscle.

toskeleton. Thin filament attachment sites are **dense bodies** scattered throughout the myoplasm and **membrane dense areas** at intervals along the sarcolemma (Figure 14-4). These structures are analogous to Z-disks. Dense bodies and membrane dense areas are linked by cytoskeletal **intermediate filaments.**

The cells must be coupled to each other and linked by an extracellular matrix in order to generate force. A variety of junctions (see Figure 14-3, *B*) provide for electrical and chemical communication and mechanical linkages between smooth muscle cells. Junctions essential to organ function are also formed with other cells, such as the vascular endothelium or airway epithelium. The type and density of junctions vary with the tissue. **Gap junctions** are most prominent in phasic smooth muscles, in which contraction follows trains of action potentials that are propagated from cell to cell. *In smooth muscle the functional units equivalent to skeletal muscle motor units are bundles or layers of cells. The contractile system is anatomically coupled across the plasma membranes of smooth muscle cells* (see Figure 14-4).

Smooth muscle cells synthesize and secrete elastin and collagen fibrils. The highly extensible elastin and inextensible collagen in the extracellular matrix act to limit the volume of hollow organs. The organization of the contractile system and force-transmitting structures probably facilitates the considerable shortening capacity and transmission of forces around variable radii (one smooth muscle cell in a terminal arteriole can completely encircle the lumen two or three times (see Figure 14-2, *C*).

Blood pressure can cause the wall of a vessel to bulge in an **aneurysm** when weakened. Aneurysms may be congenital, reflecting developmental defects, or result from vascular disease such as **atherosclerosis.** Sudden death is likely on rupture of an aneurysm in a large artery. Contributing to the likelihood of rupture is the physical fact that blood pressure puts a greater distending load on an enlarged segment of a vessel than on adjacent regions (this is illustrated later in Figure 14-9 and in Chapter 22).

CONTROL SYSTEMS: INPUTS TO THE SARCOLEMMA AND Ca^{++} MOBILIZATION

Contraction of smooth muscle cells is not solely determined by a motor nerve. An integrated and coordinated response results from many inputs that can be both excitatory and inhibitory. These inputs include *autonomic nerves; circulating hormones or drugs; local hormones, ions, and metabolites; signals from other cell types such as endothelial cells; and signals from coupled smooth muscle cells.* Specific details on the key mechanisms are given in the chapters devoted to individual organ systems.

Innervation

Nerves are generally the most important input. However, a few smooth muscles receive no innervation, and virtually all continue to function more or less appropriately after the central autonomic connections (**extrinsic innervation**) have been severed. Several factors influence neural control. Most smooth muscle tissues have more than one type of innervation, typically parasympathetic and sympathetic. More and more types of nerves and neurotransmitters are being discovered, and the neural system of the gastrointestinal tract rivals that of the brainstem in complexity.

Dual innervation usually acts reciprocally, with excitatory nerves causing contraction and inhibitory nerves producing relaxation. The neurotransmitter-releasing sites (varicosities) of the nerves make intimate neuromuscular contacts with each smooth muscle cell in some tissues. However, in many tissues neural contact may be fairly indirect, with long diffusion paths and gradients for the transmitters. The action of a particular class of nerve and its neurotransmitter on a specific smooth muscle cell depends on the receptors expressed for that neurotransmitter. Some cells respond to norepinephrine by contracting, whereas others relax; such disparities reflect differences in the receptors and signal transduction mechanisms. These different responses allow coordinated, whole-body adjustments. Under conditions of stress, for example, epinephrine is released from the adrenal gland. The vascular smooth muscle of the gut will constrict and shunt blood to the cardiac and skeletal muscle, where the vascular beds dilate.

The capacity of smooth (and cardiac) muscle to maintain organ function without atrophy or other phenotypic changes after denervation is a major distinction from skeletal muscle. Contributing factors include maintenance of contractile activity in response to intrinsic neural networks, local or circulating hormones, and propagated signals between smooth muscle cells.

> This potential independence from the central nervous system is essential for successful organ transplantation. A good example is the transplanted heart with its endogenous coronary vasculature that functions successfully without innervation.

Calcium Mobilization

The myoplasmic Ca^{++} concentration ($[Ca^{++}]$) in smooth muscle regulates crossbridge interactions, as it also does in skeletal muscle. However, the mechanisms that regulate the myoplasmic $[Ca^{++}]$ are much more

complex in smooth muscle. One factor involves the multiplicity of inputs, as already discussed. Another factor is the necessity for smooth muscle to regulate precisely the myoplasmic $[Ca^{++}]$ at submicromolar values during tonic contractions. Regulation of Ca^{++} in smooth muscle involves the sarcolemma as well as the sarcoplasmic reticulum. It also involves another Ca^{++} pool: the extracellular fluid, where the $[Ca^{++}]$ is about 1.6 mM.

Ca^{++} mobilization and regulation are schematized for phasic and tonic contractions in Figure 14-5. Phasic contractions last a few seconds or minutes and are elicited by brief periods of activation, which could be a burst of action potentials or a short period of receptor occupancy by an activating neurotransmitter or hormone without action potentials. The result is a transient increase in myoplasmic $[Ca^{++}]$ and a small contraction, or a series of transient increases in $[Ca^{++}]$ with a larger contraction, analogous to a tetanus (Figure 14-5, A). Ca^{++} release from the sarcoplasmic reticulum and influx through the sarcolemma may both contribute to the transient increase. After the stimulus the myoplasmic Ca^{++} is pumped out of the cell and resequestered in the sarcoplasmic reticulum (Figure 14-6).

If the stimulus is prolonged, the conductance of Ca^{++} channels in the sarcolemma remains elevated, and the myoplasmic $[Ca^{++}]$ remains above threshold values after an initial transient increase (see Figure 14-5, B). The resulting contraction is fast, and it may be sustained at or near peak levels, despite falling $[Ca^{++}]$. The initial transient increase occurs in the absence of extracellular Ca^{++}, which shows that it is induced by release of Ca^{++} from the sarcoplasmic reticulum. On the other hand, the sustained modest elevation in Ca^{++} is totally dependent on extracellular Ca^{++} (dashed line in Figure 14-5, B). The same force can be generated with reduced crossbridge cycling and ATP splitting in a much slower contraction in the absence of an initial Ca^{++} transient from the sarcoplasmic reticulum.

Signal Transduction Mechanisms in the Sarcolemma

How is the myoplasmic $[Ca^{++}]$ regulated by the inputs just discussed? Five mechanisms are involved: (1) membrane potential–dependent Ca^{++} influx through Ca^{++} channels from the extracellular space, (2) receptor-activated Ca^{++} channels in the sarcolemma, (3) control of sarcoplasmic reticulum Ca^{++} release, (4) sequestration and extrusion of Ca^{++} by pumps in both membranes, and (5) **Na^+/Ca^{++} exchange** across the sarcolemma, where the concentration gradients usually lead to the removal of Ca^{++} and an associated Na^+ movement into the cell (see Figure 14-6 and Chapter 18).

The membrane potential in smooth muscle is the re-

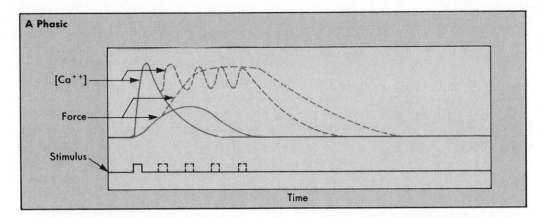

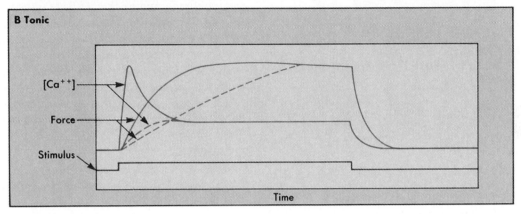

FIGURE 14-5 Patterns of Ca^{++} mobilization in phasic and tonic contractions of smooth muscle. **A,** Phasic contractions result from brief-acting stimuli that elicit only a transient increase in Ca^{++} and a small contraction *(solid lines)*. Repeated brief stimuli can lead to summation of individual contractions and greater force, much like tetanization in skeletal muscle *(dashed lines)*. **B,** In tonic contractions activation is maintained with continued receptor occupancy by neurotransmitters or other agents. An initial Ca^{++} peak, representing release from the sarcoplasmic reticulum plus influx from the extracellular pool, induces rapid force development *(solid lines)*. However, the peak $[Ca^{++}]$ is not maintained. During the sustained, tonic contraction the $[Ca^{++}]$ remains somewhat elevated and depends on extracellular Ca^{++}. If the initial high Ca^{++} peak is blocked, the same force will be developed, although the rates of the contraction are much slower *(dashed lines)*. Thus $[Ca^{++}]$ can affect the rate as well as the force of a contraction. This phenomenon, in which crossbridge cycling rates as manifested in the rate of force development are regulated in a Ca^{++}-dependent manner, does not occur in skeletal muscle. Differences in the mechanisms by which Ca^{++} regulates crossbridges in smooth muscle are responsible.

sult of two processes. A major factor is the **Donnan potential,** which reflects the relative permeabilities and concentration gradients for K^+ and Na^+ across the sarcolemma (as in skeletal muscle, Chapter 2). However, the Na^+-K^+ pump extrudes three Na^+ ions in exchange for two K^+ ions, and it thereby results in the transfer of one positive charge out of the cell for each cycle. This process may add as much as 20 mV to the Donnan potential, to give a normal membrane potential of -50 to -70 mV in the absence of depolarizing stimuli. Changes in the activity of the Na^+ K^+ pump are responsible for slow oscillations in the membrane

potential (Figure 14-7). The ion channels in the sarcolemma that open in response to depolarization in smooth muscle are primarily permeable to Ca^{++} ions **(potential-dependent Ca^{++} channels).** Increases in the Na^+-K^+ pump activity also lower myoplasmic $[Ca^{++}]$ directly. Because the Na^+ concentration gradient into the cell is increased, more Na^+ will enter in exchange for Ca^{++}.

Some smooth muscles (mainly phasic) generate action potentials at some critical level of depolarization (Figure 14-7, *A* and *B*). Ca^{++} is the current-carrying ion in these action potentials. Other smooth muscles (functionally

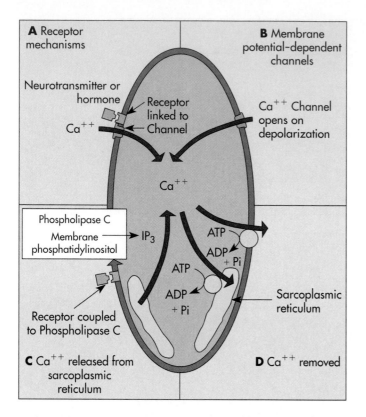

A Receptor mechanisms

Neurotransmitter or hormone

Receptor linked to Channel

Ca^{++}

B Membrane potential-dependent channels

Ca^{++} Channel opens on depolarization

Ca^{++}

Phospholipase C

Membrane phosphatidylinositol

IP_3

ATP

ADP + Pi

ATP

ADP + Pi

Sarcoplasmic reticulum

Receptor coupled to Phospholipase C

C Ca^{++} released from sarcoplasmic reticulum

D Ca^{++} removed

FIGURE 14-6 Mechanisms regulating myoplasmic $[Ca^{++}]$ in smooth muscle. **A,** Receptor-operated channels allow Ca^{++} influx through the sarcolemma. **B,** Membrane potential–dependent Ca^{++} channels also lead to Ca^{++} influx. Active transport of 3 Na^+ ions out of the cell in exchange for 2 K^+ ions lowers the membrane potential and closes some Ca^{++} channels. **C,** Ca^{++} release from the sarcoplasmic reticulum is probably caused by the second messenger, inositol 1,4,5-trisphosphate (IP_3) generated at the plasma membrane. **D,** Ca^{++} pumps in the sarcoplasmic reticulum and plasma membrane reduce the $[Ca^{++}]$. Na^+/Ca^{++} exchange also lowers myoplasmic $[Ca^{++}]$. *ATP,* Adenosine triphosphate; *ADP,* adenosine diphosphate; *P_i,* inorganic phosphate.

tonic) do not generate action potentials. Nevertheless, these cells have potential-dependent Ca^{++} channels. Graded depolarization caused by reduced electrogenic Na^+-K^+ pumping will increase Ca^{++} influx (Figure 14-7, *C*).

Receptors mediate two paths for Ca^{++} mobilization (see Figure 14-6, *A* and *C*). The binding of an agent to a receptor can increase the Ca^{++} permeability of the sarcolemma by opening **receptor-activated channels.** This allows influx of Ca^{++} from the extracellular pool. Receptor occupancy can also cause Ca^{++} release from the sarcoplasmic reticulum. Such receptor-mediated mechanisms can increase (or decrease) myoplasmic $[Ca^{++}]$ to alter tone without detectable changes in the membrane potential (see Figure 14-7, *D*). **Pharmaco-mechanical coupling** refers to activation mechanisms

that do not involve changes in the membrane potential, in contrast to **excitation-contraction coupling.**

A chemical messenger is implicated in Ca^{++} release from the sarcoplasmic reticulum. This differs from skeletal muscle, where charge coupling between the closely associated T-tubular system and the sarcoplasmic reticulum causes Ca^{++} release. Phospholipase C is activated by occupancy of specific receptors. Phospholipase C hydrolyzes membrane phosphatidylinositol to yield **inositol 1,4,5-trisphosphate (IP$_3$)** and diacylglycerol. IP_3 diffuses to receptors in the sarcoplasmic reticulum to induce Ca^{++} release.

The mechanisms that reduce myoplasmic $[Ca^{++}]$ are the membrane pumps that actively transport Ca^{++} back into the sarcoplasmic reticulum or extrude it into the extracellular space, and the Na^+/Ca^{++} exchangers in the sarcolemma. Drugs, neurotransmitters, or hormones that inhibit contraction (induce relaxation) in smooth muscle act by various mechanisms, including reduction of Ca^{++} influx or Ca^{++} release from the sarcoplasmic reticulum, as well as enhancement of Ca^{++} pump activity.

Pathological excitatory signals that increase cell Ca^{++} are responsible for a variety of life-threatening diseases. **Asthma** is one example; airborne irritants or antigens can act on the airway epithelial cells to activate pathways that lead to increases in airway smooth muscle cell $[Ca^{++}]$, airway constriction, and difficulty in breathing. Some **myocardial infarctions** (death of cardiac muscle cells because of impaired blood flow) are the result of vasospasm of coronary arteries, and some **strokes** result from cerebral arterial vasospasm.

Ca^{++} AND CROSSBRIDGE REGULATION IN SMOOTH MUSCLE

Smooth muscle lacks troponin—the Ca^{++}-binding, thin-filament, regulatory protein of skeletal and cardiac muscle. Although smooth muscle thin filaments contain other proteins of potential regulatory significance, the available evidence suggests that regulation occurs at the crossbridge itself (Figure 14-8). The mechanism is not allosteric (i.e., reversible binding of Ca^{++} to a regulatory site to induce conformational changes), but involves phosphorylation of the crossbridge at a specific serine-residue. This **covalent regulatory mechanism** (the crossbridge is chemically altered) uses ATP as the phosphate donor.

In smooth muscle, the crossbridges cannot attach to the thin filament and cycle unless they are phosphory-

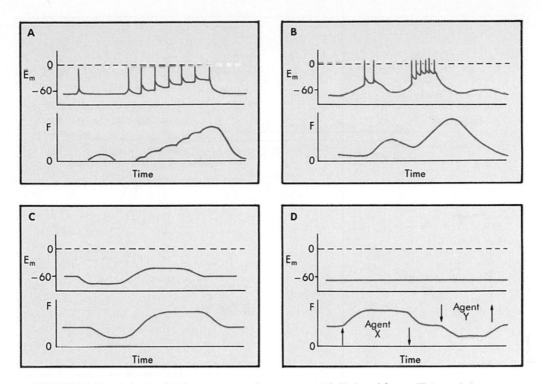

FIGURE 14-7   Relationship between membrane potential *(E_m)* and force *(F)* in various types of smooth muscle. **A,** Action potentials may be generated by pacemaker cells and propagated through the tissue. **B,** Slow waves reflecting oscillations in the Na^+-K^+ pump activity can trigger bursts of action potentials and rhythmic contractile activity. **C,** Tone will vary with E_m in tonic tissues. Action potentials are not normally generated in tonic tissues such as most vascular smooth muscle. **D,** Receptor-mediated mechanisms can alter cell $[Ca^{++}]$ without detectable changes in the membrane potential.

lated by a specific enzyme, **myosin kinase.** The active form of myosin kinase is a complex with Ca^{++}-calmodulin (Figure 14-8). **Calmodulin** is a cytoplasmic protein that has four high-affinity Ca^{++}-binding sites, and it participates in the activation of several Ca^{++}-dependent enzymes. The phosphorylated crossbridge can cycle with no further need for Ca^{++} until it is dephosphorylated by **myosin phosphatase.** The regulatory scheme illustrated in Figure 14-8 is incomplete, because it postulates that phosphorylation is a simple switch to turn on a crossbridge. The model cannot explain the phenomenon illustrated in Figure 14-5, *B,* in which crossbridge cycling rates depend on $[Ca^{++}]$. It was only recently discovered that crossbridge function in smooth muscle has features that are not present in skeletal muscle. These reflect a more complex regulatory process.

Contractile System Function

The same two steady-state relationships that describe contractile system function in skeletal muscle also apply to smooth muscle: the force-length and velocity-load re-

lationships. Force varies with tissue or cell length, as in skeletal muscle, and the maximal force at the optimal length is similar in smooth and skeletal muscle (Figure 14-9). A significant passive force, which reflects the connective tissue component, is present at the optimal length. Although smooth muscle shortens more than do the muscle cells tethered to the skeleton, other factors such as wall thickening contribute to large changes in organ volume (Figure 14-9).

When smooth muscle becomes overstretched, it tends to lose its ability to fully contract. This can occur in the urinary bladder in men when the urethra is obstructed by an enlarged prostate, as in **benign prostatic hypertrophy.** Inability to empty the bladder, with urine retention, often accompanies this ailment.

Although smooth muscle can generate a high force, contraction velocities are very low (Figure 14-10). In fact, the ATPase activity of phosphorylated smooth muscle myosin is more than 100-times lower than that of the

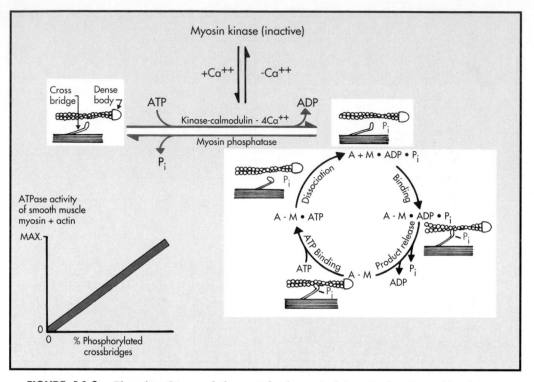

FIGURE 14-8 Phosphorylation of the crossbridges regulates attachment and cycling in smooth muscle. Smooth muscle myosin undergoes the same cyclic interaction with the thin filaments, powered by the hydrolysis of ATP *(black arrows),* as does skeletal muscle myosin (Figure 12-3). However, cycling cannot occur until the crossbridges are phosphorylated *(colored arrows),* a reaction whereby inorganic phosphate *(P_i)* derived from ATP hydrolysis is covalently attached to the head of the myosin molecule, forming a phosphorylated crossbridge. Phosphorylation requires ATP in addition to the ATP used to power crossbridge cycling. Phosphorylation is proportional to the $[Ca^{++}]$, which determines the activity of myosin kinase. If the myoplasmic $[Ca^{++}]$ is lowered, Ca^{++} and calmodulin dissociate from myosin kinase, and the crossbridges are dephosphorylated by myosin phosphatase. This scheme explains the dependence of smooth muscle actomyosin ATPase activity in vitro on myosin phosphorylation *(inset).* Note that the P_i group attached to the crossbridge by myosin kinase *(red)* is at a different site from the P_i group released during the crossbridge cycle *(black).*

fast skeletal muscle myosin isoenzyme. However, the velocity-force curves exhibit the characteristic hyperbolic dependence of velocities on load. Furthermore, the power output is maximal at a moderate load, and the ability to resist an imposed load of 1.6 times the load that can be generated is the same as that illustrated for skeletal muscle in Figure 12-5. The similarity of the force-length and velocity-load relationships provides strong evidence that the sliding filament-crossbridge mechanism also operates in smooth muscle.

Figure 14-10 shows that covalent regulation by crossbridge phosphorylation is associated with new properties. The number of attached crossbridges that determine the developed force depends on phosphorylation, but the relationship is not linear (Figure 14-10, *B*). Fairly low $[Ca^{++}]$ and phosphorylation values support high levels of force. These steady-state relationships explain

why force can be sustained in a tonic contraction when $[Ca^{++}]$ and crossbridge phosphorylation fall from initial values (Figure 14-11). Shortening velocities, which reflect average crossbridge cycling rates, increase in proportion to crossbridge phosphorylation for a given load (see Figure 14-10, *C*). *In smooth muscle both load and crossbridge phosphorylation levels determine crossbridge cycling rates and thereby ATP consumption and shortening velocities.*

Crossbridge Regulation

The properties illustrated in Figures 14-10 and 14-11 confer advantages for muscle in the walls of hollow organs with diverse functions. With appropriate modulation of the myoplasmic $[Ca^{++}]$, comparatively rapid phasic contractions are possible. However, tonic contrac-

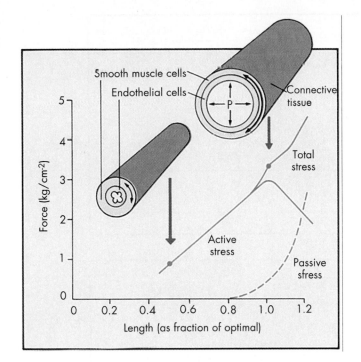

FIGURE 14-9 Force-length and pressure-volume behavior of smooth muscle. The active and passive force-length curves for a strip of smooth muscle dissected from the wall of a hollow organ are similar to those of skeletal muscle (graph). Diagram representing a small arteriole shows how shortening of the smooth muscle cells by only 50% can give a 36-fold reduction in lumen area and stop red cell flow. The smooth muscle cells cannot develop as much force at the shortened length. However, if the blood pressure remains the same, the force on the smooth muscle cells tending to distend the vessel also falls for geometric reasons. This force, indicated by the double-headed arrows, equals $P \cdot r$, where P = pressure and r = radius (Laplace's law). Bulging of the compressed endothelium also contributes to luminal reduction. As the vessel radius increases, the distending wall force rises and the load is increasingly resisted by the connective tissue (also see Chapter 22).

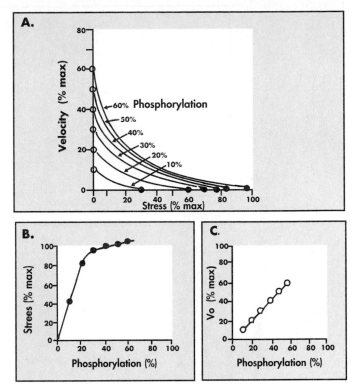

FIGURE 14-10 **A,** The velocity-load curve in smooth muscle varies with the level of crossbridge phosphorylation. **B,** Active force development at the optimum cell length (F_0 = the intercepts on the abscissa in **A**) rises rapidly with phosphorylation. Near-maximal force may be generated with only 25% to 30% of the crossbridges in the phosphorylated state. **C,** Maximal shortening velocities (intercepts on the ordinate of **A**) are directly dependent on phosphorylation.

tions are also possible with reduced crossbridge turnover rates (Figure 14-11). Much current research seeks to explain this behavior.

Covalent regulation by crossbridge phosphorylation may explain why smooth muscle exhibits properties that are lacking in striated muscle (Figure 14-12). Two crossbridge cycles are possible with four crossbridge states. One is a slow cycle (state 2 to 3 to 2) that is characteristic of phosphorylated crossbridges and in which one ATP is consumed. A very slow cycle can occur via states 1, 2, 3, and 4 to 1. Two ATP molecules are hydrolyzed in this cycle: one for chemomechanical transduction and one for crossbridge phosphorylation. At high Ca^{++} and phosphorylation values, most crossbridges will be phosphorylated in states 2 and 3. At reduced Ca^{++} and phosphorylation levels, crossbridges in state 4 predominate because their detachment is rate limiting. Shortening velocity depends on whether most crossbridges cycle slowly (high Ca^{++} and phosphorylation) or very slowly (reduced Ca^{++} and phosphorylation), as well as on the load that opposes crossbridge movement.

ATP used for phosphorylation and dephosphorylation is not captured as mechanical work. Thus covalent regulation reduces the efficiency of work performed while smooth muscle shortens. This disadvantage is offset by the gains in the economy of a tonic contraction when no shortening or work is done and crossbridge turnover is reduced. Total ATP consumption rates are low in smooth muscle, and requirements of the contractile system are met by oxidative phosphorylation. Smooth

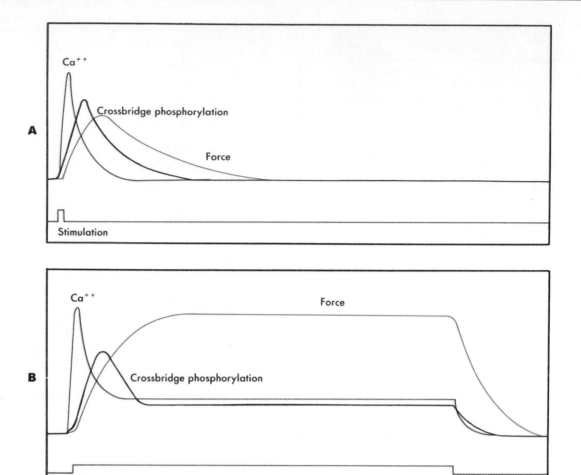

FIGURE 14-11   Time course of events in activation and contraction of smooth muscle. **A,** Phasic contraction following a brief period of stimulation. The initial event is a transient increase in Ca^{++}, followed by crossbridge phosphorylation by myosin kinase, cycling, and force development. These changes are reversed as myoplasmic $[Ca^{++}]$ is restored to resting levels. **B,** In a tonic contraction produced by sustained stimulation, the $[Ca^{++}]$ and phosphorylation levels remain somewhat elevated after an initial peak (allowing rapid force development). In this case, the force (tone) is maintained with reduced crossbridge cycling rates, as manifested by lower shortening velocities, lower ATP consumption, and an overall increase in economy (despite the ATP used for phosphorylation).

muscles do not exhibit fatigue as long as the blood supply is intact.

CARDIAC MUSCLE

The muscle in the heart has unique characteristics (see Chapter 18). The cells are striated, and their thin filaments possess a troponin-based regulatory system. The myosin isoenzymes differ from striated muscle. However, the cells have metabolic and contractile properties comparable to those of slow skeletal muscle, and ATP consumption is met by oxidative phosphorylation. Nevertheless, the heart is a hollow organ. Like smooth muscle cells, cardiac muscle cells are small, with one central nucleus, and are connected to one another by specialized junctions that provide both electrical and mechanical coupling. Action potentials are propagated from cell to cell, and their contractions are thereby synchronized.

Cardiac muscle is well suited for the special function of the heart. This organ is a pump that must contract and relax fairly rapidly. The heart must generate substantial power during shortening to eject the blood, and the efficiency of chemomechanical transduction is important. Force is generated only briefly for each heartbeat (a twitch), and the economy of force maintenance is irrelevant.

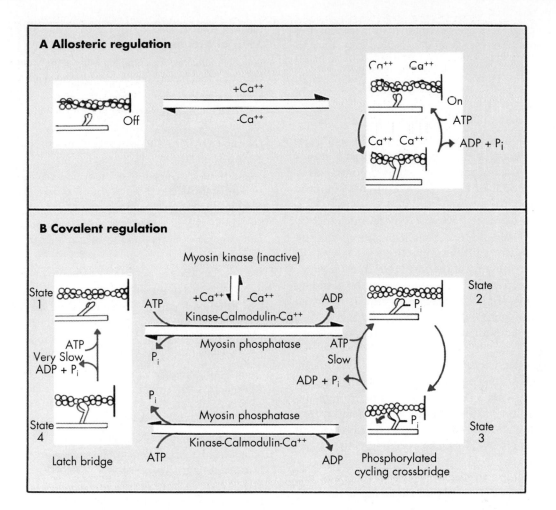

FIGURE 14-12 **A,** Allosteric regulation characteristic of skeletal and cardiac muscle allows two crossbridge states: free and attached, or force-generating (Figure 13-5). **B,** Covalent regulation can produce four crossbridge states, as both free and attached crossbridges are substrates for myosin kinase and myosin phosphatase. (Compare with the incomplete scheme in Figure 14-8, noting that the phosphorylated crossbridge cycle is simplified here to show only free and attached states.) If phosphorylated (state 3) and dephosphorylated (state 4) crossbridges are identical except for a slowed detachment rate of the latter, the behavior illustrated in Figures 14-5, 14-10, and 14-11 can be explained.

Table 14-1 Mechanisms for Grading Contractile Force

| | Occurrence | | |
General Mechanism	Skeletal	Cardiac	Smooth*
Recruit more cells (motor units)	+	−	(+)
Sum twitches by increasing stimulation frequency (tetanus)	+	−	+
Alter filament overlap by stretch	(+)	+	+
Vary twitch by changing Ca^{++} transient	−	+	+
Alter Ca^{++} sensitivity of regulatory systems	−	+	+
Tonic depolarization and activation of potential-dependent Ca^{++} channels without action potentials	−	−	+
Receptor-activated channels (pharmacomechanical coupling)	−	−	+

*The relative importance of these mechanisms varies greatly with the type of smooth muscle. Parentheses indicate the possibility of little physiological importance.

Table 14-1 compares the mechanisms for grading the force of contraction in skeletal, cardiac, and smooth muscle.

SUMMARY

1. Smooth (involuntary) muscle is primarily a component of the walls of hollow organs, where it serves to stabilize organ dimensions in tonic contractions, or to mix, propel, or expel organ contents in phasic contractions.

2. Smooth muscle cells have one central nucleus, and are anatomically discrete, but they must contract synchronously. A variety of junctions between cells coordinate communication and force transmission.

3. Contraction of smooth muscle is based on a sliding filament/crossbridge mechanism, as is skeletal muscle, although the thick (myosin) and thin (actin) filaments of smooth muscle are not organized into sarcomeres.

4. Smooth muscle is controlled by various systems, including autonomic nerves (both excitatory and inhibitory, involving a large number of neurotransmitters), circulating hormones, locally generated hormones or metabolites from associated cell types, and electrical or chemical signals coupling cells via gap junctions.

5. The sarcolemma in smooth muscle regulates cell [Ca^{++}] in response to extracellular inputs that determine Ca^{++} influx via receptor-operated or voltage-gated channels, and Ca^{++} efflux via Ca^{++} pumps and Na^{+}/Ca^{++} exchanges.

6. The sarcoplasmic reticulum is an intracellular Ca^{++} compartment that generates Ca^{++} transients.

7. Bursts of action potentials typically trigger phasic contractions. Action potentials are characteristically absent in tonic smooth muscles, where force (tone) is proportional to the membrane potential (excitation-contraction coupling) and agonist-induced changes in activation (pharmacomechanical coupling).

8. Ca^{++} regulates contraction in smooth muscle by binding to calmodulin, followed by the formation of an active myosin kinase–calmodulin–Ca^{++} complex. Activated myosin kinase uses ATP to phosphorylate crossbridges, which enables the crossbridges to attach to the thin filament and cycle.

9. Dephosphorylation of attached crossbridges by myosin phosphatase slows their detachment rate, reducing crossbridge cycling rates and ATP consumption in sustained contractions.

10. Relaxation is caused by lowering of cell Ca^{++} to levels that inactivate myosin kinase and thus lead to cessation of myosin phosphorylation.

BIBLIOGRAPHY
Journal Articles

Blaustein MP et al: Physiological roles of the sodium-calcium exchanger in nerve and muscle, *Ann NY Acad Sci* 639:254, 1991

Chen Q, van Breeman C: Function of smooth muscle sarcoplasmic reticulum, *Adv Second Messenger Phosphoprotein Res* 26:335, 1992

Gabella G: Hypertrophy of visceral smooth muscle, *Anat Embryol* 182:409, 1990

Gerthoffer WT: Regulation of the contractile element of airway smooth muscle, *Am J Physiol* 261 (*Lung Cell Mol Physiol* 5):L15, 1991

McDonald TF et al: Regulation and modulation of calcium channels in cardiac, skeletal, and smooth muscle cells, *Physiol Rev* 74:365, 1994

Murphy RA: What is special about smooth muscle? The significance of covalent crossbridge regulation, *FASEB J* 8:311, 1994

Paul RJ: Smooth muscle energetics, *Annu Rev Physiol* 51:331, 1989

Pozzan T et al: Molecular and cellular physiology of intracellular calcium stores, *Physiol Rev* 74:595, 1994

Somlyo AP: Myosin isoforms in smooth muscle: how they affect function and structure, *J Muscle Res Cell Motility* 14:557, 1993

Somlyo AV, Somlyo AP: Intracellular signaling in vascular smooth muscle, *Adv Exper Med Biol* 346:31, 1993

Stull JT et al: Vascular smooth muscle contractile elements: cellular regulation, *Hypertension* 17:723, 1991.

Trybus KM: Regulation of smooth muscle myosin, *Cell Motil Cytoskeleton* 18:81, 1991

Books and Monographs

Coburn RF: *Airway smooth muscle in health and disease,* New York, 1989, Plenum Press.

Daniel EE et al, eds: *Sphincters: normal function changes in diseases,* Boca Raton, Fla, 1992, CRC Press, Inc.

Fozzard HA et al, eds: *The heart and cardiovascular system: scientific foundations,* ed 2, vols 1 and 2, New York, 1991, Raven Press.

Frank GB et al, eds: *Excitation-contraction coupling in skeletal, cardiac and smooth muscle,* New York, 1992, Plenum Press.

Moreland RS, ed: *Regulation of smooth muscle contraction,* New York, 1991, Plenum Press.

Motta PM, ed: *Ultrastructure of smooth muscle,* Norwell, Mass, 1990, Kluwer Academic Publishers.

Ryan US, Rubanyi GM, eds: *Endothelial regulation of vascular tone,* New York, 1992, Marcel Dekker, Inc.

Sperelakis N, Wood JD, eds: *Frontiers in smooth muscle research,* New York, 1990, Wiley-Liss.

Wood JD, ed: *Handbook of physiology,* section 6: The gastrointestinal system, vol 1: Motility and circulation, Bethesda, Md, 1989, American Physiological Society.

CARDIOVASCULAR SYSTEM

ROBERT M. BERNE
MATTHEW N. LEVY

15

Blood and Hemostasis

BLOOD

The main function of the circulating blood is to carry oxygen and nutrients to the tissues and to remove carbon dioxide and waste products. However, blood also transports other substances (e.g., hormones) from their sites of formation to their sites of action and white blood cells and platelets to where they are needed. In addition, blood aids in the distribution of water, solutes, and heat, and thus contributes to **homeostasis,** a constancy of the body's internal environment.

Blood is a suspension of red cells, white cells, and platelets in a complex solution **(plasma)** of gases, salts, proteins, carbohydrates, and lipids. The circulating blood volume is about 7% of body weight. Approximately 55% of the blood is plasma, whose protein content is 7 g/dl (about 4 g/dl of albumin and 3 g/dl of immunoglobulins).

Blood Components

Erythrocytes The erythrocytes (red cells) are anuclear, flexible, biconcave disks that transport oxygen to the body tissues (Figure 15-1). The erythrocytes average 7 μm in diameter and 5 million per deciliter. They arise from stem cells in the bone marrow, and during maturation they lose their nuclei before entering the circulation, where their average life span is 120 days.

The main protein in erythrocytes is hemoglobin (about 15g/dl of blood) which consists of **heme,** an iron containing tetrapyrrole, linked to **globin,** a protein composed of four polypeptide chains (two α and two β in the normal adult). The iron moiety of hemoglobin binds loosely and reversibly to oxygen to form **oxyhemoglobin.** The affinity of hemoglobin for oxygen is affected by pH, temperature, and 2,3-diphosphoglycerate concentration. These factors facilitate O_2 uptake in the lungs and its release in the tissues (see Chapter 30).

Changes in the polypeptide subunits of globin can also affect the affinity of hemoglobin for O_2 (e.g., fetal hemoglobin has two γ chains instead of two β chains and has greater affinity for O_2), or they can result in disease states, such as **sickle cell anemia** or **thalassemia.**

The number of circulating red cells is fairly constant under normal conditions. The production of erythrocytes **(erythropoiesis)** is regulated by the glycoprotein **erythropoietin,** which is secreted mainly by the kidneys. Erythropoietin acts by accelerating the differentiation of stem cells in the bone marrow.

Anemia and chronic hypoxia (e.g., as a result of living at high altitudes) stimulate erythrocyte production and can produce **polycythemia** (increased number of red cells). When the hypoxic stimulus is removed in subjects with altitude polycythemia, the high red cell concentration in the blood inhibits erythropoiesis. The red cell count is also greatly increased in **polycythemia vera,** a disease of unknown cause. The elevated erythrocyte concentration increases blood viscosity, often to a degree that blood flow to vital tissues becomes impaired.

Leukocytes There are normally 4000 to 10,000 leukocytes (white blood cells) per microliter of blood. The leukocytes include granulocytes (65%), lymphocytes (30%), and monocytes (5%). Of the granulocytes, about 95% are neutrophils, 4% eosinophils, and 1% basophils. White blood cells originate from the primitive stem cells in the bone marrow (Figure 14-1). After birth the granulocytes and monocytes continue to originate in the bone marrow, whereas the lymphocytes take origin in lymph nodes, spleen, and thymus.

The granulocytes and the monocytes are motile, nucleated cells that contain **lysosomes,** which in turn contain enzymes capable of digesting foreign material, such as microorganisms, damaged cells, and cellular debris. Thus the leukocytes constitute a major defense mechanism against infections. Microorganisms or the products of cell destruction release **chemotactic substances** that attract granulocytes and monocytes. When the migrating leukocytes reach the foreign agents, they engulf them **(phagocytosis)** and then destroy them by action of enzymes that form **oxygen-derived free radicals** and **hydrogen peroxide.**

Lymphocytes The lymphocytes vary in size, they have large nuclei, and most lack cytoplasmic granules (Figure 15-1). The two main types are **B lymphocytes,** which confer humoral immunity, and **T lymphocytes,** which confer cell-mediated immunity. When stimulated by an **antigen,** the B lymphocytes are transformed into **plasma cells,** which synthesize and secrete antibodies (gamma globulin), which are carried by the bloodstream to their site of action.

> The main T lymphocytes are cytotoxic and are responsible for long-term protection against some viruses, bacteria, and cancer cells. They are also responsible for the rejection of transplanted organs.

Other T lymphocytes are **helper T cells,** which activate B cells, and **suppressor T cells,** which inhibit B cell activity. Special B and T lymphocytes, called **memory cells,** "remember" specific antigens. These cells can quickly generate an immune response when subsequently exposed to the same antigen.

> Protection against several infectious diseases has been achieved by injection of the appropriate antigen. Also, **vaccines** have been developed for certain diseases by injection of killed or attenuated organisms (antigens) into suitable hosts (horses, sheep).

Platelets The platelets are small (3 mm) anuclear cell fragments of **megakaryocytes.** The megakaryocytes reside in the bone marrow, and when mature they break up into platelets, which enter the circulation. The platelets are important in hemostasis, as discussed later.

Blood Groups

In humans there are four principal blood groups, designated O, A, B, and AB. The plasma of group O

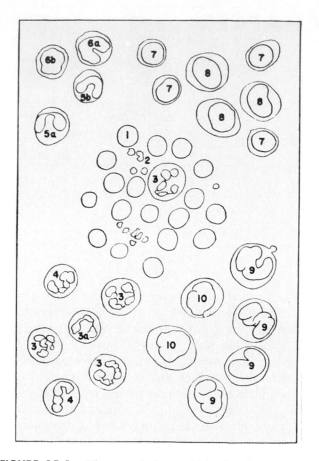

FIGURE 15-1 The morphology of blood cells. *1,* Normal red cells. *2,* Platelets. *3,* Neutrophils. *4,* Neutrophil, band form. *5a,* Eosinophil, two lobes. *5b,* Eosinophil, band form. *6a,* Basophil, band form. *6b,* Metamyelocyte, basophilic. *7,* Lymphocyte, small. *8,* Lymphocyte, large. *9,* Monocyte, mature. *10,* Monocyte, young. (From Daland GA: *A color atlas of morphologic hematology,* Cambridge, Mass, 1951, Harvard University Press.)

Continued.

blood contains antibodies to red cells of group A, group B, and group AB. Group A plasma contains antibodies to red cells of group B, and group B plasma contains antibodies to red cells of group A. Group AB plasma has no antibodies to red cells of group O, A, or B. In blood transfusions, crossmatching is necessary to prevent agglutination of donor red cells by antibodies in the plasma of the recipient. Because plasma of groups A, B, and AB has no antibodies to group O red cells, people with group O blood are called **universal donors.** Conversely, persons with AB blood are called **universal recipients,** because their plasma has no antibodies to red cells of the other three groups.

In addition to the ABO blood grouping, there are **Rh (rhesus factor)-positive** and **Rh-negative groups.**

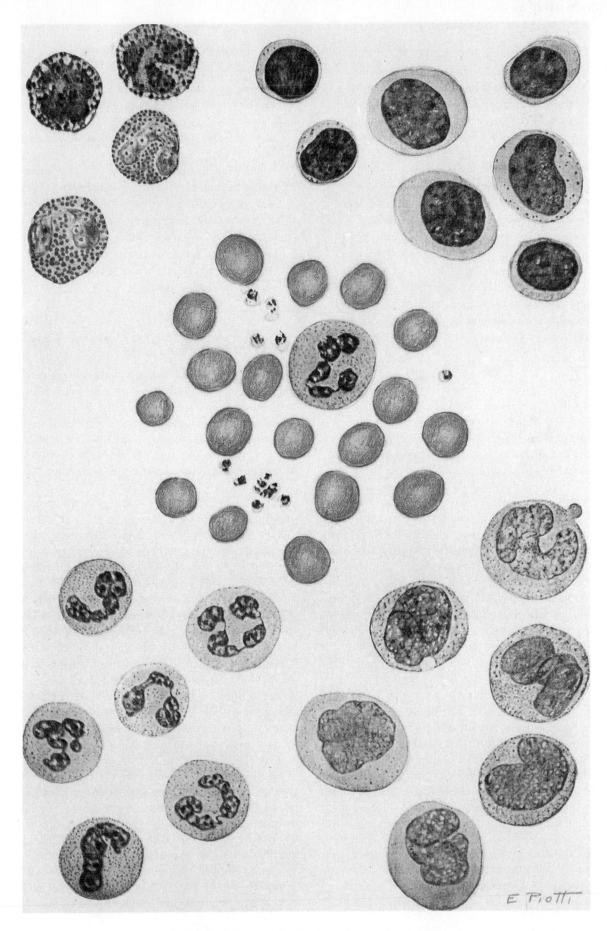

FIGURE 15-1, cont'd For legend see opposite page.

An Rh-negative person can develop antibodies to Rh-positive red cells if exposed to Rh-positive blood. This can occur during pregnancy if the mother is Rh-negative and the fetus is Rh-positive (inherited from the father). Rh-positive red cells from the fetus can enter the maternal bloodstream at the time of placental separation and induce Rh-positive antibodies in the mother's plasma. The Rh-positive antibodies from the mother can reach the fetus also via the placenta and agglutinate and hemolyze fetal red cells (**erythroblastosis fetalis,** a hemolytic disease of the newborn.) Red cell destruction can also occur in Rh-negative individuals who have previously been transfused with Rh-positive blood and have developed Rh antibodies. If these individuals are given a subsequent transfusion of Rh-positive blood, the transfused red cells will be destroyed by the Rh antibodies in their plasma.

HEMOSTASIS

When blood vessels are damaged, *three processes act to stem the flow of blood: vasoconstriction, platelet aggregation, and blood coagulation.*

Vasoconstriction

Physical injury to a blood vessel elicits a contractile response of the vascular smooth muscle and thus a narrowing of the vessel. Vasoconstriction in severed arterioles or small arteries can completely obliterate the lumen of the vessel and stop the flow of blood. The contraction of the vascular smooth muscle is probably caused by direct mechanical stimulation by the penetrating object, as well as by mechanical stimulation of the perivascular nerves.

Platelet Aggregation

Damage to the endothelium of a blood vessel engenders platelet adherence at the site of injury. The adherent platelets release **adenosine diphosphate** and **thromboxane A$_2$,** which produce adherence of additional platelets. The aggregation of platelets may continue in this manner until some of the small blood vessels become blocked by the mass of aggregated platelets. Extension of the platelet aggregate along the vessel is prevented by the antiaggregation action of **prostacyclin.** This substance is released from the normal endothelial cells in the adjacent, uninjured part of the vessel. Platelets also release **serotonin (5-hydroxytryptamine),** which enhances vasoconstriction, as well as **thromboplastin,** which hastens blood coagulation.

Bleeding of one form or another is an important clinical problem. Trauma is the most common cause of bleeding. Gastrointestinal bleeding can also occur and cause severe anemia or even cardiovascular shock, and occult blood in the stools can be the first clue in cancer of the bowel or peptic ulcer.

When the platelet count is low, as in **thrombocytopenic purpura,** tiny hemorrhages **(petechiae)** or **ecchymoses** (larger hemorrhages) may appear in the skin and mucous membranes.

Bleeding occurs into the tissues (especially joints) in **hemophilia,** a hereditary disease that afflicts humans. The disease occurs only in males, but the genetic abnormality is carried by females.

Blood Coagulation

The clotting of blood is a complex process consisting of sequential activation of various factors that are present in an inactive state in the blood. The cascade of reactions in which one activated factor activates another, and so on, is depicted in Figure 15-2. Several of the factors are synthesized in the liver, as is vitamin K, which is essential for synthesis of these liver-derived clotting factors.

The key step in blood clotting is the conversion of fibrinogen to fibrin by thrombin. The clot that is formed by this reaction consists of a dense network of fibrin strands in which blood cells and plasma are trapped (Figure 15-3). The two blood coagulation pathways, the **extrinsic pathway** and the **intrinsic pathway,** converge on the activation of factor X, which catalyzes the cleavage of prothrombin to thrombin (Figure 15-2). Blood clotting via the extrinsic pathway is initiated by tissue damage and the release of tissue thromboplastin. Blood clotting via the intrinsic pathway is initiated by exposure of the blood to a negatively charged surface. This can occur within blood vessels when the endothelium is damaged and blood comes in contact with collagen. Alternatively, it can occur outside the body when blood comes in contact with negatively charged surfaces, such as glass. If blood is carefully drawn into a syringe coated with silicone, clotting is greatly delayed.

After a clot is formed, the actin and myosin of the platelets trapped in the fibrin mesh interact in a manner similar to that in muscle. The resultant contraction pulls the fibrin strands toward the platelets, and thereby extrudes the **serum** (plasma without fibrinogen) and shrinks the clot. The process is called **clot retraction.** The function of clot retraction is not clear, but it may serve to approximate the edges of severed blood vessels.

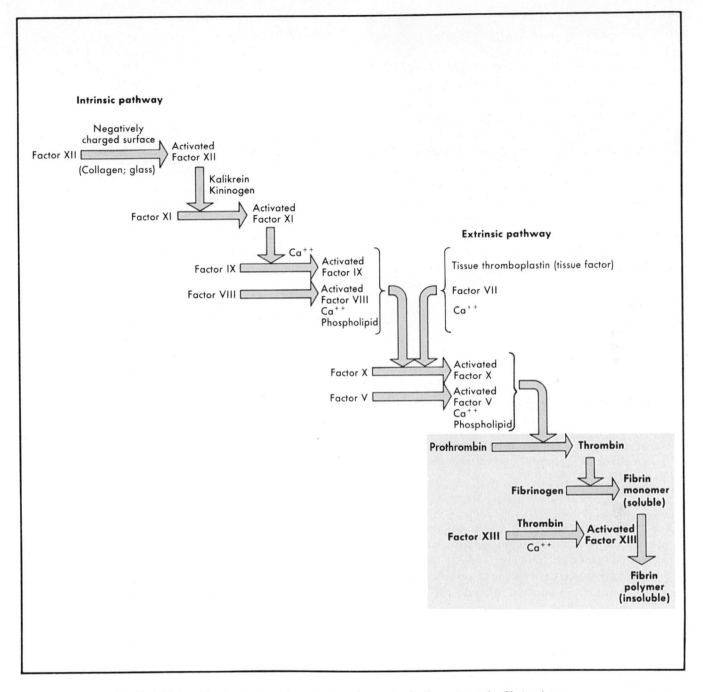

FIGURE 15-2   The intrinsic and extrinsic pathways in the formation of a fibrin clot.

Several cofactors are required for blood coagulation (Figure 15-2); the most important is calcium. If the calcium ions in blood are removed or bound, coagulation will not occur.

Clot Lysis Blood clots may be liquified (**fibrinolysis**) by a proteolytic enzyme called **plasmin.** Normal blood contains **plasminogen,** an inactive precursor of plasmin. Activators of the conversion of plasminogen to plasmin are found in tissues, plasma, and urine (**urokinase**).

Exogenous plasminogen activators, such as **streptokinase** and **tissue plasminogen activator (tPA),** are used clinically to dissolve intravascular clots. This treatment is used especially to dissolve clots in the coronary arteries of patients with **acute myocardial infarction** (damage to the heart muscle, most frequently caused by a clot in a major coronary artery).

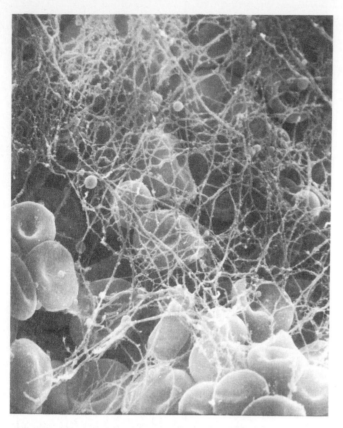

FIGURE 15-3   Human blood clot showing red blood cells immobilized within a network of fibrin threads. The small spheres are platelets. Scanning electron micrograph (×9000). (From Shelly WB: *JAMA* 249:3089, 1983.)

Anticoagulants Blood coagulation can be prevented in vitro by the addition of citrate or oxalate, which removes the calcium ions from solution. For rapid in vivo anticoagulation, **heparin,** a sulfated polysaccharide produced by mast cells, is injected intravenously.

Heparin is used in extracorporeal circuits during open-heart surgery and for prevention of intravascular clot extension. For prolonged anticoagulation, **dicumarol** is used. This drug inhibits the synthesis of vitamin K–dependent factors and is used in treating such conditions as **thrombophlebitis** (inflammation of a vein associated with an intravascular blood clot).

SUMMARY

1. Blood consists of red cells (erythrocytes), white cells (leukocytes and lymphocytes), and platelets, all suspended in a solution containing salts, proteins, carbohydrates, and lipids.
2. There are four major blood groups, O, A, B, and AB. Type O blood can be given to persons with any of the blood groups because the plasma of all of the blood groups lacks antibodies to type O red cells. Hence people with type O blood are referred to as universal donors. By the same token, people with AB blood are referred to as universal recipients because their plasma lacks antibodies to red cells of all of the blood groups. In addition to O, A, B, and AB blood groups, there are Rh-positive and Rh-negative blood groups.
3. A cascade of reactions that constitute an intrinsic pathway and an extrinsic pathway is involved in blood coagulation. The final steps where the two pathways join are (a) the conversion of prothrombin to thrombin and (b) the conversion of fibrinogen to fibrin, a reaction catalyzed by thrombin.
4. Blood clots may be liquefied by plasmin, a proteolytic enzyme whose formation from plasminogen is catalyzed by tissue activators (e.g., urokinase) or by exogenous activators (e.g., streptokinase, tissue plasminogen activator [tPA]).

BIBLIOGRAPHY
Journal Articles

Jackson CM, Nemerson Y: Blood coagulation, *Annu Rev Biochem* 49:765, 1980.

Shattil SJ, Bennett JS: Platelets and their membranes in hemostasis: physiology and pathophysiology, *Ann Intern Med* 94:108, 1981.

Books and Monographs

Babior BM, Stossel TP: *Hematology: a pathophysiological approach,* New York, 1984, Churchill Livingstone, Inc.

Eastham RD: *Clinical haematology,* ed 6, Bristol, 1984, John Wright/PSG Inc.

Erslev AJ, Gabuzda TG: *Pathophysiology of blood,* ed 3, Philadelphia, 1985, WB Saunders Co.

Ogston D: *The physiology of hemostasis,* Cambridge, 1983, Harvard University Press.

Ratnoff OD, Forbes CD, eds: *Disorders of hemostasis,* Orlando, Fla, 1984, Grune & Stratton, Inc.

16

Overview of the Circulation

The cardiovascular system is made up of a pump, a series of distributing and collecting tubes, and an extensive system of thin-walled vessels that permit rapid exchange of substances between the tissues and the vascular channels. The heart consists of two pumps in series: (1) the right ventricle, to propel blood through the lungs for exchange of oxygen and carbon dioxide, and (2) the left ventricle, to propel blood to all other tissues of the body. Unidirectional flow through the heart is achieved by the appropriate arrangement of effective flap valves.

Although the cardiac output is intermittent, continuous flow to the periphery is accomplished by distension of the aorta and its branches during ventricular contraction **(systole)** and elastic recoil of the walls of the large arteries with forward propulsion of the blood during ventricular relaxation **(diastole).** Blood moves rapidly through the aorta and its arterial branches. The branches become narrower and their walls become thinner and change histologically toward the periphery. The aorta is predominantly an elastic structure. However, the peripheral arteries are more muscular, and in the arterioles the muscular layer predominates (Figure 16-1).

In the aorta and the large arteries, frictional (viscous) resistance to blood flow is relatively small, and the pressure drop from the root of the aorta to the small arteries is also relatively small (Figure 16-2). However, in the small arteries and arterioles, resistance to blood flow is large and the pressure drop across these vessels is also large. The greatest resistance is in the arterioles, which are sometimes referred to as the "stopcocks" of the circulatory system. The degree of contraction in the circular muscle of these small vessels regulates tissue blood flow and aids in controlling the arterial blood pressure.

In addition to a sharp reduction in pressure across the arterioles, flow changes from pulsatile to steady. The pulsatile arterial blood flow, caused by the intermittent cardiac ejection, is damped at the capillary level by the combination of distensibility of the large arteries and frictional resistance in the small arteries and arterioles. Many capillaries arise from each arteriole. Hence the total cross-sectional area of the capillary bed is very large, despite the fact that the cross-sectional area of each capillary is less than that of each arteriole. As a result of the large total cross-sectional area, blood flow velocity slows considerably in the capillaries, just as the flow velocity decreases at the wide regions of a river. Because the capillaries consist of short tubes with walls only one cell thick, and because flow velocity is slow, conditions in the capillaries are ideal for the exchange of diffusible substances between blood and tissue.

On its return to the heart from the capillaries, blood passes through venules and then through veins of increasing caliber and decreasing number. The thickness and composition of the vein walls change (see Figure 16-1), the total cross-sectional area of the veins diminishes, and the velocity of blood flow in the veins increases (see Figure 16-2). Also, most of the blood in the systemic circulation is located in the venous vessels (see Figure 16-2). Conversely, the blood in the pulmonary vascular bed is about equally divided among the arterial, capillary, and venous vessels.

Blood entering the right ventricle from the right atrium is pumped through the pulmonary arterial system at a mean pressure about one-seventh that in the systemic arteries. The blood then passes through the lung capillaries, where carbon dioxide is released and oxygen is taken up. The oxygen-rich blood returns via the pulmonary veins to the left atrium and ventricle to complete the cycle. The systemic and pulmonary circulation systems are diagrammed in Figure 16-3.

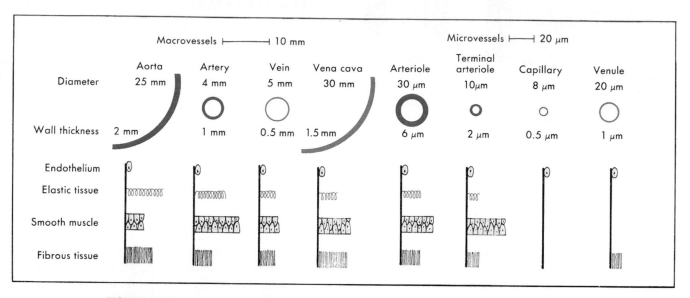

FIGURE 16-1 Internal diameter, wall thickness, and relative amounts of the principal components of the vessel walls of the various blood vessels that compose the circulatory system. Cross sections of the vessels are not drawn to scale because of the huge range from aorta and venae cavae to capillary. (Redrawn from Burton AC: *Physiol Rev* 34:619, 1954.)

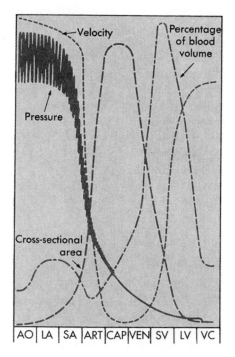

FIGURE 16-2 Pressure, velocity of flow, cross-sectional area, and capacity of the blood vessels of the systemic circulation. The important features are the inverse relationship between velocity and cross-sectional area, the major pressure drop across the small arteries and arterioles, the maximal cross-sectional area and minimal flow velocity in the capillaries, and the large capacity of the venous system. The small but abrupt drop in pressure in the venae cavae indicates the point of entrance of these vessels into the thoracic cavity and reflects the effect of the negative intrathoracic pressure. To permit schematic representation of velocity and cross-sectional area on a single linear scale, only approximations are possible at the lower values. *AO,* Aorta; *LA,* large arteries; *SA,* small arteries; *ART,* arterioles; *CAP,* capillaries; *VEN,* venules; *SV,* small veins; *LV,* large veins; *VC,* venae cavae.

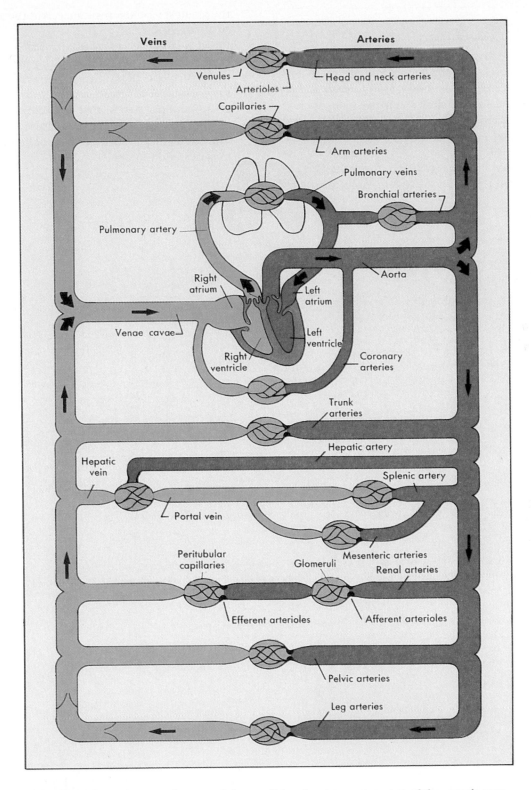

FIGURE 16-3   Schematic diagram of the parallel and series arrangement of the vessels composing the circulatory system. The capillary beds are represented by thin lines connecting the arteries (on the right) with the veins (on the left). The black crescent-shaped thickenings proximal to the capillary beds represent the arterioles (resistance vessels). (Redrawn from Green HD: In Glasser O, ed: *Medical physics,* vol 1, Chicago, 1944, Year Book Medical Publishers, Inc.)

SUMMARY

1. The aorta and large arteries are predominantly elastic tubes, whereas the smaller arteries are more muscular. The arterioles consist of smooth muscle and endothelium, and the capillaries consist solely of endothelium. The veins are thinner than their corresponding arteries and contain relatively less smooth muscle and elastic tissue.

2. Blood pressure falls progressively from the aorta to the vena cavae; the greatest pressure drop, and hence the greatest vascular resistance, is at the arterioles and small arteries.

3. Blood velocity decreases from aorta to capillaries, and then increases from capillaries to vena cavae. The velocity of blood flow in each segment of the circulatory system is inversely related to the total vascular cross-sectional area of that segment.

4. Most of the blood in the vasculature lies in the veins.

5. The circulatory system consists of conduits arranged in series and in parallel.

17

Electrical Activity of the Heart

TRANSMEMBRANE POTENTIALS OF CARDIAC CELLS

The electrical behavior of cardiac cells differs considerably from that of nerve cells or of smooth or skeletal muscle cells (Chapters 3, 13, and 14). In general, the durations of the action potentials are much longer in cardiac cells than in nerve cells or in smooth or skeletal muscle cells. Furthermore, the action potentials differ substantially among various types of cardiac cells, depending upon the function and location of those cells.

Figure 17-1, *A,* shows the potential changes recorded from a ventricular muscle cell immersed in an *electro-* lyte solution. When a microelectrode and a reference electrode are placed in the solution near the quiescent cell, no measurable potential difference *(a)* exists between the two electrodes. At *b* the microelectrode is inserted into the interior of the cell. Immediately a potential difference is recorded across the cell membrane; the potential of the cell interior is about 90 mV lower than that of the surrounding medium. Such electronegativity of the cell interior is also characteristic of skeletal and smooth muscle, of nerve, and indeed of most cells within the body (Chapter 2).

At *c* the cell is stimulated and the cell membrane rapidly **depolarizes;** that is, the potential difference across the cell membrane tends to disappear. Actually, however, the potential difference across the cell membrane becomes reversed, such that the potential of the interior of the cell exceeds that of the exterior by about 20 mV. The rapid upstroke of the **action potential** is designated **phase 0.** A brief period of partial repolarization **(phase 1)** occurs immediately after the upstroke, and it is followed by a plateau **(phase 2)** that persists for about 0.2 sec. The internal potential then becomes negative again, and it becomes progressively more negative **(phase 3)** until the resting potential is again attained (at *e*). This **repolarization** (phase 3) proceeds more slowly

than does the **depolarization** (phase 0). The interval from the completion of repolarization until the beginning of the next action potential is designated the **resting membrane potential,** or **phase 4.**

The temporal relationships between the electrical and mechanical behavior of cardiac muscle are shown in Figure 17-2. Note that rapid depolarization (phase 0) precedes force development and that repolarization is completed at about the same time that peak force is attained; the subsequent decline in force denotes relaxation. Thus, the duration of force generation parallels the duration of the action potential.

Principal Types of Cardiac Action Potentials

Two main types of action potentials may be recorded in the heart (see Figure 17-1). One type, the **fast response** (panel *A*), occurs in atrial and ventricular myocardial fibers and in specialized conducting fibers **(Purkinje fibers)** that exist mainly in the endocardial surfaces of the ventricles. The other type of action potential, the **slow response** (panel *B*), is found in the **sinoatrial (SA) node,** the natural pacemaker region of the heart, and the **atrioventricular (AV) node,** the specialized tissue that conducts the cardiac impulse from atria to ventricles.

Fast responses may change to slow responses under certain pathological conditions. For example, in patients with coronary artery disease, when a region of cardiac muscle is deprived of its normal blood supply, the K^+ concentration in the interstitial fluid bathing the affected muscle cells rises because K^+ is lost from the inadequately perfused **(ischemic)** cells. The action potentials in some of these cells may then be converted from fast to slow responses. An experimental conversion from a fast to a slow response is illustrated later in this chapter (see Figure 17-8).

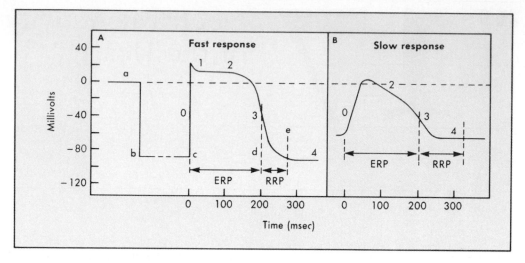

FIGURE 17-1   Changes in transmembrane potential recorded from a fast-response (**A**) and a slow-response (**B**) cardiac fiber, in isolated cardiac tissue immersed in an electrolyte solution. **A,** At time *a,* the microelectrode was in the solution surrounding a cardiac fiber. At time *b,* the microelectrode was introduced into the fiber. At time *c* an action potential was initiated in the impaled fiber. Time *c* to *d* represents the effective refractory period *(ERP),* and time *d* to *e* represents the relative refractory period *(RRP).* **B,** An action potential recorded from a slow-response cardiac fiber. Note that, compared with the fast-response fiber, the resting potential of the slow fiber is less negative, the upstroke (phase 0) of the action potential is less steep, the amplitude of the action potential is smaller, phase 1 is absent, and the relative refractory period (RRP) extends well into phase 4, after the fiber has fully repolarized.

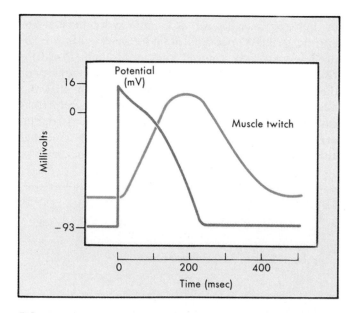

FIGURE 17-2   Time relationships between the mechanical force developed by a thin strip of ventricular muscle and the changes in transmembrane potential. (Redrawn from Kavaler F, Fisher VJ, Stuckey JH: *Bull NY Acad Med* 41:592, 1965.)

As shown in Figure 17-1, the resting membrane potential of the slow response is considerably less negative than that of the fast response. Also the slope of the upstroke (phase 0) and the amplitude and overshoot of the slow response action potentials are less than the corresponding values for the fast response action potentials. The amplitude of the action potential and the rate of rise of the upstroke are important determinants of the conduction velocity, as described below. Hence, in tissues composed of slow response fibers, conduction velocity is diminished and impulses more frequently fail to be conducted than in tissues composed of fast response fibers.

The fast and slow fiber types are affected differentially by certain types of drugs. For example, conduction in fast-response fibers is inhibited by Na^+ channel antagonists, whereas conduction in slow-response fibers is inhibited by Ca^{++} channel antagonists, as explained in the next section. Many of the drugs that are used clinically to treat cardiac rhythm disturbances are Na^+ or Ca^{++} channel antagonists.

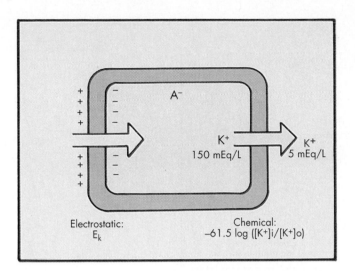

FIGURE 17-3 The balance of chemical and electrostatic forces acting on a resting cardiac cell membrane, based on 30:1 ratio of the intracellular to extracellular K^+ concentrations, and the presence of a nondiffusible anion *(A⁻)* inside but not outside the cell.

IONIC BASIS OF THE MEMBRANE POTENTIAL

The various phases of the cardiac action potential are associated with changes in the conductance of the cell membrane, mainly to sodium (Na^+), potassium (K^+), and calcium (Ca^{++}) ions. The conductance to an ion is an index of the permeability of the membrane to that ion, as explained in Chapter 2. Each phase of the action potential is associated with a change in conductance to one or more specific ions.

Resting Potential

The resting potential (phase 4) depends mainly on the conductance of the cell membrane to K^+. Just as with all other cells in the body, the concentration of potassium ions inside a cardiac muscle cell, $[K^+]_i$, greatly exceeds the concentration outside the cell, $[K^+]_o$ (Figure 17-3). The concentration gradients for Na^+ and Ca^{++} are opposite to that for K^+. Estimates of the extracellular and intracellular concentrations of Na^+, K^+, and Ca^{++}, and of the **Nernst equilibrium potentials** (Chapter 2) for these ions, are shown in Table 17-1. The ability of the resting cell membrane to conduct K^+ greatly exceeds its ability to conduct Na^+ or Ca^{++}. Thus, *the resting membrane potential will be determined mainly by the ratio of the intracellular to the extracellular K^+ concentration (i.e., by $[K^+]_i/[K^+]_o$).*

This assertion has been verified experimentally. Changes in the extracellular concentrations of Na^+ or

Table 17-1 Intracellular and Extracellular Ion Concentrations and Equilibrium Potentials in Cardiac Muscle Cells

Ion	Extracellular Concentrations (mM)	Intracellular Concentrations (mM)*	Equilibrium Potential (mV)
Na^+	145	10	70
K^+	4	135	−94
Ca^{++}	2	10^{-4}	132

Modified from Ten Eick RE et al: *Prog Cardiovasc Dis* 24:157, 1981.
*The intracellular concentrations are estimates of the free concentrations in the cytoplasm.

Ca^{++} scarcely affect the resting potential (V_m). However, when $[K^+]_i/[K^+]_o$ is decreased experimentally by raising $[K^+]_o$, the measured value of the resting potential (V_m) approximates that predicted by the Nernst equation (E_K) for K^+ (Figure 17-4). For extracellular K^+ concentrations above 5 mM, the measured values of V_m correspond closely with the predicted values. The measured values are slightly less negative than those predicted by the Nernst equation because of the small but finite conductance (g_{Na}) of the resting cell membrane to Na^+.

Fast-Response Action Potential

Genesis of the Upstroke Fast-response cardiac fibers, notably myocardial fibers, do not ordinarily initiate cardiac impulses. Such impulses usually originate in nodal tissue, especially in the SA node. Therefore, a given myocardial fiber is generally excited when an action potential, originating from some distant automatic cell, arrives at that fiber by the process of cell-to-cell conduction. The action potentials in fibers adjacent to the given fiber cause V_m to be less negative in the given fiber. When this change in V_m attains a threshold value, certain voltage-sensitive channels, called **fast Na^+ channels,** open quickly; they are said to be **activated.** Na^+ then quickly enters the myocardial cell; the process is so rapid because (1) the fast Na^+ channels are abundant in the membranes of fast response fibers, and therefore the conductance (g_{Na}) of the cell membrane to Na^+ increases substantially, (2) the interior of the cell is negatively charged, and therefore Na^+ is pulled into the cell by the electrostatic attraction, and (3) $[Na^+]_o$ greatly exceeds $[Na^+]_i$ (Table 17-1), and therefore the net diffusional forces favor the inward movement of Na^+.

The characteristics of the upstroke of the action potential in fast-response cardiac fibers depend almost entirely on the influx of Na^+; other ions are unimportant. As shown in Figure 17-5, the amplitude of the cardiac action

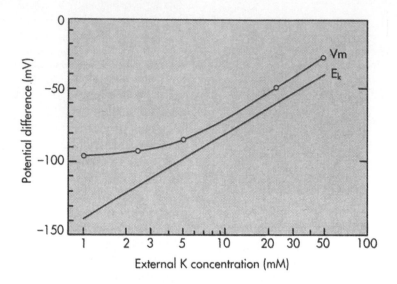

FIGURE 17-4   Transmembrane potential of a cardiac muscle fiber varies inversely with the potassium concentration of the external medium *(red curve)*. The oblique black line represents the change in transmembrane potential predicted by the Nernst equation for E_K. (Redrawn from Page E: *Circulation* 26:582, 1962.)

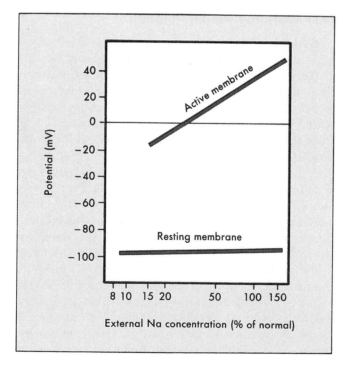

FIGURE 17-5   The concentration of sodium (logarithmic scale) in the external medium is the main determinant of the action potential amplitude *(upper curve)* in cardiac muscle, but it has relatively little influence on the resting potential *(lower curve)*. (Redrawn from Weidmann S: *Elektrophysiologie der Herzmuskelfaser,* Bern, 1956, Verlag Hans Huber.)

potential (i.e., the magnitude of the potential change during phase 0) varies linearly with the logarithm of $[Na^+]_o$. *Conversely, changes in $[Na^+]_O$ have very little effect on the resting membrane potential.*

The inrush of Na^+ into the myocardial cell ceases within 1 or 2 msec after the excitation of the cell, for two principal reasons. First, as V_m becomes less negative and approaches the Nernst equilibrium potential for Na^+ (Table 17-1), the electrostatic force that tends to pull Na^+ into the cell progressively diminishes and then it tends to repel the influx of Na^+ as V_m becomes positive. Second, and even more important, the fast Na^+ channels close (that is, they are **inactivated**) very soon after they open. Hence, g_{Na} quickly returns to its low, preactivation value (Figure 17-6). The Na^+ channels do not recover from inactivation, and the cardiac cell remains inexcitable until the cell membrane has almost fully repolarized, as explained below in the section on cardiac excitability.

Genesis of Phase 1   In fast-response cardiac cells, phase 1 is a brief period of limited repolarization that occurs immediately after the action potential upstroke; this limited repolarization may or may not be accompanied by a notch between the action potential upstroke and the plateau. Two mechanisms are mainly responsible for phase 1. In those cells (e.g., ventricular endocardial cells) in which a distinct notch is not evident, phase 1 mainly reflects the initial inactivation of the fast sodium

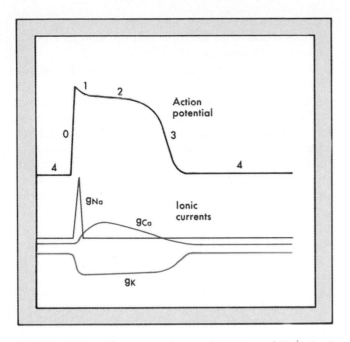

FIGURE 17-6 Changes in the conductances of Na$^+$ (g_{Na}), Ca^{++} (g_{Ca}), and K$^+$ (g_K) during the various phases of the action potential of a fast-response cardiac cell. The conductance diagrams indicate directional changes only.

is positive than when V_m is negative. This dependence of the conductance on membrane polarity is referred to as **rectification.** Such reduction in g_K during the action potential plateau (Figure 17-6) protects the cell from an excessive loss of K$^+$.

During the plateau, the efflux of K$^+$ from the cell is balanced electrically by the influx of Ca^{++}, which enters the cell through specific Ca^{++} channels. The Ca^{++} channels are activated when V_m reaches the threshold voltage of about -35 mV. Opening of these channels is reflected by an increased Ca^{++} conductance (g_{Ca}), which begins shortly after the upstroke of the action potential (Figure 17-6). Because of the increase in g_{Ca}, and because the Ca^{++} concentration is much less inside than outside the cardiac cell (Table 17-1), Ca^{++} enters the cardiac cell throughout the plateau. This Ca^{++} influx is involved in **excitation-contraction coupling,** as described in Chapters 12 and 18.

channels (Figure 17-6); that is, the small deflection from the peak of the upstroke to the action potential plateau is caused by the closing of some of the fast Na$^+$ channels. In those cells (e.g., ventricular epicardial cells and Purkinje fibers) that are characterized by a prominent notch (see Figure 17-8, *A*), phase 1 also reflects the activation of a specific K$^+$ current (the so-called **transient outward current**). These K$^+$ channels open briefly, and the transient efflux of K$^+$ produces the notch at the very beginning of the plateau.

Genesis of the Plateau During the plateau (phase 2) of the action potential, V_m is slightly positive and it remains fairly constant for about 100 to 300 msec, depending on the type of fiber. The relative constancy of the membrane potential indicates that any efflux of cations has been balanced electrically by an influx of cations. The principal cations that cross the cell membrane during phase 2 are K$^+$ and Ca^{++}; K$^+$ leaves the cell, whereas Ca^{++} enters it.

In the resting cell, the conductance of K$^+$ exceeds that for any other relevant cation, as stated above. Relatively little K$^+$ leaves the resting cell, because the electrical and chemical forces across the cell membrane are almost balanced (see Figure 17-3). During the plateau, however, V_m is positive, and therefore both chemical and electrical forces act to expel K$^+$ from the cell. The efflux of K$^+$ is minimized, however, because the conductance of the relevant K$^+$ channels is much less when V_m

Various drugs and neurotransmitters may influence the Ca^{++} current. This current may be increased by **catecholamines,** such as **epinephrine** and **norepinephrine.** An increase in Ca^{++} current is a crucial step in the mechanism by which neurally released or circulating catecholamines strengthen myocardial contraction. Conversely, **Ca^{++} channel antagonists,** such as **verapamil, nifedipine,** and **diltiazem,** impede the Ca^{++} current. By reducing the amount of Ca^{++} that enters the myocardial cells, these drugs weaken the cardiac contraction (Figure 17-7), diminish the firing frequency of SA node cells (see Figure 17-11), and retard conduction in AV nodal fibers (see Figure 17-15). Also, by altering the balance between Ca^{++} influx and K$^+$ efflux, the Ca^{++} channel antagonists diminish the level of V_m during the plateau and abridge the duration of the plateau (see Figure 17-7). The Ca^{++} channel antagonists are widely used clinically in the treatment of cardiac **rhythm disturbances, heart failure,** and **hypertension.**

Genesis of Repolarization Final repolarization (phase 3) is achieved by a much greater efflux than influx of relevant cations across the cardiac cell membrane. This imbalance is achieved by two main processes: (1) an increase in K$^+$ efflux, by virtue of the return of g_K back toward the level that prevails in the resting cell membrane, and (2) a decrease in Ca^{++} influx, by virtue of a reduction in g_{Ca} (see Figure 17-6). The reduction in g_{Ca} principally reflects the inactivation of the Ca^{++} channels. The increase in g_K is mediated by changes in at least three different types of specific K$^+$ channels in the cell mem-

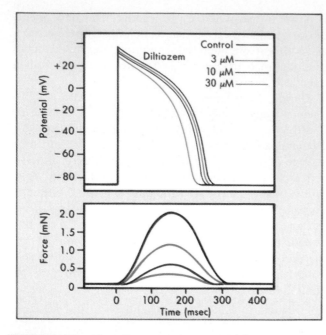

FIGURE 17-7   The effects of diltiazem, a Ca^{++} channel blocking drug, on the action potentials (in millivolts) and isometric contractile forces (in millinewtons) recorded from an isolated papillary muscle of a guinea pig. The tracings were recorded under control conditions and in the presence of diltiazem, in concentrations of 3, 10, and 30 μmol/L. (Redrawn from Hirth C, Borchard U, Hafner D: *J Mol Cell Cardiol* 15:799, 1983.)

brane. An important component of the change in g_K is the channel rectification cited above. As repolarization proceeds and V_m changes from positive to negative, the conductance of certain K^+ channels suddenly increases substantially, which reflects the rectification. The efflux of K^+ during phase 3 rapidly restores the resting level of the membrane potential (phase 4).

Any alterations in the intracellular concentrations of Na^+ and K^+ produced by the ionic fluxes during the various phases of the action potential are corrected principally by the activity of the Na^+-K^+ pump (ATPase), as described in Chapter 1. Similarly, alterations in the intracellular concentration of Ca^{++} are corrected mainly by the Na^+/Ca^{++} exchanger.

Slow-Response Action Potential

Fast-response action potentials (see Figure 17-1, *A*) consist of three principal components, a spike (a very steep upstroke, often accompanied by a notch), a plateau (phase 2), and a repolarization (phase 3). The upstroke is produced by activation of fast Na^+ channels. In contrast, in slow-response fibers (see Figure 17-1, *B*), such as those in the SA and AV nodes, the resting membrane potential is much less negative than in fast-response fibers, the upstroke rises much more gradually to the pla-

teau, the action potential amplitude is smaller, and the notch is absent. The upstroke and the plateau are produced by activation of Ca^{++} channels.

When the fast Na^+ channels are blocked in fast-response fibers by specific antagonists, such as the puffer fish toxin, **tetrodotoxin,** slow responses may be generated in those same fibers under appropriate conditions; this toxin does not affect Ca^{++} channels. The Purkinje fiber action potentials shown in the first four panels of Figure 17-8 clearly exhibit a steep upstroke and notch prior to the plateau. With progressively greater concentrations of tetrodotoxin (panels *B* to *E*) to block more and more fast Na^+ channels, the steep upstroke and notch become progressively less prominent. Thus, tetrodotoxin has a pronounced effect on the upstroke and notch, but only a negligible effect on the plateau. In panel *E,* the steep upstroke and notch are absent, and the action potential is typical of the slow response.

CONDUCTION IN CARDIAC FIBERS

An action potential traveling down a myocardial fiber is propagated by local circuit currents, just as in nerve and skeletal muscle fibers (see Figure 3-13). The electrical potential at the surface of the depolarized zone differs from that in the polarized (resting) region of the myocardial fiber. The interstitial fluid between myocardial fibers is an electrolyte solution and thus conducts electricity well. Therefore, electrical currents flow through the interstitial fluid between the polarized and depolarized zones. At the border between the polarized and depolarized zones, these local currents will act to depolarize the region of the resting fiber adjacent to the depolarized zone. Hence, the margin between the polarized and depolarized zones moves in the direction of the polarized zone. *This movement constitutes the propagation of the action potential.*

In fast-response fibers, the fast Na^+ channels will be activated when the transmembrane potential is suddenly brought to the threshold value of about −70 mV. The inward Na^+ current will then depolarize the marginal cells very rapidly, and this repetitive process will move rapidly down the fiber as a wave of depolarization (Figure 3-13, *B*). The fast-response conduction velocities are about 0.3 to 1 m/sec for myocardial cells and about 1 to 4 m/sec for the specialized conducting fibers in the atria and ventricles.

Local circuits also propagate the cardiac impulse in slow-response fibers. However, the characteristics of the conduction process differ from those of the fast response. The threshold potential is about -40 mV for exciting the adjacent polarized region in a slow-response fiber, and the conduction velocity is much less. The con-

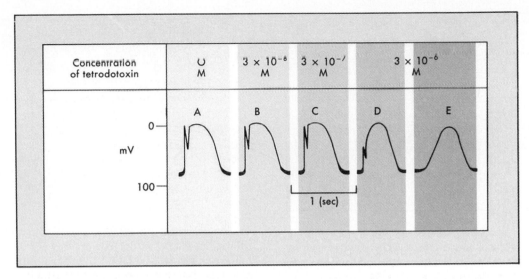

FIGURE 17-8   Effect of tetrodotoxin on the action potential recorded in a calf Purkinje fiber perfused with a solution containing epinephrine and 10.8 mM K^+. The concentration of tetrodotoxin was 0 M in *A*, 3×10^{-8} M in *C*, and 3×10^{-6} M in *D* and *E; E* was recorded later than *D*. (Redrawn from Carmeliet E, Vereecke J: *Pflugers Arch* 313:300, 1969.)

duction velocities of the slow responses in the SA and AV nodes are only about 0.02 to 0.1 m/sec. *Slow-response fibers are more likely to be blocked by certain drugs (such as **digitalis** and Ca^{++} channel antagonists) and by certain pathological processes (such as those induced by an inadequate blood supply) than are fast-response fibers.* Also, slow-response fibers cannot conduct as many impulses per second as can fast-response fibers.

CARDIAC EXCITABILITY

The excitability of a cardiac cell is the ease with which it can be activated. One way to measure the excitability of a cardiac cell is to measure how much electrical current is necessary to induce an action potential. *Changes in cardiac excitability are important because they generate many cardiac rhythm disturbances, and they must be considered in the design and use of artificial pacemakers and other electrical devices for correcting life-threatening rhythm abnormalities.* Fast- and slow-response fibers differ considerably in excitability.

Fast-Response Fibers

Once a fast-response action potential has been initiated, the depolarized cell will no longer be excitable until the middle of the period of final repolarization (phase 3). The interval from the beginning of the action potential until the time the fiber can conduct another action potential is called the **effective refractory period.** In the fast response, this period extends from the beginning of phase 0 to the time in phase 3 at which V_m has reached about -50 mV (period *c* to *d* in Figure 17-1, *A*). At this value of V_m some of the fast Na^+ channels have begun to recover from inactivation.

Full excitability is not regained until the cardiac fiber has been fully repolarized (point *e* in Figure 17-1, *A*). During period *d* to *e* in the figure, an action potential may be evoked, but only when the stimulus is stronger than that which elicits a response during phase 4. Period *d* to *e* is called the **relative refractory period.**

Premature depolarizations of the atria or ventricles are commonly occurring disturbances of cardiac rhythm. They occur occasionally in most normal individuals, and they occur much more frequently in various cardiac diseases. When a premature depolarization arises during the relative refractory period of an antecedent excitation, its characteristics vary with the membrane potential that exists at that time.

The dependency of the premature depolarization on the prevailing transmembrane potential is illustrated in Figure 17-9. As a myocardial fiber is stimulated later and later in its relative refractory period, the amplitude and the rate of rise of the upstroke of the premature action potential increase progressively. Presumably, the number of fast Na^+ channels that have recovered from inactivation increases as repolarization proceeds during phase

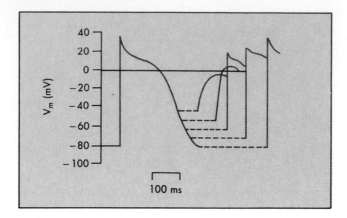

FIGURE 17-9 The changes in action potential amplitude and slope of the upstroke as premature action potentials are initiated at different stages of the relative refractory period of the preceding excitation. (Redrawn from Rosen MR, Wit AL, Hoffman BF: *Am Heart J* 88:380, 1974.)

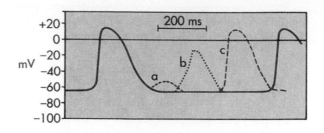

FIGURE 17-10 The effects of excitation at various times after the initiation of an action potential in a slow-response fiber. In this fiber, excitation very late in phase 3 (or early in phase 4) induces a small, nonpropagated (local) response (*a*). Later in phase 4, a propagated response (*b*) may be elicited; its amplitude is small and the upstroke is not very steep. This response (*b*) will be conducted very slowly. Still later in phase 4, full excitability will be regained, and the response (*c*) will display its normal characteristics. (Modified from Singer DH et al: *Prog Cardiovasc Dis* 24:97, 1981.)

3. Therefore, the propagation velocity of the cardiac impulse increases. Once the fiber is fully repolarized, its excitability has been fully restored, and an evoked response is constant no matter when in phase 4 the stimulus is applied.

Slow-Response Fibers

The relative refractory period in slow-response fibers frequently extends well beyond the time that full repolarization has been restored (Figure 17-1, *B*). This characteristic has been termed **post-repolarization refractoriness.** Even after the cell has completely repolarized (phase 4), a relatively strong stimulus may be required to evoke a propagated response. Hence, the recovery of full excitability is much slower than it is for the fast response.

Until excitability is fully restored, the characteristics of the evoked action potentials and the velocity of the propagated impulses vary with the excitability (Figure 17-10). Action potentials that are induced in a slow-response fiber early in its relative refractory period (e.g., *b* in Figure 17-10) are smaller and have a more gradual upstroke than those (e.g., *c* in Figure 17-10) induced later in the period of full repolarization. Furthermore, such early action potentials will be propagated much more slowly than those arriving late in that period. The lengthy refractory periods also explain why conduction tends to be blocked in slow-response fibers. For example, a strong stimulus delivered very early in the relative refractory period may elicit only a feeble depolarization (such as *a* in Figure 17-10) that will not be propagated at all. Even after the fiber has fully repolarized, an impulse that arrives early in the period of full repolarization (phase 4) may not be conducted unless that impulse is well above the threshold value for an impulse that arrives later in phase 4.

Effects of Cycle Length

Changes in cycle length alter the action potential duration of cardiac cells and thus change their refractory periods. Consequently, *the time between consecutive depolarizations is often an important factor in initiating or terminating certain arrhythmias.* The magnitude of the change in action potential duration evoked by a given change in cycle length varies substantially in different types of cardiac cells. The decreases in action potential duration produced by various reductions in cycle length in a Purkinje fiber are shown in Figure 17-11.

The mechanism responsible for the correlation between action potential duration and cycle length is not fully understood. Changes in K^+ conductance (g_K) that involve two specific types of K^+ channels, namely those that conduct the delayed rectifier current and the transient outward currents, appear to be involved. The **delayed rectifier current** activates and inactivates very slowly. Thus, the shorter the time between consecutive depolarizations, the earlier a given depolarization falls in the inactivation period of the delayed rectifier current from the preceding depolarization. Hence, the diminished inactivation of this specific K^+ current equates to a greater g_K during the next depolarization. The increased g_K then induces an earlier repolarization, and hence the action potential is abbreviated.

The second K^+ current that contributes to the relation between cycle length and action potential duration is the **transient outward current.** This is the same current that produces the notch in the action potentials of

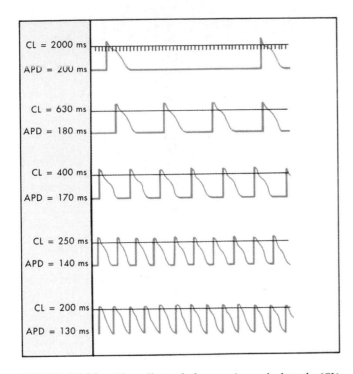

FIGURE 17-11   The effect of changes in cycle length *(CL)* on the action potential duration *(APD)* of canine Purkinje fibers. (Modified from Singer D, Ten Eick RE: *Am J Cardiol* 2:381, 1971.)

certain types of cardiac cells, notably Purkinje fibers and epicardial ventricular cells (e.g., see Figure 17-8, *A* to *C*). A strong correlation exists between the prominence of the transient outward current and the rate dependency of action potential duration in cardiac cells. However, the mechanism responsible for the suspected causal relationship remains to be established.

NATURAL EXCITATION OF THE HEART

The properties of **automaticity** (the ability to initiate a heart beat) and of **rhythmicity** (the frequency and regularity of such pacemaking activity) are intrinsic to cardiac tissue. *The heart will continue to beat for some time even when it is completely removed from the body, and therefore devoid of any influence of the central nervous system.* If the coronary vessels are perfused with an appropriate solution, the heart will contract rhythmically for many hours. At least some cells in each cardiac chamber can initiate beats; such automatic cells reside mainly in the nodal and specialized conducting tissues. The nervous system affects the frequency at which the heart will beat, and it influences other important cardiac functions as well. However, the large number of patients who have had cardiac transplants and who lead relatively normal lives have established irrefut-

ably that intact nervous pathways are not essential for effective cardiac function.

In the mammalian heart the automatic cells that ordinarily fire at the highest frequency are located in the **sinoatrial (SA) node**; this structure is the natural pacemaker of the heart. Other regions of the heart that can initiate beats under special circumstances are called **ectopic pacemakers.** Ectopic pacemakers may become dominant when (1) their own rhythmicity is enhanced, (2) the more rhythmic pacemakers are depressed, or (3) all conduction pathways between the ectopic focus and the more rhythmic foci are blocked.

When the SA node is destroyed, automatic cells in the **AV node** usually have the next highest level of rhythmicity, and they become the pacemakers for the entire heart. After some time, which may vary from minutes to days, automatic cells in the atria usually then become dominant. In the dog, the most common site for impulse initiation is at the junction of the inferior vena cava and the right atrium.

Purkinje fibers in the specialized conduction system of the ventricles are also automatic. Characteristically, these **idioventricular pacemakers** fire at a very slow rate (about 35 beats/min). Ordinarily, they do not fire at all, because impulses that originate in the SA node depolarize the Purkinje fibers at a frequency that is much greater than their intrinsic frequency. Thus, the automaticity of the Purkinje fibers is inhibited by the process of **overdrive suppression,** which is explained below.

Various processes may interfere temporarily or permanently with the conduction of the cardiac impulse from atria to ventricles. Such processes include intense neural activity in the vagus nerves, the action of certain drugs (such as **digitalis, adenosine,** and **Ca^{++} channel antagonists),** and certain pathological processes (such as coronary artery occlusion and degeneration of the conducting fibers). When the AV junction fails to conduct the cardiac impulse from the atria to the ventricles, the Purkinje fibers in the ventricles serve as idioventricular pacemakers that initiate ventricular contractions. However, they ordinarily generate impulses at such a low rate that the heart cannot pump enough blood to support normal body function. An artificial pacemaker may then be required to correct this deficiency.

Sinoatrial Node

The SA node is the phylogenetic remnant of the sinus venosus of lower vertebrate hearts. In humans it is

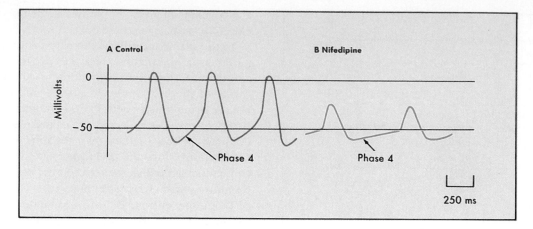

FIGURE 17-12   The effects of nifedipine (5.6×10^{-7}; M), a Ca^{++} channel antagonist, on the transmembrane potentials recorded from a rabbit's SA node cell. (From Ning W, Wit AL: *Am Heart J* 106:345, 1983.)

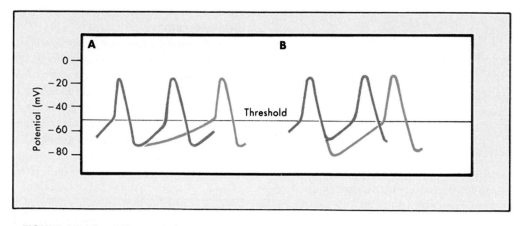

FIGURE 17-13   Effects of changes in the slope of the pacemaker potential **(A)** or in the maximum diastolic potential **(B)** on the firing period of an SA nodal cell.

about 15 mm long, 5 mm wide, and 2 mm thick. It lies in the terminal sulcus on the posterior aspect of the heart, at the junction of the superior vena cava and the right atrium.

Typical transmembrane action potentials recorded from an SA node cell are depicted in Figure 17-12, *A*. Compared with an action potential recorded from a ventricular myocardial cell (Figure 17-1, *A*), the minimum potential of the SA node cell is less negative, the upstroke is less steep, the plateau is abridged, and repolarization is more gradual. These are all characteristic of the slow response. Tetrodotoxin does not influence the SA node action potential, because the upstroke of the action potential is not produced by an influx of Na^+ through fast Na^+ channels.

The principal distinguishing feature of an automatic fiber resides in phase 4. In nonautomatic cells (see Figure 17-1), the transmembrane potential remains constant during phase 4, regardless of whether the cell is a fast-

or a slow-response fiber. However, an automatic fiber displays a slow **diastolic depolarization** (also called the **pacemaker potential**) during phase 4 (Figure 17-12). Depolarization proceeds at a steady rate until the threshold for firing is attained, and then an action potential is triggered.

The firing frequency of an automatic cell is usually varied by changing either the slope of the slow diastolic depolarization or the maximum negativity during phase 4. When the slope of the diastolic depolarization decreases (Figure 17-13, *A*), more time is required for the transmembrane potential to reach threshold, and therefore the firing frequency diminishes. Similarly, if a more negative transmembrane potential is achieved at the beginning of phase 4 (i.e., if the membrane becomes **hyperpolarized**), more time will again be required for the slow diastolic depolarization to reach threshold (Figure 17-13, *B*), and therefore the firing frequency also will diminish. Of course, a combination

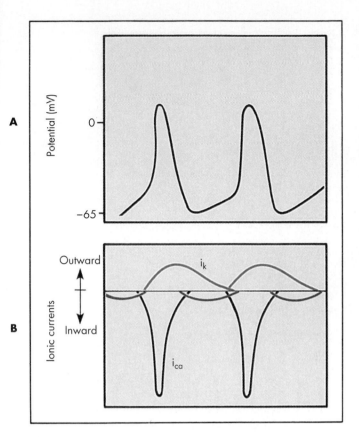

A

B

FIGURE 17-14   The transmembrane potential changes **(A)** that occur in SA node cells are produced by three principal currents **(B):** (1) the Ca^{++} current, i_{Ca}, (2) a hyperpolarization-induced inward current, i_f; and (3) an outward K^+ current, i_K. (Redrawn from Brown HF: *Physiol Rev* 61:644, 1981.)

of these two mechanisms will also alter the firing frequency.

Ionic Basis of Automaticity   Several ionic currents contribute to the slow diastolic depolarization that occurs in automatic cardiac cells. In the automatic cells of the SA node, diastolic depolarization is implemented by changes in at least three ionic currents: (1) an inward "funny" current, i_f; (2) an inward Ca^{++} current, i_{Ca}; and (3) an outward K^+ current, i_K (Figure 17-14, *B*).

The inward current, i_f, is carried mainly by Na^+. It is referred to as a funny current, because the investigators who so named this current had not expected to find such an inward current. This current is activated during repolarization (phase 3) of the action potential, as the membrane potential becomes more negative than about -50 mV. The more negative the membrane potential becomes at the end of repolarization, the greater will be the magnitude of the i_f current.

The inward Ca^{++} current begins toward the end of phase 4, as the transmembrane potential becomes about -55 mV (Figure 17-14). The influx of Ca^{++} accelerates

depolarization, which soon leads to the upstroke of the action potential. A decrease in the external Ca^{++} concentration or the addition of a calcium channel antagonist (e.g., nifedipine) diminishes the amplitude of the action potential and the slope of the slow diastolic depolarization in SA node cells (see Figure 17-12, *B*).

The progressive diastolic depolarization mediated by the two inward currents, i_f and i_{Ca}, is opposed by a third, outward current, i_K. This outward current is activated during the plateau, and it slowly inactivates during repolarization and throughout phase 4. The efflux of K^+ tends to oppose the depolarizing effects of i_f and i_{Ca} (and also of a slow inward "leak" of Na^+ through nonselective channels). However, the outward K^+ current decays steadily throughout phase 4 (Figure 17-14) because of the gradual inactivation, and hence the opposition to the depolarizing effects of the inward currents gradually diminishes. This diminishing opposition thus contributes to the slow diastolic depolarization.

The ionic basis for automaticity in the AV node pacemaker cells is probably identical to that in the SA node cells. Similar mechanisms probably also account for automaticity in Purkinje fibers, except that the Ca^{++} current does not contribute. Hence, the slow diastolic depolarization in Purkinje fibers is mediated principally by the balance between the hyperpolarization-induced inward current, i_f, the inward "leak" of Na^+, and the gradually diminishing outward K^+ current, i_K.

Effects of Autonomic Nerves   The transmitters released by the autonomic nerves affect automaticity by altering the ionic currents across the pacemaker cell membranes. Increased sympathetic nervous activity, through the release of **norepinephrine,** raises the heart rate by increasing the slope of slow diastolic depolarization (see Figure 17-13, *A*). The increase in slope is achieved mainly by augmenting i_f and i_{Ca} (see Figure 17-14) in the membranes of the SA node cells.

Increased parasympathetic activity, through the release of **acetylcholine,** diminishes the heart rate by increasing the maximum negativity (Figure 17-13, *B*) and by reducing the slope (Figure 17-13, *A*) of the slow diastolic depolarization. The neurally released acetylcholine increases the maximum negativity during phase 4 by interacting with **cholinergic receptors (muscarinic type)** that activate specific potassium channels in the membranes of the automatic cells. The acetylcholine also decreases the slope of slow diastolic depolarization by diminishing the ionic currents through the i_f and i_{Ca} channels.

Overdrive Suppression   The automaticity of pacemaker cells is suppressed temporarily after these cells are driven at a high frequency. This phenomenon is known as **overdrive suppression.** Because the SA node cells usually fire at a greater frequency than do the

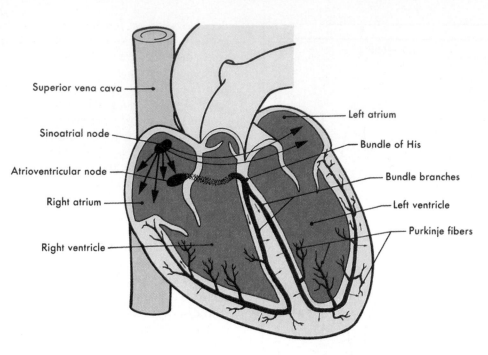

Superior vena cava

Sinoatrial node

Atrioventricular node

Right atrium

Right ventricle

Left atrium

Bundle of His

Bundle branches

Left ventricle

Purkinje fibers

FIGURE 17-15   The conduction system of the heart.

automatic cells in the other latent pacemaking sites in the heart, the firing of the SA node cells at their greater frequency tends to suppress the automaticity in the other **(ectopic)** sites.

> In certain disturbances, such as the **sick sinus syndrome,** the SA node cells periodically cease firing. The automatic cells in the various ectopic sites in the heart often do not generate impulses immediately, but instead they might remain quiescent for several seconds. Hence, the subject might lose consciousness whenever the SA node stops functioning, even though automatic cells are abundant in various ectopic sites in the heart.

The mechanism responsible for overdrive suppression involves the membrane pump (Na^+-K^+-ATPase) that actively extrudes Na^+ from the cell, in partial exchange for K^+; the ratio is 3 Na^+ : 2 K^+ (Chapter 1). During each depolarization of an automatic Purkinje fiber, for example, a certain quantity of Na^+ enters the cell during phase 0 of the action potential. The more frequently the cell is depolarized, therefore, the more Na^+ that enters the cell per minute. During the period of overdrive, the Na^+ pump extrudes this larger quantity of Na^+ more actively from the cell interior. The quantity of Na^+ extruded by the Na^+-K^+-ATPase exceeds the quantity of K^+ that enters the cell. This enhanced activity of the

pump hyperpolarizes the cell membrane, because there is a net loss of cations from the cell interior; because of the change in V_m, the pump is said to be **electrogenic.** As a consequence of the hyperpolarization, the transmembrane potential requires more time to reach the threshold (Figure 17-13, *B*). Furthermore, when the overdrive suddenly ceases, the Na^+ pump usually does not decelerate instantaneously, but it continues to operate more actively for some time. This excessive extrusion of Na^+ opposes the gradual depolarization of the pacemaker cell during phase 4, and thereby suppresses its automaticity temporarily.

Atrial Conduction

From the SA node, the cardiac impulse spreads radially throughout the right atrium (Figure 17-15) along ordinary atrial myocardial fibers, at a conduction velocity of approximately 1 m/sec. A special pathway, the **anterior interatrial band** (or **Bachmann's bundle**), conducts the impulse most directly from the SA node to the left atrium. However, even if this direct pathway is destroyed experimentally, conduction proceeds expeditiously from the right to the left atrium along ordinary myocardial fibers. Some of the action potentials that proceed inferiorly through the right atrium ultimately reach the atrioventricular (AV) node, which is normally the sole source of entry of the cardiac impulse from the atria to the ventricles.

In some individuals, an extraneous **bypass tract** serves in parallel with the AV node as a functional pathway between the atria and the ventricles. Such a bypass tract may lead to serious cardiac rhythm disturbances, because it may constitute a component of a reentry loop (which will be described below).

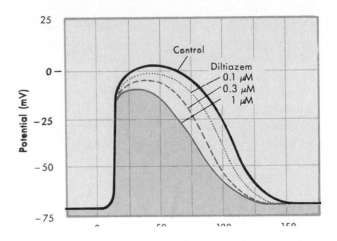

FIGURE 17-16 Transmembrane potentials recorded from a rabbit AV node cell under control conditions and in the presence of the calcium channel antagonist diltiazem in concentrations of 0.1, 0.3, and 1 μmol/L. (Redrawn from Hirth C et al: *J Mol Cell Cardiol* 15:799. 1983.)

Atrioventricular Conduction

The AV node in adult humans is approximately 22 mm long, 10 mm wide, and 3 mm thick. It is situated posteriorly, on the right side of the interatrial septum near the ostium of the coronary sinus. The AV node has been divided into three functional regions: (1) the A-N region, which is the transitional zone between the atrium and the remainder of the node; (2) the N region, which is the midportion of the AV node; and (3) the N-H region, in which the nodal fibers gradually merge with the **bundle of His,** which is the beginning of the specialized conducting system for the ventricles.

Several features of AV conduction are physiologically and clinically significant. The principal delay in the passage of the impulse from the atrial to the ventricular myocardial cells occurs in the A-N region of the node. The conduction velocity is actually less in the N region than in the A-N region. However, the path length is substantially greater in the A-N than in the N region, which accounts for the greater delay in the A-N than in the N region.

The conduction times through the A-N and N zones account for a considerable fraction of the **P-R interval** (see Figure 17-19), which signifies the delay between atrial and ventricular excitation in the electrocardiogram. Functionally, this delay permits atrial contraction to contribute optimally to ventricular filling.

The cells in the N region are characterized by slow-response action potentials. The resting potential is about −60 mV, the upstroke is not very steep, and the conduction velocity is about 0.05 m/sec. Tetrodotoxin, which blocks the fast Na$^+$ channels, has almost no effect on the action potentials in this region. Conversely, diltiazem, a calcium channel antagonist, decreases the amplitude and duration of the action potentials (Figure 17-16) and retards AV conduction. The action potentials of cells in the A-N region are intermediate in shape between those of cells in the N region and the atria. Similarly, the action potentials of cells in the N-H region are transitional between those of cells in the N region and the bundle of His. The entire conduction system between atria and ventricles, including the various transitional zones, is often referred to as the **AV junction.**

Cells in the N region display post-repolarization refractoriness (see Figure 17-10). As the time between successive atrial depolarizations is decreased, the conduction time through the AV junction becomes prolonged. For example, when the atria were paced electrically in a group of human subjects, the conduction time (AH interval) from the atria to the His bundle increased progressively as the interval between pacing stimuli was decreased (Figure 17-17).

Impulses tend to be blocked in the AV junction at cycle lengths that are easily conducted in other regions of the heart. This often has a beneficial effect in certain clinical rhythm disturbances. For example, in **atrial tachycardia** (which is an abnormally fast heart rate that originates in the atria), the atria might be depolarized at a frequency of about 200 times per minute. Under such circumstances, some fraction of the atrial impulses is usually blocked in the AV node. For example, if only alternate atrial impulses were conducted through the AV node, the ventricles would be depolarized only 100 times per minute. This AV nodal conduction pattern, which is referred to as **2:1 AV block,** tends to protect the ventricles from excessively high contraction frequencies. At frequencies as high as 200 per minute, the time available for ventricular filling between contractions would not be adequate, and therefore the heart would not be able to pump a sufficient volume of blood per minute (see Chapter 24). Thus, when the atria contract at an abnormally rapid rate, the ventricles can pump more blood per minute when some fraction of the atrial depolarizations is blocked in the AV junction than when all the atrial impulses do reach the ventricles. When the ventricles respond at too high a frequency, the physician usually attempts to reduce the number of atrial impulses that are conducted to the ventricles. He or she may attempt to block AV conduction partially by increasing vagal activity reflexly or by administering certain drugs (e.g., **adenosine** or **digitalis**) that inhibit AV conduction.

Autonomic Effects The vagus nerves release **acetylcholine,** which inhibits AV conduction. Moderate vagal activity may simply prolong the AV conduction time. In the study of the effects of atrial pacing cycle length on AV conduction time shown in Figure 17-17, the vagal activity to the heart was increased reflexly in human subjects by infusing **phenylephrine** (an adrenergic vasoconstrictor drug) to raise the subjects' arterial blood pressure. The figure shows that for any given pacing cycle length, the AH conduction time was greater when vagal activity was augmented than under control conditions. Stronger vagal activity may cause some or all of the atrial impulses to fail to be conducted through the AV node. The delayed conduction or block occurs largely in the N region of the node.

The cardiac sympathetic nerves, on the other hand, facilitate AV conduction. They decrease the AV conduction time and enhance the rhythmicity of the latent pacemakers in the AV junction. The **norepinephrine** released at the sympathetic nerve terminals increases the amplitude and slope of the upstroke of the AV nodal action potentials, principally in the N region of the node.

Ventricular Conduction

The bundle of His is the beginning of the **specialized conduction system** for the ventricles. It passes subendocardially down the right side of the interventricular septum for approximately 12 mm and then divides into the right and left **bundle branches** (see Figure 17-15). The right bundle branch, a direct continuation of the bundle of His, proceeds down the right side of the interventricular septum. The left bundle branch is considerably thicker than the right. It arises almost perpendicularly from the bundle of His and perforates the interventricular septum.

The bundle branches ultimately subdivide into a complex network of conducting fibers, called Purkinje fibers, which ramify over the subendocardial surfaces of both ventricles. **Purkinje fibers** are the broadest cells in the heart, 70 to 80 μm in diameter, compared with 10 to 15 μm for ventricular myocardial cells. Their large diameter accounts in part for the greater conduction velocity in Purkinje than in myocardial fibers. The conduction velocity of the cardiac impulse over the Purkinje fiber system is the fastest of any tissue within the heart; estimates vary from 1 to 4 m/sec. This permits a rapid activation of the entire endocardial surface of the ventricles. The endocardial surfaces of both ventricles are activated rapidly, but the wave of excitation spreads from endocardium to epicardium more slowly (about 0.3 to 0.4 m/sec).

The action potentials recorded from Purkinje fibers (see Figure 17-8, *A*) differ slightly from those obtained from ordinary ventricular myocardial fibers (see Figure 17-2). In general, Purkinje fiber action potentials have a prominent notch (phase 1). This feature resembles action potentials from ventricular epicardial fibers, but differs from action potentials recorded from endocardial fibers in that the latter type of action potentials do not have prominent notches. The duration of the plateau (phase 2) is longer in Purkinje fibers than in myocardial action potentials. The prolonged plateau confers a long refractory period on the Purkinje fibers. Hence, many premature atrial depolarizations may be conducted through the AV junction only to be blocked by the Purkinje fibers. This function of protecting the ventricles against the effects of premature atrial depolarizations is especially pronounced at slow heart rates, because the action potential duration and hence the effective refractory period of the Purkinje fibers vary inversely with the heart rate (see Figure 17-11). Similar inverse relationships between refractory period and heart rate occur in most other cardiac cells. However, in the AV node, the effective refractory period does not change appreciably over the normal range of heart rates, but it actually increases with heart rate at very rapid rates. Therefore at high atrial rates, it is the AV node, rather than the Purkinje fibers, that protects the ventricles from excitation at excessive frequencies.

The first portions of the ventricles to be excited are the interventricular septum and the papillary muscles. The wave of activation spreads into the substance of the septum from both its left and its right endocardial surfaces. Early contraction of the septum makes it more rigid and allows it to serve as an anchor point for the contraction of the remaining ventricular myocardium. Furthermore, early contraction of the papillary muscles prevents eversion of the AV valves into the atria during ventricular systole (Chapter 18). Because the right ventricular wall is appreciably thinner than the left, the epicardial surface of the right ventricle is activated earlier than that of the left ventricle. Also, apical and central epicardial regions of both ventricles are activated earlier

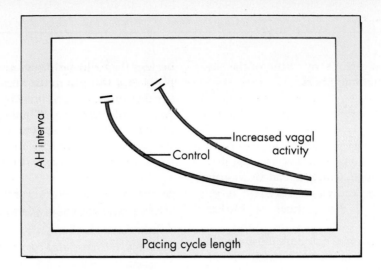

FIGURE 17-17   The changes in atrial-His (AH) intervals induced by pacing the atria at various cycle lengths in a group of 8 human subjects under control conditions and during increased vagal activity produced by intravenous infusions of phenylephrine. (Redrawn from Page RL et al: *Circ Res,* 68:1614, 1991.)

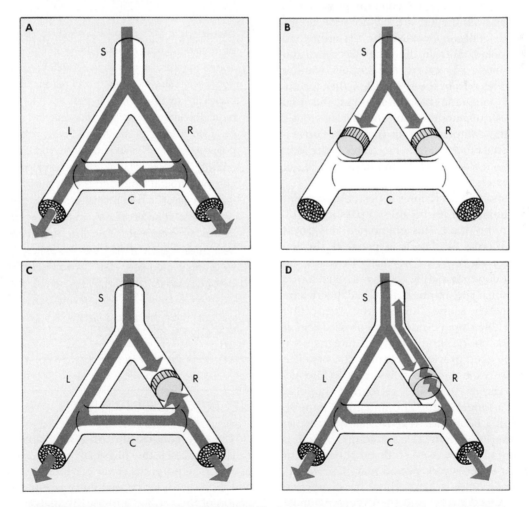

FIGURE 17-18   The role of unidirectional block in reentry. **A,** An excitation wave traveling down a single bundle *(S)* of fibers continues down the left *(L)* and right *(R)* branches. The depolarization wave enters the connecting branch *(C)* from both ends and is extinguished at the zone of collision. Reentry does not occur. **B,** The wave is blocked in the *L* and *R* branches. Reentry does not occur. **C,** The antegrade and retrograde impulses in branch *R* are both blocked. Reentry does not occur. **D,** Unidirectional block exists in branch *R.* The antegrade impulse is blocked. The retrograde impulse arrives later, after the previously refractory segment of tissue in branch *R* has regained its excitability. The retrograde impulse travels through branch *R* and reexcites bundle *S,* to complete the reentry circuit.

than their respective basal regions, by virtue of the distribution pattern of the Purkinje fibers.

REENTRY

Under appropriate conditions, a cardiac impulse may reexcite some region through which it had previously passed. This phenomenon, known as **reentry,** is responsible for many clinical disturbances of cardiac rhythm.

The conditions necessary for reentry are illustrated in Figure 17-18. In each of the four panels, a single bundle (S) of cardiac fibers splits into a left (L) and a right (R) branch. A connecting bundle (C) runs between the two branches. Normally, the impulse coming down bundle S is conducted along the L and R branches (panel A). As the impulse reaches connecting link C, it enters from both sides and becomes extinguished at the point of collision. The impulse from the left side cannot proceed beyond the point of collision because the tissue beyond has just been depolarized from the other direction, and hence it is absolutely refractory. The impulse cannot pass through bundle C from the right side either, for the same reason. If antegrade block exists in the two branches (L and R) of bundle S, the impulse obviously cannot traverse the complete circuit (panel B). Furthermore, if bidirectional block exists at any point in the loop (for example, branch R in panel C), the impulse will not be able to reenter.

A necessary condition for reentry is that at some point in the loop, the impulse must be able to pass in one direction but not in the other. This phenomenon is called **unidirectional block.** As shown in panel D, the impulse may travel down branch L normally. However, the impulse traveling down branch R may be blocked in the antegrade direction. The impulse that had been conducted down branch L and through the connecting branch C may be able to penetrate the depressed region retrogradely in branch R, even though the antegrade impulse had been blocked previously at this same site. The antegrade impulse in branch R will arrive at the depressed region earlier than the retrograde impulse, which traverses a longer path. The antegrade impulse may be blocked simply because it arrives early at the depressed region, during its effective refractory period. If the retrograde impulse is delayed sufficiently because of the longer path, the refractory period may have ended, and the retrograde impulse can then be conducted through the depressed region and then reenter bundle S, to complete the circuit.

Unidirectional block is a necessary condition for reentry, but not a sufficient one. It is also essential that the effective refractory period of the reentered region

be less than the propagation time around the loop. In panel D, if the retrograde impulse is conducted through the depressed zone in branch R and if the tissue just beyond is still refractory from the antegrade depolarization, branch S will not be reexcited. Therefore *the conditions that promote reentry are those that prolong conduction time or shorten the effective refractory period.*

The **Wolff-Parkinson-White syndrome** is a common cause of reentry in humans. In this congenital syndrome, a bypass tract constitutes a secondary conduction pathway between atria and ventricles; this bypass tract parallels the AV node and bundle of His. Ordinarily, atrial impulses traverse the normal and bypass pathways concomitantly, and no functional problems are evident to the patient. The presence of such a bypass tract is ordinarily manifested only by electrocardiographic abnormalities (a short P-R interval and a characteristic QRS complex). At times, however, the atrial impulse may travel to the ventricles exclusively by one of the parallel pathways (usually the normal route), and the impulse may then travel retrograde by the second pathway (usually the bypass tract) to reexcite the atria. The impulse may then continue to travel around this reentry loop for minutes or hours; one conduction pathway mediates antegrade conduction, and the other pathway mediates the retrograde conduction. Conduction time around the reentry loop is usually much less than the duration of a normal cardiac cycle. Hence, the heart beats at an excessively fast rate, which is not conducive to the optimal mechanical performance of the heart. Interruption of the reentry loop may be achieved by giving a drug (e.g., **Ca^{++} channel blockers, adenosine**) that will inhibit the normal AV node pathway (which mainly includes slow-response fibers) but will not inhibit the bypass tract (which comprises only fast-response fibers).

ELECTROCARDIOGRAPHY

The **electrocardiograph** is a valuable instrument because it enables the physician to record the variations in electrical potential at various loci on the body surface and thereby to derive vital information about the propagation of the cardiac impulse. By analyzing the details of these fluctuations in potential, the physician gains valuable insight concerning (1) the anatomical orientation of the heart, (2) the relative sizes of its chambers, (3) a variety of disturbances of rhythm and conduction, (4) the extent, location, and progress of injury to the myocardium, (5) the effects of altered electrolyte concentrations, and (6) the influence of certain drugs (notably digitalis and its derivatives). The science of electrocardiography is extensive and complex, but only the elementary features of the electrocardiogram will be presented here.

The electrocardiogram reflects the temporal changes in the electrical potential between pairs of points on the skin surface. The cardiac impulse progresses through the heart in a complex three-dimensional pattern. Hence, the precise configuration of the electrocardiogram varies from person to person, and in any given individual the pattern varies with the anatomical location of the recording electrodes.

In general, the pattern consists of **P, QRS,** and **T waves** (Figure 17-19). The **P-R interval** is the time from the beginning of atrial activation to the beginning of ventricular activation; it normally ranges from 0.12 to 0.20 sec. Most of this time involves the passage of the impulse through the AV conduction system. *Pathological prolon-gations of the P-R interval are associated with disturbances of AV conduction produced by inflammatory, circulatory, pharmacological, or nervous mechanisms.*

The configuration and amplitude of the **QRS complex** vary considerably among individuals. The duration is usually between 0.06 and 0.10 sec. *Abnormal prolongation may indicate a block in the normal conduction pathways through the ventricles (such as a block of the left or right bundle branch).*

During the **S-T interval** the entire ventricular myocardium is depolarized. Because all of the myocardial cells are at about the same potential, the S-T segment lies on the **isoelectric line** (which is the line that reflects that virtually all regions of the cardiac surface are at the same electrical potential).

The **Q-T interval** is sometimes referred to as the period of **electrical systole** of the ventricles; it reflects the action potential duration of the ventricular myocardial cells. The duration of the Q-T interval is about 0.4 sec, but it varies inversely with the heart rate, mainly because the action potential duration varies inversely with the heart rate (see Figure 17-11).

The **T wave** reflects the repolarization of the ventricular myocardial cells. The T wave is usually deflected in the same direction from the isoelectric line as the major component of the QRS complex. When the T wave and QRS complex deviate in the same direction from the isoelectric line, it indicates that the propagation of the repolarization process does not follow the same route as the propagation of the depolarization process. Normally, depolarization proceeds from endocardium to epicardium, whereas repolarization proceeds in the oppo-

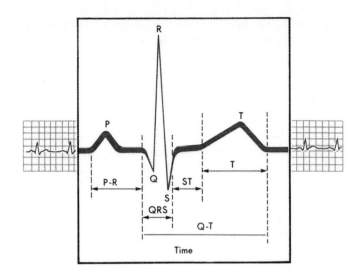

FIGURE 17-19   The important deflections and intervals of a typical scalar electrocardiogram.

site direction. The reason is related to disparities in duration of the action potentials generated in different regions of the heart.

SUMMARY

1. The transmembrane action potentials that can be recorded from cardiac myocytes consist of five phases (0 to 4):

   Phase 0, upstroke. A suprathreshold stimulus rapidly depolarizes the membrane by activating the fast Na^+ channels.

   Phase I, early partial repolarization. Achieved by the efflux of K^+ through channels that conduct the transient outward current, i_{to}.

   Phase 2, plateau. Achieved by a balance between the influx of Ca^{++} through Ca^{++} channels and the efflux of K^+ through several types of K^+ channels.

   Phase 3, final repolarization. Initiated when the efflux of K^+ exceeds the influx of Ca^{++}. The resulting partial repolarization rapidly increases the K^+ conductance and rapidly restores full repolarization.

   Phase 4, resting potential. The transmembrane potential of the fully repolarized cell is determined mainly by the conductance of the cell membrane to K^+.

2. Two principal types of action potentials may be recorded from cardiac cells:

   Fast-response action potential. Recorded from atrial and ventricular myocardial fibers and from specialized conducting (Purkinje) fibers. The upstroke of the action potential is steep, it has a large amplitude, and it is produced by the activation of fast Na^+ channels. The effective refractory period extends from the up-

---

stroke to about the middle of phase 3. The fiber is relatively refractory during the remainder of phase 3, but it regains full excitability as soon as it is fully repolarized (phase 4).

Slow-response action potential. Recorded from normal SA and AV nodal cells, and from abnormal myocardial cells that have been partially depolarized. The action potential is characterized by a less negative resting potential, a smaller amplitude, and a more gradual upstroke than is the fast-response action potential. The upstroke is produced by the activation of Ca^{++} channels. The fiber becomes absolutely refractory at the beginning of the upstroke, but partial excitability may not be regained until very late in phase 3 or after the fiber is fully repolarized. The fiber does not regain full excitability until well after the fiber has completely repolarized.

3. Automaticity is characteristic of certain cells in the SA and AV nodes and in the specialized conducting system. Automaticity is achieved by a slow depolarization of the membrane during phase 4. Ultimately, the transmembrane potential reaches threshold; a rapid influx of Na^+ leads to the upstroke of the action potential, and the remainder of the action potential ensues.

4. Normally, the SA node initiates the impulse that induces cardiac contraction. This impulse is propagated from the SA node to the atria, and the wave of excitation ultimately reaches the AV node.

5. The cardiac impulse travels very slowly through the slow-response fibers in the AV node. The consequent delay between atrial and ventricular depolarization provides adequate time for atrial contraction to help fill the ventricles.

6. Automatic cells in the atrium, AV node, or His-Purkinje system may initiate propagated cardiac impulses either because the normal pacemaker cells in the SA node are suppressed or because the firing rate of the ectopic focus is abnormally enhanced.

7. Distubances of impulse conduction consist mainly of simple conduction block and reentry.

Simple conduction block. Propagation of a cardiac impulse may fail as the result of a disease process (ischemia, inflammation) or a drug (digitalis, Na^+ or Ca^{++} channel antagonist).

Reentry. A cardiac impulse may traverse a loop of cardiac fibers and reenter previously excited tissue when (a) the impulse is conducted slowly around the loop and (b) the impulse is blocked unidirectionally in some section of the loop.

8. The electrocardiogram is recorded from the surface of the body, and it traces the conduction of the cardiac impulse through the heart. The component waves of the electrocardiogram are:

P wave. Spread of excitation over the atria.
QRS interval. Spread of excitation over the ventricles.
T wave. Spread of repolarization over the ventricles.

BIBLIOGRAPHY
Journal articles

Antzelevitch C et al: Heterogeneity within the ventricular wall: electrophysiology and pharmacology of epicardial, endocardial, and M cells, *Circ Res* 69:1427, 1991.

Armstrong CM: Voltage-dependent ion channels and their gating, *Physiol Rev* 72: S5, 1992.

Balke CW et al: Biophysics and physiology of cardiac calcium channels, *Circulation* 87: VII-49, 1993.

Delmar M: Role of potassium currents on cell excitability in cardiac ventricular myocytes, *J Cardiovasc Electrophys* 3:474, 1992.

DiFrancesco D, Zaza A: The cardiac pacemaker current i_f, *J Cardiovasc Electrophys* 3:334, 1992.

Irisawa H et al: Cardiac pacemaking in the sinoatrial node, *Physiol Rev* 73:197, 1993.

Levy MN et al: Assessment of beat-by-beat control of heart rate by the autonomic nervous system: molecular biology techniques are necessary, but not sufficient, *J Cardiovasc Electrophysiol* 4:183, 1993.

Liu D-W et al: Ionic bases for electrophysiological distinctions among epicardial, midmyocardial, and endocardial myocytes from the free wall of the canine left ventricle, *Circ Res* 72:671, 1993.

Opthof T: Mammalian sinoatrial node, *Cardiovasc Drugs Therap* 1:573, 1988.

Pallotta BS, Wagoner PK: Voltage-dependent potassium channels since Hodgkin and Huxley, *Physiol Rev* 72: S49, 1992.

Spach MS, Josephson ME: Initiating reentry: role of nonuniform anisotropy in small circuits, *J Cardiovasc Electrophysiol* 5:182, 1994.

Waldo AL, Wit AL: Mechanisms of cardiac arrhythmias, *Lancet* 341:1189, 1993.

Watanabe Y, Watanabe M: Impulse formation and conduction of excitation in the atrioventricular node, *J Cardiovasc Electrophysiol* 5:517, 1994.

Books and Monographs

Bouman LN, Jongsma HJ: *Cardiac rate and rhythm*, The Hague, 1982, Martinus Nijhoff Publishers.

Kulbertus HE, Franck G: *Neurocardiology* Mt Kisco, NY, 1988, Futura Publishing Co, Inc.

Levy MN, Schwartz PJ: *Vagal control of the heart: experimental basis and clinical implications*, Mt Kisco, NY, 1994, Futura Publishing Co., Inc.

Mazgalev T et al: *Electrophysiology of the sinoatrial and atrioventricular nodes*, New York, 1988, Alan R Liss, Inc.

Nathan RD: *Cardiac muscle: regulation of excitation and contraction*, Orlando, 1986, Academic Press, Inc.

Noble D, Powell T: *Electrophysiology of single cardiac cells,* Orlando, 1987, Academic Press, Inc.

Rosen MR et al: *Cardiac electrophysiology: a textbook,* Mt Kisco, NY, 1990, Futura Publishing Co, Inc.

Sperelakis N: Physiology and pathophysiology of the heart, ed 2, Boston, 1989, Kluwer Academic Publishers.

Wit AL, Janse MJ: *Ventricular arrhythmias of ischemia and infarction: electrophysiological mechanisms,* Mt Kisco, NY, 1993, Futura Publishing Co, Inc.

Zipes DP, Jalife J: *Cardiac electrophysiology: from cell to bedside,* Philadelphia, 1990, WB Saunders Co.

18

The Cardiac Pump

The heart exhibits a wide range of activity and functional capacity, and performs a staggering amount of work over the lifetime of an individual. To understand how the heart accomplishes its important task, it is first necessary to consider the relationships between the structure and the function of its components.

ANATOMICAL BASIS OF CARDIAC FUNCTION

Myocardial Cell

Several important morphological and functional differences exist between myocardial and skeletal muscle cells (see Chapters 13 and 14). However, the contractile elements within the two types of cells are quite similar; each skeletal or cardiac muscle cell is made up of sarcomeres that contain thick filaments composed of myosin and thin filaments composed of actin. As in skeletal muscle, shortening of the cardiac sarcomere occurs by the sliding filament mechanism. Actin filaments slide along adjacent myosin filaments by cycling of the intervening crossbridges, and thereby the Z lines are brought closer together.

Skeletal muscle and cardiac muscle show similar length-force relationships. The sarcomere length has been determined with electron microscopy in regions of the ventricles that have been rapidly fixed during systole or diastole. The developed force is maximal when cardiac muscle begins its contraction at resting sarcomere lengths of 2.0 to 2.4 μm. At such lengths the thick and thin filaments overlap, and the number of crossbridge attachments is maximal. Developed force of cardiac muscle is less than the maximal value when the sarcomeres are stretched beyond the optimal length, because the overlap of the filaments is less, and thus the cycling of the crossbridges is less. At resting sarcomere lengths shorter than the optimal value, the thin filaments that extend from adjacent Z lines overlap each other in

the central region of the sarcomere. This arrangement of the thin filaments diminishes contractile force.

The length-force relationship for the intact heart may be expressed graphically, as in the upper curve in Figure 18-1. Developed force (the force attained during contraction) may be expressed as ventricular systolic pressure, and myocardial resting fiber length may be expressed as end-diastolic ventricular volume. The lower curve in Figure 18-1 depicts the ventricular pressure produced by increments in ventricular volume during diastole (at rest). The upper curve represents the peak pressure developed by the ventricle during systole at each filling volume. *The graph illustrates the relationship of force (or pressure) development by the ventricle as a function of initial fiber length (or initial volume).* This is known as the **Frank-Starling** relationship, named after the scientists who first described it.

Note that the pressure-volume curve in diastole is quite flat at low volumes. Thus large increases in volume can be accommodated with only small increases in pressure (ventricle is very compliant). Nevertheless, systolic pressure development is considerable at the lower filling pressures. The ventricle becomes much less compliant with greater filling, however, as evidenced by the sharp rise of the diastolic curve at large intraventricular volumes. The normal heart operates only on the ascending portion of the Frank-Starling curve depicted in Figure 18-1 *(upper curve).*

A striking difference in the appearance of cardiac and skeletal muscle is that cardiac muscle appears to be a **syncytium** (a single multinucleated cell formed from many fused cells), with branching interconnecting fibers, whereas skeletal muscle cells do not interconnect. However, the myocardium is not a true anatomical syncytium, because (1) the myocardial fibers are separated laterally from adjacent fibers by their respective **sarcolemmas,** and (2) the end of each fiber is separated from its neighbor by dense structures, **intercalated disks,** that are continuous with the sarcolemma (Figures 18-2 and

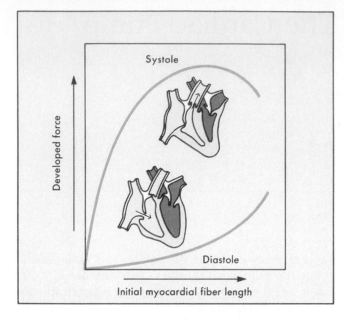

FIGURE 18-1 Relationship of myocardial resting fiber length (sarcomere length) or end-diastolic volume to developed force or peak systolic ventricular pressure during ventricular contraction in the intact dog heart. (Redrawn from Patterson SW et al: *J Physiol* 48:465, 1914.)

18-3). Nevertheless, cardiac muscle functions as a syncytium, because a wave of depolarization, followed by contraction of the atria or ventricles (an **all-or-none response**), occurs when a suprathreshold stimulus is applied (see also Chapter 17).

As the wave of excitation approaches the end of a cardiac cell, the spread of excitation to the next cell depends on the electrical conductance of the boundary between the two cells. **Gap junctions (nexi)** with high conductances are present in the intercalated disks between adjacent cells (Figures 18-2 and 18-3). These gap junctions facilitate the conduction of the cardiac impulse from one cell to the next. Consequently impulse conduction in cardiac tissues progresses more rapidly in a direction parallel to, rather than perpendicular to, the long axes of the constituent fibers.

Another difference between cardiac and fast skeletal muscle fibers is in the number of mitochondria **(sarcosomes)** in the two tissues. Fast skeletal muscle (1) has relatively few mitochondria, (2) is called on for relatively short periods of repetitive or sustained contraction, and (3) can metabolize anaerobically and build up a substantial oxygen debt. In contrast, cardiac muscle (1) is richly endowed with mitochondria (Figures 18-2 and 18-3), (2) must contract repetitively for a lifetime, and (3) is incapable of developing a significant oxygen debt. Rapid oxidation of substrates, with the synthesis of adenosine triphosphate (ATP), can keep pace with the myocardial energy requirements because of the large numbers of mi-

tochondria, which contain the respiratory enzymes necessary for oxidative phosphorylation (see also Chapter 14).

To provide adequate oxygen and substrate for its metabolic machinery, the myocardium is also endowed with a rich capillary supply, about one capillary per fiber. Thus diffusion distances are short, and oxygen, carbon dioxide, substrates, and waste material can move rapidly between myocardial cell and capillary. With respect to such exchanges, electron micrographs of myocardium show deep invaginations of the sarcolemma into the fiber at the Z lines (Figures 18-2 and 18-3). These sarcolemmal invaginations constitute the **transverse-tubular,** or **T-tubular, system.** The lumina of these T-tubules are continuous with the bulk interstitial fluid, and they play a key role in excitation-contraction coupling.

A network of **sarcoplasmic reticulum** consists of small-diameter sarcotubules that surround the myofibrils (Figure 18-4). The sarcoplasmic reticulum releases and takes up calcium, and thereby plays a key role in myocardial contraction and relaxation.

Excitation-Contraction Coupling The heart requires optimal concentrations of Na^+, K^+, and Ca^{++} to function normally. In the absence of Na^+ the heart is not excitable and will not beat because the action potential of myocardial fibers depends on extracellular Na^+. In contrast, the resting membrane potential is independent of the Na^+ gradient across the membrane (see Figure 17-5). Under normal conditions the extracellular K^+ concentration is about 4 mM. An increase in extracellular K^+, if great enough, produces depolarization, loss of excitability of the myocardial cells, and cardiac arrest in diastole. Ca^{++} is also essential for cardiac contraction. Removal of Ca^{++} from the extracellular fluid decreases contractile force and eventually causes arrest in diastole. Conversely, an increase in extracellular Ca^{++} concentration enhances contractile force, but very high Ca^{++} concentrations induce cardiac arrest in systole (rigor). *The level of the free intracellular Ca^{++} concentration is mainly responsible for the contractile state of the myocardium.*

Initially a wave of excitation spreads rapidly along the myocardial sarcolemma from cell to cell via gap junctions, and graded depolarization spreads into the interior of the cells via the T-tubules. During the plateau (phase 2) of the action potential, Ca^{++} permeability of the sarcolemma increases (see Chapter 17). Ca^{++} flows down its electrochemical gradient and enters the cell through Ca^{++} channels in the sarcolemma and in the invaginations of the sarcolemma, the T-tubules (see Figures 18-2 and 18-3). The primary source of extracellular Ca^{++} is the interstitial fluid (2 mM Ca^{++}).

The amount of calcium that enters the cell from the

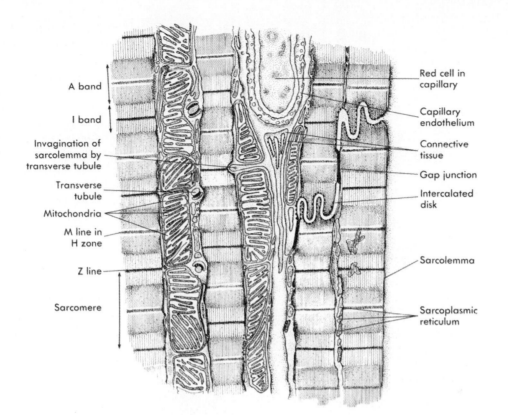

A band

I band

Invagination of
sarcolemma by
transverse tubule

Transverse
tubule

Mitochondria

M line in
H zone

Z line

Sarcomere

Red cell in
capillary

Capillary
endothelium

Connective
tissue

Gap junction

Intercalated
disk

Sarcolemma

Sarcoplasmic
reticulum

FIGURE 18-2   Diagram of an electron micrograph of cardiac muscle. Note the large number of mitochondria and the intercalated disks with nexi (gap junctions), transverse tubules, and longitudinal tubules.

extracellular space is not sufficient to induce contraction of the myofibrils, but it serves as a trigger **(trigger Ca^{++})** to release Ca^{++} from the intracellular Ca^{++} stores, the sarcoplasmic reticulum (Figure 18-4). The cytosolic free Ca^{++} concentration increases from a resting level of less than 0.1 μM to levels of 1.0 to 10 μM during excitation, and the Ca^{++} binds to the protein **troponin C** (see Chapter 12). The Ca^{++}-troponin complex interacts with **tropomyosin** to unblock active sites between the actin and myosin filaments (Figure 18-5). This unblocking action allows crossbridge cycling and thus contraction of the myofibrils (systole). *Mechanisms that raise the cytosolic Ca^{++} concentration increase the developed force, and mechanisms that lower the Ca^{++} concentration decrease the developed force.*

During emotional excitement the force of cardiac contractions is enhanced because the norepinephrine released at the terminals of the sympathetic nerve fibers to the heart increases calcium uptake by phosphorylation of the calcium channels via cyclic AMP formation (Figure 18-5).

Digitalis, a drug used in the treatment of heart failure, also increases contractile force by elevating the level of intracellular calcium. Digitalis inhibits the Na$^+$-K$^+$-ATPase; hence less sodium is pumped out of the myocytes. This results in a decrease in the sodium gradient across the cell membrane so that less sodium can enter the cell and therefore less calcium can leave the cell by Na$^+$/Ca^{++} exchange (Figure 18-5).

At the end of systole the Ca^{++} influx ceases and the sarcoplasmic reticulum is no longer stimulated to release Ca^{++}. In fact, the sarcoplasmic reticulum avidly takes up Ca^{++} by means of an ATP-energized calcium pump that is stimulated by **phospholamban** after the phospholamban is phosphorylated by cAMP-dependent protein kinase. Phosphorylation of troponin I inhibits the Ca^{++} binding of troponin C, which permits tropomyosin to again block the sites for interaction between the actin and myosin filaments, and relaxation (diastole) occurs (Figure 18-5).

The Ca^{++} that enters the cell to initiate contraction must be removed during diastole. The removal is primar-

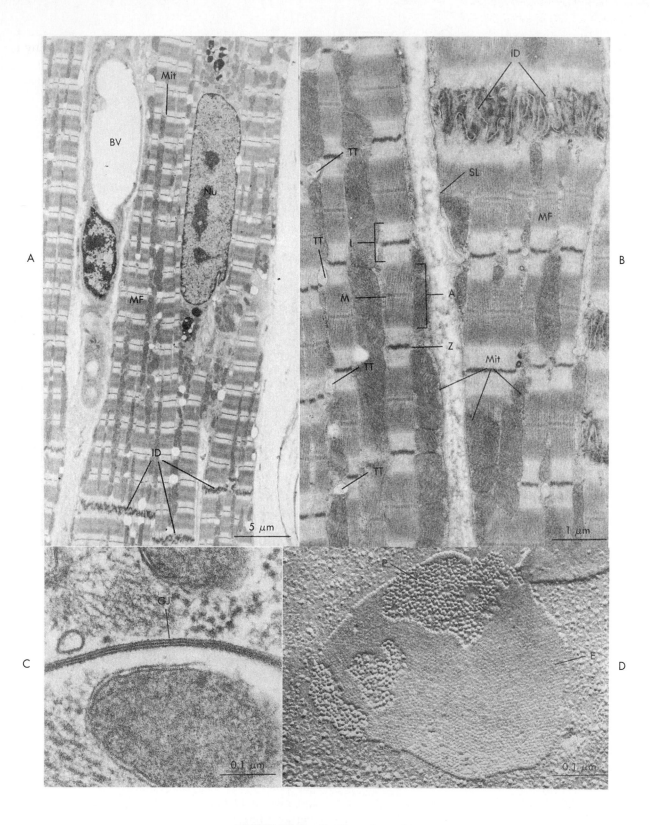

FIGURE 18-3   For legend see facing page.

ily accomplished by an electroneutral exchange of three Na^+ for one Ca^{++} (Figure 18-5). Ca^{++} is also removed from the cell by an electrogenic pump that uses energy to transport Ca^{++} across the sarcolemma (Figure 18-5).

Myocardial Contractile Machinery and Contractility The sequence of events in a preloaded and afterloaded isotonic contraction of a papillary muscle is illustrated in Figure 18-6. Point A represents the resting state, in which the **preload** is responsible for the existing initial stretch. With stimulation the contractile element begins to shorten, and at point B the elastic element has been stretched, but the load has not yet been lifted because the overall length of the muscle has not changed; the muscle fibers have shortened at the expense of the elastic element.

Stretch of the elastic element is represented in the diagram (Figure 18-6) as a progressive rise in force with no external shortening. At point C the force developed by the contractile element has equaled the load (the **afterload**), and the load begins to rise without further stretch of the elastic element. This is represented in the diagram (Figure 18-6) as external shortening of the muscle without further increase in force. *Velocity and force of contraction are functions of the intracellular concentration of free Ca^{++} ions. Force and velocity are inversely related, so that with no load, velocity is maximal but force is negligible. In an isometric contraction, where no external shortening occurs, force is maximal but velocity is zero.*

When these observations on papillary muscle are applied to the whole heart, the preload refers to the stretch of the left ventricle just before the onset of contraction (the end-diastolic volume) and the afterload refers to the aortic pressure during the period when the aortic valve is open.

The preload can be increased by greater filling of the left ventricle during diastole (Figure 18-1). At the lower end-diastolic volumes, increments in filling pressure during diastole elicit a greater systolic pressure during the subsequent contraction, until a maximum systolic pressure is reached at the optimum preload. Further diastolic filling beyond this point results in no further increase in developed pressure; at very high filling pressures, peak pressure development in systole is reduced (Figure 18-1).

At a constant preload, a higher systolic pressure can be reached during ventricular contractions by raising the afterload (e.g., increasing aortic pressure by restricting the runoff of blood to the periphery during diastole). Increments in afterload will produce progressively higher peak systolic pressures until the afterload is so great that the ventricle can no longer generate enough force to open the aortic valve. At this point ventricular systole is totally isometric; there is no ejection of blood, and hence no change in volume of the ventricle during systole. The maximal pressure developed by the left ventricle under these conditions is the maximal isometric force of which the ventricle is capable at a given preload. Of course, at preloads below the optimal filling volume an increase in preload can yield a greater maximal isometric force (see Figure 18-1).

The preloads and afterloads depend on the characteristics of the vascular system and the behavior of the heart. With respect to the vasculature, venomotor tone and peripheral resistance influence preload and afterload. With respect to the heart, a change in rate or stroke volume can also alter preload and afterload. Hence, the cardiac and vascular factors are interactive in their effect on preload and afterload (see Chapters 19 and 24 for a full explanation).

FIGURE 18-3 **A,** Low-magnification electron micrograph of a monkey heart (ventricle). Typical features of myocardial cells include the elongated nucleus *(Nu)*, striated myofibrils *(MF)* with columns of mitochondria *(Mit)* between the myofibrils, and intercellular junctions (intercalated disks, *ID*). A blood vessel *(BV)* is located between two myocardial cells. **B,** Medium-magnification electron micrograph of monkey ventricular cells, showing details of ultrastructure. The sarcolemma *(SL)* is the boundary of the muscle cells and is thrown into multiple folds where the cells meet at the intercalated disk region *(ID)*. The prominent myofibrils *(MF)* show distinct banding patterns, including the A band *(A)*, dark Z lines *(Z)*, I band regions *(I)*, and M lines *(M)* at the center of each sarcomere unit. Mitochondria *(Mit)* occur either in rows between myofibrils or masses just underneath the sarcolemma. Regularly spaced transverse tubules *(TT)* appear at the Z line levels of the myofibrils. **C,** High-magnification electron micrography of a specialized intercellular junction between two myocardial cells of the mouse. Called a gap junction *(GJ)* or nexus, this attachment consists of very close apposition of the sarcolemmal membranes of the two cells and appears in thin section to consist of seven layers. **D,** Freeze-fracture replica of mouse myocardial gap junction, showing distinct arrays of characteristic intramembranous particles. Large particles *(P)* belong to the inner half of the sarcolemma of one myocardial cell, whereas the "pitted" membrane face *(E)* is formed by the outer half of the sarcolemma of the cell above.

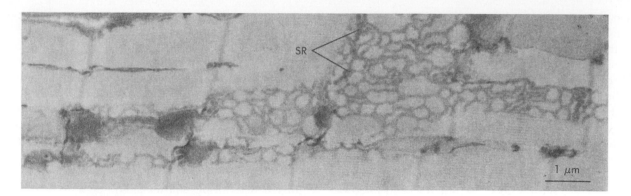

FIGURE 18-4   Mouse cardiac muscle. Tissue treated with ferrocyanide-reduced osmium tetroxide so as to identify the internal membrane system (sarcoplasmic reticulum, *SR*). Specific staining of the SR reveals its architecture as a complex network of small-diameter tubules that are closely associated with myofibrils and mitochondria.

> In **heart failure** the preload can be substantially increased because of the poor ventricular ejection and an increased blood volume caused by fluid retention. In **essential hypertension** the high peripheral resistance augments the afterload by decreasing the peripheral runoff of the blood from the arterial system.

Contractility represents the performance of the heart at a given preload and afterload. One index of contractility is the change in peak isometric force (isovolumic pressure) at a given initial fiber length (end-diastolic volume). Thus, contractility reflects the activity taking place at the interacting crossbridges, rather than the number of crossbridges that are interacting. Augmentation of contractility is observed with certain drugs, such as norepinephrine or digitalis, and with an increase in contraction frequency **(tachycardia).** The increase in contractility **(positive inotropic effect)** produced by any of these interventions is reflected by increments in developed force.

Another reasonable index of myocardial contractility can be derived from the contour of ventricular pressure curves (Figure 18-7). A hypodynamic heart is characterized by an elevated end-diastolic pressure, a slowly rising ventricular pressure, and a somewhat reduced ejection phase (curve *C*, Figure 18-7). A normal ventricle under adrenergic stimulation shows a reduced end-diastolic pressure, a fast rising ventricular pressure, and a brief ejection phase (curve *B*, Figure 18-7). The slope of the ascending limb of the ventricular pressure curve indicates the maximal rate of force development by the ventricle (maximum rate of change in pressure with time; **maximum dP/dt,** as illustrated by the tangents to the steepest portion of the ascending limbs of the ventricu-

lar pressure curves in Figure 18-7). The slope is maximal during the isovolumic phase of systole (p. 252). At any given degree of ventricular filling, the maximum slope provides an index of the initial contraction velocity and hence of contractility. Similarly, one can obtain an indication of contractility of the myocardium from the initial velocity of blood flow in the ascending aorta (the initial slope of the aortic flow curve) (see Figure 18-11).

> The **ejection fraction,** which is the ratio of the volume of blood ejected from the left ventricle per beat **(stroke volume)** to the volume of blood in the left ventricle at the end of diastole, is widely used clinically as an index of contractility.

Cardiac Chambers

The atria are thin-walled, low-pressure chambers that function more as large reservoirs and conduits of blood for their respective ventricles than as important pumps for ventricular filling. The ventricles are formed by a continuum of muscle fibers that originate from the fibrous skeleton at the base of the heart, primarily around the aortic orifice. These fibers sweep toward the apex at the epicardial surface. They also pass toward the endocardium as they gradually undergo a 180-degree change in direction to lie parallel to the epicardial fibers and form the endocardium and papillary muscles. At the apex of the heart the fibers twist and turn inward to form papillary muscles. At the base of the heart and around the valve orifices, the myocardial fibers form a thick, powerful muscle that decreases the ventricular circumference

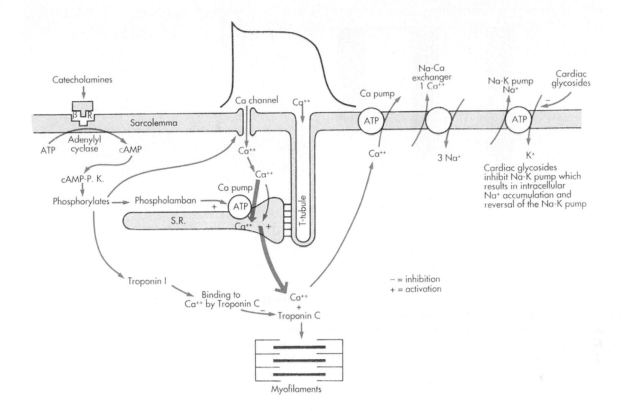

FIGURE 18-5   Schematic diagram of the movements of calcium in excitation-contraction coupling in cardiac muscle. The influx of Ca^{++} from the interstitial fluid during excitation triggers the release of Ca^{++} from the sarcoplasmic reticulum *(SR)*. The free cytosolic Ca^{++} activates contraction of the myofilaments (systole). Relaxation (diastole) occurs as a result of uptake of Ca^{++} by the sarcoplasmic reticulum and extrusion of intracellular Ca^{++} by Na^+-Ca^{++} exchange and to a limited degree by the Ca pump. *βR*, Beta-adrenergic receptor; *cAMP*, cyclic adenosine monophosphate; *cAMP-PK*, cyclic AMP–dependent protein kinase.

to aid in the ejection of blood, and narrow the atrioventricular (AV) valve orifices as an aid to valve closure. Ventricular ejection is implemented not only by a reduction in circumference but also by a decrease in the longitudinal axis; the decrease is accomplished by a descent of the base of the heart. The early contraction of the ventricular apex coupled with approximation of the ventricular walls propels the blood toward the outflow tracts.

Cardiac Valves

The cardiac valves consist of thin flaps of flexible, tough, endothelium-covered fibrous tissue firmly attached at the base to the fibrous valve rings. Movements of the valve leaflets are essentially passive, and the orientation of the cardiac valves is responsible for the unidirectional flow of blood through the heart. There are two types of valves in the heart: the atrioventricular (AV) and semilunar valves (Figures 18-8 and 18-9).

Atrioventricular Valves The **tricuspid valve** lies between the right atrium and right ventricle and is made up of three cusps, whereas the **mitral valve** lies between the left atrium and left ventricle and has two cusps. The total area of the cusps of each AV valve is approximately twice that of the respective AV orifice, and thus the leaflets overlap considerably in the closed position (Figures 18-8 and 18-9). Attached to the free edges of these valves are fine, strong filaments (**chordae tendineae**), which arise from the powerful papillary muscles of the respective ventricles and prevent eversion of the valves during ventricular systole.

In the normal heart the valve leaflets are relatively close to one another during ventricular filling and provide a funnel for the transfer of blood from atrium to ventricle. This partial approximation of the valve surfaces during diastole is caused primarily by eddy currents behind the leaflets. Also, the chordae tendineae and the papillary muscles are stretched by the filling ventricle and exert tension on the free edges of the valve leaflets.

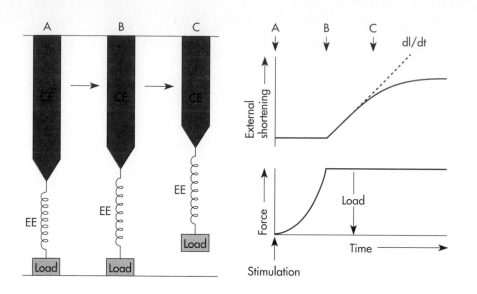

FIGURE 18-6 Model for a preloaded and afterloaded contraction of papillary muscle. **A,** Muscle at rest. Preload is represented by partial stretch of the elastic element *(EE)*. **B,** Partial contraction of the contractile element *(CE)* with stretch of the EE elastic element and no external shortening (the isometric phase of the contraction). **C,** Further contraction of the contractile element with external shortening and lifting of the afterload. The tangent *(dl/dt)* to the initial slope of the shortening curve on the right is the velocity of initial shortening. (Redrawn from Sonnenblick EH: *The myocardial cell,* Philadelphia, 1966, University of Pennsylvania Press.)

Movements of the mitral valve leaflets throughout the cardiac cycle are shown in an echocardiogram (Figure 18-10). **Echocardiography** consists of sending short pulses of high-frequency sound waves (ultrasound) through the chest tissues and the heart, and recording the echoes reflected from the various cardiac structures.

> The timing and the pattern of the reflected waves provide important clinical information, such as the diameter of the heart, the ventricular wall thickness, and the magnitude and direction of the movements of various components of the heart, including the valves. Echocardiography is a very useful diagnostic tool for the cardiologist.

In Figure 18-10 the echocardiograph transducer is positioned to depict movement of the anterior leaflet of the mitral valve. The posterior leaflet moves in a pattern that is a mirror image of the anterior leaflet, except that in the projection shown in Figure 18-10 the excursions of the leaflet appear to be much smaller. At point *D* the mitral valve opens, and during rapid filling (*D* to *E*) the anterior leaflet moves toward the ventricular septum. During the reduced filling phase (*E* to *F*) the valve leaf-

lets float toward each other, but the valve does not close. The ventricular filling contributed by atrial contraction (*F* to *A*) forces the leaflets apart, and a second approximation of the leaflets follows (*A* to *C*). At point *C* the valve is closed by ventricular contraction. The valve leaflets, which bulge toward the atrium, stay pressed together during ventricular systole (*C* to *D*).

Semilunar Valves The valves between the right ventricle and the pulmonary artery and between the left ventricle and the aorta consist of three cuplike cusps attached to the valve rings (see Figures 18-8 and 18-9). At the end of the reduced ejection phase of ventricular systole, blood flow reverses briefly toward the ventricles (shown as a negative flow in the phasic aortic flow curve in Figure 18-11). This flow reversal snaps the cusps together and prevents regurgitation of blood into the ventricles. During ventricular systole the cusps do not lie back against the walls of the pulmonary artery and aorta; rather, they float in the bloodstream approximately midway between the vessel walls and their closed position. Behind the semilunar valves are small outpocketings (**sinuses of Valsalva**) of the pulmonary artery and aorta. Eddy currents develop in these sinuses and keep the valve cusps away from the vessel walls. The orifices of the right and left coronary arteries are located behind the right and left cusps, respectively, of the aortic valve. Were it not for the presence of the sinuses of Valsalva

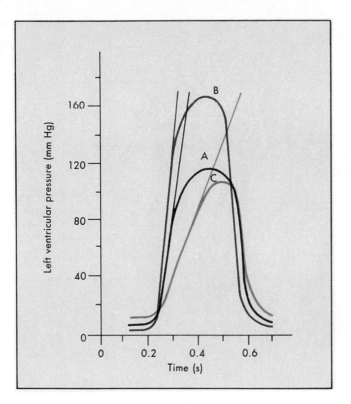

FIGURE 18-7   Left ventricular pressure curves with tangents drawn to the steepest portions of the ascending limbs to indicate the maximal dP/dt value. *A,* Control; *B,* hyperdynamic heart, as with norepinephrine administration; *C,* hypodynamic heart, as in cardiac failure (see text for details).

and the eddy currents developed therein, the coronary ostia could be blocked by the valve cusps.

Pericardium

The pericardium is an epithelialized fibrous sac. It closely invests the entire heart and the cardiac portion of the great vessels and is reflected onto the cardiac surface as the epicardium. The sac normally contains a small amount of fluid, which provides lubrication for the continuous movement of the enclosed heart. The pericardium is not very distensible and thus strongly resists a large, rapid increase in cardiac size. Therefore the pericardium helps prevent sudden overdistension of the heart chambers.

In contrast to an acute change in intracardiac pressure, progressive and sustained enlargement of the heart (as can occur in **cardiac hypertrophy**) or a slow progressive increase in pericardial fluid (as can occur with **pericardial effusion**) gradually stretches the intact pericardium.

HEART SOUNDS

Four sounds are usually produced by the heart, but only two are ordinarily audible through a stethoscope. With electronic amplification the heart sounds, even the less intense sounds can be detected and recorded graphically as a **phonocardiogram.**

The first heart sound is initiated at the onset of ventricular systole (Figure 18-11) and consists of a series of vibrations of mixed, unrelated, low frequencies (a noise). It is the loudest and longest of the heart sounds, and has a crescendo-decrescendo quality. The first heart sound is primarily caused by the oscillation of blood in the ventricular chambers and vibration of the chamber walls. The vibrations are engendered in part by the abrupt rise in ventricular pressure with acceleration of blood back toward the atria. However, the sound is produced mainly by sudden tension and recoil of the AV valves and adjacent structures when the blood is decelerated by closure of the AV valves.

The second heart sound, which occurs with closure of the semilunar valves (Figure 18-11), is composed of higher-frequency vibrations (higher pitch), is of shorter duration and lower intensity, and has a more snapping quality than the first heart sound. The second sound is caused by abrupt closure of the semilunar valves, which initiates oscillations of the columns of blood and the tensed vessel walls by the stretch and recoil of the closed valves.

The third heart sound is usually not audible, but it is sometimes heard in children with thin chest walls or in patients with left ventricular failure. This sound consists of a few low-intensity, low-frequency vibrations heard best in the region of the apex. It occurs in early diastole and is believed to be the result of vibrations of the ventricular walls caused by abrupt cessation of ventricular distension and deceleration of blood entering the ventricles.

A fourth, or atrial, sound, consisting of a few low-frequency oscillations, is occasionally heard in normal individuals. It is caused by oscillation of blood and cardiac chambers resulting from atrial contraction (Figure 18-11).

Asynchronous valve closures can produce **split sounds** over the apex of the heart for the AV valves and over the base for the semilunar valves. Deformities of the valves can produce cardiac **murmurs.** Valve lesions (stenosis or incompetence) may be congenital or produced by disease (e.g., **rheumatic fever**), and the timing (systolic or diastolic) and the character of the murmur provide clues regarding the type of valve damage.

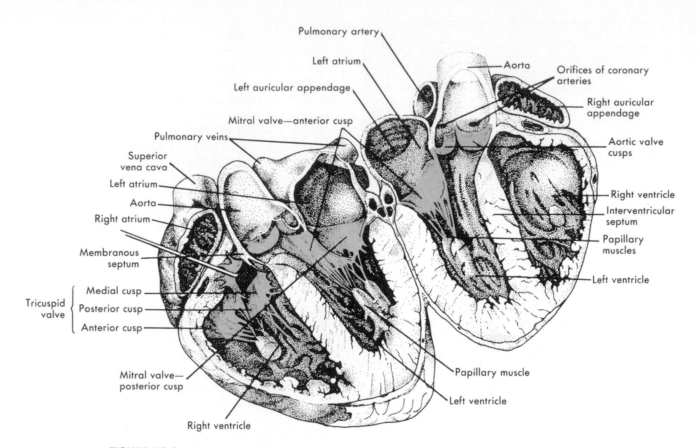

FIGURE 18-8 Drawing of a heart that is split perpendicular to the interventricular septum to illustrate the anatomical relationships of the leaflets of the AV and aortic valves.

CARDIAC CYCLE

Ventricular Systole

Isovolumic Contraction The onset of ventricular contraction coincides with the peak of the R wave of the electrocardiogram and the initial vibration of the first heart sound. It is indicated on the ventricular pressure curve as the earliest rise in ventricular pressure after atrial contraction (Figure 18-11). The interval between the start of ventricular systole and the opening of the semilunar valves (when ventricular pressure rises abruptly) is called **isovolumic contraction,** because ventricular volume is constant during this brief period (Figure 18-11).

Ejection Opening of the semilunar valves marks the onset of the **ejection phase,** which may be subdivided into an earlier, slightly shorter phase **(rapid ejection)** and a later, longer phase **(reduced ejection).** The rapid ejection phase is characterized by (1) the sharp rise in ventricular and aortic pressures that terminates at the peak ventricular and aortic pressures, (2) an abrupt decrease in ventricular volume, and (3) a large aortic blood flow (Figure 18-11). During the reduced ejection period,

runoff of blood from the aorta to the periphery exceeds ventricular output, and therefore aortic and ventricular pressures decline. Throughout ventricular systole the blood returning to the atria progressively increases the atrial pressure.

Note that during rapid ventricular ejection, left ventricular pressure slightly exceeds aortic pressure and flow accelerates (continues to increase), whereas during reduced ventricular ejection the reverse holds true. This reversal of the ventricular/aortic pressure gradient in the presence of continued flow of blood from the left ventricle to the aorta (caused by the momentum of the forward blood flow) results from the storage of potential energy in the stretched arterial walls, which decelerates the flow of blood into the aorta.

The effect of ventricular systole on left ventricular diameter is shown in an echocardiogram (Figure 18-10). During ventricular systole (Figure 18-10, *C* to *D*) the interventricular septum and the free wall of the left ventricle become thicker and move closer to each other.

At the end of ejection a volume of blood approximately equal to that ejected during systole remains in the ventricular cavities. This **residual volume** is fairly constant in normal hearts. However, it is smaller when heart

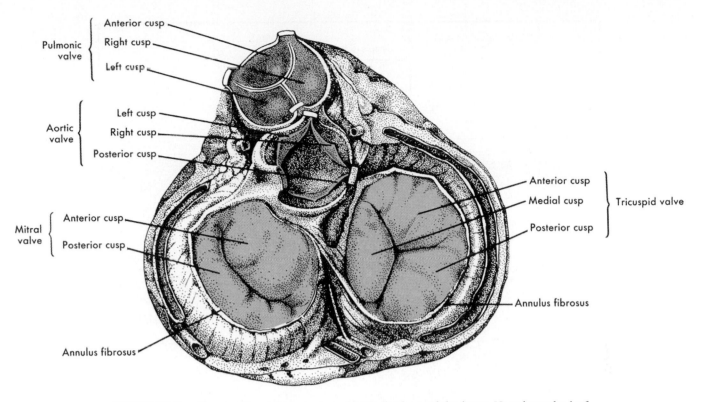

FIGURE 18-9   Four cardiac valves as viewed from the base of the heart. Note how the leaflets overlap in the closed valves.

rate increases or when outflow resistance is reduced, and it is larger when the opposite conditions prevail.

> An increase in myocardial contractility may decrease residual volume, especially in the depressed heart. In severely hypodynamic and dilated hearts, as in **heart failure,** the residual volume can become much greater than the stroke volume.

Ventricular Diastole

Isovolumic Relaxation   Closure of the aortic valve produces the **incisura** (a notch) on the descending limb of the aortic pressure curve; it marks the end of ventricular systole. The period between closure of the semilunar valves and opening of the AV valves is called isovolumic relaxation. It is characterized by a precipitous fall in ventricular pressure without a change in ventricular volume (Figure 18-11).

Rapid Filling Phase   Most of the ventricular filling occurs immediately after the AV valves open. The blood that had returned to the atria during the previous ventricular systole is abruptly released into the relaxing ventricles. This period of ventricular filling is called the **rapid filling phase** (Figure 18-11). The atrial and ventricular pressures decrease despite the increase in ventricular volume because the relaxing ventricles are exerting less and less force on the blood in their cavities.

Diastasis   The rapid filling phase is followed by a phase of slow filling called **diastasis.** During diastasis blood returning from the periphery flows into the right ventricle, and blood from the pulmonary circulation flows into the left ventricle. This small, slow addition to ventricular filling is indicated by gradual increases in atrial, ventricular, and venous pressures and in ventricular volume (Figure 18-11).

Atrial Systole   The onset of atrial systole occurs soon after the beginning of the P wave of the electrocardiogram (curve of atrial depolarization). The transfer of blood from atrium to ventricle accomplished by the peristalsis-like wave of atrial contraction completes the period of ventricular filling (Figure 18-11). Throughout ventricular diastole, atrial pressure barely exceeds ventricular pressure. This small pressure gradient indicates that the resistance of the pathway through the open AV valves during ventricular filling is normally very low.

Because there are no valves at the junctions of the venae cavae and right atrium or junctions of the pulmonary veins and left atrium, atrial contraction can force blood in both directions. Little blood is pumped back

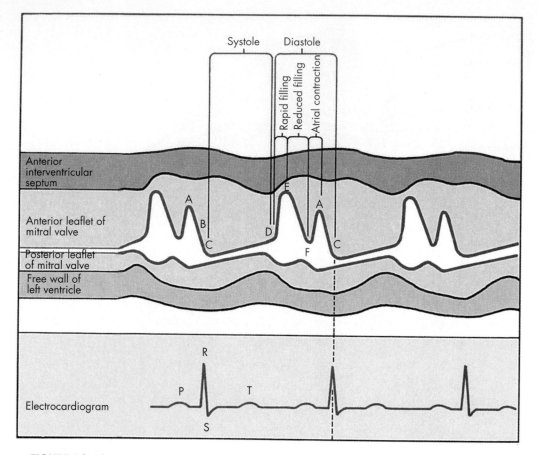

FIGURE 18-10 Drawing made from an echocardiogram showing movements of the mitral valve leaflets (particularly the anterior leaflet) and the changes in the diameter of the left ventricular cavity and the thickness of the left ventricular walls during cardiac cycles in a normal person. *D* to *C*, ventricular diastole; *C* to *D*, ventricular systole; *D* to *E*, rapid filling; *E* to *F*, reduced filling (diastasis); *F* to *A*, atrial contraction. Mitral valve closes at *C* and opens at *D*. Simultaneously recorded electrocardiogram at bottom.

into the venous tributaries during the brief atrial contraction, mainly because of the inertia of the inflowing blood.

> Atrial contraction is not essential for ventricular filling. Adequate filling is often observed in patients with **atrial fibrillation** or **complete heart block,** despite the absence of an effective atrial contribution.

The contribution of atrial contraction is governed to a great extent by the heart rate and the structure of the AV valves. At slow heart rates, filling practically ceases toward the end of diastasis, and atrial contraction contributes little additional filling. When the heart rate is rapid, diastasis is abbreviated and the atrial contribution can become substantial, especially if the atrium contracts immediately after the rapid filling phase, when the AV pressure gradient is maximal.

> When heart rates become so rapid that the period of ventricular relaxation becomes markedly abbreviated, ventricular filling is seriously impaired, despite the contribution of atrial contraction.
>
> In certain disease states, the AV valves may be severely narrowed (e.g., **mitral stenosis**) and atrial contraction can become more important to ventricular filling than it is in the normal heart.

Pressure-Volume Relationship The changes in left ventricular pressure and volume throughout the cardiac cycle are summarized in Figure 18-12. The element of time is not considered in this pressure-volume loop. Diastolic filling starts at *A* and terminates at *C*, when the mitral valve closes. The initial decrease in left ventricular pressure (*A* to *B*), despite the rapid inflow of blood from the atrium, results from progressive ventricular relaxation and increased distensibility. During the r[e]-der of diastole (*B* to *C*) the increase in ventricular pre[s]-sure reflects ventricular filling and the passive elastic characteristics of the ventricle. Note that after the initial phase of ventricular diastole, only a small increase in pressure occurs with the increase in ventricular volume (*B* to *C*) and that atrial systole (the small upward deflection just to the left of point *C*) contributes to ventricular volume and pressure. With isovolumic contraction (*C* to *D*), pressure rises steeply and ventricular volume remains constant. At *D* the aortic valve opens. During the first phase of ejection (rapid ejection, *D* to *E*) the large reduction in volume is associated with progressive increase in ventricular pressure that is less than the increase occurring during isovolumic contraction. This phase is followed by reduced ejection (*E* to *F*) and a small decrease in ventricular pressure. The aortic valve closes at *F*; this is followed by isovolumic relaxation (*F* to *A*), which is characterized by a sharp drop in pressure and no change in volume. The mitral valve opens at *A* to complete one cardiac cycle.

MEASUREMENT OF CARDIAC OUTPUT

Fick Principle

Adolph Fick contrived the first method for measuring cardiac output in intact animals and humans. The basis for this method, called the **Fick principle,** is simply an application of the law of conservation of mass. It is derived from the fact that the quantity of O_2 delivered to the pulmonary capillaries via the pulmonary artery plus the quantity of O_2 that enters the pulmonary capillaries from the alveoli must equal the quantity of O_2 that is carried away by the pulmonary veins.

This is depicted schematically in Figure 18-13. The rate, $\dot{q}_1$, of O_2 delivery to the lungs equals the O_2 concentration in the pulmonary arterial blood, $[O_2]_{pa}$, times the pulmonary arterial blood flow, Q, which equals the cardiac output; that is,

$$\dot{q}_1 = Q[O_2]_{pa} \qquad \text{(1)}$$

Let $\dot{q}_2$ be the net rate of O_2 uptake by the pulmonary capillaries from the alveoli. At equilibrium, $\dot{q}_2$ equals the O_2 consumption of the body. The rate, $\dot{q}_3$, at which O_2 is carried away by the pulmonary veins equals the O_2 concentration in the pulmonary venous blood, $[O_2]_{pv}$, times the total pulmonary venous flow, which is virtually equal to the pulmonary arterial blood flow, Q; that is,

$$\dot{q}_3 = Q[O_2]_{pv} \qquad \text{(2)}$$

From the conservation of mass,

$$\dot{q}_1 + \dot{q}_2 = \dot{q}_3 \qquad \text{(3)}$$

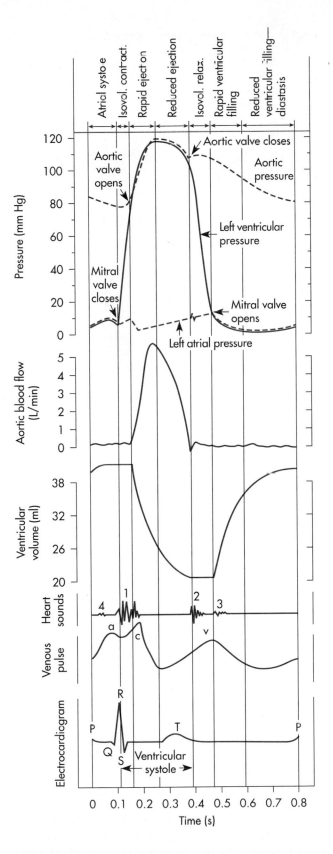

FIGURE 18-11 Left atrial, aortic, and left ventricular pressure pulses correlated in time with aortic flow, ventricular volume, heart sounds, venous pulse, and the electrocardiogram for a complete cardiac cycle in the dog (see text for details).

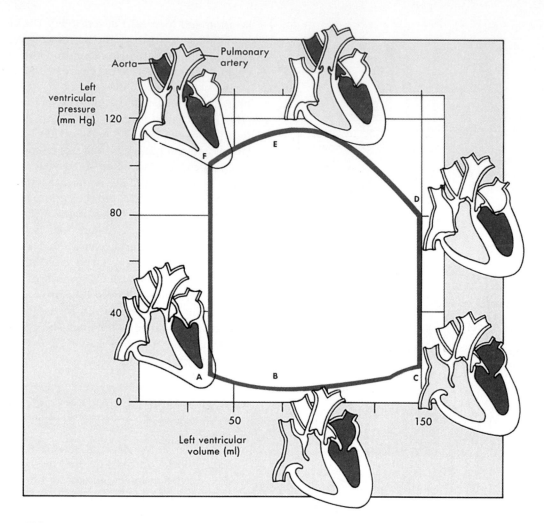

FIGURE 18-12 Pressure-volume loop of the left ventricle for a single cardiac cycle (*A to F*) (see text for details).

Therefore, from equations 1 to 3

$$Q[O_2]_{pa} + \dot{q}_2 = Q[O_2]_{pv} \quad \textbf{(4)}$$

Solving for cardiac output,

$$Q = \dot{q}_2 / ([O_2]_{pv} - [O_2]_{pa}) \quad \textbf{(5)}$$

Equation 5 is the statement of the Fick principle.

In the clinical determination of cardiac output, O_2 consumption is computed from measurements of the volume and O_2 content of expired air over a given interval. Because the O_2 concentration of peripheral arterial blood is essentially identical to that in the pulmonary veins, $[O_2]_{pv}$, it is determined on a sample of peripheral arterial blood withdrawn by needle puncture. Pulmonary arterial blood actually represents mixed systemic venous blood. Samples for O_2 analysis are obtained from the pulmonary artery or right ventricle through a cardiac catheter.

An example of the calculation of cardiac output in a normal, resting adult is illustrated in Figure 18-13. With an O_2 consumption of 250 ml/min, an arterial (pulmonary venous) O_2 content of 0.20 ml O_2/ml blood, and a mixed venous (pulmonary arterial) O_2 content of 0.15 ml O_2/ml blood, the cardiac output would be 250/(0.20 − 0.15) = 5000 ml/min.

The Fick principle is also used for estimating the O_2 consumption of organs in situ, when blood flow and the O_2 contents of the arterial and venous blood can be determined. Algebraic rearrangement of equation 5 reveals that O_2 consumption equals the blood flow times the arteriovenous O_2 concentration difference. For example, if the blood flow through one kidney is 700 ml/min, arterial O_2 content is 0.20 ml O_2/ml blood, and renal venous O_2 content is 0.18 ml O_2/ml blood, then the rate of O_2 consumption by that kidney must be 700(0.20 − 0.18) = 14 ml/min.

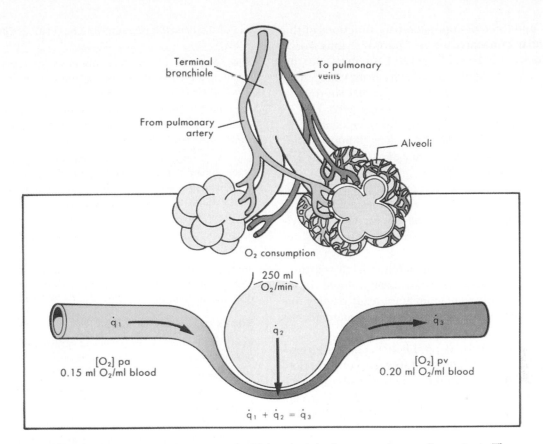

FIGURE 18-13 Schema illustrating the Fick principle for measuring cardiac output. The change in color from pulmonary artery to pulmonary vein represents the change in color of the blood as venous blood becomes fully oxygenated.

Indicator Dilution Technique

The **indicator dilution technique** has been widely used to estimate cardiac output in humans. A measured quantity of some indicator (a dye or isotope that remains within the circulation) is injected rapidly into a large central vein or into the right side of the heart through a catheter. Arterial blood is continuously drawn through a detector (densitometer or isotope rate counter), and a curve of indicator concentration is recorded as a function of time. The greater the blood flow (cardiac output), the greater is the dilution of the injected dye. Currently the most common indicator is a bolus of cold saline injected into the pulmonary artery via a cardiac catheter. The cardiac output can be calculated from the change in the temperature of the blood flowing past the temperature detector at the tip of the catheter.

SUMMARY

1. An increase in myocardial fiber length, as occurs with an augmented ventricular filling during diastole (pre-load), produces a more forceful ventricular contraction. This relationship between fiber length and strength of contraction is known as the Frank-Starling relationship or Starling's law of the heart.

2. Although the myocardium is made up of individual cells with discrete membrane boundaries, the cardiac myocytes that constitute the ventricles contract almost in unison, as do those of the atria. The myocardium functions as a syncytium with an all-or-none response to excitation. Cell-to-cell conduction occurs through gap junctions that connect the cytosol of adjacent cells.

3. During the upstroke of the action potential, voltage-gated calcium channels open to admit extracellular Ca^{++} into the cell. The influx of Ca^{++} triggers the release of Ca^{++} from the sarcoplasmic reticulum. The elevated intracellular Ca^{++} produces contraction of the myofilaments.

4. Relaxation of the myocardial fibers is accomplished by restoration of the resting cytosolic Ca^{++} level by pumping it back into the sarcoplasmic reticulum and exchanging it for extracellular Na^+ across the sarcolemma.

5. Velocity and force of contraction are functions of the intracellular concentration of free Ca^{++} ions. Force and velocity are inversely related, so that with no load, force is negligible and velocity is maximal. In an isometric contraction, where no external shortening occurs, force is maximal and velocity is zero.

6. In the ventricles the preload is the stretch of the fibers by the blood during ventricular filling, and the afterload is the aortic pressure against which the left ventricle ejects the blood.

7. Contractility is an expression of cardiac performance at a given preload and afterload. Contractility is increased mainly by interventions that increase intracellular Ca^{++} and decreased by interventions that decrease intracellular Ca^{++}

8. Simultaneous recording of the left atrial, left ventricular, and aortic pressures, ventricular volume, heart sounds, and electrocardiogram graphically portray the sequential and related electrical and cardiodynamic events throughout a cardiac cycle.

9. Cardiac output can be determined, according to the Fick principle, by dividing the oxygen consumption of the body by the difference between the oxygen content of arterial and mixed venous blood. It can also be measured by dye dilution or thermodilution techniques.

BIBLIOGRAPHY

Journal Articles

Bers DM, Lederer WJ, Berlin JR: Intracellular Ca transients in rat cardiac myocytes: role of Na-Ca exchange in excitation-contraction coupling, *Am J Physiol* 258:C944, 1990.

Brady AJ: Mechanical properties of isolated cardiac myocytes, *Physiol Rev* 71:413, 1991.

Brutsaert DL, Sys SU: Relaxation and diastole of the heart, *Physiol Rev* 69:1228, 1989.

Carafoli E: Calcium pump of the plasma membrane, *Physiol Rev* 71:129, 1991.

Chapman RA: Control of cardiac contractility at the cellular level, *Am J Physiol* 245:H535, 1983.

Elzinga G, Westerhof N: Matching between ventricle and arterial load, *Circ Res* 68:1495, 1991.

Fabiato A, Fabiato F: Calcium and cardiac excitation-contraction coupling, *Ann Rev Physiol* 41:473, 1979.

Gilbert JC, Glantz SA: Determinants of left ventricular filling and of the diastolic pressure-volume relation, *Circ Res* 64:827, 1989.

Katz AM: Cyclic adenosine monophosphate effects on the myocardium: a man who blows hot and cold with one breath, *J Am Coll Cardiol*, 2:143, 1983.

Katz AM: Interplay between inotropic and lusitropic effects of cyclic adenosine monophosphate on the myocardial cell, *Circulation* 82:I-7, 1990.

Luo W, Grupp IL, Harrer J, Ponniah S, Grupp G, Duffy JJ, Doetschman T, Kranias EG: Targeted ablation of the phospholamban gene is associated with markedly enhanced myocardial contractility and loss of β-agonist stimulation, *Circ Res* 75:401, 1994.

Sagawa K: The ventricular pressure-volume diagram revisited, *Circ Res* 43:677, 1978.

Smith JS, Rousseau E, Meissner G: Single sarcoplasmic reticulum Ca^{2+}-release channels from calmodulin modulation of cardiac and skeletal muscle, *Circ Res* 64:352, 1989.

Books and Monographs

Brady AJ: Mechanical properties of cardiac fibers. In *Handbook of physiology,* section 2: The cardiovascular system—the heart, vol 1, Bethesda, Md, 1979, American Physiological Society.

Gibbons WR, Zygmunt AC: Excitation-contraction coupling in the heart. In Fozzard HA et al, eds: *The heart and cardiovascular system,* ed 2, New York, 1991, Raven Press.

Lakatta EG: Length modulation of muscle performance: Frank-Starling law of the heart. In Fozzard HA et al, eds, *The heart and cardiovascular system,* ed 2, New York, 1991, Raven Press.

Lytton J, MacLennan DH: Sarcoplasmic reticulum. In Fozzard HA et al, eds, *The heart and cardiovascular system,* ed 2, New York, 1991, Raven Press.

Parmley WW, Talbot L: Heart as a pump. In *Handbook of physiology,* section 2: The cardiovascular system—the heart, vol 1, Bethesda, Md, 1979, American Physiological Society.

Ruegg JC: *Calcium in muscle activation,* Heidelberg, 1988, Springer-Verlag.

Sheu SS, Blaustein MP: Sodium/calcium exchange and control of cell calcium and contractility in cardiac muscle and vascular smooth muscle. In Fozzard HA et al, eds, *The heart and cardiovascular system,* New York, 1991, Raven Press.

Sommer JR, Johnson EA: Ultrastructure of cardiac muscle. In *Handbook of physiology,* section 2: The cardiovascular system—the heart, vol 1, Bethesda, Md, 1979, American Physiological Society.

Regulation of the Heartbeat

The quantity of blood pumped by the heart each minute **(cardiac output)** equals the volume of blood pumped each beat **(stroke volume)** multiplied by the number of heartbeats per minute **(heart rate).** *Thus, the cardiac output may be varied by changing the heart rate or the stroke volume.* A discussion of the control of cardiac activity may therefore be subdivided into a consideration of the regulation of pacemaker activity and the regulation of contractile strength. The control of pacemaker activity is mediated almost exclusively by the autonomic nervous system. The cardiac nerves also regulate contractile strength, but a number of mechanical and humoral factors are also important.

CONTROL OF HEART RATE

In normal adults the average heart rate at rest is about 70 beats/min, but the rate is significantly greater in children. During sleep the heart rate diminishes by 10 to 20 beats/min, but during exercise or emotional excitement it may accelerate to rates considerably above 100. In many infectious diseases, especially in those associated with fever, the heart rate is elevated. In various types of heart failure, the heart rate also is usually high. In well-trained athletes at rest the heart rate is only about 50.

The SA node is usually under the tonic influence of both divisions of the autonomic nervous system. *Stimulation of the sympathetic system increases heart rate, whereas stimulation of the parasympathetic system decreases it.* Changes in heart rate usually involve a reciprocal action of the two divisions of the autonomic nervous system. Thus an increased heart rate is usually achieved by a waning of parasympathetic activity and a concomitant increase in sympathetic activity; deceleration is usually accomplished by the opposite changes in neural activity.

Ordinarily, in healthy, resting individuals, parasympa-

thetic activity predominates. Abolition of parasympathetic influences by the drug **atropine** (a **muscarinic receptor antagonist**) usually increases the heart rate substantially (Figure 19-1). Conversely, abolition of sympathetic effects by the drug **propranolol** (a **β-adrenergic receptor antagonist**) usually slows the heart only slightly (Figure 19-1). When the effects of both divisions of the autonomic nervous system are blocked by the combination of these two drugs, the heart rate of adults averages about 100 beats/minute. The rate that prevails after complete autonomic blockade is called the **intrinsic heart rate.**

Nervous Control

Sympathetic Pathways *The cardiac sympathetic fibers (see also Chapter 10) originate in the upper five or six thoracic and lower one or two cervical segments of the spinal cord.* These preganglionic fibers emerge from the spinal column through the white communicating branches and enter the paravertebral chains of ganglia. Most of the preganglionic fibers ascend the paravertebral chains and synapse with postganglionic neurons, mainly in the stellate and middle cervical ganglia. Postganglionic sympathetic fibers then join with parasympathetic fibers to form the **cardiac plexus,** which is a complex network of nerve trunks that contain sympathetic and parasympathetic efferent nerves to the heart, as well as afferent nerves from sensory receptors in the heart and great vessels.

Sympathetic fibers from the right and left sides of the body are distributed asymmetrically to the various structures in the heart. In the dog, for example, right cardiac sympathetic nerve stimulation increases the heart rate more than does equivalent stimulation of sympathetic fibers on the left side; the asymmetry is reversed for the control of ventricular contractile force (Figure 19-2). In some dogs left cardiac sympathetic

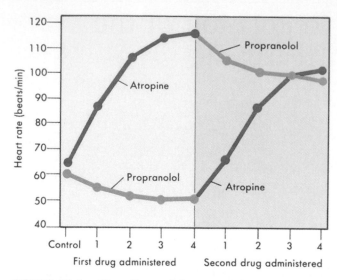

FIGURE 19-1 The effects of four equal doses of atropine (0.04 mg/kg total) and of propranolol (0.2 mg/kg total), given sequentially, on the heart rate of 10 healthy young men (mean age, 21.9 years). In half of the trials, atropine was given first *(top curve)*; in the other half, propranolol was given first *(bottom curve)*. (Redrawn from Katona PG et al: *J Appl Physiol* 52:1652, 1982.)

nerve stimulation may not affect heart rate at all, even though it may strengthen the ventricular contraction profoundly, as shown in Figure 19-2. The available evidence indicates that this bilateral asymmetry also prevails in humans.

The effects of sympathetic stimulation decay gradually after the cessation of stimulation. In Figure 19-2, for example, about 2 minutes was required for the peak left ventricular pressure generated on consecutive heartbeats to return to its control value after sympathetic stimulation was discontinued. Most of the **norepinephrine** released from the sympathetic nerve endings during stimulation is taken up again by those terminals, and much of the remaining neurotransmitter is carried away by the bloodstream. Both of these removal processes are slow.

The adrenergic receptors in the cardiac tissues are predominantly of the β type; that is, they are responsive to β-**adrenergic receptor agonists,** such as **isoproterenol,** and are inhibited by specific β-**adrenergic receptor antagonists,** such as **propranolol.**

Parasympathetic Pathways *The preganglionic parasympathetic fibers (see also Chapter 10) to the heart originate in the* **medulla oblongata,** *in cells that lie in the* **dorsal motor nucleus of the vagus** *or in the* **nucleus ambiguus.** The precise location varies from species to species. Centrifugal fibers from these nuclei pass inferiorly through the neck via the vagus nerves (the tenth cranial nerves), which lie close to the common carotid arteries. The nerve fibers then travel

through the mediastinum to synapse with postganglionic cells located on the epicardial surface or within the walls of the heart itself. Many of the cardiac ganglion cells are located near the SA and AV nodes.

The right and left vagi are usually distributed differentially to the various cardiac structures. *The vagal effects are usually inhibitory.* The right vagus nerve affects the SA node predominantly; stimulation decreases the firing rate. The left vagus nerve mainly retards AV conduction, and may actually block impulse conduction from atria to ventricles. However, the innervation overlaps considerably; left vagal stimulation inhibits the SA node, and right vagal stimulation impedes AV conduction.

The effects of vagal activity are mediated mainly by the neurotransmitter **acetylcholine,** which is released from the postganglionic vagus nerve endings in the cardiac tissues. The acetylcholine interacts with **cholinergic receptors (muscarinic type)** in the membranes of the various types of cardiac cells. The action of the released acetylcholine can be blocked by the muscarinic receptor antagonist **atropine.** The cardiac tissues are rich in **cholinesterase,** which rapidly hydrolyzes the neurally released acetylcholine. Hence, after vagal activity ceases, the effects decay quickly.

The parasympathetic effects preponderate over sympathetic effects at the SA node, as shown in Figure 19-3. As the frequency of sympathetic stimulation in an anesthetized dog was increased from 0 to 4 Hz in the absence of any concomitant vagal stimulation *(top curve),* the heart rate increased by 80 beats/min. However, during concurrent vagal stimulation at 8 Hz *(bottom curve),* the same increase in sympathetic stimulation had scarcely any effect on heart rate. The mechanisms responsible for this vagal predominance will be discussed below, in relation to the neural control of myocardial contractility.

Higher Centers A number of higher cerebral centers help regulate cardiac rate, rhythm, and contractile strength (see also Chapter 10). Stimulation of specific nuclei in the **thalamus** or **hypothalamus** will alter the heart rate. Hypothalamic centers are also involved in the circulatory responses to fluctuations in environmental temperature. Experimentally induced temperature changes in the anterior hypothalamus markedly alter heart rate and peripheral resistance (as described in the section on hypothalamic functions in Chapter 9). Stimuli applied to the H_2 fields of Forel in the **diencephalon** elicit cardiovascular responses that resemble those observed during muscular exercise. In the **cerebral cortex** the centers that influence cardiac function are located mostly in the anterior half of the brain, principally in the frontal lobe, the orbital cortex, the motor and premotor cortex, the temporal lobe, the insula, and the cingulate gyrus.

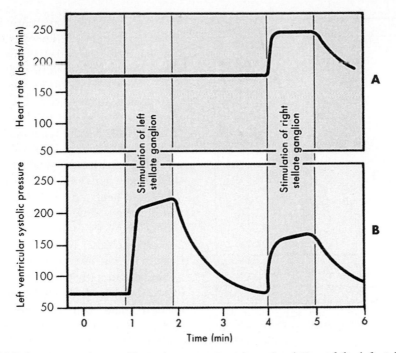

FIGURE 19-2 In a canine total heart bypass preparation, stimulation of the left stellate ganglion did not affect heart rate at all (panel **A**), but it enhanced left ventricular performance substantially (panel **B**). In most experiments, left stellate stimulation does increase heart rate, but not as much as does right stellate stimulation. Left ventricular performance was assessed by measuring the peak pressure generated by the left ventricle on successive heartbeats; the values attained during individual heartbeats are not shown. (Levy, M.N. Unpublished observations.)

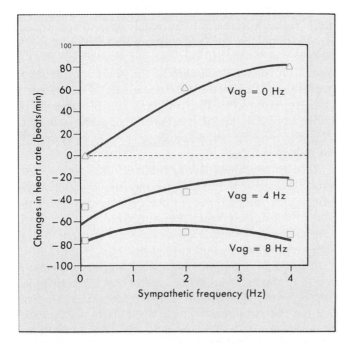

FIGURE 19-3 The changes in heart rate in an anesthetized dog when the vagus and cardiac sympathetic nerves were stimulated simultaneously. The sympathetic nerves were stimulated at 0, 2, and 4 Hz; the vagus nerves at 0, 4 and 8 Hz. The symbols represent the observed changes in heart rate; the curves were derived from the computed regression equation. (Modified from Levy MN, Zieske H: *J Appl Physiol* 27:465, 1969.)

Cortical and diencephalic centers initiate the cardiovascular reactions that occur during excitement, anxiety, and other emotional states, and during certain febrile disease states.

Reflex Control

Baroreceptor Reflex Acute changes in blood pressure reflexly alter heart rate. Such changes in heart rate are mediated mainly by the baroreceptors located in the carotid sinuses and aortic arch (see Chapter 23). An example of the changes in heart rate elicited by drug-induced changes in arterial blood pressure in a group of normal human subjects is shown in Figure 19-4. Blood pressure was elevated by infusing **phenylephrine,** which is a potent vasoconstrictor, whereas blood pressure was reduced by infusing **nitroprusside,** a vasodilator. Over the induced range of blood pressures, the cardiac cycle length (reciprocal of heart rate) varied linearly with the arterial blood pressure.

When the arterial blood pressure is in the normal range, moderate alterations in baroreceptor stimulation change heart rate by evoking reciprocal changes in autonomic neural activity. For example, a moderate in

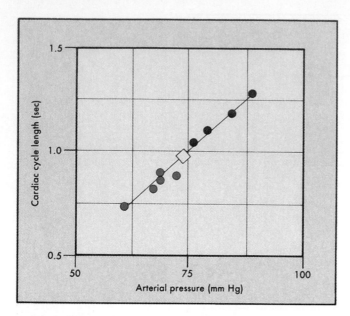

FIGURE 19-4   The relation between cardiac cycle length and diastolic arterial blood pressure in a group of healthy human subjects. The pressure changes were produced by infusions of nitroprusside *(blue circles)* and phenylephrine *(red circles)*. The diamond symbol represents the mean values of cardiac cycle length and arterial blood pressure in these subjects prior to the drug infusions. (Modified from Eckberg DL et al: *J Clin Invest* 78:366, 1986.)

crease in arterial blood pressure will lower the heart rate by increasing efferent vagal activity and by decreasing efferent sympathetic activity concurrently. However, when blood pressure is increased by more than about 25 mm Hg, cardiac sympathetic tone is completely suppressed. Thereafter the additional reduction in heart rate produced by any further rise in blood pressure is evoked entirely by increased vagal activity.

> The converse applies in response to **hypotension,** such as that induced by **hemorrhage.** In response to the loss of a moderate amount of blood, vagal tone diminishes and sympathetic activity increases. As blood continues to be lost, however, vagal activity finally ceases after the blood pressure has declined to about 20 or 30 mm Hg below the normal level. Any further acceleration of the heart in response to further declines in blood pressure is mediated exclusively by progressive increases in sympathetic activity.

Bainbridge Reflex and Atrial Receptors   In 1915 Bainbridge reported that infusions of blood or saline solution increased the heart rate, regardless of whether the infusions did or did not raise the arterial blood pressure. Cardiac acceleration was observed whenever central ve-

nous pressure rose sufficiently to distend the right side of the heart, and the effect was abolished by cutting both vagi. Bainbridge postulated that increased cardiac filling raised the heart rate reflexly, and that the afferent impulses were conducted by the vagi.

Many investigators have confirmed that the heart may accelerate in response to the intravenous administration of fluid. However, the magnitude and direction of the response depend on a number of factors, especially the prevailing heart rate. When the heart rate is relatively slow, intravenous infusions usually accelerate the heart. When the heart rate is more rapid, however, infusions will ordinarily slow the heart. Acute increases in blood volume not only evoke the Bainbridge reflex, but they also activate other reflexes (notably the baroreceptor reflex) that tend to change heart rate in the opposite direction (Figure 19-5). The actual change in heart rate induced by an intravenous infusion is therefore the result of these antagonistic reflex effects.

Sensory receptors that influence heart rate exist in both atria. The receptors are located principally in the venoatrial junctions—in the right atrium at its junctions with the venae cavae, and in the left atrium at its junctions with the pulmonary veins. Distension of these receptors sends impulses centrally in the vagi. The efferent impulses are carried by sympathetic and parasympathetic fibers to the SA node.

Stimulation of the atrial receptors also increases urine flow. A reduction in renal sympathetic nerve activity might be partially responsible for this diuresis. However, the principal mechanisms appear to be (1) a neurally mediated reduction in the secretion of **vasopressin (antidiuretic hormone)** by the posterior pituitary gland (Chapter 44) and (2) the release of another peptide, **atrial natriuretic peptide,** which is released from the atrial tissues in response to stretch (Chapter 37).

Respiratory Sinus Arrhythmia   Cardiac cycle length often fluctuates rhythmically at the frequency of respiration. Such fluctuations are detectable in most resting adults, and they are more pronounced in children. Typically the cycle length decreases during inspiration and increases during expiration (Figure 19-6).

Recordings of action potentials from the autonomic nerves to the heart in animals reveal that the activity increases in the sympathetic nerve fibers during inspiration but that it increases in the vagal fibers during expiration (Figure 19-7). The acetylcholine released at the vagal endings quickly alters the firing of the pacemaker cells in the SA node, and the acetylcholine is also removed very rapidly. Therefore, periodic bursts of vagal activity can cause heart rate to vary rhythmically. Conversely, the norepinephrine released at the sympathetic endings affects the pacemaker cells much more gradually than does acetylcholine, and norepinephrine is also

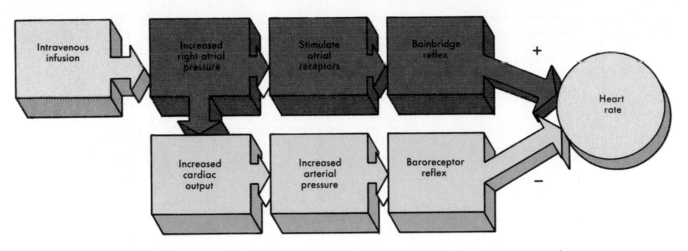

FIGURE 19-5   Intravenous infusions of blood or electrolyte solutions tend to increase heart rate via the Bainbridge reflex and to decrease heart rate via the baroreceptor reflex. The actual change in heart rate induced by an such infusions is the result of these two opposing effects.

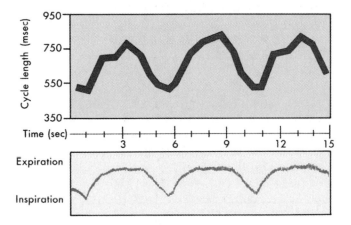

FIGURE 19-6   Respiratory sinus arrhythmia in a resting, unanesthetized dog. Note that the cardiac cycle length increases during expiration and decreases during inspiration. (Modified from Warner MR et al: *Am J Physiol* 251:H1134, 1986.)

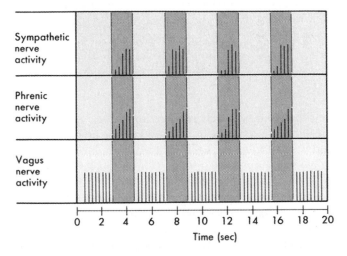

FIGURE 19-7   The respiratory fluctuations in efferent neural activity in the cardiac nerves of an anesthetized dog. Note that the sympathetic nerve activity occurs synchronously with the phrenic nerve discharges (which initiate diaphragmatic contraction), whereas the vagus nerve activity occurs between the phrenic nerve discharges. The phrenic discharges mark the inspiratory phase of respiration, whereas the periods between the phrenic discharges denote the expiratory phase. (From Kollai M, Koizumi K: *J Auton Nerv Syst* 1:33, 1979.)

removed from the cardiac tissues more slowly than is acetylcholine. Thus the effects of rhythmic variations in sympathetic activity on heart rate are damped out. Hence, *the rhythmic changes in heart rate associated with respiration are ascribable almost entirely to the oscillations in vagal activity.* Respiratory sinus arrhythmia is exaggerated when vagal tone is enhanced.

Reflex and central factors both contribute to the genesis of the respiratory cardiac arrhythmia (Figure 19-8). During inspiration the lung volume increases and the intrathoracic pressure decreases (see Chapter 28). Lung distension stimulates pulmonary stretch receptors and can reflexly increase heart rate. The reduction in intrathoracic pressure during inspiration increases venous return to the right side of the heart (see Figure 24-12). The resulting distension of the right atrium elicits the

Bainbridge reflex (Figure 19-8). After the time delay required for the increased systemic venous return to reach the left side of the heart, left ventricular stroke volume increases and thereby raises systemic arterial blood pressure. This in turn reduces heart rate reflexly through baroreceptor stimulation (Figure 19-8).

The respiratory center in the medulla oblongata directly influences the nearby cardiac autonomic centers (Figure 19-8). This influence has been established in anesthetized animals that have been placed on a heart-

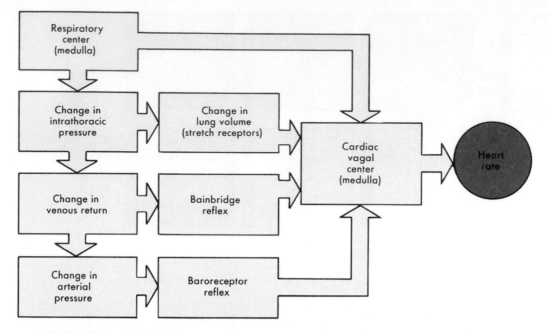

FIGURE 19-8   Respiratory sinus arrhythmia is generated by a direct interaction between the respiratory and cardiac centers in the medulla, and also by reflexes originating from stretch receptors in the lungs, stretch receptors in the right atrium (Bainbridge reflex), and baroreceptors in the carotid sinuses and aortic arch.

lung machine. In such preparations, the chest is open, the lungs are collapsed; and the arterial blood pressure and central venous pressure do not fluctuate rhythmically. Nevertheless, respiratory movements of the rib cage and diaphragm demonstrate that the medullary respiratory center is still active. Cyclic heart rate changes accompany the respiratory movements. These rhythmic changes in heart rate are almost certainly induced by a direct interaction between the respiratory and cardiac centers in the medulla.

Chemoreceptor Reflex The cardiac response to peripheral chemoreceptor stimulation (see Chapter 23) merits special consideration, because it illustrates the complexity that may be introduced when one stimulus excites two organ systems simultaneously. In intact animals, stimulation of the carotid chemoreceptors consistently increases ventilatory rate and depth (see Chapter 31) but ordinarily has little effect on heart rate. The small, directional changes in heart rate are related to the enhancement of pulmonary ventilation, as shown in Figure 19-9. When chemoreceptor stimulation augments respiration only slightly, heart rate usually decreases; when the increment in pulmonary ventilation is more pronounced, heart rate usually increases.

The cardiac response to peripheral chemoreceptor stimulation is the result of primary and secondary reflex mechanisms (Figure 19-10). The **primary reflex effect** of carotid chemoreceptor excitation on the SA node is inhibitory; this primary effect becomes evident when the

usual respiratory response is absent. The **secondary effects** of respiratory excitation tend to increase heart rate, and therefore they tend to mask the primary inhibitory effects on the SA node.

A dramatic example of the primary inhibitory influence of chemoreceptor stimulation on heart rate in a human subject is displayed in Figure 19-11. This figure is a segment of an electrocardiogram recorded from a quadriplegic patient who was unable to breathe naturally, but required tracheal intubation and artificial respiration. When the tracheal catheter was disconnected briefly to permit removal of excess tracheal secretions, the patient developed a marked bradycardia within seconds. The bradycardia could be prevented by injecting the muscarinic receptor antagonist **atropine,** and the onset of the bradycardia could be delayed substantially by hyperventilating the patient before disconnecting the tracheal cannula.

Ventricular Receptor reflexes Sensory receptors located near the ventricular endocardium initiate reflex effects similar to those elicited by the arterial baroreceptors. Excitation of these endocardial receptors diminishes the heart rate and peripheral resistance. The receptors discharge in a pattern that parallels the changes in ventricular pressure. Impulses that originate in these re-

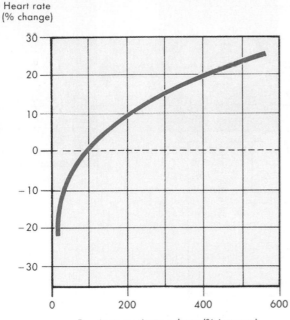

FIGURE 19-9   Relationship between the change in heart rate and the change in respiratory minute volume during carotid chemoreceptor stimulation in spontaneously breathing cats and dogs. When respiratory stimulation was relatively slight, heart rate usually diminished; when respiratory stimulation was more pronounced, heart rate usually increased. (Modified from Daly M deB, Scott MJ: *J Physiol* 144:148, 1958.)

ceptors are transmitted to the medulla oblongata via the vagus nerves.

Other sensory receptors have been identified in the epicardial regions of the ventricles. These receptors discharge in patterns that are not related to the changes in ventricular pressure. These ventricular receptors are excited by various mechanical and chemical stimuli, but their physiological functions are not clear.

REGULATION OF MYOCARDIAL PERFORMANCE

Intrinsic Regulation

Just as the heart can initiate its own beat in the absence of any nervous or hormonal control, so also can the myocardium adapt to changing hemodynamic conditions by mechanisms that are intrinsic to cardiac muscle itself. Experiments on animals with denervated hearts as well as observations in human subjects with cardiac transplants reveal that this organ adjusts remarkably well to stress even in the absence of any innervation. For example, racing greyhounds with denervated

hearts perform almost as well as those with intact innervation. Their maximal running speed is only 5% less after complete cardiac denervation. In these dogs, the fourfold increase in cardiac output that occurs when they run is achieved principally by an increase in stroke volume. In normal dogs the increase in cardiac output with exercise is accompanied by a proportionate increase in heart rate; stroke volume does not change much. The cardiac adaptation in the denervated animals is not achieved entirely by intrinsic mechanisms, however; circulating catecholamines contribute significantly. If the β-adrenergic receptors are blocked by propranolol in greyhounds with denervated hearts, their racing performance is severely impaired.

The intrinsic cardiac adaptation that has received the greatest attention involves changes in the resting length of the myocardial fibers. This adaptation is designated **Starling's law of the heart** or the **Frank-Starling mechanism.** The mechanical and structural bases for this mechanism have been explained in Chapters 12 and 18. However, certain other intrinsic mechanisms that do not necessarily involve any changes in resting length also help to regulate myocardial performance.

Frank-Starling Mechanism

Isolated Hearts   In 1895 the German physiologist Otto Frank described the response of the isolated heart of the frog to alterations in the stretching force (**preload**) on the myocardial fibers just prior to contraction. He observed that as the preload was increased, the heart responded with a more forceful contraction. About 20 years later, the English physiologist Ernest Starling described the intrinsic response of the canine heart to changes in right atrial and aortic pressure in the isolated heart-lung preparation.

In this preparation the right ventricular filling pressure is varied by altering the height of a reservoir connected to the right atrium; the filling pressure just prior to ventricular contraction constitutes the preload for the myocardial fibers in the ventricular wall. The right ventricle then pumps this blood through the pulmonary vessels to the left atrium. The lungs are artificially ventilated. Blood is pumped by the left ventricle into the aortic arch, and then through some external tubing back to the right atrial reservoir. A resistance device in the external tubing allows the investigator to control the aortic pressure; this pressure constitutes the **afterload** for left ventricular ejection.

One of Starling's recordings of the changes in ventricular volume evoked by a sudden increase in right atrial pressure is shown in Figure 19-12. Aortic pressure in this experiment was permitted to increase only slightly. In the top tracing, an increase in ventricular volume is registered as a downward deflection. Hence the upper border of the tracing represents the systolic ventricular volume, the lower border indicates the diastolic

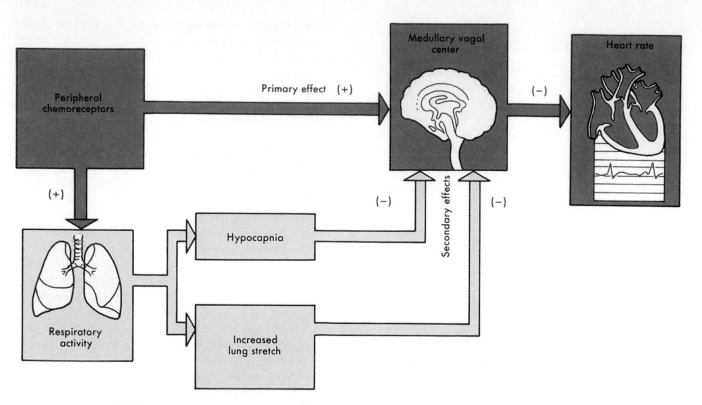

FIGURE 19-10   The primary effect of stimulation of the peripheral chemoreceptors on heart rate is to excite the cardiac vagal center in the medulla, and thus to decrease heart rate. Peripheral chemoreceptor stimulation also excites the respiratory center in the medulla. This effect produces hypocapnia and increases lung inflation, both of which secondarily inhibit the medullary vagal center. Thus, these secondary influences attenuate the primary reflex effect of peripheral chemoreceptor stimulation on heart rate.

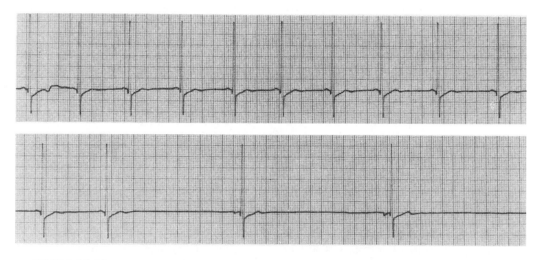

FIGURE 19-11   Electrocardiogram of a 30-year-old man who could not breathe spontaneously and required tracheal intubation and artificial respiration. The two strips are continuous. The tracheal catheter was temporarily disconnected from the respirator at the beginning of the top strip, at which time his heart rate was 65 beats/min. In less than 10 seconds, his heart rate decreased to about 20 beats/min. (Modified from Berke JL, Levy MN: *Eur Surg Res* 9:75, 1977.)

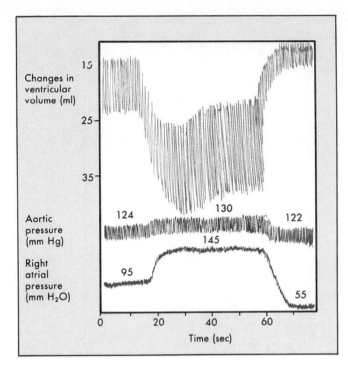

FIGURE 19-12 Changes in ventricular volume in a heart-lung preparation when the venous reservoir was suddenly raised (right atrial pressure increased from 95 to 145 mm H_2O) and subsequently lowered (right atrial pressure decreased from 145 to 55 mm H_2O). Note that an increase in ventricular volume is registered as a downward shift in the volume tracing. (Redrawn from Patterson SW, Piper H, Starling EH: *J Physiol* 48:465, 1914.)

ventricular volume, and the width of the tracing reflects the stroke volume.

For several beats after the rise in preload, the ventricular volume progressively increased. This indicates that a disparity must have existed between ventricular inflow during diastole and ventricular outflow during systole; that is, during a given systole the ventricles did not expel as much blood as had entered them during the preceding diastole. This progressive accumulation of blood dilated the ventricles and lengthened the individual myocardial fibers in the walls of the ventricles.

The increased diastolic fiber length somehow facilitates ventricular contraction and enables the ventricles to pump a greater stroke volume. Diastolic fiber length continues to increase on successive heartbeats in response to a sustained increase in preload until, at equilibrium, the cardiac output exactly matches the augmented filling volume. However, an optimum fiber length exists, beyond which contraction is actually impaired (Chapters 12 and 18). Therefore, excessively high preloads may depress rather than enhance the pumping capacity of the ventricles by overstretching the myocardial fibers. The principal way in which

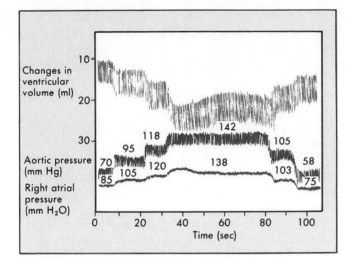

FIGURE 19-13 Changes in ventricular volume, aortic pressure, and right atrial pressure in a heart-lung preparation when peripheral resistance was raised and subsequently lowered in several steps. Note that an increase in ventricular volume is registered as a downward shift in the volume tracing. (Redrawn from Patterson SW, Piper H, Starling EH: *J Physiol* 48:465, 1914.)

changes in preload affect the pumping capacity of the ventricles is by altering the alignment of the thick and thin filaments in the component myocardial cells. Moderate increases in preload align these filaments such that more crossbridges are able to interact, whereas excessive increases in preload reduce the number of crossbridges that can interact.

Most commonly, changes in preload occur as a consequence of changes in blood volume. A common clinical condition characterized by a reduction in preload is **hemorrhage.** The blood loss is accompanied by reductions in pressure in the great central veins, and hence the cardiac filling pressure (preload) is also diminished. Even though blood loss also usually leads to a concomitant reduction in arterial blood pressure (afterload), the influence of the preload usually predominates over the influence of the afterload, and cardiac output tends to decrease in response to hemorrhage (see also Chapters 24 and 26). Large blood transfusions, on the other hand, will increase the cardiac filling pressure, and therefore they act to increase cardiac output.

Changes in diastolic fiber length also permit the isolated heart to compensate for an increase in afterload. In Starling's experiment depicted in Figure 19-13, the arterial pressure (afterload) was abruptly raised in three steps, whereas venous return to the right ventricle was held fairly constant. With each abrupt elevation of afterload, the left ventricle was at first unable to pump a normal stroke volume. Because venous return to the right atrium was held constant, the ventricular filling volume exceeded the diminished stroke volume. This disparity between ventricular filling and emptying augmented the ventricular diastolic volume (and hence the preload), and therefore increased the length of the myocardial fibers. This change in end-diastolic fiber length finally enabled the ventricle to pump a stroke volume equal to the control stroke volume, despite the greater afterload.

When cardiac compensation involves ventricular dilation, the force required by each myocardial fiber to generate a given intraventricular systolic pressure must be appreciably greater than that developed by the fibers in a ventricle of normal size. The relationship between wall tension and cavity pressure (**Laplace's law**, Chapter 22) resembles that for cylindrical tubes, in that for a constant internal pressure, wall tension varies directly with the radius (see Figure 22-2). As a consequence, the myocardial fibers in a dilated heart must develop considerably more tension than do those in a normal-sized heart. Therefore, these fibers require considerably more oxygen to perform a given amount of external work than do those in a normal-sized heart.

The most common clinical condition that is characterized by a chronic increase in afterload is **essential hypertension,** which is a sustained increase in arterial blood pressure of unknown cause. The principal hemodynamic change responsible for the elevated blood pressure is a generalized arteriolar vasoconstriction. The heart adapts initially to this increased afterload by an increase in diastolic ventricular volume, as explained above. Ultimately, however, the mass of ventricular muscle cells also increases; that is, the heart **hypertrophies.** This constitutes an additional mechanism by which the heart can adapt to an increase in afterload.

Studies in vivo A major problem in assessing the role of the Frank-Starling mechanism in intact animals and humans is the difficulty of measuring end-diastolic myocardial fiber length. The Frank-Starling mechanism has been represented graphically by plotting some index of ventricular performance (ordinate) as a function of some index of fiber length (abscissa). The most commonly used indices of ventricular performance are cardiac output, stroke volume, and stroke work. The indices of fiber length include ventricular end-diastolic volume, ventricu-

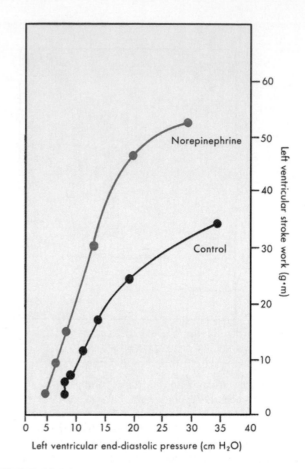

FIGURE 19-14 A constant infusion of norepinephrine in a dog shifts the ventricular function curve to the left. This shift signifies an enhancement of ventricular contractility. (Redrawn from Sarnoff SJ et al: *Circ Res* 8:1108, 1960.)

lar circumference, ventricular end-diastolic pressure, and mean atrial pressure.

Because ventricular performance depends not only on myocardial fiber length at end-diastole but also on afterload and contractility, the regulation of ventricular performance is better represented by a family of so-called **ventricular function curves,** rather than by a single curve. To construct a basal ventricular function curve, blood volume is altered over a range of values, and stroke work and ventricular end-diastolic pressure are measured at each step. Similar observations are then made during some desired experimental intervention. For example, the ventricular function curve obtained during a norepinephrine infusion lies above and to the left of a control ventricular function curve (Figure 19-14). The graph shows that, for a given level of left ventricular end-diastolic pressure, the left ventricle performs more work during a norepinephrine infusion than during control conditions. Hence a shift of the ventricular function curve to the left usually signifies an improvement in ventricular contractility; a shift to the right usu-

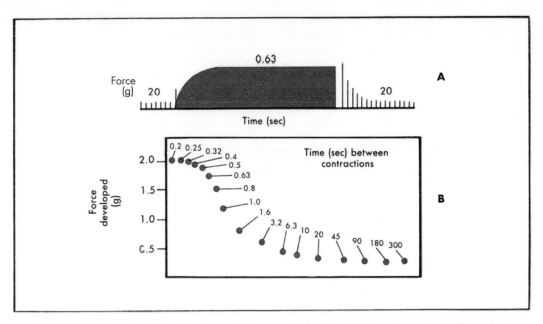

FIGURE 19-15 **A,** Changes in force development in an isolated papillary muscle from a cat as the interval between contractions was changed from 20 sec to 0.63 sec, and then back to 20 sec. **B,** The steady-state forces developed at the indicated intervals (in sec). (Redrawn from Koch-Weser J, Blinks JR: *Pharmacol Rev* 15:601, 1963.)

ally indicates an impairment of contractility and a consequent tendency toward **cardiac failure.**

Contractility is a measure of cardiac performance at a given level of preload and afterload. Thus, ***contractility is an index of the biochemical and biophysical processes taking place at the interacting crossbridges between the thick and thin filaments, preload is an index of the alignment of the thick and thin filaments and hence the number of crossbridges that are available to interact, and afterload is the mechanical force that is stored in the arteries and that opposes the ejection of blood from the ventricles during systole.*** Preload and afterload are valid determinants of ventricular performance, but they are often confusing at first. One of the principal reasons for confusion is that preload and afterload are not only *determinants of* ventricular performance, but also they are *determined by* ventricular performance. This problem is addressed in considerable detail in Chapter 24.

The Frank-Starling mechanism is ideally suited for matching the cardiac output to the venous return. Any sudden, excessive output by one ventricle soon results in a greater venous return to the other ventricle. The consequent increase in diastolic fiber length serves as the stimulus to increase the output of the second ventricle to correspond with that of its mate. For this reason *it is the Frank-Starling mechanism that maintains a precise balance between the outputs of the right and left ventricles.* Because the two ventricles are arranged

in series in a closed circuit, even a small, but maintained, imbalance in the outputs of the two ventricles would be catastrophic.

Frequency-Induced Regulation The effects of contraction frequency on the force developed in an isometrically contracting cat papillary muscle are shown in Figure 19-15, *A.* Initially the strip of cardiac muscle was stimulated to contract only once every 20 sec. When the muscle was made to contract once every 0.63 sec, the developed force increased progressively over the next several beats. This progressive increase in developed force induced by a change in contraction frequency is known as the **staircase** (or **Treppe**) **phenomenon.** At the new steady state, the developed force was more than five times as great as it was at the lower contraction frequency. A return to the slower rate had the opposite influence on developed force.

The effect of the interval between contractions on the steady-state level of developed force is shown in Figure 19-15, *B,* for a wide range of intervals. As the interval was diminished from 300 sec down to about 10 to 20 sec, developed force increased only slightly. However, as the interval was reduced further, to a value of about 0.5 sec, force increased sharply. Further reduction of the interval to 0.2 sec had little additional effect on developed force.

The progressive rise in developed force as the interval between contractions was suddenly decreased (e.g., from 20 to 0.63 sec) is mediated by a gradual rise

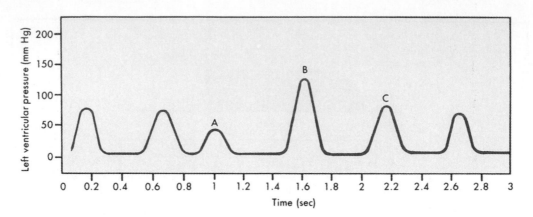

FIGURE 19-16 In an isovolumic canine left ventricle preparation a premature ventricular systole (beat *A*) is typically feeble, whereas the postextrasystolic contraction (beat *B*) is characteristically strong. The enhanced contractility may persist to a diminishing degree over a few beats (for example, contraction *C*). (MN Levy, unpublished tracing.)

in intracellular Ca^{++} content. Ca^{++} enters the cell during each action potential plateau (Chapter 17). Hence, when the time between contractions is reduced (i.e., when contraction frequency is increased), the Ca^{++} influx per minute increases. As contraction frequency increases, the plateau shortens (see Figure 17-11), and therefore less Ca^{++} enters per contraction. However, the increment in the number of beats per minute exceeds the decrement in Ca^{++} influx per beat. Therefore, the intracellular Ca^{++} content rises and thereby augments the contractile force, as shown in Figure 19-15.

Postextrasystolic Potentiation Another influence of the elapsed time between beats has been termed **postextrasystolic potentiation.** When the ventricles contract prematurely, the premature contraction (**extrasystole**) itself is feeble (e. g., beat *A* in Figure 19-16). However, the next beat *(B)*, which usually occurs after a short pause, is very strong. In the intact circulatory system, this behavior is partly mediated by the Frank-Starling mechanism. For most premature beats, the time available for ventricular filling is not adequate, and the consequently small preload could be mainly responsible for the feeble premature contraction (beat *A*). Similarly, the augmented filling associated with the subsequent pause could largely explain the vigorous postextrasystolic contraction (beat *B*).

Although the Frank-Starling mechanism is certainly involved in the usual ventricular adaptation to a premature beat, it is not the exclusive mechanism. The ventricular pressure curves illustrated in Figure 19-16 were recorded from a total heart bypass preparation, in which the left ventricle neither fills nor ejects; that is, its volume remains constant throughout the cardiac cycle. Nevertheless, the premature beat *(A)* is feeble, and the succeed-

ing contraction *(B)* is supernormal. Such enhanced contractility in contraction *B* is an example of postextrasystolic potentiation, and the enhancement may persist for one or more additional beats (for example, contraction *C*). The mechanism responsible for this phenomenon has not been adequately explained, but it may be related to a substantial delay between the time that Ca^{++} is taken up by the sarcoplasmic reticulum after one cardiac contraction and the time that it is available to be released again from that organelle during the next contraction.

Extrinsic Regulation of Contractility

Although the heart possesses effective intrinsic mechanisms of adaptation, various extrinsic mechanisms are also important in regulating myocardial contractility. Under many natural conditions the extrinsic mechanisms dominate the intrinsic mechanisms. The extrinsic regulatory factors may be subdivided into nervous and humoral components.

Nervous Control
Sympathetic Influences Sympathetic neural activity enhances atrial and ventricular contractility. The density of the sympathetic innervation of the atria and of the SA and AV nodes is about three times that of the ventricles.

The alterations in ventricular contraction evoked by electrical stimulation of the cardiac sympathetic nerves in a total heart bypass preparation are shown in Figure 19-17. In this preparation, a balloon filled with a fixed volume of saline is placed in the left ventricle; ventricular volume remains constant throughout the cardiac cycle. During sympathetic stimulation, the peak ventricular pressure and the maximum rate of pressure rise (dP/

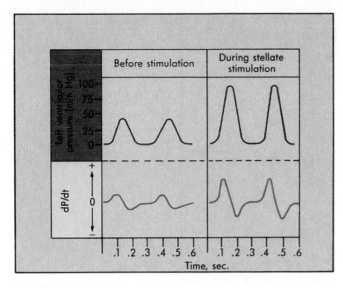

FIGURE 19-17 In an isovolumic canine left ventricle preparation, stimulation of the cardiac sympathetic nerves increases the peak left ventricular pressure and increases the maximum rates of intraventricular pressure rise and fall *(dP/dt).* (Unpublished tracing from experiments of Levy MN et al: *Circ Res* 19:5, 1966.)

dt) during systole are markedly increased. Also, the duration of systole is reduced and the rate of ventricular relaxation (as indicated by the minimum value of dP/dt) is increased.

The shortening of systole and the more rapid ventricular relaxation assist ventricular filling. For a given cardiac cycle length, the abridgement of systole allows more time for diastole and hence for ventricular filling. In the experiment shown in Figure 19-18, for example, the animal's heart was paced at a constant rapid rate. Sympathetic stimulation *(right panel)* shortened systole, which allowed substantially more time for ventricular filling. These factors gain importance at fast heart rates, and of course the rate is rapid when sympathetic activity is increased.

When the resting heart rate is slow (about 60 or 70 beats/min), a substantial increase in heart rate induced by sympathetic activity is associated with a much greater abridgement of diastole than of systole. The shortening of diastole may, of course, impede ventricular filling somewhat (see Figure 24-11). However, the concomitant shortening of systole diminishes the extent of this impediment.

Sympathetic nervous activity enhances myocardial performance. Neurally released norepinephrine interacts with β-adrenergic receptors on the cardiac cell membranes (see Chapter 47). This reaction activates **adenylyl cyclase,** which raises the intracellular levels of **cyclic AMP.** As a consequence, protein kinases are activated and promote the phosphorylation of various pro-

teins within the myocardial cells. Phosphorylation of specific sarcolemmal proteins augments the opening of the calcium channels in the myocardial cell membranes. Hence Ca^{++} influx increases during each action potential plateau, and more Ca^{++} is released from the sarcoplasmic reticulum in response to each cardiac excitation. The contractile strength of the heart is thereby increased. The sympathetically induced acceleration of relaxation is mediated by the phosphorylation of a specific protein that facilitates the reuptake of cytosolic Ca^{++} by the sarcoplasmic reticulum.

The overall effect of increased cardiac sympathetic activity on ventricular performance in intact animals can best be appreciated in terms of families of ventricular function curves. When sympathetic activity increases, the ventricular function curves shift progressively to the left. The changes parallel those produced by norepinephrine infusions (see Figure 19-14). Hence, for any given left ventricular end-diastolic pressure, ventricular performance improves as the sympathetic nervous activity increases.

During cardiac sympathetic stimulation the increase in performance is usually accompanied by a reduction in left ventricular end-diastolic pressure. An example of the response to stellate ganglion stimulation in a paced heart is shown in Figure 19-18. In this experiment, stroke volume and stroke work increased substantially (not shown in the figure), despite a 7 cm H_2O reduction in the left ventricular end-diastolic pressure (which is the preload). The reason for the reduction in the preload is explained in Chapter 24.

Parasympathetic Influences The vagus nerves strongly inhibit the SA node, atrial myocardium, and AV conduction tissue. The vagus nerves also depress the ventricular myocardium, but the effects are less pronounced. In the total heart bypass preparation, vagal stimulation decreases the peak left ventricular pressure, the maximum rate of pressure development (dP/dt), and the maximum rate of pressure decline during diastole. The effects are opposite to those elicited by sympathetic stimulation (see Figure 19-17).

The effects of increased vagal activity on the ventricular myocardium are achieved largely by antagonizing the facilitatory effects of any concurrent sympathetic activity. This antagonism takes place at two levels. At the level of the autonomic nerve endings in the heart, the acetylcholine released from vagal endings inhibits the release of norepinephrine from nearby sympathetic endings. At the level of the cardiac cell membranes, the rise in intracellular cyclic AMP that would ordinarily be produced in response to a given concentration of neurally released norepinephrine is attenuated by the acetylcholine released from nearby vagal endings.

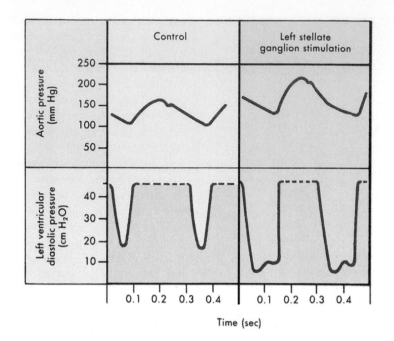

FIGURE 19-18 Stimulation of the left stellate ganglion of an anesthetized dog increases arterial pressure *(top tracing)*. The stroke volume and stroke work increased (not shown), despite a concomitant reduction in left ventricular end-diastolic pressure *(bottom tracing)*. Note also the abridgment of systole, which allows more time for ventricular filling; the heart was paced at a constant rate. In the ventricular pressure tracings the pen excursion is limited *(dashed horizontal lines)* at 45 mm Hg; actual ventricular pressures during systole can be estimated from the aortic pressure tracings. (Redrawn from Mitchell JH et al: *Circ Res* 8:1100, 1960.)

These antagonistic interactions between the sympathetic and vagal effects on the ventricular myocardium also take place in other cardiac structures. For example, they probably account for the responses of sinus node pacemaker cells shown in Figure 19-3. In the absence of vagal stimulation, sympathetic stimulation at a frequency of 4 Hz increased heart rate substantially. However, vagal stimulation at a frequency of 8 Hz attenuated the sympathetic influence so markedly that increasing the frequency of concurrent sympathetic stimulation from 0 Hz to 4 Hz had virtually no effect on heart rate.

Humoral Control

Hormones Various hormones influence cardiac function. The principal hormone secreted by the adrenal medulla is **epinephrine,** although some norepinephrine is also released (see Chapter 47). The rate of catecholamine secretion by the adrenal medulla is regulated by essentially the same mechanisms that control the activity of the sympathetic nervous system, and the effects of those catecholamines on the heart are qualitatively similar to those released from the sympathetic nerve endings. However, the concentrations of circulating catecholamines rarely rise sufficiently high to affect cardiac function appreciably.

Thyroid hormones have pronounced effects on cardiac function (see Chapter 43). Cardiac activity is sluggish in patients with inadequate thyroid function **(hypothyroidism);** that is, the heart rate is slow and cardiac output is diminished. The converse is true in patients with overactive thyroid glands **(hyperthyroidism).** Characteristically, such patients exhibit tachycardia, high cardiac output, palpitations, and arrhythmias.

Numerous studies on intact animals and humans have demonstrated that thyroid hormones enhance myocardial contractility. The rates of Ca^{++} uptake and of ATP hydrolysis by the sarcoplasmic reticulum are increased in experimental hyperthyroidism, and the opposite effects occur in hypothyroidism. Thyroid hormones increase protein synthesis in the heart, which leads to cardiac hypertrophy. These hormones also affect the composition of myosin isoenzymes in cardiac myscle. They increase principally those isoenzymes with the greatest ATPase activity, which thereby enhances myocardial contractility.

Insulin has a prominent, direct, positive inotropic ef-

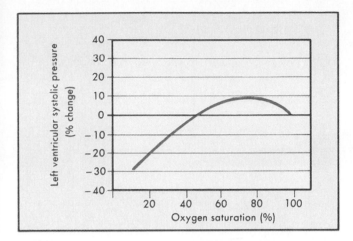

FIGURE 19-19 In an isovolumic canine left ventricle preparation, a reduction in the O_2 saturation of coronary arterial blood to between 45% and 100% stimulates ventricular contractility (as assessed by the left ventricular systolic pressure), whereas an O_2 saturation below 45% depresses ventricular contractility. (Redrawn from Ng ML et al: *Am J Physiol* 211:43, 1966.)

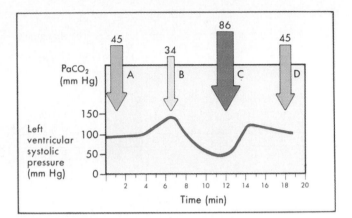

FIGURE 19-20 Decrease in Pa_{CO_2} from 45 to 34 mm Hg increases left ventricular systolic pressure (arrow *B*) in an isovolumic canine left ventricle preparation. A subsequent rise in Pa_{CO_2} to 86 mm Hg has the reverse effect. When the Pa_{CO_2} is returned to the control level (45 mm Hg), left ventricular systolic pressure returns to its original value. (From experiments by Ng ML et al: *Am J Physiol* 213:115, 1967.)

fect on the heart in several mammalian species. The effect of insulin is evident even when hypoglycemia is prevented by glucose infusions and when the β-adrenergic receptors are blocked. In fact, the positive inotropic effect of insulin is potentiated by β-adrenergic receptor blockade. The enhancement of contractility cannot be explained satisfactorily by the concomitant augmentation of glucose transport into the myocardial cells.

Glucagon has potent positive inotropic and chronotropic effects on the heart. The endogenous hormone is probably not involved in the normal regulation of the cardiovascular system, but it has been used pharmacologically to treat various cardiac conditions. The effects of glucagon on the heart closely resemble those of the catecholamines, and certain of the metabolic effects are similar.

Blood Gases Changes in oxygen tension (Pa_{O_2}) of the blood perfusing the brain and the peripheral chemoreceptors affect the heart through nervous mechanisms, as described earlier in this chapter. These indirect effects of hypoxia are usually prepotent. Moderate degrees of hypoxia characteristically increase heart rate, cardiac output, and myocardial contractility by increasing sympathetic nervous activity. These changes are largely abolished by β-adrenergic receptor blockade.

The Pa_{O_2} of the blood perfusing the myocardium also influences myocardial performance directly. The effect of hypoxia is biphasic; moderate degrees are stimulatory and more severe degrees are depressant. As shown in Figure 19-19, when the O_2 saturation is reduced to lev-

els below 50 per cent in isolated hearts, the peak left ventricular pressures are less than the control levels. However, with less severe degrees of hypoxia (O_2 saturation >50%), the peak pressures exceed the control level.

Changes in CO_2 tension (Pa_{CO_2}) in the blood may also affect the myocardium directly and indirectly. The indirect, neurally mediated effects produced by increased Pa_{CO_2} are similar to those evoked by a decrease in Pa_{O_2}.

The direct effects on myocardial performance elicited by changes in Pa_{CO_2} in the coronary arterial blood are illustrated in Figure 19-20. In this experiment on an isolated left ventricle preparation, the control Pa_{CO_2} was 45 mm Hg (arrow *A*). Decreasing the Pa_{CO_2} to 34 mm Hg (arrow *B*) was stimulatory, whereas increasing Pa_{CO_2} to 86 mm Hg (arrow *C*) was depressant. In intact animals, systemic hypercapnia activates the sympathoadrenal system, which tends to compensate for the direct depressant effect of the increased Pa_{CO_2} on the heart.

Neither the Pa_{CO_2} nor the blood pH is a primary determinant of myocardial behavior; the induced change in intracellular pH is the critical factor. The reduced intracellular pH diminishes the influx of Ca^{++} into the cell via the Ca^{++} channels and the Na^{+}/Ca^{++} exchanger, and it decreases the amount of Ca^{++} released from the sarcoplasmic reticulum in response to excitation. The intracellular acidosis also affects the myofilaments directly. When they are exposed to a given concentration of Ca^{++}, the myofibrils develop less force as the prevailing intracellular pH decreases.

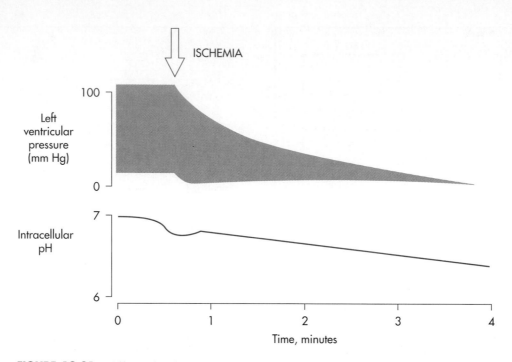

FIGURE 19-21   Effect of ischemia on left ventricular pressure and intracellular pH in an isolated perfused rabbit heart. (Modified from Mohabir R et al: *Circ Res* 69:1525, 1991.)

In patients with coronary artery disease, a narrowed region of a major coronary artery may suddenly become occluded by a blood clot; this is the most common cause of a "heart attack." The consequent inadequate blood flow to the myocardial tissue **(myocardial ischemia)** leads to progressive impairment of the contractile function of the deprived myocardial cells. This impairment of contractility is mediated by a combination of extracellular and intracellular changes in the blood gases and pH in the ischemic region. These changes include reductions in Po_2 and pH and increases in Pco_2.

Figure 19-21 shows the relationship between the changes in contractile performance and intracellular pH in an animal model of myocardial ischemia. When perfusion was suddenly halted in an isolated, perfused rabbit heart, the left ventricular pressure diminished rapidly until the contraction virtually ceased within about 4 minutes. These changes were accompanied by a progressive reduction in intracellular pH, from a control value of 7.0 to a value of about 6.3 after 4 minutes of ischemia.

SUMMARY

1. Cardiac function is regulated by various intrinsic and extrinsic mechanisms.

2. Heart rate is regulated mainly by the autonomic nervous system. Sympathetic activity increases heart rate, and parasympathetic (vagal) activity decreases heart rate. When both systems are active, the vagal effects tend to dominate.

3. The baroreceptor, chemoreceptor, pulmonary inflation, atrial receptor (Bainbridge), and ventricular receptor reflexes all serve to regulate heart rate.

4. The principal intrinsic mechanisms that regulate myocardial contraction are the Frank-Starling and rate-induced mechanisms.
   a. Frank-Starling mechanism: a change in the resting length of the muscle influences the subsequent contraction by altering the number of interacting cross-bridges between the thick and thin filaments and by altering the affinity of the myofilaments for calcium.
   b. Rate-induced regulation: a sustained change in contraction frequency affects the strength of contraction by altering the influx of Ca^{++} into the cell per minute, whereas a transient change in contraction frequency alters contractile strength because an appreciable delay exists between the time that Ca^{++} is taken up by the sarcoplasmic reticulum and the time that it becomes available again for release.

5. The autonomic nervous system regulates myocardial performance mainly by varying the Ca^{++} conductance of the cell membrane via the adenylyl cyclase system.

6. Various hormones, including epinephrine, adrenocortical steroids, thyroid hormones, insulin, and glucagon, participate in the regulation of myocardial performance.

7. Changes in the blood concentrations of O_2, CO_2, and H^+ alter cardiac function directly and, via the chemoreceptors, reflexly.

BIBLIOGRAPHY

Journal articles

Brodde O-E: β_1- and β_2-adrenoceptors in the human heart: properties, function, and alterations in chronic heart failure, *Pharmacol Rev* 43:203, 1991.

Cooper MW: Postextrasystolic potentiation: do we really know what it means and how to use it? *Circulation* 88:2962, 1993.

Elzinga G: Starling's "law of the heart": rise and fall of the descending limb, *News Physiol Sci* 7:134, 1992.

Hainsworth R: Reflexes from the heart, *Physiol Rev* 71:617, 1991.

Hartzell HC: Regulation of cardiac ion channels by catecholamines, acetylcholine, and second messenger systems, *Prog Biophys Mol Biol* 52:165, 1988.

Koizumi K, Kollai M: Multiple modes of operation of cardiac autonomic control: development of the ideas from Cannon and Brooks to the present, *J Auton Nerv Syst* 41:19, 1992.

Kollai M et al: Relation between tonic sympathetic and vagal control of human sinus node function, *J Auton Nerv Syst* 46:273, 1994.

Levy MN: Autonomic interactions in cardiac control, *Ann NY Acad Sci* 601:209, 1990.

Löffelholz K, Pappano AJ: The parasympathetic neuroeffector junction of the heart, *Pharmacol Rev* 37:1, 1985.

Mohabir R, et al: Effects of ischemia and hypercarbic acidosis on myocyte calcium transients, contraction, and pH_i in perfused rabbit hearts, *Circ Res* 69:1525, 1991.

Polikar R: Thyroid and the heart, *Circulation,* 87:1435, 1993.

Schouten VJA, ter Keurs HEDJ: Role of I_{ca} and Na^+/Ca^{2+} exchange in the force-frequency relationship of rat heart muscle, *J Mol Cell Cardiol* 23:1039, 1991.

Spyer KM: Central nervous mechanisms contributing to cardiovascular control, *J Physiol [London]* 474:1, 1994.

Walley KR et al: Effects of hypoxia and hypercapnia on the force-velocity relation of rabbit myocardium, *Circ Res* 69:1616, 1991.

Books and monographs

Fozzard HA, Haber E, Jennings RB, Katz AM, Morgan HE, eds: *Heart and cardiovascular system:* scientific foundations, ed 2, New York, 1991, Raven Press.

Garfein OB, ed: *Current concepts in cardiovascular physiology,* San Diego, 1990, Academic Press.

Katz AM: *Physiology of the heart,* New York, 1991, Raven Press.

Kulbertus HE, Franck G, eds: *Neurocardiology,* Mt Kisco, NY, 1988, Futura Publishing Co.

Levy MN, Schwartz PJ, ed: *Vagal control of the heart: experimental basis and clinical implications,* Armonk, 1993, Futura Publishing Co.

Opie L, ed: *The heart,* ed 2, New York, 1991, Raven Press.

Randall WC, ed: *Nervous control of cardiovascular function,* New York, 1984, Oxford University Press.

Sperelakis N, ed: *Physiology and pathophysiology of the heart,* ed 2, Boston, 1989, Kluwer Academic Publishers.

Zucker IH, Gilmore JP: *Reflex control of the circulation,* Boca Raton, Fla, 1990, CRC Press.

CHAPTER 20

Hemodynamics

The fluid mechanics of the circulatory system are very complicated, and therefore difficult to analyze precisely. The heart is an intermittent pump, and its behavior is regulated by many physical and chemical factors. The blood vessels are branched, distensible conduits of continuously varying dimensions. The blood is a suspension mainly of erythrocytes, but also of leukocytes, platelets, and lipid globules, all dispersed in a colloidal solution of proteins. Despite this complexity, however, an understanding of the relevant, elementary principles of fluid mechanics provides considerable insight into the physical behavior of the cardiovascular system. Certain basic principles will be expounded in this chapter in an attempt to illuminate the interrelationships among vascular geometry, blood velocity, blood flow, and blood pressure.

VELOCITY OF THE BLOODSTREAM

The relationship between the velocity of the bloodstream and the dimensions of the vascular bed is illustrated by the hydraulic system in Figure 20-1; **velocity (v)** refers to the displacement of a particle of the blood per unit time. Consider that the conduit is rigid and that it has a wide section (area $A_1 = 5$ cm^2) and a narrow section (area $A_2 = 1$ cm^2). Also, let an incompressible fluid enter the wide end of the tube at a flow, Q_1, of 5 cm^3/sec; **flow (Q)** refers to the volume of fluid that passes a given cross-section of the conduit per unit time. Then, the velocity, v_1, of a fluid particle as it passes cross-section A_1 would be

$$v_1 = Q_1/A_1$$
$$= \frac{5 \text{ cm}^3/\text{sec}}{5 \text{ cm}^2} \qquad \textbf{(1)}$$
$$= 1 \text{ cm/sec}$$

Thus, a particle of fluid advances a distance (Δl_1) of 1

cm each second (Figure 20-1). When the fluid enters the narrow section of the tube, the volume (Q_2) of fluid that passes cross-section A_2 each second must equal the volume (Q_1) that had passed cross-section A_1 each second; that is, $Q_2 = Q_1$. The velocity, v_2, in the narrow section would be

$$v_2 = Q_2/A_2$$
$$= \frac{5 \text{ cm}^3/\text{sec}}{1 \text{ cm}^2} \qquad \textbf{(2)}$$
$$= 5 \text{ cm/sec}$$

Thus, each particle of fluid must move past section A_2 five times faster than it did past section A_1. Each particle would have to move a distance (Δl_2) of 5 cm each second.

By the law of conservation of mass, the flow (Q_1) of fluid past A_1 must equal the flow (Q_2) past A_2; that is,

$$Q_1 = Q_2 \qquad \textbf{(3)}$$

From equations 1 to 3,

$$v_1 A_1 = v_2 A_2 \qquad \textbf{(4)}$$

Therefore,

$$v_1/v_2 = A_2/A_1 \qquad \textbf{(5)}$$

Hence, when the caliber of a tube varies with the axial location along the tube, the fluid velocities at these axial sites are inversely proportional to the corresponding cross-sectional areas. This relationship also holds for more complex hydraulic systems, such as the circulatory system, which are composed of large numbers of conduits, arranged both in series and in parallel.

Note that in Figure 16-2 the velocity decreases progressively as the blood traverses the aorta, its primary and secondary branches, the arterioles, and finally the capillaries. As the blood then passes through the venules and continues centrally through the larger veins toward the venae cavae, the velocity increases progressively

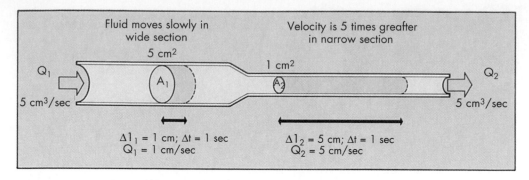

FIGURE 20-1   In a conduit that contains a wide segment and a narrow segment, the fluid velocities in the two segments are inversely proportional to the cross-sectional areas of the segments.

again. The velocities in the various serial sections of the circulatory system are inversely proportional to the total cross-sectional areas of the respective sections. The total cross-sectional area of all the parallel systemic capillaries greatly exceeds the total cross-sectional area of any other serial section of the systemic vascular bed. Hence, the velocity of the bloodstream in the capillaries is much less than that in any other vascular segment. *The very slow movement of the blood through the capillaries allows ample time for exchange of materials between the tissues and the blood.*

Localized changes in vascular cross-sectional area may be associated with pronounced localized changes in the velocity of the bloodstream. In the congenital disorder **coarctation of the aorta,** for example, a short segment of the aorta (usually just distal to the origin of the left subclavian artery) is much narrower than the more proximal or more distal aortic segments. Consequently, the velocity of the bloodstream in this narrow segment is much greater than in the aortic segments just proximal and just distal to it. The combination of high velocity and sharp change in lumen diameter causes the flow to be **turbulent** (see Figure 20-8), for reasons that will be discussed later in this chapter. The turbulent flow generates **murmurs** that can be heard through the stethoscope.

RELATIONSHIP BETWEEN PRESSURE AND FLOW

The most useful equation that defines the relationships among some of the physical factors that govern pressure and flow in hydraulic systems (including the circulatory system) was derived by the French physician Poiseuille over a century ago. This equation, known as **Poiseuille's law,** applies to the flow of fluids through cylindrical tubes, but it applies precisely only under restricted conditions. The equation applies specifically to the steady, laminar flow of newtonian fluids. The term **steady flow** signifies the absence of variations of flow in time. **Laminar flow** is the type of motion in which the fluid moves as a series of infinitesimally thin layers, with each layer moving at a velocity different from that of its neighboring layers (see Figure 20-7). A **newtonian fluid** has certain critical physical properties that will be described below; such a fluid is essentially a homogeneous fluid, such as an electrolyte solution, in contrast to a suspension, such as blood.

Effects of Pressure Difference

Pressure is a salient determinant of flow. The pressure, P, in dynes/cm^2, at a distance h cm below the surface of a liquid is

$$P = h\rho g \qquad (6)$$

where ρ is the density of the liquid in g/cm^3, and g is the acceleration of gravity in cm/sec^2. For convenience, however, pressure is frequently expressed in terms of the height, h, of the column of liquid above an arbitrary reference level.

Consider the tube that connects reservoirs R_1 and R_2 in Figure 20-2. Let reservoir R_1 be filled with liquid to height h_1, and let reservoir R_2 be empty, as in panel *A*. The outflow pressure, P_o, is therefore equal to the atmospheric pressure, which shall be designated as the zero, or reference, level. The inflow pressure, P_i, is then equal to the same reference level plus the height, h_1, of the column of liquid in reservoir R_1. Under these conditions let the flow, Q, through the tube be 5 ml/sec. If reservoir R_1 is filled to height h_2, which is twice h_1, and reservoir R_2 is again empty (as in panel *B*), the flow will be twice as great (that is, 10 ml/sec) as it is in panel *A*. Thus with reservoir R_2 empty, the flow will be directly proportional to the inflow pressure, P_i.

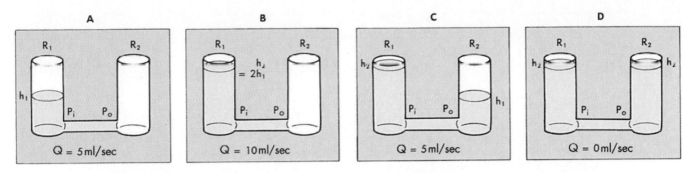

FIGURE 20-2   The flow, Q, of fluid through a tube connecting two reservoirs, R_1 and R_2, is proportional to the difference between the pressure at the inflow end (P_i) and the pressure at the outflow end (P_o) of the tube. **A,** When R_2 is empty, fluid flows from R_1 to R_2 at a rate proportional to the pressure in R_1. **B,** When the fluid level in R_1 is increased twofold, the flow increases proportionately. **C,** Flow from R_1 to R_2 is proportional to the difference between the pressures in R_1 and R_2. **D,** When pressure in R_2 rises to equal the pressure in R_1, flow ceases.

If reservoir R_2 is now allowed to fill to height h_1, and the fluid level in R_1 is maintained at h_2 (as in panel C), the flow will again become 5 ml/sec. If the fluid level in R_2 attains the same height as in R_1, flow will cease; that is, $Q = 0$ ml/sec (panel D). *Thus flow is directly proportional to the difference between the inflow and outflow pressures:*

$$Q \propto P_i - P_o \qquad (7)$$

The blood flow through specific vascular beds is affected by the difference between the inflow (arterial) and outflow (venous) pressures that prevail for that vascular bed. Such pressure differences may be affected substantially by gravitational forces and by the competence of the venous valves, as explained in Chapter 24.

The blood flow through the legs and feet may be entirely different in standing than in recumbent people. In standing subjects, the arterial blood pressure in the legs will be considerably higher than the blood pressure in the thoracic arteries; this difference will, of course, depend on the subjects' height. In subjects with normal venous valves, the venous blood pressure in the legs and feet may be only slightly above the atmospheric pressure (see Chapter 24). However, in patients with **varicose veins** (abnormally dilated veins) in the legs, the venous valves are incompetent, and the venous blood pressure in the legs in standing subjects will be elevated by the same amount as is the arterial blood pressure in the legs. Thus, the arteriovenous pressure difference in the legs will be substantially greater in standing people with normal venous valves than in those with varicose veins. Blood flow to the legs may be affected proportionately in such subjects.

Effects of Tube Dimensions

For any given pressure difference between the two ends of a tube, the flow will depend on the dimensions of the tube. Consider the tube connected to the reservoir in Figure 20-3, A. If the tube's length is l_1 and its radius is r_1, the flow Q_1 is observed to be 10 ml/sec.

The tube connected to the reservoir in panel B has the same radius, but is twice as long. Under these conditions the flow Q_2 is found to be 5 ml/sec, or only half as great as Q_1. Conversely, for a tube half as long as l_1, the flow would be twice as great as Q_1. In other words, flow is inversely proportional to the length of the tube:

$$Q \propto 1/l \qquad (8)$$

The length, l_3, of the tube connected to the reservoir in Figure 20-3, C, is the same as l_1, but the radius is twice as great as r_1. Under these conditions, the flow Q_3 increases to a value of 160 ml/sec, which is 16 times greater than Q_1. The precise measurements of Poiseuille revealed that flow varies directly as the fourth power of the radius (just as in the above example):

$$Q \propto r^4 \qquad (9)$$

Thus, in the example above, because $r_3 = 2r_1$, then Q_3 will be proportional to $(2r_1)^4$, or $16r_1^4$; therefore, Q_3 will equal $16Q_1$.

Effect of Viscosity

Finally, for a given pressure difference across a cylindrical tube of given dimensions, the flow will be affected by the nature of the fluid itself. This flow-determining property of fluids is termed **viscosity**, η. Consider that the fluid level in the reservoir in panel D of Figure 20-3 equals that in panel A, and that the tubes connected to the bottoms of both reservoirs are identical. However, if

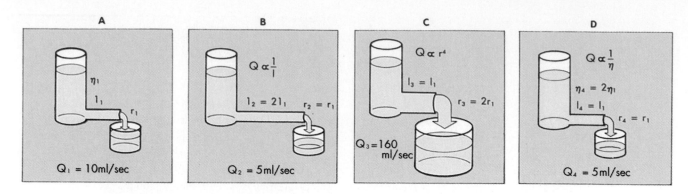

FIGURE 20-3   The flow, Q, of fluid through a tube is inversely proportional to the length, l, and the viscosity, η, and is directly proportional to the fourth power of the radius, r. **A,** Reference condition: for a given pressure, length, radius, and viscosity, let the flow (Q_1) equal 10 ml/sec. **B,** If the tube length doubles, flow decreases by 50%. **C,** If the tube radius doubles, flow increases sixteen-fold. **D,** If viscosity doubles, flow decreases by 50%.

the viscosity, η_4, of the fluid in the reservoir in panel D is twice the viscosity, η_1, of the fluid in that of panel A, then the flow, Q_4, through the tube in panel D will be only half the flow, Q_1, through the tube in panel A. Thus,

$$Q \propto 1/\eta \qquad (10)$$

For most homogeneous liquids, such as water itself or true solutions in water, this inverse proportionality prevails during laminar flow. Such fluids are said to be **newtonian.** For heterogeneous liquids, notably suspensions such as blood, this precise inverse proportionality does not apply. Such fluids are said to be **nonnewtonian.**

Poiseuille's Law

Poiseuille's law takes into account the various factors that influence the flow of a fluid through a tube under restricted conditions. This law states that for the steady, laminar flow of a newtonian fluid through a cylindrical tube, the flow, Q, varies directly as the difference between the inflow and outflow pressures, $P_i - P_o$, and the fourth power of the radius, r, of the tube, and it varies inversely as the length, l, of the tube and the viscosity, η, of the fluid. The full statement of Poiseuille's law is

$$Q = \pi(P_i - P_o)r^4/8\eta l \qquad (11)$$

where $\pi/8$ is the constant of proportionality.

RESISTANCE TO FLOW

In electrical theory, resistance, R, is defined as the ratio of voltage drop, E, to current flow, I. By analogy, a hydraulic resistance, R, may be defined as the ratio of pressure drop, $P_i - P_o$, to flow, Q. For the steady, laminar flow of a newtonian fluid through a cylindrical tube, the physical components of hydraulic resistance may be appreciated by rearranging Poiseuille's law to yield the hydraulic resistance equation:

$$R = (P_i - P_o)/Q = 8\eta l/\pi r^4 \qquad (12)$$

Thus, when Poiseuille's law applies, the resistance to flow depends only on the dimensions (l and r) of the tube and on the viscosity (η) of the fluid.

The principal determinant of the resistance to blood flow through any individual vessel within the circulatory system is its caliber, because resistance varies inversely as the fourth power of the radius. The resistance to flow through small blood vessels in the cat mesentery has been measured, and the resistance per unit length of vessel (R/l) is plotted against the vessel diameter in Figure 20-4. The resistance is highest in the individual capillaries (diameter, 7 μm), and it diminishes as the vessels increase in diameter on the arterial and venous sides of the capillaries. The values of R/l were found to be virtually proportional to the fourth power of the diameter for the larger vessels on both sides of the capillaries.

When the various blood vessels are considered as groups of vessels of specific types, the distribution of resistances is different from that shown in Figure 20-4. For example, for the systemic circulation as a whole, the resistance to flow through the arterioles exceeds that through the capillaries, despite the larger caliber of individual arterioles than of individual capillaries. This apparent contradiction is explained by the relative numbers of arterioles and capillaries in the systemic circulation, as described below in the discussion of resistances in parallel.

Figure 16-2 shows that, among the various types of

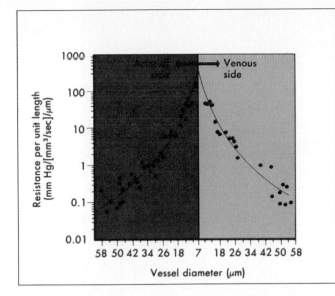

FIGURE 20-4   The resistance per unit length (R/l) of individual small blood vessels in the cat mesentery. The capillaries, diameter 7 μm, are denoted by the vertical line between the red and blue panels. Resistances of the arterioles are plotted to the left and resistances of the venules are plotted to the right of that vertical line. The solid circles represent the actual data. The two curves through the data represent the following regression equations for the arteriole and venule data, respectively: arterioles, $R/l = 1.02 \times 10^6 D^{-4.04}$, and venules, $R/l = 1.07 \times 10^6 D^{-3.94}$. Note that for both types of vessels, the resistance per unit length is inversely proportional to the fourth power (within 1%) of the vessel diameter (D). (Redrawn from Lipowsky HH et al: *Circ Res* 43:738, 1978.)

blood vessels aligned in series, the greatest pressure drop occurs across the very small arteries and arterioles. This indicates that the greatest resistance resides in these vessels, for the following reasons. The same amount of blood flows per minute through each of these serial components of the circulatory system; that is, if the left ventricle pumps 5 liters of blood per minute into the aorta, then 5 L/min must flow through the arteries, through the arterioles, through the capillaries, through the venules, and so on. Hence, because flow through each group of vessels is equal, equation 12 states that the pressure drop across any specific group of vessels is proportional to its resistance. It follows, therefore, that the greatest resistance to flow resides in the small arteries and arterioles.

The small arteries and arterioles possess a thick coat of circularly arranged smooth muscle fibers, by means of which the lumen radius may be varied. Changes in vascular resistance are induced mainly by nervous and humoral factors that alter the contractile state of the arteriolar smooth muscle cells. The control of vascular resistance is described in Chapter 23.

In severe **arteriosclerosis**, a lipid deposit in the intima of a major artery may protrude into the lumen and severely reduce it. In this event, the major resistance to flow in the vascular bed supplied by that diseased artery may reside in that large artery itself, rather than in the small arteries and arterioles of that vascular bed. Such occlusive lesions in important large arteries, such as the coronary arteries, are often treated by balloon dilatation **(angioplasty)** or by a **surgical bypass procedure.**

Resistances in Series and in Parallel

In the cardiovascular system the various types of vessels listed along the horizontal axis in Figure 16-2 lie in **series** with one another; in a series arrangement a red blood cell could travel sequentially from one component of the series to the next component and then on to the next component, as illustrated in Figure 20-5. Furthermore, the individual members within each category of vessels are ordinarily arranged in **parallel** with one another (Figure 16-3). In a parallel arrangement, a red blood cell that arrives at the junction of a number of parallel vessels would have the immediate option of traveling through just one of these parallel channels, as illustrated in Figure 20-6. The capillaries throughout the lungs are in parallel with one another, and similarly the capillaries throughout the systemic circulation are also in most instances in parallel with one another. Notable exceptions are the capillaries in the renal vasculature (wherein the peritubular capillaries are in series with the glomerular capillaries) and those in the splanchnic vasculature (wherein the hepatic capillaries are in series with the intestinal capillaries). Formulas for the total hydraulic resistance of conduits arranged in series and in parallel can be derived in the same manner as for electrical resistances.

Three hydraulic resistances, R_1, R_2, and R_3, are aligned in series in Figure 20-5. The pressure drop across the entire system (that is, the difference between inflow pressure, P_i, and outflow pressure, P_o) consists of the sum of the pressure drops across each of the individual resistances (Figure 20-5, equation 1). Under steady-state conditions, the flow, Q, through any given cross-section must equal the flow through any other cross-section. By dividing each component in equation 1 by Q (equation 2), it becomes evident from the definition of resistance (i.e., $R = (P_i - P_o)/Q$) that the total resistance, R_t, of the entire system of resistances in series equals the sum of the individual resistances; that is,

$$R_t = R_1 + R_2 + R_3 \qquad \textbf{(13)}$$

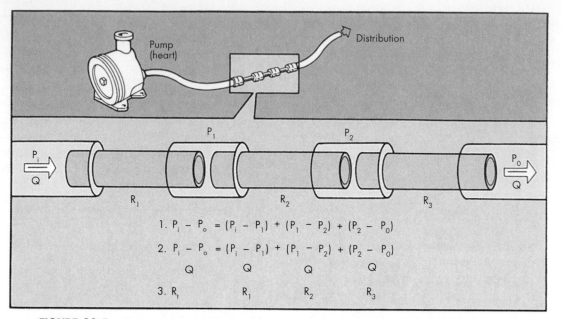

FIGURE 20-5 For resistances $(R_1, R_2,$ and $R_3)$ arranged in series, the total resistance, R_t, equals the sum of the individual resistances.

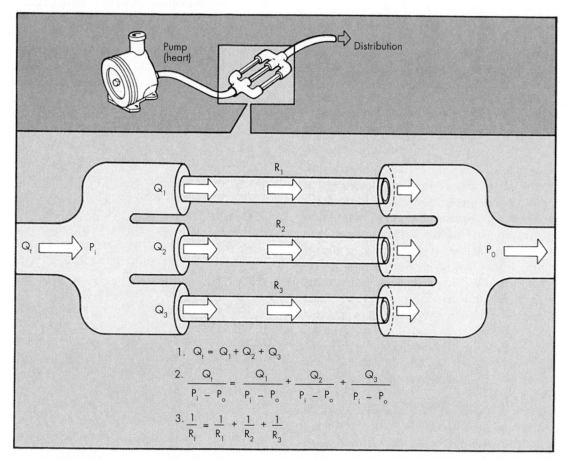

FIGURE 20-6 For resistances $(R_1, R_2,$ and $R_3)$ arranged in parallel, the reciprocal of the total resistance, R_t, equals the sum of the reciprocals of the individual resistances.

For resistances in parallel (Figure 20-6), all tubes have the same inflow pressures and the same outflow pressures. The total flow, Q_t, through the system equals the sum of the flows through the individual parallel elements (Figure 20-6, equation 1). Because the pressure difference ($P_i - P_o$) is identical for all parallel elements, each term in equation 1 may be divided by that pressure difference to yield equation 2. From the definition of resistance, equation 3 may be derived. This equation states that the reciprocal of the total resistance, R_t, equals the sum of the reciprocals of the individual resistances; that is,

$$1/R_t = 1/R_1 + 1/R_2 + 1/R_3 \qquad \textbf{(14)}$$

Stated in another way, if we define hydraulic **conductance** as the reciprocal of resistance, it becomes evident that, *for tubes in parallel, the total conductance is the sum of the individual conductances.*

If we consider a few simple illustrations, some of the fundamental properties of parallel hydraulic systems become apparent. For example, if the resistances of the three parallel elements in Figure 20-6 were all equal, then

$$R_1 = R_2 = R_3 \qquad \textbf{(15)}$$

Therefore

$$1/R_t = 3/R_1 \qquad \textbf{(16)}$$

and hence

$$R_t = R_1/3 \qquad \textbf{(17)}$$

Thus the total resistance is less than any of the individual resistances. Furthermore, for any parallel arrangement, the total resistance must be less than that of any of the individual parallel tubes. For example, consider a system in which a very high-resistance tube is added in parallel to a low-resistance tube. The total resistance must be less than that of the low-resistance component by itself, because the high-resistance component affords an additional pathway, or conductance, for fluid flow.

Similarly, the total resistance across a set of parallel tubes diminishes as the number of tubes increases. This accounts for the greater resistance through the set of arterioles than through the set of capillaries in the systemic circulation, despite the smaller caliber of the individual capillaries than of the individual arterioles. The number of parallel capillaries far exceeds the number of parallel arterioles; this is documented in Figure 16-2 by the much greater cross-sectional area of the capillary bed than of the arteriolar bed. The much greater number of systemic capillaries than of systemic arterioles accounts for the lower resistance to flow through the capillaries than through the arterioles.

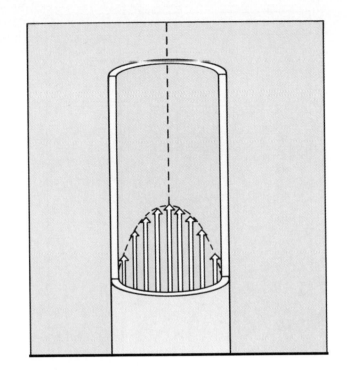

FIGURE 20-7 In laminar flow, all elements of the fluid move in streamlines that are parallel to the axis of the tube; no fluid moves in a radial or circumferential direction. The layer of fluid in contact with the wall is motionless; the fluid that moves along the axis of the tube has the maximum velocity.

LAMINAR FLOW AND TURBULENT FLOW

Under certain conditions, the flow of a fluid in a cylindrical tube will be **laminar,** as illustrated in Figure 20-7. The thin layer of fluid in contact with the inner lining of the tube adheres to the lining and hence is motionless. The thin layer of fluid just central to this external lamina must shear against this motionless layer. Therefore this adjacent layer moves slowly, but with a finite velocity. Similarly, the next more central layer travels still faster. The longitudinal velocity profile is a parabola. The velocity of the fluid adjacent to the wall is zero, whereas the velocity at the center of the stream is maximum. The maximum velocity is twice the mean velocity of flow across the entire cross-section of the tube. In laminar flow, fluid elements remain in one lamina, or streamline, as the fluid progresses longitudinally along the tube. Flow occurs only in an axial direction—that is, parallel to the axis of the tube. No particles of fluid move in either a radial or a circumferential direction.

Irregular motions of the fluid elements may develop in the flow of fluid through a tube; this irregular flow is called **turbulent flow** (Figure 20-8). Under such condi-

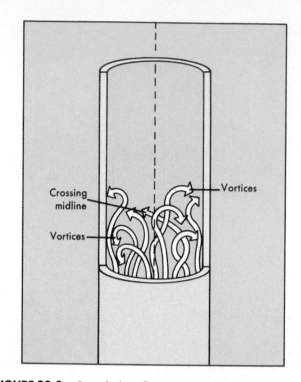

FIGURE 20-8 In turbulent flow, various elements of the fluid move irregularly in axial, radial, and circumferential directions. Vortices frequently develop.

tions fluid elements do not remain confined to definite laminae, but rapid radial and circumferential mixing occurs, and vortices may develop. More pressure is required to force a given flow of fluid through the same tube when the flow is turbulent than when it is laminar. In turbulent flow the pressure drop is approximately proportional to the square of the flow, whereas in laminar flow, the pressure drop is proportional to the first power of the flow. *Hence, to produce a given flow, a pump, such as the heart, must do considerably more work if turbulence develops.*

Whether the flow through a tube will be turbulent or laminar may be predicted by computing a dimensionless number called **Reynold's number, N_R,** which is defined as follows:

$$N_R = \rho D \bar{v} / \eta \qquad \textbf{(18)}$$

where ρ is the fluid density, D is the tube diameter, $\bar{v}$ is the mean velocity over the cross-section of the tube, and η is the fluid viscosity. For $N_R <2000$, the flow will usually be laminar, and for $N_R >3000$, turbulence will usually prevail. Various flow conditions may develop in the transition range of N_R between 2000 and 3000. Because flow tends to be laminar at low N_R and turbulent at high N_R, it is evident from equation 18 that large diameters, high velocities, and low viscosities predispose to the development of turbulence. In addition to these factors,

abrupt variations in tube dimensions or irregularities in the tube walls may produce turbulence. Turbulence is usually accompanied by vibrations of the fluid and surrounding structures. Some of these vibrations within the cardiovascular system are in the auditory frequency range, and they may be detected as a **murmur.**

The factors cited above that predispose to turbulence may account for some of the **cardiac murmurs** that are heard clinically. In certain disorders of the cardiac valves, the valves are **stenotic** (narrowed). As the blood passes through such valves, blood flow becomes turbulent and a cardiac murmur can be detected with the stethoscope. For similar reasons, murmurs may be heard in patients with aortic coarctation, as described earlier in this chapter. In severe anemia, **functional cardiac murmurs** (murmurs not caused by structural abnormalities) are often detectable. Such murmurs are caused by the reduced viscosity of the blood (because of the low red blood cell content) and the high flow velocities that usually prevail in severely anemic patients.

RHEOLOGICAL PROPERTIES OF BLOOD

The viscosity of a newtonian fluid, such as water, may be determined by measuring the rate of flow of the fluid at a given pressure difference through a cylindrical tube of known length and radius. As long as the fluid flow is laminar, the viscosity may be computed by substituting these values into Poiseuille's equation. The calculated viscosity of a given newtonian fluid at a specified temperature will be constant, regardless of the tube dimensions and flows. However, for a nonnewtonian fluid, the viscosity calculated from Poiseuille's equation may vary considerably when different tube dimensions and flows are used. Therefore, in considering the rheological (flow related) properties of a suspension such as blood, the term viscosity does not have a unique meaning. The term **apparent viscosity** is frequently applied to the value of viscosity obtained for blood under the particular conditions of measurement.

Rheologically, blood is a suspension, principally of erythrocytes in a relatively homogeneous liquid, the blood plasma. For this reason the apparent viscosity of blood varies as a function of the **hematocrit ratio** (ratio of volume of red blood cells to volume of whole blood). In Figure 20-9 the upper curve represents the ratio of the apparent viscosity of whole blood to that of plasma over a range of hematocrit ratios from 0% to 80%. The data were derived from measurements of flow

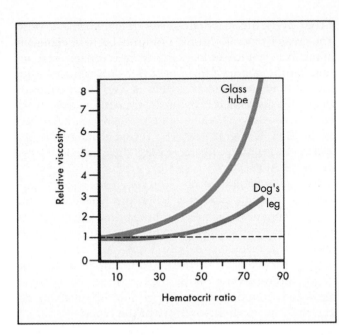

FIGURE 20-9 The viscosity of whole blood, relative to that of plasma, increases progressively as the hematocrit ratio rises. For any given hematocrit ratio the apparent viscosity of blood is less when measured in a biological viscometer (such as the tissues of an anesthetized dog) than in a glass capillary tube with a 1 mm lumen diameter. (Redrawn from Levy MN, Share L: *Circ Res* 1:247, 1953.)

through a glass tube 1 mm in internal diameter. The viscosity of plasma is 1.2 to 1.3 times that of water.

Figure 20-9 *(upper curve)* shows that blood, with a normal hematocrit ratio of 45%, has an apparent viscosity 2.4 times that of plasma. In severe anemia, blood viscosity is low. With increasing hematocrit ratios the slope of the curve increases progressively; it is especially steep at the upper range of erythrocyte concentrations. If the hematocrit ratio rises to about 70%, which it may in patients with **polycythemia vera** (abnormally high erythrocyte counts), the apparent viscosity increases more than twofold, and the resistance to blood flow increases proportionately. The effect of such a change in hematocrit ratio on peripheral resistance may be appreciated when it is recognized that in patients with severe **essential hypertension** (the most common cause of a chronic elevation of arterial blood pressure), the total peripheral resistance (ratio of the systemic arteriovenous pressure difference to the cardiac output) rarely increases more than twofold. In essential hypertension, the increase in peripheral resistance is usually achieved by arteriolar vasoconstriction.

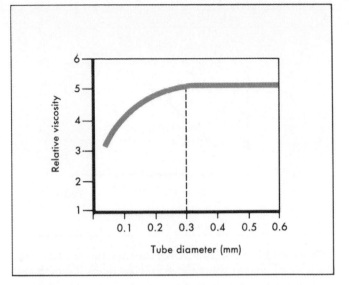

FIGURE 20-10 The viscosity of blood, relative to that of water, increases as a function of tube diameter, up to a diameter of about 0.3 mm. (Redrawn from Fahraeus R, Lindqvist T: *Am J Physiol* 96:562, 1931.)

For any given hematocrit ratio the apparent viscosity of blood depends on the dimensions of the tube used to estimate the viscosity. Figure 20-10 demonstrates that the apparent viscosity of blood is not affected appreciably by changes in tube diameter when the diameters exceed 0.3 mm, but the apparent viscosity does diminish progressively as the tube diameter is decreased to values below about 0.3 mm. The major resistance to blood flow in vascular beds normally resides in the very small arteries and arterioles, which have diameters substantially less than 0.3 mm. Thus, these small vessels would be the principal determinants of the apparent viscosity of the blood flowing through living tissues. The tendency for such small tubes to diminish the apparent viscosity of the blood, as shown in Figure 20-10, explains why the apparent viscosity is less when a biological tissue is used as a viscometer (*lower curve*, Figure 20-9) than when a glass tube with a lumen diameter of 1 mm is used as a viscometer (*upper curve*, Figure 20-9). This tendency for apparent viscosity to decrease with tube diameter confers a lower resistance to blood flow in biological tissues than would otherwise prevail if resistance were determined exclusively by vascular dimensions.

The influence of tube diameter on apparent viscosity is explained in part by the difference in the actual composition of the blood as it flows from large tubes into small tubes. The composition changes because in the small tubes the red blood cells tend to accumulate in the faster axial stream, whereas the plasma is mainly consigned to the slower marginal layers of the bloodstream. Because the red blood cells traverse the tube more

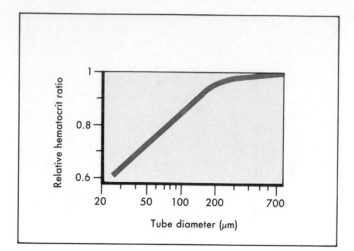

FIGURE 20-11 The "relative hematocrit ratio" of blood flowing from a feed reservoir through capillary tubes of various calibers, as a function of the tube diameter. The relative hematocrit ratio is the ratio of the hematocrit of the blood in the tubes to that of the blood in the feed reservoir. (Redrawn from Barbee JH, Cokelet GR: *Microvasc Res* 3:6, 1971.)

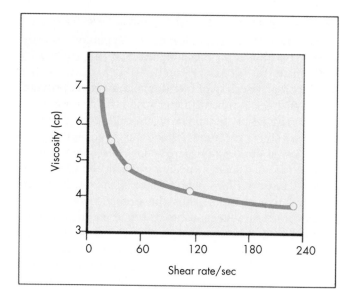

FIGURE 20-12 The viscosity of blood (centipoise) as a function of the shear rate, which is the ratio of the velocity of one layer of fluid to that of the adjacent layers. The shear rate is directionally related to the flow. (Redrawn from Amin TM, Sirs JA: *Q J Exp Physiol* 70:37, 1985.)

quickly than does the plasma, the hematocrit ratio of the blood in the tube is actually less than the hematocrit ratio of the blood in the reservoir to which the tube is connected (Figure 20-11).

The apparent viscosity of blood diminishes as the shear rate is increased (Figure 20-12), a phenomenon called **shear thinning.** The **shear rate** is the rate at which one layer of fluid moves with respect to the ad-

jacent layers; the shear rate varies directly with the flow. The greater tendency of the erythrocytes to accumulate in the axial laminae at higher flow rates is partly responsible for the nonnewtonian behavior of blood. However, a more important factor is that at very slow rates of shear, the suspended cells tend to aggregate, which increases viscosity. This tendency to aggregate decreases as the flow is augmented. The resultant diminution in apparent viscosity with increasing shear rate is shown in Figure 20-12.

The deformability of the erythrocytes is also a factor in shear thinning, especially when the hematocrit ratio is high. The mean diameter of human red blood cells is about 7 μm, yet these cells are able to pass through openings with a diameter of only 3 μm. As blood that is densely packed with erythrocytes is caused to flow at progressively greater rates, the erythrocytes become more and more deformed. The greater deformability diminishes the apparent viscosity of the blood.

The flexibility of human erythrocytes is enhanced when the concentration of fibrinogen in the plasma is elevated. Conversely, the erythrocytes are misshapen and inflexible in patients with **sickle cell anemia.** This may result in serious disturbances in regional blood flow.

SUMMARY

1. The vascular system is composed of two major subdivisions, the systemic and pulmonary circulations, which are in series with one another.
2. Each subdivision comprises a number of types of vessels (e.g., arteries, arterioles, and capillaries) that are aligned in series with one another. In general, most vessels of a given type (e.g., capillaries) are arranged in parallel with each other.
3. The mean velocity ($\bar{v}$) of the bloodstream in a given type of vessel is directly proportional to the total blood flow (Q_t) through all the vessels of that type, and it is inversely proportional to the cross-sectional area (A) of all the parallel vessels of that type; that is, $\bar{v} = Q_t/A$.
4. When blood flow is steady and laminar in vessels larger than arterioles, the flow (Q) is proportional to the difference between the inflow and outflow pressures ($P_i - P_o$) and to the fourth power of the radius (r), and it is inversely proportional to the length (l) of the vessel and to the viscosity (η) of the fluid; that is, $Q = \pi (P_i - P_o)r^4/8\eta l$ (Poiseuille's law).

5. For resistances aligned in series, the total resistance equals the sum of the individual resistances.

6. For resistances aligned in parallel, the reciprocal of the total resistanc equals the sum of the reciprocals of the individual resistances.

7. Flow tends to become turbulent when flow velocity is high, when fluid viscosity is low, when tube diameter is large, or when the lumen of the vessel is very irregular.

8. Blood flow is nonnewtonian in very small vessels; that is, Poiseuille's law is not applicable. The apparent viscosity of the blood diminishes as shear rate (flow) increases and as the tube dimensions decrease.

BIBLIOGRAPHY
Journal articles

Badeer HS, Hicks JW: Hemodynamics of vascular "waterfall": is the analogy justified? *Resp Physiol* 87:205, 1992.

Cokelet GR, Goldsmith HL: Decreased hydrodynamic resistance in the two-phase flow of blood through small vertical tubes at low flow rates, *Circ Res* 68:1, 1991.

Goldsmith HL: The microrheology of human blood, *Microvasc Res* 31:121, 1986.

Jonsson V et al: Significance of plasma skimming and plasma volume expansion, *J Appl Physiol* 72:2047, 1992.

Maeda N, Shiga T: Velocity of oxygen transfer and erythrocyte rheology, *News Physiol Sci* 9:22, 1994.

McKay CB, Meiselman HJ: Osmolality-mediated Fahraeus and Fahraeus-Lindqvist effects for human RBC suspensions, *Am J Physiol* 254:H238, 1988.

Peacock JA: An in vitro study of the onset of turbulence in the sinus of Valsalva, *Circ Res* 67:448, 1990.

Pries AR et al: Blood flow in microvascular networks: experiments and simulation, *Circ Res* 67:826, 1990.

Reinhart WH et al: Influence of endothelial surface on flow velocity in vitro, *Am J Physiol* 265:H523, 1993.

Secomb TW: Flow-dependent rheological properties of blood in capillaries, *Microvasc Res* 34:46, 1987.

Sutera SP et al: Vascular flow resistance in rabbit hearts: "apparent viscosity" of RBC suspensions, *Microvasc Res* 36:305, 1988.

Books and Monographs

Chien S, Usami S, Skalak R: Blood flow in small tubes. In Renkin EM, Michel CC, eds: *Handbook of physiology,* section 2: The cardiovascular system—Microcirculation, vol IV, Bethesda, Md, 1984, American Physiological Society.

Fung YC: *Biodynamics: circulation,* New York, 1984, Springer-Verlag New York, Inc.

Lowe GDO: *Clinical blood rheology,* vol 1, Boca Raton, Fla, 1988, CRC Press, Inc.

Milnor WR: *Hemodynamics,* Baltimore, 1982, Williams & Wilkins.

Taylor DEM, Stevens AL, eds: *Blood flow: theory and practice,* New York, 1983, Academic Press, Inc.

CHAPTER 21

The Arterial System

HYDRAULIC FILTER

The principal function of the systemic and pulmonary arterial systems is to distribute blood to the capillary beds throughout the body. The arterioles, which are the terminal components of the arterial system, regulate the distribution of blood flow to the various capillary beds. The aorta and pulmonary artery and their major branches constitute a system of conduits that lie between the heart and the arterioles. These conduits have a substantial volume, and in normal individuals they are very compliant.

Because the normal arteries are so compliant and the arterioles present such a high resistance to blood flow, the arterial system constitutes a **hydraulic filter,** so called because *the arterial system converts the intermittent flow generated by the heart to a virtually steady flow through the capillaries* (Figure 21-1). The entire ventricular stroke volume is discharged into the arterial system during ventricular systole, which usually occupies approximately one third of the cardiac cycle duration. In fact, most of the stroke volume is pumped during the rapid ejection phase of ventricular systole (see Figure 18-10), which constitutes about half of total systole. Part of the energy released by the cardiac contraction is kinetic energy; it is dissipated as forward capillary flow during ventricular systole. The remainder is stored as potential energy, in that much of the stroke volume is retained by the distensible arteries (Figure 21-1, *A*). During diastole the elastic recoil of the arterial walls converts this potential energy into capillary blood flow, which continues throughout diastole (Figure 20-1, *B*). Thus, capillary flow is normally continuous, but it does vary slightly throughout the cardiac cycle.

In a patient with severe **arteriosclerosis** (rigid arterial walls), capillary flow would be much more pulsatile than normal. Ventricular ejection would lead to an appreciable increase in capillary flow (Figure 21-1, *C*), and flow would virtually cease during diastole (Figure 21-1, *D*). Similarly, the hydraulic filter is much less effective in subjects with abnormally large stroke volumes than in normal individuals. For example, in patients with **aortic valve regurgitation** (an aortic valve deformation that allows much of the stroke volume ejected during systole to regurgitate back into the left ventricle during diastole), the volume of blood ejected during systole may greatly exceed the normal stroke volume. Such a supernormal stroke volume would induce a pulsatile flow in all the systemic capillaries. In such patients, a **capillary pulse** could be perceived in the patient's nailbeds during the physical examination. Application of a slight force at the distal tip of a fingernail will blanch the adjacent region of the nail. In a normal person, the margin between the blanched and undisturbed region of the nail will remain quiescent. However, in a patient with aortic regurgitation, the margin will pulsate perceptibly with each heartbeat, because the flow through the capillaries in the nailbed increases substantially with each ventricular contraction.

ARTERIAL COMPLIANCE

The elastic properties of the arterial wall may be appreciated by considering first the static pressure-volume relationship for the aorta. To obtain the curves shown in Figure 21-2, aortas were obtained at autopsy from people in different age groups. All branches of the aorta were tied and successive volumes of liquid were injected

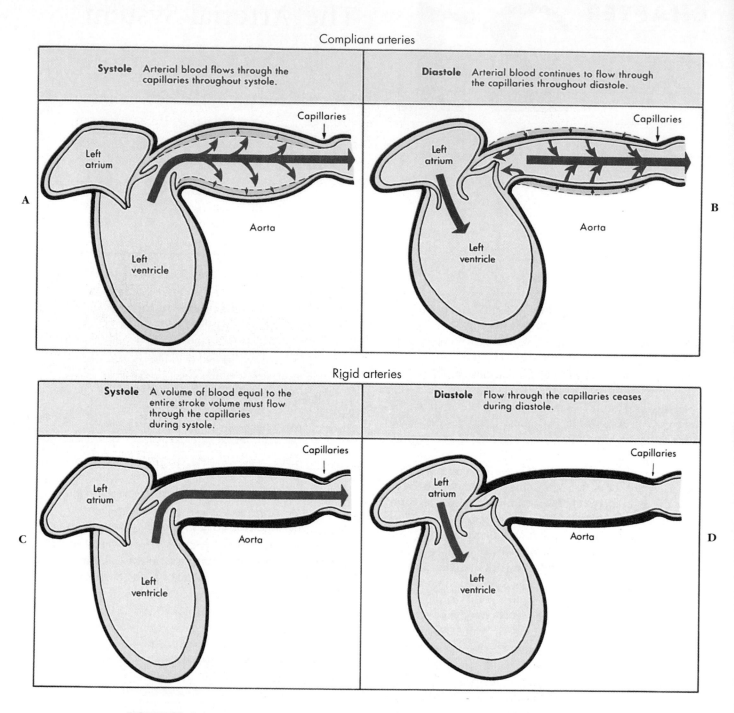

Compliant arteries

Systole   Arterial blood flows through the capillaries throughout systole.

Diastole   Arterial blood continues to flow through the capillaries throughout diastole.

Rigid arteries

Systole   A volume of blood equal to the entire stroke volume must flow through the capillaries during systole.

Diastole   Flow through the capillaries ceases during diastole.

FIGURE 21-1   When the arteries are normally compliant, blood flows through the capillaries throughout the cardiac cycle. When the arteries are rigid, blood flows through the capillaries during systole, but it ceases to flow during diastole. **A,** When the arteries are normally compliant, a substantial fraction of the stroke volume is stored in the arteries during ventricular systole. The arterial walls are stretched. **B,** During ventricular diastole the previously stretched arteries recoil. The volume of blood that is displaced by the recoil ensures continuous capillary flow throughout diastole. **C,** When the arteries are rigid, virtually none of the stroke volume can be stored in the arteries; almost all of the stroke volume must flow through the capillaries during systole. **D,** Rigid arteries cannot recoil appreciably during diastole; hence, capillary flow virtually ceases during diastole.

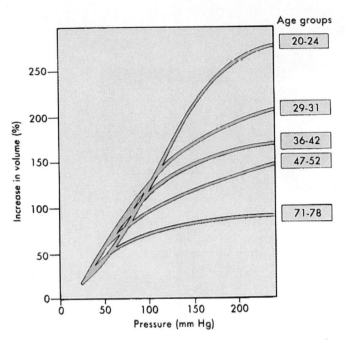

FIGURE 21-2   Pressure-volume relationships for aortas obtained at autopsy from humans in different age groups (denoted by the numbers at the right end of each of the curves). (Redrawn from Hallock P, Benson IC: *J Clin Invest* 16:595, 1937.)

into this closed elastic system, just as successive small volumes of water might be introduced into a balloon. After each increment of volume (V) was injected, the internal pressure (P) in the aorta was measured.

In Figure 21-2 the curve that relates pressure to volume for the youngest age group (20 to 24 years) is sigmoidal. The curve is approximately linear over most of its extent, but the slope decreases at the upper and lower ends. At any given point, the slope (dV/dP) represents the **arterial compliance.** Thus, in normal young people the arterial compliance is least at very high and at very low pressures, and it is greatest over the normal range of pressure variations. These compliance changes resemble the familiar changes encountered in inflating a balloon. Introducing air into the balloon is more difficult (i.e., the balloon is less compliant) at the beginning of inflation and again at near-maximum volume, just prior to rupture of the balloon. At intermediate volumes, however, the balloon is easier to inflate; that is, its compliance is greater.

> Figure 21-2 shows that the aortic pressure-volume curves become displaced downward and the slopes diminish as a function of advancing age; thus, the compliance decreases with age. The diminished compliance is a manifestation of the progressive increase in the collagen

> and decrease in the elastin contents of the arterial walls. The heart cannot eject a given stroke volume into a rigid arterial system as readily as It can Into a more compliant system.

DETERMINANTS OF THE ARTERIAL BLOOD PRESSURE

The factors that determine the arterial blood pressure cannot be evaluated with great precision. Yet the arterial blood pressure is routinely measured in most patients who come to the physician's office, and it provides some useful clues to the patient's cardiovascular status. We will therefore take a simplified approach in an attempt to understand the principal determinants of the arterial blood pressure. To accomplish this, we will first analyze the determinants of the mean arterial pressure (defined in the following section). The systolic and diastolic arterial pressures will then be considered to be the upper and lower limits of periodic oscillations about this mean pressure. Secondarily, we will analyze the determinants of these periodic oscillations.

The determinants of the arterial blood pressure will be arbitrarily subdivided into physical and physiological factors (Figure 21-3). For the sake of simplicity, the arterial system will be assumed to be a static, elastic system. The only two **physical factors** to be considered will be the **blood volume** within the arterial system and the elastic characteristics **(compliance)** of the system. The **physiological factors** to be considered are **cardiac output** (the product of heart rate and stroke volume) and **peripheral resistance.** Such physiological factors will be shown to operate through one or both of the physical factors.

Mean Arterial Pressure

The **mean arterial pressure** is the pressure in the arteries, averaged over time. It may be obtained from an arterial pressure tracing by measuring the area under the curve and dividing this area by the appropriate time interval, as shown in Figure 21-4. The mean arterial pressure, $\bar{P}_a$, usually can be estimated satisfactorily from the measured values of the systolic (P_s) and diastolic (P_d) pressures by means of the following formula:

$$\bar{P}_a \approx P_d + (P_s - P_d)/3 \qquad \textbf{(1)}$$

Just as the pressure in a balloon depends on the fluid volume in the balloon and on the compliance of its walls, so also does *the mean pressure in the arterial system depend physically on the mean volume of blood in the*

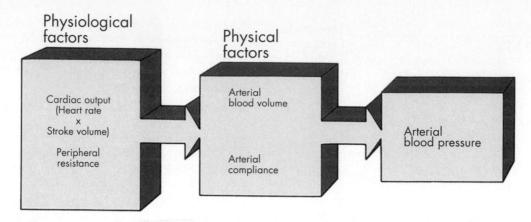

FIGURE 21-3 The arterial blood pressure is determined directly by two major physical factors, the arterial blood volume and the arterial compliance. These physical factors are affected, in turn, by certain physiological factors, mainly the heart rate, stroke volume, and peripheral resistance.

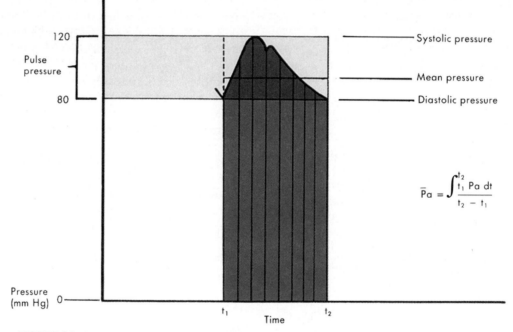

$$\bar{P}a = \frac{\int_{t_1}^{t_2} Pa\ dt}{t_2 - t_1}$$

FIGURE 21-4 Arterial systolic, diastolic, pulse, and mean pressures. The mean arterial pressure $(\bar{P}_a)$ represents the area under the arterial pressure curve *(red shaded area)* divided by the cardiac cycle duration $(t_2 - t_1)$.

arterial system and on the compliance of the arterial walls (Figure 21-3). The arterial blood volume, V_a, in turn, depends on the balance between the rate (Q_h) at which the heart pumps blood into the arteries **(cardiac output)** and the rate (Q_r) at which the blood flows out of the arteries through the resistance vessels and capillaries **(peripheral runoff)** and into the veins. Expressed mathematically:

$$dV_a/dt = Q_h - Q_r \qquad (2)$$

This equation is an expression of the **law of conser-** **vation of mass.** The equation states that any change in arterial blood volume simply reflects the difference in the rates at which blood enters and leaves the arterial system. If arterial inflow exceeds peripheral runoff, then arterial volume increases, the arterial walls are distended, and pressure rises. The converse happens when peripheral runoff exceeds arterial inflow. Finally, if arterial inflow equals peripheral runoff, arterial pressure remains constant.

Cardiac Output The change in pressure evoked by an alteration in cardiac output (Q_h) can be better appreciated by considering some simple examples. Under control conditions, let cardiac output be 5 L/min and mean arterial pressure ($\overline{P}_a$) be 100 mm Hg (Figure 21-5, A). Under steady-state conditions $Q_h = Q_r$. The **total peripheral resistance** (R_t) is the resistance to blood flow in the systemic vascular bed; that is:

$$R_t = (\overline{P}_a - \overline{P}_{ra})/Q_r \qquad (3)$$

Because $\overline{P}_{ra}$ (mean right atrial pressure) is usually much less than $\overline{P}_a$, then

$$R_t \approx \overline{P}_a/Q_r \qquad (4)$$

Therefore, in the example shown in Figure 21-5, A, R_t is 100/5, or 20 mm Hg/L/min.

Now let cardiac output, Q_h, suddenly increase to 10 L/min (Figure 21-5, B). Thus, the inflow (Q_h) of blood to the arteries will exceed the outflow (Q_r) of blood from the arteries. Initially, however, the increment in blood volume in the arteries will be negligible, and $\overline{P}_a$ will be unchanged. Because the outflow (Q_r) from the arteries depends on $\overline{P}_a$ and R_t, then Q_r also will not change appreciably at first. Therefore Q_h, now 10 L/min, will exceed Q_r, still only 5 L/min. This will progressively increase the arterial blood volume (V_a). From equation 2, when $Q_h > Q_r$, then $dV_a/dt > 0$; that is, volume increases.

The increase in blood volume in the arteries will raise the blood pressure in the arteries, just as an increase in volume in any hollow elastic structure, such as a balloon, will increase the pressure. In the above example of a sudden increase in cardiac output, blood will continue to accumulate in the arteries until the pressure rises sufficiently high to force through the peripheral resistance a runoff, Q_r, that will equal the elevated cardiac output, Q_h (Figure 21-5, C). When equation 4 is solved for Q_r, it becomes evident that Q_r will not attain a value of 10 L/min until $\overline{P}_a$ reaches a level of 200 mm Hg (if R_t remains constant at 20 mm Hg/L/min). Hence, as $\overline{P}_a$ approaches 200, Q_r will almost equal Q_h, and $\overline{P}_a$ will rise very slowly. By contrast, when Q_h is first augmented, Q_h greatly exceeds Q_r, and therefore $\overline{P}_a$ will rise sharply. The pressure-time tracing in Figure 21-6 indicates that, regardless of the arterial compliance (C_a), the slope of the pressure tracing diminishes progressively as pressure rises, and the final steady-state pressure is achieved asymptotically.

Furthermore, the height to which $\overline{P}_a$ will rise is independent of the compliance of the arterial walls. At equilibrium, $\overline{P}_a$ must rise to a level such that $Q_r = Q_h$. Rearrangement of equation 3 shows that Q_r depends only on pressure difference and resistance to flow:

$$Q_r = (\overline{P}_a - P_{ra})/R_t \qquad (5)$$

Hence C_a determines only the rate at which the new equilibrium value of $\overline{P}_a$ will be attained, but not the value of $\overline{P}_a$, as illustrated in Figure 21-6. When C_a is small (rigid vessels), a relatively slight increment in V_a (caused by a

transient excess of Q_h over Q_r) increases $\overline{P}_a$ greatly. Hence $\overline{P}_a$ attains its new equilibrium level quickly. Conversely, when C_a is large (distensible vessels), then considerable volumes can be accommodated with relatively small pressure changes. Therefore the new equilibrium value of $\overline{P}_a$ is reached at a slower rate.

Peripheral Resistance Similar reasoning may now be applied to explain the changes in $\overline{P}_a$ that accompany alterations in total peripheral resistance (R_t). Let the control conditions be identical to those of the preceding example, that is, $Q_h = 5$, $\overline{P}_a = 100$, and $R_t = 20$ (see Figure 21-5, A). Then let R_t suddenly be increased to 40 (Figure 21-5, D). Initially $\overline{P}_a$ will be unchanged, because sufficient time has not passed to allow a substantial change in V_a. While $\overline{P}_a$ is still equal to 100 mm Hg, the peripheral runoff (Q_r) will suddenly decrease to 2.5 L/min when R_t is increased to 40. If the cardiac output, Q_h, remains constant at 5 L/min, Q_h will exceed Q_r, and V_a will increase; hence $\overline{P}_a$ will rise. $\overline{P}_a$ will continue to rise until it reaches 200 mm Hg (see Figure 21-5, E). At this pressure level $Q_r = 200/40 = 5$ L/min, which equals the cardiac output, Q_h. $\overline{P}_a$ will then remain at this new equilibrium level of 200 mm Hg as long as Q_h and R_t do not change.

It is clear, therefore, that *the mean arterial pressure depends only on cardiac output and total peripheral resistance* (Figure 21-7). It is immaterial whether the change in cardiac output is accomplished by an alter-

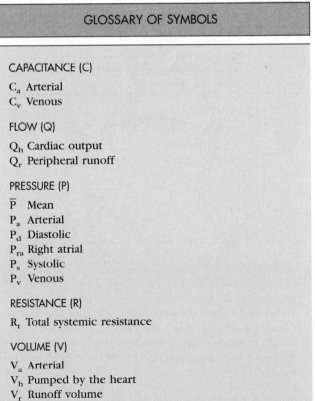

GLOSSARY OF SYMBOLS

CAPACITANCE (C)

C_a Arterial
C_v Venous

FLOW (Q)

Q_h Cardiac output
Q_r Peripheral runoff

PRESSURE (P)

$\overline{P}$ Mean
P_a Arterial
P_d Diastolic
P_{ra} Right atrial
P_s Systolic
P_v Venous

RESISTANCE (R)

R_t Total systemic resistance

VOLUME (V)

V_a Arterial
V_h Pumped by the heart
V_r Runoff volume
V_v Venous

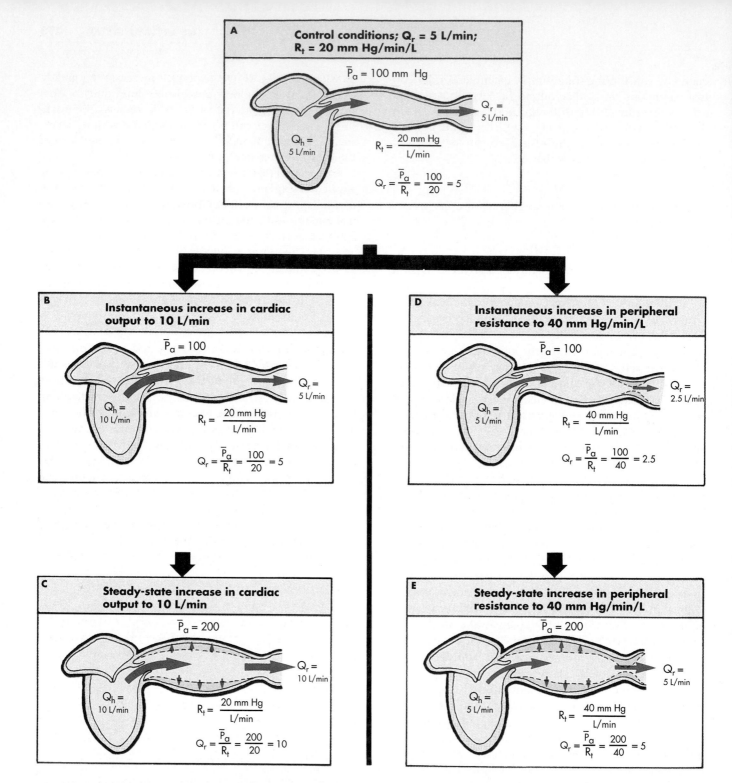

FIGURE 21-5 The relationship of mean arterial blood pressure ($\overline{P}_a$) to cardiac output (Q_h), peripheral runoff (Q_r), and total peripheral resistance (R_t) under control conditions (**A**), in response to an increase in cardiac output (**B** and **C**), and in response to an increase in total peripheral resistance (**D** and **E**). **A,** Under control conditions, cardiac output (Q_h) = 5 L/min, mean arterial pressure ($\overline{P}_a$) = 100 mm Hg, and total peripheral resistance (R_t) = 20 mm Hg/L/min. At equilibrium, the peripheral runoff (Q_r) must equal the cardiac output (Q_h), and therefore the mean blood volume ($\overline{V}_a$) in the arteries will remain constant from heartbeat to heartbeat. **B,** If Q_h suddenly increases to 10 L/min, instantaneously the increment in arterial volume is negligible, and therefore $\overline{P}_a$ and Q_r are essentially unchanged. Initially, the arterial blood volume (V_a) will increase at a rate of $Q_h - Q_r$, or 5 L/min. Therefore, $\overline{P}_a$ will begin to rise rapidly. **C,** The disparity between Q_h and Q_r results in a progressive increase in V_a. V_a continues to increase until $\overline{P}_a$ reaches a level (200 mm Hg) that will generate a peripheral runoff (Q_r) of 10 L/min through an unchanged peripheral resistance (R_t) of 20 mm Hg/L.min. **D,** If R_t is abruptly increased to 40 mm Hg/L/min, Q_r suddenly decreases from 5 to 2.5 L/min, and therefore Q_h exceeds Q_r. Initially, the increment in V_a is negligible; hence, $\overline{P}_a$ still equals 100 mm Hg. However, the flow into the arterial system exceeds the flow out of the system, and therefore $\overline{P}_a$ will rise progressively. **E,** The excess of Q_h over Q_r results in a rise in V_a. V_a continues to increase until $\overline{P}_a$ reaches a level sufficient to force a peripheral runoff (Q_r) of 5 L/min (equal to Q_h) through the high peripheral resistance.

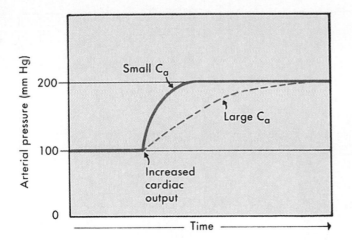

FIGURE 21-6   When cardiac output is suddenly increased, the arterial compliance (C_a) determines the rate at which the mean arterial pressure will attain its new, elevated value, but C_a will not determine the magnitude of the new pressure.

ation in heart rate, stroke volume, or both. Because cardiac output equals heart rate times stroke volume, any change in heart rate that is balanced by an inverse change in stroke volume will not alter cardiac output, and therefore $\overline{P}_a$ will not be affected.

Pulse Pressure

The two **physical factors** that mainly determine the arterial pressure (P_a) at any given time are the **arterial blood volume** (V_a) and the **arterial compliance** (C_a) that prevail at that time (see Figure 21-3). As used here, P_a refers to pressure at any given time in the cardiac cycle, whereas $\overline{P}_a$ refers to the mean pressure over one or more cardiac cycles. The **physiological factors (heart rate, stroke volume,** and **peripheral resistance)** that affect P_a operate through these physical factors to affect the arterial **pulse pressure** (the amplitude of the pressure wave), just as they do to influence the mean arterial pressure (see Figure 21-3). As shown in Figure 21-7, the pulse pressure equals the difference between the **systolic pressure** (P_s) and the **diastolic pressure** (P_d) for that heartbeat; P_s and P_d are the maximum and minimum values, respectively, of the arterial pressure during that heartbeat.

Stroke Volume   The effect of a change in stroke volume on pulse pressure may be analyzed more clearly under conditions in which the arterial compliance (C_a) remains constant. C_a is constant over any linear region of a pressure-volume curve (see Figure 21-2). If the arterial volume is plotted along the vertical axis, and the arterial pressure is plotted along the horizontal axis, the slope, dV_a/dP_a, of the volume-pressure curve is the arterial compliance, C_a, by definition.

To appreciate the effect of a change in stroke volume on the arterial pulse pressure, let us consider a normal person who initially has a heart rate (HR) of 100 beats/min, a stroke volume (SV) of 50 ml, and a peripheral resistance (R_t) of 20 mm Hg/L/min. Because cardiac output equals HR times SV, this person's cardiac output is 5 L/min. As shown in Figure 21-5, A, this person would have a mean arterial pressure $(\overline{P}_a)$ of 100 mm Hg. His systolic pressure, P_s, will of course be greater than 100 mm Hg, and his diastolic pressure, P_d, will be less than 100 mm Hg. Let the values of P_s and P_d for this person be 120 and 90 mm Hg, respectively; these values for $\overline{P}_a$, P_s, and P_d satisfy equation 1. Note that the average normal diastolic pressure is about 80 mm Hg; the value of 90 mm Hg is used here only to simplify certain computations below.

Let us review the hemodynamic events that take place just before and during the rapid ejection phase of systole of a normal heartbeat (see Chapter 18). Throughout the ventricular diastole of the preceding heartbeat and during the brief period of isovolumic contraction, the heart had pumped no blood into the arteries, but blood had continuously flowed out of the arteries and through the capillaries. Hence, the arterial volume, V_a, had been declining and consequently the arterial pressure, P_a, had been falling continuously throughout these phases of the cardiac cycle. Given an individual whose diastolic pressure was 90 mm Hg, this minimum pressure would have been attained at the end of the isovolumic contraction phase of systole (see Figure 18-10).

In healthy people, most of the stroke volume is ejected during the rapid ejection phase of ventricular systole (see Figure 18-10). Let us assume that 80% of the stroke volume (50 ml) in our hypothetical subject is expelled during rapid ejection. Thus, during this phase, the blood volume (V_h) propelled by the heart into the arterial system was 0.80×50 ml, or 40 ml (Figure 21-8, A).

Blood exits from the arteries throughout the cardiac cycle. At equilibrium, the volume of blood that leaves the arteries and flows through the capillaries during each cardiac cycle equals the stroke volume. The fractional volume that flows out of the arteries during rapid ejection is approximately equal to the fraction of the cardiac cycle duration that is occupied by the rapid ejection phase. Let us assume that during rapid ejection, 16% of the stroke volume exits the arteries through the peripheral resistance; that is, the peripheral runoff volume (V_r) during rapid ejection is 0.16×50 ml, or 8 ml.

Hence, the **volume increment** (ΔV_a) that prevails in the arteries during the rapid ejection phase of the cardiac cycle equals $V_h - V_r$, or 32 ml; that is, the heart has propelled a greater volume of blood into the arteries during this phase than the volume that has exited through the peripheral resistance and into the veins. In this hy-

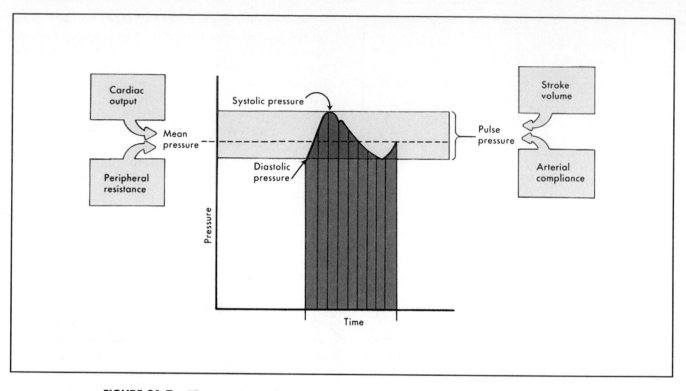

FIGURE 21-7   The mean arterial pressure is determined by the cardiac output and peripheral resistance, whereas the arterial pulse pressure is determined mainly by the stroke volume and the arterial compliance.

pothetical person, the volume increment causes P_a to rise from the diastolic level of 90 mm Hg to the systolic level of 120 mm Hg. In other words, the volume increment (ΔV_a) of 32 ml produced a *pressure increment* (ΔP_a) of 30 mm Hg; this pressure increment is, by definition, the **pulse pressure**. Thus, the pressure increment caused by the disparity between the volume of blood (V_h) ejected by the heart during rapid ejection and the volume (V_r) that exits through the peripheral resistance during rapid ejection represents the rise in arterial pressure from its minimum to its maximum value. The extent of that pressure rise, ΔP_a, which is the pulse pressure, is determined simply by the values of the volume increment (ΔV_a) and of the arterial compliance (C_a). The relationship is evident from a rearrangement of the equation that defines the arterial compliance; that is,

$$\Delta P_a = \Delta V_a / C_a \qquad \textbf{(6)}$$

Thus, the pulse pressure equals the arterial volume increment during the rapid ejection phase of ventricular systole, divided by the arterial compliance. In the hypothetical example analyzed above, the arterial compliance was equal to $\Delta V_a/\Delta P_a = 32/30$, or 1.07 ml/mm Hg.

Let us now suppose that the cardiovascular status changes, such that the heart rate decreases to 50 beats/min and the stroke volume increases to 100 ml (Figure

21-8, *B*). However, let the peripheral resistance (20 mm Hg/L/min) and arterial compliance (1.07 ml/mm Hg) remain unchanged. Because the cardiac output (HR × SV) still equals 5 L/min and the peripheral resistance is also unchanged, the mean arterial pressure ($\overline{P}_a$) remains at 100 mm Hg (see equation 4).

To determine the new pulse pressure, let us assume that the fraction of the stroke volume (100 ml) expelled during rapid ejection is still 80%. Hence, V_h during rapid ejection is 0.80 × 100 ml, or 80 ml. Also, let us assume again that 16% of the stroke volume exits the arterial system through the peripheral resistance during rapid ejection. Hence, the peripheral runoff volume (V_r) during rapid ejection is 0.16 × 100 ml, or 16 ml. Thus, the volume increment (ΔV_a) during rapid ejection is 80 − 16 ml, or 64 ml.

A volume increment of 64 ml in an arterial system with a compliance of 1.07 ml/mm Hg would produce a pressure increment (i.e., a pulse pressure) of 60 mm Hg, which is twice the pulse pressure that prevailed under the preceding control conditions (Figure 21-8, *A*). If we assume also that the relation among the mean, systolic, and diastolic pressures satisfies equation 1, this person's systolic and diastolic pressures would equal 140 and 80 mm Hg, respectively. Hence, if cardiac output and arterial compliance remain constant, an increase in stroke

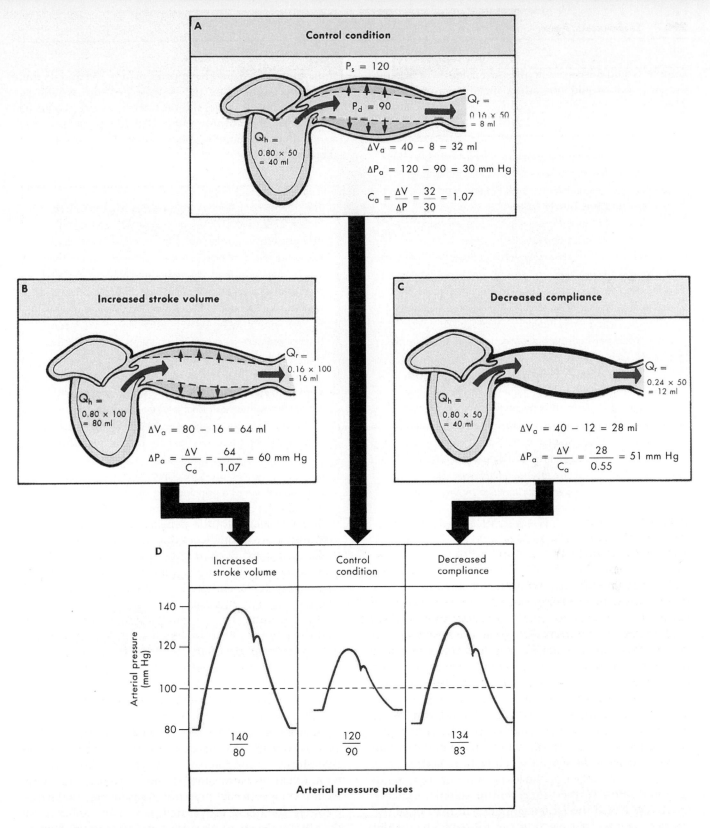

FIGURE 21-8   The effects of an increase in stroke volume and of a decrease in arterial compliance on the arterial pulse pressure. **A,** In a hypothetical subject, the heart pumps a volume (V_h) of 40 ml of blood into the arterial system with each heartbeat during rapid ejection. Concomitantly, a volume (V_r) runs out of the arteries through the peripheral resistance. Thus, a volume increment (ΔV_a) of 32 ml accumulates in the arteries during rapid ejection. This produces a pressure increment (ΔP_a), from the diastolic to the systolic level, of 30 mm Hg; ΔP_a is the pulse pressure. Because $C_a = \Delta V_a / \Delta P_a$, the arterial compliance, C_a, equals 32/30, or 1.07 ml/mm Hg. **B,** If the heart now pumps 80 ml with each beat into the arteries during rapid ejection, and if 16 ml runs out of the arteries during the same period, then ΔV_a is now 64 ml. If C_a is still 1.07, ΔP_a will then be 60 mm Hg. **C,** Assume that the heart can still pump 80% of the stroke volume (50 ml) during rapid ejection, despite a decreased arterial compliance (0.55 ml/mm Hg). Assume also that, because of the decreased compliance, 24% (rather than 16%) of the stroke volume flows out of the arteries during rapid ejection. The volume increment of 28 ml will produce a pulse pressure of 51 mm Hg. **D,** Under conditions that do not alter the mean arterial pressure, an increase in stroke volume or a decrease in arterial compliance will increase the arterial pulse pressure.

volume would raise systolic pressure and lower diastolic pressure, but would not affect mean pressure (Figure 21-8, *D*).

A patient's arterial pulse pressure affords valuable clues about his or her stroke volume, provided the arterial compliance is essentially normal. Patients who have severe **congestive heart failure** or who have lost a large amount of blood are likely to have very small arterial pulse pressures because their stroke volumes are abnormally small. Conversely, individuals with large stroke volumes are likely to have above-average arterial pulse pressures. For example, well-trained athletes at rest tend to have low heart rates. Thus, the prolonged ventricular filling times induce the ventricles to pump more blood per heartbeat. Consequently, the pulse pressures of such athletes tend to be greater than average. Similarly, in patients with **aortic valve regurgitation,** blood leaks back into the left ventricle from the aorta during diastole. This backflow of blood into the ventricle diminishes the aortic diastolic pressure and increases the blood volume in the ventricle during diastole. The augmented ventricular filling volume results in a greater stroke volume during ventricular systole. Characteristically, patients with aortic valve regurgitation have a low arterial diastolic pressure, an elevated arterial systolic pressure, and consequently a greatly increased arterial pulse pressure.

Arterial Compliance Now let us examine the effect of a change in arterial compliance (C_a) on the pulse pressure (Figure 21-8, *C*). Assume that heart rate, stroke volume, and peripheral resistance are all the same as in Figure 21-8, *A*. However, consider that the arteries are about half as compliant as previously (i.e., C_a now equals 0.55). Because the cardiac output and peripheral resistance are the same as in the two preceding examples (panels *A* and *B*), the mean arterial pressure will still be 100 mm Hg.

Let us also assume that even though the arteries are much less compliant, the heart can still pump 80% of the stroke volume during rapid ejection (i.e., $V_h = 40$ ml). Assume also that, because the less compliant arteries will accommodate a smaller increment in volume, a greater fraction (24%) of the stroke volume will run out of the arteries during rapid ejection ($V_r = 12$ ml) than it did under the control conditions (16%). In arteries whose compliance is only 0.55 ml/mm/Hg, the volume increment ($40 - 12 = 28$ ml) will produce a pressure increment (pulse pressure) of 51 mm Hg (Figure 21-8, *C*). If we assume that the relation among systolic, diastolic, and mean pressures satisfies equation 1, the systolic and dia-

stolic pressures would be approximately 134 and 83 mm Hg, respectively. Thus, the systolic pressure would exceed the control value, the diastolic pressure would be lower than the control value, and the pulse pressure would be substantially greater than the control value (Figure 21-8, *D*).

The arterial compliance of old people in general (see Figure 21-2) and of younger people with substantial arteriosclerosis (hardening of the arteries) is low, and this condition is reflected by their pulse pressures. Their systolic pressures tend to be abnormally high (so-called **systolic hypertension**), but their diastolic pressures tend to be substantially lower than the average (80 mm Hg) for normal young people. Estimation of the mean arterial pressure by means of equation 1 indicates that most of the subjects with pure systolic hypertension do not have the most common type of hypertension (**essential hypertension),** which is characteristically associated with a high total peripheral resistance.

The characteristic changes in arterial blood pressure in patients with essential hypertension are a moderate increase in diastolic pressure and a substantially greater increase in systolic pressure. Thus, their pulse pressures are augmented. The principal reason for the increase in arterial pulse pressure is the diminished arterial compliance that prevails when the arterial pressure is elevated. The reduced compliance is reflected by the decrease in the slope of the volume-pressure curve as the pressure is raised (see Figure 21-2). Thus if a normal stroke volume is ejected into an arterial system with diminished compliance, the blood ejected each systole will produce an augmented arterial pulse pressure.

BLOOD PRESSURE MEASUREMENT IN HUMANS

Needles or catheters may be introduced into peripheral arteries of patients in the cardiac catheterization laboratories or the intensive care units of hospitals. Arterial blood pressure can then be measured directly by means of electronic pressure transducers. Ordinarily, however, the blood pressure of patients is estimated indirectly by means of a **sphygmomanometer.** This instrument consists of an inextensible cuff and an inflatable bag. The cuff is wrapped around the arm above the elbow, and the inflatable bag, which lies between the cuff and the skin, is positioned over the brachial artery. To measure the patient's blood pressure, a rubber squeeze bulb is used to raise the pressure in the bag to

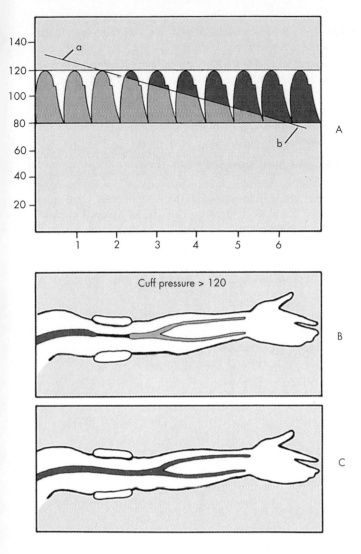

FIGURE 21-9   Measurement of arterial blood pressure with a sphygmomanometer. **A,** Consider that the patient's arterial blood pressure is 120/80 mm. Hg. The pressure (represented by the oblique line) in a cuff around the patient's arm is allowed to fall from over 120 mm Hg (point *a*) to below 80 mm Hg (point *b*) in about 6 sec. **B,** When the cuff pressure exceeds 120 mm Hg, no blood passes through the arterial segment under the cuff, and no sounds can be detected via a stethoscope bell placed on the arm distal to the cuff. **C,** When the cuff pressure falls below 80 mm Hg, arterial flow through the region surrounded by the cuff is continuous, and no sounds are audible. When the cuff pressure is between 120 and 80 mm Hg, spurts of blood traverse the artery segment under the cuff during each heartbeat, and the Korotkoff sounds are heard through the stethoscope.

a value in excess of the patient's arterial systolic pressure. This occludes the brachial artery. Pressure is then released from the bag at a rate of 2 or 3 mm Hg per second by means of a needle valve in the inflation bulb.

The physician or nurse listens with a stethoscope applied to the skin in the antecubital space, over the brachial artery. While the pressure in the bag exceeds the systolic pressure, the brachial artery is occluded and no sounds are heard. When the inflation pressure falls just below the arterial systolic pressure (*upper horizontal line* in Figure 21-9, *A*), small spurts of blood pass through the artery each time the arterial pressure exceeds the cuff pressure. Consequently, slight tapping sounds (called **Korotkoff sounds**) are heard with each heartbeat. The pressure at which the first sound is detected represents the **systolic pressure.** As inflation pressure continues to fall, more blood escapes under the cuff per beat and the sounds become louder. As the inflation pressure approaches the diastolic pressure (*lower horizontal line*), the Korotkoff sounds become muffled. As they fall just below the diastolic pressure, the sounds disappear; this indicates the **diastolic pressure.**

The Korotkoff sounds are generated by the the impact of the spurt of blood that passes under the cuff and meets the static column of blood beyond the cuff; the impact creates turbulence and generates audible vibrations. Once the inflation pressure is less than the diastolic pressure, flow is continuous in the brachial artery and sounds are no longer heard.

SUMMARY

1. The arteries serve not only to conduct blood from the heart to the capillaries, but also to store some of the ejected blood during each cardiac systole. Consequently, flow can continue through the capillaries throughout cardiac diastole as well.
2. The aging process diminishes the compliance of the arteries.
3. The less compliant the arteries, the more work the heart must do to pump a given cardiac output.
4. The mean arterial pressure varies directly with the cardiac output and the total peripheral resistance.
5. The arterial pulse pressure varies directly with stroke volume but inversely with the arterial compliance.
6. When blood pressure is measured by a sphygmomanometer: (a) The systolic pressure is manifested by the occurrence of a tapping sound that originates in the artery distal to the cuff as the cuff pressure falls below the peak arterial pressure. The sounds are produced by the spurts of blood that pass through the compressed artery during the periods in which the arterial pressure exceeds the cuff pressure. (b) The diastolic pressure is manifested by the disappearance of the sound as the cuff pressure falls below the minimum arterial pressure, because flow through the artery then becomes continuous.

BIBLIOGRAPHY
Journal Articles

Blank SG et al: Wideband external pulse recording during cuff deflation: a new technique for evaluation of the arterial pressure pulse and measurement of blood pressure, *Circulation* 77:1297; 1988

Burattini R et al: Total systemic arterial compliance and aortic characteristic impedance in the dog as a function of pressure: a model based study, *Computers Biomed Res* 20:154, 1987.

Farrar DJ et al: Anatomic correlates of aortic pulse wave velocity and carotid artery elasticity during atherosclerosis progression and regression in monkeys, *Circulation* 83:175, 1991.

Folkow B, Svanborg A: Physiology of cardiovascular aging, *Physiol Rev* 73:725, 1993.

Kenner T: Arterial blood pressure and its measurement, *Basic Res Cardiol* 83:107, 1988.

Laskey WK et al: Estimation of total systemic arterial compliance in humans, *J Appl Physiol* 69:112, 1990.

McIlroy MB, Targett RC: Model of the systemic arterial bed showing ventricular-systemic arterial coupling, *Am J Physiol* 254:H609, 1988.

Mulvany MJ, Aalkjaer C: Structure and function of small arteries, *Physiol Rev* 70:921, 1990

Perloff D et al: Human blood pressure determined by sphygmomanometry, *Circulation* 88:2460-2470; 1993.

Piene H: Pulmonary arterial impedance and right ventricular function, *Physiol Rev* 66:606, 1986.

Reneman RS et al: Noninvasive assessment of arterial flow patterns and wall properties in humans, *News Physiol Sci* 4:185, 1989

Simon AC et al: Role of arterial compliance in the physiopharmacological approach to human hypertension, *J Cardiovasc Pharmacol* 19:S11, 1992.

Books and Monographs

Dobrin PB: Vascular mechanics. In *Handbook of physiology; section 2: The cardiovascular system—peripheral circulation and organ blood flow,* vol III, Bethesda, Md, 1983, American Physiological Society.

Fung YC: *Biodynamics: circulation,* Heidelberg, 1984, Springer-Verlag.

Milnor WR: *Hemodynamics,* Baltimore, 1982, Williams & Wilkins.

O'Rourke M et al: *Arterial pulse.* Baltimore, 1992, Williams and Wilkins Co.

Westerhof N ed: *Vascular dynamics. physiological perspectives,* New York, 1989, Plenum Press.

22

The Microcirculation and Lymphatics

FUNCTIONAL ANATOMY

The entire circulatory system is geared to supply the body tissues with blood in amounts commensurate with their requirements for oxygen and nutrients. The capillaries, consisting of a single layer of endothelial cells, permit rapid exchange of water and solutes with interstitial fluid. The muscular arterioles, which are the major **resistance vessels**, regulate regional blood flow to the capillary beds, and the venules and veins serve primarily as collecting channels and **storage**, or **capacitance**, **vessels.**

Arterioles

The arterioles, which range in diameter from about 5 to 100 mm, have a thick smooth muscle layer, a thin adventitial layer, and an endothelial lining (Figure 16-1). The arterioles give rise directly to the **capillaries** (5 to 10 μm diameter) or in some tissues to **metarterioles** (10 to 20 μm diameter), which then give rise to capillaries (Figure 22-1). The metarterioles can serve either as thoroughfare channels to the venules, bypassing the capillary bed, or as conduits to supply the capillary bed. There are often cross connections from arteriole to arteriole and from venule to venule, as well as in the capillary network. Arterioles that give rise directly to capillaries regulate flow through their cognate capillaries by constriction or dilation. The capillaries form an interconnecting network of tubes of different lengths; the average length is 0.5 to 1 μm.

The diameter of the resistance vessels is determined by the balance between the contractile force of the vascular smooth muscle and the distending force produced by the intraluminal pressure. The greater the contractile activity of the vascular smooth muscle of an arteriole, the smaller will be its diameter. At some point, in the case

of small arterioles, complete occlusion of the vessel will occur, partly because of infolding of the endothelium. With reduction in the intravascular pressure, vessel diameter decreases, as does tension in the vessel wall (Laplace's law, see next page). When perfusion pressure is progressively reduced, a point is reached at which blood flow ceases, even though a positive pressure gradient may still exist. This phenomenon has been referred to as the **critical closing pressure.**

Capillaries

Capillary distribution varies from tissue to tissue. In metabolically active tissues, such as cardiac and skeletal muscle and glandular structures, capillaries are numerous, whereas less active tissues, such as subcutaneous tissue or cartilage, have few capillaries. Also, not all capillaries have the same diameter. Because some capillaries have diameters less than that of erythrocytes, the red cells must become temporarily deformed in order to pass through these capillaries. Fortunately, normal red cells are quite flexible and readily change their shape to conform to that of the small capillaries.

Blood flow in the capillaries is not uniform and depends chiefly on the contractile state of the arterioles. The average velocity of blood flow in the capillaries is approximately 1 mm/sec; however, it can quickly vary from zero to several millimeters per second in the same vessel. The capillary blood flow may vary randomly or it may oscillate rhythmically at different frequencies, as determined by contraction and relaxation (**vasomotion**) of the precapillary vessels. This vasomotion is partly an intrinsic contractile behavior of the vascular smooth muscle and is independent of external influence. Furthermore, changes in **transmural pressure** (intravascular minus extravascular pressure) affect the contractile state of the precapillary vessels. An increase in transmural pressure, whether produced by an increase

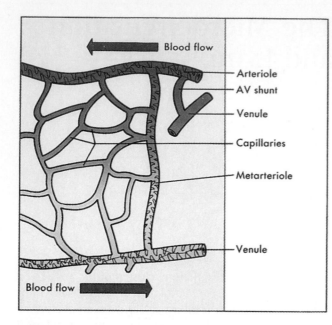

Blood flow

Arteriole

AV shunt

Venule

Capillaries

Metarteriole

Venule

Blood flow

FIGURE 22-1 Schematic drawing of the microcirculation. The circular structures on the arteriole and venule represent smooth muscle fibers, and the branching solid lines represent sympathetic nerve fibers. The arrows indicate the direction of blood flow.

in venous pressure or by dilation of arterioles, elicits contraction of the terminal arterioles at the points of origin of the capillaries. Conversely, a decrease in transmural pressure elicits precapillary vessel relaxation. In addition, humoral and neural factors also affect vasomotion.

Although reduced transmural pressure will relax the terminal arterioles, blood flow through the capillaries obviously cannot increase if the reduced intravascular pressure is caused by severe constriction of the parent arterioles or metarterioles. Large arterioles and metarterioles also exhibit vasomotion. However, in the contraction phase they usually do not completely occlude the lumen of the vessel and arrest blood flow, whereas contraction of the terminal arterioles may arrest blood flow. Because blood flow through the capillaries provides for exchange of gases and solutes between blood and tissue, it has been termed **nutritional flow,** whereas blood flow that bypasses the capillaries in traveling from the arterial to the venous side of the circulation has been termed **nonnutritional,** or **shunt, flow** (Figure 22-1). In some areas of the body (fingertips), true arteriovenous (AV) shunts exist (see Chapter 25). In many tissues such as muscle, however, evidence of anatomical shunts is lacking.

The true capillaries are devoid of smooth muscle and are therefore incapable of active constriction. Nevertheless, the endothelial cells that form the capillary wall contain actin and myosin and can alter their shape in response to certain chemical stimuli. However, such changes in endothelial cell shape do not regulate blood

flow through the capillaries. *Changes in capillary diameter are passive and are caused by alterations in precapillary and postcapillary resistance.*

For many years it was believed that endothelial cells were inert and merely served as barriers to blood cells and large molecules, such as plasma proteins. However it is now known that the endothelium can synthesize substances that affect the contractile state of the arterioles (Figure 22-2). One such substances that causes vasodilation is **endothelium-derived relaxing factor** (EDRF), which has been shown to be **nitric oxide.** It is formed and released in response to stimulation of the endothelium by various agents (acetylcholine, ATP, serotonin, bradykinin, histamine, substance P). **Prostacyclin** is another substance that is synthesized by endothelial cells and produces vasodilation. However, its major function is to inhibit platelet adherence to the endothelium and platelet aggregation, and thereby it aids in the prevention of intravascular thrombosis. A vasoconstrictor substance, **endothelin** has also been isolated from endothelial cells.

Because of their narrow lumen, the thin-walled capillaries can withstand high internal pressures without bursting. This can be explained in terms of Laplace's law:

$$T = Pr \tag{1}$$

where

T = Tension in the vessel wall (dynes/cm)
P = Transmural pressure (dynes/cm^2)
r = Radius of the vessel (cm)

Wall tension is the force per unit length tangential to the vessel wall. This force opposes the distending force (Pr) that tends to pull apart a theoretical longitudinal slit in the vessel (Figure 22-3). Transmural pressure is essentially equal to intraluminal pressure, because extravascular pressure is usually negligible.

At normal aortic (100 mm Hg) and capillary (25 mm Hg) pressures, the wall tension of the aorta is about 12,000 times greater than that of the capillary (100 mm Hg × radius of 1.5 cm for the aorta, versus 25 mm Hg × radius of 5 × 10^{-4} cm for the capillary). In a person standing quietly, capillary pressure in the feet may reach 100 mm Hg. Under such conditions capillary wall tension increases to a value that is only one three-thousandth that of the wall tension in the aorta at the same internal pressure.

According to the Laplace equation, wall tension increases as vessels dilate, even when internal pressure remains constant. Such is the case in **aneurysm** (local widening) **of the aorta,** where wall tension may become high enough to rupture the vessel.

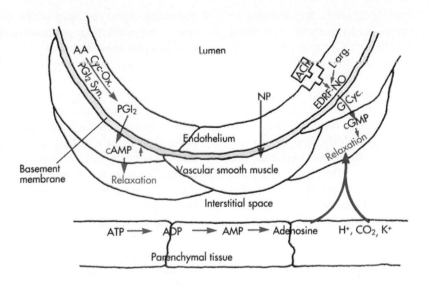

FIGURE 22-2 Diagram of an arteriole illustrating endothelium and nonendothelium mediated vasodilation. Prostacyclin *(PGI₂)* is formed from arachidonic acid *(AA)* by the action of cyclooxygenase *(Cyc Ox)* and prostacyclin synthetase *(PGI₂ Syn)* in the endothelium and elicits relaxation of the adjacent vascular smooth muscle via increases in cyclic adenosine monophosphate *(cAMP)*. Stimulation of the endothelial cells with acetylcholine *(Ach)* or other agents (see text) results in the formation and release of an endothelium-derived relaxing factor *(EDRF)*, which is nitric oxide (NO). The EDRF stimulates guanylyl cyclase *(G Cyc)* to increase cyclic guanosine monophosphate *(cGMP)* in the vascular smooth muscle to produce relaxation. The vasodilator agent nitroprusside *(NP)* acts directly on the vascular smooth muscle. Substances such as adenosine, hydrogen ions (H^+), CO_2, and potassium ions (K^+) can arise in the parenchymal tissue and elicit vasodilation by direct action on the vascular smooth muscle.

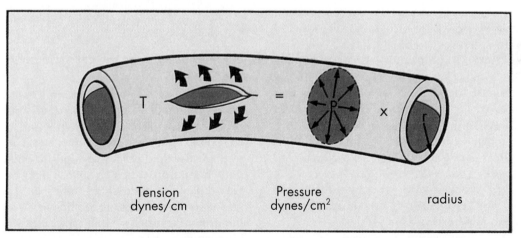

FIGURE 22-3 Diagram of a small blood vessel to illustrate Laplace's law: $T = P \cdot r$, where P = intraluminal pressure, r = radius of the vessel, and T = wall tension as the force per unit length tangential to the vessel wall, tending to pull apart a theoretical longitudinal slit in the vessel.

Capillary Pores

The permeability of the capillary endothelial membrane is not the same in all body tissues. For example, the liver capillaries are very permeable, and albumin escapes at a rate several-times greater than from the less permeable muscle capillaries. Also, permeability is not uniform along the whole capillary; the venous ends are more permeable than the arterial ends, and permeability is greatest in the venules. The greater permeability at the venous end of the capillaries and in the venules is caused by the greater number of pores. In most tissues the **clefts (pores)** are sparse and represent only about 0.02% of the capillary surface area (Figures 22-4 and 22-5). In brain, clefts are absent in the capillaries, and a **blood-brain barrier** to many small molecules exists.

In addition to clefts, some of the more porous capillaries (e.g., in the kidney and the intestine) contain fenestrations 20 to 100 nm wide, whereas others (e.g., in the liver) have a discontinuous endothelium (Figure 22-5). The fenestrations appear to be sealed by a thin diaphragm, but they are quite permeable. Large molecules can penetrate capillaries with fenestrations or gaps caused by discontinuous endothelium, but only small molecules can pass through the intercellular clefts of the endothelium.

TRANSCAPILLARY EXCHANGE

Solvent and solute move across the capillary endothelial wall by three processes: diffusion, filtration, and pinocytosis.

Diffusion

Under normal conditions only about 0.06 ml of water per minute moves back and forth across the capillary wall per 100 g of tissue as a result of filtration and absorption (movement of fluid across the capillary wall by hydrostatic and osmotic forces). By contrast, 300 ml of water per minute per 100 g of tissue does so by diffusion, a 5000-fold difference (see Chapter 1). Relating filtration and diffusion to blood flow, we find that only about 2% of the plasma passing through the capillaries is filtered. In contrast, the rate that water diffuses across the endothelium is 40 times greater than the rate at which it is delivered to the capillaries by blood flow. The transcapillary exchange of solutes is also governed by diffusion. Thus *diffusion is the key factor in the exchange of gases, substrates, and waste products between the capillaries and the tissue cells.* However, net transfer of fluid across the capillary endothelium is primarily attributable to filtration and absorption.

Lipid-Insoluble Substances For small molecules, such as water, NaCl, urea, and glucose, the capillary pores do not restrict diffusion. Diffusion proceeds so rapidly that the mean concentration gradient across the capillary endothelium is extremely small. Water passes through the capillary pores between endothelial cells. As the size of lipid-insoluble molecules is increased, diffusion through muscle capillaries becomes progressively more restricted. Diffusion of molecules with a molecular weight greater than about 60,000 becomes minimal. With small molecules the only limitation to net move-

ment across the capillary wall is the rate at which blood flow transports the molecules to the capillary; transport is said to be **flow limited.**

When transport across the capillary is flow limited, a small-molecule solute (e.g. an inert tracer) diffuses from the blood near the origin of the capillary from the cognate arteriole into the interstitial fluid and parenchymal cells (Figure 22-6). A somewhat larger molecule moves farther along the capillary before its concentration in the blood becomes insignificant, and a still larger molecule cannot pass at all through the capillary pores (Figure 22-6, *A*). An increase in blood flow velocity extends the detectable concentration of small molecules farther down the capillary and increases the capillary diffusion capacity (rate of tissue uptake of the solute).

With large molecules, diffusion across the capillaries becomes the factor that limits exchange (**diffusion limited**). In other words, capillary permeability to a large-molecule solute limits the transport of the solute across the capillary wall (Figure 22-6, *A*).

Under normal conditions the rate of diffusion of small lipid-insoluble molecules is so rapid that exchange of such molecules between blood and tissue is never a problem. Such exchange can become a problem when long distances exist between capillaries and tissue cells, as can occur when capillary density is very low (e.g., after arterial occlusions) or when tissue edema is extensive (Figure 22-6, *B*).

Lipid-Soluble Molecules In contrast to lipid-insoluble molecules, movement of lipid-soluble molecules across the capillary wall is not limited to capillary pores, because such molecules can pass directly through the lipid membranes of the entire capillary endothelium. Consequently, lipid-soluble molecules move very rapidly between blood and tissue. The lipid solubility (oil-to-water partition coefficient) is a good index of the ease of transfer of lipid molecules through the capillary endothelium.

Oxygen and carbon dioxide are both lipid soluble, and readily pass through the endothelial cells. Calculations based on (1) the diffusion coefficient for O_2, (2) capillary density and diffusion distances, (3) blood flow, and (4) tissue O_2 consumption indicate that the O_2 supply of normal tissue at rest and during activity is not limited by diffusion or by the number of open capillaries.

Measurements of oxygen tension and oxygen saturation of hemoglobin in the microvessels indicate that in many tissues blood O_2 content at the entrance of the

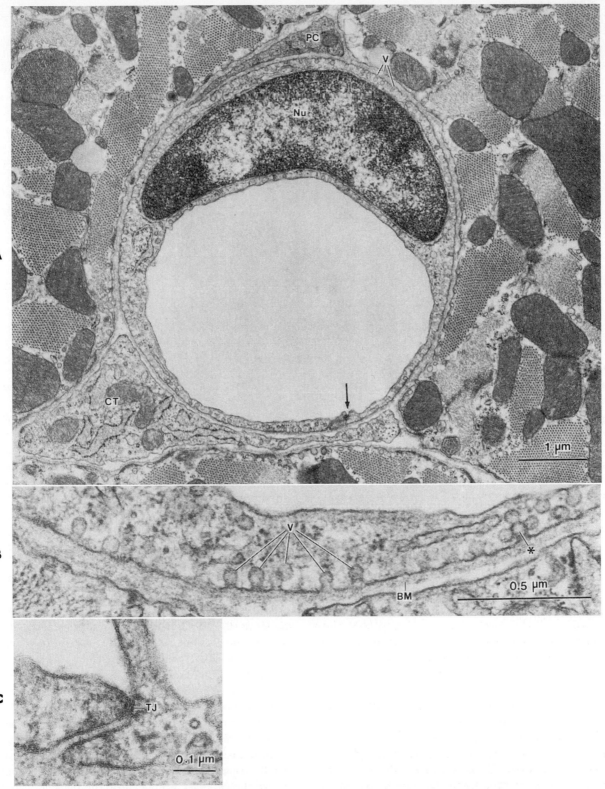

FIGURE 22-4 **A,** Cross-sectioned capillary in mouse ventricular wall. Luminal diameter is approximately 4 μm. In this thin section, the capillary wall is formed by a single endothelial cell (*Nu,* endothelial nucleus), which forms a junctional complex *(arrow)* with itself. The thin pericapillary space is occupied by a pericyte *(PC)* and a connective tissue *(CT)* cell (fibroblast). Note the numerous endothelial vesicles *(V).* **B,** Detail of endothelial panel, **A,** showing plasmalemmal vesicles *(V)* that are attached to the endothelial cell surface. These vesicles are especially prominent in vascular endothelium and are involved in transport of substances across the blood vessel wall. Note the complex alveolar vesicle *(*). BM,* Basement membrane. **C,** Junctional complex in a capillary of mouse heart. Tight junctions *(TJ)* typically form in these small blood vessels and appear to consist of fusions between apposed endothelial cell surface membranes.

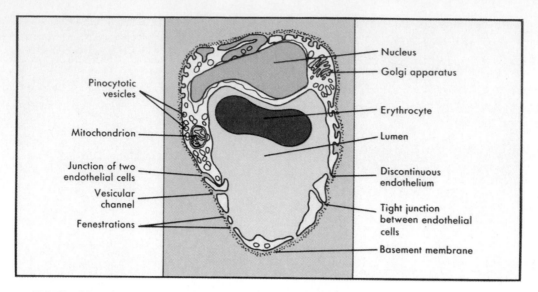

Pinocytotic vesicles

Mitochondrion

Junction of two endothelial cells

Vesicular channel

Fenestrations

Nucleus

Golgi apparatus

Erythrocyte

Lumen

Discontinuous endothelium

Tight junction between endothelial cells

Basement membrane

FIGURE 22-5   Diagrammatic sketch of an electron micrograph showing a composite capillary in cross section.

capillaries has already decreased to about 80% of that in the aorta as a result of diffusion of O_2 from the arterioles. Such studies also have shown that CO_2 loading and the resultant intravascular shifts in the oxyhemoglobin dissociation curve occur in the precapillary vessels. These findings indicate not only that gas does move to respiring tissue at the precapillary level, but also that O_2 and CO_2 pass directly between adjacent arterioles, venules, and possibly arteries and veins (countercurrent exchange, see Chapter 36). This exchange of gas represents a diffusional shunt of gas around the capillaries; at low blood flow rates, it may limit the supply of O_2 to the tissue.

Capillary Filtration

The direction and the magnitude of the movement of water across the capillary wall are determined by the algebraic sum of the hydrostatic and osmotic pressures that exist across the membrane. An increase in intracapillary hydrostatic pressure favors movement of fluid from the vessel to the interstitial space. Conversely, an increase in the concentration of osmotically active particles within the vessels favors movement of fluid into the vessels from the interstitial space.

Hydrostatic Forces The hydrostatic pressure (blood pressure) within the capillaries is not constant; it depends on the arterial pressure, the venous pressure, and the precapillary (arteriolar) and postcapillary (venules and small veins) resistances. A rise in arterial or venous pressure increases capillary hydrostatic pressure (P_c), whereas a reduction in each has the opposite effect. An increase in arteriolar resistance reduces capillary pres-

sure, whereas a greater venous resistance increases capillary pressure.

Capillary hydrostatic pressure is the principal force in capillary filtration, and it varies from tissue to tissue and even within the same tissue. Average values, obtained from many direct measurements in human skin, are about 32 mm Hg at the arterial end of the capillaries and about 15 mm Hg at the venous end of the capillaries at the level of the heart (Figure 22-7). The hydrostatic pressure in capillaries of the lower extremities will be higher and that of capillaries in the head will be lower when an individual is standing.

Tissue pressure, or more specifically **interstitial fluid pressure** (P_i), outside the capillaries, opposes capillary filtration. Hydrostatic pressure minus interstitial fluid pressure ($P_c - P_i$) constitutes the driving force for filtration. In the absence of edema, the interstitial fluid pressure is essentially zero.

Osmotic Forces *The key factor that restrains fluid loss from the capillaries is the osmotic pressure of the plasma proteins, usually termed the* **colloid osmotic pressure,** *or* **oncotic pressure (π_p).** The total osmotic pressure of plasma is about 6000 mm Hg, whereas the oncotic pressure is only about 25 mm Hg. However, this small oncotic pressure is important in fluid exchange across the capillary wall because the plasma proteins are essentially confined to the intravascular space. The electrolytes that are mainly responsible for the total osmotic pressure of plasma are practically equal in concentration on both sides of the capillary endothelium. The relative permeability of solute to water influences the actual magnitude of the osmotic pressure. The **reflection coefficient** is the relative impediment to the passage of a

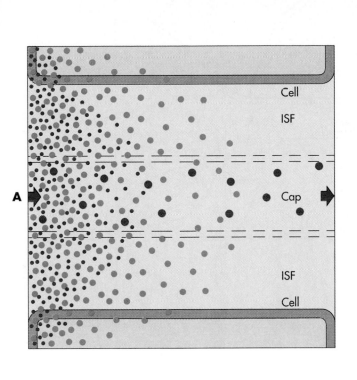

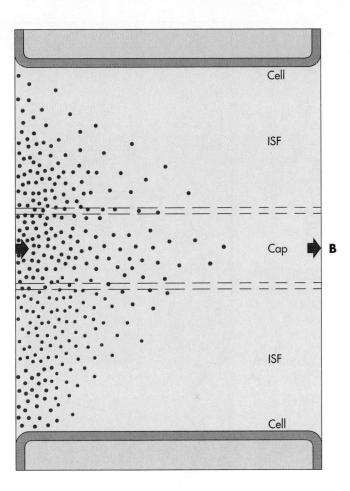

FIGURE 22-6   Flow- and diffusion-limited transport from capillaries *(Cap)* to tissue. **A,** Flow-limited transport. The smallest water-soluble inert tracer particles *(red dots)* reach negligible concentrations after passing only a short distance down the capillary. Larger particles *(blue dots)* with similar properties travel farther along the capillary before reaching insignificant intracapillary concentrations. Both substances cross the interstitial fluid *(ISF)* and reach the parenchymal tissue. Because of their size, more of the smaller particles are taken up by the tissue cells. The largest particles *(purple dots)* cannot penetrate the capillary pores and thus do not escape from the capillary lumen except by pinocytotic vesicle transport. An increase in the volume of blood flow or an increase in capillary density will increase tissue supply for the diffusible solutes. Note that capillary permeability is greater at the venous end of the capillary (also in the venule, not shown) because of the larger number of pores in this region. **B,** Diffusion-limited transport. When the distance between the capillaries and the parenchymal tissue is large, as a result of edema or low capillary density, diffusion becomes a limiting factor in the transport of solutes from capillary to tissue, even at high rates of capillary blood flow.

substance through the capillary membrane. The reflection coefficient of water is 0 and that of albumin (to which the endothelium is almost impermeable) is 1. Filterable solutes have reflection coefficients between 0 and 1.

Of the plasma proteins, albumin preponderates in determining oncotic pressure. The average albumin molecule (molecular weight 69,000) is approximately half the size of the average globulin molecule (MW 150,000), and it is present in a greater concentration than the globulins (4.0 versus 3.0 g/dl of plasma). Albumin also exerts a greater osmotic force than can be accounted for solely on the basis of the number of molecules dissolved in the plasma. Therefore albumin cannot be completely replaced by inert substances of the same molecular size, (e.g., dextran). This additional osmotic force becomes disproportionately greater at high concentrations of albumin (as in plasma), and it is

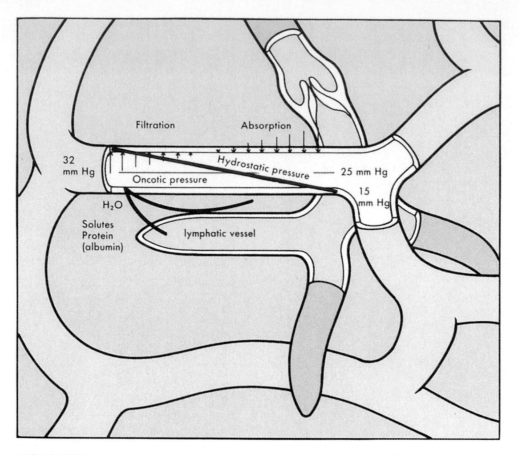

FIGURE 22-7   Schematic representation of the factors responsible for filtration and absorption across the capillary wall and the formation of lymph.

weak to absent in dilute solutions of albumin (as in interstitial fluid).

One reason for this behavior of albumin is its negative charge at the normal blood pH and the attraction and retention of cations (principally Na^+) in the vascular compartment (the Gibbs-Donnan effect, Chapter 2). Furthermore, albumin binds a small number of chloride ions, which increases its negative charge and thus its ability to retain more sodium ions inside the capillaries. The small increase in electrolyte concentration of the plasma over that of the interstitial fluid produced by the negatively charged albumin enhances its osmotic force to that of an ideal solution containing the same concentration of a solute with a molecular weight of 37,000.

Small amounts of albumin escape from the capillaries and enter the interstitial fluid. The albumin exerts a very small osmotic force (0.1 to 5 mm Hg), because the concentration of albumin is low in the interstitial fluid, and because at low albumin concentrations, the osmotic force of albumin simply becomes a function of the number of albumin molecules per unit volume of interstitial fluid; the additional osmotic force caused by albumin's negative charge is absent.

Balance of Hydrostatic and Osmotic Forces—The Starling Hypothesis   The relationship between hydrostatic pressure and oncotic pressure, and the role of these pressures in regulating fluid passage across the capillary endothelium, were expounded by Starling in 1896. The Starling hypothesis is expressed by the equation

$$Q_f = k[(P_c + \pi_i) - (P_i + \pi_p)] \qquad \textbf{(2)}$$

Where,

Q_f = Fluid movement
P_c = Capillary hydrostatic pressure
P_i = Interstitial fluid hydrostatic pressure
π_p = Plasma oncotic pressure
π_i = Interstitial fluid oncotic pressure
k = Filtration constant for the capillary membrane

Net filtration occurs when the algebraic sum is positive, and net absorption occurs when the sum is negative.

Classically, filtration has been thought to occur at the arterial end of the capillary and absorption at its venous end because of the gradient of hydrostatic pressure along the capillary. This is true for the idealized capillary, as depicted in Figure 21-5. However, direct observations have revealed that many capillaries filter along their entire length, whereas others only absorb. In some vascular beds (e.g., the renal glomerulus) hydrostatic pressure in the capillary is high enough to result in filtration along the entire length of the capillary. In other vascular beds (e.g., in the intestinal mucosa) the hydrostatic and oncotic forces are such that absorption occurs along the whole capillary.

In the normal steady state, arterial pressure, venous pressure, postcapillary resistance, interstitial fluid hydrostatic and oncotic pressures, and plasma oncotic pressure remain relatively constant, and changes in precapillary resistance determine the movement of fluid across the capillary wall. Because water moves so quickly across the capillary endothelium, the hydrostatic and osmotic forces are nearly in equilibrium along the entire capillary. Thus filtration and absorption normally occur with very small imbalances of pressure across the capillary wall. Only about 2% of the plasma flowing through the vascular system is filtered, and of this about 85% is absorbed in the capillaries and venules. The remainder returns to the vascular system in the lymph, along with the albumin that escapes from the capillaries.

In the lungs the mean capillary hydrostatic pressure is only about 8 mm Hg. Because the plasma oncotic pressure is 25 mm Hg and interstitial fluid oncotic pressure is approximately 15 mm Hg, the net force slightly favors reabsorption. Pulmonary lymph is formed, however, and it consists of fluid that is osmotically drawn out of the capillaries by the small amount of plasma protein that escapes through the capillary endothelium.

In pathological conditions, such as **left ventricular failure** or **mitral valve stenosis,** pulmonary capillary hydrostatic pressure may exceed plasma oncotic pressure. When this occurs, it may produce **pulmonary edema,** a condition that can seriously interfere with gas exchange in the lungs.

Capillary Filtration Coefficient The rate of fluid movement (Q_f) across the capillary membrane depends not only on the algebraic sum of the hydrostatic and osmotic pressures (ΔP) across the endothelium, but also on the area (A_m) of the capillary wall available for filtration, the distance (Δx) across the capillary wall (i.e., the thickness), the viscosity (η) of the filtrate, and the filtration constant (k) of the membrane. The equation is:

$$Q_f = \frac{kA_m \Delta P}{\eta \Delta x} \tag{3}$$

Because the thickness and area of the capillary wall and the viscosity of the filtrate are relatively constant for a given preparation, they can all be included with k in a total filtration constant, k_t, which is expressed per unit weight of tissue. Hence the equation can be simplified to

$$Q_f = k_t \Delta P \tag{4}$$

where k_t is the total capillary filtration coefficient, and the units for Q_f are milliliters per minute per 100 g of tissue per millimeter of mercury pressure difference.

In any given tissue the filtration coefficient per unit area of capillary surface, and thus capillary permeability, is not changed by certain physiological conditions, such as arteriolar dilation and capillary distension, nor by such adverse conditions as hypoxia, hypercapnia, or acidosis.

Capillary injury (toxins, severe burns) increases capillary permeability greatly (as indicated by an increased filtration coefficient), and significant amounts of fluid and protein leak out of the capillaries into the interstitial space. One of the important therapeutic measures used in the treatment of extensive **burns** is replacement of lost fluid and plasma proteins.

Disturbances in Hydrostatic-Osmotic Balance
Modest changes in arterial pressure per se may have little effect on filtration, because the change in pressure may be countered by adjustments of the precapillary resistance vessels (autoregulation, Chapter 23).

However, in a condition such as **hemorrhage,** in which arterial and venous pressures are severely reduced, capillary hydrostatic pressure falls. Furthermore, the low arterial blood pressure in hemorrhage decreases the blood flow (and thus O_2 supply) to the tissues, and therefore vasodilator metabolites accumulate and induce relaxation of arterioles. The reduced transmural pressure also induces precapillary vessel relaxation. As a consequence of these several factors, absorption predominates over filtration and constitutes one of the body's compensatory mechanisms to restore blood volume (see also Chapter 26).

An increase in venous pressure, as occurs in the feet when one changes from the lying to the standing position, would elevate capillary pressure and enhance filtration. However, the increase in transmural pressure causes precapillary vessel closure (myogenic mechanism, Chapter 23), and therefore the capillary filtration coefficient actually decreases. This reduction in capillary surface available for filtration protects against the extravasation of large amounts of fluid into the interstitial space.

However, the elevation of venous pressure (e.g., in **pregnancy** or **congestive heart failure,** combined with standing), enhances filtration beyond the capacity of the lymphatic system in the legs to remove the capillary filtrate from the interstitial space. **Edema** of the ankles and lower legs results.

The protein concentration in plasma may also change in pathological states and thus may alter the osmotic force and movement of fluid across the capillary membrane.

The plasma protein concentration is increased in **dehydration** (e.g., from water deprivation, prolonged sweating, severe vomiting, or diarrhea), and water moves by osmotic forces from the tissues to the vascular compartment. In contrast, the plasma protein concentration is reduced in **nephrosis** (a renal disease in which protein is lost in the urine), and edema may occur. When capillaries are injured, as in burns, plasma protein escapes into the interstitial space along with fluid and increases the oncotic pressure of the interstitial fluid. This greater osmotic force outside the capillaries causes additional fluid loss from the vascular system and may lead to severe intravascular **hypovolemia.**

Pinocytosis

Some transfer of substances across the capillary wall can occur in tiny vesicles; this process is called **pinocytosis.** The pinocytotic vesicles, formed by pinching off a section of the surface membrane, can take up substances on one side of the capillary endothelial cell, move by thermal kinetic energy across the cell, and deposit their contents at the other side of the endothelial cell (see Figure 22-4). The amount of material that can be transported in this way is much less than that moved by diffusion. However, pinocytosis may move large lipid-insoluble molecules (30 nm) between blood and interstitial fluid. The number of pinocytotic vesicles in the endothelium varies with the tissue

(muscle>lung>brain) and increases from the arterial to the venous end of the capillary.

LYMPHATICS

The terminal lymphatic vessels consist of a widely distributed closed-end network of highly permeable lymph capillaries that resemble blood capillaries in appearance. However, they generally lack tight junctions between endothelial cells, and they possess fine filaments that anchor them to the surrounding connective tissue. During skeletal muscle contraction, these fine strands may distort the lymphatic vessel and open spaces between the endothelial cells. This will permit protein, large particles, and cells in the interstitial fluid to enter the lymphatic capillaries.

The blood capillary filtrate and the protein and cells that have passed from the intravascular compartment to the interstitial fluid compartment are returned to the circulation by virtue of tissue pressure, facilitated by intermittent skeletal muscle contractions, contractions of the lymphatic vessels, and an extensive system of one-way valves. The lymph flows through thin-walled vessels of progressively larger diameter and finally enters the right and left subclavian veins at their junctions with the respective internal jugular veins. Only cartilage, bone, epithelium, and tissues of the central nervous system are devoid of lymphatic vessels.

The volume of fluid that flows through the lymphatic system in 24 hours is about equal to an animal's total plasma volume. The protein returned by the lymphatics to the blood in a day is about one fourth to one half of the circulating plasma proteins. Lymphatic return is the only means whereby protein (mainly albumin) that leaves the vascular compartment can be returned to the blood, because back diffusion into the capillaries is negligible against the large albumin concentration gradient. If the protein was not removed from the interstitial spaces by the lymph vessels, it would accumulate in the interstitial fluid and act as an oncotic force to draw fluid from the blood capillaries and produce edema.

In addition to returning fluid and protein to the vascular bed, the lymphatic system filters the lymph at the lymph nodes and removes foreign particles such as bacteria. Thus, the lymphatic system is an important component of the body's defense against bacterial invasion.

The largest lymphatic vessel, the thoracic duct, drains the lower extremities, returns protein lost through the permeable liver capillaries, and carries substances (principally fat in the form of chylomicrons) that are absorbed from the gastrointestinal tract to the circulating blood.

Lymph flow varies considerably; it is almost nil in resting skeletal muscle but increases during exercise in proportion to the degree of muscular activity. Lymph flow is increased by any mechanism that enhances the rate of blood capillary filtration; such mechanisms include increased capillary pressure, increased capillary permeability, and decreased plasma oncotic pressure.

> If the volume of interstitial fluid exceeds the drainage capacity of the lymphatics, or if the lymphatic vessels become blocked as in **elephantiasis** (caused by **filariasis,** a worm infestation), interstitial fluid accumulates (edema), chiefly in the more compliant tissues (e.g., subcutaneous tissue).

SUMMARY

1. Blood flow through the capillaries is chiefly regulated by contraction and relaxation of the arterioles (resistance vessels).
2. The capillaries, which consist of a single layer of endothelial cells, can withstand high transmural pressure by virtue of their small diameter. According to the law of Laplace, T (wall tension) = P (transmural pressure) $\times$ r (radius of capillary).
3. The endothelium is the source of endothelium-derived relaxing factor (EDRF, shown to be nitric oxide) and prostacyclin, which relax vascular smooth muscles.
4. Movement of water and small solutes between the vascular and interstitial fluid compartments occurs through capillary pores mainly by diffusion, but also by filtration and absorption.
5. Because the rate of diffusion is about 40 times greater than the blood flow in the tissue, exchange of small lipid-insoluble molecules is flow limited. The larger the molecules the slower the diffusion, until the lipid insoluble molecules become diffusion limited. Molecules larger than about 60,000 MW are essentially confined to the vascular compartment.
6. Lipid-soluble substances, such as carbon dioxide and oxygen, pass directly through the lipid membranes of the capillary endothelial cells, and the ease of transfer is directly proportional to the degree of lipid solubility of the substance.
7. Capillary filtration and absorption are described by the Starling equation: $Q_f = k[(P_c + \pi_i) - (P_i + \pi_p)]$, where Q_f = fluid movement; P_c = capillary hydrostatic pressure; P_i = interstitial fluid hydrostatic pressure; π_i = interstitial fluid oncotic pressure; and π_p = plasma protein oncotic pressure. Net filtration occurs when the algebraic sum is positive, and net absorption occurs when it is negative.
8. By a process called pinocytosis, large molecules can move across the capillary wall in vesicles formed from the lipid membrane of the capillaries.
9. Fluid and protein that have escaped from the blood capillaries enter the lymphatic capillaries and are transported via the lymphatic system back to the blood vascular compartment.

BIBLIOGRAPHY
Journal Articles

Aukland K: Why don't our feet swell in the upright position? *News Physiol Sci* 9:214, 1994.

Aukland K, Reed RK: Interstitial-lymphatic mechanisms in the control of extracellular fluid volume, *Physiol Rev* 73:1, 1993.

Curry FRE: Regulation of water and solute exchange in microvessel endothelium: studies in single perfused capillaries, *Microcirculation* 1:11, 1994.

Duling BR, Klitzman B: Local control of microvascular function: role in tissue oxygen supply, *Ann Rev Physiol* 42:373, 1980.

Feng Q, Hedner T: Endothelium-derived relaxing factor (EDRF) and nitric oxide. II. Physiology, pharmacology, and pathophysiological implications, *Clin Physiol* 10:503, 1990.

Furchgott RF, Vanhoutte PM: Endothelium-derived relaxing and contracting factors, *FASEB J* 3:2007, 1989.

Krogh A: The number and distribution of capillaries in muscles with calculation of the oxygen pressure head necessary for supplying the tissue, *J Physiol* 52:409, 1919.

Lewis DH, ed: Symposium on lymph circulation, *Acta Physiol Scand [Suppl]* 463:9, 1979.

Pries AR, Secomb TW, Gessner T, Sperandio MB, Gross JF, Gaehtgens P: Resistance to blood flow in microvessels in vivo, *Circ Res* 75:904, 1994.

Rippe B, Haraldsson B: Transport of macromolecules across microvascular walls: the two-pore theory, *Physiol Rev* 74:163, 1994.

Rosell S: Neuronal control of microvessels, *Ann Rev Physiol* 42:359, 1980.

Starling EH: On the absorption of fluids from the connective tissue spaces, *J Physiol* 19:312, 1896.

Books and Monographs

Bert JL, Pearce RH: The interstitium and microvascular exchange. In *Handbook of physiology,* section 2: The cardiovascular system—microcirculation, vol IV, Bethesda, Md, 1984, American Physiological Society.

Crone C, Levitt DG: Capillary permeability to small solutes. In *Handbook of physiology,* section 2: the cardiovascular system—microcirculation, vol IV, Bethesda, Md, 1984, American Physiological Society.

Hudlicka O: Development of microcirculation: capillary growth and adaptation. In *Handbook of physiology,* section 2: the cardiovascular system—microcirculation, vol IV, Bethesda, Md, 1984, American Physiological Society.

Krogh A: *The anatomy and physiology of capillaries,* New York, 1959, Hafner Co.

Luscher TF, Vanhoutte PM: *The endothelium: modulator of cardiovascular function,* Boca Raton, Fla, 1990, CRC Press.

Michel CC: Fluid movements through capillary walls. In *Handbook of physiology,* section 2: the cardiovascular system—microcirculation, vol IV, Bethesda, Md, 1984, American Physiological Society.

Mortillaro NA: *Physiology and pharmacology of the microcirculation,* vol 1, New York, 1983, Academic Press, Inc.

Renkin EM: Control of microcirculation and blood-tissue exchange. In *Handbook of physiology,* section 2: the cardiovascular system—microcirculation, vol IV, Bethesda, Md, 1984, American Physiological Society.

Shepro D, D'Amore PA: Physiology and biochemistry of the vascular wall endothelium. In *Handbook of physiology,* section 2: the cardiovascular system—microcirculation, vol IV, Bethesda, Md, 1984, American Physiological Society.

Simionescu M, Simionescu N: Ultrastructure of the microvascular wall: functional correlations. In *Handbook of physiology,* section 2: the cardiovascular system—microcirculation, vol IV, Bethesda, Md, 1984, American Physiological Society.

Taylor AE, Granger DN: Exchange of macromolecules across the microcirculation. In *Handbook of physiology,* section 2: the cardiovascular system—microcirculation, vol IV, Bethesda, Md, 1984, American Physiological Society.

Wiedeman MP: Architecture. In *Handbook of physiology,* section 2: the cardiovascular system—microcirculation, vol IV, Bethesda, Md, 1984, American Physiological Society.

Zweifach BW, Lipowsky HH: Pressure-flow relations in blood and lymph microcirculation. In *Handbook of physiology,* section 2: The cardiovascular system—microcirculation, vol IV, Bethesda, Md, 1984, American Physiological Society.

CHAPTER 23

The Peripheral Circulation and Its Control

The peripheral circulation is essentially under dual control: centrally by the nervous system, and locally in the tissues by the conditions in the immediate vicinity of the blood vessels. The relative importance of these two control mechanisms is not the same in all tissues. In some areas of the body, such as the skin and the splanchnic regions, neural regulation of blood flow predominates, whereas in others, such as the heart and brain, local factors are dominant.

The vessels that regulate the blood flow throughout the body are called the **resistance vessels** (very small arteries and arterioles). These vessels offer the greatest resistance to flow of blood pumped to the tissues by the heart and thereby are important in the maintenance of arterial blood pressure. Smooth muscle fibers are the main component of the walls of the resistance vessels (see Figure 16-1). Therefore the vessel lumen can be varied from complete obliteration, by strong contraction of the smooth muscle, with infolding of the endothelial lining, to maximal dilation, by full relaxation of the smooth muscle. Some resistance vessels are closed at any given time by partial contraction (or **tone)** of the arteriolar smooth muscle. If all the resistance vessels in the body dilated simultaneously, blood pressure would fall precipitously.

VASCULAR SMOOTH MUSCLE

Vascular smooth muscle is responsible for the control of total peripheral resistance, arterial and venous tone, and the distribution of blood flow throughout the body. The smooth muscle cells are small, mononucleate, and spindle shaped. They are generally arranged in several helical or circular layers around the larger blood vessels and in a single circular layer (many cells thick) around the arterioles (see Chapter 14).

INTRINSIC OR LOCAL CONTROL OF PERIPHERAL BLOOD FLOW

Autoregulation and the Myogenic Mechanism

In certain tissues the blood flow is adjusted to the existing metabolic activity of that tissue. Furthermore, at constant levels of tissue metabolism, imposed changes in the perfusion pressure are met with vascular resistance changes that maintain a constant blood flow. This mechanism is commonly referred to as **autoregulation of blood flow** and is illustrated graphically in Figure 23-1. In the skeletal muscle preparation from which these data were gathered, the muscle was completely isolated from the rest of the animal and was in a resting state. The pressure was abruptly increased or decreased from a control pressure of 100 mm Hg. The blood flows observed immediately after the changing of the perfusion pressure are represented by the black curve. Maintenance of the pressure at each new level was followed within 60 seconds by a return of flow to or toward the control level; the red curve represents these steady-state flows. Over the pressure range of 20 to 120 mm Hg, the steady-state flow is relatively constant. Calculation of resistance (pressure/flow) across the vascular bed during steady-state conditions indicates that with elevation of perfusion pressure, the resistance vessels constricted, whereas with reduction of perfusion pressure, they dilated.

The mechanism that appears to be responsible for this constancy of blood flow in the presence of an altered perfusion pressure, is called the **myogenic mechanism.** *According to the myogenic mechanism the vascular smooth muscle contracts in response to stretch and relaxes with a reduction in stretch.* An abrupt increase in perfusion pressure initially distends the bloodvessels. This passive vascular distension is followed by

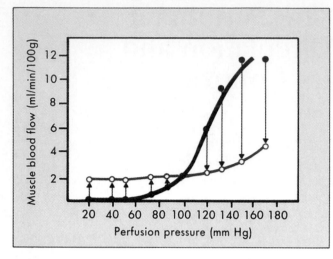

FIGURE 23-1 Pressure-flow relationship in the skeletal muscle vascular bed of the dog. The closed circles represent the flows obtained immediately after abrupt changes in perfusion pressure from the control level (point where lines cross). The open circles represent the steady-state flows obtained at the new perfusion pressure. (Redrawn from Jones RD, Berne RM: *Circ Res* 14:126, 1964.)

contraction of the smooth muscles of the resistance vessels and a return of flow to the previous control level.

Because blood pressure is reflexly maintained at a fairly constant level under normal conditions, we would expect the operation of a myogenic mechanism to be minimized. However, when a person changes from a lying to a standing position, a large increase in transmural pressure occurs in the vessels of the lower extremities. The precapillary vessels constrict in response to this imposed stretch. The constriction diminishes capillary filtration until the increase in plasma oncotic pressure and the increase in interstitial fluid pressure balance the elevated capillary hydrostatic pressure associated with the vertical position.

If arteriolar resistance did not increase with standing, the hydrostatic pressure in the lower parts of the legs would reach such high levels that large volumes of fluid would pass from the capillaries into the interstitial fluid compartment and produce **edema.** In patients with elevated venous pressure, as in **heart failure,** the additional hydrostatic pressure in the standing position produces edema of the feet, ankles, and lower legs.

Endothelium-Mediated Regulation

In isolated coronary arterioles perfused at constant transmural pressure, high velocity of blood flow elicits vasodilation. The high flow velocity is achieved by increasing the perfusion pressure. Constant transmural pressure is achieved by increasing extravascular pressure the same amount as the increase in perfusion (intravascular) pressure. The vasodilation is caused by **endothelium-derived relaxing factor (nitric oxide),** which is released by the endothelial cells in response to the shear stress that the rapid flow exerts on the vascular endothelium. Removal of the endothelium from the arterioles abolishes this dilator response to enhanced flow velocity.

Metabolic Regulation

According to the metabolic hypothesis, *the blood flow is governed by the metabolic activity of the tissue. Any intervention that results in an inadequate oxygen supply for the tissue requirements releases vasodilator metabolites from the tissue.* When the metabolic rate of the tissue increases or the oxygen delivery to the tissue decreases, more vasodilator substance is formed and blood flow increases. If perfusion pressure is constant, a decrease in metabolic activity will decrease the concentration of the vasodilator in the tissue, and the change in concentration will increase precapillary resistance. Similarly, if metabolic activity is constant, an increase in perfusion pressure, and consequently in blood flow, will decrease the tissue concentration of the vasodilator agent (metabolite washout) and increase precapillary resistance. An attractive feature of the metabolic hypothesis is that in most tissues, blood flow closely parallels metabolic activity. Thus, although blood pressure is kept fairly constant, metabolic activity and blood flow in the different tissues vary together under physiological conditions.

Many substances have been proposed as mediators of metabolic vasodilation (lactic acid, carbon dioxide, hydrogen ions, potassium ions, inorganic phosphate ions, interstitial fluid osmolarity, adenosine, and nitric oxide). However, none of these agents alone fulfills all of the criteria for a physiological vasodilator in skeletal muscle.

Metabolic control of vascular resistance by the release of a vasodilator is predicated on the existence of **basal tone,** which is the partial contraction (tonic activity) of vascular smooth muscle. In contrast to tone in skeletal muscle, basal tone in vascular smooth muscle is independent of the nervous system, and the factor responsible is not known. It could be myogenic and/or a vasoconstrictor substance in the blood.

If arterial inflow to a vascular bed is stopped for a few seconds to several minutes, the blood flow immediately after release of the occlusion (**reactive hyperemia**) exceeds the flow before occlusion, and it returns only gradually to the control level. This is illustrated in Figure 23-2, where blood flow to the leg was stopped by

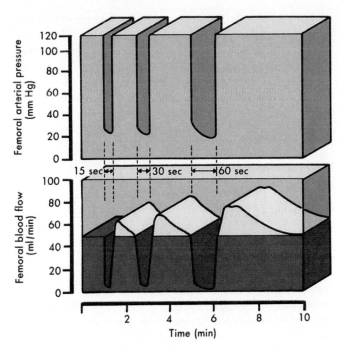

FIGURE 23-2 Reactive hyperemia in the hind limb of the dog after 1-, 30-, and 60-second occlusions of the femoral artery. (Berne, R.M. Unpublished observations.)

clamping the femoral artery for 15, 30, and 60 seconds. Release of the 60-second occlusion resulted in a peak blood flow 70% greater than the control flow, with a return to control flow within about 110 seconds.

When this same experiment is done in humans by inflating a blood pressure cuff on the upper arm, dilation of the resistance vessels of the hand and forearm, immediately after release of the cuff, is evident from the bright red color of the skin and the fullness of the veins. Within limits the peak flow, and particularly the duration of the reactive hyperemia, are proportional to the duration of the occlusion (Figure 23-2). If the arm is exercised during the occlusion period, reactive hyperemia is increased. These observations and the close relationship that exists between metabolic activity and blood flow in the unoccluded limb are consonant with metabolic regulation of tissue blood flow.

EXTRINSIC CONTROL OF PERIPHERAL BLOOD FLOW

Neural Sympathetic Vasoconstriction

Several regions in the medulla oblongata influence cardiovascular activity (Chapter 19). Some of the effects of stimulation of the dorsal lateral medulla are vasoconstriction, cardiac acceleration, and enhanced myocardial contractility. Caudal and ventromedial to the pressor region is

a zone that decreases blood pressure on stimulation. This **depressor area** exerts its effect by direct inhibition of spinal neurons and by inhibition of the medullary **pressor region.** These areas do not constitute a center in an anatomical sense because no discrete group of cells is discernible. However, they do constitute a center in a physiological sense, in that stimulation of the pressor region produces the responses mentioned previously.

From the vasoconstrictor regions, fibers descend in the spinal cord and synapse at different levels of the thoracolumbar region (T1 to L2 or L3). Fibers from the intermediolateral gray matter of the cord emerge with the ventral roots, but leave the motor fibers to join the paravertebral sympathetic chains through the white communicating branches (see Chapter 10). These preganglionic white (myelinated) fibers may pass up or down the sympathetic chains to synapse in the various ganglia within the chains or in certain outlying ganglia. Postganglionic gray branches (unmyelinated) then join the corresponding segmental spinal nerves and accompany them to the periphery to innervate the arteries and veins. Postganglionic sympathetic fibers from the various ganglia join the large arteries and accompany them as an investing network of fibers to the resistance (arterioles) and capacitance (veins) vessels.

The vasoconstrictor regions are tonically active. Reflexes or humoral stimuli that enhance this activity increase the frequency of impulses reaching the terminal branches of the vessels. A constrictor neurohumor (norepinephrine) is released from the postganglionic nerve endings and elicits constriction (α-adrenergic effect) of the resistance vessels. Inhibition of the vasoconstrictor areas diminishes the frequency of impulses in the efferent fibers, which results in vasodilation. In this manner *neural regulation of the peripheral circulation is accomplished primarily by altering the number of impulses passing down the vasoconstrictor fibers of the sympathetic nerves to the blood vessels.* The tonic activity of the vasomotor regions may vary rhythmically, which is manifested as oscillations of arterial pressure. Some oscillations occur at the frequency of respiration **(Traube-Hering waves)** and are caused by an increase in sympathetic impulses to the resistance vessels coincident with inspiration. Other oscillations **(Mayer waves)** occur at a lower frequency than respiration.

Sympathetic Constrictor Influence on Resistance and Capacitance Vessels

The vasoconstrictor fibers of the sympathetic nervous system supply the arteries, arterioles, and veins, but the neural influence on the large vessels is far less important functionally than it is on the microcirculation. Capacitance vessels (veins) are more responsive to sympathetic nerve stimulation than are resistance vessels; they are maximally constricted at a lower frequency of stimulation than are the resistance vessels. However, capacitance vessels do not respond to vasodilator metabolites. Norepinephrine is the neurotransmitter released at the sympathetic nerve terminals at the blood vessels. Many factors, such as circulating hormones and particularly locally released substances, modify the liberation of norepinephrine from the nerve terminals.

At basal tone approximately one third of the blood volume of a tissue can be mobilized on stimulation of the sympathetic nerves at physiological frequencies. The basal tone is very low in capacitance vessels, and only small increases in volume are obtained with maximal doses of the potent vasodilator acetylcholine. Therefore, at basal tone the tissue blood volume is close to its maximal value. Blood is mobilized from capacitance vessels in response to physiological stimuli. In exercise, activation of the sympathetic nerve fibers constricts veins and thus augments central venous pressure and hence the cardiac filling pressure.

In **arterial hypotension,** as induced by hemorrhage, the capacitance vessels constrict and thereby aid in overcoming the associated decrease in central venous pressure. In addition, the resistance vessels constrict in **hemorrhagic shock** and thereby assist in the restoration of arterial pressure (see Chapter 26). Furthermore, extravascular fluid is mobilized by a greater reabsorption of fluid from the tissues into the capillaries in response to the lowered capillary hydrostatic pressure caused by the low arterial pressure.

Parasympathetic Neural Influence

The efferent fibers of the cranial division of the parasympathetic nervous system supply blood vessels of the head and viscera, whereas fibers of the sacral division supply blood vessels of the genitalia, bladder, and large bowel. Skeletal muscle and skin do not receive parasympathetic innervation. Because only a small proportion of the resistance vessels of the body receives parasympathetic fibers, the effect of these cholinergic fibers on total vascular resistance is small.

Humoral Factors

Epinephrine and norepinephrine exert a profound effect on the peripheral blood vessels. In skeletal muscle epinephrine in low concentrations dilates resistance vessels (β-adrenergic effect) and in high concentrations constricts them (α-adrenergic effect). In skin, only vasocon-

striction is obtained with epinephrine, whereas in all vascular beds the main effect of norepinephrine is vasoconstriction. When stimulated, the adrenal gland releases mainly epinephrine, but also some norepinephrine, into the systemic circulation (see Chapter 47). Under physiological conditions, however, the effect of catecholamine release from the adrenal medulla is less important than norepinephrine release from the sympathetic nerves.

Vascular reflexes

Areas of the medulla oblongata that mediate sympathetic and vagal effects are under the influence of neural impulses (arising in the baroreceptors, chemoreceptors, hypothalamus, cerebral cortex, and skin) and of local carbon dioxide and oxygen concentrations.

Baroreceptors The **baroreceptors** (or **pressoreceptors**) are stretch receptors located in the **carotid sinuses** (slightly widened areas of the internal carotid arteries at their points of origin from the common carotid arteries) and in the **aortic arch** (Figures 23-3 and 23-4). Impulses arising in the carotid sinus travel up afferent fibers in the **carotid sinus nerve,** which is a branch of the glossopharyngeal nerve. Impulses arising in the baroreceptors of the aortic arch reach the medulla via afferent fibers in the vagus nerves. These fibers from both sets of baroreceptors travel to the nucleus of the **tractus solitarius** (NTS) in the medulla. The NTS is the site of central projection of the chemoreceptors and baroreceptors. Stimulation of the NTS inhibits sympathetic nerve impulses to the peripheral blood vessels and produces vasodilation **(depressor effect),** whereas experimental destruction of the NTS produces vasoconstriction **(pressor effect).**

The baroreceptor nerve terminals in the walls of the carotid sinus and aortic arch respond to the stretch and deformation of the vessel induced by the arterial pressure. The frequency of firing is enhanced by an increase in blood pressure and diminished by a reduction in blood pressure. An increase in impulse frequency inhibits the medullary vasoconstrictor regions and results in peripheral vasodilation and a lowering of blood pressure. Contributing to a lowering of the blood pressure is a bradycardia brought about by stimulation of the vagal nuclei in the medulla. The carotid sinus baroreceptors are more sensitive to pressure change than are the aortic baroreceptors. However, when the changes in blood pressure are pulsatile, the two sets of baroreceptors respond similarly.

The carotid sinus with the sinus nerve intact can be isolated from the rest of the circulation and artificially perfused. Under these conditions, changes in the pressure within the carotid sinus elicit reciprocal changes in the blood pressure of the experimenal animal. The re-

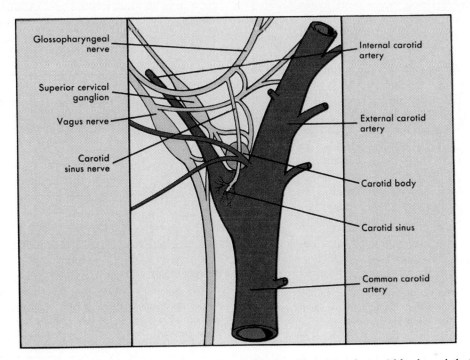

FIGURE 23-3   Diagrammatic representation of the carotid sinus and carotid body and their innervation in the dog. (Redrawn from Adams WE: *The comparative morphology of the carotid body and carotid sinus,* Springfield, Ill, 1958, Charles C Thomas, Publisher.)

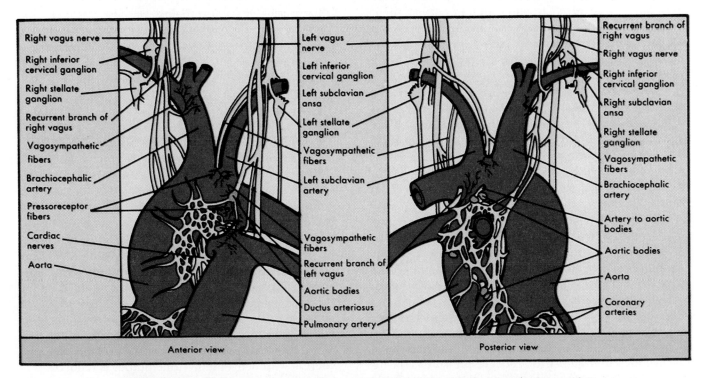
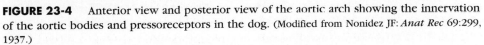

FIGURE 23-4   Anterior view and posterior view of the aortic arch showing the innervation of the aortic bodies and pressoreceptors in the dog. (Modified from Nonidez JF: *Anat Rec* 69:299, 1937.)

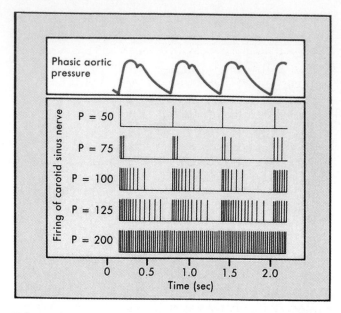

FIGURE 23-5 Relationship of phasic aortic blood pressure to the firing of a single afferent nerve fiber from the carotid sinus at different levels of mean arterial pressure.

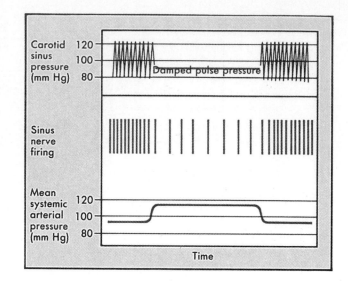

FIGURE 23-6 Effect of reducing pulse pressure in the vascularly isolated perfused carotid sinuses *(top record)* on impulses recorded from a fiber of a sinus nerve *(middle record)* and on mean systemic arterial pressure *(bottom record)*. Mean pressure in the carotid sinuses *(colored line, top record)* is held constant when pulse pressure is damped.

ceptors in the walls of the carotid sinus show some adaptation, and therefore they are more responsive to constantly changing pressures than to sustained constant pressures; this is illustrated in Figure 23-5. At normal levels of blood pressure a barrage of impulses from a single fiber of the carotid sinus nerve is initiated in early systole by the pressure; only a few spikes are observed during late systole and early diastole (Figure 23-5). At lower pressures these phasic changes are even more evident, but the overall frequency of discharge is reduced. The blood pressure threshold for eliciting sinus nerve impulses is about 50 mm Hg, and a maximal level of sustained firing is reached at approximately 200 mm Hg.

Because the baroreceptors do adapt, their response is greater with a large than with a small pulse pressure. This is illustrated in Figure 23-6, which shows the effects of damping pulsations in the carotid sinus on the frequency of firing in a sinus nerve fiber and on the systemic arterial pressure. When the pulse pressure in the carotid sinuses is reduced but mean pressure remains constant, the frequency of neural impulses recorded from a sinus nerve fiber decreases and the systemic arterial pressure increases. Restoration of the pulse pressure in the carotid sinus returns the frequency of sinus nerve discharge and systemic arterial pressure to control levels (Figure 23-6).

The resistance increases that occur in the peripheral vascular beds in response to a reduced pressure in the carotid sinus vary from one vascular bed to another and thereby redistribute blood flow. For example, the resistance changes elicited in the dog by altering carotid sinus pressure are greatest in the femoral vessels, less in the renal vessels, and least in the mesenteric and celiac vessels.

The sensitivity of the carotid sinus reflex can be altered. For example, in **hypertension,** when the carotid sinus becomes stiffer and less deformable as a result of the high intraarterial pressure, baroreceptor sensitivity decreases. In some individuals the carotid sinus is overly sensitive to pressure; tight collars, or other forms of external pressure over the region of the carotid sinus, may elicit **hypotension** and **fainting.**

The baroreceptors play a key role in short-term adjustments of blood pressure, when the changes in blood volume, cardiac output, or peripheral resistance (as in exercise) are relatively abrupt. However, *long-term control of blood pressure (i.e., over days or weeks) is determined mainly by the individual's fluid balance,* namely the balance between fluid intake and fluid output and the resulting blood volume. At constant peripheral resistance, an increase in blood volume increases blood pressure by augmenting cardiac output (see Chapter 24). *By far the single most important organ in the control of body fluid volume, and thus blood pressure, is the kidney.* With overhydration the excess fluid is excreted, whereas with dehydration urine output is reduced.

Cardiopulmonary Baroreceptors In addition to the carotid sinus and aortic baroreceptors, cardiopulmonary receptors also exist; both types of receptors are necessary for the full expression of blood pressure regulation. The cardiopulmonary receptors initiate reflexes via vagal and sympathetic afferent and efferent nerves. These reflexes are tonically active, and they can alter peripheral resistance in response to changes in intracardiac, venous, or pulmonary vascular pressures.

Peripheral Chemoreceptors The *peripheral chemoreceptors* consist of the carotid body (at the bifurcation of the carotid artery), and of several small, highly vascular bodies in the regions of the aortic arch (Figures 23-3 and 23-4). *The chemoreceptors are sensitive to changes in the arterial blood oxygen tension (Pa_{O_2}), carbon dioxide tension (Pa_{CO_2}), and pH.* Although they are mainly concerned with the regulation of respiration (Chapter 31), the peripheral chemoreceptors reflexly influence the circulatory system to a minor degree.

A reduction in Pa_{O_2} stimulates the chemoreceptors. The resultant increase in the frequency of impulses in the afferent nerve fibers from the carotid and aortic bodies stimulates the vasoconstrictor regions; this action increases tone in the resistance and capacitance vessels. The reflex vascular effect induced by increased Pa_{CO_2} and by reduced pH is much less than the direct effect of **hypercapnia** (elevated Pa_{CO_2}) and of hydrogen ions on the vasomotor regions in the medulla. When hypoxia and hypercapnia coexist **(asphyxia),** the stimulation of the chemoreceptors is greater than the sum of the two stimuli when they act alone. Stimulation of the chemoreceptors simultaneously with a reduction of arterial pressure (reduced stimulation of the baroreceptors) enhances the vasoconstrictor response of the peripheral vessels. However, when the baroreceptors and chemoreceptors are both stimulated (e.g., high carotid sinus pressure and low Pa_{O_2}), the cardiovascular effects of the baroreceptors predominate.

> Chemoreceptors with sympathetic afferent fibers exist in the heart. These cardiac chemoreceptors are activated by ischemia, and they transmit the precordial pain **(angina pectoris)** associated with an inadequate blood supply to the myocardium.

Hypothalamus Optimal function of the cardiovascular reflexes requires the integrity of pontine and hypothalamic structures. Furthermore, these structures are responsible for behavioral and emotional control of the cardiovascular system. Stimulation of the anterior hypothalamus decreases blood pressure and heart rate, whereas stimulation of the posterolateral region of the hypothalamus increases blood pressure and heart rate. The hypothalamus also contains a temperature-regulating center that affects the skin vessels. Cooling the skin or the blood perfusing the hypothalamus results in constriction of the skin vessels and heat conservation, whereas warm stimuli have the opposite effects.

Cerebrum The cerebral cortex can also affect the blood flow distribution in the body. Stimulation of the motor and premotor areas can affect blood pressure; usually a pressor response is obtained. However, vasodilation and depressor responses may be evoked (e.g., blushing or fainting) in response to an emotional stimulus.

Skin and Viscera Painful stimuli can elicit either pressor or depressor responses, depending on the magnitude and location of the stimulus. Distension of the viscera often decreases blood pressure, whereas painful stimuli on the body surface usually raise blood pressure.

Pulmonary Reflexes

Inflation of the lungs reflexly dilates systemic resistance vessels and decreases arterial blood pressure. Conversely, collapse of the lungs causes constriction of systemic vessels. Afferent fibers that mediate this reflex are carried in the vagus nerves. Stimulation of the pulmonary stretch receptors inhibits the vasomotor areas. The magnitude of the depressor response to lung inflation is directly related to the degree of inflation and to the existing level of vasoconstrictor tone; the greater the vascular tone, the greater the hypotension produced by lung inflation.

Chemosensitive Regions of the Medulla

Increases in Pa_{CO_2} stimulate the medullary vasoconstrictor regions and thereby increase peripheral resistance. Reduction in Pa_{CO_2} below normal levels (as with hyperventilation) decreases the tonic activity in these areas and thus decreases peripheral resistance. The chemosensitive regions are also affected by changes in pH. A lowering of blood pH stimulates and a rise in blood pH inhibits these areas.

Changes in oxygen tension usually have little direct effect on the vasomotor region in the medulla. The reflex effect of hypoxia is mainly mediated by the carotid and aortic chemoreceptors. Moderate reduction of Pa_{O_2} stimulates the vasomotor region, but severe reduction depresses vasomotor activity, just as very low O_2 tensions depress other areas of the brain.

> **Cerebral ischemia,** as may occur with an expanding **intracranial tumor,** results in severe peripheral vasoconstriction. The stimulation is probably caused by a local accumulation of carbon dioxide and a reduction of oxygen in the brain. With prolonged severe ischemia, extreme depression of cerebral function eventually supervenes and blood pressure falls.

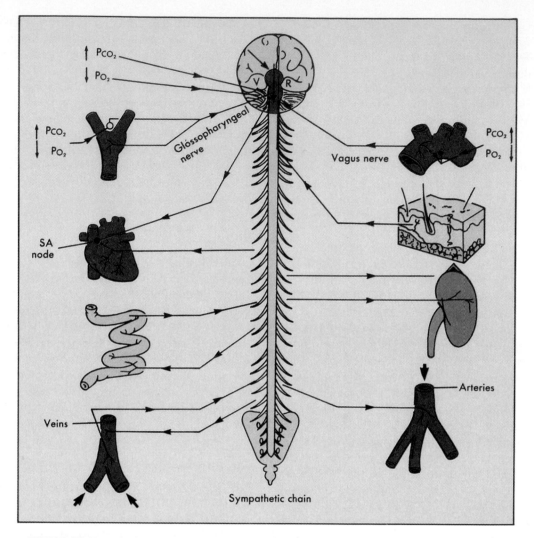

FIGURE 23-7 Schematic diagram illustrating the neural input and output of the vasomotor region *(VR)*. *SA node,* Sinoatrial node.

BALANCE BETWEEN INTRINSIC AND EXTRINSIC FACTORS IN THE REGULATION OF PERIPHERAL BLOOD FLOW

Dual control of the peripheral vessels by—intrinsic and extrinsic mechanisms—constitutes a complex system of vascular regulation. This system enables the body to direct blood flow to areas where the need is greater and to divert it away from areas where the need is less. In some tissues the relative potency of extrinsic and intrinsic mechanisms is constant. However, in other tissues the ratio is changeable, depending on the state of activity of that tissue. In the brain and the heart, which are vital structures with very limited tolerance for a reduced blood supply, intrinsic flow-regulating mechanisms are dominant.

For instance, massive discharge of the medullary vasoconstrictor region (which might occur in severe, acute **hemorrhage**) has negligible effects on the cerebral and cardiac resistance vessels, but it greatly constricts the skin, renal, and splanchnic blood vessels.

In the skin the extrinsic vascular control is dominant. Not only do the cutaneous vessels participate strongly in a general vasoconstrictor discharge, but they also respond selectively through the hypothalamic pathways that subserve body temperature regulation. However, intrinsic control can be demonstrated by local changes of skin temperature that can modify or override the central influence on the resistance and capacitance vessels.

In skeletal muscle the changing balance between extrinsic and intrinsic mechanisms can be clearly seen. In resting skeletal muscle, neural control (vasoconstrictor tone) is dominant. This can be demonstrated by the large increment in blood flow that occurs immediately after the sympathetic nerves to the muscle are cut. Just before and at the start of running, blood flow increases in the leg muscles. After the onset of exercise the intrinsic flow-regulating mechanism assumes control. Because of the local increase in metabolites, vasodilation occurs in the active muscles. Vasoconstriction occurs in the inactive muscles and other tissues as a manifestation of the general sympathetic discharge associated with exercise. However, the constrictor impulses that reach the resistance vessels of the active muscles are overridden by the local metabolic effect, which dilates the vessels. Operation of this dual-control mechanism thus provides more blood where it is required, and shunts it away from the inactive areas.

Similar effects may be achieved by a general increase in Pa_{CO_2}. Normally the hyperventilation associated with exercise keeps Pa_{CO_2} at normal levels. However, if Pa_{CO_2} does increase during exercise, a generalized vasoconstriction occurs because of stimulation of the medullary vasoconstrictor region by CO_2. In the active muscles, where the CO_2 concentration is highest, the smooth muscle of the arterioles relaxes in response to the high P_{CO_2} concentration locally. Factors that affect and that are affected by the medullary vasomotor region are summarized in Figure 23-7.

SUMMARY

1. The arterioles, often referred to as the resistance vessels, are important in the regulation of blood flow through their cognate capillaries. The smooth muscle, which constitutes a major fraction of the wall of the arterioles, contracts and relaxes in response to neural and humoral stimuli.

2. Most tissues show autoregulation of blood flow, a phenomenon characterized by a relatively constant blood flow in the face of a substantial change in perfusion pressure. A logical explanation of autoregulation is the myogenic mechanism, whereby an increase in transmural pressure elicits a direct contractile response of the vascular smooth muscle, whereas a decrease in transmural pressure directly elicits relaxation.

3. The striking parallelism between tissue blood flow and tissue oxygen consumption indicates that blood flow is largely regulated by a metabolic mechanism. A decrease in the oxygen supply/oxygen demand ratio of a tissue releases one or more vasodilator metabolites that dilate arterioles and thereby enhances the oxygen supply.

4. Neural regulation of blood flow is accomplished mainly by the sympathetic nervous system. Sympathetic nerves to blood vessels are tonically active; inhibition of the vasoconstrictor center in the medulla reduces peripheral vascular resistance. Stimulation of the sympathetic nerves constricts resistance vessels and capacitance vessels (veins).

5. In the organs and tissues supplied by the cranial and sacral divisions of the parasympathetic nervous system, the blood vessels are under parasympathetic (as well as sympathetic) control. Parasympathetic activity usually induces vasodilation, but the effect is ordinarily weak.

6. The baroreceptors (pressoreceptors) in the internal carotid arteries and aorta are tonically active and regulate blood pressure on a moment-to-moment basis. Stretch of these receptors by an increase in arterial pressure reflexly inhibits the vasoconstrictor center in the medulla and induces vasodilation, whereas a decrease in arterial pressure disinhibits the vasoconstrictor center and induces vasoconstriction.

7. The carotid baroreceptors predominate over those in the aorta, and both respond more vigorously to pulsatile pressure (stretch) than to steady (nonpulsatile) pressures.

8. Baroreceptors are also present in the cardiac chambers and large pulmonary vessels (cardiopulmonary baroreceptors); they have less influence on blood pressure but participate in blood volume regulation.

9. Peripheral chemoreceptors in the carotid bodies and aortic arch and central chemoreceptors in the medulla oblongata are stimulated by a decrease in blood oxygen tension (Pa_{O_2}) and by an increase in blood carbon dioxide tension (Pa_{CO_2}). Stimulation of these chemoreceptors primarily increases the rate and depth of respiration, but it also produces peripheral vasoconstriction.

10. Peripheral vascular resistance and hence blood pressure can be affected by stimuli arising in the skin, viscera, lungs, and brain.

11. The combined effect of neural and local metabolic factors is to distribute blood to active tissues and divert it from inactive tissues. In vital structures, such as the heart and brain, and in contracting skeletal muscle, the metabolic factors predominate over the neural factors.

BIBLIOGRAPHY
Journal Articles

Belloni FL, Sparks HV: The peripheral circulation, *Ann Rev Physiol* 40:67, 1978.

Berne RM, Knabb RM, Ely SW, Rubio R: Adenosine in the local regulation of blood flow: a brief overview, *Fed Proc* 42:3136, 1983.

Brown AM: Receptors under pressure—an update on baroreceptors, *Circ Res* 46:1, 1980.

Coleridge HM, Coleridge JCG: Cardiovascular afferents involved in regulation of peripheral vessels, *Ann Rev Physiol* 42:413, 1980.

Cowley AW, Jr: Long-term control of blood pressure, *Physiol Rev* 72:231, 1992.

Donald DE, Shepherd JT: Autonomic regulation of the peripheral circulation, *Ann Rev Physiol* 42:429, 1980.

Hainsworth R: Reflexes from the heart, *Physiol Rev* 71:617, 1991.

Hilton SM, Spyer KM: Central nervous regulation of vascular resistance, *Ann Rev Physiol* 42:399, 1980.

Hirst GDS, Edwards FR: Sympathetic neuroeffector transmission in arteries and arterioles, *Physiol Rev* 69:546, 1989.

Kuo L, Davis JJ, Chilian WM: Endothelium-dependent flow-induced dilation of isolated coronary arterioles, *Am J Physiol* 259:H1063, 1990.

Marshall JM: Peripheral chemoreceptors and cardiovascular regulation, *Physiol Rev* 74:543, 1994.

Shen Y-T, Knight DR, Thomas JX, Jr, Vatner SF: Relative roles of cardiac receptors and arterial baroreceptors during hemorrhage in conscious dogs, *Circ Res* 66:397, 1990.

Shepherd JT: Reflex control of arterial blood pressure, *Cardiovasc Res* 16:357, 1982.

Books and Monographs

Abboud FM, Thames MD: Interaction of cardiovascular reflexes in circulatory control. In *Handbook of physiology*, section 2: The cardiovascular system—peripheral circulation and organ blood flow, vol III, Bethesda, Md, 1983, American Physiological Society.

Bishop VS, Malliani A, Thoren P: Cardiac mechanoreceptors. In *Handbook of physiology*, section 2: the cardiovascular system, vol III, Bethesda, Md, 1983, American Physiological Society.

Eyzaguirre C, Fitzgerald RS, Lahiri S, Zapata P: Arterial chemoreceptors. In *Handbook of physiology*, section 2: The cardiovascular system—peripheral circulation and organ blood flow, vol III, Bethesda, Md, 1983, American Physiological Society.

Johnson PC: The myogenic response. In *Handbook of physiology*, section 2: The cardiovascular system—vascular smooth muscle, vol II, Bethesda, Md, 1980, American Physiological Society.

Kovach AGB, Sandos P, Kollii M, eds: *Cardiovascular physiology: neural control mechanisms*, New York, 1981, Academic Press, Inc.

Mancia G, Mark AL: Arterial baroreflexes in humans. In *Handbook of physiology*, section 2: The cardiovascular system—peripheral circulation and organ blood flow, vol III, Bethesda, Md, 1983, American Physiological Society.

Mark, AL, Mancia G: Cardiopulmonary baroreflexes in humans. In *Handbook of physiology*, section 2: The cardiovascular system—peripheral circulation and organ blood flow, vol III, Bethesda, Md, 1983, American Physiological Society.

Mulvany MJ, Strandgaard S, Hammersen F, eds: *Resistance vessels: physiology, pharmacology, and hypertensive pathology*, Basel, Switzerland, 1985, S Karger.

Persson PB, Kirchheim HR, eds: *Baroreceptor reflexes*, Berlin, 1991, Springer-Verlag.

Rothe CF: Venous system: physiology of the capacitance vessels. In *Handbook of physiology*, section 2: The cardiovascular system—peripheral circulation and organ blood flow, vol III, Bethesda, Md., 1983, American Physiological Society.

Sagawa K: Baroreflex control of systemic arterial pressure and vascular bed. In *Handbook of physiology*, section 2: The cardiovascular system—peripheral circulation and organ blood flow, vol III, Bethesda, Md, 1983, American Physiological Society.

Shepherd JT: Cardiac mechanoreceptors. In Fozzard HA et al, eds: *The heart and cardiovascular system: scientific foundations*, ed 2, 1991, Raven Press.

Somlyo AP, Somlyo AV: Smooth muscle structure and function. In Fozzard HA et al eds: *The heart and ardiovascular system: scientific foundations*, ed. 2, New York, 1991, Raven Press.

Sparks HV, Jr: Effect of local metabolic factors on vascular smooth muscle. In *Handbook of physiology*, section 2: The cardiovascular system—vascular smooth muscle, vol II, Bethesda, Md, 1980, American Physiological Society.

Zucker IH, Gilmore JP, eds: *Reflex control of the circulation*, Boca Raton, Fla, 1991, CRC Press.

CHAPTER 24

Control of Cardiac Output: Coupling of the Heart and Blood Vessels

CONTROLLING FACTORS

Four factors control cardiac output: heart rate, myocardial contractility, preload, and afterload (Figure 24-1). Heart rate and myocardial contractility are strictly **cardiac factors.** They are intrinsic characteristics of the cardiac tissues, although they are modulated by various neural and humoral mechanisms. As explained in Chapter 18, the **preload** is the stretching force that acts on cardiac muscle prior to contraction, whereas the **afterload** is the opposing force that acts on cardiac muscle while it shortens. The preload and afterload depend on the characteristics of both the heart and the vascular system. *Not only are preload and afterload important determinants of cardiac output, but they themselves are also determined by the cardiac output.* Hence, preload and afterload may be designated as **coupling factors** (Figure 24-1), because they constitute a functional coupling between the heart and blood vessels.

The heart pumps blood around the vascular system, which is a closed circuit. The rate at which the heart pumps the blood is an important determinant of the preload and afterload. Concomitantly, the vascular characteristics also determine the preload and afterload, and hence these coupling factors regulate the quantity of blood that the heart will pump around the circuit per unit time. To understand the regulation of cardiac output, therefore, it is important to appreciate the nature of the coupling between the heart and the vascular system.

Guyton and his colleagues have developed a graphic technique that we shall use in modified form to analyze the interactions between the cardiac and vascular components of the circulatory system. This analysis involves two independent functional relationships between the cardiac output and the pressure in the right atrium and thoracic venae cavae (that is, the **central venous pressure**).

The curve defining one of these relationships will be called the **cardiac function curve.** It is an expression of the well-known Frank-Starling relationship (see Chapters 18 and 19), and it portrays the dependence of the cardiac output on the preload. The cardiac function curve is a characteristic of the heart itself; classically, it has been studied in hearts completely isolated from the rest of the circulatory system. This curve indicates that a rise in central venous pressure ordinarily increases cardiac output; that is, *cardiac output ordinarily varies directly with central venous pressure.*

The other functional relationship between the central venous pressure and the cardiac output is defined by a second curve, which we shall call the **vascular function curve.** This relationship depends only on certain characteristics of the vascular system, namely, the peripheral resistance, the arterial and venous compliances, and the blood volume. The vascular function curve is entirely independent of the characteristics of the heart; it applies even if the heart were replaced by a mechanical pump. The vascular function curve reflects the fact that if the rate at which the blood is pumped around the body is increased, the central venous pressure decreases; that is, *the central venous pressure varies inversely with the cardiac output.*

Thus, the cardiac function curve indicates that cardiac output ordinarily varies directly with central venous pressure, whereas the vascular function curve indicates that central venous pressure varies inversely with cardiac output. On first exposure to this pair of curves, the student often believes that these assertions are contradictory. The explanations that follow are intended to avert this misconception. We will show that the pair of graphs, one with a direct and one with an inverse relationship between the same two variables, are not at all contradictory. Instead, the combination of a direct and an inverse relationship is a prerequisite for the stable operation of the cardiovascular system.

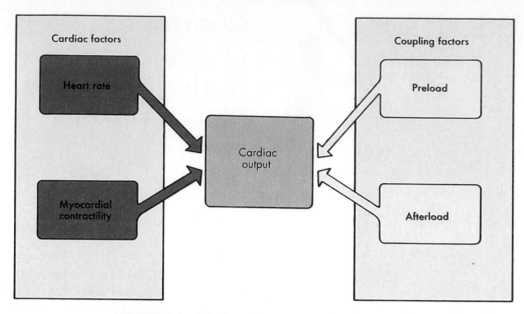

FIGURE 24-1   The four determinants of cardiac output.

We will first examine the vascular function curve, and analyze why central venous pressure varies inversely with the cardiac output. We will then examine the interactions between the cardiac and vascular systems, and we will ascertain that at equilibrium, the stable values of cardiac output and central venous pressure lie at the intersection of the characteristic cardiac and vascular function curves.

VASCULAR FUNCTION CURVE

The vascular function curve defines the change in central venous pressure that is generated by a change in cardiac output; that is, central venous pressure is the dependent variable (or response), and cardiac output is the independent variable (or stimulus).

The simplified schema of the circulation in Figure 24-2 will help elucidate how the cardiac output determines the level of the central venous pressure. The essential components of the cardiovascular system have been lumped into four elements. The right and left sides of the heart and the pulmonary vascular bed are considered simply to be a pump-oxygenator, much as that employed during open heart surgery. In Figure 24-2, the energy source is simply called a pump. The high-resistance microcirculation is designated the peripheral resistance. Finally, the entire compliance of the system is subdivided into two components, the total arterial compliance, C_a, and the total venous compli-

ance, C_v. As defined in Chapter 21, compliance is the increment of volume (ΔV) accommodated per unit change of pressure (ΔP); that is,

$$C = \Delta V/\Delta P \qquad (1)$$

The venous compliance normally is about 20 times as great as the arterial compliance. In the example to follow, the ratio of C_v to C_a will be set at $19:1$ to simplify certain calculations. Thus, if it were necessary to add x ml of blood to the arterial system in order to increase the arterial pressure by 1 mm Hg, then it would be necessary to add $19x$ ml of blood to the venous system to raise venous pressure by the same amount.

Our model (Figure 24-2) illustrates very simply why the central venous pressure varies inversely with the cardiac output. For this example, let us first endow our model with characteristics that resemble those of a normal, resting, adult person (Figure 24-2, A). Let the cardiac output (Q_h) be 5 L/min, the mean arterial pressure (P_a) be 102 mm Hg, and the central venous pressure (P_v) be 2 mm Hg. The total peripheral resistance (R_t) is the ratio of pressure difference ($P_a - P_v$) to flow (Q_r) through the peripheral resistance. At equilibrium, Q_r equals Q_h; that is, the runoff through the peripheral resistance equals the cardiac output. Hence, R_t equals 100/5, or 20 mm Hg/L/min. From heartbeat to heartbeat, the volume (V_a) of blood in the arteries and the volume (V_v) of blood in the veins remain constant, because the volume of blood transferred from the veins to the arteries each

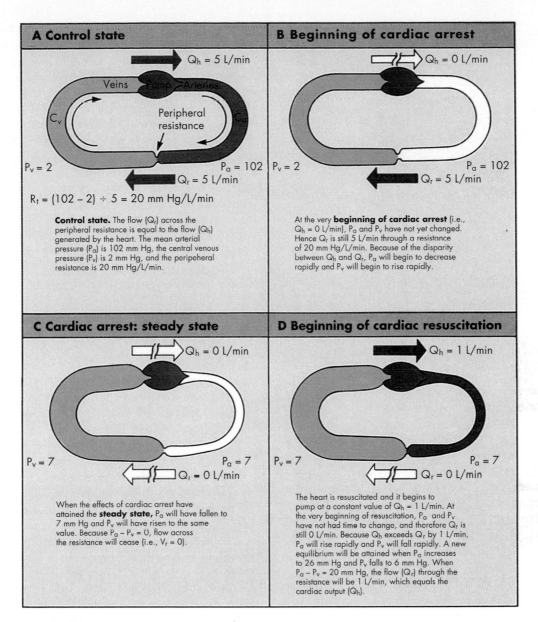

A Control state

$Q_h = 5$ L/min

Veins   Pump   Arteries

Peripheral
resistance

C_v C_a

$P_v = 2$ $P_a = 102$

$Q_r = 5$ L/min

$R_t = (102 - 2) \div 5 = 20$ mm Hg/L/min

Control state. The flow (Q_r) across the
peripheral resistance is equal to the flow (Q_h)
generated by the heart. The mean arterial
pressure (P_a) is 102 mm Hg, the central venous
pressure (P_v) is 2 mm Hg, and the peripheral
resistance is 20 mm Hg/L/min.

B Beginning of cardiac arrest

$Q_h = 0$ L/min

$P_v = 2$ $P_a = 102$

$Q_r = 5$ L/min

At the very **beginning of cardiac arrest** (i.e.,
$Q_h = 0$ L/min), P_a and P_v have not yet changed.
Hence Q_r is still 5 L/min through a resistance
of 20 mm Hg/L/min. Because of the disparity
between Q_h and Q_r, P_a will begin to decrease
rapidly and P_v will begin to rise rapidly.

C Cardiac arrest: steady state

$Q_h = 0$ L/min

$P_v = 7$ $P_a = 7$

$Q_r = 0$ L/min

When the effects of cardiac arrest have
attained the **steady state,** P_a will have fallen to
7 mm Hg and P_v will have risen to the same
value. Because $P_a - P_v = 0$, flow across
the resistance will cease (i.e., $V_r = 0$).

D Beginning of cardiac resuscitation

$Q_h = 1$ L/min

$P_v = 7$ $P_a = 7$

$Q_r = 0$ L/min

The heart is resuscitated and it begins to
pump at a constant value of $Q_h = 1$ L/min. At
the very beginning of resuscitation, P_a and P_v
have not had time to change, and therefore Q_r is
still 0 L/min. Because Q_h exceeds Q_r by 1 L/min,
P_a will rise rapidly and P_v will fall rapidly. A new
equilibrium will be attained when P_a increases
to 26 mm Hg and P_v falls to 6 mm Hg. When
$P_a - P_v = 20$ mm Hg, the flow (Q_r) through the
resistance will be 1 L/min, which equals the
cardiac output (Q_h).

FIGURE 24-2   Simplified cardiovascular system model, consisting of a pump, an arterial compliance (C_a), a peripheral resistance, and a venous compliance (C_v).

minute by the heart (Q_h) equals the volume of blood that flows each minute from the arteries through the resistance vessels and into the veins (Q_r).

The initiation of cardiac arrest in our model (Figure 24-2) will help to illustrate the basis for understanding the reason for the inverse relation between cardiac output and central venous pressure that characterizes the vascular function curves. At the very moment that the heart ceases to beat, the volumes of blood in the arteries (V_a) and veins (V_v) have not had time to change. The

arterial and venous pressures depend on V_a and V_v, respectively. Therefore, these pressures are identical to the respective pressures in Figure 24-2, *A* (i.e., $P_a = 102$ and $P_v = 2$). The arteriovenous pressure gradient of 100 mm Hg will force a flow of 5 L/min through the peripheral resistance of 20 mm Hg/L/min. Thus, although cardiac output (Q_h) now equals 0 L/min, the flow (Q_r) through the microcirculation equals 5 L/min (Figure 24-2, *B*). In other words, *the potential energy stored in the arteries by the previous contractions of the heart causes blood to continue to be transferred from arteries to veins, initially at the normal control rate, even though the heart can no longer transfer blood from the veins back into the arteries.*

Figure 24-2 also illustrates what occurs in a patient with **sudden cardiac arrest.** This dramatic and usually lethal event occurs commonly in people with severe disease of the coronary arteries. The usual basis for the sudden cessation of effective mechanical activity by the heart is the development of **ventricular fibrillation,** a state in which the ventricles are activated by multiple, highly irregular reentrant circuits (see Chapter 17). No coordinated ventricular contractions or relaxations occur, and therefore the heart pumps no blood; i.e., $Q_h = 0$. If **defibrillation** is not instituted by delivering a strong electrical shock to the patient's chest, death will ensue quickly.

After the initiation of cardiac arrest, blood continues to flow from the systemic arteries to the systemic veins as long as the arterial pressure exceeds the venous pressure (Figure 24-2, *B*). Therefore, the blood volume in the arteries progressively decreases, and the blood volume in the veins progressively increases. Because the vessels are elastic structures, the arterial pressure falls progressively and the venous pressure rises progressively. This process will continue until the arterial and venous pressures become equal (Figure 24-2, *C*). Once this condition is reached, the flow (Q_r) from arteries to veins through the resistance vessels will be zero, as is the cardiac output (Q_h).

At zero flow equilibrium (Figure 24-2, *C*), the pressure attained in the arteries and veins depends on the relative compliances of these vessels. Had the arterial (C_a) and venous (C_v) compliances been equal, the decline in P_a would have been equal to the rise in P_v, because the decrement in arterial volume equals the increment in venous volume (principle of conservation of mass). P_a and P_v would have both attained the average of P_a and P_v in panels A and B; that is, $P_a = P_v = (102 + 2)/2 = 52$ mm Hg.

However, the veins are much more compliant than the arteries; the ratio is approximately equal to the ratio ($C_v : C_a = 19$) that we have assumed for our model. Hence, the transfer of blood from arteries to veins at equilibrium would induce a fall in arterial pressure 19 times as great as the concomitant rise in venous pressure. As Figure 24-2, *C*, shows, P_v would increase by 5 mm Hg (to 7 mm Hg), whereas P_a would fall by $19 \times 5 = 95$ mm Hg (to 7 mm Hg). This equilibrium pressure that exists in the absence of flow is referred to as the **mean circulatory pressure** (or the **static pressure**). The pressure in the static system reflects the total volume of blood in the system and the overall compliance of the system.

The algebraic solution of the above problem was obtained by solving the following set of simultaneous equations:

$$\Delta V_v = -\Delta V_a \tag{2}$$

which states that the volume of blood gained by the veins equals that lost from the arteries,

$$C_v = \Delta V_v / \Delta P_v \tag{3}$$

$$C_a = \Delta V_a / \Delta P_a \tag{4}$$

which are the definitions of the venous and arterial compliances, and

$$C_v = 19 \, C_a \tag{5}$$

which is the arbitrary system characteristic that we assigned above.

Two important points on the vascular function curve have already been derived, as shown in Figure 24-3. One point (*A*) represents the normal status (depicted in Figure 24-2, *A*). Under the control conditions, when cardiac output was 5 L/min, the central venous pressure was 2

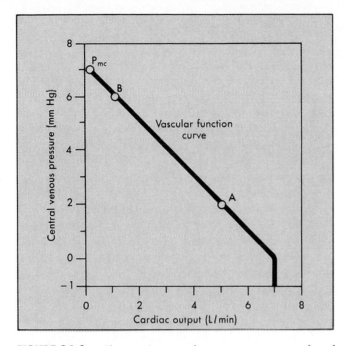

FIGURE 24-3   Changes in central venous pressure produced by changes in cardiac output. P_{mc} is the mean circulatory pressure (or static pressure), which is the equilibrium pressure throughout the cardiovascular system when cardiac output is 0. Points B and A represent the values of venous pressure at cardiac outputs of 1 and 5 L/min, respectively.

mm Hg. Then, when flow stopped (cardiac output was zero), the central venous pressure became 7 mm Hg at equilibrium; this pressure is the mean circulatory pressure, P_{mc}. This point is the P_v axis intercept.

The inverse relation between central venous pressure and cardiac output simply expresses the fact that when cardiac output is suddenly decreased, the rate (Q_r) at which blood flows from arteries to veins through the capillaries is temporarily greater than the rate (Q_h) at which the heart pumps it from the veins back into the arteries. During that transient period, a net volume of blood is translocated from arteries to veins, and hence arterial pressure falls and venous pressure rises.

In many patients who have experienced an episode of sudden cardiac arrest, **cardiac resuscitation** by the delivery of a strong electrical current to the chest wall has succeeded in restoring the heartbeat. This intense current depolarizes all of the excitable cells in the heart renders them refractory. Hence, this current will interrupt all of the irregular reentry circuits in the ventricles of a patient with ventricular fibrillation, which is probably the main cause of sudden cardiac arrest. The ventricle then

may be able to respond with a more normal wave of excitation in response to the next cardiac impulse, which may originate spontaneously in the SA node or in some other automatic focus in the heart. If resuscitation is successful and the ventricles regain the ability to pump blood, the very first ventricular contraction will begin to pump blood from the veins into the arteries. If the heart has been arrested for several minutes, its contractility will be impaired because the myocardium will have been deprived of its blood supply, and therefore the heart will not begin immediately to pump blood at the normal rate.

In the circulation model shown in Figure 24-2, let us assume that when the heart has been resuscitated after a period of cardiac arrest, it immediately begins to generate a cardiac output (Q_h) of 1 L/min. Instantaneously, virtually no blood has yet been transferred from the veins to the arteries, and therefore the arteriovenous pressure gradient is zero (Figure 24-2, D). Consequently, blood does not flow from the arteries through the capillaries and into the veins; that is, $Q_r = 0$ L/min. Hence, when pumping resumes, blood is being depleted from the veins at the rate of 1 L/min, and the arterial volume is being repleted at this very same rate. Hence, venous pressure begins to fall, and arterial pressure begins to rise. Because of the difference in compliances (equation 5), arterial pressure will rise 19 times more rapidly than venous pressure will fall.

The resulting pressure gradient will cause blood to flow through the resistance. If the pump maintains a constant output of 1 L/min, the arterial pressure will continue to rise and the venous pressure will continue to fall until the pressure gradient becomes 20 mm Hg. This gradient will force a flow of 1 L/min through a resistance of 20 mm Hg/L/min. This gradient will be achieved by a 19 mm Hg rise (to 26 mm Hg) in arterial pressure and a 1 mm Hg fall (to 6 mm Hg) in venous pressure. This equilibrium value of $P_v = 6$ mm Hg for a cardiac output of 1 L/min appears as point B on the vascular function curve of Figure 24-3. It reflects a net transfer of blood from the venous to the arterial side of the circuit, and a consequent reduction of the venous pressure.

The vascular function curve shows that as cardiac output is increased, the central venous pressure is diminished. The reduction of venous pressure that can be achieved by an increase in cardiac output is limited, however. At some critical maximum value of cardiac output, sufficient fluid will be translocated from the venous to the arterial side of the circuit to reduce the venous pressure below the ambient pressure (external to the central veins). In a system of very distensible vessels, such as the venous system, the vessels will be collapsed by the greater external pressure. This venous collapse constitutes an impediment to venous return to the heart. Hence, it will limit the maximum value of cardiac output, regardless of the strength of the pump. Note that, in Figure 24-3, cardiac output remains constant as venous pressure decreases below zero.

Blood Volume

The vascular function curve is affected by changes in total blood volume. As stated above, the mean circulatory pressure depends only on overall vascular compliance and total blood volume. Thus for a given vascular compliance the mean circulatory pressure will increase when the blood volume is expanded **(hypervolemia)** and will decrease when the blood volume is diminished **(hypovolemia).** In the family of vascular function curves shown in Figure 24-4, for example, consider that either blood was transfused into the static system until the mean circulatory pressure (P_{mc}) reached 9 mm Hg at equilibrium *(top curve),* or blood was withdrawn from the static system until P_{mc} reached 5 mm Hg at

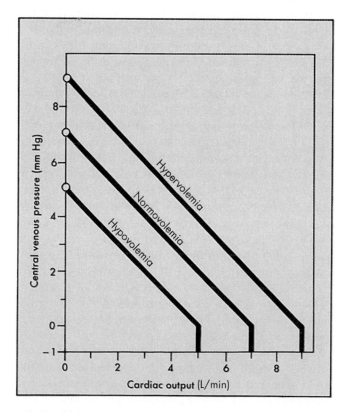

FIGURE 24-4   Effects of increased blood volume (hypervolemia) and of decreased blood volume (hypovolemia) on the vascular function curve.

equilibrium *(bottom curve).* Note that P_{mc} is the P_v-axis intercept in Figure 24-4.

Also note that the various vascular function curves in Figure 24-4 are all parallel to each other. To understand why the curves are parallel, let us consider the example of hypervolemia *(top curve),* in which the mean circulatory pressure had been raised to 9 mm Hg. When the system is static, arterial and venous pressures would both be 9 mm Hg. If cardiac output were then increased suddenly to 1 L/min (as in Figure 24-2, *D*), and if the peripheral resistance were still 20 mm Hg/L/min, an arteriovenous pressure gradient of 20 mm Hg would still be necessary for 1 L/min to flow through the resistance vessels. This does not differ from the example for normovolemia. Assuming the same ratio of C_v to C_a of 19:1, the pressure gradient would be achieved by a 1 mm Hg decline in P_v and a 19 mm Hg rise in P_a.

Therefore, a change in cardiac output from 0 to 1 L/min would evoke the same 1 mm Hg reduction in P_v, irrespective of the total blood volume, as long as the $C_v:C_a$ ratio and the peripheral resistance remain constant. The slope of the vascular function curve is, by definition, the change in P_v per unit change in cardiac output. Because the change in P_v induced by a unit change in cardiac output is not affected by the blood volume, the vascular function curves that represent different blood volumes are parallel to each other, as shown in Figure 24-4.

Figure 24-4 also shows that the cardiac output at which P_v becomes zero varies directly with the blood volume. Therefore the maximum value that cardiac output can attain becomes progressively more limited as the total blood volume is reduced. However, the pressure at which the veins collapse (sharp change in slope of the vascular function curve) is not altered appreciably by changes in blood volume. This pressure depends only on the ambient pressure. When P_v falls below the ambient pressure, the veins collapse and venous return to the heart is thereby limited.

Venomotor Tone

The effects of changes in **venomotor tone** (i.e., the contractile state of the venous smooth muscle) on the vascular function curve closely resemble those for changes in blood volume. In Figure 24-4, for example, the hypervolemia curve could just as well represent the effects of increased venomotor tone, whereas the hypovolemia curve could represent the effects of decreased venomotor tone. A given increment in P_v could be achieved as readily by constriction of the smooth muscle in the venous walls as by an increase in the volume of blood in the veins.

Acute hemorrhage, either external as a consequence of injury or internal in such diseases as peptic ulcer, is a common cause of hypovolemia. *Venoconstriction is one of the body's most efficacious defenses against the effects of blood loss.* The extent of venoconstriction is considerably greater in certain regions of the body than in others. The vascular bed of the skin is one of the major blood reservoirs in humans. During blood loss, profound cutaneous venoconstriction occurs, giving rise to the characteristic pale appearance of the skin. The resultant redistribution of blood away from the skin liberates several hundred milliliters of blood to be perfused through more vital regions of the body. Hence, vascular beds that undergo appreciable venoconstriction constitute **blood reservoirs.**

Peripheral Resistance The modification of the vascular function curve introduced by changes in total peripheral resistance is shown in Figure 24-5. The arterioles contain only about 3% of the total blood volume (see Figure 16-2). Hence, changes in the contractile state of these vessels do not significantly alter the mean circulatory pressure (P_{mc}). Thus the vascular function curves that represent various peripheral resistances converge at a common point (the P_{mc}) on the P_v axis.

Inspection of Figure 24-5 indicates that for any given cardiac output, the central venous pressure decreases as the peripheral resistance is increased. The principal reason for this relationship is that for a given cardiac output, an increase in peripheral resistance will redistribute the blood volume such that a greater fraction of the blood will reside in the arteries and consequently a lesser volume will reside in the veins. Arteriolar constriction sufficient to double the peripheral resistance will cause a twofold rise in arterial pressure, as illustrated in Figure 21-5, *D* and *E;* this is achieved by sequestering a greater volume of blood in the arteries. In a closed system with a constant total blood volume, therefore, this increase in arterial blood volume will be attended by an equivalent decrease in venous blood volume. This reduction in venous volume would be attended by a proportionate fall in venous pressure.

Increases in peripheral resistance produce a clockwise rotation of the vascular function curves about a common intercept on the P_v axis, because an increase in peripheral resistance tends to decrease P_v without affecting P_{mc} (Figure 24-5). Conversely, arteriolar vasodilation produces a counterclockwise rotation. A higher maximum cardiac output is attainable when the arterioles are dilated than when they are normal or constricted (Figure 24-5).

COUPLING BETWEEN THE HEART AND THE VASCULATURE

The central venous pressure constitutes the filling pressure (essentially, the preload) for the right ventricle. In accordance with the Frank-Starling mechanism (Chapters 18 and 19), the central venous pressure is a cardinal determinant of the cardiac output. Ordinarily, cardiac output varies directly with the central venous pressure; that is, *over a wide range of venous pressures, a rise in venous pressure increases the cardiac output.* In the discussion to follow, graphs of cardiac output as a function of venous pressure will be called **cardiac function curves.** Alterations in myocardial contractility will be represented by shifts in these curves.

In order to appreciate the coupling between the heart and the blood vessels, we will examine the interrelations between the cardiac function curve and the vascular function curve (Figure 24-6). Both curves reflect the relations between cardiac output and central venous pressure. As stated in the preceding paragraph, the cardiac function curve expresses how cardiac output varies in response to a change in venous pressure. Hence, cardiac output here is the dependent variable (or response) and venous pressure is the independent variable (or stimulus). By convention, the dependent variable is scaled

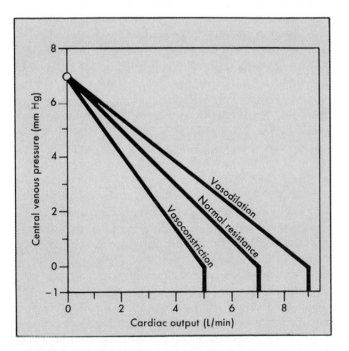

FIGURE 24-5 Effects of arteriolar vasodilation and vasoconstriction on the vascular function curve.

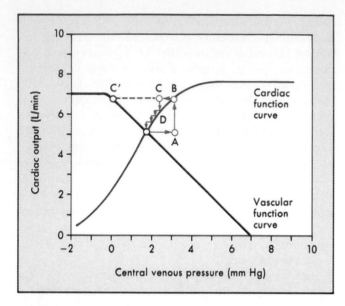

FIGURE 24-6 Typical vascular and cardiac function curves plotted on the same coordinate axes. Note that to plot both curves on the same graph, it is necessary to switch the X and Y axes for the vascular function curve. The coordinates of the equilibrium point, at the intersection of the cardiac and vascular function curves, represent the stable values of cardiac output and central venous pressure at which the system tends to operate. Any perturbation (such as when venous pressure is suddenly increased to point *A*) initiates a sequence of changes in cardiac output and venous pressure such that these variables gradually approach their equilibrium values.

along the Y axis and the independent variable is scaled along the X axis. Note that in Figure 24-6, the assignment of X and Y axes is conventional for the cardiac function curve.

The vascular function curve, conversely, reflects how central venous pressure is affected by a change in cardiac output. For the vascular function curve, venous pressure is the dependent variable (or response) and cardiac output is the independent variable (or stimulus). By convention, venous pressure should be scaled along the Y axis and cardiac output should be scaled along the X axis. Note that this convention was observed for the vascular function curves displayed in Figures 24-3 to 24-5.

However, in order to include the vascular function curve on the same set of coordinate axes with the cardiac function curve (Figure 24-6), it is necessary to violate the plotting convention for one of these curves. We have arbitrarily violated the convention for the vascular function curve. Note that the vascular function curve in Figure 24-6 reflects how the central venous pressure (scaled along the X axis) varies in response to a change of cardiac output (scaled along the Y axis).

Simultaneous examination of the two curves, one that characterizes the heart and the other that characterizes the vessels, provides some insight about the coupling between the heart and the vessels. Theoretically the heart can operate at all combinations of venous pressure and cardiac output that fall on the appropriate cardiac function curve. Similarly, the vascular system can operate at all combinations of venous pressure and cardiac output that fall on the appropriate vascular function curve. At equilibrium, therefore, the entire cardiovascular system (i.e., the combination of heart and vessels) must operate at the point of intersection of these two curves. Only at this point of intersection will the prevailing venous pressure evoke the cardiac output defined by the cardiac function curve, and, simultaneously, only at this point of intersection will the prevailing cardiac output evoke the venous pressure defined by the vascular function curve.

The tendency for the cardiovascular system to operate about such an equilibrium point may best be illustrated by examining its response to a sudden perturbation. Consider the changes elicited by a sudden rise in venous pressure from the equilibrium point to point *A* in Figure 24-6. Such a change in P_v might be induced by the rapid injection, during ventricular systole, of a given volume of blood on the venous side of the circuit, accompanied by the rapid withdrawal of an equal volume from the arterial side; the total blood volume would remain constant.

As defined by the cardiac function curve, this elevated venous pressure would increase cardiac output (from *A* to *B*) during the very next ventricular systole. The increased cardiac output, in turn, would result in the net transfer of blood from the venous to the arterial side of the circuit, with a consequent reduction in venous pressure.

In one heartbeat, the reduction in venous pressure would be small (from *B* to *C*), because the heart would transfer only a small fraction of the total venous blood volume over to the arterial side. Because of this reduction in venous pressure, the cardiac output during the very next beat would diminish (*C* to *D*) by an amount dictated by the cardiac function curve. Because *D* is still above the intersection point, the heart will pump blood from the veins to the arteries at a rate greater than the blood will flow across the peripheral resistance from arteries to veins. Hence, central venous pressure will continue to fall. This process will continue in ever-diminishing steps until the point of intersection is reached. Only one specific combination of cardiac output and venous pressure (denoted by the coordinates of the point of intersection) will satisfy simultaneously the requirements of the cardiac and vascular function curves.

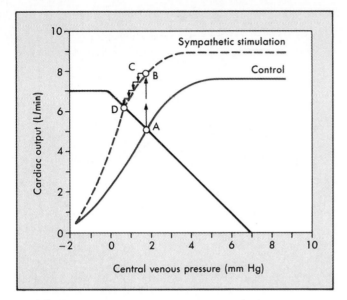

FIGURE 24-7   Enhancement of myocardial contractility, as accomplished by cardiac sympathetic nerve stimulation, causes the equilibrium values of cardiac output and central venous pressure to shift from the intersection *(A)* of the control vascular and cardiac function curves *(continuous lines)* to the intersection *(D)* of the same vascular function curve with the cardiac function curve *(dashed line)* that represents enhanced myocardial contractility.

Enhanced Myocardial Contractility

Graphs of cardiac and vascular function curves help explain the effects of alterations in ventricular contractility. **Contractility** refers to an alteration in myocardial performance based on processes that take place at the cross-bridges between the thick and thin filaments in the sarcomeres. It is distinguished from those changes in cardiac performance that are produced by changes in the number of cross-bridges that interact or from changes in the afterload that opposes the cardiac contraction. In Figure 24-7, the lower cardiac function curve represents the control contractility state, whereas the upper curve reflects an enhanced contractility. This pair of cardiac function curves is analogous to the pair of ventricular function curves shown in Figure 19-14. The change in contractility reflected by the two cardiac function curves in Figure 24-7 might be achieved by selective stimulation of the sympathetic nerves only to the heart. Such selective stimulation would not directly affect the vasculature, and so only one vascular function curve need be included in Figure 24-7.

During the control state the equilibrium values for cardiac output and venous pressure in Figure 24-7 are designated by point *A*. With the onset of cardiac sympathetic nerve stimulation (assuming the effects to be instantaneous and constant), the prevailing level of P_v

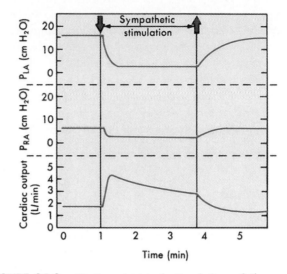

FIGURE 24-8   During electrical stimulation of the cardiac sympathetic nerves, aortic blood flow increased while pressures in the left atrium (P_{LA}) and right atrium (P_{RA}) diminished. These data support the graphic analysis shown in Figure 24-7, in which, at equilibrium, cardiac output is predicted to increase and venous pressure is predicted to decrease during stimulation of the cardiac sympathetic nerves. (Redrawn from Sarnoff SJ et al: Circ *Res* 8:1108, 1960.)

would abruptly raise cardiac output to point *B* because of the enhanced contractility. However, this high cardiac output would increase the net transfer of blood from the venous to the arterial side of the circuit, and consequently venous pressure will begin to fall (to point *C*). Cardiac output will continue to fall until it reaches a new equilibrium point *(D)*, which is located at the intersection of the vascular function curve with the new cardiac function curve. The new equilibrium point *(D)* lies above and to the left of the control equilibrium point *(A)*. This shift reveals that sympathetic stimulation increases cardiac output, despite the diminution of the ventricular filling pressure (i. e., central venous pressure). Such a change accurately describes the true response. In the experiment shown in Figure 24-8, stimulation of the left stellate ganglion in an anesthetized dog increased cardiac output, but decreased right and left atrial pressures (P_{RA} and P_{LA}).

Similar changes occur in patients with **congestive heart failure** when they are treated with drugs, such as digitalis, that improve myocardial contractility. Classically, patients with congestive heart failure have high central venous pressures and abnormally low cardiac outputs. Drugs that exert a **positive inotropic effect** (i.e., that enhance contractility) raise the cardiac output and decrease the central venous pressure. When such observations were first made, they were interpreted to be incompatible with Starling's law of the heart. Now we recognize that this important physiological principle is more faithfully represented by a family of cardiac function curves, and that changes in contractility are reflected by shifts from one component curve to another.

Blood Volume

Changes in blood volume do not directly affect the cardiac function curve, but they do influence the vascular function curve in the manner shown in Figure 24-4. Therefore, to understand the circulatory alterations evoked by a given change in blood volume, the appropriate cardiac function curve must be plotted along with the vascular function curves that represent the control and altered vascular states.

Figure 24-9 illustrates the immediate response to an **acute hemorrhage.** The parallel shift in the vascular function curve reflects the occurrence of a pure reduction in blood volume (such as that shown in Figure 24-4). Equilibrium point B, which denotes the values for cardiac output and central venous pressure immediately after a sudden hemorrhage, lies below and to the left of the control equilibrium point A. Thus, a pure, sudden reduction in blood volume decreases both cardiac output and venous pressure.

Peripheral Resistance

Predictions concerning the effects of changes in peripheral resistance on cardiac output and central venous pressure are complex, because the cardiac and vascular function curves both shift when peripheral resistance changes (Figure 24-10). The vascular function curve is rotated by a change in resistance, as explained previously. In the vascular function curves plotted in Figure 24-5, vasoconstriction caused a clockwise rotation of the vascular function curve. The axes are reversed in Figure 24-10, however, and therefore vasoconstriction is associated with a counterclockwise rotation of this curve. Hence, for any given value of P_v, the cardiac output is diminished by vasoconstriction. The cardiac function curve is also shifted downward, because at any given cardiac filling pressure

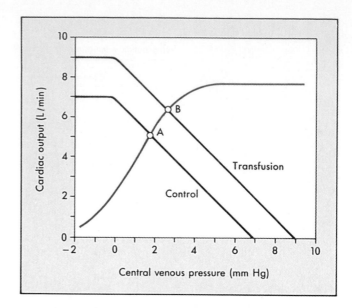

FIGURE 24-9 After a hemorrhage, the vascular function curve is shifted to the left. Therefore cardiac output and venous pressure are both decreased, as denoted by the translocation of the equilibrium point from A to B.

sure (P_v), the heart pumps less blood when the resistive load (or afterload) is increased. Because both curves are displaced downward by vasoconstriction, the new equilibrium point, B, will fall below the control point, A.

ROLE OF HEART RATE

Cardiac output is the product of stroke volume and heart rate. The preceding portion of this chapter was restricted to the control of stroke volume. This section will be devoted to the role of heart rate as a determinant of cardiac output.

Analysis of the effect of changes in heart rate on cardiac output is difficult, because a change in heart rate will alter the other factors (namely, the preload, afterload, and contractility) that determine stroke volume (see Figure 24-1). For example, an increase in heart rate decreases the duration of diastole. Hence, the time available for ventricular filling is abridged; and consequently preload is reduced. If the proposed increase in heart rate did alter cardiac output, the arterial blood pressure (afterload) would change. Finally, an increase in heart rate would augment the net influx of Ca^{++} into the cardiac myocytes, and this would enhance myocardial contractility (see Figure 19-15).

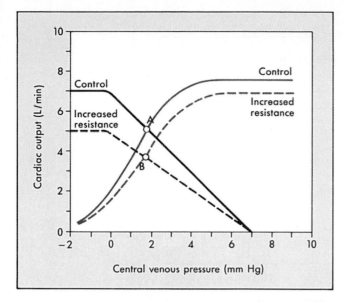

FIGURE 24-10 An increase in peripheral resistance shifts the cardiac and the vascular function curves downward. At equilibrium, the cardiac output is less *(B)* when the peripheral resistance is high than when it is normal *(A)*.

Many investigators have varied heart rate by artificial pacing in experimental animals and in humans. The effects on cardiac output resemble qualitatively the experimental results shown in Figure 24-11. In that experiment, as the atrial pacing frequency was gradually increased in an anesthetized dog, the stroke volume progressively diminished (Figure 24-11, *A*). Presumably, the curtailment of stroke volume was induced by the abridged time for ventricular filling. However, the change in cardiac output induced by a change in heart rate was influenced markedly by the actual level of the heart rate. In this experiment, for example, as the pacing frequency was increased within the range of 50 to 100 beats/min, an increase in heart rate augmented the cardiac output (Q_h). Presumably, at these lower frequencies, the reduction in stroke volume (SV) evoked by a given increase in heart rate (HR) was proportionately less than the increase in heart rate itself; that is, because $Q_h = SV \times HR$, if a given increment in HR exceeds the induced decrement in SV, then the induced Q_h will be greater than the initial Q_h.

Over the frequency range from about 100 to 200 beats/min, however, cardiac output was not affected appreciably by changes in pacing frequency (Figure 24-11, *B*). Hence, as the pacing frequency was increased, the stroke volume decreased proportionately to the increase in heart rate. Finally, at excessively high pacing frequen-

cies (above 200 beats/min), increments in heart rate induced reductions in cardiac output. Therefore, the induced decrement in SV must have exceeded increment in HR over this high range of pacing frequencies. Although the relationship of Q_h to HR is characteristically that of an inverted U, the relationship varies quantitatively among subjects and among physiological states in any given subject.

The characteristic relationship between cardiac output and heart rate explains the urgent need for treatment of patients who have excessively slow or excessively fast heart rates. Profound **bradycardias** (slow rates) may occur as the result of a very slow sinus rhythm in patients with **sick sinus syndrome** or as the result of a slow **idioventricular rhythm** in patients with **complete atrioventricular block.** In either rhythm disturbance, the capacity of the ventricles to fill during a prolonged diastole is limited (often by the noncompliant pericardium). Hence, cardiac output usually decreases substantially, because the very slow heart rate cannot be overcome by a sufficiently great stroke volume. Consequently, these rhythm disturbances often require the installation of an artificial pacemaker.

At the other end of the heart rate spectrum, excessively high heart rates in patients with **supraventricular** or **ventricular tachycardias** often require emergency treatment because their cardiac outputs may be critically low. In such patients, the filling time is so restricted at very high heart rates that small additional reductions in filling time elicit disproportionately severe reductions in filling volume. Reversion of the tachycardia to a more normal rhythm may be accomplished pharmacologically in less severe cases, but **cardioversion** by delivering a strong electrical current across the thorax or directly to the heart through an implanted device may be required in emergencies.

ANCILLARY FACTORS

In the above discussion, we have oversimplified the interrelations between central venous pressure and cardiac output. We have attempted to explain the effects on cardiac output that are elicited by changes in just one factor. However, because many feedback control mechanisms operate to regulate the cardiovascular system, an isolated change in a single variable rarely occurs. A change in blood volume, for example, reflexly alters car-

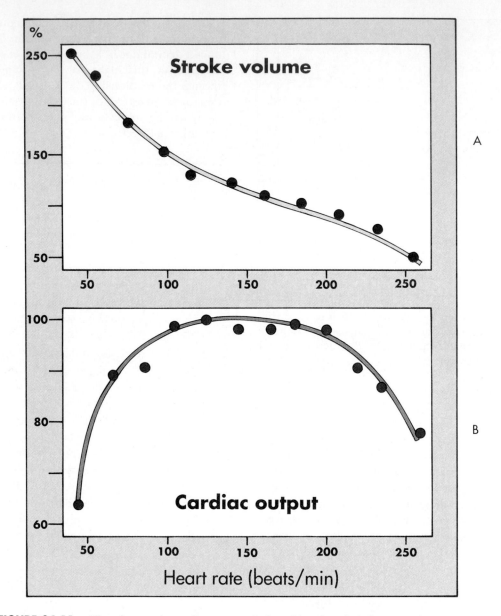

FIGURE 24-11 The changes in cardiac output induced by changing the rate of atrial pacing in an anesthetized dog. (Redrawn from Kumada M et al: *Jpn J Physiol* 17:538, 1967.)

diac function, peripheral resistance, and venomotor tone. Furthermore, several auxiliary factors also contribute to the regulation of cardiac output. Among these, some actually serve as additional energy sources to help the heart pump blood around the body.

Gravity

Gravitational forces may affect cardiac output profoundly. Among soldiers standing at attention for long periods, particularly in hot weather, some individuals may faint because their cardiac outputs decrease. Under such conditions, gravity impedes venous return from the dependent regions of the body, but it acts to promote flow on the arterial side of the same circuit. Therefore, in the dependent regions of the body, the vascular system behaves much like a U-tube, where the effects of gravity in the descending limb (arteries) and ascending limbs (veins) of the U-tube neutralize each other. Such neutralization does not take place in the vessels above the level of the heart, because the pressure in the veins at some level above the heart might fall below the ambient pressure, and these veins would collapse.

The compliance of the blood vessels accounts for the effect that gravitational forces can exert on cardiac output. When a person stands, the blood vessels below the level of the heart will be distended by the gravitational forces that act on the columns of blood in the vessels.

The distension will be more prominent on the venous than on the arterial side of the circuit, because the venous compliance is so much greater than the arterial compliance. Such venous distension is readily observed on the backs of the hands when the arms are allowed to hang below the level of the heart. The hemodynamic effects of distension of the veins (**venous pooling**) below the heart level resemble those caused by the loss of an equivalent volume of blood from the body. When a person shifts from a supine position to a relaxed standing position, from 300 to 800 ml of blood may be pooled in the legs. This may reduce cardiac output by about 2 L/min.

The compensatory adjustments to the erect position are similar to the adjustments to blood loss. For example, venous pooling and other gravitational effects tend to lower the pressure in the regions of the arterial baroreceptors. The resulting diminution in baroreceptor excitation reflexly speeds the heart, strengthens the cardiac contraction, and constricts the arterioles and veins. The baroreceptor reflex has a greater effect on the resistance vessels (arterioles) than on the capacitance vessels (veins). On hot days, the compensatory vasomotor reactions are less efficacious, and the absence of muscular activity exaggerates the pooling effects of gravity, as explained below.

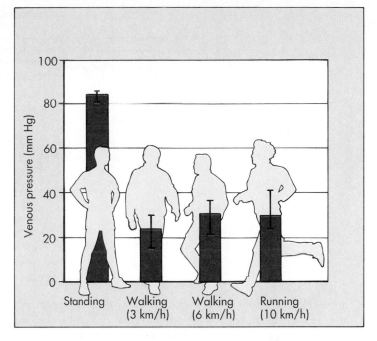

FIGURE 24-12   Mean pressures (±95% confidence intervals) in the foot veins of 18 people during quiet standing, walking, and running. (From Stick C et al: *J Appl Physiol* 72:2063, 1992.)

Many of the vasodilator drugs used to treat **essential hypertension** and other circulatory disorders also interfere with the reflex adaptation to standing. Similarly, astronauts exposed to weightlessness lose their adaptations after a few days, and they experience difficulties when they first return to a normal gravitational field. When individuals with impaired reflex adaptations stand, their blood pressures may fall dramatically. This response is called **orthostatic hypotension** which may cause lightheadedness or fainting.

Muscular Activity and Venous Valves

When a relaxed individual stands, the pressure in the veins below the heart rises. The venous pressure in the legs increases gradually and does not reach an equilibrium value until almost 1 minute after the subject stands. The slowness of the rise in venous pressure is attributable to the **venous valves,** which permit flow only toward the heart. When a person stands, the valves prevent blood in the veins from actually falling toward the feet. Hence the column of venous blood is supported at numerous levels by these valves; temporarily the venous column consists of many separate segments. However,

blood continues to enter the column from many venules and small tributary veins, and pressure continues to rise. As soon as the pressure in one segment exceeds that in the segment just above it, the valve between the two segments is forced open. Ultimately, all the valves in the dependent veins are open, and the column is continuous.

Precise measurement reveals that the final level of venous pressure in the feet during quiet standing is only slightly greater than that in a static column of blood extending from the right atrium to the feet. This indicates that the pressure drop caused by flow from foot veins to the right atrium is very small; hence, the resistance to flow is also small. This very low venous resistance justifies our lumping all the veins together as a common venous compliance in the model shown in Figure 24-2.

When an individual who has been standing quietly begins to walk or run, the venous pressure in the legs and feet decreases appreciably (Figure 24-12). Because of the intermittent venous compression produced by the contracting muscles and because of the presence of the venous valves, blood is forced from the veins toward the heart (see Figure 26-3). Hence, muscular contraction lowers the mean venous pressure in the legs and serves as an auxiliary blood pump. Furthermore, it prevents ve-

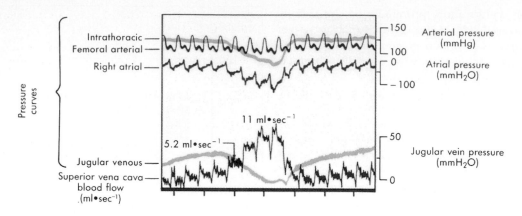

FIGURE 24-13 During inspiration in an anesthetized dog, intrathoracic, right atrial, and jugular venous pressures decrease, and flow in the superior vena cava increases (from 5.2 to 11 ml/sec). (Modified from Brecher GA: *Venous return,* New York, 1956, Grune & Stratton, Inc.)

nous pooling and lowers capillary hydrostatic pressure, and thereby reduces the tendency for edema fluid to collect in the feet during standing. This mechanism operates very effectively in normal people, in that not much motion is required for appreciable auxiliary pumping to occur. Thus, if a standing person shifts his weight periodically, the pressure in his foot veins will be considerably less than the pressure that prevails if he remains absolutely still.

> This auxiliary pumping mechanism is not very effective in people with **varicose veins** in their legs. The valves in such veins do not operate properly, and therefore when the leg muscles are contracting, the blood in the leg veins can be forced in the retrograde as well as in the antegrade direction. Thus, when an individual with varicose veins stands or walks, the venous pressure in the ankles and feet is excessively high. The consequent high capillary pressure leads to the accumulation of extracellular fluid **(edema)** in the ankles and feet.

Respiratory Effects

The normal, periodic activity of the respiratory muscles causes rhythmic variations in vena caval flow, and it constitutes an auxiliary pump to promote venous return of blood to the heart. Coughing, straining at stool, and other activities that require the respiratory muscles may affect cardiac output substantially.

The changes in blood flow in the superior vena cava during the respiratory cycle of an anesthetized dog are shown in Figure 24-13. During respiration the changes in intrathoracic pressure are transmitted to the lumina of the thoracic blood vessels. The reduction in central venous pressure during inspiration increases the pressure gradient between the extrathoracic and intrathoracic veins. The consequent acceleration of venous return to the right atrium is displayed in Figure 24-13 as an increase in superior vena caval blood flow from 5.2 ml/sec during expiration to 11 ml/sec during inspiration.

During expiration, flow into the central veins decelerates. However, the mean rate of venous return during normal respiration exceeds the venous flow that occurs in the temporary absence of respiration. Hence normal inspiration facilitates venous return more than normal expiration impedes it. This facilitation is partly attributable to the valves in the veins of the extremities and neck. These valves prevent any reversal of flow during expiration. Thus the respiratory muscles and venous valves constitute an auxiliary pump for venous return.

SUMMARY

1. Two important relationships between cardiac output (Q_h) and central venous pressure (P_v) prevail in the cardiovascular system. One applies to the heart and the other to the vascular system.
2. With respect to the heart, Q_h varies directly with P_v (preload) over a wide range of P_v. This relationship is represented by the cardiac function curve, and it expresses the Frank-Starling mechanism.
3. With respect to the vascular system, P_v varies inversely with Q_h. This relationship is represented by the vascular function curve, and it reflects the fact that as Q_h increases, for example, a greater fraction of the total blood volume resides in the arteries and hence a smaller volume resides in the veins.
4. The principal mechanisms that govern the cardiac function curve are the changes in numbers of crossbridges that interact and in the affinity of the con-

tractile proteins for calcium. These mechanisms are evoked by changes in the cardiac filling pressure (preload).

5. The principal factors that govern the vascular function curve are the arterial and venous compliances, the peripheral vascular resistance, and the total blood volume.

6. The equilibrium values of Q_h and P_v that prevail under a given set of physiological conditions are determined by the intersection of the cardiac and vascular function curves.

7. At very low and very high heart rates, the heart is unable to pump an adequate Q_h. At the very low rates, the increment in filling during diastole cannot compensate for the small number of cardiac contractions per minute. At the very high rates, the larger number of contractions per minute cannot compensate for the inadequate ventricular filling time.

8. Gravity influences Q_h because the veins are so compliant, and substantial quantities of blood tend to pool in the veins of the dependent portions of the body.

9. Respiration changes the pressure gradient between the intrathoracic and extrathoracic veins. Hence, respiration serves as an auxiliary pump, which may affect the mean level of Q_h and may produce the rhythmic changes in stroke volume that occur throughout the respiratory cycle.

BIBLIOGRAPHY
Journal Articles

Geddes LA et al: Cardiac output, stroke volume, and pacing rate: a review of the literature and a proposed technique for selection of the optimum pacing rate for an exercise responsive pacemaker, *J Cardiovasc Electrophysiol* 2:408, 1991.

Hainsworth R: The importance of vascular capacitance in cardiovascular control, *News Physiol Sci* 5:250, 1990.

Lacolley PJ et al: Microgravity and orthostatic intolerance: carotid hemodynamics and peripheral responses, *Am J Physiol* 264:H588, 1993.

Rothe CF, Gaddis ML: Autoregulation of cardiac output by passive elastic characteristics of the vascular capacitance system, *Circulation* 81:360, 1990.

Rothe CF: Mean circulatory filling pressure: its meaning and measurement, *J Appl Physiol* 74:499, 1993.

Seymour RS et al: The heart works against gravity, *Am J Physiol* 265:R715, 1993.

Sheriff DD et al: Dependence of cardiac filling pressure on cardiac output during rest and dynamic exercise in dogs, *Am J Physiol* 265:H316, 1993.

Stick, C et al: Measurements of volume changes and venous pressure in the human lower leg during walking and running, *J Appl Physiol* 72:2063, 1992.

Tyberg JV: Venous modulation of ventricular preload, *Am Heart J* 123:1098, 1992.

Books and Monographs

Guyton AC, Jones CE, Coleman TG: *Circulatory physiology: cardiac output and its regulation,* ed 2, Philadelphia, 1973, WB Saunders Co.

Sagawa K et al.: *Cardiac contraction and the pressure-volume relationship,* New York, 1988, Oxford University Press.

Shepherd JT, Vanhoutte PM: *Veins and their control,* Philadelphia, 1975, WB Saunders Co.

Smith JJ, editor: *Circulatory response to the upright posture,* Boca Raton, Fla, 1990, CRC Press.

Yin FCP, editor: *Ventricular/vascular coupling,* New York, 1987, Springer-Verlag.

CHAPTER 25

Special Circulations

CUTANEOUS CIRCULATION

The oxygen and nutrient requirements of the skin are relatively small. The supply of these essential materials is not the chief governing factor in the regulation of cutaneous blood flow, in contrast to the regulation in most other body tissues. *The primary function of the cutaneous circulation is maintenance of a constant body temperature.* Consequently blood flow to the skin fluctuates widely, depending on the need for loss or conservation of body heat. Mechanisms responsible for alterations in skin blood flow are mainly activated by changes in ambient and internal body temperatures.

Regulation of Skin Blood Flow

Essentially two types of resistance vessels are present in skin: **arterioles** and **arteriovenous (AV) anastomoses.** The arterioles are similar to those found elsewhere in the body. AV anastomoses shunt blood from arterioles to venules and venous plexuses; thus they bypass the capillary bed. AV anastomoses are found mainly in the fingertips, palms of the hands, toes, soles of the feet, ears, nose, and lips. AV anastomoses differ morphologically from the arterioles, in that they are either short and straight or long and coiled vessels, about 20 to 40 μm in lumen diameter. They have thick muscular walls, richly supplied with nerve fibers (Figure 25-1). These vessels are almost exclusively under sympathetic neural control, and they dilate maximally when their nerve supply is interrupted. Conversely, reflex stimulation of the sympathetic fibers to these vessels may constrict them to the point that the vascular lumen is completely obliterated. Although AV anastomoses do not exhibit basal tone (tonic activity of the vascular smooth muscle independent of innervation), they are highly sensitive to vasoconstrictor agents, such as epinephrine and norepineph-

rine. Furthermore, AV anastomoses are not under metabolic control, and they fail to show reactive hyperemia or autoregulation of blood flow. Thus the regulation of blood flow through these anastomotic channels is governed mainly by the nervous system in response to reflex activation by temperature receptors or by higher centers of the central nervous system.

The bulk of the skin resistance vessels exhibit some basal tone. Vascular resistance in the skin is under the dual control of the sympathetic nervous system and local regulatory factors, in much the same manner as are other vascular beds. *In the skin, however, neural control is more important than local factors.* Stimulation of sympathetic nerve fibers to skin blood vessels (arteries and veins, as well as arterioles) induces vasoconstriction, and severance of the sympathetic nerves induces vasodilation. After chronic denervation of the cutaneous blood vessels, the degree of tone that existed before denervation is gradually regained over several weeks. This is accomplished by an enhancement of basal tone that compensates for the degree of tone previously contributed by sympathetic nerve fiber activity. Epinephrine and norepinephrine elicit only vasoconstriction in cutaneous vessels.

Parasympathetic vasodilator nerve fibers do not supply the cutaneous blood vessels. However, stimulation of the sweat glands, which are innervated by cholinergic fibers of the sympathetic nervous system, results in dilation of the skin resistance vessels. Sweat contains an enzyme that acts on a protein moiety in the tissue fluid to release bradykinin, a polypeptide with vasodilator properties. Bradykinin formed in the tissue can act locally to dilate the arterioles and increase blood flow to the skin.

The skin vessels of certain body regions, particularly the head, neck, shoulders, and upper chest, are under the influence of the higher centers of the central nervous system. Blushing, caused by embarrassment or anger, and blanching, caused by fear or anxiety, are examples of ce-

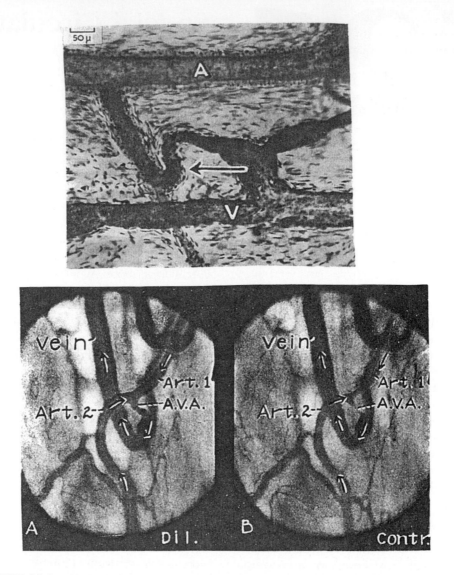

FIGURE 25-1   *Top,* Arteriovenous *(AV)* anastomosis in the human ear injected with Berlin blue. *A,* Artery; *V,* vein; arrow points to AV anastomosis. The walls of the AV anastomosis in the fingertips are thicker and more cellular. (From Pritchard MML, Daniel PM: *J Anatomy* 90:309, 1956.) *Bottom,* Two frames from a motion picture record of the same relatively large arteriovenous anastomosis *(A.V.A.)* in a stable rabbit ear chamber installed 3½ months previously. *Frame A,* A.V.A. dilated; *Frame B,* contracted. On this day the lumen of the A.V.A. measured 51 μm dilated and 5 μm contracted at its narrowest point. (From Clark ER, Clark EL: *Am J Anat* 54:229, 1934).

rebral inhibition and stimulation, respectively, of the sympathetic nerve fibers to the affected regions.

In contrast to AV anastomoses in the skin, the cutaneous resistance vessels show autoregulation of blood flow and reactive hyperemia. If the arterial inflow to a limb is stopped by inflating a blood pressure cuff briefly, the skin reddens greatly below the point of vascular occlusion when the cuff is deflated. This increased cutaneous blood flow **(reactive hyperemia)** is also manifested by the distension of the superficial veins in the erythematous extremity.

Ambient and Body Temperature The primary function of the skin is to preserve the internal milieu and protect it from adverse changes in the environment. Also, the ambient temperature is one of the most important external variables with which the body must contend. Therefore, it is not surprising that the vasculature of the skin is chiefly influenced by environmental temperature. Exposure to cold elicits a generalized cutaneous vasoconstriction that is most pronounced in the hands and feet. This response is chiefly mediated by the nervous system, because arrest of the circulation to a hand with a pressure cuff and immersion of that hand in cold water result in vasoconstriction in the skin of the other extremities that are exposed to room temperature. When the circulation to the chilled hand is not occluded, the reflex vasoconstriction is caused partly by the cooled blood returning to the general circulation and stimulating the temperature-regulating center in the anterior hypothalamus. Direct application of cold to this region of the brain produces cutaneous vasoconstriction.

The skin vessels of the cooled hand also respond directly to cold. Moderate cooling or exposure of the hand to severe cold (0° C to 15° C) for brief periods constricts the resistance and capacitance vessels, including AV anastomoses. However, prolonged exposure of the hand to severe cold has a secondary vasodilator effect. Prompt vasoconstriction and severe pain are elicited by immersion of the hand in water near 0° C, but they are soon followed by dilation of the skin vessels, reddening of the immersed part, and alleviation of the pain. With continued immersion of the hand, alternating periods of constriction and dilation occur, but the skin temperature rarely drops as low as it did during the initial vasoconstriction. Prolonged severe cold damages the tissue.

> The rosy faces of people outdoors in the cold are examples of cold vasodilation. However, the blood flow through the skin of the face may be very low, despite the flushed appearance. The red color of the slowly flowing blood is largely the result of the reduced oxygen uptake by the cold skin and the cold-induced shift to the left of the oxyhemoglobin dissociation curve.

Direct application of heat produces not only local vasodilation of resistance and capacitance vessels and AV anastomoses, but also reflex dilation in other parts of the body. The local effect is independent of the vascular nerve supply, whereas the reflex vasodilation is a combination of anterior hypothalamic stimulation by the returning warmed blood and of stimulation of receptors in the heated part.

The proximity of the major arteries and veins to each other permits considerable heat exchange (**countercurrent**) between artery and vein. Cold blood that flows in veins from a cooled hand toward the heart takes up heat from adjacent arteries; this warms the venous blood and cools the arterial blood. Heat exchange occurs in the opposite direction when the extremity is exposed to heat. Thus heat conservation is enhanced during exposure of the extremities to cold, and heat gain is minimized during exposure of the extremities to warmth.

> In patients with **Raynaud's disease,** exposure to cold or emotional stimuli may initiate ischemic attacks in the extremities (especially the fingers). The response is characterized by blanching, followed by cyanosis, and then by redness. The attacks are often associated with numbness, tingling, pain, and burning sensations.

Skin Color The color of the skin is determined in large part by pigment. In all but very dark skin, however, the pallor or ruddiness is primarily a function of the amount of blood in the skin. With little blood in the venous plexus the skin appears pale, whereas with larger quantities of blood in the venous plexus, the skin has more color. Whether this color is bright red, blue, or some intermediate shade is determined by the degree of oxygenation of the blood in the subcutaneous vessels. For example, a combination of vasoconstriction and reduced hemoglobin can produce an ashen gray color of the skin, whereas a combination of venous engorgement and reduced hemoglobin can result in a dark purple hue. Skin color provides little information about the rate of cutaneous blood flow. Rapid blood flow and pale skin may coexist when the AV anastomoses are open, and slow blood flow and red skin may coexist when the extremity is exposed to cold.

SKELETAL MUSCLE CIRCULATION

The blood flow to skeletal muscle varies directly with contractile activity of the tissue and the type of muscle. Blood flow and capillary density in red (slow-twitch, high-oxidative) muscle are greater than in white (fast-twitch, low-oxidative) muscle. In resting muscle the arterioles exhibit asynchronous intermittent contractions and relaxations. Thus, at any given moment, a large percentage of the capillary bed is not perfused. Consequently total blood flow through quiescent skeletal muscle is low (1.4 to 4.5 ml/min/100 g). During exercise the resistance vessels relax and the muscle blood flow may increase manyfold (up to 15 to 20 times the resting level); the magnitude of the increase depends largely on the severity of the exercise.

Regulation of Skeletal Muscle Blood Flow

Control of muscle circulation is achieved by neural and local factors; the relative contribution of these factors is dictated by muscle activity. *At rest neural and myogenic regulations are predominant, whereas dur-*

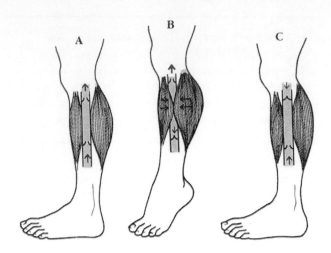

FIGURE 25-2 Action of the muscle pump in venous return from the legs. **A,** Standing at rest. The venous valves are open and blood flows upward toward the heart by virtue of the pressure generated by the heart and is transmitted through the capillaries to the veins from the arterial side of the vascular system (vis a tergo). **B,** Contraction of the muscle compresses the vein so that the increased pressure in the vein drives blood toward the thorax through the upper valve and closes the lower valve in the uncompressed segment of the vein just below the point of muscular compression. **C,** Immediately after muscle relaxation the pressure in the previously compressed venous segment falls, and the reversed pressure gradient causes the upper valve to close. The valve below the previously compressed segment opens because pressure below it exceeds that above it, and the segment fills with blood from the foot. As blood flow continues from the foot, the pressure in the previously compressed segment rises. When it exceeds the pressure above the upper valve, this valve opens and continuous flow occurs as in part **A.**

ing exercise metabolic control supervenes. As with all tissues, physical factors such as arterial pressure, tissue pressure, and blood viscosity influence blood flow to muscle. However, another physical factor plays a role during exercise: the squeezing effect of the active muscle on the vessels (see also Chapter 24) . With intermittent contractions inflow is restricted and venous outflow is enhanced during each brief contraction. The presence of the venous valves prevents backflow of blood in the veins between contractions, thereby aiding in the forward propulsion of the blood (Figure 25-2). With strong sustained contractions, the vascular bed can be compressed to the point at which blood flow actually ceases temporarily.

Neural Factors Although the resistance vessels of muscle have a high basal tone, they also display tone attributable to continuous low-frequency activity in the sympathetic vasoconstrictor nerve fibers.

The tonic activity of the sympathetic nerves is greatly influenced by the baroreceptor reflex. An increase in carotid sinus pressure dilates the vascular bed of the muscle, and a decrease in carotid sinus pressure elicits vasoconstriction. Because muscle is the major body component on the basis of mass and thereby represents the largest vascular bed, the reflex participation of resistance vessels in the muscles is important in maintaining a constant arterial blood pressure.

A comparison of the vasoconstrictor and vasodilator effects of the sympathetic nerves to blood vessels of muscle and skin is provided in Figure 25-3. Note the lower basal tone of the skin vessels, their greater constrictor response, and the absence of active cutaneous vasodilation.

Local Factors It already has been stressed that neural regulation of muscle blood flow is superseded by metabolic regulation (see Chapter 23) when the muscle changes from the resting to the contracting state. However, local control also occurs in innervated resting skeletal muscle when the vasomotor nerves are not active. Thus autoregulation can be observed in innervated as well as in denervated muscle, and in both conditions it is characterized by a low venous blood oxygen saturation.

CORONARY CIRCULATION

Factors That Influence Coronary Blood Flow

Physical Factors The main factor responsible for perfusion of the myocardium is the aortic pressure, which is generated by the heart itself. Changes in aortic pressure generally shift coronary blood flow in the same direction. However, alterations of cardiac work, produced by an increase or decrease in aortic pressure, have a considerable effect on coronary resistance. Increased metabolic activity of the heart decreases coronary resistance, and a reduction in cardiac metabolism increases coronary resistance. Under normal conditions blood pressure is kept within relatively narrow limits by the baroreceptor reflex. Therefore *changes in coronary blood flow are primarily caused by caliber changes of the coronary resistance vessels in response to metabolic demands of the heart.* When the rate of myocardial metabolism is unchanged and coronary perfusion pressure is raised or lowered, coronary blood flow remains relatively constant (**autoregulation of blood flow**).

In addition to providing the head of pressure to drive blood through the coronary vessels, the heart also influences its blood supply by the squeezing effect of the contracting myocardium on the blood vessels that course through it (**extravascular compression** or **extracoro-**

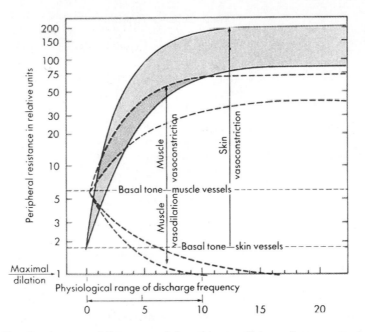

FIGURE 25-3   Basal tone and the range of the response of the resistance vessels in muscle and skin to sympathetic nerve stimulation. Peripheral resistance is plotted on a logarithmic scale. (Redrawn from Celander O, Folkow B: *Acta Physiol Scand* 29:241, 1953.)

nary resistance). This force is so great during early ventricular systole that blood flow in a large coronary artery supplying the left ventricle is briefly reversed. Left coronary inflow is maximal in early diastole, when the ventricles have relaxed and extravascular compression of the coronary vessels is virtually absent. This flow pattern is seen in the phasic coronary flow curve for the left coronary artery (Figure 25-4). After an initial reversal in early systole, left coronary blood flow parallels the aortic pressure until early diastole, when it rises abruptly. It then declines slowly as aortic pressure falls during the remainder of diastole.

The minimum extravascular resistance and absence of left ventricular work during diastole are used to advantage clinically to improve myocardial perfusion in patients with damaged myocardium and low blood pressure. The method is called **counterpulsation** and consists of the insertion of an inflatable balloon into the thoracic aorta through a femoral artery. The balloon is inflated during ventricular diastole and deflated during systole. This procedure enhances coronary blood flow during diastole by raising diastolic pressure at a time when coronary extravascular resistance is lowest. Furthermore, it reduces cardiac energy requirements by lowering aortic pressure during ventricular ejection.

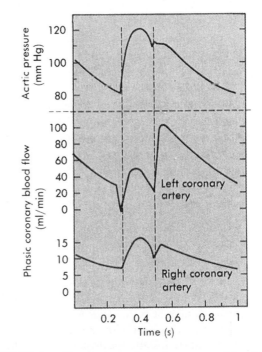

FIGURE 25-4   Comparison of phasic coronary blood flow in the left and right coronary arteries.

Left ventricular myocardial pressure (pressure within the wall of the left ventricle) is greatest near the endocardium and lowest near the epicardium. However, under normal conditions this pressure gradient does not impair endocardial blood flow, because a greater blood

Myocardial oxygen balance

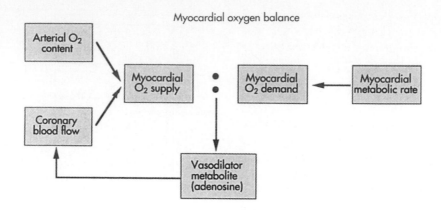

FIGURE 25-5 Imbalance in the oxygen supply/oxygen demand ratio alters coronary blood flow by changing the rate of release of a vasodilator metabolite from the cardiomyocytes. A decrease in the ratio elicits an increase in vasodilator release, whereas an increase in the ratio has the opposite effect.

flow to the endocardium during diastole compensates for the greater blood flow to the epicardium during systole. In fact, blood flow is slightly higher in the endocardium than in the epicardium under normal conditions. Because extravascular compression is greatest at the endocardial surface of the ventricle, equality of epicardial and endocardial blood flow must mean that the tone of the endocardial resistance vessels is less than that of the epicardial vessels.

> Under abnormal conditions, when diastolic pressure in the coronary arteries is low, such as in **severe hypotension, partial coronary artery occlusion,** or **severe aortic stenosis,** blood flow to the endocardial regions is more severely impaired than is that to the epicardial regions of the ventricle. For this reason, the myocardial damage after occlusion of the anterior descending branch of the left coronary artery is usually greatest in the inner wall of the left ventricle.

Flow in the right coronary artery shows a similar pattern (Figure 25-4), but because of the lower pressure developed by the thin right ventricle during systole, blood flow does not reverse in early systole. Systolic blood flow in the right corinary artery constitutes a much greater proportion of total coronary inflow than it does in the left coronary artery.

Tachycardia and bradycardia have dual effects on coronary flow. A change in heart rate is accomplished chiefly by shortening or lengthening diastole. During tachycardia the proportion of time spent in systole, and consequently during the period of restricted inflow, increases. However, this mechanical reduction in mean coronary flow is overridden by the coronary dilation associated with the increased metabolic activity of the more rapidly beating heart. During bradycardia the opposite is true; restriction of coronary inflow is less (more time in diastole) but so are the metabolic (O_2) requirements of the myocardium.

Neural and Neurohumoral Factors *The primary effect of stimulation of the sympathetic nerves to the coronary vessels is vasoconstriction. However, the observed effect is a great increase in coronary blood flow.* The increase in flow is associated with cardiac acceleration and a more forceful systole. The stronger myocardial contractions and the tachycardia (with consequent greater proportion of time spent in systole) tend to restrict coronary flow. However, the increase in myocardial metabolic activity, as evidenced by the rate and contractility changes, tends to dilate the coronary resistance vessels. The increase in coronary blood flow elicited by cardiac sympathetic nerve stimulation is the algebraic sum of these factors.

Metabolic Factors One of the most striking characteristics of the coronary circulation is the close parallelism between the level of myocardial metabolic activity and the magnitude of the coronary blood flow. This relationship is also found in the denervated heart. The link between cardiac metabolic rate and coronary blood flow remains unsettled. However, it appears that *a decrease in the ratio of oxygen supply to oxygen demand* (whether produced by a reduction in oxygen supply or by an increment in oxygen demand) *releases a vasodilator substance from the myocardium into the interstitial fluid, where the substance can relax the coronary resistance vessels.*

As diagrammed in Figure 25-5, a decrease in arterial blood oxygen content, coronary blood flow, or both, or an increase in metabolic rate decreases the oxygen supply/demand ratio. This causes the release of a vasodilator substance such as adenosine, which dilates the arterioles and thereby adjusts oxygen supply to demand. A decrease in oxygen demand would reduce the release of the vasodilator substance and permit greater expression of basal tone.

Numerous agents, generally referred to as metabolites, have been suggested as mediators of the vasodilation evoked by increased cardiac work. Among the substances implicated are CO_2, O_2 (reduced O_2 tension), hydrogen ions (lactic acid), potassium ions, and adenosine. Of these agents, adenosine comes closest to satisfying the criteria for the physiological mediator. Accumulation of vasoactive metabolites also may be responsible for reactive hyperemia in the heart, because the duration of coronary flow after release of the briefly occluded vessel is, within certain limits, proportional to the duration of the period of occlusion.

Cardiac Oxygen Consumption and Work

The volume of O_2 consumed by the heart is determined by the amount and the type of activity the heart performs. Under basal conditions myocardial O_2 consumption is about 8 to 10 ml/min/100 g of heart. It can increase several-fold with exercise and decrease moderately under such conditions as hypotension and hypothermia. The cardiac venous blood normally has a low O_2 content (about 5 ml/dl), and the myocardium can therefore receive little additional O_2 by further O_2 extraction from the coronary blood.

Left ventricular work per beat **(stroke work)** is approximately equal to the product of the stroke volume and the mean aortic pressure against which the blood is ejected by the left ventricle. At resting levels of cardiac output, the kinetic energy component is negligible (see Chapter 20). However, at high cardiac outputs, as in severe exercise, the kinetic component can account for up to 50% of total cardiac work. One can simultaneously halve the aortic pressure and double the cardiac output, or vice versa, and still arrive at the same value for cardiac work. However, the O_2 requirements are greater for any given increment of cardiac work when it is achieved by a rise in pressure than when it is achieved by an increase in stroke volume. Pumping an increase in cardiac output at a constant aortic pressure **(volume work)** is accomplished with a small increase in left ventricular O_2 consumption. Conversely, pumping against an increased arterial pressure at a constant cardiac output **(pressure work)** is accompanied by a large increment in myocardial O_2 consumption.

The greater energy demand of pressure work over volume work is clinically important. For example, in **aortic stenosis** (a narrowed aortic valve), left ventricular O_2 consumption is increased because of the high intraventricular pressure developed during systole to overcome the resistance of the stenotic valve, whereas coronary perfusion pressure is normal or reduced because of the pressure drop across the narrowed orifice of the diseased aortic valve. The result is a greater oxygen need by the myocardium in the face of a reduced oxygen supply. This can produce **angina pectoris** (chest pain) and eventually left ventricular failure.

Coronary Collateral Circulation

In the normal human heart there are virtually no functional intercoronary channels, and an abrupt occlusion of a coronary artery or one of its branches leads to **ischemic necrosis** (tissue death caused by inadequate blood flow) and eventual fibrosis of the areas of myocardium supplied by the occluded vessel.

However, if narrowing of a coronary artery occurs slowly and progressively over a period of weeks, months, or years, as often occurs in **coronary atherosclerosis,** collateral vessels develop and may furnish sufficient blood to the ischemic myocardium to prevent or reduce the extent of myocardial injury.

Collateral vessels develop between branches of occluded and nonoccluded arteries. They originate from preexisting small vessels that undergo proliferative changes of the endothelium and smooth muscle, possibly in response to wall stress and chemical agents released by the ischemic tissue.

When disease causes discrete occlusions or severe narrowing in coronary arteries (lumen diameters as small as 1 mm), the lesions can be bypassed with an artery or vein graft, or the narrow segment can be dilated by inserting a balloon-tipped catheter into the diseased vessel via a peripheral artery and inflating the balloon. Distension of the vessel by balloon inflation **(angioplasty)** can produce a lasting dilation of a narrowed coronary artery.

CEREBRAL CIRCULATION

Blood reaches the brain through the internal carotid and vertebral arteries. The latter join to form the basilar artery, which, in conjunction with branches of the internal carotid arteries, forms the **circle of Willis.** A unique feature of the cerebral circulation is that it all lies within a rigid structure, the cranium. Because intracranial contents are incompressible, any increase in arterial inflow, as occurs with arteriolar dilation, must be associated with a comparable increase in venous outflow. The volume of blood and of extravascular fluid can vary considerably in most tissues. In brain the volume of blood and extravascular fluid is relatively constant; changes in either of these fluid volumes must be accompanied by a reciprocal change in the other. In contrast to most other organs, the total cerebral blood flow is held within a relatively narrow range; in humans it averages 55 ml/min/100 g of brain.

Regulation of Cerebral Blood Flow

Of the various body tissues, the brain is the least tolerant of ischemia. Interruption of cerebral blood flow for as little as 5 seconds results in loss of consciousness. Ischemia lasting just a few minutes results in irreversible tissue damage. Fortunately, regulation of the cerebral circulation is mainly under direction of the brain itself. Local regulatory mechanisms and reflexes that originate in the brain tend to maintain cerebral circulation relatively constant. This constancy prevails even in the presence of possible adverse extrinsic effects, such as sympathetic vasomotor nerve activity, circulating humoral vasoactive agents, and changes in arterial blood pressure. Under certain conditions the brain also regulates its blood flow by initiating changes in systemic blood pressure.

> For example, elevation of intracranial pressure, as may occur with a brain tumor, results in an increase in systemic blood pressure. This response, called **Cushing's phenomenon,** is apparently caused by ischemic stimulation of vasomotor regions of the medulla. It aids in maintaining cerebral blood flow in the face of an elevated intracranial pressure.

Neural Factors The cerebral vessels are innervated by the cervical sympathetic nerve fibers that accompany the internal carotid and vertebral arteries into the cranial cavity. Relative to the control of other vascular beds, the sympathetic control of the cerebral vessels is weak, and the contractile state of the cerebrovascular smooth muscle depends mainly on local metabolic factors. There are no known sympathetic vasodilator nerves to the cerebral vessels. However, the vessels do receive parasympathetic fibers from the facial nerve, which produce a slight vasodilation on stimulation.

Local Factors *Generally, total cerebral blood flow is constant. However, regional cortical blood flow is associated with regional neural activity.* For example, movement of one hand results in increased blood flow only in the hand area of the contralateral sensory-motor and premotor cortex. Also, talking, reading, and other stimuli to the cerebral center are associated with increased blood flow in the appropriate regions of the cortex (Figure 25-6). The mediator of the link between cerebral activity and blood flow has not been established, but nitric oxide and adenosine may be involved.

It is well known that the cerebral vessels are very sensitive to carbon dioxide tension (Pco_2). Increases in arterial blood CO_2 tension ($Paco_2$) elicit marked cerebral vasodilation; inhalation of 7% CO_2 results in a twofold increment in cerebral blood flow. By the same token, decreases in $Paco_2$, which may be elicited by hyperventilation, decrease the cerebral blood flow. Carbon dioxide changes arteriolar resistance by altering the perivascular pH and probably the intracellular pH of the vascular smooth muscle. By independently changing the $Paco_2$ and the bicarbonate concentration, it has been demonstrated that pial vessel diameter (and presumably blood flow) and pH are inversely related, regardless of the level of the $Paco_2$.

The cerebral circulation shows reactive hyperemia and excellent autoregulation between pressures of about 60 and 160 mm Hg. Mean arterial pressures less than 60 mm Hg result in reduced cerebral blood flow and syncope, whereas mean pressures greater than 160 may increase the permeability of the blood-brain barrier and cause cerebral edema.

SPLANCHNIC CIRCULATION

The splanchnic circulation consists of the blood supply to the gastrointestinal tract, liver, spleen, and pancreas. The most noteworthy feature of the splanchnic circulation is that two large capillary beds are partly in series with one another. The small splanchnic arterial branches supply the capillary beds in the gastrointestinal tract, spleen, and pancreas. From these capillary beds the venous blood ultimately flows into the portal vein, which normally provides most of the blood supply to the

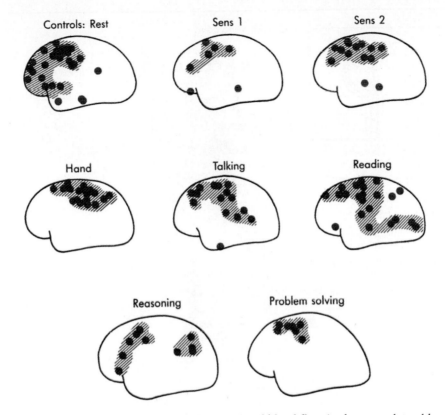

FIGURE 25-6   Effects of different stimuli on regional blood flow in the contralateral human cerebral cortex. *Sens 1,* Low-intensity electrical stimulation of the hand; *Sens 2,* High-intensity electrical stimulation of the hand (pain). (Redrawn from Ingvar DG: *Brain Res* 107:181, 1976.)

liver. However, the hepatic artery also supplies blood to the liver.

Intestinal Circulation

Neural Regulation   The neural control of the mesenteric circulation is almost exclusively sympathetic. Increased sympathetic activity constricts the mesenteric arterioles and capacitance vessels. These responses are mediated by α-receptors, which are prepotent in the mesenteric circulation; however, β-receptors are also present. Infusion of a β-receptor agonist, such as isoproterenol, causes vasodilation.

Autoregulation   Autoregulation in the intestinal circulation is not as well developed as it is in certain other vascular beds, such as those in the brain and kidney. The principal mechanism responsible for autoregulation is metabolic, although a myogenic mechanism probably also participates.

Functional Hyperemia   Food ingestion increases intestinal blood flow. The secretion of the gastrointestinal hormones, gastrin and cholecystokinin, augments intestinal blood flow. The absorption of food also increases intestinal blood flow; the principal mediators of mesenteric hyperemia are glucose and fatty acids.

Hepatic Circulation

Regulation of Flow   Blood flow in the portal venous and hepatic arterial systems varies reciprocally. When blood flow is curtailed in one system, the flow increases in the other system. However, the resultant increase in flow in one system usually does not fully compensate for the initiating reduction in flow in the other system.

The portal venous system does not autoregulate. As the portal venous pressure and flow are raised, resistance either remains constant or decreases. However, the hepatic arterial system does autoregulate.

The sympathetic nerves constrict the presinusoidal resistance vessels in the portal venous and hepatic arterial systems. However, neural effects on the capacitance vessels are more important. The effects are mediated mainly by α-receptors.

Capacitance Vessels   The liver contains about 15% of the total blood volume of the body. Under appropriate conditions, such as in response to hemorrhage, about

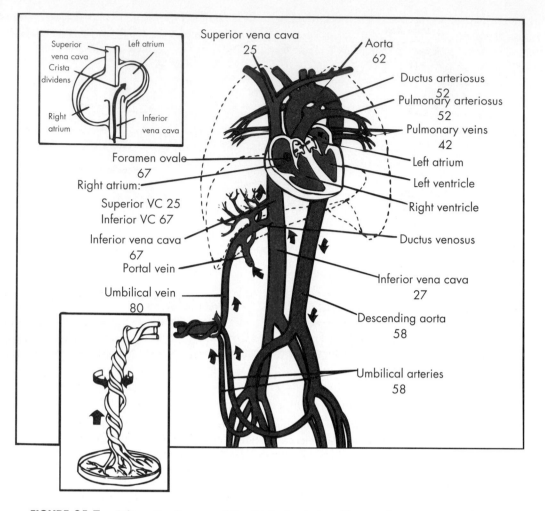

FIGURE 25-7 Schematic diagram of the fetal circulation. The numbers represent the percentage of O_2 saturation of the blood flowing in the indicated blood vessel. The insert at the upper left illustrates the direction of flow of a major portion of the inferior vena cava blood through the foramen ovale to the left atrium. (Values for O_2 saturations are from Dawes GS et al: *J Physiol* 126:563, 1954.)

half of the hepatic blood volume can be rapidly expelled. Thus the liver constitutes an important blood reservoir in humans. In certain species, notably the dog, the spleen also may serve as an effective blood reservoir. It does not play an important role in humans, however.

FETAL CIRCULATION

Before Birth

The circulation of the fetus differs from that of the postnatal infant. The fetal lungs are functionally inactive, and the fetus depends completely on the placenta for oxygen and nutrient supply. Oxygenated fetal blood from the placenta passes through the umbilical vein to the liver. Approximately half passes through the liver, and

the remainder bypasses the liver to the inferior vena cava through the **ductus venosus** (Figure 25-7). In the inferior vena cava, blood from the ductus venosus joins blood returning from the lower trunk and extremities, and this combined stream is in turn joined by blood from the liver through the hepatic veins.

The streams of blood tend to maintain their identity in the inferior vena cava, but they are divided into two streams of unequal size by the edge of the interatrial septum **(crista dividens).** The larger stream, which is mainly blood from the umbilical vein, is shunted to the left atrium through the **foramen ovale,** which lies between the inferior vena cava and the left atrium (*inset,* Figure 25-7). The other stream passes into the right atrium, where it is joined by superior vena caval blood returning from the upper parts of the body and by blood from the myocardium.

In contrast to the adult, in whom the right and left ventricles pump in series, the ventricles in the fetus operate essentially in parallel. Because of the large pulmonary resistance, only one tenth of the right ventricular output goes through the lungs. The remainder passes through the **ductus arteriosus** from the pulmonary artery to the aorta, at a point distal to the origins of the arteries to the head and upper extremities. Blood flows from the pulmonary artery to the aorta because the pulmonary artery resistance is high and the diameter of the ductus arteriosus is as large as the descending aorta.

The large volume of blood coming through the foramen ovale into the left atrium is joined by blood returning from the lungs, and it is pumped out by the left ventricle into the aorta. Most of the blood in the ascending aorta goes to the head, upper thorax, and arms, and the remaining amount joins blood from the ductus arteriosus and supplies the rest of the body and the placenta. The amount of blood pumped by the left ventricle is about half of that pumped by the right ventricle. The major fraction of the blood that passes down the descending aorta comes from the ductus arteriosus and right ventricle and flows by way of the two umbilical arteries to the placenta.

Figure 25-7 indicates the O_2 saturations of the blood at various points of the fetal circulation. Fetal blood leaving the placenta is 80% saturated, but the saturation of the blood passing through the foramen ovale is reduced to 67% by mixing with desaturated blood returning from the lower part of the body and the liver. Addition of the desaturated blood from the lungs reduces the O_2 saturation of left ventricular blood to 62%, which is the level of saturation of the blood reaching the head and upper extremities.

The blood in the right ventricle, a mixture of desaturated superior vena caval blood, coronary venous blood, and inferior vena caval blood, is only 52% saturated with O_2. When the major portion of this blood traverses the ductus arteriosus and joins that pumped out by the left ventricle, the resulting O_2 saturation of blood traveling to the lower part of the body and back to the placenta is 58%. Thus it is apparent that the tissues receiving blood of the highest O_2 saturation are the liver, heart, and upper parts of the body, including the head.

At the placenta the chorionic villi dip into the maternal sinuses, and O_2, CO_2, nutrients, and metabolic waste products exchange across the membranes. The barrier to exchange is large, and the equilibrium of O_2 tension (Po_2) between the two circulations is not reached at normal rates of blood flow. Therefore, Po_2 of the fetal blood leaving the placenta is very low. If fetal hemoglobin did not have a greater affinity for O_2 than adult hemoglobin, the fetus would not receive an adequate O_2 supply. The fetal oxyhemoglobin dissociation curve is shifted to the left. Therefore, at equal O_2 pressures, fetal blood carries significantly more O_2 than does maternal blood.

> If the mother is subjected to hypoxia, the reduced blood Po_2 is reflected in the fetus by tachycardia and an increase in blood flow through the umbilical vessels. If the hypoxia persists or if flow through the umbilical vessels is impaired, fetal distress occurs; it is first manifested as bradycardia.

In early fetal life the high cardiac glycogen levels that prevail may protect the heart from acute periods of hypoxia. The glycogen levels gradually decrease to adult levels by term.

Circulatory Changes That Occur at Birth

The umbilical vessels have thick muscular walls that are very reactive to trauma, tension, sympathomimetic amines, bradykinin, angiotensin, and changes in Po_2. In animals in which the umbilical cord is not tied, hemorrhage of the newborn is prevented because these large vessels constrict in response to one or more of the stimuli just listed. Closure of the umbilical vessels increases total peripheral resistance and blood pressure. When blood flow through the umbilical vein ceases, the ductus venosus, a thick-walled vessel with a muscular sphincter, closes. The factor initiating closure of the ductus venosus is still unknown.

The asphyxia that starts with constriction or clamping of the umbilical vessels, plus the cooling of the body, activate the respiratory center of the newborn infant. After the lungs fill with air, pulmonary vascular resistance decreases to about one tenth of the value that existed before lung expansion. This resistance change is not caused by the presence of O_2 in the lungs, because the change is just as great if the lungs are filled with nitrogen.

The left atrial pressure is raised above that in the inferior vena cava and right atrium by (1) the decrease in pulmonary resistance, with the resulting large flow of blood through the lungs to the left atrium; (2) the reduction of flow to the right atrium caused by occlusion of the umbilical vein; and (3) the increased resistance to left ventricular output produced by occlusion of the umbilical arteries. This reversal of the pressure gradient across the atria abruptly closes the valve over the foramen ovale, and the septal leaflets fuse over several days.

With the decrease in pulmonary vascular resistance, the pressure in the pulmonary artery falls to about half its previous level (to about 35 mm Hg). This change in pressure, coupled with a slight increase in aortic pressure, reverses the flow of blood through the ductus arteriosus. However, within several minutes the large ductus arteriosus begins to constrict, producing turbulent flow, which manifests as a murmur in the newborn. Constriction of the ductus arteriosus is progressive and usually is complete within 1 to 2 days after birth. Closure of the ductus arteriosus appears to be initiated by the high Pao_2 of the arterial blood passing through it; pulmonary ventilation with O_2 closes the ductus, whereas ventilation with air low in O_2 opens this shunt vessel. Whether O_2 acts directly on the ductus or mediates the release of a vasoconstrictor substance is not known.

At birth the walls of the two ventricles are about equally thick, or the right ventricle is slightly thicker. Also, in the newborn, the pulmonary arterioles are thick, which is partly responsible for the high pulmonary vascular resistance of the fetus. After birth the thickness of the walls of the right ventricle diminishes, as does the muscle layer of the pulmonary arterioles; the left ventricular walls become thicker. These changes are progressive for several weeks after birth.

The **ductus arteriosus** occasionally fails to close after birth. This constitutes a common congenital cardiac abnormality that is now amenable to surgical correction.

SUMMARY

Skin Circulation

1. Most of the resistance vessels in the skin are under the dual control of the sympathetic nervous system and local vasodilator metabolites, but the arteriovenous anastomoses found in the skin of the hands, feet, and face are solely under neural control.
2. The main function of skin blood vessels is to aid in the regulation of body temperature by constricting, which conserves heat, and by dilating, which loses heat.
3. Skin blood vessels dilate directly and reflexly in response to heat and constrict directly and reflexly in response to cold.

Skeletal Muscle Circulation

1. Skeletal muscle blood flow is regulated centrally by the sympathetic nerves and locally by the release of vasodilator metabolites.
2. At rest, neural regulation of blood flow is paramount, but it yields to metabolic regulation during muscle contractions.

Coronary Circulation

1. The physical factors that influence coronary blood flow are the viscosity of the blood, the frictional resistance of the vessel walls, the aortic pressure, and the extravascular compression of the vessels within the walls of the left ventricle. Left coronary blood flow is restricted during ventricular systole as a result of extravascular compression, and it is greatest during diastole, when the intramyocardial vessels are not compressed.
2. Neural regulation of coronary blood flow is much less important than is metabolic regulation. Activation of the cardiac sympathetic nerves directly constricts the coronary resistance vessels. However, the enhanced myocardial metabolism caused by the associated increase in heart rate and contractile force produces vasodilation, which overrides the direct constrictor effect of sympathetic nerve stimulation. Stimulation of the cardiac branches of the vagus nerves slightly dilates the coronary arterioles.
3. A striking parallelism exists between metabolic activity of the heart and coronary blood flow. A decrease in oxygen supply or an increase in oxygen demand apparently releases a vasodilator that decreases coronary resistance.
4. In response to gradual occlusion of a coronary artery, collateral vessels develop from adjacent unoccluded arteries, and they supply blood to the compromised myocardium distal to the point of occlusion.

Cerebral Circulation

1. Cerebral blood flow is predominantly regulated by metabolic factors, expecially CO_2, K^+, and adenosine.
2. Increased regional cerebral activity produced by stimuli such as touch, pain, hand motion, talking, reading, reasoning, and problem solving are associated with enhanced blood flow in the activated areas of the cerebral cortex.

Intestinal Circulation

1. The microcirculation in the intestinal villi constitutes a countercurrent exchange system for O_2. This places the villi in jeopardy in states of low blood flow.
2. The splanchnic resistance and capacitance vessels are very responsive to changes in sympathetic neural activity.

Hepatic Circulation

1. The liver receives about 25% of the cardiac ouput; about three fourths of this comes via the portal vein and about one fourth via the hepatic artery.

2. When flow is diminished in either the portal or the hepatic system, flow in the other system usually increases, but not proportionately.

3. The liver tends to maintain a constant O_2 consumption, in part because its mechanism for extracting O_2 from the blood is so efficient.

4. The liver normally contains about 15% of the total blood volume. It serves as an important blood reservoir for the body.

Fetal Circulation

1. In the fetus a large percentage of the right atrial blood passes through the foramen ovale to the left atrium, and a large percentage of the pulmonary artery blood passes through the ductus arteriosus to the aorta.

2. At birth, the umbilical vessels, ductus venosus, and ductus arteriosus close by contraction of their muscle layers.

3. The reduction in the pulmonary vascular resistance caused by lung inflation is the main factor that reverses the pressure gradient between the atria, thereby closing the foramen ovale.

BIBLIOGRAPHY
Journal Articles

Abboud FM, ed: Regulation of the cerebral circulation (symposium), *Fed Proc* 40:2296, 1981.

Baron JF, Vicaut E, Hou X, Duvelleroy M: Independent role of arterial O_2 tension in local control of coronary blood flow, *Am J Physiol* 258:H1388, 1990.

Belardinelli L, Linden J, Berne RM: The cardiac effects of adenosine, *Prog Cardiovasc Dis* 32:73, 1989.

Berne RM: Role of adenosine in the regulation of coronary blood flow, *Circ Res* 47:807, 1980.

Berne RM, Winn HR, Rubio R: The local regulation of cerebral blood flow, *Prog Cardiovasc Dis* 24:243, 1981.

Brunner JJ, Greene AS, Frankle AE, Skoukas AA: Carotid sinus baroreceptor control of splanchnic resistance and capacity, *Am J Physiol* 255:H1305, 1988.

Chou CC: Contribution of splanchnic circulation to overall cardiovascular and metabolic homeostasis, *Fed Proc* 42:1656, 1983.

Clyman RI, Saugstad OD, Mauray F: Reactive oxygen metabolites relax the lamb ductus arteriosus by stimulating prostaglandin production, *Circ Res* 64:1, 1989.

Feigl EO: Coronary physiology, *Phys Rev* 63:1, 1983.

Granger DN, Kvietys PR: The splanchnic venous system: intrinsic regulation, *Annu Rev Physiol* 43:409, 1981.

Greenway CV: The role of the splanchnic venous system in overall cardiovascular homeostasis, *Fed Proc* 42:1678, 1983.

Greenway CV, Lautt WW: Distensibility of hepatic venous resistance sites and consequences on portal pressure, *Am J Physiol* 254:H452, 1988.

Gregg DE: The natural history of coronary collateral development, *Circ Res* 35:335, 1974.

Heymann MA, Iwamoto HS, Rudolf AM: Factors affecting changes in the neonatal systemic circulation, *Ann Rev Physiol* 43:371, 1981.

Heymann MA, Rudolph AM: Control of the ductus arteriosus, *Physiol Rev* 55:62, 1975.

Hoffman JIE, Spaan JAE: Pressure-flow relations in coronary circulation, *Physiol Rev* 70: 331,1990.

Klocke FJ, Ellis AK: Control of coronary blood flow, *Ann Rev Med* 31:489, 1980.

Kontos HA: Regulation of the cerebral circulation, *Ann Rev Physiol* 43:397, 1981.

Laughlin MH: Skeletal muscle blood flow capacity: role of muscle pump in exercise hyperemia, *Am J Physiol* 253:H993, 1987.

Lautt WW: Hepatic nerves: a review of their functions and effects, *Can J Physiol Pharmacol* 58:105, 1980.

Mohri M, Tomoike H, Noma M, Inoue T, Hisano K, Nakamura M: Duration of ischemia is vital for collateral development: repeated brief coronary artery occlusions in conscious dogs, *Circ Res* 64:287,1989.

Olsson RA: Local factors regulating cardiac and skeletal muscle blood flow, *Ann Rev Physiol* 43:385,1981.

Olsson RA, Bunger R: Metabolic control of coronary blood flow, *Prog Cardiovasc Dis* 29:369, 1987.

Rubio R, Berne RM: Regulation of coronary blood flow, *Prog Cardiovasc Dis* 18:105, 1975.

Schwartz LM, McKenzie JE: Adenosine and active hyperemia in soleus and gracilis muscle of cats, *Am J Physiol* 259:H1295, 1990.

Shepherd AP, Riedel GL: Intramural distribution of intestinal blood flow during sympathetic stimulation, *Am J Physiol* 255:H1091, 1988.

Symons JD, Firoozmand E, Longhurst JC: Repeated dipyridamole administration enhances collateral-dependent flow and regional function during exercise: a role for adenosine, *Circ Res* 73:503,1993.

Wearn JT, Mettier SR, Klumpp TG, Zschiesche LJ: The nature of the vascular communications between the coronary arteries and the chambers of the heart, *Am Heart J* 9:143, 1933.

Books and Monographs

Berne RM, Rubio R: Coronary circulation. In *Handbook of physiology,* section 2: The cardiovascular system—the heart, vol I, Bethesda, Md, 1979, American Physiological Society.

Berne RM, Winn HR, Rubio R: Metabolic regulation of cerebral flood flow. In *Mechanisms of vasodilation—second symposium,* New York, 1981, Raven Press.

Donald DE: Splanchnic circulation. In *Handbook of physiology,* section 2: The cardiovascular system—peripheral circulation and organ blood flow, vol III, Bethesda, Md, 1983, Raven Press.

Faber JJ, Thornburg KL: *Placental physiology,* New York, 1983, Raven Press.

Gootman N, Gootman PM, eds: *Perinatal cardiovascular function,* New York, 1983, Marcel Dekker, Inc.

Granger DN, Kvietys PR, Korthuis RJ, Premen AJ: Microcirculation of the intestinal mucosa. In *Handbook of physiology,* Gastrointestinal system, vol I, Betheseda, Md, 1989, American Physiological Society.

Greenway CV, Lautt WW: Hepatic circulation. In *Handbook of physiology,* Gastrointestinal system: motility and circulation, vol I, Bethesda, Md, 1989, American Physiological Society.

Gregg DE: *Coronary circulation in health and disease,* Philadelphia, 1950, Lea & Febiger.

Guth PH, Leung FW, Kauffman GL, Jr: Physiology of gastric circulation. In *Handbook of physiology,* Gastrointestinal system, vol I, Bethesda, Md, 1989, American Physiological Society.

Heistad DD, Kontos HA: Cerebral circulation. In *Handbook of physiology,* section 2: The cardiovascular system—peripheral circulation and organ blood flow, vol III, Bethesda, Md, 1983, American Physiological Society.

Hellon R: Thermoreceptors. In *Handbook of physiology,* section 2: The cardiovascular system—peripheral circulation and organ blood flow, vol III, Bethesda, Md, 1983, American Physiological Society.

Lautt WW, ed: *Hepatic circulation in health and disease,* New York, 1981, Raven Press.

Longo LD, Reneau DD, eds: *Fetal and newborn cardiovascular physiology,* vol 1, Developmental aspects, New York, 1978, Garland STPM Press.

Marcus ML: *The coronary circulation in health and disease,* New York, 1983, McGraw-Hill Book Co.

Mott JC, Walker DW: Neural and endocrine regulation of circulation in the fetus and newborn. In *Handbook of physiology,* section 2: The cardiovascular system—peripheral circulation and organ blood flow, vol III, Bethesda, Md, 1983, American Physiological Society.

Olsson RA, Bunger R, Spaan JAE: Coronary circulation.. In Fozzard HA et al, eds: *The heart and cardiovascular system,* ed 2, New York, 1991, Raven Press.

Owman C, Hardebo JE, eds: *Neural regulation of brain circulation,* Amsterdam, 1985, Elsevier.

Phillis JW, ed: *The regulation of cerebral blood flow,* Boca Raton, Fla, 1993, CRC Press.

Roddie EC: Circulation to skin and adipose tissue. In *Handbook of physiology,* section 2: The cardiovascular system—peripheral circulation and organ blood flow, vol III, Bethesda, Md, 1983, American Physiological Society.

Schaper W: *The collateral circulation of the heart,* New York, 1971, North-Holland Publishing Co.

Schaper W, Bernotat-Danielowski S, Niennaber C, Schaper J: Collateral circulation. In Fozzard HA et al, eds: *The heart and cardiovascular system,* ed 2, New York, 1991, Raven Press.

Shepherd AP, Granger DN: *Physiology of the intestinal circulation,* New York, 1984, Raven Press.

Shepherd JT: Circulation to skeletal muscle. In *Handbook of physiology,* section 2: The cardiovascular system—peripheral circulation and organ blood flow, vol III, Bethesda, Md, 1983, American Physiological Society.

Sparks HV, Jr, Wangler RD, Gorman MW: Control of the coronary circulation. In Sperelakis N, ed: *Physiology and pathophysiology of the heart,* ed 2, Boston, 1989, Wolters-Kluwer Publishers.

CHAPTER 26

Interplay of Central and Peripheral Factors in the Control of the Circulation

The primary function of the circulatory system is to deliver the supplies needed for tissue metabolism and growth and to remove the products of metabolism. To explain how the heart and blood vessels serve this function, it has been necessary to analyze the system morphologically and functionally and to discuss how the component parts contribute to maintaining adequate tissue perfusion under different physiological conditions.

Once the functions of the various components are understood, it is essential that their interrelationships in the overall role of the circulatory system be considered. Tissue perfusion depends on arterial pressure and local vascular resistance; and arterial pressure, in turn, depends on cardiac output and total peripheral resistance (TPR). Arterial pressure is maintained within a relatively narrow range in the normal individual, a feat that is accomplished by reciprocal changes in cardiac output and TPR. However, cardiac output and peripheral resistance are each influenced by a number of factors, and it is the interplay among these factors that determines the level of these two variables. The autonomic nervous system and the baroreceptors play the key role in regulating blood pressure. However, from the long-range point of view, the control of fluid balance by the kidney, adrenal cortex, and central nervous system is crucial to maintaining a constant blood volume.

In a well-regulated system one way to study the extent and sensitivity of the regulatory mechanism is to disturb the system and observe its response to restore the preexisting steady state. Disturbances in the form of physical exercise and hemorrhage will be used to illustrate the effects of the various factors that regulate the circulatory system.

EXERCISE

The cardiovascular adjustments in exercise represent a combination and integration of neural and local (chemical) factors. The neural factors consist of (1) central command, (2) reflexes originating in the contracting muscle, and (3) the baroreceptor reflex. **Central command** is the cerebrocortical activation of the sympathetic nervous system that produces cardiac acceleration, increased myocardial contractile force, and peripheral vasoconstriction. Reflexes can be activated by stimulation of intramuscular mechanoreceptors (stretch, tension) and chemoreceptors (products of metabolism) in response to muscle contraction. Impulses from these receptors travel centrally via small myelinated (group III) and unmyelinated (group IV) afferent nerve fibers. The central connections of this reflex are unknown, but the efferent limb consists of the sympathetic nerve fibers to the heart and peripheral blood vessels. The baroreceptor reflex has been described on p. 316, and the local factors that influence skeletal muscle blood flow (metabolic vasodilators) are described on pp. 314 and 341.

Mild to Moderate Exercise

In humans or in trained animals, anticipation of physical activity inhibits the vagal nerve impulses to the heart and increases sympathetic discharge. The concerted inhibition of parasympathetic control areas and activation of sympathetic control areas in the medulla oblongata elicit an increase in heart rate and myocardial contractility. The tachycardia and the enhanced contractility increase cardiac output.

Peripheral Resistance At the same time that cardiac stimulation occurs, the sympathetic nervous system also changes vascular resistance in the periphery. In skin, kidneys, splanchnic regions, and inactive muscle, sympathetic-mediated vasoconstriction increases vascular resistance, which diverts blood away from these areas (Figure 26-1). This increased resistance in vascular beds of inactive tissues persists throughout the period of exercise.

As cardiac output and blood flow to active muscles

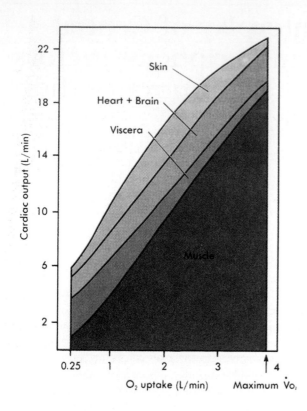

FIGURE 26-1 Approximate distribution of cardiac output at rest and at different levels of exercise up to the maximum O_2 consumption (VO_{2max}) in a normal young man. Viscera refers to splanchnic and renal blood flow. (Redrawn from Ruch HP, Patton TC: *Physiology and biophysics,* ed 12, Philadelphia, 1974, WB Saunders.)

increase with progressive increments in the intensity of exercise, blood flow to the splanchnic and renal vasculatures decreases. Blood flow to the myocardium increases, whereas that to the brain is unchanged. Skin blood flow initially decreases during exercise and then increases as body temperature rises with increments in duration and intensity of exercise. Skin blood flow finally decreases when the skin vessels constrict as the total body O_2 consumption nears maximum (Figure 26-1).

The major circulatory adjustment to prolonged exercise involves the vasculature of the active muscles. Local formation of vasoactive metabolites markedly dilates the resistance vessels; this dilation progresses as the intensity level of exercise is raised. The local accumulation of metabolites relaxes the terminal arterioles. Blood flow through the muscle may increase to fifteen to twenty times above the resting level. This metabolic vasodilation of the precapillary vessels in active muscles occurs very soon after the onset of exercise, and the decrease in TPR enables the heart to pump more blood at a lesser load and more efficiently (less pressure work, p. 345) than if TPR were unchanged.

Only a small percentage of the capillaries are perfused in resting muscle, whereas in actively contracting muscle all or nearly all of the capillaries contain flowing blood (**capillary recruitment**). The surface available for exchange of gases, water, and solutes is increased manyfold. Furthermore, the hydrostatic pressure in the capillaries is increased because of the relaxation of the resistance vessels. Hence, there is a net movement of water and solutes into the muscle tissue. Tissue pressure rises and remains elevated during exercise, as fluid continues to move out of the capillaries and is carried away by the lymphatics. Lymph flow is increased as a result of the rise in capillary hydrostatic pressure and the massaging effect of the contracting muscles on the valve-containing lymphatic vessels.

The contracting muscle avidly extracts O_2 from the perfusing blood (increased $AV-O_2$ difference, Figure 26-2), and the release of O_2 from the blood is facilitated by the nature of oxyhemoglobin dissociation. The reduction in pH caused by the high concentration of CO_2 and the formation of lactic acid, and the increase in temperature in the contracting muscle contribute to shifting the oxyhemoglobin dissociation curve to the right (see Chapter 27). At any given partial pressure of O_2, less O_2 is held by the hemoglobin in the red cells; consequently O_2 removal from the blood is more effective. Oxygen consumption may increase as much as sixtyfold with only a fifteenfold increase in muscle blood flow. However, Pao_2 and $Paco_2$ are normal during exercise. Muscle myoglobin may serve as a limited O_2 store in exercise, and it can release bound O_2 at very low partial pressures. It also facilitates O_2 transport from capillaries to mitochondria by serving as an O_2 carrier.

Cardiac Output The enhanced sympathetic drive and the reduced parasympathetic inhibition of the sinoatrial node continue during exercise; consequently, tachycardia persists. If the work load is moderate and constant, the heart rate will reach a certain level and remain there throughout the period of exercise. However, if the work load increases, a concomitant increase in heart rate occurs until a plateau is reached in severe exercise at about 180 beats/min (Figure 26-2). In contrast to the large increment in heart rate, the increase in stroke volume is only about 10% to 35% (see Figure 26-2), the larger values occurring in trained individuals. (In very well-trained distance runners, whose cardiac outputs can reach six to seven times the resting level, stroke volume reaches about twice the resting value.)

If the baroreceptors are denervated, the cardiac output and heart rate responses to exercise are sluggish when compared to the changes in animals with normally innervated baroreceptors. However, in the absence of autonomic innervation of the heart, as occurs experimentally after total cardiac denervation, exercise still elicits

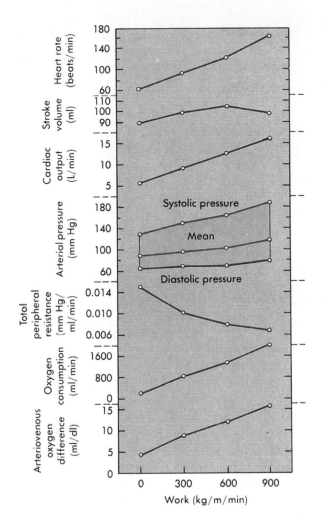

FIGURE 26-2 Effect of different levels of exercise on several cardiovascular variables. (Data from Carlsten A, Grimby G: *The circulatory response to muscular exercise in man,* Springfield, Ill, 1966, Charles C Thomas, Publisher.)

an increment in cardiac output comparable to that observed in normal animals, but then it is associated with an elevated stroke volume.

Venous Return In addition to the contribution made by sympathetically mediated constriction of the capacitance vessels in both exercising and nonexercising parts of the body, venous return to the heart is aided by the working skeletal muscles and the muscles of respiration (see also Chapter 24). The intermittently contracting muscles compress the vessels that course through them. The valves in the compressed veins permit blood flow only toward the right atrium (the valves prevent backflow). The flow of venous blood to the heart is also aided by the increase in the pressure gradient developed by the more negative intrathoracic pressure produced by deeper and more frequent respirations.

The effects of exercise in a conscious dog with in-

duced atrioventricular block and trained to run on a treadmill are shown in Figure 26-3. With the onset of exercise, heart rate, cardiac output, and central venous pressure (right atrial pressure) increase and reach a steady state as long as the ventricles (in this dog with AV node block) are paced at the same rate as the atrium, whose rate is under normal SA node control. However, when ventricular pacing is slowed in a stepwise manner and thereby dissociated from the atrial rate, cardiac output falls and central venous pressure rises in a concomitant stepwise manner. Only when the ventricles are once again paced at the atrial rate do cardiac output and central venous pressures return to their previous exercise levels. The relationship between central venous pressure (indicative of venous return) and cardiac output holds for different levels of exercise and is independent of the autonomic nervous system (**autonomic blockade** with **hexamethonium** and **atropine**). This experiment illustrates the importance of venous return (increased by lowered vascular resistance and muscle pumping in the active muscles) and ventricular rate in producing the increase in cardiac output observed in exercise.

In a normal resting person an increase in heart rate does not usually increase cardiac output appreciably (see Figure 24-11). However, in whole-body exercise, the enhanced venous return caused by a decreased TPR in concert with the skeletal muscle pump (Figure 25-2) is mainly associated with an increase in heart rate. If heart rate cannot increase normally in reponse to exercise (for example, in a patient with complete AV block), the patient's ability to exercise will be severely limited. *Hence, an increase in heart rate is necessary to enable the reduced TPR and the skeletal muscle pump to augment cardiac output appropriately during exercise.*

In humans, little evidence exists that blood reservoirs contribute much to the circulating blood volume, with the exception of the skin, lungs, and liver. In fact, blood volume is usually reduced slightly during exercise, as evidenced by a rise in the hematocrit ratio, because of water loss externally by sweating and enhanced ventilation, and by fluid movement into the contracting muscle. The fluid loss from the vascular compartment into the interstitium of contracting muscle reaches a plateau as interstitial fluid pressure rises and opposes the increased hydrostatic pressure in the capillaries of the active muscle. The fluid loss is partially offset by movement of fluid from the splanchnic regions and inactive muscle into the bloodstream. This influx of fluid occurs as a result of a decrease in hydrostatic pressure in the capillaries of these tissues and an increase in plasma osmolarity, because of movement of osmotically active particles into the blood from the contracting muscle. In addition, reduced urine formation by the kidneys helps to conserve body water.

In light to moderate exercise, blood returning to the

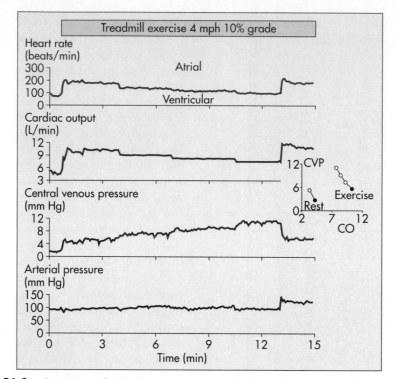

FIGURE 26-3   Response of central venous pressure (CVP) to changes in cardiac output (CO) produced by reduction in ventricular pacing rate in an atrioventricular (AV)-blocked dog. Step reductions in CO during exercise are accompanied by step increases in CVP. *Inset:* Plot of CVP versus CO from four steady-state levels of CO during exercise. (From Sheriff DD et al: *Am J Physiol* 265:316, 1993).

heart is so rapidly pumped through the lungs and out into the aorta that central venous pressure (diastolic filling pressure) increases only slightly. In fact, *chest x-ray films of individuals at rest and during exercise reveal a decrease in heart size in exercise,* which is in harmony with the observations of a constant ventricular diastolic volume. However, in maximum or near-maximum exercise, right atrial pressure and end-diastolic ventricular volume do increase. Thus, the Frank-Starling mechanism contributes to the enhanced stroke volume in vigorous exercise.

Arterial Pressure If the exercise involves a large proportion of the body musculature, such as in running or swimming, the reduction in total vascular resistance can be considerable (see Figure 26-2). Nevertheless, arterial pressure starts to rise with the onset of exercise, and the increase in blood pressure roughly parallels the severity of the exercise performed (see Figure 26-2). Therefore, the increase in cardiac ouput is proportionally greater than the decrease in TPR. The vasoconstriction produced in the inactive tissues by the sympathetic nervous system (and to some extent by the release of catecholamines from the adrenal medulla) is important for maintaining normal or increased blood pressure; sympathectomy or drug-induced block of the adrenergic sympathetic nerve fibers results in a decrease in arterial pressure (**hypotension**) during exercise.

Sympathetic-mediated vasoconstriction also occurs in active muscle when additional muscles are activated after about half of the total skeletal musculature is contracting. In experiments in which one leg is working at maximum levels and then the other leg starts to work, blood flow decreases in the first working leg. Furthermore, blood levels of norepinephrine rise significantly in exercise, and most of it comes from sympathetic nerves in the active muscles.

As body temperature rises during exercise, the skin vessels dilate in response to thermal stimulation of the heat-regulating center in the hypothalamus, and TPR decreases further. This would result in a decline in blood pressure were it not for the increasing cardiac output and constriction of arterioles in the renal, splanchnic, and other tissues.

In general, mean arterial pressure rises during exercise as a result of the increase in cardiac output. However, the effect of enhanced cardiac output is offset by the overall decrease in TPR so that the mean blood pressure increase is relatively small (Figure 26-2). Vasoconstriction in the inactive vascular beds contributes to the maintenance of a normal arterial blood pressure for adequate perfusion of the active tissues. The actual pressure attained represents a balance between cardiac output and TPR. Systolic pressure usually increases more

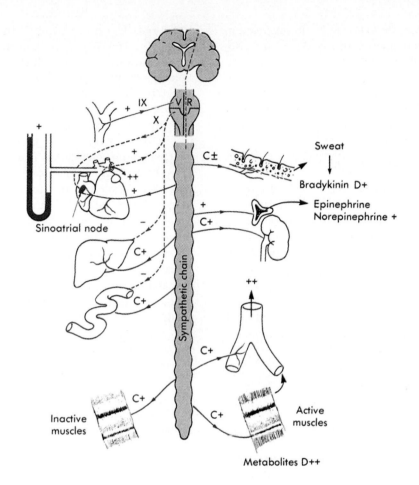

FIGURE 26-4   Cardiovascular adjustments in exercise. *VR,* Vasomotor region; *C,* vasoconstrictor activity; *D,* vasodilator activity; *IX,* glossopharyngeal nerve; *X,* vagus nerve; +, increased activity, −, decreased activity.

than diastolic pressure, which results in an increase in pulse pressure (Figure 26-2). The larger pulse pressure is primarily attributable to a greater stroke volume and to a lesser degree to a more rapid ejection of blood by the left ventricle with less peripheral runoff during the brief ventricular ejection period.

Severe Exercise

In severe exercise taken to the point of exhaustion, the compensatory mechanisms begin to fail. Heart rate attains a maximum level of about 180 beats/minute, and stroke volume reaches a plateau and often decreases, resulting in a fall in blood pressure. Dehydration occurs. Sympathetic vasoconstrictor activity supersedes the vasodilator influence on the cutaneous vessels, and its effect on the capacitance vessels produces a slight increase in effective blood volume. However, cutaneous vasocon-striction also decreases the rate of heat loss. Body temperature is normally elevated in exercise, and reduction in heat loss through cutaneous vasoconstriction can, under these conditions, lead to very high body temperatures with associated feelings of acute distress **(heat exhaustion** or **heat stroke).** The tissue and blood pH decrease, as a result of increased lactic acid and CO_2 production. The reduced pH is probably the key determinant of the maximum amount of exercise an individual can tolerate because of muscle pain and the subjective feeling of exhaustion.

A summary of the neural and local effects of exercise on the cardiovascular system is presented in Figure 26-4.

Postexercise Recovery

When exercise stops, an abrupt decrease in heart rate and cardiac output occurs—the sympathetic drive to the heart is essentially removed. In contrast, TPR remains low for some time after the exercise is ended, presumably because of the accumulation of vasodilator metabolites in the muscles during the exercise period. As a result of the reduced cardiac output and persistence of vasodilation in the muscles, arterial pressure may fall briefly below preexercise levels. Blood pressure is then stabilized at normal levels by the baroreceptor reflexes.

Limits of Exercise Performance

The two main forces that could limit skeletal muscle performance are the rate of O_2 utilization by the muscles and the O_2 supply to the muscles. Muscle O_2 usage is probably not critical, because during exercise maximum O_2 consumption (Vo_{2max}) by a large percentage of the body muscle mass is not increased when additional muscles are activated. If muscle O_2 utilization were limiting, recruitment of more contracting muscle would use additional O_2 to meet the enhanced O_2 requirements and would thereby increase total body O_2 consumption. Therefore O_2 supply to the active muscles must be at fault. Limitation of O_2 supply could be caused by inadequate oxygenation of blood in the lungs or by limitation of the supply of O_2-laden blood to the muscles. Failure of the lungs to fully oxygenate blood can be excluded, because even with the most strenuous exercise at sea level, arterial blood is fully saturated with O_2. Therefore O_2 delivery (blood flow) to the active muscles appears to be the limiting factor in muscle performance.

This limitation in muscle performance translates into the inability of the heart to increase its output beyond a certain level. Cardiac output equals heart rate times stroke volume, and heart rate reaches maximum levels before Vo_{2max} is reached. Hence, stroke volume must be the limiting factor. However, blood pressure provides the energy for skeletal muscle perfusion, and blood pressure depends on peripheral resistance as well as on cardiac output. During intense exercise at peak Vo_{2max} and peak cardiac output, blood pressure would fall as more muscle vascular beds dilate in response to locally released vasodilator metabolites if some centrally mediated vasoconstriction (via the baroreceptor reflex) did not occur in the resistance vessels of the active muscles. Hence, the adjustment of resistance in the active muscles appears to be an important factor in the limitation of whole-body exercise (swimming and running), but *the major factor is the pumping capacity of the heart.* When the exercise involves only a small group of muscles (e.g., those of the hand), the cardiovascular system is not taxed, and the limiting factor, although unknown, lies within the muscle.

Physical Training and Conditioning

The response of the cardiovascular system to regular exercise is to increase its capacity to deliver O_2 to the active muscles and to improve the ability of the muscle to use O_2. The Vo_{2max} is quite reproducible in a given individual, and it varies with the level of physical conditioning. Training progressively increases the Vo_{2max}, which reaches a plateau at the highest level of conditioning. Highly trained athletes have a lower resting heart rate, greater stroke volume, and lower peripheral resistance than they had before training or after deconditioning (becoming sedentary). The low resting heart rate is caused by a higher vagal tone and a lower sympathetic tone. With exercise, the maximum heart rate of the trained individual is the same as that in untrained persons, but it is attained at a higher level of exercise. The trained person also exhibits a low vascular resistance that is inherent in the muscle. For example, if an individual exercises one leg regularly over an extended period and does not exercise the other leg, the vascular resistance of the resting individual is lower and the Vo_{2max} is higher in the "trained" leg than in the "untrained" leg. Also, the well-trained athlete has a lower resting sympathetic outflow to the viscera than does a sedentary counterpart.

Physical conditioning is also associated with greater extraction of O_2 from the blood (greater AV-O_2 difference) by the muscles, but not with an improvement in cardiac contractility. With long-term training, capillary density in skeletal muscle increases, as do the oxidative enzymes in the mitochondria. Also, the ATPase activity, myoglobin, and enzymes involved in lipid metabolism appear to increase with physical conditioning.

Endurance training, such as running or swimming, increases left ventricular volume without increasing left ventricular wall thickness. In contrast, strength exercises, such as weight lifting, increase left ventricular wall thickness **(hypertrophy),** but have little effect on ventricular volume. However, this increase in wall thickness is small relative to that observed in **chronic hypertension,** in which the elevation of afterload persists because of the high peripheral resistance.

HEMORRHAGE

The principal physical findings in an individual who has lost a large quantity of blood are related to the cardiovascular system. The arterial systolic, diastolic, and pulse pressures are reduced and the arterial pulse is rapid and feeble. The cutaneous veins are collapsed and fill slowly when they are compressed centrally. The skin is pale, moist, and slightly **cyanotic** (blue). Respiration is rapid, but the depth of respiration may be shallow or deep.

Course of Arterial Blood Pressure Changes

Cardiac output decreases as a result of blood loss (see Chapter 24). The changes in mean arterial pressure evoked by an acute hemorrhage in experimental animals are illustrated in Figure 26-5. If the animal is bled rapidly to bring mean arterial pressure to 50 mm Hg, the arterial pressure tends to rise spontaneously toward control over the subsequent 20 to 30 minutes. In some animals (curve *A*, Figure 26-5) this trend continues, and normal pressures are regained within a few hours; these animals tend to survive even if their shed blood is not returned to them. In other animals (curve *B*), the pressure begins to decline after the initial rise, and it continues to fall at an accelerating rate until death ensues. If their shed blood is transfused early in the phase of declining arterial pressure, the animals will usually survive. However,

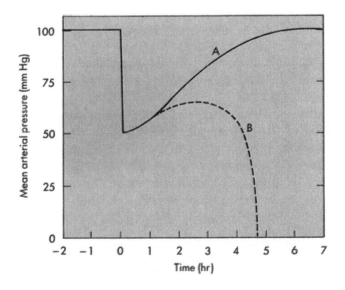

FIGURE 26-5 The changes in mean arterial pressure after a rapid hemorrhage. At time zero, the animal is bled rapidly to a mean arterial pressure of 50 mm Hg. The arterial pressure tends to return toward normal at first in most animals. In some animals (curve *A*), the pressure will continue to improve until the normal pressure is attained, even though none of the shed blood has been transfused back into the animal. In other animals (curve *B*), after a temporary period of apparent improvement, the arterial blood pressure will begin to decline, and it will fall at a steeper and steeper rate until death ensues.

at some point later in this phase of declining pressure, the deterioration becomes irreversible.

This state, known as **hemorrhagic shock,** has a lethal outcome, regardless of whether they receive a transfusion of their shed blood, a transfusion of matched donor blood of a volume substantially greater than the shed blood volume, or any other treatment presently available.

Compensatory Mechanisms

The prominent tendency for arterial blood pressure to rise toward normal levels immediately after an acute blood loss (Figure 26-5) indicates that potent compensatory mechanisms must be invoked. Any mechanism that acts to restore normal levels of arterial blood pressure in response to blood loss may be designated a **negative feedback mechanism.** It is termed "negative" because the induced change in pressure is opposite to its initiating change. The following negative feedback responses are evoked by hemorrhage: (1) the baroreceptor reflexes, (2) the chemoreceptor reflexes, (3) cerebral ischemia responses, (4) reabsorption of tissue fluids, (5) release of endogenous vasoconstrictor substances, and (6) renal conservation of salt and water.

Baroreceptor Reflexes The reduction in mean arterial pressure and in pulse pressure during hemorrhage decreases the stimulation of the baroreceptors in the carotid sinuses and aortic arch (see Chapter 23). Several cardiovascular responses are thus evoked, all of which act to restore the normal arterial blood pressure. Unloading of the arterial baroreceptors decreases vagal tone and augments sympathetic tone, and both of these changes in efferent neural activity increase heart rate and enhance myocardial contractility (see Chapter 19).

The increased sympathetic discharge also produces generalized venoconstriction, which has the same hemodynamic consequences as a transfusion of blood (see Chapter 24). Sympathetic activation constricts certain blood reservoirs, which provides an autotransfusion of blood into the circulating bloodstream. In the dog, considerable quantities of blood are mobilized by contraction of the spleen. In humans, the spleen is not an important blood reservoir. Instead, the cutaneous and hepatic vasculatures probably constitute the principal blood reservoirs.

Generalized arteriolar vasoconstriction is a prominent response to the diminished baroreceptor stimulation during hemorrhage. The reflex increase in peripheral resistance minimizes the decline in arterial pressure that results from the reduction of cardiac output. Figure 25-6 shows the effect of graded blood loss on mean arterial pressure, heart rate, hindquarter blood flow, and hindquarter vascular resistance in a conscious, chronically instrumented rabbit. Under control conditions (Figure 26-6, *A*), removal of up to 25% of the estimated blood volume was attended by substantial reductions in hindquarter blood flow and increases in heart rate and hindquarter vascular resistance, but mean arterial pressure was only slightly reduced. However, when blood loss exceeded about 70% of the estimated blood volume, the reductions in arterial blood pressure and in blood flow became much more pronounced.

After the blood volume had been restored and then the sinoaortic baroreceptors were subsequently denervated (Figure 26-6, *B*), the hindquarter blood flow was substantially less and the heart rate and vascular resistance were much greater than were the respective values that prevailed prior to bleeding when the

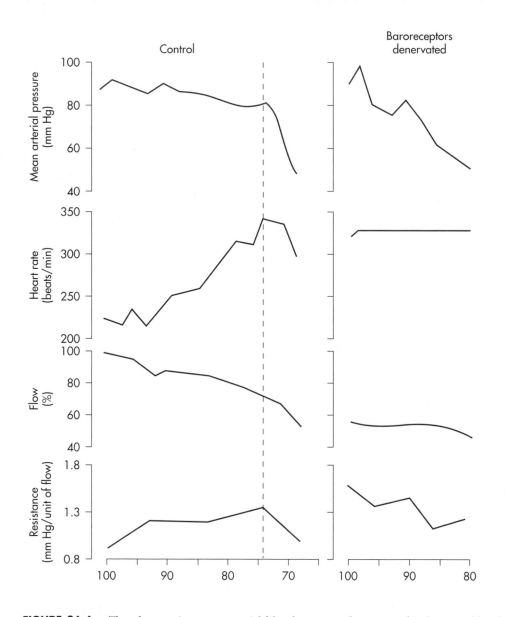

FIGURE 26-6 The changes in mean arterial blood pressure, heart rate, hindquarter blood flow, and hindquarter vascular resistance in a chronically instrumented, unanesthetized rabbit. Blood was withdrawn from a peripheral vein under control conditions **(A),** and several days later, after sino-aortic baroreceptor denervation **(B).** Hindquarter blood flow is expressed as a percentage of the flow that prevailed prior to blood withdrawal and prior to baroreceptor denervation. Hindquarter vascular resistance is the ratio of mean arterial pressure to hindquarter blood flow. (Redrawn from Courneya et al: *Am J Physiol* 261:H380, 1991.)

baroreceptor innervation was intact. Removal of up to 20% of the blood volume in this animal after baroreceptor denervation caused a profound reduction in blood pressure, but it had little influence on the heart rate or hindquarter blood flow. The hindquarter vascular resistance tended to decrease as blood was gradually withdrawn. The reduction in vascular resistance was probably mediated by the release of vasodilator metabolites (adenosine, lactate, etc.) from the tissues consequent to the reductions in blood flow. From this experiment, therefore, it is evident that the arterial baroreceptor reflexes are very important in sustaining arterial blood pressure under conditions of small to moderate losses of blood.

Probably the principal hemodynamic mechanism for preventing severe reductions in arterial blood pressure in response to hemorrhage is generalized peripheral vasoconstriction. Although the arteriolar vasoconstriction is widespread throughout the systemic vascular bed, it is by no means uniform. Hemorrhage-induced vasoconstriction is most severe in the cutaneous, skeletal muscle, and splanchnic vascular beds, and it is slight or absent in the cerebral and coronary circulations. In many instances the cerebral and coronary vascular resistances are diminished. Thus the reduced cardiac output is redistributed to favor flow through the brain and the heart, which are certainly among the most crucial of the body's organs.

The severe cutaneous vasoconstriction accounts for the characteristic pale, cold skin of patients suffering from blood loss. Warming the skin of such patients improves their appearance considerably, much to the satisfaction of well-meaning individuals who might render first aid. However, it also inactivates an effective, natural compensatory mechanism, to the patient's possible detriment.

In the early states of mild to moderate hemorrhage, the changes in renal resistance are usually slight. The tendency for increased sympathetic activity to constrict the renal vessels is counteracted by autoregulatory mechanisms (see Chapter 33). With more prolonged and more severe hemorrhages, however, renal vasoconstriction becomes intense. The reductions in renal circulation are most severe in the outer layers of the renal cortex; the inner zones of the cortex and outer zones of the medulla are spared.

The severe renal and splanchnic vasoconstriction during hemorrhage helps to minimize the arterial hypotension and favors the circulation to the heart and brain.

If the renal and splanchnic vasoconstriction persists too long, it may be harmful. Frequently patients survive the acute hypotensive period only to die several days later from **kidney failure** resulting from renal ischemia.

Intestinal ischemia also may have dire effects. In the dog, for example, intestinal bleeding and extensive sloughing of the mucosa occur after only a few hours of hemorrhagic hypotension. Furthermore, the low splanchnic flow causes the centrilobular cells in the liver to swell. The resulting obstruction of the hepatic sinusoids raises the portal venous pressure and thereby intensifies the intestinal blood loss. Fortunately the pathological changes in the liver and intestine are usually much less severe in humans than in dogs.

Chemoreceptor Reflexes Hemorrhage-induced reductions in arterial pressure below about 60 mm Hg do not evoke any additional responses through the baroreceptor reflexes, because this pressure level constitutes the threshold for baroreceptor stimulation (see Chapter 23). However, low arterial pressure may lead to peripheral chemoreceptor stimulation, because the inadequate blood flow to the aortic and carotid bodies will diminish the Po_2 in the chemoreceptor tissues. Chemoreceptor excitation augments the prevailing vasoconstriction evoked by the baroreceptor reflexes. Also, the respiratory stimulation induced by chemoreceptor excitation assists venous return by the auxiliary pumping mechanism described in Chapter 24.

Cerebral Ischemia When hemorrhage lowers the arterial blood pressure below about 40 mm Hg, the resulting cerebral ischemia activates the sympathoadrenal system. The resulting sympathetic nervous discharge is intense; it is several times greater than the maximum activity that occurs when the baroreceptors cease to be stimulated. Therefore, the vasoconstriction and facilitation of myocardial contractility evoked by cerebral ischemia may be pronounced. When cerebral ischemia is severe, the vagal centers also become activated. The resulting bradycardia may aggravate the hypotension that initiated the cerebral ischemia.

Reabsorption of Tissue Fluids The arterial hypotension, arteriolar constriction, and reduced central venous pressure induced by hemorrhage lower the hydrostatic pressure in the capillaries. Consequently, the balance of forces across the capillary endothelium promotes the net reabsorption of interstitial fluid into the vascular compartment. The rapidity of this response is displayed in Figure 26-7. In a group of cats, 45% of the estimated blood volume was removed over a 30-minute period. The mean arterial blood pressure declined rapidly to about 45 mm Hg during the hemorrhage. When bleeding was

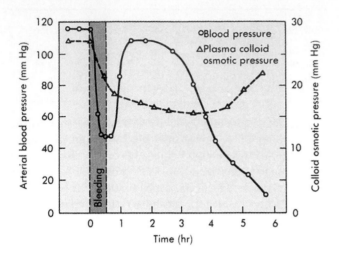

FIGURE 26-7 The changes in arterial blood pressure and plasma colloid osmotic pressure in response to withdrawal of 45% of the estimated blood volume over a 30-minute period, beginning at time zero. (Redrawn from Zweifach BW: *Anesthesiology* 41:157, 1974.)

terminated, the arterial pressure rose rapidly, but only temporarily, to near the control level. About 2.5 hours after cessation of the hemorrhage, the arterial pressure began to fall, and it declined progressively until the animals died. The time course of this response to hemorrhage resembles that of curve *B* in Figure 26-5.

Associated with these changes in arterial blood pressure, the plasma colloid osmotic pressure declined markedly during the bleeding and it continued to decrease more gradually for several hours (Figure 26-7). The reduction in colloid osmotic pressure reflects the dilution of the blood by tissue fluids. Considerable quantities of fluid thus may be drawn into the circulation during hemorrhage; about 0.25 ml of fluid per minute per kilogram of body weight may be reabsorbed. Approximately 1 liter of fluid per hour might be autoinfused into the circulatory system of an average individual from the interstitial spaces after an acute blood loss.

Considerable quantities of fluid also may be slowly shifted from the intracellular to the extracellular spaces. This fluid exchange is probably mediated by the secretion of cortisol from the adrenal cortex in response to hemorrhage. Cortisol appears to be essential for a full restoration of the plasma volume after hemorrhage.

Endogenous Vasoconstrictors The **catecholamines** (mainly epinephrine, but also norepinephrine) are released from the adrenal medulla in response to those types of stimuli, such as acute blood loss, that evoke widespread sympathetic nervous discharge (see Chapter 47). The concentrations of epinephrine and norepinephrine increase substantially in the arterial blood within the first minute of hemorrhage, and remain el-

evated (up to fifty times normal) throughout the period of hemorrhage. Retransfusion of the shed blood will reduce the blood catecholamine levels to normal. The epinephrine that appears in the blood in response to hemorrhage comes almost exclusively from the adrenal medulla, whereas the norepinephrine is derived both from the adrenal medulla and the peripheral sympathetic nerve endings throughout the body. The catecholamines released from the adrenal medulla reinforce the effects of the reflexly induced sympathetic nervous activity that has been described above.

Vasopressin, a potent vasoconstrictor, is actively secreted by the posterior pituitary gland in response to hemorrhage (see Chapter 44). Removal of about 20% of the blood volume in experimental animals increases vasopressin secretion about fortyfold. The sensory receptors responsible for the augmented release are the sino-aortic baroreceptors and the stretch receptors in the left atrium.

The diminished renal perfusion during hemorrhagic hypotension leads to the secretion of **renin** from the juxtaglomerular apparatus (see Chapters 37 and 46). Consequently, the concentration of renin in the peripheral blood rises as a function of the volume of shed blood. Renin is an enzyme that acts on the plasma protein **angiotensinogen** to form **angiotensin,** which is a very powerful vasoconstrictor.

Renal Conservation of Water The diminution in arterial blood pressure induced by the loss of blood decreases the glomerular filtration rate, and thus curtails the excretion of water and electrolytes (see Chapter 37). During hemorrhage, fluid and electrolytes are also conserved by the kidneys in response to the release of various hormones, including vaspressin and renin, as noted in the preceding section. The peptide angiotensin, formed by the action of renin, accelerates the release of **aldosterone** from the adrenal cortex. Aldosterone, in turn, stimulates sodium reabsorption by the renal tubules, and water accompanies the sodium that is actively reabsorbed (see Chapters 36 and 37).

Decompensatory Mechanisms In contrast to the negative feedback mechanisms just described, latent **positive feedback mechanisms** are also invoked by hemorrhage. Such mechanisms exaggerate any primary change initiated by the blood loss. Specifically, positive feedback mechanisms aggravate the hypotension induced by blood loss and tend to initiate vicious cycles, which may lead to death. The operation of positive feedback mechanisms is manifested in the accelerating and eventually lethal decline in blood pressure reflected by curve *B* of Figure 26-5.

Whether a positive feedback mechanism will lead to a vicious cycle depends on the gain of that mechanism; **gain** is the ratio of the secondary change evoked by a

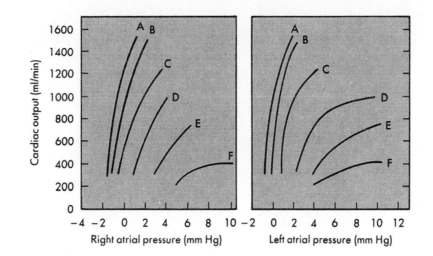

FIGURE 26-8 Ventricular function curves for the right and left ventricles during the course of hemorrhagic shock. Curve *A* represents the control curve. Curves *B* through *F* represent the curves at 117, 247, 280, 295, and 310 minutes, respectively, after the initial hemorrhage. (Redrawn from Crowell JW, Guyton AC: *Am J Physiol* 203:248, 1962.)

given feedback mechanism to the initiating change itself. A gain greater than 1 induces a vicious cycle; a gain less than 1 does not. For example, consider a positive feedback mechanism with a gain of 2. If some intervention decreases the mean arterial blood pressure by 10 mm Hg, the positive feedback mechanism (with a gain of 2) would then evoke a secondary pressure reduction of 20 mm Hg. This secondary change would in turn cause a further decrement of 40 mm Hg; that is, each change would induce a subsequent change of twice the magnitude. Hence mean arterial pressure would decline at an ever-increasing rate until death supervened, much as is depicted by curve *B* in Figure 26-5.

A positive feedback mechanism with a gain less than 1 would also amplify the effect of any initiating intervention, but it would not necessarily generate a vicious cycle. For example, if an intervention suddenly decreased the arterial blood pressure by 10 mm Hg, a positive feedback mechanism with a gain of 0.5 would initiate a secondary, additional pressure decline of 5 mm Hg. This, in turn, would provoke a decrease of 2.5 mm Hg. The process would continue in ever-diminishing steps, and the arterial blood pressure would approach an equilibrium value asymptotically.

> Some of the important positive feedback mechanisms that are invoked in response to hemorrhage include (1) cardiac failure, (2) acidosis, (3) inadequate cerebral blood flow, (4) aberration of blood clotting, and (5) depression of the reticuloendothelial system.

Cardiac Failure The role of cardiac failure in the progression of shock during hemorrhage is controversial. All investigators agree that the heart fails terminally, but the importance of cardiac failure during earlier stages of hemorrhagic hypotension remains to be established. Shifts to the right in ventricular function curves (Figure 26-8) constitute experimental evidence of a progressive depression of myocardial contractility during hemorrhage.

The hypotension induced by hemorrhage reduces the coronary blood flow, which tends to depress ventricular function. The consequent reduction in cardiac output leads to a further decline in arterial pressure, a classical example of a positive feedback mechanism. Furthermore, the reduced tissue blood flow leads to a local accumulation of vasodilator metabolites, which decreases peripheral resistance and therefore serves to intensify the decline in arterial pressure.

Acidosis The inadequate blood flow during hemorrhage affects the metabolism of all cells in the body. The resulting stagnant anoxia accelerates the production of lactic acid and other acid metabolites by the tissues. Furthermore, impaired kidney function prevents adequate excretion of the excess H^+, and generalized metabolic acidosis ensues (Figure 26-9). The depressant effect of the metabolic acidosis on the heart further reduces tissue perfusion and thus aggravates the acidosis. The reactivity of the heart and resistance blood vessels to neurally released and circulating catecholamines is diminished by the acidosis, and therefore the hypotension is intensified.

Central Nervous System Depression The hypotension in shock reduces cerebral blood flow. Moderate de-

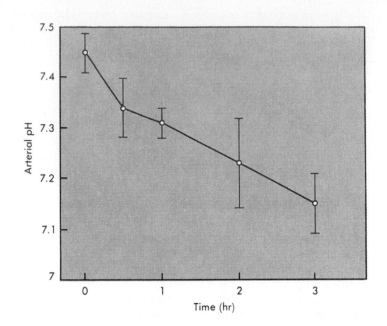

FIGURE 26-9   The reduction in arterial blood pH (mean ± SD) in a group of eleven dogs whose blood pressure had been held at a level of 35 mm Hg by bleeding into a reservoir, beginning at time zero. (Modified from Markov AK, Oglethorpe N, Young DB, Hellems HK: *Circ Shock* 8:9, 1981).

grees of cerebral ischemia induce a pronounced sympathetic nervous stimulation of the heart, arterioles, and veins. Hence, moderate cerebral ischemia invokes negative feedback mechanisms, as explained above. When hypotension is severe, however, the cardiovascular centers in the brainstem are depressed, because the blood flow to the brain is inadequate. The resulting reduction of sympathetic neural activity then reduces cardiac output and peripheral resistance. The consequent decline in mean arterial pressure intensifies the inadequate cerebral perfusion.

Various endogenous **opioids,** such as **enkephalins** and β-**endorphin,** may be released into the brain substance or into the circulation in response to the same stresses that provoke circulatory shock. Enkephalins exist along with catecholamines in secretory granules in the adrenal medulla, and they are released together in response to stress. Similar stimuli release β-endorphin from the anterior pituitary gland. These opioids depress the centers in the brainstem that mediate some of the compensatory autonomic adaptations to blood loss and other shock-provoking stresses. Conversely, the opioid antagonist **naloxone** improves cardiovascular function and survival in various forms of shock.

Aberrations of Blood Clotting *The alterations of blood clotting after hemorrhage are typically biphasic—an initial phase of hypercoagulability is followed by a secondary phase of hypocoagulability and fibrinolysis* (see Chapter 15). In the initial phase, intravascu-

lar clots, or **thrombi,** develop within a few minutes of the onset of severe hemorrhage, and coagulation may be extensive throughout the microcirculation.

Thromboxane A$_2$, which is released from various ischemic tissues, aggregates platelets, The trapped platelets release more thromboxane A$_2$, which then serves to trap additional platelets. This form of positive feedback intensifies the clotting tendency. The mortality from certain standard shock-provoking procedures has been reduced considerably by the administration of certain anticoagulants such as **heparin.**

In the later stages of hemorrhagic hypotension, the clotting time is prolonged and fibrinolysis is prominent. Hemorrhage into the intestinal lumen is common after several hours of hemorrhagic hypotension in the dog and certain other species. Blood loss into the intestinal lumen would, of course, aggravate the hemodynamic effects of the original hemorrhage.

Reticuloendothelial System Hemorrhage depresses the phagocytic activity of the reticuloendothelial system (RES). Consequently, the antibacterial and antitoxic defense mechanisms of the body are impaired. Endotoxins from the normal bacterial flora of the intestine constantly enter the circulation. Ordinarily they are inactivated by the RES, principally in the liver. When RES function is depressed, these endotoxins invade the general circulation. *Endotoxins tend to lower arterial blood pressure by inducing the synthesis of nitric oxide synthase, mainly in the vascular smooth muscle.* The nitric oxide generated by the activity of this enzyme is a potent vasodilator. Therefore, the reduction of arterial blood pressure induced by the generation of nitric oxide intensifies the hypotension caused by diminished blood volume itself.

Interactions of Positive and Negative Feedback Mechanisms

Hemorrhage alters a multitude of circulatory and metabolic activities. Some of these alterations are compensatory; others are decompensatory. For those activities that are involved in feedback systems, some of those systems possess a high gain; others, a low gain. Furthermore, the gain of any specific system usually varies with the severity of the hemorrhage. For example, when the loss of blood is minimal, mean arterial pressure is usually within the normal range, and the gain of the baroreceptor reflexes is high. When the hemorrhage is more severe, mean arterial pressure may be below about 60 mm Hg (that is, below the threshold for the baroreceptors), and therefore further reductions of pressure will have no additional influence via the baroreceptor reflexes. Hence, below this critical pressure the baroreceptor reflex gain will be near zero.

When blood loss is minor, the gains of the negative feedback mechanisms are usually high, whereas those of the positive feedback mechanisms are usually low. The converse is true when hemorrhage is more severe. The gains of the various mechanisms are additive algebraically. Therefore, the generation of a vicious cycle depends on whether the sum of the various positive and negative gains exceeds 1. The development of a vicious cycle is, of course, more likely when blood losses are severe. Therefore, to avert a vicious cyle, serious hemorrhages must be treated quickly and intensively, mainly by blood replacement, before the process becomes irreversible.

SUMMARY

Exercise

1. In anticipation of exercise the vagus nerve impulses to the heart are inhibited and the sympathetic nervous system is activated by central command. The result is an increase in heart rate, myocardial contractile force, and regional vascular resistance.
2. With exercise, vascular resistance increases in skin, kidneys, splanchnic regions, and inactive muscles and decreases in active muscles. The increase in cardiac output is mainly caused by the increase in heart rate. Stroke volume increases only slightly. Total peripheral resistance decreases, oxygen consumption and blood oxygen extraction increase, and systolic and mean blood pressure increase slightly.
3. As body temperature rises during exercise, the skin blood vessels dilate. However, when heart rate becomes maximal during severe exercise, the skin vessels constrict. This increases the effective blood volume but causes greater increases in body temperature and a feeling of exhaustion.
4. The limiting factor in exercise performance is the delivery of blood to the active muscles.

Hemorrhage

1. Acute blood loss induces the following hemodynamic changes: tachycardia, hypotension, generalized arteriolar vasoconstriction, and generalized venoconstriction.
2. Acute blood loss invokes various negative feedback (compensatory) mechanisms, such as baroreceptor and chemoreceptor reflexes, responses to moderate cerebral ischemia, reabsorption of tissue fluids, release of endogenous vasoconstrictors, and renal conservation of water and electrolytes.
3. Acute blood loss also induces various positive feedback (decompensatory) mechanisms, such as cardiac failure, acidosis, central nervous system depression, aberrations of blood coagulation, and depression of the reticuloendothelial system.
4. The outcome of an acute blood loss depends on the gains of the various feedback mechanisms and on the interactions between the positive and negative feedback mechanisms.

BIBLIOGRAPHY
Journal Articles

Abboud FM et al: Role of vasopressin in cardiovascular and blood pressure regulation, *Blood Vessels* 27:106, 1990.

Abel FI: Myocardial function in sepsis and endotoxin shock, *Am J Physiol* 257:R1265, 1989.

Blomqvist CG, Saltin B: Cardiovascular adaptations to physical training, *Ann Rev Physiol* 15:169. 1983.

Booth FW, Thomason DB: Molecular and cellular adaptation of muscle in response to exercise: perspectives of various models, *Physiol Rev* 71:541, 1991.

Briand R, Yamaguchi N, Gagne J: Plasma catecholamine and glucose concentrations during hemorrhagic hypotension in anesthetized dogs, *Am J Physiol* 257:R317, 1989.

Cameron JD, Dart AM: Exercise training increases total systemic arterial compliance in humans, *Am J Physiol* 266:H693, 1994.

Cheng K-P, Igarashi Y, Little WC: Mechanism of augmented rate of left ventricular filling during exercise, *Circ Res* 70:9,1992.

Christensen NJ, Galbo H: Sympathetic nervous activity during exercise, *Ann Rev Physiol* 45:139, 1983.

Courneya C-A et al: Afferent vascular resistance control during hemorrhage in normal and autonomically blocked rabbits, *Am J Physiol* 261:H380, 1991.

Eldridge FL et al: Stimulation by central command of locomotion, respiration, and circulation during exercise *Resp Physiol* 59:313, 1985.

Geerdes BP et al: Carotid baroreflex control during hemorrhage in conscious and anesthetized dogs, *Am J Physiol* 265:R195, 1993.

Herd JA: Cardiovascular response to stress, *Physiol Rev* 71:305, 1991.

Hosomi H et al: Interactions among reflex compensatory systems for posthemorrhage hypotension, *Am J Physiol* 250:H944, 1986.

Laughlin MH, Armstrong RB: Muscle blood flow during locomotory exercise, *Exerc Sport Sci Rev* 13:95, 1985.

Mitchell JH, Kaufman MP, Iwamoto GA: The exercise pressor reflex: its cardiovascular effects, afferent mechanisms, and central pathways, *Ann Rev Physiol* 45:229, 1983.

Rea RF et al: Relation of plasma norepinephrine and sympathetic traffic during hypotension in humans, *Am J Physiol* 258:R982, 1990.

Roarty TP, Raff H: Renin response to graded haemorrhage in conscious rats, *Clin Exp Pharmacol Physiol* 15:373, 1988.

Saltin B, Rowell LB: Functional adaptations to physical activity and inactivity, *Fed Proc* 39:1506, 1980.

Sanders JS, Mark AL, Ferguson DW: Importance of aortic baroreflex in regulation of sympathetic responses during hypotension, *Circulation* 79:83, 1989.

Schadt JC, Gaddis RR: Renin-angiotensin system and opioids during acute hemorrhage in conscious rabbits, *Am J Physiol* 258:R543, 1990.

Schadt JC, Ludbrook J: Hemodynamic and neurohumoral responses to acute hypovolemia in conscious mammals, *Am J Physiol* 260:H305, 1991.

Share L:Role of vasopressin cardiovascular regulation, *Physiol Rev* 68:1248, 1988.

Shen Y-T et al: Relative roles of cardiac and arterial baroreceptors in vasopressin regulation during hemorrhage in conscious dogs, *Circ Res* 68:1422, 1991.

Sheriff DD, Zhou XP, Scher AM, Rowell LB: Dependence of cardiac filling pressure on cardiac output during rest and dynamic exercise in dogs, *Am J Physiol* 265:H316, 1993.

Triedman JK et al: Mild hypovolemic stress alters autonomic modulation of heart rate, *Hypertension* 21:236, 1993.

Vissing SF, Scherrer U, Victor RG: Stimulation of skin sympathetic nerve discharge by central command: differential control of sympathetic outflow to skin and skeletal muscle during static exercise, *Circ Res* 69:229,1991.

Books and Monographs

Altura BM, Lefer AM, Schumer W: *Handbook of shock and trauma,* vol 1, Basic Science, New York, 1983, Raven Press.

Bond RF, Adams HR, Chaudry IH, eds: *Perspectives in shock research,* New York, 1988, Alan R Liss.

Brooks BA, Fahey TD: *Exercise physiology: human bioenergetics and its applications,* New York, 1984, John Wiley & Sons, Inc.

Janssen HF, Barnes CD, eds: *Circulatory shock: basic and clinical implications,* New York, 1985, Academic Press.

Lind AR: Cardiovascular adjustments to isometric contractions: static effort. In *Handbook of physiology,* section 2: The cardiovascular system—peripheral circulation and organ blood flow, vol III, Bethesda, Md, 1983, American Physiological Society.

Mitchell JH, Schmidt RF: Cardiovascular reflex control by afferent fibers from skeletal muscle receptors. In *Handbook of physiology,* section 2: The cardiovascular system—peripheral circulation and organ blood flow, vol III, Bethesda, Md, 1983, American Physiological Society.

Roth BL, Nielsen TB, McKee AE, eds: Molecular and cellular mechanisms of septic shock. In *Progress in clinical and biological research,* vol 286, New York, 1988, Alan R Liss.

Rowell LB: *Human cardiovascular control,* New York, 1993, Oxford University Press.

RESPIRATORY SYSTEM

NORMAN C. STAUB

PART V

27

An Overview of the Respiratory System

NECESSITY OF RESPIRATION

There are two essential physiological functions necessary to life: breathing and circulation of blood. One may live for several days without liver, kidney, or higher (cerebral) brain function. However, five minutes without breathing or without circulation for the distribution of oxygenated blood to all the body's cells is usually fatal.

> **Cardiopulmonary resuscitation** includes both breathing and **blood circulation.** Contrary to media presentations, street-corner attempts at resuscitation by well-meaning citizens infrequently succeed, because chest compression alone may move blood but does not adequately ventilate the lungs. That is why modern emergency resuscitation demands one 'kiss of life' (mouth-to-mouth breath) after every 5 or 6 chest compressions. Hospital emergency teams and paramedics are trained and equipped to do resuscitation properly.

Breathing

Breathing is an automatic, rhythmic, and centrally-regulated mechanical process (Chapters 28 and 31) by which the contraction of the skeletal muscles of the diaphragm and rib cage move gas in and out of the alveoli.

Respiration

Respiration includes breathing, but it also includes the circulation of blood to and from the tissue capillaries, so that oxygen (O_2) may reach every cell and be used to oxidize metabolites and produce useful energy. Carbon dioxide (CO_2), the spent fuel of cellular respiration, is carried away by venous blood to the lungs for exhalation. (See Chapters 22 and 23 for details about the regulation of the peripheral circulation).

Lung Function

The principal function of the lungs is to provide an adequate distribution of inspired air and pulmonary blood flow, such that the exchange of O_2 and CO_2 between the gas in the alveoli and the pulmonary capillary blood is accomplished with minimal expenditure of energy (work of breathing and work of the right ventricle).

Ventilation and Perfusion

Ventilation is measured as the rate of breathing multiplied by the volume of each breath. Ventilation maintains the normal concentrations of O_2 and CO_2 in the alveolar gas and maintains the normal O_2 and CO_2 partial pressures in the blood flowing into the pulmonary veins from the capillaries by the process of **diffusive gas exchange.**

Perfusion refers to pulmonary blood flow, which equals the heart rate multiplied by the right ventricular stroke volume.

Ventilation and perfusion are normally matched in the lung so that gas transport from the outside air (ventilation) to the systemic arterial blood (perfusion) is nearly optimal. Ideally, the ventilation/perfusion ratios of all parts of the lungs are identical. That does not occur in real life, but the match is surprisingly good, even when there is considerable lung dysfunction. The differences between the O_2 and CO_2 partial pressures in the expired gas and the systemic arterial blood are useful in determining the efficiency of overall lung function.

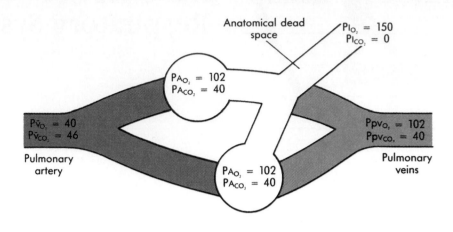

FIGURE 27-1   A model of the normal ventilation/perfusion process. The normal matching of ventilation to perfusion is simplified by showing only two parallel lung units, which are arranged vertically so that you will not automatically think of right and left lungs. Each unit receives equal quantities of fresh air (ventilation) and blood flow (perfusion) for its size. The numbers indicate normal human adult resting values for the gas partial pressures (P) in inspired air (I), in alveolar gas (A), and in mixed venous ($\bar{v}$) blood arriving at the capillaries from the right ventricle via the pulmonary artery. Note that in the ideal lung the alveolar and pulmonary venous (pv) oxygen and carbon dioxide partial pressures are equal.

The following extreme example illustrate the importance of ventilation and perfusion matching. A 2-year-old child is brought to the hospital emergency room because he inhaled a peanut into the main air tube (bronchus) of the left lung, and ventilation of that lung was blocked. Unfortunately, the child was born with a very narrow right pulmonary artery, and therefore nearly all pulmonary blood flow goes to the left lung. This congenital defect did not seriously bother the child before he aspirated the peanut, but now it threatens his life. All of the fresh air goes to the right lung, but nearly all of the lung's blood flow goes to the left lung. Little or no useful gas exchange occurs.

Figure 27-1 shows the normal ventilation and perfusion matching process in the lungs with respect to the partial pressures of O_2 and CO_2 in gas and blood. Your understanding of breathing will be adequate when you are able explain how these gas tensions are maintained in the lung and in the circulation.

BLOOD GAS TRANSPORT AND TISSUE GAS EXCHANGE

Oxygenated blood leaves the lungs via the pulmonary veins and is pumped by the left ventricle through the systemic arteries to the capillaries associated with all of the respiring cells of the body. Likewise, the principal waste product of metabolism, CO_2, is transported away from the respiring cells via the systemic veins to the lungs for elimination.

Oxygen Transport; Role of Hemoglobin

Because oxygen is rather insoluble in water, it is critical to know something about the special protein, **hemoglobin**, within the erythrocytes (see Chapter 30). The remarkable property of hemoglobin is its ability to combine rapidly and reversibly with oxygen, so as to dramatically increase its solubility in blood manyfold. The normal **oxygen-hemoglobin equilibrium curve** (Figure 27-2) shows the relationship between the partial pressure of oxygen in blood and the relative amount (percent saturation) bound to normal hemoglobin. The normal blood hemoglobin concentration of 150 g/L accounts for the normal arterial blood oxygen concentra-

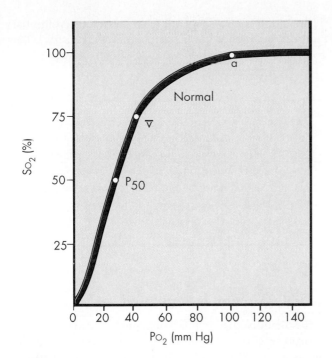

FIGURE 27-2   The ideal normal human oxyhemoglobin equilibrium curve. It gives the percent of hemoglobin bound to oxygen (saturation, So_2) as a function of the partial pressure of O_2. In the systemic arteries, *point a*, the blood is 97.5% saturated at Po_2 of 100 mm Hg, representing an O_2 concentration of 200 ml/L of blood (including a small amount of dissolved oxygen). In the venous blood returning to the lungs, *point $\bar{v}$*, the blood is 75% saturated at Po_2 of 40 mm Hg, representing an O_2 concentration of 150 ml/L of blood. The half-saturation point, P_{50}, is at Po_2 of 26 mm Hg.

tion of 200 ml O_2/L at the normal arterial oxygen partial pressure of 100 mm Hg.

The cardiac output in a resting human adult is about 5 L/min (see Chapter 18). Only about 25% of the oxygen bound to hemoglobin is exchanged between the blood and systemic tissues during each circulation; that is, the arterial-venous O_2 difference = 50 ml/L.

Even during the heaviest steady-state exercise sustainable by average normal humans, cardiac output is unlikely to increase to more than three times the resting level (15 L/min). Because oxygen consumption by the body in steady-state exercise may increase sixfold from the resting value (from 250 ml O_2/min to 1500 ml O_2/min), an additional 25% of the O_2 bound to hemoglobin in systemic arterial blood must be unloaded in the capillaries. The threefold increase in blood flow multiplied by the doubling of O_2 removal from blood accounts for the sixfold increase in O_2 consumption. Highly trained athletes can increase their exercise cardiac output or oxygen consumption to more than twice the values of a normal person (see also Chapter 26).

Carbon Dioxide Transport; Role of Bicarbonate

Carbon dioxide is transported from tissue to lungs mainly in solution and as sodium bicarbonate. Fortunately, CO_2 dissolves in water much better than does oxygen. Even more important, the CO_2 reacts with water to form carbonic acid, which dissociates to release the bicarbonate ion (HCO_3^-), the principal constituent of blood CO_2 transport.

Diffusion: Roadblock on the Oxygen Transport Pathway

Total O_2 transport from ambient (room) air to cells is accomplished by two efficient convective (bulk flow) processes: (alveolar ventilation and pulmonary blood flow); hence the central importance of ventilation/perfusion matching in lung physiology. However, between alveolar gas and blood in the pulmonary capillaries, and between blood in the systemic capillaries and the mitochondria of the respiring cells, diffusion is the only gas transport process available. Diffusion is the passive thermodynamic flow of molecules between regions with different partial pressures (chemical activities) (see Chapters 1 and 22). One would expect diffusion of O_2 to be rate-limiting at these two roadblocks on the oxygen transport highway. However, evolution has resulted in various strategies (see p. 373) to reduce resistance to O_2 diffusion.

CONTROL SYSTEM

The Act of Breathing

Inspiration (inhaling) is the active phase of breathing. It is initiated by neural impulses from the respiratory control centers in the brainstem (medulla). These neural impulses stimulate the diaphragm and intercostal muscles (between the ribs) to contract. The muscle contraction causes the thoracic cavity to expand, which lowers the pressure in the pleural space surrounding the lungs. As the pressure falls, the distensible lungs expand passively, which causes the pressure in the alveoli (terminal air spaces) to decrease. As the alveolar pressure decreases, fresh air flows along the branching airways into the alveoli until alveolar pressure equals the pressure at the airway opening (usually atmospheric pressure). During expiration, which occurs passively by elastic recoil of the lungs, the process is reversed. Pleural and alveolar pressures rise, and gas flows out of the lung. Normal breathing at rest is completely automatic and uses little energy.

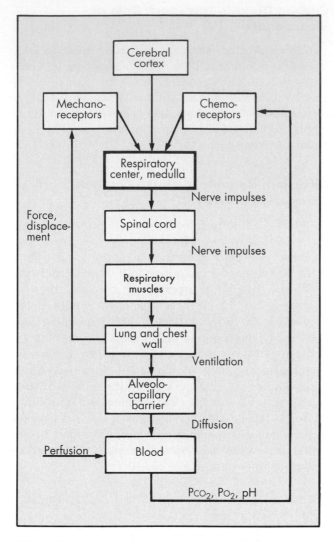

FIGURE 27-3 Block diagram showing overall regulation of breathing. The respiratory center neurons, dispersed into several groups in the medulla, show spontaneous cyclic activity, but are strongly influenced by stimuli descending from the cerebral cortex (volitional control) and from two sensory loops, mechanoreceptors and chemoreceptors (automatic control).

Regulation of Breathing

Breathing is regulated automatically by a variety of strategically placed sensors (Figure 27-3). **Mechanoreceptors** (stretch, position, or muscle force generated) located within the chest wall and lungs monitor muscular effort and rate of lung volume change. **Chemoreceptors** (CO_2, O_2) located at the bifurcation of the carotid artery in the neck and in the aortic arch sense arterial blood oxygenation, and those near the ventrolateral surface of the brainstem (medulla) sense the carbon dioxide tension within the brain tissue.

Breathing is also susceptible to conscious (volitional) control from higher brain centers (cerebral cortex). This control is required when we talk, cough, or vomit.

Some patients in deep coma may be clinically brain dead (no cerebral cortical electroencephalographic activity), but they continue to breathe and to regulate the cardiovascular system. Their automatic control centers in the medulla and adjacent brainstem structures may continue to function for long periods, if the body is fed intravenously and otherwise supported.

How Respiration is Regulated

Respiration, the cellular demand for oxygen, is not normally regulated by blood flow. Thus, steady-state oxygen consumption is not usually affected by changes in breathing, inspired gas composition, cardiac output, or blood composition. However, the delivery of oxygen from the ambient air to the mitochondria of the cells is regulated on demand through very complex local and central events.

STRUCTURE-FUNCTION RELATIONSHIPS

In normal humans the lungs completely fill and conform to the shape of the pleural cavities in such a manner as to minimize the structural stress on the lung tissue elements. During inspiration, the lung expands in all directions as the chest cavity expands.

In adults the lung weight represents about 1.5% of body weight (1 kg in a 70 kg adult); the lung tissue accounts for 60%, and blood accounts for the remainder.

What one sees in a normal chest x-ray film is mainly the larger airways (bronchi) and the blood in the arteries and veins (Figure 27-4). Because lung roentgenograms are usually taken after a maximum inspiration, the alveolar walls and capillaries are too thin to affect the penetration of the x-rays. That makes it easier for the radiologist to see changes that occur in many lung diseases.

Figure 27-5 is a light-microscopic picture of a thin slice of an expanded normal lung. The alveolar tissue occupies only a tiny fraction of the total lung volume. Fig-

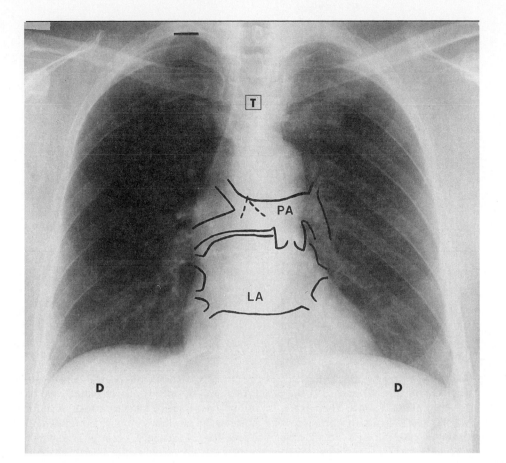

FIGURE 27-4   A normal posterior-to-anterior chest roentgenogram taken at the end of a normal expiration (functional residual capacity, FRC). One can the see chest wall (ribs and diaphragm, *D*). The lung fills all of the thoracic cavity, except for the heart and the mediastinum, which contains the trachea *(T)*, dividing into the right and left mainstem bronchi *(inverted dashed V)*. Other mediastinal structures not labeled include the aorta, vena cavae, and the esophagus. Some of the larger distributing pulmonary vessels in the lung can be seen because they contain blood, in contrast to the air-filled lung. One cannot normally see any detail of the alveolar structures. The radiologist outlined the main pulmonary artery *(PA)* and left atrium *(LA)*, into which the pulmonary veins empty, so that the distribution of lung below and above the heart would be clear. (Modified from Staub NC: *Basic respiratory physiology*, New York, 1991, Churchill-Livingstone.)

ure 27-6 is an electron micrograph of several alveolar walls. The average distance between the alveolar gas and the hemoglobin in the red blood cells is 1.5 μm. The diffusion pathway is exceedingly short and is one evolutionary solution to the alveoli-to-blood diffusion problem. Diffusion, which is very rapid over short distances, efficiently exchanges the respiratory gases across the normal alveolar-capillary interface.

The huge number of alveoli give the human lung a total internal surface area of some 70 m^2, when the lung is at **functional residual capacity** (FRC) (the volume of gas in the lung at the end of a normal expiration). This vast area is necessary to distribute the pulmonary blood flow (cardiac output) into a film one red blood cell thick,

so that the time spent by each erythrocyte flowing along the capillaries is sufficient to permit equilibration of O_2 and CO_2 between blood and alveolar gas. This is a second major evolutionary solution to the alveoli-to-blood diffusion problem. The lungs are extremely efficient in terms of gas exchange.

> In spite of the best engineering efforts, the heart-lung machines used in the operating room to substitute temporarily for the patient's heart and lungs are large and inefficient.

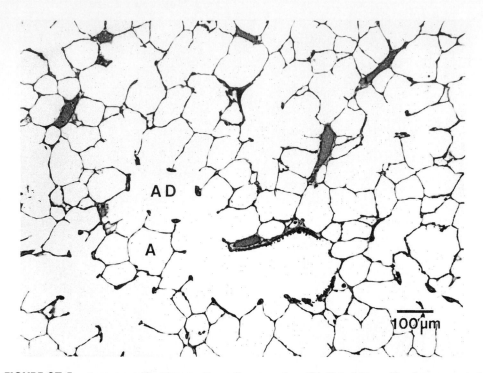

FIGURE 27-5 Low-magnification section of a normal, well-inflated lung. The larger central openings are the alveolar ducts, *AD,* the final branches of the airways. Surrounding each duct are the anatomical alveoli, *A.* There is a great deal of alveolar gas (includes both alveolar ducts and alveoli) and little tissue. The main value in having an enormous number of tiny alveoli is to greatly increase the alveolocapillary surface area for gas exchange. (Courtesy KH Albertine, University of Utah.)

Airways

The conducting airways (cartilaginous **bronchi** and the larger membranous **bronchioles)** do not participate in gas exchange. Thus, their portion of each breath is wasted. This is referred to as the **anatomical dead space,** about 30% of each normal breath.

Bronchi The bronchi (Figure 27-7) are lined by columnar epithelium, which rests on a layer of smooth muscle. The bronchi are not directly attached to the lung tissue; therefore they can dilate or constrict independently of lung volume. Among the numerous cell types in the epithelium are a large number of ciliated cells, whose rhythmic beating in a thin surface liquid layer effectively transports secreted mucus and inhaled particles out of the lung by way of the trachea.

The bronchi decrease in diameter and length with each successive branching, and the cartilaginous support gradually disappears until it is absent in tubes smaller than about 1 mm diameter.

Bronchioles The bronchioles (Figure 27-8) are the continuation of the bronchi. They constitute all of the airways less than 1 mm diameter, contain no cartilage, and have a simple cuboidal epithelium. The bronchioles

are embedded directly into the connective tissue framework of the lung; their diameter is dependent on lung volume.

Eventually, the bronchioles develop outpouchings—the alveoli. The first bronchioles with alveoli are called **respiratory bronchioles,** because they participate in gas exchange. With further branchings, the number and size of the alveoli increase until the walls of the bronchioles are almost completely replaced by the mouths of the alveoli (Figure 27-8). These final airway branches are called **alveolar ducts.**

Blood Supply The airways receive their nutritive blood supply via the **bronchial arteries,** which are small branches of the aorta, a systemic source. Bronchial blood flow is normally about 1% of cardiac output, and bronchial vascular resistance is high.

Innervation The airways are innervated by both motor and sensory nerves of the sympathetic and parasympathetic branches of the autonomic nervous system. The various nerve fibers participate in the reflex regulation of breathing, airway caliber, glandular secretion, and bronchial vasomotor control. When the airway smooth muscle contracts (usually by activity in

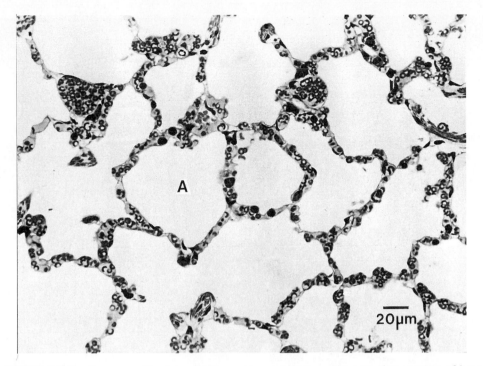

FIGURE 27-6   Low-magnification electron microscopic view of a very thin section of lung tissue. The lung was at low volume when it was fixed, which permits numerous alveolar walls to be included. The bulk of the alveolar walls is capillaries, the other tissue being reduced so as to optimize oxygen diffusion from the alveoli, *A*, to the erythrocytes *(black objects)* in the capillaries. (Courtesy KH Albertine, University of Utah.)

the parasympathetic nerves), the lumen is narrowed.

Sensory fibers are located beneath and within the intercellular junctions of the epithelial cells. There are two main types of large airway receptors, those sensitive to physical distortion (stretch) and those sensitive to chemical substances (irritants). All along the airways, particularly in the bronchioles but also in the alveolar walls, are small, nonmyelinated, slowly conducting C-fibers. These fibers occur in great numbers and can be stimulated by various chemical mediators by which certain pulmonary-cardiac reflex responses are elicited. The control of breathing, including the sensory and motor nerves of the airways, will be discussed in more detail in Chapter 31.

Pulmonary Circulation

Perfusion Matches Ventilation The pulmonary artery (perfusion) accompanies and branches with the airways (ventilation). Thus, the physiological theme of **ventilation/perfusion matching** is reflected in the anatomical pattern of the bronchovascular relations. The pulmonary veins do not follow the airways. They are situated within the interlobular loose connective tissue septa and receive blood from many lung units. They connect to the left atrium.

Pulmonary arteries and veins are thin walled and elastic, and they contain nearly 10% (500 ml) of the total body blood volume 5000 ml).

Although the pulmonary circulation normally has a low pressure and low resistance, the smaller arteries (less than about 500 μm diameter) are muscular and can actively regulate their diameters and alter resistance to blood flow.

In the fetus the pulmonary vessels are constricted, and therefore only a small fraction of right ventricular output flows through the lungs (Chapters 25 and 30). The pulmonary vascular resistance is high. At the onset of air breathing, the vessels dilate and vascular resistance falls. In a small percentage of babies a congenital opening in the septum between the right and left ventricles persists **(interventricular septal defect),** so

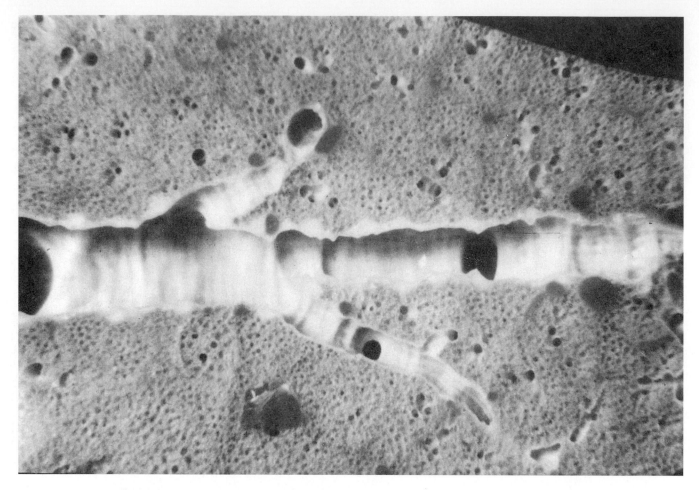

FIGURE 27-7 Cartilaginous bronchi cut in longitudinal section in a block of frozen lung. There are frequent irregular branchings in all directions, with each daughter bronchus smaller than the parent bronchus. The gas in the bronchi represents wasted ventilation (anatomical dead space). The design of the airways keeps the anatomical dead space as small as possible, commensurate with keeping airflow resistance as low as possible. (From Staub NC: *Basic respiratory physiology,* New York, 1991, Churchill Livingstone.)

that some of the left ventricular blood at high pressure flows into the right ventricle. This results in abnormally high pressures and flow in the pulmonary arteries, which respond by constricting, and increasing their smooth muscle coat. This eventually results in a permanent increase in pulmonary vascular resistance. If the congenital septal defect can be corrected early, the long-term effects on the pulmonary vessels can be prevented or reversed.

Innervation The motor nerve supply to the arterial and venous smooth muscle is through the sympathetic branch of the autonomic nervous system. In contrast to the systemic circulation, the normal pulmonary circulation shows little evidence of active external regulation (see Chapter 30).

The extensive sensory innervation, located in the adventitia (outer connective tissue layer) of the blood vessels, can be stimulated by vascular pressure changes (stretch) and by various chemical substances.

Pulmonary Capillaries The pulmonary capillaries form an extensive interdigitating network that is continuous over several alveoli. When the capillaries are well filled with blood, most of the surface area (70% to 80%) of the alveolar wall overlies red cells, as Figure 27-6 suggests.

In a resting human, the effective volume of blood in the capillaries is about 70 ml, although the maximal cap-

FIGURE 27-8 Bronchioles cut in longitudinal section in a block of frozen lung. Only the final generations of bronchioles have names (terminal bronchioles, respiratory bronchioles, alveolar ducts). Shown here is a respiratory bronchiole, *RB*, with primitive small alveoli coming into view from the right edge of the photograph, followed by numerous branches with more and larger alveoli evolving into the alveolar ducts, *AD*. The latter are seen mainly as right-angle branches above and below the row of respiratory bronchioles. Although the bronchioles are part of the airway system, they are also part of the gas exchange volume. (From Staub NC: *Basic respiratory physiology,* New York, 1991, Churchill Livingstone.)

illary volume of the human adult lung is about 200 ml. Capillary volume can be increased by opening (recruiting) closed or compressed capillary segments. This regularly occurs during exercise as cardiac output rises. The capillaries can also be distended as their internal pressure rises (see Chapter 30). Capillary blood volume at any instant is about equal to the stroke volume of the right ventricle. This means that the average erythrocyte remains in the capillaries a sufficient time (0.8 sec) for the diffusive exchange of O_2 and CO_2 to reach equilibrium across the thin alveolar-capillary barrier.

THE PHYSIOLOGICAL LUNG UNIT

The **physiological (functional) lung unit** is defined as the largest lung unit (usually at the respiratory bronchiole) in which the partial pressures of O_2 and CO_2 are uniform. The physiological unit is much larger than any anatomical alveolus. The human adult lung has 60,000 of these units. Each unit contains 5000 anatomical alveoli and 250 alveolar ducts. Figure 27-9 shows the relationships among the various structural elements of the lung.

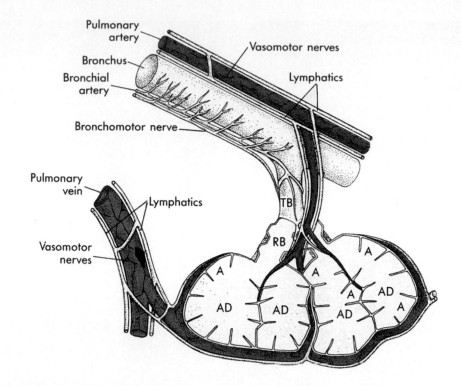

FIGURE 27-9 This model helps one to understand normal lung structure-function relations. It shows the correct anatomical relation between the pulmonary artery and the airways. The veins have no relation to the airways. Two physiological lung units are represented by the two respiratory bronchioles *(RB)*. Note that the small pulmonary arterial (resistance) vessels are surrounded by the alveolar gas of the unit they perfuse (see Chapter 30). *TB,* Terminal bronchiole; *AD,* alveolar ducts; *A,* alveoli.

Why do humans have so many small anatomical alveoli, if the functional unit is so large? It is evolution's answer to the problem of the high metabolic demand for oxygen in mammals. In order to get enough oxygen to the mitochondria in the systemic tissue cells, cardiac output must be high. If the lungs had a few huge alveoli (as do frogs and snakes), the resistance to blood flow through the alveolar wall capillaries would be enormous and would necessitate a powerful right ventricle that used as much oxygen as the normal left ventricle. Furthermore, the blood would flow through the capillaries so rapidly that the time for O_2 to diffuse from the alveolar gas into the red cells would be insufficient to achieve normal systemic arterial oxyhemoglobin saturation.

SUMMARY

1. The main function of the lung is to bring fresh air into close contact with blood flowing in the pulmonary capillaries, so that the exchange of oxygen and carbon dioxide by passive diffusion will take place efficiently.

2. Ventilation is measured as the volume of each breath times the frequency of breathing. Perfusion is measured as heart rate times right ventricular stroke volume (pulmonary blood flow).

3. Ventilation and perfusion are normally well matched, so that gas transport from the outside air to the systemic arterial blood is nearly optimal.

4. Blood gas transport and tissue oxygen exchange complete the oxygen transport highway.

5. Hemoglobin within red blood cells vastly increases the ability of blood to transport oxygen.

6. Carbon dioxide is transported from the tissue capillaries back to the lung, mainly as bicarbonate ion in plasma.

7. Passive diffusion of oxygen is normally not a rate-limiting step (a) at the alveolar-capillary barrier or (b) at the systemic capillary-tissue barrier, but it can become limiting under extremely stressful conditions or in severe lung disease.

8. The active phase of breathing is inspiration, during which the muscles of the chest wall, mainly the diaphragm, contract and lower the pressure in the lung's alveoli so that air can flow into the lungs.

9. Breathing is controlled from the respiratory center in the medulla (brainstem) by sensory information from mechanoreceptors that monitor chest wall motion and lung volume changes, and by chemoreceptors that detect the carbon dioxide and oxygen partial pressures in brain interstitial fluid and systemic arterial blood, respectively.

10. Anatomically, the main lung function theme, which is the matching of ventilation to perfusion, is reflected by the coordinated branching pattern of the airways and the pulmonary arteries.

11. The airways are of two types, the bronchi (cartilaginous) and the bronchioles. The alveolar ducts (final branches of the bronchioles), together with the alveoli in their walls, form the gas exchange part of the lung.

12. One pulmonary artery branch accompanies each airway and branches with it. To maximize the rate of O_2 and CO_2 diffusional exchange with the pulmonary blood flow, the lung has the most extensive capillary network of any organ; this network occupies 70% to 80% of the alveolar surface area. The erythrocytes remain in the capillaries long enough for gas exchange to reach equilibrium, even in strenuous exercise.

13. Groups of alveolar ducts and their alveoli, together with their supplying arteries, are combined into small functional lung units. These physiological lung units, like the leaves on a tree, are arranged in parallel, thus ensuring the efficient distribution of inspired air and mixed venous blood.

BIBLIOGRAPHY
Journal Articles

Motoyama EK, Brody JS, Colten HR, Warshaw JB: Postnatal lung development in health and disease, *Am Rev Resp Dis* 137:742, 1988.

Staub NC: The interdependence of pulmonary structure and function, *Anesthesiology* 24:831, 1963.

Staub NC: Pulmonary structure as related to its function. *Basics Resp Dis* 1:1, 1972.

Tyler WS, Julian MD: Gross and subgross anatomy of lungs, pleura, connective tissue septa, distal airways and structural units. In Parent RA, ed: *Comparative biology of the normal lung.* Boca Raton, Fla, 1991, pp 37-48.

Books and Monographs

Miller WS: *The lung,* ed 2, Springfield, Ill, 1947, Charles C Thomas.

Staub NC: *Basic respiratory physiology,* New York, 1991, Churchill-Livingstone.

Staub NC, Albertine KH: The structure of the lung relative to its principal function. In: Murray JF, Nadel JA, eds: *Textbook of respiratory medicine,* ed 2, Philadelphia, 1994, WB Saunders pp 3-35.

von Hayek H: *The human lung,* New York, 1960, Hafner.

Weibel ER: *Morphometry of the human lung,* New York, 1963, Academic Press.

28

Mechanical Aspects of Breathing

VENTILATION

Lung Volumes

Figure 28-1 shows some of the important lung volumes, as might be measured in the pulmonary function laboratory. The volume of gas moved during normal quiet breathing is the **tidal volume** (TV), about 0.5 liter. The **functional residual capacity** (FRC) is the volume of gas that remains in the lung at the end of a passive expiration; normally, it is 2 to 2.4 liters (40% of maximal lung volume) in adults. The FRC is the most stable position of the breathing system under any given circumstance. The **total lung capacity** (TLC) is the maximal lung volume that can be achieved voluntarily, 5 to 6 liters in normal adults. If one inspires to TLC, then breathes out as much as possible, one can decrease lung volume below FRC to **residual volume** (RV), normally 1 to 1.2 liters. The volume of air moved between TLC and RV is the **vital capacity** (VC) (normally, 4 to 5 liters in adults). It is the largest possible tidal volume in an individual.

The Components of Ventilation

The main purpose of ventilation is to maintain an optimal composition of alveolar gas. Half of the process of ventilation/perfusion matching involves getting fresh air to the alveoli. Ventilation raises the alveolar P_{O_2} above that of mixed venous blood, as was shown in Figure 28-1. Oxygen molecules then diffuse along their partial pressure gradient into the pulmonary capillary blood and increase the HbO_2 concentration. Ventilation lowers the alveolar P_{CO_2} below that in mixed venous blood. Carbon dioxide molecules then diffuse along their partial pressure gradient into the alveolar gas and reduce the CO_2 concentration in the pulmonary capillary blood.

When the demand for oxygen is increased, as in ex-

ercise, the P_{O_2} of venous blood decreases and its P_{CO_2} increases. Ventilation must be increased in some regulated manner, so as to maintain the partial pressures of both gases at normal levels in the arterial blood.

Minute Ventilation and Alveolar Ventilation
Minute ventilation is the total volume of air entering or leaving the lungs each minute. It is measured in the pulmonary function laboratory with an electronic flowmeter (pneumotachograph) to integrate airflow at the mouth, as shown in Figure 28-2.

Alveolar ventilation, $\dot{V}_A$, is the volume of fresh air that enters the alveoli each minute. Alveolar ventilation is always less than total ventilation; how much less depends on the anatomical dead space and the tidal volume.

Anatomical Dead Space, Tidal Volume, and Alveolar Ventilation Inspired air does not go directly to the alveoli. It first flows through the conducting airways (nose or mouth to terminal bronchioles), which form the **anatomical dead space** (no gas exchange), about 0.15 liter (2 ml/kg of body weight) in adults.

Alveolar ventilation equals the breathing frequency (12/min at rest) times the difference between tidal volume (0.5 liter) and the anatomic dead space (0.15 liter). In the normal resting adult alveolar ventilation is $12 \times (0.5 - 0.15) = 4.2$ liters/min.

Wasted Ventilation The concept of **wasted ventilation** is used clinically to describe the deviation of the overall lung ventilation/perfusion ratio from ideal. Figure 28-3 shows an extreme example, blockage of a branch pulmonary artery by a blood clot (embolus). The alveolar ventilation to the portion of lung with the obstructed blood flow is wasted (gas tensions are the same as in the inspired air) because it does not participate in gas exchange. Wasted ventilation is the sum of anatomical dead space and that portion of alveolar ventilation going to lung units that receive no blood flow.

Alveolar P_{CO_2} and Ventilatory Control It is not the

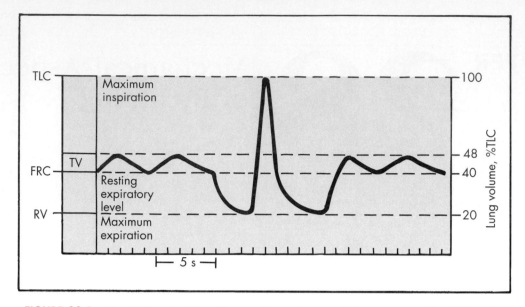

FIGURE 28-1 A real-time tracing of lung volume changes during two normal tidal breaths *(TV)*, after which the subject made a maximal expiration to residual volume, *RV,* an inspiration to total lung capacity *(TLC)*, then an expiration to RV (the volume exhaled between TLC and RV is the vital capacity, VC). The tracing ends with two more tidal breaths. The left-hand Y axis shows the main named components of lung volume, while the right-hand axis shows their relative volumes. Note that normal tidal breathing is near the middle of total lung capacity.

oxygen supply to the alveoli but the partial pressure of CO_2 in arterial blood, Pa_{CO_2}, that is sensed by the brainstem respiratory center and used to regulate alveolar ventilation (see Chapter 31).

The alveolar ventilation equation, the most important quantitative relationship in pulmonary physiology, describes the exact relation between alveolar ventilation and arterial P_{CO_2} for any given metabolic state (CO_2 production, $\dot{V}_{CO_2}$).

$$\dot{V}_A \text{ (L/min)} \times P_{aCO_2} \text{ (mm Hg)} = \dot{V}_{CO_2} \text{ (ml/min)} \times K$$

The $\dot{V}_A$ is given in L/min and the CO_2 production, $\dot{V}_{CO_2}$, is in ml/min (see metabolism, Chapter 41). Usually, the unit conversion constant, K, is 0.863 mm Hg × L/ml. When ventilation rises, Pa_{CO_2} decreases; when ventilation falls, Pa_{CO_2} increases.

The adequacy of alveolar ventilation is measured in terms of arterial P_{CO_2}. Normal alveolar ventilation means that Pa_{CO_2} equals 40 mm Hg. **Hyperventilation** (overventilation) means that Pa_{CO_2} is less than 40 mm Hg. **Hypoventilation** (underventilation), which is the more common condition encountered in patients with lung diseases, means that Pa_{CO_2} is more than 40 mm Hg.

Alveolar Oxygen Partial Pressure If Pa_{CO_2} is known, one can calculate the effective alveolar P_{O_2} by

the **alveolar gas equation.** However, the equation is complex, so pulmonary physiologists use specific electrodes to simultaneously measure arterial P_{O_2}, P_{CO_2}, and hydrogen ion concentration ($[H^+]$, pH).

When arterial P_{CO_2} rises (hypoventilation), alveolar P_{O_2} must decrease, and when arterial P_{CO_2} falls (hyperventilation), P_{O_2} must increase, because the total pressure of all alveolar gases cannot exceed atmospheric pressure, P_B. Ideally, when arterial P_{CO_2} is 40 mm Hg, arterial P_{O_2} is 100 mm Hg, presuming the subject is breathing room air at sea level (see Figure 27-1). The reason why Pa_{O_2} is not used to estimate the adequacy of ventilation is that inspired P_{O_2} can be changed readily by inspiring special gas mixtures, such as 100% O_2, or by traveling to different altitudes from sea level. Inspired P_{CO_2}, on the other hand, is rarely increased appreciably above 0.

THE CHEST WALL IS THE BREATHING PUMP

The Structure of the Chest Wall

The functional chest wall includes the **diaphragm** and the **rib cage,** as is apparent in the normal chest roentgenogram in Figure 27-4. In addition to those obvious structures, the chest wall includes the abdominal contents and anterior abdominal muscles because these structures lie between the external (abdominal) surface of the diaphragm and the atmosphere.

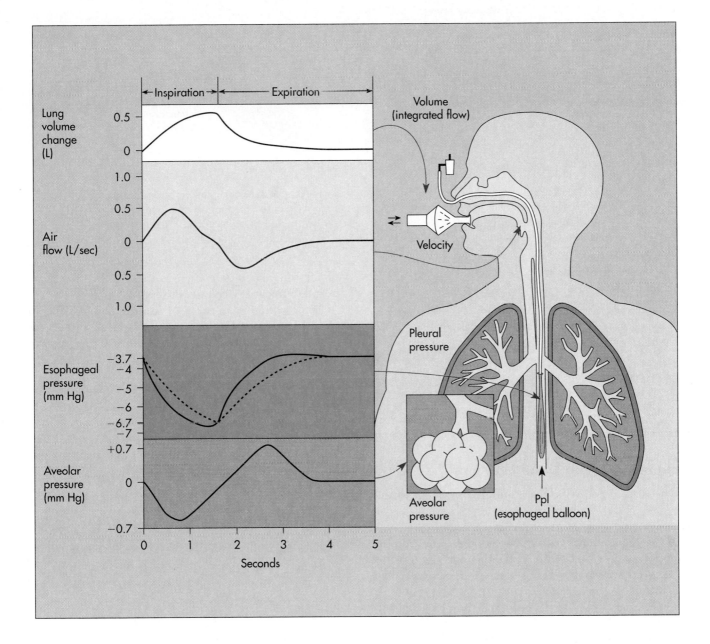

FIGURE 28-2 Dynamics of a normal tidal breath, showing the relationships among changes in lung volume, air flow, esophageal (pleural) pressure, and alveolar pressure. In modern pulmonary physiology, flow is measured at the mouth using a device called a pneumotachograph. The flow signal is integrated (added up over time) to generate the lung volume change. Pleural pressure is approximated by pressure in the intrathoracic portion of the esophagus, measured with a long flaccid balloon. The duration of the breath is 5 seconds (frequency = 12 breaths/min). Inspiration lasts 2 seconds. The curve for pleural pressure includes the pressure required to change volume *(dashed line)* plus the pressure needed to generate airflow; total pressure is the *solid line*.

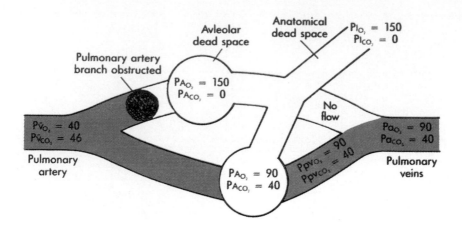

FIGURE 28-3 Wasted ventilation includes the anatomical dead space plus any portion of the alveolar ventilation that does not exchange O_2 or CO_2 with the pulmonary blood. This simple sketch shows a pulmonary artery completely obstructed by a blood clot. However, in most subjects the anatomical definition of the wasted ventilation is not easily made.

The last sentence may seem incredible, but consider a normal woman near the end of pregnancy. The enlarged uterus fills the lower abdomen and raises the intraabdominal pressure, which is evident in the outward bulge of the abdominal wall. The increased intraabdominal pressure also pushes the diaphragm into the thoracic cavity, and thereby limits both end-expiratory lung volume (FRC) and maximal lung volume (TLC).

Movements of the Breathing Pump

The diaphragm is the main muscle of the chest wall. In Figure 27-4 the diaphragm appeared as a dome-shaped structure that separates the thoracic and abdominal cavities. The diaphragm receives its blood supply from intercostal arteries and sends its venous outflow into the inferior vena cava. The diaphragm is innervated by the two phrenic nerves, which arise at the third to fifth cervical segments of the spinal cord, and then pass caudally in the mediastinum to the right and left halves of the diaphragm (Figure 28-4, *A*).

When the diaphragm contracts, it moves caudally and displaces the abdominal contents caudally (downward in the upright posture) or ventrally (outward). At its attachments to the lower several ribs, the contracting diaphragm rotates the ribs toward the horizontal plane, and thereby further expands the chest cavity (Figure 28-4, *A* and *B*).

In humans the 12 ribs on each side articulate with the thoracic vertebrae. The only rib motion possible is rotation toward (inspiration) or away from (expiration) the horizontal plane (pail handle effect). This motion

thereby increases or decreases the cross-sectional area of the thoracic cavity, respectively.

The inspiratory muscles of the rib cage are the **external intercostals;** the expiratory muscles are the **internal intercostals.** The rib cage muscles are supplied by intercostal arteries and veins and are innervated by intercostal motor and sensory nerves.

Some of the muscles of the neck (**sternocleidomastoids** and the **scalenes**) are called **accessory muscles** of breathing because when they contract they pull up on the upper ribs and assist inspiration. These accessory muscles are activated in exercise or when inspiratory airflow is limited, as in asthma; see below.

In accidents that block or sever the spinal cord low in the neck, diaphragmatic breathing continues because the phrenic nerves arise above the injury site. The rib cage and abdominal muscles, of course, are paralyzed, as are all other skeletal muscles whose motor nerves leave the spinal cord caudal to the injury site.

An important characteristic of the rib cage is its stiffness, which prevents any inward (paradoxical) movement of the thoracic wall as pleural pressure becomes more subatmospheric during inspiration.

Normally expiration is caused by the passive elastic recoil of the lungs, generated by inspiratory lung expansion. However, when a great deal of air has to be moved quickly, as in exercise, or when the airways narrow excessively during expiration, as in asthma, the internal intercostal muscles and the anterior abdominal muscles

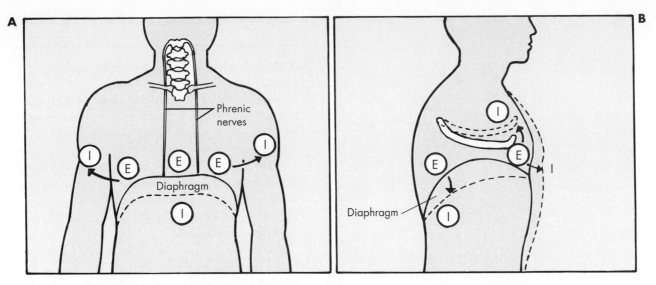

FIGURE 28-4   **A,** Schematic front view in an upright human to show the descent of the diaphragm and the flaring of the lower ribs caused by the contraction of the diaphragm at its rib attachments during inspiration. The phrenic motor nerves to the diaphragm are also shown from their origin in the neck. **B,** Schematic lateral view to show the outward bulge of the abdominal wall as the contracting diaphragm shortens and moves caudally and the limited rotatory motion (pail-handle action) permitted for the rib cage. (**B** redrawn from Staub NC: *Basic respiratory physiology,* New York, 1991, Churchill Livingstone.)

contract, thereby accelerating expiration by raising pleural pressure.

Coupling of the Lungs to the Chest Wall

The lungs change volume or shape when the thoracic cavity changes volume or shape. And yet, the lungs are not directly attached to the chest wall. The **visceral** and **parietal pleuras** cover the surfaces of the lungs and thoracic cavity, respectively, and form between them the **pleural space.** The lung and the chest wall pleuras are coupled together by a thin layer (about 20 μm thick) of liquid. The liquid coupling (molecular cohesion) allows the lungs to slide across the chest wall during breathing and thereby to accommodate themselves to changes in thoracic configuration. **Pleural pressure changes** refer to the pressure in the thin liquid layer that couples the lungs to the chest wall.

THE BREATHING CYCLE

Relative or Absolute Pressures

In pulmonary physiology one uses the term absolute pressure only when referring to atmospheric pressure,

P_B; for example, the absolute atmospheric pressure at sea level is 760 mm Hg. The pressures and pressure differences within and across the components of the respiratory system (alveolus-to-pleura, pleura-to-body surface, pleura-to-abdominal cavity; always read from inside to outside) are measured relative to atmospheric pressure. When we say that alveolar pressure is zero, we mean it is not different from atmospheric pressure (760 mm Hg absolute). That is why pleural pressure can be a negative number (subatmospheric) (Figure 28-5). Normal translung pressure is +3.7 mm Hg, when the lung is at functional residual capacity. It does not matter what alveolar and pleural pressures are in absolute terms. As long as their difference ($P_{alv} - P_{pl}$) is +3.7 mm Hg, lung volume will be at FRC.

Pressure-Volume Relationships

The **compliance** of the lung or chest wall, singly or together, refers to the ease with which either or both can be distended. The standard procedure for measuring compliance in humans is to determine the pressure-volume relationship during a passive (no breathing muscle activity) expiration from total lung capacity. If the respiratory system deflates slowly, alveolar pressure is essentially equal to atmospheric pressure, and pleural pres-

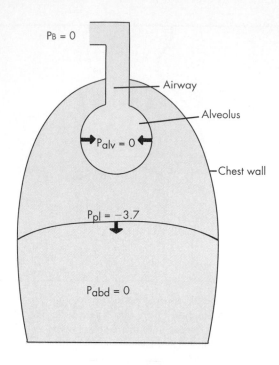

FIGURE 28-5 A simple drawing useful in thinking about the pressures and pressure differences across the respiratory system. The numbers given for the alveolar (P_{alv}) and pleural (P_{pl}) pressures are normal values in humans at end-expiration (FRC). The small arrows show the inward recoil of the lung and the outward recoil of the chest wall. The pressure differences are always read from inside to outside. P_L, the translung pressure, is $P_{alv} - P_{pl} = 0 - (-3.7) = +3.7$ mm Hg. Similarly, $P_{CW} = P_{pl} - P_B = -3.7 - 0 = -3.7$ mm Hg; the minus sign indicates that the chest wall is under compression. At FRC the pressure across the respiratory system, RS, is $P_{alv} - P_B = 0 - 0 = 0$; there is no pressure difference, indicating that lung recoil exactly balances chest wall recoil.

sure is nearly the same as the pressure in the esophagus, which is usually measured with a thin-walled balloon attached via a plastic tube to a pressure-sensing device (see Figure 28-2).

Lung Compliance Lung compliance is defined as the slope ($\Delta V/\Delta P$) of the straight line joining any two points on the deflation pressure-volume curve. For the respiratory system there are three different compliance curves—lung, chest wall, and combined—as shown in Figure 28-6. Lung compliance is the change in lung volume divided by the change in translung pressure: $\Delta V/\Delta P_L$. **Chest wall compliance** is the change in lung volume divided by the change in trans–chest wall pressure: $\Delta V/\Delta P_{CW}$. **Respiratory system compliance** is the change in lung volume divided by the change in trans–respiratory system pressure: $\Delta V/\Delta P_{RS}$.

Lung Compliance May Be Affected By Disease

In **chronic obstructive pulmonary disease** (COPD, emphysema) the alveolar walls progressively degenerate, which increases lung compliance (distensibility). Small changes in transpulmonary pressure evoke large changes in lung volume (Figure 28-7, *upper curved line*).

In **chronic restrictive lung disease** (lung fibrosis), characterized by an increased stiffness of the lung (low compliance), large changes in transpulmonary pressure evoke small changes in lung volume (Figure 28-7, *lower curved line*).

In **asthma** (hyperactive airway smooth muscle) the lung compliance is normal, although FRC may be much increased because of excessive expiratory airway narrowing (see p.390).

Babies have a lower lung compliance than do adults, not because there is anything abnormal about their lungs, but simply because their lungs are small. Neonatologists use tables of normal values for babies or measure compliance per unit of lung volume, which ratio is called **specific compliance,** in order to correct for lung size.

Chest Wall Compliance May Be Affected By Disease

Changes in chest wall compliance are less common than changes in lung compliance. In **kyphoscoliosis** (a distorted spinal column) the rib cage cannot move normally; hence chest wall compliance is decreased. When abdominal pressure is high, chest wall compliance is also decreased, not because the diaphragm is stiffer but because the abdomen interferes with descent of the diaphragm.

In newborns the rib cage is not very stiff (ribs are cartilage, not bone), so that when pleural pressure decreases during diaphragmatic contraction, the rib portion of the chest wall retracts (is pulled inward).

This can be a serious problem, if the lungs are stiffer than normal, as in **respiratory distress syndrome.** A related adult problem is the **flail chest** that may occur when several ribs are broken.

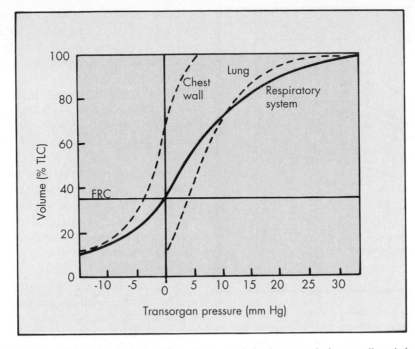

FIGURE 28-6   The distensibilities (compliances) of the lungs and chest wall and their sum, the respiratory system. These are passive static (no movement) curves. From below FRC up to about 75% TLC, lung compliance is high, fairly linear, and parallel to that of the chest wall. The lungs become stiffer near total lung capacity, TLC, as the noncompliant collagen fibers of the lung become taut. Over the normal useful range of breathing, the respiratory system curve has a low slope (lower compliance) than either the lungs or the chest wall. At FRC, the lung and chest wall recoil pressures are equal and opposite. The thin vertical line at transorgan pressure of 0 shows that if the lungs and chest wall are uncoupled (by opening the pleural space), the thoracic volume will expand and the lungs will collapse to their minimal volume.

The Balance Between the Lung and the Chest Wall

As shown in Figure 28-5, at normal end-expiration, FRC, the pressure across the respiratory system, RS, must be zero because the system is at rest. This stable normal condition is caused by the inward recoil (collapsing tendency) of the lung (translung pressure at FRC is +3.7 mm Hg), which exactly balances the outward (expanding) recoil of the chest wall (the transchest wall pressure at FRC is −3.7 mm Hg). That means the lungs are above their minimal or unstressed volume (volume at which translung pressure = 0), and the chest wall is below its rest volume. This important relationship is shown both in Figure 28-5 and in Figure 28-6 (compare the pressures along the line labeled FRC).

When the chest wall is opened during thoracic surgery, air enters the pleural space because the pleural pressure is less than atmospheric pressure. The lungs tend to collapse, whereas the thoracic cavity gets larger. The condition is called a **pneumothorax** (air in the thoracic cavity outside the lung). A traumatic or spontaneous pneumothorax may be a life-threatening emergency, if the lungs are completely uncoupled from the chest wall pump, so that they do not move when the diaphragm contracts. This occurs when the air leak is large and continuous.

When lung volume exceeds FRC by about 25% of TLC (normally FRC + 1 to 1.5 liters), the transchest wall pressure, P_{CW}, is 0 (see Figure 28-6); hence, pleural pressure must be equal to the pressure on the body surface (usually, atmospheric pressure). The total pressure across the respiratory system equals translung pressure: $P_{RS} = P_L$. The compliance lines of the lung and respiratory system cross (see Figure 28-6).

Surface Tension; the Problem of Alveolar Stability

The alveoli are lined by a thin film of liquid. At the interface between the liquid and the alveolar gas, strong

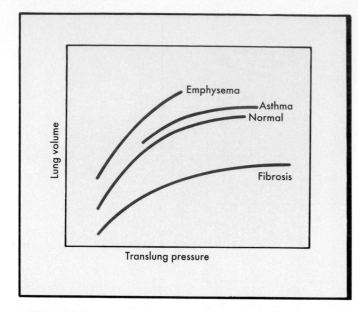

FIGURE 28-7   Compliance (distensibility) curves representing lungs with various chronic diseases. The emphysema line lies above and to the left of the normal line, indicating that the lungs are more distensible (larger) at any translung pressure. The lung in a patient with pulmonary fibrosis is stiffer (less distensible) than normal. The lungs of a person with asthma (hyperirritable airways) have normal compliance, although they are hyperinflated; FRC *(left end of line)* is at a higher volume and translung pressure than normal.

intermolecular forces in the liquid tend to cause the area of the lining to shrink (the alveoli tend to get smaller). These forces contribute more to the elastic recoil of the lung than do the lung tissue components (elastic and collagen fibers). The effects of surface forces on lung compliance become evident when the pressure-volume curves of air-filled and saline-filled lungs are compared (Figure 28-8).

Ventilating the lung with saline eliminates the air-liquid interface and abolishes the surface forces without affecting the lung tissue. At any given volume, the transpulmonary pressure of the liquid-inflated lung is less than that of the air-inflated lung. In the air-filled lung distensibility (compliance) is less during inflation than during deflation, as can be seen by the higher pressure required at any given volume in Figure 28-8.

If one blows air through a straw into some liquid dishwashing compound, fairly stable bubbles are formed. Since the bubble walls have essentially no tensile strength, they exist solely because the air-liquid surface tension is much lower than that of plain water. The walls between bubbles of equal diameter are flat because the pressure within the adjacent bubbles is equal. However, the outside bubbles of

a foam have a curved surface because that gives the least surface area for a given air volume, while it allows a component of the surface tension to act centripetally to oppose the internal pressure of the bubble.

Soaps and detergents are surface-active (prefer to be at the air-liquid surface), because they are partially hydrophobic (hate water). Therefore they prefer the surface, where they reduce the interfacial tension.

Alveolar Surfactant: Nature's Special Detergent

The most important component of the liquid film lining the alveolar walls is **surfactant.** It is produced by type 2 alveolar epithelial cells (granular pneumonocytes), and its major constituent is **dipalmitoyl phosphotidylcholine** (DPPC), a phospholipid with detergent properties.

Surfactant is special because it allows alveolar surface tension to vary with lung volume. Alveolar surface tension rises (approaching that of blood plasma, 50 milliNewtons/m) as the lung is inflated towards TLC. That is why the inflation limb of the pressure-volume curve in Figure 28-8 is displaced to the right (higher pressure at any volume).

As deflation begins, alveolar surface tension decreases rapidly, which is why the deflation limb of the pressure-volume curve of the air-filled lung moves to the left (less translung pressure needed to maintain a given volume compared to inflation). By reducing surface tension during deflation, surfactant promotes stability among the alveoli, which anatomically are of different sizes. Thus, the smaller alveoli do not collapse **(atelectasis)** at end-expiration (FRC). Alveoli are not like soap bubbles; each is connected via the alveolar ducts and airways to all the other alveoli (Figure 28-9).

The Laplace Equation and Alveolar Surface Tension   In Chapter 22 you learned the Laplace equation, which relates transmural pressure to wall tension and radius of a cylindrical blood vessel: $P_{tm} = T/r$. The Laplace equation also applies to the alveoli, because in the air-filled lung, wall tension is almost entirely surface tension, as illustrated in Figure 28-8. The component of translung pressure, P_L, caused by surface tension (st) is: $P_{Lst} = 2T/r$.

At end-inspiration, the surface tension in all the expanding, interconnected alveoli is about the same as that of blood plasma (50 milliNewtons/m). As deflation begins, the translung pressure decreases rapidly because surface tension falls to less than 10 mN/m. The low surface tension is important for the maintenance of alveolar stability as lung volume decreases toward FRC. See the description accompanying Figure 28-9.

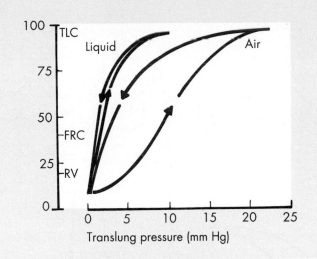

FIGURE 28-8 Pressure-volume loops of a normal human lung for the liquid-filled and air-filled states. Conventionally, each loop begins and ends at translung pressure equal to 0 (minimal lung volume). The solid line shows the liquid-filled lung. The curve is steep and the inflation and deflation lines are nearly the same. Total lung capacity is achieved at a translung pressure of 11 mm Hg. The air-filled pressure-volume curve is shown by the dashed lines. The inflation limb is displaced far to the right of the curve of the liquid-filled lung, whereas the deflation limb is closer to it. This means that the pressure-volume curve of the lung is dependent on something that is different between inflation and deflation—namely, alveolar surface tension.

In **respiratory distress syndrome** (RDS), a leading cause of morbidity and mortality in babies with immature lungs, the key defect is failure of the type 2 alveolar epithelial cells to secrete adequate quantities of surfactant. The lungs are somewhat more difficult to inflate, but the main problem is that during deflation the alveoli collapse readily, because surface tension does not fall. The deflation limb of the P-V curve is similar to the inflation curve. Thus, each lung inflation is like the first breath of air after birth, when the liquid-filled lungs have to be inflated with air. The increased work required to inflate the lungs fatigues the diaphragm.

Distensibility of the Entire Respiratory System

The solid line in Figure 28-6 shows the combined passive-deflation pressure-volume curve of the lungs and the chest wall—that is, the respiratory system, RS. When the lungs are at FRC, alveolar pressure (P_{alv}) is equal to atmospheric pressure (P_B). Hence, no pressure difference exists across the respiratory system. To passively maintain the system at volumes above FRC, one must apply pressure to the alveolar gas (positive pressure inflation); to force volume below FRC one must apply pressure on the body surface (compressive pressure). Note in Figure 28-6 that the slope of the P-V line of the respiratory system is less than that of either the lung or the chest wall; in other words, the compliance of the respiratory system is less than either of its components.

Measurement of the respiratory system compliance requires that the muscles of breathing be completely relaxed. In normal breathing, of course, the diaphragm and other chest wall muscles are doing all of the work. In active breathing it is impossible to measure the respiratory system compliance, because for every change of lung volume the P_{alv} equals P_B; that is, compliance of the respiratory system or of the chest wall appears to be infinite: $\Delta V / \Delta P_{RS} = V/0 = \infty$. In Figure 28-6 active breathing is represented by a vertical line at trans-organ pressure equal to zero. Lung compliance, however, can still be determined because the lung is always passive during breathing.

Resistance to Airflow

In addition to the pressure necessary to hold the lungs and chest wall at volumes above FRC, a pressure differ-

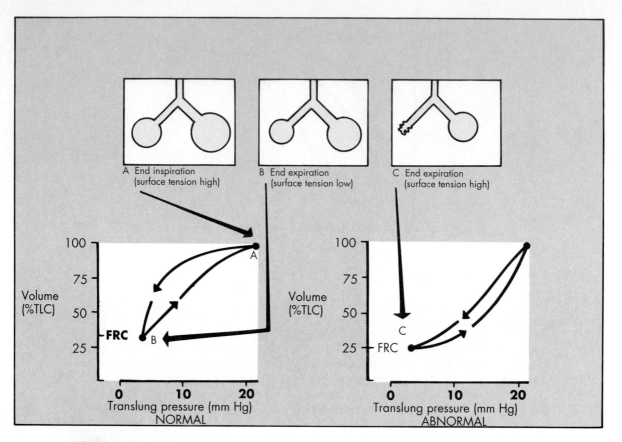

FIGURE 28-9   Effect of alveolar surface tension. **A,** At the end of a normal inspiration the surface tension in the lung's alveoli is essentially uniform and high. **B,** Surface tension decreases rapidly as lung volume starts to decrease. The decrease is greater in the more rapidly deflating alveoli. Consequently, all alveoli deflate relatively uniformly in proportion to their volumes. This maintains even the smallest alveoli inflated at FRC and makes the next lung inflation much less work. **C,** If there is a deficiency in surfactant production or secretion by the type 2 alveolar epithelial cells (respiratory distress syndrome of immature babies), surface tension remains high throughout the breathing cycle. The inflation curve is shifted to the right (more work required to inflate the lungs). The deflation limb follows the inflation limb. The smaller alveoli collapse to minimal volume; even the large ones are smaller; FRC is lower.

ence between the atmosphere and the alveoli is required to overcome viscous resistance (friction) during airflow in and out of the lungs. Although air has a very low viscosity compared to water, peak airflow is fast, about 0.5 L/sec, even during quiet breathing, and, of course, it is much faster during the increased ventilation of exercise.

Active Regulation of the Airways   The submucosal smooth muscle bands that encircle the airways are able to change the caliber of the bronchi and bronchioles independently of lung volume or translung pressure, which allows individual airways to alter their resistance to air flow. In the normal lung, airway smooth muscle tone is continuously modulated by airway reflexes.

Stimulation of the vagus nerves (parasympathetic, cholinergic), which are the motor nerves to the airway smooth muscles, leads to increased airway resistance. Stimulation of pulmonary sympathetic nerves (or adrenergic cell surface receptors) inhibits airway constriction.

A number of agents act reflexly via the vagus nerves to constrict the airways. Inhalation of smoke, dust, cold air, or irritant substances may have this effect. For example, exercise-induced asthma (cold or dry air) is common among people with hyperirritable airways.

Some substances affect the airway smooth muscle directly. For example, histamine-like compounds are used as provocative constrictor agents in testing patients who may have hypersensitive airways (asthma). On the other

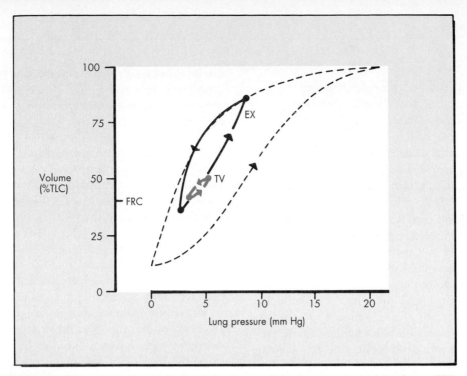

FIGURE 28-10 Dynamic pressure-volume loops of the lung; resting tidal volume (TV) and exercise (EX). The maximum air-filled lung P-V loop from Figure 28-8 is shown, too. The dynamic compliance of the tidal breath (slope of the line connecting end-inspiration with end-expiration) is less than that of the lung during exercise.

hand, agents that inhibit the parasympathetic nerves or stimulate the adrenergic nerves or cell surface receptors tend to dilate the airways.

The Time Dependence of Breathing Breathing is a dynamic event, hence time is a critical factor. We must ventilate our lungs adequately to supply oxygen to and remove carbon dioxide from the alveolar gas. Figure 28-2 shows some of the dynamic events associated with normal breathing. Inspiration lasts about 2.0 seconds and expiration about 3.0 seconds. If the duration of a breath is 5 seconds, the breathing frequency must be 12 breaths/min. In order to breathe more frequently, each breath must be shorter. Breathing faster than 30 breaths/min—2 seconds per breath—is uncommon, except in babies; it is inefficient (wasted ventilation is higher) and costly in terms of the energy required.

To achieve airflow one must generate a difference between ambient pressure (atmosphere) and alveolar pressure. Thus, during inspiration alveolar pressure is subatmospheric; it is higher than atmospheric pressure during expiration.

The changes in alveolar pressure during airflow are generated by changes in the pleural pressure. Thus, trans-lung pressure is slightly greater during inspiration and slightly less during expiration than necessary to sustain the lung volume changes, as is shown in the pleural pressure curve (second curve from the bottom) in Figure 28-2.

Calculation of Airway Resistance Physiologists use the simplified relationship that airway resistance (R_{aw}) equals driving pressure (ΔP) divided by flow ($\dot{V}$):

$R_{aw} = \Delta P/\dot{V}$, where ΔP is the difference between atmospheric pressure, P_B, and alveolar pressure, P_{alv}. During tidal inspiration the average decrease in P_{alv} is about 0.4 mm Hg and the average airflow rate is 0.25 L/sec. Thus, $R_{aw} = 0.4/0.25 = 1.6$ mm Hg $\times$ scc/L. During expiration the airways are a bit narrower, as will be discussed below, so that the average alveolar pressure is increased above P_B by about 0.4 mm Hg and the average airflow velocity is 0.2 L/sec (expiration is slower). Thus, $R_{aw} = 0.4/0.2 = 2.0$ mm Hg/L/sec, slightly more than during inspiration.

The Site of Airway Resistance From the trachea to the alveolar ducts the total cross-sectional area of the airways progressively increases. Thus, the velocity of airflow diminishes rapidly. In the trachea and main bronchi, airflow is turbulent (noisy), which is why one can hear breath sounds and why in normal people 80% of total resistance to airflow is in the upper airways (see also Chapter 20). At the low airflow velocities in the small airways, flow is laminar and silent.

In **asthma** (airway smooth muscle is hyperactive to a variety of internal or environmental stimuli) most airways are involved. The tiny bronchioles are constricted and may be further narrowed by edema (swelling) of the mucosa and excessive secretions from the airway epithelial cells. The loud wheezing sounds emitted by asthmatics during an attack are caused by turbulent airflow in the bronchioles.

Status asthmaticus and allergic laryngo-spasm are medical emergencies that demand swift treatment. The work of overcoming the high airway resistance rapidly fatigues the diaphragm and other breathing muscles. In these conditions the accessory muscles of breathing are very active.

Effect of Breathing on Airway Dimensions Because the bronchi are not directly attached to the lung and are distensible and collapsible, they are passively affected by translung pressure. During inspiration, as pleural pressure becomes more negative, the bronchi tend to distend (increased transmural pressure). The bronchioles also get wider because of the increase in lung volume. In normal tidal breathing the change of lung volume is only about 0.5 liter, so the effect on bronchiolar diameters is trivial.

During expiration pleural pressure rises and the bronchi are compressed; this explains why airway resistance is slightly increased. The bronchioles are not compressed as long as lung volume does not change much.

During a forced expiration pleural pressure may exceed pressure within the bronchi and, thus, tends to compress them. The site of the bronchial compression is usually in the mainstem bronchi because they are less well supported by their cartilaginous plates. Coughing is the best example of forced expiration; it causes a mainstem bronchus or the intrathoracic portion of the trachea to be compressed at one point. Airflow velocity at the point of compression is high and turbulent (coughing makes noise); both effects help to expel irritants from the upper airways.

People with advanced **chronic obstructive pulmonary disease** not only lose lung tissue, but lung compliance is also markedly increased (like blowing up a damp paper bag, instead of a rubber balloon). Also, the large airways may become flimsy. During expiration some of the bronchi collapse, and thereby increase airway resistance. A more serious consequence is that, if the sufferer tries to forcibly expire, pleural pressure rises more and further increases the airway compression. The correct strategy is to breathe in rapidly and breathe out slowly, which is what people with **emphysema** do automatically.

The Dynamic Pressure-Volume Curve of the Lung
During normal tidal breathing, the pressure-volume loop is much different from the static P-V loop shown in Figure 28-8, because neither volume nor translung pressure change much, as is shown in Figure 28-10. The small tidal breathing loop is located approximately in the center of the static P-V loop.

The slope of the line that joins the points of no flow (end-expiration and end-inspiration) is used to calculate the **dynamic lung compliance** (an oxymoron, because compliance is not a dynamic characteristic). In Figure 28-10 notice that the tidal P-V loop is tilted toward the X axis slightly more than either the inflation or deflation limbs of the static loop at the same lung volumes. Thus, dynamic lung compliance is less than static lung compliance, which is attributed to the fact that alveolar surface tension is higher during normal tidal breathing, because the lung volume changes are slight (0.5 L).

When one exercises (larger solid loop in Figure 28-10), the slope of the dynamic compliance line becomes steeper, which indicates greater lung distensibility because the larger tidal volumes stretch the alveolar walls and recruit more surfactant, which lowers the surface tension. Tidal volumes in heavy exercise may reach 50% of TLC (2.5 to 3 liters).

Limitation of Expiratory Airflow One of the most useful pulmonary function tests to evolve during the last 20 years is the **flow-volume curve,** which is achieved by simultaneously measuring and plotting flow at the mouth (obtained with a pneumotachograph; see Figure 28-2) and lung volume (obtained by integrating the flow signal). The results are shown in Figure 28-11 for tidal breathing and for a forced vital capacity.

The most important aspect of the forced flow–vital capacity loop is the expiratory portion. This portion shows the maximal flow at any given lung volume that can be achieved, no matter how much effort is expended. Contracting the powerful anterior abdominal muscles and the intercostal muscles (active expiration) generates a high flow velocity when lung volume is at or near TLC. The velocity of flow decreases rapidly (approximately linearly) as volume decreases, but not because of decreasing effort. Once the linear portion of the expiratory flow-volume curve is reached, it is effort-independent; that is, no matter how hard the subject tries, he cannot exceed the maximum flow at the given volume.

This occurs because the high pleural pressure causes dynamic compression of the airways, which limits flow more and more as lung volume decreases. Older people with more distensible lungs, and especially people with chronic obstructive pulmonary disease, generate less flow during expiration because of dynamic airway compression.

The Use of Muscular Energy in Breathing The **work** (energy cost) of breathing is done on the lung by the inspiratory muscles. The work done is low under normal conditions (about 1% to 2% of resting oxygen consumption), so that the respiratory muscles have a

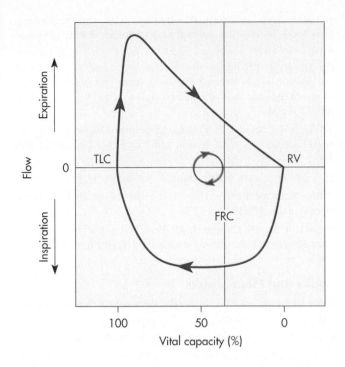

FIGURE 28-11   The flow-volume loop, the modern way to assess the dynamics of breathing. The small loop in the center represents normal tidal breathing. The large loop represents a maximal forced (as fast as possible) inspiration from residual volume to total lung capacity, followed by a maximal forced expiration back to RV (forced vital capacity maneuver). The most important information is contained in the expiratory curve, which is independent of effort once a certain flow is achieved. The effort-independent flow velocity is a function of airway collapsibility (dynamic compression).

large metabolic reserve. However, in certain diseases the diaphragm (or other chest wall muscles) may become fatigued (insufficient supply of metabolic substrates), and respiratory failure with hypoventilation ensues.

Most of the work used to expand the chest wall and lungs during inspiration is stored as elastic energy, which is used to restore the system to FRC during expiration. In fact, expiration could not occur passively without the potential energy stored during inspiration. However, the work expended to cause airflow is wasted (unrecoverable frictional heat).

During expiration the potential energy stored in the expanded lungs is used to pull the chest wall back to its FRC position and to overcome expiratory airway resistance.

HOW THE LUNG IS SUPPORTED IN THE THORAX

The lungs in the upright human are not supported by the trachea (which is rather extensible), nor do they rest on the diaphragm (there is a normal pleural space

interposed). One explanation of lung support is that each level of lung is supported by the lung above it; the lower parts of the lung are "hanging" from the upper parts. What happens when one reaches the visceral pleural surface of the lung? The force holding the lung up is transmitted through the pleural liquid layer to the chest wall. The result is that at higher levels in the lung, pleural pressure becomes more subatmospheric; the pressure varies from 2 to 6 mm Hg from the bottom to top when the lung is at FRC. Toward the top the lung must be more expanded because regional P_L ($P_{alv} - P_{pl}$) is increased. The average pleural pressure for the normal lung at FRC, however, is still 3.7 mm Hg for the entire lung.

SUMMARY

1. Functional residual capacity, FRC, is the volume of gas in the lungs at end-expiration. Total lung capacity, TLC, is the maximal lung volume that can be achieved.

2. The alveolar ventilation equation states that alveolar ventilation times arterial P_{CO_2} is a constant at any given level of aerobic metabolism (carbon dioxide production).

3. The main muscle of the chest wall is the diaphragm, but the intercostal muscles aid expansion, especially during exercise.

4. Functionally, the chest wall includes not only the diaphragm and rib cage but also the abdomen. Coordinated muscle contraction in the chest wall enlarges the thoracic cavity during inspiration, and the lungs expand passively in all directions to fill the cavity.

5. At end-expiration the lungs try to recoil to a smaller volume, while the chest wall tries to recoil in the opposite direction toward a larger volume, because the chest wall is normally under compression, except at very high volumes. FRC is the condition that occurs when these opposite recoil tendencies are equal.

6. The lungs and chest wall must move together. The total compliance of the respiratory system is less than either part alone.

7. Lung compliance (distensibility) during inflation is less than its compliance during deflation, because of the variable air-liquid surface tension in the alveoli. The reason for the variable surface tension is a special detergent (surfactant) secreted by type 2 alveolar epithelial cells; its chief component is dipalmitoyl phosphatidylcholine (DPPC).

8. Lung compliance is determined from the slope of the pressure-volume curve during passive slow deflation from TLC.

9. Because of resistance to airflow, alveolar pressure is

either slightly above or slightly below atmospheric (ambient) pressure, except at points of no air flow—that is, end-inspiration and end-expiration. During breathing the pressure difference across the lung is the sum of the compliance and resistive pressures. Breath sounds are caused by turbulence in the upper airways.

10. An important physical factor that affects airway resistance is the transmural pressure across the bronchi, whose walls are distensible and collapsible. In a forced expiratory maneuver, pleural pressure may rise above large airway pressure and dynamically compress the airways and thereby limit expiratory flow velocity. The forced flow–volume loop clearly shows expiratory flow limitation in health and disease.

11. The work of breathing at rest is small. In various lung or chest wall diseases, work may be increased to such an extent that respiratory failure occurs (increased Pa_{CO_2}).

BIBLIOGRAPHY

Journal Articles

Clements JA: Surface phenomena in relation to pulmonary function, *Physiologist* 5:11, 1962.

Dayman H: Mechanics of air flow in health and emphysema, *J Clin Invest* 30:1175, 1951.

Derenne JPH, Macklem PT, Roussos CS: The respiratory muscles. I. Mechanics, control and pathophysiology, *Am Rev Resp Dis* 48:119, 1978.

Fry DL, Hyatt RE: Pulmonary mechanics: a unified analysis of the relationship between pressure, volume and gas flow in the lungs of normal and diseased human subjects, *Am J Med* 29:672, 1960.

Macklem PT, Mead J: Resistance of central and peripheral airways measured by a retrograde catheter, *J Appl Physiol* 22:395, 1962.

Otis AB: The work of breathing, *Physiol Rev* 34:449, 1954.

Pattle RE: Properties, function and origin of the alveolar lining layer, *Nature* 175:1125, 1955.

Rahn H, Otis AB, Chadwick LE, Fenn WO: The pressure volume diagram of the thorax and lung, *J Appl Physiol* 146:161, 1946.

Books and Monographs

Bates DV, Macklem PT, Christie RV: *Respiratory function in disease*, ed 2, Philadelphia, 1971, WB Saunders.

Forgars P: *Lung sounds*, London, 1978, Bailliere Tindall.

Forster RE, Dubois AB, Briscoe WA, Fisher AB: *The lung*, ed 3, Chicago, 1984, Year Book Medical Publishers, Inc.

Murray JF: *The normal lung*, ed 2, Philadelphia, 1986, WB Saunders.

Nunn JF: *Applied respiratory physiology*, ed 3, London, 1987, Butterworths.

CHAPTER 29

Pulmonary and Bronchial Circulations and the Distribution of Ventilation and Perfusion

OVERVIEW

The work of pumping all of the cardiac output through the lungs is less than 10% of that required for the systemic circulation. The difference is attributed to the enormous *parallel* array of pulmonary resistance arteries that are normally dilated (relaxed), so that pulmonary vascular resistance is very low.

Under appropriate conditions the pulmonary vascular bed exerts considerable vasomotor control, so that blood flow (perfusion) to the myriad lung units supplied by the small resistance arteries matches the ventilation (fresh air inspired) of those units (ventilation/perfusion matching) and thereby maintains arterial P_{O_2} and P_{CO_2} near their ideal levels. Normally, flow in the pulmonary circulation is regulated by potent local mechanisms, of which alveolar oxygen tension is the most important.

Another remarkable feature of the pulmonary circulation is the network of capillaries in the alveolar walls. In resting adults it contains about 75 ml of blood, spread out in a vast array, generally one red cell thick. During exercise the capillary blood volume increases and approaches the maximum anatomical capillary volume, which is about 200 ml. The average thickness of the alveolocapillary wall is less than 1 μm, which helps to optimize oxygen diffusion between the alveolar gas and the hemoglobin in the red blood cells, as was shown in Figure 27-6. The total capillary surface area for gas exchange is about 40 times the body surface area.

One of the principal design features of the lung is to allow the erythrocytes to remain in the capillaries long enough to ensure equilibration between alveolar gas and blood; that is, the diffusive exchange of O_2 and CO_2 is not rate limiting, except under extreme conditions.

Because airplanes routinely fly at altitudes above 30,000 feet ($P_B \leq 230$ mm Hg), the airlines must pressurize the cabins to about 7000 feet ($P_B = 600$ mm Hg; the altitude of Santa Fe, New Mexico), so that the passengers and flight crew will not lose consciousness because of oxygen lack **(hypoxemia).** In spite of the pressurized cabin, people with severe chronic lung disease may require additional oxygen during flight.

The blood volume of the lung (main pulmonary artery to left atrium) in a normal adult is about 500 ml, which is about 10% of the total circulating blood volume. The relatively large volume serves as a reservoir (buffer) for filling of the left atrium. During normal breathing, as pleural pressure falls during inspiration (Chapter 28), venous return into the right ventricle rises, while the stroke volume of the left ventricle is reduced. During expiration the opposite occurs.

The lung has its own metabolic needs. Although the alveolar wall tissue can get all of the metabolites it requires from the pulmonary circulation and oxygen from the alveolar gas, the large conduit or support structures (bronchi, arteries and veins, pleura, and interlobular connective tissue) cannot. Thus, these structures receive a systemic blood supply, consisting of small bronchial arteries, which branch from the aorta or intercostal arteries.

PRESSURE AND RESISTANCE

Low Pressure and High Flow

Figure 29-1 shows the normal pressures in the human pulmonary and systemic circulations. The data are for a resting adult lying supine. The pressure in the pulmo-

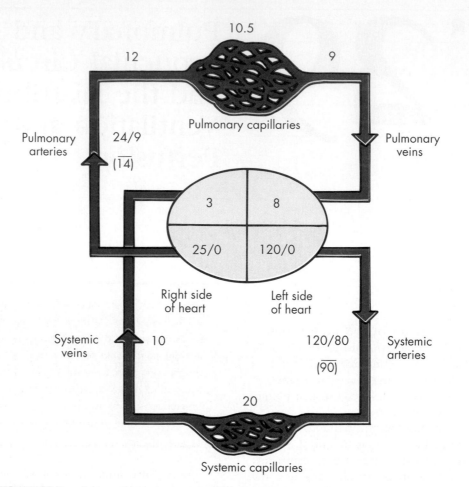

FIGURE 29-1 Schematic representation of the phasic or average pressure distribution within the systemic and pulmonary circulations in a normal, resting human adult lying in the dorsal recumbent position (supine); all pressures shown in millimeters of mercury (mm Hg). Note that the driving pressure in the systemic circuit ($P_{ao} - P_{ra}$) is $90 - 3$ is 87 mm Hg, compared to the driving pressure in the pulmonary circuit ($P_{pa} - P_{la}$) : $14 - 8 = 6$ mm Hg. Since cardiac output must be the same in both circuits in the steady state because they are in series, the resistance to flow through the lungs is less than 10% that of the rest of the body. Furthermore, the pressures in the left heart chambers are higher than those of the right heart. Thus, any congenital openings between the right and left sides of the heart favor left-to-right flow.

nary artery (P_{pa}) is about one-seventh that in the aorta. Left atrial pressure (P_{la}) is usually about 5 mm Hg higher than right atrial pressure. Pulmonary vascular resistance (PVR) at the normal resting cardiac output (Q) of 5 L/min and at mean pulmonary arterial pressure (P_{pa}) of 14 mm Hg and mean left atrial pressure (P_{la}) of 8 mm Hg is, according to the standard formula: PVR = ($P_{pa} - P_{la}$)/Q = $(14 - 8)/5 = 1.2$ mm Hg/(L×min).

Pressure-Flow Curves

A comprehensive view of pulmonary hemodynamics under a given set of conditions can be obtained by mea-

suring the changes in driving pressure ($P_{pa} - P_{la}$) as cardiac output varies. The resulting graphs are called **pressure-flow curves.** A normal curve is shown in Figure 29-2.

Vascular resistance is represented on a pressure-flow curve by the slope of the line drawn from the origin of the graph to a specified point on the curve, *not* by the slope of the curve itself. The most important aspect of the pressure-flow curve is its nonlinear shape. The curve bends toward the flow axis as flow increases (Figure 29-2). That means that pulmonary vascular resistance decreases as flow increases, because of the passive distensibility of the resistance vessels.

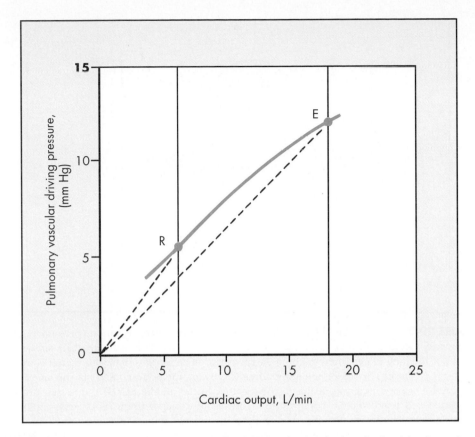

FIGURE 29-2   A representative pressure-flow curve that might be obtained in the pulmonary circulation of a dorsally recumbent (supine) normal human. R is the resting condition, and E is submaximal steady-state exercise. The slopes of the dashed lines from the origin to the two points on the curve represent pulmonary vascular resistance ($\Delta P/Q$). The key feature is that the pressure-flow line bends toward the abscissa (flow) as flow rises, meaning resistance decreases because of recruitment of capillaries and passive distention of resistance vessels.

When someone with lung cancer has half of his lung removed, pulmonary vascular resistance at rest may be only modestly elevated. This is so because the pulmonary vascular bed of the remaining lung can take up the doubled flow by recruitment of more alveolar wall capillaries. However, when he tries to exercise, there is much less vascular recruitment available. Hence, the pulmonary artery pressure rises during exercise and puts a limit on exercise ability.

PULMONARY BLOOD FLOW

Vessels Inside and Outside the Lung

In terms of lung anatomy, not everything is what it seems. For example, only the alveolar wall capillaries and the smallest venules and arterioles are functionally in the lung, because only these vessels are exposed on their external surfaces to the total pressure of the alveolar gas. These vessels are usually called **alveolar vessels.**

The outer surfaces of all other arteries and veins are exposed to pleural pressure (subatmospheric), because as stated in Chapter 27, the conduit vessels are not attached directly to the lung connective tissue structure. These vessels are usually called **extra-alveolar vessels.**

In the section on lung zones, which follows, the importance of the distinctions of these vessels is discussed.

The lung is wrapped around the extra-alveolar vessels. Thus, the conducting blood vessels and the bronchi are easily separated from the lung tissue by liquid (as in **pulmonary edema**) or air (as in **interstitial emphysema**). This arrangement is also important to the tho-

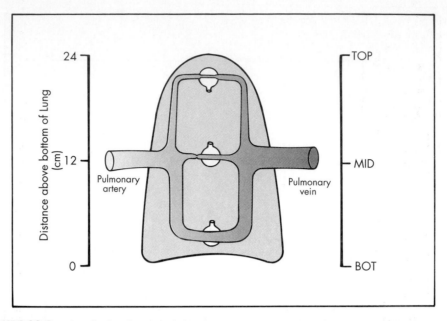

FIGURE 29-3   An idealized view of the distribution of vascular distention (pressure) in the pulmonary circulation, as it relates to the height of the lung. The ordinate scale shows the height of the lung in centimeters, about 24 cm at FRC in an upright man, although it is true in any body position. The vessels are more distended toward the bottom *(BOT)* of the lung (zone 3) because gravity increases all vascular pressures. In the middle *(MID)* the vessels are less distended and may even be compressed at the venous end of the alveolar wall capillaries when left atrial pressure falls below alveolar pressure (see zone 2 in Figure 29-4). Near the top *(TOP)* the alveolar wall capillaries may even be completely compressed, if the arterial pressure falls below alveolar pressure.

racic surgeon, who carefully separates the lung tissue from the main conduit vessels before removing a portion of a lung lobe. If the separation could not be done, the entire lobe would have to be removed, even though only a small portion is cancerous.

Figure 29-3 is a simplified model of the lung's rather complicated arrangement of blood vessels. The figure shows vessels at three heights (top, middle, and bottom) in the upright human lung. Because all of the pulmonary vessels, including the resistance vessels, are relatively thin-walled soft tubes, their diameters are sensitive to the distending pressure—that is, to the pressure difference from the inside to the outside of each vessel.

The pressure outside the alveolar vessels is alveolar gas pressure, which is ordinarily atmospheric pressure. The pressure outside the extra-alveolar vessels, however, is below alveolar pressure, being similar to pleural pressure (−3.7 mm Hg at FRC).

In normal breathing, when the lung expands and pleural pressure falls, the pressure outside the extra-alveolar vessels decreases. This causes the extra-alveolar vessels to enlarge during inspiration, which increases pulmonary blood volume. On the other hand, alveolar pressure does not vary much with breathing.

The anesthetist is sometimes faced with a dilemma. In order to ventilate the lung, it may be necessary to use a machine that inflates the lungs by delivering positive pressure at the nose and mouth. That raises alveolar pressure, which compresses the capillaries. Pulmonary vascular resistance is increased, but if there is lung disease, the pressure effect may not be uniformly distributed. This nonuniformity may markedly affect the matching of perfusion to ventilation and disturb not only the work of the right ventricle but also systemic arterial oxygen concentration.

How do the transmural pressures of the various blood vessels affect pulmonary blood flow? Because of gravity, the pressure of blood in all of the pulmonary vessels varies from top to bottom. That affects the perfusion of the various alveolar units and the ventilation/perfusion matching. Any deviation of ventilation/perfusion matching from the ideal always decreases the arterial Po_2.

Distribution of Blood Flow

Gravity affects venous return (cardiac output) and the distribution of systemic blood volume (see Chapters 23 and 24). Recall that a systemic arterial pressure of 120/80 mm Hg refers to pressure at the level of the heart or in the dorsal recumbent (supine) position (see Figure 29-1). Luminal pressure rises in both arteries and veins by 0.7 mm Hg for each centimeter below the heart and falls by a comparable amount above the heart.

The effects of gravity are greater in the pulmonary circulation than in the systemic circuit because the vascular pressures are much lower. Furthermore, the heart is about halfway up the lung in the upright human posture (see Figure 27-4). At functional residual capacity (FRC; normal lung volume at end-expiration) the bottom of the lung is about 12 cm below the lung hilum (level of left atrium).

Although the higher pressure toward the bottom of the lung does not change the driving pressure along the vessels, it does change the transmural distending pressure, which tends to decrease the resistance to flow. *Flow is greater toward the bottom of the lung.*

The opposite effect occurs for the 12 cm of lung above the heart level. When the lung is at FRC, the mean pulmonary arterial pressure at the top will be 5 mm Hg, and the pulmonary venous (outflow) pressure will be −1 mm Hg—that is, less than alveolar (atmospheric) pressure. When the pressure inside a distensible vessel is less than the pressure outside, the vessel is compressed and thereby increases resistance to flow at the point of compression. The pressure differences resulting from gravity affect the distribution of blood flow up and down the lung, as is illustrated in Figure 29-4. There are three zones, depending on the relationship among pulmonary arterial, venous, and alveolar pressures.

In zone 1 there is no blood flow, because pulmonary arterial pressure is less than alveolar pressure (capillaries compressed). In zone 3 flow is high because all vessels are distended. In zone 2 the alveolar pressure exceeds the venous outflow pressure, so that capillaries or venules exposed to alveolar pressure are compressed at the outflow end of the alveolar compartment (see Figure 29-3).

Although all three zones can exist in the human lung, pulmonary arterial pressure is normally high enough so that the zone 1 condition of no flow does not occur, and the condition described for zone 2 is limited to the top of the lung because left atrial pressure is normally well above alveolar pressure.

The intensive care physician often uses a positive-pressure machine to ventilate the patient with diseased lungs. If the pressure that inflates the lungs is too high, some of the lung may be put into zone 1, where there is no blood flow. In fact, the healthiest part of the lung may be most affected; thus, blood flow may be shifted to the more diseased parts of the lung. Hence, the ventilator may make ventilation/perfusion mismatching worse, instead of better.

Venous Admixture

Venous blood returning to the lungs that does *not* pass by air-filled alveoli is called **venous admixture,** because the venous blood mixes with arterialized blood from the lung. Venous admixture always lowers systemic arterial oxygen tension and hemoglobin saturation.

Venous admixture includes anatomical shunts and blood from mismatched ventilation/perfusion lung units. A small amount of venous admixture is normal (bronchopulmonary venous anastomoses [see later] and the intracardiac Thebesian veins). This amounts to less than 1% of the cardiac output normally, but in some diseases the percentage may be much higher.

In the fetus it is normal for nearly all of the cardiac output to bypass the lungs. Some blood flows through a hole **(foramen ovale)** between the atria, but most of the blood flows through a vessel **(ductus arteriosus)** that connects the pulmonary artery to the ascending aorta (see Chapter 25). After birth, as the blood is oxygenated by the lungs and pulmonary vascular resistance falls toward the low adult levels, these fetal shunts quickly close. However, in certain congenital defects they not only remain patent but allow venous blood to keep flowing through them.

After birth, left atrial pressure is higher than right atrial pressure, and aortic pressure is higher than pulmonary arterial pressure normally (see Figure 29-1). Even if the foramen ovale or the ductus arteriosus remains open after birth, blood flow is from *left to right,* so that systemic arterial blood is normally oxygenated. In a few uncommon anomalies (abnormal developmental anatomy) flow goes in the other direction, and it causes **cyanotic heart disease** ("blue babies," who have low hemoglobin saturation of their arterial blood).

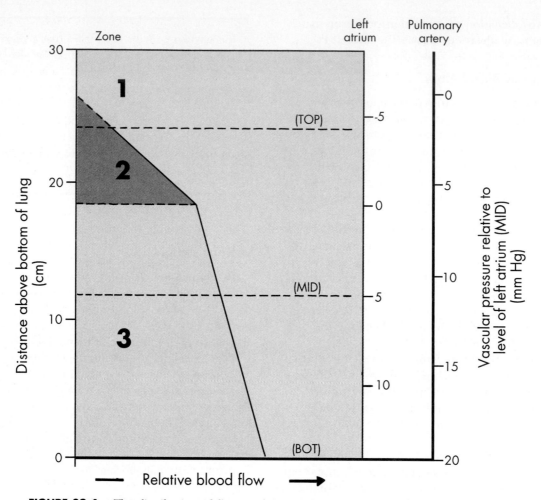

FIGURE 29-4   The distribution of flow in the normal upright human lung at rest. At FRC, the *TOP* of the lung is at 24 cm (left-hand ordinate) above the bottom *(BOT)*, the same as in Figure 29-3. At the level of the left atrium *(MID)*, left atrial and pulmonary arterial mean pressures are 5 and 11 mm Hg, respectively, as shown on the two parallel scales along the right-hand ordinate. Note that these pressures are each about 3 mm Hg less than in the supine resting individual (Figure 29-1). This is the result of gravity affecting the whole body in the upright posture. Clearly, most of the lung is in zone 3 because left atrial pressure does not fall below 0 until some 18 cm above the *BOT.* Zone 2 occurs in the upper 6 centimeters. Arterial pressure never falls below alveolar pressure, meaning there is no zone 1. In zone 2 flow is regulated by alveolar pressure compressing the outflow from the alveolar microvessels, as Figure 29-3 shows. In zone 3, although the driving pressure is constant, flow (abscissa shows relative flow only) continues to increase toward the bottom of the lung because of passive distention of the vessels by the increasing vascular pressure.

A left-to-right shunt is always *in addition to* the systemic flow; that is, pulmonary blood flow exceeds systemic blood flow and left ventricular stroke volume exceeds right ventricular stroke volume by the amount of blood flowing through the shunt.

If one thinks of venous admixture as a stream of blood being pumped around and around the body without delivering any O_2 or picking up any CO_2, then venous admixture reduces the efficiency of blood gas exchange. Venous admixture is the blood flow equivalent of wasted ventilation. Wasted ventilation gives a ventilation/perfusion ratio = V/Q = ∞.

In pulmonary diseases venous admixture occurs commonly because ventilation is low relative to perfusion in many lung units. Low ventilation/perfusion units effectively shunt some of the blood around the alveolar gas (wasted blood flow). On the other hand, high ventilation/perfusion does not cause venous admixture, only wasted ventilation.

Liquid Filtration in the Lung

All capillaries throughout the body are leaky in varying degrees to liquid, electrolytes, and proteins of blood plasma (Chapter 22). Net outward filtration continues throughout life. Most of the filtrate is returned to the circulation as lymph by the efficient lymphatic system.

The lung is no exception. The hydrostatic pressure favoring liquid filtration from the microvessels is higher near the bottom of the lung than near the top. Accumulation of excess extracellular liquid in the lung (**pulmonary edema**) may be life-threatening because it may cause serious venous admixture (ventilation/perfusion mismatching). Alveoli full of liquid cannot be adequately ventilated.

In patients with **congestive heart failure,** the left atrial pressure is elevated and edema fluid tends to accumulate, especially in the lower half of the lung. The physician generally finds reduced breath sounds or bubbling **rales** (noises in the small airways) near the bases (bottom) of the lung. The chest roentgenogram shows more density (more fluid) toward the bottom. Patients with left ventricular failure often complain that they must sit up in bed to sleep (**orthopnea**). When they lie flat, pulmonary vascular pressures rise and fluid filtration rises throughout the lung. The patient has a feeling of suffocation. The accumulating edema fluid in the alveoli interferes more with O_2 exchange and causes systemic arterial oxygen desaturation.

REGULATION OF THE PULMONARY CIRCULATION

Passive Regulation

In exercise most humans can increase their cardiac output at least threefold. As mentioned earlier, the large increase in blood flow is accommodated by the pulmonary circulation without an equivalent rise in the pulmonary vascular driving pressure because of recruitment and distention of microvessels. For example, during exercise the volume of blood in the pulmonary capillaries may more than double, so that the resistance to blood flow in the capillaries decreases substantially.

Active Regulation

Although the passive effects of pressure on the distensible pulmonary vascular bed are generally predominant, active regulation occurs under various physiological and pathological conditions. Contrary to what is often written, the smooth muscle associated with the small muscular pulmonary arteries, arterioles, and veins is adequate to alter pulmonary vascular resistance substantially. The smooth muscle may hypertrophy remarkably in pathological conditions.

Pulmonary hypertension, although not common, is a very serious problem to treat. Many of the remedies used to alleviate left ventricular high output failure are not effective for the right ventricle. **Primary pulmonary hypertension** (that is, high pressure in the pulmonary arteries of no known cause) strikes young women especially and may progress rapidly. Causes of **secondary pulmonary hypertension** include massive destruction of alveolar wall capillaries. Such destruction may occur in advanced chronic obstructive lung disease or after multiple pulmonary emboli. Surgical resection of more than about 60% of the total lung mass leads to pulmonary hypertension, even when the patient is at rest.

Autonomic Nerve Regulation The pulmonary vascular bed is innervated by parasympathetic (cholinergic) nerves (vagus nerves) and sympathetic (adrenergic) nerves. However, external autonomic stimulation does not greatly affect vascular resistance, but it seems chiefly to modulate lung vascular distensibility (compliance). Efferent vagal stimulation releases acetylcholine, which acts via endothelial cell receptors to cause the release of EDRF (endothelium-derived relaxing factor—nitric oxide, NO), which increases vascular compliance. It also relaxes the resistance vessels, but the effect is normally minimal because the vessels are already relaxed.

When humans (and other mammals) go to high altitude, pharmacological blockade of the release of endothelium-derived relaxing factor tends to further increase pulmonary vascular resistance. This response indicates that nitric oxide production is increased normally at high altitude. This, however, is probably a local effect and not an autonomic reflex, because carotid chemoreceptor stimulation (by hypoxemia; see Chapter 31) tends to constrict the pulmonary vascular resistance vessels.

Sympathetic nerves are the main motor nerves to pulmonary vascular smooth muscle. Baroreceptor stimulation may dilate pulmonary resistance vessels, most likely because of withdrawal of sympathetic tone. The sympathetic nerves release norepinephrine, as in systemic vascular beds. Norepinephrine acts principally via α-adrenergic receptors on the smooth muscle cells to increase their tone, and thereby to reduce vascular distensibility. Norepinephrine also tends to increase vascular resistance. The main circulating catecholamine, however, is epinephrine, which stimulates β-adrenergic receptors on the smooth muscle cells to relax them.

Local Hormones A number of naturally occurring substances affect the vasomotor tone of pulmonary arterial or venous vessels. Some are normally constrictors. The more important of these include: thromboxane A_2, α-adrenergic catecholamines, several arachidonic acid metabolites, neuropeptides, endothelin, and increased P_{CO_2}. Some are normally dilators. The more important of these include increased alveolar P_{O_2}, β-adrenergic catecholamines, prostacyclin, and nitric oxide.

Alveolar Oxygen Tension The factor of overriding importance that governs minute-to-minute normal regulation of the pulmonary circulation is the alveolar oxygen tension (P_{AO_2}).* The P_{AO_2} is critical because the small muscular pulmonary arteries are surrounded by the alveolar gas of the terminal respiratory units they subserve. Ordinarily, alveolar P_{O_2} is about 100 mm Hg.

The response of pulmonary vascular smooth muscle to low P_{O_2} is different from that of the systemic circulation. In the lung, low alveolar oxygen tension leads to *constriction* of nearby arterioles, whereas in the systemic circulation low oxygen tension *relaxes* the resistance vessels.

In the lung the low P_{O_2} appears to act directly on the vascular smooth muscle cells, although the exact molecular physiological mechanism is elusive.

As alveolar oxygen tension falls in any region of the lung, the adjacent arterioles constrict. This decreases local blood flow and shifts it to other regions of the lung. Because alveolar hypoxia usually results from inadequate ventilation, the local hypoxic vasoconstriction reduces perfusion, so that ventilation/perfusion is restored toward normal; this is an efficient self-control mechanism.

On the other hand, a global reduction in alveolar oxygen tension, such as occurs at high altitude, causes *all* of the resistance vessels to constrict and thereby raises pulmonary vascular resistance. Pulmonary vascular resistance in people traveling or living at altitudes above 10,000 feet may be more than twice normal.

An important clinical test that is used to assess the possible reversibility of a pulmonary hypoxic component is to have the subject breathe 100% O_2. Any hypoxic component of the hypertension ought to be reversed. Recently, other potent pulmonary vasodilators have begun to supplant the oxygen test. These include breathing nitric oxide or infusing β-agonists.

Thromboxane and Prostacyclin The most important of the vasoconstrictors is probably **thromboxane A_2**, a product of arachidonic acid metabolism. Many cells, including leucocytes, macrophages, platelets, and possibly endothelial cells, can produce and release thromboxane. Thromboxane A_2 is one of the most powerful constrictors of pulmonary arterial and venous smooth muscle. It constricts these pulmonary vessels mainly locally, because thromboxane is metabolized in seconds.

Prostacyclin (prostaglandin I_2), another product of arachidonic acid metabolism, is a potent vasodilator. Endothelial cells are the chief source of this substance, whose principal function is to prevent platelet and leucocyte adherence to endothelium (see Chapter 22).

BRONCHIAL CIRCULATION

The bronchial arteries supply water and nutrients to the airways down to and including the terminal bronchioles. They also nourish the pulmonary arteries and veins, pleura, and interlobular septa.

The pressure in the main bronchial arteries is essentially the same as in the aorta; hence the driving pressure is high. Although bronchial blood flow is less than 1% of cardiac output, it is appropriate for the portion of the lung tissue it serves.

About half of the bronchial blood flow returns to the right side of the heart via the bronchial veins. The remainder flows through small bronchopulmonary anastomoses (<100 μm diameter) into the pulmonary veins and thereby contributes to the small, normal venous admixture (right-to-left shunt).

In certain inflammatory diseases of the airways (**bronchitis, bronchiectasis, bronchogenic carcinoma,** the bronchial circulation expands dramatically and may contribute as much as 10% to 20% venous admixture.

---

*P_{AO_2} and P_{ACO_2} refer to the partial pressures of O_2 and CO_2 in the alveoli, whereas P_{aO_2} and P_{aCO_2} refer to the partial pressures of O_2 and CO_2 in arterial blood.

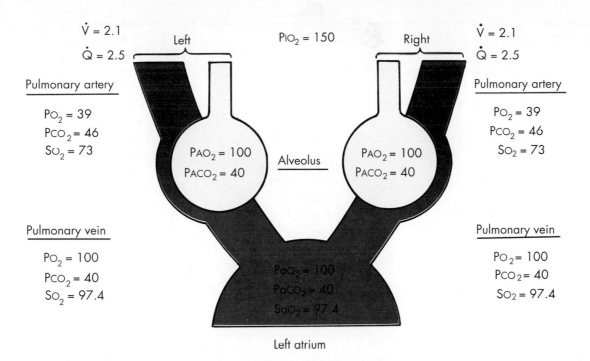

FIGURE 29-5   A two-compartment schema showing the effect of uniform alveolar ventilation/perfusion (V/Q) ratios of 0.84 on the partial pressures of oxygen (Po_2, mm Hg) and carbon dioxide (Pco_2, mm Hg), and the oxygen saturation (So_2, %), of the pulmonary venous blood leaving each compartment and after mixing in the left atrium (systemic arterial blood). Each lung receives half of the total ventilation (L/min) and the blood flow (L/min), so that the left atrial blood has the same So_2, 97.4%, as the outflowing blood from each lung.

The bronchial circulation participates in warming and humidifying the inspired air, especially when the inspired air bypasses the nose, as in mouth breathing during exercise. The inhaled air is completely warmed and humidified in the upper airways, and this process prevents any evaporation from the alveolar surfaces.

MATCHING VENTILATION TO PERFUSION

In the ideal lung the ventilation/perfusion ratios (V/Q) are uniform at about 0.8 (no units). This value yields the normal arterial Po_2 of 100 mm Hg and Pco_2 of 40 mm Hg, as shown in Figure 29-5. However, in the real lung, neither ventilation nor blood flow is uniformly distributed, even in a healthy adult.

In patients with cardiopulmonary disease, the most common cause of systemic arterial hypoxemia is uneven matching of alveolar ventilation to alveolar blood flow.

Ventilation-perfusion mismatching is generally expressed in terms of its effect on the alveolar-arterial difference in the partial pressure of O_2. Note that in Figure 29-5, the ideal lung has no A-a Po_2 difference.

The two extreme examples of ventilation-perfusion mismatching are **wasted ventilation, V/Q = ∞** (Chapter 26), and **venous admixture, V/Q = 0**.

The Normal Ventilation/Perfusion Distribution

Between totally wasted ventilation and totally venous admixture lie all possible ventilation/perfusion ratios. The V/Q distribution throughout the normal lung is not homogeneous, although the average value is about 0.8. Some lung units are overventilated and some are underventilated.

Figure 29-6 shows the distribution of V/Q in a normal human adult, which in this individual is 0.84, as shown by the thin vertical line extending up from the abscissa. The ordinate shows the person's ventilation or perfusion (L/min). The range of ventilation/perfusion ratios (derived by dividing each ventilation

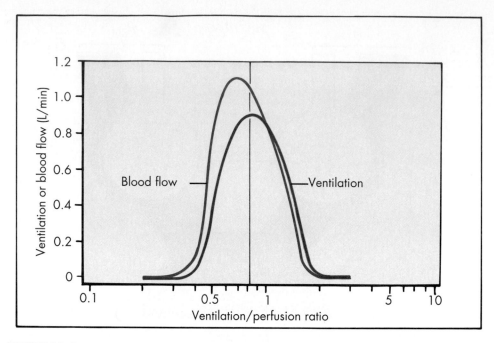

FIGURE 29-6   Normal ventilation and blood flow (perfusion) distribution curves in an adult human. On the abscissa is the V/Q ratio distribution, shown on a logarithmic scale ranging from 0.1 to 10. The overall lung normal value of 0.84 in this man is shown by the vertical line. The ordinate gives the absolute value of ventilation or blood flow for each V/Q ratio found.

value by its associated perfusion value) is shown on the X axis; the scale is logarithmic in order to encompass the wide range of V/Q ratios that may be encountered.

The conclusion to be drawn from the normal V/Q distribution graph is that the lung functions exceedingly well in spite of gravity and uneven lung expansion (Chapter 28) and uneven distribution of blood flow. The V/Q distribution is narrow.

DISTRIBUTION OF VENTILATION

Regional Distribution

In Figure 29-7 the middle inset illustration shows that the end-expiratory volume (FRC) of alveoli toward the bottom of the lung is less than that for the alveoli toward the top. This is a direct consequence of the fact that in a lung at FRC translung pressure is greater as one moves up the lung because of the way the lung is supported within the pleural space. However, as long as the compliance (slope of the P-V line in Figure 29-7) is linear, the inspired fresh air goes proportionally to all al-

veoli; that is, each alveolus (terminal lung unit) receives fresh air in proportion to its volume. Thus, regional maldistribution of ventilation is a minimal problem in the healthy young person.

One can increase the regional maldistribution of ventilation by temporarily reducing end-expiratory lung volume to residual volume (RV) or by inspiring to total lung capacity (TLC). Figure 29-7 shows both of these effects, which depend entirely on the shape of the pressure-volume curve.

Increased regional maldistribution is a normal part of the aging process, because the lung as a whole becomes more distensible (increased compliance).

Local Distribution

Local maldistribution of ventilation is probably more important than regional maldistribution in the normal adult. Resistance, R, of the small airways and compliance, C, differences among terminal respiratory units (alveoli) vary substantially on both functional and anatomical grounds. *The product of R × C (called the time constant) regulates not only how quickly a lung unit changes volume but also what the final volume will be.*

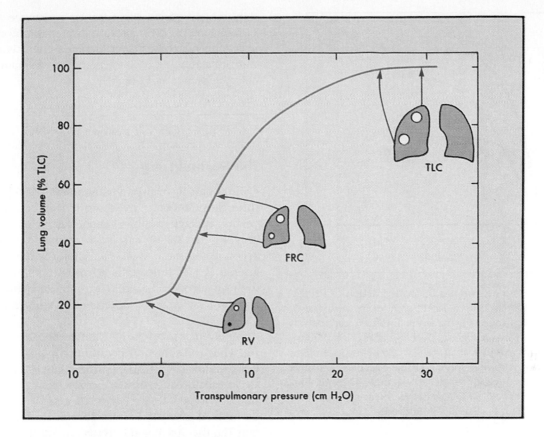

FIGURE 29-7   Regional distribution of lung volume at end-expiration. Because of the gradually more subatmospheric pleural pressure, (greater translung distending pressure [X axis]), as one goes from the bottom to the top of the lung, the end expiratory volume of individual lung units is less near the bottom of the lung *(FRC)*. As long as the various lung ventilation units are on the nearly linear part of the lung pressure-volume (compliance) curve, each unit is ventilated during normal tidal breathing in proportion to its volume, so that the regional distribution range of ventilation is fairly narrow. However, if end-expiratory lung volume goes below FRC, say to residual volume *(RV)*, some lower ventilation units are compressed. During the next tidal breath they are less well ventilated because their compliance is less. If one takes a maximal inspiration to total lung capacity *(TLC)*, all ventilation units are fully expanded, accentuating uneven ventilation among units that began the inspiration at different relative volumes.

In **chronic obstructive lung disease** (COPD, emphysema or **chronic bronchitis),** now the fourth leading cause of death in the United States (after heart disease, cancer, and accidents), maldistribution of ventilation may become so severe that the patient requires supplemental oxygen at rest. Some lung units may fill or empty very slowly. When chest physicians speak of "fast" or "slow" alveoli, they are referring to the time constants of various lung ventilation units.

Figures 29-8 shows how three different units will inflate during a normal inspiration: one unit is normal, one has twice the airway resistance of the others, and one has reduced compliance. At the normal breathing rate of 12 breaths/min, inspiratory time is about 2 seconds, which may not be long enough for every functional lung unit to achieve a new steady state. The normal unit, N, and the low compliance unit both attain new steady states, although the volume of the stiff (low-compliance) unit is reduced. The unit with increased R fills to only 80% of its steady-state volume. If inspiration is prolonged, the high-resistance unit will eventually fill.

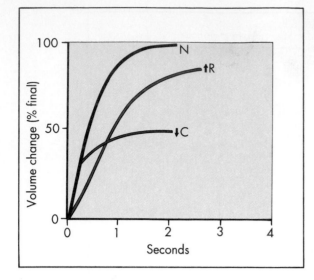

FIGURE 29-8 Three examples of local variation of ventilation among lung units caused by differences in airway resistance *(R)* or compliance *(C)*. The normal unit, *N*, reaches 97% of its equilibrium volume during a 2-second normal inspiration. The unit with increased resistance (↑R) fills slowly, reaching only 80% of its equilibrium volume in the 2 seconds available; it is underventilated relative to its FRC volume. The unit with decreased compliance (↓C) fills as fast as the normal unit, but, because it is stiffer, it is also underventilated.

DISTRIBUTION OF PERFUSION

Regional Distribution

We have already shown how gravity affects the regional distribution of blood flow in the lung by its effect on the transmural distending pressure of the blood vessels and on the relative arterial, venous, and alveolar pressures (lung zones) (see Figure 29-4). In the range of normal breathing at rest, gravitational effects on the blood flow distribution over the height of the lung may not be very large. Figure 29-4 shows that most of the lung is in zone 3, where flow changes do occur but not as dramatically as in zone 2.

Local Distribution

The local uneven distribution of blood flow among alveoli may be substantial. There are two components.

The first is anatomical (geometrical); that is, the diameter and length of the vessels to each functional unit affect the resistance to flow, according to Poiseuille's equation (Chapter 20). Local uneven flow on this congenital basis is not affected by large changes in total lung blood flow.

The second component is the alveolar oxygen tension, which affects the vascular resistance within the various lung units by altering local vasomotor tone. Normally, the hypoxic vasoconstrictor effect is not significant; inhalation of 100% oxygen changes pulmonary vascular resistance only slightly.

THE EFFECT OF V/Q MISMATCHING

The Normal Lung

Although in healthy young adults right-to-left shunts (bronchopulmonary anastomotic flow and intracardiac venous leakage through Thebesian veins) account for only 1% to 2% of cardiac output, V/Q mismatching accounts for some 4% to 5%. Thus, the normal systemic arterial Po_2 in humans is 85 to 90 mm Hg, compared with an alveolar Po_2 of 100 mm Hg, which gives an alveolar-to-arterial Po_2 difference (A-a gradient) of 10 to 15 mm Hg.

An alveolar-to-arterial Po_2 difference less than 20 mm Hg, while the subject is breathing room air, is considered normal. However, if a patient breathes 100% O_2, the physician usually can estimate the percentage venous admixture at roughly 1% for every 20 mm Hg difference. For example, if alveolar Po_2 is 550 mm Hg but Pao_2 is 150 mm Hg, the A-a Po_2 difference is 400 mm Hg, and the venous admixture is 20%.

V/Q Mismatching Depends on Oxygen Saturation or Concentration, Not Po_2

In healthy resting adults the average alveolar ventilation is 4.2 L/min and the average pulmonary blood flow is about 5.0 L/min. Thus, the V/Q ratio normally is about 0.84 (see Figure 29-6).

In Figure 29-5 half the alveolar ventilation and half the blood flow go to each lung. Thus, the V/Q ratio of each lung is also 0.84, and the pulmonary venous blood leaving each has the same oxygen saturation, namely 97.5%. When the two streams join in the left atrium, the So_2 remains at 97.5%.

Figure 29-9 shows that the lung on the right side is overventilated but has normal blood flow. Thus, the V/Q is high and the Pao_2 is increased. This, however, only raises the So_2 in the outflowing blood to 98.0% because the HbO_2 curve has nearly plateaued (see Figure 27-2).

On the other hand, the left lung is underventilated, but it has a normal blood flow. Thus, its V/Q is low and the Pa_{O_2} is decreased, which reduces the saturation of the outflowing blood to 89%. When the venous blood mixes in the left atrium, the saturation will be 93.5% and the Po_2 will be 68 mm Hg.

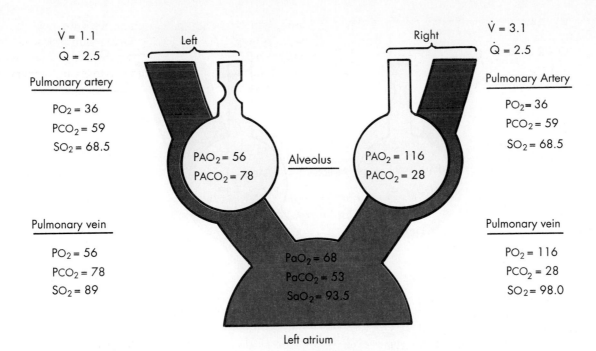

FIGURE 29-9   This two-compartment schema shows the effect of uneven alveolar ventilation/perfusion ratios (V/Q) on the P_{O_2}, P_{CO_2} and S_{O_2} of the pulmonary venous blood leaving each compartment and after mixing in the left atrium (systemic arterial blood). The right lung receives 3 L/min ventilation instead of 2 in the ideal lung (Figure 29-5), but it still receives half of the normal blood flow, 2.5 L/min. Its V/Q is 3/2.5 = 1.2. Clearly, it is overventilated. Its P_{CO_2}, which must be calculated using the alveolar ventilation equation (Chapter 28), is 28 mm Hg—hyperventilation, by definition. Its P_{AO_2} is increased to 116 mm Hg. However, the saturation, S_{O_2}, of the outflowing venous blood is only slightly increased, to 98.0%. The left lung, of course, is underventilated by 1 L/min but still receives half of the total blood flow. Its V/Q is 1.1/2.5 = 0.44. Its P_{CO_2}, calculated using the alveolar ventilation equation again, is 78 mm Hg—hypoventilation, by definition. Its P_{AO_2} is decreased to 56 mm Hg, which gives an S_{O_2} equal to 89% in the outflowing venous blood. When the two bloodstreams mix in the left atrium, the resultant systemic arterial blood is hypoxemic (S_{O_2} = 93.5%; P_{O_2} = 68 mm Hg.) Clearly, the hyperventilated lung compartment cannot compensate for the hypoventilated compartment because of the markedly alinear shape of the HbO_2 equilibrium curve (Figure 27-2). Note that in this uncompensated example the Pa_{CO_2} is increased.

V/Q Mismatching and P_{CO_2}

V/Q mismatching also affects the systemic arterial P_{CO_2}, as shown in Figure 29-9. Usually, compensation is effective, so that the Pa_{CO_2} is only slightly above normal (see Figure 27-10).

Compensation for V/Q Mismatching

A number of compensatory mechanisms come into play to partially alleviate V/Q mismatches. Initially, these may involve increasing the overall ventilation, because the Pa_{CO_2} is increased. However, increasing total ventilation is only a stopgap measure, because much of the increased ventilation goes to the lung that is already overventilated.

Local regulatory factors are much more important. As already mentioned, hypoxic vasoconstriction (in hypoventilated units) is the single most effective mechanism to shift flow away from underventilated units, as Figure 29-10 demonstrates. Likewise, when ventilation is wasted (hyperventilated unit), the local Pa_{CO_2} falls. This lowers the hydrogen ion concentration (raises pH) in and around the associated airway smooth muscle. The fall in $[H^+]$ increases in local airway resistance and shifts ventilation to units with higher V/Q ratios (Figures 29-7 and 29-10).

If blood flow is sufficiently reduced to a lung unit, the local alveolar cell metabolism (notably, surfactant production) may be decreased. The increase in the air-liquid interfacial tension reduces the unit's compliance, and its FRC volume decreases (see Figure 29-8).

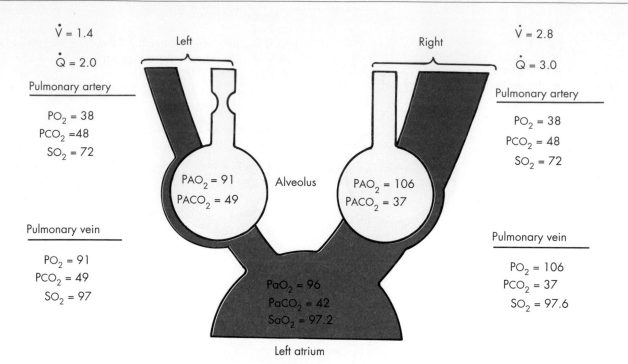

Left

$\dot{V} = 1.4$

$\dot{Q} = 2.0$

Pulmonary artery

$PO_2 = 38$
$PCO_2 = 48$
$SO_2 = 72$

$PAO_2 = 91$
$PACO_2 = 49$

Alveolus

$PAO_2 = 106$
$PACO_2 = 37$

Pulmonary vein

$PO_2 = 91$
$PCO_2 = 49$
$SO_2 = 97$

Right

$\dot{V} = 2.8$

$\dot{Q} = 3.0$

Pulmonary artery

$PO_2 = 38$
$PCO_2 = 48$
$SO_2 = 72$

Pulmonary vein

$PO_2 = 106$
$PCO_2 = 37$
$SO_2 = 97.6$

$PaO_2 = 96$
$PaCO_2 = 42$
$SaO_2 = 97.2$

Left atrium

FIGURE 29-10   The two-compartment schema showing the marked improvement of the systemic arterial blood gases after modest local compensatory mechanisms have occurred. Hypoxic vasoconstriction has occurred in the left lung compartment, shifting 0.5 L/min of blood flow to the right, overventilated lung. Increased airway resistance in the right lung has shifted 0.3 L/min to the left lung. These shifts bring the V/Q of the left and right lungs to 0.7 and 0.94, respectively. The compensation brings the systemic arterial blood gases to near normal.

SUMMARY

1. Pulmonary vascular resistance is normally much less than systemic vascular resistance. The pressure-flow curve is an excellent way to assess changes in pulmonary hemodynamics under various conditions.

2. Blood flowing from the systemic venous to the systemic arterial system without being fully oxygenated (right-to-left shunt) decreases systemic arterial oxygen tension and concentration. In left-to-right shunts, pulmonary blood flow is increased, but the systemic arterial oxygen concentration is normal.

3. The distribution of pulmonary blood flow is sensitive to gravity over the height of the air-filled lung. Gravity leads to three flow conditions: zone 1, no flow; zone 2, flow is regulated by compression of microvessels at the outflow from the alveolar walls; zone 3, flow is dependent on driving pressure, vascular geometry and smooth muscle tone. The normal human lung is mostly in zone 3.

4. Although the normal passive regulation of pulmonary blood flow distribution predominates, active regulation may become very important. The main regulator of flow distribution is alveolar oxygen tension, such that a decrease in alveolar oxygen tension (al-

veolar hypoxia) leads to a local increase in vascular resistance. Thromboxane and prostacyclin are powerful constrictors and dilators, respectively.

5. The bronchial circulation nourishes the walls of the airways and blood vessels; it warms and humidifies the incoming air.

6. The ventilation/perfusion ratios among functional lung units are not uniform, even in the normal lung, and they may deviate markedly from normal in disease. The limiting ratios are wasted ventilation (V/Q = ∞) and venous admixture (V/Q = 0).

7. Ventilation is distributed nonuniformly in the lung on regional (gravitational) and local (nongravitational) bases. Local factors are probably more important under normal conditions by affecting the ventilation distribution to each lung unit on the basis of its resistance and compliance.

8. Blood flow is also distributed nonuniformly in the lung on regional and local bases. Regional distribution predominates in the normal upright human. Acute alveolar hypoxia effectively regulates local blood flow.

9. The overall effect of ventilation/perfusion distribution can be judged by determining the alveolar-arterial oxygen tension difference, which is normally 10 to 15 mm Hg.

REFERENCES
Journal Articles

Bhattacharya J, Staub NC: Direct measurement of microvascular pressures in the isolated perfused dog lung. *Science* 210:327, 1980.

Deffenbach ME et al: The bronchial circulation: small, but a vital attribute of the lung, *Am Rev Resp Dis* 135:463, 1987.

Marshall C, Marshall B: Sitc and sensitivity for stimulation of hypoxic pulmonary vasoconstriction, *J Appl Physiol* 55:711, 1983.

Milic-Emili J et al: Regional distribution of inspired gas in the lung; *J Appl Physiol* 21:749, 1966.

Mitzner W: Resistance of the pulmonary circulation, *Clin Chest Med* 4:127, 1983.

West JB, Dollery CT, Naimach A: Distribution of blood flow in isolated lung: relation to vascular and alveolar pressures, *J Appl Physiol* 19:713, 1964.

West JF: Ventilation-perfusion relationships, *Am Rev Resp Dis* 116:919, 1977.

Books and Monographs

Grover RF ct al: Pulmonary circulation. In: *Handbook of physiology*, section 2: The cardiovascular system, vol III. Bethesda, Md, 1984, American Physiological Society, pp 103-136.

Harris P, Heath D: *The human pulmonary circulation*. ed 2, Edinburgh, 1978, Livingstone.

Staub NC: *Basic respiratory physiology*, New York, 1991, Churchill-Livingstone.

30

Transport of Oxygen and Carbon Dioxide Between Lungs and Cells of the Body

ARTERIAL AND VENOUS STREAMS

The cardiovascular system transports O_2 from the lungs to all of the systemic capillaries and transports CO_2 from the systemic capillaries back to the lungs. However, the cardiovascular system is not directly under the control of the respiratory system, except for the number of erythrocytes ($5 \times 10^6/\mu L$; 5 trillion per liter) and the concentration of hemoglobin, the oxygen transport protein in each red cell. These blood parameters are regulated in large part by the renal tissue hormone **erythropoietin,** whose production and secretion are inversely related to kidney tissue Po_2.

Red blood cells also contribute in a major way to carbon dioxide transport because hemoglobin is an important CO_2-binding protein and because the enzyme **carbonic anhydrase,** which greatly accelerates the reaction of CO_2 with water, is present in erythrocytes but not in extracellular liquids, including plasma.

An interesting feature of respiratory gas transport in blood is shown in Figure 30-1. Oxygen is transported for metabolic purposes only in the systemic arterial blood from the lungs to the tissue capillaries. Carbon dioxide is transported for metabolic purposes only in the systemic venous blood from the tissue capillaries to the lungs.

OXYGEN TRANSPORT

Combination of Oxygen with Hemoglobin

As briefly outlined in Chapter 27, almost all of the oxygen transported in the systemic arterial blood is chemically bound to hemoglobin. This protein is specifically made within the bone marrow during the latter stages of red blood cell development. Normal human hemoglobin (**hemoglobin A**) consists of four O_2-binding heme molecules (iron-containing porphyrin rings), each attached to a polypeptide chain. The total molecular weight of hemoglobin is 66,500, one of the smaller proteins. Normally, the concentration of hemoglobin in blood is about 150 g/L. Each gram of hemoglobin can combine with about 1.34 ml of O_2. Thus, the hemoglobin in 1 liter of blood can combine with approximately 200 ml of O_2 (100% hemoglobin saturation).

The Hemoglobin-Oxygen Equilibrium Curve

The binding of oxygen to hemoglobin is directly dependent on the partial pressure of oxygen (Po_2) with which the hemoglobin is equilibrated. The empiric (measured) relationship is called the **hemoglobin-oxygen equilibrium curve.** The normal HbO_2 curve for human adult systemic arterial blood is shown in Figure 30-2 (see also Figure 27-2).

There are three important physiological properties of the chemical binding between hemoglobin and O_2. These properties are not obvious from looking at the HbO_2 equilibrium curve. They are: (1) Hemoglobin combines reversibly with O_2. The oxygen-bound form is called **oxyhemoglobin,** HbO_2, and the deoxygenated form is called **hemoglobin,** Hb. (2) Molecular oxygen rapidly reacts with hemoglobin. The reaction times are measured in milliseconds, even when the hemoglobin is densely packed in erythrocytes. This fast reaction is critical for O_2 transport, because blood remains in the pulmonary and systemic capillaries less than 1 second (see Chapters 27 and 29). (3) Heme group interaction affects the shape of the HbO_2 equilibrium curve. The S-shaped (sigmoid) HbO_2 equilibrium curve is caused by a change in the shape (configuration) of the hemoglobin molecule after the iron in three of a molecule's heme groups binds oxygen. The shape change increases hemoglobin's affinity for oxygen—that is, greatly increases the ability of Hb

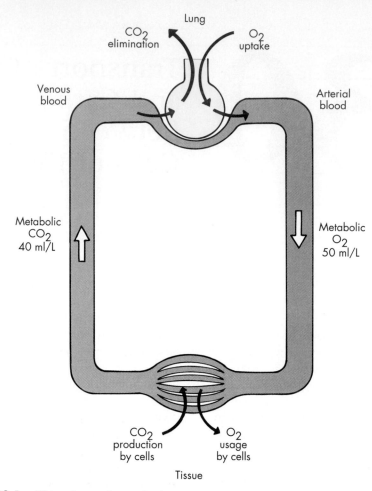

FIGURE 30-1   This schema shows the key concept about oxygen and carbon dioxide transport in the blood. Oxygen is carried from the pulmonary capillaries to the systemic capillaries by the systemic arteries. Carbon dioxide is carried from the systemic capillaries to the pulmonary capillaries by the systemic veins. The normal resting arterial-venous O_2 difference is 50 ml/L of blood, and the venous-arterial CO_2 difference is 40 ml/L. At a resting normal cardiac output of 5 L/min, [CO_2 production (Vco_2) is 200 ml/min and O_2 consumption (Vo_2) is 250 ml/min]. The respiratory exchange ratio is Vco_2/Vo_2; 200/250 = 0.8.

to bind O_2 at a given partial pressure. Functionally, the shape of the curve maximizes the quantity of O_2 taken up in the lungs even at lower-than-normal alveolar oxygen tension and increases the quantity released in the systemic capillaries at a relatively high partial pressure.

Oxygen Concentration or Hemoglobin Saturation

The amount of oxygen bound to hemoglobin is described in two ways: the actual concentration, in milliliters per liter of blood, and the relative concentration, or (percentage) of the maximum amount that can be bound (**oxygen capacity**). The relative amount is called the saturation; for example, So_2 is 50% when Po_2 is 26 mm Hg, as Figure 30-2 shows.

Both ways of expressing O_2 binding are necessary and useful. The normal saturation of systemic arterial blood is 97.5% when Pao_2 is 100 mm Hg and of systemic mixed

venous (pulmonary arterial) blood is 75% when Pvo_2 is 40 mm Hg.

Individuals with **anemia** have lower-than-normal blood hemoglobin; for example, in the anemia associated with **chronic renal failure** the circulating hemoglobin concentration may be 50 g/L, because of inadequate renal output of the red cell stimulating hormone **erythropoietin**. Each liter of blood has only one-third the normal oxygen capacity, even though the hemoglobin may be normally saturated in systemic arterial blood. Despite such O_2 transport limitations, the victims of this insidious process may complain only of tiredness or the inability to do much exercise. When the patient is treated properly with recombinant human erythropoietin, the blood oxygen capacity may improve dramatically, so that the patient feels invigorated.

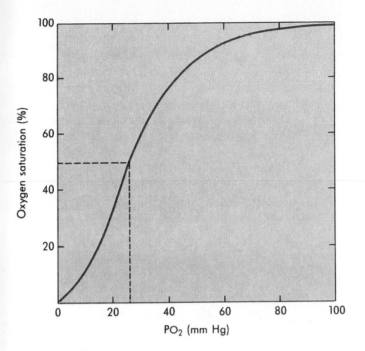

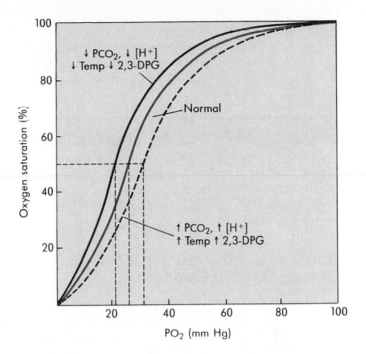

FIGURE 30-2   The normal hemoglobin-oxygen (HbO$_2$) equilibrium curve is an experimentally determined relationship (which has profound molecular physical-chemical significance) between the partial pressure of oxygen (abscissa) and the saturation (percent maximal binding of oxygen to hemoglobin.) The P$_{50}$ (26 mm Hg normally in systemic arterial blood) is used to compare HbO$_2$ curves under different conditions.

FIGURE 30-3   The position of the hemoglobin-oxygen equilibrium curve is shifted to the right when any of the four physiological factors (P$_{CO_2}$, [H$^+$], temperature, 2,3-DPG) shown is increased. A shift to the right ($\rightarrow$) decreases the ability of Hb to bind O$_2$ (decreased oxygen affinity). A shift to the left ($\leftarrow$) signifies a decrease in one or more of the four physiological factors and represents an increased oxygen affinity for hemoglobin.

Physiological Factors that Affect the HbO$_2$ Equilibrium Curve

The normal hemoglobin-oxygen equilibrium curve for human adult hemoglobin (type A) shown in Figure 30-2 is for: hydrogen ion concentration [H$^+$] = 40 nmol/L (pH = 7.40), P$_{CO_2}$ = 40 mm Hg, temperature = 37° C, and 2,3-diphosphoglycerate concentration [2,3-DPG] = 15 µmol/g Hb.

When any of these four physiological factors ([H$^+$], P$_{CO_2}$, temperature, [2,3-DPG]) increases, the affinity of hemoglobin for oxygen decreases. It is customary to describe changes in Hb affinity in terms of the P$_{O_2}$ at 50% Hb saturation (P$_{50}$). The P$_{50}$ increases (decreased affinity) when any of the factors increases (Figure 30-3, *dashed curve*). The entire HbO$_2$ equilibrium curve is shifted proportionally to the right of the standard curve.

Conversely, when the concentration of any of the physiological factors decreases, the affinity of hemoglobin for oxygen increases and the P$_{50}$ decreases. The entire HbO$_2$ equilibrium curve is shifted proportionally to the left (Figure 30-3, *solid curve*).

The effect of [H$^+$] is caused by a greater affinity of hydrogen ions for hemoglobin than for oxyhemoglobin. Carbon dioxide, which forms carbonic acid in plasma, has its principal effect by increasing the H$^+$ concentra-

tion. The shift in the position of the oxyhemoglobin equilibrium curve, caused by a change in P$_{CO_2}$, is called the **Bohr effect.** Increased temperature favors dissociation and decreased temperature favors association of O$_2$ with Hb. The compound 2,3-diphosphoglycerate (2,3-DPG) is present in erythrocytes in high concentration relative to other cells because mature red blood cells, having no mitochondria, respire by anaerobic metabolism (glycolysis), which produces 2,3-DPG as a side reaction. 2,3-DPG binds to hemoglobin more strongly than to oxyhemoglobin and reduces its affinity for oxygen. Increases in [2,3-DPG] occur in chronic hypoxemia (decreased Pa$_{O_2}$) and when blood [H$^+$] decreases (increased pH), whereas decreases in [2,3-DPG] occur in blood stored for transfusions.

One of the factors that favors oxygen diffusion across the placenta to the developing fetus is that the hemoglobin (type F) produced in utero is not affected by 2,3-DPG; therefore the fetal HbO$_2$ curve shows slightly greater O_2 affinity than does that of the mother. Furthermore, the Hb concentration in fetal blood is high (up to 200 g/L). Thus the fetus's arterial blood has nearly the same O_2 concentration as that of its mother, even though the fetus's arterial P_{O_2} is less than 40 mm Hg.

In everyday life, shifts in $[H^+]$, P_{CO_2}, temperature, and [2,3-DPG] are small. The HbO_2 curve for mixed venous blood lies only slightly to the right of the curve for arterial blood, as shown in Figure 30-4. The P_{50} increases from 26 mm Hg in arterial blood to 29 mm Hg in mixed venous blood.

Other Factors That Can Affect Oxygen Binding to Hemoglobin

Myoglobin If the hemoglobin molecule were separated into its four globin chains, each with one heme group, one would have a molecule very much akin to **myoglobin,** the oxygen-binding heme protein (molecular weight 16,500 daltons) in skeletal muscle cells, which gives meat its red color.

The important physiological feature of the myoglobin-oxygen (MbO_2) equilibrium curve is that it lies far to the left of the HbO_2 equilibrium curve, and it has a different shape (hyperbola). The two curves are compared in Figure 30-5, *A*. Because each molecule has only one heme group, there cannot be any molecular interaction. Also, because of the low P_{O_2} at which myoglobin binds O_2, it is unsuited for O_2 transport. However, myoglobin is useful for storing O_2 temporarily in skeletal muscle cells, where the P_{O_2} is normally low and in the range over which myoglobin is only partly saturated. The half-saturation partial pressure (P_{50}) of oxymyoglobin is <5 mm Hg.

Carbon Monoxide Hemoglobin Figure 30-5, *A*, also shows the equilibrium curve for the pathological **carbon monoxide–hemoglobin** (HbCO). Carbon monoxide has more than 200 times the affinity (ability to bind at a given P_{O_2}) for hemoglobin than does oxygen. Carbon monoxide associates readily with hemoglobin, but it does not dissociate unless the P_{CO} is less than 1.0 mm Hg.

One of nature's oddities is that carbon monoxide is produced normally whenever hemoglobin is catabolized in the liver. Because humans degrade 0.8% of their erythrocytes daily, they produce considerable CO. Fortunately, it is slowly exhaled in the expired air and does not accumulate.

Another oddity is that nitric oxide (NO), the endothelium-derived relaxing factor (EDRF, Chapters 22 and 29), binds to hemoglobin 200,000 times more

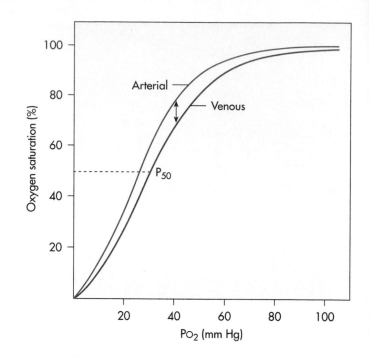

FIGURE 30-4 When blood passes through the systemic capillaries the HbO_2 equilibrium curve shifts slightly to the right *(blue line)*, which decreases the oxygen affinity of the hemoglobin. For example, in arterialized blood *(red line)* the saturation at P_{O_2} of 40 mm Hg is 80%, but in venous blood it is 75%. Thus, about 5% more oxygen is released. The effect is caused by the rise in P_{CO_2} and $[H^+]$ in the systemic capillaries. The reverse occurs in the lung, which increases the oxygen affinity of the blood, although the effect on oxygen saturation is much less than in the systemic capillaries because of the plateau of the HbO_2 curve at high P_{O_2}. This physiological shift is called the Bohr effect.

strongly than does oxygen and 1000 times more strongly than does CO. Thus, NO is strictly a local hormone, active only at its site of production, because any NO that diffuses into blood is rapidly and irreversibly bound by hemoglobin.

Carbon monoxide poisoning is a much more serious problem than is an equivalent reduction of hemoglobin O_2 capacity caused by anemia, because of the high affinity of CO for hemoglobin. Figure 30-5, *B,* shows this effect. In Figure 30-5, *A,* the ordinate is the usual saturation (relative concentration) of either CO or O_2, whereas in Figure 30-5, *B,* the ordinate is the absolute oxygen concentration in blood. HbCO has a hyperbolic equilibrium curve that is shifted far to the left, compared with the equilibrium curve of a simple hemoglobin deficiency anemia of the same degree (50% decrease in oxygen capacity).

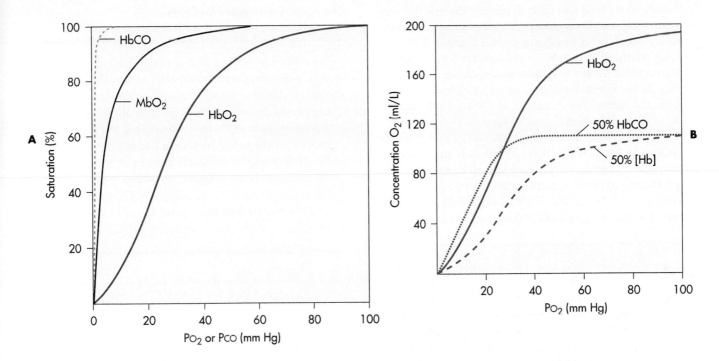

FIGURE 30-5   **A,** Comparison among the saturation curves of hemoglobin for oxygen and for carbon monoxide, and of myoglobin (the skeletal muscle oxygen storage protein) for oxygen. The P_{50} of myoglobin is <5 mm Hg, which is similar to the normal intracellular P_{O_2} of muscle. Myoglobin serves as a temporary intracellular O_2 storage depot, useful when blood flow is interrupted briefly by phasic muscle contraction. Carbon monoxide–hemoglobin has a P_{50} of 0.1 mm Hg. **B,** The full effect of the damage caused by carbon monoxide to oxygen transport is seen when the actual concentration of oxygen in blood is plotted versus P_{O_2}. This panel compares normal blood with blood with half the normal hemoglobin concentration (anemia) and with blood with 50% HbCO. The dramatic change in the shape of the oxyhemoglobin equilibrium curve in CO poisoning can be seen by the shift in the mixed venous point, if the a-v oxygen difference remains at 50 ml/L.

Methemoglobin Hemoglobin binds oxygen only when the iron atoms in the porphyrin rings are in the ferrous (Fe^{++}, reduced) state. One of the developmental advantages of having hemoglobin packaged within erythrocytes is that several reducing systems are present that keep hemoglobin functional. Certain chemicals (nitrates and sulfates) can, however, oxidize the iron to the inactive ferric (Fe^{+++}) form. Nowadays, this is an infrequent clinical problem.

Oxygen in Physical Solution

Unlike CO_2, oxygen has a low physical solubility in water or whole blood, so that normally one can disregard the contribution of dissolved O_2 to oxygen transport. However, when a person breathes 100% O_2, the high partial pressure of oxygen (up to 675 mm Hg at sea level; $P_B - P_{H_2O} - P_{ACO_2}$) permits about 20 ml O_2/L to be carried in physical solution—nearly 40% of the normal resting a-v P_{O_2} difference. Under such conditions,

one must take dissolved O_2 into account when assessing blood oxygen transport.

Oxygen Reserves

For the body as a whole, tissue P_{O_2} is low and oxygen is poorly soluble in water. Thus, very little O_2 can be stored in solution for emergencies. Fortunately, aerobic metabolism can be sustained for a few minutes, because there are three locations in the body where some oxygen is normally stored.

Some oxygen is stored in the alveoli of the lung. At an FRC of 2.4 liters and with P_{AO_2} at 100 mm Hg, there is enough O_2 in the gas phase for about 1 to 2 minutes of basal O_2 consumption.

In the circulating blood volume of 5 liters there is 150 ml/L of O_2 over and above the quantity transported in systemic arterial blood. This quantity gives another 3 minutes of oxygen consumption, provided that all of it can be utilized.

Finally, in skeletal and cardiac muscle there is the oxygen storage protein myoglobin. However, the concentration of myoglobin is low, being only 4 mg/g (0.2 mM) in cardiac ventricular muscle—sufficient for 3 to 4 seconds at the normal basal O_2 consumption of the heart. The myoglobin concentration in skeletal muscle is less than that in the heart. For example, the total body muscle mass is 43% of body weight (30 kg in a 70 kg adult human). If the myoglobin concentration is 0.1 mM, the total O_2 stored as MbO_2 is less than 10 ml! This small amount of oxygen is useful during brief muscle contractions, when the blood flow is temporarily stopped or slowed.

Overall, the O_2 in the body is sufficient to provide about 5 minutes of basal O_2 consumption.

> Every winter television or newspapers report the miracle of someone successfully resuscitated after being found in the snow. The victim has a low body temperature and no detectable heartbeat or breathing for up to 1 hour. The apparent miracle that the person can be revived without permanent brain damage is the result of the low body temperature, which drastically reduces oxygen demand by the brain and other tissues. Before the advent of practical heart-lung bypass machines, surgeons used total body hypothermia to reduce oxygen consumption so that they could stop the heart for several minutes to do quick surgical repair of congenital cardiac lesions.

Oxygen Utilization

Even in individuals at rest, the normal myocardium extracts the maximum quantity of oxygen per volume of blood it can without reducing tissue Po_2 to dangerously low levels. The coronary sinus (venous) oxygen saturation is about 50%; Po_2 is 29 mm Hg (P_{50} after Bohr shift caused by increased acidity) or even a little less. Increases in oxygen demand by the heart are met almost entirely by an increase in coronary blood flow.

During maximal steady-state exercise, the total body (a-v) O_2 concentration difference increases to about 100 ml O_2/L blood. It is uncommon for steady-state mixed venous oxygen saturation to fall below 50% ($P\bar{v}o_2 = 29$ mm Hg), although the venous blood from individual organs may do so. In fact, at maximal steady-state exercise, when cardiac output may reach 15 to 20 L/min in normal people, 70% of the cardiac output goes to the exercising muscle. Because so much flow passes through the active muscles, mixed venous So_2 during exercise is a reasonable estimate of the hemoglobin saturation of the blood leaving the muscle capillaries.

Oxygen Transport Defects

> There are four types of clinically relevant O_2 transport deficiencies: **hypoxic hypoxia** (inadequate O_2 uptake into blood in the lung caused by breathing low-oxygen atmospheres or by large amounts of venous admixture, as in individuals with marked V/Q mismatching); **stagnant hypoxia** (inadequate blood flow to an organ or the whole body as may occur in hemorrhagic shock); **anemic hypoxia** (inadequate blood oxygen-carrying capacity, as in CO poisoning); and **histotoxic hypoxia** (biochemical blockade of mitochondrial respiration, as in cyanide poisoning).

RESPIRATORY GAS DIFFUSION IN THE BODY

As stated in Chapters 27 and 29, diffusion of oxygen or carbon dioxide could be a bottleneck along the gas transport highway. The general relationship is that diffusion is favored by a large surface area, a short distance to travel from a region of high concentration to lower concentration, a large partial pressure difference (equivalent to concentration for the respiratory gases), and high diffusivity (small molecular size and high water solubility). Diffusion is a fast, efficient process over short distances but can be seriously rate-limiting when distance exceeds several micrometers (see Chapter 1).

One serious misconception about respiratory gas diffusion needs to be clarified. Although O_2 is a smaller molecule than CO_2, CO_2 is much more soluble in water. The **diffusivity** (diffusion coefficient times solubility) of CO_2 is therefore much greater than that of O_2. Thus, the initial partial pressure difference for CO_2 between mixed venous blood and alveolar gas is normally about 6 mm Hg and that for O_2 is 60 mm Hg. Nevertheless, CO_2 diffuses about twice as rapidly as O_2 across the alveolar-capillary barrier. Even that slight advantage is partly eliminated, because the uptake of oxygen and the release of carbon dioxide in pulmonary capillary blood are interactive (see Bohr and Haldane effects below). These comments are also true about the reverse process that occurs in the peripheral (systemic) capillaries.

Diffusion of Oxygen Across the Alveolar-Capillary Barrier

The respiratory (O_2, CO_2, N_2) or anesthetic gases (ether, nitrous oxide, fluothane) diffuse along their partial pressure gradients (from a region of higher to a re-

LUNG

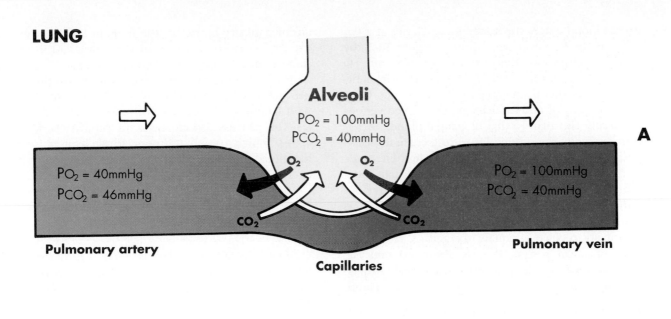

TISSUE

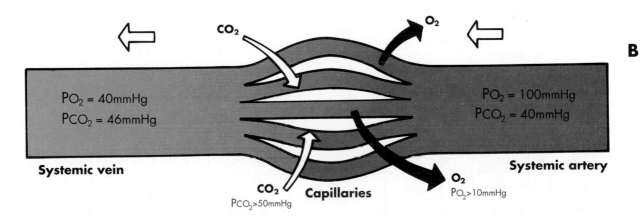

FIGURE 30-6   The respiratory gases, oxygen and carbon dioxide, generally diffuse in opposite directions, each along its partial pressure gradient. The diffusion streams are completely independent of each other. **A,** Pulmonary capillaries. **B,** Systemic capillaries.

gion of lower partial pressure) independently of the movements of any other gases. Thus, O_2 and CO_2 diffuse in opposite directions in the lungs and in the systemic capillaries, as illustrated in Figure 30-6.

Some of the adaptations of the lungs to reduce any diffusion limitation for O_2 or CO_2 are to vastly increase the alveolar capillary surface area and to make the barrier as thin as possible (Chapter 27). Furthermore, the relatively large capillary blood volume in the lung (75 ml in resting men) allows blood to remain in the exchange vessels long enough (one heart beat) to completely equilibrate.

Diffusion of Oxygen to the Systemic Tissue Mitochondria

To get sufficient oxygen from the capillary blood to the mitochondria of the tissue cells throughout the body, so that steady-state respiration is aerobic, nature has tried to follow the same principles as for the lung—namely, to maximize transit time in the systemic capillaries and to minimize the distance between the capillaries and the mitochondria. However, because all body organs, except the lung, are solid (mostly water), the distance from the capillaries to the mitochondria is much longer in those tissues than in the lung.

Arterial blood enters the systemic capillaries at a P_{O_2} of 100 mm Hg. However, according to Figure 30-2, the P_{O_2} of the capillary blood must fall to below 80 mm Hg before any significant quantity of O_2 dissociates from hemoglobin. Thus, the mean capillary P_{O_2} of 55 mm Hg is closer to the P_{O_2} of end-capillary blood (mixed venous $P_{O_2} = 40$ mm Hg) than to that of arterial blood.

As in the lungs, the driving force for oxygen diffusion is directly dependent on the partial pressure difference between the capillary blood and the most distant mitochondria. Fortunately, mitochondria are able to carry out oxidative metabolism at tissue oxygen tensions, P_{tO_2}, as low as 1 to 2 mm Hg. In the steady state, all mitochondria normally receive adequate oxygen. A reasonable estimate of mean resting P_{tO_2} is about 10 mm Hg, which in skeletal muscle means that the oxygen-storage protein myoglobin is about 75% saturated. (See Figure 30-5.)

In the left ventricular myocardium, which is a critical tissue that requires a large oxygen supply per unit mass, the capillaries are about 25 μm apart—the width of one left ventricular muscle fiber (Chapter 25). Thus, from each capillary oxygen must diffuse outward within a tissue cylinder of about 13 μm radius. While that distance seems short, it is 10 times greater than that across the alveolocapillary barrier in the lung. In brain cortex, the capillaries are some 35 to 40 μm apart, and in resting skeletal muscle they are about 80 μm apart.

The body's most effective mechanism to improve oxygen delivery to tissue cells is to decrease the diffusion path length by recruitment of more functional capillaries. This process also increases the capillary surface area across which oxygen diffuses. In skeletal muscle the functional capillary density increases threefold in heavy exercise (Figure 30-7).

CARBON DIOXIDE TRANSPORT

Metabolic CO_2 transport is solely a function of systemic venous blood, just as metabolic O_2 transport is solely a function of systemic arterial blood (see Figure 30-1).

The Respiratory Exchange Ratio

Under basal steady-state conditions, we exhale 80 molecules of CO_2 for every 100 molecules of O_2 we take up into the pulmonary capillary blood. This ratio is called the **respiratory exchange ratio**, R. It is calcu-

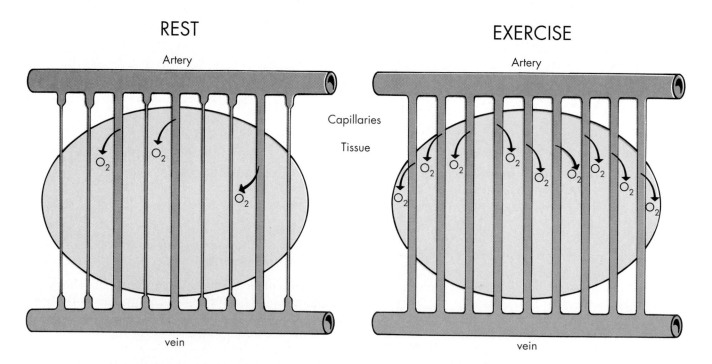

FIGURE 30-7   Tissue oxygen transport is limited by the relatively long distances over which oxygen must diffuse to reach the cell mitochondria. This is especially true in skeletal muscle, where oxygen demand may increase manyfold between rest and maximum activity. In order to supply adequate oxygen for metabolism during muscular exercise, the number of perfused capillaries increases by about three to four times, which reduces the oxygen diffusion path by half or more.

lated by dividing the CO_2 production by the O_2 consumption. R is normally 80/100, or 0.8.

In the steady state the respiratory exchange ratio, R, is the same as the **respiratory quotient**, RQ; the latter applies to the respiratory events at the mitochondria (aerobic oxidation). R can vary from 0.7 (lipid) to 1.0 (carbohydrate); amino acids are intermediate (0.85) (see Chapter 41). Lipid is the major fuel under basal conditions, because R is about 0.8. That is much closer to 0.7 than to 1.0, regardless of food fadists' propanganda. Under non–steady state conditions (for example, submaximal exercise when the heart rate exceeds 160 beats/min), one can have a respiratory exchange ratio above 1.0 because as the body becomes more acid, because of the inadequacy of the oxygen supply **(oxygen debt)**, body stores of carbonic acid (mostly extracellular bicarbonate) are converted to CO_2 gas and exhaled. One can detect the onset of anaerobic metabolism (anaerobic threshold) in exercise by measuring R breath-by-breath at the mouth.

The Chemistry of Carbon Dioxide in Whole Blood

In contrast to hemoglobin and its reversible binding of oxygen, no single, efficient, carbon dioxide transport molecule exists in blood. However, three mechanisms solve the CO_2 transport problem effectively: dissolved CO_2, carbaminohemoglobin compounds, and sodium bicarbonate ($NaHCO_3$), as shown in Figure 30-8.

Physical Solution Because CO_2 is 20 times more soluble in blood than O_2, about 10% of systemic venous CO_2 transport occurs by physical solution, even though the a-v P_{CO_2} difference is only 6 mm Hg normally.

Carbaminohemoglobin Because so much hemoglobin exists in blood (150 g/L), a great many amine (NH_2) side groups are available on the amino acids. At rest, during passage through systemic capillaries, about 25% of the hemoglobin is deoxygentated. Deoxygenated hemoglobin is a weaker acid than HbO_2; that means it is less dissociated ($HHb \leftrightarrow H^+ + Hb^-$). Some of the CO_2 diffusing into the red blood cells reacts with the hemoglobin-NH_2 side chains to form carbamino (NH-COOH) hemoglobin. These compounds instantly dissociate into $NHCOO^-$ and H^+ (Figure 30-8). As our knowl edge of the physical chemistry of protein has increased, so has the importance of the carbamino transport mechanism. About 30% of normal venous CO_2 transport occurs by that mechanism.

Sodium Bicarbonate Sodium bicarbonate ($NaHCO_3$) is the principal form by which CO_2 is carried in blood. As CO_2 is formed and diffuses into capillary

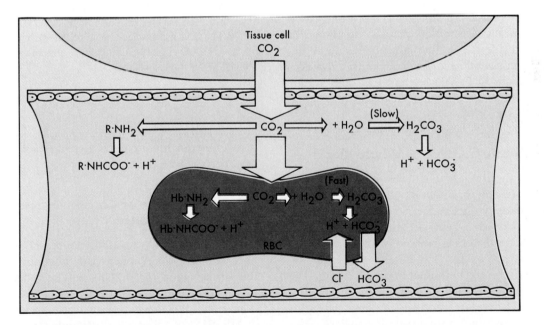

FIGURE 30-8   The three mechanisms of CO_2 transport in blood. As CO_2 diffuses into capillary blood from peripheral tissues, it may slowly react with water to form carbonic acid or with the amine side groups of plasma proteins to form carbamino compounds. However, the vast majority of CO_2 diffuses into the red blood cells (RBC), where it rapidly forms carbonic acid, a reaction catalyzed by carbonic anhydrase. The bicarbonate ions formed exchange with chloride ions in plasma because the erythrocyte membrane is readily permeable to small anions. As the hemoglobin loses its oxygen, it becomes more able to bind CO_2 to its NH_2 side groups to form carbaminohemoglobin compounds.

blood, it reacts with water to form carbonic acid (H_2CO_3), which in turn instantly dissociates to form H^+ and bicarbonate (HCO_3^-) ions:

$$CO_2 + H_2O \leftrightarrow H_2CO_3 \leftrightarrow H^+ + HCO_3^-$$

However, the formation of H_2CO_3 proceeds relatively slowly, with a half-time of many seconds, which is too slow because each unit of blood passes through capillaries in about 1 second and arrives at the lungs within 30 seconds. Circulation time (one trip around the body) averages 60 seconds, because blood volume/resting cardiac output is about 1.

The enzyme carbonic anhydrase, is present in red blood cells (and many other body cells). Carbonic anhydrase is not present in plasma. The carbonic anhydrase catalyzes the reaction shown in the equation above; it speeds up the reaction more than 10,000-fold in either direction, depending on the P_{CO_2} and $[HCO_3^-]$. Thus, the HCO_3^- concentration rises or falls much more rapidly in the erythrocytes than in the plasma in the capillaries.

The red blood cell membrane is very permeable to HCO_3^- and other anions. Hence HCO_3^- diffuses rapidly along its concentration gradient between the erythrocyte and plasma.

As with other mammalian cells, the red cell membrane is not very permeable to cations (Na^+, K^+, or H^+). However, the laws of physical chemistry require that there be electrical neutrality in the bulk phase of solutions. To achieve this, chloride (Cl^-), the main plasma anion, diffuses along its concentration gradient and counterbalances the flux of HCO_3^-. This exchange of chloride for bicarbonate in tissue or pulmonary capillaries is called the **chloride shift.**

The CO₂ Equilibrium Curve of Whole Blood

As a result of the three mechanisms described above, the CO_2 concentration in whole blood varies with P_{CO_2}, but here the analogy to the HbO_2 equilibrium curve completely breaks down. Over the range of P_{CO_2} found in life the CO_2 equilibrium curve is essentially a straight line, as Figure 30-9 shows.

In Figure 30-4, the oxyhemoglobin equilibrium curve of venous blood was slightly shifted to the right of that of arterial blood. This shift improves the dissociation of oxygen from hemoglobin in the systemic capillaries: the **Bohr effect.** In Figure 30-9, notice that the CO_2 equilibrium curve of systemic venous blood at $P_{O_2} = 40$ mm Hg is shifted to the left of that for systemic arterial blood at $P_{O_2} = 100$ mm Hg. This shift increases the transport of CO_2 in systemic venous blood: the **Haldane effect.** The Haldane effect is attributed to the fact that deoxy-

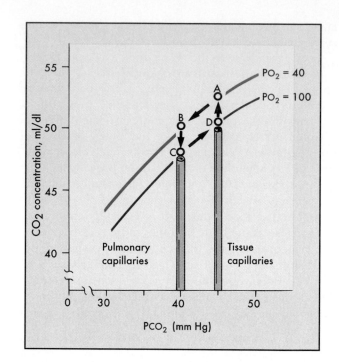

FIGURE 30-9 The CO_2 equilibrium curves for normal systemic arterial and mixed venous blood. In the pulmonary capillaries, as the P_{O_2} of the blood increases, the hydrogen ion concentration of the red cells tends to rise (HbO_2 is a stronger acid, more dissociated, than Hb). Thus, CO_2 is liberated from carbamino compounds and from HCO_3^- (left shift of the CO_2 equilibrium curve). Conversely, in the tissue capillaries, as the P_{O_2} of the blood decreases, the erythrocyte cytoplasm becomes less acidic and more CO_2 can be carried by the blood (right shift of the CO_2 equilibrium curve). This physiological shift is called the Haldane effect.

genated hemoglobin (Hb) is a weaker acid than oxygenated hemoglobin (HbO_2).

A Major Function of Carbon Dioxide is Acid-Base Balance

Normally the two strong cations (Na^+, K^+) in plasma have a considerably higher total concentration than does the strong anion (Cl^-). This deficit, called the **strong ion difference,** SID, is made up by the plasma bicarbonate ion plus the net plasma protein anions.

The $[H^+]$ of various body liquids (extracellular, intracellular, and transcellular) is exactly defined by the strong ion difference, the total protein concentration, and the P_{CO_2}. The role of P_{CO_2} and its associated products (carbonic acid and bicarbonate) as part of the various H^+ buffer systems of the body is covered more fully in the renal section (see Chapter 39).

Clearly, the alveolar ventilation equation (Chapter 28) describes how the body's P_{CO_2} regulates or is regulated by breathing. Suffice it to say here that when the systemic arterial blood $[H^+]$ rises, as it does in **metabolic acidosis** (see Chapter 39), the body's P_{CO_2} also rises, which stimulates the brain to increase alveolar ventilation to return the P_{CO_2} toward normal. On the other hand, if alveolar ventilation is impaired, as in **respiratory insufficiency** (advanced emphysema is the most common cause), alveolar P_{CO_2} rises and increases the body's P_{CO_2}, which characterizes **respiratory acidosis.**

An increase in ventilation sometimes occurs during emotional excitement. The resultant reduction in alveolar P_{CO_2} decreases the body's $[H^+]$ ($\uparrow pH$) and is called **respiratory alkalosis.** Likewise, loss of body acids, such as during prolonged vomiting, leads to **metabolic alkalosis.** This causes breathing to decrease, which raises body P_{CO_2}, bicarbonate, and the strong ion difference.

Thus, there are four distinct deviations of acid-base balance from normal, two caused primarily by changes in breathing and P_{CO_2} (respiratory acidosis and alkalosis) and two in which secondary, compensatory changes in breathing and P_{CO_2} occur (metabolic acidosis and alkalosis).

SUMMARY

1. Oxygen is transported from the lungs to the systemic capillaries in arterial blood, while carbon dioxide formed by aerobic metabolism is transported from the systemic tissues to the lungs in venous blood.

2. Hemoglobin quickly and reversibly binds with oxygen. The oxygen capacity of normal human blood (150 g Hb/L) is 200 ml O_2/L (absolute concentration) or 100% saturation (relative concentration).

3. The position of the hemoglobin-oxygen equilibrium curve is well suited for the loading of oxygen in the lungs and for unloading in systemic tissue capillaries.

4. Normal physiological factors that affect the HbO_2 curve are hydrogen ion concentration, P_{CO_2}, temperature, and the concentration of 2,3-diphosphoglycerate in erythrocytes. Increases in any of these shift the position of the HbO_2 curve to the right, thereby decreasing hemoglobin affinity for oxygen.

5. A 3 mm Hg shift to the right (decreased oxygen affinity) occurs as blood passes through systemic capillaries, because P_{CO_2} and hydrogen ion concentration increase. This shift increases O_2 unloading in

the systemic capillaries; a reverse shift occurs in the lung and favors O_2 uptake.

6. The shape of the HbO_2 equilibrium curve is such that in the systemic capillaries hemoglobin releases large quantities of oxygen as P_{O_2} falls below 70 mm Hg.

7. The shape of the curve is affected by carbon monoxide, which has an affinity for hemoglobin more than 200 times greater than does oxygen. Even in very low concentrations, carbon monoxide combines avidly with hemoglobin, interferes with oxygen binding, and destroys heme group interactions.

8. The transport of oxygen between body compartments is governed by diffusion at a rate dependent on the P_{O_2} difference. The principal diffusion limitation for oxygen is from the systemic capillaries through several micrometers of interstitial fluid to the mitochondria of the respiring cells. The number of actively perfused capillaries is an important mechanism for increasing oxygen diffusion to skeletal muscle cells during muscle activity.

9. Carbon dioxide is produced by aerobic metabolism in mitochondria. The CO_2 diffuses into systemic capillary blood and is transported to the lungs, where it diffuses into alveolar gas and is exhaled. CO_2 is transported in blood by a combination of three mechanisms (physical solution, carbaminohemoglobin, and as HCO_3^-)

10. Carbon dioxide slowly combines chemically with water to produce carbonic acid. In erythrocytes the enzyme carbonic anhydrase catalyzes this reaction. Once formed, carbonic acid dissociates instantly into hydrogen ions and bicarbonate ions. The hydrogen ions are removed by combining chemically with the enormous amount of hemoglobin within the red cells. The bicarbonate ions diffuse out of the red cells in exchange for chloride ions.

11. Carbon dioxide also plays a major role in acid-base balance. There are four deviations of acid-base balance from normal, two caused primarily by changes in breathing and P_{CO_2} (respiratory acidosis and alkalosis) and two in which secondary, compensatory changes in breathing and P_{CO_2} occur (metabolic acidosis and alkalosis).

BIBLIOGRAPHY
Journal Articles

Bidani A: Velocity of CO_2 exchanges in the lungs, *Annu Rev Physiol* 50:639, 1988.

Grant BJB: Influence of Bohr-Haldane effect on steady-state gas exchange, *J Appl Physiol* 52:1330, 1982.

Jennings ML: Kinetics and mechanism of anion transport in red blood cells, *Annu Rev Physiol* 47:519, 1985.

Books and Monographs

Baumann R: Interaction between hemoglobins, CO_2 and anions. In Bauer C et al, eds: *Biophysics and physiology of carbon dioxide,* New York, 1980, Springer.

Grote J: Tissue respiration. In Schmidt RF, Thews G, eds: *Human physiology,* New York, 1983, Springer.

Stewart, PA: *How to understand acid-base,* New York, 1981, Elsevier, New York.

Weibel ER: *The pathway for oxygen: structure and function in the mammalian respiratory system,* Cambridge, Mass, 1984, Harvard University Press.

Control of Breathing

Both the rate and depth of breathing are regulated so that arterial and alveolar P_{CO_2} are maintained close to 40 mm Hg, according to the alveolar ventilation equation (Chapter 28). When Pa_{CO_2} equals 40 mm Hg, Pa_{O_2} is automatically set to its normal value of 100 mm Hg; this adjustment depends, of course, on the atmospheric partial pressure of oxygen (effect of altitude or inspired gas composition).

Although the Pa_{CO_2}-sensitive mechanism is the main controller, it can be overridden in systemic arterial hypoxemia (for example, acclimatization to high altitude) by a Pa_{O_2}-sensitive controller. If arterial oxygen tension were to decrease below 70 mm Hg, the supply of oxygen to the systemic tissue mitochondria might be impaired because the oxygen diffusion gradient between the capillaries and the tissue cells is reduced (See Chapter 30).

CENTRAL ORGANIZATION OF BREATHING

Breathing is Mainly Controlled at the Brainstem Level

Two patterns are involved in the control of breathing: the **metabolic** (automatic) **pattern** and the **behavioral** (voluntary) **pattern.** Metabolic breathing is the basic pattern, and it is concerned with the regulation of Pa_{CO_2} through modulation of alveolar ventilation. We can override our automatic breathing pattern, at least briefly, as when a truculent child threatens to "hold my breath until I die." As parents know, such a threat cannot be successful because within a minute or so the metabolic control system reasserts its authority.

The metabolic controller (respiratory neuronal groups in Figure 31-1) resides within the brainstem. Formerly, neurophysiologists thought that the neurons responsible for inspiration and expiration were located in

specific brainstem centers, but, as now conceived, the organization is much looser.

Surrounding and interdigitating throughout the brainstem is a loose network of interneurons known as the **reticular activating system** that influences the brainstem controller by affecting the state of alertness (wakefulness) of the brain. That means that breathing is regulated somewhat differently during sleep than when we are awake.

> Control of breathing in sleep is one of the most rapidly growing aspects of pulmonary medicine. Sleep disorders vary from the tragic **sudden infant death syndrome** to **adult sleep apneas.** The former is the leading killer of babies, usually in the first 4 months after birth. Its cause is unknown, but it may be due to immaturity of the central breathing controller. The latter disorder is usually obstructive (closure of nasopharynx during sleep) but sometimes central—that is, a depression of the controller of unknown cause.

As for the behavioral controller, about all we know is that the thalamus (primitive integrating system) and cerebral cortex (higher integrating system) are required to coordinate breathing in relation to the many complex but volitional motor activities that make use of the lungs and chest walls. Several pathways carry the rostrally located (anterior) cortical and thalamic neuronal axons caudally to the pontine and medullary controllers. Some volitional activities include talking, singing, suckling, swallowing, coughing, sneezing, defecation, parturition, and responses to anxiety and fear (such as hyperventilation, which leads to respiratory alkalosis).

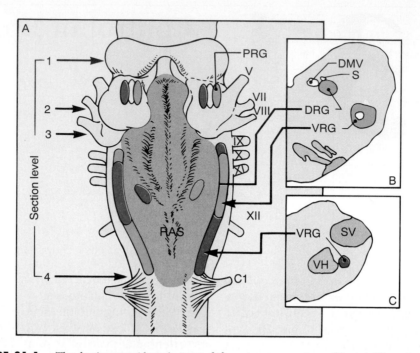

FIGURE 31-1   The brainstem (dorsal view) **(A)** and cross sections **(B** and **C)** at the mid-medulla and caudal medulla, respectively, to illustrate locations of the dorsal *(DRG)*, ventral *(VRG)*, and pontine *(PRG)* respiratory neuron groups (the metabolic controllers of breathing). The three colors roughly indicate the locations of neurons that fire during inspiration *(red)*, expiration *(blue)*, or both *(yellow)*. The reticular activating system *(gray overlay)* surrounds and interdigitates with the various neuronal groups. The reticular formation regulates wakefulness (state of alertness), which has important influences on breathing. The main difference between modern and classical views of the brainstem controller is that all respiratory neuronal groups are active during each phase of breathing; no separate inspiratory and expiratory centers exist. The roman numerals *(right side)* show the corresponding cranial nerves, some of which have both sensory and motor divisions. The arabic numerals *(left side)* show the classic levels of brainstem transections, which helped early physiologists to localize the respiratory neuron groups. The breathing patterns caused by successive ablations from rostral to caudal are shown in Figure 31-2. Other selected anatomical landmarks: *C1,* first cervical nerve; *DMV,* dorsal motor nucleus of the vagus; *S,* nucleus of the tractus solitarius; *SV,* spinal trigeminal nucleus; and *VH,* ventral horn.

Defining the brain centers involved in breathing control is one area in which "natural experiments" (patients with neurological disorders) have given important insights. For example, after **head trauma** a patient may breathe with inspiratory breathholds lasting many seconds followed by brief exhalations **(apneustic breathing).** This has been traced to damage in the pons.

Basic Organization of the Brainstem Controllers

Two regions in the brainstem contain the intrinsic breathing controllers: the **medullary respiratory area** and the **pneumotaxic center** in the pons. Thus, the earliest portion of the brainstem to evolve regulates the basic breathing pattern nearly completely, which makes evolutionary sense. The pattern of breathing is essentially normal even when the pons and medulla are separated from the rest of the brain (see Figure 31-2).

The precise anatomical and functional organization of the respiratory medullary neurons is still under investigation, but Figure 31-1 shows the current schema, in

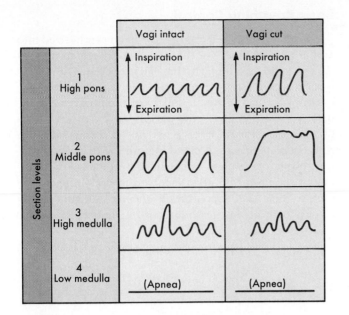

	Vagi intact	Vagi cut
1 High pons	↑ Inspiration ↓ Expiration	↑ Inspiration ↓ Expiration
2 Middle pons		
3 High medulla		
4 Low medulla	(Apnea)	(Apnea)

(Section levels)

FIGURE 31-2 Different patterns of breathing, produced by successive transections of the brainstem in anesthetized animals. These patterns may also occur in humans with selected brainstem lesions. The four levels of transection correspond to the numbers in Figure 31-1, *A*. The importance of the sensory division of the vagus nerve, carrying information about the state of lung inflation, is shown by contrasting the intact to the cut columns. *Transection level 1:* There is no effect on the basic breathing pattern, showing that the main controller lies caudal. Furthermore, it eliminates all volitional (conscious) changes in breathing (not shown here). When vagal sensory input is eliminated, breathing slows and tidal volume increases. Skipping down to *Transection level 4:* Breathing stops completely, showing that the main controller lies rostral. *Transection level 2:* As long as sensory input from the vagus nerve continues, ablation of the pneumotaxic center has only a moderate slowing effect on breathing, with tidal volume increased to maintain alveolar ventilation at approximately normal. After vagotomy, apneustic breathing (deep inspiration held for many seconds) appeared. Obviously, both the pontine pneumotaxic center and vagal sensory input are important in terminating inspiration, under the conditions of these experiments. *Transection level 3:* Irregular, gasping breathing develops, which is not affected by cutting the vagus nerves. Clearly, medullary controllers are sufficient to sustain life, but are not adequate to provide normal modulation of breathing.

which the classical notion of completely separate inspiratory and expiratory neuron groups has been replaced by regional groups of mixed respiratory neurons (dorsal, ventral, and pontine). Note also that the cranial nerves to the face, mouth, throat, and lungs, which must be coordinated in metabolic and volitional breathing activities, are close to the respiratory neuron groups.

Normal people cannot breathe and swallow simultaneously without gagging or coughing. However, the newborn can! Yes, babies are born with the correct anatomy (a large flexible epiglottis to direct incoming milk laterally into the oropharynx, down and around the glottis) and the absence of breathing interrupter reflexes that are activated when swallowing begins. We lose the ability to breathe and swallow simultaneously at about 1 year of age. Excessive alcohol intake in adults may dull the gag reflex sufficiently so that food or vomitus enters the glottis or trachea. This medical emergency is not rare. The **Heimlich maneuver** (sharp upper abdominal pressure) is designed to force the food out of the trachea under strong expiratory pressure.

The mechanism by which the medullary neuronal groups cause switching between inspiration and expiration is not completely clear. However, most neurophysiologists agree that no spontaneously excitable breathing pacemaker operates (nothing analogous to the sinoatrial node in the heart).

Currently, the main view is that *rhythmic breathing depends on a continuous (tonic) inspiratory drive from neurons in the dorsal motor group, with intermittent (phasic) excitatory inputs from various sources, but especially from the pontine respiratory group.* Thus, breathing results from the reciprocal inhibition of interconnected neuronal networks.

The pontine group of respiratory neurons (pneumotaxic center) influences the switching between inspiration and expiration. When the pneumotaxic center is inactivated, inspiration becomes greatly prolonged.

Lung Stretch Receptors

Lung stretch receptors are specialized sensory nerve receptors that are stimulated by lung inflation to send impulses up the vagus nerves (cranial nerve X) to the ventral medullary motor group (see Figure 31-1). These sensory fibers enter the medulla adjacent to the ventral respiratory motor group. The sensory impulses provide positive reinforcement to neurons that when activated send impulses rostrally to the pontine respiratory group. The pontine neurons (pneumotaxic center) then send inhibitory impulses to the dorsal medullary motor group to turn off the inspiratory neurons. These effects that originate in the lung stretch receptors constitute the **Hering-Breuer reflex.**

Input from the stretch receptors mainly influences the inspiratory time. Interruption of the stretch receptor traffic may markedly prolong the duration of inspiration (see Figure 31-2, *right panel.*) The effects of lung stretch receptors are variable, however. In awake humans, in whom there is increased reticular formation activity caused by an increase in general sensory information, input from the pulmonary stretch receptors has less influence than in most experiments on anesthetized humans or animals. Apparently these reflexes are more important when we are asleep.

During spontaneous breathing, termination of inspiration prevents overinflation of the lung. Shortening inspiratory time also facilitates the higher breathing frequencies needed for adequate increases in ventilation when breathing is stimulated, as in exercise. The regulation of expiratory time is less well understood, although it is correlated with changes in inspiratory time.

Basic Breathing Control

The facts presented above have been synthesized into an operational model of the brainstem mechanisms that generate the breathing rhythm (Figure 31-3). The system is represented by three neuron pools, which are not necessarily identical to the medullary respiratory groups. Signals from the central and peripheral chemoreceptors ($\uparrow$ Pco$_2$ or $\downarrow$ Po$_2$) reinforce activity in inspiratory neurons (pool A) that are located in both the dorsal and ventral respiratory groups. Some of these neurons send axons to the motor neurons involved in breathing and increase their activity (phrenic nerve and intercostal muscle neurons in the spinal cord). Stimulation via this pathway causes inspiration.

The central inspiratory activity stimulates neurons in

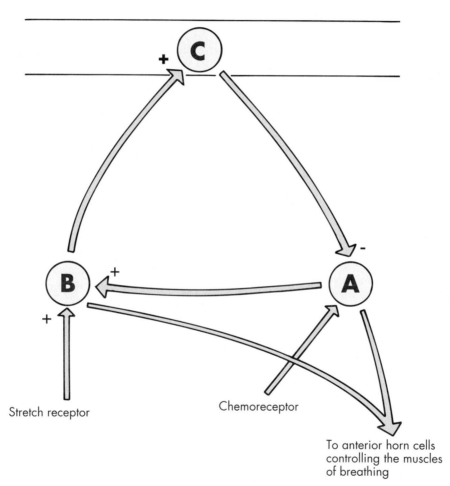

Stretch receptor

Chemoreceptor

To anterior horn cells controlling the muscles of breathing

FIGURE 31-3 The basic wiring diagram of the brainstem ventilatory controller. The signs on the main outputs *(arrows)* of the neuron pools indicate excitatory (+) or inhibitory (−). The pools are named A, B, and C because they are not anatomically identical to the respiratory groups shown in Figure 31-1. *Pool A* is tonically active, sending impulses via the phrenic nerve (cranial nerve IX) to the diaphragm or via the spinal cord to the ventral horn motor neurons that activate the intercostals and other chest wall muscles of breathing. *Pool B* neurons are stimulated by pool A and provide additional stimulation to the muscles of breathing. But pool B also sends stimuli to *pool C,* which sends inhibitory impulses back to pool A. Afferent information feeds back to the medullary neuron pools from various sensors.

pool *B* (Figure 31-3). In addition, these neurons are reinforced by impulses from the lung stretch receptors. As the lung expands, the stretch receptor signals increase more and more. Some of the pool B neurons project anteriorly to stimulate pool *C* (mainly the pneumotaxic center). The pool C neurons send inhibitory impulses to the inspiratory neurons in the dorsal medulla until their impulse activity is extinguished, inspiration ends and expiration occurs.

This scheme of breathing pattern generation—namely, tonic inspiratory activity, inhibited only when sufficient inhibitory signals are received—is attractive because no spontaneous cyclic breathing rhythm is required and none has been found. The scheme also integrates the effects of CO_2 and O_2 (chemoreceptor activity) and of other reflex stimuli into various normal and abnormal breathing patterns.

SPINAL INTEGRATION

The impulses from the inspiratory neurons pass down into the spinal cord along axons that reach various segments, where they are integrated with intrasegmental and intersegmental activity so as to modulate the membrane potentials of the motor nerve cells.

Excitation and inhibition of the breathing muscles also involve segmental interneuronal networks. For example, inhibition of antagonist muscles takes place through interneurons that travel between the motor neurons of the inspiratory and expiratory muscles. Stretch of intercostal muscles or electrical stimulation of sensory input into thoracic segments T9 to T12 excites intercostal and phrenic motor neurons to enlarge the thoracic cavity. In contrast, stimuli to segments T1 to T8 inhibit phrenic motor neuron activity and terminate inspiration. Spinal reflexes are important because they augment muscle force when the resistance to breathing is increased or when the lung compliance is decreased.

> Spinal reflexes from intercostal muscle distortion are beneficial to the newborn, whose cartilaginous rib cage is very compliant. The baby's rib cage needs to be stabilized during inspiration, so that the subatmospheric pleural pressure does not suck the rib cage inward. Intercostal muscle stretch receptors (different from the airway stretch receptors) sense the inward movement of the rib cage as pleural pressure decreases during inspiration. These receptors reflexly influence the intercostal muscle motor neurons to stimulate contraction and thereby to oppose the distortion.

CHEMORECEPTOR CONTROL OF BREATHING

Carbon Dioxide

Two main types of chemoreceptors (central and peripheral) are involved in the control of breathing. By far the more important ones are located in the central nervous system at or near the ventrolateral surface of the medulla, between the origins of the seventh and tenth cranial nerves (Figure 31-4). By comparing Figures 31-1 (dorsal view of medulla and pons) and 31-4 (ventral view of medulla and pons), you will see that the central chemoreceptors are near the ventral respiratory neuron group.

The other chemoreceptors are located peripherally—that is, outside the central nervous system. The main peripheral chemoreceptors are the **carotid bodies,** which are located at the bifurcation of the common carotid artery in the neck. There are also chemoreceptors in the aortic arch **(aortic bodies),** but these are less studied and appear to be of minimal importance.

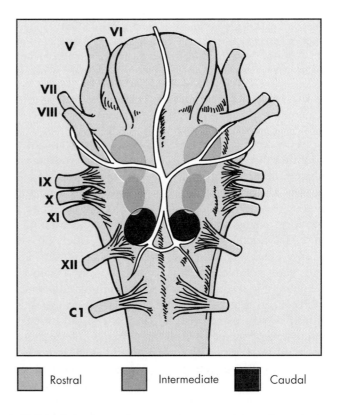

Rostral   Intermediate   Caudal

FIGURE 31-4   Ventral view of the brainstem, to show the locations of the central chemoreceptors. The chemosensitive cells are believed to lie just beneath the surface of the medulla. Hence, they are affected by the $[H^+]$ in brain interstitial and cerebrospinal fluids.

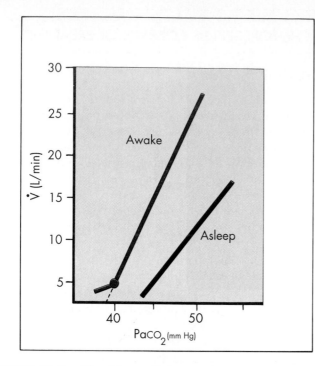

FIGURE 31-5   The CO_2-ventilation tension-response curve. Ventilation is sensitive to P_{aCO_2}, as shown by the solid line (awake). The curve is defined by the slope of the straight portion (about 2.5 L/min) and the extrapolated X axis intercept 38 mm Hg). The normal operating point is 5 L/min at 40 mm Hg *(large dot)*. In the awake (alert) state, ventilation becomes insensitive to decreases in CO_2 below 40 mm Hg. When the reticular activating system is turned off, as during sleep *(dashed line)*, the CO_2 intercept is increased (shifted to the right) and the slope is decreased (decreased sensitivity).

In awake humans the **central chemoreceptors** account for three-quarters of CO_2-induced increases in ventilation. Sensitivity to CO_2 is measured as the slope of the line relating P_{CO_2} (alveolar or arterial) to ventilation (Figure 31-5). The medullary chemoreceptor cells respond to changes in the hydrogen ion concentration of the surrounding brainstem interstitial fluid and the nearby cerebrospinal fluids. Because CO_2 is very lipid and water soluble, it easily crosses the blood-brain barrier and rapidly equilibrates with brain fluids. Conversely, increases in arterial blood acidity ($\uparrow[H^+]$) cause much slower responses because H^+ does not readily cross the blood-brain barrier or cell membranes. Elevations of P_{aCO_2} up to 100 mm Hg cause a nearly linear increase in ventilation. The effects of CO_2 are greater when the individual is awake (Figure 31-5), because of greater activity in the reticular formation (increased alertness).

In normal individuals the average ventilatory response to inspired CO_2 is about 2.5 L/min/mm Hg, but the response varies considerably among individuals because of differences in body size, age, sex, genetic makeup, or personality.

The **peripheral chemoreceptors** (carotid bodies) account for about 25% of human sensitivity to CO_2. Again, the chemoreceptors cells respond primarily to changes in the hydrogen ion concentration of the surrounding interstitial fluid, not to the CO_2 molecule itself.

Oxygen

It may not seem logical, but the effect of reduced arterial oxygen tension on breathing is small until P_{aO_2} falls to 70 mm Hg, below which ventilation increases in a non-linear fashion (Figure 31-6). However, the explanation of the insensitivity to oxygen is fairly straightforward. The HbO_2 equilibrium curve (Figure 30-2) shows that arterial oxygen saturation does not decrease much until the P_{aO_2} falls below 70 mm Hg (S_{O_2} = 94%). Thus, pulmonary physiologists generally regard low oxygen in arterial blood as an emergency mechanism to drive ventilation only when the supply of oxygen is seriously impaired, as in patients with severe ventilation/perfusion mismatching (chronic obstructive pulmonary disease) or in persons being strangled or smothered to death.

This standard view about the oxygen control of breathing is not entirely true. Even at normal P_{aO_2} breathing is slightly stimulated. If it were not so, then it would be difficult to explain how living at moderately high altitudes (say at 7000 feet at Santa Fe, New Mexico) leads to an increase in alveolar ventilation; P_{aCO_2} averages about 33 mm Hg, which indicates a 20% rise in ventilation, according to the alveolar ventilation equation (Chapter 28). Breathing stimulation at high altitude continues to develop over some weeks and is long lasting, in spite of the fact that the fall in P_{ACO_2} at the central chemoreceptors of the brainstem tends to depress ventilation. That paradox will be resolved in the next section.

Interactions Between Oxygen and Carbon Dioxide

When one first goes to a higher altitude, the ventilatory response to the decreased inspired alveolar and arterial P_{O_2} is attenuated, because the increasing ventilation decreases the P_{ACO_2}. Thus, when physiologists test a person's sensitivity to low inspired oxygen, they distinguish between tests in which the P_{aCO_2} is maintained constant and those in which it is allowed to change. Figure 31-7 shows the ventilatory response to hypoxia when P_{ACO_2} is controlled at 40 mm Hg and when it is not controlled.

The mechanism by which the carotid body is stimulated, rather than depressed, by hypoxemia has not been completely elucidated. What is known is that carotid body blood flow per gram of tissue is high, as is its meta-

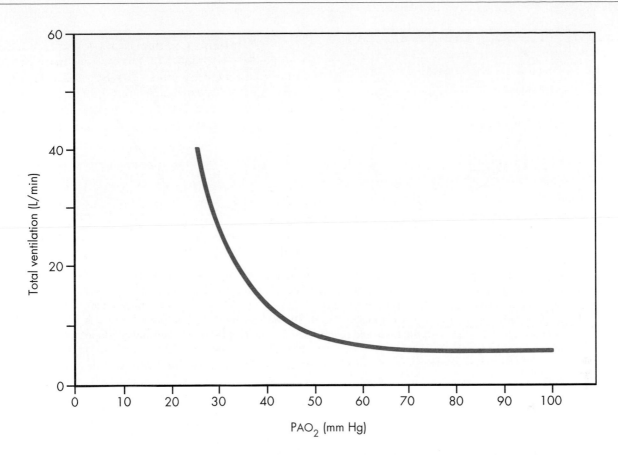

FIGURE 31-6 The O_2-ventilation dose-response curve. Ventilation is not sensitive to Pa_{O_2} until the tension falls below about 70 mm Hg. Part of the reason is that the response follows the HbO_2 equilibrium curve, which is very flat near the top. Furthermore, as ventilation starts to rise the Pa_{CO_2} decreases. For example, at Po_2 of 40 mm Hg, ventilation is about three times normal, meaning Pa_{CO_2} must be decreased to one third of normal (13 mm Hg).

bolic rate, yet the blood leaving the carotid bodies is not measurably desaturated. *It is the partial pressure of oxygen, rather than the oxygen concentration or saturation of the arterial blood, that is the primary stimulus.*

This explains why in carbon monoxide poisoning there is little stimulation to breathe, even though one might be dying of hypoxemia. While the arterial O_2 concentration is markedly decreased, the Pa_{O_2} and Sa_{O_2} are normal (Figure 30-5).

> One reason people commit suicide by sitting in a closed garage with the car motor running is that CO poisoning is a painless way to die of essentially pure hypoxemia. One becomes drowsy, nods off, and dies of **anemic hypoxia** and **histotoxic hypoxia.** Suicide by this method is less effective since the advent of catalytic converters, which remove nearly all of the carbon monoxide gas.

MECHANICAL CONTROL OF BREATHING

Receptors in the Lungs

There are three types of lung sensory receptors: **stretch receptors,** located within the smooth muscle layer of the extrapulmonary airways; **irritant receptors,** located adjacent to the epithelial cells of the nasal cavity and in the bronchi; and **C fibers** (unmyelinated; J receptors in the older literature), situated in the lung interstitium and alveolar walls. The afferent nerve fibers of all these receptors travel to the brain in the vagus nerves, and they enter the medulla adjacent to the respiratory nerve pools.

As mentioned earlier, the stretch receptors are excited by an increase in lung or airway distending pressure. One of their signal features is that they adapt very slowly to a sustained stimulus; that is, they continue to send impulses up the vagus nerve to the medulla as long as the

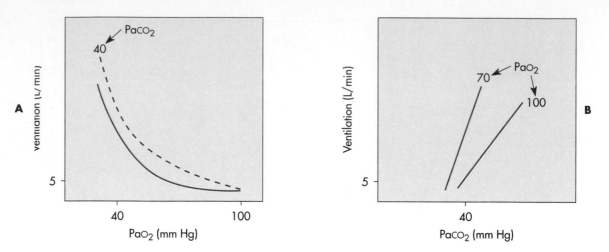

FIGURE 31-7 The effect of interactions between O_2 and CO_2 on ventilation. **A,** The normal ventilation response to decreased inspired oxygen causes Pa_{CO_2} to decrease, which opposes the hypoxic drive. When Pa_{CO_2} is maintained at 40 mm Hg, by adding CO_2 to the inspired air, the ventilatory response begins at a higher Po_2. **B,** The normal ventilatory response to CO_2 is enhanced by hypoxia; both the threshold (extrapolated X-intercept) and the sensitivity (slope of response line) are affected.

stretch stimulus continues. As the lung is inflated, they tend to inhibit inspiration and promote expiration **(Hering-Breuer reflex).**

Irritant receptors are stimulated chemically by noxious agents (sulfur dioxide, ammonia) or mechanically (particulate matter impinging on the bronchial surfaces). Irritant receptor stimulation constricts the airways and promotes rapid, shallow breathing. One likely effect of this response is to limit penetration of harmful substances deep into the lung and thereby prevent these substances from injuring the gas-exchanging surfaces. Apparently, the irritant receptors also contribute to the initiation of periodic sighs (large breaths) that occur during normal breathing, which expand the alveolar surface area and replenish the surface-active molecules (Chapter 28). Chemical mediators (histamine, leukotriene C_4, or bradykinin) released in the lung during allergic reactions also stimulate the irritant receptors. The irritant receptors adapt within a few seconds when they are subjected to a sustained stimulus; that is, these nerves stop sending impulses up the vagus nerve after the initial burst of activity.

The unmyelinated C fibers are excited by distortion of the lung's connective tissue structure, such as by fluid accumulation (edema) or by any one of several chemicals, including histamine and capsaicin (the active irritant in pepper). Activation of the C fibers initially causes **laryngeal closure (spasm)** and **apnea,** but this is quickly followed by rapid shallow breathing. Stimulation of C fibers and irritant receptors together may be respon-

sible for the **tachypnea** (rapid breathing) seen in patients with multiple pulmonary emboli, lung edema, or pneumonia. The C fibers also adapt rapidly to a sustained stimulus.

Receptors in the Chest Wall

When activated, the muscles of breathing (diaphragm, intercostals, abdominal wall) develop tension that depends on their initial length (preload; passive stretch) and their afterload (resistance to expansion of the thorax). The preload varies with posture, and the afterload varies with chest wall expansion or the resistance to inspiratory and expiratory airflow.

There are various chest wall sensory nerve receptors (joint, tendon, and muscle spindles, as in other skeletal muscles). **Joint receptors** are stimulated by the extent and speed of rib movement relative to the vertebrae and the sternum. **Tendon organs** in the intercostal muscles and the diaphragm monitor the force of muscle contraction and tend to inhibit inspiration. **Muscle spindles** are abundant in the intercostal and abdominal wall muscles, but scarce in the diaphragm. The spindles help coordinate breathing during changes in posture and speech, and stabilize the rib cage when breathing is impeded by increases in airway resistance or decreases in lung compliance. Normally, we are unaware of breathing, but sometimes we become aware that we are breathing hard, such as after running up two flights of stairs.

The feeling generated in our consciousness when chest wall movement is inappropriate is called **dyspnea** (difficult or painful breathing). Dyspnea is generally equated by chest physicians with breathlessness (the person experiencing the sensation cannot get enough air to breathe). Even healthy people who exercise to exhaustion, such as an Olympic runner or a cross-country skier, experience severe shortness of breath. However, it is normal to be short of breath after an exhausting race. Therefore, dyspnea includes the concept that the sensation is inappropriate to the effort.

RESPIRATORY FAILURE

As lung disease progresses, it eventually reaches a stage where impairment of chest wall movement is severe enough so that adequate alveolar ventilation cannot be preserved, a condition called **respiratory failure.** *Respiratory failure means that ventilation cannot keep up with O_2 demand or CO_2 production.* When respiratory failure occurs, the alveolar ventilation equation indicates that alveolar, arterial, and tissue P_{CO_2} will rise. It is clearly a non–steady state condition, leading to acidosis (metabolic and respiratory, as lactic acid and H_2CO_3 accumulate). This occurs physiologically when one exceeds the anaerobic threshold during exercise. Respiratory failure is a bad sign and requires serious attention.

ABNORMAL BREATHING PATTERNS

Breathing is normally a smoothly recurring cyclic process of which we are unaware, but in some (usually serious) conditions the drive to breathe is affected in a stereotypic fashion that is useful in diagnosis.

Cheyne-Stokes breathing is a manifestation of instability of ventilatory control in which tidal volume cyclically increases and decreases to the point of apnea. This breathing pattern, during which the systemic arterial partial pressures of oxygen and carbon dioxide fluctuate markedly, may appear in normal individuals during sleep, as a person fluctuates between various levels of sleep. In disease conditions, it is a serious sign.

Delays in information transfer within the breathing control system occur when the transit time around the circulation is prolonged. Ordinarily, changes in alveolar P_{CO_2} are sensed by the medullary (central) chemorecep-

tors within several seconds. In severe congestive heart failure, in which cardiac output may be only half of normal, the time required for the blood to circulate from the lungs to the central chemoreceptors is prolonged, so that the controller "hunts"; that is, it increases and decreases ventilation as it seeks a level that will maintain arterial P_{CO_2} in the normal range.

In **Biot's breathing,** periods of normal breathing are interrupted suddenly by periods of apnea. The mechanism for this pattern is unclear; it may be a variant of Cheyne-Stokes breathing. It occurs in patients with central nervous system diseases—for example, meningitis.

Apneustic breathing, with its prolonged inspiratory pauses, has already been mentioned on p. 424.

Grossly **irregular breathing** occurs in some patients with medullary lesions and occasionally in people who have absent responses to chemical stimuli (primary alveolar hypoventilation).

Increased breathing with hypocapnia (low P_{aCO_2}) occurs when the lung irritant receptors or C fibers are stimulated (in **asthma** or **pulmonary microembolism**), in psychiatric disturbances **(hysteria),** or in **severe metabolic acidosis.** In **diabetic coma** this form of continuous rapid deep breathing is called **Kussmaul breathing;** aggressive treatment is needed.

SLEEP

When we are awake, stimuli from the environment modulate rhythmic breathing, both reflexly via brain centers and by their effect on alertness, which is dependent on the activity of the reticular activating system. Fluctuations in alertness, such as dozing off during a boring lecture, occur in the awake state. However, breathing fluctuations (for example, Cheyne-Stokes breathing) are more pronounced during sleep. The reticular activating system is shut down during sleep.

Sleep consists of two main stages: a slow-brain-wave stage (non-REM; light) and a rapid-eye-movement (REM) stage (deep sleep during which dreaming occurs; see also Chapter 11).

Regulation of Breathing in Sleep

During the initial stage of sleep (slow-wave sleep) ventilation usually decreases. Hence arterial P_{CO_2} increases; note that the CO_2 response curve is shifted to the right (*dashed line,* Figure 31-5). However, the ventilatory response to hypoxia, but not to CO_2, is maintained in slow wave sleep.

REM sleep (dreaming) is divided into phasic and tonic

substages. In tonic REM sleep, breathing maintains its regularity, but tidal volume may decrease progressively, as the ventilatory responses to inspired CO_2 are less pronounced than in non-REM sleep (first stage). External stimuli (baby crying) and changes in blood gas tensions are less effective in producing arousal than in slow-wave sleep. Phasic REM sleep is associated with irregular breathing patterns, because the activity of higher brain centers (dreaming) modulates medullary respiratory neuronal activity.

Sleep Apneas

Sleep apneas are classified into two distinct categories: central and obstructive (Figure 31-8).

Central Sleep Apnea Central sleep apnea is characterized by a cessation of all breathing efforts; that is, the medullary centers are depressed and do not send sufficient impulses to the breathing muscle motor neurons to elicit activation. As shown in Figure 31-8, *A*, the hallmark of central sleep apnea is no attempt to breathe. In addition to no airflow, pleural pressure does not change. Many pediatricians believe central sleep apnea is the basis for the **sudden infant death syndrome.**

Obstructive Sleep Apnea When we are awake, the muscles of our tongue and throat are tonically active, and they maintain a patent oropharynx (the space behind the tongue that connects the nose to the larynx). Activity of skeletal muscle decreases during sleep; the oropharyngeal muscles are almost completely relaxed. Furthermore, the soft palate and uvula may be drawn dorsally (especially when one sleeps on one's back), because of the slight negative upper airway pressure during inspiratory airflow. This is frequently manifested by snoring, and it indicates a partial obstruction of the oropharynx.

When occlusion of the oropharynx becomes complete, the condition is called **obstructive sleep apnea.** As shown in Figure 31-8, *B*, the hallmark of this malady is that pleural pressure cycles, but there is no airflow. The apnea may last for many seconds and be associated with reductions in arterial oxygen saturation to 75% ($Pa_{O_2} \leq 40$ mm Hg). The pleural pressure swings increase as Pa_{CO_2} rises.

Obstructive apnea occurs in all sleep stages, but it is more common in slow-wave sleep. Arousal may be an important element in terminating sleep apnea; arousal occurs when chemoreceptor excitation by hypoxia and hypercapnia are sufficient. Obstructive sleep apnea is the most common sleep disorder, aside from insomnia (inability to sleep).

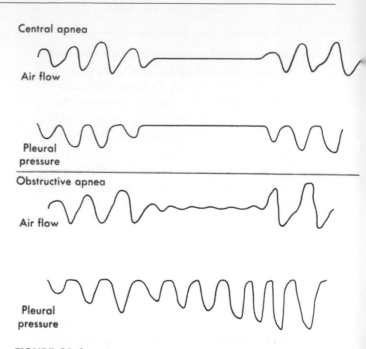

FIGURE 31-8   The two types of sleep apnea. **A,** Central apnea is characterized by no attempt to breathe, as demonstrated by no change in pleural pressure. **B,** In obstructive sleep apnea, pleural pressure swings are augmented, indicating that airflow resistance is very high because of blockage of the oropharynx by the soft palate or tongue.

ACCLIMATIZATION TO ALTITUDE

Barometric pressure and inspired P_{O_2} decrease with increasing altitude (Figure 31-9); the relationship is approximately exponential. As one ascends to higher altitudes, the resulting systemic arterial hypoxemia elicits a variety of compensatory responses. Some of these occur quickly, whereas others develop gradually during prolonged exposure (weeks).

The initial hyperventilation that occurs at high altitude is caused by stimulation of the peripheral chemoreceptors, particularly the carotid bodies (decreased Pa_{O_2} depresses central breathing). The increase in ventilation reduces the arterial P_{CO_2} and $[H^+]$. These changes decrease the excitation of central chemoreceptors and thus act as a negative feedback (opposing the hypoxia-induced hyperventilation). After a few days at altitude, ventilation increases further until a new steady state of breathing is achieved.

Acclimatization occurs, in part, by the following mechanisms: (1) the renal excretion of sodium reduces plasma HCO_3^- concentration (decreased strong ion difference), which raises blood H^+ concentration toward normal (decreased pH); (2) at the same time the HCO_3^- concentration is decreased in brain interstitial fluid, probably by an active metabolic process that moves so-

Exposure to high altitude also increases the circulating hemoglobin concentration as more erythropoietin is manufactured and secreted by the hypoxic kidney cells. The concentration of 2,3-diphosphoglycerate also rises in the red blood cells and decreases the affinity of hemoglobin for O_2 (shift to the right). The P_{50} of the blood is increased, which improves the unloading of oxygen at a higher P_{O_2} in the systemic capillaries (Chapter 30).

Occasionally, tolerance of high altitude disappears and serious symptoms develop, such as ventilatory depression, severe **polycythemia** (hematocrit may rise above 60%) and heart failure. This intolerance to altitude **(chronic mountain sickness)** is relieved by descent to a lower altitude or by administration of oxygen. People who have suffered from chronic mountain sickness should never go to high altitude again.

SUMMARY

1. There are two main components to the regulation of breathing. Metabolic (automatic) control is concerned with oxygen delivery and with acid-base balance (P_{aCO_2}). Behavioral (voluntary) control is related to coordinated activities in which breathing may be temporarily suspended or altered.

2. The control system consists of a central controller (driver) located in the brainstem (medulla and pons), an effector (the muscles of the chest wall, especially the diaphragm), and various sensory receptors, which report back to the central controller the results of the intended action.

3. The modern view of the brainstem controller is that it contains a tonically active inspiratory neuron pool that receives input from a variety of sensors. As lung volume increases, the summed sensory input from various receptors inhibits inspiratory activity acting via the pneumotaxic center (pons). The cortex, thalamus, and hypothalamus induce behavioral breathing by temporarily overriding the brainstem pattern generator.

4. The sensory components of breathing control include central chemoreceptors (near the surface of the medulla); peripheral chemoreceptors (carotid bodies); and proprioceptors (lung stretch, irritant, and C-fiber receptors; diaphragm, intercostal, and abdominal muscle spindles; and tendon and joint organs).

5. The medullary receptors are sensitive to brain and cerebrospinal P_{CO_2} which is closely related to P_{aCO_2}. The initial ventilatory response to CO_2 is large and occurs rapidly.

6. The carotid body chemoreceptors are mainly sensitive to reduced arterial oxygen tension. However, a

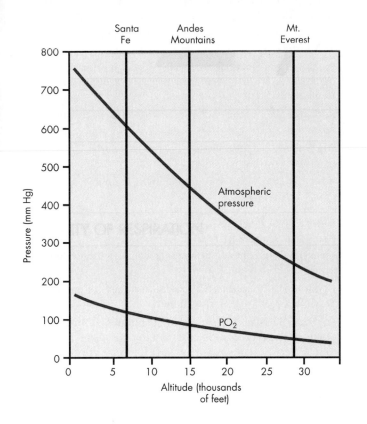

FIGURE 31-9   Barometric pressure and P_{O_2} fall exponentially as one ascends to high altitude (X axis). Vertical lines show some locations relative to human ability to live at altitude. Santa Fe, New Mexico, at 7000 feet ($P_B = 585$ mm Hg; $P_{O_2} = 118$ mm Hg), is the highest large US city. The highest permanent human community is in the Andes mountains at about 15,000 feet ($P_B = 430$ mm Hg; $P_{O_2} = 90$ mm Hg). The top of Mt Everest is at 29,000 feet ($P_B = 235$ mm Hg; $P_{O_2} = 49$ mm Hg).

dium ions from the brain interstitial fluid or cerebrospinal fluid into the blood; (3) a small amount of anaerobic metabolism may occur in the hypoxic brain and allows brain lactate anions to replace brain bicarbonate ions.

Another important acclimatization response is a shift to the left of the ventilatory response to CO_2 and a steepening of the slope of the response curve (increased sensitivity) caused by the interaction with reduced P_{aO_2} (Figure 31-7). Consequently, the threshold for central breathing stimulation occurs at a lower P_{CO_2}. Overall, *the acclimatization process decreases the negative feedback that opposes the carotid chemoreceptor-induced drive to increase ventilation—hence, the rate and depth of breathing increase.*

significant response to hypoxia comes into play only at low Pa_{O_2} (<70 mm Hg).

7. Irritant receptors in the nose and large airways protect the delicate alveolar surfaces from inhaled water, particles, and chemical vapors.

8. C-fiber receptors in the lung interstitium are stimulated by distortion of the alveolar walls (lung vascular congestion or edema).

9. Many factors influence breathing in exercise in a manner that is incompletely understood, but there can be no doubt that the controlled variable is Pa_{CO_2}.

10. Sleep is a complex phenomenon that consists of two stages (light and REM). The sensitivities to CO_2 and O_2 are both diminished, and the reticular activating system is depressed.

11. Acute hypoxia and chronic hypoxia affect breathing differently. The immediate hyperventilation that occurs at high altitude is caused by stimulation of the peripheral chemoreceptors by lower Pa_{O_2}. The chronic or slow adjustments are achieved by restoration of the pH in the brain interstitium and cerebrospinal fluid, and this change in pH affects CO_2 sensitivity.

BIBLIOGRAPHY

Journal Articles

Berger AJ: Pattern generator for the basic breathing rhythmicity, *J Appl Physiol* 55:1647, 1983.

Coleridge JCG, Coleridge HM: Afferent vagal fibre innervation of the lung and airways and its functional significance, *Rev Physiol Biochem Pharmacol* 99:2, 1984.

Eyzaguirre C, Zapata P: Perspectives in carotid body research, *J Appl Physiol* 57:931, 1984.

Lydic R: State-dependent aspects of regulatory physiology, *Faseb J* 1:6, 1987.

Pack AI: Sensory inputs to the medulla, *Annu Rev Physiol* 43:73, 1981.

Phillipson EA: Sleep disorders. In Murray J, Nadel J, eds: *Textbook of respiratory medicine,* vol 2, Philadelphia, 1994, WB Saunders, p 2310.

Schiaefke ME: Central chemosensitivity: a respiratory drive, *Rev Physiol Biochem Pharmacol* 90:171, 1981.

von Euler C: On the central pattern generator for the basic breathing rhythmicity, *J Appl Physiol* 55:1647, 1983.

Whipp BJ, SA Ward: Cardiopulmonary coupling during exercise, *J Exp Biol* 100:175, 1982.

Books and Monographs

Cherniack NS, Widdicombe JG, eds: *Handbook of physiology,* section 3: The control of breathing, vol II, Bethesda, Md, 1986, American Physiological Society.

Comroe JH Jr: *Physiology of respiration,* ed 2, Chicago, 1974, Year Book Medical Publishers, Inc.

Feldman JL: Neurophysiology of breathing in mammals. In Mountcastle VB, Bloom FE, eds: *Handbook of physiology,* section 1: Nervous system, vol IV: Intrinsic regulatory systems of the brain, Bethesda, Md, 1986, American Physiological Society.

Nunn FJ: *Applied respiratory physiology,* ed 3, London, 1987, Butterworths.

von Euler C, Lagercrantz H, eds: *Neurobiology of the control of breathing,* New York, 1987, Raven Press.

GASTROINTESTINAL SYSTEM

HOWARD C. KUTCHAI

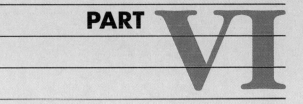

PART VI

32 Motility of the Gastrointestinal Tract

The gastrointestinal system consists of the gastrointestinal tract and certain associated glandular organs that produce secretions that function in the gastrointestinal tract. The major subdivisions of the gastrointestinal tract are the mouth, pharynx, esophagus, stomach, duodenum, jejunum, ileum, colon, rectum, and anus. The duodenum, jejunum, and ileum constitute the small intestine. Associated glandular organs include salivary glands, liver, gallbladder, and pancreas.

The major physiological functions of the gastrointestinal system are to digest foodstuffs and absorb nutrient molecules into the bloodstream. The activities by which the gastrointestinal system carries out these functions may be subdivided into motility, secretion, and digestion and absorption. **Motility** refers to gastrointestinal movements that mix and circulate the gastrointestinal contents and propel them along the length of the tract. Usually net propulsion occurs in the orthograde direction—that is, away from the mouth and toward the anus. Retrograde propulsion does occur, however; vomiting is a notable example. **Secretion** refers to the processes by which the glands associated with the gastrointestinal tract release water and substances into the tract. **Digestion** is defined as the processes by which ingested pieces of food and large molecules are chemically degraded to produce smaller molecules that can be absorbed across the wall of the gastrointestinal tract. **Absorption** refers to the processes by which nutrient molecules are absorbed by the gastrointestinal tract and enter the bloodstream.

STRUCTURE OF THE GASTROINTESTINAL TRACT

The structure of the gastrointestinal tract varies greatly from region to region, but common features exist in the overall organization of the tissue. Figure 32-1 depicts the general layered structure of the gastrointestinal tract wall.

The **mucosa** consists of an epithelium, the lamina propria, and the muscularis mucosae. The **epithelium** is a single layer of specialized cells that lines the lumen of the gastrointestinal tract. The nature of the epithelium varies greatly from one part of the digestive tract to another. The **lamina propria** consists largely of loose connective tissue containing collagen and elastin fibrils. The lamina propria is rich in several types of glands and contains lymph nodules and capillaries. The **muscularis muscosae** is the thin, innermost layer of intestinal smooth muscle. Contractions of the muscularis mucosae throw the mucosa into folds and ridges.

The **submucosa** consists largely of loose connective tissue with collagen and elastin fibrils. In some regions submucosal glands are present. The larger nerve trunks and blood vessels of the intestinal wall travel in the submucosa.

The **muscularis externa** typically consists of two substantial layers of smooth muscle cells: an inner circular layer and an outer longitudinal layer. Contractions of the muscularis externa mix and circulate the contents of the lumen and propel them along the gastrointestinal tract.

The wall of the gastrointestinal tract contains many neurons that are highly interconnected. A dense network of nerve cells in the submucosa is the **submucosal plexus (Meissner's plexus).** The prominent **myenteric plexus (Auerbach's plexus)** is located between the circular and longitudinal smooth muscle layers. The submucosal and myenteric plexuses, together with the other neurons of the gastrointestinal tract, constitute the **intramural plexuses** or the **enteric nervous system,** which helps to integrate the motor and secretory activities of the gastrointestinal system. If the sympathetic and parasympathetic nerves to the gut are cut, many motor

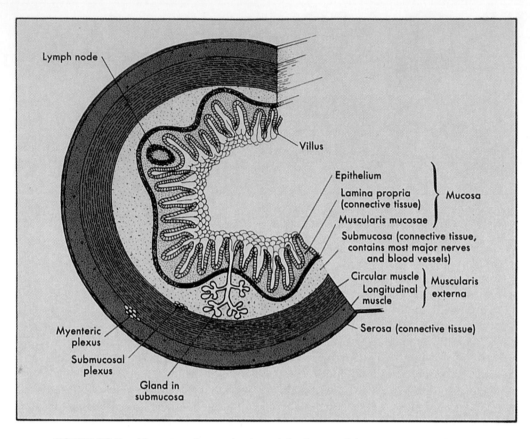

FIGURE 32-1   The general organization of the layers of the gastrointestinal tract.

and secretory activities continue because these processes are controlled by the enteric nervous system.

The **serosa,** or **adventitia,** is the outermost layer and consists mainly of connective tissue covered with a layer of squamous mesothelial cells.

INNERVATION OF THE GASTROINTESTINAL TRACT

Sympathetic Innervation

Sympathetic innervation of the gastrointestinal tract is mainly via postganglionic adrenergic fibers whose cell bodies are in prevertebral and paravertebral ganglia (Figure 32-2). The celiac, superior and inferior mesenteric, and hypogastric plexuses provide sympathetic innervation to various segments of the gastrointestinal tract. Activation of the sympathetic nerves usually inhibits the motor and secretory activities of the gastrointestinal system. *Most of the sympathetic fibers do not directly innervate structures in the gastrointestinal tract, but rather terminate on neurons in the intramural plexuses.* Some vasoconstrictor sympathetic fibers directly innervate blood vessels of the gastrointestinal

tract. Other sympathetic fibers innervate glandular structures in the wall of the gut.

Stimulation of the sympathetic input to the gastrointestinal tract inhibits motor activity of the muscularis externa, but stimulates contraction of the muscularis mucosae and certain sphincters. The inhibitory effect of the sympathetic nerves on the muscularis externa is not a direct action on the smooth muscle cells, because few sympathetic nerve endings lie in the muscularis externa. Rather, the sympathetic nerves act to influence neural circuits in the enteric nervous system, and these circuits provide input to the smooth muscle cells. The sympathetic nerves may reinforce this effect by reducing blood flow to the muscularis externa. Other fibers that travel with the sympathetic nerves may be cholinergic; still others release neurotransmitters that remain to be identified.

Parasympathetic Innervation

Parasympathetic innervation of the gastrointestinal tract down to the level of the transverse colon is provided by branches of the vagus nerves (Figure 32-2). The remainder of the colon, the rectum, and the anus receive parasympathetic fibers from the pelvic nerves.

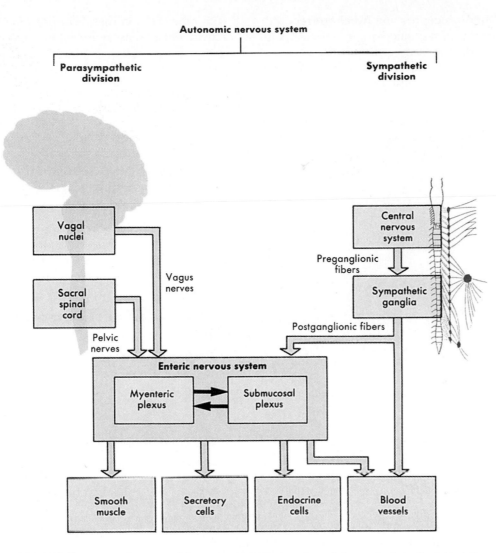

FIGURE 32-2   Major features of the autonomic innervation of the gastrointestinal tract. In most cases the autonomic nerves influence the functions of the gastrointestinal tract by modulating the activities of neurons of the enteric nervous system. (Redrawn from Costa M, Furness JB: *Br Med Bull* 38:247, 1982.)

These parasympathetic fibers are preganglionic and predominantly cholinergic. Other fibers that travel in the vagus and its branches release other transmitters, some of which have not been identified. The parasympathetic fibers terminate predominantly on the ganglion cells in the intramural plexuses. The ganglion cells then directly innervate the smooth muscle and secretory cells of the gastrointestinal tract. Excitation of parasympathetic nerves usually stimulates the motor and secretory activity of the gut.

The Enteric Nervous System

The myenteric and submucosal plexuses are the most well-defined plexuses in the wall of the gastrointestinal tract (Figure 32-3). The plexuses are networks of nerve fibers and ganglion cell bodies. Interneurons in the plexuses connect afferent sensory fibers with efferent neurons to smooth muscle and secretory cells to form reflex arcs that are wholly within the gastrointestinal tract wall. Consequently the myenteric and submucosal plexuses can coordinate activity in the absence of extrinsic innervation of the gastrointestinal tract. Axons of plexus neurons innervate gland cells in the mucosa and submucosa, smooth muscle cells in the muscularis externa and muscularis mucosae, and intramural endocrine and exocrine cells.

Reflex Control

Afferent fibers in the gastrointestinal tract provide the afferent limbs of both **local and central reflex arcs**

(Figure 32-4). **Chemoreceptor** and **mechanoreceptor endings** are present in the mucosa and muscularis externa. The cell bodies of many of these sensory receptors are located in the myenteric and submucosal plexuses. The axons of some of these receptor cells synapse with other cells in the plexuses to mediate local reflex activity. Other sensory receptors send their axons back to the central nervous system. The complex afferent and efferent innervation of the gastrointestinal tract allows for fine control of secretory and motor activities.

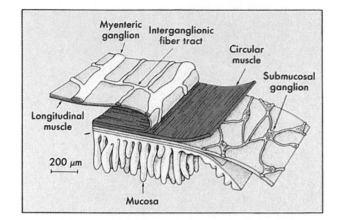

FIGURE 32-3 Enteric neurons of the submucosal and myenteric plexuses in the wall of the gastrointestinal tract. The plexuses consist of ganglia interconnected by fiber tracts. (Redrawn from Wood JD: In Johnson RL, ed: *Physiology of the gastrointestinal tract,* ed 2, New York, 1987, Raven Press.)

GASTROINTESTINAL SMOOTH MUSCLE

Properties of Gastrointestinal Smooth Muscle Cells

Smooth muscle was discussed in Chapter 14. The smooth muscle cells of the gastrointestinal tract are long (about 500 μm in length) and slender (5 to 20 μm across). The cells are arranged in bundles that are separated and defined by connective tissues.

Electrophysiology of Gastrointestinal Smooth Muscle

Resting Membrane Potential The resting membrane potential of gastrointestinal smooth muscle cells ranges from approximately -40 to -80 mV. The electrogenic Na^+-K^+ pump (see Chapter 2) contributes significantly

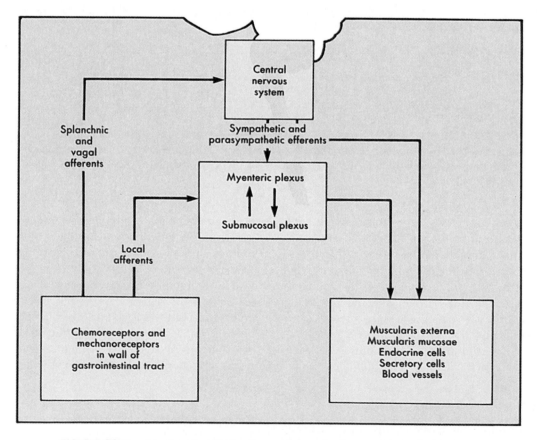

FIGURE 32-4 Local and central reflex pathways in the gastrointestinal system.

to the resting membrane potential in smooth muscle. In guinea pig taenia coli, for example, about one third of the resting membrane potential results from the electrogenicity of the Na^+-K^+-ATPase.

Slow Waves In most other excitable tissues the resting membrane potential is rather constant. In gastrointestinal smooth muscle the resting membrane potential characteristically varies in time. Oscillations of the resting membrane potential called **slow waves** (also known as the **basic electrical rhythm**) are characteristic of gastrointestinal smooth muscle (Figure 32-5). The frequency of slow waves varies from about 3 per minute in the stomach to 12 per minute in the duodenum.

Slow waves are generated by **interstitial cells,** which have properties of both fibroblasts and smooth muscle cells. A thin layer of interstitial cells is located between the longitudinal and circular layers of muscularis externa. Processes of the interstitial cells form gap junctions with longitudinal and circular smooth muscle cells. These gap junctions enable the slow waves to be rapidly conducted to both muscle layers. Because the smooth muscle cells of both longitudinal and circular layers are well-coupled electrically, the slow wave spreads throughout the smooth muscle of each segment of the gastrointestinal tract.

The amplitude and, to a lesser extent, the frequency of the slow waves can be modulated by the activity of intrinsic and extrinsic nerves and by circulating hormones. In general, sympathetic nerve activity decreases the amplitude of the slow waves or abolishes them, whereas stimulation of parasympathetic nerves increases the size of the slow waves.

If the peak of the slow wave is above threshold for the cells to fire action potentials, one or more action potentials may be triggered during the peak of the slow wave (Figure 32-5).

Action Potentials Action potentials in gastrointestinal smooth muscle are more prolonged (10 to 20 msec) than those of skeletal muscle and have little or no overshoot. The rising phase of the action potential is caused by ion flow through channels that conduct both Ca^{++} and Na^+ and are relatively slow to open. Ca^{++} that enters the cell during the action potential helps to initiate contraction.

When the membrane potential of gastrointestinal smooth muscle reaches threshold, typically near the peak of a slow wave, a train of action potentials (1 to 10 per second) occurs (Figure 32-5). The extent of depolarization of the cells and the frequency of action potentials are enhanced by certain hormones and by compounds liberated from excitatory nerve endings. Inhibitory hormones and neuroeffector substances hyperpolarize the smooth muscle cells and may abolish action potential spikes.

Relationship Between Membrane Potential and Tension In the example shown in Figure 32-5 the slow waves without action potentials elicit weak contractions of the smooth muscle layers. Much stronger contractions are evoked by the action potentials that are intermittently triggered near the peaks of the slow waves. The greater the frequency of action potentials that occur at the peak of a slow wave, the more intense is the contraction of the smooth muscle. Because smooth muscle cells contract rather slowly (about one-tenth as fast as skeletal muscle), the individual contractions caused by each action potential in a burst are not visible as distinct

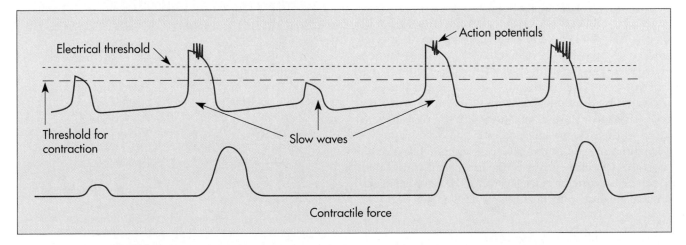

FIGURE 32-5 Contraction of small intestinal smooth muscle occurs when the depolarization caused by the slow wave exceeds a threshold for contraction. When depolarization of a slow wave exceeds the electrical threshold, a burst of action potentials occurs. The action potentials elicit a much stronger contraction than occurs in the absence of action potentials. The contractile force increases with increasing number of action potentials.

twitches; rather, they sum temporally to produce a smoothly increasing level of tension (Figure 32-5).

Between bursts of action potentials the tension developed by gastrointestinal smooth muscle falls, but not to zero. This nonzero resting, or baseline, tension developed by the smooth muscle is called **tone.** The tone of gastrointestinal smooth muscle is altered by neuroeffectors, hormones, and drugs.

Electrical Coupling Between Smooth Muscle Cells
Neighboring cells are said to be well coupled electrically if a perturbation of the membrane potential of one cell spreads rapidly, and with little decrement, to the other cell. The smooth muscle cells of the circular layer are better coupled than those of the longitudinal layer. The cells of the circular layer are joined by frequent gap junctions that allow the spread of electrical current from one cell to another (see also Chapter 4).

INTEGRATION AND CONTROL OF GASTROINTESTINAL MOTILITY

Control of the contractile activities of gastrointestinal smooth muscle involves the central nervous system, the enteric nervous system, and hormones and paracrine substances. The autonomic nervous system typically modulates the patterns of muscular and secretory activity that are controlled more directly by the enteric nervous system.

Neuromuscular Interactions

The neurons of the intramural plexuses send axons to the smooth muscle layers, and each axon may branch extensively to innervate many smooth muscle cells. Neuromuscular interactions in the gastrointestinal tract do not involve true neuromuscular junctions with specialization of the postjunctional membrane, as occurs at neuromuscular junctions in skeletal muscle. The circular smooth muscle layer of the muscularis externa is heavily innervated by excitatory and inhibitory motor nerve terminals that form close association with the plasma membranes of the smooth muscle cells.

Longitudinal smooth muscle cells are much less richly innervated by the neurons of the intrinsic plexuses than are the cells of the circular layer, and the neuromuscular contacts are not so intimate.

The Enteric Nervous System

The plexuses of the gastrointestinal tract function as a semiautonomous nervous system that controls the motor and secretory activities of the digestive system. The enteric nervous system of the large and small intestines alone contains about 10^8 neurons, about as many neu-

rons as in the spinal cord. Figure 32-3 depicts the myenteric and submucosal plexuses and their locations in the wall of the intestine. Both plexuses consist of ganglia that are interconnected by tracts of fine, unmyelinated nerve fibers. The neurons in the ganglia include sensory neurons, with their sensory endings in gastrointestinal tract wall. Sensory neurons that respond to mechanical deformation, to particular chemical stimuli, and to temperature have been identified. Some of the neurons in the enteric ganglia are effector neurons that send axons to smooth muscle cells of the circular or longitudinal layers, to secretory cells of the gastrointestinal tract, or to gastrointestinal blood vessels. Many of the neurons in the enteric ganglia are interneurons that are part of the network of neurons that integrates the sensory input to the ganglia and formulates the output of the effector neurons.

Neuromodulatory Substances Most of the neuromodulatory substances that function in the central nervous system (see Chapter 4) are also present in the gastrointestinal tract. Table 32-1 lists some of the neuroactive substances present in the gastrointestinal tract.

Functions of Myenteric Neurons
Myenteric Neurons The majority of neurons in myenteric ganglia are motor neurons, both excitatory and inhibitory, to the smooth muscle cells of the muscularis externa. In addition, there are sensory neurons and interneurons in myenteric ganglia. Excitatory motor neurons release **acetylcholine** onto muscarinic receptors on the smooth muscle cells; they also release **substance P.** Inhibitory motor neurons release VIP (**vasoactive**

Table 32-1 Neurotransmitters or Neuromodulatory Substances in the Enteric Nervous System	
Acetylcholine	Somatostatin
Norepinephrine	Vasoactive intestinal
5-Hydroxytryptamine	polypeptide (VIP)
(5-HT)	Enkephalin
Purine nucleotides	Substance P
Dopamine	Bombesin
Neurotensin	γ-Aminobutyric acid
Cholecystokinin (CCK)	(GABA)
Glycine	Gastrin
Motilin	Histamine
Angiotensin	Thyrotropin-releasing
Secretin	hormone
Galanin	Gastrin-releasing peptide (GRP)
Neuropeptide Y	Prostaglandins
	Peptide, histidine, isoleucine (PHI)

Modified from Wood JD: Physiology of the enteric nervous system. In Johnson RL, ed: *Physiology of the gastrointestinal tract.* New York, 1987, Raven Press, p. 69.

intestinal polypeptide) and NO **(nitric oxide).** About one third of neurons in myenteric ganglia are sensory. Other myenteric neurons project to neurons in submucosal ganglia. Most myenteric interneurons release acetylcholine onto nicotinic receptors on motor neurons or on other interneurons.

Submucosal Neurons *Most neurons in submucosal ganglia regulate secretion by glandular, endocrine, and epithelial cells.* Stimulatory secretomotor neurons release acetylcholine and VIP onto gland cells or epithelial cells. There are numerous sensory neurons in submucosal ganglia. These neurons respond to chemical stimuli or to mechanical deformation of the mucosa, and they are the afferent limbs of secretomotor reflexes. Submucosal interneurons release acetylcholine onto other neurons in submucosal ganglia or project to myenteric ganglia. Submucosal ganglia also contain vasodilator neurons that release acetylcholine and/or VIP onto submucosal blood vessels.

Intrinsic Reflexes An **intrinsic reflex** is a reflex whose component cells are all located in the wall of the gastrointestinal tract. Numerous intrinsic reflexes control the motor and secretory activities of each segment of the gastrointestinal tract. A well-characterized intrinsic reflex is shown in Figure 32-6. *Localized mechanical or chemical stimulation of the intestinal mucosa*

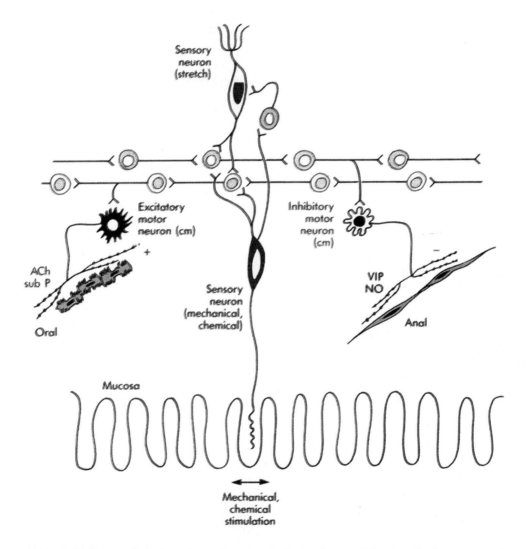

FIGURE 32-6 Localized mechanical or chemical stimulation of the intestinal mucosa typically elicits contraction above and relaxation below the point of stimulation. This figure depicts the enteric neuronal circuitry responsible for this reflex behavior. In the center of the figure are two sensory neurons: a stretch-sensitive neuron in the muscle layer *(white cytoplasm, colored nucleus)* and a mechanosensitive or chemosensitive neuron *(colored cytoplasm, white nucleus)* with its receptive ending in the mucosa. Stimulation of either of these sensory neurons results in activation of ascending (oral) excitatory pathways and descending (anal) pathways to circular muscle *(cm)*. (Courtesy Dr. Terence K. Smith.)

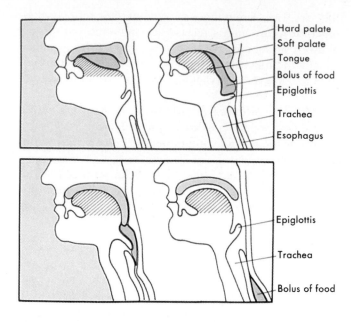

Hard palate
Soft palate
Tongue
Bolus of food
Epiglottis
Trachea
Esophagus

Epiglottis
Trachea
Bolus of food

FIGURE 32-7 Major events involved in the swallowing reflex.

elicits contraction above (oral to) and relaxation below (anal to) the point of stimulation.

CHEWING (MASTICATION)

Chewing can be carried out voluntarily, but it is more frequently a reflex behavior. Chewing lubricates the food by mixing it with salivary mucus, mixes starch-containing food with salivary amylase, and subdivides the food so that it can be mixed more readily with the digestive secretions of the stomach and duodenum.

SWALLOWING

Swallowing can be initiated voluntarily, but thereafter it is almost entirely under reflex control. The swallowing reflex is a rigidly ordered sequence of events that propels food from the mouth to the stomach. At the same time it inhibits respiration and prevents the entrance of food into the trachea (Figure 32-7). The afferent limb of the swallowing reflex begins with touch receptors, most notably those near the opening of the **pharynx.** Sensory impulses from these receptors are transmitted to certain areas in the medulla. The central integrating areas for swallowing lie in the medulla and lower pons; they are collectively called the **swallowing center.** Motor impulses travel from the swallowing center to the musculature of the pharynx and upper esophagus via various cranial nerves and to the remainder of the esophagus by vagal motor neurons.

The Oral Phase

The **oral,** or **voluntary, phase,** of swallowing is initiated by separating a bolus of food from the mass in the mouth with the tip of the tongue. The bolus to be swallowed is moved upward and backward in the mouth by pressing first the tip of the tongue and later the more posterior portions of the tongue as well against the hard palate. This forces the bolus into the pharynx, where the bolus stimulates the tactile receptors that initiate the swallowing reflex.

The Pharyngeal Phase

The **pharyngeal phase** of swallowing involves the following sequence of events, which occur in less than 1 second:

1. The soft palate is pulled upward, and the palatopharyngeal folds move inward toward one another. This prevents reflux of food into the nasopharynx and provides a narrow passage through which food moves into the pharynx.
2. The vocal cords are pulled together. The larynx is moved forward and upward against the epiglottis. These actions prevent food from entering the trachea and help to open the upper esophageal sphincter.
3. The upper esophageal sphincter relaxes to receive the bolus of food (Figure 32-8). Then the pharyngeal superior constrictor muscles contract strongly to force the bolus deeply into the pharynx.
4. A **peristaltic wave** is initiated with contraction of the pharyngeal superior constrictor muscles, and moves toward the esophagus (Figure 32-8). This forces the bolus of food through the relaxed upper esophageal sphincter.

During the pharyngeal stage of swallowing, respiration is reflexly inhibited.

The Esophageal Phase

The esophageal phase of swallowing is controlled mainly by the swallowing center. After the bolus of food passes the upper esophageal sphincter, the sphincter reflexly constricts. A peristaltic wave then begins just below the upper esophageal sphincter and traverses the entire esophagus in less than 10 seconds (Figure 32-8). This initial wave of peristalsis, called **primary peristalsis,** is controlled by the swallowing center. The peristaltic wave travels down the esophagus at 3 to 5 cm/sec. If the primary peristalsis is insufficient to clear the esophagus of food, distension of the esophagus initiates another peristaltic wave, called **secondary peristalsis,** which begins above the site of distension and moves downward. Input from esophageal sensory fibers to the

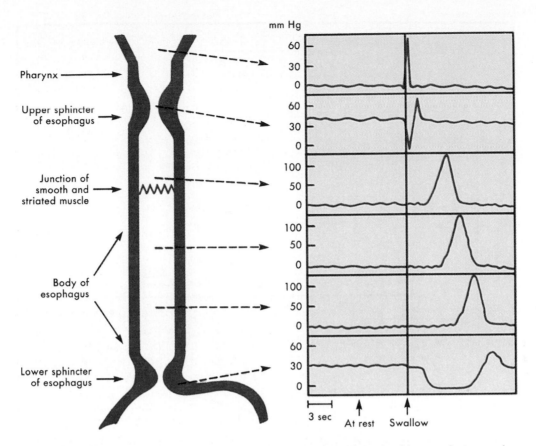

FIGURE 32-8   Pressures in the pharynx, esophagus, and esophageal sphincters during swallowing. Note the reflex relaxation of the upper and lower esophageal sphincters and the timing of the relaxation. (Redrawn from Christensen JL: In Christensen J, Wingate DL, eds: *A guide to gastrointestinal motility,* Bristol, England, 1983, John Wright and Sons.)

central and enteric nervous systems modulates esophageal peristalsis.

ESOPHAGEAL FUNCTION

After food is swallowed, the esophagus functions as a conduit to move the food from the pharynx to the stomach. In the upper third of the esophagus both the inner circular and the outer longitudinal muscle layers are striated. In the lower third, the muscle layers are composed entirely of smooth muscle cells. In the middle third, skeletal and smooth muscles coexist, with a gradient from all skeletal above to all smooth below.

The esophageal musculature, both striated and smooth, is mainly innervated by branches of the vagus nerve. Somatic motor fibers of the vagus form motor endplates on striated muscle fibers. Visceral motor nerves are preganglionic parasympathetic fibers that synapse primarily on the nerve cells of the myenteric plexus. Neurons of the myenteric plexus directly innervate the smooth muscle cells of the esophagus and communicate with one another. The neural circuits that control the esophagus are schematized in Figure 32-9.

The upper esophageal sphincter (UES) and the lower esophageal sphincter (LES) prevent the entry of air and gastric contents, respectively, into the esophagus. The LES opens when a wave of esophageal peristalsis begins (see Figure 32-8). The opening of the LES is mediated by impulses in branches of the vagus nerve. In the absence of esophageal peristalsis the sphincter must remain tightly closed to prevent reflux of the gastric contents, which would cause esophagitis and the sensation of heartburn.

Reflux is particularly problematic because the pressure in the thoracic esophagus is close to intrathoracic pressure, which is almost always less than intraabdominal pressure. The difference between intraabdominal and intrathoracic pressures increases during each inspiration (see Chapter 28). The reflux of gastric contents up into

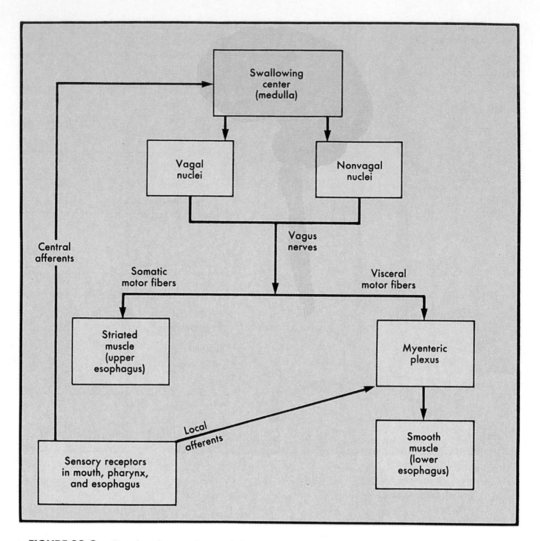

FIGURE 32-9   Local and central neural circuits involved in the control of esophageal motility.

the esophagus is opposed by the LES. In addition, because the crura of the diaphragm wrap around the esophagus at the level of the LES, contraction of the diaphragm helps to increase the pressure in the LES with each inspiration. In individuals with weakness of the diaphragm, and particularly in those with **hiatal hernia,** there is a tendency toward esophagitis caused by increased reflux.

The Lower Esophageal Sphincter

Control of LES Tone   The resting pressure in the LES is about 30 mm Hg. The tonic contraction of the circular musculature of the sphincter is regulated by nerves, both intrinsic and extrinsic, and by hormones and neuromodulators. A significant fraction of basal tone in this sphincter is mediated by vagal cholinergic nerves. Stimu-

lation of sympathetic nerves to the sphincter also causes the LES to contract.

Relaxation of the LES   The intrinsic and extrinsic innervation of the lower esophageal sphincter is both excitatory and inhibitory. A major component of the sphincter's relaxation that occurs in response to primary peristalsis in the esophagus is mediated by vagal fibers inhibitory to the circular muscle of the LES.

In some individuals the sphincter fails to relax sufficiently during swallowing to allow food to enter the stomach, a condition known as **achalasia.** Therapy for achalasia may involve mechanically dilating or surgically weakening the lower esophageal sphincter or therapy with drugs that inhibit the tone of the LES. Individuals with **diffuse esophageal spasm** have prolonged

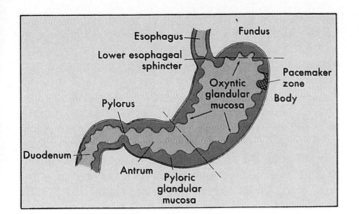

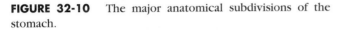

FIGURE 32-10   The major anatomical subdivisions of the stomach.

and painful contraction of the lower part of the esophagus after swallowing instead of the normal esophageal peristaltic wave. In individuals with **incompetence of the LES,** gastric juice can reflux back up into the lower esophagus and cause erosion of the esophageal mucosa.

GASTRIC MOTILITY

The major functions of gastric motility are (1) to allow the stomach to serve as a reservoir for the large volume of food that may be ingested at a single meal, (2) to fragment food into smaller particles and to mix food with gastric secretions so that digestion can begin, and (3) to empty gastric contents into the duodenum at a controlled rate. Figure 32-10 shows the major anatomical subdivisions of the stomach.

The **fundus** and the **body** of the stomach can accommodate volume increases as large as 1.5 liters without a great increase in intragastric pressure; the phenomenon is called **receptive relaxation.** Contractions of the fundus and body are normally weak, so that much of the gastric contents remains relatively unmixed for long periods. Thus the fundus and the body serve the reservoir functions of the stomach. In the antrum, however, contractions are vigorous and thoroughly mix antral chyme with gastric juice and subdivide food into smaller particles. The antral contractions serve to empty the gastric contents in small squirts into the duodenal bulb. The rate of gastric emptying is adjusted by several mechanisms so that chyme is not delivered to the duodenum too rapidly. The physiological mechanisms that underlie this behavior are discussed below.

Structure and Innervation of the Stomach

The basic structure of the gastric wall follows the scheme presented in Figure 32-1. The circular muscle layer of muscularis externa is more prominent than the longitudinal layer. The muscularis externa of the fundus and the body is relatively thin, but that of the antrum is considerably thicker and increases in thickness toward the pylorus.

The stomach is richly innervated by extrinsic nerves and by the neurons of the enteric nervous system. Axons from the cells of the intramural plexuses innervate smooth muscle and secretory cells.

Parasympathetic innervation comes via the vagus nerves and sympathetic innervation from the celiac plexus. In general, parasympathetic nerves stimulate gastric smooth muscle motility and gastric secretions, whereas sympathetic activity inhibits these functions. Numerous sensory afferent fibers leave the stomach in the vagus nerves, and some travel with sympathetic nerves. Other fibers are the afferent links of intrinsic reflex arcs via the intramural plexuses of the stomach. Some of these afferent fibers relay information about intragastric pressure, gastric distension, intragastric pH, or pain.

Responses to Gastric Filling

When a wave of esophageal peristalsis begins, the lower esophageal sphincter reflexly relaxes. This is followed by receptive relaxation of the fundus and body of the stomach. The stomach also will relax if it is directly filled with gas or liquid. The nerve fibers in the vagi are a major efferent pathway for reflex relaxation of the stomach. The vagal fibers that mediate this response release vasoactive intestinal polypeptide (VIP) as their transmitter.

In some patients with duodenal ulcer the vagus nerves to the stomach are sectioned (vagotomy) in order to diminish the rate of gastric acid secretion. This procedure also eliminates the efferent pathway of the receptive relaxation reflex, so that in response to ingestion of a meal, intragastric pressure increases to a much greater level than in normal individuals. This results in accelerated emptying, known as the **dumping syndrome,** of gastric contents. Gastric contents are emptied into the small intestine faster than they can be processed, and thus chronic diarrhea may occur in these patients.

Mixing and Emptying of Gastric Contents

The muscle layers in the fundus and body are thin; weak contractions characterize these parts of the stomach. As a result, the contents of the fundus and the body tend to form layers that are based on their density. The gastric contents may remain unmixed for as long as 1 hour after eating. Fats tend to form an oily layer on top of the other gastric contents. *Consequently, fats are emptied later than other gastric contents.* Liquids can flow around the mass of food contained in the body of the stomach and are emptied more rapidly into the duodenum. Large or indigestable particles are retained in the stomach for a longer period (Figure 32-11).

Gastric contractions, with a frequency of about 3 per minute, usually begin in the middle of the body of the stomach and travel toward the pylorus. The contractions increase in force and velocity as they approach the gastroduodenal junction. As a result, the major mixing activity occurs in the antrum, the contents of which are mixed rapidly and thoroughly with gastric secretions.

Fed Versus Fasted State

After an individual eats, regular contractions of the antrum occur about 3 per minute. As discussed later, the rate of gastric emptying is regulated by feedback mechanisms that diminish the force of antral contractions.

In a fasted animal a different pattern of antral contractions occurs. The antrum is quiescent for 75 to 90 minutes; then a brief period (5 to 10 minutes) of intense electrical and motor activity occurs. This activity is characterized by strong contractions of the antrum with a relaxed pylorus. During this period even large chunks of material that remain from the previous meal are emptied from the stomach. The period of intense contractions is followed by another 75 to 90 minutes of quiescence. This cycle of contractions in the stomach is part of a pattern of contractile activity that periodically sweeps from the stomach to the terminal ileum during fasting. This cyclic contractile activity is known as the **migrating myoelectric complex** (MMC) and is discussed later.

Electrical Activity and Gastric Contractions

The gastric peristaltic waves occur at the frequency of the gastric slow waves that are generated by a **pacemaker zone** (see Figure 32-10) near the middle of the body of the stomach. These waves are conducted toward the pylorus. In humans the frequency of slow waves is about 3 per minute.

The gastric slow wave is triphasic (Figure 32-12). Its shape resembles the action potentials in cardiac muscle. However, the gastric slow wave lasts about 10 times longer than the cardiac action potential and does not overshoot.

Gastric smooth muscle contracts when the depolarization during the slow wave exceeds the threshold for

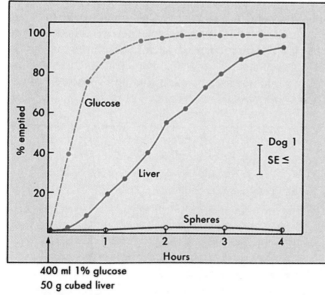

400 ml 1% glucose
50 g cubed liver
40 plastic spheres

FIGURE 32-11   Rates of emptying of different meals from dog stomach. A solution (1% glucose) is emptied faster than a digestible solid (cubed liver). An indigestible solid (7 mm plastic spheres) remains in the stomach under these conditions. (Redrawn from Hinder RA, Kelly KA: *Am J Physiol* 233:E335, 1977.)

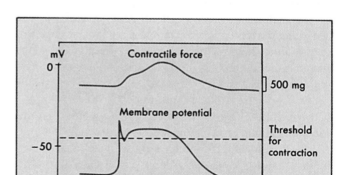
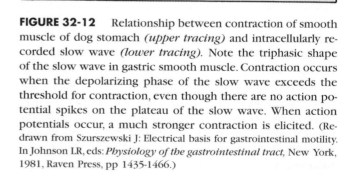

FIGURE 32-12   Relationship between contraction of smooth muscle of dog stomach *(upper tracing)* and intracellularly recorded slow wave *(lower tracing)*. Note the triphasic shape of the slow wave in gastric smooth muscle. Contraction occurs when the depolarizing phase of the slow wave exceeds the threshold for contraction, even though there are no action potential spikes on the plateau of the slow wave. When action potentials occur, a much stronger contraction is elicited. (Redrawn from Szurszewski J: Electrical basis for gastrointestinal motility. In Johnson LR, eds: *Physiology of the gastrointestinal tract*, New York, 1981, Raven Press, pp 1435-1466.)

contraction (Figure 32-12). The greater the extent of depolarization and the longer the cell remains depolarized above the threshold, the greater the force of contraction. In the gastric antrum, action potential spikes may occur during the plateau phase; when action potentials occur, the resulting contraction is much stronger than in the absence of action potentials (see also Figure 32-5). Acetylcholine and the hormone gastrin stimulate gastric contractility by increasing the amplitude and duration of the plateau phase of the gastric slow wave. Norepinephrine has the opposite effect.

Gastroduodenal Junction

The pylorus separates the gastric antrum from the first part of the **duodenum,** the duodenal bulb. The pylorus functions as a sphincter. The circular smooth muscle of the pylorus forms two ringlike thickenings that are followed by a connective tissue ring that separates pylorus from duodenum.

The duodenum has a basic electrical rhythm of 10 to 12 waves per minute, far faster than the 3 per minute of the stomach. The duodenal bulb is influenced by the basic electrical rhythms of both the stomach and the postbulbar duodenum. It thus contracts somewhat irregularly. However, the antrum and duodenum are coordinated; when the antrum contracts, the duodenal bulb relaxes.

The essential functions of the gastroduodenal junction are (1) to allow the carefully regulated emptying of gastric contents at a rate consistent with the ability of the duodenum to process the chyme and (2) to prevent regurgitation of duodenal contents back into the stomach.

> The gastric mucosa is highly resistant to acid but may be damaged by bile. The duodenal mucosa has the opposite properties. Thus too rapid gastric emptying may lead to **duodenal ulcers,** whereas regurgitation of duodenal contents may contribute to **gastric ulcers.**

The pylorus is densely innervated by both vagal and sympathetic nerve fibers. Sympathetic fibers increase the constriction of the sphincter. Vagal fibers are both excitatory and inhibitory to pyloric smooth muscle. Excitatory cholinergic vagal fibers stimulate constriction of the sphincter. Inhibitory vagal fibers release another transmitter, probably VIP, that relaxes the sphincter. The hormones **cholecystokinin, gastrin, gastric inhibitory peptide,** and **secretin** all elicit constriction of the pyloric sphincter.

Regulation of Gastric Emptying

The emptying of gastric contents is regulated by both neural and hormonal mechanisms. The duodenal and jejunal mucosa have receptors that sense acidity, osmotic pressure, and certain fats (Figure 32-13). *The presence of fatty acids or monoglycerides (products of fat digestion) in the duodenum dramatically decreases the rate of gastric emptying.* The chyme that leaves the stomach is usually hypertonic, and it becomes more hypertonic because of the action of the digestive enzymes in the duodenum. *Gastric emptying is retarded by hypertonic solutions in the duodenum, by duodenal pH below 3.5, and by the presence of amino acids and peptides in the duodenum.* As a result of these mechanisms:

1. Fat is not emptied into the duodenum at a rate greater than that at which it can be emulsified by the bile acids and lecithin of the bile.
2. Acid is not dumped into the duodenum more rapidly than it can be neutralized by pancreatic and duodenal secretions and by other mechanisms.
3. The rates at which the other components of chyme are presented to the small intestine do not exceed the rate at which the small intestine can process those components.

Mechanisms that Regulate Gastric Emptying The slowing of gastric emptying in response to components of the content of the duodenum is mediated by neural and hormonal mechanisms.

1. Acid in the duodenum. In response to the acid in the duodenum, the force of gastric contractions promptly decreases and duodenal motility increases. This response has neural and hormonal components. The presence of acid in the duodenum releases **secretin,** which diminishes the rate of gastric emptying by inhibiting antral contractions and stimulating contraction of the pyloric sphincter (Figure 32-13).
2. Fat-digestion products. The presence of fat-digestion products in the duodenum and jejunum decreases the rate of gastric emptying. This response results mainly from the release of **cholecystokinin** from the duodenum and jejunum. Cholecystokinin decreases the rate of gastric emptying. The presence of fatty acids in the duodenum and jejunum releases another hormone, **gastric inhibitory peptide,** that also decreases the rate of gastric emptying.
3. Osmotic pressure of duodenal contents. Hyperosmotic solutions in the duodenum and jejunum slow the rate of gastric emptying. This response has both neural and hormonal components. Hypertonic solutions in the duodenum release an unidentified hormone that diminishes the rate of gastric emptying.

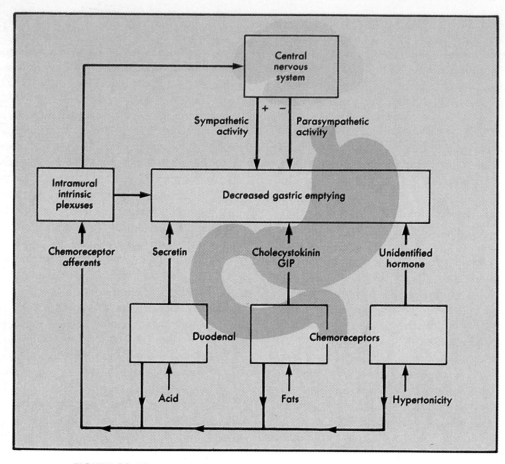

FIGURE 32-13 Neural and hormonal inhibition of gastric emptying.

4. Peptides and amino acids in the duodenum. Peptides and amino acids release **gastrin** from **G cells** located in the antrum of the stomach and the duodenum. Gastrin increases the strength of antral contractions and increases constriction of the pyloric sphincter; the net effect usually diminishes the rate of gastric emptying.

> In patients with duodenal ulcers the underlying physiological malfunction is frequently diminished effectiveness of the mechanisms by which hormones released from the duodenum decrease the rate of gastric emptying and/or gastric acid secretion. In normal individuals, experimental instillation of acid into the duodenum via a nasogastric tube causes a dramatic decrease in the rate and force of contractions of the gastric antrum. In many patients with duodenal ulcer this response to acid in the duodenum is markedly diminished.

VOMITING

Vomiting is the expulsion of gastric (and sometimes duodenal) contents from the gastrointestinal tract via the mouth. Vomiting often is preceded by a feeling of nausea, a rapid or irregular heartbeat, dizziness, sweating, pallor, and dilation of the pupils. Vomiting usually is preceded by **retching,** in which gastric contents are forced up into the esophagus but do not enter the pharynx.

Vomiting is a reflex behavior controlled and coordinated by a **vomiting center** in the medulla oblongata (Figure 32-14). Many areas in the body have receptors that provide afferent input to the vomiting center. Distension of the stomach and duodenum is a strong stimulus that elicits vomiting. Tickling the back of the throat, painful injury to the genitourinary system, dizziness, and certain other stimuli can bring about vomiting.

Certain chemicals, called **emetics,** can elicit vomiting. Some emetics do this by stimulating receptors in the stomach or more often in the duodenum. The widely used emetic **ipecac** stimulates duodenal receptors. Cer-

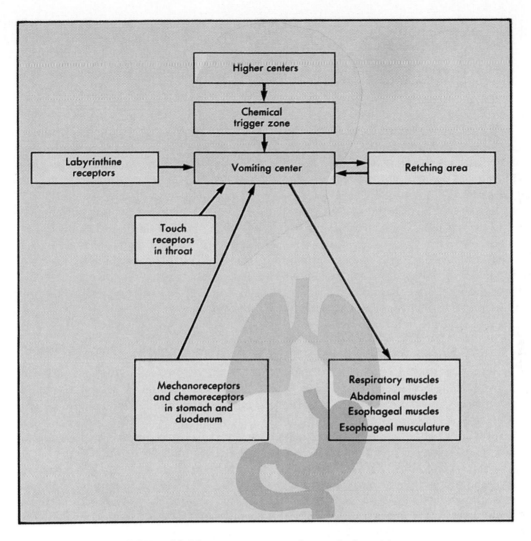

FIGURE 32-14   Some aspects of control of vomiting.

tain other emetics (e.g., **apomorphine**) act at the level of the central nervous system on receptors in the floor of the fourth ventricle, in an area known as the **chemoreceptor trigger zone.** The chemoreceptor trigger zone lies on the blood side of the blood-brain barrier and thus can be reached by most blood-borne substances.

When the **vomiting reflex** *is activated, the sequence of events is the same regardless of the stimulus that initiates the reflex.* Early events in the vomiting reflex include a wave of **reverse peristalsis** that sweeps from the middle of the small intestine to the duodenum. The pyloric sphincter and the stomach relax to receive intestinal contents. Then a forced inspiration occurs against a closed glottis. This decreases intrathoracic pressure, whereas the lowering of the diaphragm increases intraabdominal pressure. Then a forceful contraction of

abdominal muscles sharply elevates intraabdominal pressure, driving gastric contents into the esophagus. The lower esophageal sphincter relaxes reflexly to receive gastric contents, and the pylorus and antrum contract reflexly. When a person retches, the upper esophageal sphincter remains closed, preventing vomiting. When the respiratory and abdominal muscles relax, the esophagus is emptied by secondary peristalsis into the stomach. Often a series of stronger and stronger retches precedes vomiting.

When a person vomits, the rapid propulsion of gastric contents into the esophagus is accompanied by a reflex relaxation of the upper esophageal sphincter. **Vomitus** is projected into the pharynx and mouth. Entry of vomitus into the trachea is prevented by approximation of the vocal chords, closure of the glottis, and inhibition of respiration.

MOTILITY OF THE SMALL INTESTINE

The small intestine accounts for about three fourths of the length of the human gastrointestinal tract. The small intestine is about 5 meters in length, and chyme typically takes 2 to 4 hours to traverse it.

The first 5% or so of the small intestine is the duodenum, which has no mesentery and can be distinguished from the rest of the small intestine histologically. The remaining small intestine is divided into the jejunum and the ileum. The **jejunum** is more proximal and occupies about 40% of the length of the small bowel. The **ileum** is the distal part of the small intestine and accounts for its remaining length.

The small intestine, particularly the duodenum and the jejunum, is the site of most digestion and absorption. The movements of the small intestine mix chyme with digestive secretions, bring fresh chyme into contact with the absorptive surface of the microvilli, and propel chyme toward the colon.

The most frequent type of movement of the small intestine is termed **segmentation.** Segmentation (Figure 32-15) is characterized by closely spaced contractions of the circular muscle layer. The contractions divide the small intestine into small neighboring segments. In rhythmic segmentation the sites of the circular contractions alternate, so that a given segment of gut contracts and then relaxes. Segmentation effectively mixes chyme with digestive secretions and brings fresh chyme into contact with the mucosal surface.

Peristalsis is the progressive contraction of successive sections of circular smooth muscle. The contractions move along the gastrointestinal tract in an orthograde direction. Peristaltic waves occur in the small intestine, but they usually involve only a short length of intestine. As in other parts of the digestive tract, the slow waves of the smooth muscle cells determine the timing of intestinal contractions.

Electrical Activity of Small Intestinal Smooth Muscle

Regular slow waves occur all along the small intestine. The frequency is highest (11 to 13 per minute in humans) in the duodenum and declines along the length of the small bowel (to a minimum of 8 or 9 per minute in humans in the terminal part of the ileum). The slow waves may or may not be accompanied by bursts of action potential spikes. When action potentials occur, they elicit much stronger contractions of the smooth muscle that cause the major mixing and propulsive movements of the small intestine (see Figure 32-5). Action potential bursts are localized to short segments of the intestine, and they elicit the highly localized contractions of the circular smooth muscle that cause segmentation.

The basic electrical rhythm of the small intestine is independent of the extrinsic innervation. The frequency of the action potential spike bursts that elicit strong contractions depends on the excitability of the smooth muscle cells of the small intestine, which is influenced by circulating hormones, the autonomic nervous system, and enteric neurons. Excitability is enhanced by parasympathetic nerves and is inhibited by sympathetic nerves, both acting via the intramural plexuses. Even though much of the direct control of intestinal motility resides in the intramural plexuses, the parasympathetic and sympathetic innervation of the small intestine modulates contractile activity. The extrinsic neural circuits are essential for certain long-range intestinal reflexes discussed later.

Contractile Behavior of the Small Intestine

Contractions of the duodenal bulb mix chyme with pancreatic and biliary secretions, and they propel the chyme along the duodenum. Contractions of the duodenal bulb typically follow contractions of the gastric antrum. This helps prevent regurgitation of duodenal contents back into the stomach.

Segmentation is the most frequent type of movement by the small intestine (Figure 32-15). Segmental contractions occurs at about 11 or 12 per minute in the duodenum and 8 or 9 per minute in the ileum.

In the jejunum, contractile activity usually occurs in bursts, which are separated by an interval when contractions are weak or absent. Because the bursts are approximately 1 minute apart, the pattern has been termed the **minute rhythm** of the jejunum (Figure 32-16).

Short-range peristalsis also occurs in the small intestine, although much less frequently than segmentation. The relatively low rate of net propulsion of chyme in the small intestine allows time for digestion and absorption.

The importance of the slow rate of propulsion in the small intestine can be demonstrated by treatment with agents that alter small intestinal motility. For example administration of **codeine** and other **opiates** typically markedly reduces the frequency and the volume of stools. This results from a decrease in small intestinal motility that increases the transit time of jejunal contents. The longer transit time allows more complete absorption of salts and

A

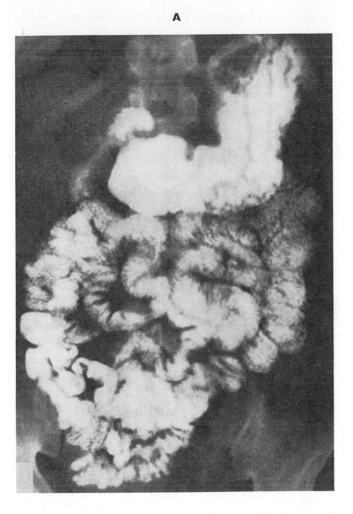

B

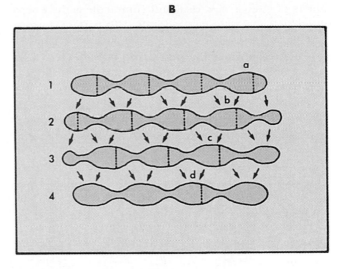

FIGURE 32-15 **A,** X-ray view showing the stomach and small intestine filled with barium contrast medium in a normal individual. Note that segmentation of the small intestine divides its contents into ovoid segments. **B,** The sequence of segmental contractions in a portion of cat's small intestine. Lines *1* through *4* indicate successive patterns in time. The dotted lines indicate where contractions will occur next. The arrows show the direction of chyme movement. (**A** from Gardner EM et al: *Anatomy, a regional study of human structure,* ed 4, Philadelphia, 1975, WB Saunders Co. **B** redrawn from Cannon WB: *Am J Physiol* 6:251, 1902.)

water and certain nutrients in the small intestine, so that less than a normal volume enters the colon. **Castor oil,** a potent laxative, contains hydroxy fatty acids that stimulate small intestinal motility and decrease small intestinal transit time. Hence, salts and water are delivered to the colon at a rate that overwhelms the ability of the colon to absorb them; this results in diarrhea.

Intestinal Reflexes

When a bolus of material is placed in the small intestine, the intestine typically contracts behind the bolus and relaxes ahead of it (see Figure 32-6), a response known as the **law of the intestine.** This may propel the bolus in an orthograde direction, similar to a peristaltic wave.

Certain intestinal reflexes can occur along a consid-

erable length of the gastrointestinal tract. These long-range reflexes depend on the function of both intrinsic and extrinsic nerves.

Overdistension of one segment of the intestine relaxes the smooth muscle in the rest of the intestine. This response is known as the **intestinointestinal reflex.**

The stomach and the terminal part of the ileum interact reflexly. Elevated secretory and motor functions of the stomach increase the motility of the terminal part of the ileum and accelerate the movement of material through the **ileocecal sphincter.** This response is called the **gastroileal reflex.**

Migrating Myoelectric Complex

The contractile behavior of the small intestine previously discussed is characteristic of the period after ingestion of a meal. In a fasted individual or some hours after the processing of the previous meal, small intestinal motility follows a different pattern, characterized by bursts of intense electrical and contractile activity separated by longer quiescent periods. This pattern is propagated from the stomach to the terminal ileum (Figure 32-17) and is known as the **migrating myoelectric complex** (MMC). The MMC in the stomach is discussed earlier in this chapter.

The MMC repeats every 75 to 90 minutes in humans (Figure 32-17). About the time that one MMC reaches the distal ileum, the next MMC begins in the stomach.

The strongest contractions of the MMC, both in the stomach and in the small intestine, are more vigorous and more propulsive than the contractions that occur in the fed individual. These intense contractions sweep the small bowel clean and empty its contents into the colon. Thus the MMC has been termed the "housekeeper of the small intestine."

> The MMC inhibits the migration of colonic bacteria into the terminal ileum. Individuals with weak or absent phase 3 contractions are frequently troubled by bacterial overgrowth in the ileum.

Contractile Activity of the Muscularis Mucosae

Sections of the muscularis mucosae contract irregularly at a rate of about 3 contractions per minute. These contractions alter the pattern of ridges and folds of the mucosa, mix the luminal contents, and bring different parts of the mucosal surface into contact with freshly mixed chyme. The villi contract irregularly, especially in the proximal part of the small intestine. This helps to empty the central lacteals of the villi and increases intestinal lymph flow.

Emptying the Ileum

The **ileocecal sphincter** separates the terminal end of the ileum from the **cecum,** the first part of the colon. Normally the sphincter is closed, but short-range peristalsis in the terminal part of the ileum relaxes the sphincter and allows a small amount of chyme to squirt into the cecum. The ileocecal sphincter normally allows ileal chyme to enter the colon at a slow enough rate so that the colon can absorb most of the salts and water of the chyme. The ileocecal sphincter

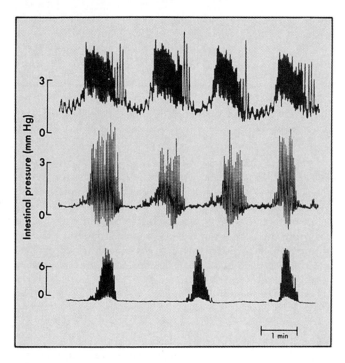

FIGURE 32-16   The minute rhythm recorded from the jejunums of three different ferrets in the fed state. Similar minute rhythms occur in the human jejunum. The pattern of the minute rhythm varies from one ferret to another and varies in time and with location in the small intestine in the same ferret. (From Collman PI et al: *J Physiol [Lond]* 345:65, 1983.)

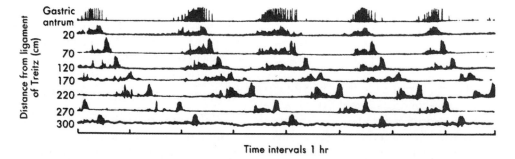

FIGURE 32-17   The occurrence of the migrating myoelectric complex in the stomach and small intestine of a fasting dog. The ligament of Treitz marks the border between the duodenum and the jejunum. (From Itoh Z, Sekiguchi T: *Scand J Gastoenterol [Suppl]* 82:121, 1983.)

is coordinated primarily by the neurons of the intramural plexuses.

MOTILITY OF THE COLON

The colon receives 500 to 1500 ml of chyme per day from the ileum. Most of the salts and water that enter the colon are absorbed; the feces normally contain only about 50 to 100 ml of water each day. Colonic contractions mix the chyme and circulate it across the mucosal surface of the colon. As the chyme becomes semisolid, this mixing resembles a kneading process. The progress of colonic contents is slow, about 5 to 10 cm/hr at most.

One to three times daily a wave of contraction, called a **mass movement,** occurs. A mass movement resembles a peristaltic wave in which the contracted segments remain contracted for some time. Mass movements push the contents of a significant length of colon in an orthograde direction.

Structure and Innervation of the Colon

As shown in Figure 32-18, the major subdivisions of the large intestine are the cecum, the ascending colon, the transverse colon, the descending colon, the sigmoid colon, the rectum, and the anal canal.

The structure of the wall of the large bowel follows the general plan presented earlier in this chapter, but the longitudinal muscle layer of the muscularis externa is concentrated into three bands called the **taenia coli.** In between the taenia coli the longitudinal muscle layer is thin. The longitudinal muscle of the rectum and anal canal is substantial and continuous.

Parasympathetic innervation of the cecum and the as-

cending and transverse colon is via branches of the vagus nerve; that of the descending and sigmoid colon, the rectum, and the anal canal is via the pelvic nerves from the sacral spinal cord. The parasympathetic fibers end primarily on neurons of the intramural plexuses. Sympathetic fibers innervate the proximal part of the large intestine via the superior mesenteric plexus, the distal part of the large intestine via the inferior mesenteric and superior hypogastric plexuses, and the rectum and anal canal via the inferior hypogastric plexus.

Stimulation of the sympathetic nerves stops colonic movements. Vagal stimulation causes segmental contractions of the proximal part of the colon. Stimulation of the pelvic nerves brings about expulsive movements of the distal colon and sustained contraction of some segments.

The anal canal usually is kept closed by the internal and external sphincters. The **internal anal sphincter** is a thickening of the circular smooth muscle of the anal canal. The **external anal sphincter** is more distal, and it consists entirely of striated muscle. The external anal sphincter is innervated by somatic motor fibers via the pudendal nerves, which allow it to be controlled both reflexly and voluntarily.

Motility of the Cecum and Proximal Colon

Most contractions of the cecum and proximal part of the large bowel are segmental and are more effective at mixing and circulating the contents than at propelling them. The mixing action facilitates absorption of salts and water by the mucosal epithelium.

Localized segmental contractions divide the colon into neighboring ovoid segments, called **haustra** (Figure 32-19). Thus, segmentation in the colon is known as **haustration.** The most dramatic difference between haustration and the segmentation that occurs in the small intestine is the regularity of the segments (haustra) produced by haustration and the large length of bowel involved in haustration at one time. Haustral contractions result in back-and-forth mixing of luminal contents.

In the proximal colon "antipropulsive" patterns occur. Reverse peristalsis and segmental propulsion toward the cecum both take place; consequently chyme is retained in the proximal colon, thus facilitating the absorption of salts and water there.

Motility of the Central and Distal Colon

Normally the distal part of the colon is filled with semisolid feces by a mass movement. Segmental contractions knead the feces, facilitating absorption of remaining salts and water. About one to three times daily, mass movements occur and sweep the feces toward the rectum.

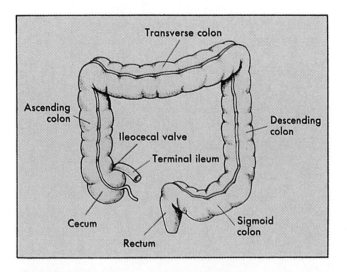

FIGURE 32-18   Major anatomic subdivisions of the colon.

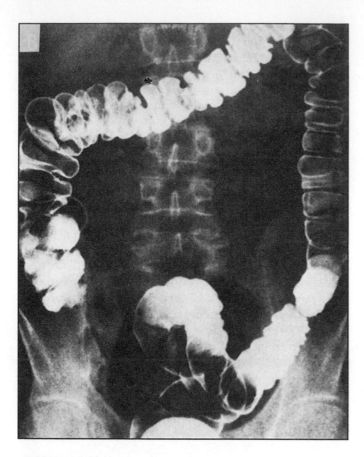

FIGURE 32-19 X-ray image showing a prominent haustral pattern in the colon of a normal individual. (From Keats TE: An atlas of normal roentgen variants, ed 2, Chicago, Year Book Medical Publishers, Inc.)

Control of Colonic Motility

As in other segments of the gastrointestinal tract, the intramural plexuses control the contractile behavior of the colon, and the extrinsic autonomic nerves to the colon are modulatory. The **defecation reflex,** discussed later, is an exception to this rule: it requires the function of the spinal cord via the pelvic nerves.

Electrophysiology of the Colon

Circular Muscle There are two classes of rhythm-generating cells in the colon. Interstitial cells near the inner border of the circular muscle produce regular slow waves with a frequency about 6 per minute. The slow waves have high amplitude, and their shape resembles that of gastric slow waves. Interstitial cells near the outer border of the circular muscle produce **myenteric potential oscillations,** which are low in amplitude and much higher in frequency than the slow waves.

The circular muscle does not usually fire action potentials. Contractile agonists, such as acetylcholine re-

leased from excitatory enteric motor neurons, enhance contractions by increasing the duration of some of the slow waves. The longer slow waves elicit contractions of the circular muscle.

Longitudinal Muscle Longitudinal colonic muscle displays the myenteric potential oscillations. In contrast with the circular smooth muscle, the longitudinal muscle cells fire occasional action potentials at the peaks of the myenteric potential oscillations. The action potentials elicit contraction of the longitudinal muscle. Contractile agonists increase the frequency of action potentials.

Reflex Control of Colonic Motility

Distension of one part of the colon reflexly relaxes other parts of the colon. This **colonocolonic reflex** is mediated partly by the sympathetic fibers that supply the colon. The motility of proximal and distal colon and the frequency of mass movements increase reflexly via the **gastrocolic reflex** after a meal enters the stomach.

The Rectum and Anal Canal

The rectum is usually empty, or nearly so. The rectum is more active in segmental contractions than is the sigmoid colon, so that the rectal contents tend to move retrogradely into the sigmoid colon. The anal canal is tightly closed by the anal sphincters. Before defecation the rectum is filled as a result of a mass movement in the sigmoid colon. Filling the rectum brings about reflex relaxation of the internal anal sphincter and reflex constriction of the external anal sphincter and causes the urge to defecate. Persons who lack functional motor nerves to the external anal sphincter defecate involuntarily when the rectum is filled. The reflex reactions of the sphincters to rectal distension are transient. If defecation is postponed, the sphincters regain their normal tone, and the urge to defecate temporarily subsides.

In **Hirschsprung's disease,** also known as **congenital megacolon,** enteric neurons are congenitally absent from part of the colon. Usually only the internal anal sphincter and a short length of colon proximal to it are involved, but larger segments of the colon may be affected. In the normal person, filling of the rectum by a mass movement leads to reflex relaxation of the distal rectum and the internal anal sphincter. In the absence of enteric neurons, this reflex relaxation cannot occur; this results in functional obstruction of the distal colon and dilation of the colon above the obstruction.

Defecation

When an individual feels the circumstances are appropriate, he voluntarily relaxes the external anal sphincter to allow defecation to proceed. Defecation is a complex behavior involving both reflex and voluntary actions. The integrating center for the reflex actions is in the sacral spinal cord and is modulated by higher centers. The principal efferent pathways are cholinergic parasympathetic fibers in the pelvic nerves. The sympathetic nervous system does not play a significant role in normal defecation.

Voluntary actions are important in defecation. The external anal sphincter is voluntarily held in the relaxed state. Intraabdominal pressure is elevated to aid in expulsion of feces. Evacuation is normally preceded by a deep breath, which moves the diaphragm downward. The glottis is then closed, and contractions of the respiratory muscles on full lungs elevates both the intrathoracic and the intraabdominal pressure. Contractions of the muscles of the abdominal wall further increase intraabdominal pressure, which may be as great as 200 cm H_2O. This helps to force feces through the relaxed sphincters. The muscles of the pelvic floor are relaxed to allow the floor to drop. This helps to straighten out the rectum and prevent rectal prolapse.

SUMMARY

1. The gastrointestinal tract has a characteristic layered structure consisting of mucosa, submucosa, muscularis externa, and serosa; this structure varies somewhat from one segment to another.
2. The gastrointestinal tract receives both sympathetic and parasympathetic innervation. Autonomic nerves influence the motor and secretory activities of the gastrointestinal tract and regulate the caliber of blood vessels of the gastrointestinal tract.
3. Contractions of the smooth muscle of the muscularis externa mix and propel the contents of the gastrointestinal tract.
4. Gastrointestinal smooth muscle cells are electrically coupled, and their resting membrane potential oscillates with a rhythm that is characteristic of each segment of the gastrointestinal tract. The membrane potential oscillations, called slow waves, control the timing and the force of contractions of gastrointestinal smooth muscle.
5. The nerve plexuses of the gastrointestinal tract, the enteric nervous system, contain about 10^8 neurons, as many as in the spinal cord. The enteric nervous system contains motor neurons, sensory neurons, and interneurons. Enteric sensory neurons function as the afferent arms of enteric reflex arcs by which the enteric nervous system controls most of the motor and secretory activities of the gastrointestinal tract. The autonomic nervous system modulates the activities of the enteric nervous system.
6. Swallowing is a reflex coordinated by a swallowing center in the medulla and pons. The swallowing reflex is initiated by touch receptors in the pharynx. This reflex involves a series of ordered and coordinated motor impulses to the muscles of the pharynx, upper esophageal sphincter, esophageal striated muscle, esophageal smooth muscle, and lower esophageal sphincter.
7. Contractions of the stomach mix food with gastric juice and mechanically subdivide the food. Gastric emptying is closely regulated, which ensures that gastric contents are not emptied into the duodenum at a rate faster than the duodenum and jejunum can neutralize the gastric acid and process the chyme.
8. Hormonal and neural mechanisms initiated by the presence of acid, fats, peptides, and hypertonicity in the duodenum regulate gastric emptying.
9. Segmentation is the major contractile activity in the small intestine. Segmental contractions mix and circulate intestinal contents, but are not very propulsive. The slow rate of transport of intestinal contents allows adequate time for digestion and absorption.
10. In a fasted individual a different pattern of motility, called the migrating myoelectric complex (MMC), occurs. The MMC is characterized by 70- to 90-minute periods of quiescence interrupted by periods of vigorous and intensely propulsive contractions that last 3 to 6 minutes. The MMC sweeps the stomach and the small intestine clear of any debris left from the previous meal.
11. In the proximal colon antipropulsive contractions predominate, which allows time for absorption of salts and water. In the transverse and descending colon haustral contractions mix and knead colonic contents to facilitate extraction of salts and water. Mass movements that occur in the colon one to three times daily sweep colonic contents toward the anus.
12. Filling the rectum with feces initiates the defecation reflex. The integrating center for the defecation reflex is in the sacral spinal cord, and the pelvic nerves are the principal motor pathway that regulates the actions of the distal colon, the rectum, the anal canal, and the internal and external anal sphincters in defecation. Both reflex and voluntary activities are involved in defecation.

BIBLIOGRAPHY
Journal Articles

Bornstein JC, Furness JB: Correlated electrophysiological and histochemical studies of submucous neurons and their contribution to understanding enteric neural circuits, *J Autonom Nerv Syst* 25:1, 1988.

Bornstein JC, Furness JB, Smith TK, Trussell DC: Synaptic responses evoked by mechanical stimulation of the mucosa in morphologically characterized myenteric neurons of the guinea-pig ileum, *J Neurosci* 11:505, 1991.

Bywater RAB, Taylor GS, Furukawa K: The enteric nervous system in the control of motility and secretion, *Digest Dis* 5:193, 1987.

Costa M, Brookes S, Steele P, Vickers J: Chemical coding of neurons in the gastrointestinal tract, *Adv Exptl Med Biol* 298:17, 1991.

Furness JB, Bornstein JC, Murphy R., Pompolo S.: Roles of peptides in the enteric nervous system, *Trends Neurosci* 15:66, 1992.

Furness JB, Bornstein JC, Smith TK, Murphy R, Pompolo, S: Correlated functional and structural analysis of enteric neural circuits, *Arch Histol Cytol* 52(Suppl):161, 1989.

Furness JB, Bornstein JC, Trussell DC: Shapes of nerve cells in the myenteric plexus of the guinea-pig small intestine revealed by the intracellular injection of dye, *Cell Tissue Res* 254:561, 1988.

Hara Y, Kubota Y, Szurszewski JH: Electrophysiology of smooth muscle of the small intestine of some mammals, *J Physiol* 372:501, 1986.

Lang IM: Digestive tract motor correlates of vomiting and nausea, *Can J Physiol Pharmacol* 68:242, 1990.

Langton P et al: Spontaneous electrical activity of interstitial cells of Cajal isolated from canine proximal colon, *Proc Nat Acad Sci USA* 86:7280, 1989.

Sanders KM: Ionic mechanisms of electrical rhythmicity in gastrointestinal smooth muscles, *Annu Rev Physiol* 54:439, 1992

Smith TK, Bornstein JC, Furness JB: Distension-evoked ascending and descending reflexes in the circular muscle of guinea-pig ileum: an intracellular study, *J Autonom Nerv Syst* 29:203, 1990.

Smith TK, Bornstein JC, Furness, JB: Interactions between reflexes evoked by distension and mucosal stimulation: electrophysiological studies of guinea-pig ileum, *J Autonom Nerv Syst* 34:69, 1991.

Smith TK, Reed JB, Sanders KM: Interaction of two electrical pacemakers in muscularis of canine proximal colon, *Am J Physiol* 252:C290, 1987.

Wood JD: Enteric neurophysiology, *Am J Physiol* 247:G585, 1984.

Books and Monographs

Christensen J, Wingate DL, eds: *A guide to gastrointestinal motility,* Bristol, England, 1983, John Wright and Sons.

Conklin JL, Christensen J: Motor functions of the pharynx and esophagus. In: Johnson RL, ed: *Physiology of the gastrointestinal tract,* ed 3, New York, 1994, Raven Press.

Davenport HW: *Physiology of the digestive tract,* ed 5, Chicago, 1985, Year Book Medical Publishers.

Furness JB, Bornstein JC: The enteric nervous system and its extrinsic connections. In Yamada T, ed: *Textbook of gastroenterology,* vol 1, Philadelphia, 1991, JB Lippincott Co.

Furness JB, Costa M: *The enteric nervous system,* Edinburgh, 1987, Churchill Livingstone.

Gabella G: Structure of muscles and nerves in the gastrointestinal tract. In Johnson RL, ed: *Physiology of the gastrointestinal tract,* ed 3, New York, 1994, Raven Press.

Grundy D: *Gastrointestinal motility: the integration of physiological mechanisms,* Lancaster, England, 1985. MTP Press.

Kamm MA, Lennard-Jones JE, eds: *Gastrointestinal transit,* Petersfield, 1991, Wrightson Biomedical Publishing Ltd.

Makhlouf GM, ed: *Handbook of physiology,* section 6: The gastrointestinal system, vol II. Neural and endocrine biology, Bethesda, Md, 1989, American Physiological Society.

Makhlouf GM: Neuromuscular function of the small intestine. In Johnson RL, ed: *Physiology of the gastrointestinal tract,* ed 3, New York, 1994, Raven Press.

Mayer EM: The physiology of gastric storage and emptying. In Johnson RL, ed: *Physiology of the gastrointestinal tract,* ed 3, New York, 1994, Raven Press.

Sanders KM, Smith TK: Electrophysiology of colonic smooth muscle. In Wood JD, ed: *Handbook of physiology,* section 6: The gastrointestinal system, vol I, part 1, Bethesda, Md, 1989, American Physiological Society.

Wood JD: Electrical and synaptic behavior of enteric neurons, In Wood JD, ed: *Handbook of physiology,* section 6: The gastrointestinal system, vol I, part 1, Bethesda, Md, 1989, American Physiological Society.

Gastrointestinal Secretions

This chapter deals with the glandular secretion of fluids and compounds that have important functions in the digestive tract. In particular, the secretions of salivary glands, gastric glands, the exocrine pancreas, and the liver are considered. In each case the nature of the secretions and their functions in digestion are discussed, and the regulation of the secretory processes is emphasized.

The secretions just mentioned are elicited by the action of specific effector substances on the secretory cells. These substances may be classified as neurocrine, endocrine, or paracrine (see Chapter 5).

A substance that stimulates a particular cell to secrete is called a **secretagogue.** Many secretagogues exist, but there are few signal transduction mechanisms (discussed in Chapter 6) by which these substances elicit secretion.

SECRETION OF SALIVA

In humans the salivary glands produce about 1 liter of saliva each day. Saliva lubricates food for greater ease of swallowing and also facilitates speaking.

> In people who lack functional salivary glands, **xerostomia** (dry mouth), **dental caries,** and infections of the buccal mucosa are prevalent. Saliva contains **secretory immunoglobulins** (antibodies) directed against microorganisms in the mouth. In the absence of these antibodies, organisms that cause buccal infections and dental caries proliferate. The basic pH of saliva also helps prevent dental caries.

Functions of Saliva

Mucins, which are glycoproteins produced by the submaxillary and sublingual glands, lubricate food so that it may be more readily swallowed. The major digestive function of saliva is the action of **salivary amylase** on starch. Salivary amylase is an enzyme with the same specificity as the α-amylase of pancreatic juice; it reduces starch to oligosaccharide molecules. The pH optimum of salivary amylase is about 7, but it is active between pH 4 and 11. Amylase action continues in the mass of food in the stomach and is terminated only when the contents of the antrum are mixed with enough gastric acid to lower the pH to less than 4. More than half the starch in a well-chewed meal may be reduced to small oligosaccharides by the action of salivary amylase. However, because of the large capacity of the pancreatic α-amylase to digest starch in the small intestine, starch is well-absorbed even in the absence of salivary amylase.

Structure of Salivary Glands

In humans the **parotid glands,** the largest salivary glands, are entirely serous. Their watery secretion lacks mucins. The **submaxillary** and **sublingual glands** are mixed mucous and serous glands, and they secrete a more viscous saliva that contains mucins. Many smaller salivary glands are present in the oral cavity. The microscopic structure of a mixed salivary gland is depicted in Figure 33-1. The **serous acinar cells,** located in the **secretory end-pieces** (also called **acini**), have apical **zymogen granules** that contain salivary amylase and perhaps certain other salivary proteins as well (Figure 33-2). **Mucous acinar cells** secrete glycoprotein mucins into the saliva. **Intercalated ducts** drain the acinar fluid into somewhat larger ducts, the **striated ducts,** which empty into still larger **excretory ducts.** A single large duct brings the secretions of each major gland into the mouth.

A **primary secretion** is elaborated in the secretory end-pieces, and the cells that line the ducts modify the primary secretion.

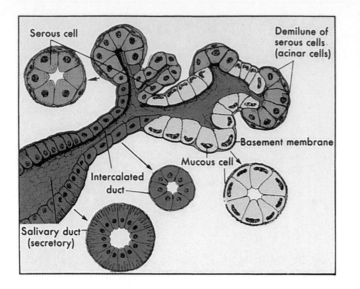

FIGURE 33-1 The structure of the human submandibular gland, as seen with the light microscope. (Redrawn from Braus H: *Anatomie des Menschen,* Berlin, 1934, Julius Springer.)

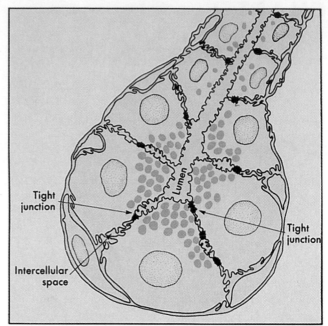

FIGURE 33-2 A schematic representation of the cellular morphology of a secretory end-piece of a serous salivary gland. Secretory canaliculi drain into the acinar lumen. Tight junctions separate the secretory canaliculi from the lateral intercellular spaces. The small colored circles represent zymogen granules. Upon stimulation the contents of zymogen granules are released by exocytosis into the lumen of the acinus. (Redrawn from Young JA, van Lennep LW: *Morphology of salivary glands,* London, 1978, Academic Press.)

Metabolism and Blood Flow of Salivary Glands

The salivary glands produce a prodigious flow of saliva. The maximal rate in humans is about 1 ml/min/g of gland; *at this rate the glands are producing their own weight in saliva each minute!* Salivary glands have a high rate of metabolism and a high blood flow; both are proportional to the rate of saliva formation. The blood flow to maximally secreting salivary glands is approximately 10 times that of an equal mass of actively contracting skeletal muscle. Stimulation of the parasympathetic nerves to salivary glands increases blood flow by dilation of the vasculature of the glands. **Vasoactive intestinal polypeptide** (VIP) and **acetylcholine** are released from parasympathetic nerve terminals in the salivary glands; both of these compounds contribute to vasodilation during secretory activity.

Secretion of Saliva

Ionic Composition of Saliva In humans saliva is always hypotonic to plasma. As shown in Figure 33-3, salivary concentrations of Na⁺ and Cl⁻ are less than those of plasma. The greater the secretory flow rate, the higher is the tonicity of the saliva; at maximal flow rates the tonicity of saliva in humans is about 70% of that of plasma. The pH of saliva from resting glands is slightly acidic. During active secretion, however, the saliva becomes basic, with the pH near 8. The increase in pH with secretory flow rate is partly caused by the increase in salivary bicarbonate concentration.

Secretion of Water and Electrolytes A two-stage model of salivary secretion (Figure 33-4) postulates that:

1. The secretory end-pieces, perhaps with the participation of intercalated ducts, produce a primary secretion that is isotonic to plasma. The amylase concentration and the rate of fluid secretion vary with the level and type of stimulation. However, the electrolyte composition of the secretion is fairly constant, and the levels of Na⁺, K⁺, and Cl⁻ are close to plasma levels.
2. The excretory ducts, and probably the striated ducts as well, modify the primary secretion by extracting Na⁺ and Cl⁻ from and adding K⁺ and HCO3⁻ to the saliva. The ducts do not add to the volume of saliva.

As saliva flows down the ducts, it becomes progressively more hypotonic. Thus the ducts remove more ions from saliva than they contribute to it. The faster the flow rate of the saliva down the striated and excretory ducts, the closer to isotonicity is the saliva.

Secretion of Salivary Amylase In their apical cytoplasm, serous acinar cells have zymogen granules (see Figure 33-2) that contain salivary amylase. When the

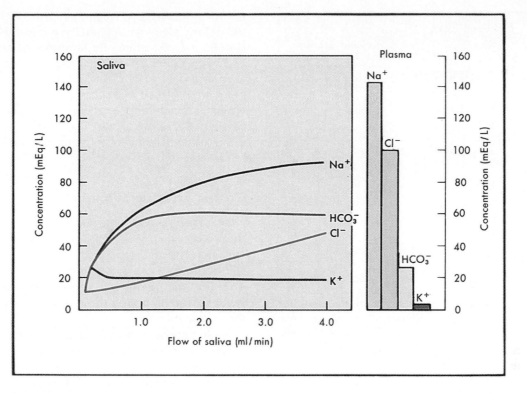

FIGURE 33-3 Average composition of the parotid saliva as a function of the salivary flow rate. Saliva is hypotonic to plasma at all flow rates. The bicarbonate level in saliva exceeds that in plasma, except at very low flow rates. (Redrawn from Thaysen JH et al: *Am J Physiol* 178:155, 1954).

gland is stimulated to secrete, the zymogen granules fuse with the plasma membrane and release their contents into the lumen of an acinus by exocytosis.

Neural Control of Salivary Gland Function

Excitation of either sympathetic or parasympathetic nerves to the salivary glands stimulates salivary secretion, but the effects of the parasympathetic nerves are stronger and more long lasting. Interruption of the sympathetic nerves causes no major defect in the function of the salivary glands. If the parasympathetic supply is interrupted, however, the salivary glands atrophy. *The essential physiological control is by way of the parasympathetic nervous system.*

Sympathetic fibers to the salivary glands come from the superior cervical ganglion. Preganglionic parasympathetic fibers come via branches of the facial and glossopharyngeal nerves (cranial nerves VII and IX, respectively), and they synapse with postganglionic neurons in or near the salivary glands. The acinar cells and ducts are supplied with parasympathetic nerve endings.

Parasympathetic stimulation increases the synthesis and secretion of salivary amylase and mucins, enhances the transport activities of the ductular epithelium, greatly increases blood flow to the glands, and stimulates glandular metabolism and growth.

The increase in salivary secretion that results from stimulation of sympathetic nerves is transient. Sympathetic stimulation constricts blood vessels, with consequent reductions in salivary gland blood flow.

Cellular Mechanisms of Salivary Secretion

Duct Cells The ducts of salivary glands respond to both cholinergic and adrenergic agonists by increasing their rates of secretion of K^+ and HCO_3^-.

Acinar Cells The neuroeffector substances that stimulate acinar cell secretions act mainly by elevating intracellular cAMP or by increasing the level of Ca^{++} in the cytosol (Figure 33-5).

Acetylcholine, norepinephrine, substance P, and VIP are released in salivary glands by specific nerve terminals. Each of these neuroeffectors may increase the secretion of salivary amylase and the flow of saliva.

Norepinephrine acting on β-receptors and VIP elevates cAMP in acinar cells. By contrast, acetylcholine, substance P, and activation of α-receptors by norepinephrine increase intracellular Ca^{++}.

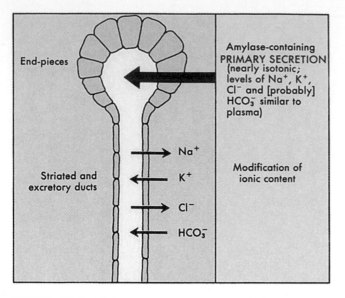

FIGURE 33-4 Schematic representation of the two-stage model of salivary secretion. The primary secretion, containing salivary amylase, is secreted by the acinar cells. The striated and excretory ducts modify the composition of saliva.

GASTRIC SECRETION

Structure of the Gastric Mucosa

The surface of the gastric mucosa (Figure 33-6) is covered by columnar **epithelial cells** that secrete mucus and an alkaline fluid that protects the epithelium from mechanical injury and gastric acid. The surface is studded with **gastric pits;** each pit is the opening of a duct into which one or more **gastric glands** empty (Figure 33-6, *A*). *The gastric pits are so numerous that they account for a significant fraction of the total surface area of the gastric mucosa.*

The gastric mucosa can be divided into three regions, based on the structures of the glands present. The small **cardiac glandular region,** just below the lower esophageal sphincter, contains primarily mucus-secreting gland cells. The remainder of the gastric mucosa is divided into the **oxyntic** (acid-secreting) **glandular region,** above the notch, and the **pyloric glandular region,** below the notch (see Figure 32-10).

The structure of a gastric gland from the oxyntic glandular region is illustrated in Figure 33-6, *B.* The surface epithelial cells extend slightly into the duct opening. In the narrow neck of the gland are the **mucous neck cells,** which secrete mucus. Deeper in the gland are **parietal** or **oxyntic cells,** which secrete HCl and intrinsic factor (discussed below), and **chief** or **peptic cells,** which secrete pepsinogens. Oxyntic cells are particularly numerous in glands in the fundus.

Mucus-secreting cells predominate in the glands of the pyloric glandular region. Pyloric glands also contain **G cells,** which secrete the hormone **gastrin.**

Surface epithelial cells are exfoliated into the lumen at a considerable rate during normal gastric function. They are replaced by mucous neck cells, which differentiate into columnar epithelial cells and migrate up out of the necks of the glands. The capacity of the stomach to repair damage to its epithelial surface in this way is remarkable.

Gastric Acid Secretion

The fluid secreted into the stomach is called **gastric juice.** Gastric juice is a mixture of the secretions of the surface epithelial cells and the secretion of gastric glands. *Among the important components of gastric juice are salts, water, HCl, pepsins, intrinsic factor, and mucus.* Secretion of all these components increases after a meal.

Ionic Composition of Gastric Juice The ionic composition of gastric juice depends on the rate of secretion. Figure 33-7 shows that the higher the secretory rate, the higher the concentration of hydrogen ion. At lower secretory rates, $[H^+]$ diminishes and $[Na^+]$ increases. $[K^+]$ is always higher in gastric juice than in plasma, and consequently prolonged vomiting may lead to hypokalemia. At all rates of secretion, Cl^- is the major anion of gastric juice. At high rates of secretion the composition of gastric juice resembles that of an isotonic solution of HCl. Gastric HCl converts pepsinogens to active pepsins (see below) and provides an acid pH at which pepsins are active.

> The high acidity of gastric juice kills most microorganisms that are ingested. Individuals who have low rates of gastric acid secretion, either because of disease or because they are taking medications that suppress HCl secretion, are more susceptible to infection by ingested pathogens.

Rate of Secretion of Gastric Acid The rate of gastric acid secretion varies considerably among individuals. Basal (unstimulated) rates of gastric acid production typically range from about 1 to 5 mEq/hr in humans. On maximal stimulation, HCl production rises to 6 to 40 mEq/hr. *On average, patients with gastric ulcers secrete less HCl but those with duodenal ulcers secrete more HCl than do normal individuals* (Figure 33-8). The reasons for this are discussed below.

Morphological Changes that Accompany Gastric Acid Secretion Parietal cells have a distinctive ultrastruc-

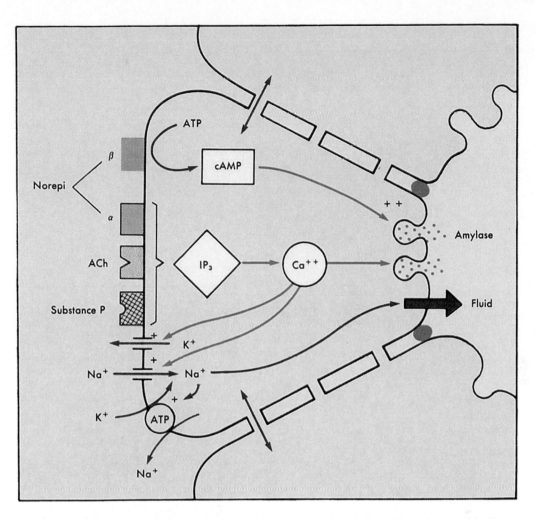

FIGURE 33-5   The cellular mechanisms whereby norepinephrine *(Norepi),* acetylcholine *(ACh),* and substance P evoke salivary secretion. ACh, substance P, and norepinephrine acting on α-adrenergic receptors increase intracellular Ca^{++}. Norepinephrine acts on β-adrenergic receptors to increase intracellular levels of cAMP. Effectors that increase cellular cAMP elicit a primary secretion that is richer in amylase than is the secretion evoked by agents that increase intracellular Ca^{++}. Substances that increase intracellular Ca^{++} produce a greater volume of acinar cell secretion than do agonists that increase intracellular cAMP. (Modified from Peterson OH. In Johnson RL, ed: *Physiology of the gastrointestinal tract,* New York, 1981, Raven Press.)

ture (Figure 33-9) and an elaborate system of branching **secretory canaliculi,** which course through the cytoplasm and are connected by a common outlet to the cell's luminal surface. Microvilli line the surfaces of the canaliculi. The cytoplasm of unstimulated parietal cells contains numerous tubules and vesicles—the **tubulovesicular system.** The membranes of the tubovesicles contain the transport proteins responsible for secretion of H^+ and Cl^- into the lumen of the gland. When parietal cells are stimulated to secrete HCl (Figure 33-9), tubulovesicular membranes fuse with the plasma membrane of the secretory canaliculi; this extensive membrane fu-

sion greatly increases the number of HCl pumping sites available at the surface of the secretory canaliculi.

Cellular Mechanisms of Gastric Acid Secretion At maximal rates of secretion H^+ is pumped against a concentration gradient that is more than one million to one. Cl^- also enters the gastric lumen against a large electrochemical potential difference. Thus energy is required for transport of both H^+ and Cl^-.

The apical membrane of the parietal cell (the membrane that lines the secretory canaliculus) contains an H^+-K^+-ATPase, which exchanges H^+ for K^+ (Figure 33-10). This ATPase is the primary H^+ pump. Both H^+ and

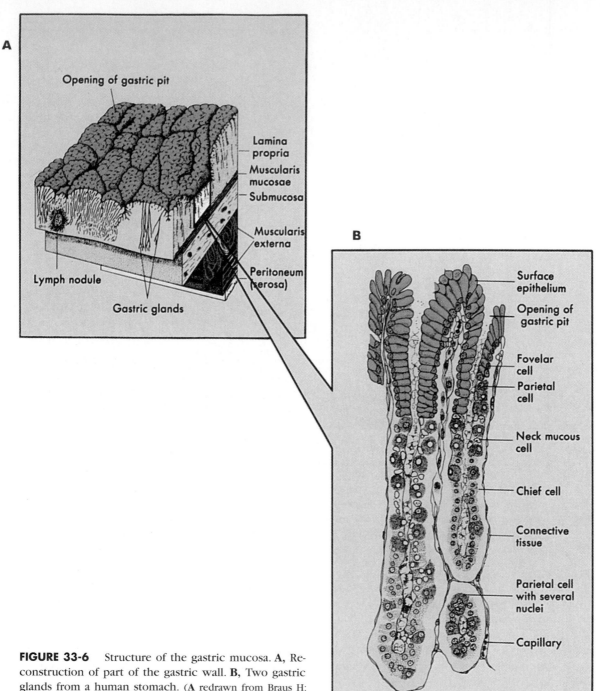

FIGURE 33-6   Structure of the gastric mucosa. **A,** Reconstruction of part of the gastric wall. **B,** Two gastric glands from a human stomach. (**A** redrawn from Braus H: *Anatomie des Menschen,* Berlin, 1934, Julius Springer. **B** redrawn From Weis L, ed: *Histology: cell and tissue biology,* ed 5, New York, 1981, Elsevier.)

K^+ are pumped against their electrochemical potential gradients.

When H^+ is pumped out of the parietal cell, an excess of HCO_3^- is left behind. HCO_3^- flows down its electrochemical gradient across the basolateral plasma membrane. The protein that mediates HCO_3^- efflux transports Cl^- in the opposite direction. Cl^- moves against its electrochemical potential gradient into the cell, and the energy for the active transport of Cl^- comes from the downhill movement of HCO_3^-.

As a result of the combined action of the H^+-K^+-ATPase and the Cl^-/HCO_3^- counter-transporter, Cl^- is

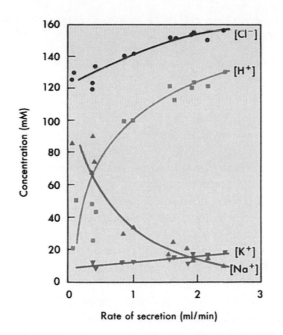

FIGURE 33-7 Concentrations of ions in gastric juice as a function of the rate of secretion in a normal young person. At low flow rates gastric juice is hypotonic to plasma. At high flow rates, gastric juice approaches isotonicity and contains predominantly H^+ and Cl^- (Adapted from Nordgren B: *Acta Physiol Scand* 58[suppl 202]:1, 1963.)

concentrated in the cytoplasm of the parietal cell. The Cl^- leaves the parietal cell at the apical membrane via an electrogenic anion channel.

Secretion of Pepsins

Pepsins, often collectively called **pepsin,** are a group of proteases secreted by the chief cells of the gastric glands. Pepsins are secreted as inactive proenzymes called **pepsinogens.** Cleavage of acid-labile linkages converts pepsinogens to active pepsins: the lower the pH, the more rapid the conversion. Pepsins also act proteolytically on pepsinogens to form more pepsins.

The pepsins have their highest proteolytic activity at pH 3 and below. Pepsins may digest as much as 20% of the protein in a typical meal. When the duodenal contents are neutralized, pepsins are inactivated irreversibly by the neutral pH.

Pepsinogens are contained in membrane-bound zymogen granules in the chief cells. The contents of the zymogen granules are released by exocytosis when the chief cells are stimulated to secrete.

Secretion of Intrinsic Factor

Intrinsic factor, a glycoprotein secreted by the parietal cells of the stomach, is required for the normal absorption of vitamin B_{12} (see Chapter 34). Intrinsic factor

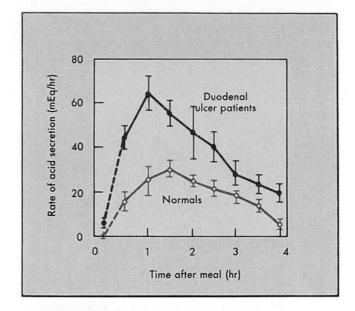

FIGURE 33-8. Rate of gastric acid secretion after a meal in six normal subjects and seven patients with duodenal ulcers. (From Fordtran JS, Walsh JH: *J Clin Invest* 52:645, 1973.)

is released in response to the same stimuli that evoke the secretion of HCl by parietal cells. *Secretion of intrinsic factor is the only gastric function that is essential for human life.*

Secretion of Mucus and Bicarbonate

Secretions that contain glycoprotein mucins are viscous and sticky and are collectively termed **mucus.** Mucins are secreted by mucous neck cells in the necks of gastric glands and by the surface epithelial cells of the stomach. Mucus is stored in large granules in the apical cytoplasm of mucous neck cells and surface epithelial cells; mucus is released by exocytosis.

Secretion of mucus is stimulated by some of the same stimuli that enhance acid and pepsinogen secretion, especially by acetylcholine released from parasympathetic nerve endings near the gastric glands.

The surface epithelial cells also secrete watery fluid with Na^+ and Cl^- concentrations similar to those of plasma, but with higher K^+ and HCO_3^- concentrations than in plasma. The high HCO_3^- concentration makes the mucus alkaline. Mucus is secreted by the resting mucosa and lines the stomach with a sticky, viscous, alkaline coat. When food is eaten, the rates of secretion of mucus and of HCO_3^- increase.

The Gastric Mucosal Barrier The mucus forms a gel on the luminal surface of the mucosa. *The mucus, and alkaline secretions entrapped within it, constitute a* **gastric mucosal barrier** *that prevents damage to the mucosa by gastric contents* (Figure 33-11). Pepsins cleave certain peptide bonds in the mucin molecules and

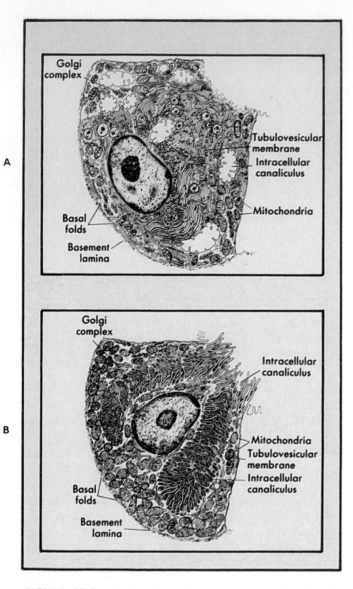

FIGURE 33-9 **A,** Drawing of a resting parietal cell with cytoplasm full of tubulovesicles and an internalized intracellular canaliculus. **B,** An acid-secreting parietal cell. Tubulovesicles have fused with the membrane of the intracellular canaliculus, which is now open to the lumen of the gland and lined with abundant, long microvilli. (Redrawn after Ito S. In Johnson RL: *Physiology of the gastrointestinal tract,* New York, 1981, Raven Press.)

thereby dissolve the gel. The gel must be replenished by synthesis of new mucin molecules.

The mucus gel layer prevents the bicarbonate-rich secretions of the surface epithelial cells from rapidly mixing with the contents of the gastric lumen. Thus, the surface of the epithelial cells can be maintained at nearly neutral pH, despite a luminal pH of about 2. The protection depends on both mucus and HCO_3^- secretion; nei-

ther mucus alone nor HCO_3^- alone can hold the pH at the epithelial cell surface near neutral.

The gastric mucosal barrier of a normal individual is capable of protecting the stomach even when rates of secretion of HCl and pepsins are elevated. If the secretion of either HCO_3^- or mucus is suppressed, however, the gastric mucosal barrier is compromised and the effects of acid and pepsin on the surface of the stomach may produce **gastric ulcers.**

α-Adrenergic agonists diminish HCO_3^- secretion. This effect may play a role in the pathogenesis of **stress ulcers:** a chronically elevated level of circulating epinephrine may suppress HCO_3^- secretion sufficiently to decrease protection of the epithelial cell surface. Aspirin and other nonsteroidal antiinflammatory agents inhibit secretion of both mucus and HCO_3^-; prolonged use of these drugs may damage the mucosal surface.

Control of Gastric Acid Secretion

Control of HCl Secretion by the Parietal Cell Acetylcholine, histamine, and gastrin are the three physiological agonists of HCl secretion. Each of these secretagogues binds to a distinct class of receptors on the plasma membrane of the parietal cell and directly stimulates the parietal cell to secrete HCl (Figure 33-12). Acetylcholine is released near parietal cells by cholinergic nerve terminals. Gastrin, a hormone, is produced by G cells in the mucosa of the gastric antrum and the duodenum and reaches parietal cells via the bloodstream. Histamine is released from cells in the gastric mucosa and diffuses to the parietal cells.

Cellular Mechanisms of Parietal Cell Agonists The receptors on the parietal cell membrane for acetylcholine, gastrin, and histamine and the intracellular second messengers by which these secretagogues act are shown in Figure 33-12.

Histamine is a major physiological mediator of HCl secretion. **Cimetidine,** a specific antagonist of H_2 receptors, blocks a large portion of the acid secretion elicited by any known secretagogue. **Enterochromaffin-like cells** (ECL cells) are present in the gastric muscosa, and these cells synthesize and store histamine. When stimulated by acetylcholine or gastrin, the ECL cells release histamine, which diffuses to nearby parietal cells to stimulate HCl secretion.

Gastrin is not as potent a direct stimulant of parietal cells as acetylcholine or histamine. The physiological response to elevated levels of gastrin in the blood is greatly attenuated by cimetidine. Thus a major component of

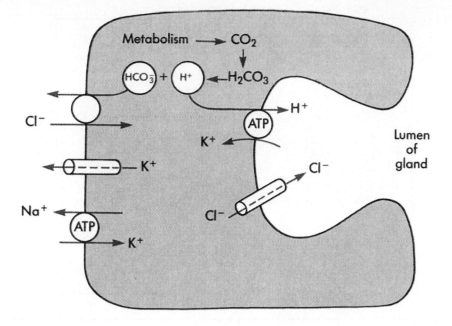

FIGURE 33-10   A simplified view of the major ionic transport processes involved in the secretion of H^+ and Cl^- by parietal cells. Cl^- enters the cell across the basolateral membrane against an electrochemical gradient. Cl^- entry is powered by the downhill efflux of HCO_3^-. The high level of HCO_3^- in the cytosol is generated by the extrusion of proteins across the luminal membrane. Protons are pumped into the secretory canaliculus by the H^+-K^+-ATPase. Cl^- enters the canalicular fluid by an electrogenic ion channel.

the physiological response to gastrin may result from gastrin-stimulated release of histamine.

The availability of cimetidine and other H_2 receptor blockers has revolutionized therapy for duodenal ulcer disease and other disorders related to hypersecretion of gastric acid. These drugs usually diminish secretion of HCl dramatically, and they have few side effects. H_2 receptor blockers are not so effective in patients with **Zollinger-Ellison syndrome.** Such patients have gastrin-secreting tumors that result in very high serum levels of gastrin. In treating these patients, **omeprazole** (a specific and irreversible inhibitor of the H^+-K^+-ATPase) is the current drug of choice.

In Vivo Control of Acid Secretion Rate When the stomach has been empty for several hours, HCl is secreted at a basal rate, which is approximately 10% of the maximal rate. After a meal the rate of acid secretion by the stomach increases promptly. There are three phases of increased acid secretion in response to food: the ce phalic phase, elicited before food reaches the stomach; the **gastric phase,** elicited by the presence of food in

the stomach; and the **intestinal phase,** elicited by mechanisms that originate in the duodenum and upper jejunum (Table 33-1).

The Cephalic Phase The cephalic phase of gastric secretion is normally elicited by the sight, smell, and taste of food. Cholinergic vagal fibers and cholinergic neurons of the intramural plexuses evoke cephalic phase secretion. Acetylcholine released from these neurons directly stimulates parietal cells to secrete HCl. Acetylcholine also stimulates acid secretion indirectly by releasing gastrin from G cells in the antrum and duodenum and histamine from ECL cells in the gastric mucosa.

Low pH in the antrum of the stomach diminishes the amount of HCl secreted during the cephalic phase. In the absence of food in the stomach to buffer the acid secreted, the pH of the antral contents falls rapidly during the cephalic phase. Low pH limits the amount of acid secreted by inhibiting the parietal cells directly and by evoking inhibitory intrinsic neural reflexes.

The Gastric Phase The gastric phase of gastric secretion is elicited by the presence of food in the stomach. The principal stimuli are distension of the stomach and the presence of amino acids and peptides resulting from the actions of pepsins. Most of the acid secreted in response to a meal is secreted during the gastric phase.

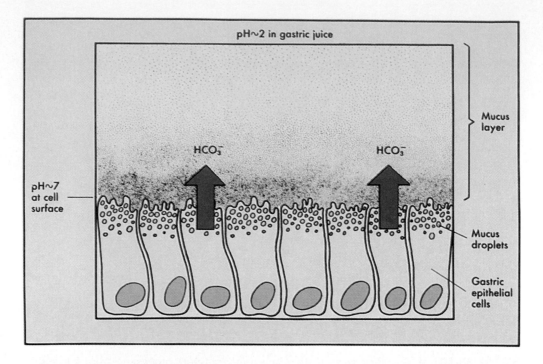

FIGURE 33-11 The protection provided to the mucosal surface of the stomach by the bicarbonate-containing mucus layer is known as the gastric mucosal barrier. Buffering by the bicarbonate-rich secretions of the surface epithelial cells and the restraint to convective mixing caused by the high viscosity of the mucus layer allow the pH at the cell surface to remain near 7, whereas the pH in the gastric juice in the lumen is 1 to 2.

Secretion of HCl is blocked effectively by bathing the mucosal surface with a solution that has a pH of 2 or less. Once the buffering capacity of the gastric contents is saturated, gastric pH falls rapidly and inhibits further acid release. In this way the acidity of gastric contents regulates itself.

The presence of amino acids and peptides in the antrum elicits HCl secretion by causing G cells in the antrum to release gastrin. Intact proteins do not have this effect. Other ingested substances that can enhance gastric acid secretion include calcium ions, caffeine, and alcohol.

The Intestinal Phase The presence of chyme in the duodenum brings about neural and endocrine responses that first stimulate and later inhibit secretion of acid by the stomach. Early in gastric emptying, when the pH of gastric chyme is greater than 3, the stimulatory influences predominate. Later, when the buffer capacity of gastric chyme is exhausted and the pH of chyme emptied into the duodenum falls to less than 3, inhibitory influences prevail. Tables 33-1 and 33-2 summarize the major mechanisms that stimulate and inhibit gastric acid secretion.

Stimulation of secretion. Gastric secretion is enhanced by distension of the duodenum and by the presence of protein digestion products (peptides and amino acids) in the duodenum. The duodenum and proximal jejunum contain G cells that release gastrin when stimulated by peptides and amino acids.

Inhibition of secretion. Several different mechanisms that operate during the intestinal phase inhibit gastric secretion (Table 33-2). These mechanisms are evoked by the presence of acid, fat digestion products, and hypertonicity in the duodenum and proximal part of the jejunum.

Acid solutions in the duodenum release the hormone **secretin** into the bloodstream. Secretin inhibits gastric acid by inhibiting gastrin release by G cells and by decreasing the response of parietal cells to secretagogues. Acid in the duodenum also inhibits gastric acid secretion via a local nervous reflex. Acid in the duodenal bulb releases another hormone, **bulbogastrone,** which inhibits acid secretion by the parietal cells.

Products of triglyceride digestion in the duodenum and proximal part of the jejunum release two hormones, **gastric inhibitory peptide** and **cholecystokinin** (CCK), that inhibit acid secretion by parietal cells.

Hyperosmotic solutions in the duodenum release a hormone that inhibits gastric acid secretion.

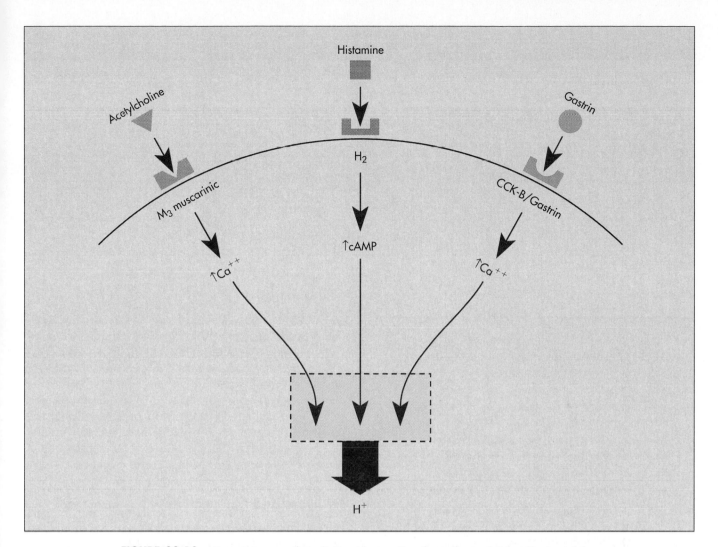

FIGURE 33-12   Secretagogues that elicit acid secretion from the parietal cells. Acetylcholine binds to M_3 muscarinic receptors. Histamine acts via H_2 histamine receptors. Gastrin binds to CCK-B/gastrin receptors. Acetylcholine and gastrin act to increase cytosolic free Ca^{++}. Histamine increases intracellular levels of cAMP.

Table 33-1   Major Mechanisms for Stimulation of Gastric Acid Secretion

Phase	Stimulus	Pathway	Stimulus to Parietal Cell
Cephalic	Chewing, swallowing, etc.	Vagus nerve to:	
		1. Parietal cells	Acetylcholine
		2. G cells	Gastrin
Gastric	Gastric distension	Local and vagovagal reflexes to:	
		1. Parietal cells	Acetylcholine
		2. G cells	Gastrin
Intestinal	Protein digestion products in duodenum	1. Intestinal G cells	Gastrin
		2. Intestinal endocrine cells	Enterooxyntin

Modified from Johnson LR, ed: *Gastrointestinal physiology,* ed 4, St Louis, 1991, Mosby.

Table 33-2 Major Mechanisms for Inhibition of Gastric Acid Secretion

Region	Stimulus	Mediator	Inhibit Gastrin Release	Inhibit Acid Secretion by Partietal Cell
Antrum	Acid (pH < 3.0)	None, direct	+	
Duodenum	Acid	Secretin	+	+
		Bulbogastrone	+	+
		Neural reflex		+
Duodenum and jejunum	Hyperosmotic solutions	Unidentified enterogastrone		+
	Fatty acids, monoglycerides	Gastric inhibitory peptide	+	+
		Cholecystokinin		+
		Unidentified enterogastrone		+

Modified from Johnson LR, ed: *Gastrointestinal physiology,* ed 4, St Louis, 1991, Mosby.

Patients with gastric ulcers frequently have subnormal rates of HCl secretion. This may be counterintuitive. Gastric ulcers are usually caused by a failure of the gastric mucosal barrier to a decrease in pH and to a bacterial infection. The decrease in pH at the mucosal surface suppresses HCl secretion.

Duodenal ulcer patients, by contrast, often have elevated rates of HCl secretion (see Figure 33-8). A common cause of the hypersecretion is a diminished sensitivity of certain mechanisms that inhibit gastric HCl secretion. As a result HCl is emptied into the duodenum more rapidly than the H^+ can be neutralized; this leads to ulceration of the duodenum.

Pepsinogen Secretion

Most of the agents that stimulate parietal cells to secrete acid also elicit release of pepsinogens from chief cells (Figure 33-13). Thus the rates of release of acid and pepsinogens from the gastric glands are highly correlated. Acetylcholine is a potent stimulus for the chief cells to release pepsinogens. Gastrin also directly stimulates chief cells. Acid in contact with the gastric mucosa stimulates pepsinogen release by a local neural reflex. Secretin and CCK released by the duodenal mucosa stimulate chief cells to secrete pepsinogens.

PANCREATIC SECRETION

The human pancreas weighs less than 100 g, yet each day it secretes 1 liter (10 times its mass) of pancreatic juice. The pancreas is unusual in having both endocrine and exocrine secretory functions. The exocrine juice is composed of an **aqueous component** that is rich in bicarbonate and that helps to neutralize duodenal contents and an **enzyme component** that contains enzymes for digesting carbohydrates, proteins, and fats. Pancreatic exocrine secretion is controlled by both neural and hormonal signals, elicited mainly by the presence of acid and digestion products in the duodenum. Secretin chiefly elicits secretion of the aqueous component, and CCK stimulates the secretion of pancreatic enzymes.

Structure and Innervation of the Pancreas

The structure of the exocrine pancreas resembles that of the salivary glands (see Figures 33-1 and 33-2). Microscopic, blind-ended tubules are surrounded by polygonal acinar cells whose primary function is to secrete the enzyme component of pancreatic juice. The acini are organized into lobules; the tiny ducts that drain the acini are called **intercalated ducts.** The intercalated ducts empty into somewhat larger **intralobular ducts.** The intralobular ducts of a particular lobule drain into a single **extralobular duct** that empties the lobule into still larger ducts. The larger ducts converge into a main duct that enters the duodenum along with the **common bile duct.**

The endocrine cells of the pancreas reside in the islets of Langerhans. Although islet cells account for less than 2% of the volume of the pancreas, their hormones are essential in regulating metabolism. **Insulin, glucagon, somatostatin,** and **pancreatic polypeptide** are hormones released from cells of the islets of Langerhans (see Chapter 42).

The pancreas is innervated by branches of the vagus nerve. Vagal fibers synapse with cholinergic neurons that lie within the pancreas and that innervate both aci-

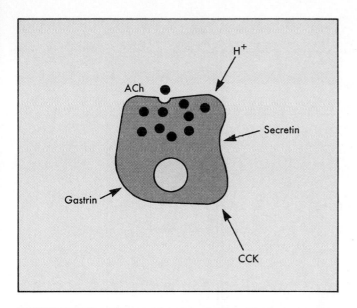

FIGURE 33-13 The agonists that elicit pepsinogen secretion by chief cells: acetylcholine *(ACh)*, H^+, secretin, cholecystokinin *(CCK)*, and gastrin.

nar and islet cells. Postganglionic sympathetic nerves from the celiac and superior mesenteric plexuses innervate pancreatic blood vessels. Secretion of pancreatic juice is stimulated by parasympathetic activity and inhibited by sympathetic activity.

Aqueous Component of Pancreatic Juice

The aqueous component of pancreatic juice is elaborated principally by the columnar epithelial cells that line the ducts. The Na^+ and K^+ concentrations of pancreatic juice are similar to those in plasma. HCO_3^- (at levels well above those in plasma) and Cl^- are the major anions. The HCO_3^- concentration varies from approximately 70 mEq/L at low rates of secretion to more than 100 mEq/L at high secretory rates (Figure 33-14). HCO_3^- and Cl^- concentrations vary reciprocally. The aqueous component secreted by the duct cells is slightly hypertonic, and its HCO_3^- concentration is high. As the secretion flows down the ducts, water equilibrates across the epithelium to make the pancreatic juice isotonic, and some HCO_3^- exchanges for Cl^- (Figure 33-15).

Under resting conditions the aqueous component is produced primarily by the intercalated and other intralobular ducts. When secretion is stimulated by secretin, however, the additional flow comes mostly from the extralobular ducts (Figure 33-15). Secretin is the major physiological stimulus for secretion of the aqueous component.

Enzyme Component of Pancreatic Juice

The secretions of the acinar cells constitute the **enzyme component** of pancreatic juice. The fluid that is secreted by the acinar cells resembles plasma in its tonicity and in the concentrations of various ions. The enzyme component contains enzymes important for the digestion of all the major classes of foodstuffs. If pancreatic enzymes are absent, lipids, proteins, and carbohydrates are malabsorbed.

Proteases of pancreatic juice are secreted in inactive zymogen form. The major pancreatic proteases are **trypsin, chymotrypsin,** and **carboxypeptidase.** They are secreted as trypsinogen, chymotrypsinogen, and procarboxypeptidase, respectively. Trypsinogen is specifically activated by **enteropeptidase** (also called **enterokinase**), which is secreted by the duodenal mucosa. Trypsin then activates **trypsinogen, chymotrypsinogen, and procarboxypeptidase. Trypsin inhibitor,** a protein present in pancreatic juice, prevents the premature activation of proteolytic enzymes in the pancreatic ducts.

Pancreatic juice contains an **α-amylase** that is secreted in active form. **Pancreatic amylase** cleaves starch molecules into oligosaccharides. Pancreatic juice also contains a number of lipid-digesting enzymes, or **lipases.** Among the major pancreatic lipases are **triacylglycerol hydrolase, cholesterol ester hydrolase,** and **phospholipase A_2.**

Cl^- enters the acinar lumen via electrogenic Cl^- channels in the apical plasma membranes of the acinar cells. The primary molecular defect in **cystic fibrosis** is a mutation in the gene that encodes this Cl^- channel; this mutation renders it ineffective in Cl^- transport. The decreased transport of Cl^- into the acinar lumen impairs transport of Na^+ and water as well. Consequently, in cystic fibrosis, the acini and intercalated ducts of the pancreas and the small airways of the lung become clogged with mucus. The pancreatic exocrine function of most infants with cystic fibrosis has been irreversibly damaged in utero. For this reason, infants with cystic fibrosis frequently have severe digestive difficulties, especially in the digestion and absorption of fats.

Regulation of Secretion of Pancreatic Juice

Stimulation of the vagal branches to the pancreas enhances secretion. Activation of sympathetic fibers inhibits pancreatic secretion, partly by decreasing blood flow to the pancreas. Secretin and CCK, hormones released from the duodenal mucosa, stimulate secretion of the aqueous and enzyme components, respectively. Because

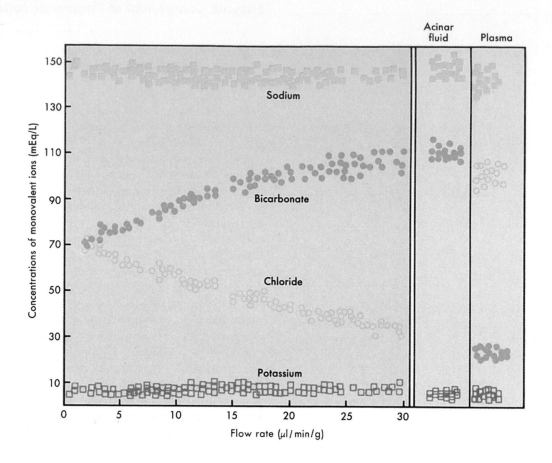

FIGURE 33-14 Concentrations of the major ions in rat pancreatic juice as functions of the secretory flow rate. The concentrations of the ions in acinar fluid and in plasma are shown for reference. At all flow rates the concentration of HCO_3^- in pancreatic juice is well above the plasma level. Secretion was stimulated by intravenous injection of secretin. (Redrawn from Mangos JA, McSherry NA: *Am J Physiol* 221:496, 1971.)

the aqueous and enzyme components of pancreatic juice are separately controlled (Figure 33-15), the composition of the juice varies from less than 1% to as much as 10% protein.

The Cephalic Phase Gastrin released from the mucosa of the gastric antrum in response to vagal impulses stimulates pancreatic secretion during the cephalic phase. Gastrin is a member of the same class of peptides as CCK, but it is much less potent as a pancreatic secretagogue than CCK.

The Gastric Phase During the gastric phase of secretion, gastrin, released in response to gastric distension and the presence of amino acids and peptides in the antrum of the stomach, enhances secretion by the pancreas. In addition, neural reflexes elicited by distension of the stomach evoke pancreatic secretion.

The Intestinal Phase In the intestinal phase of secretion, certain components of the chyme in the duodenum and upper jejunum evoke pancreatic secretion. Acid in

the duodenum and upper jejunum elicits the secretion of a large volume of pancreatic juice that is low in enzyme content. *The hormone secretin is the major mediator of this response to acid.* Secretin is released by certain cells in the mucosa of the duodenum and upper jejunum in response to acid in the lumen, and it directly stimulates the cells of the pancreatic ductular epithelium to secrete the bicarbonate-rich aqueous component of the pancreatic juice.

The presence of peptides and certain amino acids in the duodenum elicits the secretion of pancreatic juice rich in enzyme components. Fatty acids and monoglycerides in the duodenum also elicit secretion of protein-rich pancreatic juice. **CCK,** a hormone released by particular cells in the duodenum and upper jejunum in response to these digestion products, is the most important physiological mediator of the enzyme component of pancreatic juice.

CCK potentiates the stimulatory effect of secretin on

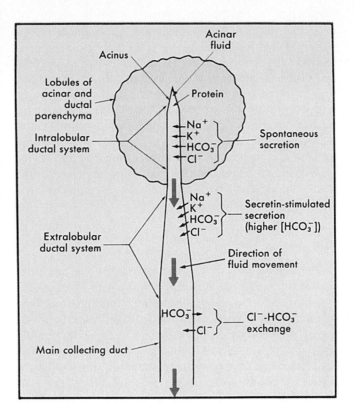

FIGURE 33-15 The locations of important transport processes involved in the elaboration of pancreatic juice. Acinar fluid is isotonic and resembles plasma in its concentrations of Na^+, K^+, Cl^-, and HCO_3^-. The secretion of acinar fluid and the proteins it contains is stimulated by CCK and acetylcholine. A spontaneous secretion that is produced by the intralobular ducts has higher concentrations of K^+ and HCO_3^- than does plasma. The hormone secretin stimulates water and electrolyte secretion by the cells that line the extralobular ducts. The secretin-stimulated secretion is still richer in HCO_3^- than the spontaneous secretion. (Adapted from Swanson CH, Solomon AK: *J Gen Physiol* 62:407, 1973.)

the ducts. Secretin potentiates the effect of CCK on acinar cells.

FUNCTIONS OF THE LIVER AND GALLBLADDER

Structure of the Liver

The histology of the liver is shown in Figure 33-16. Each liver **lobule** is organized around a **central vein.** At the periphery of the lobule blood enters the **sinusoids** from branches of the **portal vein** and the **hepatic artery.** In the sinusoids blood flows toward the center of the lobule between plates of **hepatocytes** that are one or two cells thick. Because of the large fenestrations between the endothelial cells that line the sinusoids, each hepatocyte is in direct contact with sinusoidal

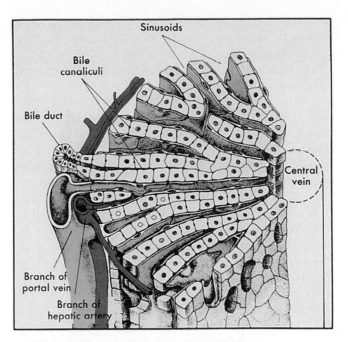

FIGURE 33-16 Diagrammatic representation of a hepatic lobule. A central vein is located in the center of the lobule, with plates of hepatocytes disposed radially. Branches of the portal vein and hepatic artery are located on the periphery of the lobule, and blood from both perfuses the sinusoids. Peripherally located bile ducts drain the bile canaliculi that run between the hepatocytes. (Adapted from Bloom W, Fawcett DW: *A textbook of histology,* ed 10, Philadelphia, 1975, WB Saunders.)

blood. The intimate contact of a large fraction of the hepatocyte surface with blood contributes to the ability of the liver to clear the blood effectively of certain classes of compounds. **Biliary canaliculi** lie between adjacent hepatocytes, and the canaliculi drain into bile ducts at the periphery of the lobule.

Functions of the Liver

The liver performs many vital functions. *The liver is essential in regulating metabolism, synthesizing many proteins and other molecules, storing certain vitamins and iron, degrading certain hormones, and inactivating and excreting many drugs and toxins.*

The liver regulates the metabolism of carbohydrates, lipids, and proteins. Liver and skeletal muscle are the two major sites of glycogen storage in the body. When the level of glucose in the blood is high, glycogen is deposited in the liver. When the blood glucose level is low, glycogen is broken down to glucose **(glycogenolysis),** and the glucose is then released into the blood. In this way the liver helps to maintain a relatively constant blood glucose level. The liver is also the major site of **gluconeogenesis,** the conversion of amino acids, lipids, or simple carbohydrates (e.g., lactate) into glucose. Carbohydrate metabolism by the liver is regulated by several hormones (see Chapters 42 and 47).

The liver is also centrally involved in lipid metabolism. As described in Chapter 34, lipids absorbed by the intestine leave the intestine in **chylomicrons** in the lymph. Lipoprotein lipase on the endothelial cell surface of blood vessels hydrolyzes some of the triglyceride in the chylomicrons and releases glycerol and fatty acids that are taken up by **adipocytes.** This results in formation of **chylomicron remnants** rich in cholesterol. Chylomicron remnants are taken up by hepatocytes and degraded. Hepatocytes synthesize and secrete **very-low-density lipoproteins (VLDL).** VLDL are then converted to the other types of serum lipoproteins. These lipoproteins are the major sources of cholesterol and triglycerides for most other tissues of the body. *Cholesterol present in bile represents the only route of excretion of cholesterol.* Hepatocytes are thus a principal source of cholesterol in the body and the major site of excretion of cholesterol. Thus, *hepatocytes play an important role in the regulation of serum cholesterol levels.* (See also Chapter 41.)

Because carbohydrate utilization is impaired in **diabetes mellitus,** β-oxidation of fatty acids provides a major source of energy for the body (Chapter 42). In the liver the oxidation of fatty acids produces acetoacetate, β-hydroxybutyrate, and acetone. These three compounds are called **ketone bodies.** Ketone bodies are released from hepatocytes and carried in the circulation to other tissues, where they are metabolized. The levels of ketone bodies in the urine and blood can indicate the severity of diabetic acidosis.

The liver is centrally involved in protein metabolism. When proteins are broken down (catabolized), amino acids are deaminated to form ammonia (NH_3). Ammonia cannot be further metabolized by most tissues and becomes toxic at levels achievable by metabolism. Ammonia is dissipated by conversion to urea, mainly in the liver. The liver also synthesizes all the nonessential amino acids.

The liver synthesizes all the major plasma proteins, including the plasma lipoproteins, albumins, globulins, fibrinogens, and other proteins involved in blood clotting.

The liver stores certain substances important in metabolism. Next to hemoglobin in red blood cells, the liver is the most important storage site for iron. Also, certain vitamins, most notably A, D, and B_{12}, are stored in the liver. Hepatic storage protects the body from limited dietary deficiencies of these vitamins.

The liver transforms and excretes many hormones, drugs, and toxins. These substances are frequently converted to inactive forms by reactions that occur in hepatocytes. The smooth endoplasmic reticulum of hepatocytes contains systems of enzymes and cofactors that are responsible for chemical transformations of many substances. Certain other enzymes in the endoplasmic reticulum catalyze the conjugation of many compounds with glucuronic acid, glycine, or glutathione. The transformations that occur in the liver render many compounds more water soluble, and thus they are more readily excreted by the kidneys. Some liver metabolites are secreted into the bile.

Bile

The hepatic function most important to the digestive tract is the secretion of **bile.** Bile, elaborated by hepatocytes, contains bile acids, cholesterol, lecithins, and bile pigments. These constituents are all synthesized and secreted by hepatocytes into the bile canaliculi, along with an isotonic fluid that resembles plasma in its electrolyte concentrations. The bile canaliculi merge into ever larger ducts and finally into a single large bile duct. The epithelial cells that line the bile ducts secrete a watery fluid that is rich in bicarbonate and contributes to the volume of bile leaving the liver.

The secretory function of the liver resembles that of the exocrine pancreas. In both organs the major parenchymal cell type elaborates a primary secretion containing the substances responsible for the major digestive function of the organ. In both the liver and the pancreas the primary secretion is isotonic to plasma and contains Na^+, K^+, and Cl^- at close to plasma levels, and the primary secretion is stimulated by CCK. In both pancreas and liver the epithelial cells that line the duct systems modify the primary secretion. When stimulated by secretin, the epithelial cells contribute an aqueous secretion with a high bicarbonate concentration.

Between meals bile is diverted into the **gallbladder.** *The gallbladder epithelium extracts salts and water from the stored bile, and the bile acids are thereby concentrated 5- to 20-fold.* After an individual has eaten, the gallbladder contracts and empties its concentrated bile into the duodenum. The most potent stimulus for emptying of the gallbladder is CCK. From 250 to 1500 ml of bile enter the duodenum each day.

Bile acids emulsify lipids, thereby increasing the surface area available to lipolytic enzymes. Bile acids then form **mixed micelles** (see Chapter 34) with the products of lipid digestion. Micelles increase the transport of the products of lipid digestion products to the brush border surface, and thus micelles enhance absorption of lipids by the epithelial cells. Bile acids are actively absorbed, mainly in the terminal ileum. A small fraction of bile acids escapes absorption and is excreted. Bile acids return-

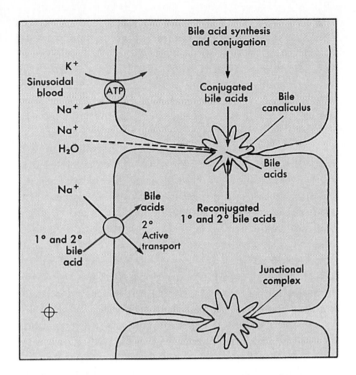

FIGURE 33-17 A representation of cellular mechanisms responsible for the uptake by hepatocytes of bile acids returning to the liver in portal blood and the resecretion of bile acids into the bile canaliculi. Bile acids are taken up by the hepatocytes by secondary active transport mechanisms. In the hepatocyte, deconjugated bile acids are largely reconjugated. Bile acids are resecreted into bile.

ing to the liver are avidly taken up by hepatocytes and are rapidly resecreted during the course of digestion (Figure 33-17). The entire bile acid pool is recirculated two or more times in response to a typical meal. The recirculation of the bile acids is known as the **enterohepatic circulation.** Approximately 20% of the bile acid pool is excreted in the feces each day and is replenished by hepatic synthesis of new bile acids. Figure 33-18 summarizes some major aspects of the enterohepatic circulation.

> Bile acids lost into the feces are a significant mechanism of excretion of cholesterol. Treatment with drugs that inhibit the reabsorption of bile acids in the ileum promotes the synthesis of new bile acids from cholesterol. Such drugs have been used to lower the level of cholesterol in the blood.

Fraction of Bile Secreted by Hepatocytes

Bile Acids Bile acids constitute about 50% of the dry weight of bile. Other important compounds secreted by the hepatocytes into the bile include lecithin, bile pigments, and proteins.

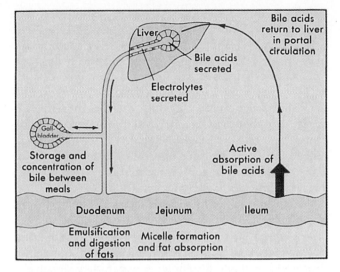

FIGURE 33-18 Overview of the enterohepatic circulation of bile. Bile is dumped into the duodenum by contractions of the gallbladder. In the small intestine, bile acids first emulsify dietary fat and then form mixed micelles with the products of fat digestion. In the terminal ileum, bile acids are reabsorbed. Bile acids return to the liver in the portal blood; they arc avidly taken up by hepatocytes and resecreted into bile.

Bile acids have a steroid nucleus and are synthesized by the hepatocytes from cholesterol. The major bile acids synthesized by the liver are called **primary bile acids.** These are **cholic acid** (3-hydroxyl groups) and **chenodeoxycholic acid** (2-hydroxyl groups). The presence of the carboxyl and hydroxyl groups make the bile acids much more water soluble than the cholesterol from which they are synthesized.

Bacteria in the digestive tract dehydroxylate bile acids to form **secondary bile acids.** The major secondary bile acids are **deoxycholic acid** (from dehydroxylation of cholic acid) and **lithocholic acid** (from dehydroxylation of chenodeoxycholic acid). Bile contains both primary and secondary bile acids.

Bile acids normally are secreted conjugated with glycine or taurine. In conjugated bile acids the glycine or taurine is linked by a peptide bond between the carboxyl group of an unconjugated bile acid and the amino group of glycine or taurine. At the near-neutral pH of the gastrointestinal tract, conjugated bile acids are more completely ionized, and thus more water soluble, than unconjugated bile acids. Conjugated bile acids are present almost entirely as salts of various cations (mostly Na^+) and are often called **bile salts.**

The steroid nucleus of bile acids is almost planar. In solution, bile acids have their polar (hydrophilic) groups—the hydroxyl groups, the carboxyl moiety of glycine or taurine, and the peptide bond—all on one surface of the molecule. This makes the bile acid molecule amphipathic—that is, having both hydrophilic and hydrophobic domains. Because they are amphipathic, bile acids tend to form molecular aggregates, called **micelles,** in which the hydrophobic side of the bile acid faces inside and away from water and the hydrophilic surface faces outward toward the water (see Chapter 34). Whenever the concentration of bile acids exceeds a certain concentration, called the **critical micelle concentration,** bile acid micelles will form. Above this concentration any additional bile acid will go into the micelles exclusively and not into molecular solution. Normally in bile, the bile acid concentration is much greater than the critical micelle concentration.

Phospholipids in Bile Hepatocytes also secrete phospholipids, especially lecithins, into bile. Cholesterol is also secreted into the bile, and this is the major route for cholesterol excretion. Although lecithin and cholesterol are essentially insoluble in water, they dissolve in the bile acid micelles. The lecithin increases the amount of cholesterol that can be solubilized in the micelles.

> If more cholesterol is present in the bile than can be solubilized in the micelles, crystals of cholesterol will form in the bile. These crystals are important in the formation of **cholesterol gallstones** (the most common gallstones) in the duct system of the liver or more often in the gallbladder.

Bile pigments When senescent red blood cells are degraded in reticuloendothelial cells, the porphyrin moiety of hemoglobin is converted to **bilirubin.** Bilirubin is released into the plasma, where it is bound to albumin. Hepatocytes efficiently remove bilirubin from blood in the sinusoids and conjugate bilirubin with one or two glucuronic acid molecules. Bilirubin glucuronides are secreted into the bile. Bilirubin is yellow and contributes to the yellow color of bile.

Secretion of the Bile Duct Epithelium The epithelial cells that line the bile ducts contribute an aqueous secretion that accounts for about 50% of the total volume of the bile. The secretion of the bile duct epithelium is isotonic and contains Na^+ and K^+ at levels similar to those of plasma. However, the concentration of HCO_3^- is greater and the concentration of Cl^- is less than in plasma. The secretory activity of the bile duct epithelium is specifically stimulated by secretin.

Bile Concentration and Storage in the Gallbladder

Between meals the tone of the **sphincter of Oddi,** which guards the entrance of the common bile duct into the duodenum, is high. Thus most bile flow is diverted into the gallbladder. The gallbladder is a small organ, having a capacity of 15 to 60 ml (average about 35 ml) in humans. Many times this volume of bile may be secreted by the liver between meals. The gallbladder concentrates the bile by absorbing Na^+, Cl^-, HCO_3^-, and water from the bile, such that the bile acids are concentrated from 5 to 20 times. The active transport of Na^+ is the primary active process in the concentrating action of the gallbladder.

Because of its high rate of water absorption, the gallbladder serves as a model for water and electrolyte transport by tight-junctioned epithelia. The **standing osmotic gradient mechanism** for fluid absorption was first proposed for the gallbladder. It was noted that during fluid reabsorption by the gallbladder, the lateral intercellular spaces between the epithelial cells were large and swollen. When fluid transport was blocked, the intercellular spaces almost disappeared. These observations suggested that the intercellular spaces are a major route of fluid flow during absorption.

The primary active transport process in the standing osmotic gradient mechanism is the active transport of Na^+ into the lateral intercellular spaces. The Na^+-K^+-ATPase molecules are especially concentrated in the basolateral membrane near the mucosal (apical) end of the intercellular channels (Figure 33-19). Cl^- and HCO_3^- are also transported into the intercellular space, probably because of the electrical potential created by electrogenic Na^+ transport. The high ion concentration near the apical end of the intercellular space causes the fluid there to be hypertonic. This produces an osmotic flow of water from the lumen via adjacent cells into the intercellular space. Water distends the intercellular channels because of increased hydrostatic pressure. As a result of water flow from adjacent cells, the fluid becomes less hypertonic as it flows down the intercellular channel, so that the fluid is essentially isotonic when it reaches the serosal (basal) end of the channel. Ions and water move across the basement membrane of the epithelium, and they are carried away by the capillaries.

Emptying of the Gallbladder

Emptying of the gallbladder begins several minutes after the start of a meal. Intermittent contractions of the gallbladder force bile through the partially relaxed sphincter of Oddi. During the cephalic and gastric phases of digestion, gallbladder contraction and relaxation of the sphincter are mediated by cholinergic fibers in branches of the vagus nerve and by gastrin released from the stomach. Stimulation of sympathetic nerves to the gallbladder and duodenum inhibits emptying of the gallbladder.

The highest rate of gallbladder emptying occurs during the intestinal phase of digestion; the strongest stimulus for the emptying is CCK. CCK reaches the gallbladder via the circulation and causes strong contractions of the gallbladder and relaxation of the sphincter of Oddi. Substances, such as gastrin, that mimic the actions of CCK in promoting gallbladder emptying are called **cholecystagogues.** Gastrin has the same sequence of five amino acids at its C-terminus as does CCK; however, gastrin is not nearly as potent a cholecystagogue as CCK. Nevertheless, gastrin helps to elicit gallbladder contractions during the cephalic and gastric phases of digestion.

Under normal circumstances the rate of gallbladder emptying is sufficient to keep the concentration of bile acids in the duodenum above the critical micelle concentration.

Intestinal Absorption of Bile Acids and their Enterohepatic Circulation

The functions of bile acids in emulsifying dietary lipid and in forming mixed micelles with the products of lipid digestion are discussed in Chapter 34. Normally, by the time chyme reaches the terminal part of the ileum, dietary fat is almost completely absorbed. Bile acids are then absorbed. Transport mechanisms are present in the brush border of the terminal ileum for uptake of both conjugated and unconjugated bile acids. Conjugated bile acids can be taken up against a large concentration gradient. Because bile acids are also lipid soluble, they can be taken up by simple diffusion as well. Bacteria in the terminal part of the ileum and colon deconjugate bile acids and also dehydroxylate them to produce secondary bile acids. Both deconjugation and dehydroxylation lessen the polarity of bile acids, and thereby enhance their lipid solubility and their absorption by simple diffusion.

Typically about 0.5 g of bile acids escapes absorption and is excreted in the feces each day. This quantity is 15% to 35% of the total bile acid pool, and normally it is replenished by synthesis of new bile acids by the liver.

Bile acids, whether absorbed by active transport or simple diffusion, are transported away from the intestine in the portal blood, mostly bound to plasma proteins. In the liver, hepatocytes avidly extract the bile acids from the portal blood. *In a single pass through the liver, the portal blood is essentially cleared of bile acids.* Bile acids in all forms, primary and secondary, both conjugated and deconjugated, are taken up by the hepatocytes. The hepatocytes reconjugate almost all the deconjugated bile acids and rehydroxylate some of the secondary bile acids. These bile acids are secreted into the bile along with newly synthesized bile acids (see Figure 33-18).

Control of Bile Acid Synthesis and Secretion

The rate of return of the bile acids to the liver affects the rate of synthesis and secretion of bile acids. Bile acids in the portal blood stimulate the uptake and resecretion of bile acids by the hepatoctyes. This is called the

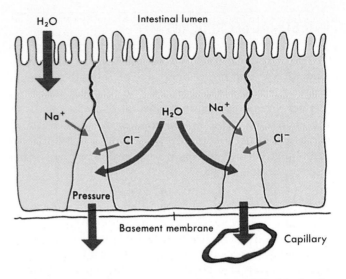

FIGURE 33-19 Water absorption from the gallbladder by the standing gradient osmotic mechanism. Na^+ is actively pumped into the lateral intercellular spaces; Cl^- follows. Water is drawn by osmosis to enter the intercellular spaces, elevating the intercellular hydrostatic pressure. Water, Na^+, and Cl^- are filtered across the porous basement membrane and enter the capillaries.

choleretic effect of bile acids; substances that enhance bile acid secretion are called choleretics. So powerful is the stimulus to resecrete the returning bile acids that the entire pool of bile acids (1.5 to 31.5 g) recirculates twice in response to a typical meal. In response to a meal with a very high fat content, the bile acid pool may recirculate five or more times.

Gallstones

Cholesterol is essentially insoluble in water. When bile contains more cholesterol than can be solubilized in the bile acid–lecithin micelles, crystals of cholesterol form in the bile. Such bile is said to be **supersaturated** with cholesterol. The greater the concentration of bile acids and lecithin in bile, the greater is the amount of cholesterol that can be contained in the mixed micelles.

The formation of cholesterol gallstones was mentioned previously. **Bile pigment gallstones** are the other major class of gallstones; their major constituent is the calcium salt of unconjugated bilirubin. Conjugated bilirubin is quite soluble and does not form insoluble calcium salts in bile. In liver disease, bile may contain elevated levels of unconjugated bilirubin because hepatocytes are deficient in forming the glucuronides of bilirubin. Individuals with liver disease have an increased likelihood of forming bile pigment stones.

INTESTINAL SECRETIONS

The mucosa of the intestine, from the duodenum through the rectum, elaborates secretions that contain mucus, electrolytes, and water. The total volume of intestinal secretions is about 1500 ml/day. The mucus in the secretions protects the mucosa from mechanical damage. The nature of the secretions and the mechanisms that control secretion vary in different segments of the intestine.

Duodenal Secretions

The duodenal submucosa contains branching glands that elaborate a secretion rich in mucus. The duodenal epithelial cells also contribute to duodenal secretions, but most of the secretions are produced by the glands. The duodenal secretion contains mucus and an aqueous component that does not differ significantly from plasma in its concentrations of the major ions.

Secretions of the Small Intestine

Goblet cells, which lie among the columnar epithelial cells of the small intestine, secrete mucus. During normal digestion an aqueous secretion is elaborated by the epithelial cells at a rate only slightly less than the rate of fluid absorption by the small intestine.

Secretions of the Colon

The secretions of the colon are smaller in volume but richer in mucus than are the small intestinal secretions. The mucus is produced by numerous goblet cells in the colonic mucosa. The aqueous component of colonic secretions is rich in K^+ and HCO_3^-. Colonic secretion is stimulated by mechanical irritation of the mucosa and by activation of cholinergic pathways to the colon. Stimulation of sympathetic nerves to the colon decreases the rate of colonic secretion.

SUMMARY

1. The epithelial cells that line the gastrointestinal tract and the cells of various glands associated with the gastrointestinal tract produce secretions that contain water, electrolytes, and proteins. These secretions have important functions in the gastrointestinal tract.

2. The physiological regulation of gastrointestinal secretions is effected by intrinsic and extrinsic neurons, hormones, and paracrine mediators.

3. Salivary glands produce a hypotonic fluid with bicarbonate and potassium concentrations in excess of plasma levels. Saliva contains an α-amylase that begins the digestion of starch. Mucus in saliva lubricates food. Parasympathetic nerves are the key regulators of salivary secretion.

4. The stomach serves as a reservoir for ingested food and empties gastric contents into the duodenum at a regulated rate. Parietal cells secrete hydrochloric acid and intrinsic factor into the stomach. Chief cells secrete pepsinogens.

5. The regulation of HCl secretion in the stomach involves extrinsic and intrinsic nerves, with acetylcholine as the major stimulatory neurotransmitter. Gastrin, a hormone released by G cells in the gastric antrum and in the duodenum, and histamine, a paracrine agonist released by ECL cells in the stomach, are also important physiological agonists of HCl secretion.

6. HCl catalyzes the conversion of pepsinogens to active pepsins. Pepsins convert a significant fraction of ingested protein to oligopeptides.

7. Mucus and bicarbonate secretions form the "gastric mucosal barrier" that protects the epithelial cells of the stomach from the effects of HCl and pepsins.

8. The pancreas produces a bicarbonate-rich fluid that contains enzymes essential for the digestion of carbohydrates, proteins, and fats. Pancreatic acinar cells produce the enzyme component of pancreatic juice; the intralobular and extralobular ducts secrete much of the aqueous component (water and electrolytes) of pancreatic juice.

9. Cholecystokinin (CCK) is the major physiological agonist of acinar cell secretion of the enzyme component. Secretin is the major stimulus for secretion of bicarbonate-rich fluid by the extralobular ducts of the pancreas. CCK and secretin are hormones released by cells in the duodenum and jejunum in response to the presence of fat digestion products and acid, respectively.

10. The liver produces and the gallbladder concentrates a secretion called bile. Bile is a bicarbonate-rich fluid that contains bile acids, bile pigments, lecithin, cholesterol, and numerous other components. Bile acids play a vital role in the digestion and absorption of lipids.

11. Hepatocytes are responsible for secreting the organic components of bile. The cells of the bile ducts secrete a bicarbonate-rich fluid. CCK is a major secretagogue for secretion by the hepatocytes. Secretin stimulates the bile ducts to produce their bicarbonate-rich fluid.

12. Bile acids are absorbed in the terminal ileum and return to the liver in the portal vein. Hepatocytes rap-

idly clear the blood of bile acids and resecrete them. Bile acids in the portal blood are a powerful stimulus to the hepatocytes to resecrete bile acids. The bile acid pool may be recirculated 2 to 5 times in response to a single meal. The secretion, return, and resecretion of bile acids is known as the enterohepatic circulation of bile acids.

BIBLIOGRAPHY
Journal Articles

Allen A, Garner A: Mucus and bicarbonate secretion in the stomach and their possible role in mucosal protection, *Gut* 21:249, 1980.

Chew CS: CCK, carbachol, gastrin, histamine, and forskolin increase [Ca]$_i$ in gastric glands, *Am J Physiol* 250:G312, 1986.

Chew CS: Intracellular mechanisms in control of acid secretion, *Curr Opinion Gastroenterol* 7:856, 1991.

Gerber JG, Payne NA: The role of gastric secretagogues in regulating gastric histamine release in vivo, *Gastroenterology* 102:403, 1992.

Hoffman AF: Current concepts of biliary secretion, *Dig Dis Sci* 34(suppl 1):16S, 1989.

Jensen RT, Gardner JD: The cellular basis of action of gastrointestinal peptides, *Adv Cyclic Nucleotide Protein Phosphorylation Res* 17:375, 1984.

Klaasen CD, Watkins III JB: Mechanisms of bile formation, hepatic uptake, and biliary excretion, *Pharmacol Rev* 36:1, 1984.

Putney Jr JW: Identification of cellular activation mechanisms associated with salivary secretion, *Annu Rev Physiol* 48:75, 1986.

Rabon EC, Reuben MA: The mechanism and structure of the gastric H,K-ATPase, *Annu Rev Physiol* 52321, 1990.

Raufman J-P: Gastric chief cells: receptors and signal transduction mechanisms, *Gastroenterology* 102:699, 1992.

Reuss L: Ion transport across gallbladder epithelium, *Physiol Rev* 69:503, 1989.

Shamburek RD, Schubert ML: Control of gastric acid secretion, *Gastroenterol Clin North Am* 21:527, 1992.

Walsh JH: Peptides as regulators of gastric acid secretion, *Annu Rev Physiol* 50:41, 1988.

Williams JA: Regulatory mechanisms in pancreas and salivary acini, *Annu Rev Physiol* 46:361, 1984.

Books and Monographs

Allen A, ed: *Mechanisms of mucosal protection in the upper gastrointestinal tract,* New York, 1984, Raven Press.

Argent BE, Case RM: Pancreatic ducts: cellular mechanisms and control of bicarbonate secretion. In Johnson LR, ed: *Physiology of the gastrointestinal tract,* ed 3, New York, 1994, Raven Press.

Arias IM et al: *The liver: biology and pathobiology,* ed 2, New York, 1988, Raven Press.

Cook DI et al: Secretion by the major salivary glands. In Johnson LR, ed: *Physiology of the gastrointestinal tract,* ed 3, New York, 1994, Raven Press.

Davenport HW: *Physiology of the digestive tract,* ed 5, Chicago, 1982, Year Book Medical Publishers, Inc.

Feldman M: Gastric secretion. In Sleisenger M, Fordtran JS, eds: *Gastrointestinal diseases,* ed 5, Philadelphia, 1993, WB Saunders.

Flemström G: Gastric and duodenal secretion of mucus and bicarbonate. In Johnson LR, ed: *Physiology of the gastrointestinal tract,* ed 3, New York, 1994, Raven Press.

Forte JG, Soll AH: Cell biology of hydrochloric acid secretion. In *Handbook of physiology,* section 6, vol 3, Washington, DC, 1989, American Physiological Society.

Go VLW et al, eds: *The pancreas: biology, pathobiology, and disease,* ed 2, New York, 1993, Raven Press.

Hernandez DE, Glavin GB, eds: *Neurobiology of stress ulcers,* Ann NY Acad Sci, vol 299, New York, 1990, New York Academy of Sciences.

Hersey SJ: Gastric secretion of pepsins. In Johnson LR, ed: *Physiology of the gastrointestinal tract,* ed 3, New York, 1994, Raven Press.

Hoffman AF: Biliary secretion and excretion: the hepatobiliary components of the enterohepatic circulation of bile acids. In Johnson LR, ed: *Physiology of the gastrointestinal tract,* ed 3, New York, 1994, Raven Press.

Sachs G: The gastric H,K-ATPase: regulation and structure/fuction of the acid pump of the stomach. In Johnson LR, ed: *Physiology of the gastrointestinal tract,* ed 3, New York, 1994, Raven Press.

Siegers C-P, Watkins JB III, eds: *Biliary excretion of drugs and other chemicals,* New York, 1991, Gustav Fischer Verlag.

Soll AH, Berglindh T: Receptors that regulate gastric acid secretory function. In Johnson LR, ed: *Physiology of the gastrointestinal tract,* ed 3, New York, 1994, Raven Press.

Tavoloni N, Berk PD, eds: *Hepatic transport and bile secretion,* New York, 1993, Raven Press.

Yule DI, Williams JA: Stimulus-secretion coupling in the pancreatic acinus. In Johnson LR, ed: *Physiology of the gastrointestinal tract,* ed 3, New York, 1994, Raven Press.

CHAPTER 34

Digestion and Absorption

In most instances nutrients cannot be absorbed by the cells that line the gastrointestinal (GI) tract in the forms in which they are ingested. **Digestion** refers to the processes by which ingested molecules are cleaved into smaller ones by reactions catalyzed by enzymes in the lumen or on the luminal surface of the GI tract. As a result of digestion, ingested molecules are converted to forms that can be absorbed from the lumen of the GI tract. **Absorption** refers to the processes by which molecules are transported through the epithelial cells that line the GI tract and then enter the blood or lymph draining that region of the tract.

DIGESTION AND ABSORPTION OF CARBOHYDRATES

Carbohydrates in the Diet

For most people, carbohydrate is usually the principal source of calories. Plant starch, **amylopectin,** is the major source of carbohydrate in most human diets. Amylopectin is a high-molecular-weight (MW $> 10^6$), branched polymer of glucose units. A smaller proportion of dietary starch is **amylose,** a smaller-molecular-weight (MW $> 10^5$), linear α-1,4-linked polymer of glucose. **Cellulose** is a β-1,4-linked glucose polymer. Intestinal enzymes cannot hydrolyze β-glycosidic linkages. Thus cellulose and other molecules with β-glycosidic linkages remain undigested and contribute to **dietary fiber.**

The amount of the animal starch **glycogen** that is ingested varies widely among cultures and among individuals within a given culture. **Sucrose** and **lactose** are the principal dietary disaccharides, and **glucose** and **fructose** are the major monosaccharides.

Digestion of Carbohydrates

The structure of a branched starch molecule is depicted in Figure 34-1. Starch is a polymer of glucose and consists of chains of glucose units linked by α-1,4 glycosidic bonds. The α-1,4 chains have branch points formed by α-1,6 glycosidic linkages.

The digestion of starch begins in the mouth with the action of **salivary amylase.** This enzyme catalyzes the hydrolysis of the internal α-1,4 links of starch, but it cannot hydrolyze the α-1,6 branching links. As shown in Figure 34-1, the principal products of α-amylase digestion of starch are maltose, maltotriose, and branched oligosaccharides known as α-limit dextrins. Considerable digestion of starch by the salivary amylase may occur normally, but this enzyme is not required for normal digestion and absorption of the starch. After salivary amylase is inactivated by the low pH of gastric contents, no further processing of carbohydrate occurs in the stomach.

Pancreatic juice contains a highly active α-amylase. The products of starch digestion by the pancreatic enzyme are the same as for the salivary amylase, but the total amylase activity in pancreatic juice is considerably greater than that in saliva, Within 10 minutes after entering the duodenum, starch is entirely converted to the oligosaccharides shown in Figure 34-1.

The further digestion of these oligosaccharides is accomplished by enzymes that reside in the brush border membrane of the epithelium of the duodenum and jejunum (Figure 34-2). The major brush border oligosaccharidases are **lactase,** which splits lactose into glucose and galactose; **sucrase,** which splits sucrose into glucose and fructose; α-**dextrinase** (also called **isomaltase**), which "debranches" the α-limit dextrins by cleaving the α-1,6 linkages at the branch points; and **glucoamylase,** which cleaves the terminal α-1,4-glycosidic bonds to break maltooligosaccharides down to glucose units. The activities of these four enzymes

481

are highest in the brush border of the upper jejunum, and they gradually decline through the rest of the small intestine.

Absorption of Carbohydrates

The duodenum and upper jejunum have the highest capacity to absorb sugars. The capacities of the lower jejunum and ileum are progressively less. *The only*

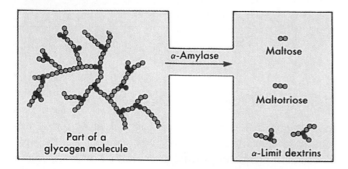

FIGURE 34-1   Structure of a branched starch molecule and the action of α-amylase. The circles represent glucose monomers. The colored circles show glucose units linked by α-1,6 linkages at the branch points. The α-1,6 links and terminal α-1,4 bonds cannot be cleaved by α-amylase.

monosaccharides that are well absorbed are glucose, galactose, and fructose.

Glucose and galactose are actively taken up across the brush border of epithelial cells through an Na^+-powered secondary active transport system (Figure 34-3). Glucose and galactose compete for entry. Na^+ and glucose or galactose are transported into the cell by the transport protein, which has 2 Na^+ binding sites and one sugar binding site. The presence of Na^+ in the intestinal lumen enhances the absorption of glucose and galactose and vice versa. The energy released by Na^+ moving down its electrochemical potential gradient is harnessed to transport glucose or galactose into the cell against a concentration gradient of the sugar. Glucose and galactose leave the intestinal epithelial cell at the basal and lateral plasma membranes via facilitated transport, and they diffuse into the mucosal capillaries.

Fructose does not compete well for the glucose-galactose transporter. However, fructose is transported almost as rapidly as glucose and galactose and much more rapidly than other monosaccharides. Fructose is taken up across the brush border membrane by a fructose-specific facilitated transporter. Fructose crosses the basolateral membrane by a different facilitated transporter that also can transport glucose, but not galactose.

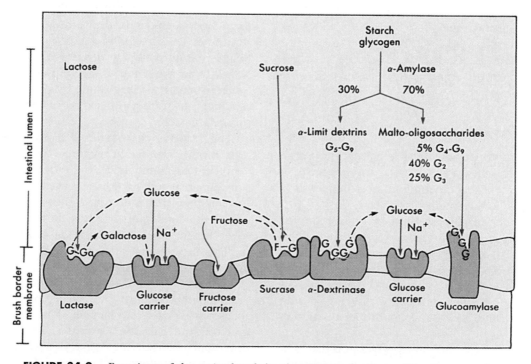

FIGURE 34-2   Functions of the major brush border oligosaccharidases. The glucose, galactose, and fructose molecules released by enzymatic hydrolysis are then transported into the epithelial cells by specific transport proteins in the brush border membrane. *G,* Glucose; *Ga,* galactose; *F,* fructose. (Redrawn from Gray GM: *N Engl J Med* 292:1225, 1975.)

In individuals with low levels of brush border lactase activity, undigested lactose is passed on to the colon. The colonic bacteria rapidly metabolize the lactose, and the bacteria produce gas and release metabolic products that enhance colonic motility and cause diarrhea. This condition is called **lactose intolerance.**

Lactose intolerance in the newborn, **congenital lactose intolerance,** is uncommon. Lactose-intolerant infants are usually fed formula with sucrose as the major source of carbohydrate. By contrast, more than 50% of the world's adults are lactose intolerant. This is genetically determined. Almost all Asian and African adults are lactose intolerant, but most Northern European adults tolerate lactose.

DIGESTION AND ABSORPTION OF PROTEINS

The amount of dietary protein varies greatly among cultures and among individuals within a culture. In poor societies it is difficult for adults to obtain the amount of protein (0.5 to 0.7 g/day/kg of body weight) required to balance the normal catabolism of proteins. It is even more difficult for children to receive the relatively greater amounts of protein required to sustain normal growth. In wealthier societies a typical individual ingests protein far in excess of the nutritional requirement.

In normal humans essentially all ingested protein is digested and absorbed. Most of the protein in digestive secretions and exfoliated cells is also digested and absorbed. The small amount of protein in the feces is derived principally from colonic bacteria, exfoliated colonic cells, and proteins in mucous secretions of the colon. Ingested protein is almost completely absorbed by the time the meal has traversed the jejunum.

Digestion of Proteins

Digestion in the Stomach Pepsinogens are secreted by the chief cells of the stomach and are converted by hydrogen ions to active **pepsins.** The extent to which pepsins hydrolyze dietary protein is significant but highly variable. At most, about 15% of dietary protein may be reduced to peptides and amino acids by pepsins. The duodenum and small intestine have such a high capacity to digest protein that the total absence of pepsins does not impair the digestion and absorption of dietary protein.

Digestion in the Small Intestine Proteases in pancreatic juice play a major role in protein digestion. The most important of these proteases are **trypsin, chymotrypsin,** and **carboxypeptidase.** Pancreatic juice contains these enzymes in inactive, proenzyme forms. The enzyme **enteropeptidase** (also known as **enterokinase**), secreted by the mucosa of the duodenum and jejunum, converts trypsinogen to active trypsin. Trypsin activates trypsinogen and also converts chymotrypsinogen and procarboxypeptidase to the active enzymes (Figure 34-4). The pancreatic proteases are very active in the duodenum and rapidly convert dietary protein to small peptides. About 50% of ingested protein is digested and absorbed in the duodenum.

The brush border of the duodenum and the small intestine contains a number of peptidases (Figure 34-4). These peptidases are integral membrane proteins whose active sites face the intestinal lumen.

The principal products of protein digestion by pancreatic proteases and brush border peptidases are small peptides and single amino acids. The small peptides (mainly dipeptides, tripeptides, and tetrapeptides) are about three or four times more concentrated than the single amino acids. As discussed next, small peptides and amino acids are transported across the brush border plasma membrane into intestinal epithelial cells. Small peptides are then hydrolyzed by peptidases in the cytosol of intestinal epithelial cells; consequently, only single amino acids appear in the portal blood. The cytosolic peptidases are particularly active against dipeptides and tripeptides, which are transported with high efficiency across the brush border membrane. The brush border peptidases, on the other hand, are mainly active against peptides of four or more amino acids.

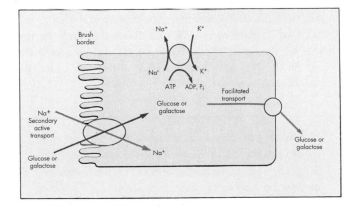

FIGURE 34-3 Major features of glucose and galactose absorption in the small intestine. Glucose and galactose enter the epithelial cell against a concentration gradient. The gradient of Na$^+$ provides the energy for sugar entry. Glucose and galactose leave the cell at the basolateral membrane by facilitated transport.

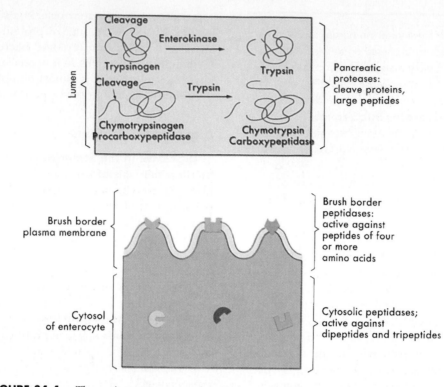

FIGURE 34-4   The major proteases and peptidases present in the lumen of the small intestine, on the brush border plasma membrane, and in the cytosol of the epithelial cells of the small intestine.

Absorption of Protein Digestion Products

Absorption of Intact Proteins and Large Peptides

Intact proteins are not absorbed by humans to an extent that is nutritionally significant, but *amounts of intact proteins sufficient to trigger an immunologic response can be absorbed.* In ruminants and rodents, but not in humans, the neonatal intestine has a high capacity for the specific absorption of immune globulins present in colostrum. This absorptive process is vital in development of normal immune competence in ruminants and rodents.

Absorption of Small Peptides Dipeptides and tripeptides are transported across the brush border membrane. The rate of transport of dipeptides or tripeptides usually exceeds the rate of transport of individual amino acids. For example, glycine is absorbed by the human jejunum less rapidly as the amino acid than it is as glycylglycine or as glycylglycylglycine. A single membrane transport system with broad specificity is responsible for absorption of small peptides. The transport system has high affinity for dipeptides and tripeptides, but very low affinities for peptides of four or more amino acid residues. Transport of dipeptides and tripeptides across the brush border plasma membrane is a secondary active transport process, powered by the electrochemical potential difference of H^+ across the membrane.

Absorption of Amino Acids Amino acids are transported across the brush border plasma membrane into the intestinal epithelial cell by way of certain specific amino acid transport systems. Transport of amino acids out of the epithelial cell across the basolateral membrane occurs by transporters different from the ones in the brush border membrane. Some of the transporters depend on the Na^+ gradient, whereas other transport systems are independent of Na^+.

Hartnup's disease is a rare hereditary disorder in which one of the major neutral amino acid transport proteins is deficient in the brush border of the small intestine and the proximal renal tubule. Individuals with Hartnup's disease have elevated urinary levels of certain neutral amino acids. However, such patients absorb neutral amino acids normally and are not malnourished, because the affected neutral amino acids are absorbed well in dipeptides and tripeptides in the small intestine.

INTESTINAL ABSORPTION OF WATER AND ELECTROLYTES

Under normal circumstances humans absorb almost 99% of the water and ions presented to them in ingested food and in GI secretions (Figure 34-5). Thus net fluxes of water and ions are normally from the lumen of the gut to the blood. In most cases the net fluxes of water and ions are the differences between much larger unidirectional flows from lumen to blood and from blood to lumen.

Absorption of Water

Typically about 2 liters of water are ingested each day, and approximately 7 L/day are contained in GI secretions (Figure 34-5). Only about 50 to 150 ml of water per day are lost in the feces. Thus the GI tract typically absorbs more than 8 L/day.

Very little net absorption occurs in the duodenum, but the chyme is brought to isotonicity in the duodenum. The chyme that is delivered from the stomach is often hypertonic. The action of digestive enzymes creates still more osmotic activity. The epithelium of the duodenum is highly permeable to water and ions, and very large fluxes of water occur from lumen to blood and from blood to lumen. Usually the net flux is from blood to lumen because of the hypertonicity of the chyme. Large net water absorption occurs in the small intestine: the jejunum is more active than the ileum in absorbing water. The net absorption that occurs in the colon is relatively small, approximately 400 ml/day.

Absorption of Na$^+$

Na$^+$ is absorbed along the entire length of the intestine. Net absorption is the result of large unidirectional fluxes of Na$^+$ from blood to lumen and from lumen to blood. The unidirectional fluxes are greater in the proximal gut than in the distal intestine. Na$^+$ crosses the brush border membrane down an electrochemical gradient, and it is actively extruded from the epithelial cells by the Na$^+$-K$^+$-ATPase in the basolateral plasma membrane. The contents of the small bowel are normally isotonic to plasma. Luminal contents have about the same Na$^+$ concentration as plasma, and therefore Na$^+$ absorption normally occurs in the absence of a significant net concentration gradient.

The net rate of absorption of Na$^+$ is highest in the jejunum. Here Na$^+$ absorption is enhanced by the presence of glucose, galactose, and neutral amino acids in the lumen. These substances and Na$^+$ cross the brush bor-

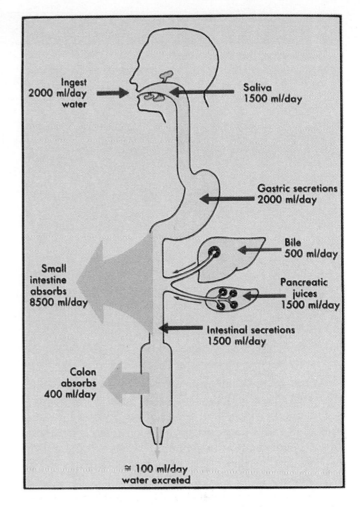

FIGURE 34-5   Overall fluid balance in the human gastrointestinal tract. Approximately 2 liters of water is ingested each day, and 7 liters of various secretions enters the GI tract. Of this total of 9 liters, about 8.5 liters is absorbed in the small intestine. Approximately 500 ml is passed on to the colon, which normally absorbs 80% to 90% of the water presented to it.

der membrane on the same transport proteins. Na$^+$ moves down its electrochemical potential gradient and provides the energy for moving the sugars (glucose and galactose) and neutral amino acids into the epithelial cells against a concentration gradient. In this way Na$^+$ enhances the absorption of sugars and amino acids, and vice versa.

The ability of glucose to enhance the absorption of Na$^+$ and hence of Cl$^-$ and water is exploited in oral rehydration therapy for *cholera* and other secretory diarrheas. When cholera patients drink a solution con-

Table 34-1 Transport of Na^+, K^+, Cl^-, and HCO_3^- in the Large and Small Intestines

Segment of Intestine	Na^+	K^+	Cl^-	HCO_3^-
Jejunum	Actively absorbed; absorption enhanced by sugars, neutral amino acids	Passively absorbed when concentration rises because of absorption of water	Absorbed	Absorbed
Ileum	Actively absorbed	Passively absorbed	Absorbed, some in exchange for HCO_3^-	Secreted, partly in exchange for Cl^-
colon	Actively absorbed	Net secretion occurs when K^+ concentration in lumen < 25 mM	Absorbed, some in exchange for HCO_3^-	Secreted, partly in exchange for Cl^-

taining glucose, NaCl, and certain other constituents, the absorption of salt and water helps to counteract the secretory fluxes of salt and water that characterize this disease.

In the ileum the net rate of Na^+ absorption is smaller than that in the jejunum. Na^+ absorption is only slightly stimulated by sugars and amino acids because the sugar and amino acid transport proteins are less concentrated in the ileum than in the jejunum. The ileum can absorb Na^+ against a larger electrochemical potential than can the jejunum.

In the colon, Na^+ is normally absorbed against a large electrochemical potential difference. Sodium concentrations in the luminal contents can be as low as 25 mM, compared with about 120 mM in the plasma.

Absorption of Cl^- and HCO_3^-

In the jejunum both chloride and bicarbonate are absorbed in large amounts. By the end of the jejunum most of the HCO_3^- of the hepatic and pancreatic secretions has been absorbed. In the ileum Cl^- is absorbed, but HCO_3^- is normally secreted. If the HCO_3^- concentration in the lumen of the ileum exceeds about 45 mM, the flux from lumen to blood exceeds that from blood to lumen, and net absorption occurs. In the colon the transport of these ions is similar to that in the ileum: Cl^- is absorbed and HCO_3^- is usually secreted.

Absorption of K^+

As with the other ions, the net movement of potassium across the intestinal epithelium is the difference between large unidirectional fluxes from lumen to blood and from blood to lumen. In the jejunum and in the il-

eum, the net flux is from lumen to blood. As the volume of intestinal contents is reduced because of the absorption of water, the increased concentration of K^+ provides a driving force for the movement of K^+ across the intestinal mucosa and into the blood.

In the colon, there is usually net secretion of K^+. Net secretion occurs when the luminal concentration is less than about 25 mM; if it is greater than 25 mM, net absorption occurs. The secretion of K^+ is powered by the negative luminal electrical potential (about -30 mV) in the colon.

Most absorption of K^+ results from its enhanced concentration in the lumen caused by the absorption of water. Significant K^+ loss may occur in diarrhea. If diarrhea is prolonged, the K^+ level in the extracellular fluid compartment of the body falls. Maintaining normal K^+ levels is important, especially for the heart and other muscles. K^+ imbalance can have life-threatening consequences, such as **cardiac arrhythmias.** Infants with prolonged diarrhea are particularly susceptible to **hypokalemia** (low plasma K^+).

Table 34-1 summarizes the transport of Na^+, K^+, Cl^-, and HCO_3^- in the small and large intestines.

Ion and Water Transport by Intestinal Epithelium

Tight Junctions The epithelial cells that line the intestine are connected to their neighbors by tight junctions near their luminal surfaces. The tight junctions are leaky; that is, they are somewhat permeable to water and ions. The tight junctions of the duodenum are the least tight and hence have the highest permeability. Tight junctions in the jejunum are somewhat tighter, those in the ileum are still tighter, and those in the colon are tightest.

Transcellular Versus Paracellular Transport Because the tight junctions are leaky, some fraction of the water and ions that traverses the intestinal epithelium passes between, rather than through, the epithelial cells. Transmucosal movement through the tight junctions and the lateral intracellular spaces is called **paracellular transport.** Passage through the epithelial cells is termed **transcellular transport.**

Because the tight junctions in the duodenum are very leaky, most of the large unidirectional fluxes of water and ions that take place in the duodenum occur via the paracellular pathway. The proportions of water or of a particular ion that pass through the transcellular and paracellular routes are determined by the relative permeabilities of the two pathways for the substance in question. Even in the ileum, where the junctions are much tighter than in the duodenum, the paracellular pathway contributes more to the total ionic conductance of the mucosa than does the transcellular pathway.

Mechanism of Water Absorption

The absorption of water depends on the absorption of nutrients, such as sugars and amino acids, and on the absorption of ions, principally Na^+ and Cl^-. Water absorption in the small intestine normally occurs in the absence of a significant osmotic pressure difference between the luminal contents and the blood in the intestinal capillaries. A significant fraction of the net water absorption occurs by a mechanism known as **standing gradient osmosis,** described in Chapter 33. As shown in Figure 33-19, the active pumping of Na^+ into the lateral intercellular spaces by the Na^+-K^+-ATPase drives the absorption of Cl^- and water.

> Any substance that cannot be absorbed in the intestine will, because of its osmotic effect, prevent an iso-osmotic equivalent of water from being absorbed. This is the basis for the action of **osmotic laxatives,** such as magnesium sulfate (epsom salts). When a nutrient is malabsorbed, as in **lactose intolerance,** for example, the osmotic effect of unabsorbed nutrient contributes to diarrhea.

Control of Intestinal Electrolyte Absorption

Electrolyte transport in the intestine is regulated by certain hormones, neurotransmitters, and paracrine substances.

Autonomic Nervous System Stimulation of sympathetic nerves to the intestine or an elevated plasma level of epinephrine increases the absorption of Na^+, Cl^-, and water. Stimulation of parasympathetic nerves to the gut decreases the net rate of ion and water absorption.

Adrenal Hormones Aldosterone strongly stimulates the secretion of K^+ and the absorption of Na^+ and water by the colon and to a much lesser extent by the ileum. Aldosterone acts by increasing the number of Na^+ channels in the luminal membrane of the colonic epithelial cells (Figure 34-6, *C*) and the number of active Na^+-K^+-ATPase molecules in the basolateral membrane. Aldosterone has similar effects on the epithelial cells of the distal tubule of the kidney (see Chapter 37). The enhanced absorption of NaCl and water induced by aldosterone in the colon and kidney is an important mechanism in the body's compensatory response to dehydration. Glucocorticoids also increase the content of Na^+-K^+-ATPase in the basolateral membrane and thereby enhance Na^+ and water absorption and K^+ secretion in the colon.

Ion Transport Processes in the Intestine

The ion transport processes that occur in the jejunum, ileum, and colon are summarized in Figure 34-6.

Ion Transport in the Jejunum In the jejunum (Figure 34-6, *A*) there is net absorption of Na^+, Cl^-, and HCO_3^-. Na^+ enters the epithelial cell at the brush border via the nutrient-coupled, Na^+-powered transporters (electrogenic) and via the Na^+/H^+ exchanger. Na^+ is extruded from the cell across the basolateral membrane by the Na^+-K^+-ATPase. Absorption of Cl^- and HCO_3^- is powered by the slight luminal electronegativity generated by Na^+ uptake at the brush border and by the Na^+-K^+-ATPase and the concentration of luminal ions by the large net absorption of water in the jejunum.

Acidification of the jejunal contents by gastric acid and by the Na^+/H^+ exchanger pushes the bicarbonate/carbonic acid equilibrium toward carbonic acid, which is in equilibrium with CO_2 and water. CO_2 is highly diffusible (see Chapter 30) and is readily absorbed across the mucosa and into the blood. Most of the HCO_3^- dumped into the duodenum in bile and pancreatic juice is absorbed by this mechanism.

Ion Transport in the Ileum In the ileum (Figure 34-6, *B*), net absorption of Na^+ and Cl^- occurs by mechanisms similar to those in the jejunum. The Na^+-powered nutrient transporters are less numerous than in the jejunum. Net secretion of HCO_3^- in exchange for absorption of Cl^- occurs via an **anion exchanger** in the brush border membrane. HCO_3^- enters the epithelial cells across the basolateral membrane by Na^+-powered secondary active transport. Note that the coupled operation

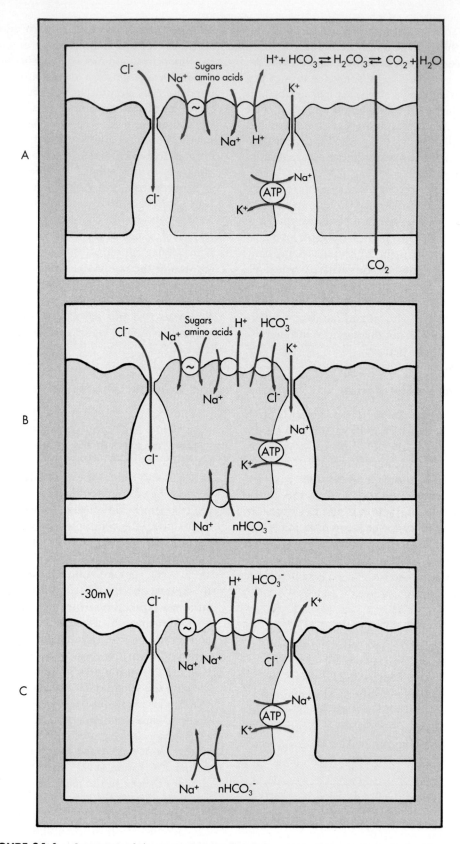

FIGURE 34-6   Summary of the major ion transport processes that occur in **A,** the jejunum; **B,** the ileum; and **C,** the colon. *ATP,* Adenosine triphosphate.

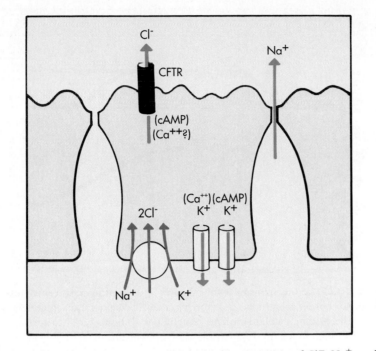

FIGURE 34-7   The ion transport processes involved in secretion of Cl^-, Na^+, and water by the epithelial cells in Lieberkühn's crypts in the small intestine. Cl^- is actively taken up into the cell at the basolateral membrane; this is powered by the Na^+ gradient. Cl^- enters the lumen via an electrogenic channel; the resulting luminal negativity drives secretion of Na^+, partly via the tight junctions. The secretion of Na^+ and Cl^- results in osmotic movement of water into the lumen.

of the Na^+/H^+ exchanger and the Cl^-/HCO_3^- transporter of the luminal membrane results in the absorption of NaCl and the secretion of H_2CO_3. In both jejunum and ileum K^+ is concentrated by absorption of water; the elevated K^+ concentration drives net absorption of K^+, predominantly via the tight junctions.

Ion Transport in the Colon   In the colon net absorption of Na^+ and Cl^- and net secretion of HCO_3^- occur by mechanisms similar to those in the ileum. In the colon, however, entry of Na^+ across the brush border membrane occurs via an electrogenic Na^+ channel. Because the tight junctions of the colon are so tight, electrogenic Na^+ transport produces an electrical potential of about 30 mV (lumen negative) across the mucosa. This potential drives the net secretion of K^+ into the lumen, mostly via the tight junctions. In the distal colon, active absorption of K^+ and secretion of H^+ are powered by an H^+-K^+-ATPase like the gastric H^+ pump.

Intestinal Secretion of Electrolytes and Water

The normal net absorption of Na^+, Cl^-, and water is the result of large unidirectional fluxes from lumen to blood and from blood to lumen. Mature intestinal epithelial cells near the tips of the villi are active in net absorption of Na^+, Cl^- and water; the processes described

above and shown in Figure 34-6 occur in the cells at the villous tips. The more immature epithelial cells in the crypts of Lieberkühn are net secretors of Na^+, Cl^-, and water (Figure 34-7). Secretion by the crypt cells is a normal physiological function and is subject to physiological regulation.

Cl^- is actively transported into the crypt cell across the basolateral membrane (Figure 34-7); Cl^- is driven into the cell by the electrochemical potential of Na^+ via a transporter that is known as the **Na^+,K^+,2 Cl^- transporter.** Cl^- is secreted into the lumen across the brush border membrane via an **electrogenic chloride channel.** Na^+ is secreted along with Cl^- to preserve electroneutrality, and water is secreted because of the osmotic pressure generated by the secretion of Na^+ and Cl^-.

In **secretory diarrheal diseases,** such as **cholera,** the secretion of Cl^-, Na^+, and water into the intestinal lumen by the cells in Lieberkühn's crypts is specifically elevated. Cholera is caused by cholera toxin produced by the bacterium *Vibrio cholerae*. Cholera toxin acts to permanently activate adenylyl cyclase, elevating the level of cAMP in the crypt cells. cAMP activates the brush border

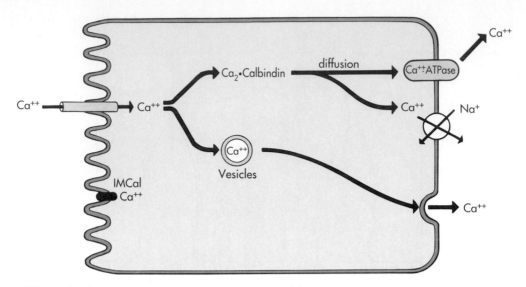

FIGURE 34-8 Cellular mechanisms involved in Ca^{++} absorption in the small intestine. Ca^{++} crosses the brush border membrane probably via a Ca^{++} channel. An integral membrane protein (intestinal membrane calcium-binding protein, IMCal) binds Ca^{++} at the inner surface of the brush border membrane. In the cytosol of the epithelial cell, Ca^{++} is bound to calbindin. Ca^{++} is extruded across the basolateral membrane by a Ca^{++}-ATPase and by an Na^+/Ca^{++} exchange mechanism. Intracellular vesicles take up some Ca^{++}, transport it through the cytosol, and release it across the basolateral membrane by exocytosis.

Cl^- channels and thereby enhances the secretion of Cl^- (and therefore also of Na^+ and water). Cholera patients may produce up to 20 L/day of watery stool! Such patients are likely to die, unless they are promptly and adequately rehydrated.

The brush border Cl^- channel is the same protein that is defective in **cystic fibrosis (CF).** CF is by far the most common autosomal recessive disorder; about 1 in 20 American adults are CF carriers. CF carriers, who have one normal and one defective copy of the gene for the Cl^- channel, may be less likely to suffer severe cholera symptoms than normal individuals. Because cholera and related secretory diarrheas are the major causes of loss of life in children in societies that lack modern sanitation, the resistance of CF carriers to secretory diarrheas may explain the unusual prevalence of this mutation.

ABSORPTION OF CALCIUM

Calcium ions are actively absorbed in all segments of the intestine. The duodenum and jejunum are especially active and can absorb Ca^{++} against a greater than tenfold concentration gradient.

The ability of the intestine to absorb Ca^{++} is regulated. Animals with a calcium-deficient diet increase their ability to absorb Ca^{++}. Animals receiving high-calcium diets have less capacity to absorb Ca^{++}. Intestinal absorption of Ca^{++} is stimulated by vitamin D and slightly stimulated by parathyroid hormone (see Chapter 43).

Brush Border Plasma Membrane

Ca^{++} moves down its electrochemical potential gradient across the brush border membrane, via a Ca^{++} channel (Figure 34-8). An integral protein of the brush border plasma membrane, called the **intestinal membrane calcium-binding protein** (IMCal) appears to be involved in binding Ca^{++} near the inner surface of the brush border membrane.

Epithelial Cell Cytosol

The cytosol of the intestinal epithelial cells contains a calcium-binding protein called **calbindin** or **CaBP.** In mammals the calbindin has a molecular weight of about 9000 and binds two calcium ions with high affinity. The level of calbindin in the epithelial cells correlates well with the capacity to absorb Ca^{++}. Calbindin allows large amounts of Ca^{++} to traverse the cytosol, but it averts concentrations of free Ca^{++} that are high enough to form insoluble salts with intracellular anions. In addition, Ca^{++} traverses the epithelial cell cytosol in membrane-bounded vesicles, with which a fraction of the calbindin is associated.

Basolateral Membrane

The basolateral plasma membrane contains two transport proteins that are capable of ejecting Ca^{++} from the cell against its electrochemical potential gradient. A **Ca^{++}-ATPase** extrudes Ca^{++} across the basolateral plasma membrane. A smaller amount of Ca^{++} is transported across the basolateral plasma membrane by an **Na^+/Ca^{++} exchanger.** The Ca^{++}-containing vesicles are believed to extrude Ca^{++} across the basolateral plasma membrane by exocytosis (Figure 34-8).

Actions of Vitamin D

Vitamin D is essential for development of the normal capacity for calcium absorption by the intestine. The actions of vitamin D are also discussed in Chapter 43.

In **rickets,** a disease caused by vitamin D deficiency, the rate of absorption of Ca^{++} is very low. In children with rickets, because of insufficient availability of Ca^{++}, the growth of bones is abnormal. Because of failure to deposit normal levels of calcium salts in the bone matrix, bones are softer and more flexible than normal. These changes contribute to the characteristic "bow-legged" appearance of children with rickets.

Vitamin D has multiple effects that enhance the absorption of Ca^{++} by the epithelium of the small intestine. Treatment with vitamin D increases transport of Ca^{++} across the brush border membrane; the mechanism of this effect is not clear. Vitamin D treatment also enhances the transport of Ca^{++} through the cytosol of the intestinal epithelial cell by dramatically increasing the level of calbindin. Vitamin D increases the rate of extrusion of Ca^{++} across the basolateral membrane of intestinal epithelial cells by increasing the level of Ca-ATPase in the membrane.

ABSORPTION OF IRON

A typical adult should ingest approximately 15 to 20 mg of iron daily. Of this amount, only 0.5 to 1 mg is absorbed by normal adult men and 1 to 1.5 mg is absorbed by premenopausal adult women. Iron depletion (e.g.,

caused by hemorrhage) increases the capacity of the intestine to absorb iron. Growing children and pregnant women absorb greater amounts of iron than do adult men.

Iron absorption is limited because iron tends to form insoluble salts with hydroxide, phosphate, bicarbonate, and other anions present in intestinal secretions. Iron also forms insoluble complexes with other substances typically present in food, such as phytate, tannins, and the fiber of cereal grains. These iron complexes are more soluble at low pH. Therefore HCl secreted by the stomach enhances iron absorption, whereas iron absorption is usually low in individuals deficient in HCl secretion.

Vitamin C effectively promotes iron absorption by forming a soluble complex with iron and by reducing Fe^{+++} to Fe^{++}. Iron complexed with ascorbate or in the form of Fe^{++} has less tendency to form insoluble complexes than Fe^{+++}, and thus is better absorbed. Individuals who take iron supplements are well advised to ingest vitamin C along with their iron tablets.

Absorption of Heme Iron

Iron is present in the diet as inorganic iron salts and as part of the heme prosthetic groups of proteins such as hemoglobin, myoglobin, and cytochromes. About 20% of ingested heme iron is absorbed. Proteolytic enzymes release heme groups from proteins in the intestinal lumen. Heme is taken up by facilitated transport by the epithelial cells that line the upper small intestine. In the epithelial cell, iron is split from the heme. No intact heme is transported into the portal blood.

Cellular Mechanism of Iron Absorption

The epithelial cells of the duodenum and jejunum release **transferrin,** an iron-binding protein, into the lumen (Figure 34-9). The intestinal transferrin is similar, but not identical, to the transferrin that is the principal iron-binding protein of plasma. In the lumen of the duodenum and jejunum, each transferrin can bind two iron ions. Receptors on the brush border surface of the duodenum and jejunum bind the transferrin-iron complex, and the complex is taken up into the epithelial cell by receptor-mediated endocytosis. In the cytosol of the intestinal epithelial cell, transferrin acts as a soluble iron carrier. Much of the transferrin, after it releases its bound iron, is resecreted into the lumen. Iron ultimately appears in plasma bound to plasma transferrin. Transport of iron

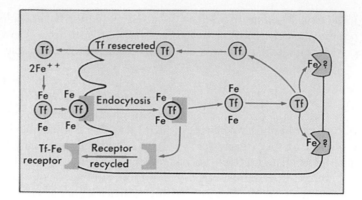

FIGURE 34-9 Current view of the mechanism of iron absorption by the epithelial cells of the small intestine. *Tf* denotes a form of transferrin that is secreted into the lumen of the intestine, where it binds Fe^{++}.

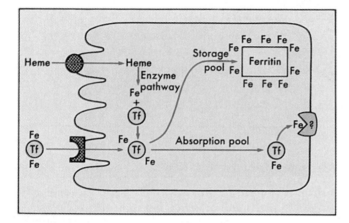

FIGURE 34-10 In the epithelial cells of the small intestine there are two pools of iron. One pool of iron is available for absorption into the blood, where iron is bound to transferrin. The other pool, iron that becomes bound to ferritin in the epithelial cell, is unavailable for transport across the basolateral membrane; this iron is lost into the lumen when the epithelial cell is sloughed off. The latter mechanism prevents absorption of more iron than is needed.

across the basolateral membrane requires metabolic energy.

Regulation of Iron Absorption

Iron absorption is regulated in accordance with the body's need for iron. *In chronic iron deficiency or after hemorrhage, the capacity of the duodenum and jejunum to absorb iron is increased.* The intestine also protects the body from the consequences of absorbing too much iron.

An important mechanism for preventing excess absorption of iron is the almost irreversible binding of iron to **ferritin** *in the intestinal epithelial cell.* Iron

bound to ferritin is not available for transport into the plasma (Figure 34-10), and it is lost into the intestinal lumen and excreted in the feces when the intestinal epithelial cell is sloughed off. The amount of apoferritin present in the intestinal cells determines how much iron can be trapped in the nonabsorbable pool. The synthesis of apoferritin is stimulated by iron, and this protects against absorption of excessive amounts of iron.

The capacity of the duodenum and jejunum to absorb iron increases 3 or 4 days after a hemorrhage. The intestinal epithelial cells require this time to migrate from their sites of formation in Lieberkühn's crypts to the tips of the villi, where they are involved in absorptive activities. The iron-absorbing capacity of the epithelial cells is programmed when the cells are in Lieberkühn's crypts. The brush border membranes of the duodenum and jejunum of an iron-deficient animal also have an increased number of receptors for the complex of iron with transferrin and thus absorb the iron-transferrin complex from the lumen more rapidly.

ABSORPTION OF OTHER IONS

Magnesium Magnesium is absorbed along the entire length of the small intestine. About half the normal dietary intake is absorbed, and the rest is excreted.

Phosphate Phosphate is absorbed, in part by active transport, all along the small intestine.

Copper Copper is absorbed in the jejunum; approximately 50% of the ingested load is absorbed. Copper is secreted in the bile bound to certain bile acids; this copper is lost in the feces.

Table 34-2 Intestinal Absorption of Vitamins

Vitamin	Species	Site of Absorption	Transport Mechanism	Maximal Absorptive Capacity in Humans (Per Day)	Dietary Requirement in Humans (Per Day)
Ascorbic acid (C)	Humans, guinea pig	Ileum	Active	> 5000 mg	< 50 mg
Biotin	Hamster	Upper small intestine	Active	?	?
Choline	Guinea pig, hamster	Small intestine	Facilitated	?	?
Folic acid					
Pteroylglutamate	Rat	Jejunum	Facilitated	> 1000 μg/dose	100-200 μg
5-Methyltetrahydrofolate	Rat	Jejunum	Diffusion		
Nicotinic acid	Rat	Jejunum	Facilitated	?	10-20 mg
Pantothenic acid		Small intestine	?	?	(?) 10 mg
Pyridoxine (B_6)	Rat, hamster	Small intestine	Diffusion	> 50 mg/dose	1-2 mg
Riboflavin (B_2)	Humans, rat	Jejunum	Facilitated	10-12 mg/dose	1-2 mg
Thiamin (B_1)	Rat	Jejunum	Active	8-14 mg	$\approx$ 1 mg
Vitamin B_{12}	Humans, rat, hamster	Distal ileum	Active	6-9 μg	3-7 μg

Data from Matthews DM: In Smyth DH, ed: *Intestinal absorption,* vol 4B: Biomembranes, London, 1974, Plenum Press, and Rose RC: *Annu Rev Physiol* 42:157, 1980.

ABSORPTION OF WATER-SOLUBLE VITAMINS

Most water-soluble vitamins can be absorbed by simple diffusion if they are taken in sufficiently high doses. Nevertheless, specific transport mechanisms are important in the normal absorption of most water-soluble vitamins (Table 34-2).

Vitamin B_{12}

A specific active transport process has been implicated in the absorption of vitamin B_{12}. The dietary requirement for B_{12} is fairly close to the maximal absorptive capacity for the vitamin (Table 34-2). Enteric bacteria synthesize vitamin B_{12} and other B vitamins, but the colonic epithelium lacks specific mechanisms for their absorption.

When the intestinal absorption of vitamin B_{12} is impaired, the resulting vitamin B_{12} deficiency retards the maturation of red blood cells and causes the disease called **pernicious anemia.** Because of the occurrence of this disorder, much attention has focused on the absorption of vitamin B_{12}.

Most patients with pernicious anemia have pronounced atrophy of gastric glands, and their stomachs are defective in secreting HCl and pepsins, as well as intrinsic factor. These individuals have circulating antibodies against parietal cells; the antibodies may cause the destruction of parietal cells.

Storage in Liver The liver contains a large store of vitamin B_{12} (2 to 5 mg). Vitamin B_{12} is normally present in the bile (0.5 to 5 μg daily), but approximately 70% of this is normally reabsorbed. Because only about 0.1% of the store is lost daily, even if absorption totally ceases, the store will last for 3 to 6 years.

Absorption of Vitamin B_{12}

Gastric Phase Most of the vitamin B_{12} present in food is bound to proteins. The low pH in the stomach and the digestion of proteins by pepsins releases free vitamin B_{12}, which is then rapidly bound to a class of glycoproteins known as **R proteins.** R proteins are present in saliva and in gastric juice, and they bind vitamin B_{12} tightly over a wide pH range.

Intrinsic factor (IF) is a vitamin B_{12}–binding protein secreted by the gastric parietal cells. IF binds vitamin B_{12} with less affinity than do the R proteins; thus in the stomach most of the vitamin B_{12} present in food is bound to R proteins.

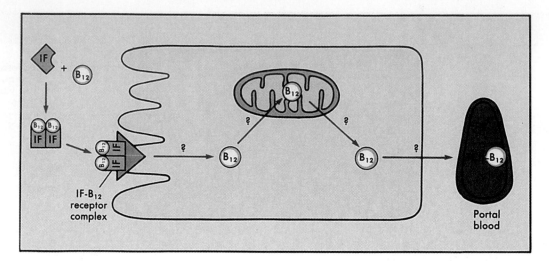

FIGURE 34-11 Mechanism of absorption of vitamin B_{12} by epithelial cells in the ileum. *IF,* Intrinsic factor. The IF-B_{12} complex is taken up across the brush border membrane, and the IF is degraded. The intracellular handling of vitamin B_{12} is poorly understood, but it causes a time lag of about 4 hours in the absorption process. Vitamin B_{12} in portal blood is bound to transcobalamin II *(TCII).*

Intestinal Phase Pancreatic proteases degrade the complex between R proteins and vitamin B_{12}, which causes vitamin B_{12} to be released. The free vitamin B_{12} is taken up by IF, which is highly resistant to digestion by pancreatic proteases.

Mechanism of Absorption The normal absorption of vitamin B_{12} depends on the presence of IF (Figure 34-11). The brush border plasma membranes of the epithelial cells of the ileum contain a receptor protein that recognizes and binds the IF-B_{12} complex. Free IF does not compete for binding, and the receptor does not recognize free vitamin B_{12}. Binding to the receptor is required for normal B_{12} uptake. After being absorbed, vitamin B_{12} appears in the portal blood bound to a protein called **transcobalamin II.**

Absorption in the Absence of Intrinsic Factor In the complete absence of IF, approximately 1% to 2% of the vitamin B_{12} ingested is absorbed. If large doses of B_{12} are taken (about 1 mg/day), enough can be absorbed to treat pernicious anemia.

DIGESTION AND ABSORPTION OF LIPIDS

The primary lipids of a normal diet are **triglycerides.** The diet contains smaller amounts of **sterols** (such as cholesterol), **sterol esters,** and **phospholipids.** Because lipids are only slightly soluble in water, they pose special problems at every stage of their processing. In the stomach lipids tend to separate out into an oily phase. In the duodenum and small intestine lipids are **emulsified** with the aid of bile acids. The emulsion consists of small droplets of lipid coated by bile acids. The large surface area of the emulsion droplets allows access of the water-

soluble lipolytic enzymes to their substrates. The digestion products of lipids form small molecular aggregates, known as **micelles,** with the bile acids. The micelles are small enough to diffuse among the microvilli and to allow absorption of the lipids from molecular solution along the entire surface of the intestinal brush border.

Lipids in the Stomach

Because fats tend to separate out into an oily phase, they usually are emptied from the stomach later than the other gastric contents. Fat in the duodenum strongly inhibits gastric emptying. This helps to ensure that the fat is not emptied from the stomach more rapidly than it can be accommodated by the duodenal mechanisms that provide for emulsification and digestion.

Digestion of Lipids and Micelle Formation

Lingual lipase is produced by serous glands in the tongue. **Gastric lipase** is secreted by chief cells. Together these two lipases constitute **preduodenal lipase.** These enzymes are specific for hydrolysis of triglycerides. In humans gastric lipase is much more abundant than lingual lipase. The amount of triglyceride hydrolyzed by preduodenal lipases varies considerably among individuals.

Lipases present in pancreatic juice are responsible for hydrolysis of most dietary lipid. The lipolytic enzymes of the pancreatic juice are water-soluble molecules and thus have access to lipids only at the surfaces of the fat droplets. *The surface available for digestion is increased many thousand times by emulsification of the lipids* (Figure 34-12). Bile acids themselves are rather

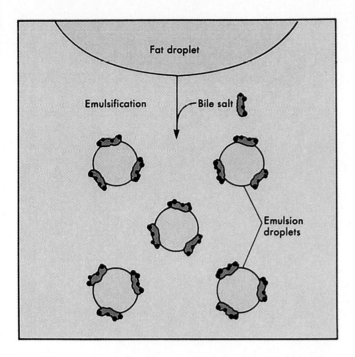

FIGURE 34-12   Emulsification of fats by bile salts and lecithin greatly increases the surface area available to the fat-digesting enzymes. (Vander AJ et al: *Human physiology*, ed 4, New York, 1985, McGraw-Hill.)

poor emulsifying agents. However, with the aid of lecithins, which are present in high concentration in bile, the bile acids emulsify dietary fats.

Action of Lipolytic Enzymes   Pancreatic juice contains the major lipolytic enzymes responsible for digestion of lipids.

Glycerol ester hydrolase, also called **pancreatic lipase,** cleaves the 1 and 1′ fatty acids preferentially from a triglyceride to produce two free fatty acids and one 2-monoglyceride. **Colipase,** a small protein present in pancreatic juice, is essential for the function of glycerol ester hydrolase. Colipase is required for glycerol ester hydrolase to bind to the surface of the emulsion droplets in the presence of bile acids.

Cholesterol esterase cleaves the ester bond in a cholesterol ester to give one fatty acid and free cholesterol.

Phospholipase A$_2$ cleaves the ester bond at the 2 position of a glycerophosphatide to yield, in the case of lecithin, one fatty acid and one lysolecithin.

Infants with **cystic fibrosis** (CF) secrete extremely low levels of pancreatic enzymes. This is because the defective Cl$^-$ channels in the apical membranes of pancreatic acinar cells make the cells unable to secrete Cl$^-$, Na$^+$, and water into the acinar lumen. Because little water is secreted, mucus obstructs the small pancreatic ducts,

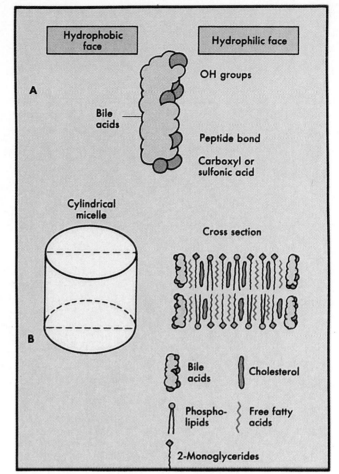

FIGURE 34-13   Structure of bile acids and micelles. **A,** A bile acid molecule is amphipathic because it has a hydrophobic face and a hydrophilic face. **B,** Model of the structure of a bile acid–lipid mixed micelle, showing the way that bile acids and the major products of lipid digestion pack into the mixed micelle.

and the pancreatic acinar cells are destroyed. (Obstruction by mucus of bronchioles is responsible for the pulmonary problems associated with CF.) Because of the deficiency of pancreatic lipases, children with CF have marked difficulties in digesting dietary lipids. Consequently, they may suffer from **steatorrhea** (fatty stool) and malnutrition.

Formation of Micelles   Bile acids form micelles with the products of fat digestion, especially 2-monoglycerides. The micelles are multimolecular aggregates (about 5 nm in diameter) containing approximately 20 to 30 molecules (Figure 34-13). Bile acids are flat molecules that have a polar face and a nonpolar face. Much of the surface of the micelles is cov-

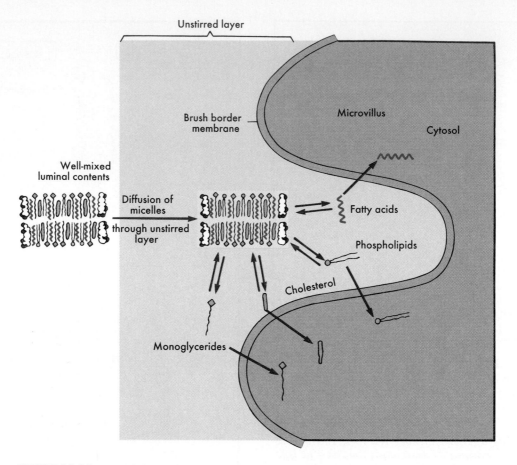

FIGURE 34-14   Lipid absorption in the small intestine. Mixed micelles of bile acids and lipid digestion products diffuse through the unstirred layer. As lipid digestion products are absorbed from free solution, more lipids partition out of the micelles. Because lipids are rapidly transported across the brush border membrane, the diffusion of the micelles through the unstirred layer is the rate-limiting step in lipid absorption.

ered with bile acids, with the nonpolar face toward the lipid interior of the micelle and the polar face toward the outside. Hydrophobic molecules, such as long-chain fatty acids, monoglycerides, phospholipids, cholesterol, and fat-soluble vitamins, tend to partition into the micelles.

Bile acids must be present at a certain minimal concentration, called the critical micelle concentration, before micelles will form. In the normal state, bile acids are always present in the duodenum at greater than the critical micelle concentration.

Absorption of Lipid Digestion Products

Transport into the Intestinal Epithelial Cell Micelles are important in the absorption of the products of lipid digestion and in the absorption of most other fat-soluble molecules (e.g., fat-soluble vitamins). The micelles diffuse among the microvilli that form the brush border, and this allows the huge surface area of the brush border membrane to participate in lipid absorption (Figure 34-14). The presence of micelles tends to keep the aqueous solution near the brush border plasma membrane saturated with fatty acids, 2-monoglycerides, cholesterol, and other micellar contents.

Because of their high lipid solubility, the fatty acids, 2-monoglycerides, cholesterol, and lysolecithin can diffuse across the brush border membrane. In addition, the brush border plasma membrane contains specific transport proteins that facilitate the transport of particular lipid digestion products. An Na^+-dependent **fatty acid transport protein** enhances the movement of long-chain fatty acids across the brush border mem-

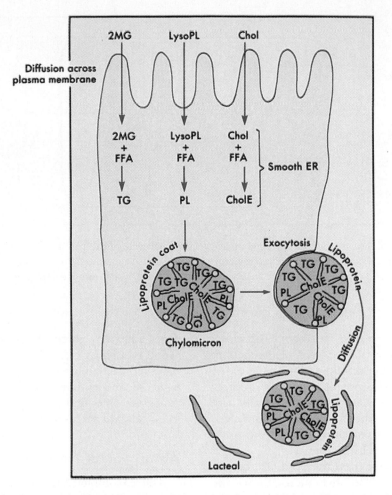

FIGURE 34-15 Resynthesis of lipids in the epithelial cells of the small intestine, formation of chylomicrons, and subsequent transport of chylomicrons into the lymphatic vessels. *FFA*, Free fatty acid; *2MG*, 2-monoglyceride; *TG*, triglyceride; *PL*, phospholipid; *lysoPL*, lysophospholipid; *Chol*, cholesterol; *CholE*, cholesterol ester; *ER*, endoplasmic reticulum.

branes. Another transport protein mediates facilitated transport of cholesterol across the brush border plasma membrane. Because lipids can be taken up across the brush border plasma membrane quite rapidly, the main limitation to the rate of lipid uptake by the epithelial cells of the upper small intestine is the diffusion of the mixed micelles through an unstirred layer on the luminal surface of the brush border plasma membrane (Figure 34-14). Partly because of the convoluted surface of the intestinal mucosa, the fluid in immediate contact with the epithelial cell surface is not readily mixed with the bulk of the luminal contents. Thus this fluid forms an effective unstirred layer with an effective thickness of 200 to 500 μm. Nutrients present in the well-mixed contents of the intestinal lumen must diffuse through the unstirred layer to reach the plasma membrane of the brush border.

The duodenum and jejunum are most active in fat ab-

sorption, and most of the ingested fat is absorbed by the time chyme reaches midjejunum. The fat present in normal stool is not ingested fat (which is completely absorbed), but it is derived from colonic bacteria and from exfoliated intestinal epithelial cells.

Lipid Processing in Intestinal Epithelial Cells
The products of lipid digestion are taken up by the smooth endoplasmic reticulum. A cytoplasmic **fatty acid–binding protein** transports fatty acids, and a **sterol-binding protein** transports cholesterol to the smooth endoplasmic reticulum. In the smooth endoplasmic reticulum, which is engorged with lipid after a meal, considerable chemical reprocessing occurs (Figure 34-15). The 2-monoglycerides are reesterified with fatty acids at the 1 and 1′ carbons to reform triglycerides. Lysophospholipids are reconverted to phospholipids. Cholesterol is reesterified to a considerable extent.

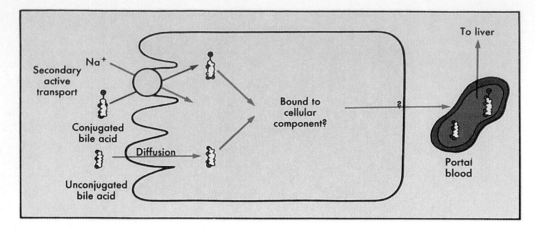

FIGURE 34-16 Absorption of bile acids by epithelial cells of the terminal ileum. Bile acids are absorbed both by simple diffusion and by Na$^+$-powered secondary active transport. Conjugated bile acids are absorbed mainly by active transport; unconjugated bile acids are absorbed chiefly by diffusion.

Chylomicron Formation and Transport The reprocessed lipids accumulate in the vesicles of the smooth endoplasmic reticulum. Phospholipids cover the external surfaces of these lipid droplets. The lipid droplets, approximately 10 nm in diameter at this point, are known as **chylomicrons.** About 10% of their surface is covered by β-lipoprotein, some of which is synthesized in the intestinal epithelial cells.

Chylomicrons are ejected from the epithelial cell by exocytosis (Figure 34-15). The chylomicrons leave the cells at the level of the nuclei and enter the lateral intercellular spaces. Chylomicrons are too large to pass through the basement membrane that invests the mucosal capillaries. However, they do enter the lacteals, which have sufficiently large fenestrations for the chylomicrons to pass through. The chylomicrons leave the intestine with the lymph, primarily via the thoracic lymphatic duct, and flow into the venous circulation.

Absorption of Bile Acids

The absorption of dietary lipids is typically complete by the time chyme reaches midjejunum. By contrast, bile acids are absorbed largely in the terminal part of the ileum. Bile acids cross the brush border plasma membrane by two routes: by an Na$^+$-powered secondary active transport process and by simple diffusion (Figure 34-16). Conjugated bile acids are the principal substrates for active absorption; deconjugated bile acids have poor affinity for the transporter. Deconjugation and dehydroxylation make bile acids less polar, and thus better absorbed by simple diffusion.

Absorbed bile acids are carried away from the intestine in the portal blood. Hepatocytes avidly extract bile

acids, and they essentially clear the bile acids from the blood in a single pass through the liver. In the hepatocytes most deconjugated bile acids are reconjugated, and some secondary bile acids are rehydroxylated. The reprocessed bile acids, together with newly synthesized bile acids, are secreted into bile.

Absorption of Fat-Soluble Vitamins

The fat-soluble vitamins (vitamins A, D, E, and K) partition into the mixed micelles formed by the bile acids and lipid digestion products. *The presence of bile acids and lipid digestion products enhances the absorption of fat-soluble vitamins.* The fat-soluble vitamins diffuse across the brush border plasma membrane into the intestinal epithelial cell. In the intestinal epithelial cell, the fat-soluble vitamins enter the chylomicrons and leave the intestine in the lymph. In the absence of bile acids, a significant fraction of the ingested load of a fat-soluble vitamin may be absorbed and leave the intestine in the portal blood.

SUMMARY

1. The α-amylases of saliva and pancreatic juice cleave branched starch molecules into maltose, maltotriose, and α-limit dextrins. These digestion products are then reduced to glucose molecules by glucoamylase and α-dextrinase on the brush border membrane.
2. Sucrase and lactase on the brush border membrane cleave sucrose and lactose into monosaccharides that can be transported into the intestinal epithelial cells by the monosaccharide transport proteins of

the brush border membrane (i.e., the glucose-galactose transporter and the fructose transporter).

3. Protein digestion begins in the stomach by pepsins. Pancreatic proteases rapidly cleave proteins in the duodenum and jejunum, primarily to oligopeptides. Peptidases on the brush border membrane reduce oligopeptides to single amino acids and to small peptides.

4. Amino acids are transported across the brush border membrane by an array of amino acid transporters. Dipeptides and tripeptides are taken up by a brush border peptide transporter with broad specificity.

5. A typical human ingests about 2 liters of water daily, and about 7 liters enters the GI tract in various secretions. About 99% of the water presented to the GI tract is absorbed; only about 100 ml of water escapes each day into the feces.

6. The absorption of water is powered by the absorption of nutrients and salts. The greatest quantity of water is absorbed in the small intestine, especially the jejunum. Mature cells at the tips of the villi are active in salt, nutrient, and water absorption. Cells in the crypts of Lieberkühn are net secretors of ions and water.

7. Calcium is actively absorbed in the small intestine. Calbindin, a Ca^{++}-binding protein, facilitates the transport of Ca^{++} through the cytosol of the intestinal epithelial cell. Ca^{++} is transported across the basolateral membrane by the Ca^{++}-ATPase and the Na^{+}/Ca^{++} exchange protein.

8. Vitamin D stimulates the absorption of Ca^{++} by enhancing the synthesis of calbindin and of the Ca^{++}-ATPase of the basolateral membrane. The capacity of the intestinal epithelial cells to absorb Ca^{++} is regulated in accordance with the body's need for Ca^{++}.

9. About 5% of inorganic iron ingested is absorbed by the small intestine; approximately 20% of heme iron is absorbed. The small intestinal epithelial cells secrete transferrin into the lumen. The complex of iron with transferrin is taken up at the small intestinal brush border by receptor-mediated endocytosis. In the epithelial cells some iron is bound to ferritin and is unavailable for absorption. The capacity to absorb iron increases in response to hemorrhage.

10. Most water-soluble vitamins are taken up by specific transporters in the small intestinal brush border membrane.

11. Vitamin B_{12} is bound to R proteins in saliva and gastric juice. When R proteins are digested, vitamin B_{12} is bound by intrinsic factor (IF). Receptors on the ileal brush border membrane recognize the IF-B_{12} complex and allow vitamin B_{12} to be absorbed by the ileal epithelial cell. Vitamin B_{12} appears in the plasma bound to transcobalamin II.

12. Triglycerides are the major dietary lipids. Lipids form droplets in the stomach and are emulsified in the duodenum by bile acids. Emulsification greatly increases the surface area available for the action of lipid-digesting enzymes of pancreatic juice.

13. The products of triglyceride digestion, 2-monoglycerides and fatty acids, form mixed micelles with bile acids. Cholesterol, fat-soluble vitamins, and other lipids partition into the micelles. Mixed micelles are small enough to diffuse among the microvilli and thus greatly enhance the brush border surface area available for lipid absorption.

14. In the epithelial cells, triglycerides and phospholipids are resynthesized and packaged along with other lipids into chylomicrons. Chylomicrons are coated with apolipoproteins and released at the basolateral membrane by exocytosis. Chylomicrons leave the intestine in the lymphatic vessels and the thoracic duct.

BIBLIOGRAPHY
Journal Articles

Buddington RK, Diamond JM: Ontogenetic development of intestinal nutrient transporters, *Annu Rev Physiol* 51:601, 1989.

Caspary WF: Physiology and pathophysiology of intestinal absorption, *Am J Clin Nutr* 55:S299, 1992.

Cheeseman CI: Molecular mechanisms involved in regulation of amino acid transport, *Progr Biophys Mol Biol* 55:71, 1991.

Eastwood MA: The physiological effect of dietary fiber: an update, *Annu Rev Nutr* 12:19, 1992.

Ferraris RP, Diamond JM: Specific regulation of intestinal nutrient transporter by their dietary substrates, *Annu Rev Physiol* 51:125, 1989.

Gray GM: Starch digestion and absorption in nonruminants, *J Nutr* 122:172, 1992.

Nemere I: Vesicular calcium transport in chick intestine, *J Nutr* 122:657, 1992.

Seetharam B, Alpers DH: Absorption and transport of cobalamin (vitamin B_{12}), *Annu Rev Nutr* 2:343, 1982.

Stremmel W: Uptake of fatty acids by jejunal mucosal cells is mediated by a fatty acid binding membrane protein, *J Clin Invest* 82:2001, 1988.

Thomson ABR et al: Intestinal aspects of lipid absorption: in review, *Can J Physiol Pharmacol* 67:179, 1989.

Thurnhofer H, Hauser H: Uptake of cholesterol by small intestinal brush border membrane is protein mediated, *Biochemistry* 29:2142, 1990.

Tso P: Gastrointestinal digestion and absorption of lipid, *Adv Lipid Res* 21:143, 1985.

Shiau YF: Mechanisms of intestinal fat absorption, *Am J Physiol* 240:G1, 1981.

Stevens BR et al: Intestinal transport of amino acids and sugars: advances using membrane vesicles, *Annu Rev Physiol* 46:417, 1984.

Turk E et al: Glucose/galactose malabsorption caused by a defect in the Na$^+$/glucose cotransporter, *Nature* 350:354, 1991.

Wasserman RH et al: Intestinal calcium transport and calcium extrusion processes at the basolateral membrane, *J Nutr* 122:662, 1992.

Wilson FA: Intestinal transport of bile acids, *Am J Physiol* 241:G83, 1981.

Young S, Bomford A: Transferrin and cellular iron exchange, *Clin Sci* 67:273, 1984.

Books and Monographs

Alpers DH: Digestion and absorption of carbohydrates and proteins. In Johnson LR, ed: *Physiology of the gastrointestinal tract,* ed 3, New York, 1994 Raven Press.

Binder HJ, Sandle GI: Electrolyte transport in the mammalian colon. In Johnson LR, ed: *Physiology of the gastrointestinal tract,* ed 3, New York, 1994 Raven Press.

Chang EB, Rao MC: Intestinal water and electrolyte transport: mechanisms of physiological and adaptive responses. In Johnson LR, ed: *Physiology of the gastrointestinal tract,* ed 3, New York, 1994 Raven Press.

Davenport HW: *Physiology of the digestive tract,* ed 5, Chicago, 1982, Year Book Medical Publishers, Inc.

Davidson NO: Cellular and molecular mechanisms of small intestinal lipid transport. In Johnson LR, ed: *Physiology of the gastrointestinal tract,* ed 3, New York, 1994, Raven Press.

Ganapathy V, Brandsch M, Leibach FH: Intestinal transport of amino acids and peptides. In Johnson LR, ed: *Physiology of the gastrointestinal tract,* ed 3, New York, 1994, Raven Press.

Hoffman AF: Intestinal absorption of bile acids and biliary constituents. In Johnson LR, ed: *Physiology of the gastrointestinal tract,* ed 3, New York, 1994, Raven Press.

Matthews DM: *Protein absorption: Development and present state of the subject,* New York, 1991, Wiley-Liss.

Wright EM: Intestinal sugar transport. In Johnson LR, ed: *Physiology of the gastrointestinal tract,* ed 3, New York, 1994, Raven Press.

RENAL SYSTEM

BRUCE M. KOEPPEN
BRUCE A. STANTON

CHAPTER 35

Elements of Renal Function

OVERVIEW OF THE KIDNEYS

The kidneys are both regulatory and excretory organs. By their excretion of water and solutes the kidneys are able to regulate the volume and composition of the body fluids within a very narrow range, despite wide variations in the intake of food and water. As a consequence of the kidneys' homeostatic role, the tissues and cells of the body are able to carry out their normal functions in a relatively constant environment.

The kidneys have several major functions, including the following:

1. *Regulation of body fluid osmolality and volume*
2. *Regulation of electrolyte balance*
3. *Regulation of acid-base balance*
4. *Excretion of metabolic products and foreign substances*
5. *Production and secretion of hormones*

The control of body fluid osmolality is important for the maintenance of normal cell volume in all tissues of the body. Control of the volume of the body fluids is necessary for normal function of the cardiovascular system. The kidneys, working in an integrated fashion with components of the cardiovascular, endocrine, and central nervous systems, accomplish these tasks by regulating the excretion of water and NaCl.

The kidneys play an essential role in regulating the amount of several important inorganic ions in the body, including Na^+, K^+, Cl^-, HCO_3^-, H^+, Ca^{++}, Mg^{++}, and $PO_4^=$. In order to maintain appropriate balance, the excretion of any one of these electrolytes must be equal to the daily intake. If intake exceeds excretion, the amount of a particular electrolyte in the body increases and the individual is in positive balance with respect to that electrolyte. Conversely, if excretion exceeds intake, the amount decreases, and the individual is in negative balance with respect to that electrolyte. For many electrolytes the kidneys are the sole or primary route for ex-

cretion from the body. Thus, electrolyte balance is achieved by carefully matching daily excretion by the kidneys with dietary intake.

Many of the metabolic functions of the body are exquisitely sensitive to pH. Thus, the pH of the body fluids must be maintained within very narrow limits. This is accomplished by buffers within the body fluids, and by the coordinated action of the lungs and kidneys.

The kidneys excrete a number of end products of metabolism that are no longer needed by the body. These so-called waste products include urea (from amino acids), uric acid (from nucleic acids), creatinine (from muscle creatine), end products of hemoglobin metabolism, and metabolites of hormones. The kidneys eliminate these substances from the body at a rate that matches their production. Thus, the kidneys regulate their concentrations within the body fluids. The kidneys also represent an important route for elimination of foreign substances from the body; such substances include drugs, pesticides, and other chemicals ingested in the food.

The kidneys are important endocrine organs that produce and secrete renin, prostaglandins, kinins, 1,25-dihydroxyvitamin D_3, and erythropoietin. Renin activates the renin-angiotensin-aldosterone system, which is important in regulating blood pressure, as well as sodium and potassium balance. Prostaglandins and kinins (e.g., bradykinin) are vasoactive, and important in the regulation and modulation of renal blood flow, and together with angiotensin II influence systemic blood pressure. 1,25-dihydroxyvitamin D_3 is necessary for normal reabsorption of Ca^{++} by the gastrointestinal tract, and for its deposition in bone. Erythropoietin stimulates red blood cell formation by the bone marrow.

FUNCTIONAL ANATOMY OF THE KIDNEYS

Structure and function are closely linked in the kidney. Consequently, an appreciation of the gross anatomi-

503

# This is

cal and histological features of the kidneys is a prerequisite for an understanding of their function.

Gross Anatomy

The gross anatomical features of the mammalian kidney are illustrated in Figure 35-1. The medial side of each kidney contains an indentation through which pass the renal artery and vein, nerves, and pelvis. On the cut surface of a bisected kidney, two regions are evident: an outer region called the **cortex** and an inner region called the **medulla.** The cortex and medulla are composed of **nephrons** (the functional units of the kidney), blood vessels, lymphatics, and nerves. The medulla in the human kidney is divided into 8 to 18 conical masses, the **renal pyramids.** The base of each pyramid originates at the corticomedullary border and the apex terminates in the papilla, which lies within the calyx. The **pelvis** represents the upper expanded region of the **ureter,** which carries urine from the pelvis to the urinary bladder. In the human kidney the pelvis divides into two or three open-ended pouches, the major **calyces,** which extend outward from the dilated end of the pelvis. Each major calyx divides into minor calyces, which collect the urine from each papilla. The walls of the calyces, pelvis, and the ureters contain smooth muscle that contracts to propel the urine toward the bladder.

The blood flow to the two kidneys is equal to about 25% (1.25 L/min) of the cardiac output in resting persons. However, the kidneys constitute less than 0.5% of the total body weight. As illustrated in Figure 35-2, the renal artery enters the kidney alongside the ureter, and it branches to form progressively the interlobar artery, the arcuate artery, the interlobular artery (cortical radial artery), and the afferent arteriole, which leads into the glomerular capillaries (i.e., **glomerulus**). The glomerular capillaries coalesce to form the efferent arteriole, which leads into a second capillary network, the peritubular capillaries, which supply blood to the nephron. The vessels of the venous system run parallel to the arterial vessels and form progressively the interlobular vein (cortical radial vein), the arcuate vein, the interlobar vein, and the renal vein, which courses beside the ureter.

Ultrastructure of The Nephron

The functional unit of the kidneys is the nephron. Each human kidney contains approximately 1.2 million nephrons, which are hollow tubes composed of a single cell layer (Figure 35-2). The nephron consists of a renal corpuscle, a proximal tubule, a loop of Henle, a distal tu-

bule, and a collecting duct system.* The **renal corpuscle** consists of the glomerulus, which is a tuft of specialized capillaries, and **Bowman's capsule.** The **proximal tubule** initially forms several coils, followed by a straight piece that descends toward the medulla. The next segment is **Henle's loop,** which is composed of the straight part of the proximal tubule, the descending thin limb (which ends at a hairpin turn), the ascending thin limb (only in nephrons with long loops of Henle), and the **thick ascending limb.** Near the end of the thick ascending limb, the nephron passes between the **afferent** and **efferent arterioles** that supply the glomerulus of the same nephron. This short segment of the thick ascending limb is called the **macula densa.** The distal tubule begins a short distance beyond the macula densa and extends to the point in the cortex where two or more nephrons join to form a **cortical collecting duct.** The cortical collecting duct enters the medulla and becomes the **outer medullary collecting duct,** and then the **inner medullary collecting duct.**

Each nephron segment is composed of cells that are uniquely suited to perform specific transport functions (Figure 35-3). Proximal tubule cells have an extensively amplified apical membrane (the urine side of the cell) called the brush border, which is only present in the proximal tubule. The basolateral membrane (the blood side of the cell) is highly invaginated. These invaginations contain many mitochondria. In contrast, the descending thin limb and ascending thin limb of Henle's loop have poorly developed apical and basolateral surfaces and few mitochondria. The cells of the thick ascending limb and the distal tubule have abundant mitochondria and extensive infoldings of the basolateral membrane. The collecting duct is composed of two cell types: **principal cells** and **intercalated cells.** Principal cells have a moderately invaginated basolateral membrane and contain few mitochondria. Intercalated cells have a high density of mitochondria. The final segment of the nephron, the inner medullary collecting duct, is composed of inner medullary collecting duct cells.

Nephrons may be subdivided into **superficial** and **juxtamedullary** types (see Figure 35-2). The renal corpuscle of each superficial nephron is located in the

*The organization of the nephron is actually much more complicated than presented here; however, for simplicity and clarity of presentation in subsequent chapters the nephron is divided into five segments. For details on the subdivisions of the five nephron segments consult the references by Kriz and Bankir, Kriz and Kaissling, and Tisher and Madsen. The collecting duct system is not actually part of the nephron. However, for simplicity, we consider the collecting duct system part of the nephron.

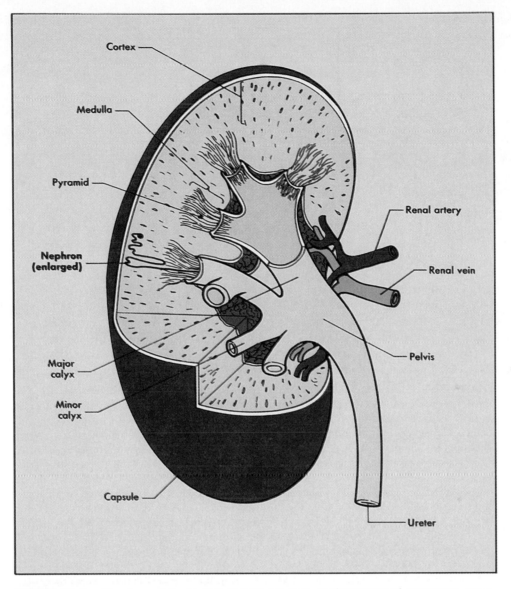

FIGURE 35-1   Structure of the human kidney, cut open to show internal structures. (Modified from Marsh DJ: *Renal physiology,* New York, 1983, Raven Press).

outer region of the cortex (see Figure 35-2). Its loop of Henle is short, and its efferent arteriole branches into peritubular capillaries that surround the tubular segments of its own and adjacent nephrons. This capillary network conveys oxygen and important nutrients to the tubular segments, delivers substances to the tubules for secretion (i.e., the movement of a substance from the blood into the tubular fluid), and serves as a pathway for the return of reabsorbed water and solutes to the circulatory system. A few species, including humans, also possess very short superficial nephrons whose loops of Henle never enter the medulla.

The renal corpuscle of each juxtamedullary nephron is located in the region of the cortex adjacent to the medulla (see Figure 35-2). In comparison with the superficial nephrons, the juxtamedullary nephrons differ anatomically in three important ways: (1) the renal corpuscle is larger, (2) the loop of Henle is longer and extends deeper into the medulla, and (3) the efferent arteriole forms not only a network of peritubular capillaries, but also a series of vascular loops called the **vasa recta.** As illustrated in Figure 35-2, the vasa recta descend into the medulla, where they form capillary networks that surround the collecting ducts and ascending limbs of Henle's loop. The blood returns to the cortex in the ascending vasa recta. *Although less than 0.7% of*

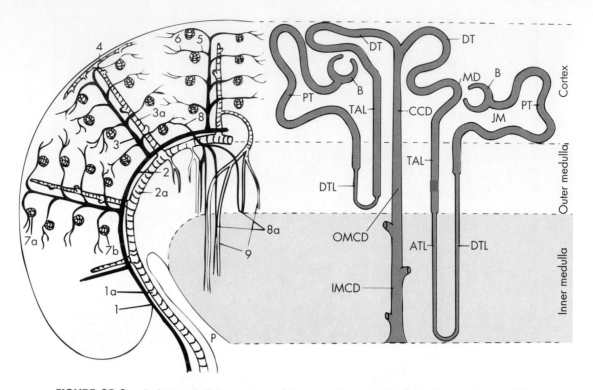

FIGURE 35-2   *Left panel,* Organization of the vascular system of the human kidney. The renal artery branches to form interlobar arteries *(1)*, which give rise to arcuate arteries *(2)*. Arcuate arteries lead to interlobular arteries *(3)*, which ascend toward the renal capsule and branch to form afferent arterioles *(5)*. Afferent arterioles branch to form glomerular capillary networks (i.e., glomeruli: *7A, 7B)*, which then coalesce to form efferent arterioles *(6)*. The efferent arterioles of the outer cortical nephrons form capillary networks (not shown) that suffuse the cells in the cortex. The efferent arterioles of the juxtamedullary nephrons divide into descending vasa recta *(8)*, which form capillary networks that supply blood to the outer and inner medulla *(8a)*. Blood from the peritubular capillaries enters, consecutively, the stellate vein *(4)*, interlobular vein *(3a)*, arcuate vein *(2a)*, and interlobar vein *(1a)*. Blood from the ascending vasa recta *(9)* enters the arcuate and interlobular veins. *P,* Pelvis. *Right panel,* Organization of the human nephron. A superficial nephron is illustrated on the left (labeled *S*), and a juxtamedullary nephron is illustrated on the right (labeled *JM*). *B,* Bowman's capsule; *DT,* distal tubule; *PT,* proximal tubule; *CCD,* cortical collecting duct; *TAL,* thick ascending limb; *DTL,* descending thin limb; *OMCD,* outer medullary collecting duct; *ATL,* ascending thin limb; *IMCD,* inner medullary collecting duct; *MD,* macula densa. The loop of Henle includes the straight portion of the PT, the DTL, ATL, and TAL.

the renal blood flow enters the vasa recta, these vessels subserve important functions, including; conveying oxygen and important nutrients to nephron segments, delivering substances to the nephron for secretion, serving as a pathway for the return of reabsorbed water and solutes to the circulatory system, and concentrating and diluting the urine (see Chapter 37).

Ultrastructure of The Renal Corpuscle

The first step in urine formation begins with the ultrafiltration of plasma across the glomerular capillaries (i.e., glomerulus). The term ultrafiltration refers to the passive movement of an essentially protein-free fluid from the glomerular capillaries into Bowman's space. To appreciate the process of ultrafiltration, it is important to describe the anatomy of the renal corpuscle. The glomerulus consists of a network of capillaries supplied by the afferent arteriole and drained by the efferent arteriole (Figure 35-4). During development, the glomerular capillaries press into the closed end of the proximal tubule, which forms Bowman's capsule of a renal corpuscle. The capillaries are covered by epithelial cells, called **podocytes,** which form the visceral layer of Bow-

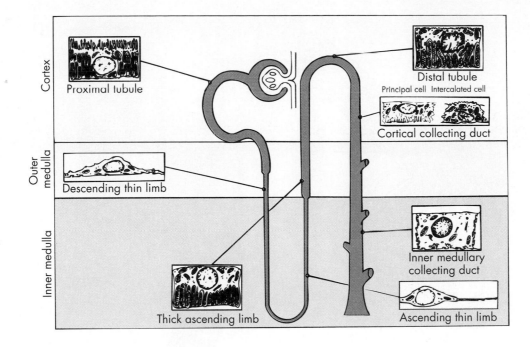

FIGURE 35-3   Diagram of a nephron including the cellular ultrastructure.

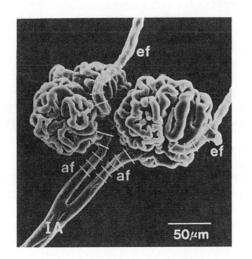

FIGURE 35-4   Scanning electron micrograph of interlobular artery *(IA)*, afferent arteriole *(af)*, efferent arteriole *(ef)*, and glomerulus. The white bars on the afferent and efferent arterioles indicate they are about 15 to 20 mm in diameter. (From Kimura K et al: *Am J Physiol* 259:F936, 1990.)

man's capsule (Figure 35-5). The visceral cells are reflected at the vascular pole to form the parietal layer of Bowman's capsule. The space between the visceral layer and the parietal layer is called **Bowman's space,** which, at the urinary pole of the glomerulus, becomes the lumen of the proximal tubule.

The endothelial cells of glomerular capillaries are cov-

ered by a **basement membrane,** which is surrounded by podocytes (Figures 35-5 to 35-7). The capillary endothelium, basement membrane, and foot processes of podocytes form the so-called **filtration barrier** (see Figures 35-5 to 35-7). *The endothelium is fenestrated (i.e., contains 700 Å holes) and is freely permeable to water, to small solutes such as sodium, urea, and glucose, and even to small protein molecules. Because the fenestrations are relatively large (700 Å), the endothelium acts as a filtration barrier only to cells.* The basement membrane is an important filtration barrier to plasma proteins. The podocytes, which are endocytic (the process of endocytosis allows materials to enter the cell without passing through the membrane; see Chapter 1), have long finger-like processes that completely encircle the outer surace of the capillaries (Figure 35-7). The processes interdigitate to cover the basement membrane and are seprated by gaps, called **filtration slits.** Each filtration slit is bridged by a thin diaphragm, which contains pores with dimensions of 40 Å × 140 Å. Therefore, the filtration slits retard the filtration of some proteins and macromolecules that pass through the endothelium and basement membrane. Because the basement membrane and the filtration slits contain negatively charged glycoproteins, some molecules are held back on the basis of size and charge. For molecules with an effective molecular radius between 18 Å and 36 Å, cationic molecules are filtered more readily than anionic molecules (see below for more details).

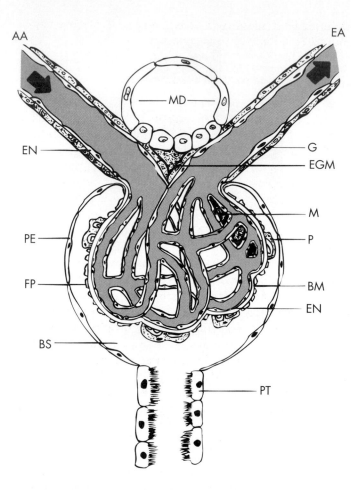

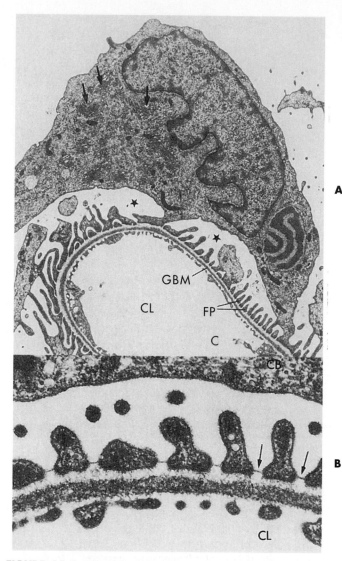

FIGURE 35-5   Anatomy of the renal corpuscle and the juxtaglomerular apparatus. The juxtaglomerular apparatus is composed of (1) the macula densa (*MD*) of the thick ascending limb, (2) the extraglomerular mesangial cells (*EGM*), and (3) the renin producing granular cells (*G*) of the afferent (*AA*) and efferent (*EA*) arterioles. *EN*, Endothelial cell; *PE*, parietal epithelium; *FP*, foot processes of podocyte; *BS*, Bowman's space; *PT*, proximal tubule cell; *BM*, basement membrane; *P*, podocyte cell; *M*, mesangial cells between capillaries. (Modified from Koushanpour E, Kriz W: *Renal physiology principles, structure and function,* ed 2, Berlin, 1986, Springer-Verlag.)

The **nephrotic syndrome** is produced by a variety of kidney disorders and is characterized by an increase in the permeability of the glomerular capillary wall to proteins. The augmented permeability results in an increase in urinary protein excretion. Thus, appearance of proteins in the urine can indicate kidney disease.

FIGURE 35-6   **A,** Electron micrograph of a podocyte surrounding a glomerular capillary *(c)*. The cell body of the podocyte contains a large nucleus with indentations. Cell processes of the podocyte form the interdigitating foot processes *(FP)*. The arrows in the cytoplasm of the podocyte indicate the well-developed Golgi apparatus. *CL*, Capillary lumen; *GMB*, glomerular basement membrane. Stars indicate Bowman's space. (Magnification ~7600×.) **B,** Electron micrograph of the filtration barrier of a glomerular capillary. *CL*, Capillary lumen; *CB*, cell body of a podocyte. The filtration barrier is composed of three layers: the endothelium with large pores, the basement membrane, and the foot processes of the podocytes. Note the diaphragm bridging the floor of the filtration slits *(arrows)*. (Magnification ~42,700×.) (From Kriz W, Kaissling B: Structural organization of the mammalian kidney. In Seldin DW, Giebisch G, eds: *The kidney: physiology and pathophysiology,* ed 2, New York, 1992, Raven Press.)

Another important component of the renal corpuscle is the **mesangium,** which consists of mesangial cells and the mesangial matrix (see Figure 35-5). Mesangial cells surround glomerular capillaries, provide structural support for the glomerular capillaries, secrete the extracellular matrix, exhibit phagocytic activity, and secrete

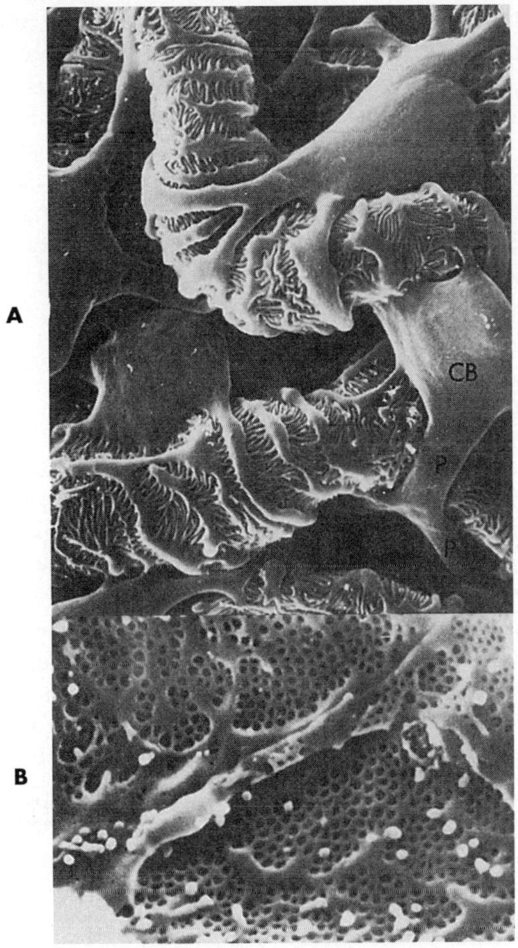

FIGURE 35-7 **A,** Scanning electron micrograph showing the outer surface of glomerular capillaries. This is the view that would be seen from Bowman's space. Processes *(P)* of podocytes run from the cell body *(CB)* toward the capillaries, where they ultimately split into foot processes. Interdigitation of the foot processes creates the filtration slits. (Magnification ~3,400×.) **B,** Scanning electron micrograph of the inner surface (blood side) of a glomerular capillary. This is the view that would be seen from the lumen of the capillary. The fenestrations of the endothelial cells are seen as small, 700 Å, holes. (Magnification ~16,000×.) (From Kriz W, Kaissling B: Structural organization of the mammalian kidney. In Seldin DW, Giebisch G eds: *The kidney: physiology and pathophysiology,* ed 2, New York, 1992, Raven Press).

prostaglandins. Mesangial cells contract and because they are adjacent to glomerular capillaries, they may influence glomerular filtration rate by regulating blood flow through glomerular capillaries, or by altering the capillary surface area. Mesangial cells located outside the glomerulus (between the afferent and efferent arterioles) are called **extraglomerular mesangial cells** (or **lacis**

cells or **Goormaghtigh cells**). Lacis cells, like mesangial cells, exhibit phagocytic activity.

The Juxtaglomerular Apparatus

The structures that compose the juxtaglomerular apparatus include (1) the **macula densa** of the thick ascending limb, (2) the **extraglomerular mesangial cells,** and (3) the renin-producing **granular cells** of the afferent and efferent arterioles (see Figure 35-5). Macula densa cells represent a morphologically distinct region of the thick ascending limb that passes through the angle formed by the afferent and efferent arterioles. The cells of the macula densa contact the extraglomerular mesangial cells and the granular cells of the afferent and efferent arterioles. Granular cells of the afferent and efferent arterioles are modified smooth muscle cells that manufacture, store, and release **renin.** Renin is involved in the formation of **angiotensin II** and ultimately in the secretion of **aldosterone** (see Chapter 37). The juxtaglomerular apparatus is one component of an important feedback mechanism (i.e., **tubuloglomerular feedback mechanism**) that is involved in the autoregulation of renal blood flow and of the glomerular filtration rate. Details of this feedback mechanism are discussed below.

Innervation of the Kidney

Renal nerves help regulate renal blood flow, glomerular filtration rate, and salt and water reabsorption by the nephron. The nerve supply to the kidneys consists of sympathetic nerve fibers that originate mainly in the celiac plexus. There is no parasympathetic innervation. Adrenergic fibers that innervate the kidneys release norepinephrine and dopamine. The adrenergic fibers lie adjacent to the smooth muscle cells of the major branches of the renal artery (interlobar, arcuate, and interlobular arteries), and the afferent and efferent arterioles. Moreover, the renin-producing granular cells of the afferent and efferent arterioles are innervated by sympathetic nerves. Renin secretion is elicited by increased sympathetic activity. Nerve fibers also innervate the proximal tubule, loop of Henle, distal tubule, and collecting duct; activation of these nerves enhances sodium reabsorption by these nephron segments (see Chapters 36 and 37).

ANATOMY AND PHYSIOLOGY OF THE LOWER URINARY TRACT

Gross Anatomy and Histology

Once urine leaves the renal calyces and pelvis, it flows through the ureters and enters the bladder, where urine is stored. The **ureters** are muscular tubes 30 cm long, and they enter the **bladder** on its posterior aspect near the base, above the bladder neck. The bladder is composed of two parts: the **fundus** or **body,** which stores urine, and the **neck,** which is funnel-shaped and connects with the urethra. The bladder neck, which is 2 to 3 cm long, is also called the posterior urethra. In females, the posterior urethra is the end of the urinary tract and the point of exit of urine from the body. In males, urine flows through the posterior urethra into the anterior urethra, which extends through the penis. Urine leaves the urethra through the external meatus.

The renal calyces, pelvis, ureter, and bladder are lined with a transitional epithelium that is composed of several layers of cells: basal columnar cells, intermediate cuboidal cells, and superficial squamous cells. This epithelium is surrounded by a mixture of spiral and longitudinal smooth muscle fibers. The bladder is also lined with a transitional epithelium that is surrounded by a mixture of smooth muscle fibers, called the **detrusor muscle.** Detrusor muscle fibers are arranged at random. They form layers except close to the bladder neck, where the fibers form three layers: inner longitudinal, middle circular, and outer longitudinal. Muscle fibers in the bladder neck form the internal sphincter, which is not a true sphincter but a thickening of the bladder wall formed by converging muscle fibers. The internal sphincter is not under conscious control. Its inherent tone prevents emptying of the bladder until appropriate stimuli initiate urination. The urethra passes through the urogenital diaphragm, which contains a layer of skeletal muscle called the external sphincter. This muscle is under voluntary control and can be used to prevent or interrupt urination, especially in males. In females the external sphincter is poorly developed; thus, it is less important in voluntary bladder control. The smooth muscle cells in the lower urinary tract are electrically coupled, exhibit action potentials, contract when stretched, and respond to parasympathetic neurotransmitters.

The walls of the ureters, bladder, and urethra are highly folded and thereby very distensible. In the bladder and urethra these folds are called **rugae.** As the bladder fills with urine the rugae flatten out and the volume of the bladder increases, with very little change in intravesical pressure. The volume of this structure can increase from a minimal volume of 10 ml following urination to 400 ml with a pressure change of only 5 cm H_2O, illustrating the highly compliant nature of the bladder.

Innervation of the Bladder

Innervation of the bladder and urethra is important in controlling urination. The smooth muscle of the bladder neck receives sympathetic innervation from the hypogastric nerves. α-adrenergic receptors, located mainly in the bladder neck and the urethra, cause contraction. Stimulation of these receptors facilitates storage of urine by inducing closure of the urethra. Sacral parasympathetic fibers (muscarinic) innervate the body of the bladder and cause a sustained bladder contraction. Sensory fibers of the pelvic nerves (visceral afferent pathway) also innervate the fundus. These sensory fibers carry input from receptors that detect bladder fullness, pain, and temperature sensation. The sacral pudendal nerves innervate the skeletal muscle fibers of the external sphincter, and excitatory impulses cause contraction.

Passage of Urine From the Kidney to the Bladder

As urine collects in the renal calyces, stretch promotes their inherent pacemaker activity. The pacemaker activity initiates a peristaltic contraction that begins in the calyces and spreads to the pelvis and along the length of the ureter, and thereby forces urine from the renal pelvis toward the bladder. Transmission of the peristaltic wave is caused by action potentials that are generated by the pacemaker, and pass along the smooth muscle syncytium. The ureters are innervated with sensory nerve fibers (pelvic nerves).

When the ureter is blocked with a kidney stone, reflex constriction of the ureter around the stone elicits severe pain.

Micturition

Micturition is the process of emptying the urinary bladder. Two processes are involved: (1) progressive filling of the bladder until the pressure rises to a critical value and (2) a neuronal reflex called the **micturition reflex,** which empties the bladder. The micturition reflex is an automatic spinal cord reflex. However, it can be inhibited or facilitated by centers in the brainstem and the cerebral cortex.

Filling of the bladder stretches the bladder wall and causes it to contract. Contractions are the result of a reflex initiated by stretch receptors in the bladder. Sensory signals from the bladder fundus enter the spinal cord via pelvic nerves and return directly to the bladder through parasympathetic fibers in the same nerves. Stimulation of parasympathetic fibers causes intense stimulation of the detrusor muscle. The smooth muscle in the bladder is a syncytium, accordingly, stimulation of the detrusor also causes the muscle cells in the neck of the bladder to contract. Because the muscle fibers of the bladder outlet are oriented both longitudinally and radially, contraction opens the bladder neck and allows urine to flow

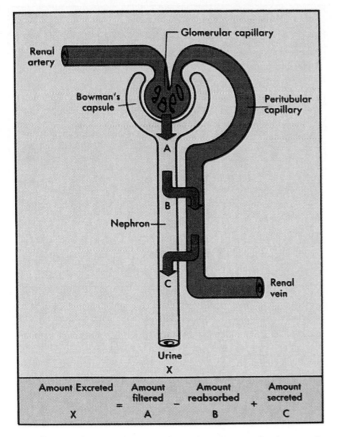

FIGURE 35-8 Schematic representation of the entire nephron population of both kidneys, depicting the three general processes that determine and modify the composition of the urine: glomerular filtration *(A)*, tubular reabsorption *(B)*, and tubular secretion *(C)*.

through the posterior urethra. A voluntary relaxation of the external sphincter, by cortical inhibition of the pudendal nerve, permits the flow of urine through the external meatus. Voluntary relaxation of the external sphincter is required and may be the event that initiates micturition. Interruption of the hypogastric sympathetic nerves and the pudendal nerves to the lower urinary tract does not alter the micturition reflex. In contrast, destruction of the parasympathetic nerves results in complete bladder dysfunction.

ASSESSMENT OF RENAL FUNCTION

Three general processes determine the composition and volume of urine:

1. *Glomerular filtration*
2. *Reabsorption of the substance from the tubular fluid into the blood*
3. *Secretion of the substance from the blood into the tubular fluid.*

These three processes are illustrated in Figure 35-8, in

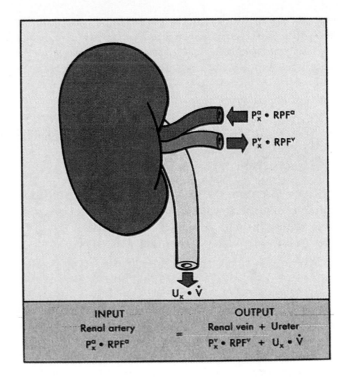

FIGURE 35-9 Mass balance relationships for the kidney. See text for definition of symbols.

which the entire nephron population of both kidneys is represented by a single nephron. This section develops the renal clearance concept, which can be used to quantitate glomerular filtration, reabsorption, and secretion. In addition, the use of clearance to measure renal blood flow is discussed.

Renal Clearance

The concept of renal clearance is based on the Fick principle (i.e., mass balance; see Chapter 18). Figure 35-9 illustrates the various factors required to describe the mass balance relationships of a kidney. The renal artery is the single input source to the kidney, whereas the renal vein and ureter constitute the two output routes. The following equation defines the mass balance relationship:

$$P^a_x \cdot RPF^a = (P^v_x \cdot RPF^v) + (U_x \cdot \dot{V}) \qquad \textbf{(1)}$$

where P^a_x and P^v_x are the concentrations of substance x in the renal artery and renal vein plasma, respectively; RPF^a and RPF^v are the renal plasma flow rates in the artery and vein, respectively; U_x is the concentration of x in the urine; and $\dot{V}$ is the urine flow rate. This relationship permits the quantitation of the amount of x excreted in the urine versus the amount returned to the systemic circulation in the renal venous blood. Thus, for any substance that is neither synthesized nor metabolized, the amount that enters the kidneys is equal to the amount that leaves the kidneys in the urine plus the amount that leaves the kidneys in the renal venous blood.

The principle of renal clearance (C_x) emphasizes the excretory function of the kidney; it considers only the rate at which a substance is excreted into the urine, and not its rate of return to the systemic circulation in the renal vein. Therefore, in terms of mass balance (equation 1), the urinary excretion rate of x ($U_x \cdot \dot{V}$) is proportional to the plasma concentration of x (P^a_x).

$$P^a_x \propto U_x \cdot \dot{V} \tag{2}$$

In order to equate the urinary excretion rate of x to its renal arterial plasma concentration, it is necessary to determine the rate at which x is removed from the plasma by the kidneys. This removal rate is the clearance (C_x).

$$P^a_x \cdot C_x = U_x \cdot \dot{V} \tag{3}$$

If equation 3 is rearranged, and if the concentration of x in the renal artery plasma (P_x) is assumed to be identical to its concentration in a plasma sample from any peripheral blood vessel, the following relationship is obtained.

$$C_x = \frac{U_x \cdot \dot{V}}{P_x} \tag{4}$$

Clearance has the dimensions of volume/time, and it represents a volume of plasma from which all the substance has been removed and excreted into the urine per unit time. This last point is best illustrated by considering the following example.

If a substance is present in the urine at a concentration of 100 mg/ml, and the urine flow rate is 1 ml/min, then the excretion rate for this substance is calculated as:

$$\text{Excretion rate} = U_x \cdot \dot{V} = (100 \text{ mg/ml}) \times (1 \text{ ml/min}) \tag{5}$$
$$= 100 \text{ mg/min}$$

If this substance is present in the plasma at a concentration of 1 mg/ml, then its clearance according to equation 4 is:

$$C_x = \frac{U_x \cdot \dot{V}}{P_x} = \frac{100 \text{ mg/min}}{1 \text{ mg/ml}} = 100 \text{ ml/min} \tag{6}$$

That is, 100 ml of plasma will be completely cleared of substance X each minute. The definition of clearance as a volume of plasma from which all the substance has been removed and excreted into the urine per unit time is somewhat misleading, because it is not a real volume of plasma; rather it is an idealized volume. The concept of clearance is important because it can be used to measure GFR and RPF and to determine whether a substance is reabsorbed or secreted along the nephron.

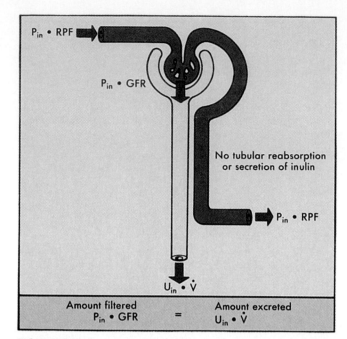

FIGURE 35-10 Renal handling of inulin. Inulin is freely filtered at the glomerulus and is neither reabsorbed, secreted, nor metabolized by the nephron. P_{in}, Plasma inulin concentration; *RPF*, renal plasma flow; *GFR*, glomerular filtration rate; U_{in}, urinary concentration of inulin; $\dot{V}$, urine flow rate. Note that all the inulin coming to the kidney in the renal artery does not get filtered at the glomerulus (normally 15% to 20% of plasma and inulin are filtered). The portion that is not filtered is returned to the systemic circulation in the renal vein.

Glomerular Filtration Rate: Clearance of Inulin

Inulin is a polymer of fructose (molecular weight, ca. 5000), and it can be used to measure the glomerular filtration rate (GFR). It is not produced by the body and therefore must be administered intravenously to measure GFR. Inulin is freely filtered at the glomerulus and is neither reabsorbed, secreted, nor metabolized by the cells of the nephron. Accordingly, the amount of inulin excreted in the urine per minute equals the amount of inulin filtered at the glomerulus each minute (Figure 35-10):

$$\text{Amount filtered} = \text{Amount excreted} \tag{7}$$

$$GFR \cdot P_{in} = U_{in} \cdot \dot{V}$$

where GFR is the glomerular filtration rate, P_{in} and U_{in} are the plasma and urine concentrations of inulin, and $\dot{V}$ is the urine flow. If equation 7 is solved for the GFR:

$$GFR = \frac{U_{in} \cdot \dot{V}}{P_{in}} \tag{8}$$

This equation is the same form as that for clearance (see equation 4). Thus, the clearance of inulin provides a means for determining the GFR.

Inulin is not the only substance that can be used to measure the GFR. Any substance that meets the following criteria will serve as an appropriate marker for the measurement of GFR. The substance must:

1. Be freely filtered by the glomerulus.
2. Not be reabsorbed or secreted by the nephron.
3. Not be metabolized or produced by the kidney.
4. Not alter GFR

Whereas inulin is used extensively in experimental studies, the fact that it must be infused intravenously limits its clinical use. Consequently, creatinine is used to estimate the GFR in clinical practice. **Creatinine** is a byproduct of skeletal muscle creatine metabolism. It is produced at a relatively constant rate, and the amount produced is proportional to the muscle mass. With regard to the measurement of GFR, creatinine has an advantage over inulin in that it is produced endogenously, and thus it obviates the need for an intravenous infusion, as is required for inulin. However, creatinine is not a perfect substance to measure GFR, because it is secreted to a small extent by the organic cation secretory system in the proximal tubule (see Chapter 36). The error introduced by this secretory component is approximately 10%. Thus, the amount of creatinine excreted in the urine exceeds the amount expected from filtration alone by 10%. In fact, the method used to quantitate the plasma creatinine concentration overestimates the true value by 10%. Consequently, the two errors cancel, and the creatinine clearance provides a reasonably accurate measure of the GFR.

As illustrated in Figure 35-10, not all the inulin (or any substance used to measure GFR) that enters the kidney in the renal arterial plasma is filtered at the glomerulus. Likewise not all of the plasma coming into the kidney is filtered.* The portion of plasma that is filtered is termed the **filtration fraction,** and is determined as:

$$\text{Filtration fraction} = \frac{\text{GFR}}{\text{RPF}} \qquad \textbf{(9)}$$

where, again, RPF is renal plasma flow. Under normal conditions the filtration fraction averages 0.15 to 0.20.

---

*Nearly all of the plasma that enters the kidney in the renal artery passes through the glomerulus. Approximately 10% does not.

This means that only 15% to 20% of the plasma that enters the glomerulus is actually filtered. The remaining 80% to 85% continues on through the glomerulus into the efferent arterioles and peritubular capillaries, and finally is returned to the systemic circulation in the renal vein.

A fall in GFR may be the first and only clinical sign of kidney disease. Thus, a measurement of GFR is important in individuals thought to have kidney disease. For example, a 50% loss of functioning nephrons will reduce the GFR by approximately 20% to 30%. This smaller-than-anticipated decline in GFR is caused by compensation of the remaining nephrons. Because measurements of GFR are cumbersome, kidney function is usually assessed in the clinical setting by measuring plasma creatinine (P_{Cr}) concentration, which is inversely related to GFR.

Renal Plasma Flow: Clearance of PAH

p-Aminohippuric acid (PAH) is an organic anion that can be used to measure renal plasma flow (RPF). As with inulin, PAH is not produced in the body, and therefore must be infused intravenously. PAH is an organic anion that is excreted into the urine by the processes of glomerular filtration and tubular secretion. For this discussion it is sufficient to recognize that the PAH secretory mechanism in the proximal tubule has a maximum rate of ~80 mg/min. Delivery of PAH to the peritubular capillaries at a rate less than this will cause virtually all of the PAH to be secreted into the tubular fluid, and thus little PAH will remain in the renal vein plasma. When the plasma PAH concentration is low and the secretory mechanism is not overwhelmed (generally at plasma [PAH] below 0.12 mg/ml), the clearance of PAH can be used to measure the renal plasma flow. Figure 35-11 depicts the renal handling of PAH in terms of whole kidney mass balance, and illustrates why, when nonsaturating concentrations of PAH are used, its clearance provides a measure of the RPF. The amount of PAH that arrives at the kidneys per minute is simply the product of the plasma PAH concentration (P^a_{PAH}) and the renal plasma flow (RPF). Because all of the PAH is excreted into the urine, and none is returned to the systemic circulation via the renal vein, the following mass balance relationship holds true:

$$RPF \cdot P^a_{PAH} = U_{PAH} \cdot \dot{V} \qquad \textbf{(10)}$$

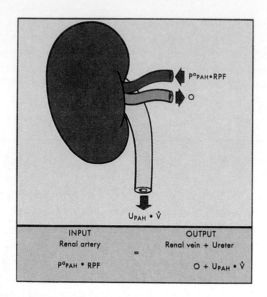

FIGURE 35-11 Mass balance relationships for the use of PAH clearance to measure the renal plasma flow (RPF). P^a_{PAH}, Plasma PAH concentration; U_{PAH}, urine PAH concentration; $\dot{V}$, urine flow rate.

where U_{PAH} is urine PAH concentration and $\dot{V}$ is urine flow. Rearranging and solving for the RPF, the following equation is obtained:

$$RPF = \frac{U_{PAH} \cdot \dot{V}}{P^a_{PAH}} \qquad (11)$$

This equation conforms to the general clearance equation (equation 4). Thus, at low plasma PAH concentrations, the PAH clearance is equal to the RPF. At high plasma PAH concentrations, however, the PAH secretory mechanism will be saturated, and a significant amount of PAH will appear in the renal venous blood. Under this condition, equations 10 and 11 do not hold, and the clearance of PAH does not equal the renal plasma flow.

The relationship between PAH clearance and renal plasma flow described here is idealized. Even at plasma PAH concentrations that do not exceed the capability of the secretory mechanism, some PAH still appears in the renal venous blood. The reason for this is related to the anatomy of the nephron and of the renal blood vessels. The PAH secretory mechanism is located in the proximal tubule. Consequently, if all of the PAH entering the renal artery were to be secreted into the tubular fluid, all of the plasma would have to flow through the peritubular capillaries surrounding the proximal tubule. Approximately 90% of plasma does in fact flow through peritubular capillaries that surround the proximal tubules. However, 10% does not (this plasma perfuses some of the medullary structures, the renal capsule, and parts of the renal hilum). Thus, the PAH in this plasma cannot be secreted, and this portion of PAH will be returned to the systemic circulation in the renal vein plasma. In recognition of the fact that the clearance of PAH does not provide a fully accurate measure of the RPF (i.e., it underestimates the true value by approximately 10%), it is more appropriate to refer to the clearance of PAH as providing a measure of the **effective renal plasma flow** (ERPF)—effective in the sense that this represents plasma flow past portions of the nephron that can effectively secrete PAH.

The clearance of PAH also can be used to estimate the renal blood flow (RBF). Normally, the plasma fraction of blood accounts for 50% to 60% of the blood volume, and the cells account for the remainder. The red cell fraction of a blood sample (the hematocrit, Hct) is measured. Normally the Hct is in the range of 0.40 to 0.50. Once the Hct is known, the renal blood flow can be calculated as:

$$RBF = \frac{RPF}{1 - Hct} \qquad (12)$$

Thus, if the Hct of an individual is 0.40 and the RPF is 700 ml/min, RBF is 1,167 ml/min (i.e., RBF = (700 ml/min)/1 − 0.4 = 1,167 ml/min.). However, measurement of RPF provides little useful information and is rarely performed in clinical situations. Kidney function is usually assessed by measuring plasma creatinine (P_{Cr}) concentration, which, as discussed above, is inversely related to GFR.

Reabsorption and Secretion: Comparison of Renal Clearance

Most substances are filtered by the kidneys and either reabsorbed or secreted. The important exceptions to this general rule are K^+ and urea, which undergo filtration as well as reabsorption and secretion. Analysis of clearance can provide information on how a particular substance is handled by the kidneys. For substances that are not filtered, or that are filtered and completely reabsorbed by the nephron, the clearance is zero. The clearance of PAH (i.e., the effective RPF) defines the upper limit of renal clearance. If it is known that a substance is filtered freely at the glomerulus, comparison of its clearance with that of inulin and PAH will indicate the net handling of the substance by the kidney. Thus:

1. If its clearance is less than that of inulin, the substance is reabsorbed by the nephron (e.g., glucose).
2. If its clearance is greater than that of inulin, the substance is secreted (e.g., PAH).

3. If its clearance equals that of inulin, the substance is only filtered.

For those substance that are both reabsorbed and secreted, the clearance will reflect the dominant transport system.

The conclusions obtained about transport mechanisms from the analysis of clearance values must be considered carefully. For example, suppose the renal handling of a substance (x) occurs solely by glomerular filtration. If substance x were filtered freely, its clearance would be equal to that of inulin. However, consider what happens if 50% of x is bound to plasma protein. Because only the unbound portion can be filtered, and thus excreted, the clearance of substance x will be less than that of inulin by 50%. If we did not know in advance that x was partially protein bound, we would conclude erroneously that substance x was reabsorbed by the nephron.*

GLOMERULAR FILTRATION

The first step in the formation of urine is the production of an ultrafiltrate of the plasma by the glomerulus. The ultrafiltrate is devoid of cellular elements and is essentially protein free. The concentrations of salts and of organic molecules, such as glucose and amino acids, are similar in the plasma and ultrafiltrate. Ultrafiltration is driven by Starling forces (see p. 516 and Chapter 22) across the glomerular capillaries, and changes in these forces and in renal plasma flow alter the glomerular filtration rate. Glomerular filtration rate and renal plasma flow normally are held within very narrow ranges by a phenomenon called **autoregulation** (see p. 518 and Chapter 23). This section will review the composition of the glomerular filtrate, the dynamics of its formation, and the relationship between renal plasma flow and glomerular filtration rate. In addition, the factors that contribute to the autoregulation of glomerular filtration rate and renal blood flow will be discussed.

Determinants of Ultrafiltrate Composition

The unique structure of the glomerular filtration barrier (capillary endothelium, basement membrane, and filtration slits of the podocytes) determines the composition of the ultrafiltrate of plasma. The glomerular filtration barrier restricts the filtration of molecules on the

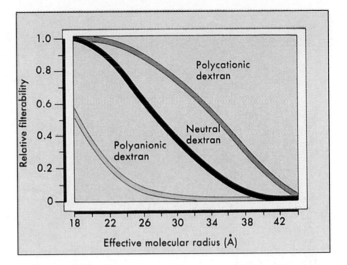

FIGURE 35-12   Influence of size and electrical charge of dextran on its filterability. A value of one indicates that it is filtered freely, whereas a value of zero indicates that it is not filtered. The filterability of dextrans between approximately 18 Å and 36 Å depends on charge. Dextrans larger than 36 Å are not filtered, regardless of charge, and polycationic dextrans and neutral dextrans smaller than 18 Å are freely filtered.

basis of size and electrical charge (Figure 35-12). In general, neutral molecules with a radius less than 18 Å are filtered freely, molecules larger than 36 Å are not filtered, and molecules between 18 and 36 Å are filtered to various degrees. For example, serum albumin, an anionic protein that has an effective molecular radius of 35.5 Å, is filtered poorly: approximately 7 g of albumin are filtered each day.* Because albumin is reabsorbed avidly by the proximal tubule, however, almost none appears in the urine.

Figure 35-12 illustrates how electrical charge affects the filtration of macromolecules (e.g., dextrans) by the glomerulus. Dextrans are a family of exogenous polysaccharides that are manufactured in various molecular weights, as well as in an electrically neutral form or with negative charges (polyanionic) or positive charges (polycationic). At constant charge, as the size (i.e., effective molecular radius) increases, filtration decreases. For any given molecular radius, cationic molecules are more readily filtered than are anionic molecules. The restriction of anionic molecules is explained by the presence of negatively charged glycoproteins on the surface of all components of the glomerular filtration barrier. These

---

*When the renal clearance of a protein-bound substance is calculated, the total plasma concentration (bound plus unbound) is used. However, only the unbound portion can be filtered, and only this portion is used to calculate the filtered load.

*Approximately 70,000 g/day of albumin passes through the glomeruli. Therefore the filtration of 7 g/day represents only 0.01%. This is well below the filtration fraction for substances that are freely filtered (15% to 20%).

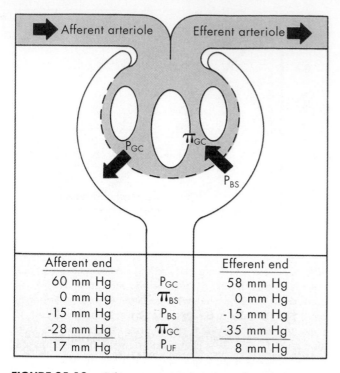

Afferent end		Efferent end
60 mm Hg	P_{GC}	58 mm Hg
0 mm Hg	π_{BS}	0 mm Hg
-15 mm Hg	P_{BS}	-15 mm Hg
-28 mm Hg	π_{GC}	-35 mm Hg
17 mm Hg	P_{UF}	8 mm Hg

FIGURE 35-13 Schematic representation of an idealized glomerular capillary and the Starling forces across the glomerular capillary. P_{UF}, Net ultrafiltration pressure; P_{GC}, glomerular capillary hydrostatic pressure; P_{BS}, Bowman's space hydrostatic pressure; π_{GC}, glomerular capillary oncotic pressure; π_{BS}, Bowman's space oncotic pressure.

charged glycoproteins repel similarly charged molecules. Because most plasma proteins are negatively charged, the negative charge on the filtration barrier restricts the filtration of proteins that have a molecular radius of 18 to 36 Å.

The importance of the negative charges on the filtration barrier in restricting the filtration of plasma proteins is illustrated in Figure 35-12. Removal of negative charges from the filtration barrier causes proteins to be filtered solely on the basis of their effective molecular radius. Hence, at any molecular radius between approximately 18 and 36 Å, the filtration of polyanionic proteins will exceed the filtration that prevails in the normal state, in which the filtration barrier has anionic charges. In a number of glomerular diseases the negative charge on the filtration barrier is lost secondarily to immunologic damage and inflammation. As a result, filtration of proteins is increased, and proteins appear in the urine **(proteinuria).**

Dynamics of Ultrafiltration

The forces responsible for the glomerular filtration of plasma are the same as those involved in fluid exchange across all capillary beds. Ultrafiltration occurs because Starling forces drive fluid from the lumen of glomerular capillaries, across the filtration barrier, into Bowman's space. As shown in Figure 35-13, the Starling forces across glomerular capillaries are similar to the forces that promote filtration across other capillary beds, and include hydrostatic and oncotic pressures (see Chapter 22). The hydrostatic pressure in the glomerular capillary (P_{GC}) is oriented to promote the movement of fluid from the glomerular capillary into Bowman's space. Because the glomerular ultrafiltrate is essentially protein-free, the oncotic pressure in Bowman's space (π_{BS}) is near zero. Therefore, P_{GC} is the only force that favors filtration, and it is opposed by the hydrostatic pressure in Bowman's space (P_{BS}) and the oncotic pressure in the glomerular capillary (π_{GC}).

As illustrated in Figure 35-13 a net ultrafiltration pressure (P_{UF}) of 17 mm Hg exists at the afferent end of the glomerulus, whereas at the efferent end the P_{UF} is 8 mm Hg (where $P_{UF} = P_{GC} - P_{BS} - \pi_{GC}$). Two additional points concerning Starling forces are important. First, P_{GC} decreases slightly along the length of the capillary because of the resistance to flow in the capillary. Second, π_{GC} increases along the length of the glomerular capillary because water is filtered and protein is retained in the glomerular capillary; accordingly, the protein concentration in the capillary rises and π_{GC} increases.

The glomerular filtration rate (GFR) is proportional to the sum of the Starling forces that exists across the capillaries ($[P_{GC} - P_{BS}] - [\pi_{GC} - \pi_{BS}]$) times the ultrafiltration coefficient, K_f:

$$GFR = K_f ([P_{GC} - P_{BS}] - [\pi_{GC} - \pi_{BS}]) \qquad (13)$$

The K_f is the product of the intrinsic permeability of the glomerular capillary and the glomerular surface area available for filtration. Although the P_{UF} in glomerular capillaries is similar to that in other capillary beds, the rate of glomerular filtration is considerably greater in glomerular capillaries, mainly because the K_f is approximately 100 times higher.

The GFR can be altered by changing K_f or by changing any of the Starling forces. Physiologically, however, GFR is affected in three main ways:

1. Changes in renal arterial pressure. An increase in blood pressure transiently increases P_{GC}, which enhances GFR, and a decrease in blood pressure depresses P_{GC}, which decreases GFR. Changes in arterial blood pressure are the most frequent cause of variations in P_{GC}.

2. Changes in afferent arteriolar resistance. A fall in resistance increases P_{GC} and GFR, whereas an increase in resistance decreases P_{GC} and GFR.
3. Changes in efferent arteriolar resistance. A fall in resistance reduces P_{GC} and GFR, whereas an increase in resistance elevates P_{GC} and GFR.

Pathological conditions and drugs may also affect GFR, mainly by changing π_{GC}, P_{BS}, and K_f. Thus, GFR may change by three additional mechanisms:

1. Changes in π_{GC}. An inverse relationship exists between π_{GC} and GFR. Alterations in π_{GC} result from changes in protein metabolism outside the kidney. In addition, protein loss in the urine caused by **glomerulonephritis** decreases π_{GC} and may increase GFR.
2. Changes in K_f. Increased K_f enhances GFR, whereas decreased K_f reduces GFR. Some kidney diseases reduce K_f by reducing the number of filtering glomeruli. Some drugs and hormones that dilate the glomerular arterioles increase K_f, and some that constrict them decrease K_f.
3. Changes in P_{BS}. Increased P_{BS} reduces GFR, whereas decreased P_{BS} facilitates GFR. Acute obstruction of the urinary tract (e.g., a **kidney stone** occluding the ureter) increases P_{BS}.

RENAL BLOOD FLOW

In resting persons the blood flow to the kidneys (about 1.25 L/min) is equal to about 25% of the cardiac output. However, the kidneys constitute less than 0.5% of the total body weight. Blood flow through the kidneys serves several important functions, including the following:

1. Indirectly determining the GFR
2. Modifying the rate of solute and water reabsorption by the proximal tubule
3. Participating in the concentration and dilution of the urine
4. Delivering oxygen, nutrients, and hormones to the cells of the nephron and returning carbon dioxide and reabsorbed fluid and solutes to the general circulation

The blood flow through any organ may be represented by the following equation:

$$Q = \Delta P/R \qquad \textbf{(14)}$$

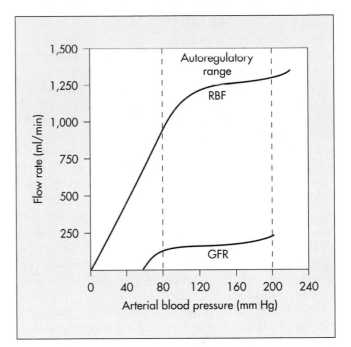

FIGURE 35-14 Relationships between arterial blood pressure and renal blood flow *(RBF)* and glomerular filtration rate *(GFR)*. Autoregulation maintains RBF and GFR relatively constant as blood pressure changes from 90 to 180 mm Hg.

where

Q = Blood flow
ΔP = Mean arterial pressure minus venous pressure for that organ
R = The resistance to flow through that organ.

Accordingly, RBF is equal to the pressure difference between the renal artery and the renal vein divided by the renal vascular resistance:

$$\text{RBF} = \frac{\text{Aortic pressure} - \text{Renal venous pressure}}{\text{Renal vascular resistance}} \qquad \textbf{(15)}$$

The afferent arteriole, the efferent arteriole, and the interlobular artery are the major resistance vessels in the kidney, and, thereby, they determine renal vascular resistance. The kidneys, like most other organs, regulate their blood flow by adjusting the vascular resistance in response to changes in arterial pressure. As illustrated in Figure 35-14, this adjustment in resistance is so precise that blood flow remains relatively constant as arterial blood pressure changes between 90 and 180 mm Hg. GFR is also regulated over the same range of arterial pressures. The phenomenon whereby RBF and GFR are maintained relatively constant is called **autoregulation.** As the term indicates, autoregulation is achieved by changes in vascular resistance exclusively within the kidney. Because both GFR and RBF are regulated over the same range of pressures, and because renal plasma flow (RPF) is an important determinant of GFR, it is not surprising that the same mechanisms regulate both flows.

Two mechanisms are responsible for autoregulation of RBF and GFR: one that responds to changes in arterial pressure, and another that responds to changes in the flow rate of tubular fluid. The pressure-sensitive mechanism, the so-called **myogenic mechanism** (see Chapter 23), is related to an intrinsic property of vascular smooth muscle—the tendency to contract when it is stretched. Accordingly, when arterial pressure rises and the renal afferent arteriole is stretched, the smooth muscle contracts. Because the increase in the resistance of the arteriole offsets the increase in pressure, RBF and therefore GFR remain constant (i.e., RBF is constant if the ratio of $\Delta P/R$ is kept constant, see equation 14, above).

The second mechanism responsible for autoregulation of GFR and RBF, the flow-dependent mechanism, is known as **tubuloglomerular feedback** (Figure 35-15). This mechanism involves a feedback loop in which the flow of tubular fluid (or some other factor, such as the rate of NaCl reabsorption, which increases in direct proportion to flow) is sensed by the macula densa of the juxtaglomerular apparatus (JGA) and converted into a signal that affects GFR. When GFR increases and causes the flow of tubular fluid at the macula densa to rise, the JGA sends a signal that causes RBF and GFR to return to normal levels. In contrast, when GFR and tubular flow past the macula densa decrease, the JGA sends a signal causing RBF and GFR to increase to normal levels. The signal affects RBF and GFR mainly by changing the resistance of the afferent arteriole, but the mediator for this effect is controversial. The major unknowns about tubuloglomerular feedback concern the variable that is sensed at the macula densa and the effector substance that alters the resistance of the afferent arteriole. It has been suggested that flow-dependent changes in NaCl reabsorption are sensed by the macula densa. The effector mechanism may be adenosine, which constricts renal arteries, in contrast to its vasodilator effect on most other vasculature beds, or it may be ATP, which selectively vasoconstricts the afferent arteriole. Nitric oxide, a vasodilator, produced by the macula densa, may also play a role in autoregulation. It may be that the macula densa releases both a vasoconstrictor and a vasodilator, such as nitric oxide, which oppose each other's action at the level of the afferent arteriole.

Because animals engage in many activities that can change arterial blood pressure, having mechanisms that

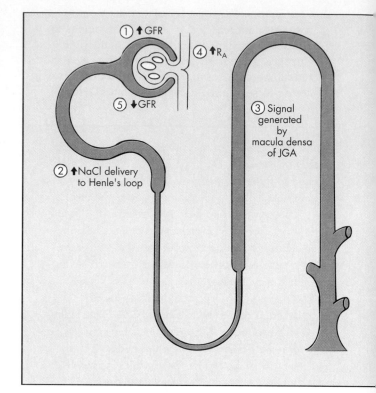

FIGURE 35-15 Tubuloglomerular feedback. An increase in GFR *(1)* increases NaCl delivery to the loop of Henle *(2)*, which is sensed by the macula densa and converted into a signal *(3)*, that increases R_A *(4)*, the resistance of the afferent arteriole), which decreases GFR *(5)*. (Adapted from Cogan MG: *Fluid and electrolytes: physiology and pathophysiology,* Norwalk, 1991, Appleton & Lange.)

maintain RBF and GFR relatively constant despite changes in arterial pressure is highly desirable. If RBF and GFR were to rise or fall suddenly in proportion to changes in blood pressure, urinary excretion of fluid and sodium would also change suddenly, because alterations in GFR influence water and solute excretion (the reason for this will be discussed in the next chapter). Such changes in water and sodium excretion, without comparable alterations in intake, would alter fluid and sodium balance. Accordingly, autoregulation of GFR and RBF provides an effective means for uncoupling renal function from arterial pressure, and ensures that fluid and solute excretion remain constant.

Three points concerning autoregulation should be made: (1) autoregulation is absent below arterial pressures of 90 mm Hg; (2) autoregulation is not perfect; RBF and GFR do change slightly as arterial blood pressure rises; and (3) despite autoregulation, GFR and RBF can be changed, under appropriate conditions, by several hormones (see below).

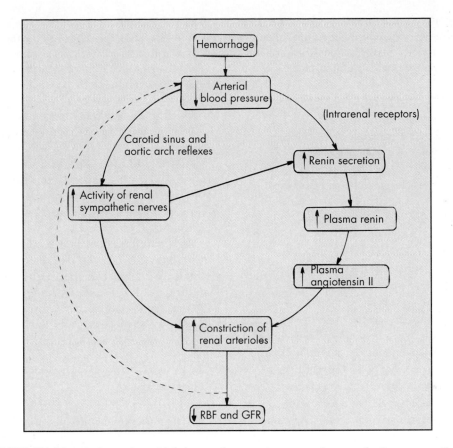

FIGURE 35-16   Pathway by which hemorrhage activates renal sympathetic nerve activity and stimulates angiotensin II production. (Modified from Vander AJ: *Renal physiology*, ed 2, New York, 1980, McGraw-Hill Book Co.)

Individuals with **renal artery stenosis** (a narrowing of the artery lumen)—caused by atherosclerosis, for example—can have an elevated systemic blood pressure mediated by stimulation of the renin-angiotensin system. The pressure in the artery proximal to the stenosis is increased, but it is normal or reduced distal to the stenosis. Autoregulation plays an important role in maintaining RBF, P_{GC}, and GFR in the presence of this stenosis. The administration of drugs to lower systemic blood pressure also lowers pressure distal to the stenosis; accordingly RBF, P_{GC}, and GFR fall.

Regulation of Renal Blood Flow

Several factors and hormones have a major effect on RBF. As discussed in the previous section, the myogenic mechanism and tubuloglomerular feedback play a key role in regulating RBF. The sympathetic nervous system and a variety of hormones also regulate RBF. Sympathetic nerves, angiotensin II, prostaglandins, and perhaps aden-

osine exert the most control over RBF. The physiological role of the other hormones discussed below remains to be determined.

Sympathetic nerves   The afferent and efferent arterioles are innervated by sympathetic neurons; however, sympathetic tone is normally minimal. However, norepinephrine, released by sympathetic nerves, and circulating epinephrine, secreted by the adrenal medulla, cause vasoconstriction by binding to α_1-adrenoceptors, which are located mainly on the afferent arterioles and thereby decrease RBF and GFR. A reduction in the effective circulating volume (ECV) or strong emotional stimuli, like fear and pain, activate sympathetic nerves and reduce RBF and GFR.

Angiotensin II   Angiotensin II is produced systemically and within the kidney, and it constricts the afferent and efferent arterioles* and decreases RBF (see Chap-

---

*The efferent arteriole is more sensitive to angiotensin II than is the afferent arteriole. Therefore with low concentrations of angiotensin II, constriction of the efferent arteriole predominates. However, with high concentrations of angiotensin II, constriction of both afferent and efferent arterioles occurs.

ter 37 for details on the renin-angiotensin system). Figure 35-16 illustrates how norepinephrine, epinephrine, and angiotensin II act together to decrease RBF and GFR, as would occur, for example, with hemorrhage.

Hemorrhage decreases arterial blood pressure, which activates the sympathetic nerves to the kidneys via the baroreceptor reflex. (See Chapters 23 and 37.) Norepinephrine elicits an intense vasoconstriction of the afferent and efferent arterioles, and thereby decreases RBF and GFR. The rise in sympathetic activity also increases the release of epinephrine and angiotensin II, which cause further vasoconstriction and a fall in RBF. The rise in the vascular resistance of the kidney and other vascular beds increases total peripheral resistance, which, by increasing blood pressure (BP = cardiac output × total peripheral resistance), offsets the fall in mean arterial blood pressure elicited by hemorrhage. Hence, this system works to preserve arterial pressure at the expense of maintaining normal RBF and GFR. This example illustrates the important point that, although autoregulatory mechanisms can prevent the effects of changes in arterial pressure on RBF and GFR, when needed, sympathetic nerves and angiotensin II have important salutary effects on RBF and GFR.

Atrial natriuretic peptide (ANP) Circulating levels of ANP rise with hypertension and ECV expansion, causing dilation of the afferent arteriole, and thereby increase RBF and GFR.

Antidiuretic hormone (ADH) High concentrations of ADH cause renal vasoconstriction and contraction of mesangial cells and thereby lower RBF and GFR.

ATP Various cell types release ATP into the interstitial fluid. ATP selectively vasoconstricts the afferent arteriole and may play a role in tubuloglomerular feedback.

Glucocorticoids Glucocorticoids also cause dilation of the afferent arteriole and thereby increase RBF and GFR.

Nitric oxide Nitric oxide, an endothelium-derived relaxing factor, elicits vasorelaxation and increases blood flow. An increase in shear force acting on endothelial cells in the glomerulus increases the production of nitric oxide, which may cause glomerular vasodilation by increasing cGMP in mesangial cells (see Chapter 23).

Endothelin Endothelin is a potent vasoconstrictor secreted by endothelial cells of arterioles. Its secretion is stimulated by stretch.

Prostaglandins Prostaglandins do not regulate RBF or GFR in healthy resting people. However, during pathophysiological conditions, such as hemorrhage, prosta-

glandins, notably PGE_2 and PGI_2, are produced locally within the kidneys. These substances modulate the afferent and efferent arterioles and thereby dampen the vasoconstrictor effects of sympathetic nerves and angiotensin II. This effect of prostaglandins is important because it prevents severe and potentially harmful renal vasoconstriction and renal ischemia. Prostaglandin synthesis is stimulated by sympathetic nerve activity and angiotensin II.

Adenosine Adenosine is a vasoconstrictor hormone in the kidney. It is produced within kidney cells in response to anoxia. Adenosine is also a metabolite of ATP and is produced inside cells as well as in the extracellular fluid (see Chapter 23).

Kinins Kallikrein is a proteolytic enzyme produced in the kidneys; it cleaves circulating kininogen to kinins, which are vasodilators that act by stimulating the release of nitric oxide and prostaglandins.

Dopamine Dopamine, a vasodilator hormone produced by the proximal tubule, has several actions, such as increasing RBF and inhibiting renin secretion.

SUMMARY

1. The functional unit of the kidneys is the nephron, which consists of a renal corpuscle, proximal tubule, loop of Henle, distal tubule, and collecting duct.

2. The renal corpuscle is composed of glomerular capillaries and Bowman's capsule.

3. The juxtaglomerular apparatus is one component of an important feedback mechanism that regulates renal blood flow and the glomerular filtration rate. The structures that compose the juxtaglomerular apparatus include the macula densa, the extraglomerular mesangial cells, and the renin-producing granular cells.

4. The lower urinary tract is composed of the ureters, bladder, and urethra. Micturition is the process of emptying the urinary bladder. The micturition reflex is an automatic spinal cord reflex; however, it can be inhibited or facilitated by centers in the brainstem and cortex.

5. The formation of urine occurs by three general processes: (a) glomerular filtration, (b) reabsorption of solutes and water from the ultrafiltrate into the peritubular capillaries, and (c) secretion of selected solutes from the peritubular capillaries into the tubular fluid.

6. The rate of glomerular filtration is calculated by measuring the clearance of inulin or creatinine.

7. Effective renal plasma flow is determined by the clearance of *p*-aminohippurate (PAH).

8. The renal clearance equation can be used to deter-

mine if a substance undergoes either net reabsorption or secretion by the nephron.

9. The first step in the production of urine is the formation of an ultrafiltrate of plasma by the glomerulus. Starling forces across the glomerular capillaries provide the driving force for the ultrafiltration of plasma from the glomerular capillaries into Bowman's space.

10. The glomerular ultrafiltrate is devoid of cellular elements and contains very little protein, but otherwise is identical to plasma. Proteins with a molecular radius smaller than 18 Å are readily filtered, proteins between 18 and 36 Å are filtered at rates that depend on size and charge (cationic proteins are more readily filtered than anionic proteins), and proteins with molecular radii greater than 36 Å are not filtered.

11. Renal blood flow (1.25 L/min) is about 25% of the cardiac output, yet the kidneys constitute less than 0.5% of the body weight. Renal blood flow serves several important functions, it determines the glomerular filtration rate, modifies solute and water reabsorption by the proximal tubule, participates in concentration and dilution of the urine, delivers oxygen, nutrients, and hormones to the cells of the nephron, and returns carbon dioxide and reabsorbed fluid and solutes to the general circulation.

12. Renal blood flow and glomerular filtration rate are maintained constant, despite changes in arterial blood pressure between 90 and 180 mm Hg, by the phenomenon of autoregulation. Autoregulation is achieved by the myogenic reflex and tubuloglomerular feedback.

BIBLIOGRAPHY
Journal Articles

Baylis C, Blantz RC: Glomerular hemodynamics, *News Physiol Sci* 1:86, 1986.

Gottschalk CW, ed: Renal and electrolyte physiology section: tubuloglomerular feedback mechanisms, *Ann Rev Physiol* 49:249, 1987.

Inscho EW, Mitchell KD, Navar LG: Extracellular ATP in the regulation of renal microvasculature. *FASEB J* 8:319, 1994.

Ito S: Role of nitric oxide in glomerular arterioles and macula densa, *News Physiol Sci* 9:115, 1994.

Kriz W, Bankir L: A standard nomenclature for structures of the kidney, *Am J Physiol* 254:F1, 1988.

Moe OW, Alpern RJ, Henrich WL: The renal proximal tubule renin-angiotensin system, *Semin Nephrol* 13:552, 1993.

Takabatake T, Thurau K, editors: Tubuloglomerular feedback system. *Kidney Int [Suppl] 32:1, 1991.*

Books and Monographs

Bradley WE: Physiology of the urinary bladder. In Walsh PC, Gittes RF, Perlmutter AD, Stamey TA, eds: *Cambell's urology,* ed 5, Philadelphia, 1986, WB Saunders.

Dworkin LD, Brenner BM: Biophysical basis of glomerular filtration. In Seldin DW, Giebisch G, eds: *The kidney: physiology and pathophysiology,* vol 1, New York, 1985, Raven Press.

Dworkin LD, Brenner BM: The renal circulations. In Brenner BM, Rector FC Jr, eds: *The kidney,* ed 4, Philadelphia, 1991, WB Saunders Co.

Kassier JP, Harrington JT: Laboratory evaluation of renal function. In Schrier RW, Gottschalk CW, eds: *Diseases of the kidney,* ed 4, Boston, 1988, Little Brown & Co.

Koeppen BM, Stanton BA: *Renal physiology,* St. Louis, 1992, Mosby.

Koushanpour E, Kriz W: *Renal physiology,* ed 2, Berlin, 1986, Springer-Verlag.

Kriz W, Kaissling B: Structural organization of the kidney. In Seldin DW, Giebisch G, eds: *The kidney: physiology and pathophysiology,* New York, 1985, Raven Press.

Maddox DA, Brenner BM: Glomerular ultrafiltration. In Brenner BM, Rector FC Jr, eds: *The kidney,* ed 4, Philadelphia, 1991, WB Saunders Co.

Ofstad J, Aukland K: Renal circulation. In Seldin DW, Giebisch G, eds: *The kidney: physiology and pathophysiology,* New York, 1985, Raven Press.

Rose BD: *Clinical physiology of acid-base and electrolyte disorders,* ed 4, New York, 1994, McGraw-Hill Information Services Co.

Tanagho EA: Anatomy of the lower urinary tract. In Walsh PC, Gittes RF, Perlmutter AD, Stamey TA, eds: *Cambell's urology,* ed 5, Philadelphia, 1986, W.B. Saunders.

Tanagho EA: Anatomy of the genitourinary tract. In Tanagho EA, McAnich JW, eds: *Smith's general urology,* ed 12, Norwalk, 1988, Appleton & Lange.

Tisher CC, Madsen KM: Anatomy of the kidney. In Brenner BM, Rector FC Jr, eds: *The kidney,* ed 4, Philadelphia, 1991, WB Saunders Co.

CHAPTER 36

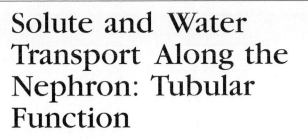

Solute and Water Transport Along the Nephron: Tubular Function

Formation of urine involves three basic processes: ultrafiltration of plasma by the glomerulus, reabsorption of water and solutes from the ultrafiltrate, and secretion of selected solutes into the tubular fluid. Although 180 liters of essentially protein-free fluid is filtered by the human glomeruli each day, less than 1% of the filtered water and Na^+, and variable amounts of the other solutes, is excreted in the urine (Table 36-1). By the processes of reabsorption and secretion, the renal tubules modulate the volume and composition of the urine (Table 36-2). Consequently, the tubules control precisely the volume, osmolality, composition, and pH of the intracellular and extracellular fluid compartments.

GENERAL PRINCIPLES OF TRANSEPITHELIAL SOLUTE AND WATER TRANSPORT

Solutes may be transported across cell membranes by passive or active mechanisms. As defined in Chapter 1, passive movement of solutes occurs down an electrochemical gradient. Active transport results in the movement of a solute against an electrochemical gradient, and requires energy derived from metabolic processes. Transport against an electrochemical gradient that is directly coupled to an energy source (hydrolysis of ATP) is termed **primary active transport.** As detailed later, the primary active transport of Na^+ by the Na^+-K^+-ATPase pump is central to the function of the kidneys. Transport against an electrochemical gradient coupled indirectly to an energy source (e.g., energy stored in an ion gradient) is termed **secondary active transport.** The reabsorption of amino acids by the kidney is an example of such a secondary active process (see later in this chapter).

In contrast to solutes, which are transported by both passive and active mechanisms in the kidney, *water is always transported by passive processes.* The driving force for water movement in the kidneys is an osmotic gradient. Water moves from an area of low osmolality to an area of high osmolality.

The nephron, like other epithelia, such as the intestine, can transport solutes and water from one side of the tubule to the other. **Reabsorption** is the net transport of a substance from the tubular lumen into the blood, whereas **secretion** is the net transport in the opposite direction.

As illustrated in Figure 36-1, renal cells are held together by **tight junctions.** Below the tight junctions, the cells are separated by lateral intercellular spaces. The tight junctions separate the apical membranes from the basolateral membranes. An epithelium can be compared to a six-pack of soda where the cans are the cells and the plastic holder represents the tight junctions.

In the nephron, a substance can be reabsorbed or secreted across cells, the so-called **transcellular pathway,** or between cells, the so-called **paracellular pathway** (Figure 36-1). Na^+ reabsorption by the proximal tubule is a good example of transport by the transcellular pathway. Na^+ reabsorption in this nephron segment depends on the operation of the Na^+-K^+-ATPase pump (Figure 36-1). The Na^+-K^+-ATPase pump, which is located exclusively in the basolateral membrane, moves Na^+ out of the cell into the blood and moves K^+ into the cell. Thus, the operation of the Na^+-K^+-ATPase pump lowers intracellular Na^+ concentration and increases intracellular K^+ concentration. Because intracellular $[Na^+]$ is low (12 mEq/L) and the $[Na^+]$ in tubular fluid is high (140 mEq/L), Na^+ moves across the apical cell membrane down a chemical concentration gradient from the tubular lumen into the cell. The Na^+-K^+-ATPase pump senses the addition of Na^+ to the cell and is stimulated to increase its rate of Na^+ extrusion into the blood and thereby returns intracellular Na^+ to normal levels (see Chapter 1). Thus, a component of Na^+ reabsorption by the proximal tubule is transcellular and is a two-step pro-

Table 36-1   Filtration, Excretion, and Reabsorption of Water, Electrolytes and Solutes*

Substance	Measure	Filtered	Excreted	Reabsorbed	% Filtered Load Reabsorbed
Water	L/day	180	1.5	178.5	99.2
Na^+	mEq/day	25,200	150	25,050	99.4
K^+	mEq/day	720	100	620	86.1
Ca^{++}	mEq/day	540	10	530	98.2
HCO_3^-	mEq/day	4,320	2	4,318	99.9+
Cl^-	mEq/day	18,000	150	17,850	99.2
Glucose	mmol/day	800	0	800	100.0
Urea	g/day	56	28	28	50.0

*The filtered amount of any substance is calculated by multiplying the concentration of that substance in the ultrafiltrate by the glomerular filtration rate; for example, the filtered load of Na^+ is calculated as: $[Na^+]_{ultrafiltrate}$ (140 mEq/L) × Glomerular filtration rate (180 L/day) = 25,200 mEq/day.

Table 36-2   Composition of the Urine

Substance	Concentration
Na^+	50-130 mEq/L
K^+	20-70 mEq/L
NH_4^+	30-50 mEq/L
Ca^{++}	5-12 mEq/L
Mg^{++}	2-18 mEq/L
Cl^-	50-130 mEq/L
$PO_4^=$	20-40 mEq/L
Urea	200-400 mM
Creatinine	6-20 mM
pH	5.0-7.0
Osmolality	500-800 mOsm/kg H_2O
Glucose*	0
Amino acids*	0
Protein*	0
Blood*	0
Ketones*	0
Leukocytes*	0
Bilirubin*	0

These values represent average ranges. Asterisks indicate that the presence of these substances in freshly voided urine is measured with dipstick reagent strips. These small strips of plastic contain reagents that change color in a semi-quantitative manner in the presence of specific compounds. Water excretion ranges between 0.5 and 1.5 liters/day. (Table modified from Valtin HV: *Renal physiology*, ed 2, Boston, 1983, Little, Brown & Co.)

cess: (1) movement across the apical membrane into the cell down an electrochemical gradient established by the Na^+-K^+-ATPase pump, and (2) movement across the basolateral membrane against an electrochemical gradient via the Na^+-K^+-ATPase pump.

The reabsorption of Ca^{++}, Mg^{++}, and K^+ across the proximal tubule is a good example of paracellular transport. Some of the water reabsorbed across the proximal tubule traverses the paracellular pathway. Some solutes dissolved in this water—in particular, Ca^{++}, Mg^{++}, and K^+—are entrained in the reabsorbed fluid, and are

thereby reabsorbed by the process of **solvent drag** (for additional details of transcellular and paracellular reabsorption, as well as solvent drag, see later in this chapter).

PROXIMAL TUBULE

The proximal tubule reabsorbs approximately 67% of the filtered water, Na^+, Cl^-, K^+, and other solutes. In addition, virtually all of the glucose and amino acids filtered by the glomerulus are reabsorbed. *The key element in proximal tubule reabsorption is the Na^+-K^+-ATPase pump in the basolateral membrane.* The reabsorption of every substance, including water, is linked to the operation of the Na^+-K^+-ATPase pump.

Na^+ Reabsorption Na^+ is reabsorbed by different mechanisms in the early (first half of the proximal tubule) and late (second half of the proximal tubule) segments of the proximal tubule. In the early segment, Na^+ is reabsorbed primarily with HCO_3^- and a number of organic molecules (e.g., glucose, amino acids, $PO_4^=$, lactate). By contrast, Na^+ is reabsorbed mainly with Cl^- in the second half of the proximal tubule. This occurs because of differences in Na^+ transport systems present in the early and late segments of the proximal tubule, as well as differences in the composition of tubular fluid at these sites.

As illustrated in Figure 36-2, in the early segment of the proximal tubule, Na^+ uptake into the cell is coupled with organic solutes, anions, and H^+. Some Na^+ is also reabsorbed with Cl^-. Na^+ entry into the cell across the apical membrane is mediated by specific **symporter** and **antiporter proteins** and not by simple diffusion.*

---

*As described in Chapter 1, symporters and antiporters are examples of coupled transport proteins. Symporters couple the movement of two or more molecules in the same direction across the membrane. Antiporters couple the movement of two or more molecules in opposite directions across the membrane.

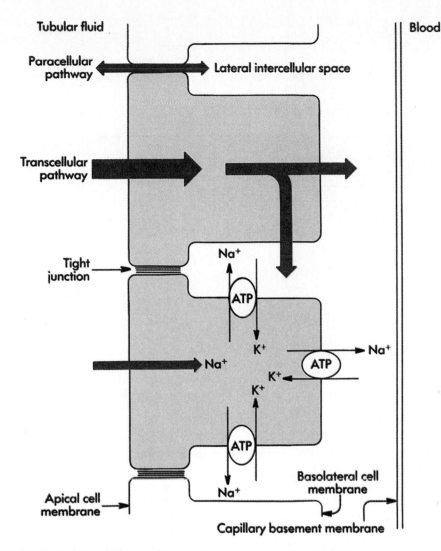

FIGURE 36-1 Paracellular and transcellular transport pathways in the proximal tubule. See text for details.

For example, Na^+ enters proximal cells by Na^+-glucose, Na^+-amino acid, Na^+-$PO_4^{\equiv}$, and Na^+-lactate symporters. Na^+ entry is also coupled with H^+ extrusion from the cell by the Na^+-H^+ antiporter (Figure 36-2). H^+ secretion, by the Na^+-H^+ antiporter, results in HCO_3^- reabsorption (see Chapter 39 for details). The Na^+ that enters the cell across the apical membrane, by either a symport or antiport mechanism, leaves the cell and enters the blood via the Na^+-K^+-ATPase. The solutes and anions that enter the cell with Na^+ (e.g., glucose, amino acids, $PO_4^{\equiv}$, and lactate) exit across the basolateral membrane by passive mechanisms. In summary, in the early-segment of the proximal tubule the reabsorption of Na^+ is coupled to that of HCO_3^-, and a number of organic molecules. Reabsorption of many of these organic molecules is so avid that these solutes are almost completely removed from the tubular fluid in the first half of the

proximal tubule (Figure 36-3). Because water is reabsorbed in excess of Cl^- in the early segment of the proximal tubule, the Cl^- concentration in tubular fluid rises along the length of the early proximal tubule (Figure 36-3).

In the second half of the proximal tubule Na^+ is reabsorbed with Cl^- across both the transcellular and paracellular pathways (Figure 36-4). Na^+ is reabsorbed with Cl^- rather than with organic anions or HCO_3^- as the accompanying anion. This occurs because the cells lining the second half of the proximal tubule have different Na^+ transport mechanisms than the early segment of the proximal tubule, and because the tubular fluid that enters the second half of the proximal tubule contains very little glucose and amino acids, and has a high concentration of Cl^- (140 mEq/L versus 105 mEq/L in the first half of the proximal tubule). The Cl^- concentration

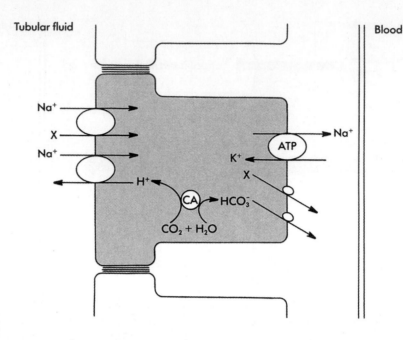

Tubular fluid

Blood

FIGURE 36-2 Na^+ transport processes in the first half of the proximal tubule. Na^+-X cotransport protein indicates the presence of four unique symporters. X represents either glucose, amino acids, phosphate, or lactate. CO_2 and H_2O combine inside the cells to form H^+ and HCO_3^- in a reaction facilitated by the enzyme carbonic anhydrase *(CA)*.

is high because, in the first half of the proximal tubule, Na^+ is preferentially reabsorbed with HCO_3^-, glucose, and organic anions, and therefore the tubular fluid becomes enriched in Cl^-.

The mechanism of transcellular Na^+ reabsorption is illustrated in Figure 36-4. Na^+ enters the cell across the luminal membrane by the parallel operation of Na^+-H^+ and one or more Cl^--$Base^-$ antiporters. Because the secreted H^+ and base combine in the tubular fluid and re-enter the cell by passive diffusion, the operation of the Na^+-H^+ and Cl^--$Base^-$ antiporters is equivalent to NaCl uptake from tubular fluid into the cell. Na^+ leaves the cell by the Na^+-K^+-ATPase pump, and Cl^- leaves the cell and enters the blood by a KCl symport protein in the basolateral membrane.

Paracellular NaCl reabsorption occurs because the rise in [Cl$^-$] in the tubule fluid of the early segment of the proximal tubule creates a gradient that favors the diffusion of Cl$^-$ from the tubular lumen across the tight junctions into the lateral intercellular space. Movement of the negatively charged Cl^- generates a positive transepithelial voltage (tubular fluid positive relative to the blood), which causes the diffusion of positively charged Na^+ out of the tubular fluid across the tight junction into the blood. Thus, in the second half of the proximal tubule some Na^+ and Cl^- is reabsorbed across the tight junctions by passive diffusion.

In summary, reabsorption of Na^+ and Cl^- in the proximal tubule occurs across the paracellular pathway and across the transcellular pathway. Approximately 17,000 mEq of the 25,200 mEq of NaCl filtered each day is reabsorbed in the proximal tubule (~67% of the filtered load). Of this, two thirds moves across the transcellular pathway, while the remaining one third moves across the paracellular pathway.

Water Reabsorption Figure 36-5 illustrates the mechanism of water reabsorption in the proximal tubule. *The driving force for water reabsorption is a transtubular osmotic gradient established by Na^+ reabsorption.* The reabsorption of Na^+ with organic solutes, HCO_3^-, and Cl^- increases the osmolality of the lateral intercellular space. This occurs because some Na^+-K^+-ATPase pumps and organic solute, HCO_3^-, and Cl^- transporters are located on the lateral cell membranes and deposit these solutes in this space. Furthermore, some NaCl also enters the lateral intercellular space by diffusion across the tight junction (i.e., paracellular pathway). Because the lateral intercellular space becomes slightly hyperosmotic (3 to 5 mOsm/kg H_2O) with respect to tubular fluid, and because the proximal tubule is highly permeable to water, water will flow by osmosis across both the tight junctions and the proximal tubular cells into this hyperosmotic compartment. Accumulation of fluid and solutes within the lateral intercellular space in-

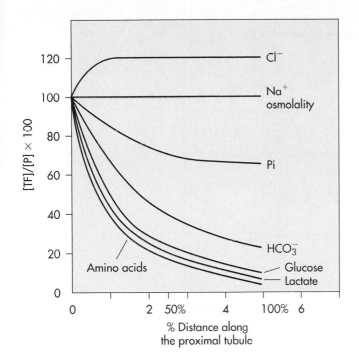

FIGURE 36-3 Concentration of solutes in tubule fluid as a function of length along the proximal tubule. *[TF]* is the concentration of the substance in tubular fluid; *[P]* is the concentration of the substance in plasma. Values above 100 indicate that relatively less of the solute than water was reabsorbed, and values below 100 indicate that relatively more of the substance than water was reabsorbed. (Modified from Vander AJ. *Renal physiology*, ed 4, New York, 1991, McGraw-Hill.)

creases the hydrostatic pressure in this compartment, which, in turn, forces fluid and these solutes to move into the capillaries. Thus, water reabsorption follows solute reabsorption in the proximal tubule. The reabsorbed fluid is essentially isosmotic to plasma. An important consequence of osmotic water flow across the proximal tubule is that some solutes, especially K^+, Ca^{++}, and Mg^{++}, are entrained in the reabsorbed fluid and are thereby reabsorbed by the process of solvent drag (Figure 36-5). Because the reabsorption of virtually all organic solutes, Cl^-, other ions, and water is coupled to Na^+ reabsorption, changes in Na^+ reabsorption will influence the reabsorption of water and the other solutes by the proximal tubule.

Protein Reabsorption Proteins that are filtered are also reabsorbed in the proximal tubule. As mentioned previously, peptide hormones, small proteins, and even small amounts of larger proteins, such as albumin, are filtered by the glomerulus. Although filtration of proteins is small (the concentration of proteins in the ultrafiltrate is only 40 mg/L), the amount of protein filtered per day

is significant because the GFR is so high (filtered protein = GFR × [protein] in the ultrafiltrate: thus, filtered protein = 180 L/day × 40 mg/L = 7.2 g/day). These proteins are partially degraded by enzymes on the surface of the proximal tubule cells, and then they are taken up into the cell by endocytosis. Once inside the cell, enzymes digest the proteins and peptides into their constituent amino acids, which leave the cell across the basolateral membrane, and are returned to the blood. Normally, this mechanism reabsorbs virtually all of the protein filtered, and hence the urine is essentially protein free. However, because the mechanism is easily saturated, protein will appear in the urine if the amount of protein filtered increases. Disruption of the glomerular barrier to proteins will increase the filtration of proteins and result in proteinuria (appearance of protein in the urine). Proteinuria is frequently seen with kidney disease.

Organic Anion and Organic Cation Secretion In addition to reabsorbing solutes and water, cells of the proximal tubule also secrete organic cations and organic anions (see Tables 36-3 and 36-4 for a partial listing). Many of these substances are end-products of metabolism, and they circulate in the plasma. Cells of the proximal tubule also secrete numerous exogenous organic compounds, including *p*-aminohippurate (PAH) and drugs such as penicillin. Because many of these organic compounds can be bound to plasma proteins, they are not readily filtered. Therefore, excretion by filtration alone eliminates only a small portion of these potentially toxic substances from the body. Excretion rates of these substances are high because they are secreted from the peritubular capillaries into the tubular fluid. Because the kidneys remove virtually all organic ions and drugs from the plasma entering the kidneys, it is evident that these secretory mechanisms are very powerful and that they serve a vital function by clearing these substances from the plasma.

Figure 36-6 illustrates the mechanism of PAH transport across the proximal tubule as an example of organic anion secretion. This secretory pathway has a maximal transport rate (T_m) and a low specificity and is responsible for the secretion of all organic anions listed in Table 36-3. The organic anion PAH, which can be used to measure RPF, has been used to unravel the details of this pathway. PAH is taken up into the cell, across the basolateral membrane, against its chemical gradient by a PAH-di- and tricarboxylate antiport mechanism. The di- and tricarboxylates accumulate inside the cell by an Na^+-di- and tricarboxylate symporter, also present in the basolateral membrane. Thus, PAH uptake into the cell against its electrochemical gradient is coupled to the exit of di- and tricarboxylates out of the cell down their chemical gradients by the antiport mechanism. The resulting high

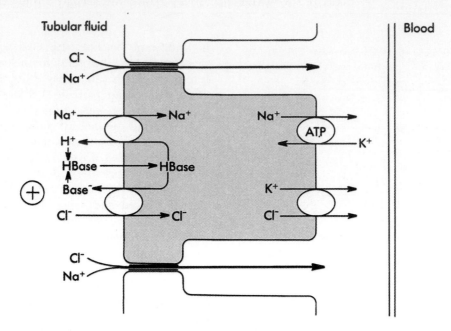

FIGURE 36-4 Na$^+$ transport processes in the second half of the proximal tubule. Na$^+$ and Cl$^-$ enter the cell across the apical membrane by the operation of parallel Na$^+$-H$^+$ and Cl$^-$-Base$^-$ antiporters. More than one base antiporter may be involved in this process, but only one is depicted here. The secreted H$^+$ and Base$^-$ combine in the tubular fluid to form an HBase complex that is permeable to the plasma membrane. Accumulation of HBase in tubular fluid establishes an HBase concentration gradient that favors HBase diffusion across the apical plasma membrane into the cell. Inside the cell H$^+$ and the Base$^-$ dissociate and recycle back across the apical plasma membrane. The net result is NaCl uptake across the apical membrane. Base may be OH$^-$, formate (HCO$_2^-$), oxalate$^-$, or HCO$_3^-$. The lumen positive transepithelial voltage, indicated by the plus sign inside the circle in the tubular lumen, is generated by the diffusion of Cl$^-$ (lumen-to-blood) across the tight junction. The high [Cl$^-$] of tubular fluid provides the driving force for Cl$^-$ diffusion.

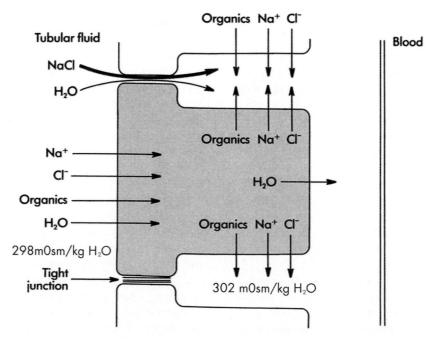

FIGURE 36-5 Routes of water reabsorption across the proximal tubule. Transport of Na$^+$, Cl$^-$, and organic solutes into the lateral intercellular space increases the osmolality of this compartment, which establishes the driving force for osmotic water reabsorption across the proximal tubule. An important consequence of osmotic water flow across the proximal tubule is that some solutes, especially K$^+$, Ca^{++}, and Mg^{++}, are entrained in the reabsorbed fluid and are thereby reabsorbed by the process of solvent drag.

intracellular concentration of PAH provides the driving force for PAH exit across the luminal membrane into the tubular fluid via a PAH-anion antiporter (Figure 36-6).

Because all organic anions compete for the same transporter, elevated plasma levels of one anion will inhibit the secretion of the others. For example, a reduction of **penicillin** secretion by the proximal tubule can be produced by infusing PAH. Because the kidneys are responsible for eliminating penicillin from the body, the infusion of PAH into individuals receiving penicillin will reduce urinary penicillin excretion and thereby extend the biological half-life of the drug. In World War II when penicillin was in short supply, **hippurates** were given with the penicillin to extend the drug's therapeutic effect.

The details of the secretory pathway for organic cations in the proximal tubule have not been elucidated. Organic cations are taken up into the cell, across the basolateral membrane, by an organic cation transport mechanism that involves either simple facilitated diffusion (i.e, uniport mechanism) or exchange for intracellular organic cations. Organic cation (OC^+) transport across the luminal membrane into the tubular fluid is mediated by an OC^+-H^+ antiporter. The transport mechanisms for organic cation secretion are nonspecific (see Table 36-4); several cations compete for the transport pathway.

The histamine H_2 antagonist **cimetidine,** used to treat gastric ulcers, is secreted by the organic cation pathway in the proximal tubule. Cimetidine reduces the urinary excretion of the antiarrythmic drug **procainamide,** also an organic cation, by competing with procainamide for the secretory pathway. It is important to recognize that co-administration of organic cations can often increase the plasma concentration of both drugs, to levels that are much higher than the plasma concentration of the drugs when given alone, and it can often lead to drug toxicity.

HENLE'S LOOP

Henle's loop reabsorbs approximately 25% of the filtered NaCl and K^+. Ca^{++}, HCO_3^-, and Mg^{++} are also reabsorbed in the loop of Henle (see Chapter 38). This re-

Table 36-3 Some Organic Anions Secreted by the Proximal Tubule

Endogenous Anions	Drugs
cAMP	Acetazolamide
Bile salts	Chlorothiazide
Hippurates	Furosemide
Oxalate	Penicillin
Prostaglandins	Probenecid
Urate	Salicylate (aspirin)
	Hydrochlorothiazide
	Bumetanide

absorption occurs almost exclusively in the thick ascending limb. By comparison, the ascending thin limb has a much lower reabsorptive capacity, and the descending thin limb does not reabsorb significant amounts of solutes (see Chapter 37). The loop of Henle reabsorbs approximately 15% of the filtered water. This reabsorption, however, occurs exclusively in the descending thin limb. *The ascending limb is impermeable to water.*

The key element in solute reabsorption by the thick ascending limb is the Na^+-K^+-ATPase pump in the basolateral membrane (Figure 36-7). As with reabsorption in the proximal tubule, the reabsorption of every solute by the thick ascending limb is linked to the Na^+-K^+-ATPase pump. The operation of the Na^+-K^+-ATPase pump maintains a low cell $[Na^+]$. This low $[Na^+]$ provides a favorable chemical gradient for the movement of Na^+ from the tubular fluid into the cell. The movement of Na^+ across the apical membrane into the cell is mediated by the $1Na^+$-$2Cl^-$-$1K^+$ symporter, which couples the movement of $1Na^+$ with $2Cl^-$ and $1K^+$. This symport protein uses the potential energy released by the downhill movement of Na^+ and Cl^- to drive the uphill movement of K^+ into the cell. An Na^+-H^+ antiporter in the apical cell membrane also mediates Na^+ reabsorption as well as H^+ secretion (HCO_3^- reabsorption) in the thick ascending limb (Figure 36-7). Na^+ leaves the cell across the basolateral membrane via the Na^+-K^+-ATPase pump, and K^+, Cl^-, and HCO_3^- leave the cell across the basolateral membrane by separate pathways.

The voltage across the thick ascending limb is positive in the tubular fluid relative to the blood because of the unique location of transport proteins in the apical and basolateral membranes. *The important points to recognize are that increased salt transport by the thick ascending limb increases the magnitude of the positive voltage in the lumen, and that this voltage is an important driving force for the reabsorption of several cations, including Na^+, K^+, Ca^{++}, and Mg^{++}, across the*

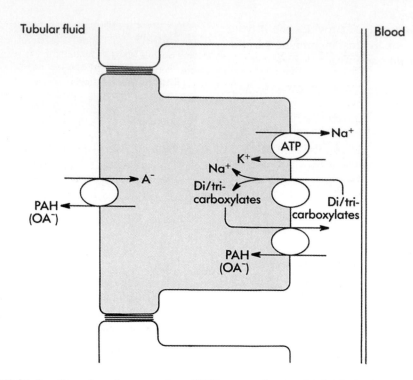

FIGURE 36-6   Organic anion secretion *(PAH)* across the proximal tubule. PAH or another organic anion *(OA⁻)* enters the cell across the basolateral membrane by a PAH-di- and tricarboxylate antiport mechanism. The uptake of di- and tricarboxylates into the cell, against their chemical gradients, is driven by the movement of Na^+ into the cell. The di- and tricarboxylates recycle across the basolateral membrane. PAH leaves the cell across the apical membrane, down its chemical gradient, by a PAH/organic anion (OA^-) antiport mechanism. The OA^- indicates one of several possible anions (e.g., urate).

Table 36-4   Some Organic Cations Secreted by the Proximal Tubule

Endogenous Cations	Drugs
Creatinine	Atropine
Dopamine	Isoproterenol
Epinephrine	Cimetidine
Norepinephrine	Morphine
	Quinine
	Amiloride
	Procainamide

Inhibition of the $1Na^+$-$2Cl^-$-$1K^+$ symporter in the thick ascending limb by loop diuretics, such as **furosemide (Lasix),** inhibits NaCl reabsorption by the thick ascending limb and, thereby, increases urinary NaCl excretion. Furosemide also inhibits K^+, Ca^{++}, HCO_3^-, and Mg^{++} reabsorption by reducing the lumen positive voltage which drives the paracellular reabsorption of these ions. Thus, furosemide also increases urinary K^+, Ca^{++}, HCO_3^-, and Mg^{++} excretion.

paracellular pathway (Figure 36-7). Thus, salt reabsorption across the thick ascending limb occurs by transcellular and paracellular pathways. Fifty percent of solute transport is transcellular and 50% is paracellular.

Because the thick ascending limb is very impermeable to water, reabsorption of NaCl and other solutes reduces the osmolality of tubular fluid to less than 150 mOsm/kg H_2O.

DISTAL TUBULE AND COLLECTING DUCT

The distal tubule and the collecting duct reabsorb approximately 7% of the filtered NaCl, secrete variable amounts of K^+ and H^+, and reabsorb a variable amount of water (~17%). Water reabsorption depends on the plasma concentration of ADH. The initial segment of the distal tubule (early distal tubule) reabsorbs Na^+, Cl^-, and

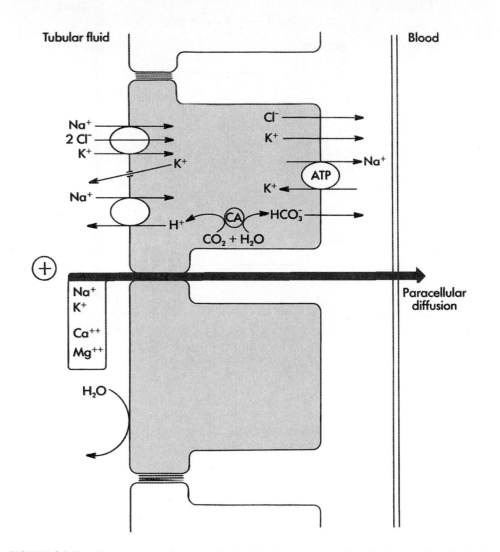

FIGURE 36-7   Transport mechanisms for NaCl reabsorption in the thick ascending limb of Henle's loop. The lumen positive transepithelial voltage results from the diffusion of K^+ from the cell into the tubular fluid, and plays a major role in driving passive paracellular reabsorption of cations.

Ca^{++}, and, like the thick ascending limb, is impermeable to water (Figure 36-8). NaCl entry into the cell across the apical membrane is mediated by an NaCl symporter (Figure 36-8). Na^+ leaves the cell via the Na^+-K^+-ATPase pump, and Cl^- leaves the cell by diffusion via channels. NaCl reabsorption is reduced by thiazide diuretics, which inhibit the NaCl symporter. Thus, the active dilution of the tubular fluid begins in the thick ascending limb and continues in the early segment of the distal tubule.

The last segment of the distal tubule (late distal tubule) and the collecting duct are composed of two cell types, principal cells and intercalated cells. As illustrated in Figure 36-9, **principal cells** reabsorb Na^+ and water and secrete K^+. **Intercalated cells** either secrete H^+ (re-

absorb HCO_3^-) or secrete HCO_3^- and thus are important in regulating acid-base balance (see Chapter 39). Intercalated cells also reabsorb K^+. Both Na^+ reabsorption and K^+ secretion by principal cells depend on the activity of the Na^+-K^+-ATPase pump in the basolateral membrane (Figure 36-9). This enzyme maintains a low cell $[Na^+]$, which provides a favorable chemical gradient for the movement of Na^+ from the tubular fluid into the cell. Because Na^+ enters the cell across the apical membrane by diffusion through channels in the membrane, the negative potential inside the cell facilitates Na^+ entry. Na^+ leaves the cell across the basolateral membrane and enters the blood via the Na^+-K^+-ATPase pump. Sodium reabsorption generates a lumen-negative voltage across the late distal tubule and collecting duct. Cells in

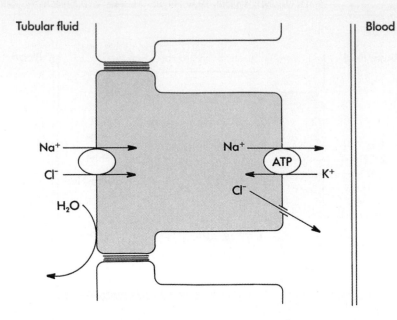

FIGURE 36-8 Transport mechanism for Na^+ and Cl^- reabsorption in the early segment of the distal tubule. This segment is impermeable to water. See the text for details.

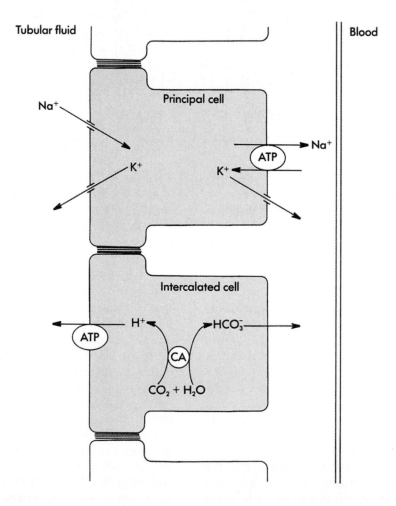

FIGURE 36-9 Transport pathways in principal cells and α intercalated cells of the distal tubule and collecting duct. See the text for details. *CA*, carbonic anhydrase. β-intercalated cells (not shown) secrete HCO_3^-.

Table 36-5 Hormones that Regulate NaCl and Water Reabsorption

Hormone	Major Stimulus	Nephron Site of Action	Effect on Transport
Angiotensin II	⇑ Renin	PT	⇑ NaCl and H_2O reabsorption
Aldosterone	⇑ Angiotensin II, ⇑ $[K^+]_p$	TAL, DT/CD	⇑ NaCl and H_2O reabsorption*
ANP	⇑ BP, ⇑ ECF	CD	⇓ H_2O and NaCl reabsorption
Sympathetic nerves	⇓ ECF	PT, TAL, DT/CD	⇑ NaCl and H_2O reabsorption*
Dopamine	⇑ ECF	PT	⇓ H_2O and NaCl reabsorption
ADH	⇑ P_{osm}, ⇓ ECF	TAL, DT/CD	⇓ NaCl and H_2O reabsorption*

PT is proximal tubule; TAL is thick ascending limb; DT/CD is the distal tubule and collecting duct. All of the hormones listed act within minutes, except aldosterone, which exerts its action on NaCl reabsorption with a delay of one hour. * indicates that the effect on H_2O reabsorption does not occur in TAL. ⇓ ECF is decrease in extracellular fluid volume; ⇑ ECF is increase in extracellular fluid volume; ⇑ BP is increase in blood pressure; ⇑ [K]p is increase in plasma $[K^+]$; and ⇑ P_{osm} is increase in plasma osmolality.

the collecting duct reabsorb significant amounts of Cl^-, most likely across the paracellular pathway. Reabsorption of Cl^- is driven by the lumen-negative transepithelial voltage.

K^+ is secreted from the blood into the tubular fluid by principal cells in two steps (Figure 36-9). K^+ uptake across the basolateral membrane is mediated by the Na^+-K^+-ATPase pump. Because the $[K^+]$ inside the cells is high (140 mEq/L) and the $[K^+]$ in tubular fluid is low (~10 mEq/L), K^+ diffuses down its concentration gradient across the apical cell membrane into the tubular fluid. Although the negative potential inside the cells tends to retain K^+ within the cell, the electrochemical gradient across the apical membrane favors K^+ secretion from the cell into the tubular fluid. Additional details of K^+ secretion and its regulation are considered in Chapter 38.

The mechanism of K^+ reabsorption by intercalated cells is not completely understood, but is thought to be mediated by an H^+-K^+-ATPase located in the apical cell membrane (see Chapter 39).

Amiloride is a diuretic that inhibits Na^+ reabsorption by the distal tubule and collecting duct by directly inhibiting Na^+ channels in the luminal cell membrane. Amiloride also inhibits Cl^- reabsorption indirectly, by its effect on inhibiting Na^+ reabsorption. Inhibition of Na^+ reabsorption reduces the lumen-negative voltage, the driving force for paracellular Cl^- reabsorption. Because of amiloride's effect on reducing the magnitude of the lumen-negative voltage, it also acts to inhibit K^+ secretion. By inhibiting K^+ secretion by the distal tubule and collecting duct, amiloride reduces the amount of K^+ excreted in the urine (see Chapter 38). Consequently, amiloride is frequently referred to as a **K^+-sparing diuretic.** It is most often used in patients who excrete too much K^+ in their urine.

REGULATION OF NaCl AND WATER REABSORPTION

Several hormones and factors regulate NaCl reabsorption (Table 36-5). Quantitatively, angiotensin II, aldosterone, ANP, and the sympathetic nerves are the most important hormones that regulate NaCl reabsorption and thereby urinary NaCl excretion. However, other hormones (including dopamine, glucocorticoids, and ADH), Starling forces, and the phenomenon of glomerulotubular balance influence NaCl reabsorption. ADH is the major hormone that regulates the amount of water excreted by the kidneys.

Angiotensin II Angiotensin II is one of the most potent hormones that stimulates NaCl and water reabsorption in the proximal tubule. A decrease in the extracellular fluid volume activates the renin-angiotensin system (discussed in Chapter 37) and thereby increases plasma angiotensin II concentration.

Aldosterone Aldosterone is synthesized in the glomerulosa cells of the adrenal cortex and stimulates NaCl reabsorption by the thick ascending limb of Henle's loop and the distal tubule and collecting duct. Aldosterone also stimulates K^+ secretion by the distal tubule and collecting duct (see Chapter 38). *The two most important stimuli to aldosterone secretion are an increase in angiotensin II concentration and an increase in plasma $[K^+]$.* By its action of stimulating NaCl reabsorption in the collecting duct, aldosterone also increases water reabsorption by this nephron segment.

Some individuals with an expanded extracellular fluid volume and an elevated blood pressure (e.g., **congestive heart failure** and **hypertension**) are treated with drugs that inhibit **angiotensin-converting enzyme (ACE inhibitors,** such as **captopril)** to lower fluid volume and blood pressure. Inhibition of angiotensin-converting enzyme blocks the degradation of angiotensin I to angiotensin II and thereby lowers plasma angiotensin II levels (see Chapter 37). The decline in plasma angiotensin II concentration has three effects: (1) NaCl and water reabsorption by the proximal tubule falls; (2) aldosterone secretion falls, which reduces NaCl reabsorption in the distal tubule and collecting duct; and (3) because angiotensin is a potent vasoconstrictor, the systemic arterioles dilate and arterial blood pressure falls. Thus, ACE inhibitors decrease the extracellular fluid volume and the arterial blood pressure by promoting renal NaCl and water excretion and by depressing total peripheral resistance.

Atrial Natriuretic Peptide (ANP) ANP is secreted by cells of the cardiac atria and the kidneys. ANP secretion is stimulated by a rise in blood pressure and an increase in the extracellular fluid volume. ANP increases urinary NaCl by increasing GFR, which increases the filtered load of NaCl, and by inhibiting NaCl reabsorption by the medullary portion of the collecting duct. ANP also increases urinary water excretion by directly inhibiting water reabsorption across the collecting duct and by inhibiting the secretion of ADH.

Sympathetic Nerves Catecholamines released from sympathetic nerves (norepinephrine) and the adrenal medulla (epinephrine) stimulate NaCl and water reabsorption from the proximal tubule and thick ascending limb of Henle's loop. Activation of sympathetic nerves—for example after hemorrhage or a decrease in the extracellular fluid volume—stimulates NaCl and water reabsorption by the proximal tubule, the thick ascending limb of Henle's loop, the distal tubule, and the collecting duct.

Dopamine Dopamine, a catecholamine, is released from dopaminergic nerves in the kidney and may also be synthesized in the kidneys. The action of dopamine is opposite that of norepinephrine and epinephrine: dopamine secretion is stimulated by an increase in extracellular fluid volume, and its secretion directly inhibits NaCl and water reabsorption in the proximal tubule.

Antidiuretic Hormone (ADH) Antidiuretic hormone is the most important hormone that regulates water balance (see Chapters 37 and 44). This hormone is secreted by the posterior pituitary in response to an increase in plasma osmolality or a decrease in the extracellular fluid volume. ADH increases the permeability of the collecting duct to water, and because an osmotic gradient exists across the wall of the collecting duct, the hormone increases water reabsorption by the collecting duct (see

Chapter 37 for details). ADH has little effect on urinary NaCl excretion.

Starling Forces Starling forces* (see also Chapter 22) regulate NaCl and water reabsorption across the proximal tubule (Figure 36-10). As described above, Na^+, Cl^-, HCO_3^-, amino acids, glucose, and water are transported into the intercellular space of the proximal tubule. Starling forces between this space and the peritubular capillaries facilitate the movement of the reabsorbate into the capillaries. Starling forces that favor movement from the interstitium into the peritubular capillaries are the capillary oncotic pressure (π_{cap}) and the hydrostatic pressure in the intercellular space(P_{IS}). The opposing Starling forces are the interstitial oncotic pressure (π_{IS}) and the capillary hydrostatic pressure (P_{cap}). Normally, the sum of the Starling forces favors movement of solute and water from the interstitium into the capillary. However, some of the solutes and fluid that enter the lateral intercellular space leak back into the proximal tubular fluid (Figure 36-10). Starling forces do not affect transport by the loop of Henle, distal tubule, and collecting duct, because these segments are less permeable to H_2O than is the proximal tubule.

Starling forces across the peritubular capillaries surrounding the proximal tubule are readily altered. Dilation of the efferent arteriole increases the hydrostatic pressure in the peritubular capillaries (P_{cap}), whereas constriction of the efferent arteriole decreases P_{cap}. An increase in P_{cap} inhibits solute and water reabsorption by increasing the back-leak of NaCl and water across the tight junction, whereas a decrease in P_{cap} stimulates reabsorption by decreasing back-leak across the tight junction (Figure 36-10).

The oncotic pressure in the peritubular capillary is determined in part by the rate of formation of the glomerular ultrafiltrate. For example, if one assumes a constant plasma flow in the afferent arteriole, as less ultrafiltrate is formed (i.e., as GFR decreases) the plasma proteins become less concentrated in the plasma that enters the efferent arteriole and peritubular capillary. Hence the peritubular oncotic pressure decreases. Thus, the peritubular oncotic pressure is directly related to the filtration fraction (FF = GFR/RPF). A fall in the FF, owing to a de-

---

*Starling forces across the wall of the peritubular capillaries are the hydrostatic pressure in the peritubular capillary (P_{cap}) and lateral intercellular space (P_{IS}) and the oncotic pressure in the peritubular capillary (π_{cap}) and the lateral intercellular space (π_{IS}). Thus, the reabsorption of water, resulting from sodium transport from tubular fluid into the lateral intercellular space, will be modified by the Starling forces. Thus:

$$Q = K_f\{(P_{IS} - P_{cap}) - (\pi_{cap} - \pi_{IS})\}$$

where Q equals flow (positive numbers indicate flow from the intercellular space into blood).

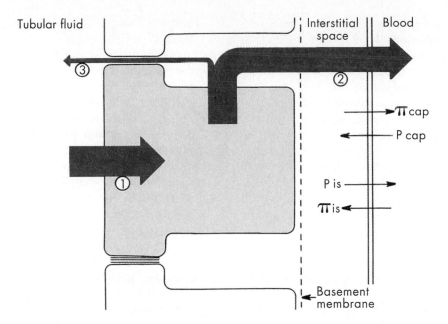

FIGURE 36-10   Routes of solute and water transport across the proximal tubule and the Starling forces that modify reabsorption. *(1)*, Solute and water are reabsorbed across the apical membrane. This solute and water then cross the lateral cell membrane. Some solute and water reenter the tubule fluid (indicate by arrow labeled *3*), and the remainder enters the interstitial space and then flows into the capillary (indicated by arrow labeled *2*). The width of the arrows is directly proportional to the amount of solute and water moving by the pathways labeled *1* to *3*. Starling forces across the capillary wall determine the amount of fluid flowing through pathway *2* versus *3*. Transport mechanisms in the apical cell membranes determine the amount of solute and water entering the cell (pathway *1*). π_{cap}, Capillary oncotic pressure; P_{cap}, capillary hydrostatic pressure; π_{is}, interstitial fluid oncotic pressure; P_{is}, interstitial hydrostatic pressure. Thin arrows across the capillary wall indicate direction of water movement in response to each force.

crease in GFR at constant RPF, decreases the peritubular capillary oncotic pressure. This, in turn, increases the backflux of NaCl and water from the lateral intercellular space into the tubular fluid and thereby decreases net solute and water reabsorption across the proximal tubule. An increase in the FF has the opposite effect.

The importance of Starling forces in regulating solute and water reabsorption by the proximal tubule is underscored by the phenomenon of **glomerulotubular balance (G-T balance).** Spontaneous changes in GFR alter the filtered load of sodium markedly (filtered load = GFR × [Na$^+$]). Unless such changes were rapidly accompanied by adjustments in Na$^+$ reabsorption, urine Na$^+$ excretion would fluctuate widely and disturb the Na$^+$ balance of the whole body. However, spontaneous changes in GFR do not alter Na$^+$ balance because of the phenomenon of G-T balance. *G-T balance refers to the fact that, when body Na$^+$ balance is normal, Na$^+$ and water reabsorption increases in parallel with an increase in GFR and filtered load of Na$^+$. Thus, a constant fraction of the filtered Na$^+$ and water is reabsorbed from the proximal tubule despite variations in*

GFR. The net result of G-T balance is to reduce the impact of GFR changes on the amount of Na$^+$ and water excreted in the urine.

Two mechanisms are responsible for G-T balance. One is related to the oncotic and hydrostatic pressures between the peritubular capillaries and the lateral intercellular space (i.e., Starling forces), and the other is related to the filtered load of glucose and amino acids. As an example of the first mechanism, an increase in GFR (at constant RPF) raises the protein concentration above normal in the glomerular capillary plasma. This protein-rich plasma leaves the glomerular capillaries, flows through the efferent arteriole, and enters the peritubular capillaries. The increased oncotic pressure in the peritubular capillaries augments the movement of solute and fluid from the lateral intercellular space into the peritubular capillaries and thereby increases net solute and water reabsorption.

The second mechanism responsible for G-T balance is initiated by an increase in the filtered load of glucose and amino acids. As discussed earlier in this chapter, the reabsorption of Na$^+$ in the early segment of the proximal tubule is coupled to that of glucose and amino acids. The rate of Na^+ reabsorption therefore depends in part on the filtered load of glucose and amino acids. As GFR and the filtered load of glucose and amino acids increase, Na^+ and water reabsorption also rise.

In addition to G-T balance, another physiological mechanism operates to minimize changes in the filtered load of Na^+. As described earlier in this chapter, an increase in GFR, and thus in the amount of Na^+ filtered by the glomerulus, activates the tubuloglomerular feedback mechanism, which returns GFR and the filtration of Na^+ to normal values. Thus, spontaneous changes in GFR—for example, caused by changes in posture—only increase the amount of Na^+ filtered for a few minutes. Until GFR returns to normal values, the mechanisms that underlie G-T balance maintain urinary sodium excretion constant and thereby maintain Na^+ homeostasis.

SUMMARY

1. The four major segments of the nephron determine the composition and volume of the urine by the processes of selective reabsorption of solutes and water and secretion of solutes.
2. Tubular reabsorption allows the kidneys to retain those substances that are essential, and to regulate their levels in the plasma by altering the degree to which they are reabsorbed. The reabsorption of Na^+, Cl^-, other anions, and organic solutes together with water constitutes the major function of the nephron. Approximately 25,000 mEq of Na^+ and 178 liters of water are reabsorbed each day. The proximal tubule cells reabsorb 67% of the glomerular ultrafiltrate, and cells of the loop of Henle reabsorb about 25% of the NaCl that was filtered and about 15% of the water that was filtered. The distal segments of the nephron (distal tubule and collecting duct system) have a more limited reabsorptive capacity. However, the final adjustments in the composition and volume of the urine

and most of the regulation by hormones and other factors occur in distal segments.
3. Secretion of substances into tubular fluid is a means for excreting various by-products of metabolism, and it also serves to eliminate exogenous organic anions and bases (e.g., drugs) from the body. Many organic compounds are bound to plasma proteins, and therefore are unavailable for ultrafiltration. Thus secretion is their major route of excretion in the urine.
4. Various hormones (including angiotensin II, aldosterone, ADH, and ANP), sympathetic nerves, dopamine, ADH, and Starling forces regulate NaCl and water reabsorption by the kidneys.

BIBLIOGRAPHY
Journal Articles

Giebisch G, Boulpaep EL, eds: Symposium on cotransport mechanisms in renal tubules, *Kidney Int* 36:333, 1989.

Groves CE, Evans KK, Dantzler WH, Wright SH: Peritubular organic cation transport in isolated rabbit proximal tubules, *Am J Physiol* 266(*Renal Fluid Electrolyte Physiol* 35):F450, 1994.

Kriz W, Bankir L: A standard nomenclature for structures of the kidney, *Am J Physiol* 254:F1, 1988.

Moe OW, Alpern RJ, Henrich WL: The renal proximal tubule renin-angiotensin system, *Semin Nephrol* 13:552, 1993.

Books and Monographs

Berry CA, Rector FC, Jr: Renal transport of glucose, amino acids, sodium, chloride and water. In Brenner BM and Rector FC Jr, editors: *The kidney*, ed 4, Philadelphia, 1991, WB Saunders Co.

Byrne JH, Schultz SG: *An introduction to membrane transport and bioelectricity*, New York, 1988, Raven Press.

Koeppen BM, Stanton BA: *Renal physiology*, St Louis, 1992, Mosby.

Koushanpour E, Kriz W: *Renal physiology*, ed 2, Berlin, 1986, Springer-Verlag.

Rose BD: *Clinical physiology of acid-base and electrolyte disorders*, ed 4, New York, 1994, McGraw-Hill Information Services Co.

CHAPTER 37

Control of Body Fluid Volume and Osmolality

The kidneys maintain the osmolality and volume of the body fluids within a very narrow range by regulating the excretion of water and NaCl, respectively. This chapter discusses regulation of renal water excretion (urine concentration and dilution) and renal NaCl excretion.

THE BODY FLUID COMPARTMENTS

Volumes of Body Fluid Compartments

In adults, water accounts for approximately 60% of the body weight. The water content of different individuals varies with the amount of adipose tissue; the greater the amount of adipose tissue, the smaller the fraction of body weight attributable to water.

The **total body water** (TBW) is contained within two major compartments, which are divided by the cell membranes. The larger of these compartments is termed the **intracellular fluid** (ICF), and represents the water contained within cells. The ICF accounts for approximately 40% of body weight. The other fluid compartment is the **extracellular fluid** (ECF), which represents water outside of cells, and accounts for 20% of the body weight. The ECF is subdivided into several compartments. The largest of these is the **interstitial fluid** (ISF), which is the fluid surrounding the cells in the various tissues of the body. Also included within this compartment is the water contained within lymph, bone, and dense connective tissue. Three quarters of the ECF is located within the interstitial space, and the remaining one quarter is **plasma** within the circulatory system. Figure 37-1 summarizes the relationships among these various fluid compartments and provides estimates of their volumes in a normal adult.

Composition of Body Fluid Compartments

Sodium (Na^+) and its anions, chloride (Cl^-) and bicarbonate (HCO_3^-), are the major ions of the ECF. The compositions of the two major compartments of the ECF (ISF and plasma) are very similar. Because these two compartments are separated only by the capillary endothelium, and because this barrier is freely permeable to small ions, *the major difference between the ISF and plasma is that plasma contains significantly more protein.* The presence of protein in the plasma can also affect the ionic composition of the ISF and plasma by the Gibbs-Donnan effect (see Chapter 2). The Gibbs-Donnan effect is normally quite small, and for most purposes the ionic composition of the plasma and ISF can be considered identical.

Na^+ and its attendant anions (Cl^- and HCO_3^-) are the major determinants of the osmolality of the ECF. A rough estimate of the ECF osmolality, which is a measure of the number of solute particles in a kilogram of water (see Chapter 1), can be obtained simply by multiplying the plasma sodium concentration $[Na^+]$ by two, which accounts for the anions that are associated with Na^+. For example, if the plasma $[Na^+]$ is 142 mEq/L, the osmolality of the plasma and ECF can be estimated as:

$$\text{Plasma osmolality} = 2 \,(\text{plasma } [Na^+])$$
$$= 284 \text{ mOsm/kg } H_2O$$

The normal plasma osmolality ranges from approximately 285 to 295 mOsm/kg H_2O. The difference between the actual osmolality and the value estimated from the plasma $[Na^+]$ is caused by the presence of other solutes in the plasma, such as K^+ salts, glucose, and urea.

In the ICF, the $[Na^+]$ is extremely low, whereas potassium (K^+) is the predominant intracellular cation. As explained in Chapter 2, this distribution of Na^+ and K^+ is maintained by the Na^+-K^+-ATPase. The anion composi-

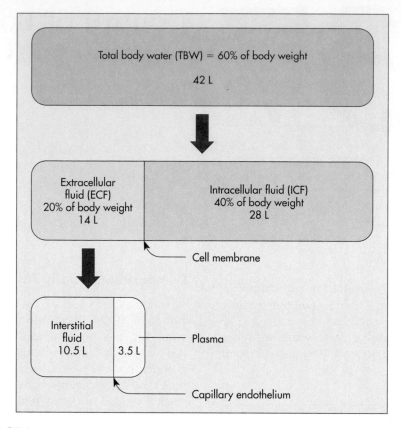

FIGURE 37-1   Volumes of the major body fluid compartments calculated for a 70 kg individual.

tion of the ICF also differs greatly from that of the ECF. The major intracellular anions are phosphates and organic anions, whereas the Cl^- concentration is relatively low.

Fluid Exchange Between Compartments

Water moves readily between the various body fluid compartments. *Two forces determine this movement: hydrostatic pressure and osmotic pressure.* Hydrostatic pressure and the fraction of the osmotic pressure generated by proteins (oncotic pressure) are important determinants of fluid movement across the capillary endothelium (see Chapter 22), whereas osmotic pressure differences between the ICF and ECF are responsible for water movement across the cell membranes. Because the plasma membranes of cells are highly permeable to water, a change in the osmolality of either the ICF or the ECF will shift water rapidly between these compartments until the osmolality of both compartments is the same. Thus, except for transient changes, the ICF and ECF compartments have the same osmolality (285 to 295 mOsm/kg H_2O).

While water can freely cross most cell membranes, the movement of ions is more restricted, and is carried out by specific ion transport mechanisms in the plasma membrane. Thus, fluid shifts between the ICF and ECF compartments primarily by the movement of water and not ions.

Shifts of fluid between the ICF and ECF compartments are important when intravenous solutions are administered to patients. For example, the infusion of 1 liter of an NaCl solution, which is isoosmotic to plasma, will increase the volume of the ECF by 1 liter, because there will be no osmotic gradient to cause fluid to move into or out of the ICF. In contrast, if the NaCl solution is either hypoosmotic or hyperosmotic with respect to the plasma, fluid will move between the ICF and ECF. With the infusion of a hypoosmotic NaCl solution, the osmolality of the ECF will be lower than that of the ICF and fluid will move into the cells. Thus, the increase in ECF volume will be less than the volume of the infused solution. Conversely, infusion of a hyperosmotic NaCl solution will shift fluid out of the cells into the ECF, and the increase in volume of the ECF will be greater than the infused volume of solution.

Table 37-1 Normal Routes of Water Gain and Loss in Adults (at 23° C)

Route	ml/day
WATER INTAKE	
Fluid*	1200
In food	1000
Metabolically produced from food	300
WATER OUTPUT	
Insensible	700
Sweat	100
Feces	200
Urine	1500

* Fluid intake varies widely for both social and cultural reasons.

CONTROL OF BODY FLUID OSMOLALITY: URINE CONCENTRATION AND DILUTION

The kidneys are the major organs responsible for regulating water balance, and under most conditions are the major route for the elimination of water from the body (Table 37-1). Other routes of water loss from the body include evaporation from the cells of the skin and the respiratory passages. Collectively, water loss by these routes is termed insensible water loss, because the individual is unaware of its occurrence. Additional water can be lost by the production of sweat. Water loss by this mechanism can increase dramatically in a hot environment, with exercise, or in the presence of fever (Table 37-2). Finally, water can be lost from the gastrointestinal tract. Fecal water loss is normally small, but increases with diarrhea. Gastrointestinal losses can also increase with vomiting.

Water loss in sweat, feces, and evaporation from the lungs and skin is not regulated. In contrast, the renal excretion of water is tightly regulated to maintain the osmolality of the body fluids constant. The maintenance of a constant body fluid osmolality requires that water intake and loss from the body are precisely matched. If intake exceeds losses, positive water balance exists and the osmolality of the body fluids decreases. Conversely, when intake is less than losses, negative water balance exists and the osmolality of the body fluids increases.

When water intake is low, or when water losses increase, the kidneys conserve water by producing a small volume of urine that is hyperosmotic with respect to plasma. When water intake is high, a large volume of hypoosmotic urine is produced. In a normal individual the urine osmolality can vary from approximately 50 to 1200 mOsm/kg H_2O, and the urine volume can vary from 0.5 to near 20 L/day.

It is important to recognize that disorders of water balance are manifested by alterations in the body fluid osmolality (e.g., plasma osmolality). Because the major determinant of the plasma osmolality is Na^+ (with its anions Cl^- and HCO_3^-), disorders of water balance will alter the plasma $[Na^+]$. When one evaluates an abnormal plasma $[Na^+]$ in an individual, it is tempting to suspect a problem in Na^+ balance. However, the problem relates not to Na^+ balance, but to water balance. Changes in Na^+ balance alter the volume of the extracellular fluid, and not its osmolality (see below).

In the clinical setting, **hypoosmolality** (a reduction in plasma osmolality, P_{osm}) shifts water into cells, and this process results in cell swelling. Symptoms associated with hypoosmolality are related primarily to swelling of brain cells. For example, a rapid fall in P_{osm} can alter neurologic function and thereby cause nausea, malaise, headache, confusion, lethargy, seizures, and coma. When P_{osm} is increased (i.e., **hyperosmolality**), water is lost from cells. The symptoms of an increase in P_{osm} are also primarily neurologic, and they include lethargy, weakness, seizures, coma, and even death.

The kidneys can control water excretion independently of their ability to control the excretion of a number of other physiologically important substances (e.g., Na^+, K^+, H^+, urea). Indeed, this ability is necessary for survival, because it allows water balance to be achieved without perturbing the other homeostatic functions of the kidneys.

Antidiuretic Hormone

Antidiuretic hormone (ADH), or **vasopressin,** acts on the kidneys to regulate the excretion of water. When plasma ADH levels are low, a large volume of urine is excreted **(diuresis)** and the urine is dilute (hypoosmotic to plasma). When plasma levels of ADH are elevated, a small volume of urine is excreted **(antidiuresis)** and the urine is concentrated (hyperosmotic to plasma).

ADH is a small peptide, nine amino acids in length. It is synthesized in cells of the hypothalamus (Figure 37-2), and it is stored and released from axon terminals located in the neurohypophysis (posterior lobe of the pituitary gland).

The secretion of ADH from the posterior lobe of the pituitary is mainly regulated by the osmolality of the plasma, but both blood volume and pressure also have an influence. Cells **(osmoreceptors)** within the hypo-

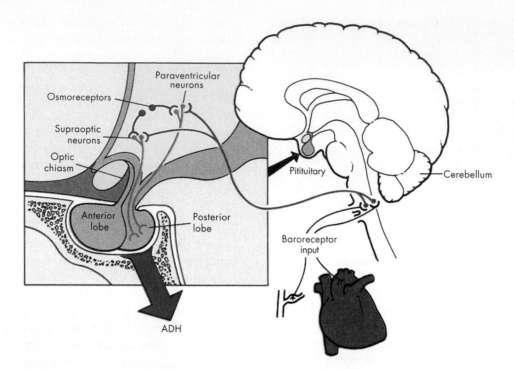

FIGURE 37-2   Anatomy of the hypothalamus and pituitary gland (midsagittal section) depicting the pathways for vasopressin (ADH) secretion. Also shown are pathways involved in regulating ADH secretion. Afferent fibers from the baroreceptors are carried in the vagus and glossopharyngeal nerves. Closed box illustrates expanded view of the hypothalamus and pituitary.

Table 37-2   Effect of Environmental Temperature and Excercise on Water Loss and Intake in Adults (in ml/day).*

	Normal Temperature	Hot Weather	Prolonged Heavy Exercise
WATER LOSS			
Insensible Loss:			
Skin	350	350	350
Lungs	350	250	650
Sweat	100	1400	5000
Feces	200	200	200
Urine	1500	1200	500
TOTAL LOSS	2500	3400	6700
WATER INTAKE TO MAINTAIN WATER BALANCE	2500	3400	6700

*In hot weather and during prolonged heavy exercise water balance is maintained only if the individual increases water intake to match the increased loss of water in sweat. Decreased water excretion by the kidneys alone is insufficient to maintain water balance.

thalamus sense changes in the osmolality of the plasma. These cells are distinct from those that produce ADH. They appear to behave as osmometers and sense changes in plasma osmolality by either shrinking or swelling. The osmoreceptors are very sensitive, and a change in plasma osmolality of only 1% changes ADH secretion significantly. When plasma osmolality is elevated, ADH secretion increases. Conversely, when plasma osmolality is reduced, secretion is inhibited. Because ADH is rapidly degraded in the plasma, circulating levels can be reduced to zero within minutes after secretion is inhibited. As a result the ADH system can respond rapidly to fluctuations in plasma osmolality. Figure 37-3, *A*, illustrates the effect of changes in plasma osmolality on plasma ADH levels.

A decrease in blood volume or pressure also stimulates ADH secretion. The receptors responsible for this response are located in both the low-pressure (left atrium) and the high-pressure (aortic arch and carotid sinus) sides of the circulatory system (see also Chapters 23 and 44). Signals from these receptors are relayed to the ADH secretory cells via afferent fibers in the vagus and glossopharyngeal nerves. The sensitivity of this baroreceptor system is less than that of the osmoreceptors, and a 5% to 10% change in volume is required to alter ADH secretion. Figure 37-3, *B,* illustrates the effect of changes in blood volume or pressure on plasma levels of ADH.

The osmolality and the blood volume and pressure systems for ADH secretion interact. Alterations in blood volume and pressure also affect the response to changes in body fluid osmolality (Figure 37-3, *C*). When blood volume or pressure decreases, the osmotic set point (plasma osmolality at which ADH secretion begins) is shifted to lower values and the slope of the relationship is steeper. When blood volume increases, the opposite occurs; the osmotic set point is shifted to higher values and the slope of the relationship is decreased.

As already noted, the primary action of ADH on the kidneys is to decrease the excretion of water. At the level of the nephron, ADH has the following actions:

1. Increases the permeability of the collecting duct to water
2. Stimulates the active reabsorption of NaCl by the thick ascending limb of Henle's loop and by the collecting duct
3. Increases the permeability of the inner medullary collecting duct to urea

How these actions result in the excretion of a small volume of hyperosmotic urine is considered in the following sections.

Inadequate release of ADH from the posterior pituitary results in the excretion of large volumes of dilute urine **(polyuria).** To compensate for this loss of water, the individual must ingest large volumes of water **(polydipsia)** in order to maintain plasma osmolality constant. If the patient is deprived of water, the plasma osmolality will increase. This condition is called **central diabetes insipidus** or **pituitary diabetes insipidus.** Central diabetes insipidus can be inherited, although this is rare. It occurs more commonly after head trauma, with brain neoplasms, or with brain infections. Patients with central diabetes insipidus have a concentrating defect that is corrected by the administration of exogenous ADH.

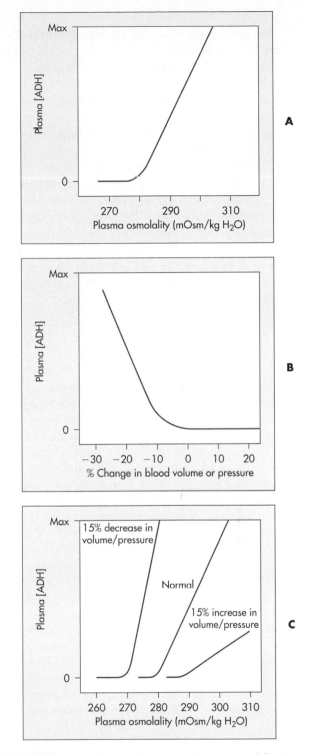

FIGURE 37-3   **A,** Effect of changes in plasma osmolality (constant blood volume and pressure) on plasma ADH levels. **B,** Effect of changes in the blood volume or blood pressure (constant plasma osmolality) on plasma ADH levels. **C,** Interactions between osmolar and blood volume and pressure stimuli on ADH secretion.

The **syndrome of inappropriate ADH secretion** (i.e., **SIADH**) is a common clinical problem that is characterized by inappropriately high ADH levels, which results in plasma hypoosmolality. SIADH can be caused by infections or neoplasms of the central nervous system, drugs (e.g., antitumor drugs), pulmonary diseases, or carcinoma of the lung, and may occur in post-operative patients.

The collecting ducts of some individuals do not respond normally to ADH. These individuals cannot maximally concentrate their urine, and they also suffer from **polyuria** and **polydipsia.** This entity is termed **nephrogenic diabetes insipidus** to distinguish it from central diabetes insipidus. Although inherited forms of nephrogenic diabetes insipidus exist, most cases are secondary to other factors, such as metabolic disorders (e.g., hypercalcemia) or certain drugs (e.g., lithium).

Thirst

In addition to affecting the secretion of ADH, changes in plasma osmolality or blood volume and pressure influence the thirst center of the brain. When plasma osmolality is increased, or when the blood volume or pressure is decreased, the person experiences the desire to drink. Hyperosmolality is the more potent stimulus, because a 2% to 3% change in osmolality can produce a strong desire to drink, whereas a 10% to 15% decrease in blood volume or pressure is required for the same response. The sensation of thirst is suppressed when the osmolality of the body fluids is decreased. The brain centers that control the thirst response are not completely defined. They appear to be located within the hypothalamus, near to but distinct from the osmoreceptor cells.

Note that the ADH and thirst systems work together to maintain body fluid osmolality. An increase in body fluid osmolality evokes drinking and, via ADH action on the kidneys, the conservation of water. Conversely, when plasma osmolality is decreased, thirst is suppressed, and in the absence of ADH, renal water excretion is enhanced.

Urine Concentration and Dilution

Under normal circumstances the excretion of water is regulated separately from the excretion of solute (e.g., NaCl). In order for this to occur, the kidneys must be able to excrete urine that is either hypoosmotic or hyperosmotic with respect to plasma, which in turn requires that solute be separated from water at some point along the nephron. As discussed in Chapter 36, reabsorption of solute in the proximal tubule results in the reabsorption of a proportional amount of water; hence there is no separation of solute and water in this portion of the nephron. *Henle's loop, and in particular the thick ascending limb, is the main nephron site where this separation of solute and water occurs.* As a consequence of its ability to separate solute and water, Henle's loop serves two important functions. First, it dilutes the tubular fluid, which in the absence of ADH allows the excretion of a dilute urine. Second, it renders the interstitial fluid within the medulla hyperosmotic to plasma. The hyperosmotic medullary interstitium is essential for the excretion of a concentrated urine, because it establishes an osmotic gradient across the wall of the collecting duct, and thereby allows water to be reabsorbed when ADH increases the water permeability of the collecting duct cells.

Figure 37-4 summarizes the essential features of the mechanisms whereby the kidneys excrete either a dilute (Figure 37-4, *A*) or a concentrated urine (Figure 37-4, *B*). First, we will consider how the kidneys excrete a dilute urine (water diuresis), when P_{osm} and plasma ADH levels are low. The following numbers 1 to 7 refer to those encircled in Figure 37-4, *A*.

1. Fluid entering the thin descending limb of Henle's loop from the proximal tubule is isosmotic with respect to plasma. This reflects the essentially isosmotic nature of solute and water reabsorption in the proximal tubule (see Chapter 36).

2. The thin descending limb is highly permeable to water and much less so to NaCl and urea. Consequently, as the fluid descends deeper into the hyperosmotic medulla, water is reabsorbed. By this process, fluid at the bend of the loop will have an osmolality equal to that of the surrounding interstitial fluid. Although the osmolalities of the tubular and interstitial fluids are similar at the bend of the loop, their compositions are markedly different. The tubular fluid [NaCl] is greater than that of the surrounding interstitial fluid. However, the [urea] of the tubular fluid is less than that of the interstitial fluid (see p. 545).

3. The thin ascending limb is impermeable to water but permeable to NaCl and urea. Consequently, as tubular fluid moves up the ascending limb, NaCl is passively reabsorbed (luminal [NaCl] > interstitial [NaCl]), while urea passively diffuses into the tubular fluid (luminal [urea] < interstitial [urea]). The net effect is that the volume of the tubular fluid remains unchanged along the length of the thin ascending limb, but the [NaCl] decreases and the [urea] increases.

4. The thick ascending limb of Henle's loop is impermeable to water and urea. This portion of the nephron actively reabsorbs NaCl and thereby dilutes the tubular

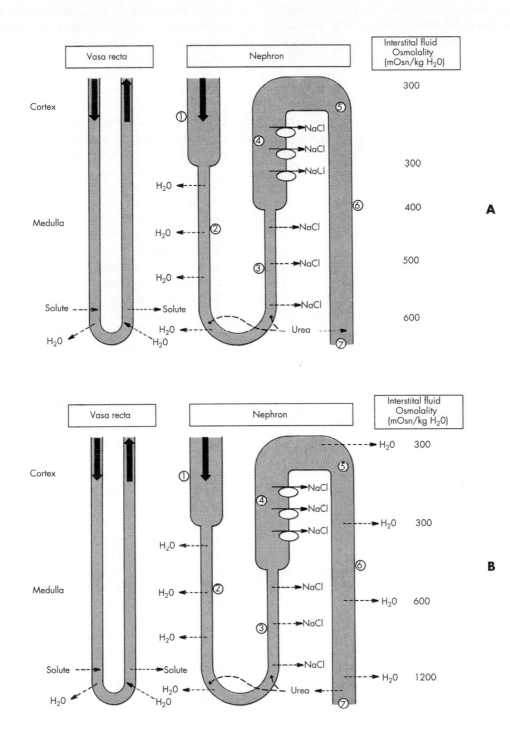

FIGURE 37-4   **A,** Mechanism for the excretion of dilute urine (water diuresis). ADH is absent, and the collecting duct is essentially impermeable to water. Note also that during a water diuresis the osmolality of the medullary interstitium is reduced as a result of increased vasa recta blood flow, and the entry of some urea into the medullary collecting duct. **B,** Mechanism for the excretion of a concentrated urine (antidiuresis). Plasma ADH levels are maximal, and the collecting duct is highly permeable to water. Under this condition the medullary interstitial gradient is maximal. See text for details.

fluid. Dilution occurs to such a degree that this segment is often referred to as the diluting segment of the kidney. Fluid leaving the thick ascending limb is hypoosmotic with respect to plasma (approximately 150 mOsm/kg H_2O).

5. The distal tubule and cortical collecting duct actively reabsorb NaCl, and are impermeable to urea. In the absence of ADH the water permeability of these segments is low. Thus, when ADH is absent, or present at low levels (i.e., decreased P_{osm}), the distal tubule and the cortical collecting duct are impermeable to water. The osmolality of tubule fluid in these segments is reduced further because NaCl is reabsorbed without water. Fluid entering the cortical portion of the collecting duct is hypoosmotic with respect to plasma (approximately 100 mOsm/kg H_2O).

6. The medullary collecting duct actively reabsorbs NaCl. Even in the absence of ADH this segment is slightly permeable to water and urea. Consequently, some urea enters the collecting duct from the medullary interstitium, and a small volume of water is reabsorbed.

7. The urine will have an osmolality of ~50 mOsm/kg H_2O and will contain low concentrations of NaCl and urea.

Next, we will consider how the kidneys excrete a concentrated urine (antidiuresis), when P_{osm} and plasma ADH levels are high. The following numbers 1 to 7 refer to those encircled in Figure 37-4, *B*.

1-4. Steps 1 to 4 are similar when either a dilute or a concentrated urine is produced. It is important to recognize that while reabsorption of NaCl by the thick ascending limb of Henle's loop dilutes the tubular fluid, the reabsorbed NaCl accumulates in the medullary interstitium. *The accumulation of NaCl and urea (see below) in the medullary interstitium is critically important for the production of urine hyperosmotic to plasma, because it provides the osmotic driving force for water reabsorption from the collecting duct.* The overall process by which the loop of Henle, and in particular the thick ascending limb, generates the hyperosmotic medullary interstitial gradient is termed **countercurrent multiplication.***

*The term countercurrent multiplication derives from both the form and the function of the loop of Henle. The loop of Henle consists of two parallel limbs with tubular fluid flowing in opposite directions (countercurrent flow). Fluid flows into the medulla in the descending limb, and out of the medulla in the ascending limb. The ascending limb is impermeable to water and reabsorbs solute from the tubular fluid. Thus, fluid within the ascending limb becomes diluted. This separation of solute and water by the ascending limb is termed the single effect of the countercurrent multiplication process. The solute removed from the ascending limb tubular fluid accumulates in the surrounding interstitial fluid, and raises its osmolality. Because

5. Because of NaCl reabsorption by the thick ascending limb of Henle's loop, fluid reaching the collecting duct is hypoosmotic with respect to the surrounding interstitial fluid. Thus, there is an osmotic gradient across the collecting duct. In the presence of ADH, which increases the water permeability of the collecting duct, water diffuses out of the tubule lumen and the tubule fluid osmolality increases. This begins the process of urine concentration. The maximum osmolality that the fluid in the cortical collecting duct can attain is approximately 300 mOsm/kg H_2O, which is the osmolality of the surrounding interstitial fluid and plasma. Although the fluid at this point has the same osmolality as that which entered the descending thin limb, its composition has been altered dramatically. Because of reabsorption by the preceding nephron segments, NaCl accounts for a much smaller portion of the total tubular fluid osmolality. Instead, the tubular fluid osmolality reflects the presence of urea (filtered urea, plus urea added in the descending thin and ascending thin limbs of Henle's loop) and other nonreabsorbed solutes (e.g., creatinine).

6. The osmolality of the interstitial fluid in the medulla progressively increases from the corticomedullary junction, where it is 300 mOsm/kg H_2O, to the papilla, where it is 1200 mOsm/kg H_2O. Thus, there is an osmotic gradient between tubule fluid and the interstitial fluid along the entire medullary collecting duct. In the presence of ADH, which makes the medullary collecting duct permeable to water, the osmolality of tubule fluid increases. Because most of the NaCl has been reabsorbed and because the initial portion of the collecting duct is impermeable to urea, even in the presence of ADH, urea remains in tubule fluid and its concentration increases as water is reabsorbed. In the presence of ADH the urea permeability of the last portion of the medullary collecting duct is increased. Because the urea concentration of the tubular fluid has been increased by water reabsorption, some urea will diffuse out of the tubule lumen and into the medullary interstitium, thereby increasing the urea concentration of the medullary interstitial fluid. The maximal osmolality that the fluid in the

the descending limb is highly permeable to water, the increased osmolality of the medullary interstitium causes water to be absorbed and thereby concentrates the tubular fluid. The countercurrent flow within the descending and ascending limbs of Henle's loop magnifies, or multiplies, the osmotic gradient between the tubular fluid in the descending and ascending limbs of Henle's loop.

Table 37-3 Transport and Permeability Properties of Nephron Segments Involved in Urine Concentration and Dilution

Tubule Segment	Active Transport	Passive Permeability*			Effect of ADH
		NaCl	**Urea**	**H₂O**	
Henle's loop					
Thin descending limb	0	+	+	+++	None
Thin ascending limb	0	+++	+	0	None
Thick ascending limb	+++	+	0	0	↑NaCl absorption
Distal tubule	++	+	0	0	None on early portion
Collecting duct					
Cortex	+	+	0	+	↑H_2O permeability and NaCl absorption
Medulla	+	+	++	+	↑H_2O and urea permeability

* Permeability is proportional to the number of + indicated: +, low permeability; +++, high permeability; 0, impermeable.

medullary collecting duct can attain is equal to that of the surrounding interstitial fluid. The major components of the fluid are substances that have either escaped reabsorption or have been secreted into the tubular fluid. Of these, urea is the most abundant.

7. The urine will have an osmolality of 1200 mOsm/kg H_2O and contain high concentrations of urea and other nonreabsorbed solutes. Because urea in the tubular fluid tends to equilibrate with the interstitial urea, its concentration in the urine will not exceed that of the interstitium (approximately 600 mmol/L).

Table 37-3 summarizes the transport and passive permeability properties of the nephron segments involved in the process of concentrating and diluting the urine.

Medullary Interstitial Fluid: Importance of Urea

As already described, the generation of a hyperosmotic medullary interstitium, which is dependent on NaCl reabsorption by the ascending limb of Henle's loop, is critically important for the ability of the kidneys to excrete urine that is hyperosmotic with respect to plasma. Measurements of the composition of the medullary interstitial fluid have shown that its principal components are NaCl and urea, and that the distribution of these solutes is not uniform throughout the medulla. At the junction of the medulla and the cortex, the interstitial fluid has an osmolality of approximately 300 mOsm/kg H_2O, with virtually all osmoles attributable to NaCl. The concentrations of both NaCl and urea increase progressively with increasing depth into the medulla, and at the papilla the osmolality of the interstitial fluid

is approximately 1200 mOsm/kg H_2O. Of this value, 600 mOsm/kg H_2O are attributed to NaCl and 600 mOsm/kg H_2O to urea.

The medullary gradient for NaCl results from the accumulation of NaCl reabsorbed by the nephron segments in the medulla. The most important segment in this regard is the ascending limb of Henle's loop.

Urea accumulation within the medullary interstitium is more complex, and occurs most effectively when a hyperosmotic urine is excreted. Urea is generated by the liver as a result of protein metabolism, and enters the tubular fluid by glomerular filtration. As indicated in Table 37-3, the permeability to urea of most nephron segments is relatively low, with the exception of the medullary collecting duct, especially in the presence of ADH. As fluid moves along the nephron, and especially as water is reabsorbed in the collecting duct, the urea concentration in the tubular fluid increases. When this urea-rich tubular fluid reaches the medullary collecting duct, where the permeability to urea is not only high but is increased by ADH, urea diffuses down its concentration gradient into the medullary interstitial fluid, where it accumulates. Some urea enters the thin descending and thin ascending limbs of Henle's loop and thereby causes the recycling of urea between the collecting duct and the loop of Henle. This process also serves to trap urea and thus facilitates its accumulation in the medulla.

As already described, the hyperosmotic medullary interstitium is essential for concentrating the tubular fluid within the collecting duct. Because water reabsorption is a passive process driven by an osmotic gradient, the maximal concentration that the urine can attain is equal to that of the medullary interstitium at the papilla (approximately 1200 mOsm/kg H_2O). Because a hyperosmotic medullary interstitium is essential for urine concentration, any condition that reduces this gradient will impair the ability of the kidney to concentrate the urine maximally.

Individuals on a protein-deficient diet can exhibit a defect in urine-concentrating ability. This inability to maximally concentrate the urine reflects decreased urea levels within the medullary interstitial fluid. When protein intake is inadequate, urea production in the body is decreased. Therefore the urea content, and thus the osmolality of the medullary interstitium, is reduced. This in turn will reduce water reabsorption from the collecting duct and thereby reduce the concentrating ability of the kidneys. Feeding of adequate amounts of protein will correct this defect.

Function of the Vasa Recta

The **vasa recta,** the capillary networks that supply blood to the medulla, are highly permeable to solute and water. As with Henle's loops, the vasa recta form a parallel set of hairpin loops within the medulla (see Chapter 35). The vasa recta function not only to bring nutrients and oxygen to the tubules within the medulla, but more importantly to remove excess water and solute, which is continuously added to the medullary interstitium by the nephron segments located in this region. It should be emphasized that *the ability of the vasa recta to maintain the medullary interstitial gradient is flow dependent.* Despite the countercurrent arrangement, a substantial increase in blood flow through the vasa recta will ultimately dissipate the medullary gradient.

The intake of large volumes of fluid results in the excretion of large volumes of urine **(water diuresis).** During water diuresis vasa recta blood flow increases, and the gradient within the medullary interstitium is dissipated (Figure 37-4, A). If individuals who have drunk a large volume of water are abruptly deprived of water, they will not be able to maximally concentrate their urine. Normal concentrating ability will be restored once the medullary interstitial gradient is reestablished. The time needed to reestablish the gradient is variable, depending on the magnitude and duration of the water diuresis.

Because water intake is not determined solely by thirst, but is influenced by social and behavioral factors, certain individuals will drink large volumes of liquids over extended periods of time, and thus undergo a prolonged water diuresis. Approximately 5% to 15% of patients with chronic psychiatric illness will drink excessive quantities of water **(psychogenic water drinkers),** and thus will demonstrate a concentrating defect if acutely challenged with water deprivation.

Quantitating the Urine Dilution and Concentration Processes

Assessment of the dilution and concentration processes involves measurements of urine osmolality and the volume of urine excreted. As noted previously, the range of urine osmolality is from approximately 50 mOsm/kg H_2O to 1200 mOsm/kg H_2O. The corresponding range in urine volume is approximately 20 L/day to as little as 0.5 L/day, although this can vary depending on the need for solute excretion (see below).

In order for the kidneys to excrete water maximally, the following conditions must exist:

1. Adequate delivery of tubular fluid to those nephron segments in which separation of solute and water occurs. Most important in this regard is the thick ascending limb of Henle's loop. Delivery of tubular fluid to the loop of Henle is in turn dependent upon the GFR and proximal tubule reabsorption.
2. Normal reabsorption of NaCl by the nephron segments. Again the most important segment is the thick ascending limb of Henle's loop.
3. Absence of ADH

If each of these conditions is met, large volumes of dilute urine will be excreted. However, the volume of urine will also depend on the amount of solute excreted. For example, if 600 mmole of solute must be excreted to maintain steady-state balance, then the maximum volume of urine excreted at an osmolality of 50 mOsm/kg H_2O would be 12 liters (i.e., 600 mOsm/50 mOsm/kg H_2O, with 1 kg of water equaling 1 liter). Thus, in order to attain the maximum urine volume output of approximately 20 L/day, solute excretion would need to be 1000 mOsm/day.

Similar requirements also apply to the conservation of water by the kidneys. In order for the kidneys to conserve water maximally, the following conditions must exist:

1. Adequate delivery of tubular fluid to those nephron segments in which separation of solute and water occurs. Most important in this regard is the thick ascending limb of Henle's loop. Delivery of tubular fluid to the loop of Henle is in turn dependent upon the GFR and proximal tubule reabsorption.

2. Normal reabsorption of NaCl by the nephron segments. Again the most important segment is the thick ascending limb of Henle's loop.

3. Presence of a hyperosmotic medullary interstitium. The interstitial osmolality will be maintained by NaCl reabsorption from the Henle's loop (conditions 1 and 2) and by effective accumulation of urea. Urea accumulation in turn depends on adequate dietary protein intake.

4. Maximum levels of ADH and responsiveness of the collecting duct to ADH.

If these conditions are met, a small volume of hyperosmotic urine will be excreted. The actual volume again depends on solute excretion. Thus, with a solute excretion of 600 mmole/day and a maximum urine osmolality of 1200 mOsm/kg H_2O, the minimum urine volume will be 0.5 L/day.

CONTROL OF EXTRACELLULAR FLUID VOLUME

The major solute of the ECF is NaCl. Because NaCl is also the major determinant of ECF osmolality, it is often assumed that alterations in NaCl balance disturb ECF osmolality. However, under normal circumstances this is not the case. Changes in NaCl balance do not normally alter the ECF osmolality, because the ADH and thirst systems maintain the osmolality of the ECF within a very narrow range. For example, addition of NaCl (without water) to the ECF will increase the osmolality of this compartment. This increase in osmolality will in turn stimulate ADH secretion and thirst. The ADH-induced decrease in urinary water excretion, together with the increased water intake, will restore the ECF osmolality to normal. However, the volume of this compartment will be increased in proportion to the amount of added NaCl. Thus, in the new steady-state, addition of NaCl is equivalent to adding an isosmotic solution to the body. Conversely, a decrease in the ECF NaCl content will reduce the volume of this compartment.

The maintenance of a constant ECF volume depends on the body's ability to regulate the amount of NaCl in this compartment. This is accomplished by the kidneys, which regulate their excretion of NaCl to match the amount ingested in the diet.

The typical diet contains approximately 150 mEq/day of Na^+ ($\approx$8 g of NaCl), and thus daily Na^+ excretion is also about 150 mEq/day. However, the kidneys can vary the excretion of Na^+ over a wide range. Excretion rates as low as 10 mEq/day can be attained when individuals are placed on a low-salt diet. Conversely, the kidneys can increase their excretion rate to more than 1000 mEq/day when challenged by the ingestion of a high-salt diet. These changes in Na^+ excretion occur with only modest changes in the steady-state total Na^+ content of the body.

The response of the kidneys to abrupt changes in Na^+ intake typically takes from 3 to 5 days. During this transition period, intake and excretion of Na^+ are not matched as they are in the steady state. Thus, the individual will experience either positive Na^+ balance (intake > excretion) or negative Na^+ balance (intake < excretion). However, by the end of the transition period, intake once again equals excretion.

Provided the ADH and thirst systems are intact and normal, changes in Na^+ balance result in changes in the volume of the ECF, not in changes of the serum [Na^+]. These changes are detected clinically as changes in body weight (1 liter of ECF = 1 kg body weight).

To maintain ECF volume, the body monitors the volume of this compartment and in response to changes signals the kidneys to make appropriate adjustments in NaCl excretion.

Effective Circulating Volume

To understand the role of the kidneys in regulating the ECF volume, it is necessary to consider the concept of **effective circulating volume** (ECV). ECV is not a measurable and distinct body fluid compartment; rather, it is related to the adequacy of tissue perfusion. Thus, it is related to the "fullness" of and "pressure" within the vascular tree. In the normal individual, ECV varies in parallel with the volume of the ECF. However, this relationship is not maintained under some pathological conditions.

Patients with **heart failure** frequently have an increase in their ECF volume. This reflects the fact that reduced cardiac output, secondary to the heart disease, is sensed by the body as a decrease in the ECV. As a result, the

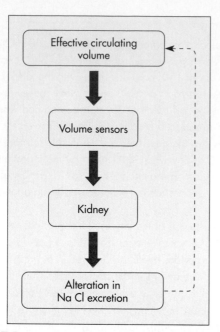

FIGURE 37-5   General scheme for monitoring and controlling the effective circulating volume.

kidneys retain NaCl and the volume of the ECF is expanded (manifested as tissue edema) in an attempt to increase the ECV. Thus, in **low-cardiac-output heart failure** the ECF volume is increased but the ECV is decreased.

An important point regarding ECV is that the kidneys will alter their excretion of NaCl in response to changes in ECV. When ECV is decreased, renal NaCl excretion is reduced. This adaptive response restores ECV to its normal value and maintains adequate tissue perfusion. Conversely, an increase in ECV enhances renal NaCl excretion, termed **natriuresis.** Figure 37-5 illustrates the components of the ECV regulatory system, each of which is discussed in detail below.

In the normal individual, the terms ECV and ECF can be, and quite often are, interchanged. However, it is the ECV that determines renal Na^+ excretion, especially under certain pathological conditions. Consequently, to provide a framework for understanding the pathophysiological basis of some clinically significant conditions, the remaining sections of this chapter will refer primarily to the ECV.

Volume Sensors

Volume sensors have been identified in both the low-pressure and high-pressure sides of the vascular tree. These receptors act as stretch receptors and respond to both volume and pressure. In general, 5% to 15% changes in blood volume and pressure are necessary to evoke a response.

The low-pressure receptors are found in the pulmonary vasculature and cardiac atria (see Chapters 19 and 23). Afferent fibers from these sensors are carried in the vagus nerve to the central nervous system. Activity of these sensors modulates both sympathetic nerve outflow and ADH secretion. For example, a decrease in filling of the pulmonary vessels and atria increases sympathetic nerve activity and stimulates ADH secretion. Conversely, distention of these structures decreases sympathetic nerve activity.

The cardiac atria also synthesize and secrete **atrial natriuretic peptide** (ANP). This constitutes an additional mechanism for the control of renal NaCl excretion, as described later in this chapter (p. 550). When the atria are distended, this peptide is released, and acts on the kidneys to increase the excretion of NaCl and H_2O.

High-pressure baroreceptors have been identified in the aortic arch, carotid sinus, and the afferent arterioles of the kidneys. Afferent fibers from the aortic arch and carotid sinus sensors are carried in the vagus and glossopharyngeal nerves, respectively (see Chapter 23). Like the low-pressure sensors, activity of the high-pressure sensors modulates sympathetic nerve activity and ADH secretion. Thus, a decrease in blood pressure will increase sympathetic nerve activity and ADH secretion. An increase in pressure reduces sympathetic nerve activity.

The afferent arterioles of the kidneys respond directly to changes in perfusion pressure. If pressure is decreased, renin is secreted. Renin secretion is suppressed with an increase in pressure.

As summarized in Figure 37-5, changes in the ECV are detected by these sensors, which in turn signal the kidneys to adjust NaCl excretion (also water) appropriately, and thus return the ECV toward normal. Accordingly, when the ECV is expanded, renal NaCl and water excretion are increased. Conversely, when the ECV is contracted, renal NaCl and water excretion are reduced. The signals involved in coupling the volume receptors to the kidney are both neural and hormonal.

Renal Sympathetic Nerves

As described in Chapter 35, sympathetic fibers innervate the afferent and efferent arterioles, as well as the cells of the renal tubule. These fibers can be activated by a decrease in the ECV and blood pressure, which results in the following responses:

1. Constriction of the afferent and efferent arterioles, which decreases the glomerular filtration rate.
2. Renin secretion by cells of the afferent and efferent arterioles is increased.

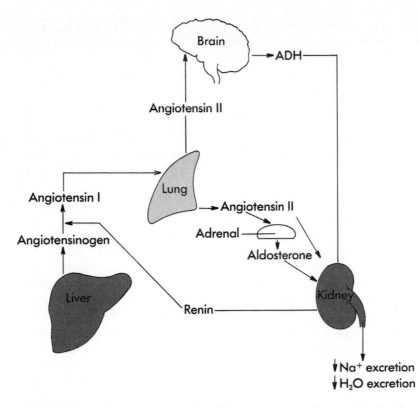

FIGURE 37-6   The renin-angiotensin-aldosterone system. See text for details.

3. NaCl reabsorption by the proximal tubule and Henle's loop is increased.

These responses are mediated via activation of α-adrenergic and β-adrenergic receptors. The net result is to reduce the excretion of NaCl. The mechanisms responsible for this decrease in renal NaCl excretion are considered below (see p. 552).

Other neural pathways are also affected by a change in ECV. One of these pathways involves the control of ADH secretion. As already described, a 5% to 10% reduction in ECV stimulates ADH secretion. ADH acting on the collecting duct increases water reabsorption. This, together with the reduced excretion of NaCl, restores the ECV to normal.

When the ECV is increased, sympathetic nerve activity is reduced. This generally reverses the effects just described, with the result that NaCl and water excretion are enhanced and the ECV is restored to normal.

Renin-Angiotensin-Aldosterone System

Modified smooth muscle cells in the afferent and efferent arterioles of the glomerulus are the site of synthesis, storage, and release of renin. Three factors are important in stimulating **renin** secretion. These are:

1. Perfusion pressure. The afferent arteriole behaves as a high-pressure baroreceptor. When perfusion pressure to the kidney is reduced, renin secretion is stimulated. Conversely, an increase in perfusion pressure inhibits renin release.
2. Sympathetic nerve activity. Activation of the sympathetic nerve fibers that innervate the afferent and efferent arterioles increases renin secretion. Renin secretion will be decreased as renal sympathetic nerve activity is decreased.
3. Delivery of NaCl to the macula densa. When NaCl delivery is decreased, renin secretion is enhanced. Conversely, an increase in NaCl delivery inhibits renin secretion.

Figure 37-6 illustrates the essential components of the renin-angiotensin-aldosterone system. Renin alone does not have a physiological function; it functions solely as a proteolytic enzyme. Its substrate is a circulating protein, **angiotensinogen,** which is produced by the liver. Angiotensinogen is cleaved by renin to yield a 10-amino acid peptide, **angiotensin** I. Angiotensin I also has no known physiological function and is further cleaved to an 8-amino acid peptide, **angiotensin II,** by a converting enzyme (**angiotensin converting enzyme;** ACE) found on the surface of vascular endothelial cells. The pulmonary endothelial cells are an important site for the conversion of angiotensin I to angiotensin II. Angiotensin II has several important physiological functions, including (1) stimulation of aldosterone secretion by the adrenal cortex, (2) arteriolar vasoconstriction, which increases blood pressure, (3) stimulation of ADH secretion and thirst, and (4) enhancement of NaCl reabsorption by the proximal tubule. Aldosterone is a steroid hormone with many important actions in the kidney (see Chapters 36, 38 and 46). With regard to the regulation of the ECV, aldosterone reduces NaCl excretion by stimulating its reabsorption primarily by the collecting duct and to a lesser degree by the thick ascending limb of Henle's loop.

Diseases of the adrenal cortex can alter aldosterone levels, and thereby impair the ability of the kidneys to maintain an adequate ECV (see also Chapter 46). With decreased secretion of aldosterone **(hypoaldosteronism),** there is reduced reabsorption of Na^+ by the collecting duct, resulting in the loss of Na^+ in the urine. Because urinary Na^+ loss can exceed the amount ingested in the diet, the ECV will decrease. In response to the decreased ECV, there is increased sympathetic tone, and elevated levels of renin, angiotensin II, and ADH. With increased aldosterone secretion **(hyperaldosteronism),** the opposite effects are seen. Na^+ reabsorption by the collecting duct is enhanced, resulting in reduced excretion of Na^+. Consequently, the ECV is increased, sympathetic tone is decreased, and the levels of renin, angiotensin II, and ADH are decreased. As described below, atrial natriuretic peptide levels are also elevated in this setting.

Atrial Natriuretic Peptide

The myocytes of the cardiac atria produce a peptide hormone that promotes water and NaCl excretion by the kidneys. This hormone, which is 28 amino acids in length, is called **atrial natriuretic peptide** (ANP). ANP is released in response to atrial stretch, such as occurs when the ECV is increased, and elicits various effects throughout the body. In general, ANP antagonizes the effects of the renin-angiotensin-aldosterone system. ANP increases renal NaCl and water excretion by renal and extrarenal actions. Actions of ANP include the following:

1. ANP dilates the afferent arterioles and thereby increases the glomerular filtration rate and thus the filtered load of NaCl.
2. ANP directly inhibits NaCl reabsorption by the collecting duct and indirectly inhibits NaCl reabsorption by the collecting duct through its inhibitory actions on the renin-angiotensin-aldosterone system.

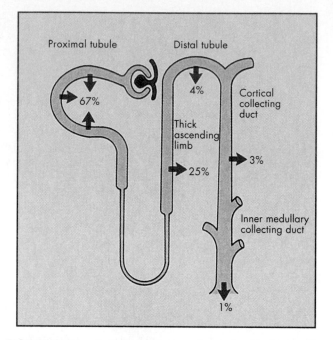

FIGURE 37-7   Sodium transport along the nephron. The values represent the percentage of the filtered load (i.e., GFR $\times$ P_{Na^+}) reabsorbed by each segment.

3. ANP reduces plasma aldosterone levels by inhibiting the release of renin and directly inhibiting the secretion of aldosterone from the adrenal cortex.
4. ANP promotes the excretion of water, because it inhibits the secretion of ADH from the posterior pituitary and inhibits ADH action on the collecting duct.

Control of Na^+ Excretion with Normal Effective Circulating Volume

As already noted, the maintenance of a normal ECV (euvolemia) requires a precise balance between the amount of NaCl ingested and the amount lost from the body. Because the kidneys are the major route for NaCl excretion, they adjust the amount of NaCl excreted in the urine to the amount ingested in the diet during euvolemia. To understand how renal NaCl excretion is regulated, the general features of renal Na^+ handling must be understood. Figure 37-7 summarizes the contribution of each nephron segment to the handling of the filtered load of Na^+ (see Chapter 36). The following discussion considers only the renal handling of Na^+. Although not specifically addressed, Cl^- reabsorption is regulated in parallel.

In a normal adult the filtered load of Na^+ is approximately 25,000 mEq/day. With a typical diet, 1% or less of this filtered load is excreted in the urine ($\approx$150 mEq/day). Because of this large filtered load of Na^+, it is important to recognize that small changes in Na^+ reabsorption by the nephron can have a large effect on the ECF volume (and ECV). For example, an increase in the excretion of Na^+ from 1% to 3% of the filtered load would represent an additional Na^+ loss of approximately 500 mEq/day. Because the Na^+ concentration in the ECF is approximately 140 mEq/L, such an Na^+ loss would decrease the ECF volume by more than 3 liters. (A loss of 500 mEq of Na^+ would also result in an increase in water excretion so as to maintain P_{osm} constant. Thus, the decrease in the ECF volume would be: 500 mEq Na^+/140 mEq/L = 3.6 liters).

During euvolemia the collecting duct is the main nephron segment where Na^+ reabsorption is adjusted to maintain excretion at a level appropriate for dietary intake. This does not mean, however, that the other portions of the nephron are not important in this process. Because the reabsorptive capacity of the collecting duct is limited, these other portions of the nephron must reabsorb the bulk of the filtered load of Na^+. As indicated in Figure 37-7, Na^+ reabsorption by the proximal tubule, Henle's loop, and distal tubule results in the delivery of only about 4% of the filtered load of Na^+ to the collecting duct. Importantly, this delivery is maintained relatively constant despite small variations in Na^+ intake. Two mechanisms operate to ensure this constant delivery. First, the glomerular filtration rate, and thus the filtered load of Na^+, is maintained relatively constant by the process of autoregulation (see Chapters 23 and 35). Second, when the filtered load of Na^+ changes slightly, such as during transient changes in blood pressure, reabsorption of NaCl by the proximal tubule is adjusted in parallel. The mechanism responsible for this phenomenon is termed **glomerulotubular balance** (see Chapter 36).

With a constant delivery of Na^+, small adjustments in collecting duct reabsorption are sufficient to balance excretion with intake. Aldosterone is the primary regulator of collecting duct Na^+ reabsorption under this condition (see earlier discussion). Other hormones also influence collecting duct Na^+ reabsorption. For example, ANP inhibits Na^+ reabsorption. Na^+ reabsorption is also inhibited by prostaglandins and bradykinin, but their roles in regulating Na^+ excretion during euvolemia are not established. Finally, ADH not only stimulates water reabsorption, but also stimulates Na^+ reabsorption by the collecting duct.

As long as variations in the dietary intake of NaCl are small, the mechanisms just described can regulate renal Na^+ excretion appropriately and thereby maintain a constant ECV. Large changes in NaCl intake, however, cannot be handled effectively by these mechanisms. Consequently the ECV will be altered. When this occurs, additional factors act on the kidney to adjust Na^+ reabsorp-

tion to reestablish the euvolemic state. These factors will be discussed in the next section.

Control of Na$^+$ Excretion With Increased Effective Circulating Volume

When the ECV is increased, the various signals just described act on the kidney to increase Na^+ excretion. The integrated response to an increase in the ECV is illustrated in Figure 37-8. The important difference between this condition and that described earlier for the euvolemic state is that the renal response is not limited to the collecting duct; rather, it involves the entire nephron. Three general responses to an increase in ECV include:

1. The glomerular filtration rate increases. This is achieved mainly by decreased activity of the sympathetic nerve fibers innervating the afferent arterioles. The decreased sympathetic activity leads to vasodilation, and an increase in the glomerular capillary pressure, and thereby in GFR. ANP may also contribute to this response by dilating afferent arterioles. Because the glomerular filtration rate is increased, the filtered load of Na^+ increases in parallel.

2. Reabsorption of Na^+ decreases in the proximal tubule. Various mechanisms are thought to contribute to this inhibition of reabsorption. Because activation of the sympathetic fibers that innervate this portion of the nephron stimulate Na^+ reabsorption, the decreased sympathetic activity engendered by the expansion of the ECV may contribute to the observed decrease in Na^+ reabsorption. Angiotensin II stimulates proximal tubule reabsorption. Because angiotensin II levels are reduced under this condition, it is possible that proximal tubule Na^+ reabsorption may decrease as a result. Also, dilation of the afferent arteriole will result in a parallel increase in the hydrostatic pressure within the peritubular capillaries surrounding the proximal tubule (see Chapter 36). This increase in pressure inhibits the uptake of NaCl and water by the capillary and thereby inhibits their reabsorption. Note that proximal tubule reabsorption is reduced even though the glomerular filtration rate is increased. This shows that glomerulotubular balance only occurs during euvolemia, not when the ECV is altered.

3. Na^+ reabsorption decreases in the collecting duct. Both the increase in filtered load and the decrease in proximal tubule reabsorption enhance the delivery of Na^+ to the loop of Henle, which in turn increases the delivery of Na^+ to the collecting duct. Thus the delivery of Na^+ to the collecting duct is no longer a constant fraction of the filtered load, as is the case during euvolemia. Instead, Na^+ delivery varies in parallel with the degree of ECV expansion. This large

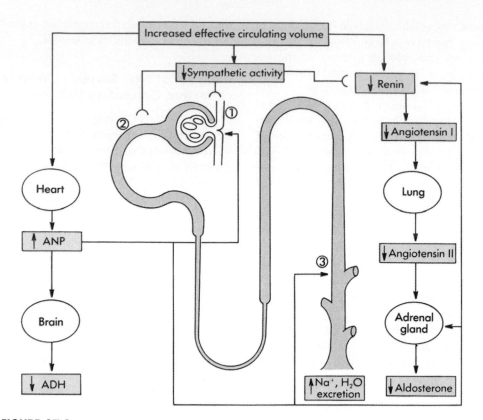

FIGURE 37-8 Integrated response to an increase in the effective circulating volume. The numbers along the nephron indicate sites where sodium handling is regulated. See text for details.

load of Na^+ overwhelms the reabsorptive capacity of the collecting duct; the capacity is reduced by the action of ANP and by the diminution in the circulating levels of aldosterone. Thus, by altering Na^+ reabsorption along the entire length of the nephron, the capacity to excrete Na^+ is increased greatly above that seen during euvolemia.

One final component to be considered in the response to ECV expansion is the excretion of water. As Na^+ excretion is increased, plasma osmolality begins to fall. This inhibits the secretion of ADH and thereby increases water excretion. An additional mechanism by which ADH secretion and action at the collecting duct are suppressed is through the action of ANP (see earlier section). Because of the sensitivity of the ADH system, the excretion of Na^+ parallels that of water. Thus the ECV will be restored to normal, and its osmolality will be unchanged.

Control of Na^+ Excretion With Decreased Effective Circulating Volume

When the ECV is decreased, Na^+ and water excretion by the kidneys is reduced. The mechanisms involved are

essentially the opposite of those just described and are summarized in Figure 37-9. Again, the entire nephron contributes to this response. The three general responses to a decrease in the ECV are:

1. The glomerular filtration rate, and therefore the filtered load of Na^+, is reduced. This results from afferent and efferent arteriolar constriction mediated by the sympathetic nerves.
2. Na^+ reabsorption by the proximal tubule is stimulated (note that glomerulotubular balance does not occur). This stimulation is the result of increased sympathetic nerve activity and the direct action of angiotensin II.
3. Na^+ reabsorption by the collecting duct is enhanced. The decreased filtered load of Na^+ together with enhanced reabsorption by the proximal tubule, and additional reabsorption by the loop of Henle, result in a decrease in Na^+ delivery to the collecting duct. Na^+ reabsorption by the collecting duct is stimulated mainly by the action of aldosterone. Other hormones may also contribute to this response. In particular, the circulating levels of ANP are reduced, and ADH levels are elevated. ADH not only stimulates Na^+ reabsorption but also increases the reabsorption of water

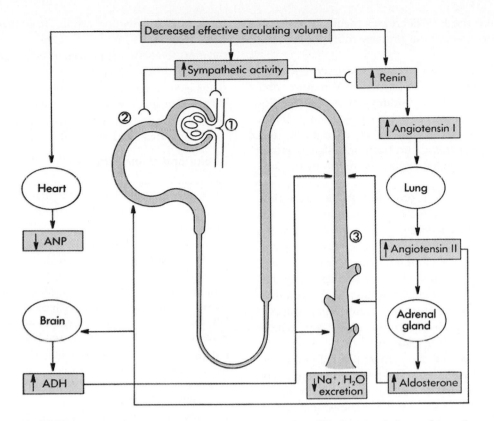

FIGURE 37-9   Integrated response to a decrease in the effective circulating volume. Symbols are the same as in Figure 37-8.

by the collecting duct. As a result, the urine can be rendered virtually Na^+ free. This retention of Na^+, and the concurrent conservation of water, will restore the ECV.

SUMMARY

1. The osmolality and volume of the body fluids are maintained within a narrow range, despite wide variation in water and solute intake. The kidneys play the central role in this regulatory process by virtue of their ability to vary the excretion of water and solutes.
2. Regulation of body fluid osmolality requires that water intake and loss from the body be equal. This involves the integrated interaction of the ADH secretory and thirst centers of the hypothalamus, and the ability of the kidneys to excrete urine that is either hypoosmotic or hyperosmotic with respect to the body fluids.
   a. When body fluid osmolality increases, ADH secretion and thirst are stimulated. ADH acts on the kidneys to increase the permeability of the collecting duct to water. Hence, water is reabsorbed from the

lumen of the collecting duct, and a small volume of hyperosmotic urine is excreted. This renal conservation of water, together with increased water intake, restores body fluid osmolality to normal.
   b. When body fluid osmolality decreases, ADH secretion and thirst are suppressed. In the absence of ADH the collecting duct is impermeable to water, and a large volume of hypoosmotic urine is excreted. With this increased excretion of water, and with a decreased intake of water caused by suppression of thirst, the osmolality of the body fluids is restored to normal.
3. Central to the process of concentrating and diluting the urine is the loop of Henle. The transport of NaCl by the loop allows the separation of solute and water, which is essential for the elaboration of hypoosmotic urine. By this same mechanism, the interstitial fluid in the medullary portion of the kidney is rendered hyperosmotic. This hyperosmotic medullary interstitial fluid in turn provides the osmotic driving force for the reabsorption of water from the lumen of the collecting duct when ADH is present.
4. The ECV is determined by Na^+ balance. When intake of Na^+ exceeds excretion, the ECV increases (positive Na^+ balance). Conversely, when excretion of Na^+

exceeds intake, the ECV decreases (negative Na$^+$ balance). The kidneys are the primary route for Na$^+$ excretion.

5. The coordination of Na$^+$ intake and excretion, and thus the maintenance of a normal ECV, requires the integrated action of the kidneys with the cardiovascular and sympathetic nervous systems. Cardiovascular volume receptors detect changes in the ECV, and by sympathetic and hormonal signals effect appropriate adjustments in Na$^+$ excretion by the kidneys.

6. Under normal conditions (euvolemia), Na$^+$ excretion by the kidneys is matched to the amount of Na$^+$ ingested in the diet. The kidneys accomplish this by reabsorbing virtually all of the filtered load of Na$^+$ (typically less than 1% of the filtered load is excreted). During euvolemia, the collecting duct is responsible for making small adjustments in urinary Na$^+$ excretion to effect Na$^+$ balance. The major factor regulating collecting duct Na$^+$ reabsorption is aldosterone, which acts to stimulate Na$^+$ reabsorption.

7. When the volume of the extracellular fluid is increased, low-pressure and high-pressure volume sensors initiate a response that ultimately leads to increased excretion of Na$^+$ by the kidneys, and the return of the ECV to normal. The components of this response include a decrease in sympathetic outflow to the kidney, a suppression of the renin-angiotensin-aldosterone system, and release from the cardiac atria of atrial natriuretic peptide. By the actions of these effectors, glomerular filtration rate is enhanced, which increases the filtered load of Na$^+$, and Na$^+$ reabsorption by the proximal tubule and collecting duct is reduced. Together, these changes in renal Na$^+$ handling enhance Na$^+$ excretion.

8. When the volume of the extracellular fluid is decreased, the above sequence of events is reversed (increased sympathetic outflow to the kidney, activation of the renin-angiotension-aldosterone system, and suppression of atrial natriuretic peptide secretion). This decreases the glomerular filtration rate, and enhances reabsorption of Na$^+$ by the proximal tubule and collecting duct, and thus reduces Na$^+$ excretion.

BIBLIOGRAPHY
Journal Articles

Bayliss PH: Osmoregulation and control of vasopressin secretion in healthy humans, *Am J Physiol* 253:R671, 1987.

Breyer MD, Ando Y: Hormonal signaling and regulation of salt and water transport in the collecting duct, *Annu Rev Physiol* 56:711, 1994.

De Wardner HE: The control of sodium excretion, *Am J Physiol* 235:F163, 1978.

Funder JW: Aldosterone action, *Annu Rev Physiol* 55:115, 1993.

Gillin AG, Sands JM: Urea transport in the kidney, *Semin Nephrol* 13:146, 1993.

Gunning ME, Brenner BM: Natriuretic peptides and the kidney: current concepts, *Kidney Int* 42 [suppl 38]:S127, 1992.

Jamison RL, Maffly RH: The urinary concentrating mechanism, *New Engl J Med* 295:1059, 1976.

Books and Monographs

Ballerman BJ, Zeidel ML: Atrial natriuretic hormone. In Seldin DW, Giebisch G, eds: *The kidney: physiology and pathophysiology,* ed 2, New York, 1992, Raven Press.

Fitzsimons JT: Physiology and pathophysiology of thirst and sodium appetite. In Seldin DW, Giebisch G, eds: *The kidney: physiology and pathophysiology,* ed 2, New York, 1992, Raven Press.

Gonzalez-Campoy JM, Knox, FG: Integrated responses of the kidney to alterations in extracellular fluid volume. In Seldin DW, Giebisch G, eds: *The kidney: physiology and pathophysiology,* ed 2, New York, 1992, Raven Press.

Hall JE, Brands MW: The renin-angiotensin-aldosterone systems: renal mechanisms and circulatory homeostasis. In Seldin DW, Giebisch G, eds: *The kidney: physiology and pathophysiology,* ed 2, New York, 1992, Raven Press.

Hays RM: Cell biology of vasopressin. In Brenner BM, Rector FC Jr, eds: *The kidney,* ed 4, Philadelphia, 1991, WB Saunders Co.

Kirk KL, Schaffer JA: Water transport and osmoregulation by antidiuretic hormone in terminal nephron segments. In Seldin DW, Giebisch G, eds: *The kidney: physiology and pathophysiology,* ed 2, New York, 1992, Raven Press.

Knepper MA, Rector FC Jr: Urinary concentration and dilution. In Brenner BM, Rector FC Jr, eds: *The kidney,* ed 4, Philadelphia, 1991, WB Saunders Co.

Koeppen BM, Stanton BA: *Renal physiology,* St Louis, 1992, Mosby.

Moe GW, Legault L, Skorecki KL: Control of extracellular fluid volume and pathophysiology of edema formation. In Brenner BM, Rector FC Jr, eds: *The kidney,* ed 4, Philadelphia, 1991, WB Saunders Co.

Robertson GL: Regulation of vasopressin secretion. In Seldin DW, Giebisch G, eds: *The kidney: physiology and pathophysiology,* ed 2, New York, 1992, Raven Press.

Robertson GL, Berl T: Pathophysiology of water metabolism. In Brenner BM, Rector FC Jr, eds: *The kidney,* ed 4, Philadelphia, 1991, WB Saunders Co.

Rose BD: *Clinical physiology of acid-base and electrolyte disorders,* ed 4, New York, 1994, McGraw-Hill Inc.

Roy DR et al: Countercurrent mechanism and its regulation. In Seldin DW, Giebisch G, eds: *The kidney: physiology and pathophysiology,* ed 2, New York, 1992, Raven Press.

Seldin DW, Giebisch G, eds: *The regulation of sodium and chloride balance,* New York, 1990, Raven Press.

CHAPTER 38

Renal Regulation of Potassium, Calcium, Magnesium, and Phosphate Balance

POTASSIUM (K⁺)

Potassium (K^+) is one of the most abundant cations in the body. It is critical for many cell functions, and its concentration in cells and extracellular fluid remains constant despite wide fluctuations in dietary K^+ intake. Two sets of regulatory mechanisms safeguard K^+ homeostasis. First, several mechanisms regulate the potassium concentration ($[K^+]$) in the extracellular fluids. Second, another set of mechanisms maintains the amount of K^+ in the body constant by adjusting renal K^+ excretion to match dietary K^+ intake. It is the kidneys that regulate K^+ excretion. This section will focus on the hormones and factors that influence the $[K^+]$ in the extracellular fluid compartments and the hormones and factors that regulate the amount of K^+ excreted in the urine.

Overview of K⁺ Homeostasis

The distribution of K^+ in the body is shown in Figure 38-1. Total body K^+ has been estimated at 50 mEq/kg of body weight, or 3500 mEq for a 70 kg individual. Ninety-eight percent of the K^+ in the body lies within cells, where its average concentration is 150 mEq/L. A high intracellular concentration of K^+ is required for many cell functions, including cell growth and division and volume regulation. Only 2% of total body K^+ is located in the ECF, where its normal concentration is 4 mEq/L. When the $[K^+]$ of the ECF exceeds 5.5 mEq/L, **hyperkalemia** exists. Conversely, **hypokalemia** exists when the $[K^+]$ of the ECF is less than 3.5 mEq/L.

The large concentration difference of K^+ across cell membranes (146 mEq/L) is maintained by the operation of the Na^+-K^+-ATPase. This gradient is important in maintaining the potential difference across cell membranes. Thus, K^+ is critical for the excitability of nerve and muscle cells, as well as for the contractility of car-

diac, skeletal, and smooth muscle cells (see Chapters 2, 3, 17 and 18).

After a meal the K^+ absorbed by the gastrointestinal tract rapidly (minutes) enters the ECF (Figure 38-1). If the K^+ ingested during a normal meal ($\approx$33 mEq) were to remain in the ECF compartment, plasma $[K^+]$ would increase by a potentially lethal 2.4 mEq/L (33 mEq added to 14 L of ECF; therefore, 33 mEq/L ÷ 14 L = Δ2.4 mEq/L). This rise in plasma $[K^+]$ is prevented by the rapid uptake of K^+ into cells. *Because the excretion of K^+ by the kidneys after a meal is relatively slow (hours), the buffering of K^+ by cells is essential to prevent life-threatening hyperkalemia.* To maintain total body K^+ constant, all of the K^+ absorbed by the gastrointestinal tract must eventually be excreted by the kidneys. This K^+ is slowly excreted, such that after 6 hours it is eliminated from the body, thereby maintaining a constant amount of K^+ in the body.

Several hormones promote the uptake of K^+ into cells after a rise in plasma $[K^+]$ and thereby prevent dangerous hyperkalemia. As illustrated in Figure 38-1, and summarized in Table 38-1, these hormones include epinephrine, insulin, and aldosterone. They all increase K^+ uptake into skeletal muscle, liver, bone, and red blood cells by stimulating the Na^+-K^+-ATPase pump, either directly or indirectly. A rise in plasma $[K^+]$, subsequent to K^+ absorption by the gastrointestinal tract, stimulates insulin secretion from the pancreas, aldosterone release from the adrenal cortex, and epinephrine secretion from the adrenal medulla. In contrast, a decrease in plasma $[K^+]$ inhibits release of these hormones. Whereas insulin and epinephrine act within a few minutes, aldosterone requires about 1 hour to stimulate K^+ uptake into cells.

Epinephrine Catecholamines affect the distribution of K^+ across cell membranes by activating α- and β_2-adrenergic receptors. Stimulation of α receptors increases plasma $[K^+]$, whereas stimulation of β_2 receptors decreases plasma $[K^+]$.

555

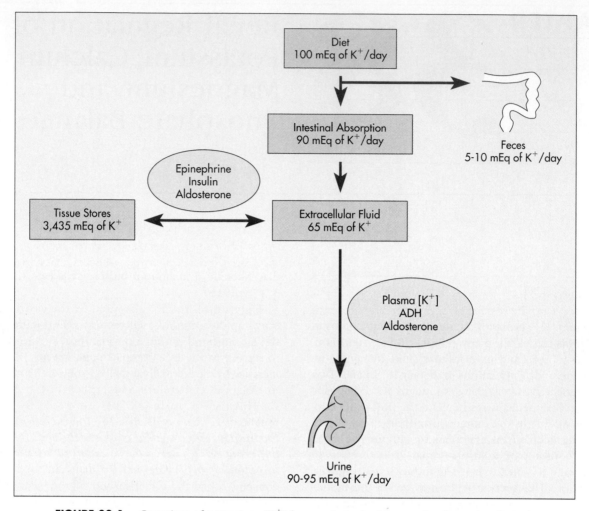

FIGURE 38-1 Overview of potassium (K^+) homeostasis. An increase in plasma insulin, epinephrine, and aldosterone stimulates K^+ movement into cells and decreases plasma [K^+], whereas a fall in the plasma concentration of these hormones increases plasma [K^+]. The amount of K^+ in the body is determined by the kidneys. An individual is in K^+ balance when dietary intake and urinary output (plus output by the gastrointestinal tract) are equal. The excretion of K^+ by the kidneys is regulated by plasma [K^+], aldosterone, and ADH.

α-adrenoceptor activation after exercise is important in preventing hypokalemia. The importance of β₂ receptors is illustrated by two observations. First, the rise in plasma [K^+] after a K^+-rich meal is greater if the subject has been pretreated with **propranolol,** a β-adrenergic blocker. Second, the release of **epinephrine** during stress (e.g., coronary ischemia) can rapidly lower plasma [K^+].

Insulin Insulin also stimulates K^+ uptake into cells. The importance of insulin is illustrated by two observations. First, the rise in plasma [K^+] after a K^+-rich meal is greater in patients with diabetes mellitus (i.e., insulin deficiency) than in normal people. Second, infusion of

insulin and glucose constitutes acute therapy for hyperkalemia. *Insulin is the most important hormone that shifts K^+ into cells after ingestion of K^+ in a meal.*

Aldosterone Aldosterone, like catecholamines and insulin, also promotes K^+ uptake into cells (see Chapter 46). A rise in aldosterone levels (e.g., **primary aldosteronism**) causes hypokalemia, and a fall in aldosterone levels (e.g., **Addison's disease**) causes hyperkalemia. As discussed below, aldosterone also stimulates urinary K^+ excretion. Thus, aldosterone alters plasma [K^+] by acting on K^+ uptake into cells and by altering urinary K^+ excretion.

Thus far, the discussion has focused on hormones that maintain the distribution of K^+ across cell membranes constant. Other factors, however, influence K^+ movements across the cell membrane, but they are not ho-

Table 38-1 Major Factors and Hormones Influencing the Distribution of K^+ between the ICF and the ECF

PHYSIOLOGICAL: KEEP PLASMA [K^+] CONSTANT

Epinephrine
Insulin
Aldosterone

PATHOPHYSIOLOGICAL: DISPLACE PLASMA [K^+] FROM NORMAL

Acid-base balance
Plasma osmolality
Cell lysis
Exercise

meostatic mechanisms because they displace plasma [K^+] from normal levels. Table 38-1 summarizes these factors.

Acid-base balance In general, metabolic acidosis increases and metabolic alkalosis decreases plasma [K^+]. In contrast, respiratory acid-base disorders have little or no effect on plasma [K^+]. A metabolic acidosis produced by the addition of inorganic acids (e.g., HCl, H_2SO_4) increases plasma [K^+] to a much greater extent than does a similar acidosis produced by the accumulation of organic acids (e.g., lactic acid, acetic acid, keto acids). The reduced pH promotes movement of H^+ into cells, and the reciprocal movement of K^+ out of cells. Metabolic alkalosis has the opposite effect; plasma [K^+] decreases as K^+ moves into cells and H^+ leaves cells. The mechanism responsible for this shift is not fully understood. It has been proposed that the movement of H^+ occurs as the cells buffer changes in the [H^+] of the extracellular fluid. As H^+ moves across the cell membranes, K^+ moves in the opposite direction, and thus cations are neither gained nor lost across the cell membranes. Although organic acids produce a metabolic acidosis, they do not cause a significant degree of hyperkalemia. Two possible explanations have been suggested for the reduced effect of organic acids to cause hyperkalemia. First, the organic anion may enter the cell with H^+, thereby eliminating the need for K^+ for H^+ exchange across the membrane. Second, organic anions may stimulate insulin secretion, which moves K^+ into cells, counteracting the direct effect of the acidosis that moves K^+ out of cells.

Plasma osmolality The osmolality of the plasma also influences the distribution of K^+ across cell membranes. An increase in the osmolality of the extracellular fluid enhances K^+ release by cells and thus increases extracellular [K^+]. The plasma K^+ level may increase by 0.4 to 0.8 mEq/L for a 10 mOsm/kg H_2O elevation in plasma osmolality. Hypoosmolality has the opposite action.

As plasma osmolality increases, water will leave cells because of the osmotic gradient across the plasma membrane. Water leaves cells until the intracellular osmolality becomes equal to that of the extracellular fluid osmolality. This loss of water shrinks cells, and causes cell [K^+] to rise. The rise in intracellular [K^+] provides a driving force for K^+ efflux from the cells. This sequence increases plasma [K^+]. A fall in plasma osmolality has the opposite effect.

Cell lysis

Cell lysis causes hyperkalemia. Severe trauma (e.g., burns) and some diseases such as **tumor lysis syndrome** and **rhabdomyolysis** (i.e., destruction of skeletal muscle) cause cell destruction and release of K^+ (and other cell solutes) into the extracellular fluid. In addition, gastric ulcers may cause seepage of red blood cells into the gastrointestinal tract. The blood cells are digested, and the K^+ released from the cells is absorbed and can cause hyperkalemia.

Exercise During exercise more K^+ is released from skeletal muscle cells than during rest. Release of K^+, during the recovery phase of the action potential, and the ensuing hyperkalemia, depends on the degree of exercise. Plasma [K^+] increases by 0.3 mEq/L with slow walking and may increase by up to 2.0 mEq/L above normal with exercise.

Exercise-induced changes in plasma [K^+] usually do not produce symptoms and are reversed after several minutes of rest. However, in individuals who have certain endocrine disorders that affect the release of insulin, epinephrine, or aldosterone, or whose ability to excrete K^+ is impaired (e.g., **renal failure),** or who are on certain medications, exercise can lead to potentially life-threatening hyperkalemia. For example, during exercise plasma [K^+] may increase by 2 to 4 mEq/L in individuals taking β-**adrenergic blockers** for hypertension.

Acid-base balance, plasma osmolality, cell lysis, and exercise do not maintain plasma [K^+] at a normal value, and therefore do not contribute to K^+ homeostasis. The extent to which these pathophysiological states alter plasma [K^+] depends on the integrity of the homeostatic mechanisms that regulate plasma [K^+] (e.g., secretion of epinephrine, insulin, and aldosterone).

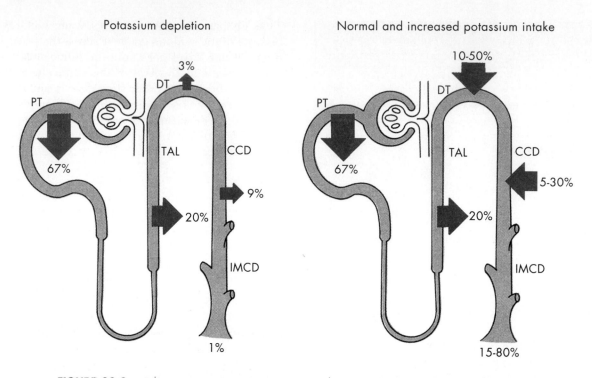

FIGURE 38-2 K^+ transport along the nephron. K^+ excretion depends on the rate and direction of K^+ transport by the distal tubule and the collecting duct. Percentages refer to the amount of filtered K^+ reabsorbed or secreted by each nephron segment. *Left,* Dietary K^+ depletion. An amount of K^+ equal to 1% of the filtered load of K^+ is excreted. *Right,* Normal and increased dietary K^+ intake. An amount of K^+ equal to 15% to 80% of the filtered load is excreted. *PT,* Proximal tubule; *TAL,* thick ascending limb; *DT,* distal tubule; *CCD,* collecting duct; *IMCD,* inner medullary collecting duct.

K^+ Excretion by the Kidneys

The kidneys play the major role in maintaining K^+ balance. As illustrated in Figure 38-1, the kidneys excrete 90% to 95% of the K^+ ingested in the diet. Excretion equals intake even when intake increases by as much as 10-fold. This equality between urinary excretion and dietary intake underscores the importance of the kidneys in maintaining K^+ homeostasis. Although small amounts of K^+ are lost each day in the stool and sweat (~5% to 10% of the K^+ ingested in the diet), this amount is essentially constant, is not regulated, and therefore is relatively much less important than is the K^+ excreted by the kidneys.* *The primary event in determining urinary K^+ excretion is K^+ secretion from the blood into the tubular fluid by the cells of the distal tubule and collecting duct system.* The transport pattern of K^+ by the major nephron segments is illustrated in Figure 38-2.

Because K^+ is not bound to plasma proteins, it is freely filtered by the glomerulus. Normally, when indi-

viduals ingest a normal diet, urinary K^+ excretion is 15% of the amount filtered. Accordingly, K^+ must be reabsorbed along the nephron. When dietary K^+ intake is augmented, however, K^+ excretion can exceed the amount filtered, indicating K^+ can also be secreted.

The proximal tubule reabsorbs 67% of the filtered K^+ under most conditions. Approximately 20% of the filtered K^+ is reabsorbed by Henle's loop and, as with the proximal tubule, reabsorption is a constant fraction of the amount filtered. In contrast to these segments, which are capable of only reabsorbing K^+, the distal tubule and the collecting duct have the dual capacity to reabsorb and secrete K^+. The rate of K^+ reabsorption or secretion by the distal tubule and the collecting duct depends on a variety of hormones and factors. When K^+ intake is normal (100 mEq/day), K^+ is secreted. A rise in dietary K^+ intake increases K^+ secretion, such that the amount of K^+ appearing in the urine may approach 80% of the amount filtered (Figure 38-2). In contrast, a low potassium diet activates K^+ reabsorption along the distal tubule and collecting duct, such that urinary excretion falls to 1% of the K^+ filtered by the glomerulus (Figure 38-2). The kidneys are not able to reduce K^+ excretion to the same low levels as

---

*Loss of K^+ in the feces can become significant during periods of diarrhea.

they can for Na^+ (0.2%). Therefore, hypokalemia can develop in individuals placed on a K^+-deficient diet.

Because the magnitude and direction of K^+ transport by the distal tubule and collecting duct are variable, the overall rate of urinary K^+ excretion is determined by these tubular segments.

In individuals with **chronic renal failure,** the kidneys are unable to eliminate K^+ from the body and plasma $[K^+]$ rises. The resulting hyperkalemia reduces the resting membrane potential (i.e., the voltage becomes less negative) and decreases the excitability of neurons, cardiac cells, and muscle cells by inactivating fast Na^+ channels in the membrane (see Chapter 3). Severe, rapid increases in plasma $[K^+]$ can lead to cardiac arrest and death. In contrast, in patients taking diuretic drugs for hypertension, urinary K^+ excretion often exceeds dietary K^+ intake. Accordingly, body K^+ balance falls and hypokalemia develops. This decline in extracellular $[K^+]$ hyperpolarizes the resting cell membrane potential (i.e., the voltage becomes more negative) and reduces the excitability of neurons, cardiac cells, and muscle cells (see Chapter 3). Severe hypokalemia can lead to paralysis, cardiac arrhythmia, and death. Hypokalemia can also impair the ability of the kidneys to concentrate the urine, and can stimulate renal production of NH_4^+. Therefore, maintenance of a high intracellular $[K^+]$ and a low extracellular $[K^+]$, as well as a high K^+ concentration gradient across cell membranes, is essential for a number of cellular functions.

Cellular Mechanisms of K^+ Transport by the Distal Tubule and Collecting Duct

Figure 38-3 illustrates the cellular mechanism of K^+ secretion by principal cells in the distal tubule and collecting duct. Secretion from blood into tubular fluid is a two-step process involving: (1) K^+ uptake across the basolateral membrane by Na^+-K^+-ATPase and (2) diffusion of K^+ from the cell into the tubular fluid. The operation of the Na^+-K^+-ATPase creates a high intracellular $[K^+]$, which provides the chemical driving force for K^+ exit across the apical membrane through K^+ channels. Although K^+ channels are also present in the basolateral membrane, K^+ preferentially leaves the cell across the apical membrane and enters the tubular fluid for two reasons. First, the electrochemical gradient of K^+ across the apical membrane favors the downhill movement into the tubular fluid. Second, the permeability of the apical membrane to K^+ is greater than that of the basolateral mem-

brane. Therefore K^+ preferentially diffuses across the apical membrane into the tubular fluid. The three major factors that control the rate of K^+ secretion by the distal tubule and the collecting duct are (see Figure 38-3):

1. The activity of the Na^+-K^+-ATPase
2. The driving force (electrochemical gradient) for K^+ movement across the apical membrane
3. The permeability of the apical membrane to K^+

Every change in K^+ secretion results from an alteration in one or more of these factors.

In contrast, the cellular pathways and mechanisms of K^+ reabsorption in the distal tubule and collecting duct are not completely understood. Intercalated cells may reabsorb K^+ by an H^+-K^+-ATPase transport mechanism located in the apical membrane. This transporter mediates K^+ uptake in exchange for H^+. However, the pathway of K^+ exit from intercalated cells into the blood is unknown. Reabsorption of K^+ is activated by a low K^+ diet.

Regulation of K^+ Secretion by the Distal Tubule and Collecting Duct

Regulation of K^+ excretion is achieved mainly by alterations in K^+ secretion by principal cells of the distal tubule and collecting duct. Plasma $[K^+]$ and aldosterone are the major physiological regulators of K^+ secretion. ADH also stimulates K^+ secretion; however, it is less important than plasma $[K^+]$ and aldosterone. Other factors, including the flow rate of tubular fluid and acid-base balance, influence K^+ secretion by the distal tubule and collecting duct. However, they are not homeostatic mechanisms because they disturb K^+ balance.

Plasma $[K^+]$ *Plasma $[K^+]$ is an important determinant of K^+ secretion by the distal tubule and collecting duct.* Hyperkalemia (e.g., high K^+ diet or **rhabdomyolysis**) rapidly (in minutes) stimulates secretion. Several mechanisms are involved. First, hyperkalemia stimulates the Na^+-K^+-ATPase pump, and thereby increases K^+ uptake across the basolateral membrane. This raises intracellular $[K^+]$ and increases the electrochemical driving force for K^+ exit across the apical membrane. Second, hyperkalemia also increases the permeability of the apical membrane to K^+. Third, hyperkalemia stimulates aldosterone secretion by the adrenal cortex, which, as discussed below, acts synergistically with plasma $[K^+]$ to stimulate K^+ secretion.

Hypokalemia (e.g., caused by a low K^+ diet or diarrhea) decreases K^+ secretion by actions opposite to those described for hyperkalemia. Hence, hypokalemia inhibits the Na^+-K^+-ATPase pump, decreases the electrochemical driving force for K^+ efflux across the apical membrane, reduces the permeability of the apical membrane to K^+, and causes a reduction in plasma aldosterone levels.

Principal Cell

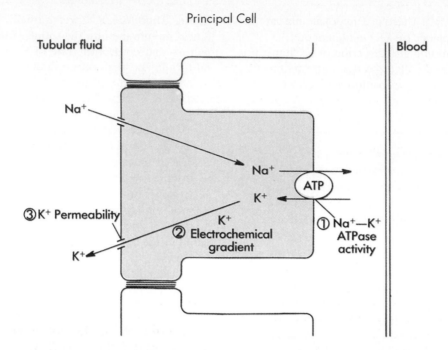

FIGURE 38-3   Cellular mechanism of K^+ secretion by the principal cell in the distal tubule and collecting duct. The numbers indicate the sites where K^+ secretion is regulated: *(1)* Na^+-K^+-ATPase; *(2)* electrochemical gradient of K^+ across the apical membrane; and *(3)* the K^+ permeability of the apical membrane.

Aldosterone   Aldosterone, in addition to stimulating Na^+ reabsorption, enhances K^+ secretion across the distal tubule and collecting duct by increasing the amount of Na^+-K^+-ATPase in principal cells. This elevates cell $[K^+]$. Aldosterone also increases the driving force for K^+ exit across the apical membrane and increases the permeability of the apical membrane to K^+. Aldosterone secretion is increased by hyperkalemia and by angiotensin II (following activation of the renin-angiotensin system), and it is decreased by hypokalemia and atrial natriuretic peptide. Stimulation of K^+ secretion by aldosterone occurs after a 1-hour lag period and attains its highest level after 1 day (see also Chapter 46).

Antidiuretic hormone   Although antidiuretic hormone (ADH) increases the electrochemical driving force for K^+ exit across the apical membrane of principal cells in the distal tubule and collecting duct, this hormone does not change K^+ secretion by these nephron segments. The reason for this relates to the effect of ADH on tubular fluid flow. ADH decreases tubular fluid flow by stimulating water reabsorption. The decrease in tubular flow in turn decreases K^+ secretion (see below for a discussion). As illustrated in Figure 38-4, changes in ADH levels do not alter K^+ secretion by the distal tubule or collecting duct or urinary K^+ excretion because the stimulatory effect of ADH on the electrochemical driving force for K^+ exit across the apical membrane is offset by the inhibitory effect of decreased tubular fluid flow.

If ADH did not increase the electrochemical gradient favoring K^+ secretion, urinary K^+ excretion would fall as ADH levels increase (decreasing urine flow) and K^+ balance would change in response to alterations in water balance. Thus, these effects of ADH on the electrochemical driving force for K^+ exit across the apical membrane and tubule flow enable urinary K^+ excretion to be maintained constant despite wide fluctuations in water excretion.

Flow of tubular fluid   A rise in the flow of tubular fluid (e.g., diuretic treatment, extracellular fluid volume expansion, excess water intake) rapidly (in minutes) stimulates K^+ secretion, whereas a fall in flow (e.g., extracellular fluid volume contraction caused by hemorrhage, or severe vomiting or diarrhea) reduces K^+ secretion by the distal tubule and collecting duct. Increments in tubular fluid flow are more effective in stimulating K^+ secretion as dietary K^+ intake is increased from a low K^+ diet to a high K^+ diet. Alterations in tubular fluid flow influence K^+ secretion by changing the driving force for K^+ exit across the apical membrane. As K^+ is secreted into the tubular fluid, the $[K^+]$ of the fluid increases. This will reduce the electrochemical driving force for K^+ exit across the apical membrane and thereby reduce the rate of secretion. An increase in tubular fluid flow minimizes the rise in tubular fluid $[K^+]$ as the secreted K^+ is washed downstream. As a result, K^+ secretion is stimulated by an increase in the flow of

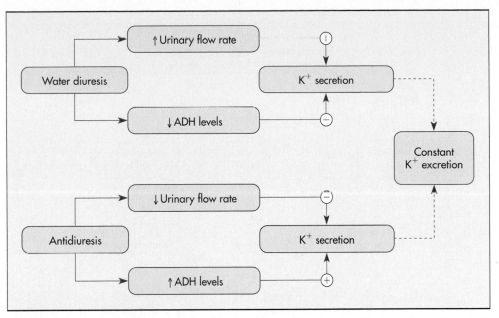

FIGURE 38-4   Opposing effects of ADH on K^+ secretion by the distal tubule and cortical collecting duct. Because the two effects oppose each other, net K^+ secretion and urinary K^+ excretion are not affected by ADH.

tubular fluid. *Because diuretic drugs increase the flow of tubular fluid through the distal tubule and collecting duct, they also enhance urinary K^+ excretion.* In contrast, a decline in tubular fluid flow inhibits K^+ secretion. A decline in the tubular fluid flow facilitates the rise in tubular fluid [K^+], and thereby reduces secretion.

Acid-base balance   Another factor that modulates K^+ secretion is the [H^+] of the extracellular fluid. Acute alterations (over a period of minutes to hours) in the pH of the plasma influence K^+ secretion by the distal tubule and collecting duct. **Alkalosis** (i.e., a plasma pH above a normal value of 7.4) increases H^+ secretion, whereas **acidosis** (i.e., a plasma pH below 7.4) decreases K^+ secretion. An acute acidosis reduces K^+ secretion by two mechanisms: (1) it inhibits the Na^+-K^+-ATPase pump and thereby reduces cell [K^+] and the electrochemical driving force for K^+ exit across the apical membrane, and (2) it reduces the permeability of the apical membrane to K^+. An alkalosis has the opposite effects.

The effect of metabolic acidosis on K^+ excretion is time dependent. When a metabolic acidosis is prolonged for several days, urinary K^+ excretion is stimulated (Figure 38-5). This occurs because chronic metabolic acidosis decreases water and NaCl reabsorption by the proximal tubule, by inhibiting the Na^+-K^+-ATPase. Hence, the flow of tubular fluid is augmented through the distal tubule and collecting duct. The inhibition of proximal tubular water and NaCl reabsorption also causes a decrease in ECV and thereby stimulates aldosterone secretion. In addition, chronic acidosis, caused by inorganic acids, in-

creases plasma [K^+], which stimulates aldosterone secretion. The rise in tubular fluid flow, plasma [K^+], and aldosterone offsets the effects of acidosis on cell [K^+] and apical membrane permeability, such that K^+ secretion rises. Thus, metabolic acidosis may either inhibit or stimulate potassium excretion, depending upon the duration of the disturbance.

Chronic hypokalemia, a plasma [K^+] < 3.5 mEq/L, occurs most often in individuals who receive diuretics for hypertension. Hypokalemia also occurs in individuals who vomit, have nasogastric suction, have diarrhea, abuse laxatives, or have **hyperaldosteronism.** Hypokalemia occurs because the excretion of K^+ by the kidneys exceeds dietary intake of K^+. Vomiting, nasogastric suction, diuretics, and diarrhea all decrease the ECV, which stimulates secretion of aldosterone, a hormone that stimulates K^+ excretion by the kidneys.

Chronic hyperkalemia, a plasma [K] > 5.5 mEq/L, occurs most frequently in individuals with a reduced urine flow or a low plasma aldosterone level or whose GFR falls to less than 20% of normal. In these individuals, hyperkalemia occurs because the excretion of K^+ by the kidneys is less than dietary intake of K^+. Less common causes for hyperkalemia are deficiencies of insulin, epinephrine, or aldosterone secretion, or metabolic acidosis caused by inorganic acids.

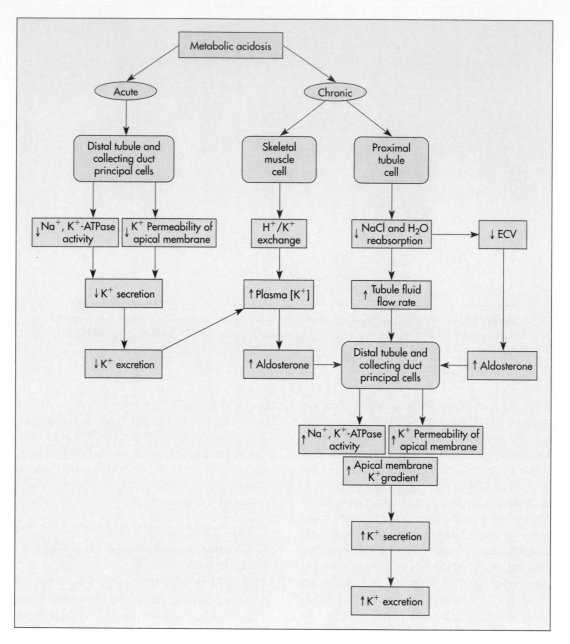

FIGURE 38-5   Acute versus chronic effect of metabolic acidosis on K^+ excretion. See text for details.

MULTIVALENT IONS

Ca^{++}, Mg^{++}, and inorganic phosphate $(PO_4^=)$* are multivalent ions that subserve many complex and vital functions. In a normal adult, the renal excretion of these ions is balanced by gastrointestinal absorption. If body stores decline substantially, gastrointestinal absorption, bone resorption, and renal tubular reabsorption increase and return body stores to normal levels. During growth and pregnancy, intestinal absorption exceeds urinary excretion, and these ions accumulate in newly formed fetal tissue and bone. In contrast, bone disease (e.g., osteoporosis) or a decline in lean body mass increases urinary multivalent ion loss without a change in intestinal absorption. In these conditions there is a net loss of Ca^{++}, Mg^{++}, and $PO_4^=$ from the body.

---

*At physiological pH inorganic phosphate exists as $HPO_4^=$ and $H_2PO_4^-$ (pK = 6.8). For simplicity we collectively refer to these ion species as $PO_4^=$.

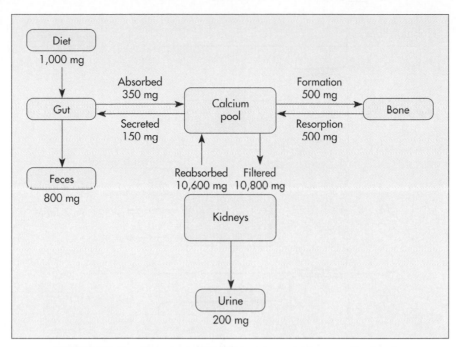

FIGURE 38-6   Overview of Ca^{++} homeostasis. See text for details.

This brief introduction reveals that *the kidneys, in conjunction with the gastrointestinal tract and bone, play a major role in maintaining Ca^{++}, Mg^{++} and $PO_4^=$ homeostasis.* Accordingly, this section will discuss Ca^{++}, Mg^{++}, and $PO_4^=$ handling by the kidneys, with an emphasis on the hormones and factors that regulate urinary excretion.

Calcium (Ca^{++})

Calcium ions play a major role in many processes, including bone formation, cell division and growth, blood coagulation, hormone-response coupling, and electrical stimulus-response coupling (such as muscle contraction and neurotransmitter release). Ninety-nine percent of Ca^{++} is stored in bone, 1% is found in the intracellular fluid, and 0.1% is located in the extracellular fluid. The total $[Ca^{++}]$ in plasma is about 2.5 mM, and its concentration is normally maintained within very narrow limits. A low $[Ca^{++}]$ increases the excitability of nerve and muscle cells. Diseases that lower plasma $[Ca^{++}]$ cause **hypocalcemic tetany** (intermimittent muscular contraction), which is characterized by skeletal muscle spasms that if severe can cause death by asphyxiation. Hypercalcemia can also be lethal by causing cardiac arrhythmia and decreased neuromuscular excitability.

Overview of Ca^{++} homeostasis   The maintenance of Ca^{++} homeostasis depends on (1) the total amount of Ca^{++} in the body and (2) the distribution of Ca^{++} between the intracellular and extracellular fluid compartments (see Chapter 43). Total body Ca^{++} is deter-

mined by the relative amounts of Ca^{++} absorbed by the gastrointestinal tract and excreted by the kidneys Figure (38-6). Ca^{++} absorption by the gastrointestinal tract occurs by an active, carrier-mediated transport mechanism that is stimulated by 1,25-dihydroxyvitamin D_3 (1,25$[OH]_2D_3$) (see Chapter 43). Net Ca^{++} absorption is normally 200 mg/day, but it can increase to 600 mg/day when 1,25$(OH)_2D_3$ levels rise. Ca^{++} excretion by the kidneys is equal to the amount absorbed by the gastrointestinal tract (200 mg/day), and it changes in parallel with reabsorption of Ca^{++} by the gastrointestinal tract. Thus, Ca^{++} balance is maintained because the amount of Ca^{++} ingested in an average diet (1000 mg/day) is equal to the amount lost in the feces (800 mg/day: the amount that escapes reabsorption by the gastrointestinal tract) plus the amount excreted in the urine (200 mg/day).

The second factor that controls Ca^{++} homeostasis is the distribution of Ca^{++} between bone and the extracellular fluid. Two hormones, parathyroid hormone (PTH) and 1,25$(OH)_2D_3$, are the most important hormones that control this factor and thereby regulate plasma $[Ca^{++}]$ (see Chapter 43). PTH is secreted by the parathyroid glands, and its secretion is stimulated by a decline in plasma $[Ca^{++}]$ (i.e., hypocalcemia). PTH increases plasma $[Ca^{++}]$ by:

1. Stimulating bone resorption
2. Increasing Ca^{++} reabsorption by the kidneys
3. Stimulating the production of 1,25$(OH)_2D_3$, which, in turn, increases Ca^{++} absorption by the gastrointestinal tract and stimulates bone resorption.

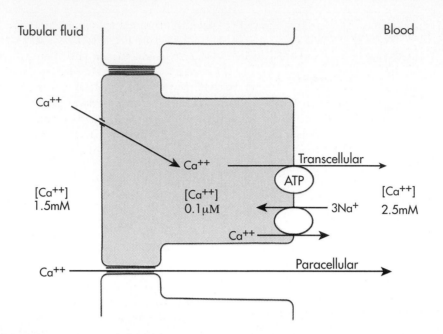

FIGURE 38-7   Cellular mechanisms of Ca^{++} reabsorption by the proximal tubule. Ca^{++} is reabsorbed by transcellular and paracellular routes. The mechanism of Ca^{++} diffusion into the cell across the apical membrane has not been characterized, but is likely to occur via Ca^{++} channels.

Hypercalcemia reduces PTH secretion, which leads to actions opposite to those described above. Thus, by controlling the amount of Ca^{++} in the body, the kidneys also regulate plasma $[Ca^{++}]$.

Ca^{++} transport along the nephron Approximately 50% of the Ca^{++} in plasma is ionized, 45% is bound to plasma proteins (mainly albumin), and 5% is complexed to several anions, including HCO_3^-, citrate, $PO_4^=$, and $SO_4^=$. The pH of the plasma influences this distribution. Acidosis increases the percentage of ionized calcium at the expense of Ca^{++} bound to proteins, whereas alkalosis decreases the percentage of ionized calcium, again by altering Ca^{++} bound to proteins. Thus, individuals with alkalosis are susceptible to tetany, whereas individuals with acidosis are less susceptible to tetany, even when total plasma Ca^{++} levels are reduced. Ca^{++} available for filtration consists of the ionized fraction and that complexed with anions. Thus, about 55% of the Ca^{++} in the plasma is available for glomerular filtration.

Normally, 99% of the filtered Ca^{++} is reabsorbed by the nephron. The proximal tubule reabsorbs 70% of the filtered Ca^{++}. Another 20% is reabsorbed in the loop of Henle (mainly the thick ascending limb), 5% to 10% is reabsorbed by the distal tubule, and <5% is reabsorbed by the collecting duct. About 1% (200 mg/day) is excreted in the urine. This fraction is equal to the net amount absorbed daily by the gastrointestinal tract.

Cellular mechanisms of Ca^{++} reabsorption Ca^{++}

reabsorption by the proximal tubule occurs by two pathways: transcellular and paracellular (Figure 38-7). Ca^{++} reabsorption across the cellular pathway (i.e., transcellular) accounts for one third of proximal reabsorption. Ca^{++} reabsorption through the cell is an active process, and it occurs in two steps. Ca^{++} diffuses across the apical membrane into the cell down its electrochemical gradient. This gradient is exceptionally steep, because the Ca^{++} concentration in the cell is only $0.1\mu M$, about 10,000-fold less than that in the tubular fluid (1.5 mM). The cell interior is electrically negative with respect to the luminal side of the apical membrane, and this also favors Ca^{++} entry into the cell. Ca^{++} is extruded across the basolateral membrane against its electrochemical gradient. The mechanism for the extrusion of Ca^{++} is a Ca^{++}-ATPase and a $3Na^+$-Ca^{++} antiporter.

Two thirds of Ca^{++} is reabsorbed between cells across the tight junctions (i.e., paracellular pathway). This passive paracellular reabsorption of Ca^{++} occurs by solvent drag along the entire length of the proximal tubule and is also driven by the positive luminal voltage in the second half of the proximal tubule (see Chapter 36).

Ca^{++} reabsorption by Henle's loop is restricted to the thick ascending limb. Ca^{++} is reabsorbed via a cellular and a paracellular route by mechanisms similar to those described for the proximal tubule, except that Ca^{++} is not reabsorbed by solvent drag in this segment (recall that the thick ascending limb is impermeable to water).

Table 38-2 Hormones and Factors Influencing Urinary Ca^{++} Excretion

Increase Excretion	Decrease Excretion
Decrease in [PTH]	Increase in [PTH]
ECF expansion	ECF contraction
$PO_4^=$ depletion	$PO_4^=$ loading
Metabolic acidosis	Metabolic alkalosis

In the thick ascending limb, Ca^{++} and Na^+ reabsorption parallel each other because of the significant component of Ca^{++} reabsorption that occurs by passive, paracellular mechanisms secondary to Na^+ reabsorption and by the presence of the lumen-positive transepithelial voltage (see Chapter 36). Therefore, *changes in Na^+ reabsorption will also result in parallel changes in Ca^{++} reabsorption by the proximal tubule and the thick ascending limb of Henle's loop.*

In the distal tubule and collecting duct, where the voltage in the tubule lumen is electrically negative with respect to the blood, Ca^{++} reabsorption is entirely active because Ca^{++} is reabsorbed against its electrochemical gradient. Ca^{++} reabsorption by these segments is transcellular, and the mechanism is similar to that in the proximal tubule and thick ascending limb: uptake across the apical membrane by a Ca^{++}-permeable ion channel and extrusion across the basolateral membrane by a Ca^{++}-ATPase and a $3Na^+$-Ca^{++} antiporter. *Because the reabsorption of Ca^{++} and Na^+ by the distal tubule and the collecting duct are independent and are differentially regulated, urinary Ca^{++} and Na^+ excretion do not always change in parallel.*

Regulation of Ca^{++} excretion Table 38-2 summarizes the hormones and factors that influence urinary Ca^{++} excretion. *Parathyroid hormone (PTH) exerts the most powerful control on renal Ca^{++} excretion and is responsible for maintaining Ca^{++} homeostasis.* Overall, this hormone stimulates Ca^{++} reabsorption by the kidneys. Although PTH inhibits the reabsorption of fluid, and therefore of Ca^{++} by the proximal tubule, it dramatically stimulates Ca^{++} reabsorption by the thick ascending limb of Henle's loop and the distal tubule. As a result, urinary Ca^{++} excretion declines. All of the other hormones and factors listed in Table 38-2 disturb Ca^{++} homeostasis. An increase in plasma $[PO_4^=]$ (e.g., increased dietary intake of $PO_4^=$ or ingestion of large amounts of $PO_4^=$-containing antacids) elevates PTH levels and thereby decreases Ca^{++} excretion, whereas a decline in plasma $[PO_4^=]$ (e.g., dietary $PO_4^=$ depletion) has the opposite effect.

Changes in the extracellular fluid volume alter Ca^{++} excretion mainly by affecting Na^+ and fluid reabsorption in the proximal tubule. Contraction of the extracellular fluid volume increases Na^+ and water reabsorption by the proximal tubule and thereby enhances Ca^{++} reabsorption. Accordingly, urinary Ca^{++} excretion declines. Expansion of the extracellular fluid volume has the opposite effect. Acidosis increases Ca^{++} excretion, whereas alkalosis decreases excretion. The regulation of Ca^{++} reabsorption by pH occurs in the distal tubule by an unknown mechanism.

Clinically, the most common causes of hypercalcemia are primary hyperparathyroidism and malignancy-associated hypercalcemia. **Primary hyperparathyriodism** results from the overproduction of PTH caused by a tumor of the parathyroid glands. In contrast, **malignancy-associated hypercalcemia** is caused by secretion of parathyroid hormone-related peptide (PTHRP), a PTH-like hormone secreted by carcinomas in a variety of organs. Increased levels of PTH and PTHRP cause hypercalcemia, hypercalcuria, and hypophosphatemia.

Magnesium (Mg^{++})

Mg^{++} is the second most abundant intracellular multivalent electrolyte, and it has many biochemical roles, including activation of enzymes and regulation of protein synthesis. It is also important for bone formation. Fifty-four percent of Mg^{++} is located in bone, 45% is located in the intracellular fluid, and 1% is located in the extracellular fluid. The plasma $[Mg^{++}]$ is about 1 mM. Approximately 30% is protein-bound and therefore unavailable for ultrafiltration by the glomerulus. The Mg^{++} that is filtered consists of an ionized fraction (55%) and a nonionized component (15%) that is complexed to HCO_3^-, citrate, $PO_4^=$, and $SO_4^=$. Accordingly, the Mg^{++} concentration in the glomerular ultrafiltrate is 30% less than the Mg^{++} concentration in the plasma.

Overview of Mg^{++} homeostasis A general scheme of Mg^{++} homeostasis is illustrated in Figure 38-8. As with Ca^{++}, the maintenance of $[Mg^{++}]$ in body fluids depends on two factors: (1) the total amount of Mg^{++} in the body and (2) the distribution of Mg^{++} between the intracellular and extracellular fluid compartments. Total body Mg^{++} is determined by the relative amount of net Mg^{++} absorption by the gastrointestinal tract and excretion by the kidneys. Body Mg^{++} balance is maintained by the kidneys because of their ability to excrete in the urine an amount of Mg^{++} equal to the amount absorbed by the gastrointestinal tract (i.e., 75 mg/day). Although the gastrointestinal absorption of Mg^{++} is not regulated as closely as that of Ca^{++}, when

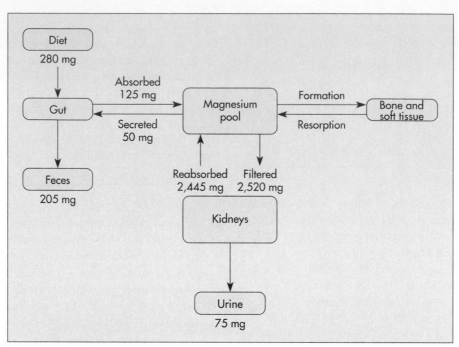

FIGURE 38-8   Overview of Mg^{++} homeostasis. See text for details.

dietary intake is restricted, intestinal absorption rises. The hormones and other factors responsible for the regulation of intestinal Mg^{++} absorption are unknown. The second factor that regulates Mg^{++} homeostasis is the distribution of Mg^{++} between the intracellular fluid and the extracellular fluid. The hormones and factors that regulate this distribution have not been identified.

Mg^{++} transport along the nephron The kidneys are vital in Mg^{++} homeostasis, and they excrete 3% of the filtered Mg^{++}. Approximately 25% of the filtered Mg^{++} is reabsorbed by the proximal tubule, and some 65% of the filtered Mg^{++} is reabsorbed in the loop of Henle, primarily by the thick ascending limb of Henle's loop. *Alterations in urinary Mg^{++} excretion usually arise from changes in Mg^{++} reabsorption by the thick ascending limb of Henle's loop.* Relatively little Mg^{++} is reabsorbed by the distal tubule and collecting duct (<7%). The mechanisms of Mg^{++} transport along the nephron are poorly understood. However, reabsorption appears to be passive and to be driven by the lumen-positive transepithelial voltage in the second half of the proximal tubule and the thick ascending limb.

Regulation of Mg^{++} excretion Table 38-3 summarizes the major hormones and other factors that influence Mg^{++} excretion. These factors influence urinary Mg^{++} excretion by altering Mg^{++} reabsorption by the thick ascending limb of Henle' loop. An increase in Mg^{++} excretion is caused by hypercalcemia, hypermagnesemia, extracellular fluid volume expansion, acidosis, or a decrease in plasma concentration of PTH, glucagon, calcitonin, or ADH. A decrease in excretion is caused by hypocalcemia, hypomagnesemia, extracellu-

Table 38-3   Hormones and Factors Influencing Urinary Mg^{++} Excretion

Increase Excretion	Decrease Excretion
Hypercalcemia	Hypocalcemia
Hypermagnesemia	Hypomagnesemia
ECF expansion	ECF contraction
Acidosis	Alkalosis
Decrease in [PTH], [calcitonin], [glucagon], or [ADH]	Increase in [PTH], [calcitonin], [glucagon], or [ADH]

lar volume contraction, alkalosis, or an increase in plasma concentration of PTH, glucagon, calcitonin, or ADH. All of these regulatory influences have a direct effect on Na^+ reabsorption and, thus, affect the transepithelial voltage across the thick ascending limb of Henle's loop. By virtue of changing the transepithelial voltage, these hormones and factors modulate Mg^{++} reabsorption across the paracellular pathway of the thick ascending limb.

Phosphate ($PO_4^=$)

$PO_4^=$ is an important component of many organic molecules, including DNA, RNA, ATP, and intermediates of metabolic pathways. It is also a major constituent of bone. Its concentration in plasma is an important determinant in bone formation and resorption. In addition, uri-

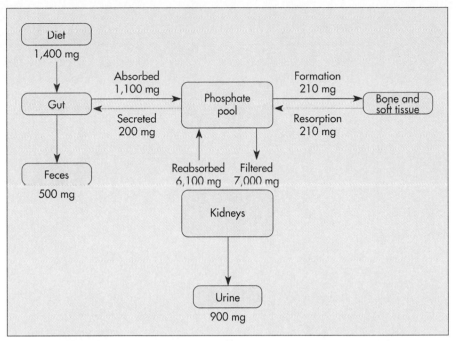

FIGURE 38-9   Overview of $PO_4^=$ homeostasis. See text for details.

nary $PO_4^=$ is an important buffer (titratable acid) for the maintenance of acid-base balance (see Chapter 39). Eighty-six percent of $PO_4^=$ is located in bone, 14% is located in the intracellular fluid, and 0.03% is located in the extracellular fluid. The plasma $[PO_4^=]$ is 1.3 mM. Approximately 10% of the $PO_4^=$ in the plasma is protein-bound and therefore unavailable for ultrafiltration by the glomerulus. Accordingly, the $[PO_4^=]$ in the ultrafiltrate is 10% less than that in plasma.

Overview of $PO_4^=$ homeostasis A general scheme of $PO_4^=$ homeostasis is shown in Figure 38-9. The maintenance of $PO_4^=$ homeostasis depends on two factors: (1) the amount of $PO_4^=$ in the body, and (2) the distribution of $PO_4^=$ between the intracellular fluid and the extracellular fluid compartments. Total body $PO_4^=$ is determined by the relative amount of $PO_4^=$ absorbed by the gastrointestinal tract versus the amount excreted by the kidneys. $PO_4^=$ absorption by the gastrointestinal tract occurs by active and by passive mechanisms; it increases as dietary $PO_4^=$ rises, and it is stimulated by $1,25(OH)_2D_3$. Despite changes in $PO_4^=$ intake between 800 and 1500 mg/day, total body $PO_4^=$ balance is maintained by the kidneys, which excrete an amount of $PO_4^=$ in the urine equal to the amount absorbed by the gastrointestinal tract. Thus, *the kidneys play a vital role in maintaining $PO_4^=$ homeostasis.*

The second factor that maintains $PO_4^=$ homeostasis is the distribution of $PO_4^=$ between bone and the intracellular fluid and extracellular fluid compartments. The release of $PO_4^=$ from intracellular stores is stimulated by the same hormones (PTH and $1,25[OH]_2D_3$) that release Ca^{++} from this pool. Thus, the release of $PO_4^=$ is always

accompanied by a release of Ca^{++}. The kidneys also contribute importantly to the regulation of plasma $[PO_4^=]$. This can be illustrated by considering Figure 38-10. A small increase in plasma $[PO_4^=]$ increases the filtered load of $PO_4^=$ such that the maximum transport rate (T_m) is exceeded and urinary $PO_4^=$ excretion increases above the amount absorbed by the gastrointestinal tract. This in turn causes plasma $[PO_4^=]$ to fall. Accordingly, *the kidneys regulate plasma $[PO_4^=]$.* The T_m for $PO_4^=$ is variable and is regulated by dietary $PO_4^=$ intake. A high $PO_4^=$ diet decreases the T_m, and a low $PO_4^=$ diet increases the T_m. This effect of dietary $PO_4^=$ intake on the T_m is independent of changes in PTH levels.

$PO_4^=$ transport along the nephron The proximal tubule reabsorbs 80% of the $PO_4^=$ filtered by the glomerulus, and the distal tubule reabsorbs 10%. In contrast, the loop of Henle and the collecting duct reabsorb negligible amounts of $PO_4^=$. Therefore, 10% of the filtered load of $PO_4^=$ is excreted.

$PO_4^=$ reabsorption by the proximal tubule occurs mainly, if not exclusively, by a transcellular route. As shown in Figure 38-11, $PO_4^=$ uptake across the apical membrane occurs by a $2Na^+$-$HPO_4^=$/$H_2PO_4^-$ symport mechanism. $HPO_4^=$/$H_2PO_4^-$ exits across the basolateral membrane, most likely by a $HPO_4^=$/$H_2PO_4^-$-anion antiporter. The cellular mechanism of $PO_4^=$ reabsorption by the distal tubule and collecting duct has not been characterized.

Regulation of $PO_4^=$ excretion Table 38-4 summarizes the major hormones and factors that regulate $PO_4^=$ excretion. All act on the proximal tubule and either stimulate or inhibit $PO_4^=$ reabsorption. *Parathyroid hor-*

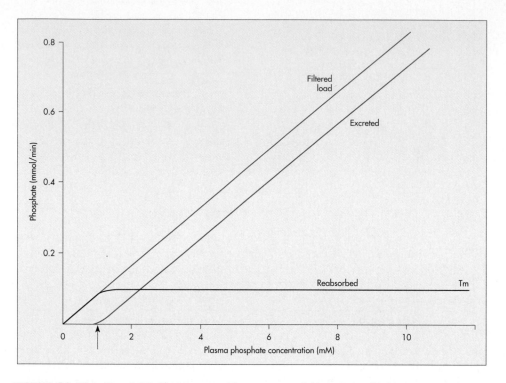

FIGURE 38-10 Renal $PO_4^=$ transport. The amount of filtered $PO_4^=$ that is reabsorbed is calculated as the difference between the amount filtered (i.e., filtered load) minus the amount excreted. The arrow pointing to the X axis indicates the normal plasma $[PO_4^=]$. T_m, Transport maximum.

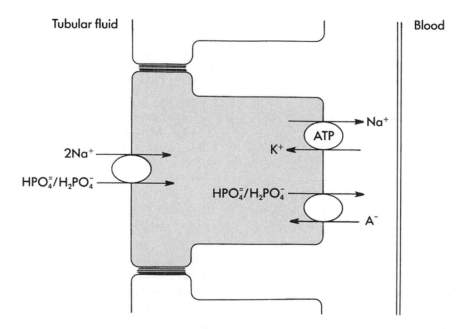

FIGURE 38-11 Cellular mechanism of $PO_4^=$ reabsorption by the proximal tubule. The apical transport pathway may operate primarily as a $2Na^+/HPO4^=$ symporter. $PO_4^=$ leaves the cell across the basolateral membrane by an $HPO_4^=/H_2PO_4^-$-anion antiporter and possibly by an $Na^+/PO_4^=$ symporter (not shown). A^- indicates an anion.

Table 38-4 Hormones and Factors Influencing Urinary $PO_4^=$ Excretion

Increase Excretion	Decrease Excretion
Increase of [PTH]	Decrease of [PTH]
$PO_4^=$ loading	$PO_4^=$ depletion
ECF expansion	ECF contraction
Glucocorticoids	
Acidosis	Alkalosis

mone (PTH) is the most important hormone that controls $PO_4^=$ excretion. PTH increases cAMP production and inhibits $PO_4^=$ reabsorption by the proximal tubule. Dietary $PO_4^=$ intake also regulates $PO_4^=$ excretion by mechanisms unrelated to changes in PTH levels. $PO_4^=$ loading increases excretion, whereas $PO_4^=$ depletion decreases excretion. Changes in dietary $PO_4^=$ intake modulate $PO_4^=$ transport by altering the transport rate of each $2Na^+\text{-}HPO_4^=/H_2PO_4^-$ symporter without changing the number of transporters.

Extracellular fluid volume also affects $PO_4^=$ excretion: volume expansion increases excretion, and volume contraction decreases excretion. Acid-base balance also influences $PO_4^=$ excretion: acidosis increases $PO_4^=$ excretion, and alkalosis decreases $PO_4^=$ excretion. Glucocorticoids increase the excretion of $PO_4^=$ (see also Chapter 46). Glucocorticoids increase the delivery of $PO_4^=$ to the distal tubule and collecting duct, by inhibiting proximal tubular $PO_4^=$ reabsorption. This inhibition enables the distal tubule and collecting duct to secrete more H^+, and to generate more HCO_3^-. $PO_4^=$ is an important urinary buffer, and as described in Chapter 39 is termed titratable acid.

In the absence of glucocorticoids (e.g., **Addison's disease**), $PO_4^=$ excretion is depressed, as is the ability of the kidneys to excrete titratable acid and to generate new HCO_3^-.

SUMMARY

Potassium

1. K^+ is one of the most abundant cations in the body and it is crucial for many cellular functions, including cell growth and division and the excitability of nerve and muscle.
2. K^+ homeostasis is maintained by hormones that regu-late plasma $[K^+]$, and by the kidneys, which adjust K^+ excretion to match dietary K^+ intake. Plasma $[K^+]$ is maintained by insulin, epinephrine, and aldosterone. In contrast, cell lysis and exercise and changes in acid-base balance and in plasma osmolality perturb plasma $[K^+]$.
3. K^+ excretion by the kidneys is determined by the rate of K^+ secretion by the distal tubule and collecting duct. K^+ secretion by these tubular segments is regulated by plasma $[K^+]$, aldosterone, and ADH. In contrast, changes in tubular fluid flow and acid-base disturbance perturb K^+ excretion by the kidneys.

Multivalent Ions

1. Ca^{++}, Mg^{++}, and inorganic phosphate ($PO_4^=$) are multivalent ions that subserve many vital functions. The kidneys, in conjunction with the gastrointestinal tract and bone, play a vital role in regulating Ca^{++}, Mg^{++}, and $PO_4^=$ homeostasis.
2. Plasma Ca^{++} is regulated by parathyroid hormone and $1,25(OH)_2D_3$. Ca^{++} excretion by the kidneys is determined by the rate of Ca^{++} reabsorption by the distal tubule and thick ascending limb (TAL) of Henle's loop. Ca^{++} reabsorption by the TAL is regulated mainly by PTH, which stimulates Ca^{++} reabsorption.
3. Although Mg^{++} has many important biological roles, relatively little is known about the factors that regulate plasma $[Mg^{++}]$ and urinary Mg^{++} excretion.
4. Plasma $[PO_4^=]$ is regulated by PTH and $1,25(OH)_2D_3$. A fall in $[PO_4^=]$ indirectly inhibits release of PTH and directly stimulates the production of $1,25(OH)_2D_3$. This decreases renal $PO_4^=$ excretion and stimulates intestinal $PO_4^=$ absorption. $PO_4^=$ excretion by the kidneys is determined mainly by the rate of reabsorption in the proximal tubule. PTH inhibits $PO_4^=$ reabsorption by the proximal tubule and enhances urinary $PO_4^=$ excretion. The opposite occurs with an increase in plasma $[PO_4^=]$.

BIBLIOGRAPHY
Journal Articles

De Rouffignac C, Quamme G: Renal magnesium handling and its hormonal control, *Physiol Rev* 74:305-322, 1994.

Friedman PA: Renal calcium transport: sites and insights, *News Physiol Sci* 3:17, 1988.

Wingo CS, Cain BD: The renal H-K-ATPase: Physiological significance and role in potassium homeostasis, *Ann Rev Physiol* 55:323, 1993.

Books and Monographs

Berndt TJ, Knox FG: Renal regulation of phosphate excretion. In Seldin DW, Giebisch G, eds: *The kidney: physiology and pathophysiology,* ed 2, New York, 1992, Raven Press.

Costanzo LS, Windhager EE: Renal regulation of calcium balance. In Seldin DW, Giebisch G, eds: *The kidney: physiology and pathophysiology,* ed 2, New York, 1992, Raven Press.

Murer H, Biber J: Renal tubular phosphate transport: cellular mechanisms. In Seldin DW, Giebisch G, eds: *The kidney: physiology and pathophysiology,* ed 2, New York, 1992, Raven Press.

Quamme GA: Magnesium: cellular and renal exchanges. In Seldin DW, Giebisch G, eds: *The kidney: physiology and pathophysiology,* ed 2, New York, 1992, Raven Press.

Stanton BA, Giebisch G: Renal potassium transport. In

Windhager EE, ed: *Handbook of physiology: renal physiology,* ed 2, New York, 1991, Oxford University Press.

Stewart AF: Hypercalcemic and hypocalcemic states. In Seldin DW, Giebisch G, eds: *The kidney: physiology and pathophysiology,* ed 2, New York, 1992, Raven Press.

Suki WN, Rouse D: Renal transport of calcium, magnesium. In Brenner BM, Rector FC Jr, eds: *The kidney,* ed 4, Philadelphia,1991, WB Saunders Co.

Wright FS, Giebisch G. Regulation of potassium excretion. In Seldin DW, Giebisch G, eds: *The kidney: physiology and pathophysiology,* ed 2 New York, 1992, Raven Press.

Role of the Kidney in Acid-Base Balance

The concentration of H^+ in the body fluids is low compared to that of other vital ions. For example, Na^+ is present at a concentration approximately 1 million times greater than that of H^+ ($[Na^+] = 140$ mEq/L; $[H^+] = 40$ nEq/L). Because of its low concentration in the body fluids, $[H^+]$ is commonly expressed as the negative logarithm, or pH. The normal pH of the body fluids is between 7.35 and 7.45.

Acid-base balance is accomplished through the coordinated functions of the liver, lungs, and kidneys. Through metabolism, the liver adds various acid and base equivalents to the ECF (other cells, such as skeletal muscle, can also add acid to the body via metabolism, especially under anaerobic conditions). In a normal adult the amount of acid or base added to the body fluids by metabolism is determined primarily by the diet.

With a normal diet, the body produces approximately 15 to 20 moles of acid per day. Virtually all of this acid is derived from CO_2 ($CO_2 + H_2O \leftrightarrow H_2CO_3$) and is therefore called volatile acid. The lungs handle this potential acid load by excreting CO_2 (see Chapter 30). In addition, the metabolism of food produces acids and alkali that cannot be excreted by the lungs. In people on a typical meat-containing diet, acid production exceeds alkali production. This acid, termed nonvolatile acid, is produced at a small fraction of the rate of volatile acid production (50-100 mmole/day vs. 15 to 20 moles/day). The kidneys, together with various buffers, play an important role in the handling of nonvolatile acid; they minimize the effect on body fluid pH and maintain a steady-state balance.

Many of the metabolic functions of the body are exquisitely sensitive to pH, and normal function can occur only over a very narrow pH range. The pH range that is generally compatible with life is 6.8 to 7.8 (160-16 nEq/L of H^+). This chapter discusses the renal and buffer mechanisms involved in the maintenance of acid-base balance. The role of the lungs is described in Chapter 30.

THE CO_2/HCO_3^- BUFFER SYSTEM

Bicarbonate (HCO_3^-) is the primary buffer of the ECF. Given a plasma $[HCO_3^-]$ of 23 to 25 mEq/L, and a volume of 14 liters, the ECF can potentially buffer 350 mEq of H^+. The CO_2/HCO_3^- buffer system differs from the other buffer systems of the body because it is regulated by both the lungs and the kidneys. This is best appreciated by considering the following reaction:

$$CO_2 + H_2O \overset{CA}{\leftrightarrow} H_2CO_3 \leftrightarrow H^+ + HCO_3^- \qquad (1)$$

The first reaction (hydration/dehydration of CO_2) is the rate-limiting step. This reaction, which is normally slow, is greatly accelerated in the presence of the enzyme carbonic anhydrase (CA). The ionization of H_2CO_3 to H^+ and HCO_3^- is virtually instantaneous.

The dissociation constant for the previous reaction can be written as:

$$K' = \frac{[H^+]\,[HCO_3^-]}{[CO_2]\,[H_2CO_3]} \qquad (2)$$

Because the reaction includes not only the hydration and dehydration of CO_2 but also the ionization of H_2CO_3, K' is not a true dissociation constant. For plasma at $37°$ C, this apparent dissociation constant equals $10^{-6.1}$ ($pK' = 6.1$).

The terms in the denominator of equation 2 represent the total amount of CO_2 dissolved in solution. Most of this CO_2 is in the gaseous form, with only 0.3% being H_2CO_3. Because the amount of CO_2 in solution depends on its partial pressure (P_{CO_2}) and on its solubility (α), equation 2 can be rewritten as:

$$K' = \frac{[H^+]\,[HCO_3^-]}{\alpha P_{CO_2}} \qquad (3)$$

For plasma at $37°$ C, $\alpha = 0.03$.

A more useful form of this equation is obtained by solving for $[H^+]$:

$$[H^+] = \frac{K'\alpha P_{CO_2}}{[HCO_3^-]} \qquad \textbf{(4)}$$

Taking the negative logarithm of both sides of this equation yields:

$$-\log[H^+] = \frac{-\log[K'] + -\log\alpha P_{CO_2}}{-\log[HCO_3^-]} \qquad \textbf{(5)}$$

$$pH = pK' + \log\frac{[HCO_3^-]}{\alpha P_{CO_2}} \text{ or} \qquad \textbf{(6)}$$

$$pH = 6.1 + \log\frac{[HCO_3^-]}{0.03 P_{CO_2}}$$

Equation 6 is the Henderson-Hasselbalch equation. Inspection of this equation clearly shows that the pH varies when either the $[HCO_3^-]$ or the P_{CO_2} is altered. Disturbances of acid-base balance that result from a change in the $[HCO_3^-]$ of the ECF are termed **metabolic acid-base disorders,** whereas those resulting from a change in the P_{CO_2} are termed **respiratory acid-base disorders.** As discussed shortly, the kidneys mainly control the $[HCO_3^-]$, whereas the lungs control the P_{CO_2}.

PRODUCTION OF NONVOLATILE ACID

As already mentioned, a tremendous amount of potential acid is produced each day in the form of CO_2. This is generated from the metabolism of carbohydrates and fats. In addition, the metabolism of certain other foods produces nonvolatile acids. The bulk of these nonvolatile acids is produced from the metabolism of certain amino acids. The sulfur-containing amino acids, cysteine and methionine, yield sulphuric acid when metabolized, whereas hydrochloric acid results from the metabolism of lysine, arginine, and histidine. In addition to the nonvolatile acid produced by these amino acids, the presence of phosphate $(H_2PO_4^-)$ in ingested food increases the dietary acid load.

A portion of this nonvolatile acid load is offset by the production of HCO_3^- through the metabolism of the amino acids aspartate and glutamate as well as by metabolism of certain organic anions (e.g., citrate). On balance, in individuals ingesting a meat-containing diet, acid production exceeds HCO_3^- production. During the process of digestion some HCO_3^- is normally lost in the feces. This HCO_3^- loss is equivalent to the addition of nonvolatile acid to the body. Together, dietary intake, cellular metabolism, and fecal HCO_3^- loss result in the addition of approximately 1 mEq/kg body weight of nonvolatile acid to the body each day (50 to 100 mEq/day for most adults).

Nonvolatile acids do not circulate throughout the body but are immediately buffered. For example:

$$H_2SO_4 + 2NaHCO_3 \leftrightarrow Na_2SO_4 + 2CO_2 + 2H_2O \qquad \textbf{(7)}$$

$$HCl + NaHCO_3 \leftrightarrow NaCl + CO_2 + H_2O \qquad \textbf{(8)}$$

This titration process yields the sodium salts of the strong acid, and removes HCO_3^- from the ECF. The kidneys must excrete these sodium salts and replenish the HCO_3^- lost by titration.

When insulin levels are normal, carbohydrates and fats are completely metabolized to $CO_2 + H_2O$. However, if insulin levels are abnormally low (e.g., **diabetes mellitus**), the metabolism of carbohydrates leads to the production of several organic ketoacids (e.g., β-hydroxybutyric acid).

Anaerobic metabolism by cells can also lead to the production of organic acids (e.g., lactic acid) rather than $CO_2 + H_2O$. This frequently occurs in normal individuals during vigorous exercise. Clinically, poor tissue perfusion, as can occur with reduced cardiac output, can also lead to anaerobic metabolism by cells, and thus to acidosis.

In all of the above-mentioned conditions, the pH of the body fluids will decrease (metabolic acidosis). Treatment, by the administration of insulin in the case of diabetes, or improved delivery of adequate levels of O_2 to the tissues in the case of poor tissue perfusion, will result in the further metabolim of these organic acids to $CO_2 + H_2O$, and thereby help to correct the acid-base disorder.

RENAL ACID EXCRETION

To maintain acid-base balance, the kidneys must excrete an amount of acid equal to the nonvolatile acid production. In addition, they must prevent the loss of HCO_3^- in the urine. This latter task is quantitatively more important, because the filtered load of HCO_3^- is approximately 4320 mEq/day (24 mEq/L $\times$ 180 L/day), compared to the 70 mEq/day of nonvolatile acid excretion. *Both the reabsorption of the filtered load of HCO_3^- and the excretion of acid are accomplished through the process of H^+ secretion by the nephrons.* Thus in a single day the nephrons secrete approximately 4390 mEq of H^+. Most of this H^+ is not excreted in the urine, but it serves to reclaim the filtered load of HCO_3^-. Only 70 mEq/day of H^+ are excreted. As a result of this acid excretion, the urine is normally acidic (i.e., pH < 7.0).

Theoretically the kidneys could excrete the nonvolatile acids and replenish the HCO_3^- lost during their titration by reversing the reactions shown in equations 7 and 8. However, because the pK's of these acids are so low, this process would require a urine pH of 1.0. The minimal urine pH attainable by the kidneys is only 4.0 to 4.5, therefore the kidneys must instead excrete the sodium salts of these acids (e.g., Na_2SO_4, $NaCl$), while the H^+ is excreted with other buffers. The two major urinary buffers are ammonia (NH_3/NH_4^+) and phosphate ($HPO_4^=/H_2PO_4^-$). The urinary phosphate and some other buffer species (e.g., creatinine) are collectively termed **titratable acid.** The overall process of acid excretion by the kidney can be quantitated as follows:

Net acid excretion
$$\text{rate} = (U_{NH_4^+} \times \dot{V}) + (U_{TA} \times \dot{V}) - (U_{HCO_3^-} \times \dot{V}) \quad \textbf{(9)}$$

where $U_{NH_4^+} \times \dot{V}$ and $U_{TA} \times \dot{V}$ are the rates of H^+ excretion (mEq/day) as NH_4^+ and titrable acid (TA), and

$U_{HCO_3^-} \times \dot{V}$ is the amount of HCO_3^- lost in the urine (equivalent to adding H^+ to the body). To maintain acid-base balance, the net acid excretion must equal the nonvolatile acid production.

BICARBONATE REABSORPTION ALONG THE NEPHRON

Glomerular filtration delivers 4320 mEq/day of HCO_3^- to the proximal tubule. Approximately 85% of this HCO_3^- is reabsorbed by this segment. The general processes by which this reabsorption of HCO_3^- occurs is illustrated in Figure 39-1 (see also Chapter 36):

1. The apical membrane of the proximal tubule cell contains an Na^+-H^+ antiporter that secretes H^+ into the tubular fluid, using the energy from the lumen-to-cell Na^+ gradient; this antiporter secretes H^+ into the tubule lumen.

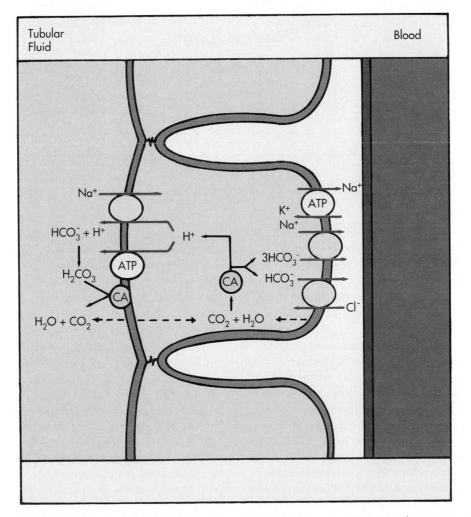

FIGURE 39-1   Cellular mechanism for H^+ secretion by the proximal tubule. H^+ secretion at this site results in reabsorption of 85% of the filtered load of HCO_3^-. See text for details. *CA,* Carbonic anhydrase.

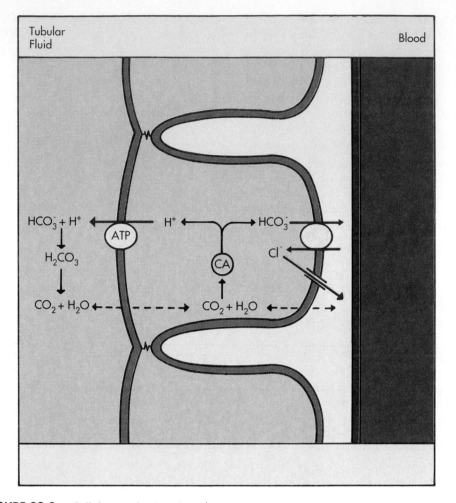

FIGURE 39-2 Cellular mechanism for H^+ secretion by the intercalated cell of the distal tubule and collecting duct. See text for details. *CA*, Carbonic anhydrase.

2. A small portion of H^+ secretion is also mediated by an H^+-ATPase.

3. Within the cell H^+ and HCO_3^- are produced in a reaction catalyzed by carbonic anhydrase (equation 1). The H^+ is secreted into the tubule lumen, whereas the HCO_3^- exits the cell across the basolateral membrane.

4. Although the electrochemical gradient for HCO_3^- would allow passive exit from the cell across the basolateral membrane, HCO_3^- movement is coupled to that of other ions (e.g., Na^+-$3HCO_3^-$ symporter and Cl^-/HCO_3^- antiporter).

5. Within the tubule lumen the secreted H^+ combines with the filtered HCO_3^- to form H_2CO_3. This is rapidly converted to CO_2 and H_2O by the carbonic anhydrase present in the apical membrane of the cell. Because the tubule is highly permeable to both CO_2 and H_2O, these substances are rapidly reabsorbed. The net effect of this process is that for each HCO_3^- removed from the tubule lumen, one HCO_3^- appears in the peritubular blood.

An additional 10% of the filtered load of HCO_3^- is reabsorbed by Henle's loop. Most of this is reabsorbed by the thick ascending limb, which, as with the proximal tubule, has an Na^+-H^+ antiporter in its apical membrane, and an Na^+-$3HCO_3^-$ symporter in the basolateral membrane to effect the net reabsorption of tubular fluid HCO_3^-.

The distal tubule and collecting duct reabsorb the small amount of HCO_3^- that escapes reabsorption by the proximal tubule and Henle's loop ($\approx$5% of the filtered load). This reabsorption is accomplished by an H^+-ATPase, and does not depend on Na^+.

Another difference between the collecting duct and the more proximal nephron segments is that not all cells of the collecting duct are involved in H^+ secretion. Only the intercalated cells secrete H^+. The mechanism by which this occurs is illustrated in Figure 39-2 (see also Chapter 36):

1. Within the intercalated cell, H^+ and HCO_3^- are produced by the hydration of CO_2. This reaction is catalyzed by carbonic anhydrase.

Table 39-1 Factors Regulating H$^+$ Secretion by the Nephron

Factor	Nephron Site of Action
INCREASING H$^+$ SECRETION	
Increase in filtered load of HCO$_3^-$	Proximal tubule
Decrease in ECF volume	Proximal tubule
Decrease in plasma [HCO$_3^-$] ($\downarrow$ pH)	Proximal tubule, thick ascending limb and collecting duct
Increase in blood P$_{CO_2}$	Proximal tubule, thick ascending limb and collecting duct
Aldosterone	Collecting duct
DECREASING H$^+$ SECRETION	
Decrease in filtered load of HCO$_3^-$	Proximal tubule
Increase in ECF volume	Proximal tubule
Increase in plasma [HCO$_3^-$] ($\uparrow$ pH)	Proximal tubule, thick ascending limb and collecting duct
Decrease in blood P$_{CO_2}$	Proximal tubule, thick ascending limb and collecting duct

2. The H$^+$ is secreted into the tubule lumen by the H$^+$-ATPase.
3. The HCO$_3^-$ exits the cells across the basolateral membrane via a Cl$^-$/HCO$_3^-$ antiporter.

A second population of intercalated cells exists within the collecting duct; these cells secrete HCO$_3^-$ rather than H$^+$ into the tubular fluid, and appear to have the H$^+$-ATPase located in the basolateral membrane and a Cl$^-$/HCO$_3^-$ antiporter in the apical membrane. Their activity can be increased during metabolic alkalosis when the kidneys must excrete excess HCO$_3^-$. However, under normal conditions H$^+$ secretion predominates in the collecting duct.

The tubule cells of the collecting duct are highly impermeable to H$^+$; thus the pH of the luminal fluid can be rendered quite acidic. Indeed, the most acidic fluid along the nephron (pH = 4.0 to 4.5) is produced at this site. In comparison, the permeability of the proximal tubule to H$^+$ and HCO$_3^-$ is much higher, and thus prevents the establishment of a significant tubular fluid-to-blood pH gradient; the minimum pH of proximal tubular fluid is 6.5.

REGULATION OF BICARBONATE REABSORPTION

HCO$_3^-$ reabsorption is regulated by several factors that act at both the proximal tubule and the collecting duct (Table 39-1). Because of the phenomenon of glomerulotubular balance (see Chapter 36), any change in the filtered load of HCO$_3^-$ is matched by an appropriate change in HCO$_3^-$ reabsorption by the proximal tubule. Recall that the bulk of proximal tubule HCO$_3^-$ reabsorption occurs by the Na$^+$-H$^+$ antiporter. Consequently, factors that affect Na$^+$ reabsorption will alter HCO$_3^-$ reabsorption secondarily. Thus expansion of the ECF volume inhibits HCO$_3^-$ reabsorption, and the opposite occurs when the ECF volume is decreased.

Changes in systemic acid-base balance also affect HCO$_3^-$ absorption in the proximal tubule. Systemic acidosis, whether produced by a decrease in the plasma bicarbonate concentration (metabolic) or by an increase in the plasma P$_{CO_2}$ (respiratory), stimulates H$^+$ secretion by the proximal tubule cells. This stimulation is the result of acidification of the intracellular fluid (ICF) of the tubule cell. When the ICF is acidic, a more favorable cell-to-lumen H$^+$ gradient exists, and H$^+$ secretion is stimulated. Conversely, metabolic and respiratory alkalosis inhibit H$^+$ secretion in the proximal tubule.

The reabsorption of HCO$_3^-$ by the thick ascending limb and collecting duct is also modulated by systemic acid-base balance; acidosis stimulates and alkalosis inhibits this process. An additional factor that regulates HCO$_3^-$ reabsorption in the collecting duct is aldosterone (see also Chapters 36 and 37). When aldosterone levels are elevated, H$^+$ secretion by the intercalated cell is stimulated. Conversely, when aldosterone levels are reduced, H$^+$ secretion is decreased. As already noted, during metabolic alkalosis HCO$_3^-$ secretion occurs by a subset of collecting duct intercalated cells.

FORMATION OF NEW BICARBONATE

As discussed previously, the reabsorption of the filtered load of HCO$_3^-$ is important for the maintenance of acid-base balance. HCO$_3^-$ loss in the urine would decrease the plasma [HCO$_3^-$], and thus it would be equivalent to the addition of acid to the body. However, HCO$_3^-$ reabsorption alone does not replenish the HCO$_3^-$ that was lost during the titration of the nonvolatile acids. For acid-base balance to be maintained, the kidneys must replace this lost HCO$_3^-$. The production of new HCO$_3^-$ is critically dependent on the availability of urinary buffers. Figure 39-3 illustrates how the titration of urinary

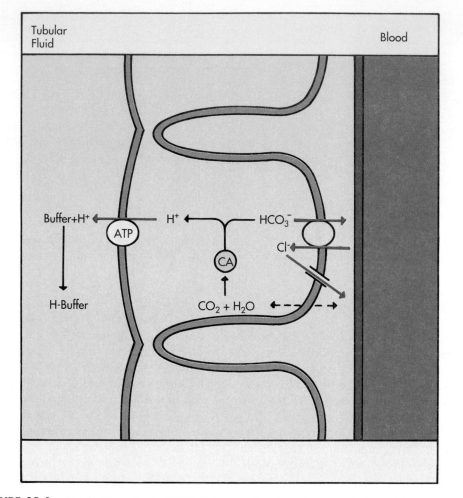

FIGURE 39-3   Production of new HCO_3^- by the titration of urinary buffers. The primary urinary buffers are $HPO_4^=/H_2PO_4^-$ (titratable acid) and NH_3/NH_4^+. *CA,* Carbonic anhydrase.

buffers results in the formation of new HCO_3^-. In the collecting duct when tubular fluid is HCO_3^- free because of HCO_3^- reabsorption in upstream tubular segments, H^+ secreted into tubular fluid combines with a urinary buffer. Thus, H^+ secretion results in the excretion of the H^+ with a buffer, and the HCO_3^- produced in the cell from the hydration of CO_2 is added back to the blood.

As mentioned, the two major urinary buffers are ammonia (NH_3/NH_4^+) and phosphate ($HPO_4^=/H_2PO_4^-$). Phosphate is derived from the diet. The amount excreted as titratable acid, therefore, depends on the filtered load minus the amount reabsorbed by the nephron. *Ammonia is produced by the kidney, and its synthesis and subsequent excretion can be regulated in response to the acid-base requirements of the body. Because of this, ammonia is the more important urinary buffer.*

Ammonia is produced in the kidneys by the metabolism of glutamine. The kidneys receive the glutamine (primarily from the liver), and the Na^+ salts of the nonvolatile acids, in the renal arterial plasma. The kidneys

metabolize the glutamine, excrete NH_4^+ with the acid salts, and return $NaHCO_3$ to the body in the renal vein plasma. The details of the steps involved in this process are illustrated in Figure 39-4 and considered below.

Glutamine is metabolized primarily in the cells of the proximal tubule. Each molecule of glutamine produces two molecules of NH_4^+ and a divalent anion. Ultimately it is the complete metabolism of this divalent anion that provides two molecules of new bicarbonate.

$$Glutamine \leftrightarrow 2NH_4^+ + Anion^= \leftrightarrow 2HCO_3^- + 2NH_4^+ \quad \textbf{(10)}$$

As already noted, the NH_4^+ is excreted into the urine, whereas the HCO_3^- is returned to the ECF to replace that which was lost during titration of the nonvolatile acids.

Figure 39-4 summarizes the mechanisms by which NH_3/NH_4^+ is handled by the various segments of the nephron:

1. NH_4^+ produced in the cells of the proximal tubule is secreted into the tubular fluid mainly by substituting for H^+ on the Na^+-H^+ antiporter.
2. A large portion of this NH_4^+ is reabsorbed by the

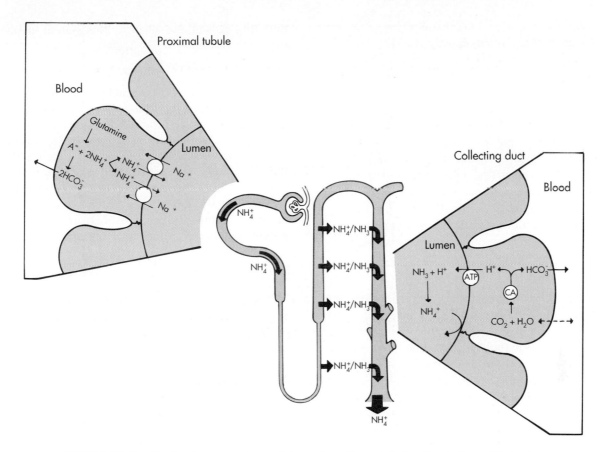

FIGURE 39-4   Production, transport, and excretion of ammonia by the nephron. Glutamine is metabolized in proximal tubule cells to NH_4^+ and HCO_3^-. The NH_4^+ is secreted into the lumen, and the HCO_3 enters the blood. The secreted NH_4^+ is reabsorbed in Henle's loop primarily by the thick ascending limb, and accumulates in the medullary interstitium by countercurrent multiplication, where it exists as both NH_4^+ and NH_3. NH_3 diffuses into the tubular fluid of the collecting duct, and H^+ secretion by the collecting duct leads to accumulation of NH_4^+ in the lumen by the process of nonionic diffusion and diffusion trapping.

thick ascending limb and accumulates in the interstitial fluid of the renal medulla, where it exists in equilibrium with NH_3. The countercurrent flow in both the loops of Henle and the vasa recta facilitate this accumulation of NH_4^+/NH_3 in the medullary interstitium (see also countercurrent multiplication in Chapter 37).

3. The NH_3 diffuses from the medullary interstitium into the lumen of the collecting duct **(nonionic diffusion),** where it is protonated to NH_4^+ by the acidic tubular fluid. Because the collecting duct is less permeable to NH_4^+ than to NH_3, NH_4^+ is trapped within the lumen of the tubule **(diffusion trapping)** and eliminated from the body in the urine.

As already indicated, an important feature of the ammonia buffer system is that it can be regulated. With systemic acidosis, the enzymes responsible for the metabolism of glutamine are stimulated. This stimulation involves the synthesis of more enzyme, a step that requires

several days. Because of this adaptive response, the kidneys can increase the excretion of H^+ and produce more new bicarbonate to defend against the acidosis.

Renal tubule acidosis (RTA) refers to conditions in which urine acidification is impaired. Under these conditions the kidneys are unable to excrete a sufficient amount of net acid to balance nonvolatile acid production, and metabolic acidosis results. RTA can occur by a defect either in proximal tubule H^+ secretion /HCO_3^- reabsorption **(proximal RTA)** or in distal tubule H^+ secretion **(distal RTA).**

Proximal RTA can be caused by a variety of hereditary and acquired conditions (e.g., **cystinosis, Fanconi syndrome,** administration of carbonic anhydrase inhibitors). H^+ secretion by proximal tubule cells is impaired and results in a decrease in the reabsorption of the filtered load of HCO_3^-. Consequently, HCO_3^- is lost in the urine, plasma $[HCO_3^-]$ decreases, and metabolic acidosis ensues.

Distal RTA also occurs in a number of hereditary and acquired conditions (e.g., **medullary sponge kidney,** certain drugs such as **amphotericin,** and secondary to urinary obstruction). Depending on the cause, secretion of H^+ by collecting duct intercalated cells is impaired or the permeability of the collecting duct to H^+ is increased. In either case, the ability to acidify the tubular fluid is impaired. Consequently, trapping of NH_4^+ is reduced. This in turn decreases net acid excretion, with the subsequent development of metabolic acidosis.

RESPONSE TO ACID-BASE DISORDERS

The pH of the body fluids is maintained within a very narrow range (7.35 to 7.45). Acidosis exists when the blood pH falls below this range, whereas alkalosis exists when the blood pH exceeds this range. When the acid-base disorder results from a primary change in the $[HCO_3^-]$, it is termed a metabolic disorder. When the primary disturbance is an alteration in the blood P_{CO_2}, it is termed a respiratory disorder.

When an acid-base disturbance develops, the body employs a series of mechanisms to defend against the change in the pH of the ECF. It is important to recognize that these defense mechanisms do not correct the acid-base disturbance, but merely minimize the change in pH imposed by the disturbance. Restoration of the blood pH to its normal value requires correction of the underlying process or processes that produced the acid-base disorder. The body has three general mechanisms to defend against changes in body fluid pH produced by acid-base disturbances. These mechanisms are (1) extracellular and intracellular buffering, (2) adjustments in blood P_{CO_2} by alterations in the ventilatory rate of the lungs, and (3) adjustments in renal acid excretion.

Extracellular and Intracellular Buffering

The first line of defense to acid-base disorders is extracellular and intracellular buffering. The response of the extracellular buffers is virtually instantaneous, whereas cellular buffering is somewhat slower and can take several minutes to reach completion.

Metabolic disorders that result from the addition of nonvolatile acids or alkali to the body fluids are buffered in both the extracellular and intracellular fluids. The CO_2/HCO_3^- buffer system is the primary ECF buffer.

Buffering of acid and alkali by this system involves the following reaction:

$$H^+ + HCO_3^- \leftrightarrow H_2CO_3 \leftrightarrow H_2O + CO_2 \quad \textbf{(11)}$$

When nonvolatile acid is added to the body fluids or alkali is lost from the body, the above reaction is driven to the right, HCO_3^- is consumed during the process of buffering the acid load, and the plasma $[HCO_3^-]$ is reduced. Conversely, when nonvolatile alkali is added to the body fluids or acid is lost from the body, the reaction is driven to the left. As a consequence, $[HCO_3^-]$ increases.

Although the CO_2/HCO_3^- buffer system is the primary extracellular buffer, phosphate and plasma protein provide additional extracellular buffering.

$$H^+ + HPO_4^= \leftrightarrow H_2PO_4^- \quad \textbf{(12)}$$
$$H^+ + Protein^- \leftrightarrow H\text{-}Protein$$

The combined action of the CO_2/HCO_3^-, phosphate, and plasma protein buffering processes accounts for 50% of the buffering of a nonvolatile acid load and 70% of a nonvolatile alkali load. The remainder of the buffering under these two conditions occurs intracellularly. Intracellular buffering involves the movement of H^+ into cells (during buffering of nonvolatile acid), or the movement of H^+ out of cells (during buffering of nonvolatile alkali). H^+ is titrated inside the cell by HCO_3^-, phosphate, and the histidine groups on protein.

Bone represents an additional source of extracellular buffer ($NaHCO_3$, $KHCO_3$, $CaCO_3$, $CaHPO_4$). With chronic acidosis, buffering by bone results in demineralization (i.e., Ca^{++} is released from bone as Ca^{++}-containing buffers bind H^+).

With respiratory acid-base disorders, body fluid pH changes as a result of alterations in $[H_2CO_3]$, which is determined directly by the P_{CO_2} (see equation 11). Virtually, all ($\sim$99%) buffering in respiratory acid-base disorders occurs intracellularly. When P_{CO_2} rises (respiratory acidosis), CO_2 moves into the cell, where it combines with H_2O to form H_2CO_3. This dissociates to H^+ and HCO_3^-. The H^+ is buffered by cellular proteins, and HCO_3^- exits the cell and raises the plasma $[HCO_3^-]$.

This process is reversed when P_{CO_2} is reduced (respiratory alkalosis). Under this condition, the hydration reaction ($H_2O + CO_2 \leftrightarrow H_2CO_3$) is shifted to the left by the decrease in P_{CO_2}. This in turn shifts the dissociation reaction ($H_2CO_3 \leftrightarrow H^+ + HCO_3^-$) to the left, thereby reducing the plasma $[HCO_3^-]$.

Respiratory Defense

The lungs are the second line of defense against acid-base disorders. As indicated by the Henderson-Hasselbalch equation, changes in the P_{CO_2} will alter the

Table 39-2 Mechanisms of Defense Against Acid-Base Disturbances

Disorder	Plasma pH	Primary Alteration	Defense Mechanisms
Metabolic acidosis	↓	↓ Plasma $[HCO_3^-]$	ICF and ECF buffers; hyperventilation (↓ P_{CO_2})
Metabolic alkalosis	↑	↑ Plasma $[HCO_3^-]$	ICF and ECF buffers; hypoventilation (↑ P_{CO_2})
Respiratory acidosis	↓	↑ P_{CO_2}	ICF buffers; ↑ renal H^+ excretion
Respiratory alkalosis	↑	↓ P_{CO_2}	ICF buffers; ↓ renal H^+ excretion

blood pH; an increase in P_{CO_2} decreases pH, and a decrease in P_{CO_2} increases pH.

The ventilatory rate determines the P_{CO_2} (see also Chapters 30 and 31). Increased ventilation decreases P_{CO_2}, whereas P_{CO_2} increases with decreased ventilation. The blood P_{CO_2} and pH are important regulators of the ventilatory rate. Chemoreceptors located in the brain (ventral surface of medulla) and in the periphery (carotid and aortic bodies) sense changes in P_{CO_2} and $[H^+]$, and alter the ventilatory rate. With metabolic acidosis, an increase in $[H^+]$ (decrease in pH) increases the ventilatory rate. Conversely, during metabolic alkalosis, a decrease in $[H^+]$ (increase in pH) leads to a decrease in the ventilatory rate. The respiratory response to metabolic acid-base disturbances may take place in several minutes, but can require several hours to complete.

Renal Defense

The third and final line of defense is the kidneys. In response to an alteration in the plasma pH and P_{CO_2}, the kidneys make appropriate adjustments in the excretion of HCO_3^- and net acid. The renal response requires several days to complete.

In the case of acidosis (increase in $[H^+]$ or P_{CO_2}), secretion of H^+ by the nephron is stimulated, and the entire filtered load of HCO_3^- is reabsorbed. The production and excretion of NH_4^+ are also stimulated, thus increasing net acid excretion by the kidneys (see equation 9). The new HCO_3^- generated during the process of net acid excretion is returned to the body, and the plasma $[HCO_3^-]$ increases.

With alkalosis (decrease in $[H^+]$ or P_{CO_2}), secretion of H^+ by the nephron is inhibited and as a result, net acid excretion and HCO_3^- reabsorption are reduced. HCO_3^- will appear in the urine, and the plasma $[HCO_3^-]$ is thereby reduced.

The primary acid-base disorders and the appropriate defense mechanisms are listed in Table 39-2. As shown in the table, the lungs compensate for metabolic disorders, and the kidneys compensate for respiratory disorders. Note again that *these compensatory mechanisms do not correct the underlying disorder, but simply reduce the magnitude of the change in blood pH. Com-*

plete recovery from the acid-base disorder requires correction of the underlying cause.

SIMPLE ACID-BASE DISORDERS

Table 39-2 summarizes the primary alteration and the subsequent defense mechanisms for the various simple acid-base disorders. The respiratory and renal defense mechanisms in simple acid-base disorders are commonly referred to as compensatory responses. Accordingly, the lungs compensate for metabolic disorders and the kidneys compensate for respiratory disorders.

Metabolic Acidosis

Metabolic acidosis is characterized by a low plasma $[HCO_3^-]$ and a low plasma pH. This condition can develop by the addition of nonvolatile acid to the body (e.g., with diabetic ketoacidosis), by the loss of nonvolatile alkali (e.g., with diarrhea), or by failure of the kidneys to excrete enough net acid to replenish the HCO_3^- used to titrate nonvolatile acids (e.g., with renal tubular acidosis or renal failure). As described above, H^+ will be buffered in both the extracellular and intracellular fluid. The fall in pH will stimulate the respiratory centers, and the ventilatory rate will increase (**respiratory compensation**). This reduces the P_{CO_2}, which further helps to minimize the fall in plasma pH. Lastly, renal excretion of net acid increases. This occurs by eliminating all HCO_3^- from the urine (enhanced reabsorption of filtered HCO_3^-), and by increasing ammonia excretion (enhanced production of new HCO_3^-). If the process that initiated the acid-base disturbance is corrected, the enhanced excretion of acid by the kidneys will ultimately return the pH and $[HCO_3^-]$ to normal. Correction of the pH will also return the ventilatory rate to normal.

Metabolic Alkalosis

Metabolic alkalosis is characterized by an elevated plasma $[HCO_3^-]$ and elevated plasma pH. This can occur by the addition of nonvolatile alkali to the body (e.g., ingestion of antacids), after a decrease in the ECV, or more commonly from loss of nonvolatile acid (e.g., loss of gastric HCl with vomiting). Buffering occurs in the extracellular and intracellular fluid compartments. The increase in pH inhibits the respiratory centers, the ventilatory rate is reduced, and the P_{CO_2} rises (**respiratory compensation**). The renal compensatory response to metabolic alkalosis is to increase the excretion of HCO_3^- by reducing its reabsorption along the nephron. Normally, this occurs quite rapidly and effectively. However, when the alkalosis occurs in conjunction with a reduced ECV (e.g., vomiting), Na^+ and HCO_3^- reabsorption are enhanced in the proximal tubule (see Chapter 37). Renal excretion of HCO_3^- and correction of the alkalosis will occur with restoration of a normal effective circulating volume. As is the case with metabolic acidosis, renal excretion of HCO_3^- will eventually return the pH and $[HCO_3^-]$ to their normal values, provided that the underlying cause of the initial disturbance is corrected. Correction of the pH also returns the ventilatory rate to normal.

Respiratory Acidosis

Respiratory acidosis is characterized by an elevated plasma P_{CO_2} and reduced plasma pH. It results from decreased gas exchange across the alveoli, as a result of either inadequate ventilation (e.g., drug-induced depression of the respiratory centers) or impaired gas diffusion (e.g., pulmonary edema as may occur in cardiovascular disease). In contrast to the metabolic disorders, buffering during respiratory acidosis occurs almost entirely in the intracellular compartment (see above). Both the increase in P_{CO_2} and the decrease in pH stimulate HCO_3^- reabsorption by the kidneys, and stimulate ammonium excretion (**renal compensation**). Together these responses increase net acid excretion and generate new HCO_3^-. However, the renal compensatory response takes several days. Consequently, respiratory acid-base disorders are commonly divided into acute and chronic phases. In the acute phase, the renal compensatory response has not had sufficient time to occur, and the body relies on intracellular buffering to minimize the change in pH. In the chronic phase, renal compensation occurs. Correction of the underlying disorder returns the P_{CO_2} to normal; secondarily the renal excretion of acid decreases to its initial level.

Respiratory Alkalosis

Respiratory alkalosis is characterized by a reduced plasma P_{CO_2} and elevated plasma pH. It results from hyperventilation and excessive pulmonary washout of CO_2, or from increased ventilation elicited by stimulation of the respiratory centers (e.g., by drugs or CNS disor-

ders). Hyperventilation can occur as a response to anxiety, pain, or fear. As already noted, buffering is mainly intracellular. The elevated pH and reduced P_{CO_2} inhibit HCO_3^- reabsorption by the nephron and reduce ammonium excretion (**renal compensation**). These two effects reduce net acid excretion. Correction of the underlying disorder returns the P_{CO_2} to normal; secondarily the renal excretion of acid increases to its initial level.

ANALYSIS OF ACID-BASE DISORDERS

Analysis of a clinical acid-base disorder is directed at identification of the underlying cause so that appropriate therapy can be initiated. Oftentimes the patient's medical history and associated physical findings provide valuable clues about the nature and origin of an acid-base disorder. An analysis of an arterial blood sample is frequently required. For example, consider the following data:

$$pH = 7.35 \qquad [HCO_3^-] = 16 \text{ mEq/L} \qquad P_{CO_2} = 30 \text{ mmHg}$$

The acid-base disorder represented by values such as these can be determined by the following three-step approach (see Figure 39-5).

1. Examination of the pH: By first considering the plasma pH, the underlying disorder can be classified as either an acidosis (pH < 7.40), or an alkalosis (pH > 7.40). Note that the defense mechanisms of the body, by themselves, cannot correct an acid-base disorder. Thus, even if the defense mechanisms are completely operative, the pH will still indicate the origin of the initial disorder. In the example shown above, the pH of 7.35 indicates an acidosis.

2. Determination of metabolic vs. respiratory disorder: Simple acid-base disorders are either metabolic or respiratory. To determine which disorder is present, the $[HCO_3^-]$ and P_{CO_2} must next be examined. As indicated by the Henderson-Hasselbalch equation, a pH value below 7.40 (i.e., acidosis) could be the result of a decrease in the $[HCO_3^-]$ (metabolic) or an increase in the P_{CO_2} (respiratory). Alternatively, a pH value above 7.40 (alkalosis) could be the result of an increase in the $[HCO_3^-]$ (metabolic) or a decrease in the P_{CO_2} (respiratory). For the above example, the $[HCO_3^-]$ is reduced from normal (normal = 23 to 25 mEq/L), as is the P_{CO_2} (normal = 40 mm Hg). The disorder must therefore be a metabolic acidosis; it cannot be a respiratory acidosis, because the P_{CO_2} is reduced.

3. Analysis of compensatory response: Metabolic disorders result in compensatory changes in ventila-

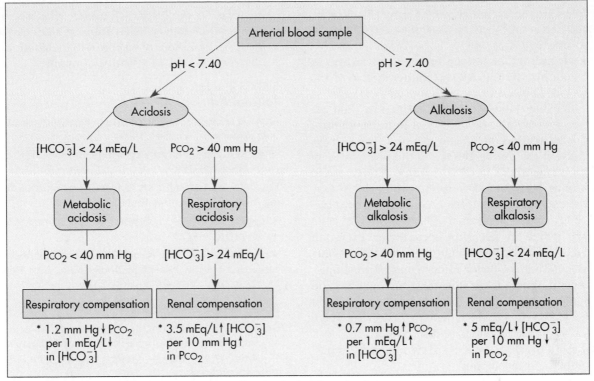

FIGURE 39-5 Approach for analysis of simple acid-base disorders.

tion, and thus in the P_{CO_2}. Respiratory disorders elicit compensatory changes in renal acid excretion, and thus in the plasma $[HCO_3^-]$. In an appropriately compensated metabolic acidosis, the P_{CO_2} will be decreased, whereas in a compensated metabolic alkalosis the P_{CO_2} will be elevated. With respiratory acidosis, complete compensation elevates the $[HCO_3^-]$, whereas with respiratory alkalosis complete compensation reduces the $[HCO_3^-]$. In the above example, the P_{CO_2} is reduced. Therefore, the acid-base disorder is a simple metabolic acidosis with appropriate respiratory compensation.

If the compensatory response is not appropriate, a mixed acid-base disorder should be suspected (see below). A mixed acid-base disorder simply reflects the presence of two or more underlying causes for the acid-base disturbance. A mixed acid-base disorder is suspected when analysis of the arterial blood sample indicates that compensation has not been appropriate. For example, consider the following data:

pH = 6.96 $[HCO_3^-]$ = 12 mEq/L P_{CO_2} = 55 mmHg

Following the three-step approach outlined above, it is evident that the disturbance is an acidosis that has both a metabolic component ($[HCO_3^-] < 24$ mEq/L) and a respiratory component ($P_{CO_2} > 40$ mm Hg). Thus, this

disorder is mixed. Such a disorder would be manifested in an individual with a history of chronic pulmonary disease such as **emphysema** (chronic respiratory acidosis), in conjunction with an acute gastrointestinal illness with diarrhea. Because diarrhea fluid contains large quantities of HCO_3^-, the loss of this fluid from the body induces a metabolic acidosis.

A mixed acid-base disorder is also indicated when a patient has abnormal plasma P_{CO_2} and plasma $[HCO_3^-]$, but the plasma pH is normal. Such a situation can be seen in a patient who has ingested a large quantity of aspirin. The salicylic acid (active ingredient in aspirin) produces a metabolic acidosis, and at the same time stimulates the respiratory centers, causing hyperventilation and a respiratory alkalosis. Thus, the patient has a reduced plasma $[HCO_3^-]$ and a reduced P_{CO_2}.

SUMMARY

1. The pH of the body fluids is maintained within a narrow range by the coordinated function of the lungs and kidneys. Volatile (CO_2 derived) and nonvolatile acids, together with any acid or alkali ingested in the diet, must be excreted in order for acid-base balance to be maintained.

2. The lungs are the excretory route for the volatile acid, whereas the kidneys are the route for excretion of the nonvolatile acid.

3. The body uses buffer systems to minimize changes in body fluid pH; the CO_2/HCO_3^- buffer system of the extracellular fluid is the most important.

4. The kidneys maintain acid-base balance by their excretion of an amount of acid equal to the amount of nonvolatile acid produced and ingested. The kidneys also prevent the loss of HCO_3^- in the urine by reabsorbing virtually all the HCO_3^- that is filtered at the glomerulus. Both the reabsorption of filtered HCO_3^- and the excretion of acid are accomplished by secretion of H^+ by the nephrons.

5. Urinary buffers are necessary for effective excretion of acid because the minimum pH of the urine is only 4.0 to 4.5. The two major urinary buffers are ammonia and phosphate (titratable acid). Ammonia is the more important urinary buffer, because its production by the kidneys and subsequent excretion are regulated in response to acid-base disturbances.

6. Acid-base disturbances are of two general types: respiratory and metabolic.

7. Respiratory acid-base disorders result from primary alterations in the blood P_{CO_2}. Elevation of P_{CO_2} produces acidosis, and the kidneys respond by an increase in excretion of acid. Conversely, reduction of P_{CO_2} produces alkalosis, and renal acid excretion is reduced. The kidneys respond to respiratory acid-base disorders over several hours to days.

8. Metabolic acid-base disorders result from primary alterations in the plasma $[HCO_3^-]$, which in turn results from addition of acid to, or loss of alkali from, the body. In response to metabolic acidosis, pulmonary ventilation is increased, which decreases the P_{CO_2}. An increase in the $[HCO_3^-]$ causes alkalosis. This decreases pulmonary ventilation, which elevates the P_{CO_2}. The pulmonary response to metabolic acid-base disorders occurs in a matter of minutes.

BIBLIOGRAPHY
Journal Articles

Capasso G et al: Acidification in mammalian cortical distal tubule, *Kidney Int* 45:1543, 1994.

Gluck SL: Cellular and molecular aspects of renal H^+ transport, *Hosp Pract* 149, 1989.

Hamm LL, Hering-Smith KS: Acid-base transport in the collecting duct, *Semin Nephrol* 13:246, 1993.

Knepper MA: NH_4^+ transport in the kidney, *Kidney Int* 40[suppl 33]:S95, 1991.

Wingo CS, Cain BD: The renal H-K-ATPase: physiological significance and role in potassium homeostasis, *Ann Rev Physiol* 55:323, 1993.

Books and Monographs

Gennari FJ, Maddox DA: Renal regulation of acid-base homeostasis: integrated response. In Seldin DW, Giebisch G, eds: *The kidney: physiology and pathophysiology*, ed 2, New York, 1992, Raven Press.

Halperin ML, Goldstein MB: *Fluid, electrolyte, and acid-base physiology*, Philadelphia, 1994, WB Saunders Co.

Hamm LL, Alpern RJ: Cellular mechanisms of renal tubular acidification. In Seldin DW, Giebisch G, eds: *The kidney: physiology and pathophysiology*, New York, 1992, Raven Press.

Lowenstein J: *Acids and bases*, New York, 1993, Oxford University Press.

Rose BD: *Clinical physiology of acid-base and electrolyte disorders*, New York, 1994, McGraw-Hill Inc.

Seldin DW, Giebisch G, eds: *The regulation of acid-base balance*, New York, 1990, Raven Press.

ENDOCRINE SYSTEM

SAUL M. GENUTH

PART

CHAPTER 40

General Principles of Endocrine Physiology

The endocrine system is a key component in the adaptation of the human organism to alterations in the internal and external environment. *This system acts to maintain a stable internal milieu in the face of changes in inflow or outflow of substrates, minerals, water, environmental molecules, heat, and so on.* Specific endocrine cells, usually grouped in glands, sense the disturbance and respond by secreting chemical substances called **hormones** into the bloodstream. These special molecules are carried via the circulation to various tissues, where they signal and act on their target cells. As a result the target cells respond in a manner that usually opposes the direction of change that evoked the secretion of hormone, and thereby restores the organism toward its original state. In addition to this fundamental role in maintaining **homeostasis,** the endocrine system also helps to initiate, mediate, and regulate the process of growth, development, maturation, reproduction, and senescence.

A hormone was originally defined as a substance that was elaborated by one type of cell, and that carried a signal through the bloodstream to distant target cells. However, this sophisticated method of signalling probably evolved from a more primitive one (Figure 40-1). Hormone molecules secreted by endocrine cells can also reach and act on target cells within the same locale simply by diffusing through the interstitial fluid separating them; this is called **paracrine function.** Hormone molecules can even act back on their cells of origin to modulate their own secretion or other intracellular processes; this is called **autocrine function.**

The endocrine system may act independently of, or it may be integrated with, the nervous system, which is the other major component in the organism's adaptability to change (Figure 40-2). These two signaling systems have several characteristics in common:

1. Both neurons and endocrine cells are capable of secreting.

2. Both endocrine cells and neurons generate electrical potentials and can be depolarized.
3. Some molecules serve as both a neurotransmitter and a hormone.
4. The mechanism of action of both hormones and neurotransmitters requires interaction with specific receptors in target cells.

Although the endocrine system responds more often to chemical stimuli and the nervous system more often to physical or mechanical stimuli, considerable overlap exists. For example, changes in the quantity of light and changes in plasma substrate concentrations may evoke responses by both systems. Interaction between the two systems takes several forms:

1. Some stimuli to hormone release are first sensed by the nervous system, which in turn signals an appropriate endocrine cell to respond.
2. Some neurons extend their axons in bundles or tracts that terminate adjacent to capillaries. Stimulation causes release of their neurotransmitters into the bloodstream. This hybrid form of signal transmission is called **neurocrine function** (see Figure 40-1), and the signaling molecules are called neurohormones.

   For example, **antidiuretic hormone** (ADH) is synthesized in the cell body of a hypothalamic neuron, but is released from the end of the neuron's axon into blood bathing the posterior pituitary gland. ADH then acts on distant cells in the kidney and causes retention of free water. The stimulus for axonal release of ADH is water deprivation, sensed in the hypothalamus as an increase in plasma osmolality and/or a decrease in volume.
3. Some stimuli evoke integrated endocrine and nervous system responses that augment each other in restoring homeostasis.

The principle of chemical homeostasis and the fundamental relationship of the endocrine system to the

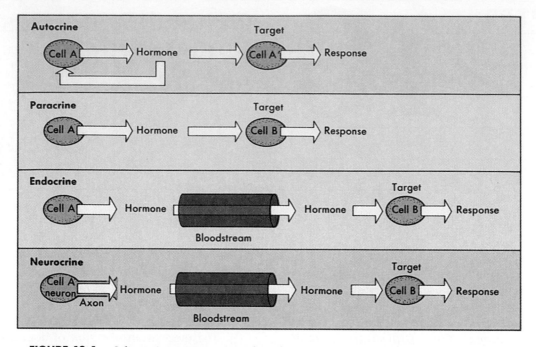

FIGURE 40-1   Schematic representation of mechanisms for cell-to-cell signaling via hormone molecules. In autocrine function the hormone signal acts back on the cell of origin or adjacent identical cells. In paracrine function the hormone signal is carried to an adjacent target cell over short distances via the interstitial fluid. In endocrine function the signal is carried to a distant target via the bloodstream. In neurocrine function the hormone signal originates in a neuron and, after axonal transport to the bloodstream, is carried to a distant target cell.

nervous system are well illustrated by the response of the organism to a lowering of the plasma concentration of glucose (hypoglycemia), such as could occur with very strenuous and prolonged exercise. (Figure 40-3). Because a supply of glucose is absolutely required to sustain brain function, hypoglycemia cannot be tolerated for long. Endocrine cells in the pancreas respond to hypoglycemia by secreting a hormone, called **glucagon,** that stimulates the release of stored glucose from the liver. Other endocrine cells in the pancreas respond in the opposite way to hypoglycemia by diminishing the secretion of the hormone **insulin,** and thereby reduce utilization of glucose by tissues other than the brain.

Certain neurons in the hypothalamus sense hypoglycemia and augment the release of stored glucose directly by transmitting sympathetic neural impulses to liver cells and indirectly by transmitting sympathetic nervous system impulses to the adrenal medulla. This neuroendocrine gland secretes the hormone **epinephrine,** which acts on the liver to release stored glucose and on other tissues to reduce glucose utilization. Finally, other neurons in the hypothalamus also sense hypoglycemia and, via combined neurocrine and endocrine pathways, stimulate the adrenal cortex to secrete the hormone **cortisol.** This hormone augments synthesis of glucose in

the liver to maintain the supply in case initial stores become depleted. Cortisol also inhibits the insulin-stimulated utilization of glucose by tissues other than the brain. Together these endocrine and neural responses to hypoglycemia promptly raise plasma glucose levels back to normal.

PATTERNS OF HORMONE SYNTHESIS, STORAGE, AND SECRETION

Hormones are synthesized, stored, and secreted in a variety of ways. **Peptide** and **protein hormones** are synthesized by a general process that characterizes the synthesis of all secreted proteins (Figure 40-4). The gene or deoxyribonucleic acid (DNA) molecule that directs hormone synthesis transcribes a messenger ribonucleic acid (mRNA) molecule. The latter traverses the nuclear membrane to the cytoplasm, where it translates its message on ribosomes by directing the assembly of the correct sequence of amino acids into a primary gene product. This is larger than the hormone itself and is called a **preprohormone.** At the N terminal, a signal peptide directs transfer of the preprohormone from the ribosome into the **endoplasmic reticulum.** During this

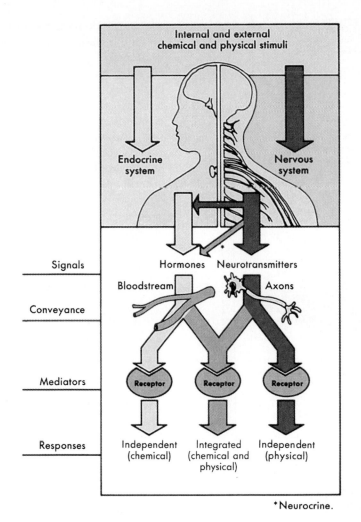

FIGURE 40-2 Overview of the relationship between the endocrine system and the nervous system. Similar stimuli may elicit activity of both systems. Hormones secreted by endocrine cells and conveyed via the bloodstream are analogous to neurotransmitters released by neurons after being conveyed by their axons. Neurotransmitters may also stimulate hormone release and themselves act as hormones. Responses are mediated by receptors in each system and may consist of either chemical or physical changes.

process the signal peptide is degraded, leaving a **prohormone.** This molecule contains the hormone as well as other peptide sequences.

The prohormone is transferred to the Golgi apparatus, where it undergoes further processing. This may include cleavage, the addition of carbohydrate units, or the combination of separate subunits derived from different genes. In the Golgi apparatus the hormone and its peptide coproducts are packaged together within a secretory granule.

On stimulation of the endocrine cell the contents of secretory granules are released by the process of exocytosis into the extracellular fluid and then into adjacent bathing capillaries (Figure 40-5). By contraction of microfilaments and guidance by microtubules, the secretory granules move to the plasma membrane of the cell and fuse with it. A **guanosine triphosphate (GTP)–binding protein** helps attach the granules to specific sites. The mechanisms of granule release require an increase in intracytoplasmic calcium (Ca^{++}) concentration; the Ca^{++} is derived from both extracellular fluid and intracellular stores within the endoplasmic reticulum and other organelles. Exocytosis is often preceded by increases in **cyclic adenosine monophosphate** (cAMP) concentration.

Catecholamine hormones (epinephrine, norepinephrine, dopamine) are synthesized from the amino acid tyrosine through a series of enzymatic reactions. However, they are stored in secretory granules and secreted in a manner similar to that of peptide hormones.

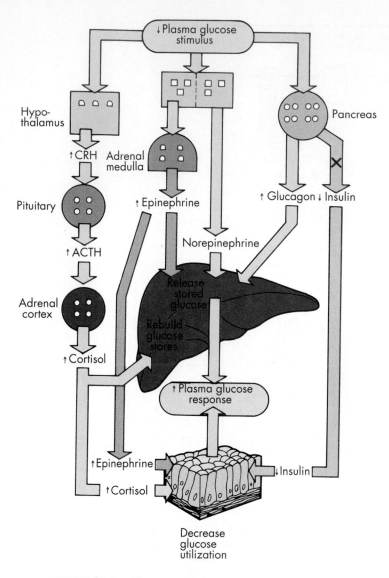

FIGURE 40-3 The integrated endocrine and neural response to hypoglycemia. The anterior pituitary gland, the adrenal cortex, the adrenal medulla, and the pancreatic islets participate in the main endocrine components of the response. The hypothalamus and the sympathetic nervous system participate in the neural components of the response. See text for details on how each component acts to restore the plasma glucose concentration to normal. *CRH,* Corticotropin-releasing hormone; *ACTH,* adrenocorticotropic hormone.

Thyroid hormones (thyroxine, triiodothyronine) are synthesized from tyrosine and iodide in a series of reactions that occur with the amino acid already incorporated via peptide linkage into a large protein molecule. The hormones are then sequestered within the protein molecule in a storage space (follicle) shared by a group of surrounding endocrine cells. Secretion of thy-

roid hormone requires retrieval from the follicle and enzymatic release from its protein storage form.

Steroid hormones (cortisol, aldosterone, androgens, estrogens, progestins, vitamin D) are synthesized from cholesterol by a series of enzymatic reactions. However, they are not stored to any appreciable extent within the gland of origin. Thus, to increase the secretion of a steroid hormone, the entire biosynthetic sequence from cholesterol must be activated. In effect, the storage form of all steroid hormones is the intracellular depot of cholesterol.

> Genetic diseases causing deficient or abnormal synthesis of peptide or protein hormones usually involve the hormone gene itself. In the case of thyroid or steroid hormones, the product of the mutant gene is usually an enzyme that catalyzes one of the many separate reactions in the biosynthetic sequence for the affected hormone.

REGULATION OF HORMONE SECRETION

The secretion of hormones is related to their roles in maintaining homeostasis. Therefore the dominant mechanism of regulation is **negative feedback** (Figure 40-6). If hormone A acts to raise the plasma concentration of substrate B, a decrease in substrate B will stimulate hormone A secretion, whereas an increase in substrate B will suppress hormone A secretion. In essence, *physiological conditions that require the action of a hormone also stimulate its release; conditions or products resulting from prior hormone action suppress further hormone release.* This homeostatic partnership may exist between a hormone and one or more substrates, minerals, other hormones, or even physical factors such as fluid volume.

Occasionally, positive feedback is observed. In such circumstances a product of hormone action initially stimulates further hormone secretion. When the product eventually reaches appropriate concentrations, it may then exert negative feedback on hormone secretion. This mechanism of regulation obtains when a biological process begins at a very low level, yet must reach rather high levels in the course of normal physiological function.

Feedback regulation may be exerted at all levels of endocrine cell function—that is, at transcription of the hormone gene, at translation of the gene message, and at release of stored hormone.

Engrafted on homeostatic feedback are patterns of

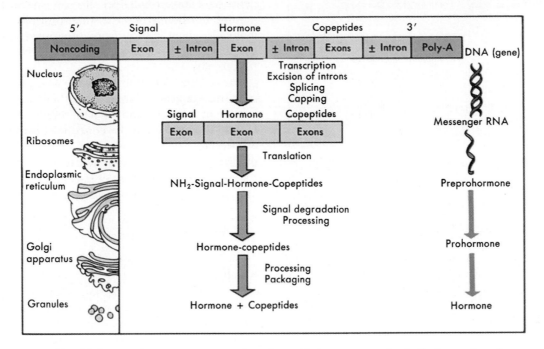

FIGURE 40-4   A schematic representation of peptide hormone synthesis. In the nucleus the primary gene transcript undergoes excision of introns (noncoding regions), splicing of the exons (coding regions), and capping. The resultant mature messenger RNA enters the cytoplasm, where it directs the synthesis of a precursor peptide sequence (preprohormone) on ribosomes. In this process the N-terminal signal is removed, and the resultant prohormone is transferred into the endoplasmic reticulum. The prohormone undergoes further processing and is then packaged into secretory granules in the Golgi apparatus. After final cleavage of the prohormone within the granules, the hormone and copeptides are secreted by exocytosis.

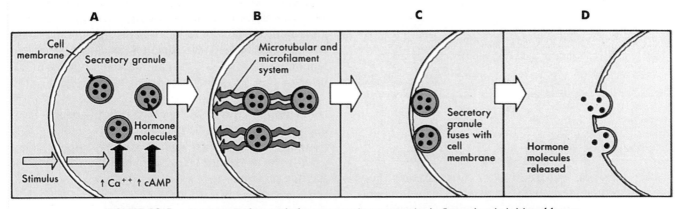

FIGURE 40-5   Secretion of peptide hormones via exocytosis. **A,** Secretion is initiated by application of a stimulus that raises cytosolic Ca^{++} and also usually raises intracellular cyclic adenosine monophosphate (cAMP) levels. **B,** The secretory granules are lined up and translocated to the plasma membrane via activation of a microtubular and microfilament system. **C,** The membrane of the secretory granule fuses with that of the cell. **D,** The common membrane is lysed, releasing the hormone into the interstitial space.

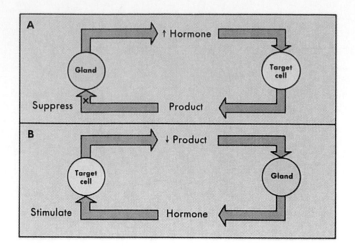

FIGURE 40-6 Negative feedback principle. **A,** If an increase in hormone secretion stimulates a greater output of product from the target cell, the product feeds back on the gland to suppress further hormone secretion. In this way hormone excess is limited or prevented. **B,** If a decrease in output of product from the target cell stimulates hormone secretion, the hormone in turn will stimulate output of product by the target cell. In this way product deficiency is limited or corrected.

hormone release dictated by diurnal (daily) or ultradian (within a day) rhythms, stages of sleep, seasonal variation, and stages of development (fetal, neonatal, pubertal, senescent). In addition, pain, emotion, fright, injury, and sexual arousal may evoke or shut off the release of hormones via complex neural pathways.

> Individuals undergoing major medical or surgical stress demonstrate a pattern of hormonal release that stimulates mobilization of fuels such as glucose or free fatty acids and augments their delivery to the heart and musculature. By contrast, growth and reproductive processes are suppressed.

HORMONE TURNOVER

After secretion into the blood, catecholamine, peptide, and protein hormones usually circulate unbound to other plasma constituents. In contrast, thyroid and steroid hormones circulate largely bound to specific globulins, as well as to albumin. The extent of this protein binding greatly influences the rates at which hormones exit from plasma into interstitial fluid, where they interact with target cells. The plasma half-life of a hormone (the time required for a 50% decrease in concentration) is increased by strong protein binding.

Larger protein hormones with carbohydrate components have longer half-lives than do smaller protein and peptide hormones. After exiting from plasma, hormone molecules may return from other compartments via lymphatic channels, sometimes after dissociation from target cells.

Irreversible removal of hormone from the body results from target cell uptake, metabolic degradation, and urinary or biliary excretion. The sum of all removal processes is expressed in the concept of **metabolic clearance rate** (MCR). This is defined as the volume of plasma cleared of hormone per unit of time. In a steady state this equals the mass of hormone removed per unit of time, divided by its plasma concentration:

$$MCR = \frac{mg/min \text{ (removed)}}{mg/ml \text{ (plasma)}} = \frac{ml \text{ plasma (cleared)}}{min} \quad \textbf{(1)}$$

MCR is an expression of the overall efficiency with which a hormone is removed from plasma, irrespective of the mechanism (just as renal clearance rate is an expression of the efficiency with which a substance is specifically removed from plasma by urinary excretion (see Chapter 35). MCR is most conveniently determined by infusing a hormone at a constant rate until a new steady-state plasma concentration is reached. At this point the rate of removal from plasma must equal the rate of infusion. Therefore,

$$MCR = \frac{Hormone \text{ infused (mg/min)}}{Hormone \text{ concentration (mg/ml)}} = ml/min \quad \textbf{(2)}$$

The kidney and liver are the major sites of the metabolic degradation of hormones. Renal clearance of a hormone, (e.g., thyroid hormones) is extremely low if it is bound to specific plasma globulins. Although peptide and smaller protein hormones are filtered to some degree by the renal glomeruli, they usually undergo tubular reabsorption and subsequent degradation within the kidney so that only a minute amount appears in the urine.

Metabolic degradation of hormones occurs by enzymatic processes that include proteolysis, oxidation, reduction, hydroxylation, decarboxylation, and methylation. Hormones or their metabolites may also be conjugated to water-soluble molecules and then excreted in the bile or urine.

Quantitation of Hormone Secretion

By isotopic techniques, the total amount of hormone secreted into the bloodstream per unit time can be measured. For clinical purposes, measurement of plasma levels or urine excretion of a hormone must usually suffice. However, they are valid indices of hormone production if certain conditions exist. In a steady state, the amount of hormone entering the plasma is equal to the amount of hormone exiting the plasma.

Secretion rate = Disposal rate =
Metabolic clearance × Plasma concentration   **(3)**

If the MCR is within normal limits and can be taken as a constant, then the secretion rate is proportional to the plasma level. This is the theoretical basis for employing plasma hormone measurement alone as an index of secretory rate. However, the secretion of many hormones is characterized by diurnal variation and episodic spurts. In such instances, multiple plasma samples may be necessary for a valid estimate.

Similarly, hormone secretion can sometimes be assessed by measuring its urinary excretion in accurately timed collections. This serves to average out plasma fluctuations over the collection interval. The urinary excretion is a valid index of secretion rate when kidney function and kidney handling of the hormone are normal.

HORMONE ACTION

Three major sequential steps are involved in eliciting responses to hormones (Figure 40-7):
1. The hormone must be recognized by the target cell.
2. An intracellular signal must then be generated.
3. One or more intracellular processes must be increased or decreased (e.g., enzyme reactions, ion movements, cytoskeletal rearrangements, gene transcription).

Hormone Recognition: Receptors

Recognition takes place by binding of the hormone to a specific receptor, which may be located within the plasma membrane, the cytoplasm, the nucleus, or possibly other organelles of the target cell. The receptor has a specific binding site with high affinity for the hormone (see Chapter 5). The two molecules associate in reversible fashion to form a hormone-receptor complex. The receptor confers specificity to the interaction of a hormone with its target cells. Only cells that have the receptor can respond to the hormone; only hormones for which the cell possesses receptors can affect the cell.

Receptors are large protein molecules that may also contain carbohydrate units. Those receptors incorporated into plasma membranes have extracellular portions that bind their hormones; transmembrane and intracellular portions interact with a signal-generating mechanism that initiates intracellular actions (see Figure 6-5).

Some plasma membrane receptors resemble immunoglobulins in structure. Certain individuals susceptible to **autoimmune diseases** develop antibodies to their own hormone receptors. When such antibodies react with the receptor molecules, they may simply block access of the hormone to the receptor and cause biological deficiency (e.g., **adrenocortical insufficiency,** or **Addison's disease;** see Chapter 46). Alternatively, the antibody-receptor combination mimics the hormone-receptor interaction and causes hyperfunction of the gland (e.g., hyperthyroidism caused by **Graves' disease;** see Chapter 45).

The reaction between hormone and receptor is the initial determinant of the rate of hormone action. This reaction can be expressed in classical chemical terms.

$$[H] + [R] \rightleftarrows [HR] \qquad (4)$$

$$K = \frac{[HR]}{[H][R]} \qquad (5)$$

$$[HR] = K[H][R] \qquad (6)$$

where [H] equals the free hormone concentration; [R] equals the free or unoccupied receptor concentration; [HR] equals the hormone-receptor complex or bound hormone concentration; and K equals the affinity (association) constant.

The amount of receptor occupied by hormone, [HR], is the critical component that governs the magnitude of hormone action at this first step. As can be seen from equation 6, [HR] is increased when the receptor has a high affinity (K) for the hormone, when the cell is exposed to high hormone concentrations [H], or when the receptor number [R] is high.

In some cases [HR] is the rate-limiting step in the whole sequence of hormone action; therefore the maximum biological response to these hormones is directly proportional to and limited by the number of receptors. In cases where [HR] is not rate limiting, the biological response can be maximum when only a small proportion of the available receptors is occupied by hormone.

Receptor molecules are continually synthesized, translocated to sites of association with hormone molecules, and degraded. These processes can be influenced by their respective hormone partners. Some hormones decrease or "down-regulate" the number of their own receptors, and thereby prevent excess hormone action on the cell. Other hormones recruit their own receptors, and thereby amplify hormone action on the cell.

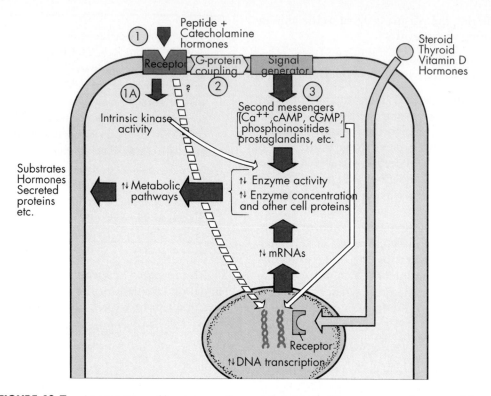

FIGURE 40-7   An overview of hormone actions on target cells. Hormones may interact with either plasma membrane or intracellular receptors. Hormones may generate second messengers within the cytoplasm or the nucleus. Metabolic pathways may be regulated by altering the activities or the concentrations of enzymes. Cell growth and architecture may also be modulated.

Signal Generation

Signal generation is the next step in hormone action. When hormone receptor association occurs within the plasma membrane of the cell, the resultant complex is often coupled to other plasma membrane components (see Chapter 5). These act to generate within the cell a variety of signal molecules, or "second messengers," which then influence metabolic and other processes (see Figures 5-1 and 40-7). In this situation the essential information for triggering the cell's response actually resides in the receptor molecule; this information is transmitted to the cytoplasm when the hormone occupies and changes the conformation of the extracellular domain of the membrane receptor. The hormone is essentially an extracellular signal.

In contrast, when hormone receptor association occurs within the cytoplasm or nucleus, the hormone-receptor complex ultimately interacts with specific DNA molecules and alters gene expression (Figures 40-7 and 40-9). Here the second messengers are transcribed RNA molecules that direct the synthesis of protein molecules. In this situation essential information for triggering the cell's response resides in the hormone molecule itself as well as in the receptor. The hormone is an intracellular signal.

Plasma Membrane–Generated Second Messengers A class of protein molecules known as **G-proteins** acts to couple hormone-receptor complexes with second-messenger generation (see Chapter 5). G-proteins contain 3 subunits (see Figure 5-6). When activated by a hormone-occupied receptor, the α-subunit effectively exchanges guanosine diphosphate (GDP) for guanosine triphosphate (GTP). The GTP-linked α-subunit dissociates from the β-δ units of the G-protein and interacts with a nearby effector enzyme or membrane carrier. As a result, the effector is either stimulated or inhibited; the α-subunit reexchanges GTP for GDP and reassociates with the β-δ subunit; the reconstituted G-protein can then begin another cycle (see Figure 5-2). This system of coupling the hormone-occupied receptor with a signal-generating enzyme or transport molecule greatly amplifies the effect of the hormone because numerous second-messenger molecules can be rapidly created.

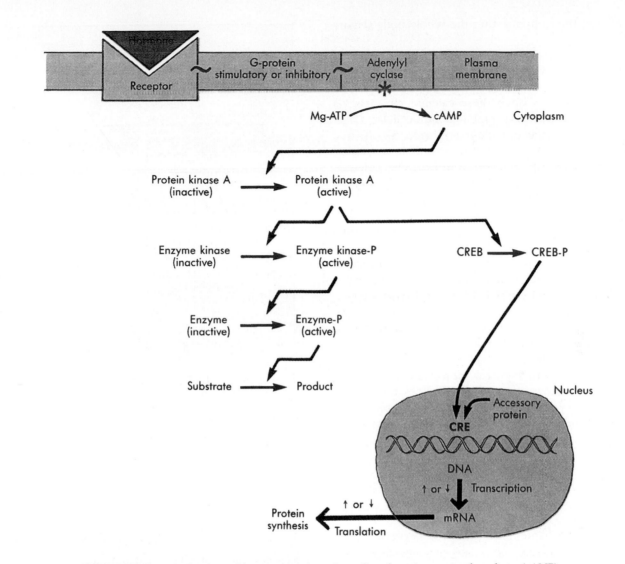

FIGURE 40-8   Mechanism of hormone action via cyclic adenosine monophosphate (cAMP) as a second messenger. The hormone-receptor complex interacts through a G-protein with adenylyl cyclase. If the G-protein is stimulatory, cAMP levels increase and activate protein kinase A, which then phosphorylates various enzyme kinases. These kinases activate enzymes, and ultimately this cascade of effects leads to increases in the intracellular levels of various substances. If the G-protein is inhibitory or if the target enzymes of the hormone are inactivated by phosphorylation, inhibition rather than stimulation of metabolic pathways will result. Activated protein kinase A also phosphorylates a cAMP-binding protein (CREB), which acts as a nuclear transcription factor that interacts with a cAMP regulatory unit (CRE) on target DNA molecules.

1. Cyclic adenosine monophosphate (cAMP). Several peptide hormone receptors are often coupled via G-proteins to the plasma membrane enzyme **adenylyl cyclase.** This enzyme complex consists of a stimulating regulatory subunit, a catalytic subunit, and an inhibiting regulatory subunit (see Figures 5-6 and 5-7). The formation of a hormone-receptor complex causes coupling of G-proteins to stimulate or inhibit the catalytic subunit. This increases or decreases the rate of forma-

tion of cAMP from magnesium–adenosine triphosphate (Mg-ATP) (see Figures 5-3 and 40-8). Cytoplasmic cAMP, the second messenger, activates the enzyme **protein kinase A** in the cytoplasm. Protein kinase A then phosphorylates various enzyme kinases, ultimately either increasing or decreasing the activity of numerous enzymes in various metabolic pathways. The final result of changing cAMP levels is a cascade of reactions that increases, decreases, or alters the direction of substrate flux within the cell and then often within the whole body (Figure 40-8).

2. Calcium-calmodulin. Plasma membranes contain channels that allow graded influx of extracellular Ca^{++} into the cytoplasm. Hormone occupancy of plasma membrane receptors can open these calcium channels via G-proteins and can also mobilize intracellular Ca^{++} bound to endoplasmic reticulum (see Figure 5-9). As free cytoplasmic Ca^{++} concentration rises, it is associated with its binding protein, **calmodulin.** The calcium-calmodulin complex is a powerful second-messenger regulator of the activity of enzymes.

3. Phospholipid products. Formation of a hormone-receptor complex activates via G-protein a membrane-bound **phospholipase-C.** This enzyme splits phosphatidylinositol 4,5-bisphosphate into two products: a **diacylglycerol,** often containing arachidonic acid as a major fatty acid component, and **inositol trisphosphate** (see Figures 5-9 and 5-10). These two second messengers, together with Ca^{++}, activate the enzyme **protein kinase C** (see Figure 5-10). This kinase in turn phosphorylates other enzymes in the cytoplasm and thereby alters flow in the metabolic pathways that they catalyze.

4. Receptor kinases. In some instances the plasma membrane receptor itself acts as a second messenger. After binding to hormone, such receptors undergo autophosphorylation of specific tyrosine residues (see Figure 5-11). The altered receptor molecule now possesses **tyrosine kinase** activity toward other proteins and may itself initiate a cascade of phosphorylations—first at tyrosine, then at serine and threonine, sites—that regulate metabolic processes and cellular growth.

As noted in Chapter 5, a group of **protein phosphatases,** which dephosphorylate enzymes, also exists. These too are subject to hormonal regulation, and they provide still another mechanism whereby hormones may modulate metabolic pathways and mitogenic processes by covalent modification of target proteins.

None of the membrane-generated second messengers is unique to any particular hormone, and a single hormone may operate through multiple messengers. In addition, messengers such as cAMP can also modulate gene expression by interacting with protein transcription factors and regulatory elements on DNA molecules (Figure 40-8). Thus, peptide and protein hormones can also increase or decrease the synthesis of target enzymes.

Nuclear Second Messengers Hormones that directly enter the cell (steroids, vitamin D, thyroid hormones) combine with receptor proteins located in the cytoplasm and nucleus (Figure 40-9). A specific C-terminus domain of the receptor molecule binds the hormone. This hormone-receptor complex undergoes

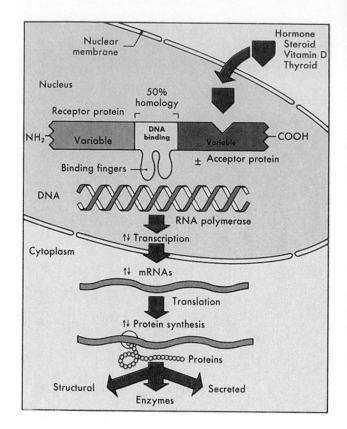

FIGURE 40-9 Mechanism of action of vitamin D, steroid and thyroid hormones. The hormone combines with a nuclear protein receptor. The carboxy-terminal portion of the receptor varies for each hormone. The midportion of the receptor molecule has considerable similarity among hormones. The midportion contains DNA-binding fingers. Binding of the hormone-receptor complex to hormone regulatory elements in DNA molecules either stimulates or suppresses transcription of target genes. The result is increased or decreased synthesis of cell proteins.

an activation process during which the hormone ligand displaces an inactivating or blocking protein from the receptor. After transformation, the complex can then associate with specific DNA molecules.

Another domain in the middle portion of the receptor binds to DNA. This second domain exhibits considerable homology among the various receptors and is coded for by a superfamily of genes related to oncogenes (growth-regulating genes). The DNA site with which the hormone-receptor complex interacts is termed the hormone regulatory site; it is usually upstream from the basal promotor site at the 5′ end of the gene. After the hormone-receptor complex has been bound, transcription of the primary gene message by RNA polymerase is either induced or repressed. Thus, by raising or lower-

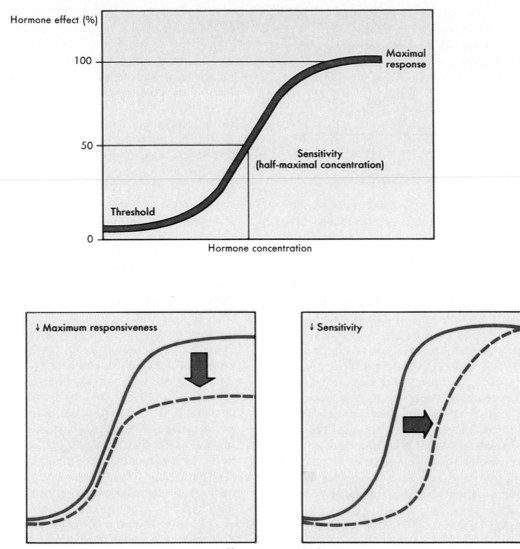

FIGURE 40-10 The general shape of a hormone dose-response curve. Alterations in this curve can take the form of a change in maximal responsiveness *(left lower panel)* or a change in sensitivity *(right lower panel).*

ing the levels of specific RNA molecules, the hormone increases or decreases the concentration of specific cell proteins. When the latter are enzymes, the rates of specific metabolic reactions are likewise increased or decreased by the hormone.

The onset of hormone actions mediated by nuclear second messengers is generally slower than that mediated by cytoplasmic second messengers. There is also a more graded response to increasing hormone concentrations, rather than amplification, because each hormone-receptor unit engages a single DNA molecule. The magnitude of action may be influenced by the concentrations of RNA polymerase, of the enzymes of protein syn-

thesis or processing, of transfer RNA, or of amino acids, or by the number and activity of ribosomes.

Although rare, diseases caused by mutant genes for receptors or G-proteins have provided much insight into mechanisms of hormone action. For example, the reduced activity of a mutant α-subunit of a stimulatory G-protein leads to diminished cAMP levels, resulting in **deficient action of parathyroid hormone** and **hypocalcemia** (see Chapter 43). Another mutant G-protein, which is constitutively overactive, is associated with continuous **hypersecretion of growth hormone** (see Chapter 44). Mutant nuclear receptors for thyroid hormone lead to **hypothyroidism,** caused by the inability of thyroid target cells to respond normally to the hormone (see Chapter 45).

Outcome of Hormone Action

The final outcome, quantitatively, of the interaction of a hormone with its target cells depends on several factors. These include hormone concentration; receptor number; duration of exposure to hormone; intervals between consecutive exposures; intracellular conditions such as concentrations of rate-limiting enzymes, cofactors, or substrates; and the concurrent effects of antagonistic or synergistic hormones. *Hormonal effects are not usually "all-or-none" phenomena.*

The dose-response curve for the action of a hormone is generally complex and often exhibits a sigmoidal shape (Figure 40-10). An intrinsic basal level of cell activity may be observed independent of the hormone. A certain minimal threshold concentration of hormone then is required to elicit a measurable response. The effect produced by saturating doses of hormone is defined as the **maximal responsiveness** of the target cells. The concentration of hormone required to elicit a half-maximal response is an index of the *sensitivity* of the target cells to that hormone.

Alterations in the dose-response curve in vivo can take two general forms (Figure 40-10):

1. A decrease in maximal responsiveness may be caused (a) by a decrease in the number of functional target cells, in the total number of receptors per cell, or in the concentration of an enzyme activated by the hormone; or (b) by an increase in the concentration of a noncompetitive inhibitor.
2. A decrease in hormone sensitivity may be caused (a) by a decrease in hormone receptor affinity or number, (b) by an increase in the rate of hormone degradation, or (c) by an increase in the concentration of antagonistic hormones.

Obesity is a good example of a condition in which sensitivity to a hormone, insulin, is considerably diminished. In **type 2 diabetes** (non–insulin dependent diabetes), both sensitivity and maximal responsiveness to insulin are reduced, and this plays a major role in causing high plasma glucose levels.

SUMMARY

1. The function of the endocrine system is to regulate metabolism, fluid status, growth, and sexual development. The endocrine and nervous systems work conjointly to maintain homeostasis.
2. Hormones are signaling molecules that are conveyed by the bloodstream (endocrine), by neural axons and the bloodstream (neurocrine), or by local diffusion (paracrine, autocrine). Hormones may be proteins, peptides, catecholamines, steroids, or iodinated tyrosine derivatives.
3. Protein or peptide hormone synthesis begins with generation of a primary gene product called a prohormone. The latter undergoes processing by proteolytic cleavage and glycosylation to yield the hormone.
4. Thyroid hormone and catecholamines are synthesized from tyrosine and steroid hormones from cholesterol, by multiple enzyme reactions.
5. Peptide and protein hormones and catecholamines are stored in granules and secreted by exocytosis. Thyroid hormone is stored within protein molecules in large quantities; steroid hormones are not stored at all. Both are released by diffusion.
6. Protein, peptide, and catecholamine hormones act on target cells via specific protein receptors located in the plasma membranes. Stimulatory or inhibitory G-proteins link the hormone-receptor complex to membrane mechanisms that generate cAMP, Ca^{++}, diacylglycerols, and inositol phosphate. These molecules act intracellularly as second messengers to increase or decrease enzyme activities.
7. Thyroid and steroid hormones act via specific protein receptors located in the nucleus. The hormone-receptor complex interacts with hormone-regulatory units in DNA molecules to induce or repress expression of target genes. In turn, this increases or decreases the concentration of enzymes and other proteins.
8. Plasma levels and urinary excretion rates of hormones are used clinically as indirect indices of hormonal secretion rates. These indices are valid as long as metabolic or renal clearance of the hormone is normal. Binding of some hormones to plasma proteins also influences their availability to target cells.
9. The sensitivity of an organism to hormone action is expressed as the hormone concentration that produces half-maximum activity. The sensitivity can be influenced by changes in receptor number, affinity, hormone degradation rate, or competitive antagonists.
10. The maximum effect produced by saturating con-

centrations of hormone can be influenced by the number of target cells, number of receptors, concentration of target enzymes, or noncompetitive antagonists.

BIBLIOGRAPHY
Journal Articles

Alford FP et al: Temporal patterns of circulating hormones as assessed by continuous blood sampling, *J Clin Endocrinol Metab* 36:108, 1973.

Birnbaumer L et al: Molecular basis of regulation of ionic channel by G proteins, *Rec Prog Horm Res* 45:121, 1989.

Carson-Jurica MA et al: Steroid receptor family: structure and function, *Endocr Rev* 11:201, 1990.

Chambon P et al: Promoter elements of genes coding for proteins and modulation of transcription by estrogens and progesterone, *Rec Prog Horm Res* 40:1, 1984.

Combarnous Y: Molecular basis of the specificity of binding of glycoprotein hormones to their receptors, *Endocr Rev,* 13:670, 1992.

Freedman LP: Anatomy of the steroid receptor zinc finger region, *Endocr Rev,* 13:129, 1992.

Glass CK: Differential recognition of target genes by nuclear receptor monomers, dimers, and heterodimers, *Endocr Rev,* 15:391, 1994.

Gordon P et al: Internalization of polypeptide hormones: mechanism, intracellular locations and significance, *Diabetologia* 18:263, 1980.

Lacy PE: Beta cell secretion: from the standpoint of a pathobiologist, *Diabetes* 19:895, 1970.

Lefkowitz R et al: Mechanisms of membrane-receptor regulation: biochemical, physiological, and clinical insights derived from studies of the adrenergic receptors, *N Engl J Med* 310:1570, 1984.

Orti E, Bodwell JE, Munck A: Phosphorylation of steroid hormone receptors, *Endocr Rev,* 13:105, 1992.

Schuchard M, Landers JP, Punkay Sandhu N, Spelsberg TC: Steroid hormone regulation of nuclear proto-oncogenes, *Endocr Rev* 14:659, 1993.

Spaulding SW: The ways in which hormones change cyclic adenosine 3'-5'-monophosphate–dependent protein kinase subunits, and how such changes affect cell behavior, *Endocr Rev* 14:632, 1993.

Spiegel AM, Shenker A, Weinstein LS: Receptor-effector coupling by G proteins: implications for normal and abnormal signal transduction, *Endocr Rev* 13:536, 1992.

Tait JF: The use of isotopic steroids for the measurement of production rates in vivo, *J Clin Endocrinol* 23:1285, 1963.

Walters MR: Steroid hormone receptors and the nucleus, *Endocr Rev* 6:512, 1985.

Zor U: Role of cytoskeletal organization in the regulation of adenylate cyclase-cyclic adenosine monophosphate by hormones, *Endocr Rev* 4:1, 1984.

Books and Monographs

Exton JH, Blackmore PF: Calcium-mediated hormonal responses. In DeGroot LJ, ed: *Endocrinology,* ed 2, Philadelphia, 1989, WB Saunders Co.

Habener JF: Genetic control of hormone formation. In Wilson JD, Foster DF, eds: *Textbook of endocrinology,* ed 8, Philadelphia, 1992, WB Saunders Co.

Kahn CR, Smith RJ, Chin WW: Mechanism of action of hormones that act at the cell surface. In Wilson JD, Foster DF, eds: *Textbook of endocrinology,* ed 8, Philadelphia, 1992, WB Saunders Co.

Whole Body Metabolism

Metabolism may be broadly defined as the sum of all the chemical (and physical) processes involved (1) in producing and expending energy from exogenous and endogenous sources, (2) in synthesizing and degrading structural and functional tissue components, and (3) in disposing of resultant waste products. *Regulating the rate and direction of the various components of metabolism is one of the major functions of the endocrine system.* Therefore, a firm grasp of the fundamentals of metabolism is essential in order to understand the important influence of hormones on body functions.

ENERGY METABOLISM

Energy Balance

The laws of thermodynamics require that total energy balance be constantly maintained in living organisms. However, energy may be obtained in various forms, may be stored in other forms, and may be expended in many different ways. Therefore numerous interconversions of chemical, mechanical, and thermal energy are possible within the basic rule that *in the steady state, energy input must always equal energy output.* Figure 41-1 illustrates this overall flow of energy through the human organism.

Energy Input

Energy input consists of foodstuffs, which are classified into three major chemical categories: carbohydrate, fat, and protein. The complete combustion of each chemical type yields characteristic amounts of energy, expressed as joules or kilocalories per gram (1 kcal = 4184 joules). Combustion of each type also requires characteristic amounts of oxygen per gram, depending on the proportions of carbon, hydrogen, and

oxygen in the substance. However, for each class of foodstuff, the energy yield per liter of oxygen used is quite similar because the ratio of carbon to hydrogen atoms is quite similar in each class. *Within the body the carbon skeletons of carbohydrate and protein can be converted to fat, and their potential energy can be stored more efficiently in that manner. The carbon skeletons of protein can also be converted to carbohydrate when that energy source is specifically needed. However, in the human, carbon atoms from fat are not converted to carbohydrate to any significant extent.*

Energy Output

Energy output can be divided into several distinct, measurable components.

1. In individuals at rest, energy is expended in a myriad of synthetic and degradative chemical reactions; in generating and maintaining chemical and electrical gradients of ions and other molecules across cell and organelle membranes; in the creation and conduction of signals, particularly in the nervous system; in the mechanical work of respiration and circulation of the blood; and in obligate heat loss to the environment. This absolute minimal energy expenditure is called the **basal** or **resting metabolic rate** (BMR or RMR). In the adult human, BMR amounts to an average daily expenditure of 20 to 25 kcal (84 to 105 kjoules)/kg body weight (or 1.0 to 1.2 kcal/min), and this uses approximately 200 to 250 ml oxygen/min.

The BMR is linearly related to lean body mass and body surface area. The central nervous system and skeletal muscle together account for 60% to 70% of the BMR. The BMR declines in the elderly, partly because lean body mass declines with age. BMR increases when environmental temperature rises. During sleep BMR falls 10% to 15%. Studies in identical twins and families suggest that some of the variation in BMR is genetically determined.

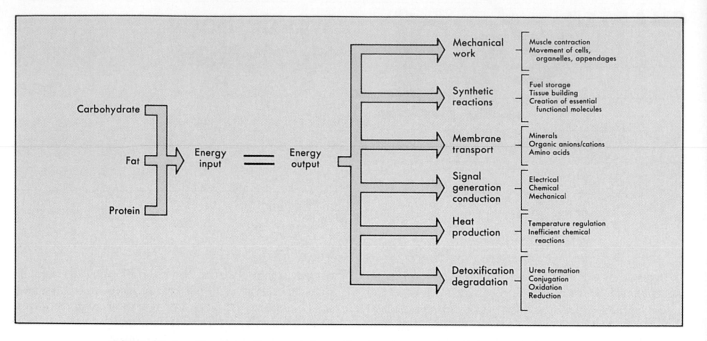

FIGURE 41-1   Overview of energy balance. In a steady state, input as caloric equivalents of food equals output as caloric equivalents of various forms of mechanical and chemical work and heat.

2. Ingestion of food causes a small obligate increase in energy expenditure referred to as **diet-induced thermogenesis.** This is explained by the increased rate of reactions involved in the disposition of the ingested calories, such as storage of glucose in the large molecule, glycogen.

3. **Facultative thermogenesis,** or nonshivering thermogenesis, comes into play during prolonged exposure to cold, when energy may be expended specifically to produce heat and maintain the core temperature of the body. Impulses from the sympathetic nervous system are important mediators of this response to a fall in temperature (see Chapter 10). Facultative thermogenesis may also be used to limit weight gain from prolonged excess of energy intake.

4. Energy is also expended by sedentary individuals in spontaneous physical activity, such as "fidgeting," at least some of which is unconscious and seemingly purposeless.

5. The additional energy expended in occupational labor and purposeful exercise varies greatly among individuals as well as from day to day and from season to season. This component generates the greatest need for variation in daily caloric intake and underscores the importance of energy stores to buffer temporary discrepancies between energy output and intake.

Of a total average daily expenditure of 2300 kcal (9700 kjoules) in a typical sedentary human, basal me-tabolism accounts for 75%, dietary thermogenesis for 7%, and spontaneous physical activity for 18%. As many as 3000 additional kcal may be used in daily physical work. During short periods of high-intensity exercise, energy expenditure can increase more than tenfold over basal levels.

ENERGY GENERATION

Chemical Pathways

The basic chemical currency of energy in all living cells consists of the two high-energy phosphate bonds contained in **adenosine triphosphate** (ATP). To a much lesser extent, other purine and pyrimidine nucleotides (guanosine triphosphate, cytosine triphosphate, uridine triphosphate, inosine triphosphate) also serve as energy sources after the energy from ATP is transferred to them. In muscle, creatine phosphate is a high-energy molecule of particular importance.

The two terminal P-O bonds of ATP each contain about 12 kcal of potential energy per mole under physiological conditions of temperature and pH. These bonds are in constant flux. They are generated by oxidative reactions and are consumed as the energy is either (1) transferred into other high-energy bonds involved in synthetic reactions (e.g., amino acid + ATP → amino acyl

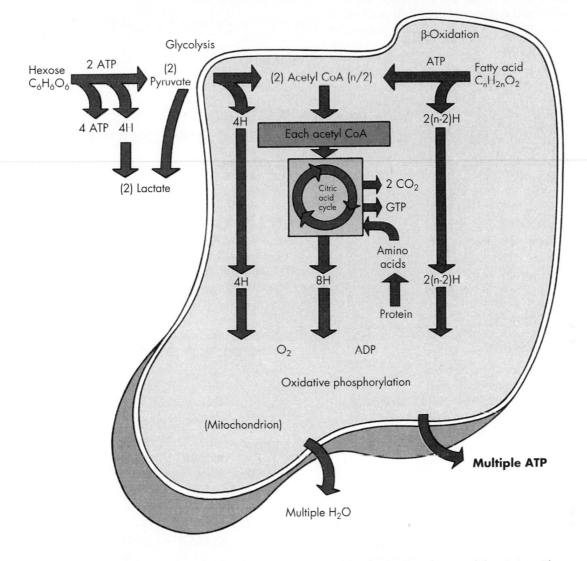

FIGURE 41-2   An overview of energy production. Glycolysis and each turn of the citric acid cycle supply only 2 ATP equivalents apiece. ATP is generated mainly when hydrogens removed from carbohydrate, fat, or protein substrates are oxidized in the mitochondria. *ATP,* Adenosine triphosphate; *ADP,* adenosine diphosphate; *GTP,* guanosine triphosphate.

AMP), (2) expended in creating lower-energy phosphorylated metabolic intermediates (e.g., glucose + ATP → glucose-6-phosphate), or (3) converted to mechanical work (e.g., propulsion of spermatozoa). Because the production and transfer of energy are not 100% efficient, about 18 kcal of substrate is required to generate each terminal P-O bond of ATP. If 2300 kcal is turned over in a typical day, about 128 moles, or 63 kg, of ATP (a mass approximating body weight) is generated and expended. An overview of energy production with generation of ATP from the major substrates is shown in Figure 41-2.

The combustion of carbohydrates, chiefly glucose with lesser amounts of fructose and galactose, includes two major phases:

1. At the end of an anaerobic phase known as **glycolysis** (Embden-Meyerhof pathway), each glucose molecule has been degraded to two molecules of pyruvate but has yielded only 8% of its energy content. Glycolysis can serve as a sole source of energy only briefly because (a) the supply of glucose is limited, and (b) the accumulated pyruvate must be syphoned off by reduction to lactate, a metabolite that if accumulated is ultimately noxious.

2. During an aerobic phase the two pyruvate molecules are decomposed to CO_2 via the **citric acid cycle** (Krebs cycle), and the remaining energy is liberated. In this pathway **acetyl coenzyme A** (acetyl CoA), initially formed by oxidative decarboxylation of pyruvate, is condensed with oxaloacetate to form citrate. Through a cyclic series of reactions the carbons of acetyl CoA appear as CO_2, and oxaloacetate is regenerated.

The combustion of fatty acids, the major energy component of fats, first requires their transfer from the cytoplasm to the mitochondria via **carnitine** esters generated from and then reconverted to CoA derivatives. The fatty acyl CoA's then enter a mitochondrial repetitive biochemical sequence known as **β-oxidation.** This process releases two carbons at a time as acetyl CoA until the entire fatty acid molecule is broken down. The resultant acetyl CoA is disposed of via the citric acid cycle, as already described. A variable portion of fatty acid oxidation in the liver stops at the last four carbons and yields **acetoacetic** and **β-hydroxybutyric acids.** When the rate of fatty acid delivery is high, more of these water-soluble ketoacids are generated and released by the liver to be oxidized in other tissues.

The condition known as **ketosis** occurs when fasting is prolonged beyond the usual overnight period or when carbohydrate intake is low and ketoacids accumulate. It develops to an extreme degree in **diabetes mellitus** when the hormone insulin is very deficient.

The combustion of protein first requires hydrolysis to its component amino acids. Each of these undergoes degradation by individual pathways, which ultimately lead to intermediate compounds of the citric acid cycle and then to acetyl CoA and CO_2.

The combustion of all substrates yields large numbers of hydrogen atoms. These hydrogens are oxidized to H_2O in the mitochondrion in linkage with phosphorylation of ADP to ATP (Figure 41-2). In this process, 3 high energy P-O bonds are formed for each atom of oxygen used. This usually yields an overall efficiency of 60% to 65% for the recovery of usable chemical energy.

Respiratory Quotient

In the process of oxidizing substrates, the proportion of carbon dioxide produced ($\dot{V}_{CO_2}$) to oxygen used ($\dot{V}_{O_2}$) varies according to the fuel mix. The ratio of $\dot{V}_{CO_2}$ to $\dot{V}_{O_2}$ is known as the **respiratory quotient** (RQ). As indicated by the following equations, RQ equals 1.0 for oxidation of carbohydrate (e.g., glucose), whereas RQ equals 0.70 for oxidation of fat (e.g., palmitic acid).

For carbohydrates:

$$C_6H_{12}O_6 + 6O_2 \rightarrow 6CO_2 + 6H_2O \qquad (1)$$

Glucose

$$RQ = \frac{6O_2}{6CO_2} = 1.0$$

For fats:

$$C_{15}H_{31}COOH + 23O_2 \rightarrow 16CO_2 + 16H_2O \qquad (2)$$

Palmitic acid

$$RQ = \frac{16CO_2}{23O_2} = 0.70$$

The RQ for protein reflects that of the individual RQs of the amino acids and averages 0.80. Ordinarily, protein is a minor energy source. The small contribution of protein oxidation to the overall RQ can be corrected for by measuring the urinary excretion of the nitrogen that results from the metabolism of amino acids.

Normal humans can vary their fuel mix and RQ from 0.7 to 1.0 without difficulty. Patients with very poor pulmonary function who cannot excrete CO_2 efficiently benefit from a low RQ. They are therefore given a higher proportion of fat calories as an energy source so that the least amount of CO_2 is produced for the quantity of O_2 used.

ENERGY STORAGE AND TRANSFERS

The intake of energy in the form of food is periodic. Its time course does not match either the constant rate of energy expenditure in the basal state or the variable rate expended during intermittent muscle work. Therefore, the organism must have mechanisms for storing ingested energy for future use. The greatest part of these energy reserves (75%) is in the form of fat, stored as triglycerides, in adipose tissue. In normal-weight humans, fat constitutes 10% to 30% of body weight, but it can reach 80% in very obese individuals. Fat is a particularly efficient storage fuel because of its high caloric density (i.e., 9 kcal/g) and because it engenders little additional weight as intracellular water.

Triglycerides are formed by esterification of free fatty acids with α-glycerol phosphate. Free fatty acids arise from digestion of dietary fat, but they can also be synthesized from acetyl CoA derived from oxidation of glucose (Figure 41-3). Thus, dietary carbohydrate can be

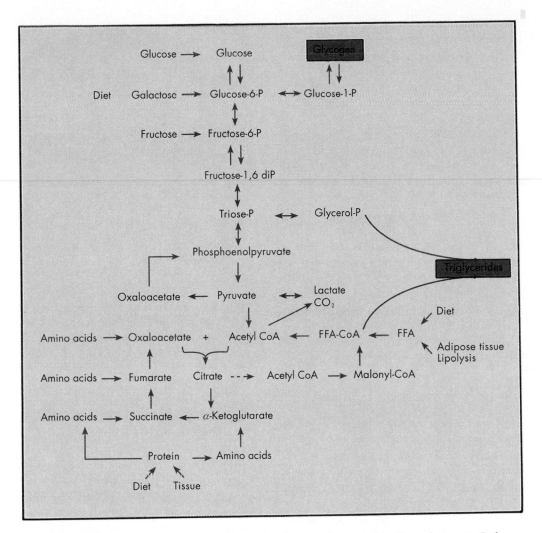

FIGURE 41-3   An overview of the chemical pathways of energy transfer and storage. Carbohydrates funnel through glucose-6-phosphate to be stored as glycogen. Alternatively, they can undergo glycolysis to pyruvate and be used for synthesis of fatty acids. The latter, regardless of source, are esterified with glycerol-phosphate and stored as triglycerides. Amino acids from endogenous or exogenous sources are converted to glucose via oxaloacetate or pyruvate.

converted to fat in the liver and transferred to adipose tissue for storage in that more efficient form.

Protein (4 kcal/g) constitutes almost 25% of the potential energy reserves, and the component amino acids can contribute to the glucose supply. However, because virtually all proteins serve some vital structural and functional role, their use as a major source of energy is not desirable.

Carbohydrate (4 kcal/g) in the form of a glucose polymer, **glycogen,** forms less than 1% of total energy reserves. However, this portion is critical for support of central nervous system metabolism and for short bursts of intense muscle work. Approximately one fourth of the glycogen stores (75 to 100 g) is in the liver, and about three fourths (300 to 400 g) is in the muscle mass. Liver glycogen can be made available to other tissues by the process of **glycogenolysis** and glucose release. Muscle glycogen can only be used by muscle, because this tissue lacks the enzyme glucose-6-phosphatase, which is required for dephosphorylation of glucose prior to its entrance into the bloodstream.

Glycogen can be formed from all three major dietary sugars (Figure 41-3). In addition, in the liver (and to a much lesser extent in the kidney) glucose itself can also be synthesized de novo from the three carbon precursors pyruvate, lactate, and glycerol, and from parts of the carbon skeleton of all 20 amino acids in protein, except leucine. This process, known as **gluconeogenesis,** converts two pyruvate molecules to glucose.

Gluconeogenesis is not a simple reversal of all the reactions of glycolysis (Figure 41-3). The chemical free energy change is too large to permit efficient backward flow of the glycolytic reactions at three steps: (1) pyruvate to phosphoenolpyruvate, (2) fructose-1,6-diphosphate to fructose-6-phosphate, and (3) glucose-6-phosphate to glucose. The first step requires energy input in the form of ATP and guanosine triphosphate (GTP). Substitution of simple phosphatase reactions reverses the last two steps.

It is also important to realize that net glucose synthesis cannot occur from acetyl CoA, even though carbon atoms from acetyl CoA can become part of glucose molecules via oxaloacetate and the citric acid cycle. Thus fat can only contribute to carbohydrate stores by way of the 3-carbon glycerol moiety of triglycerides, a minor source.

The processes of energy storage and transfer themselves expend energy. This partly accounts for the stimulation of oxygen utilization after a meal (i.e., diet-induced thermogenesis). The cost of storing dietary fatty acids as triglycerides in adipose tissue is only 3% of the original calories, and the cost of storing glucose as glycogen is only 7% of the original calories. In contrast, conversion of carbohydrate to fat uses up 23% of the original calories, and a similar amount is expended in storing dietary amino acids as protein or in converting them to glycogen.

Because glucose and fatty acids are alternative and in effect competing energy substrates, some relationship between their use and their synthesis and storage within cells could be expected. For example, when dietary glucose is plentiful, glycolysis is augmented, more acetyl CoA is generated from pyruvate, and more citrate is formed (Figure 41-3). Citrate is a potent activator of the first step in the synthesis of fatty acids (acetyl CoA → malonyl CoA). In addition, glycolysis will produce more glycerol phosphate from triose phosphates. The combination of increased fatty acid synthesis and glycerol phosphate availability results in accentuated synthesis of triglycerides and reduced oxidation of fat. Thus, *increased carbohydrate use shifts fat metabolism from oxidation to storage.* Conversely, under circumstances where fatty acid supply is augmented, β-oxidation increases. Several of its products then retard glycolysis and enhance gluconeogenesis and glycogen synthesis. Thus *increased use of fat as fuel shifts carbohydrate metabolism from oxidation to storage.* Many of these intrinsic chemical checks and balances are also reinforced by hormonal signals.

In addition to these intracellular relationships, transfer of energy between organs also occurs (Figure 41-4). The stored energy contained within adipose tissue triglycerides is transported as free fatty acids to the liver. There part of the energy (not the carbon atoms) is effectively transferred to glucose molecules. This is be-

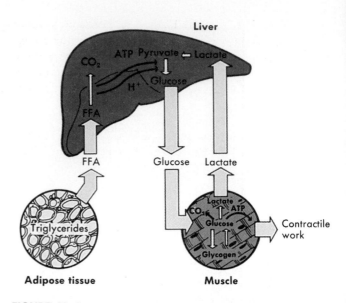

FIGURE 41-4 Interorgan energy transfers. Energy contained in free fatty acids *(FFA)* can be transferred to energy contained in glucose in the liver. Lactate released from muscle glycogen by glycolysis can carry energy back to the liver, where the lactate is built back into new glucose molecules (and glycogen).

cause as fatty acids are oxidized and yield ATP, gluconeogenesis is stimulated concurrently and uses the ATP that was generated. The newly synthesized glucose molecules in turn can then be transported to muscle tissue, where their energy is released during glycolysis and applied to muscle contraction. Furthermore, if the lactate produced exceeds the ability of the muscle to oxidize it rapidly enough in the citric acid cycle, the lactate can be returned to the liver, where it may again be built back up into glucose molecules. From this viewpoint *the liver is a flexible and versatile organ that can transmute and transfer energy from fuel depots to working tissues.* Hormones help to regulate this process.

When liver disease causes severe hepatic insufficiency (e.g., **alcoholic hepatitis** or **cirrhosis**), energy metabolism can be markedly distorted. Inability to store glucose as glycogen leads to hypoglycemia during fasting. Inability to take up lactate produced by peripheral anaerobic glycolysis can cause high plasma lactic acid levels and serious metabolic acidosis.

CARBOHYDRATE METABOLISM

Dietary carbohydrates give rise to various sugars (hexoses), the most important of which are glucose, fructose, and galactose. In addition to serving as energy sources, sugars are also components of glycoproteins, glycopeptides, and glycolipids that have structural and functional roles. Examples include basement membrane collagen, mucopolysaccharides, nerve cell myelin, hormones, and hormone receptors.

Glucose is the central molecule in carbohydrate metabolism. Other sugars are metabolized through the glucose pathways. Postabsorptive plasma glucose concentration averages 80 mg/dl (4.5 mM/L), with a range of 60 to 115 mg/dl.

> When the plasma glucose level falls below 60 mg/dl, as may occur with an overdose of insulin, uptake of the sugar and utilization of oxygen by the brain decrease in parallel. Central nervous system function becomes progressively impaired. Convulsions or coma may occur, and death may ensue.

The major products of glycolysis, lactate and pyruvate, circulate at average concentrations of 0.7 and 0.07 mM, respectively. This 10:1 ratio of lactate to pyruvate prevails even when the glycolytic rate changes. However, when tissues are deprived of oxygen, the equilibrium between the two shifts toward lactate, the reduced molecule. Plasma concentration ratios as high as 30:1 may then be observed. Ischemia may produce very high concentrations of lactate that result in metabolic acidosis.

In the basal state, glucose turnover is about 2 mg/kg/min (11 mmole/kg/min), which is equivalent to about 9 g/hr or 225 g/day in adults. Approximately 55% of glucose utilization results in terminal oxidation to CO_2, of which the brain accounts for the greatest part. Another 20% is caused by glycolysis; the resulting lactate then returns to the liver for resynthesis into glucose (Cori cycle). Reuptake by the liver and other splanchnic tissues accounts for the remaining 20% of glucose use. Most of the glucose use (about 70%) in the basal state is independent of insulin, a hormone with otherwise important regulatory effects on glucose metabolism.

The circulating pool of glucose is only slightly larger than the liver output in 1 hour, and it can maintain brain oxidation for only 3 hours. This emphasizes the crucial importance of continuous hepatic production of glucose in the fasting state. About 80% of this production results from glycogenolysis and 20% from gluconeogenesis. Hepatic uptake and utilization of lactate accounts for more than half the glucose supplied by gluconeogenesis. The remainder is largely accounted for by amino acids, especially alanine. The supply of lactate comes from glycoly-

sis in muscle, red blood cells, white blood cells, and a few other tissues. The amino acid precursors come from proteolysis of muscle.

When an individual ingests glucose after overnight fasting, approximately 70% of the load is assimilated by peripheral tissues, mainly muscle, and about 30% by splanchnic tissues, mainly liver. Only 25% is oxidized during the 3 to 5 hours required for its absorption from the gastrointestinal tract. The remainder is stored as glycogen in muscle and liver. Glucose initially stored as muscle glycogen can later be transferred to the liver by undergoing glycolysis to lactate, which is released into the circulation; the lactate is then taken up by the liver, rebuilt into glucose, and stored as glycogen in that organ. During the period of peak absorption of exogenous glucose, hepatic output of the sugar is largely unnecessary and is greatly reduced from basal levels. These metabolic adaptations are facilitated by coordinated secretion of the pancreatic islet hormones, insulin and glucagon.

PROTEIN METABOLISM

The average adult body contains 10 kg of protein, of which about 6 kg is metabolically active. Approximately 50 g of amino acids is released daily by proteolysis from muscle, the main endogenous repository. A portion of this amount is reused for protein synthesis, but much is degraded. A daily dietary protein intake of 0.8 g/kg is ordinarily sufficient for an adult human to remain in balance. When accretion of lean body mass is taking place (for example, in growing children, pregnant women, persons recovering from prior weight loss), daily protein requirements increase to 1.5 to 2.0 g/kg.

All proteins are composed of the same 20 amino acids. Half of these are called **essential amino acids** because their carbon skeletons, the corresponding α-ketoacids, cannot be synthesized by humans. Once present, however, these ketoacids can be converted to the essential amino acids by transamination. The other half, the **nonessential amino acids,** can be synthesized endogenously, because the appropriate carbon skeletons can be built from glucose metabolites in the citric acid cycle. The essential amino acids must be supplied in the diet in amounts that range from 0.5 to 1.5 g/day. All 20 amino acids are required for normal protein synthesis; therefore, a deficiency of even one essential amino acid disrupts this process.

Protein sources vary greatly in their biologic effectiveness, depending in part on the ratio of the essential to nonessential amino acids. Milk and egg proteins are of the highest quality in this regard, but properly combined vegetable sources can provide all essential amino acids.

About 40% of the protein intake of infants and children should consist of essential amino acids in order to support growth. In adults this requirement falls to 20%. In addition to their incorporation into proteins, many of the amino acids, including some essential ones, are precursors for important molecules, such as purines, pyrimidines, polyamines, phospholipids, creatine, carnitine, methyl donors, thyroid and catecholamine hormones, and neurotransmitters.

Deficiency of protein intake is a worldwide problem. When people are chronically deprived of both protein and calories, marked loss of muscle mass and adipose tissue results. When calories are sufficient but protein is deficient over relatively short periods of time, a syndrome known as **Kwashiorkor** occurs. This is characterized by low plasma albumin levels, edema caused by low plasma oncotic pressure, fragile hair, depressed immune function, lymphopenia, reduced wound healing, increased infections, and fatty infiltration of the liver.

All 20 amino acids, after removal of the amino group, are completely oxidized to CO_2 and H_2O. Each amino acid traverses a specific degradative pathway. (Refer to standard biochemistry textbooks for details.) However, all these pathways converge into three general metabolic processes: gluconeogenesis, ketogenesis, and ureagenesis. Except for leucine, all the amino acids can contribute carbon atoms for the synthesis of glucose. Five ketogenetic amino acids give rise to acetoacetate. In the degradation of all amino acids, ammonia is released. Ammonia, incorporated mainly into glutamine and alanine molecules, is then transported to the liver. In the liver ammonia is "detoxified" by incorporation into urea, a metabolically inert molecule. The synthesis of urea via the Krebs-Henseleit cycle is depicted in Figure 41-5. The urea that results from protein degradation is excreted by the kidney (see Chapter 35).

In the healthy adult under steady-state conditions, the total daily nitrogen excreted in the urine as urea plus ammonia, along with minor losses of nitrogen in the feces (0.4 g/day) and skin (0.3 g/day), is equal to the nitrogen released during metabolism of exogenous and endogenous protein. Such an individual is said to be in **nitrogen balance.** When there is no dietary protein intake, the sum of urea plus ammonia nitrogen in the urine reflects almost quantitatively the rate of endogenous protein degradation. When protein breakdown is greatly accelerated by tissue trauma or disease (for example, after major gastrointestinal surgery or with sepsis), urinary urea plus ammonia nitrogen may exceed protein nitro-

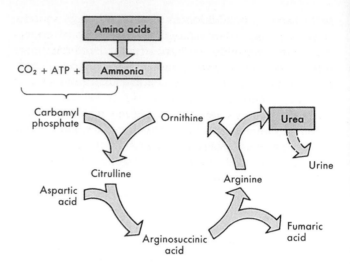

FIGURE 41-5   The Krebs-Henseleit urea cycle for disposal of amino acid ammonia. Two ammonia molecules are contributed by aspartic acid. These are effectively combined with one CO_2 molecule to form metabolically inert urea, which is excreted in the urine.

gen intake. In these cases the individual is said to be in **negative nitrogen balance.** In a growing child or in a previously malnourished individual undergoing protein repletion with a gain in lean body mass, urinary urea plus ammonia nitrogen excretion is less than the intake of protein nitrogen. This individual is said to be in **positive nitrogen balance.**

Healthy adults who receive isocaloric diets containing adequate protein and who are in nitrogen balance synthesize and degrade body protein at a rate of 3 to 4 g/kg/day. Approximately 5% of this total is accounted for by hepatic synthesis of albumin, a protein that turns over rapidly. When the diet is severely deficient in energy, in total protein, or in one of the essential amino acids, the rate of total body protein synthesis diminishes. In compensation, protein degradation also diminishes, but not to the same extent as synthesis, so that net loss of body protein results.

FAT METABOLISM

Fat represents almost half the total daily substrate for oxidation (about 100 g, or 900 kcal). As noted previously, fat is the major and most efficient form of stored fuel. The usual daily intake in the United States is also approximately 100 g, or 40% of total calories. The major component of both dietary and storage fat is triglycerides. These largely consist of long-chain saturated and monounsaturated fatty acids (chiefly palmitic, stearic, and oleic acids) esterified to glycerol. Because these fatty acids can also be synthesized in the liver and adipose tissue, in an overall sense, no strict dietary requirement exists for fat. *However, about 3% to 5% of fatty acids are polyunsaturated and cannot be synthesized in the body.* These are termed **essential fatty acids** (linoleic, linolenic, and arachidonic), because they are required as precursors for certain membrane phospholipid and glycolipid substances, as well as for important intracellular mediators known as **prostaglandins.** Patients who receive all nutrition parenterally must be provided these fatty acids to prevent dermatitis. Another component of fat is the steroid molecule **cholesterol,** which serves a variety of specific functions in membranes and is the precursor for bile acids and steroid hormones. Cholesterol is both ingested and synthesized by most cells.

> The typical fat intake cited above is now deemed excessive for good health, particularly because it promotes **atherosclerosis.** In this common condition, plaques laden with lipid form in the walls of arteries and are often the sites of thrombosis, which obstructs the flow of blood and causes necrosis of vital tissue, such as heart muscle or cerebral cortex. Current nutritional recommendations are for intake of fat not to exceed 30% and saturated fat (usually animal fat) not to exceed 10% of total calories. Monounsaturated fat should modestly exceed polyunsaturated fat. Cholesterol intake should be less than 600 mg/day and reduced to below 300 mg/day if plasma cholesterol is elevated.

Transport of the nonpolar lipids in plasma requires that they be incorporated into various complex lipoprotein particles, which turn over at different rates and also interact with each other. The protein portion of each particle is derived from several **apoproteins** synthesized in the liver and intestine. These apoproteins serve catalytic functions and interact with specific cell receptors. Figure 41-6 summarizes the metabolic pathways and interactions of the lipoprotein particles.

Chylomicrons, formed from dietary fat (see Chapter 34), are the lowest-density lipoproteins. After their absorption from the gastrointestinal tract, they disappear from plasma rapidly. Their major component is triglycerides, which are partly hydrolyzed by the key enzyme **lipoprotein lipase** on capillary endothelial surfaces. This enzyme is activated by apoprotein C-II and transferred to the chylomicron by **high-density lipoprotein** (HDL) particles (see below). The resulting free fatty acids are taken up both by adipose cells for resynthesis into triglycerides and storage and by other cells for oxidation. The residual lipoprotein particles, now relatively higher in cholesterol content and known as **chylomicron**

remnants, are taken up by the liver for further degradation. This uptake is directed by apoproteins E and B-48, which react with specific hepatic receptors.

In contrast, **very-low-density lipoprotein** (VLDL) particles are formed in the postabsorptive state by endogenous synthesis in the liver, and to a lesser extent in the intestine. They are denser, contain somewhat more cholesterol than chylomicrons, and have a longer plasma half-life. Their rate of formation varies from as little as 15 to as much as 90 g/day. The initial metabolism of VLDL is the same as that of chylomicrons. The product of lipoprotein lipase action consists of particles called **intermediate-density lipoprotein** (IDL). About half the IDLs return to the liver and are taken up, as occurs with chylomicron remnants. The other half of the IDLs are further enriched with cholesterol to form **low-density lipoprotein** (LDL) particles. Circulating LDL is responsible for transferring cholesterol into other cells. Uptake of LDL, IDL, and remnant particles takes place by initial interaction between apoproteins E and B100 and specific cell receptors, which is then followed by endocytosis.

The uptake of LDL cholesterol by cells has important regulatory actions on intracellular cholesterol metabolism. Cholesterol uptake from plasma downregulates the LDL receptor, and thereby reduces further entrance of the sterol. Uptake of cholesterol also suppresses its own de novo synthesis.

HDL particles are synthesized in the liver and intestine and have a long half-life. These particles facilitate the major steps in fat metabolism just described. HDL particles exchange key apoproteins with the other lipoprotein particles. The smaller, denser HDL_3 particles accept free-cholesterol molecules from chylomicrons, from VLDL, IDL, and remnant particles, and from peripheral cells. The cholesterol is esterified via the enzyme lecithin-cholesterol acyltransferase (LCAT). The resulting cholesterol esters are then exchanged for triglycerides on other particles via a cholesterol-ester transfer protein. The HDL_3 is transformed to the larger, more buoyant HDL_2 particle in this exchange. The triglycerides are removed from this particle by the action of hepatic lipase, and this restores the HDL_3. The net effect of such HDL cycling is to accelerate clearance of triglycerides from plasma and regulate the ratios of free to esterified cholesterol.

Free fatty acids circulate in an average concentration of 400 μmol/L bound to albumin molecules. They undergo a very rapid turnover of approximately 8 g/hr. Half of this represents oxidation and half reesterification to triglycerides. The plasma concentration of total cholesterol (average, 185 mg/dl), and especially of LDL cholesterol (average, 120 mg/dl), is a very important risk factor for atherosclerosis and death from cardiovascular events. On the other hand, a higher plasma HDL cholesterol (average, 50 mg/dl) exerts a protective effect against cardiovascular disease.

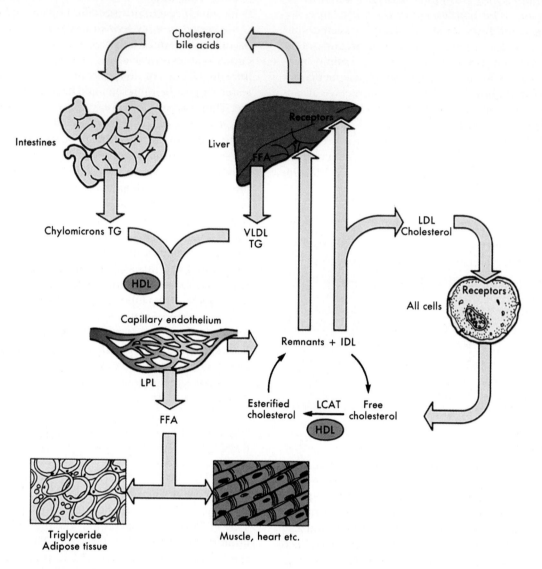

FIGURE 41-6 A schematic overview of lipoprotein metabolism and major aspects of lipid turnover in humans. Exogenous triglycerides *(TG)* in the form of chylomicrons absorbed from the intestine and endogenous triglycerides (very-low-density lipoproteins, *VLDL*) produced in the liver both give rise to free fatty acids *(FFA)* for storage in adipose tissue and oxidation in muscle. High-density lipoprotein particles *(HDL)* facilitate, and the enzyme lipoprotein lipase *(LPL)* directly catalyzes, liberation of the free fatty acids from triglycerides. The resultant particles, called remnants from chylomicrons and intermediate-density lioproteins *(IDL)* from VLDL, undergo further change in the circulation, which also is facilitated by HDL. The ratio of esterified cholesterol to free cholesterol is increased in the remnant and IDL particles by the enzyme lecithin-cholesterol acyltransferase *(LCAT)*. The remnant particles are then taken up by the liver for further metabolism. The IDL particles are partly taken up by the liver and partly converted to cholesterol-rich low-density lipoprotein *(LDL)* particles. The latter are then taken up by virtually all cells after interaction with specific LDL receptors. Cholesterol, either synthesized in the liver or extracted from remnant and IDL particles, is also excreted into the intestine, partly as bile acids.

Genetic abnormalities in apoprotein and lipid receptor particles account for a substantial proportion of **dyslipidemias.** Premature development of coronary artery disease is often the consequence. For example, mutant apoprotein E molecules cause familial forms of **hyperlipidemia,** with elevations of both triglyceride and cholesterol levels caused by accumulation of IDL and remnant particles. Excessive apoprotein B-100 content of VLDL and LDL particles characterizes **familial combined hyperlipidemia,** in which cholesterol and often triglyceride levels are high. In **familial hypercholesterolemia,** mutant LDL receptor types prevent normal cellular uptake of cholesterol. This results in extremely high plasma concentrations in homozygotes (<700 mg/dl), visible and palpable deposits of cholesterol in skin and tendons, and even coronary thrombosis in children. Attempts are underway to transplant a normal LDL receptor gene into homozygotes to prevent early death.

METABOLIC ADAPTATIONS

Fasting

In the fasting state the individual totally depends on endogenous substrates for energy. Mobilization of glucose provides essential fuel for the central nervous system; release of free fatty acids provides for the oxidative needs of the other tissues. An increase in protein degradation to amino acids is also a fundamental feature of this response. *The fasting individual is said to be in a state of catabolism because carbohydrate, fat, and protein stores are all decreasing.*

The liver supplies glucose to the circulation initially by augmenting glycogenolysis. After 12 to 15 hours of fasting, however, hepatic glycogen stores are almost depleted, and a rapid enhancement of gluconeogenesis fills the void. To supply glucose precursors, 75 to 100 g of muscle protein is broken down daily during the first few days. This is reflected in a rising excretion of nitrogen in the urine. Gluconeogenesis is also supported by the provision of 15 to 20 g of glycerol daily, which is released during the accelerated lipolysis of triglycerides in adipose tissue. Glucose oxidation in muscle and liver is spared as increasing quantities of free fatty acids become available. Their oxidation in the liver yields the ketoacids, which can also be oxidized by muscle cells, and further spares utilization of glucose. The net shift away from glucose and toward fatty acid oxidation lowers the respiratory quotient.

These adaptations are also reflected in changing plasma concentrations of substrates. The concentrations of glucose and of alanine, the major gluconeogenic amino acid, decrease, whereas the concentrations of free fatty acids, glycerol, and branch-chain amino acids, such as leucine, increase. High levels of the strong ketoacids produce a slight reduction in plasma bicarbonate and blood pH.

As fasting is prolonged beyond a few days, other important adaptations occur. Total energy expenditure, reflected in the BMR, decreases 10% to 20%, limiting the drain on energy stores. The central nervous system no longer depends entirely on glucose as an energy source, and much of its needs are eventually met by the ketoacids. Gluconeogenesis, therefore, diminishes; and protein breakdown declines to 25 g/day. In long-term fasting, body weight diminishes by an average of 300 g/day, of which two thirds is accounted for by fat and one third by lean tissue, 25% of which is protein.

Circumstances such as very prolonged hunger strikes or accidental isolation in areas totally lacking in food sources demonstrate the above adaptations to fasting. As long as sufficient water is available to prevent dehydration, the adipose stores of a normal human (about 10 kg) can sustain the reduced basal energy needs and greatly limited physical activity (about 1400 kcal/day) for up to 60 days. Likewise, the mobilizable protein stores (about 6 kg) can supply the diminished requirements for glucose oxidation. However, the loss of protein leads to progressive muscle weakness, apathy, organ dysfunction, and ultimately to death.

Exercise

The metabolic response to exercise resembles the response to fasting, in that the mobilization and generation of fuels for oxidation are dominant factors. The types and amounts of substrate vary with the intensity and duration of the exercise (Figure 41-7). For very intense, short-term exercise (e.g., a 10- to 15-second sprint), stored creatine phosphate and ATP provide the energy at a rate of approximately 50 kcal/min. When these stores are depleted, additional intensive exercise for up to 2 minutes can be sustained by breakdown of muscle glycogen to glucose-6-phosphate, with glycolysis yielding the necessary energy (at a rate of 30 kcal/min). This anaerobic phase is not limited by depletion of muscle

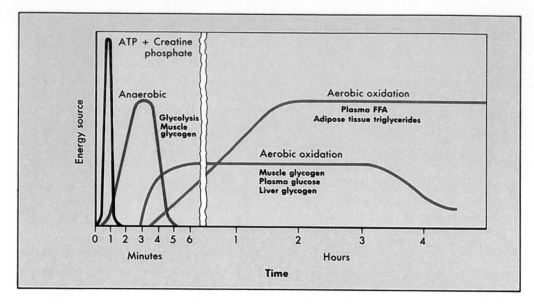

FIGURE 41-7   Energy sources during exercise. Note the sequential use of stored high-energy phosphate bonds as creatine phosphate and glycogen for short-term intensive exercise. Circulating glucose and free fatty acids *(FFA)* become increasingly important with time, and the latter dominate in sustained exercise.

glycogen at this point, but rather by the accumulation of lactic acid in the exercising muscles and the circulation.

> Several muscle diseases result from defects in glycolysis caused by deficiency of an enzyme in the pathway from glycogen to pyruvate/lactate. An example is **McArdle's disease,** or **muscle phosphorylase deficiency.** Such patients experience pain and weakness after exercise. The glycolytic defect is diagnosed by demonstrating a failure of lactate levels to increase in the antecubital vein after brief forearm muscle exercise with arterial inflow excluded by a cuff.

After several minutes of exhaustive anaerobic exercise, an oxygen debt of 10 to 12 liters can be built up. This must be repaid before the exercise can be repeated: (1) accumulated lactic acid must be oxidized or rebuilt into glucose; (2) muscle ATP and creatine phosphate contents must be restored; and (3) oxygen normally present in the lungs, body fluids, myoglobin, and hemoglobin must be replenished.

For less intense but longer periods of exercise, aerobic oxidation of substrates is required to produce the necessary energy (at about 12 kcal/min). Substrates from the circulation are added to muscle glycogen. Glucose uptake from the plasma increases dramatically, up to

thirty-fold in some muscle groups. To meet this need, hepatic glucose production increases up to five-fold. Initially this production is largely from glycogenolysis, but gluconeogenesis becomes increasingly important as liver glycogen stores become depleted. However, endurance can be improved by high-carbohydrate feedings for several days before prolonged exercise (e.g., a marathon run), because this increases both liver and muscle glycogen stores. To support gluconeogenesis, amino acids are increasingly released by muscle proteolysis. Eventually, fatty acids, liberated from adipose tissue triglycerides, form the predominant substrate, which supplies two thirds of the energy needs during sustained exercise. Except for the increases in circulating pyruvate and lactate that result from enhanced glycolysis, the pattern of change in plasma substrates is similar to that of fasting, only telescoped in time.

> Genetic defects in oxidation of fatty acids lead to reduced exercise capacity and muscle pain or even progressive muscle weakness. Cardiac muscle dysfunction may occur. **Deficiencies of carnitine or carnitine palmitoyl transferase** impair transfer of fatty acids into the mitochondria. Deficiencies of beta oxidation enzymes cause even more severe consequences in infants who cannot tolerate even short fasts because overdependence on glucose for fuel leads to hypoglycemia.

REGULATION OF ENERGY STORES

As stated, the preponderance of stored energy consists of fat. What determines the proper quantity of this energy reserve, and what regulates it? Does an ideal relationship exist between fat mass and either total body weight or lean body mass? Clear-cut answers to these questions are not yet available.

A genetic influence on fat mass is suggested by (1) the greater similarity of adipose stores in identical twins than in fraternal twins; (2) the tendency for body mass of adopted children to resemble that of their biological parents more than that of their adoptive parents; and (3) the existence of obesity genes in animal models. One of the latter genes, the function of which is unknown, has recently been located on a human chromosome and is expressed specifically in adipose tissue. Environmental influences, especially the quality and quantity of the food available, are suggested by the excessive weight gain of certain laboratory animals in response to presentation of high-fat or "junk food" diets, as well as by the much greater prevalence of obesity in relatively affluent westernized societies than in other populations. Also, the human species has more energy-storage adipose cells per unit body mass than most other species. This excess may contribute to the propensity for obesity.

Some data suggest the existence of a particular set point for energy stores in each individual. Once adult weight is reached, it tends to be constant, at least until middle age. At that point, most humans incur at least a modest weight gain, and fat becomes a higher proportion of body weight. Normal-weight and genetically obese animals subjected to overfeeding or underfeeding experiments will return to their original weight and degree of fatness when again allowed free access to food. They will do this not only by adjusting food intake, but also by adjusting energy expenditure in the appropriate direction. Decreases in energy stores caused by excessive expenditure are compensated for by increased caloric intake. Control of appetite appears to reside in the hypothalamus. In rodents, evidence exists for both a hunger center in the lateral hypothalamus and a satiety center in the ventromedial hypothalamus.

A large number of neurotransmitters and neuropeptides (both hypothalamic and gastrointestinal) influence food intake. For example, serotonin inhibits and neuropeptide-Y stimulates feeding. Synthesis of the latter is in turn suppressed by insulin. Thus a complex set of signals interact to regulate caloric intake.

Increases in energy stores resulting from excessive food intake can be compensated for by increased energy expenditure. This may occur via thermogenic processes, such as "futile cycles," that are wasteful of

ATP (e.g., glucose → glucose-6-phosphate → glucose).

A specialized form of adipose tissue, **brown adipose tissue,** is present in rodents and to some extent in humans, especially newborns. Its large mitochondria are regulated by **thermogenin,** a protein that uncouples ATP production from oxygen utilization and thus produces only heat from substrate use. The sympathetic nervous system and various hormones also stimulate energy expenditure. In some combination, these factors at least limit weight gain when normal humans are subjected to overfeeding.

Obesity is a major and common health problem in this and many other countries. Diabetes, hypertension, hyperlipidemias, and associated cardiovascular disease are all increased in frequency, as is the mortality rate. No one cause for human obesity has been found, and the condition is likely heterogenous in origin. Some individuals behave as though appetite is not physiologically regulated. Others appear to have an elevated set point for energy stores in adipose tissue; when weight is lost by enforced caloric restriction, energy expenditure declines below the level that would be appropriate for their newly reduced body mass. This contributes to an inexorable regain of weight once caloric restriction is lessened.

A profile of hormones conducive to fat deposition (increased plasma insulin, decreased plasma growth hormone, increased cortisol secretion) is often present in commonplace obesity, but such a profile does not appear to be a primary defect.

Treatment with low-calorie diets, exercise, behavior modification techniques, drugs, and support groups all have short-term benefits in the management of obesity. However, long-term success in eliminating or even significantly decreasing the obese state is frustratingly poor. The radical approach of surgically reducing the functioning volume of the stomach is more effective, but it has attendant morbidity and mortality, which diminish its appeal for widespread use. A true cure for obesity awaits the identification of a fundamental reversible metabolic or regulatory defect.

SUMMARY

1. Energy input as carbohydrate, fat, and protein calories must equal energy expenditure. The latter is composed of basal, diet-induced, and nonshivering thermogenesis and sedentary activity components plus exercise. Basal metabolic rate (75% of total) is proportional to body mass, is in part genetically determined, and declines with aging.

2. Fatty acids are the major fuel in most tissues except for the central nervous system and red blood cells, where glucose is the major and obligatory oxidative substrate. Depending on availability, fatty acids and glucose are competitive substrates in muscle mass and liver.

3. Metabolism of glucose to pyruvate (anaerobic glycolysis), beta oxidation of fatty acids, and disposal of acetyl CoA by the Krebs cycle are the biochemical mechanisms that lead to mitochondrial generation of ATP by oxidative phosphorylation. The overall efficiency of energy yield is 65%.

4. Energy is mainly stored as adipose tissue triglycerides, with lesser amounts as protein. Carbohydrate stores as glycogen are trivial; hence the need for efficient glucose production by the liver (gluconeogenesis).

5. During long-term fasting, gluconeogenesis from amino acids and glycerol is required to sustain central nervous system metabolism and other critical functions that are dependent on glucose. The pathway of gluconeogenesis is partly a reversal of glycolysis, but it requires special steps from pyruvate. Increased use of fatty acids during fasting greatly increases the production of the ketoacids, β-hydroxybutyrate, and acetoacetate.

6. Endogenous protein turnover obligates a daily ingestion of protein—in particular, essential amino acids, such as leucine. Such amino acids are irreversibly degraded, and their carbon skeletons cannot be synthesized. The carbon skeletons of nonessential amino acids, such as alanine, are degraded and resynthesized daily.

7. Fat metabolism involves a variety of circulating lipoprotein particles that transfer triglycerides and cholesterol, either originating in the diet or from hepatic synthesis, to and from various tissues. Specific apoproteins and tissue receptors are required. Low-density lipoprotein particles, rich in cholesterol, play a role in the development of atherosclerosis, whereas high-density lipoprotein particles have a protective effect.

8. Energy needs during exercise are met in sequence by stored muscle creatine phosphate plus ATP, stored muscle glycogen, anaerobic glycolysis, and, finally, aerobic oxidation of glucose and fatty acids taken up from the plasma. These substrates are supplied by hepatic glycogenolysis and gluconeogenesis and by adipose tissue lipolysis.

9. The balance between energy intake and expenditure and their correlation with energy stores constitute a complex process, probably controlled in the hypothalamus and involving numerous amine and peptide neurotransmitters and neuromodulators.

10. Obesity can result from an altered set point of energy stores, from unregulated caloric intake, or from decreased energy use.

BIBLIOGRAPHY
Journal Articles

Bogardus C et al: Familial dependence of the resting metabolic rate, *N Engl J Med* 315:96, 1986.

Bouchard C, Despres JP, Mauriege P: Genetic and nongenetic determinants of regional fat distribution, *Endocr Rev* 14:72, 1993.

Cahill GF: Starvation in man, *N Engl J Med* 282:668, 1970.

Eckel RH: Lipoprotein lipase: a multifunctional enzyme relevant to metabolic diseases, *N Engl J Med* 320:1060, 1989.

Ferrannini E et al: The disposal of an oral glucose load in healthy subjects: a quantitative study, *Diabetes* 34:580, 1985.

Groop LC et al: Role of free fatty acids and insulin in determining free fatty acid and lipid oxidation in man, *J Clin Invest* 87:83, 1991.

Harris RBS: Role of set-point theory in regulation of body weight, *FASEB J* 4:3310, 1990.

Klein S, Goran M: Energy metabolism in response to overfeeding in young adult men, *Metab* 42:1201-1205, 1993.

Ravussin E et al: Determinants of 24-hour energy expenditure in man: methods and results using a respiratory chamber, *J Clin Invest* 78:1568, 1986.

Reeds P, James W: Nutrition: the changing scene—protein turnover, *Lancet* 1:571, 1983.

Roberts SB, Heyman MB, Evans WJ, Fuss P, Tsay R, Young VR: Dietary energy requirements of young adult men, determined by using the doubly labeled water method, *Am J Clin Nutr* 54:499-505, 1991.

Schaefer E et al: Pathogenesis and management of lipoprotein disorders, *N Engl J Med* 312:1300, 1985.

Sims EAH, Danforth E Jr: Expenditure and storage of energy in man, *J Clin Invest* 79:1019, 1987.

Tall AR: Plasma high density lipoproteins. *J Clin Invest* 86:379, 1990.

Wolfe BM et al: Effect of elevated free fatty acids on glucose oxidation in normal humans, *Metab* 37:323, 1988.

Yiying Zhang, Proenca R, Maffel M, Barone M, Leopold L, Friedman JM: Positional cloning of the mouse obese gene and its human homologue, *Nature* 372:425, 1994.

Books and Monographs

Kimball SR, Flaim KE, Peavy DE, Jefferson LS: Protein metabolism: In Rifkin H, Porte D, eds: *Diabetes mellitus,* ed 4, New York, 1990, Elsevier Scientific Publishing Co.

Leibowitz SF: Brain neurotransmitters and hormones in relation to eating behavior and its disorders. In Bjorntorp P, Brodoff BN, eds: *Obesity,* Philadelphia, 1992, JP Lippincott.

McGarry JD and Foster DW: Ketogenesis. In Rifkin H, Porte D, eds: *Diabetes mellitus,* ed 4, New York, 1990, Elsevier Scientific Publishing Co.

Seifter S, England S: Carbohydrate metabolism. In Rifkin H, Porte D, eds: *Diabetes mellitus,* ed 4, New York, 1990, Elsevier Scientific Publishing Co.

CHAPTER 42

Hormones of the Pancreatic Islets

The major pancreatic islet hormones, **insulin** and **glucagon,** are rapid and powerful regulators of metabolism. Their secretion is primarily determined by plasma substrate levels. *Together they coordinate the disposition of nutrient input from meals as well as the flow of endogenous substrates by actions on the liver, adipose tissue, and muscle mass.* Their cells of origin are intimately interspersed in anatomical islets that constitute 1% to 2% of the mass of the pancreas and are scattered throughout that organ. These islets are composed of 60% β-cells, the source of insulin, and 25% α-cells, the source of glucagon. The remaining islet cells secrete the peptides **somatostatin** (δ-cells) and **pancreatic polypeptide** (F-cells).

The strategic location of the islets (Figure 42-1) reflects their functional role. Insulin and glucagon are secreted in response to nutrient inflow and gastrointestinal secretagogues, as are the enzymes of the acinar pancreas (see Chapter 33). The islet hormones may have paracrine effects on nearby acinar cells, as well as on each other, through tight junctions and gap junctions between the endocrine cells. The location of the islets dictates secretion of insulin and glucagon into the pancreatic veins and then into the portal vein. This arrangement permits the liver, the central organ in nutrient traffic, to be exposed to higher concentrations of these hormones than are peripheral tissues. It also permits the liver to modulate the availability of insulin and glucagon to peripheral tissues by extracting variable amounts of these hormones during first passage through that organ.

Insulin and glucagon are often secreted reciprocally and act reciprocally. When one is needed, the other usually is not. Therefore the ratio of insulin to glucagon concentrations may be more critical than the absolute concentrations of each hormone. The consequences of isolated insulin deficiency—the common disease **diabetes mellitus**—are so devastating that insulin has dominated our physiological thinking. In contrast, isolated glucagon deficiency is virtually unknown in medicine; moreover, it can be compensated for by other mechanisms.

INSULIN

Synthesis and Secretion

Insulin is a peptide hormone with a molecular weight of 6000. It is composed of two straight chains linked by disulfide bridges. The B chain contains the core of biological activity, whereas the A chain contains most of the species-specific sites. Human, bovine, porcine, and fish insulins have equivalent biological activity on a molar basis. Although all have been used for treatment of diabetes mellitus, human insulin produced by recombinant DNA techniques has largely replaced animal insulins for therapeutic use.

The synthesis of insulin by the β-cell follows the general pattern for peptide hormones described in Chapter 40. The gene directs the synthesis of a preprohormone, from which the signal peptide is cleaved to yield the single-chain **proinsulin.** Establishment of the disulfide linkages is followed by excision of a connecting peptide, known as **C-peptide.** Insulin and C-peptide are packaged together in the secretory granules by the Golgi apparatus. These granules also contain zinc, which acts to join six insulin molecules into hexamers. Crystalline zinc insulin is the basic pharmaceutical preparation for treatment of diabetes mellitus.

Insulin is secreted by exocytosis (see Figure 40-5) of its granules, which are arrayed in parallel with microtubules in the β-cell cytoplasm. The microtubules are associated with a web of microfilaments that contain myosin and actin near the plasma membrane. On application of a stimulus, contraction of microfilaments draws the granules to the plasma membrane, where they fuse with

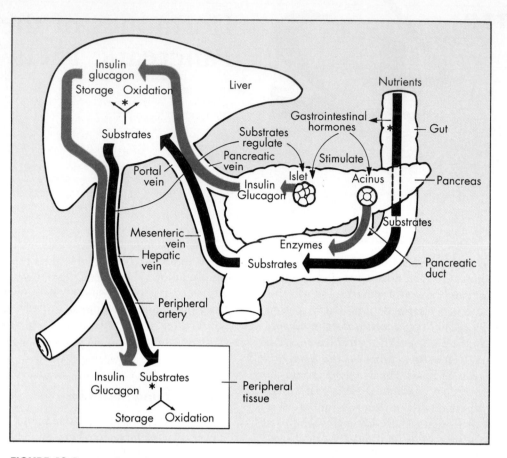

FIGURE 42-1 A schematic view of the pivotal location of the pancreatic islets. Secretion of the islet hormones insulin and glucagon is coordinated with secretion of exocrine pancreatic enzymes. Both are stimulated by entry of nutrients into the gastrointestinal tract and by gastrointestinal hormones. Pancreatic enzymes reach the intestinal lumen via the pancreatic duct. Islet hormones are secreted into the portal vein and thereby reach the liver together with the substrates resulting from nutrient digestion. Within the liver, these hormones direct storage or oxidation of the ingested substrates. Islet hormones that pass through the liver with substrates similarly affect metabolism of these substrates by peripheral tissues. Asterisks indicate the action sites

it, rupture, and release equimolar amounts of insulin and C-peptide.

Many agents can stimulate insulin secretion, but the most important stimulator of insulin secretion is glucose. The mechanism involves the following sequence: (1) A specific glucose transporter (glut-2) facilitates rapid diffusion into the β cell and maintains an intracellular glucose concentration equal to that of interstitial fluid. (2) The enzyme glucokinase, with a K_m for glucose of 5mM (the average fasting concentration), acts as a glucose sensor that controls the rates of glucose utilization. (3) Products of glucose metabolism, including ATP, NADH, and NADPH, increase and an ATP-sensitive K^+ channel closes. (4) This triggers the opening of a voltage-regulated Ca^{++} channel. The intracellular Ca^{++} increases and triggers exocytosis. (5) G-proteins linked to adenylyl cyclase and

to phospholipase C mediate the stimulatory and inhibitory actions of other modulators of insulin release via alterations in levels of cAMP and phosphatidylinositol products. In addition to causing insulin release, glucose stimulates synthesis of the hormone by increasing the transcription rate of the insulin gene and the translation rate of its mature messenger ribonucleic acid (mRNA).

Insulin-dependent, or **type 1, diabetes mellitus** results from complete destruction of the β-cells and ultimate loss of all insulin. However, in rare instances, much milder diabetes mellitus is caused by defined genetic abnormalities in insulin synthesis and secretion.

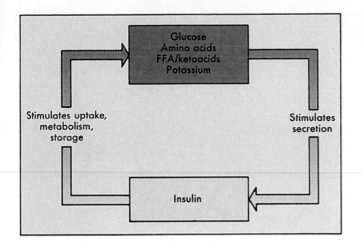

FIGURE 42-2 Feedback relationship between insulin and nutrients. Those nutrients that stimulate insulin secretion are the same nutrients whose disposal is facilitated by insulin. *FFA*, Free fatty acids.

These include mutant insulins with diminished biological activities, mutant proinsulins with impediments to processing, and mutant glucose transporters or glucokinases, which hinder rapid secretory responses to glucose.

Regulation of Secretion

In the broadest sense, insulin secretion is governed by a feedback relationship with the exogenous nutrient supply (Figure 42-2). When this supply is abundant, insulin is secreted in response; the hormone then stimulates utilization of these same incoming nutrients while it simultaneously inhibits mobilization of endogenous substrates. When nutrient supply is low or absent, insulin secretion is dampened and mobilization of endogenous fuels is enhanced.

The central regulating molecule is glucose. At plasma levels less than 50 mg/dl, little or no insulin is secreted, whereas the response is maximal at plasma levels greater than 250 mg/dl. Brief exposure of the β-cell to glucose induces a rapid but transient release of insulin. With continuous glucose exposure, this initial response fades, later to be replaced by a more prolonged second phase.

Glucose entry from the gastrointestinal tract follows digestion of the carbohydrate components of a meal. Under these circumstances insulin release is greater than can be accounted for by the degree to which plasma glucose levels rise. This results from release of gastrointestinal peptide hormones such as gastric in-

hibitory peptide (GIP), gastrin, secretin, cholecystokinin, and most notably a glucagon-like peptide, GLP-1, from intestinal cells, all of which can stimulate insulin secretion. In addition, digestion of the protein in a meal yields amino acids, some of which synergize with glucose in stimulating the β-cells. Lipids and their products contribute little directly to the β-cell response to a meal. When digestion and absorption of the nutrients are completed, plasma glucose and amino acid levels return to baseline, and insulin secretion subsides to a rate that is maintained steadily during the usual overnight fasting period.

If fasting is extended for days, insulin secretion declines below the basal rate and then resets at a lower level. In this state secretion is maintained by lower but still stimulatory plasma glucose levels, with contributions from greatly elevated ketoacid and free fatty acid levels. Insulin secretion is also modulated by cholinergic and β-adrenergic stimulatory and α-adrenergic inhibitory influences. All these factors cause physiological fluctuations in peripheral plasma insulin levels and in the average equivalent rates of insulin delivery into the peripheral circulation. These changes are summarized in Figure 42-3.

Non–insulin-dependent, or **type 2, diabetes mellitus** is the most common form of the disease. Evidence is accumulating that one important factor in type 2 diabetes is a subtle disturbance in the pattern of insulin secretion. This is characterized by altered cyclicity, diminished pulse frequency, delayed response to rising glucose levels, and/or loss of recognition of glucose as a stimulus. The primary cause of such β-cell dysfunction remains unknown.

Insulin concentration in the portal vein ranges from two to ten times higher than in the peripheral circulation. The liver extracts about half the insulin reaching it, but this varies with the nutritional state. Thus actual β-cell secretory rates are better estimated by measurement of plasma C-peptide, because this cosecreted molecule is not removed by the liver. Such estimates yield values for insulin secretion of 1.0 to 2.5 mg/day (25 to 40 units/day). These estimates are similar to the amounts of exogenous insulin required to normalize plasma glucose levels in individuals who do not produce any endogenous insulin. C-peptide and the small quantity of proinsulin secreted by β-cells have no known physiological actions.

Insulin has a short plasma half-life (6 to 8 minutes), mainly because of specific degradation in the kidney and liver. However, insulin is also degraded in conjunction

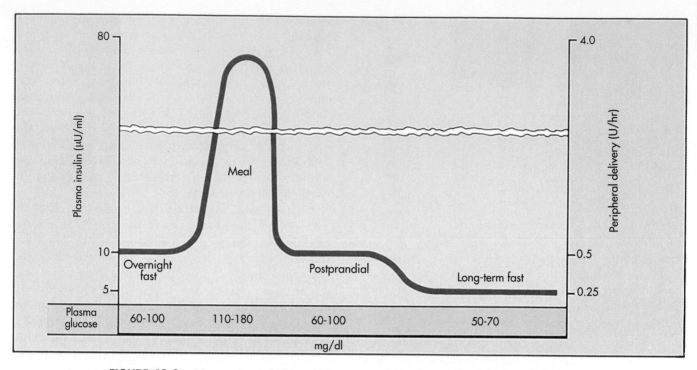

FIGURE 42-3   The pattern of plasma insulin levels and the corresponding insulin delivery rates to the peripheral circulation in various physiological states (10 μU/ml = 7 × 10⁻¹¹M). The usual prevailing plasma glucose levels are also indicated. Fasting insulin levels regulate mobilization of endogenous substrates at appropriate rates. Postmeal levels stimulate storage of substrates.

with its actions on its target cells after receptor binding and internalization of the hormone. Very little intact insulin is excreted in the urine.

Actions of Insulin

Fuel Turnover *The overall thrust of insulin action is to facilitate storage of substrates and inhibit their release* (Figure 42-4). As a result, secreted or administered insulin decreases the plasma concentrations of glucose, of free fatty acids and ketoacids, and predominantly of the essential branch-chain amino acids (leucine, isoleucine, valine). The major sites of insulin action are liver, muscle, and adipose tissue. In each target tissue, carbohydrate, lipid, and protein metabolism are regulated coordinately.

Under maximal insulin stimulation, the usual rate of glucose utilization by peripheral tissues is increased fivefold to sixfold. Simultaneously the output of glucose by the liver drops to considerably below half. Most of the extra glucose uptake occurs in muscle, with a very small fraction in adipose tissue. Approximately 75% of this glucose is converted to glycogen, and only 20% to 30% undergoes glycolysis and terminal oxidation to CO_2. The absolute rate of glucose oxidation, however, is increased threefold by insulin.

Hypersecretion of insulin by a β-cell tumor causes **hypoglycemia.** The latter produces central nervous system dysfunction, which ranges from mild difficulty in concentrating to severe behavioral disturbances, psychosis, convulsions, and coma. Symptoms are typically worse in the fasting state, and are countered by overeating, particularly carbohydrates, with resulting weight gain. The diagnosis is established by demonstrating inappropriately high plasma insulin and C-peptide levels when plasma glucose is low.

The basal rate of free fatty acid inflow to plasma from adipose tissue is also decreased markedly by insulin. (Simultaneously, inflow of glycerol, the other product of triglyceride hydrolysis, is decreased sharply.) As a result, insulin reduces the basal rate of lipid oxidation more than 90%. Because insulin also stimulates triglyceride synthesis and storage, the hormone facilitates weight gain.

The action of insulin on protein turnover can be assessed by determining the hormone's effect on the flux of the essential amino acid leucine. Maximal doses decrease the rate of inflow of leucine into plasma to almost half. Because the only source of leucine in the basal

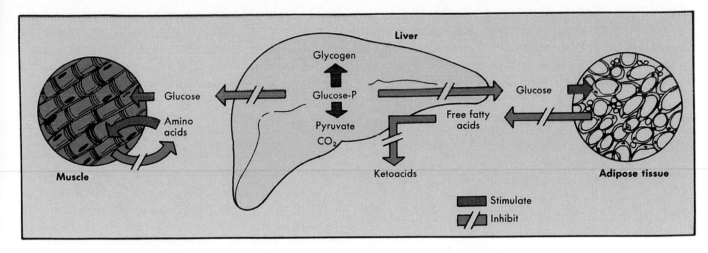

FIGURE 42-4 Effect of insulin on the overall flow of fuels. Tissue uptake of glucose and amino acids is stimulated by insulin; tissue release of glucose, amino acids, and free fatty acids is inhibited by insulin, as is ketogenesis. The net result is a decrease in plasma levels of these substrates.

state is endogenous protein, this action of insulin must be caused by inhibition of proteolysis. In addition, insulin decreases the rate of oxidation of leucine. The result of these insulin effects is a net gain in body protein.

Carbohydrate Metabolism *In muscle and adipose tissue, insulin stimulates the transport of glucose from the plasma, and into the cytoplasm, where it is rapidly phosphorylated. In muscle and liver, insulin largely stimulates glycogen formation from glucose-6-phosphate.* To a much lesser extent, insulin also stimulates glycolysis and oxidation of glucose. In adipose tissue, the most important effect of insulin is to stimulate production of α-glycerol phosphate from the triose phosphate intermediates of glycolysis. The α-glycerol phosphate is used to esterify free fatty acids and thus stores them as triglycerides.

Insulin stimulates conversion of glucose to glycogen and inhibits the reverse reaction, breakdown of glycogen to glucose, by altering the activities or concentrations (or both) of several enzymes (see Figure 41-3). The equilibrium between glucose and glucose-6-phosphate is shifted toward the latter because the phosphorylating enzyme glucokinase is increased by insulin. (In liver only, the dephosphorylating enzyme glucose-6-phosphatase is decreased by insulin.) Insulin increases the enzyme (glycogen synthase) that polymerizes phosphorylated glucose to glycogen, whereas insulin decreases the enzyme (phosphorylase) that degrades glycogen. Insulin also shifts the balance between glycolysis and gluconeogenesis toward the former and away from the latter. Glycolysis is accelerated because insulin increases the key enzymes phosphofructokinase, pyruvate kinase, and pyruvate dehydrogenase; gluconeogenesis is retarded because insulin decreases the enzymes fructose-1,6-diphosphatase, phosphoenolpyruvate carboxykinase, and pyruvate carboxylase.

In liver, glucose availability reinforces those effects of insulin that lead toward glycogen storage or glycolysis and away from glucose release. Conversely, if plasma glucose levels decline below normal, these effects of insulin become attenuated by intrahepatic autoregulatory phenomena as well as by the secretion of hormones (glucagon, epinephrine, cortisol, growth hormone) whose actions are antagonistic to insulin.

Fat Metabolism *In adipose tissue, insulin facilitates transfer of circulating fat into the adipose cell* (see Chapter 41) *by inducing the enzyme lipoprotein lipase.* More free fatty acid is thereby liberated from circulating triglyceride and is rapidly taken up into the adipose cell, where it is reesterfied. Thus, dietary fat that is not needed for immediate energy generation is stored. *Of equal or greater importance, insulin profoundly inhibits the reverse reaction (i.e., lipolysis of stored triglyceride) by inhibiting the necessary enzyme, hormone-sensitive adipose tissue lipase.* In this manner free fatty acid release and delivery to other tissues are greatly suppressed.

In liver, insulin favors shunting of incoming free fatty acids away from β-oxidation and toward esterification, again by increasing production of α-glycerol phosphate. Because β-oxidation is diminished, less β-hydroxybuturate and acetoacetate are produced. Thus, *insulin is powerfully antiketogenic.*

Virtually complete loss of insulin and its actions causes the striking manifestation of insulin-dependent (type 1) diabetes mellitus. Over a period of weeks, hyperglycemia develops to a point exceeding the renal threshold of glucose (Chapter 36). Large amounts of glucose are lost in the urine; the high urinary glucose concentration creates a continuous osmotic diuresis, **polyuria,** thirst, and dehydration. This drain of carbohydrate calories along with catabolic losses of adipose stores and lean body mass causes weight loss, despite increased food intake **(polyphagia).** Uninhibited lipolysis leads to high plasma free fatty acids, stimulation of ketogenesis, and very high levels of β-hydroxybutyrate and acetoacetate in plasma. The plasma bicarbonate and pH fall, and a profound metabolic acidosis may ensue. Coma and death follow unless treatment with insulin, intravenous fluids, and electrolytes is provided.

When insulin is administered to patients in **diabetic ketoacidosis,** increased transport into cells can cause profound drops in plasma potassium, phosphate, and magnesium levels. Potassium always, phosphate occasionally, and magnesium rarely must be given intravenously to prevent serious or even fatal consequences of hypokalemia, hypophosphatemia, or hypomagnesemia.

Insulin also stimulates de novo synthesis of free fatty acids from glucose. Cytoplasmic acetyl coenzyme A (acetyl CoA) derived from pyruvate is shunted toward fatty acid formation because insulin increases the key enzymes acetyl CoA carboxylase and fatty acid synthase. In addition, insulin stimulates synthesis of cholesterol from acetyl CoA by activating the key enzyme, hydroxymethylglutaryl CoA reductase. The net effect of insulin is, therefore, to increase the fat content of the liver and, in some circumstances, to increase the release of very-low-density lipoprotein from that organ.

Protein Metabolism In muscle, insulin stimulates the transport of certain amino acids from the plasma, across the cell membrane, and into the cytoplasm in a manner that is analogous to, but independent of, the transport of glucose. The overall synthesis of proteins from amino acids is also increased by stimulation of transcription and translation. These anabolic effects are reinforced by anticatabolic effects—that is, by inhibition of the enzymes of proteolysis and inhibition of amino acid release from the cell. Other examples of anabolic effects are observed in the liver and exocrine pancreas, where the synthesis of albumin and amylase, respectively, is increased by insulin. Moreover, in cartilage and osseous tissue, insulin and structurally related insulin growth factors enhance the general synthesis of proteins as well as of DNA, RNA, and other macromolecules. *Thus insulin is an important contributor to growth, to tissue regeneration, and to bone remodeling.*

Other Effects Both glycogen and protein synthesis require concurrent cellular uptake of potassium, phosphate, and magnesium. The translocation of all three of these electrolytes from the extracellular to the intracellular space is stimulated by insulin. Insulin also stimulates the reabsorption of potassium, phosphate, and sodium by the renal tubules. Prevention of the urinary loss of potassium and phosphate also contributes to anabolism, whereas conservation of sodium may be related to the need for additional extracellular fluid formation to accompany the expansion of lean body mass.

Insulin has other actions that are relevant to total body energy turnover. Diet-induced thermogenesis, particularly following carbohydrate ingestion, is enhanced by insulin, probably through stimulation of glycogen formation. Insulin also increases energy expenditure by stimulating Na^+-K^+-ATPase. Although most parts of the brain are unresponsive to insulin, considerable evidence points to hormonal action on the hypothalamus.

Insulin decreases the synthesis of neuropeptide Y, a hypothalamic peptide that stimulates food intake. Hence, insulin is an appetite suppressant. However, if insulin causes plasma glucose levels to drop below those needed for normal brain metabolism (i.e., less than 50 mg/dl), hunger is stimulated by other mechanisms.

Molecular Mechanisms The initial step in all actions of insulin is binding of the hormone to its plasma membrane receptor (Figure 42-5). This receptor is a glycoprotein composed of two symmetric units connected by disulfide bonds. Each of the units is made up of an α-subunit that extends externally and binds the hormone and a β-subunit that traverses the cell membrane and terminates in an intracytoplasmic tail. Receptor molecules cycle between a cytoplasmic pool and the plasma membrane and, their number is downregulated by insulin.

After insulin binding to its receptor, the following steps occur:

1. The β-subunit of the receptor undergoes autophosphorylation with ATP at specific tyrosine sites.
2. The phosphorylated receptor becomes a tyrosine kinase itself, which then phosphorylates a large protein called insulin receptor substrate (IRS).
3. Activation of IRS-1 initiates a cascade of serine and threonine phosphorylations that produce effects through multiple intermediary molecules. Metabolic and mitogenic pathways are altered in parallel though not necessarily synchronously.
4. Target enzymes are rapidly activated or deactivated, both by phosphorylation and by dephosphorylation. More slowly, the same enzymes may be induced or repressed by modulation of gene transcription.
5. Plasma membrane carrier mechanisms are activated.

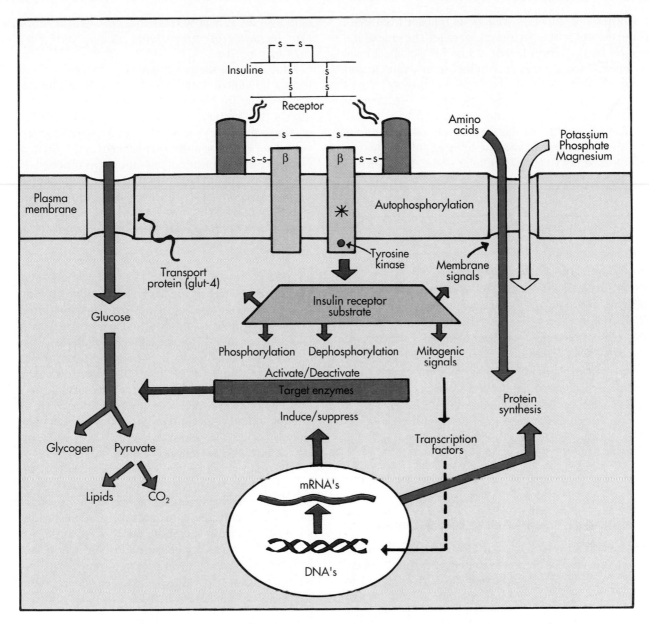

FIGURE 42-5 Insulin action on cells. Binding of insulin to its receptor causes autophosphorylation of the receptor, which then itself acts as a tyrosine kinase that phosphorylates insulin receptor substrate (IRS-1). As a result, serine and threonine residues are phosphorylated in other cell proteins and enzymes. Numerous target enzymes are ultimately activated or inactivated, and the result is to shift the metabolism of glucose toward glycogen and pyruvate. The glucose transporter, glut-4, is recruited to the plasma membrane, where it facilitates glucose entry into the cell. The transport of amino acids, potassium, magnesium, and phosphate into the cell is also facilitated. The synthesis of various enzymes is induced or suppressed and cell growth is regulated by signal molecules that modulate gene expression.

6. DNA transcription factors related to cell growth are modified.
7. In some target cells (e.g., adipose tissue) cAMP levels are lowered by insulin; and this contributes to some of the hormone's actions (e.g., the inhibition of lipolysis).

The most exclusive effect of insulin is to stimulate a specific glucose carrier system within the plasma membrane of muscle and adipose tissue cells. A glucose transporter (glut-4) facilitates diffusion (not active transport) of extracellular glucose into the cytosol, down an already existing, large concentration gradient. Insulin increases the synthesis of glut-4 as well as its transfer from cytoplasmic depots to the plasma membrane. The crucial importance of this action is that at low physiological insulin concentrations, transport of glucose into muscle and adipose tissue cells is often the rate-limiting step in its utilization. By comparison, phosphorylation of glucose is so rapid that intracellular concentrations of free glucose are usually negligible. At the high insulin concentrations that prevail after a meal, the rate-limiting step in glucose metabolism shifts to an intracellular point also governed by insulin actions.

> In rare instances, deletions as well as missense and nonsense mutations in the insulin receptor gene cause marked resistance to the action of insulin. This can lead to diabetes mellitus and disorders of growth. Impairment of insulin action (as well as inadequate insulin secretion) is also a major factor producing hyperglycemia in the very common non–insulin dependent diabetes mellitus. Candidate genes for the loci of this strongly hereditary disease include those for the insulin receptor, the glut-4 transporter, and numerous enzymes, such as glycogen synthase, that are involved in insulin action. Because most of the patients are obese and seldom exhibit increased ketogenesis, insulin resistance that is greater in muscle than in adipose tissue is suspected.

Correlation of Insulin Action and Secretion

The major actions of insulin form a hierarchy that is related to successive increases in plasma insulin concentration. The low insulin concentrations that prevail in overnight fasted humans are partly able to restrain and thereby regulate rates of endogenous release of free fatty acids and amino acids. Somewhat higher insulin concentrations elicited by incoming dietary nutrients are required to shut off completely glucose production by the liver. The peak insulin concentrations elicited by a meal greatly stimulate glucose and amino acid uptake by peripheral tissues, especially muscle, and fatty acid uptake by adipose tissue. This uptake process ensures that these substrates may be stored for future use.

GLUCAGON

Synthesis and Secretion

Although glucagon was discovered shortly after insulin, its physiological importance took much longer to

demonstrate. It is now clear that *glucagon is an important regulator of intrahepatic glucose and free fatty acid metabolism.*

Glucagon is a single straight-chain peptide with a molecular weight of 3500. The N-terminal residues 1 to 6 are essential for biological activity. The glucagon gene directs the synthesis of a preproglucagon in the α-cells of the pancreatic islets. This preproglucagon is processed to a prohormone that subsequently yields glucagon and other peptides of still unknown function. In certain cells of the intestinal tract, alternative processing of preproglucagon yields glucagon-like peptides with different functions. *In contrast to insulin, glucagon synthesis is inhibited by high glucose levels and stimulated by low glucose levels.*

The secretion of glucagon is related in feedback fashion to the principal function of the hormone—stimulating glucose output by the liver and sustaining plasma glucose levels (Figure 42-6). Thus, hypoglycemia promptly evokes a twofold to fourfold increase in plasma glucagon concentrations from basal levels of about 100 pg/ml (3×10^{-11} M), whereas hyperglycemia suppresses glucagon secretion by more than 50%. These effects of glucose are independently reinforced by insulin, possibly through a paracrine action within the islet. Thus, insulin directly inhibits glucagon secretion; conversely, when insulin is absent, the stimulatory effect of low glucose levels on glucagon secretion is exaggerated.

The other major energy substrate, free fatty acids, also suppresses glucagon release, whereas a sharp decline in plasma free fatty acid levels is stimulatory. A protein meal and amino acids, the substrates for glucose production, stimulate glucagon secretion, but this response is dampened by concurrent glucose or insulin action. As a result, the usual mixed meal produces only small and variable increases in plasma glucagon levels, in contrast to the large and consistent increases in plasma insulin levels.

> Prolonged fasting and sustained exercise—circumstances that require glucose mobilization—increase glucagon secretion. Under stressful conditions such as major infection or surgery, glucagon secretion is often greatly augmented. This probably occurs through sympathetic nervous system stimulation of the α-cells via α-adrenergic receptors.

Glucagon is extracted by the liver on the first pass, and it has a short half-life in peripheral plasma. The hormone is degraded in the kidney and liver, and very little is excreted in the urine.

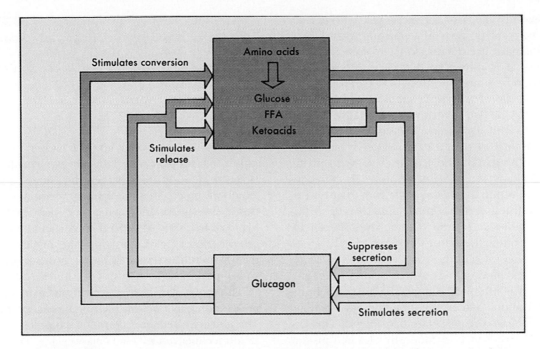

FIGURE 42-6 Feedback relationship between glucagon and nutrients. Glucagon stimulates production and release of glucose, free fatty acids (FFA), and ketoacids, which in turn suppress glucagon secretion. Amino acids stimulate glucagon secretion, and glucagon in turn stimulates amino acids' conversion to glucose.

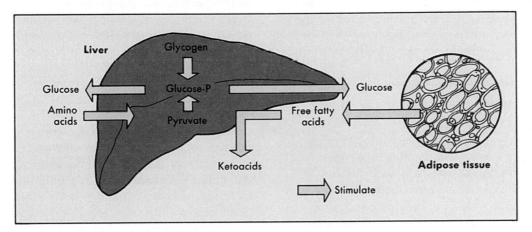

FIGURE 42-7 Effect of glucagon on the overall flow of fuels. Tissue release of glucose, free fatty acids, and ketoacids raises their plasma levels, whereas liver uptake of amino acids lowers their plasma levels.

Actions of Glucagon

In almost all respects the actions of glucagon are opposite to those of insulin. Glucagon promotes mobilization rather than storage of fuels, especially glucose (Figure 42-7). Both hormones act at numerous similar control points for glucose metabolism in the liver. Indeed, glucagon may be the primary hormone that regulates hepatic glucose production (and ketogenesis), with insulin's main role being that of a glucagon antagonist.

Glucagon exerts an immediate and profound glycogenolytic effect through activation of hepatic glycogen phosphorylase. Simultaneously, resynthesis of phosphorylated glucose molecules back to glycogen is prevented by inhibition of glycogen synthase. Glucagon also stimulates gluconeogenesis by several mechanisms. The hepatic extraction of amino acid precursors is increased. The activities of key gluconeogenic enzymes—pyruvate carboxylase, phosphoenolpyruvate carboxykinase, and fructose-1,6-diphosphatase—are increased, whereas activities of key glycolytic enzymes, phosphofructokinase and pyruvate kinase, are decreased (see Figure 41-3).

The enzyme pair, phosphofructokinase/fructose-1,6-diphosphatase, determines the flow between fructose-6-phosphate and fructose-1,6-diphosphate (see Figure 41-3). Thus, this enzyme pair determines the relative rates of gluconeogenesis and glycolysis. The activities of these two enzymes in turn are reciprocally related by the hepatic level of another metabolite, fructose-2,6-biphosphate. *Glucagon decreases the concentration of fructose-2,6-biphosphate,* an action that favors flow from fructose-1,6-diphosphate to fructose-6-phosphate, *and thereby stimulates gluconeogenesis.* (Insulin has the opposite effect, probably by inhibiting the action of glucagon.)

The importance of glucagon is shown by the sharp decline in hepatic glucose output that occurs when glucagon secretion is inhibited. Conversely, increasing glucagon concentrations powerfully stimulates glycogenolysis and rapidly raises the plasma glucose level. This occurs even in the presence of modestly elevated insulin levels. However, after an initial increase, hepatic glucose production wanes during continuous glucagon administration, probably because rising intracellular glucose levels within the liver feed back to regulate glucose synthesis. In addition, insulin release is stimulated both by glucose and glucagon. If glucagon is given in a more physiological fluctuating pattern, however, each increment in the hormone increases hepatic glucose output.

Spontaneous glucagon deficiency is virtually unknown. However, in insulin-dependent diabetes, loss of α-cell responsiveness to a drop in plasma glucose often develops. Such patients become increasingly vulnerable to the risk of severe hypoglycemia if too large a dose of insulin is administered or if a meal is missed. If hypoglycemia that is severe enough to impair consciousness occurs, glucagon can be injected to raise plasma glucose and restore central nervous function.

Another important intrahepatic action of glucagon is to direct incoming free fatty acids toward β-oxidation and away from triglyceride synthesis. Thus, *glucagon is a ketogenic hormone.* The mechanism involves the intermediate malonyl CoA, which is an inhibitor of free fatty acid transfer into the mitochondria. Glucagon suppresses the synthesis of malonyl CoA by inhibiting the enzyme acetyl-CoA carboxylase. The lower levels of malonyl CoA then allow a higher rate of influx of free fatty acids into the mitochondria for conversion to ketoacids.

In diabetic ketoacidosis (see above), high plasma glucagon levels contribute importantly to the overproduction of the ketoacids. Suppression of glucagon secretion by insulin administration helps to restore normal ketoacid levels and pH.

Glucagon actions on adipose tissue or muscle are rather insignificant unless insulin is virtually absent. Peripheral glucose use is largely unaffected by glucagon. However, glucagon can activate hormone-sensitive adipose tissue lipase and thereby increase lipolysis, the delivery of free fatty acids to the liver, and ketogenesis. Another action of glucagon, opposite to that of insulin, is to inhibit renal tubular sodium resorption and thus to cause natriuresis.

The molecular mechanism of glucagon action begins with binding to a plasma membrane receptor in the liver. The glucagon receptor complex causes a rapid increase in intracellular cAMP (see Chapter 40). This is followed by a specific enzymatic cascade (see Figure 40-8). Protein kinase A activity increases and converts inactive phosphorylase kinase to active phosphorylase kinase. The latter then converts inactive phosphorylase to active phosphorylase, and glycogenolysis results. Other enzymes in glucose metabolism whose functional state depends on addition or removal of phosphate are likewise regulated by glucagon.

INSULIN/GLUCAGON RATIO

Substrate fluxes are clearly very sensitive to the relative availability of insulin and glucagon. The usual molar ratio of insulin to glucagon in plasma is about 2.0. Under circumstances that require mobilization and increased utilization of endogenous substrates, the insulin/glucagon ratio drops to 0.5 or less. This occurs in fasting, in prolonged exercise, and in the neonatal period when the infant is abruptly cut off from maternal fuel supplies but is not yet able to assimilate exogenous fuel efficiently. The ratio usually drops because of both decreased insulin secretion and increased glucagon secretion. Conversely, when substrate storage is advantageous, as after a pure carbohydrate load or a mixed meal, this ratio rises to 10 or more, mainly because of increased insulin secretion. An interesting exception occurs when a pure protein meal is ingested, a circumstance in which the insulin/glucagon ratio changes little. Insulin secretion increases, facilitating muscle uptake of amino acids and their synthesis into proteins. But glucagon secretion also increases and prevents the decrease in hepatic glucose output and

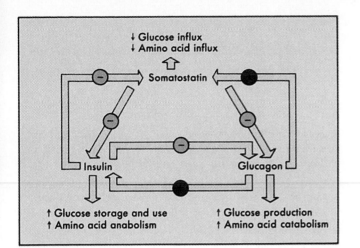

FIGURE 42-8 The interrelationships between somatostatin, insulin, and glucagon effects on each other's secretions and their effects on glucose and amino acid metabolism. (Modified from Unger RH et al: *Annu Rev Physiol* 40:307, 1978.)

hypoglycemia that would ensue if the extra insulin action had been unopposed.

ISLET SOMATOSTATIN

Somatostatin is a neuropeptide hormone originally discovered in the hypothalamus, where it serves as an inhibitor of growth hormone secretion from the anterior pituitary gland (see Chapter 44). In the δ-cells of the islets a large preprohormone is processed to two somatostatin peptides, one of 28 amino acids and one of 14 amino acids. Their secretion is stimulated by glucose, amino acids, free fatty acids, glucagon, and several gastrointestinal hormones, such as cholecystokinin and vasoactive intestinal peptide. Somatostatin secretion is inhibited by insulin. After a mixed meal, plasma somatostatin concentration increases 50% to 100%.

The overall thrust of somatostatin action is to decrease the rate of digestion and absorption of nutrients from the gastrointestinal tract and their subsequent utilization. Thus this neuropeptide inhibits gastric, duodenal, and gallbladder motility; it reduces the secretion of hydrochloric acid, pepsin, gastrin, secretin, intestinal juice, and pancreatic enzymes. Somatostatin also inhibits the absorption of glucose and triglycerides across the intestinal mucosal membrane. Finally, it powerfully inhibits insulin and glucagon secretion.

Somatostatin of islet origin probably participates in a feedback arrangement whereby entrance of food into the gut stimulates the release of the hormone so as to prevent rapid nutrient overload. The anatomical relationships between α-, β-, δ-cells, and the existence of tight junctions and gap junctions between them suggest that all three islet hormones—somatostatin, insulin, and glucagon—may influence each other's secretion by paracrine effects (Figure 42-8). This may improve coordination between bulk movement, digestion, and absorption of nutrients, with the insulin and glucagon responses necessary for proper disposition of nutrients in the liver and other tissues.

Somatostatin, in the form of a long-acting analogue, has several therapeutic uses. It decreases fluid, electrolyte, and nutrient losses in the stools of patients with secretory or pancreatic diarrhea. It also inhibits secretion of various peptide hormones from endocrine tumors and relieves the accompanying symptoms of hormone excess.

SUMMARY

1. The pancreatic islets are composed mainly of insulin-secreting β-cells, along with glucagon-secreting α-cells and somatostatin-secreting δ-cells. The microarchitecture and circulation permit paracrine and neurocrine functioning, as well as direct cell-to-cell communication.
2. Insulin is a major glucoregulatory, antilipolytic, antiketogenic, and anabolic hormone. It consists of two straight-chain peptides held together by disulfide bonds.
3. Insulin secretion is stimulated by glucose, protein, gastrointestinal peptides, and cholingeric and β-adrenergic stimuli. Release is inhibited by circumstances, such as fasting and exercise, that require fuel mobilization.
4. Insulin promotes fuel storage. It inhibits adipose tissue lipolysis, ketogenesis, hepatic glycogenolysis, gluconeogenesis and glucose release, and muscle proteolysis. It stimulates muscle glucose uptake and storage as glycogen and protein synthesis.
5. Insulin acts through a plasma membrane receptor with tyrosine kinase activity. This leads to modulation of the activities of enzymes involved in glucose and fatty acid metabolism. Insulin also affects gene expression of numerous enzymes and proteins.
6. Insulin decreases plasma levels of glucose, free fatty acids, ketoacids, and branch-chain amino acids. Insulin deficiency leads to hyperglycemia, loss of lean body and adipose tissue mass, growth retardation, and ultimately to metabolic ketoacidosis.
7. Glucagon is a single-chain peptide released in re-

sponse to hypoglycemia and to amino acids. Glucagon secretion increases during prolonged fasting and exercise.

8. Glucagon promotes mobilization of glucose by stimulating hepatic glycogenolysis and gluconeogenesis. It also increases fatty acid oxidation and ketogenesis. cAMP is its second messenger, and modification of enzyme activities by phosphorylation is the main mechanism of action. Glucagon increases the plasma levels of glucose, free fatty acids, and ketoacids, but it decreases amino acid levels.

9. The insulin/glucagon ratio controls the relative rates of glycolysis and gluconeogenesis by altering hepatic fructose-2,6-biphosphate levels. The two hormones have antagonistic effects at numerous steps in hepatic glucose and fatty acid metabolism.

10. Somatostatin is a neuropeptide that decreases the motility of the gastrointestinal tract, gastrointestinal secretions, digestion and absorption of nutrients, and the secretion of both insulin and glucagon. Somatostatin is secreted in response to meals; its actions coordinate nutrient input with substrate disposal.

BIBLIOGRAPHY
Journal Articles

Bonadonna RC et al: Dose-dependent effect of insulin on plasma free fatty acid turnover and oxidation in humans, *Am J Physiol* 259:E726, 1990.

Dupre J, Behme MT, Hramiak IM, Longo CJ: Hepatic extraction of insulin after stimulation of secretion with oral glucose or parenteral nutrients, *Metabolism* 42:921, 1993.

Exton JH: Some thoughts on the mechanism of action of insulin, *Diabetes* 40:521, 1991.

Granner DK, Andreone TL: Insulin modulation of gene expression, *Diabetes Metab Rev* 1:139, 1985.

Kahn CR: Insulin action, diabetogenes, and the cause of type II diabetes, *Diabetes* 43:1066-1084, 1994.

Katz L et al: Splanchnic and peripheral disposal of oral glucose in man, *Diabetes* 32:675, 1983.

Liljenquist JE et al: Evidence for an important role of glucagon in the regulation of hepatic glucose production in normal man, *J Clin Invest* 59:369, 1977.

Marchetti P et al: Pulsatile insulin secretion from isolated human pancreatic islets, *Diabetes* 43:827, 1994.

Nair S et al: Effect of intravenous insulin treatment on in vivo whole body leucine kinetics and oxygen consumption in insulin deprived type I diabetic patients, *Metabolism* 36:491, 1987.

Olefsky JM: The insulin receptor: a multifunctional protein, *Diabetes* 39:1009, 1991.

Philippe J: Structure and pancreatic expression of the insulin and glucagon genes, *Endocr Rev* 12:252, 1991.

Polonsky KS, Given BD, Van Cauter E: Twenty-four-hour profiles and pulsatile patterns of insulin secretion in normal and obese subjects, *J Clin Invest* 81:442, 1988.

Schwartz MW, Figlewicz DP, Baskin DG, Woods SC, Porte D Jr.: Insulin in the brain: a hormonal regulator of energy balance, *Endocr Rev* 13:387, 1992.

Seino S, Seino M, Bell GI: Human insulin receptor gene, *Diabetes* 39:129, 1990.

Thiebaud D et al: The effect of graded doses of insulin on total glucose uptake, glucose oxidation, and glucose storage in man, *Diabetes* 31:957, 1982.

Unger RH et al: Insulin, glucagon, and somatostatin secretion in the regulation of metabolism, *Annu Rev Physiol* 40:307, 1978.

Weigle DS: Pulsatile secretion of fuel regulatory hormones, *Diabetes* 36:764, 1987.

Weir GC, Bonner-Weir S: Islets of Langerhans: the puzzle of intraislet interactions and their relevance to diabetes, *J Clin Invest* 85:983, 1990.

Books and Monographs

Cook DL, Taborsky GJ: β-cell function and insulin secretion. In Rifkin H, Porte D, eds: *Diabetes mellitus,* New York, 1990, Elsevier Scientific Publishing Co.

Malaisse WJ: Insulin secretion and beta cell metabolism. In DeGroot LJ, ed: *Endocrinology,* Philadelphia, 1994, WB Saunders Co.

Polonsky, KS, O'Meara NM: Secretion and metabolism of insulin, proinsulin and C-peptide. In DeGroot LJ, ed: *Endocrinology,* Philadelphia, 1994, WB Saunders Co.

Steiner DF et al: Chemistry and biosynthesis of the islet hormones: insulin, islet amyloid polypeptide (amylin), glucagon, somatostatin and pancreatic polypeptide. In DeGroot LJ, ed: *Endocrinology,* Philadelphia, 1994, WB Saunders Co.

Unger RH, Orci L: Glucagon. In Rifkin H, Porte D, eds: *Diabetes mellitus,* New York, 1990, Elsevier Scientific Publishing Co.

CHAPTER 43

Endocrine Regulation of the Metabolism of Calcium and Related Minerals

Calcium, phosphate, and magnesium homeostasis is essential for health and life. A complex system acts to maintain normal body contents and extracellular fluid levels of these minerals in the face of environmental (e.g., diet) and internal (e.g., pregnancy) changes. The key elements in the regulatory system are **vitamin D** and **parathyroid hormone** (PTH), with subsidiary participation by **calcitonin** (CT) and other hormones. *The intestinal tract, the kidneys, the skeleton, the skin, and the liver are all involved in the homeostatic regulation of calcium, phosphate, and magnesium metabolism.*

CALCIUM, PHOSPHATE, AND MAGNESIUM TURNOVER

Calcium

The calcium ion (Ca^{++}) is of fundamental importance to all biological systems. Calcium, usually complexed to calmodulin (see Chapter 40), participates in numerous important enzymatic reactions. This ion is a vital component in the mechanisms of hormone secretion and hormone action. *Calcium is intimately involved in neurotransmission, muscle contraction, mitosis and cell division, fertilization, and blood clotting. Ca^{++} is the major cation in the crystalline structure of bone and teeth.* For these reasons, it is vital that cells be bathed with fluid in which the calcium concentration is kept within narrow limits.

Calcium metabolism may be viewed as having two parts: an intracellular microcomponent and an extracellular macrocomponent. Each is regulated differently and somewhat independently. The crucial intracellular calcium functions are carried out at an average basal cytosolic free calcium concentration of 10^{-7} molar (range 5×10^{-8} to 3×10^{-7} molar). In contrast, the free calcium concentration in extracellular fluid is approximately 10^{-3} molar, or 10,000-fold higher. This large extracellular/intracellular gradient of calcium is maintained by a low permeability of the plasma cell membrane to Ca^{++} and by the regulated activities of a Ca^{++}-ATPase pump and a Ca^{++}-Na^+ exchange system (see Chapters 1, 12, 17 and 18). Within the cell a much larger store of Ca^{++} is bound to various proteins and membranes, to the endoplasmic reticulum, and within the mitochondria. If it were in solution, this Ca^{++} content would be equivalent to an intracellular concentration of 10^{-2} molar.

The cytosolic free calcium concentration can be altered as needed, both by regulating influx from outside the cell and by mobilizing intracellular stores. When calcium is required to function as a second messenger, the free calcium concentration can rise from 10^{-7} molar to as high as 10^{-5} molar. In absolute terms, however, this represents the movement of only small amounts of extra calcium into the cytoplasmic fluid. Such changes are transient (seconds to minutes). The excess cytosolic calcium is either rapidly extruded from the cell or returned to the intracellular reservoirs. The influx and efflux of calcium are so finely balanced that this ubiquitous ion can serve as an internal cell signal with a large dynamic range of gain and sensitivity.

The concentration of calcium in the extracellular fluid and plasma normally fluctuates little. This clearly helps to maintain intracellular calcium at a proper level. When plasma and extracellular fluid calcium concentration is either greater or less than the normal range, intracellular function can be widely and severely affected. Abnormalities in neurotransmission and in the growth and renewal of the skeleton are prominent examples.

The normal range of total calcium concentration in plasma is 8.6 to 10.6 mg/dl, or 2.15 to 2.65×10^{-3} M. Because Ca^{++} is a divalent ion, this is equivalent to 4.3 to 5.3 mEq/L. Individual day-to-day variation is less than 10%. Approximately 50% of total plasma calcium is in the ionized form, Ca^{++}, which is biologically active; 40% is bound to proteins, mainly albumin; 10% is complexed in nonionic but ultrafilterable forms, such as calcium bicarbonate. The total plasma calcium concentration rises or falls with plasma albumin levels, but this has no biological consequence as long as the ionized Ca^{++} concentration remains in the normal range. The equilibrium between ionized and protein-bound calcium depends on blood pH. Alkalosis increases the protein-bound and decreases the ionized Ca^{++} concentration, whereas acidosis has the opposite effect.

When calcium concentration drops below normal, neuromuscular irritability develops. This is manifested by numbness and paresthesias ("pins and needles" sensation), and by tetanic contractions of muscles in the hands and feet **(carpopedal spasm)** and, most dangerously, in the larynx. The latter can cause **airway obstruction. Epileptic seizures** may also occur. When calcium concentration is excessive, depressed neurotransmission can cause impaired mentation, or consciousness, muscle weakness, and decreased gastrointestinal motility. Individuals who hyperventilate to the point of severe respiratory alkalosis can lower ionized calcium enough (without changing the total concentration) to produce the sensory symptoms described above.

Figure 43-1 details the normal turnover of calcium in the body. Daily dietary calcium intake may range from 200 to 2000 mg. The percentage of dietary calcium absorbed from the gut is inversely related to the intake in a curvilinear manner. Thus, in the face of dietary calcium deprivation, one important mechanism for maintaining normal plasma calcium concentration and body calcium stores is an adaptive increase in the percent of ingested calcium that is absorbed. Conversely, in the face of moderate dietary calcium excess, overload is prevented by an adaptive decrease in intestinal absorption. At a daily intake of 1000 mg, about 35% is absorbed. In a steady state, the same amount of calcium, 350 mg, is excreted. Approximately 150 mg is secreted into intestinal juices and excreted in the stools, along with the unabsorbed fraction from the diet. The remaining 200 mg is excreted in the urine. Although the kidney filters about 10,000 mg of non–protein-bound Ca^{++} per day, approximately 98% is reabsorbed by the renal tubules. Therefore alteration in renal tubular Ca^{++} transport provides a very sensitive means for maintaining calcium balance.

These control mechanisms are clinically important when conservation is needed, such as in aged people, whose calcium intakes typically fall and increase the danger of **osteoporosis,** and during pregnancy, when calcium is drained off by the fetus. In contrast, they act to protect against lethal **hypercalcemia**—for example, when metastatic cancer causes rapid destruction of bone.

The extracellular pool of calcium is only 1000 mg. The largest store of calcium, about 1.2 kg, is in the skeleton. Of this, 4000 mg is available for rapid exchange with the extracellular pool and for buffering of plasma calcium. Bone is a dynamic tissue that undergoes daily turnover. In this process approximately 500 mg of calcium is extracted from the extracellular pool as new bone is formed, and a like amount is returned to this pool as old bone is broken down.

Phosphate

The phosphate ion $(PO_4^{\equiv})$ is also of critical importance to all biological systems and is the major intracellular anion. Phosphate is a component of all intermediates in glucose metabolism. It is part of the structure of all high-energy transfer compounds, such as adenosine triphosphate (ATP); of cofactors such as nicotinic acid dinucleotide; and of lipids, such as phosphatidylcholine. Phosphate functions as a covalent modifier of numerous enzymes. *Phosphate is also an integral part of the crystalline structure of bone.*

The normal concentration of phosphate in the plasma is 2.4 to 4.5 mg/dl, or 0.81 to 1.45×10^{-3} M. Because the valence of phosphate changes with pH, it is less useful to express normal concentrations in mEq/L. The turnover of phosphate is shown in Figure 43-2. In contrast to calcium, the percentage of phosphate absorbed from the diet is relatively constant, and thus the net absorption of phosphate from the gut is more linearly related to intake. Therefore urinary excretion provides the major mechanism for regulating phosphate balance (see Chapter 38). The daily filtered load is approximately 6000 mg, but renal tubular reabsorption can vary from 70% to 100%. This variability affords the needed flexibility to compensate for fluctuation in dietary intake. Large soft tissue stores of phosphate, as in muscle, are a source for rapid regulation of the plasma concentration. Approximately 250 mg of phosphate enters and leaves the extracellular fluid daily in the course of bone turnover. Severe depletion of phosphate can result in serious cardiac and skeletal muscle dysfunction, hemolysis, and abnormal bone growth.

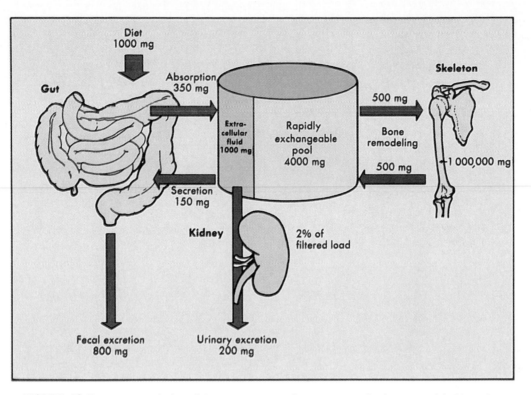

FIGURE 43-1 Average daily calcium turnover in humans. Note both external balance between intake and excretion and internal balance between entry into and exit from bone.

Magnesium

The divalent cation magnesium (Mg^{++}) is related in some metabolic respects to calcium and phosphate. Mg^{++}, *a major intracellular cation, is essential to neuromuscular transmission and is a cofactor in numerous reactions, most notably those involving energy transfers via ATP and those concerned with protein synthesis.* The normal range of magnesium concentration in plasma is 1.8 to 2.4 mg/dl, or 0.75 to 1.00×10^{-3} M, which equals 1.5 to 2.0 mEq/L. One third of plasma magnesium is bound to protein. The average daily intake is about 300 mg, of which 40% is absorbed and (in a steady state) excreted in the urine. The body content of magnesium is about 25 g, 50% of which is present in the skeleton.

Severe depletion of magnesium can result from intestinal malabsorption caused by **sprue** or **inflammatory bowel disease,** by alcoholism, or by diuretic overuse. Hypocalcemia and hypokalemia commonly accompany magnesium deficiency and can result in tetany (see above), muscle weakness, and ventricular arrhythmia.

BONE TURNOVER

As already indicated, bone is a major and dynamic reservoir for calcium and phosphate. Therefore, understanding those aspects of bone structure and function that are pertinent to endocrine regulation is essential.

Bone is broadly divided into two types, cortical and trabecular. **Cortical, or compact, bone** represents 80% of the total and is typified by the thick shafts of the appendicular skeleton (arms and legs). **Trabecular, or spongy, bone** constitutes 20%; it makes up most of the axial skeleton (vertebrae, skull, ribs) and bridges the centers of the long bones. The fivefold greater surface area of trabecular bone gives it disproportionate significance in regulation of calcium metabolism, despite its lesser mass.

Bone formation occurs on the outer surface of cortical bone, whereas bone resorption occurs on its inner surface. Both formation and resorption also take place in specialized nutrient canals within cortical bone and on the surfaces of trabecular bone. Throughout life, the processes of bone formation and resorption are tightly regulated. During growth phases, formation exceeds resorption and the skeletal mass increases. Linear growth occurs between the heads and the shafts of long bones

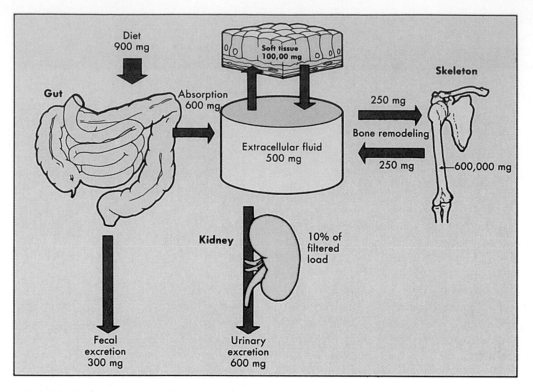

FIGURE 43-2 Average daily phosphate turnover in humans. Note both external balance between intake and excretion and internal balance between entry into and exit from bone.

in specialized areas known as **epiphyseal growth plates.** These close off at the end of puberty when adult height is reached. Bone width increases by adding bone to the outer surfaces.

Total bone mass reaches a peak between ages 20 and 30 years. Thereafter, equal rates of formation and resorption prevail until age 40 to 50 years, at which time resorption begins to exceed formation and the total bone mass slowly decreases. The process of bone turnover in the adult, known as remodeling, may involve up to 15% of the total bone mass per year. Endocrine disturbances that disrupt the coupling of formation and resorption are particularly harmful when they are superimposed on either the growth phase or the senescent phase of life.

Women have smaller bone mass than men, and during the perimenopausal period they lose bone rapidly as ovarian function declines. This is caused by estrogen deficiency, but lifelong calcium intakes that are marginally adequate may also contribute. The resultant **osteoporosis** leads to fractures of the spine and wrist. Later in life, senescent osteoporosis leads to hip fractures in both genders, but women are more affected.

Bone Formation

Three major cell types exist in bone: **osteoblasts, osteocytes,** and **osteoclasts** (Figure 43-3). The first two arise from primitive mesenchymal cells, called **osteoprogenitor cells,** within the investing connective tissue. Various bone proteins known as **skeletal growth factors** attract osteoprogenitor cells, direct their differentiation into osteoblasts, and stimulate their further growth. Osteoclasts arise from the same precursors as do circulating monocytes and tissue macrophages. Together the three major cell types form the **osteon,** or bone modeling unit.

Bone formation is carried out by active osteoblasts, which synthesize and secrete type 1 collagen. The collagen fibrils line up in regular arrays. This process creates an organic matrix known as **osteoid,** within which calcium and phosphate are subsequently deposited in amorphous masses. The slow addition of hydroxide and bicarbonate ions to the mineral phase produces mature **hydroxyapatite** crystals, which have a molar calcium-to-phosphate ratio of 1.7. As the completely mineralized bone accumulates and surrounds the osteoblast, that cell loses its synthetic activity and becomes an interior osteocyte (Figure 43-3). Osteoblastic activity therefore is observed only on the surfaces of bone, along which resting cells wait to be activated.

The mineralization process critically requires normal plasma concentrations of calcium and phosphate. The enzyme **alkaline phosphatase** and other proteins from the osteoblast also participate. **Osteocalcin,** which forms 1% to 2% of all bone protein, has a strong affinity for calcium and for uncrystallized hydroxyapatite, but its exact function is uncertain. Alkaline phosphatase and osteocalcin circulate in plasma, and their concentrations correlate well with quantitative histologic assessments of osteoblastic activity.

Within each osteon, minute fluid-containing channels, called **canaliculi,** traverse the mineralized bone; through these channels the interior osteocytes remain connected with surface osteocytes and osteoblasts via syncytial cell processes (Figure 43-3). This arrangement provides an enormous surface area for the rapid (within minutes) transfer of calcium from the interior to the exterior of the osteons and from there to the extracellular fluid. This transfer process, carried out by the osteocytes, is known as **osteocytic osteolysis.** It does not decrease mature bone mass, but simply removes calcium from the most recently formed crystals.

Bone Resorption

The process of bone resorption does not merely extract calcium; it destroys the entire organic matrix as well, thereby diminishing the bone mass. The cell responsible is the osteoclast, which is a giant multinucleated cell formed by fusion of several precursors (Figure 43-3). The osteoclast contains large numbers of mitochondria and lysosomes. It attaches to the surface of the osteon and creates at this point a ruffled border by an infolding of its plasma membrane. Within this enclosed zone the process of dissolution is carried out by collagenase and other enzymes, and protons secreted by the osteoclast. During this process the osteoclast literally tunnels its way into the mineralized bone. Calcium, phosphate, magnesium, and the constituent amino acids—including hydroxyproline and hydroxylysine, which are unique to collagen—are released into the extracellular fluid.

As already emphasized, resorption and formation of bone are closely coordinated locally. In a complex sequence, resting osteoblasts are stimulated to recruit and activate osteoclasts, probably by paracrine signaling. The resultant resorption cavity created by the osteoclast becomes the site of subsequent osteoblastic activity, which fills in the recently formed cavity with new bone. Thus bone resorption precedes and subsequently triggers bone formation.

The recruitment of osteoblasts and osteoclasts from precursors and the activity of each cell type are regu-

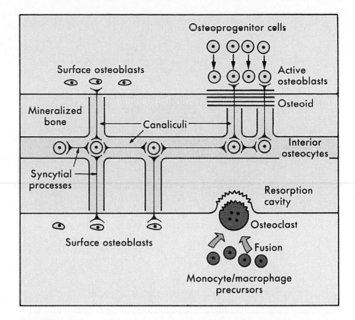

FIGURE 43-3   The relationships between bone cells and bone remodeling. Note that the canaliculi provide channels through which calcium and phosphate can be transferred from the interior to the exterior of bone. These minerals may be used for new bone formation or may be transported into the circulation. (Redrawn from Avioli LV et al: Bone metabolism and disease. In Bundy PK and Rosenberg LE: *Metabolic control and disease,* Philadelphia, 1980, WB Saunders Co.)

lated by various local factors, including lymphokines, growth factors, prostaglandins, and an array of hormones. *As a general principle, whether the primary effect of a hormone is on the formation or the resorption of bone, the phenomenon of coupling will secondarily alter the other process in the same direction.* Therefore the net effect of a hormone excess or deficiency will depend on the degree to which the coupling phenomenon defends the total bone mass.

VITAMIN D

Vitamin D, through its active metabolites, is one of two major regulators of calcium and phosphate metabolism. *It acts to sustain normal plasma concentrations of calcium and phosphate by increasing their inflow from the intestinal tract, and it also is required for normal bone formation.* Vitamin D is a hormone in the sense that it is synthesized in the body, although not by an endocrine gland; after further processing, it is transported via the circulation to act on target cells. It is also a vitamin in the sense that when it cannot be synthesized in sufficient quantities, it must be ingested for health to be maintained.

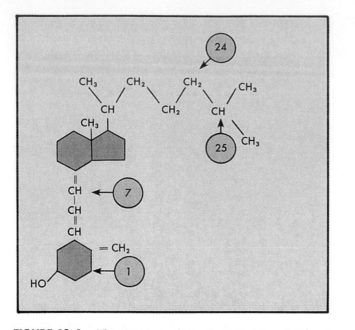

FIGURE 43-4 The structure of vitamin D. Positions 1, 24, and 25 are important sites of hydroxylation that affect biological activity.

Synthesis and Intake

The sterol structure of the synthesized form of vitamin D (D_3) is shown in Figure 43-4. The ingested form (D_2), prepared by irradiating plant or milk ergosterol, differs only slightly. Vitamins D_3 and D_2 are essentially prohormones that undergo processing that converts them to molecules with identical qualitative and quantitative actions. Henceforth, the term vitamin D will refer to both.

The minimum daily requirement of vitamin D is approximately 2.5 micrograms (100 units). Endogenously, it is synthesized in the skin by ultraviolet irradiation at specific frequencies of the precursor **7-dehydrocholesterol.** Exogenously, the fat-soluble vitamin D is available in fish, liver, and milk, and it is absorbed from the gut just as are fats. Vitamin D is stored in adipose tissue in amounts normally sufficient for several months.

Once vitamin D enters the circulation from the skin or the gut, it is concentrated in the liver. There it is hydroxylated to 25-OH-D. This molecule is transported to the kidney, where it undergoes alternative fates (Figure 43-5). Hydroxylation in the 1 position produces the metabolite 1,25-$(OH)_2$-D, which unquestionably expresses most if not all of the biologic activity of vitamin D. Alternatively, 25-OH-D may be hydroxylated in the 24 position. 24,25-$(OH)_2$-D is only one-twentieth as potent as 1,25-$(OH)_2$-D and mainly serves to dispose of excess vitamin D.

Vitamin D deficiency can occur in several ways. If individuals who live in sunlit climates (e.g., India) and are dependent on D synthesis move to cloudy countries (e.g., England), they may become deficient in vitamin D if they do not alter their dietary habits or ingest supplementary vitamin D. Black infants who are breast-fed are also at risk, because less effective ultraviolet radiation reaches sites of vitamin D synthesis. Deficiency also results from gastrointestinal diseases, such as pancreatic insufficiency, that cause malabsorption of fats when the subjects' exposure to sunlight is inadequate. Liver disease leads to diminished rates of 25-hydroxylation and hence to deficient vitamin D action. A common cause of deficient vitamin D effects is kidney failure, with virtually total loss of production of the most active metabolite, 1,25-$(OH)_2$-D.

Feedback control of vitamin D activation occurs through regulation of the renal 1-hydroxylase and 24-hydroxylase activities (Figure 43-5). 25-OH-D is preferentially directed toward the active metabolite, 1,25-$(OH)_2$-D, whenever calcium, phosphate, or vitamin D itself is lacking. Calcium deprivation leads to compensatory secretion of **parathyroid hormone.** This hormone then stimulates 1-hydroxylation. A lowering of plasma phosphate and renal phosphate content also augments 1-hydroxylase activity. In addition, 1,25-$(OH)_2$-D is a feedback inhibitor of its own synthesis; hence in vitamin D deficiency, there is compensatory enhancement of 1-hydroxylase. By contrast, 24-hydroxylase activity is stimulated by normal to elevated calcium or phosphate concentrations, and by 1,25-$(OH)_2$-D. The net result of this regulation is that the supply of active 1,25-$(OH)_2$-D is increased (and that of inactive 24,25-$(OH)_2$-D is decreased) whenever homeostasis requires increasing calcium and phosphate absorption from dietary sources or skeletal stores.

Vitamin D, 25-OH-D, and 1,25-$(OH)_2$-D circulate bound to a protein carrier. 1,25-$(OH)_2$-D has by far the lowest concentration and the shortest half-life of the three (.03 µg/L and 6 hours). However, regulation is powerful enough to maintain the appropriate concentration of 1,25-$(OH)_2$-D even when the concentrations of its precursors are quite reduced.

Actions of Vitamin D

The active form of vitamin D (1,25-$(OH)_2$-D) acts through the general mechanism outlined for steroid hormones (see Chapter 40). After binding to a cytosolic receptor, the hormone receptor complex enters the nucleus, where it stimulates or represses transcription of

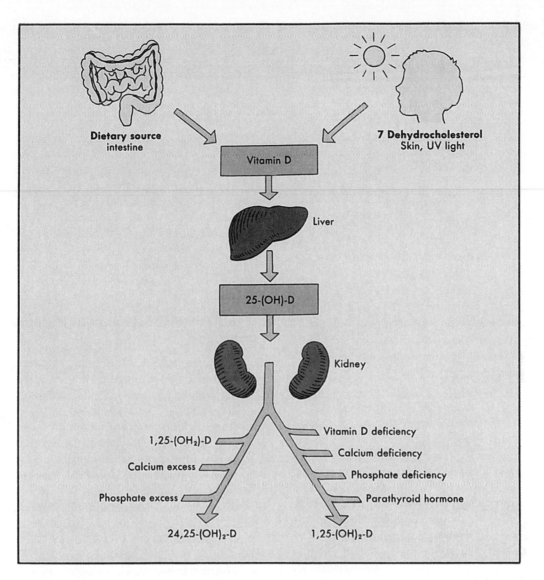

FIGURE 43-5 Vitamin D metabolism. Whether synthesized in the skin or absorbed from the diet, vitamin D undergoes 25 hydroxylation in the liver. In the kidney, it is further hydroxylated in the 1 position when more biological activity is required or in the 24 position when less biological activity is required.

messenger ribonucleic acid (mRNA) for several products. One is **calbindin,** a calcium-binding protein found in cells of the intestinal mucosa, bone, kidney, and parathyroid glands. It has significant homology with calmodulin and a high affinity for calcium. Calbindin is not essential for vitamin D stimulation of calcium transport, because the latter precedes the appearance of the protein in intestinal cells. However, calbindin protects the cells from the effect of high cytoplasmic concentrations of calcium during enhanced transport.

The major action of $1,25\text{-}(OH)_2\text{-}D$ is to stimulate absorption of calcium from the intestinal lumen against a concentration gradient (see Chapter 34). $1,25\text{-}(OH)_2\text{-}D$ localizes in the nuclei of intestinal villus and crypt cells, where it acts on the brush border. It probably stimulates

production of a plasma membrane calcium transporter protein because inhibition of protein synthesis prevents hormone action on calcium transport. $1,25\text{-}(OH)_2\text{-}D$ is responsible for the adaptation previously described whereby intestinal absorption of calcium increases in response to decreases in its dietary intake. $1,25\text{-}(OH)_2\text{-}D$ also augments active absorption of phosphate and probably of magnesium across the intestinal cell membrane.

$1,25\text{-}(OH)_2\text{-}D$ also stimulates bone resorption. Osteoclasts have no receptors for $1,25\text{-}(OH)_2\text{-}D$, but osteoblasts do; therefore the resorptive action of the osteoclast may be driven by an osteoblast-derived factor. This effect of $1,25\text{-}(OH)_2\text{-}D$ is physiologically important in sensitizing the bone to the resorptive effects of parathyroid hormone.

The normal mineralization of newly formed osteoid along a regular front is critically dependent on vitamin D. The major mechanism for this vitamin D action is augmentation of the supply of calcium and phosphate. However, osteoblasts can also respond directly to 1,25-$(OH)_2$-D, which represses transcription of the collagen gene and decreases collagen synthesis. In the hormone's absence, unmineralized osteoid accumulates from continued osteoblastic activity (Figure 43-6), and the bone so formed is weakened.

The skeletal manifestations of vitamin D deficiency vary with the stage of life. In children, the growth centers are preferentially affected and the failure of normal bone mineralization leads to abnormal epiphyses (Figure 43-6), bowing of the extremities, and collapse of the chest wall. In adults, bone pain, vertebral collapse, and fractures along stress lines occur. Plasma calcium and phosphate are decreased, whereas alkaline phosphatase is increased. Therapy with vitamin D or the appropriate metabolite is curative.

Skeletal muscle is another target tissue for vitamin D. It increases calcium transport and uptake by the sarcoplasmic reticulum, as well as cellular uptake of phosphate. Deficiency of vitamin D leads to muscle weakness, electrophysiological evidence of abnormal contraction and relaxation, and altered cytoarchitecture.

A role for vitamin D in immunomodulation has also emerged. Immune system cells such as macrophages can synthesize 1,25-$(OH)_2$-D from 25-OH-D. The hormone can decrease production of numerous lymphokines and proliferation of lymphocytes. It is likely these phenomena are part of autocrine or paracrine regulation of immunoactivity engendered by tissue injury or invasion.

In diseases, such as **sarcoidosis** or **tuberculosis,** that are characterized by formation of granulomas, the macrophages may synthesize excessive amounts of 1,25-$(OH)_2$-D. This can result in hypercalcemia and hypercalcuria.

PARATHYROID GLAND FUNCTION

The four parathyroid glands are major regulators of plasma calcium and phosphate concentrations and flux. These four glands develop from brachial pouches at 5 to 14 weeks of fetal life. They descend to lie just poste-

rior to the thyroid gland. The total weight of adult parathyroid tissue is about 130 mg.

The predominant cell of the parathyroid gland is known as the **chief cell.** These cells are present throughout life and are the source of PTH. A second and related cell type, the **oxyphil cell,** first appears at puberty and increases in number with age. Active chief cells have a large convoluted Golgi apparatus with vacuoles and vesicles, and a granular endoplasmic reticulum. During hormone secretion numerous granules undergo exocytosis.

The paramount effect of PTH is to sustain or increase the plasma calcium level. This is accomplished by stimulating entry of calcium into plasma from bone, tubular urine, and the intestinal tract. An important second effect is to decrease or prevent an undue rise in the plasma phosphate level by stimulating excretion of phosphate into the urine.

Synthesis and Secretion of Parathyroid Hormone

PTH is a single-chain protein (9600 molecular weight) that contains 84 amino acids. The biological activity of the hormone resides in the N-terminal portion of the molecule within amino acids 1 to 34. The function of the larger carboxyl portion is not known.

The gene for PTH directs the synthesis of **prepro-PTH.** 25 amino acids are enzymatically cleaved from the N-terminal end, leaving **pro-PTH.** Pro-PTH is then transported to the Golgi apparatus, where another six amino acids are cleaved. The resulting PTH is packaged for storage in secretory granules. Degradation of PTH also occurs within the gland; therefore not all synthesized molecules reach the circulation.

The dominant regulator of parathyroid gland activity is the plasma calcium level. PTH and calcium form a negative feedback pair, and secretion of PTH is inversely related to the plasma calcium concentration (Figure 43-7). Maximal secretory rates are achieved when plasma ionized calcium concentration falls below 3.5 mg/dl. PTH secretion increases within minutes if plasma Ca^{++} is selectively decreased by chelation, even though plasma total calcium concentration remains unchanged. Conversely, as ionized calcium concentration increases to 5.5 mg/dl, PTH secretion is progressively diminished. However, secretion reaches a persistent basal rate that is not suppressible by further elevation of the ambient calcium concentration.

Regulation of PTH represents an exception to the general rule that hormone secretion is stimulated by calcium. Strong evidence exists for a calcium receptor within the plasma membrane of the parathyroid cell; this receptor responds rapidly to alterations in extracellular calcium levels. Activation of the receptor by a fall in the ionized calcium level leads to exocytosis of PTH-

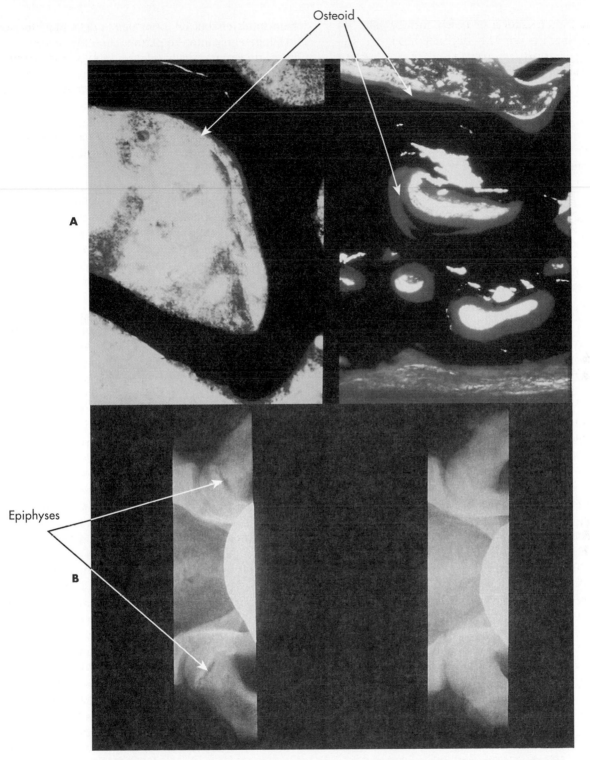

FIGURE 43-6   **A,** *Left:* Histological section of normal trabecular bone showing very low ratio of unmineralized osteoid to mineralized bone. *Right:* Similar section from an individual with vitamin D deficiency showing much higher proportion of osteoid to mineralized bone (i.e., excess osteoid). **B,** *Left:* X-ray of hip from a child with deficient vitamin D action showing widened, irregular epiphysis. *Right:* Same hip after effective treatment with vitamin D has mineralized the area normally.

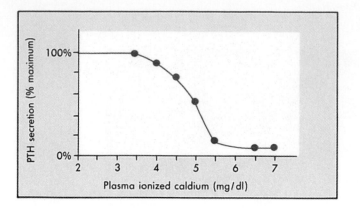

FIGURE 43-7   The inverse relationship between parathyroid hormone secretion and plasma ionized calcium concentration in humans. Note maintenance of some secretion even at high calcium levels. (Redrawn from Brent GA et al: *J Clin Endocrinol Metab* 67:944, 1988.)

containing granules. An accompanying rise in intracellular cAMP levels may partly mediate PTH secretion.

Calcium also modulates PTH turnover within the gland. Prolonged exposure to a high ambient calcium concentration lowers the rate of PTH synthesis and stimulates the intraglandular degradation of PTH. Calcium also regulates the size of parathyroid cells. The net effect of calcium excess is to decrease both the glandular stores and the release rates of PTH. Conversely, calcium deficiency increases PTH stores, secretory rates, and ultimately gland size as well.

Phosphate exerts no direct effect on PTH secretion in vitro. However, by complexing calcium and decreasing Ca^{++} concentration, a rise in plasma phosphate concentration indirectly causes a transient increase in PTH secretion. 1,25-$(OH)_2$-D directly feeds back on the parathyroid gland to inhibit transcription of the PTH gene, reduce PTH secretion, and decrease proliferation of parathyroid cells. Chronic magnesium depletion also ultimately inhibits PTH synthesis and release.

Plasma PTH concentration is about 30 pg/ml (3×10^{-12} M). Whereas the hormone has a short half-life, its C-terminal metabolites circulate with half-lives of many hours.

Actions of Parathyroid Hormone

Intracellular Effects   *The overall effect of PTH is to increase plasma calcium levels and decrease plasma phosphate levels by acting on three major target organs: kidney, bone, and, indirectly, the gastrointestinal tract.* Actions on all three targets ultimately increase calcium influx into the plasma and raise its concentration (Figure 43-8). In contrast, PTH acts on the kidney to increase phosphate exit from plasma; this action over-

whelms the effects on bone and gut, which increase phosphate entry into the plasma; therefore plasma phosphate concentration falls (Figure 43-8).

PTH action is initiated by binding to a G-protein–linked plasma membrane receptor. *In all target cells, activation of adenylyl cyclase and an increase in cAMP follow.* The second messenger then triggers a protein kinase cascade (see Chapter 40), which ultimately leads to phosphorylation of proteins necessary for expression of PTH action.

PTH also stimulates the uptake of calcium into the cytoplasm of its target cells. This may well be mediated by increases in phosphoinositide second messengers. The initial uptake of calcium is reflected in a slight transient hypocalcemia, which immediately follows PTH administration and precedes the classic hypercalcemia. The presence of 1,25-$(OH)_2$-D is required for maximal responsiveness to PTH. A sufficient intracellular concentration of magnesium is also necessary for PTH to act maximally.

Renal Effects   PTH increases the reabsorption of calcium in the ascending loop of Henle and distal tubule. PTH also decreases the reabsorption of phosphate in the proximal and distal tubules (see Chapter 38). These effects are mediated by PTH stimulation of cAMP production at the capillary surface of the renal tubular cell. The cAMP is transported to the luminal surface, where it activates protein kinases that are located in the membranes of the brush border and are involved in calcium and phosphate reabsorption. During this process cAMP is released into the tubular lumen. *Therefore the earliest observable renal effect of PTH in vivo is a dramatic increase in cAMP excretion in the urine.*

The relationship between urinary calcium excretion and plasma calcium concentration is altered by PTH (Figure 43-9). At any given plasma calcium concentration, PTH diminishes the amount of calcium lost in the urine and thus counters hypocalcemia. Conversely, suppression of PTH secretion by an excess calcium load increases calcium excretion and helps prevent hypercalcemia.

The net effect of prolonged alterations in PTH secretion on urinary calcium excretion is dominated by the influence of PTH on bone and gut. A continuous excess of PTH eventually elevates the plasma calcium level and with it the load of calcium filtered by the glomerulus. Therefore the absolute amount of calcium excreted in the urine eventually increases despite the stimulation of tubular reabsorption.

In contrast to calcium, the relationship between urinary phosphate excretion and plasma phosphate level is shifted in the opposite direction by PTH (Figure 43-9). This phosphaturic effect of PTH allows disposal of the extra phosphate that is released when the hormone stimulates bone resorption (see next section). Otherwise, PTH would simultaneously elevate plasma cal-

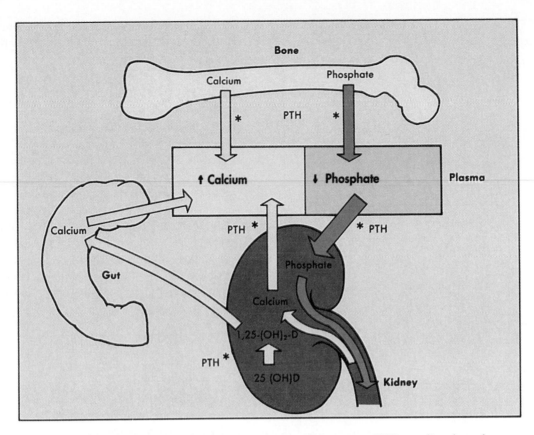

FIGURE 43-8   Overview of parathyroid hormone *(PTH)* actions. PTH acts directly on bone and kidney to increase calcium influx into plasma. By stimulating 1,25-(OH)$_2$-D synthesis, PTH indirectly also increases calcium absorption from the gut. Thus plasma calcium level increases. In contrast, PTH inhibits renal tubular resorption of phosphate, thereby increasing urinary phosphate excretion. This effect quantitatively offsets entry of phosphate from bone and gut. Therefore plasma phosphate level decreases.

cium and phosphate levels and thereby create the danger of precipitating calcium-phosphate complexes in tissue.

PTH also inhibits the reabsorption of sodium and bicarbonate in the proximal tubule (see Chapter 36). This action may prevent metabolic alkalosis, which could result from the release of bicarbonate during the dissolution of hydroxyapatite crystals in bone (see following discussion).

PTH directly stimulates the synthesis of 1,25(OH)$_2$-D from 25-(OH)$_2$-D. The decrease in plasma and renal phosphate content caused by PTH further augments this direct action on 1-hydroxylase activity. The increase in 1,25-(OH)$_2$-D stimulates calcium absorption from the gut and raises plasma calcium levels (see Figure 43-4). This important indirect effect of PTH on the intestinal tract again serves the major function of the hormone.

Skeletal Effects The major action of PTH on bone is to accelerate removal of calcium. The initial effect of PTH is to stimulate osteocytic osteolysis, which causes a transfer of calcium from the bone canalicular fluid into

the osteocyte and then out the opposite side of the cell into interstitial fluid. Replenishment of calcium in the canalicular fluid probably then occurs from the surface of partially mineralized bone.

A second, more slowly developing effect of PTH is to stimulate the osteoclasts to resorb completely mineralized bone. In this process both calcium and phosphate are released for transfer ultimately into the extracellular fluid; in addition, the organic bone matrix is hydrolyzed by PTH activation of collagenase and of lysosomal enzymes. PTH initially increases the active resorptive ruffled border of the osteoclasts. This is followed by PTH stimulation of osteoclast size, number of nuclei, fusion, and proliferation from precursors. PTH also induces increases in the enzymes acid phosphatase and carbonic anhydrase and the accumulation of an acidic environment. The resultant lowering of bone pH contributes to the resorptive process. Because of collagen degradation, PTH increases the release of hydroxyproline and other bone collagen products into plasma and urine.

The dramatic effects of PTH on osteoclasts in vitro are

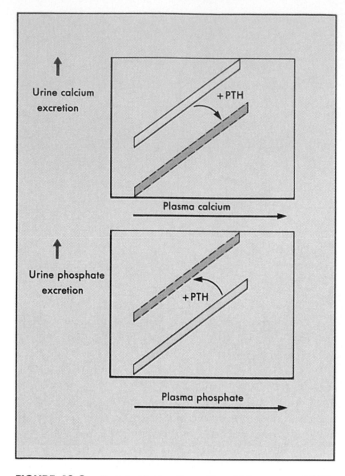

FIGURE 43-9 Renal effects of parathyroid hormone *(PTH)*. At any given level of plasma calcium, PTH decreases urinary calcium excretion. At any given level of plasma phosphate, PTH increases urinary phosphate excretion.

not evident in the absence of osteoblasts. Therefore an initial action of PTH on osteoblasts may be required to evoke a local factor that secondarily stimulates the osteoclasts. PTH receptors are present on osteoblasts, and when exposed to the hormone, these cells immediately change their shape. Later PTH inhibits the synthesis of collagen by the osteoblasts, probably at the level of transcription. Stimulation of osteoclastic bone resorption and inhibition of osteoblastic bone formation are achieved by the elevated concentrations of hormone that result from stimulation of the parathyroid glands through hypocalcemia. Thus these actions of PTH are part of its general mission to restore plasma calcium level rapidly to normal.

PTH also has anabolic actions on bone, In part, increases in bone formation may reflect coupling to enhanced resorption. However, by stimulating production of local growth factors, PTH increases the number and

activity of osteoblasts. Intermittent administration of PTH to humans increases trabecular bone mass but still decreases cortical bone.

Prolonged excess secretion of PTH occurs in primary hyperparathyroidism, usually because of a benign neoplasm in one gland. Plasma calcium concentration is high (with attendant symptoms, see above) and phosphate concentration is usually low. The increased renal excretion of calcium can cause kidney stones. Modest loss of cortical bone may result. In hyperparathyroidism secondary to kidney failure, both $1,25\text{-}(OH)_2\text{-}D$ and calcium concentrations decrease and all the parathyroid glands enlarge. Massive osteoclastic bone resorption can result, with attendant pain, fractures, and deformities. Cure of hyperparathyroidism requires removal of the excess parathyroid tissue.

CALCITONIN

Synthesis and Secretion

Another peptide hormone, **calcitonin** (CT), decreases plasma calcium levels, largely by antagonizing the actions of PTH on bone. CT is secreted by a small population of neuroendocrine cells known as **C cells,** or **parafollicular cells,** in the thyroid gland. These relatively large cells contain small secretory granules enclosed in membranes. C-cell neoplasms and other tumors of neural crest origin often secrete great amounts of CT. Although its role in normal human physiology is uncertain, CT is an important regulator of plasma calcium levels in lower animals that live in an aquatic environment high in calcium.

Calcitonin, a straight-chain peptide of 32 amino acids, has a molecular weight of 3400. Synthesis proceeds from a preprohormone through a prohormone to CT, which is packaged in granules along with N-terminal and C-terminal copeptides. The latter also has a calcium-lowering action and is present in the plasma of humans.

Calcitonin is also found in nervous tissue, where it may function as a neuromodulator and analgesic. The gene for CT illustrates well the relationship between the endocrine and nervous systems. In some cells the primary transcript of the gene is processed to a messenger RNA that directs synthesis of calcitonin. However, in other cells the same primary transcript is processed to a different RNA that directs the synthesis of an alternative peptide product. This molecule, known as **calcitonin gene-related peptide** (CGRP), circulates in human plasma and probably arises from perivascular nerve axons. It is a potent vasodilator and a cardiac inotropic agent.

The major stimulus to CT secretion is a rise in plasma calcium concentration. However, the degree of response in various species is related to their need to prevent hypercalcemia. The first vertebrates to develop CT did so after migrating from fresh water of low calcium concentration into the sea, with a high calcium concentration. When vertebrates moved to land, the emphasis in calcium economy shifted toward defense against a low calcium concentration. PTH was then developed, and the importance of CT in plasma calcium regulation probably declined.

CT circulates in humans at concentrations of 10 to 100 pg/ml (10^{-11} M). It increases considerably when plasma calcium level is raised as little as 1 mg/dl. The stimulating effect of calcium on CT secretion involves an increase in cAMP. Ingestion of food also increases CT secretion, a response that is mediated by gastrin and other gastrointestinal hormones.

Actions of Calcitonin

The major effect of CT is to decrease plasma calcium levels. Binding of CT to its plasma membrane receptors stimulates adenylyl cyclase and elevates cAMP. This second messenger initiates at least a portion of CT action in all target cells, but the subsequent intracellular events are obscure. The magnitude of the fall in plasma calcium concentration is directly proportional to the baseline rate of bone turnover. Thus young growing animals are most affected by CT, whereas adults with more stable skeletons respond minimally to the hormone. *The hypocalcemic action of CT is caused by inhibition of osteocytic osteolysis and osteoclastic bone resorption, particularly when these are stimulated by PTH.* Continued exposure to CT eventually decreases the number of osteoclasts and alters their morphology. As bone formation is also stimulated, denser bone with fewer resorption cavities eventually results.

CT is clearly a physiological antagonist to PTH with respect to calcium. However, with respect to phosphate, it has the same net effect as PTH; that is, CT decreases plasma phosphate concentration and slightly increases urinary phosphate excretion.

The importance of CT in humans is controversial. CT deficiency does not lead to overt hypercalcemia, and CT hypersecretion rarely produces hypocalcemia. It may be that abnormal CT secretion is easily compensated for by

adjustment in PTH and vitamin D levels. A role for CT in fetal bone development and in the declining bone mass of aging has been proposed.

> Calcitonin is used therapeutically to block bone resorption in situations where the resorption rate is high (e.g., in **Paget's disease** and hypercalcemia of malignancy). The hormone is also sometimes used to treat **osteoporosis.**

INTEGRATED REGULATION OF CALCIUM AND PHOSPHATE

An integrated system maintains normal concentrations of calcium and phosphate. Calcium deprivation (Figure 43-10) stimulates PTH secretion. The PTH increases urinary phosphate excretion and thereby decreases plasma and renal cortical phosphate content. Excess PTH secretion, together with the decreased phosphate concentration, stimulates the production of 1,25-$(OH)_2$-D. The sterol hormone raises the plasma calcium concentration back toward normal by increasing the absorption of calcium from the gut. PTH also increases bone resorption and calcium reabsorption from the renal tubular urine. Together then, PTH and 1,25-$(OH)_2$-D respond to calcium deprivation by increasing the flux of calcium into the plasma. Simultaneously the extra phosphate that enters with the calcium is eliminated in the urine by PTH action.

Phosphate deprivation (Figure 43-11) directly stimulates 1,25-$(OH)_2$-D production. The latter increases the flux of phosphate into plasma by stimulating bone resorption and phosphate absorption from the gut. The extra calcium that enters simultaneously will raise the plasma calcium level. This suppresses PTH secretion, and the absence of PTH causes urinary phosphate excretion to diminish and thus aids in the restoration of plasma phosphate levels back to normal. At the same time suppression of PTH diminishes renal tubular calcium reabsorption and increases urinary calcium excretion. Thus the extra calcium that was mobilized is eliminated more easily.

This combined arrangement of dual hormone regulation and dual hormone action by PTH and vitamin D permits selective defense of either plasma calcium level or plasma phosphate level, without creating a circulatory excess of the other. The same principles apply in reverse when excess loads of calcium or phosphate are imposed on the body.

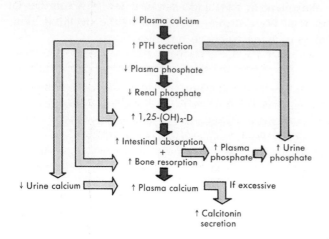

FIGURE 43-10 The compensatory response to a decrease in plasma calcium concentration. (See text for explanation.)

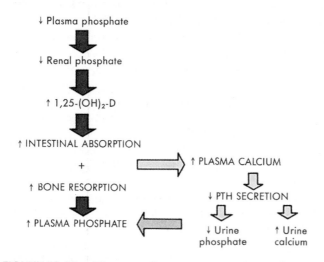

FIGURE 43-11 The compensatory response to a decrease in plasma phosphate concentration. (See text for explanation.)

The renal responses to PTH provide the most rapid (within minutes) defense against sudden changes in calcium or phosphate levels. As PTH ranges from very high to very low levels, the rate of urinary calcium excretion can rise 25-fold and that of phosphate excretion can fall to almost 0. The gastrointestinal responses of this integrated system are slower and narrower in range. Bone responses to regulation by PTH and 1,25-(OH)$_2$-D are rapid when they are produced by osteocytic osteolysis and relatively slow when caused by osteoclastic resorption. However, the capacity for calcium and phosphate uptake and release by the skeleton is enormous.

The compensatory responses of the kidney and the gut defend the total body and bone stores of calcium and phosphate against erosion. In contrast, the skeletal

mechanisms that defend the plasma calcium and phosphate levels have the important disadvantage that they eventually sacrifice the chemical and structural integrity of the bone mass.

SUMMARY

1. Calcium is critically involved in numerous functions, including neurotransmission, hormone secretion and action, enzyme activities, muscle contraction, and blood clotting. It is also the chief mineral contributing to the structural integrity of the skeleton.
2. Extracelluar Ca^{++} concentration is closely controlled, in part to help regulate the wide transient swings in intracellular concentration, which is 10,000-fold lower.
3. Phosphate ion participates in the major enzymatic pathways of energy generation, substrate flux, and protein and other macromolecule synthesis. Phosphate is also the anion partner of calcium in bone structure.
4. Calcium balance depends on dietary intake, gastrointestinal absorption, renal excretion, and internal transfers between extracellular calcium and the skeletal reservoir.
5. Phosphate balance reflects dietary intake, gastrointestinal absorption, renal excretion, and internal shifts between extracellular fluid, large soft tissue contents and the skeletal reservoir.
6. Bone is a complex organ with several cell types that are specifically devoted to a continuous process of remodeling. Mineralized bone is resorbed by osteoclasts (which release calcium and phosphate), following which new replacement bone is formed by osteoblasts (which assimilate calcium and phosphate). This coupled process is augmented during growth periods and slows with aging.
7. Vitamin D is a steroid molecule that is both synthesized from 7-dehydrocholesterol in the skin by UV light and absorbed from the diet. It undergoes successive hydroxylations in the liver and kidney to 1,25-(OH)$_2$-D, the active metabolite. The final activation steps are enhanced by calcium or vitamin D deficiency.
8. 1,25-(OH)$_2$-D acts via its intestinal epithelial nuclear receptor to increase absorption of ingested calcium (and phosphate). The hormone is therefore critical to supplying calcium for bone formation and maintaining normal plasma levels required for other calcium-dependent processes. It also enhances bone resorption. Overall, 1,25-(OH)$_2$-D increases both plasma calcium and plasma phosphate levels.
9. Parathyroid hormone (PTH) is a straight-chain pep-

tide synthesized as a preprohormone in four parathyroid glands. PTH and calcium form a classic negative feedback pair. PTH is released by exocytosis in response to a decrease in plasma calcium level, whereas its synthesis and secretion are suppressed by calcium (and $1,25$-$[OH]_2$-D).

10. PTH acts via a plasma membrane receptor and cAMP to increase osteoclastic bone resorption, to increase renal tubular reabsorption of calcium, and to increase $1,25$-$(OH)_2$-D synthesis in the kidney. PTH increases urinary excretion of phosphate by decreasing tubular reabsorption. Overall, PTH increases plasma calcium and decreases plasma phosphate.

11. Calcium deficiency evokes a synergistic sequence that increases PTH and $1,25$-$(OH)_2$-D. Their combined actions increase the inflow of calcium and restore plasma levels to normal. Simultaneously the extra phosphate entering with calcium is disposed of by renal excretion.

12. In contrast, phosphate deprivation evokes a synergistic sequence that increases $1,25$-$(OH)_2$-D secretion but decreases PTH secretion. The result is restoration of plasma phosphate to normal while excreting extra calcium into the urine.

13. Calcitonin is a peptide hormone synthesized in C cells within the thyroid gland. It is secreted in response to hypercalcemia and inhibits bone resorption. Thus calcitonin is a PTH antagonist that acts to lower the plasma level of calcium.

BIBLIOGRAPHY
Journal Articles

Brent GA et al: Relationship between the concentration and rate of change of calcium and serum intact parathyroid hormone levels in normal humans. *J Clin Endocrinol Metab* 67:994, 1988.

Brown EM: Extracellular Ca^{2+} sensing, regulation of parathyroid cell function, and role of Ca^{2+} and other ions as extracellular (first) messengers, *Physiol Rev* 71:371, 1991.

Canalis E: The hormonal and local regulation of bone formation, *Endocr Rev* 5:62, 1983.

DeLuca H, Schnoes H: Vitamin D: recent advances, *Annu Rev Biochem* 52:411, 1983.

Dempster DW, Cosman F, Parisien M, Shen V, Lindsay R: Anabolic actions of parathyroid hormone on bone, *Endocr Rev* 14:690, 1993.

Mallette LE: The parathyroid polyhormones: new concepts in the spectrum of peptide hormone action, *Endocr Rev* 12:110, 1991.

Mawer EB: Clinical implications of measurements of circulating vitamin D metabolites, *Clin Endocrinol Metab* 9:63, 1980.

Munson PL, Hirsch PF: Importance of calcitonin in physiology, clinical-pharmacology, and medicine, *Bone Miner* 16:162, 1992.

Nijweide PJ et al: Cells of bone: proliferation, differentiation and hormonal regulation, *Physiol Rev* 66:885, 1986.

Parfitt AM: The actions of parathyroid hormone on bone: relation to bone remodeling and turnover, calcium homeostasis, and metabolic bone disease. I. Mechanisms of calcium transfer between blood and bone and their cellular basis: morphologic and kinetic approaches to bone turnover, *Metabolism* 25:809, 1976.

Parfitt AM: The actions of parathyroid hormone on bone: relation to bone remodeling and turnover, calcium homeostasis, and metabolic bone disease. II. PTH and bone cells: bone turnover and plasma calcium regulation, *Metabolism* 25:909, 1987.

Raisz LG: Direct effects of vitamin D and its metabolites on skeletal tissue, *Clin Endocrinol Metab* 9:27, 1980.

Reichel H et al: The role of the vitamin D endocrine system in health and disease, *N Engl J Med* 320:980, 1989.

Stern PH: Vitamin D and bone, *Kidney Int* 29:S17, 1990.

Walters MR: Newly identified actions of the vitamin D endocrine system, *Endocr Rev* 13:719, 1992.

Books and Monographs

Avioli LV et al: Bone metabolism and disease. In Bondy PK, Rosenberg LE, eds: *Metabolic control and disease,* Philadelphia, 1980, WB Saunders Co.

Coleman DT, Fitzpatrick LA, Bilezikian J: Biochemical mechanisms of parathyroid hormone action. In Bilezikian J, ed: *The parathyroids: basic and clinical concepts,* New York, 1994, Raven Press.

Kronenberg HM, Bringhurst FR, Segre GV, Potts JT Jr.: Parathyroid hormone biosynthesis and metabolism. In Bilezikian J, ed: *The parathyroids: basic and clinical concepts,* New York, 1994, Raven Press.

Neer RM: Calcium and inorganic phosphate homeostasis. In Degroot LJ, ed: *Endocrinology,* New York, 1989, Grune & Stratton, Inc.

Rosenblatt M et al: Parathyroid hormone: physiology, chemistry, biosynthesis, secretion, metabolism, and mode of action. In Degroot LJ, ed: *Endocrinology,* New York, 1989, Grune & Stratton, Inc.

CHAPTER 44

The Hypothalamus and Pituitary Gland

The pituitary gland, once called "the master gland," retains a preeminent position in endocrinology even though it is now known to be under neural regulation by products from the hypothalamus and under feedback control by circulating products of its target glands. The pituitary gland and hypothalamus, with their associated neural and vascular connections, form a complex functional unit that epitomizes the subtle interrelationship between the endocrine system and the nervous system. *This unit regulates water metabolism, milk secretion, body growth, reproduction, lactation, and the growth and secretory activities of the thyroid, adrenal, and reproductive glands.*

The neurons of the hypothalamus synthesize and secrete neurohormones (Figure 44-1). Two of these neurohormones, **antidiuretic hormone** and **oxytocin,** are stored in secretory vesicles in terminal swellings of the axons within the posterior pituitary gland, also known as the **neurohypophysis.** From there they are released into the bloodstream to act on distant target cells (**neurocrine** function). Several other hypothalamic neurohormones are transported down axons that end in a neurovascular region known as the **median eminence,** just below the hypothalamus (Figure 44-1). From storage vesicles, these neurohormones are released into the bloodstream and stimulate or inhibit proximal endocrine target cells in the anterior pituitary gland, also known as the **adenohypophysis** (again, neurocrine function). The endocrine cells in the adenohypophysis synthesize, store, and secrete a variety of peptide and protein hormones that are released into the bloodstream to act on distant peripheral target cells (**endocrine** function). In addition, the hormones of these closely intertwined endocrine cells may act on neighboring target cells within the adenohypophysis (**paracrine** function).

ANATOMY AND EMBRYOLOGICAL DEVELOPMENT

The pituitary gland sits beneath the hypothalamus in a socket of bone called the **sella turcica,** within the skull. The gland represents a fusion of two tissues. The posterior portion, or neurohypophysis, develops as a downward outpouching of neuroectoderm from brain tissue in the floor of the third ventricle. This differentiates into the neurons of the hypothalamus. The lower part of the downward-growing neural stalk forms the bulk of the posterior pituitary. The upper part of the neural stalk expands to form the median eminence. Both the posterior pituitary and the median eminence consist largely of the terminals of various hypothalamic neurons. Both tissues are highly vascularized, and their capillaries contain fenestrations (intercellular windows) that allow influx and efflux of protein molecules.

The posterior pituitary is supplied by the inferior hypophyseal artery, whose capillary plexus invests the terminal swellings of axons from the supraoptic and paraventricular areas of the hypothalamus. These terminals are the immediate source of the peptide neurohormones, antidiuretic hormone (ADH) and oxytocin (OCT); ADH is also known as **arginine vasopressin** (AVP). These neurohormones are released into this capillary plexus, which carries them into the systemic circulation via draining veins (Figure 44-1).

The anterior pituitary develops from an upward outpouching of ectoderm from the floor of the oral cavity. After pinching off, the pouch becomes separated from the mouth by the sphenoid bone of the skull. At the junction of the anterior and posterior lobes of the pituitary gland is an intermediate zone, minuscule in humans but well developed in animals, from which another peptide hormone, **melanocyte-stimulating hormone** (MSH), is produced.

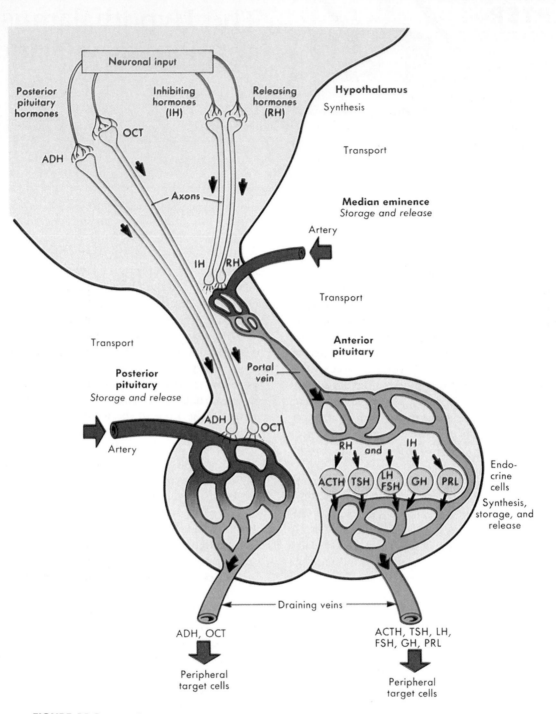

FIGURE 44-1 A schematic overview of the anatomical and functional relationships between the hypothalamus and the pituitary gland. Note that the posterior pituitary gland is an extension of neural tissue that stores neurohormones and has its own arterial blood supply. In contrast, the anterior pituitary gland is endocrine tissue with a blood supply largely derived from veins that first drain neural tissue in the median eminence. By this arrangement the endocrine cells are exposed to high concentrations of neurohormones originating in the hypothalamus and stored in the median eminence. The hormones secreted by the posterior and anterior pituitary gland reach and act on peripheral target cells. *ADH,* Antidiuretic hormone; *OCT,* oxytocin; *ACTH,* adrenocorticotropic hormone; *TSH,* thyroid-stimulating hormone; *LH,* leutinizing hormone; *FSH,* follicle-stimulating hormone; *GH,* growth hormone; *PRL,* prolactin.

The median eminence is supplied mainly by the superior hypophyseal artery (and to a lesser extent, the inferior hypophyseal artery). Its capillary plexus invests terminal swellings of axons from a variety of hypothalamic neurons. These neurons are the source of hypothalamic **releasing hormones** and **inhibiting hormones** that regulate anterior pituitary function. The capillary plexus of the median eminence forms a set of **portal veins** that descend into the anterior pituitary (Figure 44-1). These veins then give rise to a second fenestrated capillary plexus, which serves a dual role. Hypothalamic releasing and inhibiting hormones, carried down from the median eminence, exit the second plexus and regulate the secretion of the endocrine cells in the anterior pituitary. The hormone products of these cells then enter the same capillary plexus and are delivered via the circulation to distant target cells.

The anterior pituitary derives 90% of its blood supply in this manner and has little direct arterial input. Furthermore, its endocrine cells lie outside the blood-brain barrier. Reversal of flow upward in the portal veins may permit high concentrations of anterior pituitary hormones to reach the median eminence and even the hypothalamus, where they could feed back on neurons without impedance from the blood-brain barrier.

Just above the pituitary gland and sella turcica lies the crossing of the optic nerves as they course from the retina to the cerebral cortex. Any upward tumorous growth of the pituitary out of the sella turcica can compress the optic nerves and cause a characteristic loss of visual fields and acuity.

HYPOTHALAMIC FUNCTION

A comprehensive discussion of the hypothalamus is given in Chapter 10. From an endocrine standpoint, however, the hypothalamus may be viewed as a central relay station for collecting and integrating signals from diverse sources and funneling them to the pituitary gland (Figure 44-2). The hypothalamus receives input from the thalamus, the reticular activating substance, the limbic system (amygdala, olfactory bulb, hippocampus, and habenula), the eyes, and remotely from the neocortex. Through this input, *pituitary function can be influenced by sleep or wakefulness, pain, emotion, fright, smell, light, and possibly even thought*. It can be coordinated with such other behavior as mating responses. Interhypothalamic axonal connections also exist. *These*

allow the output of pituitary hormones to respond to changes in autonomic nervous system activity and to the needs of temperature regulation, water balance, and energy requirements.

The proximity of these various areas of the hypothalamus to each other has functional logic. As one example, hormones of the thyroid gland increase energy expenditure, metabolic rate, and thermogenesis. The neurons that ultimately control thyroid gland output are anatomically close to neurons that regulate energy intake via appetite control and also temperature. As the hypothalamic-pituitary unit is systematically studied, other examples become apparent.

Separation of the hypothalamus into individual nuclei or discrete anatomical centers of endocrine function is relatively imprecise, with two exceptions. The supraoptic nucleus is a collection of large neurons that secrete mainly ADH, and the paraventricular nucleus is a similar collection of neurons that secrete mainly OCT. These two neuronal pools overlap only slightly. In contrast, the small neurons that secrete the hypothalamic releasing and inhibiting hormones are more loosely aggregated in various areas, and they overlap more. As a general rule, only one cell type secretes each individual neurohormone, although in at least one instance the same hypothalamic neuron contains two hormones. In some hypothalamic neurons, monoamine neurotransmitters are also produced.

Each hypothalamic anterior pituitary releasing or inhibiting hormone can be assigned a primary target for which it has been named, such as **thyrotropin-releasing hormone** (a hormone that releases another hormone, which stimulates the thyroid gland) or **somatostatin** (soma, referring to body growth; statin, referring to halting of function). However, some of these neuropeptides act on more than one anterior pituitary cell.

In addition to those hypothalamic neurons whose axons end in the posterior pituitary and median eminence, other neurons have axons that project to different parts of the brain. In these instances the same hypothalamic peptides serve as neurotransmitters. (Furthermore, these neuropeptides have also been found in the spinal cord, the sympathetic ganglia, sensory neurons, pancreatic islets, and neuroendocrine cells of the gastrointestinal tract).

Hypothalamic neurohormones are synthesized from preprohormones (Chapter 40). Copeptide products of processing have also been identified and their functional roles investigated. Hypothalamic neurohormones are typically secreted in pulses generated by an intrinsic neural oscillator (Figure 44-3). Optimal effects on target cells result from this pulsatile pattern of signaling.

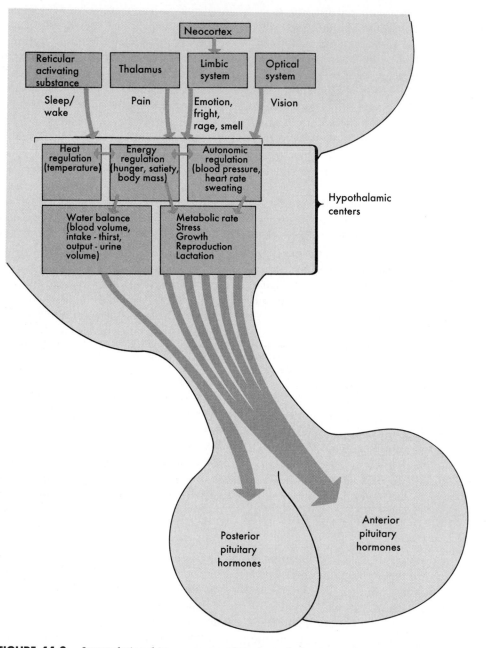

FIGURE 44-2 Interrelationships among various hypothalamic regulatory centers, their inputs from various parts of the brain, and their outputs to the pituitary gland. Note that sleep, pain, stresses, energy needs, temperature, and signals from the autonomic nervous system, as well as other factors, influence pituitary function.

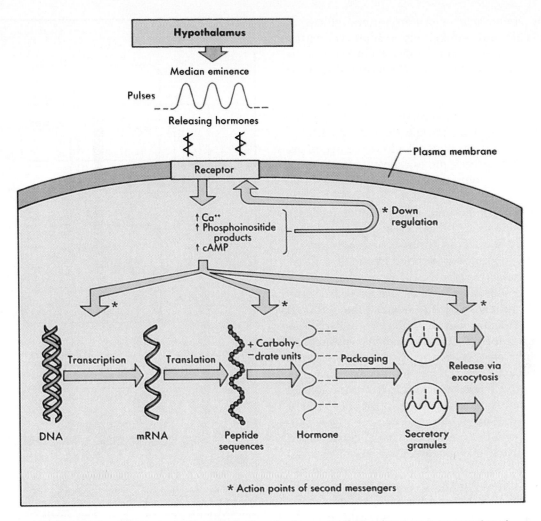

FIGURE 44-3 The action of hypothalamic releasing or inhibiting hormones on anterior pituitary cells. Characteristically the neurohormones are released in pulses, bind to plasma membrane receptors, and act through calcium ions (Ca^{++}) and other second messengers. They regulate gene expression, posttranslational processes, and secretion of anterior pituitary tropic hormones. *cAMP,* Cyclic adenosine monophosphate; *DNA,* deoxyribonucleic acid; *mRNA,* messenger ribonucleic acid.

An important clinical example is seen in reproductive function. In women who are infertile because of hypothalamic dysfunction, ovulatory menstrual cycles can only be restored if the appropriate **hypothalamic releasing hormone** is administered in pulses of the correct size and frequency throughout the day. If the releasing hormone is administered continuously, the necessary anterior pituitary response actually is ultimately lost, because of down regulation of the releasing hormone receptor. The same phenomenon is observed with regard to spermatogenesis in men.

Releasing and inhibiting hormones react with plasma membrane receptors in anterior pituitary cells; calcium, phosphatidylinositol products, and cyclic adenosine monophosphate (cAMP) are generated as second messengers. The releasing hormones all stimulate exocytosis of granules containing tropic hormone. In addition, they stimulate synthesis of the tropic hormones at the level of gene transcription and often enhance their biological activity by posttranslational modification. Inhibiting hormones have the opposite effects. Hypothalamic neurohormones can also regulate the numbers of their own receptors.

Afferent impulses to hypothalamic neurons are largely transmitted via norepinephrine, serotonin, and acetylcholine. The amino acid neurotransmitters (glutamate, aspartate, glycine, and γ-aminobutyric acid) also participate, as do a host of neuropeptides. From some hypothalamic neurons, dopamine and β-endorphin transmit signals to neighboring neurons via intrahypothalamic tracts and to the median eminence via efferent tracts. These signals directly or indirectly modulate the discharge of releasing and inhibiting hormones. In addition, neurotransmitters, such as dopamine from the hypothalamus, may themselves reach the portal vein blood and directly influence the output of anterior pituitary hormones.

The pituitary hypothalamic axis is under feedback control from its peripheral targets (Figure 44-4). Tropic hormones from the adenohypophysis increase the concentrations of (1) hormones secreted by the thyroid, adrenal, and reproductive glands; (2) peripheral peptide products; or (3) substrates such as glucose or free fatty acids. These in turn feed back to regulate the output of both the hypothalamus and the anterior pituitary. This is known as **long-loop feedback,** and it is usually negative, although it can transiently be positive. Negative feedback can also be exerted by the pituitary hormones themselves on the synthesis or discharge of the related hypothalamic releasing or inhibiting hormones. This is known as **short-loop feedback.** Because these hormones do not ordinarily cross the blood-brain barrier, short-loop feedback may occur either via fenestrated cells of the capillaries that bathe hypothalamic neurons or via retrograde flow through pituitary portal veins. Finally a hypothalamic releasing hormone may even inhibit its own synthesis and discharge or stimulate that of a paired hypothalamic inhibiting hormone. This is called **ultrashort-loop feedback.**

POSTERIOR PITUITARY FUNCTION

Antidiuretic hormone (ADH) and oxytocin (OCT), two small peptides with molecular weights of approximately 1000 and with homologous structure, are secreted by the posterior pituitary gland. *The primary role of ADH is to conserve water and regulate the tonicity of body fluids (see Chapter 37). A secondary role is to help maintain vascular volume. The primary role of OCT is to eject milk from the lactating mammary gland; a secondary role is to stimulate contraction of the uterus.* Although their functions are very different, both hormones are synthesized, stored, and secreted in similar fashion.

The genes that direct synthesis of the preprohormones for ADH and OCT are very similar and likely have a common ancestor gene. In addition to the two neuropeptides, the gene products include distinctive small

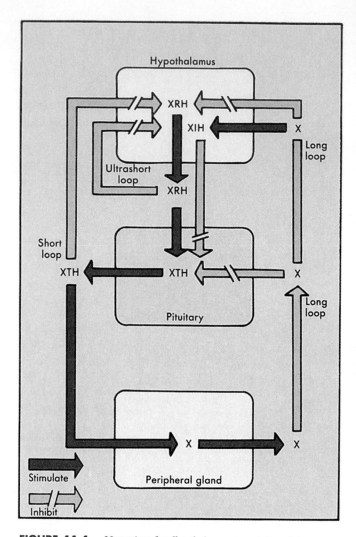

FIGURE 44-4 Negative feedback loops regulating hormone secretion in a typical hypothalamus–pituitary–peripheral gland axis. Note that feedback from the periphery can regulate both hypothalamic and pituitary function. Ultrashort loop feedback may be intra-hypothalamic. *X,* Peripheral gland hormone; *XTH,* pituitary tropic hormone; *XRH,* hypothalamic releasing hormone; *XIH,* hypothalamic inhibiting hormone.

proteins, known as **neurophysins.** Neurophysin-1 for OCT and neurophysin-2 for ADH are very similar in amino acid sequence. After processing, ADH and OCT are packaged with their respective neurophysins in neurosecretory granules. The neurophysins may serve as carrier proteins during transport of ADH and OCT down the axons to the posterior pituitary gland.

Release of ADH or OCT occurs when an electrical discharge is transmitted from the cell body in the hypothalamus down its axon, where it depolarizes the neurosecretory vesicle in the posterior pituitary. An influx of calcium into the vesicles releases the hormones by exocytosis. During this process each hormone dissociates from

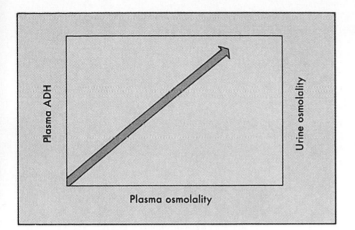

FIGURE 44-5 Positive correlation between the stimulus of plasma osmolality and the response of ADH secretion. In turn, urine osmolality rises as a result of the effect of higher plasma ADH levels on the renal tubules.

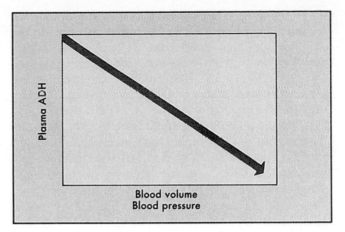

FIGURE 44-6 Negative correlation between the stimulus of blood volume or blood pressure and the response of ADH secretion. Actual plasma ADH levels evoked may be higher than those produced by osmolar stimuli.

its neurophysin, and the two molecules enter the circulation separately.

Secretion of Antidiuretic Hormone

Regulation of Secretion The secretion of ADH illustrates the homeostatic principle that the release of a hormone is stimulated by conditions that require its action (Figure 44-5). *Water deprivation raises plasma osmolality, which evokes release of ADH. In turn ADH causes retention of free water by the kidney and an increase in urine osmolality, with the result that plasma osmolality declines to normal* (see Chapter 37). Conversely, ingestion of a water load decreases plasma osmolality. This suppresses ADH release, which increases water excretion and raises plasma osmolality to normal. Thus water and ADH form a negative feedback loop, which operates to defend total body water and osmolality.

> A deficiency of ADH, known as **diabetes insipidus,** caused by disease or traumatic destruction of the ADH neurons, has dramatic consequences. Urine volume can reach 500 to 1000 ml/hr with osmolalities as low as 50 mOsm/kg. This necessitates frequent urination and requires the individual to drink equally large volumes of water to prevent collapse from volume depletion and hyperosmolality. Impairment of thirst or consciousness can, therefore, lead to death from dehydration. Treatment with ADH provides rapid relief.

The direct physiological stimulus to ADH release is an increase in the osmolality of fluids that bathe osmo-receptor neurons in the hypothalamus. This creates a gradient for water movement out of the neurons, and the consequent rise in intracellular osmolality triggers release of ADH. Any administered solute that does not readily penetrate cell membranes, such as sodium, creates the same osmotic gradient and stimulates ADH secretion. By contrast, solutes that freely enter cells, such as urea, do not stimulate ADH release.

The hypothalamic osmoreceptors respond to changes in plasma osmolality of only 1% to 2%. The osmolar threshold for ADH release is approximately 280 mOsm/kg of body weight. Plasma ADH then increases about 1 pg/ml for each 3 mOsm/kg increase in plasma osmolality. Generation of sufficient ADH to produce maximal retention of water and maximal urinary osmolality occurs when plasma osmolality reaches 294 mOsm/kg. The osmolar threshold for stimulation of thirst is close to or somewhat higher than that for ADH. Therefore, ADH secretion may precede activation of thirst in defending normal body water content and tonicity.

ADH release is also stimulated by hypovolemia. This is a much less sensitive response, because a decrease of 5% to 10% in blood volume, cardiac output, or blood pressure is required (Figure 44-6). Hemorrhage, quiet standing, and positive-pressure breathing, all of which reduce cardiac output and central blood volume, increase ADH secretion. Conversely, increasing central blood volume by administration of blood or isotonic saline solution suppresses ADH release. Hypovolemia is perceived by several pressure (rather than volume) sensors (see Chapters 19 and 23). These include carotid and aortic baroreceptors and stretch receptors in the walls of the left atrium and pulmonary veins. Normally, these pressure receptors tonically inhibit ADH release. A reduction in circulating blood volume decreases pressure on the baroreceptors and reduces the flow of inhibitory impulses to the hypothalamus. This increases ADH secretion. Hypovolemia also stimulates the generation of **renin** and **angiotensin** directly within the brain. The angiotensin augments the release of ADH and also stimulates thirst. In contrast, **atrial natriuretic peptide,** generated in response to volume or pressure overload, inhibits ADH release. Plasma ADH concentration rises much more in response to hypovolemia than in response to hyperosmolality. This correlates with the lesser sensitivity of the vascular system than of the kidney to hormone action.

The two major stimuli of ADH release interact (Figure 44-7). Increases or decreases in volume reinforce the osmolar responses by raising or lowering, respectively, the threshold for osmotic release of ADH. Thus *hypovolemia sensitizes the system to hyperosmolarity.* When hypovolemia is severe, baroregulation can override osmotic regulation. Consequently, ADH secretion is stimulated even though plasma osmolality may be below the threshold of 280 mOsm/kg.

Pain, emotional stress, nausea and vomiting, heat, and a variety of drugs also stimulate ADH release. Ethanol, on the other hand, is a frequently encountered inhibitor that causes diuresis. Cortisol and thyroid hormone restrain ADH release; when they are deficient, ADH may be secreted even though plasma osmolality is low.

> The clinical syndrome of secretion of excess ADH in amounts inappropriate to the plasma osmolality occurs in a variety of settings. These include psychiatric or cerebral disease, use of psychotropic drugs, pulmonary disease or tumor, and following major surgery. Plasma osmolality is chronically low and may reach a point where the patient becomes obtunded or has seizures. Restriction of water intake or inhibition of ADH action is required to correct this situation.

ADH circulates at basal concentrations of about 1 pg/ml (10^{-12} M). The plasma half-life is very short. During water deprivation, ADH secretion increases threefold to fivefold, and synthesis of the hormone is augmented. Plasma levels of neurophysin-2 also rise and fall in parallel with ADH.

Actions of Antidiuretic Hormone The major action of ADH is on the renal tubular mechanism for concentrating the urine—that is, for reabsorbing osmotically unencumbered water from the glomerular filtrate (see Chapter 37). ADH stimulates the two phases in the countercurrent concentrating mechanism. First, the hormone

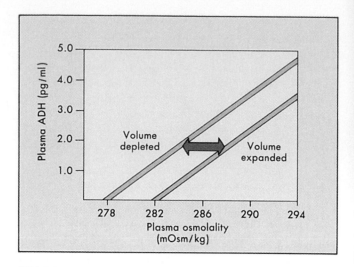

FIGURE 44-7   Regulation of ADH secretion by the interaction between plasma osmolality and blood volume. A reduction in blood volume sensitizes the hypothalamus–posterior pituitary gland so that ADH is secreted at a lower threshold of plasma osmolality. (Modified from Robertson GL et al: *J Clin Endocrinol Metab* 42:613, 1976.)

increases the transport of sodium out of the thick ascending portion of Henle's loop into the medullary interstitium, and thus helps to create the osmotic gradient for water. Second and more importantly, ADH increases the permeability of the collecting duct membranes to water, and thus facilitates back diffusion of water into the medulla. The maximal effect of ADH increases the osmolality of urine to a value four times higher than that of plasma, or about 1200 mOsm/kg. As noted in Figure 44-5, urine osmolality correlates directly with plasma ADH concentration. Without the hormone, urine osmolality falls to less than 100 mOsm/kg, and free water clearance reaches 10 to 15 mOsm/kg.

The intracellular mechanism of ADH action requires binding to a plasma membrane receptor, generation of cAMP as second messenger, and subsequent phosphorylation of proteins mediated by protein kinase A. In the collecting ducts this leads to insertion of water-conducting particles into the cell membrane and an increase in water permeability.

Several factors can blunt the action of ADH on tubular cells. These include solute diuresis, chronic water loading (which reduces medullary hyperosmolality), potassium deficiency, calcium excess, cortisol excess, and lithium administration. When any of these circumstances exist, the ineffectiveness of ADH leads to **nephrogenic diabetes insipidus.**

In addition to its major role in water metabolism, ADH may subserve other functions. It contributes in a minor way to increasing vascular tone in response to hemorrhage. When administered systemically in large doses, it elevates the blood pressure and constricts the coronary and splanchnic beds. This action requires binding to a different receptor in vascular cells and is mediated by the phosphatidylinositol–protein kinase C second-messenger system. ADH functions as a hypothalamic releasing factor via axons that project to the median eminence. ADH also serves neurotransmitter functions elsewhere within the brain—for example, at sites where it facilitates long-term memory.

Secretion of Oxytocin

OCT is required for normal nursing. Known biologically as the milk letdown factor, it is secreted within seconds in response to suckling. Sensory receptors in the nipple generate afferent impulses, which reach the hypothalamic paraventricular and supraoptic nuclei via various relays. A final cholinergic synapse causes discharge of OCT and neurophysin-1 from the posterior pituitary in a manner similar to ADH. Continued suckling further stimulates OCT synthesis and transport to the posterior pituitary. In humans there is little crossover secretion of ADH with suckling or of OCT secretion with an increase in plasma osmolality. OCT secretion can also be stimulated by vaginal distension during intercourse. Inhibition of OCT release by emotional distress can interfere with nursing.

Actions of Oxytocin OCT causes the myoepithelial cells of the alveoli in the breast to contract. This forces the milk from the alveoli into the ducts and nipple, from where it is extracted by the infant. OCT acts via plasma membrane receptors and cAMP generation in target cells. Binding of OCT to the receptor is increased by estrogen. Although basal plasma levels of OCT are similar in men and women, no role for the circulating hormone in men is known.

OCT also stimulates contraction of the uterus. Lower doses cause rhythmic contractions, whereas higher doses cause sustained tetanic contraction. In turn, reflexes inhibited by the contracting uterus stimulate oxytocin release. OCT and its receptor are also present in the human ovary and testis, where the locally produced hormone probably plays a role in reproduction.

There is controversial evidence regarding the function of OCT in normal labor in humans, but the sustained contractions OCT produces may be important in reducing blood loss from the uterus after delivery of the conceptus. OCT in large doses is often used therapeutically to induce labor or to stop excessive postpartum bleeding. Because of cross-reactivity with ADH receptors, OCT can cause water retention in this circumstance.

ANTERIOR PITUITARY FUNCTION

The **anterior pituitary gland,** or **adenohypophysis,** makes up most of the 500 mg of pituitary tissue. It contains at least five types of endocrine cells, each being the source of a different hormone with a distinct function after which it is named. These cell types, their relative proportions in the pituitary, and their major secretory products are shown in Figure 44-8. Although the five types of cells aggregate regionally on a functional basis, they do not form unique enclaves, but they are also interspersed among each other. They vary somewhat in size and in characteristics of their secretory granules, but they can only be identified with certainty by immuno-histochemical staining of the hormones within. Cells that contain no known hormone are called null cells. They do show evidence of protein hormone synthesis and contain a few secretory granules.

Each anterior pituitary cell is regulated by one or more hypothalamic neurohormones that reach them by the portal veins, as described previously. Three of the cell types produce hormones that regulate the function of the thyroid gland (**thyroid-stimulating hormone,** TSH), the adrenal glands (**adrenocorticotropic hormone,** ACTH), and the gonads (**leuteinizing** and **follicle-stimulating hormones,** LH, FSH), respectively. For purposes of better integration, the synthesis, secretion, and actions of these tropic hormones are presented in conjunction with their major peripheral target glands (see Chapters 45, 46, and 48). In this section only the function of those cells that secrete growth hormone, which acts on numerous peripheral tissues, and prolactin, which acts primarily on the mammary glands, is presented.

Growth Hormone (Somatotropin)

The major physiological effect of **growth hormone** (GH) is to stimulate postnatal somatic growth and development. Once growth and puberty have been completed, GH has a role in modulating the metabolism and body composition of adults.

Synthesis and Secretion Somatotrophs are the most numerous cells of the pituitary and are concentrated in its lateral wings (see Figure 44-8). Their product, GH, is a single, large, polypeptide chain with 191 amino acids and two disulfide bridges. The exact site(s) necessary for full biological activity is still unknown. The human genome contains multiple genes coding for a family of closely related GH molecules. Only one of these genes is expressed as normal pituitary GH. The messenger ribonucleic acid (mRNA) directs synthesis of a prehormone. After removal of a

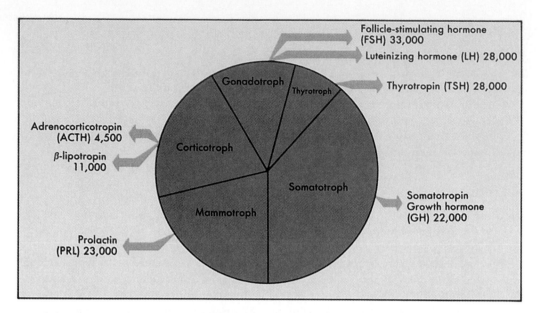

FIGURE 44-8 The relative proportions of cell types in the anterior pituitary gland, their major hormonal products, and the molecular weights of the latter. Note the preponderance of cells secreting growth hormone and prolactin.

signal peptide, the complete hormone is stored in granules. GH synthesis is increased by the specific hypothalamic releasing hormone, **growth hormone–releasing hormone** (GHRH) and by thyroid hormone and cortisol.

GH secretion by exocytosis is stimulated by GHRH, a hypothalamic peptide with 44 amino acids. GHRH interacts with its plasma membrane receptor, following which calcium, phosphatidylinositol products, and cAMP are generated as second messengers.

Somatostatin, a hypothalamic peptide with 14 or 28 amino acids, is a powerful inhibitor of GH release. Somatostatin blocks GHRH stimulation noncompetitively. The inhibitor acts through its own plasma membrane receptor, in part by decreasing both calcium entry into the cells and cAMP levels. GH is secreted in pulses caused by intermittent release of GHRH into portal pituitary vein blood. Somatostatin diminishes the frequency and amplitude of the pulses.

The secretion of GH is influenced by many factors (Figure 44-9). However, *the final common pathway for most stimulators of GH is an increase of GHRH, a decrease in somatostatin, or both. Conversely, suppressors of GH either decrease GHRH, increase somatostatin, or both.* Some agents can alter GH secretion by direct effects on the somatotroph.

GH release is regulated metabolically by the energy substrates glucose and free fatty acids and by amino acids. A sharp drop in either glucose or free fatty acid levels stimulates a large increase in plasma GH, whereas elevation of glucose or free fatty acid levels reduces plasma

GH considerably. Protein ingestion or intravenous amino acid infusion stimulates GH release. Arginine is especially effective. *Both short-term fasting and prolonged protein-calorie deprivation increase GH secretion. In contrast, obesity reduces GH responses to all stimuli, including GHRH.*

Central nervous system regulation of GH secretion takes several forms. A nocturnal surge in GH occurs 1 to 2 hours after the onset of deep sleep. Conversely, light sleep, associated with rapid eye movements (REM sleep), inhibits GH release. Various stresses, including trauma, surgery, anesthesia, fever, or even simple venipuncture, elevate plasma GH. Exercise is also a potent stimulant. These conditions influence hypothalamic GHRH and somatostatin neurons through a variety of monoamine neurotransmitters (Figure 44-9).

Age, gender, and other hormonal influences also alter GH secretion. *Children secrete somewhat more GH than adults, especially during puberty. In aged individuals GH secretion declines.* Females are usually more responsive to GH stimuli than are males.

Children who cannot secrete GH or respond to its actions grow at reduced rate and are delayed in skeletal and sexual maturation. They are short in stature and modestly obese (Figure 44-10). In some short children deficiency is easily established by failure of plasma GH to rise acutely with any stimulus. In others, a fairly spe-

Resting basal plasma levels of GH are 1 to 5 ng/ml (10^{-10} M). The hormone circulates bound to a binding protein that is identical to the extracellular domain of GH receptors. Daily GH secretion is approximately 600 mg in prepubertal children, 1800 mg in late puberty, and 300 to 500 mg in adults.

Feedback regulation of GH secretion occurs at all levels (see Figure 44-9). Long-loop negative feedback is exerted by **somatomedin,** which is a peripheral product of GH action. Somatomedin inhibits release of GHRH and the latter's action on the pituitary somatotroph, and it also stimulates somatostatin release. Short-loop negative feedback is exerted by GH itself, by stimulating somatostatin release. Ultrashort-loop negative feedback is exerted by GHRH, possibly via synapses with somatostatin neurons.

Actions of Growth Hormone GH interacts with several distinct plasma membrane receptors in target cells throughout the body. Thus far, none of the known membrane-generated second messengers appears to mediate its intracellular actions. However, much of the growth-promoting effect of GH requires the generation of an entirely different family of peptides, known as **somatomedins.** These peptides, with a molecular weight of 7000, resemble proinsulin in structure. They were originally discovered in plasma and were termed **insulin growth factors** (IGFs).

Two principal IGFs, their receptors, and the respective genes are well characterized. IGF-1 has 50% and IGF-2 has 70% amino acid homology with the A and B chains of insulin. Somatomedins or IGFs are produced by many tissues in response to GH. However, circulating somatomedins originate mainly in the liver, and the lag between administration of GH and the subsequent increase in plasma IGF-1 and IGF-2 is about 12 hours. Both growth factors circulate bound to a number of large binding proteins that regulate their availability to tissues. This accounts for their relatively stable concentration and much longer half-lives than that of GH itself. *Both somatomedins, but especially IGF-1, are greatly reduced in the plasma of GH-deficient subjects.*

Although somatomedins may function as circulating hormones in classic endocrine fashion, they also function as locally produced hormones in paracrine and even autocrine fashion. GH probably induces differentiation of precursor cells in target tissues (e.g., prechondrocytes in cartilage) into mature cells (chondrocytes), which then express the IFG-1 gene under further GH stimulation. IGF-1 then acts through its own plasma membrane receptor, which has structural similarity to the insulin receptor (see Chapter 42). The IGF-2 receptor is dissimilar to those of IGF-1 and insulin.

Somatomedins mediate the typical GH responses of cartilage, bone, muscle, adipose tissue, fibroblasts, and tumor cells in vitro. *Individuals who lack the ability to produce somatomedins show retarded growth, despite high GH levels.* Although fetal GH is not required for intrauterine growth, somatomedins generated in the placenta or other fetal tissues may participate in regulating prenatal growth. The placental GH variant may stimulate production of these growth factors.

In states of fasting and protein-calorie malnutrition, somatomedin levels in plasma are diminished, which correlates with the negative nitrogen balance in these conditions. Because GH levels are elevated in these catabolic states, factors other than GH must also regulate somatomedin production. In turn, the high GH levels most likely result from negative feedback caused by low somatomedin levels (see Figure 44-9). Somatomedin production is also diminished by cortisol and estrogens, hormones that antagonize GH action.

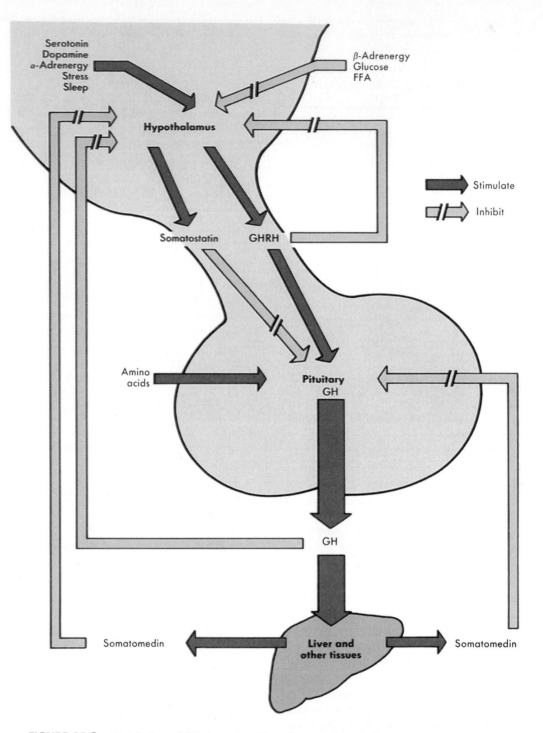

FIGURE 44-9 Regulation of GH secretion. Note that two hypothalamic peptides, one stimulatory and one inhibitory, regulate growth hormone release. Negative feedback by somatomedin, the peripheral product, is exerted at the hypothalamic and the pituitary level. There is also complex regulation by substrates and neural influences. *GHRH,* Growth hormone-releasing hormone; *FFA,* free fatty acids.

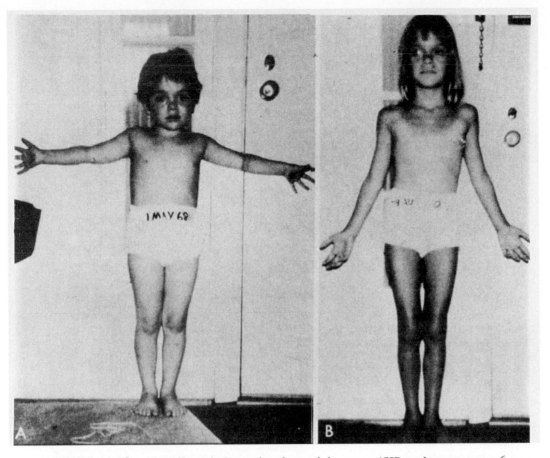

FIGURE 44-10   The effect of 15 months of growth hormone (GH) replacement on a 6-year-old child with GH deficiency. Note that GH increases linear growth and decreases adiposity. (From Foster D, Wilson J, eds: *Williams textbook of endocrinology,* Philadelphia, 1985, WB Saunders Co.)

The multiplicity of GH targets and effects is indicated in Figure 44-11. The most striking and specific effect is the acceleration of linear growth (see Figure 44-10) that results from GH action on the epiphysial cartilage growth centers of long bones (see Chapter 43). All aspects of the metabolism of chondrocytes, the cartilage-forming cells, are stimulated. This includes the synthesis of collagen and of the proteoglycan chondroitin, which together form the resilient extracellular matrix of cartilage. In addition, GH stimulates the synthesis of proteins, RNA, and DNA in these cells, as well as their proliferation. In support of the augmented protein synthesis, GH also stimulates cellular uptake of amino acids.

Many tissues share in the anabolic response to GH. The width of bones increases, as well as their length. Visceral organs (liver, kidney, pancreas, intestines), endocrine glands (adrenals, parathyroids, pancreatic islets), skeletal muscle, heart, skin, and connective tissue all enlarge. This is reflected in enhanced function of these organs.

GH affects carbohydrate and lipid metabolism. It stimulates expression of the insulin gene; without GH, insulin secretion declines. Importantly, *GH induces resistance to the action of insulin;* glucose uptake by muscle and adipose cells is inhibited and plasma glucose rises. Hyperinsulinemia results in compensation. In addition, GH enhances lipolysis and antagonizes insulin-stimulated lipogenesis. These actions increase plasma free fatty acid and ketoacid levels and decrease adipose tissue. Thus, on balance, GH is a diabetogenic hormone.

Sustained hypersecretion of GH from a slow-growing somatotroph tumor produces a unique syndrome called **acromegaly,** which reflects all the above actions. In adults, accumulation of excess soft tissue and widening of bones lead to coarse features (Figure 44-12) and spade-like digits. Thick skin, enlarged muscles such as in the tongue, and decreased subcutaneous fat are seen. Glomerular filtration and cardiac output are increased. Glucose intolerance or frank diabetes occurs in some individuals. Life expectancy is reduced by accelerated atherosclerosis. The diagnosis is confirmed by elevated plasma GH and somatomedin levels. If surgery is not curative, somatostatin analogues are effective treatment.

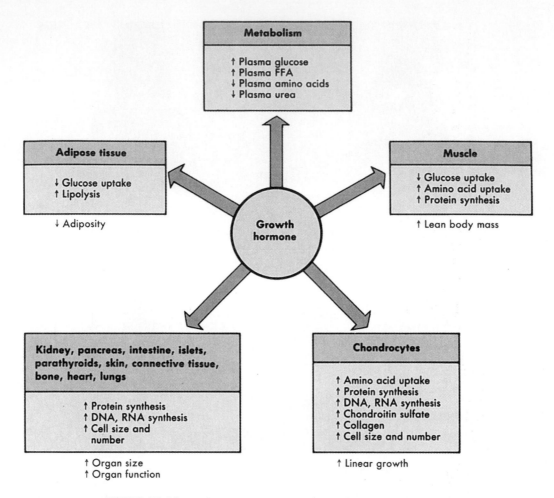

FIGURE 44-11 An overview of GH actions. *FFA,* Free fatty acids.

Correlation of Growth Hormone and Insulin Actions The secretion and actions of GH and insulin are metabolically coordinated (Figure 44-13).

1. When protein and energy intake is ample, amino acids can be used for protein synthesis and growth. Thus protein ingestion stimulates both GH and insulin secretion, and together the hormones augment the production of somatomedins. The latter in turn stimulate accretion of lean body mass. At the same time, the insulin-antagonistic effect of GH helps to prevent hypoglycemia, which might otherwise result from the increased insulin secretion in the absence of ingested carbohydrate.

2. When carbohydrate is ingested alone, insulin secretion is increased but GH secretion is suppressed. In the absence of amino acids, accelerated generation of somatomedins is not advantageous. Insulin antagonism also is not necessary; on the contrary, unrestrained expression of insulin action permits efficient storage of excess carbohydrate calories.

3. With fasting, insulin secretion falls, GH secretion rises, but somatomedins still decline. This combination seems appropriate in a situation where protein catabolism is essential and protein synthesis must decline. However, the increase in GH is beneficial, because it contributes to enhanced lipolysis and decreases peripheral tissue glucose utilization. This helps to mobilize free fatty acids for oxidative purposes and to provide glucose for central nervous system needs.

Prolactin

Prolactin (PRL) is a protein hormone principally concerned with stimulating breast development and milk production in women. In addition, it may play a role in reproductive function. The PRL-producing cells, called **mammotrophs,** are the second most prevalent in the pituitary gland (see Figure 44-8). They increase in number during pregnancy and lactation.

Synthesis and Secretion PRL is a single-chain protein with 198 amino acids and three disulfide bridges. It is structurally similar to GH, and the two genes are thought to have arisen from a common ancestor. Synthe-

A   **B**   **C**

FIGURE 44-12   Coarsening of the face in a patient who developed GH hypersecretion (acromegaly). **A,** Age 24. **B,** Age 50. **C,** Age 58.

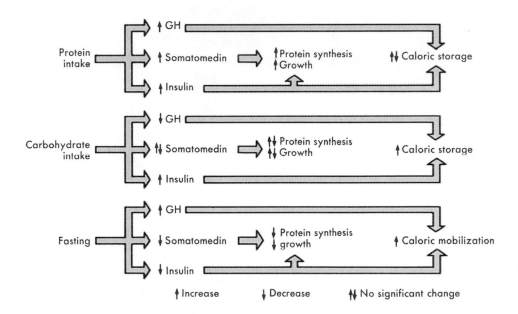

FIGURE 44-13   Complementary regulation of GH and insulin secretion. Both hormones are increased by protein intake, and both likewise stimulate anabolic processes. Insulin and GH secretion are regulated in opposite directions under circumstances where storage of ingested carbohydrate (facilitated by insulin) or caloric mobilization during fasting (facilitated by GH) is required.

sis of PRL proceeds in the manner described for GH via a prehormone. Some molecules are also glycosylated, and these are secreted constitutively. A small number of pituitary cells, called **mammosomatotrophs,** actually secrete PRL and GH. Transcription of the PRL gene is regulated by the same factors that regulate secretion of the hormone (see following discussion).

The most important influence on prolactin secretion is the combination of pregnancy, estrogens, and nurs-ing (Figure 44-14). In preparation for lactation, PRL secretion increases steadily during pregnancy to a twenty-fold plasma elevation. This is probably mediated by the high estrogen level of pregnancy, which stimulates hyperplasia of mammotrophs and transcription of the PRL gene. Although estrogen does not directly stimulate the release of prolactin, it does enhance responsiveness to other stimuli. If a new mother fails to nurse her child, the plasma PRL level declines to the range that prevails

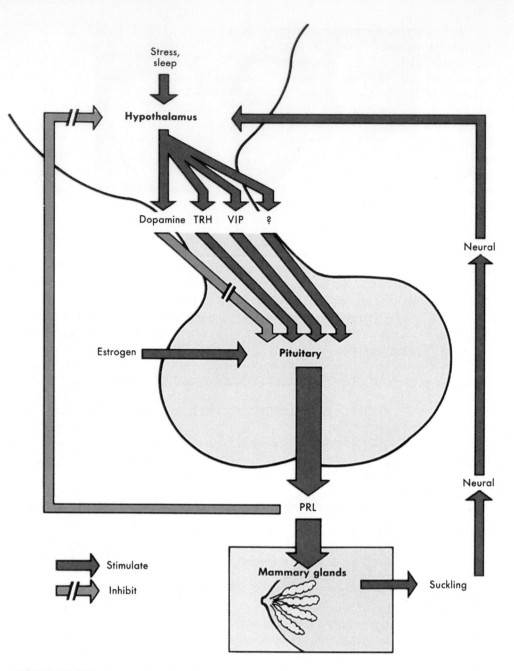

FIGURE 44-14 Regulation of prolactin *(PRL)* secretion. The predominant hypothalmic influence is normally inhibitory via dopamine. Pregnancy via high estrogen levels and lactation via suckling are the major physiological stimulators. *TRH,* Thyrotropin-releasing hormone; *VIP,* vasoactive intestinal peptide.

in nonpregnant women by 6 weeks after delivery. Suckling, however, maintains elevated PRL levels for 8 to 12 weeks.

PRL secretion rises at night and in conjunction with major stresses. The functional significance of increased PRL in these situations is unclear.

Unique among the anterior pituitary hormones, *secretion of PRL is predominantly under inhibition by hy-* *pothalamic factors* (Figure 44-14). Disruption of the connections to the hypothalamus leads to a great increase in PRL secretion, whereas the secretion of all other anterior pituitary hormones decreases. **Dopamine** released from the median eminence into the portal veins is the major hypothalamic inhibitory factor. This catecholamine neurohormone dramatically suppresses the release and synthesis of PRL. An additional prolactin-inhibiting factor may be a copeptide synthesized with the hypothalamic peptide, LH-releasing hormone. Short-loop negative feedback also operates as PRL inhibits its own secretion by stimulating the synthesis and release of hypothalamic dopamine (Figure 44-14).

The hypothalamus is also the source of PRL-releasing factors. Thyrotropin-releasing hormone (TRH) strongly stimulates PRL synthesis and release by acting through its receptors in mammotrophs. However, TRH probably does not primarily mediate the PRL response to nursing. A number of other hypothalamic peptides found in the hypothalamus, such as vasoactive intestinal peptide (VIP), have PRL-releasing activity; but their roles are unknown.

Actions of Prolactin *PRL participates in stimulating the original differentiation of breast tissue and its further expansion during pregnancy. It is the principal hormone responsible for* **lactogenesis** *(milk production).* PRL, together with estrogen, progesterone, cortisol, and GH, stimulates proliferation and branching of the breast ducts. During pregnancy, PRL, estrogen, and progesterone cause development of glandular tissue (alveoli), within which milk production will occur. After parturition, milk synthesis and secretion require PRL, along with cortisol and insulin.

The action of PRL begins by combination with a plasma membrane receptor, similar to that of GH. This receptor neither links with G-proteins nor possesses tyrosine kinase activity. Subsequent to binding, PRL rapidly induces transcription of RNAs for the milk proteins, casein and lactalbumin, and for enzymes necessary for the synthesis of lactose, the major sugar in milk.

A second area of PRL action may be on the reproductive axis. An excess of PRL blocks the synthesis and release of LH-releasing hormone, which inhibits gonadotropin secretion. This prevents ovulation and spermatogenesis. PRL also has variable effects on the synthesis of gonadal steroid hormones in both women and men. Certain behavioral effects of PRL have been described, including inhibition of libido in humans and stimulation of parental protective behavior toward the newborn in animals.

> Although the exact role(s) of PRL in normal human reproduction is uncertain, an excess of PRL from a pituitary tumor has major consequences. Ovulation and menstruation are prevented in women and spermatogenesis is blocked in men because secretion of pituitary gonadotropins is suppressed. Nonpregnant women secrete milk, and men may have breast enlargement. Surgical removal and treatment with dopamine agonists reverse these effects.

SUMMARY

1. The hypothalamic-pituitary unit regulates water metabolism, growth, lactation, and the functions of the thyroid gland, adrenal glands, and gonads.

2. Peptide hormones synthesized in some hypothalamic neurons pass down their axons to be stored in and released into the circulation from the posterior pituitary gland. Other hypothalamic peptides travel down axons to the median eminence, from which they are released into a portal venous circulation that carries them to the anterior pituitary gland. There they stimulate or inhibit release of target hormones.

3. Hypothalamic releasing and inhibiting peptides are secreted in pulses and induce effects via Ca^{++}, cAMP, and phosphatidyl inositol products as messengers. They stimulate or inhibit transcription, modulate translation, and stimulate or inhibit secretion of the target anterior pituitary hormones.

4. The anterior pituitary gland contains five functional cell types in close proximity, which suggests paracrine interactions. These are thyrotrophs, adrenocorticotrophs, gonadotrophs, somatotrophs, and mammotrophs. Each secretes a hormone(s) in response to hypothalamic stimulation. The function of each cell is also regulated by negative feedback from target glands or tissues.

5. Antidiuretic hormone (ADH) is a small peptide that is synthesized in the hypothalamus and is secreted from the posterior pituitary in response to an increase in plasma osmolality or to a decrease in plasma volume or blood pressure.

6. ADH acts on renal tubule cells via cAMP as a second messenger. The hormone increases reabsorption of free water and thus the final urine osmolality. ADH deficiency leads to polyuria and plasma hyperosmolality, whereas excess produces water retention and plasma hypo-osmolality.

7. Oxytocin (OCT) is structurally very similar to ADH, but it acts specifically on the mammary gland to cause release of milk. It is secreted in response to suckling. OCT also causes contraction of the uterus and plays a role in the overall process of parturition.

8. Growth hormone (GH) is a protein hormone with anabolic effects. It stimulates cartilage development, bone growth, and accretion of lean body mass. It acts largely via a peptide mediator (somatomedin) produced in the liver and many other cells. GH is responsible for linear growth in children.

9. GH also has insulin-antagonistic actions, such as stimulation of lipolysis and inhibition of peripheral glucose uptake, which are diabetogenic.

10. GH secretion is stimulated by growth hormone-releasing hormone and inhibited by somatostatin. These hypothalamic peptides reach the anterior pituitary via the portal veins. Glucose, free fatty acids, and the peripheral mediator somatomedin inhibit GH secretion.

11. GH excess produces the disease acromegaly. GH deficiency in childhood leads to short stature and delayed maturation.

12. Prolactin (PRL) is structurally similar to GH, but it specifically stimulates growth of the mammary glands and production of milk. PRL is normally tonically inhibited by dopamine from the hypothalamus. Its synthesis is markedly increased during pregnancy and is augmented by estrogens. Its release is stimulated by suckling.

13. An excess of PRL inhibits pituitary gonadotropin secretion and thereby results in loss of ovulation and menses in women and impairment of spermatogenesis and sexual functioning in men.

BIBLIOGRAPHY
Journal Articles

Amato G et al: Body composition, bone metabolism, and heart structure and function in growth hormone (GH)-deficient adults before and after GH replacement therapy at low doses, *J Clin Endocrinol Metab* 77:1671, 1993.

Argente J et al: Relationship of plasma growth hormone-releasing hormone levels to pubertal changes, *J Clin Endocrinol Metab* 63:680, 1986.

Brixen K et al: A short course of recombinant human growth hormone treatment stimulates osteoblasts and activates bone remodeling in normal human volunteers, *J Bone Miner Res* 5:609, 1990.

Charlton JA, Baylis PH: Mechanisms responsible for mediating the antidiuretic action of vasopressin (editorial), *J Endocrinol* 118:3, 1988.

Corpas E, Harman SM, Blackman MR: Human growth hormone and human aging, *Endocr Rev* 14:20, 1993.

Daughaday WH, Rotwein P: Insulin-like growth factors I and II: peptide, messenger, ribonucleic acid and gene structures, serum, and tissue concentrations, *Endocrinol Rev* 10:68, 1989.

Denef C: Paracrine interactions in the anterior pituitary, *Clin Endocrinol Metab* 15:1, 1986.

Hall K, Sara V: Somatomedin levels in childhood, adolescence and adult life, *Clin Endocrinol Metab* 13:91, 1984.

Hirsch AT et: Vasopressin-mediated forearm vasodilation in normal humans: evidence for a vascular vasopressin V_2 receptor, *J Clin Invest* 84:418, 1989.

Kelly PA et al: The prolactin/growth hormone receptor family, *Endocr Rev* 12:235, 1991.

Kerrigan JR, Rogol AD: The impact of gonadal steroid hormone action on growth hormone secretion during childhood and adolescence, *Endocr Rev* 13:281, 1992.

Lamberts SW, Macleod RM: Regulation of prolactin secretion at the level of the lactotroph, *Physiol Rev* 70:279, 1990.

Miller N et al: Short-term effects of growth hormone on fuel oxidation and regional substrate metabolism in normal man, *J Clin Endocrinol Metab* 70:1179, 1990.

Norsk P, Epstein M: Effects of water immersion on arginine vasopressin release in humans, *J Appl Physiol* 64:1, 1988.

Salomon F et al: The effects of treatment with recombinant human growth hormone on body composition and metabolism in adults with growth hormone deficiency, *N Engl J Med* 321:1797, 1989.

Sklar A, Schrier R: Central nervous system mediators of vasopressin release, *Physiol Rev* 63:1243, 1983.

Snyder SH: Brain peptides as neurotransmitters, *Science* 209:976, 1980.

Theill LE, Karin M: Transcriptional control of growth hormone expression and anterior pituitary development, *Endocr Rev* 14:670, 1993.

Thissen JP, Ketelslegers JM, Underwood LE: Nutritional regulation of the insulin-like growth factors, *Endocr Rev* 15:80, 1994.

Thompson CJ et al: Reproducibility of osmotic and nonosmotic tests of vasopressin secretion in men, *Am J Physiol* 260:R533, 1991.

Weitzman RE et al: The effect of nursing on neurohypophyseal hormone and prolactin secretion in human subjects, *J Clin Endocrinol Metab* 41:836, 1980.

Books and Monographs

Frohman LA et al: The physiological and pharmacological control of anterior pituitary hormone secretion. In Dunn A, Nemeroff C, eds: *Behavioral neuroendocrinology,* New York, 1983, Spectrum Publications, Inc.

Guillemin R: Neuroendocrine interrelations. In Body P, Rosenberg LE, eds: *Metabolic control and disease,* Philadelphia, 1980, WB Saunders Co.

Reeves WB, Anderoli TE: The posterior pituitary and water metabolism. In Foster D, Wilson J, eds: *Williams textbook of endocrinology,* Philadelphia, 1992, WB Saunders Co.

Reichlin S: Neuroendocrinology. In Foster D, Wilson J, eds: *Williams textbook of endocrinology,* Philadelphia, 1992, WB Saunders Co.

Riskind PN, Martin JB: Functional anatomy of the hypothalamic-anterior pituitary complex. In Degroot LJ, ed: *Endocrinology,* Philadelphia, 1994, WB Saunders Co.

Seo H: Growth hormone and prolactin: chemistry, gene organization, biosynthesis, and regulation of gene expression. In Imura H: *The pituitary gland,* New York, 1985, Raven Press.

Thorner MO, Vance ML, Horvath E, Kovacs K: The anterior pituitary. In Foster D, Wilson J, eds: *Williams textbook of endocrinology,* Philadelphia, 1992, WB Saunders Co.

45

The Thyroid Gland

The thyroid gland was the first endocrine gland to be recognized as such. Observations that its absence or enlargement was correlated with altered biology at distant body sites provided the clue that the gland produced a substance that reached tissue targets via the bloodstream. Extracts of the thyroid gland were subsequently shown to correct the striking disease state that resulted from its absence.

The thyroid gland produces two hormones, **thyroxine** and **triiodothyronine,** at a rather steady pace. *These hormones increase the rate of basal oxygen utilization and metabolism and the consequent rate of heat production, so as to adjust them to alterations in energy need, caloric supply, and thermal environment.* Thyroid hormones concordantly modulate the delivery of substrates and oxygen needed to sustain the appropriate metabolic rate. Their actions are critical for normal growth and maturation of the fetus and the child.

FUNCTIONAL ANATOMY

The thyroid gland develops from endoderm of the pharyngeal gut. The gland descends to the anterior neck, where it overlies the trachea and can often be palpated (Figure 45-1, *A*). It can also be visualized by several imaging techniques, such as ultrasonography. By 12 weeks of human gestation, the gland synthesizes and secretes thyroid hormones under the stimulus of the fetal hypothalamus and pituitary gland. *This entire axis is required for subsequent normal intrauterine development of the central nervous system and skeleton,* because neither maternal thyroid hormone nor its pituitary-stimulating hormone can cross the placenta in sufficient quantities after the first trimester.

The thyroid gland in adults weighs approximately 20 g. The histological structure is shown schematically in Figure 45-1, *B.* The endocrine cells are surrounded by a basement membrane, and they form single-layered circular **follicles.** The lumina of the follicles contain thyroid hormones stored in the form of a **colloid** material. When stimulated, the endocrine cells enlarge and assume a columnar shape, with their nuclei at the base. The colloid material in the lumen appears scalloped because it is undergoing resorption. Also scattered within the thyroid gland are the parafollicular cells, or C cells, which secrete calcitonin (see Chapter 43).

SYNTHESIS AND SECRETION OF THYROID HORMONES

Thyroid hormones are unique in that they incorporate an inorganic element, iodine, into an organic structure made up of two molecules of the amino acid, tyrosine. The secretory products of the thyroid gland are known as **iodothyronines.** *The major product is 3,5,3',5'-tetraiodothyronine, known as thyroxine and referred to as* T_4. *This molecule functions largely as a circulating prohormone. Secreted in much less quantity is 3,5,3'-triiodothyronine, known simply as triiodothyronine and referred to as* T_3. *This molecule, which provides almost all thyroid hormone activity in target cells, is actually produced mostly in various tissues from the circulating supply of the prohormone* T_4. A trivial secretory product with no identified hormonal action is 3,3',5'-triiodothyronine. This is known as reverse T_3, or rT_3, because it differs from T_3 only in the location of one of the three iodine atoms. This inactive molecule is an alternative product of the prohormone T_4, produced when less thyroid hormone action is needed. The structures of T_4, T_3, and rT_3 are shown in Figure 45-2.

Three major steps are involved in the synthesis of thyroid hormones: (1) uptake and concentration of iodide within the gland, (2) oxidation and incorporation of the

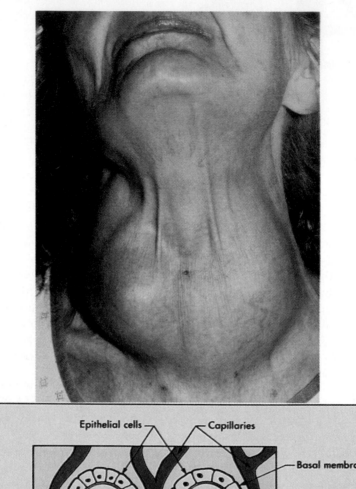

A

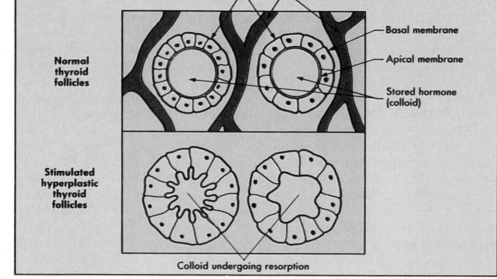

B

FIGURE 45-1   **A,** The anterior location of the thyroid gland permits visualization and palpation when it is enlarged and forms a goiter. **B,** Schematic representation of the basic thyroid unit. A normal follicle consists of a central core of colloid material surrounded by a single layer of cuboidal cells. When stimulated by thyrotropin, the cells elongate and the central core becomes scalloped because of resorption of the colloid.

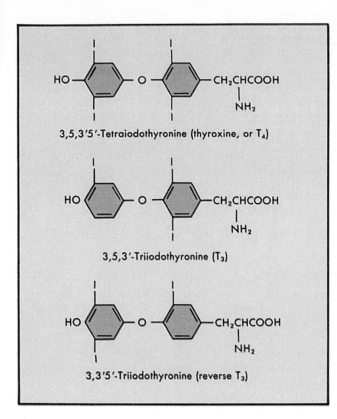

FIGURE 45-2 The structures of thyroxine (T_4), triiodothyronine (T_3), and reverse T_3 (rT_3). Note that T_3 and rT_3 differ only in the position from which an iodine atom was removed from T_4.

iodide into the phenol ring of tyrosine, and (3) coupling of two iodinated tyrosine molecules to form either T_4 or T_3 (Figure 45-3).

Prior to iodination and coupling, the tyrosine molecules must first be incorporated by standard peptide linkages into a protein known as **thyroglobulin.** Thyroglobulin is the substance actually iodinated on specific constituent tyrosines, and the latter are brought into proximity for coupling by the three-dimensional structure of the protein. The thyroid hormones formed remain in peptide linkage within thyroglobulin, and their release into the circulation requires proteolytic cleavage.

Step 1: Iodination

Iodide is an essential dietary element because of its thyroid role. The minimal daily iodide requirement for hormone synthesis is about 75 micrograms. In the United States the average daily intake is 300 to 400 μg, and almost the same amount is excreted in the urine. About 80 μg, or 20% of the extrathyroidal pool, is taken up daily by the gland. With iodide deficiency the

extrathyroidal pool size shrinks; however, the gland can increase the daily percentage uptake to 80% to 90%, and thereby still acquire sufficient iodide for hormone synthesis. A decrease in urinary excretion helps conserve iodide, as does preferenial synthesis of T_3, over T_4. Under steady-state conditions, about 80 μg of iodide are released from the gland daily, 90% in the form of T_4. The content of iodide within the thyroid gland is 100 times greater than the amount needed daily for hormone production. Because all this is stored in the form of iodinated thyroglobulin, the human is protected for approximately 2 months from the effects of iodide deficiency.

Iodide is actively transported into the thyroid gland against chemical and electrical gradients. This process, known as the **iodide trap,** maintains a high ratio of free iodide concentration in the gland to plasma iodide. The trapping mechanism for iodide requires energy generation via oxidative phosphorylation. Although a specific carrier remains to be identified, some evidence links it to an Na^+-K^+-ATPase. Various anions, such as thiocyanate (CNS), perchlorate ($HClO_4$), and pertechnetate (TcO_4), act as competitive inhibitors of iodide transport.

Small increases in dietary iodide intake lead to increases in the rate of thyroid hormone synthesis. However, as the daily dosage of iodide exceeds 2000 μg, the intraglandular concentration of free iodide or of some iodinated product reaches a level that inhibits the iodide trap and the biosynthetic mechanism, and thus the hormone production declines back to normal. A severe lack of dietary iodide (endemic in certain parts of the world) ultimately leads to thyroid hormone deficiency, despite maximum operation of the iodide trap.

Step 2: Iodination of Tyrosine within Thyroglobulin

Thyroglobulin is a large glycoprotein that is synthesized as two separate peptide units. These combine and are then glycosylated in transit to the Golgi apparatus. The completed protein, incorporated in small vesicles, moves to the apical membrane and then into the adjacent lumen of the follicle (Figure 45-3).

Just inside the follicle lumen, iodide is incorporated into thyroglobulin. An enzyme complex, known as **thyroid peroxidase,** is bound to the apical membrane. This enzyme catalyzes simultaneously the oxidation of iodide and its substitution for a hydrogen in the benzene ring of tyrosine. The immediate oxidant of iodide is hydrogen peroxide (H_2O_2). This is probably generated via the reduction of O_2 by reduced nicotinamide adenosine dinucleotide phosphate (NADPH) and flavoproteins. Ei-

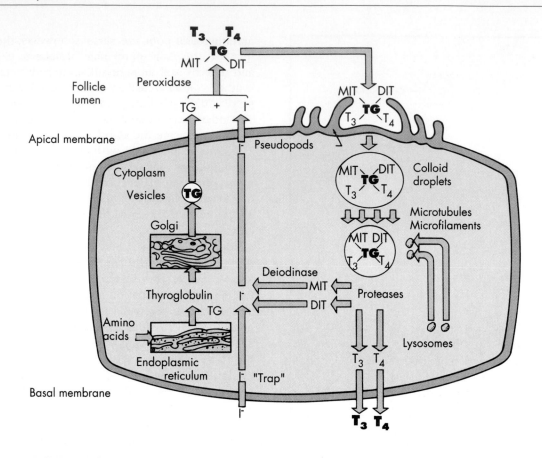

FIGURE 45-3 Overall schema of thyroid hormone synthesis and release. T_4 and T_3 synthesis occurs within the protein molecule thyroglobulin *(TG)* at the border of the cytoplasm and the follicle lumen. Retrieval of stored hormone requires endocytosis of the colloid followed by intracytoplasmic proteolysis by lysosomes. Iodide in the precursor molecules monoiodotyrosine *(MIT)* and diiodotyrosine *(DIT)* is recovered by the action of the enzyme deiodinase.

ther monoiodotyrosine (MIT) or diiodotyrosine (DIT) results from iodination.

Step 3: Coupling

The coupling step is also carried out within thyroglobulin by the enzyme peroxidase. One DIT molecule is juxtaposed either with another DIT molecule to form T_4 or with an MIT molecule to form T_3. The usual ratio of T_4 to T_3 in the gland is 10:1. When iodide availability is restricted or when the thyroid gland is hyperstimulated, the formation of T_3 is favored, thus providing relatively more active hormone.

Retrieval

Once thyroglobulin has been iodinated, it is stored within the follicle as **colloid.** Release of the peptide-linked T_4 and T_3 into the bloodstream requires retrieval of the thyroglobulin. The latter is transferred from the lumen of the follicle into the endocrine cell by **micropinocytosis** or by **endocytosis** (Figure 45-3). In the latter, the cell membrane forms pseudopods that engulf a pocket of colloid. This is pinched off by the cell membrane and becomes a colloid droplet within the cytoplasm. The droplet moves in a basal direction, probably as a result of microtubule and microfilament function. At the same time, lysosomes move from the base toward the apex of the cell and fuse with the colloid droplets. Lysosomal proteases then release free T_4 and T_3, which leave the cell through the basal membrane and enter the adjacent capillary blood (Figure 45-3).

The MIT and DIT molecules, which are also released from thyroglobulin, are rapidly deiodinated within the cell by the enzyme **deiodinase** (see Figure 45-2). Because these compounds are metabolically useless and, if secreted, would be lost in the urine, their deiodination conserves iodide for recycling into hormone synthesis. Normally, only minor amounts of intact thyroglobulin leave the cell.

Any step in the sequence from iodide trapping to thyroglobulin proteolysis may be defective in congenital biosynthetic disorders, and these defects result in thyroid hormone deficiency. A group of drugs, known as **thiouracils,** block the enzyme peroxidase and are very useful in treating states of thyroid hyperfunction. A large excess of iodide itself, its competitive anion perchlorate, or lithium (widely used in treatment of manic-depressive disorders) also inhibits T_4 synthesis. Iodide is sometimes used to treat **hyperthyroidism** for short periods until more definitive therapy takes hold.

REGULATION OF THYROID GLAND ACTIVITY

The thyroid gland is the effector component of a classic hypothalamic–anterior pituitary–peripheral gland axis (Figure 45-4) (see Chapter 44). The major stimulator of thyroid hormone secretion is **thyrotropin,** or **thyroid-stimulating hormone** (TSH), which is secreted by the anterior pituitary gland. The direct stimulator of TSH secretion is **thyrotropin-releasing hormone** (TRH) from the hypothalamus. The thyroid hormones T_4 and T_3, by negative feedback, inhibit the synthesis and release of TSH from the pituitary gland as well as TRH synthesis and release from the hypothalamus.

Thyrotropin-Releasing Hormone

TRH is a tripeptide, pyroglutamine-histadine-prolineamide. Its synthesis in the hypothalamus is directed by a gene that codes for a large precursor molecule that contains the small sequence of glutamine-histadine-proline-glycine. After translation, the glutamic acid undergoes cyclization, and the terminal glycine is replaced with an amino group. TRH is stored in the median eminence (Chapter 44) and reaches its target cells via the pituitary portal vein. There TRH interacts with specific plasma membrane receptors on the thyrotroph cell. This triggers an influx of calcium and increases in phosphatidylinositol products, which act as second messengers. TSH is then released by exocytosis. Prolonged stimulation with TRH also increases TSH synthesis and its bioactivity. TRH eventually down-regulates its own receptors; thus the releasing hormone loses effectiveness.

Thyrotropin

TSH is a glycoprotein hormone of 28,000 molecular weight. It is composed of two peptide subunits, each of which is coded for by separate genes on two different chromosomes. The α-subunit is "nonspecific" because it is also part of three unrelated hormones with reproductive function (luteinizing and follicle-stimulating hormones from the pituitary gland and chronic gonadotropin from the placenta). In contrast, the β-subunit of TSH is completely different and contains the specific biologically active sites of the hormone. Nonetheless, by noncovalent forces, the β-subunit must be combined with the α-subunit for TSH to stimulate thyroid cells. Specific glycosylation of TSH increases its activity.

TSH circulates in concentrations of about 10^{-11} M. For technical reasons, these are usually reported in units of biological activity; the normal range is approximately 0.5 to 6.0 μU/ml. The α-subunit also circulates.

TSH acts on the follicular cells of the thyroid gland to produce many effects, which are summarized in Figure 45-5. *The processes of iodide trapping and of each step in T_4 and T_3 synthesis, as well as the endocytosis of colloid and the proteolytic release of T_4 and T_3 from the gland, are all rapidly stimulated by TSH.* Sustained exposure to TSH leads to hyperplasia of the follicular cells (see Figure 45-1, *B*), accompanied by increases in endoplasmic reticulum, ribosomes, the size and complexity of the Golgi apparatus, and DNA synthesis. In the absence of TSH the gland atrophies, although it still maintains a low basal level of thyroid hormone secretion. The trophic effects of TSH on the thyroid gland may be mediated by local generation of insulin growth factors 1 and 2 or of epidermal growth factor.

The initial step in TSH action is binding to a plasma membrane receptor. This transmembrane molecule is functionally linked via a stimulatory G-protein to adenylyl cyclase. Cyclic AMP then mediates stimulation of iodide uptake by the cell. The phosphatidylinositol system may participate with cAMP in rapidly stimulating the subsequent steps in thyroid hormone secretion. Concurrently, TSH also increases glucose oxidation, which may provide NADPH needed for the peroxidase reaction. After several hours, TSH increases nucleic acid, protein, and phospholipid synthesis. These actions underlie the growth-promoting effects of TSH.

The trophic effects of TSH are commonly expressed pathophysiologically. Any genetic biosynthetic defect, an acquired impairment of thyroid hormone synthesis caused by inflammation or drugs, and iodide deficiency all increase TSH secretion via negative feedback. Chronic stimulation of the thyroid gland by TSH hypersecretion then produces (sometimes spectacular) enlargement of the gland, known as **goiter** (Fig. 45-1, *A*).

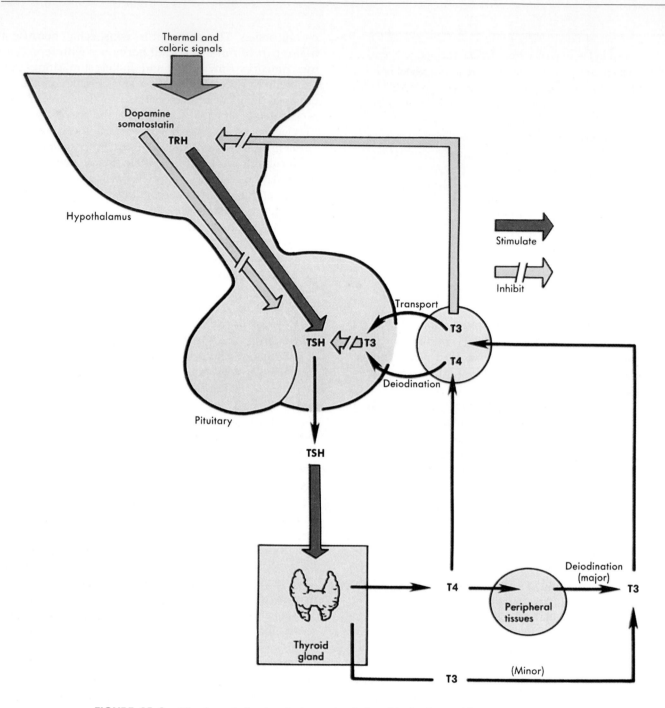

FIGURE 45-4 The hypothalamic–pituitary gland–thyroid gland axis. Thyrotropin-releasing hormone *(TRH)* stimulates thyrotropin *(TSH)* release from the pituitary gland. TSH stimulates T_4 and to a minor degree T_3 secretion by the thyroid gland. T_3 arising from T_4 in peripheral tissues or within the pituitary gland itself blocks the effect of TRH and suppresses TSH release by negative feedback. Dopamine and somatostatin also tonically inhibit TSH release.

Feedback Control

Thyroid hormone output is under sensitive feedback control, which keeps plasma T_4, and T_3, relatively constant. *Changes in thyroid hormone levels of only 10% to 30% are enough to change TSH levels in the opposite direction. Negative feedback is exerted predomi-*

nantly at the pituitary level (see Figure 45-4). This is well demonstrated by the results of repeated stimulation of the axis with TRH injections. The initially brisk increase in TSH is progressively dampened as T_4 levels rise in response to the TSH. These slightly elevated T_4 levels feed back to inhibit further responsiveness of the pi-

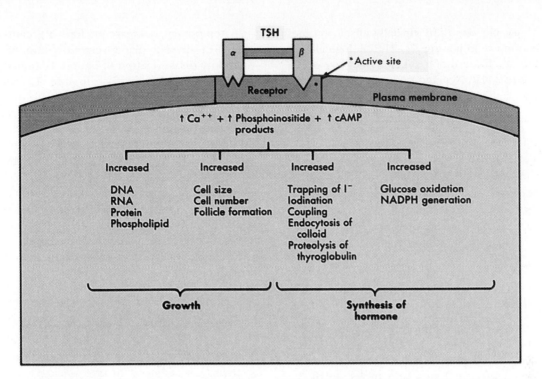

FIGURE 45-5   TSH actions on the thyroid cell. Cyclic adenosine monophosphate *(cAMP)* along with calcium ions *(Ca⁺⁺)* and phosphoinositol products act as second messengers generated by TSH binding to its receptor. All steps in thyroid hormone production, as well as many aspects of thyroid cell metabolism and growth, are stimulated by TSH.

tuitary thyrotroph to later TRH injections. With maneuvers that mainly lower plasma T_4 and T_3 levels, plasma TSH response to TRH stimulation is enhanced.

> Individuals with long-standing deficiency of thyroid hormone from thyroid gland disease have high plasma TSH levels as well as enlarged pituitary glands that contain increased numbers of thyrotroph cells and an elevated TSH content. Conversely, a pathological excess of thyroid hormone causes very low plasma TSH levels and atrophy of the thyrotroph cells.

The effector molecule of negative feedback is T_3. Although T_3 can enter the thyrotroph cell from the plasma, that T_3 which is generated within the pituitary gland by deiodination of T_4 taken up from the plasma is more important (see Figure 45-4). T_3 suppresses not only TSH release but also its synthesis by inhibiting expression of the TSH gene. T_3 further blunts TSH release by decreasing the number of TRH receptors.

TSH secretion is also tonically inhibited by dopamine and somatostatin from the hypothalamus. Cortisol and growth hormone reduce TSH secretion as well, the latter probably by stimulating somatostatin release (see Chapter 44).

The operation of the hypothalamic-pituitary-thyroid axis results in a slightly pulsatile plasma TSH level and steady plasma T_4 and T_3 levels. This befits hormones, whose actions on metabolism are gradual and wax and wane slowly. Physiological conditions that alter TSH levels, and therefore T_4 and T_3 levels, are consonant with the action of thyroid hormones on energy utilization and thermogenesis. During total fasting, TSH responsiveness to TRH stimulation and possibly TRH release itself are diminished; T_3 levels also fall. This coincides with an advantageous decrease in resting metabolic rate (see Chapter 41). In contrast, ingestion of excess calories, especially carbohydrate, tends to increase T_3 availability. In animals, exposure to cold increases TSH and thyroid hormone secretion. In humans, this is observed shortly after birth, when the change in temperature from the maternal to external environment is accompanied by a sharp rise in plasma TSH and T_4 levels. The latter remains above adult levels for some weeks.

METABOLISM OF THYROID HORMONES

T_4, the dominant secreted and circulating form, serves largely as a prohormone for T_3, but T_4 also provides some intrinsic intracellular action of its own. Average daily secretion of T_4 is 90 μg. The circulating storage function of plasma T_4 is reflected in its large pool size and long half-life of 6 days. In contrast, the major portion of T_3, 35 μg per day (and virtually all rT_3) comes from deiodination of circulating T_4. The biologically active metabolite, T_3, has a much smaller pool size and a shorter half-life of 1 day. Average plasma concentrations are: T_4, 8 μg/dl; T_3, 0.12 μg/dl; rT_3, 0.04 μg/dl.

> Replacement of thyroid hormone to deficient individuals is almost always carried out with the prohormone T_4 and not with the more active metabolite T_3. This is done to mimic the physiological situation. The biochemical end points of treatment are a level of T_4 in the normal range and a reduction (by negative feedback) of the elevated TSH level to the normal range.

Protein Binding

T_4 and T_3 circulate almost entirely bound to proteins. The major binding protein is **thyroxine-binding globulin** (TBG), a glycoprotein that is synthesized in the liver. Each TBG molecule binds one molecule of T_4. About 70% of T_4 and T_3 is bound to TBG. The remainder is bound to **transthyretin** (thyroxine-binding prealbumin) and albumin. Transthyretin has a lower affinity for T_4 than does TBG and therefore more readily transfers the hormone to target cells by dissociation.

Two biological functions can be ascribed to TBG and transthyretin. First, by creating a circulating reservoir of T_4, they buffer against acute changes in thyroid gland function. Even the sudden addition to the plasma of an entire day's thyroid gland output would cause only a 10% increase in the circulating T_4 concentration. After removal of the gland, it would take nearly 1 week for the plasma T_4 concentration to fall 50%. Second, binding T_4 and T_3 prevents their glomerular filtration and urinary excretion.

Only 0.03% of total T_4 and 0.3% of total T_3 are in the free state. However, these are the critical biologically active fractions. They not only exert the thyroid hormone effects on target tissues, but are also responsible for pituitary feedback. Therefore, the chemical equilibrium between T_4 and TBG governs the distribution of the hormone between the free T_4 and bound $T_4 \cdot$ TBG fractions:

$$T_4 + TBG = T_4 \cdot TBG \tag{1}$$

$$K_{eq} = \frac{[T_4 \cdot TBG]}{[T_4][TBG]} \tag{2}$$

$$\frac{Free\ T_4}{Bound\ T_4} = \frac{[T_4]}{[T_4 \cdot TBG]} = \frac{1}{K_{eq}[TBG]} \tag{3}$$

where K_{eq} is the equilibrium constant.

A temporary decrease in free T_4, caused by a decrease in thyroid gland secretion, can be rapidly reversed by disassociation of bound T_4 (equation 1). Likewise, a temporary increase in free T_4 can be rapidly compensated for by association of the excess T_4 with TBG, which has additional unoccupied binding sites. Sustained decreases or increases in daily T_4 supply resulting from thyroid disease, however, eventually lead to sustained decreases or increases in both the bound and free fractions.

A primary change in TBG concentration itself disturbs the ratio of free to bound T_4 (equation 3). In this situation the normal thyroid gland must increase or decrease its rate of hormone secretion appropriately until the new equilibrium state restores the absolute free T_4 level to normal.

> Acute hepatic disease, pregnancy, or estrogen therapy raises serum TBG levels. This would initially decrease free T_4 (equation 1), but negative feedback would increase TSH secretion. The latter would stimulate sufficiently more T_4 release from the thyroid gland to raise the bound T_4 to the point where the equilibrium in equation 1 would restore free T_4 to normal. In severe chronic hepatic disease (such as **cirrhosis**) or kidney disease (such as the **nephrotic syndrome**), serum TBG falls either because of reduced synthesis or loss in the urine. The sequence opposite to that just described would decrease T_4 release by the thyroid gland and restore free T_4 to normal.

Metabolic Pathways

The liver, kidney, and skeletal muscle are the major sites of degradation of thyroid hormones. The rate of disposal of T_4 is proportional to the free T_4 concentration in plasma.

Because T_4 is hormonally only 25% as active as T_3, the initial step of converting it either to the active metabolite T_3 (by outer ring deiodination; Figure 45-2) or to the inactive metabolite rT_3 (by inner ring deiodination) is an important means of adjusting thyroid hormone action on tissues. Normally the split between T_3 and rT_3 is equal. When it is physiologically desirable to have more thyroid hormone action, as in exposure to cold, more T_3 and less rT_3 are generated. The opposite, less T_3 and more rT_3, commonly occurs in critically ill persons and portends a poor outcome. The activity of the enzyme **5' monodeiodinase,** which catalyzes the conversion of T_4 to T_3, is an important regulator of this distribution.

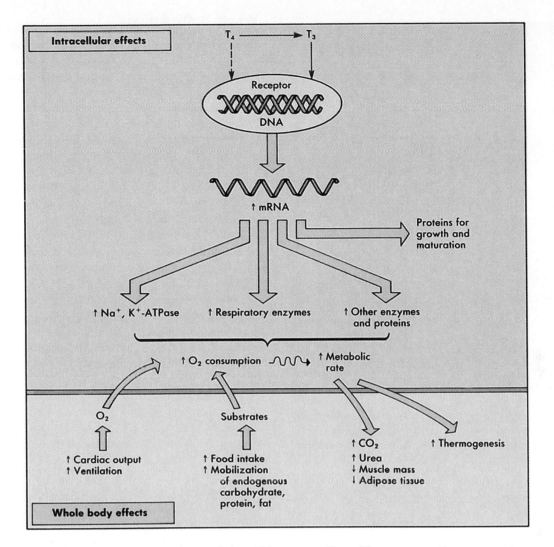

FIGURE 45-6   Overall schema of thyroid hormone effects. The upper portion represents intracellular actions; the lower portion, whole body effects.

ACTIONS OF THYROID HORMONE

Intracellular Mechanism

T_4 and T_3 enter target cells by carrier-mediated energy-dependent transport, where most of the T_4 undergoes deiodination to T_3 (Figure 45-6). Both are transferred to the nucleus, where T_3 binds to a nuclear receptor with much greater affinity than does T_4. Two distinct forms of the receptor are expressed in a tissue-specific manner. The T_3 receptor complex interacts with DNA to stimulate or inhibit transcription of numerous messenger RNAs, as described in Chapter 40. The latter then direct increased or decreased synthesis of many specific proteins in different tissues. Examples include enzymes, growth hormone, myosin

chains, TSH, and even the T_3 receptors, which are down-regulated.

The critical importance of the T_3 receptor is illustrated clinically by individuals with hypothyroidism caused by resistance to thyroid hormone. They can have mutant receptors, which are unable to tranduce the hormone signal, or a single allele for a mutant receptor that blocks T_3 binding to the normal receptor.

The quantitative responses of tissues to T_3 correlates well with their nuclear receptor content and with the degree of receptor occupancy (Chapter 40, Figure 40-12). Normally about half the available receptor sites are occupied by T_3. Because T_3 acts largely through gene

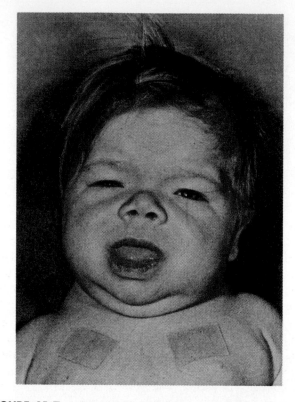

FIGURE 45-7 The typical appearance of an individual suffering from cretinism. Note the dull, apathetic appearance, puffy features, and swollen tongue in this 7-month-old infant who was born without a thyroid gland. (Andersen HJ: Hypothyroidism. In Gardner LI, ed: *Endocrine and genetic diseases of childhood and adolescence,* ed 2, Philadelphia, 1975, WB Saunders Co.)

transcription, a 12- to 48-hour delay occurs before its effects become evident in vivo. Several weeks of hormone replacement are required before all the consequences of a deficiency state are corrected.

The most obvious effect of thyroid hormone is to stimulate oxygen consumption and substrate use (Figure 45-6). A number of mechanisms are likely involved. T_3 increases the number, size, membrane areas, and certain key respiratory enzymes of mitochondria. T_3 also stimulates the activity of the Na^+-K^+-ATPase pump, which is responsible for membrane cation transport (see Chapter 1). Because large amounts of adenosine triphosphate (ATP) are thereby consumed and much adenosine diphosphate (ADP) is correspondingly generated by Na^+-K^+-ATPase, the extra ADP could be one "messenger" by which thyroid hormone stimulates mitochondrial O_2 utilization. Another possibility is that thyroid hormone simultaneously stimulates synthesis and oxidation of fatty acids and/or glucose—in effect, futile, wasteful cycles that require energy input and generate heat. In brain tissue, O_2 consumption is not stimulated by T_3, but the hormone increases the synthesis of specific structural or functional proteins.

Whole Body Actions

In humans, O_2 use at rest is approximately 225 to 250 ml/min (Chapter 41). It falls to about 150 ml/min in the absence of thyroid hormone and can increase to 400 ml/min with thyroid hormone excess. *Thus, the basal metabolic rate ranges from −40% to +80% of normal at the extremes of thyroid function.* Of necessity, thermogenesis increases or decreases concomitantly with O_2 use. In turn, increases and decreases in body temperature parallel fluctuations in thyroid hormone effect. These changes, however, are moderated by thyroid hormone–induced increases or decreases in heat loss through appropriate changes in cutaneous blood flow, sweating, and ventilation.

Thyroid hormone could not stimulate O_2 utilization for long without augmenting O_2 supply to the tissues (Figure 45-6). Thus thyroid hormone increases the resting rate of ventilation sufficiently to maintain a normal arterial oxygen pressure (Po_2) despite increased O_2 utilization, and a normal carbon dioxide pressure (Pco_2) in the face of increased carbon dioxide production. In addition, the O_2-carrying capacity of the blood is enhanced by a small increase in red cell mass.

Another important action of thyroid hormone is to increase cardiac output, which ensures a sufficient O_2 delivery to the tissues. The resting heart rate and stroke volume are both increased, and the speed and force of myocardial contractions are enhanced (see Chapter 19). These effects are partly indirect, via adrenergic stimulation. However, thyroid hormone directly increases myocardial Ca^{++} uptake, adenylyl cyclase activity, and the active form of myosin-stimulated ATPase. Systolic blood pressure rises and diastolic blood pressure falls, reflecting the combined effects of the increased stroke volume with a substantial reduction in peripheral vascular resistance. The latter results from blood vessel dilation produced by the increased tissue metabolism (see Chapter 23).

Stimulation of O_2 utilization also requires provision of substrates for oxidation. Thyroid hormone potentiates the stimulatory effects of other hormones on glucose absorption from the gastrointestinal tract, on gluconeogenesis, on lipolysis, on ketogenesis, and on proteolysis of the labile protein pool. *The overall metabolic effect of thyroid hormone has therefore aptly been described as accelerating the metabolic response to starvation.*

Thyroid hormone also stimulates the biosynthesis of cholesterol, its oxidation, its conversion to bile acids, and its biliary secretion. The net effect is to decrease the body pool and plasma level of cholesterol. The rate of metabolic disposal of steroid hormones, B vitamins, and many administered drugs is increased. Therefore, to maintain effective plasma levels of these substances in the presence of increased thyroid hormone, their endogenous production or their exogenous administration must be increased.

Thyroid Hormone and the Sympathetic Nervous System

A major intermediary in some thyroid hormone actions is the sympathetic nervous system. Although the activity of the sympathetic nervous system is diminished by thyroid hormone, as evidenced by decreased plasma levels and urinary excretion of the specific neurotransmitter, norepinephrine, the sensitivity of tissues to certain effects of the catecholamine hormones (see Chapter 47) is enhanced. These include the thermogenic, lipolytic, glycogenolytic, and gluconeogenic effects of epinephrine and norepinephrine. With regard to cardiovascular responses to catecholamines, a modest reinforcing effect of thyroid hormone may exist. The mechanism of enhanced sensitivity to catecholamines appears to lie in the ability of thyroid hormone to increase the number of β-adrenergic receptors, couple them to adenylyl cyclase, and thereby increase the catecholamine second messenger, cAMP.

Because of these effects, **hyperthroidism** presents a striking clinical picture. In most instances, the thyroid gland is enlarged. The increase in metabolic rate leads to weight loss, which is characteristically accompanied by increased intake of food. Excessive generation of heat causes discomfort in warm environments, fever if the condition is severe, excessive sweating, thirst, and increased ventilation. Muscle weakness and atrophy and even osteoporosis can result from increased protein degradation. The increase in β-adrenergic responsivity is manifested by tremor, nervousness, insomnia, and an anxious stare. The heart rate is rapid, and a high-cardiac-output form of heart failure can occur in extreme cases. The use of β-adrenergic antagonists ameliorates the sympathetic nervous system manifestations.

Effects on Growth and Development

In humans, thyroid hormone stimulates linear growth, development, and maturation of bone. A direct effect of T_3 on the activity of chondrocytes in the growth plate of bone may initiate this process. T_3 also accelerates growth by stimulating secretion of growth hormone. Although thyroid hormone is not required for linear growth until after birth, it is already essential for maturation of the growth centers in the bones of the fetus. The regular progression of tooth development and eruption is dependent on thyroid hormone, as is the normal cycle of renewal of the epidermis and hair follicles. Because thyroid hormone stimulates degradative processes in structural and integumentary tissues, elevated levels cause resorption of bone and accelerated shedding of skin and hair. The synthesis of mucopolysaccharides that form the intercellular ground substance is inhibited by thyroid hormone.

Normal skeletal muscle function also requires thyroid hormone. This may be related to the regulation of energy production and storage in this tissue. The muscle content of creatine phosphate is reduced by an excess of thyroid hormone; the inability of muscle to take up and phosphorylate creatine leads to an increase in its urinary excretion.

Thyroid hormone has critical effects on the development of the central nervous system. If thyroid hormone is deficient in utero, growth of the cerebral and cerebellar cortex, proliferation of axons and branching of dendrites, and myelinization are all impaired. Irreversible brain damage results when the deficiency of thyroid hormone is not recognized and treated immediately after birth. These anatomical defects are paralleled by biochemical abnormalities. Without thyroid hormone, RNA and protein content, protein synthesis, the enzymes necessary for DNA synthesis, protein and lipid content of myelin, neurotransmitter receptors, and neurotransmitter synthesis are decreased in various areas of the brain. In children and adults, thyroid hormone enhances the speed and amplitude of reflexes, wakefulness, alertness, responsiveness to various stimuli, awareness of hunger, memory, and learning capacity. Normal emotional tone also depends on appropriate thyroid hormone levels.

The clinical effects of hypothyroidism may be severe, especially in a newborn, in whom the condition is known as **cretinism** (Figure 45-7). The central nervous manifestations can include mental retardation (if not treated promptly at birth) and delayed developmental milestones such as sitting, standing, and walking. Later, lethargy, growth retardation, and poor school performance occur. In children and adults the decreased metabolic rate causes intolerance of cold, decreased sweating, dry skin, a low cardiac output, and weight gain. The latter results from both excess adipose tissue and fluid that accumulates in association with ground substance mucopolysaccharides. All these abnormalities vanish with thyroid hormone replacement (except any signs resulting from irreversible central nervous system damage).

Thyroid hormone contributes to the regulation of reproductive function in both genders. The normal process of sperm production; the ovarian cycle of follicular development, maturation, and ovulation; and the maintenance of a healthy pregnant state are all disrupted by significant deviations of thyroid hormone levels from normal. In part, these may be caused by alterations in the metabolism of sex steroid hormones.

SUMMARY

1. The basic endocrine unit of the thyroid gland is a follicle consisting of a circular layer of epithelial cells surrounding a central lumen that contains colloid or stored thyroid hormones. These hormones are tetraidothyronine (thyroxine, T_4) and triiodothyronine (T_3)

2. T_4 and T_3 are synthesized from tyrosine and iodine by the enzyme complex, peroxidase. The tyrosine is incorporated in peptide linkages within the protein thyroglobulin. After iodination, two iodotyrosine molecules are coupled to yield the iodothyronines.

3. Secretion of stored T_4 and T_3 requires retrieval of thyroglobulin from the follicle lumen by endocytosis. To support hormone synthesis, iodide is both actively concentrated by the gland and conserved within it by recovery from the iodotyrosine that escapes secretion.

4. Thyrotropin (TSH) acts on the thyroid gland largely via cyclic AMP to stimulate all steps in hormone production, including iodide uptake, iodination and coupling, and retrieval from thyroglobulin. TSH also stimulates growth of the epithelial cells.

5. More than 99.5% of the T_4 and T_3 circulate bound to various proteins. Only the free fractions of T_4 and T_3 are biologically active. Changes in thyroid-binding globulin levels require corresponding changes in thyroid hormone secretion in order to maintain normal concentrations of free T_4 and T_3.

6. T_4 functions largely as a prohormone. Peripheral monodeiodination of the outer ring yields most of the T_3, which is the principal active hormone. Alternatively, monodeiodination of the inner ring yields reverse T_3, which is biologically inactive. Proportioning of T_4 between T_3 and reverse T_3 helps to regulate the availability of active thyroid hormone.

7. Thyroid hormone increases the basal metabolic rate. T_3 combines with its receptor in target cell nuclei; the T_3-receptor complex interacts with many target DNA molecules to induce or suppress synthesis of a variety of enzymes and other proteins. The result is to increase oxygen utilization and thermogenesis by mechanisms that include increases in the size and number of mitochondria, in Na^+-K^+-ATPase activity, and in the rates of substrate use.

8. Additional important actions of thyroid hormone are to increase heart rate, cardiac output and ventilation and to decrease peripheral resistance. These subserve the increased tissue oxygen demand.

9. Other thyroid hormone effects on the central nervous system and skeleton are crucial to normal growth and development. In the absence of the hormone, brain development is retarded, and cretinism results. Linear growth is restricted and the bones fail to mature normally.

BIBLIOGRAPHY
Journal Articles

Acheson K et al: Thyroid hormones and thermogenesis: the metabolic cost of food and exercise, *Metabolism* 33:262, 1984.

Bantle JP et al: Common clinical indices of thyroid hormone action: relationships to serum free 3,5,3'-triiodothyronine concentration and estimated nuclear occupancy, *J Clin Endocrinol Metab* 50:286, 1980.

Chin WW: Hormonal regulation of thyrotropin and gonadotropin gene expression, *Clin Res* 36:484, 1988.

Danforth E Jr: The role of thyroid hormone and insulin in the regulation of energy metabolism, *Am J Clin Nutr* 38:1006, 1983.

Dillmann WH: Biochemical basis of thyroid hormone action in the heart, *Am J Med* 88:626, 1990.

Izumo S et al: All members of the MHC multigene family respond to thyroid hormone in a highly tissue-specific manner, *Science* 231:597, 1986.

Larsen PR et al: Relationships between circulating and intracellular thyroid hormones, physiological and clinical implications, *Endocr Rev* 2:87, 1981.

Lazar MA et al: Nuclear thyroid hormone receptors, *J Clin Invest* 86:1777, 1990.

Lazar MA: Thyroid hormone receptors: multiple forms, multiple possibilities, *Endocr Rev* 14:184, 1993.

Misrahi M et al: Cloning, sequencing and expression of human TSH receptor, *Biochem Biophys Res Commun* 166:394, 1990.

Mutvei A et al: Thyroid hormone and not growth hormone is the principal regulator of mammalian mitochondrial biogenesis, *Acta Endocrinol (Copenh)* 121:223, 1989.

Nelson BD: Thyroid hormone regulation of mitochondrial function. Comments on the mechanism of signal transduction, *Biochem Biophys Acta* 1018:275, 1990.

Oppenheimer JH et al: Advances in our understanding of thyroid hormone action at the cellular level, *Endocr Rev* 8:288, 1987.

Piolino V et al: Thermogenic effect of thyroid hormones: interactions with epinephrine and insulin, *Am J Physiol* 259:E305, 1990.

Polikar R, Burger AG, Scherrer U, Nicod P: The thyroid and the heart, *Circulation* 87:1435-1441, 1993.

Porterfield SP and Hendrich CE: The role of thyroid hormones in prenatal and neonatal neurological development—current perspectives, *Endocr Rev* 14:94, 1993.

Schimmel M et al: Thyroidal and peripheral production of thyroid hormones, *Ann Intern Med* 87:760, 1977.

Taylor T and Weintraub B: Thyrotropin (TSH)-releasing hormone regulation of TSH subunit biosynthesis and glycosylation in normal and hypothyroid rat pituitaries, *Endocrinology* 116:1968, 1985.

Books and Monographs

Galton VA: Thyroid hormone action in amphibian metamorphosis. In Oppenheimer JH and Samuels HH: *Molecular basis of thyroid hormone action,* New York, 1983, Academic Press, Inc.

Greer MA et al: Thyroid secretion. In *Handbook of physiology,* section 7: Endocrinology, vol III: Thyroid, Baltimore, 1974, American Physiological Society.

Scanlon MF: Neuroendocrine control of thyrotropin secretion. In Ingebar SH and Braverman LE, editors: *Werner's The Thyroid,* Philadelphia, 1991, JB Lippincott Co.

Schwartz HL: Effect of thyroid hormone on growth and development. In Oppenheimer JH and Samuels HH: *Molecular basis of thyroid hormone action,* New York, 1983, Academic Press, Inc.

Taurog A: Hormone synthesis: thyroid iodine metabolism. In Ingbar SH and Braverman LE, editors: *Werner's The Thyroid,* Philadelphia, 1991, JB Lippincott Co.

46

The Adrenal Cortex

The adrenal glands are multifunctional endocrine organs that secrete a variety of hormones. Abundant experimental and clinical evidence has demonstrated that the adrenal glands are essential to life. *Their secretions subserve a wide variety of physiological functions, including blood glucose regulation, protein turnover, fat metabolism, sodium, potassium, and calcium balance, modulation of tissue response to injury or infection, and, most importantly, survival in the face of stress.*

Each adrenal gland is located just above the ipsilateral kidney (Figure 46-1), and their combined weight is 6 to 10 g. Each gland is a combination of two separate functional entities (Figure 46-2). The outer zone, or **cortex**, comprises 80% to 90% of the weight. It is derived from mesodermal tissue and is the source of corticosteroid hormones. The inner zone, or **medulla**, comprises the other 10% to 20%. It is derived from neuroectodermal cells of the sympathetic ganglia and is the source of catecholamine hormones. Small clusters of adrenocortical cells are found within the medulla, and catecholamine-secreting cells likewise penetrate the cortex. Paracrine interactions between the two cell types are possible. The adrenal glands have one of the highest rates of blood flow per gram of tissue. Arterial blood enters the outer cortex and breaks up into capillaries; the venous drainage is into the medulla. This exposes the inner cells of the cortex and the cells of the medulla to high concentrations of steroid hormones from the outer cortex.

The outermost **zona glomerulosa** of the adrenal cortex is only a few cells thick (Figure 46-2). The middle **zona fasciculata** is the widest and consists of long cords of columnar cells. The innermost **zona reticularis** contains networks of interconnecting cells. Typical steroid-secreting cells are rich in lipid droplets and contain numerous large mitochondria with vesicles in their membranes.

The major hormones of the cortex are (1) the **glucocorticoid, cortisol,** which has critical roles in carbohydrate and protein metabolism and in the adaptation to stress; (2) the **mineralocorticoid, aldosterone,** which is vital to maintaining normal extracellular fluid volume and potassium levels; and (3) **sex steroid precursors,** which contribute to maintaining secondary sexual characteristics.

The discovery and synthesis of cortisol were medical landmarks. Replacement therapy with the hormone was lifesaving for patients whose adrenal glands were destroyed by disease or removed surgically, and it dramatically reversed their debilitation. Subsequently, cortisol was found to have potent antiinflammatory and antiimmune effects, which have been exploited therapeutically. In pharmacological doses, cortisol and synthetic glucocorticoid analogues are used to treat a large number of diseases in which autoimmunity plays an important pathogenetic role. They are also used to prevent rejection of organ transplants.

SYNTHESIS OF CORTICOSTEROID HORMONES

The precursor for all adrenocortical hormones is cholesterol, which is taken up from the plasma via a specific plasma membrane receptor for low-density lipoproteins (see Chapter 41). After transfer into the cell, the cholesterol is largely esterified and stored in vacuoles within the cytoplasm. Under basal conditions cholesterol just taken up from plasma is immediately used for hormone synthesis. However, when hormone production is stimulated, stored cholesterol is rapidly mobilized and transferred to the mitochondria for the first step in synthesis of all corticosteroid hormones.

Most of the reactions in corticosteroid synthesis are catalyzed by **cytochrome P-450 enzymes.** The genes

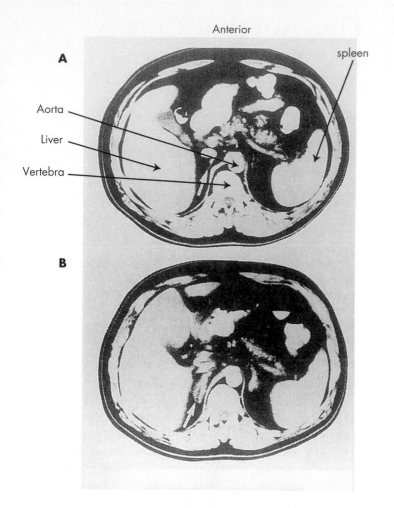

Anterior

A

spleen

Aorta

Liver

Vertebra

B

FIGURE 46-1 Computerized axial tomographic (CAT) scan of the abdomen. **A** shows the small size of the adrenal gland *(white arrow)* relative to other organs. **B** shows the marked hyperplasia that occurred after 1 week of stimulation by an excess of endogenous ACTH *(white arrow).* (From Mastorakos G et al: *J Clin Endocrinol Metab* 77: 1690, 1993.)

that direct their synthesis have considerable similarity, even though they may be located on different chromosomes. A single P-450 enzyme may catalyze more than one reaction, depending on its location in the cortex and on substrate availability. These enzymes catalyze hydroxylations of the steroid nucleus. The reactions require molecular oxygen, NADPH, a flavoprotein, and an iron-containing protein called **adrenoxin.**

Cortisol

The synthesis of cortisol, the major glucocorticoid in humans (Figure 46-3), occurs largely in the zona fasciculata. The initial and rate-limiting reaction converts cholesterol to **pregnenolone** and is catalyzed by the mitochondrial side-chain cleavage enzyme complex P-450 scc (also

known as 20,22-desmolase). The pregnenolone is then converted to **progesterone,** following which hydroxyls are successively added at the 17 and 21 positions. These reactions take place within the endoplasmic reticulum. The resultant 11-deoxycortisol is transferred to the mitochondria and hydroxylated in the 11 position, the final and critical step in creating a glucocorticoid molecule.

Neither the final product, cortisol, nor its precursors are stored in the adrenocortical cell. Thus an acute need for increased cortisol secretion requires rapid activation of the initial controlling step: side-chain cleavage of stored cholesterol.

Aldosterone

The synthesis of aldosterone, the major mineralocorticoid (Figure 46-3), is carried out exclusively in the zona glomerulosa. The reactions from cholesterol to corticosterone occur as in the zona fasciculata. In the subsequent key step, the C_{18} methyl group of corticosterone is oxidized to yield aldosterone (by the same or similar mitochondrial enzyme that catalyzes 11-hydroxylation). Deoxycorticosterone and its 18-hydroxy derivative are other steroids that have mineralocorticoid activity and that are synthesized in small quantities in the zona fasciculata.

Androgens and Estrogens

The synthesis of the sex steroids occurs largely in the zona reticularis. The potent androgen **testosterone** and the potent estrogen **estradiol** are normally secreted only in trace amounts by the adrenal cortex. However, substantial amounts of precursor steroids with weak intrinsic androgenic activity are secreted and converted to testosterone and estradiol by peripheral tissues. These precursors, **androstenedione, dehydroepiandrosterone** (DHEA), and **DHEA-sulfate** (DHEA-S) are synthesized from 17-OH-progesterone or 17-OH-pregnenolone, respectively, as shown in Figure 46-3.

In women the adrenal precursors supply 50% of the androgenic hormone requirements. In men they are unimportant because the testes produce testosterone. After menopause the estrogens that arise directly or indirectly from the adrenal cortex become the only source for this biological activity in women.

Genetic defects in cortisol biosynthesis have important and varied consequences for infants. A defect in either the 21- or 11-hydroxylase enzyme gene will lead to overproduction of androgenic steroids from the accumulated precursors, 17-OH progesterone and 17-OH

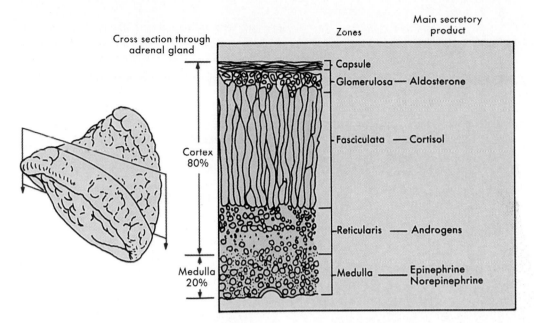

FIGURE 46-2 Schematic representation of the adrenal gland and its main secretory products.

pregnenolone (Figure 46-3). This will cause masculinization of female fetuses in utero and early secondary sexual changes in male infants and young boys. Severe deficiency of 21-hydroxylase activity may also cause manifestations of cortisol (glucocorticoid) and aldosterone (mineralocorticoid) deficiency. Deficiency of 11-hydroxylase leads to overproduction of 11-deoxycorticosterone (Figure 46-3), which causes manifestations of excess mineralocorticoid activity, such as hypertension.

CORTICOSTEROID HORMONE METABOLISM

Basal morning plasma cortisol concentrations are 5 to 20 μg/dl; by evening they are often less than 5. The hormone circulates 90% bound to a specific corticosteroid-binding globulin called **transcortin.** The concentrations of transcortin and likewise total plasma cortisol are increased during pregnancy and estrogen administration. However, because bound cortisol is biologically inactive, the physiological effects of an increase in transcortin are determined by principles similar to those discussed with regard to thyroxine binding (see Chapter 45). Free cortisol is filtered by the kidney, and when kidney function is normal, the small amount of daily urinary cortisol excretion (10-100 μg) is a valid index of cortisol secretion.

Cortisol is in equilibrium with its biologically inactive 11-keto analogue, **cortisone,** via the enzyme 11β-OH dehydrogenase. This enzyme, which is present in many tissues, renders exogenous cortisone an effective source of cortisol activity. The conversion of cortisol to cortisone in the kidney is important in preventing cortisol from exerting mineralocorticoid activity via aldosterone receptors to which it binds (see below). Almost all of the cortisol and cortisone is metabolized in the liver; the metabolites are conjugated and excreted in the urine as glucuronides. The measurement of these urinary metabolites, known generally as **17-hydroxycorticoids,** also provides an index of cortisol secretion as long as hepatic and renal functions are intact.

Aldosterone circulates bound to a specific aldosterone-binding globulin, to transcortin, and to albumin. Aldosterone and its liver-generated metabolites are excreted in the urine as glucuronides.

Adrenal androgen precursors are also metabolized in the liver and excreted in the urine in a fraction known as **17-ketosteroids.** However, these products are not specific for the adrenal gland because they also arise from gonadal androgens.

In adrenal insufficiency (**Addison's disease,** see below), measurements of plasma and urine cortisol and urine 17-hydroxycorticoids will be low. Conversely, in states of adrenocortical hyperfunction, all these values will be high.

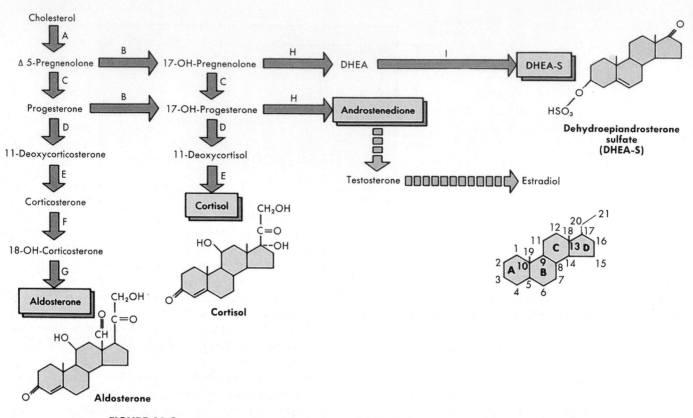

FIGURE 46-3 Sequence of reactions in the synthesis of adrenocorticosteroid hormones from the common precursor cholesterol. Step *A* is overall rate-limiting. Step *E* is critical to glucocorticoid activity. Step *G* is critical to mineralocorticoid activity. Step *B* is essential to formation of sex steroid precursors. *DHEA*, Dehydroepiandrosterone. *A,* 20,22-desmolase (P-450 scc); *B,* 17-hydroxylase (P-450 c17); *C,* 3β-ol-dehydrogenase, Δ4,5-isomerase; *D,* 21-hydroxylase (P-450 c21); *E,* 11-hydroxylase (P-450 c11); *F,* 18-hydroxylase; *G,* 18-01-dehydrogenase; *H,* 17,20-desmolase (P-450 c17); *I,* sulfotransferase.

Zona glomerulosa and zona reticularis dysfunction will likewise be reflected by deviations from normal of plasma and urine aldosterone and of DHEA-S, respectively.

REGULATION OF CORTISOL SECRETION

The pattern of cortisol secretion is very complex (Figure 46-4). The immediate stimulator of cortisol secretion is **adrenocorticotropin** (ACTH) from the anterior pituitary gland. The most important immediate stimulator of ACTH secretion is the neuropeptide **corticotropin-releasing hormone** (CRH) from the hypothalamus. Thus a hypothalamic—anterior pituitary—adrenal cortex axis exists, and the three hormones just listed form

a classic negative feedback loop (Figure 46-4). Cortisol (or any synthetic glucocorticoid analogue (e.g., dexamethasone, prednisone):

1. Feeds back within minutes on the pituitary gland to inhibit the release of ACTH by blocking the stimulatory action of CRH on the corticotroph cells
2. Feeds back more slowly (within hours) to inhibit the synthesis of ACTH by blocking transcription of its gene
3. Feeds back on the hypothalamus to block release and probably synthesis of CRH

In addition to this long-loop feedback, short-loop feedback also exists as ACTH inhibits CRH release (Figure 46-4). The neuropeptide arginine vasopressin (AVP) (see Chapter 44) also augments ACTH secretion and therefore cortisol secretion in certain situations, and cortisol feeds back to restrain AVP release.

When given therapeutically for long periods and in large doses, synthetic glucocorticoids profoundly suppress the function of the hypothalamic CRH neurons, the pituitary corticotroph cells, and consequently the cells of the zona fasciculata. Atrophy of the ACTH-dependent adrenal cells occurs. After such therapy is withdrawn, full recovery of the inactivated hypothalamic–pituitary–adrenal axis can take up to 1 year. During this time, patients must be protected against deficient responses to stress with supplemental cortisol.

Corticotropin-Releasing Hormone

CRH is a 41-amino acid peptide synthesized from a prepro-CRH. CRH enters the pituitary portal veins and travels to the corticotroph cells (see Figure 44-1). After binding to a plasma membrane receptor, CRH stimulates the release of ACTH from its secretory granules via calcium and cyclic adenosine monophosphate (cAMP) as second messengers, and also stimulates ACTH synthesis. In addition to this endocrine role, and possibly functionally related to it, CRH exhibits diverse other actions in the central nervous system. These include stimulating sympathetic nervous system activity, decreasing fever, suppressing reproductive function and sexual activity, suppressing growth hormone release, and altering behavior. Peripheral plasma CRH levels are very low, but they mirror negative feedback because they are slightly increased by cortisol deficiency and decreased by glucocorticoid administration.

Adrenocorticotropin

ACTH is a 39 amino acid peptide that increases the synthesis and immediate release of cortisol, adrenal androgens, their precursors, and aldosterone. However, only cortisol feeds back negatively on the hypothalamus and the pituitary gland. ACTH is synthesized via a large precursor called **preproopiomelanocortin,** which gives rise to a number of cosecreted products, including β-endorphin. After binding to its adrenal plasma membrane receptor, ACTH stimulates the generation of cAMP, which is the major second messenger for its actions (Figure 46-5). Calcium and phosphatidylinositol products may play adjunctive roles. The ultimate effects of ACTH are mediated by protein products that result from a cascade of enzyme phosphorylations catalyzed by protein kinases A and C. These products include enzyme activators, transcription factors, and growth factors. ACTH acutely stimulates cholesterol uptake by the cell, choles-

terol ester hydrolysis, cholesterol transfer to the mitochondria, the rate-limiting P-450 scc desmolase reaction, and the critical 11-hydroxylation step in cortisol synthesis. ACTH also alters the shape of the adrenocortical cell by affecting its cytoskeleton and bringing the cholesterol vacuoles into contact with the mitochondria. Continuous stimulation with ACTH causes hyperplasia of the adrenal cortex (see Figure 46-1).

Patterns of Cortisol Secretion

Cortisol is secreted in pulses and in a diurnal pattern (Figure 46-6). The bursts of cortisol secretion are induced by pulses of ACTH, which in turn are caused by the pulsatile release of CRH. These episodes form a circadian cortisol pattern. The peak plasma ACTH and cortisol levels are achieved about 2 hours before awakening, at 4 to 6 AM; the nadir of plasma ACTH and cortisol is reached just before falling asleep. The morning peak of cortisol constitutes 50% of the total daily secretion. The clock time of this peak can be altered by systematically shifting the sleep-wake cycle. *This phenomenon is of occupational significance, such as in transoceanic airline flights.* The circadian rhythm is intrinsic and generated within the hypothalamus, probably by the suprachiasmatic nucleus. Negative feedback can affect the setting of this center: exogenous glucocorticoid suppresses, and prior cortisol deficiency accentuates, the early-morning ACTH peak. Loss of consciousness and constant exposure to either dark or light also blunts the circadian rhythm.

Cortisol is required for survival of the "stressed" organism. Severe pain and prolonged exercise also cause release of cortisol, whereas the state of analgesia induced by endorphins blocks the cortisol response. Stress can override the diurnal pattern of cortisol secretion as well as the suppressive effects of negative feedback. Several neurotransmitters mediate the stressful inputs that stimulate CRH (plus AVP) release (see Figure 46-4).

Patients with serious medical illnesses (such as **sepsis**) or major fractures, those undergoing surgery or electroconvulsive therapy, and those experiencing hypoglycemia all secrete extra cortisol. In patients in intensive care units, plasma cortisol levels are elevated twofold to fivefold; the highest levels are associated with increased mortality.

Activation of the process of cell-mediated immunity also increases ACTH and cortisol release. Lymphokines,

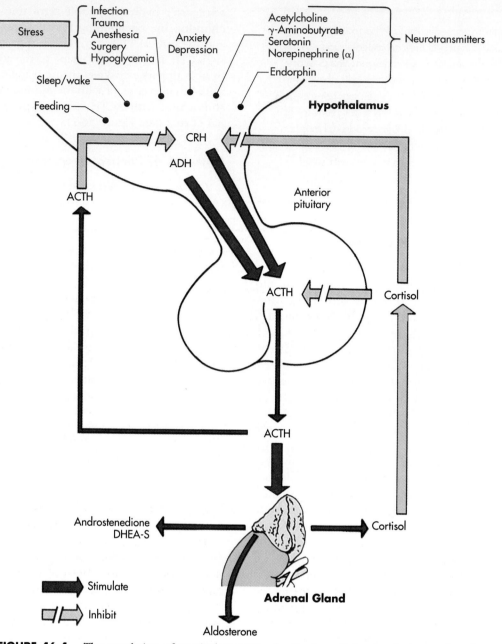

FIGURE 46-4   The regulation of cortisol secretion by the hypothalamic–pituitary–adrenal axis. A variety of inputs to the hypothalamus stimulate corticotropin-releasing hormone *(CRH)* secretion, which in turn increases adrenocorticotropin *(ACTH)* and thence cortisol secretion. ADH has an auxiliary stimulating effect on ACTH secretion. Cortisol exerts negative feedback at both the hypothalamic and the pituitary levels. *ADH,* Antidiuretic hormone; *DHEA-S,* dehydroepiandrosterone sulfate.

such as various interleukins, stimulate ACTH secretion. Because stresses, such as infection and tissue trauma, are accompanied by cell-mediated immune responses, and because cortisol is an important modulator of those responses (see following discussion), a significant feedback relationship between the immune and endocrine systems exits.

ACTIONS OF CORTISOL (GLUCOCORTICOIDS)

Despite knowledge of many important effects of cortisol, the exact reasons for its life-preserving action are not certain. *Most clearly, the hormone is required to sustain glucose production from protein and to support*

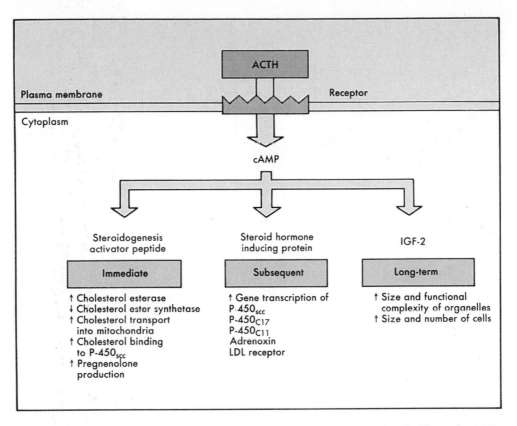

FIGURE 46-5   An overview of ACTH actions on target adrenocortical cells. Through cAMP as second messenger, ACTH activates and induces steroidogenic enzymes and also stimulates adrenocortical cell growth. See text for details. *cAMP,* Cyclic adenosine monophosphate; *IGF-2,* insulin growth factor 2; *LDL,* low-density lipoprotein.

vascular responsiveness. In addition, the hormone affects fat metabolism, central nervous system function, skeletal turnover, hematopoiesis, muscle function, renal function, and immune responses. The term **permissive** has been used to describe cortisol's action, implying that the hormone may not directly initiate, so much as allow, critical processes to occur. For example:

1. Cortisol does not itself directly stimulate glycogenolysis. However, if cortisol is present, glycogenolysis stimulated by glucagon is enhanced.
2. Cortisol does not itself alter the activity of the enzyme tyrosine transaminase. However, the enzyme's substrate, tyrosine, induces tyrosine transaminase only if cortisol is present.

Intracellular Mechanisms

Almost all effects of cortisol are mediated via transcriptional mechanisms (see Figure 40-11). Cortisol enters target cells by facilitated diffusion and binds to a type 1 or type 2 receptor in the cytoplasm and/or nucleus. The cortisol-receptor complex must undergo an activation process before it can then bind to a specific

DNA molecule. Hormonal action is directly proportional to the degree of DNA binding, and the final response is an increase or decrease in gene transcription of specific messenger ribonucleic acids (mRNAs). Although other steroids can bind to a cortisol receptor and other steroid receptors may bind to a similar site on the same DNA molecule, *the specific combination of cortisol, one of its receptors, and a responsive DNA molecule is required to elicit a cortisol action.*

Effects on Metabolism

The most important overall effect of cortisol is to stimulate the conversion of protein to glucose and the storage of glucose as glycogen; thus the term **glucocorticoid** (Figure 46-7). The type 2 or glucocorticoid receptor (GR) mediates such actions. All phases of this process are augmented, including mobilization of protein from muscle stores, entrance of the released amino acids into the hepatic gluconeogenetic pathway, conversion of pyruvate to glycogen, and disposition of the ammonia released from metabolism of the precursor amino acids. Cortisol increases the activity of enzymes involved

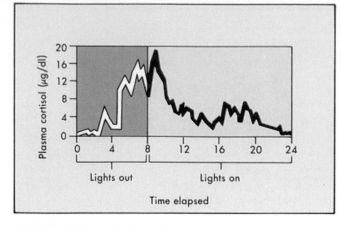

FIGURE 46-6 Pulsatile and diurnal nature of cortisol secretion. Peak cortisol levels are reached just prior to awakening. (Redrawn from Weitzman ED et al: *J Clin Endocrinol Metab* 33:14, 1971.)

in each of these steps. In some of these instances, the hormone "permits" substrate induction of its enzyme; in others, cortisol directly increases transcription of the target enzyme gene.

If the glucocorticoid effect is excessive (known as **Cushing's syndrome**) and continues beyond the point where it is beneficial, the continuous drain on body protein produces serious deleterious effects. Most notably, muscle, bone, connective tissue, and skin lose mass. This is further exacerbated by the inhibitory effects of cortisol on the synthesis of constitutive proteins such as collagen.

> The clinical expression of the negative nitrogen balance that results from **Cushing's syndrome** is striking (Figure 46-8). Skin is so thin that the capillaries show through and, because of fragile walls, rupture spontaneously, causing bruises. Muscle weakness and atrophy are prominent. Osteoporosis also occurs and can lead to atraumatic fractures and bone necrosis.

The presence of cortisol is essential for maintenance of plasma glucose levels and for survival during prolonged fasting. Without the hormone, death may occur from hypoglycemia once glycogen stores are gone. However, only a minor increase in cortisol secretion occurs with fasting, and thus ordinary levels of the hormone can effectively mobilize amino acids and promote gluconeogenesis. On the other hand, plasma cortisol levels increase sharply in response to acute hypoglycemia. In this situation cortisol amplifies the glycogenolytic actions of glucagon and epinephrine and synergizes with them in rebuilding liver glycogen stores.

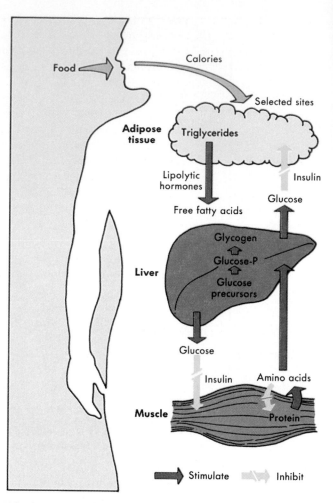

FIGURE 46-7 Effect of cortisol on the flow of fuels. Cortisol stimulates mobilization of amino acids and their conversion to glucose. The glucose is preferentially but not exclusively stored as glycogen. Insulin-mediated glucose uptake by peripheral tissues is inhibited. Cortisol facilitates storage of fat in selected adipose tissue sites but also facilitates release of free fatty acids.

Consonant with its role in preventing hypoglycemia, cortisol is a strong antagonist to insulin (see Figure 46-7). Cortisol inhibits insulin-stimulated glucose uptake by muscle and adipose tissue and blocks the suppressive effect of insulin on hepatic glucose output. The interaction between cortisol and insulin is complex. Both hormones favor hepatic glycogen storage by increasing glycogen synthase activity (see Figure 41-3). However, they have opposite effects on expression of the genes for the gluconeogenetic enzyme, phosphoenolpyruvate carboxykinase, and the glucose-releasing enzyme, glucose-6-phosphatase. Thus, cortisol favors glucose output by the liver, whereas insulin inhibits it. The net result of cortisol excess is a rise in plasma glucose concentration and a compensatory increase in plasma insulin levels. *When the rise in insulin is insufficient, diabetes mellitus can develop or, if already present, be greatly worsened.*

Cortisol also plays a complex role in fat metabolism (see Figure 46-7). The presence of cortisol "permits" maximal stimulation of fat mobilization by growth hormone, epinephrine, and other lipolytic factors during fasting. However, the hormone also greatly increases appetite and stimulates lipogenesis in certain adipose tissue depots. *Therefore, in Cushing's syndrome an excess of cortisol also results in accumulation of fat, but the obesity has a peculiar distribution, favoring the face and trunk but sparing the extremities* (see Figure 46-8).

Thus cortisol is a catabolic, antianabolic, and diabetogenic hormone. In stress situations cortisol accentuates hyperglycemia produced by other hormones and it greatly accelerates the loss of body protein. These actions are amplified if insulin secretion is simultaneously deficient.

Effects On Tissues and Organs

Muscle Basal levels of cortisol are required for maintenance of normal contractility and maximal work output of skeletal and cardiac muscle. The inotropic effects are probably exerted at the myoneural junctions. In contrast, excess cortisol produces muscle atrophy and weakness through protein wastage (see Figure 46-9).

Bone The major effect of cortisol is to decrease bone formation. Less prominently, cortisol increases bone resorption. *The net outcome of cortisol excess can be a profound reduction in bone mass and, in children with Cushing's syndrome, a reduction in linear growth as well.* Several actions contribute to this outcome. Cortisol decreases the synthesis of $1,25\text{-}(OH)_2$ vitamin D and blocks its action; therefore calcium absorption from the gastrointestinal tract is defective (see Chapter 43). At the same time, urinary calcium excretion is increased. Thus, less calcium is available for mineralization of bone. Cortisol also inhibits the differentiation of mesenchymal precursors into osteoblasts and the synthesis of collagen by these cells.

Vascular System Cortisol is required for the maintenance of normal blood pressure. The hormone permits enhanced responsiveness of arterioles to the constrictive action of adrenergic stimulation and also optimizes myocardial performance. Cortisol helps to maintain blood volume by decreasing the permeability of the vascular endothelium. *In Cushing's syndrome, hypertension frequently is present.*

Kidney Cortisol increases the rate of glomerular filtration. The hormone is also essential for the rapid excretion of a water load because it inhibits both secretion of antidiuretic hormone (ADH) and the latter's action on the collecting duct tubules (see Chapter 44). This diminishes free water clearance. *Clinically, the absence of cortisol may lead to water retention and resultant hyponatremia.*

Central Nervous System The type 1 receptor for cortisol (which is identical to the mineralocorticoid receptor) is present throughout the brain and is concentrated in the hippocampus, reticular activating substance, and autonomic nuclei of the brainstem. Cortisol modulates perceptual and emotional functioning. A deficiency of cortisol accentuates auditory, olfactory, and gustatory acuity; these actions suggest that the hormone may normally have a damping effect. The diurnal increase in CRH pulses and cortisol levels just before awakening is important for normal arousal and initiation of daytime activity. *Clinically, an excess of cortisol can cause insomnia and either euphoria or depression. Frank psychosis can even result.*

Fetus Cortisol has important permissive effects that facilitate in utero maturation of the gastrointestinal tract, the lungs, the central nervous system, the retina, and the skin. Cortisol facilitates timely preparation of the fetal lung to permit satisfactory breathing immediately after birth. The rate of development of the pulmonary alveoli, flattening of the lining cells, and thinning of the lung septa are increased. Most importantly, the synthesis of surfactant, a phospholipid vital for maintaining alveolar surface tension, is increased. Cortisol also facilitates maturation of the enzyme capacity of the intestinal mucosa from a fetal to an adult pattern. This permits the newborn to digest the disaccharides present in milk.

Inflammatory and Immune Responses

Cortisol has a profound influence on the complex set of reactions evoked by tissue trauma, chemical irritants, foreign proteins, and infection. The overriding effect is to inhibit important steps in the response to tissue injury. The clinical consequences of cortisol administration are to impede the ability of tissues either to eliminate immediately noxious substances and invaders or to wall them off from the rest of the body. Thus, long-term treatment with pharmacological doses of any glucocorticoid increases susceptibility to opportunistic infections, allows their dissemination, and masks them. Normal wound healing after injury may also be prevented.

The mechanisms by which cortisol suppresses these responses (see box) are shown by the following examples:

1. Cortisol induces a phosphoprotein called **lipocortin** that inhibits the enzyme phospholipase A_2.

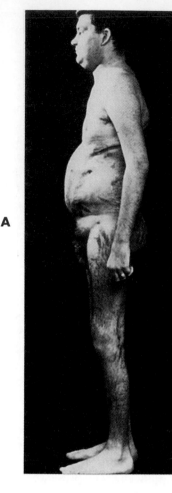

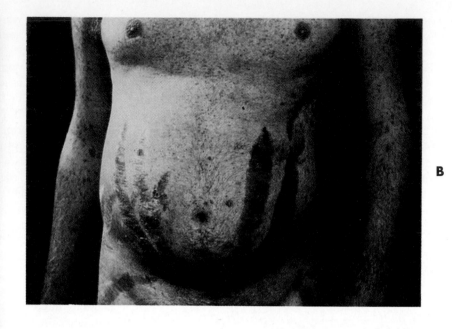

FIGURE 46-8   An individual suffering from Cushing's syndrome, an excess of cortisol. **A,** Note the selective accumulation of abdominal fat and the loss of musculature in the extremities. **B,** The extreme thinness of the skin reveals blood flowing through underlying blood vessels.

This enzyme generates arachidonic acid. Because the latter serves as the precursor for synthesis of prostaglandins and related compounds, the production of these mediators of inflammation is reduced.

2. Cortisol decreases the production of interleukin-1 by repressing expression of the lymphokine gene. In this way cortisol can block the entire cascade of cell-mediated immunity as well as the generation of fever.

3. Cortisol stabilizes lysosomes, and thereby reduces the release of enzymes capable of degrading foreign substances.

4. Cortisol blocks recruitment of neutrophils by inhibiting their ability to bind chemotactic peptides. The hormone further impairs the phagocytic and bacterial capacity of neutrophils.

5. Cortisol decreases proliferation of fibroblasts and their ability to synthesize and deposit tissue fibrils, and thus prevents encapsulation of invaders.

At this point a paradox is evident. On the one hand, augmentation of cortisol secretion is essential to the survival of severely stressed, traumatized, or infected individuals through its metabolic actions. On the other hand, many of the tissue defense mechanisms evoked by such conditions are inhibited by elevated cortisol levels. This suggests that permissive lower levels of cortisol may be required for the initial responses to stress. Subsequently, higher levels of cortisol may be secreted to limit cellular and tissue reactions so that they do not themselves seriously damage the individual.

The therapeutic use of glucocorticoids, which may require administration of high doses for some time, represents a two-edged sword. Glucocorticoids are dramatically beneficial when inflammatory reactions are so severe as to be

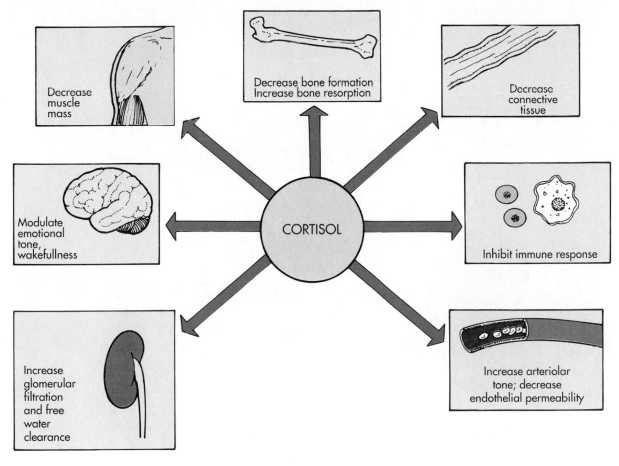

FIGURE 46-9   Cortisol effects on various tissues and organs.

functionally disabling or life threatening (for example, during a severe asthma attack), or when the rejection of transplanted tissues must be prevented. However, their adverse effects, of vulnerability to serious infection, diabetes, osteoporosis, and psychiatric disorders, require physicians to *prescribe glucocorticoids cautiously and only when no safer form of treatment exists.*

This injunction does not apply to the use of replacement doses of cortisol in patients who lack adrenocortical function **(Addison's disease).** Cortisol deficiency leads to anorexia, weight loss, fatigue, poor tolerance of stress, fever, hypoglycemia, and, in females, loss of sexual hair. Loss of negative feedback causes hypersecretion of ACTH and darkening of the skin through its melanocyte-stimulating activity. Suitable replacement therapy reverses these findings without adverse effects.

REGULATION OF ALDOSTERONE SECRETION

Aldosterone, the major product of the zona glomerulosa, has two principal functions: (1) to sustain extracellular fluid volume by conserving body sodium and (2) to prevent overload of potassium by accelerating its excretion. (See also Chapters 37 and 38.) Thus aldosterone is largely secreted in response to a reduction in circulating fluid volume and increases in plasma potassium (Figure 46-10).

When sodium is depleted, the fall in extracellular fluid and plasma volume causes a decrease in arterial blood pressure and renal blood flow. The juxtaglomerular cells in the kidney respond by secreting the enzyme **renin** into the peripheral circulation (see Chapter 37). Renin into acts on its substrate, angiotensinogen, to form **angiotensin I.** The latter is then further cleaved by a converting enzyme to the potent vasoconstrictors **angiotensin II** and **angiotensin III.** These bind to specific receptors in the zona glomerulosa and stimulate the key enzymatic steps in the synthesis and release of aldosterone (Figure 46-10). Calcium and the phosphatidylinositol messenger system are mediators.

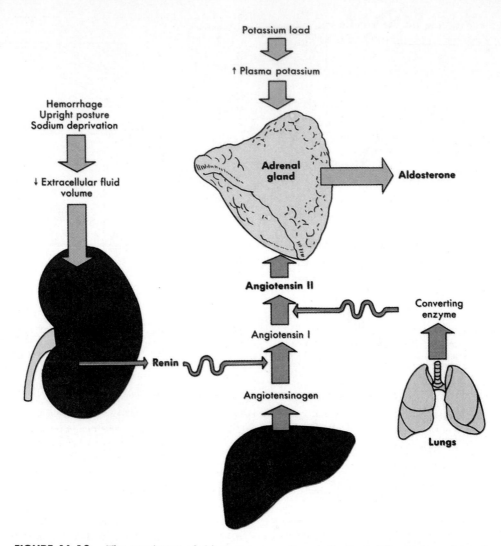

FIGURE 46-10 The regulation of aldosterone secretion. Activation of the renin-angiotensin system in response to hypovolemia is the predominant stimulus to aldosterone production. The kidney, liver, and lungs are required for the production of angiotensin II, the direct stimulator. Elevation of plasma potassium is the other major stimulus of aldosterone secretion.

Basal plasma aldosterone ranges from 5 to 15 ng/dl. When hypovolemia is produced rapidly by hemorrhage or by acute diuresis, or slowly by chronic sodium deprivation, aldosterone concentration increases markedly. Conversely, when excess sodium is ingested and extracellular fluid volume expands, renin release, angiotensin generation, and aldosterone secretion are all suppressed. Thus, *the juxtaglomerular cells and the zona glomerulosa form a physiological feedback system to defend extracellular fluid volume.* Sodium loss induces hypersecretion of renin and then aldosterone. When the additional aldosterone has caused sufficient sodium retention to restore extracellular fluid volume to normal, renin and aldosterone hypersecretion ceases.

This system plays an important role in disease states where effective blood flow to the kidney is reduced. These include cardiac failure, hepatic failure, renal artery stenosis, and hypoalbuminemia with transudation of fluid out of the plasma space. In each of these conditions, hypersecretion of aldosterone is stimulated and sodium retention occurs; this causes or augments edema.

The atrial natriuretic peptide hormones, synthesized and released by atrial myocytes in response to changes in vascular volume, decrease aldosterone secretion.

They do so directly by acting on the zona glomerulosa through specific receptors and indirectly by reducing renin release.

Aldosterone also participates in a vital feedback relationship with potassium (Figure 46-10). Aldosterone facilitates the clearance of potassium from the extracellular fluid, and concordantly potassium is an important stimulator of aldosterone secretion. In humans, raising the plasma potassium concentration only 0.5 mEq/L immediately increases plasma aldosterone. Conversely, potassium depletion lowers aldosterone secretion. Potassium acts by depolarizing the zona glomerulosa cell membrane, allowing influx of calcium and activation of aldosterone biosynthesis.

Other factors that influence renin release or angiotensin II formation secondarily affect aldosterone secretion. β-Adrenergic stimulation of the kidney in response to hypovolemia increases the output of renin and aldosterone. Certain prostaglandins produced within the kidney also increase renin release. Conversely, inhibitors of angiotensin-converting enzyme decrease aldosterone secretion.

Therefore, β-adrenergic blocking agents (e.g., **propranolol)** used in treatment of hypertension or angina, inhibitors of prostaglandin synthesis (e.g., **indomethacin)** used in treatment of inflammatory conditions, and angiotensin converting enzyme inhibitors (e.g., **captopril)** used in the treatment of hypertension or congestive heart failure can all depress aldosterone secretion and cause elevated plasma potassium levels (see below).

Aldosterone secretion is also stimulated by ACTH. However, this effect of ACTH wanes after several days because of renin and atrial natriuretic hormone responses as sodium is retained and extracellular fluid volume rises. The physiological role of ACTH in regulating aldosterone output appears to be limited to a tonic one; when ACTH is deficient, the response to the primary stimulus of sodium depletion is modestly diminished.

ACTIONS OF ALDOSTERONE (MINERALOCORTICOIDS)

The kidney is the major site of mineralocorticoid activity. In renal tubular cells aldosterone binds to the mineralocorticoid receptor (MR), which is identical to the type 1 cortisol receptor. Messenger RNAs and proteins of still undetermined nature are induced and apparently

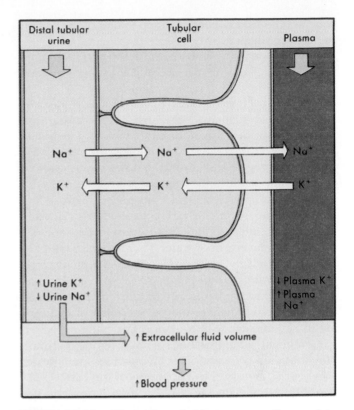

FIGURE 46-11 The action of aldosterone on the renal tubule. Sodium reabsorption from tubular urine is stimulated. Simultaneously, potassium secretion into the tubular urine is increased by responses to the electronegative gradient created by sodium movement. The net result is expansion of extracellular fluid and kaliuresis.

mediate the hormone's actions. A lag of hours is required between exposure to aldosterone and onset of its effect. Aldosterone stimulates active reabsorption of sodium from the distal tubular urine; the sodium is transported through the tubular cell and back into the capillary blood (Figure 46-11) (see Chapter 36). Thus net urinary sodium excretion is diminished, and the vital extracellular cation is conserved. Because water is passively reabsorbed with the sodium, plasma sodium concentration increases only slightly, and extracellular fluid volume expands isotonically. *Although only a small fraction of total sodium reabsorption is regulated by aldosterone, deficiency of the hormone produces a critical negative sodium balance.* Hypovolemia and hypotension result unless a large intake of sodium and water is maintained.

Aldosterone acts at various loci in distal renal tubular and collecting duct cells: (1) at the apical (luminal) surface, to increase the number of membrane channels through which sodium enters the cell along an electrochemical gradient; (2) at the basal (capillary) surface of the cell, to activate Na^+- K^+-ATPase, which pumps the sodium back into the interstitial fluid; and (3) in the mitochondria, to stimulate Kreb's cycle reactions that help generate the energy needed for operation of the sodium pump (see Chapter 37).

Aldosterone also stimulates the active secretion of potassium out of the tubular cell and into the urine concurrently with sodium reabsorption (Figure 46-11). The latter creates in the tubular lumen an electronegative condition that facilitates the transfer of potassium into the tubular urine. Therefore the extent to which aldosterone affects potassium secretion is greatly dependent on the delivery of sodium to the distal tubule. Aldosterone cannot significantly increase potassium excretion in a sodium-depleted subject; conversely, a high sodium intake exaggerates the urinary potassium loss caused by aldosterone. Unlike sodium flux, potassium flux does not entrain the movement of water. Therefore *potassium retention caused by aldosterone deficiency or by drugs that block aldosterone action (e.g., spironolactone) can result in a dangerous rise in plasma potassium.* Aldosterone also enhances tubular secretion of hydrogen ion in conjunction with sodium reabsorption.

Continued administration of aldosterone produces only a limited retention of sodium, which then ceases. This escape is caused by expansion of the extracellular fluid and is mediated in part by atrial natriuretic hormones. In contrast, the potassium loss induced by aldosterone continues because sodium delivery to the distal tubule is maintained.

The net clinical effect of **primary aldosterone excess** is modest fluid retention without detectable edema. Hypertension, hypokalemia, and metabolic alkalosis are the dominant signs. This situation can be ameliorated by aldosterone antagonists.

In contrast, **aldosterone deficiency** leads to natriuresis, dehydration, hypotension, hyperkalemia, hyponatremia, and hyperchloremic acidosis. These findings are present in **Addison's disease** caused by adrenocortical destruction.

Aldosterone significantly affects sodium and potassium exchange across muscle cells. The net result is to increase potassium content of the intracellular space, another effect that helps prevent hyperkalemia. Aldosterone also modestly stimulates sodium reabsorption from the gastrointestinal tract and enhances potassium excretion in the feces.

SUMMARY

1. The adrenal cortex secretes three types of steroid hormones: cortisol, a glucocorticoid; aldosterone, a mineralocorticoid; and androgen precursors, largely dehydroepiandrosterone. The adrenal glands are richly vascularized and essential to survival because of the cortisol they produce.

2. All adrenocorticosteroids are synthesized from cholesterol by sequential enzymatic steps consisting of side-chain cleavage and hydroxylation of key sites in the steroid molecule. Cortisol specifically requires an 11-hydroxyl group, aldosterone an 18-hydroxyl group, and androgens a 17-hydroxyl group for their respective activities. The mitochondrial and microsomal enzymes involved are P-450 mixed oxygenases. Steroid hormones are not stored directly; increased secretory demands require rapid synthesis from stored cholesterol.

3. Cortisol and androgen secretion are stimulated by adrenocorticotropin (ACTH) from the pituitary gland. ACTH secretion is stimulated by corticotropin-releasing hormone (CRH) from the hypothalamus. Cortisol feeds back negatively on the anterior pituitary and hypothalamus to suppress ACTH and CRH release.

4. ACTH, via cAMP as second messenger, stimulates cellular uptake of cholesterol, its movement from storage vacuoles into mitochondria, and all subsequent biosynthetic steps to cortisol.

5. Cortisol has major effects on protein, glucose, and fat metabolism. It acts via a nuclear receptor and modulates gene expression of numerous enzymes and proteins. Cortisol inhibits muscle proteolysis and increases hepatic conversion of the liberated amino acids into glucose and storage as glycogen. Cortisol also inhibits insulin-stimulated glucose uptake by muscle. Cortisol stimulates caloric intake and favors deposition of fat in selected sites. By inhibiting collagen synthesis, cortisol reduces bone formation and causes thinning of skin and capillary walls.

6. Cortisol strongly inhibits the entire process of inflammation, including the recruitment and function of neutrophils and the release of prostaglandin and leukotriene mediators. It also inhibits the immune system and prevents proliferation of thymus-derived lymphocytes and production of some lymphokines. All these actions underlie the broad therapeutic use of synthetic analogues of cortisol as antiinflammatory and immune suppressant agents.

7. Aldosterone is a major regulator of sodium, potassium, and fluid balance. It acts on the renal tubule via a specific nuclear receptor and gene expression. Sodium reabsorption is increased, with concomitant expansion of extracellular fluid and maintenance of blood pressure. Renal potassium excretion is concurrently increased, and plasma potassium concentration is lowered.

8. Aldosterone secretion is primarily regulated by the

renin-angiotensin system. In response to sodium deprivation, an increase in angiotensin stimulates aldosterone secretion. It is also directly stimulated by potassium.

BIBLIOGRAPHY
Journal Articles

Allison AC, Lee SW: The mode of action of anti-rheumatic drugs. I. Antiinflammatory and immunosuppressive effects of glucocorticoids, *Prog Drug Res* 33:63, 1989.

Burnstein KC, Cidlowski JA: Regulation of gene expression by glucocorticoids, *Annu Rev Physiol* 51:603, 1989.

Chrousos GP, Gold PW: The concepts of stress and stress system disorders, *JAMA* 267:1244, 1992.

Darmaun D et al: Physiological hypercortisolemia increases proteolysis, glutamine, and alanine production, *Am J Physiol* 255:E366, 1988.

DeFeo P et al: Contribution of cortisol to glucose counterregulation in humans, *Am J Physiol* 257:E35, 1989.

Gustafsson J et al: Biochemistry, molecular biology and physiology of the glucocorticoid receptor, *Endocr Rev* 8:185, 1987.

Horrocks PM et al: Patterns of ACTH and cortisol pulsatility over twenty-four hours in normal males and females, *Clin Endocrinol (Oxf)* 32:127, 1990.

Jackson R et al: Synthetic ovine corticotropin-releasing hormone: simultaneous release of propiolipomelanocortin peptides in man, *J Clin Endocrinol Metab* 58:740, 1984.

Mastorakos G, Chrousos GP, Weber JS: Recombinant interleukin-6 activates the hypothalamic-pituitary-adrenal axis in humans, *J Clin Endocrinol Metab* 77:1690, 1993.

Miller WL: Molecular biology of steroid hormone synthesis, *Endocr Rev* 9:295, 1988.

Munck A et al: Physiological functions of glucocorticoids in stress and their relation to pharmacological actions, *Endocr Rev* 5:25, 1984.

Prummel MF et al: The course of biochemical parameter of bone turnover during treatment with corticosteroids, *J Clin Endocrinol Metab* 72:382, 1991.

Quinn SJ, Williams GH: Regulation of aldosterone secretion, *Annu Rev Physiol* 50:409, 1988.

Rizza RA et al: Cortisol-induced insulin resistance in man: impaired suppression of glucose production and stimulation of glucose utilization due to a postreceptor defect of insulin action, *J Clin Endocrinol Metab* 54:131, 1982.

Simpson ER, Waterman MR: Regulation of the synthesis of steroidogenic enzymes in adrenal cortical cells by ACTH, *Annu Rev Physiol* 50:427, 1988.

Wick G, Hu Y, Schwarz S, Kroemer G: Immunoendocrine communication via the hypothalamo-pituitary-adrenal axis in autoimmune diseases, *Endocr Rev* 14:539, 1993.

Young DB: Quantitative analysis of aldosterone's role in potassium regulation, *Am J Physiol* 255:F811, 1988.

Books and Monographs

Crabbe J: Mechanism of action of aldosterone. In Degroot LJ, ed: *Endocrinology,* Philadelphia, 1989, WB Saunders Co.

Keith LD, Kendall JW: Regulation of ACTH secretion. In Imura H: *The pituitary gland,* New York, 1985, Raven Press.

Meikle AW: Secretion and metabolism of the cortiosteroids and adrenal function and testing. In Degroot LJ, ed: *Endocrinology,* Philadelphia, 1989, WB Saunders Co.

Numa S, Imura H: ACTH and related peptides: gene structure and biosynthesis. In Imura H: *The pituitary gland,* New York, 1985, Raven Press.

Orth DN, Kovacs WJ, Debold CR: The adrenal cortex. In Wilson JD, Foster DW, eds: *Williams textbook of endocrinology,* Philadelphia, 1992, WB Saunders Co.

47

The Adrenal Medulla

The adrenal medulla forms the inner core of the adrenal gland. It is the source of the circulating catecholamine hormone **epinephrine.** The medulla also secretes **norepinephrine,** primarily a neurotransmitter, which can also function as a hormone. The catecholamine hormones are important mediators of rapid fuel mobilization; they increase both glucose and free fatty acids, especially during acute stress. They also stimulate the cardiovascular system and cause contraction or relaxation of smooth muscles in the respiratory, gastrointestinal, and genitourinary tracts.

The adrenal medulla is essentially a specialized sympathetic ganglion. However, the neuronal cells of the medulla do not have axons; instead, they discharge their products directly into the bloodstream, and thus they function in true endocrine fashion. The medulla is usually activated in association with the sympathetic portion of the autonomic nervous system and acts in concert with it in the "fight-or-flight" reaction. Some of the neurotransmitter actions of norepinephrine are duplicated and amplified by the hormone actions of epinephrine, which reaches similar targets via the circulation. However, epinephrine has other effects of its own, some of which modulate those of norepinephrine.

The adrenal medulla is formed in parallel with the peripheral sympathetic nervous system. At about 7 weeks of gestation, neuroectodermal cells invade the adrenal cortex, where they develop into the medulla. The development of this tissue and induction of hormone synthesis are stimulated by nerve growth factor.

The adult adrenal medulla weighs about 1 g and is composed of **chromaffin cells.** These are organized in cords in intimate relationship with venules that drain the adrenal cortex. *The chromaffin cells are innervated by cholinergic preganglionic fibers of the sympathetic nervous system.* Within these cells are numerous granules similar to those found in postganglionic sympathetic nerve terminals. These contain catecholamines, adeno-sine triphosphate (ATP), proopiomelanocortin products (see Chapter 46), and other neuropeptides. Approximately 85% of the chromaffin granules store epinephrine, and 15% store norepinephrine.

SYNTHESIS, STORAGE, AND SECRETION OF MEDULLARY HORMONES

The catecholamines are synthesized by a series of reactions shown in Figure 47-1. The intermediates move back and forth in sequence between the cytoplasm and the storage granules. The first, rate-limiting step occurs in the cytoplasm. Here the conversion of tyrosine to dihydroxyphenylalanine (DOPA) requires molecular oxygen, a tetrahydropteridine, and NADPH. The subsequent decarboxylation of DOPA to dopamine in the cytoplasm employs pyridoxal phosphate as a cofactor. The dopamine must then be taken up into the chromaffin granules, where the next enzyme in the sequence, dopamine β-hydroxylase, exclusively resides. This enzyme catalyzes the formation of norepinephrine from dopamine, molecular oxygen, and a hydrogen donor. In a few granules the sequence ends here and the norepinephrine remains stored.

In most granules norepinephrine diffuses back into the cytoplasm, where it is N-methylated to epinephrine, with S-adenosylmethionine as the methyl donor. The epinephrine is then taken back up into the chromaffin granules for storage. Granule uptake of catecholamines and their storage at high concentrations require ATP. One mole of the nucleotide complexes with 4 moles of catecholamine hormone and a specific protein known as **chromogranin.**

The synthesis of epinephrine and norepinephrine is regulated by several factors. Acute sympathetic stimulation of the medulla activates the initial rate-limiting step. Chronic stimulation induces an increased concentration

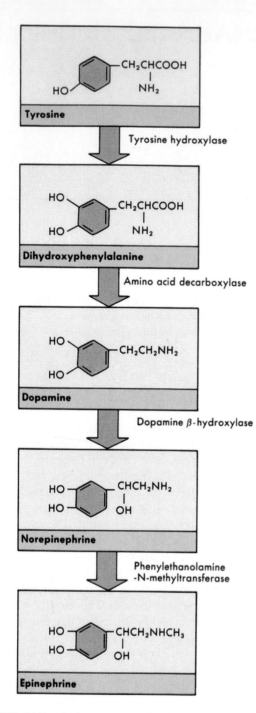

FIGURE 47-1 Pathway of catecholamine hormone synthesis in the adrenal medulla.

of tyrosine hydroxylase, and thereby catecholamine output is maintained in the face of continuous demand. Cyclic AMP mediates both of these effects. Cortisol, by specifically inducing the last enzyme in the sequence, N-methyltransferase, selectively stimulates epinephrine synthesis. The perfusion of the adrenal medulla with cortisol-enriched blood from the cortex facilitates this induction.

The effector pathway for release of adrenal medullary hormones consists of cholinergic preganglionic fibers in the splanchnic nerves. On nerve stimulation, acetylcholine released from the nerve terminals depolarizes the chromaffin cell membrane by increasing its permeability to sodium. In turn this induces an influx of calcium ions, which causes aggregation of the chromaffin granules. Exocytosis follows, with secretion of epinephrine, norepinephrine, ATP, dopamine β-hydroxylase, neuropeptides, and chromogranin.

METABOLISM OF CATECHOLAMINE HORMONES

All the circulating epinephrine is derived from adrenal medullary secretion. Basal plasma epinephrine levels are 25 to 50 pg/ml. In contrast, almost all the circulating norepinephrine is derived from sympathetic nerve terminals and from the brain; this represents norepinephrine that escaped local reuptake from the synaptic clefts. Basal norepinephrine levels are 100 to 350 pg/ml. Both catecholamines have plasma half-lives of about 2 minutes, which allows rapid turnoff of their dramatic effects. Only 2% to 3% of catecholamines, the majority of which is norepinephrine, are excreted unchanged in the urine.

Epinephrine and norepinephrine are metabolized by O-methylation and oxidative deamination, predominantly in the liver and kidney. The major end products, **vanillylmandelic acid** (VMA) and **metanephrines,** are excreted in the urine. They serve as indices of sympathetic nervous system activity or of pathological hypersecretion of epinephrine.

REGULATION OF ADRENAL MEDULLARY SECRETION

As previously noted, secretion from the adrenal medulla is part of the fight-or-flight reaction (Figure 46-2). Thus *perception or even anticipation of danger, fear, excitement, trauma, or pain, hypovolemia, hypotension, anoxia, hypothermia, hypoglycemia, and intense exercise cause rapid secretion of epinephrine and norepinephrine* (see Chapter 26). These stimuli are sensed at various levels in the sympathetic nervous system, and responses are initiated in the hypothalamus and brainstem (see Chapter 10). Epinephrine secretion may follow activation of the sympathetic nervous system by more intense stimuli. However, epinephrine secretion specifically increases in response to mild hypoglycemia, moderate hypoxia, and fasting, even though sympathetic nervous system activity remains constant or decreases.

Mild hypoglycemia causes a fivefold to tenfold increase in plasma epinephrine concentration but little

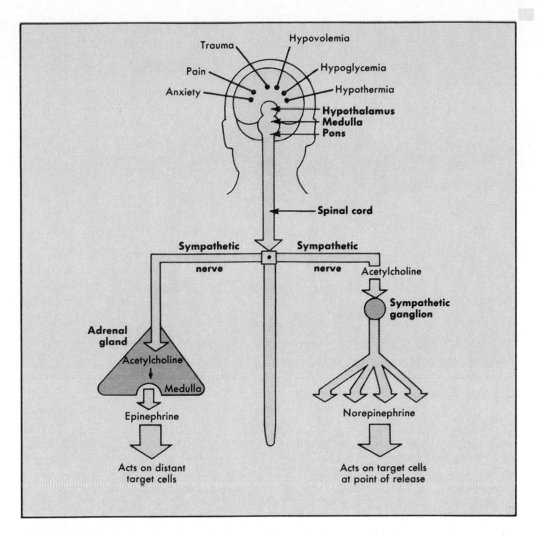

FIGURE 47-2 Activation of catecholamine effects via the sympathetic nervous system and the adrenal medulla. Note that the adrenal medulla is homologous to a sympathetic ganglion, but the adrenal medulla releases its catecholamines into the bloodstream rather than into a synaptic cleft.

change in norepinephrine concentration. The resultant epinephrine concentration can stimulate a compensatory increase in the plasma glucose level; the norepinephrine level cannot. The reduction in central venous pressure produced by assuming the upright position (see Chapter 24) increases both plasma epinephrine and norepinephrine concentrations twofold. However, only the epinephrine concentration is high enough to increase heart rate and blood pressure. Thus epinephrine functions as a true hormone in both situations, whereas norepinephrine does not. Norepinephrine, however, contributes as a neurotransmitter to the compensatory responses to hypovolemia and to more severe hypoglycemia, because the higher concentration necessary for its action is generated locally at the effector site (Figure 47-2). *In states of major metabolic decompensation, such as diabetic ketoacidosis, the circulating concentrations of both catecholamines rise high enough to evoke responses.*

ACTIONS OF CATECHOLAMINE HORMONES

Intracellular Mechanisms

Epinephrine and norepinephrine exert their many effects via several plasma membrane receptors, designated β_1, β_2, β_3, α_1, and α_2. The β_1-, β_2-, and α_2- receptors are structurally similar glycoproteins. Each winds in and out of the plasma membrane so that more than one surface is presented extracellularly for hormone binding and intracellularly for signal generation. The β_1, and β_2, receptors are coupled to the stimulating G-protein unit of adenylyl cyclase, and hormone binding increases cyclic

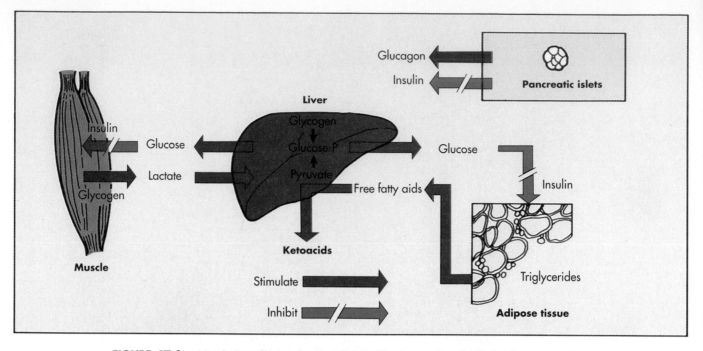

FIGURE 47-3 Metabolic effects of epinephrine. The hormone stimulates glucose production and inhibits glucose uptake. Lipolysis and ketogenesis are stimulated. The net result is an increase in plasma glucose, free fatty acids, and ketoacids.

adenosine monophosphate (cAMP) levels. In contrast, the α_2-receptor is coupled to the inhibiting G-protein of adenylyl cyclase, and hormone binding decreases cAMP levels. Catecholamine hormones therefore either trigger (β_1 or β_2) or suppress (α_2) a cascade of protein phosphorylations catalyzed by protein kinase A. The α_1-receptor is structurally different and is coupled to calcium and phosphatidylinositol products as second messengers.

Continuous exposure to catecholamines eventually down-regulates the number of receptors and induces partial refractoriness to hormone action. A distinctly different phenomenon—acute desensitization to successive doses of catecholamine hormones—is caused by phosphorylation of the receptor molecules themselves, mediated by protein kinase A or C. This desensitization process constitutes a form of rapid intracellular negative feedback that almost immediately limits hormone actions.

Whole Body Effects

The major metabolic effect of catecholamines is fuel mobilization (Figure 47-3). Glycogenolysis in the liver is stimulated via cAMP-mediated activation of phosphorylase, and glucose output increases. Glycogenolysis in muscle is similarly stimulated. This increases

muscle glucose supply and glycolysis. Lactate is released and serves as a substrate for hepatic gluconeogenesis. In addition, gluconeogenesis is directly stimulated by catecholamines, through α_1-receptors in the liver. Plasma glucose levels also rise because catecholamines inhibit insulin secretion as well as insulin-stimulated glucose uptake by muscle tissue. The availability of free fatty acids is increased by activating the enzyme, adipose tissue lipase. The enhanced lipolysis in turn leads to increased free fatty acid oxidation in the liver and ketogenesis.

> Thus the catecholamine hormones are diabetogenic. They contribute materially to the development of hyperglycemia and ketonemia in diabetic ketoacidosis, particularly when an intercurrent stress has provoked this metabolic emergency.

Catecholamines increase the basal metabolic rate by stimulating facultative, or nonshivering, thermogenesis (see Chapter 41). It is an important part of the response to cold exposure. In neonates of many species, brown adipose tissue is an important site where catecholamines increase heat production. Here the hormones uncouple ATP synthesis from oxygen utilization in the mitochondria. They do so by inducing transcription of the gene

Table 47-1   Actions of Catecholamine Hormones

	β	α
Metabolic	↑ Glycogenolysis	↑ Gluconeogenesis (α_1)
	↑ Glucose utilization	
	↑ Lipolysis and ketosis (β_1)	
	↑ Calorigenesis (β_1)	
	↑ Insulin secretion (β_2)	↓ Insulin secretion (α_2)
	↑ Glucagon secretion (β_2)	
	↑ Muscle K^+ uptake (β_2)	
Cardiovascular	↑ Cardiac contractility (β_1)	
	↑ Heart rate (β_1)	
	↑ Conduction velocity (β_1)	
	↑ Arteriolar dilation (β_2) (muscle)	↑ Arteriolar vasoconstriction (α_1) (splanchnic, renal, cutaneous, genital)
	↓ Blood pressure	↑ Blood pressure
Visceral	↑ Muscle relaxation (β_2)	↑ Sphincter contraction (α_1)
	Gastrointestinal	Gastrointestinal
	Urinary	Urinary
	Bronchial	
Other		Sweating (adrenergic)
		Dilation of pupils
		Platelet aggregation (α_2)

↑, Increased; ↓, decreased.

for an uncoupling protein known as **thermogenin.** Catecholamines also increase diet-induced thermogenesis and thereby they help to regulate overall energy balance and stores.

Sympathetic nervous system activity decreases during fasting and increases after feeding. In this way norepinephrine adapts total energy utilization to energy availability, and helps maintain balance between them. In contrast, epinephrine secretion increases slightly during prolonged fasting and also 4 to 5 hours after a meal, when plasma glucose is declining. This response serves the different purpose of sustaining glucose production for use by the central nervous system.

> The activity of the sympathetic nervous system may be decreased in obese individuals. Such a characteristic would favor storage of energy as fat when dietary calories are plentiful.

The cardiovascular and visceral effects of epinephrine are consonant with its metabolic actions (Table 47-1). For example, during exercise epinephrine increases cardiac output by increasing cardiac contractile force and heart rate (see Chapters 19 and 26). At the same time, muscle arterioles dilate, whereas renal, splanchnic, and cutane-

ous arterioles constrict. Systolic blood pressure increases. The net effect is to shunt blood to exercising muscles and away from other tissues, while maintaining essential coronary and cerebral blood flows (see Chapter 26). This guarantees delivery of oxygen and substrate for energy production to the critical tissues in situations of danger or during whole body exercise.

> In severe or prolonged states of schock, however, compensatory hypersecretion of catecholamines can eventually contribute to fatal ischemic kidney and hepatic failure and to lactic acidosis (see Chapter 26). The catecholamine response to exercise may also be disadvantageous in patients who have coronary artery disease and cannot increase myocardial blood flow adequately. β-adrenergic antagonists are used to good therapeutic advantage in this situation; by decreasing heart rate, cardiac contractility, and systolic blood pressure, the balance between myocardial work and oxygen supply is improved and **angina pectoris** (chest pain) is prevented.

During exposure to cold, the constriction of cutaneous vessels helps to conserve heat and reinforces epi-

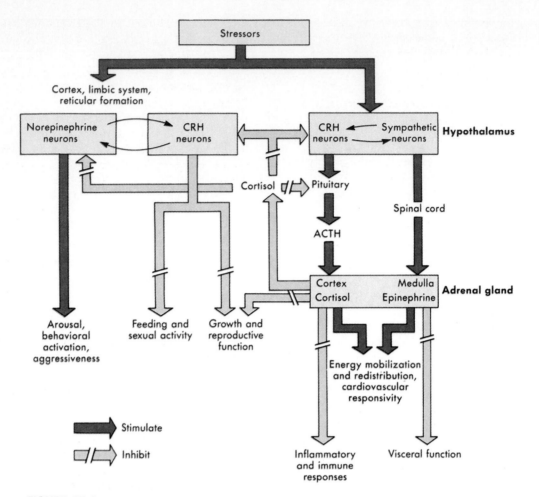

FIGURE 47-4 Integrated responses to stress mediated by the sympathetic nervous system and the hypothalamic-pituitary-adrenocortical axis. The responses are mutually reinforcing, at both the central and peripheral levels. Negative feedback by cortisol also can limit an overresponse that might be harmful to the individual. *Red arrows,* stimulation; *blue arrows,* inhibition; *CRH,* corticotropin-releasing hormone; *ACTH,* adrenocorticotropic hormone.

nephrine's thermogenic action. Other actions useful to the threatened individual are: relaxation of bronchioles, which improves alveolar gas exchange; dilation of pupils, which permits better distant vision; and inhibition of temporarily unneeded gastrointestinal and genitourinary motor activity.

The catecholamines also have significant actions on mineral metabolism. They increase sodium reabsorption

by the kidney by stimulating renal tubular sodium transport and by stimulating renin release and consequently aldosterone secretion. They also stimulate influx of potassium into muscle cells via β_2-receptors and help to prevent hyperkalemia.

During acute asthma attacks, constriction of bronchioles increases airway resistance (Chapter 28) and causes wheezing and hypoxia. Synthetic β-adrenergic agonists of epinephrine, administered through inhalers, relax the bronchioles and provide critical relief to patients in respiratory distress.

Pathological hypersecretion of epinephrine and/or norepinephrine from a tumor of the chromaffin cells **(pheochromocytoma)** results in a distinctive and dangerous syndrome. Bursts of catecholamine release can cause sudden tachycardia, extreme anxiety with a sense of impending death, cold perspiration, skin pallor resulting from vasoconstriction, blurred vision, headache, and chest pain. Blood pressure may rise greatly and cause stroke or heart failure. If epinephrine only is secreted, the heart rate increases; if norepinephrine only is secreted, the heart rate decreases because of the baroreceptor reflex in response to the hypertension. In addition to such episodes, chronic catecholamine excess may produce weight loss, as a result of the increased metabolic rate, and hyperglycemia. Prompt surgical removal of the tumor is mandatory.

INTEGRATION OF THE RESPONSE TO STRESS

The adrenal medulla and adrenal cortex are both major participants in the adaptation to stress. Their intimate anatomical juxtaposition mirrors a fundamental functional relationship between the sympathetic nervous system and the corticotropin-releasing hormone (CRH)–adrenocorticotropic hormone (ACTH)–cortisol axis. Although our knowledge about stress and the human body's adaptation to it is still incomplete, recent advances justify presenting an integrated overview (Figure 47-4).

Stress can be perceived by many areas of the brain, from the cortex down to the brainstem. Major stresses almost simultaneously activate CRH neurons and adrenergic neurons in the hypothalamus. The activation is mutually reinforcing, because norepinephrine input increases CRH release and CRH increases adrenergic discharge (Figure 47-4). Release of CRH elevates plasma cortisol levels; adrenergic stimulation elevates plasma catecholamine levels. Together these hormones increase glucose production; epinephrine does so rapidly by activating glycogenolysis, and cortisol more slowly by providing amino acid substrate for gluconeogenesis. Together they shift glucose utilization toward the central nervous system and away from peripheral tissues. Epinephrine also rapidly augments free fatty acid supply to the heart and to muscles, and cortisol facilitates this lipolytic response. Both hormones raise blood pressure and cardiac output and improve delivery of substrates to tissues that are critical to the immediate defense of the organism. If the stress involves tissue trauma or invasion by microorganisms, high cortisol levels eventually act to restrain the initial inflammatory and immune responses so that they do not lead to irreparable damage.

The same signaling molecules, the neurotransmitter norepinephrine and the neuropeptide CRH, can produce other adaptive responses to stress. A general state of arousal and vigilance, an activation of defensively useful behavior, and appropriate aggressiveness result from adrenergic stimuli to the pertinent brain centers. At the

same time, CRH input to other hypothalamic neurons inhibits growth hormone and gonadotropin release, presumably because growth and reproduction are not useful functions during stress. This is reinforced by the excess of cortisol, which also suppresses growth and ovulation. In addition, CRH inhibits sexual activity and eating, again inappropriate activities when the organism perceives itself to be in immediate serious danger. Thus the adaptation to stress represents a prime example of the integration between the nervous system and the endocrine system.

SUMMARY

1. The adrenal medulla is an enlarged, specialized sympathetic ganglion that synthesizes epinephrine and norepinephrine from tyrosine and stores them in granules. These catecholamines are released in response to activity in preganglionic cholinergic sympathetic nervous system fibers that are stimulated by hypoglycemia, hypovolemia, hypotension, exercise, or stress.
2. Circulating epinephrine increases plasma glucose, free fatty acids, and ketoacids by stimulating glycogenolysis and lipolysis and by inhibiting glucose uptake by muscle. Metabolic rate also rises. cAMP and Ca^{++} are second messengers.
3. Cardiovascular actions include increases in heart rate and cardiac contractility and variable effects on different vascular beds.
4. Circulating norepinephrine contributes to the effects, but more often epinephrine effects are reinforced by concurrent activation of the sympathetic nervous system, with norepinephrine as the neurotransmitter.

BIBLIOGRAPHY
Journal Articles

Clutter W et al: Epinephrine plasma metabolic clearance rates and physiologic thresholds for metabolic and hemodynamic actions in man, *J Clin Invest* 66:94, 1980.

Cryer PE: Physiology and pathophysiology of the human sympathoadrenal neuroendocrine system, *N Engl J Med* 303:436, 1980.

Landsberg L, Young JB: The role of the sympathetic nervous system and catecholamines in the regulation of energy metabolism, *Am J Clin Nutr* 36:1018, 1983.

Lefkowitz RJ, Caron MG: Adrenergic receptors: molecular mechanisms of clinically relevant recognition, *Clin Res* 33:395, 1985.

Matthews DE et al: Effect of epinephrine on amino acid and energy metabolism in humans, *Am J Physiol* 258:E948, 1990.

Santiago JV et al: Epinephrine, norepinephrine, glucagon, and growth hormone release in association with physiological decrements in the plasma glucose concentration in normal and diabetic man, *J Clin Endodrinol Metab* 51:877, 1980.

Silverberg A et al: Norepinephrine: hormone and neurotransmitter in man, *Am J Physiol* 234:E252, 1978.

Wortsman J et al: Adrenomedullary response to maximal stress in humans, *Am J Med* 77:779, 1984.

Books and Monographs

Landsberg L, Young JB: Catecholamines and the adrenal medulla. In Wilson JD, Foster DW, eds: *Williams textbook of endocrinology,* Philadelphia, 1992, W.B. Saunders Co.

CHAPTER 48

Overview of Reproductive Function

The endocrine glands in general are essential to maintenance of the life and well-being of the individual. In contrast, the endocrine function of the gonads is additionally concerned with the perpetuation and well-being of the species. Human reproduction requires highly complex patterns of hypothalamic-pituitary-gonadal function. These ensure the development and maintenance of mature gametes—**ova** and **spermatozoa**—from primordial germ cells, their subsequent successful union (**fertilization**), and finally the growth and development of the conceptus within the body of the mother. Although certain fundamental differences exist between male and female gonadal function, important conceptual similarities and operational homologies are also present. The gender differences are better appreciated if the common aspects of gonadal function are first understood.

The gonad, whether ovary or testis, consists of two distinct anatomical and functional parts. One part encloses the developing germ cell line with specialized membrane and cytoplasmic barriers that prevent its indiscriminate exposure to the general constituents of plasma and interstitial fluid. In the ovary the germ cell enclosure is the **follicle;** in the testis it is the **spermatogenic (seminiferous) tubule.** The other part is composed of surrounding endocrine cells that secrete sex steroid hormones, protein hormones, and other products necessary for germ cell development. The most important sex steroids are **estradiol** and **progesterone** in the female and **testosterone** in the male. Protein hormones produced by the gonads include inhibin, activin, follistatin, anti-müllerian hormone, and oocyte meiosis inhibitor, as well as various derivatives of proopiomelanocortin (see Chapter 46).

Acting locally in paracrine and autocrine fashion, the gonadal hormones stimulate the development of the respective germ cells into ova and spermatozoa. Acting peripherally in endocrine fashion, these hormones:

1. Stimulate the development and function of the sec-

ondary sexual organs essential for support and delivery of the ova and spermatozoa to the site of fertilization
2. Regulate the secretion of hypothalamic-pituitary hormones essential to gonadal function
3. Modify somatic shape and regulate certain physiological functions within each gender
4. Support the conceptus in the early phase of pregnancy in the female

There are two principal types of endocrine cells in the gonads. Those immediately encompassing the germ cells are called **granulosa cells** in the ovary and **Sertoli cells** in the testis. Those more distant from the germ cells and separated from them by a basement membrane are called **theca,** or **interstitial, cells** in the ovary and **Leydig cells** in the testis. The homologous granulosa and Sertoli cells mainly secrete estrogens, whereas the homologous theca and Leydig cells mainly secrete androgens. Progesterone is secreted in large amounts only in females by transformed granulosa and theca cells, known as **luteal cells.** The protein products come mostly from granulosa and Sertoli cells.

SYNTHESIS OF SEX STEROID HORMONES

Biosynthesis of gonadal steroid hormones follows a common pathway in both genders (Figure 48-1). *The enzymes, their organelle localization, and cofactor requirements are those described for the adrenal cortex* (see Chapter 46). Furthermore, the gonadal enzymes are identical to the adrenal enzymes, and the same genes direct their synthesis. Cholesterol, either synthesized in situ from acetylcoenzyme A (acetyl CoA) or taken up from the plasma low-density lipoproteins (LDLs), is the starting compound. P-450 scc (20,22-desmolase) catalyzes side-chain cleavage of cholesterol and is the rate-limiting step for synthesis of progester-

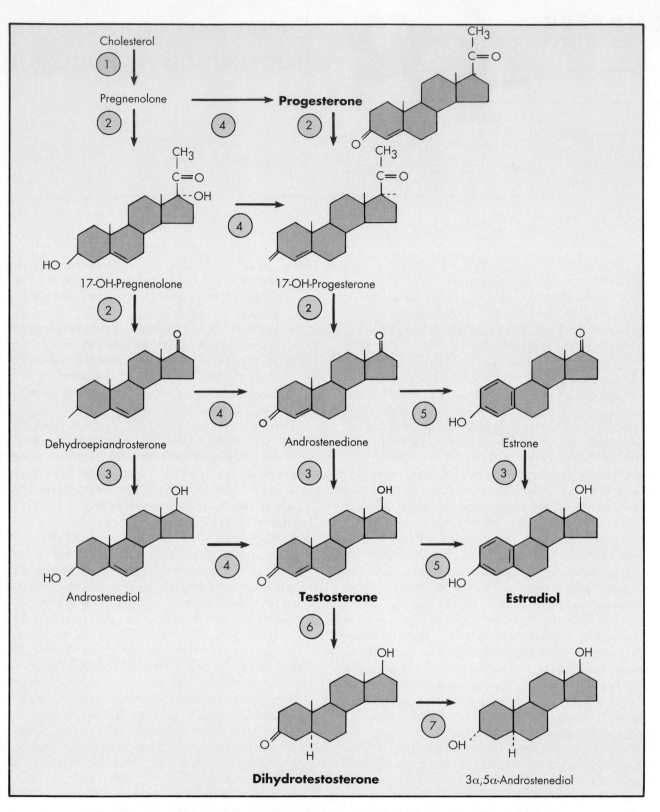

FIGURE 48-1 Pathways of synthesis of steroid hormones in the gonads. Testosterone is the major product of the testis. Estradiol and progesterone are the major products of the ovary. The enzymes are: *1,* 20,22-desmolase (P-450 scc); *2,* 17-hydroxylase/17,20-desmolase; *3,* 17β-OH-steroid dehydrogenase; *4,* 3β-ol-dehydrogenase/δ4,5-isomerase; *5,* aromatase; *6,* 5α-reductase; *7,* 3α-reductase.

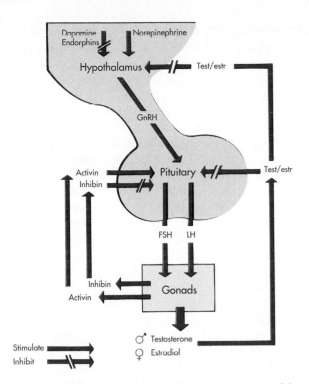

FIGURE 48-2 The hypothalamic-pituitary-gonadal axis. The hypothalamic peptide gonadotropin-releasing hormone *(GnRH)* stimulates release of two gonadotropins, luteinizing hormone *(LH)* and follicle-stimulating hormone *(FSH)*, from the pituitary gland. The gonadotropins, in turn, stimulate gonadal secretion of primarily testosterone in males and primarily estradiol in females. These feed back negatively at both the pituitary and hypothalamic levels to inhibit LH and FSH secretion. In addition, FSH stimulates release from the gonad of inhibin, which feeds back negatively to block FSH release preferentially. By contrast, estradiol (in women) and the gonadal protein activin have positive feedback effects on pituitary secretion.

one, androgens, and estrogens. Within the testes a small quantity of testosterone undergoes 5α-reduction to **dihydrotestosterone** (DHT), a potent androgen. However, a much larger and more important conversion of testosterone to DHT occurs in target tissues, catalyzed by the enzyme 5α-reductase. Estradiol and estrone are synthesized from their respective androgen precursors by the P-450 aromatase enzyme complex. This sequentially catalyzes the hydroxylation and oxidation of the 19-methyl group, the creation of a 1-2 double bond, the decarboxylation of position 19, and the formation of the characteristic benzene ring of estrogens.

REGULATION OF GONADAL STEROID HORMONE SECRETION

A hypothalamic–anterior pituitary–gonadal axis, analogous to that involved in thyroid and adrenal function, is the basis for gonadal regulation (Figure 48-2). The

components are **gonadotropin-releasing hormone** (GnRH) and two pituitary gonadotropins designated **luteinizing hormone** (LH) and **follicle-stimulating hormone** (FSH). A single pituitary cell type, the gonadotroph, generally produces both LH and FSH, although occasionally the gonadotroph contains only one or the other.

Gonadotropin-Releasing Hormone

GnRH, also known as **luteinizing hormone–releasing hormone** (LHRH), stimulates both LH and FSH secretion, but LH more so. The existence of a separate FSH-releasing hormone is suspected but unproved. GnRH, a decapeptide synthesized from a much larger preprohormone, is produced in two clusters of hypothalamic neurons in the arcuate and preoptic areas. From here the hormone is transported axonally for storage in the median eminence. Input from other areas of the brain permits reproduction to be influenced by light-dark cycles, by olfactory stimuli via airborne molecules known as pheromones, and by stress. Dopaminergic and endorphinergic tracts within the hypothalamus and the median eminence transmit important inhibitory influences on GnRH release (Figure 48-2). GnRH is released into the pituitary portal veins in pulses driven by a primary generator. Men have 8 to 10 pulses per day, whereas in women the frequency and periodicity of pulses vary with the menstrual cycle. In children pulsatility is greatly reduced.

GnRH binds to its plasma membrane receptor and triggers an influx of extracellular calcium into the gonadotroph cells. Complexed to calmodulin, the calcium acts as the major second messenger, and phosphatidylinositol products play a subsidiary role. GnRH stimulates the simultaneous release of LH and FSH from their secretory granules. The ratio of FSH to LH increases when the frequency of GnRH pulses declines. GnRH also stimulates the transcription of genes that direct the synthesis of the two gonadotropins and subsequent prohormone processing via glycosylation. Thus a GnRH infusion typically produces a biphasic LH response.

Prolonged stimulation by GnRH causes down-regulation of its receptor, desensitization of the gonadotroph to GnRH, and profound inhibition of gonadotropin secretion. Long-acting GnRH superagonists are commonly used therapeutically in clinical situations where a cessation of gonadotropin secretion and gonadal production of androgens or estrogens is beneficial. These include carcinoma of the prostate in men, **endometri-osis** in women, and in some in vitro fertilization programs. In the last circumstance, it may be easier to produce ova at a specified time with an external program of FSH and LH administration (see Chapter 50) if endogenous FSH and LH secretion is eliminated.

Gonadotropins

LH and FSH are glycoproteins that resemble thyroid-stimulating hormone (TSH). The α-subunit of all three hormones is identical, whereas their respective β-subunits are different and are determined by unique genes. The α- and β-subunits in each gonadotropin are required for binding to its gonadal receptor, and proper carbohydrate components are necessary for full biological activity of the β-submits.

LH mainly stimulates the theca cells of the female and the Leydig cells of the male to synthesize and secrete androgens and, to a much lesser extent, estrogens. LH also stimulates granulosa cells, once LH receptors have been expressed by these cells during the female cycle. cAMP is the major second messenger for LH actions. Continuous stimulation by LH down-regulates its receptor and reduces responsitivity to the hormone.

Analogous to adrenocorticotropic hormone (ACTH), LH stimulates cholesterol transfer to the mitochondria and its conversion to pregnenolone. Subsequently the concentrations of steroidogenic enzymes and of adrenoxin are increased by stimulating transcription of their genes. Most importantly, *LH increases the level of 17-hydroxylase/17,20-desmolase, the essential step in androgen synthesis* (see Figure 48-1).

FSH stimulates granulosa and Sertoli cells to secrete estrogens. Acting via its plasma membrane receptor and cAMP as a second messenger, *FSH increases transcription of the gene for aromatase, the enzyme specific to estradiol synthesis.* Another important effect of FSH is to increase the number of LH receptors in target cells, and thereby to amplify their sensitivity to LH. FSH also stimulates the secretion of inhibin and other protein products of the granulosa and Sertoli cells.

Feedback Regulation

Regulation of gonadotropin secretion, sex steroid hormone production, and other aspects of gonadal function is complex, and those aspects distinctive to each gender are described in later sections. However, certain common principles exist (see Figure 48-2). *Testosterone in men and estradiol in women inhibit secretion of LH and FSH.* In this basic negative feedback loop, the sex steroids act at the pituitary level by blocking the actions of GnRH on gonadotropin release and synthesis. They also act at the hypothalamic level via endorphin neurons to decrease GnRH. Both the frequency and the amplitude of LH and FSH pulses are diminished. In women a specific positive feedback effect of estradiol on LH secretion is included in the basic framework (see Chapter 50). This effect depends on the dose, time, and duration of exposure to estradiol.

Another negative feedback loop relates **inhibin** from granulosa and Sertoli cells to FSH secretion. *Inhibin reduces GnRH release, FSH β-subunit synthesis, and the stimulatory effect of GnRH on FSH secretion.* In contrast, from the same gonadal cells, **activin** exerts a positive feedback effect to stimulate FSH secretion. Thus the output of LH and FSH from the pituitary gland can be exquisitely and differentially regulated by interactions among hypothalamic and gonadal products. At various times the critical influence may be from either site. In this sense the gonad can be viewed as a much more self-regulatory gland than either the adrenal cortex or the thyroid. This is most apparent in women, as discussed later.

AGE-RELATED CHANGES IN REPRODUCTION

The hypothalamic-pituitary-gonadal axis is unique in that it undergoes extreme changes throughout the human life span. Although the patterns of females and males differ, again certain common aspects bear emphasis (Figure 48-3).

Intrauterine and Childhood Pattern

In humans GnRH is present in the hypothalamus at 4 weeks and FSH and LH are present in the pituitary gland by 10 to 12 weeks of gestation. A peak of gonadotropin concentrations occurs in fetal plasma at midgestation. The concentrations drop to low levels before birth, and then transiently increase (more prolonged in females) again at about 2 months of age. For the rest of childhood, FSH and LH are secreted at very low levels. These changes are mirrored by fluctuations of plasma testosterone in males and estradiol in females.

Puberty

The transition from a nonreproductive to a reproductive state requires the pubertal maturation of the entire hypothalamic-pituitary-gonadal axis. Before the child reaches age 10 years, plasma LH and FSH levels are low despite very low concentrations of gonadal hormones. Therefore, either the negative feedback system

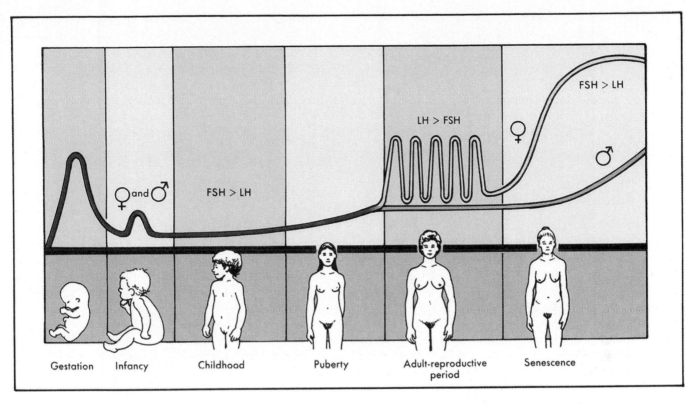

FIGURE 48-3   The pattern of gonadotropin secretion throughout life. Note transient peaks during gestation and early infancy and low levels thereafter in childhood. Women subsequently develop monthly cyclic bursts, with luteinizing hormone *(LH)* exceeding follicle-stimulating hormone *(FSH)*; men do not. Both genders show increased gonadotropin production after age 50 years, with FSH exceeding LH

is inoperative or the hypothalamus and pituitary gland are exquisitely sensitive to testosterone, estradiol, and inhibin. One factor in puberty may thus be the gradual maturing of hypothalamic neurons, and this process leads to an increased synthesis and release of GnRH. The time of onset of this maturational process ranges from age 9 to 17 and may be genetically preprogrammed, because familial patterns are apparent. As puberty approaches, a pulsatile pattern of LH and FSH secretion appears. The ratio of plasma LH to FSH rises as the pulse frequency increases. Furthermore, during early and middle puberty, but not usually thereafter, a distinct nocturnal peak in LH secretion is observed (Figure 48-4). This coincides with but is not proven to be the result of a decrease in nocturnal melatonin secretion. Notably, these changes in GnRH and gonadotropins occur even in the absence of the gonads.

During early puberty, the responsiveness of the pituitary gland to GnRH changes qualitatively such that LH exceeds FSH output. This may result from increased synthesis and storage of LH in response to pulsatile GnRH secretion, because the latter allows better maintenance of GnRH receptors. Although the gonadal target cells re-

spond to LH in childhood, their responsiveness is augmented during puberty. Therefore, plasma levels of estradiol in females, testosterone in males, and inhibin in both genders increase sharply during these years. Early and midpuberty can be viewed as a cascade of increasing maturation from the hypothalamic to the pituitary to the gonadal level.

It is often difficult to distinguish with certainty by clinical endocrine testing between a late onset of what will eventually be normal puberty and disease of the hypothalamus or pituitary gland that prevents the expected increased secretion of LH and FSH. Because failure to show physical signs of puberty by age 13 or 14 (see Chapters 49 and 50) is psychologically distressing to the child, treatment with sufficient testosterone or estradiol to induce such changes and a growth spurt may be warranted. Such hormonal support can be withdrawn after an appropriate period to determine if normal puberty has finally begun.

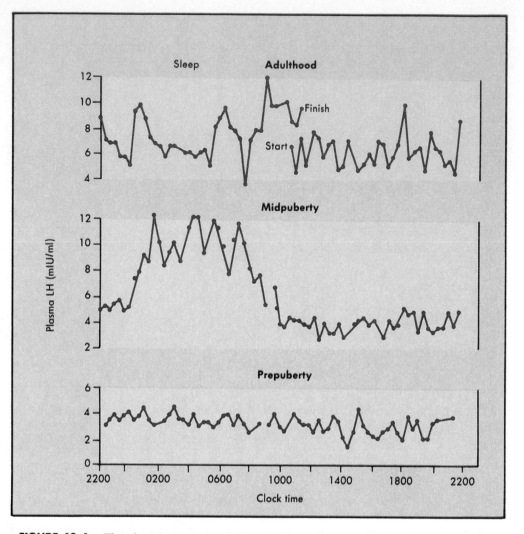

FIGURE 48-4   The changing pattern of diurnal LH secretion from childhood to adulthood. During puberty LH secretion becomes much more pulsatile. In addition, a nocturnal peak appears early in puberty and then disappears when puberty is completed. Males and females both show these changes. Each day's sampling started and finished at approximately 10 AM. (Redrawn from Boyar RM et al: *N Engl J Med* 287:582, 1972.)

Once the adult pattern of gonadotropin secretion is established, the basal plasma concentrations of LH and FSH (approximately 10^{-11} molar) are similar in men and women. An important distinguishing feature between the genders is the additional establishment of a dramatic monthly gonadotropin cycle in females only (see Chapter 50), with the LH bursts greatly exceeding the FSH bursts (see Figure 48-3).

Climacteric

In both genders, a loss of gonadal responsiveness to gonadotropin stimulation develops around the fifth decade of life. In males this is gradual, and some reproductive capacity usually persists even into the ninth decade. In females reproductive capacity is eventually lost completely, and menopause occurs. In both genders, negative feedback leads to elevated plasma gonadotropin levels. The FSH level rises more than the LH level, and the increase is more distinct in females (see Figure 48-3).

SEXUAL DIFFERENTIATION

The most fundamental and obvious difference between the genders lies in the anatomy and consequent physiology of their reproductive tracts. During the first 5 weeks of gestation, however, the gonads of males and females are indistinguishable, and their genital tracts are

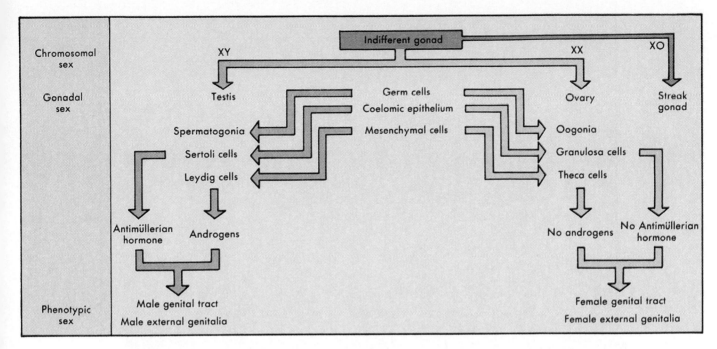

FIGURE 48-5   An overview of the development of the cells of the ovary and testis from the primitive indifferent gonad. Androgens and antimüllerian hormone from the testis induce the male pattern of differentiation of the genital tract and external genitalia. It is the absence of these products from the ovary that determines differentiation of the genital tract and the external genitalia into the female pattern.

unformed. From this stage of the "indifferent gonad" to that of the completed normal individual of either gender lies the process of sexual differentiation (Figures 48-5 and 48-6). The final maleness or femaleness is best characterized in terms of differences in genetic sex, in gonadal sex, and in genital (phenotypic) sex.

Genetic Sex

Maleness is determined positively and predominantly by the presence of the Y chromosome. The normal male has a chromosome complement of 44 autosomes and two sex chromosomes, XY. *Without the Y chromosome* (or in rare instances DNA translocated from the Y to the X chromosome), *neither testicular development nor masculinization of the genital tracts and external genitalia can occur.* The organization of the indifferent gonad into the characteristic spermatogenic tubules of the male is directed by a 14-kilobase segment known as the sex-determining region of the Y-chromosome (SRY gene) or the **testis determining factor (TDF).** This gene is located on the short arm of the Y chromosome (see Figures 48-5 and 48-6, *B*). Either identical to or closely linked to the SRY or TDF gene is one which codes for the **H-Y antigen.** This glycoprotein is present on the surface of all male cells

except diploid germ cells, and it is involved in rejection of male tissues by female recipients. The H-Y antigen causes virilization of the indifferent gonad or of disaggregated early ovarian cells in vitro. Even though it is essential, by itself the Y chromosome is not sufficient for maleness. Located on the X chromosome is the gene for the androgen receptor, which sensitizes the genital ducts and the external genitalia to the masculinizing effects of testosterone and DHT. Autosomal genes may also participate in directing the initial organization of the primitive gonad into a functioning reproductive gland.

In contrast, femaleness is partly determined positively by the presence of an X chromosome but also negatively by the absence of a Y chromosome. The normal female chromosome complement is 44 autosomes and two sex chromosomes, XX. Both X chromosomes are active in germ cells and are essential for the genesis of a normal ovary (see Figures 48-5 and 48-6, *A*). However, *female differentiation of the genital ducts and external genitalia requires that only a single X chromosome be active in directing transcription within their constituent cells.* The second X chromosome of a normal XX female is inactivated in all tissues outside the gonad. Thus, if an abnormality in meiosis or in early mitosis produces an individual with only a single sex chromosome (an X), that

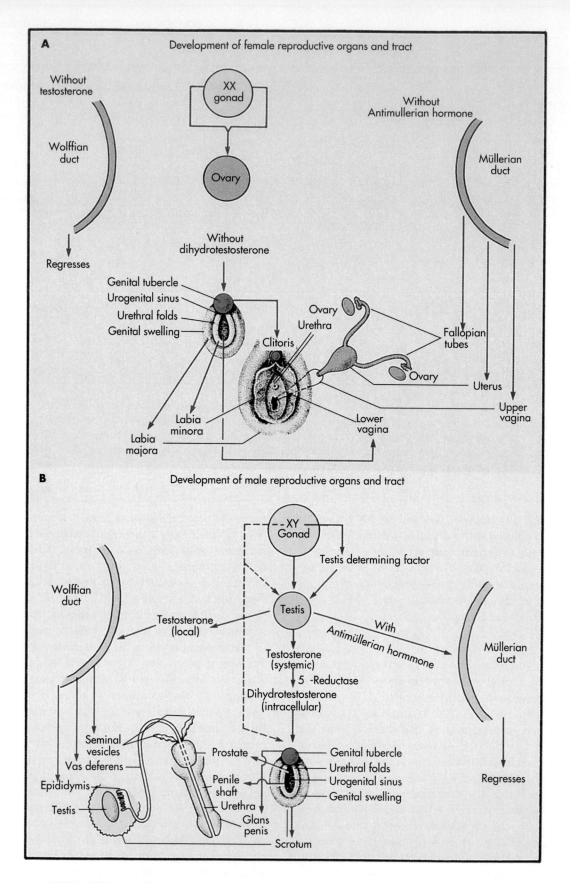

FIGURE 48-6 **A,** Development of the female reproductive organs. Note that this development does not require hormonal products from the ovary. Therefore, in the absence of gonads, the female pattern results. **B,** Development of the male reproductive organs. Note that the complete male pattern requires secretion and local action of testosterone on the wolffian duct, reduction of testosterone to dihydrotestosterone in the cells of the anlage of the external genitalia, and the secretion of antimüllerian hormone to suppress müllerian duct development.

individual will undergo normal female genital development even though her gonad is abnormal and without function (see Figure 48-5).

Gonadal Sex

The indifferent gonad of 5 weeks' gestation consists of a primordial mesonephric ridge with several components: coelomic epithelium, the precursor of granulosa and Sertoli cells; mesenchymal stromal cells, the precursors of theca and Leydig cells; and germ cells that have migrated there from the yolk sack endoderm (see Figure 48-5). This assembly is organized as an outer cortex and an inner medulla.

In a normal male fetus, the spermatogenic tubules begin to form at 6 weeks, followed by differentiation of the Sertoli cells at 7 weeks and the Leydig cells at 8 to 9 weeks. At that point the testis is recognizable and testosterone secretion has begun. The germ cells have become enclosed within the medulla, whereas the cortex has regressed. No known hormonal influences are required for this differentiation of the indifferent gonad into a testis.

In a normal female fetus, differentiation of the indifferent gonad into an ovary does not start until 9 weeks of age. At this time, activity of both X chromosomes within the germ cells is essential. The germ cells begin to undergo mitosis, giving rise to daughter cells called oogonia, which continue to proliferate. Shortly thereafter, meiosis is initiated in some oogonia, and each becomes surrounded by differentiating granulosa cells and precursor theca cells to form a follicle. The germ cells, now known as primary oocytes, remain in the first stage, or prophase, of meiosis until activated many years later. In contrast to the male, the cortex (which contains the follicles) predominates in the developed ovary, whereas the medulla regresses. The primitive ovary begins to synthesize estrogenic hormones concurrent with these developments, and these hormones may contribute to later ovarian differentiation.

Genital (Phenotypic) Sex

Up to this point in fetal development, sexual differentiation is largely independent of known hormonal products. However, *differentiation of the genital ducts and the external genitalia requires specific hormonal signals from the gonad to produce the masculine format. Without such input, the feminine format will result.*

During the sexually indifferent stage, from 3 to 7 weeks, two different genital ducts develop on each side. In the male, at about 9 to 10 weeks, the wolffian, or mesonephric, duct on each side begins to grow. Together they give rise to the **epididymis,** the **vas deferens,** the

seminal vesicles, and the **ejaculatory duct** by 12 weeks (see Figure 48-6, *B*). This constitutes the system for delivering sperm from the testis to the female. The growth and differentiation of each wolffian duct in the male is induced by testosterone, which is secreted by the ipsilateral testis and acts locally. Testosterone is not converted to its active metabolite, DHT, before acting on the wolffian duct cells, as it must be in other genital tissues. In the female the wolffian ducts regress at 10 to 11 weeks, because the ipsilateral ovary does not secrete testosterone.

Each müllerian duct arises parallel to the wolffian duct on its side. In the male the müllerian ducts begin to regress at 7 to 8 weeks, about the same time that the Sertoli cells of the testis appear. These cells produce a glycoprotein, **antimüllerian hormone** (AMH), which causes atrophy of the müllerian ducts. AMH belongs to a superfamily of growth-regulating factors that are coded for by similar genes; these include transforming growth factors, epidermal growth factor, and inhibin. AMH also initiates descent of the testes into the inguinal area. Although the homologous granulosa cells of the ovary produce AMH, they do not do so until after the müllerian ducts have already developed to the point where AMH can no longer cause their regression. Therefore in the female these ducts grow and differentiate into fallopian tubes at their upper ends and join at their lower ends to form a single uterus, cervix, and upper vagina (see Figure 48-6, *A*). This process does not require any known ovarian hormone. The early critical secretion of AMH in the male may be initiated by a factor expressed by the Y chromosome.

The external genitalia of both genders begin to differentiate at 9 to 10 weeks. They are derived from the same primitive structures: the genital tubercle, the genital swelling, the urethral or genital folds, and the urogenital sinus (see Figure 48-6). *In the male testosterone must be secreted into the fetal circulation and subsequently converted to DHT* within these tissues in order for masculine differentiation of the external genitalia to occur. With DHT stimulation the genital tubercle grows into the glans penis, the genital swellings fold and fuse into the scrotum, the urethral folds enlarge and enclose the penile urethra and corpora spongiosa, and the urogenital sinus gives rise to the prostate gland (see Figure 48-6, *B*). In the normal female, in an individual with an XO chromosome karotype, XX, or in the absence of gonads, the external genitalia develop without significant positive hormonal influence into the clitoris, labia majora, labia minora, and lower vagina (see Figure 48-6, *A*). The critical importance of androgen molecules themselves to the development of masculine external genitalia is emphasized by the presence of adequate androgen receptors in female urogenital tract cells. The estrogen molecules in the female may play a role by offsetting possible virilizing actions of normal adrenal androgens in that gender.

Ambiguity in the genital phenotype and discordance between gonadal and genital sex constitute male and female **pseudohermaphroditism.** XY individuals with normal testes may secrete inadequate amounts of testosterone or DHT because of various enzyme deficiencies (see Figure 48-1); they will exhibit partially male and partially female characteristics—for example, a scrotum but an incompletely fused penis. If the androgen receptor is absent, they will lack normal male wolffian duct derivatives (e.g., the spermatic cord) and have completely female external genitalia. However, no müllerian duct derivatives (e.g., a uterus) will be present, because of normal AMH action.

An XX individual with normal ovaries who is exposed to high levels of androgen from the maternal circulation (see Chapter 46) will have variable degrees of masculinization of the external genitalia. The degree of masculinization will depend on when the exposure began and its magnitude. Thus, the clitoris may be almost penile in size; or the labia majora may only be partially fused. The presence of the extra X chromosome in the XXY individual does not prevent the normal male external pattern but does essentially abolish spermatogenesis at puberty and modestly diminishes testosterone synthesis. It is extremely important to detect, define the cause, and, when appropriate, to treat pseudohermaphroditism early in life. Surgical reconstruction of the external genitalia may be required. Delay can cause lifelong disturbances in feelings of gender identity and in sexual functioning in the affected person, as well as great parental distress.

The initial androgen production necessary for male sexual differentiation does not depend on fetal pituitary gonadotropins. An LH-like hormone, **chorionic gonadotropin** from the placenta, stimulates early testosterone production by the Leydig cells of the testis. On the other hand, the continued growth of the male genitalia in the last 6 months requires fetal pituitary LH to support the necessary testicular androgen production. Similarly, the later molding of the female genitalia in utero may be modulated by ovarian estrogen production, which also depends on pituitary gonadotropins.

Other aspects of phenotypic sexual differentiation are not evident until long after birth. These include differences between the constant pattern of gonadotropin se-

cretion in the male versus the cyclic pattern in the female, the degree of breast development, and psychological identification with one gender. It is not certain what factors imprint or regulate these traits in humans. Evidence from rodents suggest that circulating androgens program the fetal hypothalamus to set the ultimate noncycling pattern of gonadotropin secretion in the postpubertal male. (To do so, testosterone may paradoxically require conversion to estradiol within the target neurons.) Without androgens, the ultimate cyclic pattern of the female results. This would constitute another instance in which the female pattern was the "neutral pattern," whereas the male pattern required an action derived from the Y chromosome.

Mammary gland development in the rodent embryo also is clearly regulated by androgen. In its absence a normal female breast develops; in its presence the elaborated ductal system is suppressed. In the human, however, male/female differences in breast tissue are not apparent until puberty. At that time the hormonal milieu in the female induces growth and differentiation of breast tissue, whereas that in the male suppresses it.

Evidence suggests that psychological gender identification is mostly independent of hormonal regulation or even inherently of the phenotype of the genitalia. Instead, it appears to depend on rearing cues, which of course can be influenced by parental perceptions of genital sex. A few individuals with XY male pseudohermaphroditism resulting from deficiency of the enzyme 5α-reductase who were raised as girls experienced significant growth of the penis at puberty and reversed their psychosocial gender from female to male.

SUMMARY

1. Female and male gonadal structure and function have important homologous characteristics. In each gender, cells develop within sheltered and hormonally conditioned environments provided by granulosa cells (estrogens) and thecal cells (adrogens) in the female and by Sertoli cells (estrogens) and Leydig cells (androgens) in the male.

2. Gonadal steroids are synthesized from cholesterol by the same enzymatic pathways as are adrenal steroids. The predominant products are estradiol in the female and testosterone in the male. An important protein hormone product is inhibin.

3. Testosterone production is stimulated by luteinizing hormone (LH), and estradiol and inhibin by follicle-stimulating hormone (FSH), from the anterior pituitary gonadotrophs. LH and FSH are secreted in response to a hypothalamic gonadotropin-releasing hormone (GnRH).

4. Estradiol and testosterone feed-back negatively on the hypothalamus and pituitary to decrease LH and FSH release, and inhibin feeds back negatively on FSH release.

5. Gonadotropin and sex steroid production have fetal peaks, are low during childhood, and rise to adult levels during puberty. Sex steroid levels decline during senescence, while gonadotropins increase because of negative feedback.

6. Gender differences in reproductive function derive from the process of sexual differentiation. The Y chromosome is a positive determinant of the development of the indifferent gonad into a testis and of spermatogenesis. Two active X chromosomes are required for normal development of the ovary and for oogenesis.

7. Regardless of karyotype, masculinization of the internal genital ducts into a delivery system for sperm and of the external genitalia requires normal testosterone production and action. Suppression of the development of the female internal ducts requires antimüllerian hormone from the testis.

8. In the absence of these testicular hormones, the internal ducts develop into organs for receiving sperm and housing a conceptus and the external genitalia are feminized.

BIBLIOGRAPHY

Journal Articles

Ascoli M, Segaloff DL: On the structure of the luteinizing hormone/chorionic gonadotropin receptor, *Endocr Rev* 10:27, 1989.

Belchetz PE et al: Hypophysial responses to continuous and intermittent delivery of hypothalamic gonadotropin-releasing hormone, *Science* 202:631, 1978.

Conn PM, Crowley WF, Jr: Gonadotropin-releasing hormone and its analogues, *N Engl J Med* 324:93, 1991.

Friedman RC, Downey J: Neurobiology and sexual orientation: current relationships, *J Neuropsychiatry Clin Neurosciences* 5:131, 1993.

George FW, Wilson JD: Hormonal control of sexual development, *Vitam Horm* 43:145, 1986.

Gharib SD et al: Molecular biology of the pituitary gonadotropins, *Endocr Rev* 11:177, 1990.

Haqq CM et al: Molecular basis of mammalian sexual determination: activation of müllerian inhibiting substance gene expression by SRY, *Science* 266:1494, 1994.

Hawkins JR: The SRY gene, *Trends Endocrinol Metab* 4:328, 1993.

Josso N: Anti-Müllerian hormone: new perspectives for a sexist molecule, *Endocr Rev* 7:421, 1986.

LaBarbera AR, Rebar RW: Reproductive peptide hormones: generation, degradation, reception, and action, *Clin Obstet Gynecol* 33:576, 1990.

Lee MM, Donahoe PK: Müllerian inhibiting substance: a gonadal hormone with multiple functions, *Endocr Rev* 14:152, 1993.

Matsumoto A, Bremner W: Modulation of pulsatile gonadotropin secretion by testosterone in man, *J Clin Endocrinol Metab* 58:609, 1984.

Muller U, Lattermann U: H-Y antigens, testis differentiation, and spermatogenesis, *Exp Clin Immunogenet* 5:176, 1988.

Naftolin F, Butz E: Sexual dimorphism, *Science* 211:1263, 1981.

Scott R, Burger H: An inverse relationship exists between seminal plasma inhibin and serum follicle-stimulating hormone in man, *J Clin Endocrinol Metab* 52:796, 1981.

Southworth MB et al: The importance of signal pattern in the transmission of endocrine information: pituitary gonadotropin responses to continuous and pulsatile gonadotropin-releasing hormone, *J Clin Endocrinol Metab* 72:1286, 1991.

Veldhuis J et al: Endogenous opiates modulate the pulsatile secretion of biologically active luteinizing hormone in man, *J Clin Invest* 72:2031, 1983.

Veldhuis JD: The hypothalamic pulse generator: the reproductive core, *Clin Obstet Gynecol* 33:538, 1990.

Books and Monographs

Chin W: Organization and expression of glycoprotein hormone genes. In Imura H: *The pituitary gland,* New York, 1985, Raven Press.

Knobil E, Neill JD: *The physiology of reproduction,* New York, 1988, Raven Press.

Reichlin S: Neuroendocrinology. In Foster D, Wilson J, eds: *Williams textbook of endocrinology,* ed 8, Philadelphia, 1992, WB Saunders.

Savoy-Moore RT et al: Differential control of FSH and LH secretion. In Greep RO, ed: *Reproductive physiology,* Baltimore, 1980, University Park Press.

Yen, SSC: The hypothalamic control of pituitary hormone secretion regulation. In Yen SSC, Jaffe RB, eds: *Reproductive Endocrinology,* Philadelphia, 1991, WB Saunders Co.

CHAPTER 49

Male Reproduction

ANATOMY

The testes are situated in the scrotum, where they are maintained at 1° to 2° below the body core temperature, a situation that facilitates sperm production. Each adult testis weighs about 40 g and has a long diameter of 4.5 cm. Eighty percent of the testis is made up of the **spermatogenic** or **seminiferous tubules;** the remaining 20% is composed of connective tissue containing the **Leydig cells.** The spermatogenic tubules, a coiled mass of loops, empty into a ductal system that eventually drains into the **epididymis,** a maturation and storage site for spermatozoa. From there, spermatozoa are carried via the **vas deferens** and **ejaculatory duct** into the penis for emission.

The structure of the spermatogenic tubule is shown schematically in Figure 49-1. Each tubule is bounded by a basement membrane that separates it from the Leydig cells, the **peritubular (myoid) cells,** and adjacent capillaries. Beneath this membrane are **Sertoli cells** and immature germ cells, the **spermatogonia.** As the spermatogonia divide and develop around the circumference of the tubule, columns of maturing germ cells are formed below them. These columns reach from the basement membrane to the lumen and culminate in the spermatozoa. The columns lie between the cytoplasms of two adjoining Sertoli cells, each of which extends from the basement membrane to the lumen (Figure 49-1). *Special processes of the Sertoli cell cytoplasms fuse into tight junctions that create two compartments of intercellular space between the basement membrane and the lumen.* The spermatogonia lie within the proximal or basal compartment, whereas their descendants that arise from subsequent stages in spermatozoan development lie in the distal adluminal compartment. This compartmentalization accomplishes two important functions. *The basement membrane, the overlapping peritubular cells,*

and the Sertoli cell cytoplasm together form a blood-testis barrier. This barrier can exclude harmful circulating substances from the intercellular fluid that bathes the maturing germ cells and from the tubular fluid surrounding the spermatozoa. Conversely, products from the later stages of spermatogenesis are prevented from diffusing back into the bloodstream and producing antibodies. This arrangement also provides germ cells with high local concentrations of testosterone from the Leydig cells and protein products (and probably estradiol) from the Sertoli cells. Such high concentrations are essential for spermatogenesis.

THE BIOLOGY OF SPERMATOGENESIS

Sperm production continues throughout the male's reproductive life. At peak, 100 to 200 million sperm can be produced daily. To generate this large number, the spermatogonia must renew themselves by cell division in an ongoing manner. This differs from the situation in the female, who at birth has a fixed number of germ cells, which continually decreases throughout her life.

The descendants of the spermatogonia undergo an extraordinary metamorphosis to spermatozoa as they move from the basement membrane to the tubule. Within the basal compartment (Figure 49-1) a spermatogonium undergoes two mitotic divisions, which give rise to three active cells and a single resting cell; the latter will serve as the ancestor of a later generation of spermatozoa. The active cells divide further to yield type B spermatogonia, which then generate a number of **primary spermatocytes** (Figure 49-2). These enter the prophase of meiosis, the first reduction division, in which they remain for about 20 days.

The complex process of chromosomal reduplication, synapsis, crossover, division, and separation completes meiosis. Subsequently, within the adluminal compart-

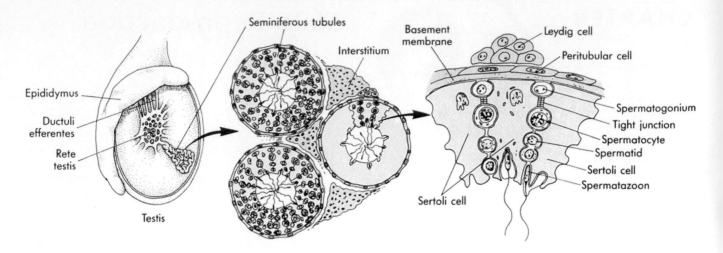

FIGURE 49-1   Schematic representation of the architecture of the testis. Note that the Leydig cells and peritubular cells are separated from the spermatogenic tubules. Within the latter, the germ cell line is completely invested by cytoplasm of the surrounding Sertoli cells. In addition, tight junctions between adjacent Sertoli cells separate the ancestral spermatogonia from their descendant spermatocytes, spermatids, and spermatozoa. Thus, a blood-testis barrier effectively filters plasma, permitting only selected substances to reach the developing germ cells from Sertoli cell cytoplasm. (Redrawn from Skinner MK: Cell-cell interactions in the testis, *Endocr Rev* 12:45, 1991.)

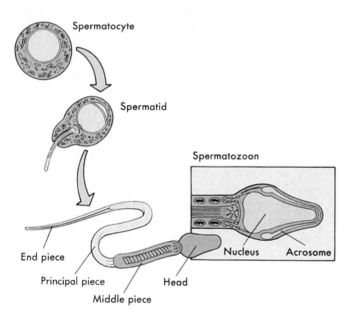

FIGURE 49-2   Schematic representation of the morphological alterations in the development of the spermatozoon from the spermatocyte. The final spermatozoan is almost devoid of cytoplasm. The acrosomal head contains enzymes necessary for penetration of the ova. The middle piece supplies energy and the principal piece contractile machinery for motility.

ment (see Figure 49-1) the daughter cells, **secondary spermatocytes,** divide again. The products, called **spermatids** (see Figure 49-2), now contain 22 autosomes and either an X or a Y chromosome. The spermatids lie near the lumen of the tubule, attached to the

abutting Sertoli cells by specialized junctions. They also remain connected to spermatocytes through intercellular bridges. The spermatids undergo nuclear condensation, shrinkage of cytoplasm, formation of an **acrosome,** development of a tail, and then emerge as flagellated spermatozoa (see Figure 49-2). In the end 64 spermatozoa arise from each spermatogonium. The spermatozoa are then extruded into the tubular lumen, and most of their cytoplasm is imbedded in Sertoli cells, where it is phagocytized and degraded. Movement of the spermatozoa into the epididymus is facilitated by fluid currents generated by the peritubular myoid cells.

The spermatozoa are now linear structures with several components (see Figure 49-2). The head contains the nucleus and an acrosomal cap in which are concentrated hydrolytic and proteolytic enzymes that will facilitate penetration of the ovum. The middle piece, or body, contains mitochondria, which generate the motile energy of the spermatozoa. The chief, or principal, piece contains stored adenosine triphosphate (ATP) and pairs of contractile microtubules down its entire length. Cross-bridging arms contain **dynein,** an ATPase that transfers the stored energy into a sliding movement between the microtubules. This imparts flagellar motion to the spermatozoa. Both Ca^{++} and cyclic adenosine monophosphate (cAMP) are involved in regulating sperm motility.

Approximately 70 days are required for this entire sequence of development. However, *individual resting spermatogonia do not begin the process of spermatogenesis randomly.* Groups of adjacent spermatogonia initiate a cycle of development about every 16 days, thus constituting one "generation." At about the same time that the primary spermatocytes of one cycle enter prophase, a second cycle of spermatogonia is activated. A third cycle begins about the time spermatids from the first cycle appear. When these spermatids have completed their transformation into spermatozoa, a fourth cycle of spermatogonia has begun. The individual descendants of any one type B spermatogonium that lie within the adluminal compartment of the tubule may not be totally separated. Continuity of cytoplasm and possibly cell-to-cell intercommunication may exist. Because of this and because of the regular topographic association of particular stages of spermatogenesis in neighboring cycles around the circumference of the tubule, products of germ cells in one stage of spermatogenesis may possibly modulate events in other stages.

The spermatozoa traverse the epididymis in 2 to 4 weeks. During this time they lose their remaining cytoplasm and become increasingly motile. The epididymis is lined by specialized epithelial cells whose function includes progressive modulation of the chemical and osmotic environment in which the sperm advance. Proteins produced by these cells bind to sperm membranes and enhance their forward mobility and their ultimate ability to fertilize an ovum. After reaching the vas deferens, sperm may be stored viably for several months.

DELIVERY OF SPERMATOZOA

The process of ejaculation delivers spermatozoa from the vas deferens out of the penile urethra; and for reproductive purposes, into the female genital tract. *Erection of the penis, a process that results from filling of its venous sinuses, is accomplished through simultaneous dilation of arterioles and constriction of veins and is under parasympathetic control. Ejaculation is then effected by sympathetic activation.* Just before ejaculation, successive fluids are added to the contents of the vas deferens. The initial secretions are from the prostate gland, the alkalinity of which helps neutralize the acid pH of the female genital secretions. The terminal portion of the ejaculate is composed of secretions from the seminal vesicles. These contain fructose, an important oxidative substrate for the spermatozoa, and prostaglandins. Within the female tract, the prostaglandins may stimulate contractions that help propel the spermatozoa toward the ovum. Seminal fluid also contains calcium, zinc, luteinizing and follicle-stimulating hormones (LH, FSH), prolactin, testosterone, estradiol, inhibin, oxytocin, endorphins, and a variety of enzymes. Their exact source and role in fertilization remain to be determined.

A typical seminal emission contains 200 to 400 million spermatozoa in a volume of 3 to 4 ml. Once within the vagina, the spermatozoa rapidly move inward (at a

rate up to 44 mm/min). Their life span in the female genital tract is approximately 2 days. Transport of sperm to the ovum requires mechanical assistance by smooth muscle contractions of the female reproductive organs.

In vivo, human sperm cannot fertilize an ovum until they have been in contact with the female reproductive tract for several hours; the process is termed **capacitation.** In vitro, however, fertilization can occur after the ejaculated sperm have been washed free of seminal fluid. This observation suggests that washing has removed inhibitory substances. Although the process of capacitation is poorly understood, it increases motility and enhances the ability of sperm to penetrate the ovum. This involves a reaction in which the acrosomal membrane and the outer sperm membrane fuse to create pores through which the enclosed enzymes can escape.

MALE PUBERTY

Beginning at an average age of 10 to 11 years and ending at about 15 to 17 years, males develop adult levels of androgenic hormones and full reproductive function. Activation of the testes results in adult size and function of the accessory organs of reproduction, complete secondary sexual characteristics, and adult musculature. Boys undergo a linear growth spurt, and the epiphyseal growth centers close when adult height is attained. A composite picture of the sequence is shown in Figure 49-3.

Enlargement of the testes is the first physical sign of puberty. This principally represents an increase in the volume of the spermatogenic tubules, and it is preceded by small increases of plasma FSH. Leydig cells appear and testosterone secretion rises secondary to increases of the plasma LH level. The plasma testosterone level climbs rapidly over a 2-year period, during which time pubic hair appears, the penis enlarges, and peak velocity in linear growth is achieved (Figure 49-3). When the boy is about age 13, sperm production begins. Growth ceases 1 to 2 years after adult testosterone levels are reached. In about one third of boys a transient stimulation of breast growth occurs, probably reflecting increased production of estradiol. As testosterone levels become dominant, the breast tissue regresses.

REGULATION OF SPERMATOGENESIS

For various reasons hormonal regulation of spermatogenesis is less completely understood than that of oogenesis. Numerous hormones and hormonal products are likely involved, and the possibilities include paracrine and autocrine actions among Leydig, Sertoli, and germ

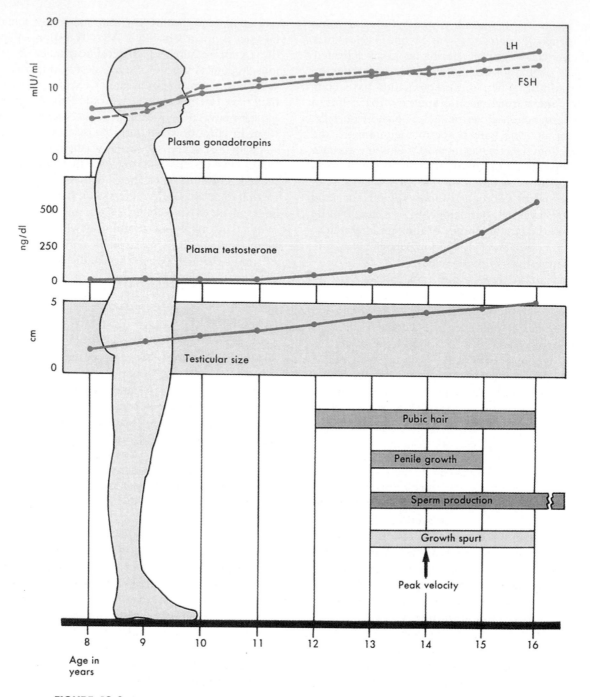

FIGURE 49-3   Average chronologic sequence of hormonal and biological events in normal male puberty. Secretion of FSH and LH initiate testicular enlargement and testosterone secretion, respectively. The latter then stimulates somatic growth as well as maturation of the organs of reproduction. (Data from Marshall WA, Tanner JM: *Arch Dis Child* 45:13, 1970; and Winter JSD et al: *Pediatr Res* 6:126, 1972.)

cells. *It is, however, essential that there be normal functioning of the adult GnRH–LH/FSH–testosterone axis. This includes pulsatility of GnRH release and of FSH/LH actions on their target cells as well as the production of very high intratesticular concentrations of testosterone.*

FSH, LH, and testosterone may also possibly coordinate with estradiol, inhibin, prolactin, and growth hormone in the regulation of spermatogenesis. For example, growth hormone deficiency delays the onset of reproductive function, probably because of a resulting lack of local insulin-like growth factor (somatomedin) produc-

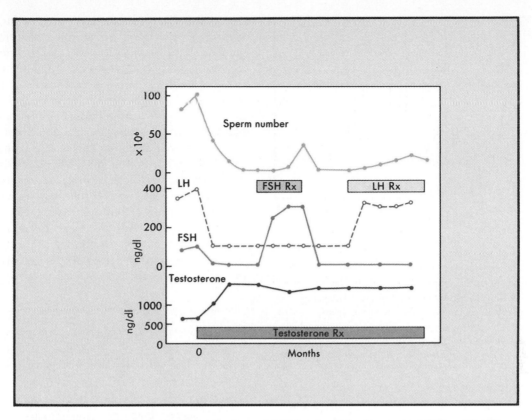

FIGURE 49-4   The individual contributions of follicle-stimulating hormone *(FSH)* and luteinizing hormone *(LH)* to sperm production in normal men. When endogenous FSH and LH secretion is suppressed by administration of exogenous testosterone (negative feedback), sperm production declines to very low levels. Selective restoration of either FSH or LH by exogenous administration raises sperm production, but neither gonadotropin alone returns sperm counts to normal. (Redrawn from Matsumoto AM et al: *J Clin Invest* 72:1005, 1983; *J Clin Endocrinol Metab* 59:882, 1984.)

tion. In normal adults experimental production of simultaneous FSH and LH deficiency by total suppression of the pituitary gonadotrophs with high doses of testosterone almost completely inhibits sperm production (Figure 49-4). Selective replacement of either FSH or LH then reinitiates sperm production, but not to normal levels. Other studies suggest that a suitable period of pubertal exposure to FSH is essential to spermatogenesis, but after that it can sometimes be maintained adequately by LH and local testosterone alone.

During intrauterine development of the testis, testosterone stimulated by the surge of fetal gonadotropins (see Figures 48-3 and 49-5) may condition the transformation of primordial germ cells into resting spermatogonia. From then until puberty the spermatogonia normally remain dormant, presumably because gonadotropin secretion and testosterone levels in the testis are low. Activation of the spermatogonia, which possess FSH receptors, then starts shortly after FSH secretion begins to undergo its pubertal increase. Shortly thereafter the concentrations of LH and testosterone also rise. Testoster-

one reaches levels many times higher in the testis than in the plasma, because of the specific action of LH on the Leydig cells. (In men who lack LH, the substitution of testosterone in amounts only sufficient to raise *plasma* levels to normal is unable to promote spermatogenesis.) This critical action of LH may also be facilitated by prolactin, which increases LH receptors on the Leydig cells. To what extent the local testosterone regulates spermatogenesis itself or via intratesticular conversion to dihydrotestosterone (DHT) or estradiol is uncertain. How or whether the sex steroids directly act on the spermatocytes and spermatids is also uncertain, since these germ cells lack receptors for these hormones. Nonetheless, testosterone somehow regulates the maturation of these precursors to spermatozoa.

FSH acts critically on the Sertoli cells, whose role is vital and complex. Their only known function until puberty is to secrete antimüllerian hormone in early fetal life. After puberty, each Sertoli cell is in contact with numerous germ cells at various stages of their development through ectoplasmic invaginations within their respec-

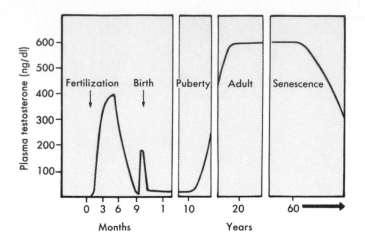

FIGURE 49-5   Plasma testosterone profile during the life span of a normal male. The intrauterine surge corresponds with masculinization of the external genitalia in the fetus. The adult plateau is achieved rapidly during puberty. The senescent decline is relatively modest, and there is slight if any loss of androgen effects. (Data from Griffin JE et al. In Bondy PK, Rosenberg LE: *Metabolic control and disease,* Philadelphia, 1980, WB Saunders Co; and Winter JSD et al: *J Clin Endocrinol Metab* 42:679, 1976.)

tive plasma membranes. In association with the cycle of spermatogenesis, the Sertoli cells show regular changes in the activity and shape of the nucleus; in the size, shape, and branching of the cytoplasmic processes; in concentrations of lipid and glycogen; in mitochondrial function; and in enzyme content. These changes are dependent on FSH and testosterone. The Sertoli cell cytoplasm also acts as a conduit through which the germ cells move from the basal to the adluminal compartment; thus their tight junctions open regularly to permit maturing primary spermatocytes to pass, and then the junctions close again behind the spermatocytes. This maintains the blood-testis barrier (see Figure 49-1).

Sertoli cells are stimulated by FSH to synthesize a wide variety of products, including estradiol from testosterone, which is provided by the Leydig cells. Conceivably, both sex steroids then have access to the developing spermatocytes. FSH also stimulates synthesis of **androgen-binding protein,** which is a unique Sertoli cell product that complexes with high affinity to testosterone, DHT, and estradiol. This protein may serve to concentrate these sex steroids in the Sertoli cells and thereby create a storage form for controlled release during appropriate stages of spermatogenesis. Androgen-binding protein is also secreted into the tubular fluid, where it prevents reabsorption of the sex steroids from the epididymis and thus ensures their availability to the spermatozoa during transit. FSH induces production of binding proteins for iron, copper, and vitamin A. The binding proteins allow these and other substances necessary for spermatogenesis to be more readily extracted

from plasma and transferred to germ cells. Other products of FSH actions on Sertoli cells provide energy sources to the germ cells and facilitate spermiation, the expulsion of spermatozoa into the lumen of the tubule.

Local feedback loops operate within and between the Sertoli, Leydig, and peritubular cells. FSH stimulates inhibin and estradiol production, but inhibin feeds back to block the critical aromatase reaction in estradiol synthesis. Testosterone from the Leydig cells stimulates inhibin secretion by the Sertoli cells, whereas activin and estradiol from the Sertoli cells block testosterone synthesis in the Leydig cells. Testosterone also stimulates differentiation and proliferation of peritubular cells. Protein products from Sertoli and peritubular cells modulate each other's function. The timing of these actions must somehow be coordinated to produce an optimal balance of the molecules that foster spermatogenesis.

Approximately 10% of otherwise normal males are completely or relatively infertile because of anatomical abnormalities (e.g. **varicocoele**), inadequate spermatogenesis, or rejection of their sperm by elements of the female genital tract, including the ova of their partners. The ejaculate may contain no sperm **(azospermia);** a subnormal number of sperm (less than 10,000,000/ml) **(oligospermia);** a high percentage of sperm with reduced mobility; or a high percentage of sperm with immature or abnormal morphology. With azospermia, serum FSH is elevated secondary to loss of negative feedback by inhibin, which is deficient. However, in the other situations basal measurements of plasma gonadotropin and testosterone levels and their responsiveness to GnRH administration are often within normal limits. Yet testicular biopsies can show spermatogenic arrest at various stages from spermatogonia to spermatids, with few normal-appearing spermatozoa. Failure of Sertoli cells to form properly functioning junctional complexes with germ cells is another abnormality that can be seen on testicular biopsy. It is not clear whether hormonal stimulation is at fault (e.g., the timing, frequency, or amplitude of FSH/LH pulses may be abnormal) or whether the production of local paracrine and autocrine regulatory factors may be defective. Even when the sperm count and morphology are normal, other potential causes of male infertility could exist; these include absence of a necessary protein in the sperm (e.g., an acrosomal enzyme or surface binding protein), inadequate contents of prostatic or seminal vesicle secretions, and the presence of antibodies in the female directed at a normal or mutant sperm surface protein.

SECRETION AND METABOLISM OF ANDROGENS

Testosterone, the major androgenic hormone, is synthesized as described in Chapter 48. In adults plasma testosterone levels show small pulses throughout the day that correspond to pulses of LH. Testosterone is in part only a circulating prohormone. Much of androgen action is supplied by the reduction of testosterone to DHT in target tissues. In addition, circulating testosterone and androstenedione are the major sources of systemic estradiol and estrone, respectively, in men. They are produced by aromatization in such sites as adipose tissue and liver. In certain instances, estradiol may even be the actual mediator of an apparent testosterone action. Only 1% to 2% of circulating testosterone is in a free form. Testosterone and DHT circulate mostly bound to a sex steroid–binding globulin (SSBG), which is identical in amino acid sequence to the androgen-binding protein of Sertoli cell origin. The remainder is bound to albumin. Only the free and loosely bound albumin fractions of the androgens are biologically active. Thus SSBG-bound fractions serve as circulating androgen reservoirs, similar to those of thyroid hormone and cortisol. SSBG concentration is itself decreased by androgens and increased by estrogens. Thus, androgens increase their own biological availability by increasing the percentage of available circulating unbound hormone. Most testosterone is metabolized to products oxidized at the 17-position and excreted in the urine; these products constitute 30% of the 17-ketosteroid fraction (see Chapter 46). The rest arises from adrenal sources.

Plasma testosterone levels vary throughout life. As shown in Figure 49-5, the plasma testosterone level rises to adult values in the fetus at the same time as plasma gonadotropins (see Figure 48-3) and when the external genitalia are undergoing differentiation. By birth, however, the levels have declined greatly. After a brief postnatal surge, plasma testosterone and LH fall to low values throughout childhood, and Leydig cells cannot even be identified in the testis. At the age of about 11 years, Leydig cells reappear, and plasma testosterone concentration begins a steep rise to approximately 600 ng/dl at about age 17 (see Figure 49-5). This plateau is sustained for some 50 years. During the late decades of life, the plasma testosterone level gradually declines, this time because the Leydig cells lose their responsiveness to LH stimulation. Because of negative feedback, plasma LH levels rise slowly. Although decreasing testosterone levels may be associated with a decline in libido and sperm production, spermatogenesis still occurs in most octogenarians.

ACTIONS OF ANDROGENS

The effects of androgens on tissues outside the testis can be divided into two major categories: those pertaining to reproductive function and secondary sexual characteristics, and those pertaining to stimulation of somatic growth and maturation.

Testosterone diffuses into target cells, where it is usually reduced to DHT. That testosterone itself can initiate androgen effects is demonstrated by the absence of 5α-reductase activity and DHT production in certain responsive cells. A single receptor binds androgens with a greater affinity for DHT than for testosterone. The androgen receptor complex interacts with DNA molecules (see Figure 40-11), probably assisted by nuclear accessory proteins. This results in stimulation of RNA polymerase, induction of messenger RNAs, and their translation into proteins. Virtually all actions of androgens are blocked by inhibitors of either RNA or protein synthesis. Androgens stimulate growth and differentiation of the epididymis and of the male accessory organs of reproduction. These effects are manifested by hypertrophy and hyperplasia of the epithelial cells, stromal components, and blood vessels.

The major androgen effects, classified according to the probable effector molecule, are shown in Figure 49-6. DHT is specifically required in the fetus for the differentiation of the penis, scrotum, penile urethra, and prostate (Figure 48-6). DHT is required again during puberty for growth of the scrotum and prostate and stimulation of other prostatic secretions. DHT stimulates the hair follicles to produce the typical masculine beard growth, diamond-shaped pubic hair, and recession of the temporal hairline. Growth of the sebaceous glands and their production of sebum also results from DHT action. Testosterone stimulates fetal differentiation of the epididymis, vas deferens, and seminal vesicles (Figure 48-6). During puberty, testosterone and DHT cause enlargement of the penis and seminal vesicles and stimulate the latter to secrete. Although spermatozoa can be produced by adults who secrete testosterone but who lack DHT because of 5α-reductase deficiency, this observation does not preclude participation of DHT normally in spermatogenesis.

Testosterone first stimulates the growth spurt of puberty but ultimately halts linear growth by closure of the epiphyseal growth centers. Estradiol may also participate in these actions in males. The androgen both potentiates growth hormone secretion and synergizes with its actions by stimulating local production of transforming growth factor in osteoblasts. Testosterone causes enlargement of the muscle mass in boys during puberty by increasing the size of muscle fibers. It enlarges the lar-

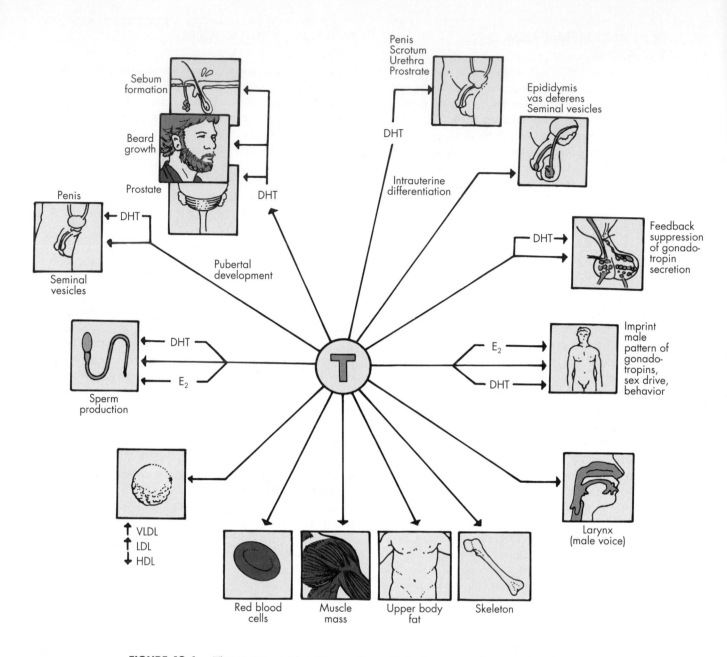

FIGURE 49-6   The spectrum of androgen effects. Note that some effects, such as increase in muscle mass, result from the action of testosterone *(T)* itself. Others such as prostate growth and development are mediated by dihydrotestosterone *(DHT).* The role of estradiol *(E$_2$)* produced from testosterone in certain target tissues such as the testis itself is likely but unproven. *VLDL, LDL,* and *HDL* are very-low-density, low-density, and high-density lipoproteins.

ynx, thickens the vocal cords, and thereby deepens the voice. In adult life, administration of testosterone to either gender causes nitrogen retention, which reflects protein anabolism.

Feedback suppression of gonadotropin secretion is largely an effect of testosterone, because the hypothalamus cannot produce DHT. Circulating DHT and estradiol

generated from testosterone within the hypothalamus may also participate to some extent in negative feedback. Libidinous drives, the ability to sustain an erection, and aggressive behavior are likely fostered by androgens, but they are also influenced by other factors that may partially sustain them when testosterone is deficient.

Androgens also increase red blood cell mass by stimulating erythropoietin synthesis (see Chapter 15) and by a direct effect on maturation of erythroid precursors. Androgens regulate the synthesis of many hepatic proteins; in particular, they decrease all hormone-binding globulins. Importantly, plasma levels of very-low-density lipoproteins are increased, whereas plasma levels of high-density lipoproteins are decreased, by androgen action. This may be partly responsible for the much higher risk of coronary artery disease in men. On the other hand, androgens create a greater bone mass in men than women, a protective effect against osteoporosis.

The clinical consequences of androgen deficiency in a male depend on when the deficiency develops. As already noted, absence of testosterone in utero results in feminization of the external genitalia. If testosterone is deficient during the chronological years of puberty, then enlargement of the penis, growth of the beard and sexual hair, acne, and voice change will not occur and sperm production will not be initiated. Short stature relative to agemates will initially be an additional problem for the child to cope with. However, because the epiphyses remain open, linear growth continues for many years, so that eventually the arms and legs will be disproportionately long. The end result is a relatively tall individual who appears physically to be sexually immature, the so-called **eunuchoidal habitus.** In adults, testosterone deficiency leads to infertility but variable loss of libido and potency. Beard growth diminishes, but it may be partially sustained by adrenal androgens. Osteoporosis and mild anemia are other consequences.

SUMMARY

1. The anatomical arrangement of the testis permits spermatogenesis to occur in a protective and conditioned environment within the spermatogenic tubules behind a blood-testis barrier.
2. Sertoli cells are stimulated by FSH to provide androgen-binding protein, growth factors, inhibin, mineral- and vitamin-binding proteins, and enzymes required for the development, sustenance, and transit of spermatozoa. Sertoli cell cytoplasm forms a conduit within which these processes occur.
3. Leydig cells in the interstitium of the testis are stimulated by LH to secrete testosterone, which in high local concentrations is essential for spermatogenesis.
4. Testosterone and its active product, dihydrotestoster-

one (DHT), are required for pubertal masculinization to occur, including enlargement of the sexual organs, function of accessory glands of reproduction (such as the prostate), growth of the beard, and deepening of the voice.
5. Testosterone also increases muscle mass and linear body growth, but ultimately halts further increase in height by closing the epiphyseal growth centers of bones.

BIBLIOGRAPHY
Journal Articles

De Kretser DM, McLachlan RI, Robertson DM, Wreford NG: Control of spermatogenesis by follicle stimulating hormone and testosterone, *Baillieres Clin Endocrinol Metab*, 6:335, 1992.

Forest MG et al: Kinetics of human chorionic gonadotropin-induced steroidogenic response of the human testis. II. Plasma 17α-hydroxyprogesterone, δ⁴-androstenedione, estrone and 178, β-estradiol: evidence for the action of human chorionic gonadotropin on intermediate enzymes implicated in steroid biosynthesis, *J Clin Endocrinol Metab* 49:284, 1979.

Harman SM et al: Reproductive hormones in aging men. I. Measurement of sex steroids, basal luteinizing hormone, and Leydig cell response to human chorionic gonadotropin, *J Clin Endocrinol Metab* 51:35, 1980.

Johnson MD: Genes related to spermatogenesis: molecular and clinical aspects, *Semin Reproduct Endocrinol* 9:72, 1991.

Kierszenbaum AL: Mammalian spermatogenesis in vivo and in vitro: a partnership of spermatogenic and somatic cell lineages, *Endocr Rev* 15:116, 1994.

Krzanowski, JJ: Regulating spermatogenesis: mechanisms of reproductive hormones, *J Fla Med Assoc*, 80:193, 1993.

Mooradian AD et al: Biological actions of androgens, *Endocr Rev* 8:1, 1987.

Saez JM: Leydig cells: endocrine, paracrine, and autocrine regulation, *Endocr Rev* 15:574, 1994.

SanFilippo S, Imbesi RM: Is the spermatogonium an androgen target cell? An histochemical, immunocytochemical and ultrastructural study in the rat, *Prog Clin Biol Res* 296:177, 1989.

Skinner MK: Cell-cell interactions in the testis, *Endocr Rev* 12:45, 1991.

Spiteri-Grech J, Nieschlag E: The role of growth hormone and insulin-like growth factor I in the regulation of male reproductive function, *Horm Res* 38 (suppl 1):22, 1992.

Weinbauer GF, Nieschlag E: Peptide and steroid regulation of spermatogenesis in primates, *Ann New York Acad Sci* 637:107, 1991.

Books and Monographs

Fawcett DW: Ultrastructure and function of the Sertoli cell. In Hamilton DW, Greep RO, eds: *Handbook of physiology,* section 7, vol 5, Bethesda, Md, 1975, The American Physiological Society.

Griffin JE, Wilson DJ: Disorders of the testes and male reproductive tract. In Wilson DJ and Foster DW, eds: *Textbook of endocrinology,* Philadelphia, 1992, WB Saunders Co.

Steinberger E, Steinberger A: Hormonal control of spermatogenesis. In Degroot LJ et al, eds: *Endocrinology,* vol 3, New York, 1989, Grune & Stratton, Inc.

Veldhuis JD: The hypothalamic-pituitary-testicular axis. In Yen SSC, Jaffe RB, eds: *Reproductive endocrinology,* Philadelphia, 1991, WB Saunders.

Yamamoto M, Turner TT: Epididymis, sperm maturation, and capacitation. In Lipshultz LI, Howards SS, eds: *Infertility in the male,* St Louis, 1991, Mosby.

Female Reproduction

The ovaries, along with the fallopian tubes and uterus, are situated in the pelvis. In adults each ovary weighs approximately 15 g and consists of three zones. The dominant zone is the **cortex,** which is lined by germinal epithelium and contains all the oocytes, each enclosed in a follicle. Follicles in various stages of development and regression are present throughout the cortex (Figure 50-1). The surrounding stroma is composed of connective tissue elements and interstitial cells. The other two zones of the ovary, the **medulla** and the **hilum,** contain scattered steroid-producing cells whose function is unknown. The ovaries and follicle development can be visualized by ultrasonography.

The **granulosa** and **theca cells** of the ovary produce hormones, and other substances act locally, to modulate the development of the ovum and its extrusion from the follicle. These hormones are also secreted into the blood and act on the fallopian tubes, uterus, vagina, breasts, hypothalamus, pituitary gland, adipose tissue, liver, kidney, and bones. Many of these endocrine effects also subserve the process of reproduction. *The fundamental reproductive unit in the ovary is the follicle, which consists of one oocyte surrounded by a cluster of granulosa and theca cells.* When fully developed, the follicle will (1) maintain, nurture, and mature the oocyte and release it at the proper time and (2) provide hormonal support for the fetus until the placenta can assume this function.

BIOLOGY OF OOGENESIS

Oogonia arise from primordial germ cells that migrate to the genital ridge at 5 to 6 weeks of gestation. There, in the developing ovary, they undergo mitosis until 20 to 24 weeks, when the total number of oogonia has reached a maximum of 7 million. Beginning at 8 to 9 weeks and continuing until 6 months after birth, oogonia start into the prophase of meiosis and become pri-

mary oocytes. The latter grow from 10 to 25 μm in diameter when meiosis begins and to 50 to 120 μm at maturity. The process of meiosis is kept suspended in the prophase by an inhibitory hormonal milieu, at least until sexual maturation of the individual, and in some primary oocytes, until menopause. Thus primary oocytes have life spans of up to 50 years.

From the start of oogenesis a process of oocyte attrition occurs so that by birth only 2 million primary oocytes exist, and by the onset of puberty only 400,000 remain. *This constitutes the entire supply of potential ova for the woman's reproductive life, because no new oogonia can be formed.* With continuing attrition, very few oocytes are left when menopause begins, and reproductive capacity ends. This contrasts sharply with the male, in whom the supply of spermatogonia is continually being renewed.

DEVELOPMENT OF THE OVARIAN FOLLICLE

First Stage

The follicle develops in distinct stages. The first stage parallels the prophase of the oocyte and occurs very slowly. It begins in utero and ends at any time during reproductive life. As an oocyte enters meiosis, it induces a single layer of spindle-shaped cells from the stroma to surround it completely. Their cytoplasmic processes attach to the plasma membrane of the oocyte. Simultaneously a membrane called the **basal lamina** forms outside the spindle cells. This delimits the **primordial follicle,** which has a diameter of 25 μm, from the surrounding stroma (Figure 50-1).

At 5 to 6 months of gestation, the spindle-shaped cells in some of the primordial follicles are transformed into cuboidal granulosa cells, thereby forming a **primary follicle.** As these granulosa cells divide and create

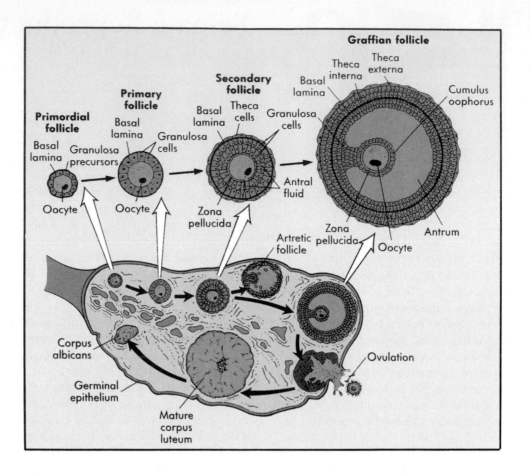

FIGURE 50-1   Schematic representation (not to scale) of the structure of the ovary, showing the various stages in the development of the follicle and its successor structure, the corpus luteum. The follicle grows from a primordial size of 25 μm to an ovulatory size of 10 to 20 mm. The oocyte is shielded from indiscriminate exposure to interstitial fluid contents by the basal lamina and the cytoplasm of the surrounding granulosa cells. The development of primary follicles is gonadotropin-independent. Subsequent stages through ovulation require FSH and LH. The hormones and other constituents of the antral fluid are important local regulators of follicular development. (Redrawn from Ham AW, Leeson TS: *Histology*, ed 4, Philadelphia, 1968, JB Lippincott Co.)

several layers around the oocyte, the complex beomes a **secondary follicle.** The granulosa cells secrete mucopolysaccharides, which form a protective halo, the **zona pellucida,** around the oocyte (Figure 50-1). However, the cytoplasmic processes of the granulosa cells continue to penetrate the zona pellucida and provide nutrients and hormonal signals to the enclosed maturing primary oocytes. The cytoplasm of the granulosa cells also forms a filter through which plasma substances must pass before reaching the germ cell (compare to Sertoli cells and spermatocytes in Figure 49-1).

The secondary follicle grows to a diameter of about 150 μm. Concurrently, a new layer of cells from the stroma is recruited outside the basal lamina and forms the **theca interna.** The granulosa cells now begin to extrude small collections of fluid around and among them.

This completes the first, or preantral, stage and is the maximal degree of follicular development ordinarily found in the prepubertal ovary.

Second Stage

The second stage of follicular development begins after the onset of menstrual cycling. It may require up to 70 to 85 days and span 3 menstrual cycles till completion. Past the midpoint of each cycle, a small number of secondary follicles are recruited for further development. The small collections of follicular fluid coalesce into a single area called the **antrum** (Figure 50-1). The fluid of the antral follicle contains a complex of substances, some secreted by the granulosa and theca cells and some transferred from the plasma through the granulosa cell cytoplasm. Included are mucopolysaccharides, plasma proteins, electrolytes, enzymes of steroid synthesis, steroid hormones, follicle-stimulating hormone (FSH) and luteinizing hormone (LH), inhibin, oxytocin and arginine vasopressin, proopiomelanocortin derivatives, and other granulosa cell products. The steroid hormones reach the antrum by secretion from granulosa cells and by diffusion from theca cells. A nonsteroidal substance (possibly antimüllerian hormone) capable of inhibiting oocyte meiosis is also secreted into the antrum.

As the granulosa cells proliferate, they form a syncytium with electrical and chemical intercommunication. The oocyte is displaced into an eccentric position on a stalk, where it is surrounded by a distinctive layer, the **cumulus oophorus,** which is two to three cells thick (Figure 50-1). The cells of the theca interna also proliferate and are transformed into cuboidal steroid-secreting cells. Additional layers of spindle cells from the stroma form an outer vascularized layer, called the **theca externa.** At the end of this stage the entire complex, called a **preovulatory** or **graafian follicle** (Figure 50-1), has reached an average diameter of 2 to 5 mm.

Third Stage

In the final stage of follicular development one of the graafian follicles is "selected" by day 5 to 7 of a cycle, and it dominates the other second-stage follicles. This dominant follicle now undergoes rapid expansion by cellular growth and augmented production of antral fluid. The colloid osmotic pressure of this fluid increases because of depolymerization of the mucopolysaccharides. However, the total pressure remains unchanged at 16 to 20 mm Hg. The granulosa cells spread apart, the cumulus oophorus loosens, and the vascularity of the theca layers increases greatly. With exponential growth the total size of the **dominant follicle** reaches 10 to 20 mm during the last 48 hours before the midpoint of the cycle, when **ovulation** (release of the ovum) occurs. At a critical point the basal lamina adjacent to the surface of the ovary undergoes proteolysis. The follicle gently ruptures, releasing the oocyte with its adherent cumulus oophorus into the peritoneal cavity. At this point, the initial meiotic division of the oocyte is completed. The resultant secondary oocyte is drawn into the closely approximated fallopian tube. The other daughter cell receives very little cytoplasm. It is called the first polar body and is discarded. In the fallopian tube, sperm penetration causes completion of the second meiotic division, resulting in the haploid (23-chromosome) ovum and a second polar body. The remaining unsuccessful follicles from that cycle undergo **atresia** (described below) within the ovary (Figure 50-1).

CORPUS LUTEUM FORMATION

The residual elements of the ruptured dominant follicle next form a new endocrine unit, the **corpus luteum** (Figure 50-1). *The corpus luteum provides the necessary steroid hormone balance that optimizes conditions for implantation of a fertilized ovum and for subsequent maintenance of the zygote until the placenta is able to do so.* The corpus luteum is made up mainly of granulosa cells. These hypertrophy and form rows, and numerous lipid droplets appear within their cytoplasm. This process, called **luteinization,** begins just before ovulation and is greatly accelerated by the exit of the oocyte from the follicle. The rest of the corpus luteum consists of somewhat luteinized theca cells, arranged in folds along its outer surface. Importantly, the basal lamina disappears, allowing ingrowth of blood vessels that supply the granulosa cells directly.

The corpus luteum regresses after a 14-day life span if conception does not follow ovulation. In this process of regression, known as **luteolysis,** the granulosa and theca cells undergo necrosis, and after the structure is invaded by leukocytes, macrophages, and fibroblasts, the corpus luteum degenerates to an avascular scar (Figure 50-1).

ATRESIA OF FOLLICLES

During an average woman's reproductive life span, only 400 to 500 oocytes (usually one per month) will undergo the complete sequence that culminates in ovulation. The remaining millions disappear in a process called atresia, which begins almost with the appearance of the initial primordial follicles. In first-stage follicles, the oocyte simply becomes necrotic and the granulosa cells degenerate. This accounts for almost all of the oocytes. In second-stage follicles (Figure 50-1), necrosis of the granulosa cells farthest from the oocyte may precipitate a resumption of meiosis in the oocyte to the point of extrusion of the first polar body. However, the granulosa cells in the cumulus oophorus also eventually die, the unsupported oocyte degenerates, and everything inside the basal lamina collapses into a scar. The theca cells dedifferentiate and return to the stroma.

HORMONAL PATTERNS DURING THE MENSTRUAL CYCLE

The menstrual cycle is divided physiologically into three phases (Figure 50-2) that correspond with the dominant events in the monthly development of each ovum. The follicular phase begins with the onset of men-

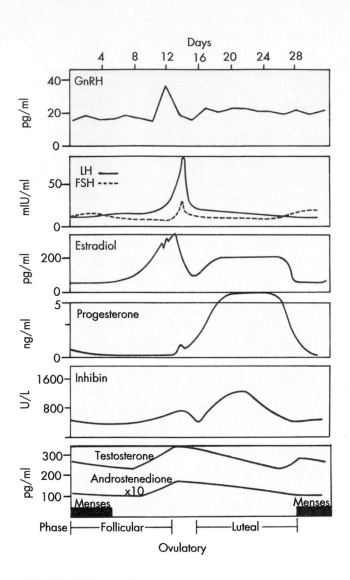

FIGURE 50-2 Profile of plasma hormone levels throughout the menstrual cycle. The dominant follicle is the source of the rising estradiol level in the latter part of the follicular phase. Note ovulatory surges of luteinizing hormone *(LH)* and follicle-stimulating hormone *(FSH)*, preceded by increases in estradiol and gonadotropin-releasing hormone *(GnRH)*. The broad peaks of progesterone and estradiol in the luteal phase result from secretion by the corpus luteum.

strual bleeding and averages 15 days (range, 9 to 23 days). The succeeding **ovulatory phase** lasts only 1 to 3 days. The final **luteal phase** lasts 13 to 14 days and ends with the onset of menstrual bleeding. A normal menstrual cycle may range from 21 to 35 days, depending mainly on the length of the follicular phase.

Normal reproductive function is characterized by a series of cyclic changes in ovarian steroid hormone and inhibin production, consequent to cyclic changes in pituitary LH and FSH secretion and in hypothalamic gonadotropin-releasing hormone (GnRH) pulses. Both

negative and positive feedback loops are involved in creating this complex pattern.

The critical regulators of the ovarian cycle are FSH and LH. Just before the start of the follicular phase, plasma FSH and LH concentrations are at their lowest levels (Figure 50-2), and the LH/FSH ratio is slightly greater than 1. The FSH level begins to rise gradually 1 day before menses begins, and it continues to do so through the first half of the follicular phase. The level of LH rises later. Then, during the second half of the follicular phase, the FSH level falls slightly, whereas the LH level continues to rise so that the LH/FSH ratio reaches about 2. Stimulated by the early rise in FSH, the plasma estradiol concentration also increases gradually during the critical first 6 to 8 days. Later in the follicular phase the plasma estradiol level increases much more sharply and reaches a peak just before the ovulatory phase (Figure 50-2). *This estradiol is secreted by the granulosa cells of the dominant follicle.* The higher estradiol level, along with increased ovarian secretion of inhibin, then feeds back to decrease the plasma FSH concentration during the second half of the follicular phase. LH levels continue to rise slowly, as do androgens produced by theca cells.

The succeeding ovulatory phase is uniquely characterized by a very large but transient spike in the plasma LH level, with a lesser spike in the FSH level (Figure 50-2). This surge in gonadotropin is preceded first by the "sawtooth" estradiol peak of the late follicular phase and then by an increase in GnRH pulses (Figure 50-2). At the same time, the plasma progesterone level rises slightly. Together these changes suggest that both the ovary and the hypothalamus contribute to the ovulatory surge of LH and FSH.

After ovulation, negative feedback from the corpus luteum causes the LH and FSH levels to decline during the luteal phase to reach their nadirs toward its end (Figure 50-2). The pulse frequency of gonadotropin secretion is also diminished. The most distinctive feature of the luteal phase is a tenfold increase in plasma progesterone of corpus luteum origin. Estradiol, also secreted by the corpus luteum, increases again. If pregnancy does not occur and the corpus luteum degenerates, progesterone and estradiol levels decrease dramatically to their lowest levels at the end of the luteal phase, FSH secretion increases, and menstrual bleeding starts.

HORMONAL REGULATION OF OOGENESIS

Primary Follicle Formation

The initial growth of the primordial follicle appears to be a local phenomenon that is independent of gonadotropins and one in which factors from the oocyte stimu-

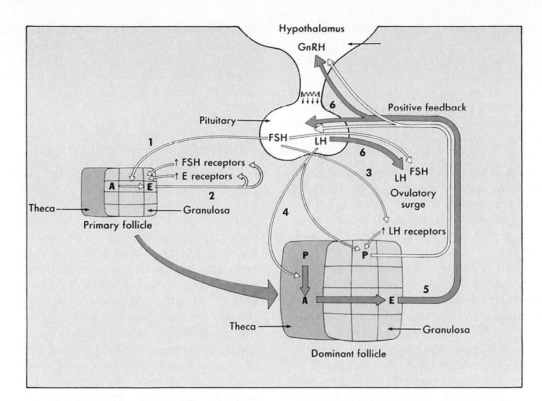

FIGURE 50-3   Hormonal regulation of follicular development. *1*, FSH stimulates granulosa cell growth and estradiol *(E)* synthesis in a small cohort of primary follicles. *2*, The local estradiol increases its own receptors and FSH receptors, amplifying both hormones' effects. Thus, a self-propelling mechanism is set into motion. *3*, FSH later increases LH receptors, initiating granulosa responsiveness to LH. *4*, LH stimulates theca cell growth and androgen *(A)* production. Androgen is then converted to estradiol in the granulosa cells. LH also stimulates progesterone *(P)* production in the granulosa cells and adds to the FSH effects by increasing cyclic AMP levels. *5*, As a result of theca and granulosa cell synergism, the dominant follicle emerges as a very efficient secretor of estradiol. *6*, Rising estradiol, with late potentiation by progesterone, feeds back positively on the pituitary gland and hypothalamus to evoke the preovulatory surge of LH and FSH.

late granulosa cell development. In turn, granulosa cell products initiate formation of the theca and then stop maturation of the oocyte once it reaches 80 μm in diameter. The transient surge of FSH and LH in midgestation and even the low levels of gonadotropins secreted during childhood are necessary for an adequate rate of follicular growth throughout the rest of life. Nonetheless, the first stage, from primordial to primary follicle, continues to occur until menopause, apparently independent of the presence or of the state of reproductive cycling.

Follicular Development

The second stage of follicular development begins in the early luteal phase of one cycle and continues gradually till the late luteal phase two cycles later. A small number of follicles that have acquired FSH and LH receptors are recruited and slowly grow over this period. From each such group of approximately 20, a single dominant follicle is "selected" by the fifth to seventh day of the ensuing follicular phase, usually in only one ovary each month. The early follicular phase rise in FSH levels stimulates a much more rapid rate of growth of the granulosa cells and of aromatase activity in that particular follicle.

The result is that estradiol synthesis from androgens is markedly enhanced (Figure 50-3). The high local estradiol concentration then induces increases in its own receptors as well as in FSH receptors. This further sensitizes the granulosa cells to both hormones, and even more estradiol is produced. In addition, granulosa cell hypertrophy and hyperplasia take place, mediated in part by local generation of insulin-like growth factors. Thus, *once started, second-stage follicular development becomes a self-propelling mechanism that combines endocrine, autocrine, and paracrine effects and that requires fine coordination between the pituitary gland*

and the ovary. The outcome is exponential follicular growth.

Two further actions contribute to continuing follicular development (Figure 50-3):

1. FSH induces LH receptors on the granulosa cells.
2. The slowly rising plasma estradiol level conditions the GnRH-gonadotropin axis so as to decrease FSH secretion but still to permit a slight increase in LH secretion.

Pituitary stores of LH are also built up, thereby creating a supply for the coming surge of LH in the ovulatory phase. The estradiol effect partly occurs within the gonadotroph cells and is partly mediated by interaction with dopaminergic and endorphinergic neurons that inhibit GnRH release.

The unique role of LH in the second half of the follicular phase is to stimulate the theca cells to produce increasing amounts of androstenedione and testosterone. These androgens diffuse across the basal lamina into the granulosa cells, where they serve as essential precursors to estradiol (see Figure 50-1). In addition, LH stimulates more granulosa cell production of progesterone. The maturation of the dominant follicle depends on a complex set of interactions between its component theca and granulosa layers (see Figure 50-3). These interactions underlie the critical goal: a high enough output of estradiol to trigger ovulation. The FSH-specific granulosa cells are greatly dependent on the LH-specific thecal cells for a sufficient supply of androgens to be converted to estrogens. This reaction is catalyzed by the action of aromatase, which is up-regulated by FSH-stimulated cAMP levels. The FSH-induced recruitment of LH receptors on granulosa cells also allows LH to augment cAMP levels and thus also to contribute to estradiol production directly in those cells as well. Another granulosa cell product that is induced by FSH, namely inhibin, stimulates androgen production by the theca cells. In turn, androgens stimulate inhibin production by granulosa cells. This local positive feedback loop, as well as other paracrine effects, contributes to the striking momentum of the dominant follicle and its production of estradiol.

FSH also stimulates enhanced graulosa cell production of trace metal and vitamin-binding proteins, protein growth factors such as IGF-1, and substrates for energy generation by germ cells. Molecules involved in the mechanism of ovulation (such as plasminogen activator) are also increased by FSH during the preovulatory period.

Both ovaries presumably receive similar inputs of FSH and LH. Therefore *the emergence of one dominant follicle in each cycle may result from its possessing more FSH receptors and greater aromatase activity at the outset.* Such characteristics would permit this particular follicle to exceed the others in early estradiol produc-

tion. Conversely, the other second-stage follicles undergoing atresia have relatively low ratios of estradiol to androgen concentrations in the antral fluid. This likely reflects declining FSH availability as FSH secretion is reduced by estradiol and inhibin released by the dominant follicle.

Ovulation

The ovulatory surge of LH and FSH is triggered by a positive feedback effect of estradiol (see Figure 50-3). A critical plasma estradiol level of at least 200 pg/ml, sustained for at least 2 preceding days, is required. The proportionally smaller preovulatory increase in plasma progesterone (see Figure 50-3) synergizes with estradiol to amplify and prolong the gonadotropin surge. This positive feedback effect takes place at both pituitary and hypothalamic levels. Estradiol and progesterone augment the flow of GnRH pulses from the hypothalamus to the pituitary gland. The pituitary gland, appropriately primed by the preceding pattern of ovarian steroid exposure, now responds to these repetitive GnRH pulses in exaggerated fashion. Furthermore, the secreted LH molecules are more biologically active.

Thus, the hypothalamic-pituitary unit is conditioned by ovarian steroids to provide a sudden increase in gonadotropin (mainly LH) stimulation of the dominant follicle. This triggers ovulation 12 hours later by a multicomponent mechanism:

1. LH neutralizes the action of a peptide oocyte maturation inhibitor, allowing completion of meiosis. At the same time, further replication of granulosa cells is halted.
2. LH stimulates progesterone synthesis; the increased progesterone augments proteolytic enzyme activity, which loosens the wall and increases distensibility of the follicle.
3. A pseudoinflammatory response ensues, characterized by local synthesis of prostaglandins, leukotrienes, and thromboxanes, some of which are required for follicular rupture.
4. FSH stimulates production of glycosaminoglycans, which mucify the environment and disperse the cumulus oophorus. FSH also induces proteolytic enzymes, which catalyze final breakdown of the follicular wall.
5. Immediately after ovulation, there is a rapid fall in estradiol production, which also contributes to loss of integrity of the follicle.

Corpus Luteum Function

The development and functioning of the corpus luteum are also under hormonal control. The ovulatory LH surge stimulates the luteinization of the granulosa cells. Lower levels of LH secreted in appropriate pulses can then maintain a very high rate of progesterone production by the corpus luteum, as well as a substantial rate of estradiol production (see Figure 50-2). Exposure to proper amounts of FSH in the preceding follicular phase ensures the presence of sufficient corpus luteum receptors for LH action. The vascular ingrowth into the corpus luteum is also important to deliver the LH and cholesterol necessary to sustain progesterone secretion. The lower levels of FSH and LH during the luteal phase withdraw support from the other follicles of this cohort and hasten their atresia.

If pregnancy does not ensue and the declining LH levels of the late luteal phase are not replaced by the equivalent placental hormone, **human chorionic gonadotropin** (HCG), the corpus luteum begins to regress after the eighth postovulatory day and its secretion of progesterone and estradiol ceases completely by the fourteenth day. By then, corpus luteum secretion has fallen low enough to release the pituitary gland from feedback inhibition and allow the FSH rise of the next cycle to begin. The process of luteolysis is partly mediated by prostaglandins.

Extraordinary coordination between the various elements of the female hypothalamic-pituitary-ovarian axis is required for ovulation and conception. This creates numerous possiblilites for failure, and infertility arising from dysfunction of this system is common. Disease or conditions that disrupt GnRH release or impair the gonadotroph responsiveness will prevent the necessary initial FSH pattern to recruit a dominant follicle and may result in complete loss of menses (**amenorrhea**). A dominant follicle may produce enough estrogen for uterine bleeding to occur (see below) but not enough to induce a midcycle peak of LH; this causes **anovulatory cycles.** On the other hand, an elevated ratio of LH to FSH in the follicular phase is associated with excessive theca cell production of androgens and the formation of numerous atretic and cystic follicles, the **polycystic ovary syndrome.** Even if ovulation occurs, an **inadequate luteal phase,** either too short or substandard in progesterone production, may lead to poor preparation of the reproductive tract for either fertilization or implantation. Various manipulative medical therapies are available for female infertility, in contrast to the situation in men. For example, the drug **clomiphene** is an estrogen receptor antagonist that acts in the hypothalamus. By simulating estrogen deficiency and producing negative feedback,

clomiphene produces an increase in GnRH and gonadotropin secretion in women with a hypothalamic origin of infertility. Alternatively, endogenous pituitary function can be suppressed with a long-acting GnRH superagonist and ovulation can be induced by carefully timed doses of exogenous FSH and LH.

Origin of the Menstrual Cycle

Substantial evidence supports the thesis that, in humans, *the monthly cycle of the LH/FSH surge and consequent ovulation is mainly a self-recruited ovarian rhythm rather than the result of an inherent rhythm generated within the central nervous system.* No cycle of LH/FSH release is observed in the absence of functional ovaries. The gonadotropin surge only occurs when the dominant follicle has reached the receptive preovulatory stage of development, irrespective of the number of days required for this to occur. Estradiol itself, administered in a proper fashion, can induce an LH surge. Finally, if the pituitary gland is severed from the hypothalamus and GnRH pulses of appropriate frequency and amplitude are provided externally to the pituitary gland in a fixed pattern, a preovulatory surge of LH and ovulation occur without abruptly altering the profile of the GnRH input.

The close coordination between the emergence of a single dominant follicle and the ovulatory signal it recruits makes multiple pregnancies unlikely in humans. For example, the natural rate of occurrence of **dizygotic twins** is less than 1% of live births. This is further emphasized by the much higher rate (15%) of multiple ova produced during cycles in which follicular development and ovulation are produced artificially "from above" by superimposed profiles of stimulation with exogenous FSH and LH. Multiple pregnancies are also more common (5%) when endogenous FSH and LH are released in response to clomiphene administered on the fifth day of the cycle to infertile women.

The ovarian signals to induce ovulation can be overridden by other influences. Loss of cyclic gonadotropin secretion can occur with caloric deprivation, habitual strenuous exercise, stress, and emotional disturbance such as depression. The inhibitory influences on GnRH and/or LH and FSH secretion may be mediated by endorphins, dopamine, or corticotropin-releasing hormone (CRH), and in some instances by changing the levels of cortisol, androgens, or thyroid hormone.

> Well-known examples of anovulation or even complete amenorrhea occur in women with **anorexia nervosa,** in ballet dancers, or in marathon runners. The ovarian dysfunction can be so serious as to cause profound estrogen deficiency with consequent **osteoporosis.**

HORMONAL REGULATION OF REPRODUCTIVE TRACT FUNCTION

The cyclic changes in ovarian estradiol and progesterone secretion affect all the reproductive tract tissues involved in conception.

Fallopian Tubes

Fertilization normally occurs in the fallopian tubes. Each tube ends in fingerlike projections called **fimbriae,** which lie close to the adjacent ovary. The tubes consist of a muscular layer enclosing an epithelial lining that contains secretory and ciliated cells. During the follicular phase of the cycle, estradiol increases the number of cilia and their rate of beating. At ovulation the fimbriae undulate so as to draw the shed ovum into the tube, and tubal contractions move the ovum toward incoming sperm. During the luteal phase, progesterone maximizes this ciliary beat, facilitating movement of any fertilized ovum toward the uterus. Estradiol and progesterone also regulate tubal secretion of mucoid fluids, ions, and substrates that facilitate movement of the ovum and sperm and help sustain a zygote.

> When the fallopian tubes are dysfunctional or blocked by disease, infertility usually results. Conversely, they can sustain a conceptus for some time within the lumen if a trapped zygote attaches to the epithelial lining. Such **ectopic pregnancies** are very dangerous because of the risk of rupture of the tube, with accompanying hemorrhage.

Uterus

The function of the uterus is to house and nurture the developing conceptus and ultimately to evacuate the mature fetus. This muscular organ encloses a cavity that is lined with a mucous membrane, called the **endometrium.** At the start of each menstrual cycle, the endometrium is thin and its glands are sparse and straight, with a narrow lumen (Figure 50-4); it exhibits few mitoses and is incapable of receiving a conceptus. After menstruation has ceased, the rise in plasma estradiol concentration during the follicular phase increases endometrial thickness threefold to fivefold. Mitoses appear in the glands and stroma, the glands become tortuous, and the spiral arteries that supply the endometrium elongate. This is the characteristic appearance of the **proliferative phase** of the endometrium. Estradiol also changes the mucus elaborated by the cervix (the opening to the uterus) from a scant, very viscous material to a copious, more watery but more elastic substance. This mucus can be stretched into a long, fine thread, and it produces a characteristic fernlike pattern when dried. Such cervical mucus creates channels that facilitate entrance of the sperm into the uterine cavity.

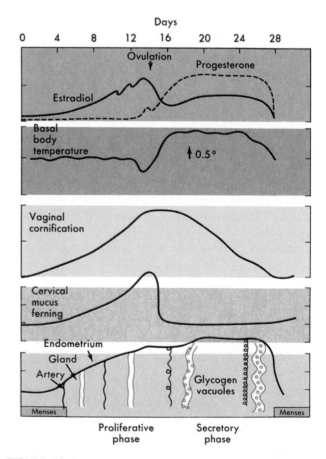

FIGURE 50-4   Correlation of changes in body temperature, vaginal cytology, and endometrial structure and function with the profiles of plasma estradiol and progesterone concentrations. (Redrawn from Odell WD: The reproductive system in women. In DeGroot LJ et al, editors: *Endocrinology,* vol 3, New York, 1989, Grune & Stratton, Inc.)

Shortly after a woman ovulates, the rise in plasma progesterone concentration greatly alters the endometrium and produces the characteristic appearance of the **secretory phase** (Figure 50-4). The rapid growth and mitotic activity of the endometrium are inhibited. The glands become much more tortuous and accumulate glycogen. As the luteal phase of the cycle progresses, the glycogen vacuoles move from the base toward the lumen, and the glands greatly increase their secretion. The stroma of the endometrium becomes edematous; the spiral arteries elongate further and coil. These changes enable the endometrium to accept, implant, and nourish a conceptus. At the same time, progesterone decreases the quantity of cervical mucus and returns it to its thick, non-elastic, and nonferning state.

If conception does not occur, the abrupt loss of progesterone and estradiol from the corpus luteum causes spasmodic contractions of the spiral arteries, mediated by increased prostaglandins. The resultant loss of blood supply produces tissue death, and the superficial endometrial cells are shed along with clotted blood. This constitutes the **menstrual flow.**

Vagina

The vaginal canal is lined with a stratified squamous epithelium that is highly sensitive to estradiol. In its absence, only a layer of basal cells is present. As estradiol levels rise during the follicular phase, more layers of epithelium are added, and the maturing vaginal cells accumulate glycogen. They become large and cornified, and their nuclei shrink or disappear. The percentage of such cells is a quantitative index of estrogenic activity (Figure 50-4). In the luteal phase progesterone reduces the percentage of cornified cells. Vaginal secretions that enhance the prospect for fertilization are also increased by estradiol.

Reproductive Sexual Functioning

Several processes combine to accomplish acceptance and inward transmission of sperm. In some women the desire for sexual activity is increased just before ovulation by the midcycle rise in plasma androgens (see Figure 50-2). With sexual stimulation, vascular erectile tissue beneath the clitoris is activated by parasympathetic impulses. During heterosexual intercourse, this causes the vagina to be tightened around the penis. Simultaneously the glands beneath the labia and in the vaginal entrance secrete copious amounts of mucus. The secretions lubricate the vagina and help it produce a massaging effect on the penis. These glands are maintained by estradiol action. Orgasm results from spinal cord reflexes that are similar to those involved in male ejaculation. Orgasm consists of involuntary contractions of the skeletal muscle of the perineum; of the musculature of the vagina, uterus, and fallopian tubes; and of the rectal sphincter.

Many spermatozoa are trapped and within a few hours are destroyed in the vagina. The remainder reach the cervix, where they dwell in storage crypts formed by the estrogen-stimulated convoluted mucosa and its mucus. From this reservoir, capacitated spermatozoa migrate into the uterine cavity and fallopian tubes over 24 to 48 hours. Of these, as few as 50 to 100 spermatozoa eventually reach an ovum, but they are sufficient for fertilization.

Breasts

The mammary glands consist of lobular ducts lined by an epithelium capable of secreting milk. These ducts empty into larger conduits that converge at the nipple. The glandular structures are embedded in supporting adipose and connective tissue. Before puberty, the breasts grow only in proportion to the rest of the body. The development of adult breasts depends on estradiol, but progesterone, insulin, growth hormone, IGF-1, cortisol, epidermal growth factor, and prolactin have synergistic effects. After puberty, estradiol stimulates growth of the lobular ducts in the area around the nipple. Estradiol also selectively increases the adipose tissue, giving the breast its distinctive female shape. Progesterone stimulates outpouching of the lobular ducts to form numerous alveoli capable of milk secretion. Cyclic changes in the breast occur in conjuction with the fluctuations in estradiol and progesterone levels during the menstrual cycle.

Ovarian Steroid Effects on Other Tissue

During puberty, estradiol is to the female what testosterone is to the male. Estradiol causes almost all the somatic changes that result in the female adult appearance. In addition to stimulating growth of the internal reproductive organs and breasts, estrogens cause pubertal enlargement of the labia majora and labia minora. Linear body growth is accelerated by estradiol; however, because the epiphyseal growth centers are more sensitive to estradiol than to testosterone, they close sooner. For this reason, the average height of women is less than that of men. The hips enlarge and the pelvic inlet widens, facilitating future accommodation of pregnancy. The predominance of estradiol over testosterone is responsible for the total weight of adipose mass of women being twice as great as that of men, whereas muscle and bone mass are only two-thirds that of men.

The adult skeleton, the kidney, and the liver are also target tissues of estrogens. Estradiol inhibits bone resorption; loss of this important action can contribute to a declining bone mass and an increased fracture rate. Estradiol stimulates reabsorption of sodium from the renal tubules, and this may contribute to cyclic fluid retention. Estradiol increases hepatic synthesis of binding proteins for thyroid and steroid hormones, of the renin substrate angiotensinogen, of clotting factors, and of very-low-density lipoproteins. The latter actions can lead to hypertension, venous thrombosis, and hyperlipidemia in estrogen-treated women.

Progesterone produces the 0.5 C° rise in body temperature that occurs shortly after ovulation (see Figure 50-4). Central nervous system actions of progesterone include an increase in appetite, a decrease in wakefulness, and a heightened sensitivity of the respiratory center to carbon dioxide.

MECHANISMS OF ACTION OF OVARIAN STEROIDS

Estrogens and progesterone enter cells freely and bind to cytoplasmic/nuclear receptors of the superfamily described in Chapter 40. The sex steroid–receptor complex undergoes an activation step that enhances its binding to specific DNA molecules. The receptors can also be phosphorylated by a protein kinase dependent on cyclic adenosine monophosphate (cAMP), and can dimerize, thereby increasing their binding activity. By stimulating transcription of the respective genes, estradiol and progesterone increase the synthesis of numerous proteins that have reproductive functions.

Because spare receptors generally are not present, the responsiveness of various tissues to ovarian steroids is proportional to receptor concentration. Estradiol and progesterone can fortify or inhibit each other's actions through receptor recruitment. Estradiol increases its own receptor and that of progesterone in the uterus during the latter part of the proliferative phase. Conversely, progesterone decreases estradiol receptors and therefore estrogen action on the endometrium during the secretory phase.

METABOLISM OF OVARIAN STEROIDS

Estradiol and estrone bind to sex steroid–binding globulin, but their affinities are lower than that of testosterone. They also circulate largely bound to albumin, and this fraction, along with the free steroids, is biologically active. In women who menstruate, most of the circulating estrogen is estradiol from the ovaries. Estrogens are excreted in the urine as sulfate and glucuronate conjugates. Progesterone circulates largely bound to albumin. It is reduced to pregnanediol and excreted in the urine.

FEMALE PUBERTY

The general process of initiation of puberty is described in Chapter 48. Reproductive function begins after gonadotropin secretion increases from the low levels of childhood (Figure 50-5). Budding of the breasts is the first physical sign of puberty and coincides with the initial increase in plasma estradiol concentration. The onset of menses occurs approximately 2 years later, at 11 to 15 years of age, after LH levels have risen more sharply. This timing may depend on achieving a critical body weight, a critical ratio of adipose mass to lean body mass, and a certain level of skeletal maturation.

The positive feedback effect of estradiol on gonadotropin secretion is the last step in the maturation of the hypothalamic-pituitary-ovarian axis; thus ovulation usually does not occur in the first few menstrual cycles. These are irregular in length because the bleeding is induced by withdrawal of estrogen secretion from graafian follicles undergoing atresia.

The growth spurt and the peak velocity of growth are characteristically attained earlier in girls than in boys. Height increase usually stops 1 to 2 years after the onset of menses. The development of pubic hair precedes menses and correlates best with rising levels of adrenal dehydroepiandrosterone sulfate (DHEA-S). The time of onset of puberty is influenced by race and individual heredity, and occurs earlier in tropical zones.

MENOPAUSE

The reproductive capacity of women wanes usually in the fifth decade of life, and menses terminate at an average age of 50. For several years before, the frequency of ovulation decreases. The menses occur at variable intervals, and the decreased menstrual flow is caused by irregular peaks of estradiol secretion and inadequate secretion of progesterone in the luteal phase. With the disappearance of almost all follicles, ovarian secretion of estradiol virtually ceases and estrone produced from theca cell and adrenal androgens becomes the predominant estrogen.

As menopause approaches, follicular sensitivity to gonadotropin stimulation diminishes, and plasma FSH and LH levels gradually increase. Once menopause occurs, loss of negative feedback from estradiol and inhibin increase plasma gonadotropins to levels 4 to 10 times those that are characteristic of the follicular phase, and FSH exceeds LH levels (see Figure 48-3). Although the cycle of gonadotropin secretion is lost, pulsatility persists.

The manifestions of ovarian insufficiency, most particularly of estradiol deficiency, depend on the stage of female life. **Intrauterine estradiol deficiency**—even caused by complete absence of the ovaries—does not prevent expression of the basic feminine phenotype (see Chapter 48), though the external genitalia may appear somewhat undersized. During puberty, estrogen deficiency causes lack of breast development and of menses. The uterus and ovaries remain infantile in size. In an **XX individual,** instead of a growth spurt, there is slow but prolonged growth until the epiphyses close late.

In adult women whose reproductive function is terminated early by disease and in normal menopause, estrogen deficiency causes thinning of the vaginal epithelium, loss of its secretions, and discomfort during intercourse. A decrease in breast mass and thinning of the skin also occur. Vascular flushing and emotional lability are disturbing symptoms. Of great importance is a sharp increase in the incidence of coronary artery disease. In women with relatively low bone mass caused by other factors such as poor calcium intake in earlier life, accelerated further bone loss from estrogen deficiency causes **osteoporosis,** with fractures of the wrist, spine, and hips.

PREGNANCY

Fertilization

After the ovum enters the widened proximal end of the fallopian tube (ampulla), it is transported down to the junction with the isthmus. There it must encounter sperm within 12 to 24 hours for fertilization to occur. The sperm in turn must reach the ovum within 48 hours after entering the vagina. Contact between the sperm and ovum is facilitated by a mixing motion of the fallopian tube. Access of the sperm to the ovum begins with dispersal of the granulosa cells of the cumulus oophorus (see Figure 50-1). Dispersal is achieved through the action of **hyaluronidase** and a **corona-dispersing enzyme,** both of which are contained in the acrosomal cap of the sperm (see Figure 49-2). The underlying zona pellucida of the ovum (see Figure 50-1) contains species-specific receptors for sperm. The single fertilizing sperm penetrates this barrier by releasing acrosin, a proteolytic enzyme. Penetration then releases materials contained in granules within the ovum, which block entrance of other sperm. This prevents **polyploidy,** which is the production of an individual with more than two sets of homologous chromosomes. The polar body resulting from the second reduction division is then ejected from the ovum, leaving a female pronucleus with 23 chromosomes. After fusion of their respective membranes, the DNA of the sperm head is engulfed by the ovum and forms the male pronucleus with 23 chromosomes. The two pronuclei then generate a spindle on which the chromosomes are arranged, and a zygote with 46 chromosomes is created.

Implantation

The zygote develops into a **blastocyst,** which traverses the fallopian tube in about 3 days. Within another 2 or 3 days, implantation in the uterus begins. *Implantation consists of three consecutive processes: adhesion, penetration and invasion.* The requisite dissolution of the zona pellucida is initiated by alternate contraction and expansion of the blastocyst, as well as by the action of lytic substances in the uterine secretions. These and other maternal factors necessary for implantation depend on adequate maternal levels of progesterone and on paracrine signals from the zygote. From the initial solid mass of cells, a layer of **trophoblasts** separates. Microvilli of these cells interdigitate with those of endometrial cells, and junctional complexes form between the cell membranes. Adhesion is aided by a variety of endometrially produced molecules, such as laminin and fibronectin. Once firmly attached, trophoblasts penetrate between and beneath endometrial cells, lyse the intercellular matrix with a variety of enzymes, and phagocytize and digest dead endometrial cells.

The depth of penetration by the trophoblasts is limited by changes in the endometrium. Late in the luteal phase, uterine stromal cells, stimulated by progesterone, enlarge and accumulate glycogen and lipid. Now called **decidual cells,** they disappear unless pregnancy supervenes and the corpus luteum is maintained. In the latter case, continuing progesterone and estrogen stimulation rapidly changes the entire stroma into a sheet of decidual cells. This **decidua** functions initially as a source of nutrients for the embryo, until trophoblastic invasion establishes vascular connections between the fetus and the mother. Thereafter, the decidua provides a mechanical and an immunological barrier to further invasion of the uterine wall. The decidua also secretes prolactin, relaxin, prostaglandins, and other molecules with paracrine effects on the uterine muscle and on the fetal membranes (**chorion** and **amnion).**

Implantation is even more susceptible to mishap than is conception. Approximately 70% of all conceptions result in **miscarriage.** The majority occur within 14 days and

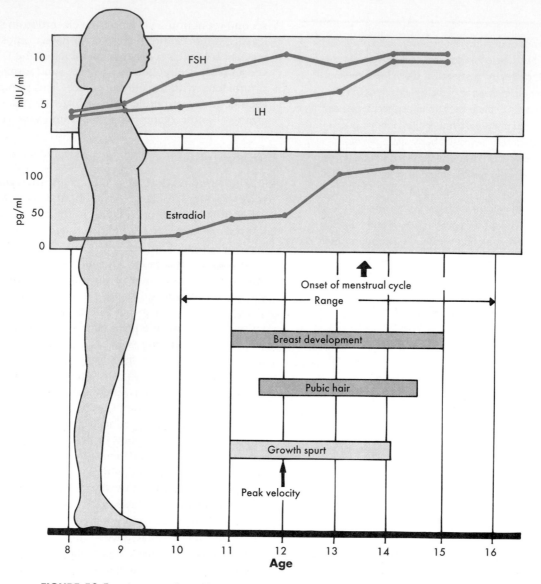

FIGURE 50-5 Average chronologic sequence of hormonal and biological events in female puberty. The growth spurt starts earlier and is shorter than in males. (Redrawn from Lee PA et al: *J Clin Endocrinol Metab* 43:775, 1976, and Marshall WA, Tanner JM: *Arch Dis Child* 45:13, 1970.)

are unrecognized by the woman, who may only have a slightly delayed menstrual period. Miscarriages later in the first trimester may still reflect suboptimal maternal-fetal attachment, but are also caused by fetal anomalies.

Functions of the Placenta

Pregnancy is marked by the development of a unique organ, the **placenta,** which has a limited life span. This organ serves as the fetal gut and nutrient supply, as the fetal lung in exchanging oxygen and carbon dioxide, and as the fetal kidney in regulating fluid volumes and disposing of waste metabolites. In addition, the placenta is an extraordinarily versatile endocrine gland, capable of synthesizing and secreting numerous protein and steroid hormones that affect maternal and fetal metabolism. These hormones can be found in fetal plasma and amniotic fluid and exhibit characteristic concentration profiles in maternal plasma (Figure 50-6).

Trophoblasts differentiate into an inner layer of **cytotrophoblasts** and an outer layer of **syncytiocytotrophoblasts,** which are fused. The inner cytotrophoblasts secrete hypothalamic-like stimulatory and inhibitory

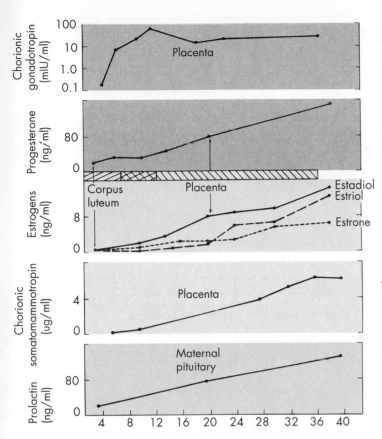

FIGURE 50-6 Profile of maternal plasma hormone changes in human pregnancy. Note logarithmic scale for human chorionic gonadotropin (HCG) in top panel. Between 6 and 12 weeks, the source of estrogens and progesterone shifts from the corpus luteum to the placenta. The latter also secretes large amounts of chorionic somatomammotropin. The maternal pituitary contributes the prolactin excess. (Redrawn from Goldstein DP et al: *Am J Obstet Gynecol* 102:110, 1968; Rigg LA et al: *Am J Obstet Gynecol* 129:454, 1977; Selenkow HA et al: In Pecile A, Frinzi C: *The foetoplacental unit*, Amsterdam, 1969, Excepta Medica; and Tulchinski D et al: *Am J Obstet Gynecol* 112:1095, 1972.)

peptides, such as corticotropin-releasing hormone, which likely regulate in paracrine manner the secretion of pituitary-like peptides, such as ACTH, by the outer syncytiocytotrophoblasts. The latter also secrete increasingly large amounts of sex steroid hormones as pregnancy progresses.

Human Chorionic Gonadotropin This hormone (HCG) is the first key hormone of pregnancy. Secreted by the placental syncytiocytotrophoblast cells, HCG can be detected in maternal plasma and urine within 9 days of conception and serves as a reliable pregnancy test. HCG is a glycoprotein with two subunits. The α-subunit is identical to that of thyroid-stimulating hormone (TSH), FSH, and LH. The β-subunit is closely homologous to that of LH, and the two hormones have indistinguishable biological actions. Plasma HCG concentration increases at

an exponential rate, reaches a peak at 9 to 12 weeks of gestation, and then declines to a stable plateau for the remainder of pregnancy (Figure 50-6).

HCG maintains the function of the corpus luteum, which would otherwise degenerate in the absence of pregnancy. It stimulates the corpus luteum to secrete progesterone and estradiol by mechanisms identical to those of LH. Later, when the placenta itself synthesizes these steroids in adequate amounts, HCG secretion declines and the corpus luteum regresses. HCG also stimulates essential DHEA-S production by the fetal zone of the adrenal gland (see following discussion). In males HCG stimulates the early secretion of testosterone by the Leydig cells, which is critical to masculine genital tract differentiation.

Progesterone *Progesterone is essential for successful implantation, initial sustenance, and long-term maintenance of the fetus.* It stimulates the endometrial glands to secrete nutrients on which the early zygote depends. Thereafter, progesterone maintains the decidual lining of the uterus, where it induces prolactin synthesis. The latter helps inhibit maternal immune responses to fetal male parental antigens and thus helps prevent rejection of the fetus. Progesterone transferred to the fetus is the substrate for synthesis of cortisol and aldosterone by the fetal adrenal cortex (Figure 50-7). The latter cannot itself synthesize progesterone because it lacks 3-β-OL dehydrogenase $\delta^{4,5}$ isomerase activity (see Figure 46-3).

Progesterone quiets uterine muscle activity and prevents premature expulsion of the fetus. Also, it stimulates mammary gland development and greatly enhances the eventual capacity to secrete milk. Finally, progesterone increases maternal ventilation, which is needed for the removal of the increased load of carbon dioxide created by pregnancy.

The placenta begins to synthesize progesterone at about 6 weeks, and by 12 weeks it produces enough to replace the corpus luteum source (see Figure 50-6). Cholesterol extracted from maternal plasma serves as the major precursor for placental progesterone. The synthetic pathway is identical to that of the adrenal gland and ovary. By term, placental progesterone production reaches a level that is tenfold greater than peak production by the corpus luteum.

Estrogens Progressive increases in estradiol, estrone, and estriol occur throughout pregnancy. Estrogens stimulate continuous growth of the uterine muscles necessary for labor. They foster relaxation and softening of the pelvic ligaments and junction of the pelvic bones; this allows better accommodation of the expanding uterus. In addition, estrogens augment growth of the ductal system of the breast to prepare for lactation.

Estrogens are initially produced by the corpus luteum. The placenta subsequently assumes this role, but be-

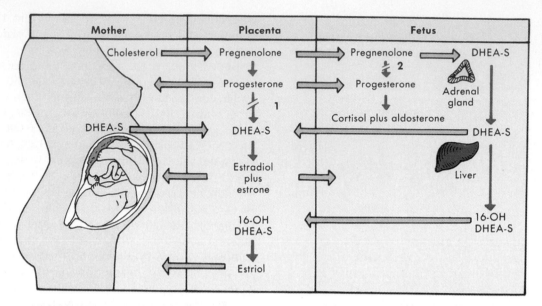

FIGURE 50-7   The maternal-fetal-placental unit in steroid hormone synthesis. Progesterone is synthesized in the placenta from maternal cholesterol. In turn, this progesterone acts on the mother and also serves as the precursor to fetal cortisol and aldosterone synthesis. Estradiol and estrone are synthesized in the placenta from maternal and fetal adrenal DHEA-S and estriol from 16-OH-DHEA-S which is synthesized in the fetal liver. *DHEA-S,* Dehydroepiandrosterone sulfate; *16-OH-DHEA-S,* 16-α-hydroxydehydroepiandrosterone sulfate; *1,* 17-hydroxylase/17,20-desmolase; *2,* 3-β-OL-dehydrogenase-Δ^{4,5} isomerase.

cause it lacks 17-hydroxylase/17,20-desmolase activity, it requires an androgen substrate from the maternal and fetal compartments. This exemplifies coordinated maternal-placental-fetal function. Thus the placenta extracts DHEA-S derived from the maternal and fetal adrenal glands, removes the sulfate, and aromatizes the androgen to estradiol and estrone (see Figure 50-7). In the case of **estriol,** the fetal liver must first 16-hydroxylate DHEA-S before the placenta acts on the precursor androgen (see Figure 50-7).

Human Chorionic Somatomammotropin   A protein hormone, unique to pregnancy, is **human chorionic somatomammotropin** (HCS), also called **human placental lactogen.** Its structure is determined by a gene in the growth hormone family.

Synthesized by placental trophoblasts within 4 weeks, maternal plasma HCS concentration rises steadily throughout pregnancy. The peak HCS production rate far exceeds that of any other human protein hormone. Although its growth-promoting activity is only a fraction of that of growth hormone, the high maternal plasma concentration of HCS makes it capable of contributing to anabolism in the mother.

HCS stimulates lipolysis and is an insulin antagonist. Thus HCS raises maternal free fatty acid and glucose levels. As discussed later, a major function of HCS is to direct maternal metabolism to shunt these substrates to the fetus.

Other Placental Hormones   The placenta produces hypothalamic- or pituitary-like peptides, including GnRH, TRH, CRH, somatostatin, ACTH, TSH, and a unique placental growth hormone variant. Placental ACTH and TSH may augment maternal adrenal and thyroid gland activity, and placental growth hormone acts on maternal target tissues. The placenta also synthesizes 1,25-(OH)$_2$-vitamin D, which helps regulate calcium homeostasis and skeletal formation in the fetus.

Hormones of Maternal Origin

Prolactin   Prolactin secretion from the maternal pituitary gland increases greatly (see Figure 50-6), in response to the high maternal estrogen levels. The prolactin is in the nonglycosylated active form, and it specifically stimulates the lactogenic apparatus of the breast (see Chapter 44). During pregnancy, however, lactation itself is inhibited by the great excess of estrogen and progesterone. After delivery of the fetus, true milk synthesis is initiated by the precipitous drop in steroid hormone levels. Thereafter, milk synthesis is maintained in a nursing mother by prolactin, facilitated by insulin and cortisol. Although basal prolactin concentrations gradually decline by 8 weeks after delivery, they are transiently elevated during each period of suckling. This helps to sustain milk secretion.

Prolactin also suppresses reproductive function in the

nursing mother. During the first 7 to 10 days after delivery, plasma FSH and LH levels remain low. FSH levels then rise, but LH levels do not; this pattern simulates the situation in early puberty. In the nursing mother this persists because of the inhibitory effects of prolactin on GnRH secretion. A decrease in circulating prolactin that follows either cessation of nursing or administration of a dopaminergic agonist (see Figure 44-12) triggers LH release and initiates menstrual cycling.

Relaxin Relaxin is a peptide hormone that is structurally similar to proinsulin. It is produced by the corpus luteum, the decidua, and placenta under HCG stimulation. Maternal plasma relaxin levels rise to a peak in the first trimester, and then decline somewhat. This hormone relaxes the mother's pelvic outlet and softens the cervix by increasing collagenase activity and decreasing tissue collagen content. Relaxin also decreases uterine muscle contractility by reducing the activity of myosin kinase. Thus, relaxin acts to maintain uterine quiescence and prevent early abortion, but later it facilitates easier passage of the fetus into the birth canal once labor has begun.

Other Hormonal Changes

The pregnant state induces important other changes in maternal endocrine function. Insulin secretion increases after the third month in response to glucose challenge or meals. It peaks during the last trimester and acts to compensate for the insulin resistance caused by HCS, placental GH, and cortisol.

Aldosterone secretion increases throughout pregnancy because of estrogen-induced augmentation of renin and angiotensinogen levels. The greater amounts of angiotensin-II generated stimulate the adrenal zona glomerulosa. This induces a positive sodium balance, which is needed to support a high maternal plasma volume and build the extracellular fluid of the fetus. Another mineralocorticoid, deoxycorticosterone (Chapter 46), is synthesized by maternal kidneys during pregnancy and contributes to sodium retention.

Total plasma thyroxine and cortisol levels are elevated because of estrogen-induced increases in their respective binding globulins. Free T_4 and T_3 may be increased in the first trimester. The level of plasma free cortisol also rises modestly and may contribute to maternal adipose tissue gain and to mammary gland development.

Parathyroid hormone (PTH) secretion also increases. PTH augments maternal plasma levels of 1,25-$(OH)_2$ vitamin D, which in turn increases dietary calcium absorption. This enhances the supply of calcium for the growing fetal skeleton.

Maternal FSH and LH are suppressed by high concentrations of estrogen, progestrone, and inhibin from the corpus luteum and placenta. Similarly, maternal pituitary

GH secretion is decreased by feedback from HCS and placental variant GH actions.

MATERNAL-FETAL METABOLISM

During pregnancy, the average gain in maternal weight is 11 kg. Approximately half of this can be attributed to changes in maternal tissues and half to the fetus and placenta. The mother must ingest approximately 300 extra kilocalories and 30 extra grams of protein daily to support fetal development, enlarge maternal energy stores, and sustain growth of certain tissues.

During the first half of pregnancy, the mother is in an anabolic state, and the conceptus represents an insignificant nutritional drain. This phase is characterized by normal or even increased maternal sensitivity to insulin. Maternal plasma levels of glucose, free fatty acids, glycerol, and amino acids are normal or slightly decreased. Dietary carbohydrate and protein loads are rapidly used. Maternal lipogenesis is favored, glycogen stores are expanded, and protein synthesis is enhanced. This supports early growth of the breasts and uterus and prepares the mother to withstand the later metabolic demands of the enlarging fetus.

During the second half of pregnancy the mother shifts into a catabolic state aptly described as "accelerated starvation." Insulin sensitivity is replaced by insulin resistance. This resistance causes elevation of postprandial plasma levels of glucose and amino acids as the uptake of dietary carbohydrate, protein, and fat by maternal tissues is reduced. Consequently, the diffusion of glucose and the facilitated transport of amino acids across the placenta into the fetus are accelerated. During maternal fasting intervals, plasma glucose and amino acid levels fall more rapidly than in nonpregnant women, because the fetus continues to siphon off these substances. Maternal lipolysis is excessively stimulated, ensuring alternate oxidative fuels for the mother and even for the fetus, to whom ketoacids and free fatty acids can be transferred across the placenta. HCS is the key hormone responsible for maternal insulin resistance and for lipid mobilization during fasting in this later stage of pregnancy. Elevated estrogen, progesterone, and cortisol levels also antagonize insulin action.

The insulin resistance of pregnancy, when added to an underlying vulerability to diabetes, produces **gestational diabetes** in 4% of pregnancies. Hyperglycemia appears around week 24 to 28 of gestation and may have serious consequences for the fetus. Maternal

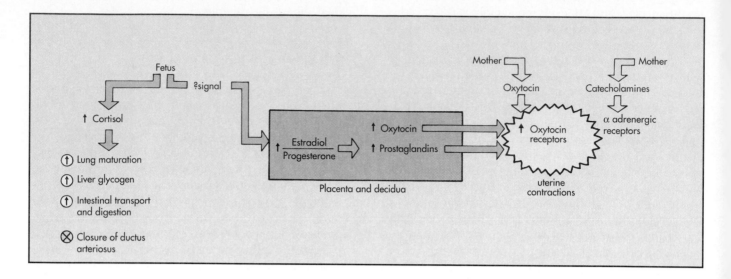

FIGURE 50-8 Endocrine regulation of parturition. The fetus probably initiates signals that increase the ratio of estrogen to effective progesterone in the myometrium. This leads to increased prostaglandins, which are the main stimulus to uterine labor contractions. Oxytocin (OCT) produced locally in the placenta and decidua and a small maternal component coupled with increased myometrial OCT receptors, may contribute to labor but is not essential. However, OCT sustains uterine contractions after expulsion of the fetus so as to minimize maternal loss of blood. Maternal catecholamines may add to the hormonal cascade of contractions. Cortisol prepares the fetus to maintain its own supply of oxygen and substrates after birth.

hyperglycemia stimulates fetal hyperinsulinemia; the latter causes fat babies that are harder to deliver and a tendency to hypoglycemia in the newborn infant. Immature lungs that lack surfactant can cause respiratory distress, and sudden death caused by heart muscle abnormalities may occur in utero.

PARTURITION

The exact mechanisms for initiation of the terminal phase of human pregnancy and their endocrine mediators are still unclear. Studies in numerous animal models suggest possible roles for cortisol, estrogens, progesterone, prostaglandins, oxytocin, relaxin, and catecholamines in the initiation, maintenance, and termination of labor. Because much species variation exists, it is uncertain what can be applied to humans. Figure 50-8 shows some current notions of the endocrine regulation of parturition.

Once the contents of the uterus reach some critical size, stretching of the uterine muscle fibers increases their contractility. Thus, uncoordinated uterine contractions begin at least 1 month before the end of gestation.

However, some signal from the fetus that indicates readiness may initiate the process of labor. In sheep, fetal cortisol has been strongly implicated. In humans, fetal cortisol production rises sharply during the last few weeks of gestation. However, the evidence for a rapid surge of cortisol immediately preceding the onset of labor in humans is contradictory. Nonetheless, the late gestational increase in cortisol secretion is important for preparing the human fetus for the abrupt transition to extrauterine life by stimulating lung maturation, increasing stores of glycogen in the liver, inducing intestinal transport and digestive mechanisms, and facilitating closure of the ductus arteriosus.

Even if the precise initiating signal in human parturition is unidentified, current concepts favor a multicomponent process of human parturition involving paracrine and endocrine mechanisms. *A large increase in local prostaglandins increases myometrial cell Ca^{++} and triggers uterine contractions. A drop in the ratio of intrauterine progesterone to estrogen levels appears to be responsible for augmenting prostaglandins and for abolishing uterine quiescence.* Increased activity at term of the enzyme, 17β,20α-hydroxysteroid dehydrogenase in the uterine tissue lowers this progesterone-to-estrogen ratio. This single enzyme inactivates progesterone by reducing the ketone group at position 20 and augments the estrogen effect by reducing estrone to estradiol at position 17.

Another major stimulator of myometrical contractions is oxytocin (OCT). Although the concentration of OCT in maternal plasma does not increase consistently just before labor, the frequency of OCT pulses does increase. Furthermore, myometrial OCT receptor content rises dramatically at term, as does the local synthesis of OCT by the decidua and the fetal membranes. OCT may, therefore, reinforce labor contractions and probably maximizes the contractions immediately after delivery so as to minimize maternal blood loss.

Uterine contractions may also be modulated by catecholamines; α-adrenergy is stimulatory and β-adrenergy is inhibitory. Maternal stress may release circulating catecholamines that participate in the final hormonal cascade of parturition.

In addition to uterine contractions, an important component of labor is the rapid changes that occur in placental and cervical tissue. At term, a sharp rise in inflammatory cytokines, such as interleukin 8, is seen in amniotic fluid. These cytokines are produced by decidual and fetal membrane tissues, probably under hormonal paracrine stimulation. They attract neutrophils, which release collegenase that loosens the attachments between the maternal and fetal tissue planes and decreases the cervical resistance to pressure from the fetal head.

Once labor has begun, it proceeds in three stages. In the first, lasting a variable number of hours, the uterine contractions originate at the fundus, sweep downward, and force the head of the fetus against the cervix. This progressively widens and thins the opening to the vaginal canal In the second stage, lasting less than 1 hour, the fetus is forced out of the uterine cavity and through the cervix and is delivered from the vagina. In the third stage, lasting 10 minutes or less, the placenta is separated from the decidual tissue of the uterus and is forcefully evacuated. Myometrial contractions at this point act to constrict the uterine vessels and to prevent excessive bleeding. Once the placenta has been removed, all its hormonal products disappear from the maternal plasma by 48 to 72 hours.

LACTATION

Maternal provision of nutrients to the newborn begins within 48 hours of delivery. First, thin fluid containing lactose and proteins but little fat, and known as colostrum, is secreted in very small quantities. True milk delivery follows shortly. Human breast milk contains 1% protein, largely as casein, lactalbumin, and lactglobulin. In addition, it contains 7% lactose and 3.5% fat, equivalent to about 70 kcal/100 ml. By 1 week, 550 ml/day is produced; later, maximum rates of up to 2000 ml/day may occur. Large quantities of calcium and phosphorus are also needed by and provided to the infant. Milk contains as well immunoglobulins, which protect against infection. Over 160 other constituents, which include many peptide hormones and growth factors, may act directly on the infant's gastrointestinal tract or even be absorbed to act systemically. Typically, infants nurse for 6 to 12 months.

Mammary cells package proteins, lactose, calcium, and phosphate in secretory vesicles and fat in droplets. Prolactin is essential to these processes (Chapter 44). Immunoglobulins enter the cells combined with membrane receptors in vesicles. All these products are then secreted into the alveoli. Suckling and even anticipatory signals such as the infant's cry stimulate oxytocin release via neural sensory pathways and the central **nucleus tractus solitarius.** Oxytocin causes contraction of myoepithial cells around the alveoli and smooth muscle cells in the duct walls. This "lets the milk down" into the areolar area, where the infant obtains it through holes in the nipple by positive pressure, not by creating a vacuum.

SUMMARY

1. In the ovary, oocytes that are suspended in the prophase of meiosis seclude themselves in primordial follicles. These follicles undergo slow, hormonally independent development to primary follicles, which are composed of nurturing, surrounding granulosa cells.
2. Monthly cohorts of follicles are stimulated sequentially by FSH and LH to progress in development. From each cohort, a single dominant follicle emerges and grows exponentially. This dominant follicle produces enough estradiol to inhibit its cohort follicles, to prepare reproductive organs for fertilization, and to condition the hypothalamus and pituitary to provide a timed surge of LH and FSH to cause ovulation.
3. Estradiol production by the follicular granulosa cells is dependent on androgen substrate, which is provided by neighboring theca cells.
4. The monthly cyclicity of ovulation is mainly determined by the ovary. A surge of LH and FSH secretion occurs when the dominant follicle secretes sufficient estradiol in an appropriate temporal pattern.
5. After ovulation, the granulosa and theca cells form a corpus luteum. This endocrine structure secretes sufficient progesterone and estradiol to condition the reproductive organs so that they can receive and implant a zygote.
6. Estradiol and progesterone in sequence and together induce cyclic changes in the structure and secretory function of the vagina, the endometrium of the

uterus, and the fallopian tubes. Estrogens also have important actions on bone remodelling, hepatic synthesis of proteins, and other systemic target tissues.

7. After implantation of a conceptus, the placenta is formed from fetal trophoblasts, which initially secrete human chorionic gonadotropin that sustains the corpus luteum. Eventually the placenta itself produces the following hormones of pregnancy: estrogens, progesterone, and a variety of proteins and peptides resembling hypothalamic and pituitary hormones.

8. Early in pregnancy, the mother is in a metabolic state that facilitates growth of reproductive tissues and energy stores. Later, she becomes insulin resistant, which allows her to shunt substrates to the growing fetus.

9. The endocrine mechanism of parturition is not yet completely defined. An increased ratio of estradiol to progesterone in uterine tissue augments production of prostaglandins, which are the main stimulators of uterine contractions. This action may be supplemented by that of locally produced and maternal oxytocin and by maternal catecholamines.

BIBLIOGRAPHY
Journal Articles

Adashi EY: Intraovarian pepetides: stimulators and inhibitors of follicular growth and differentiation, *Endocr Metab Clin* 21:1, 1992.

Apter D, Butzow TL, Laughlin GA, Yen SSC: Gonadotropin-releasing hormone pulse generator activity during pubertal transition in girls: pulsatile and diurnal patterns of circulating gonadotropins, *J Clin Endocrinol Metab* 76:940, 1993.

Batista MC, Cartledge TP, Zellmer AW, Nieman LK, Merriam GR, Loriaux DL: Evidence for a critical role of progesterone in the regulation of the midcycle gonadotropin surge and ovulation, *J Clin Endocrinol Metab* 74:565, 1992.

Bryant-Greenwood GD, Schwabe C: Human relaxins: chemistry and biology, *Endocr Rev* 15:5, 1994.

Casey ML, MacDonald PC: Human parturition: distinction between the initiation of parturition and the onset of labor, *Semin Reprod Endocrinol* 11:272, 1993.

Couzinet B, Brailly S, Bouchard P, Schaison G: Progesterone stimulates luteinizing hormone secretion by acting directly on the pituitary, *J Clin Endocrinol Metab* 74:374, 1992.

Crisp TH: Organization of the ovarian follicle and events in its biology: oogenesis, ovulation or atresia, *Mutation Research* 296:89, 1992.

Crowley WR, Armstrong WE: Neurochemical regulation of oxytocin secretion in lactation, *Endocr Rev* 13:33, 1992.

Das K, Tagatz GE, Hensleigh HC, Leung BS, Phipps WR: Paracrine interactions among insulin-like growth factor-I, insulin-like growth factor binding protein-1, and steroid hormones in follicular fluid, *Int J Fertil* 38:360, 1993.

DeRidder CM, Thijssen JHH, Bruning PF, Van Den Brande, JL, Zonderland ML, Erich WBM: Body fat mass, body fat distribution, and pubertal development: a longitudinal study of physical and hormonal sexual maturation of girls, *J Clin Endocrinol Metab* 75:442, 1992.

Findlay JK: An update on the roles of inhibin, activin, and follistatin as local regulators of folliculogenesis, *Biol Reprod* 48:15, 1993.

Flamigni B et al: Factors regulating interaction between trophoblast and human endometrium, *Ann NY Acad Sci* 622:176, 1991.

Fritz MA, McLachlan RI, Cohen NI, Dahl KD, Bremner WJ, Soules MR: Onset and characteristics of the midcycle surge in bioactive and immunoactive luteinizing hormone secretion in normal women: influence of physiological variations in periovulatory ovarian steroid hormone secretion, *J Clin Endocrinol Metab* 75:489, 1992.

Fuchs AR et al: Oxytocin secretion and human parturition: pulse frequency and duration increase during spontaneous labor in women, *Am J Obstet Gynecol* 165:1515, 1991.

Goebelsmann U: Protein and steroid hormones in pregnancy, *J Reprod Med* 23:166, 1979.

Hall JE, Schoenfeld DA, Martin KA, Crowley WF Jr: Hypothalamic gonadotropin-releasing hormone secretion and follicle-stimulating hormone dynamics during the luteal-follicular transition, *J Clin Endocrinol Metab* 74:600, 1992.

Hillier SG: Current concepts of the roles of follicle stimulating hormone and luteinizing hormone in folliculogenesis, *Human Reprod* 9:188, 1994.

Hillier SG, Whitelaw PF, Smyth CD: Follicular oestrogen synthesis: the 'two-cell, two-gonadotrophin' model revisited, *Mol Cell Endocrinol* 100:51, 1994.

Hsueh AJW, Billig H, Tsafriri A: Ovarian follicle atresia: a hormonally controlled apoptotic process, *Endocr Rev* 15:707, 1994.

Irianni F, Hodgen GD: Mechanism of ovulation, *Reprod Endocrinol* 21:19, 1992.

Keyes PL, Wiltbank MD: Endocrine regulation of the corpus luteum, *Annu Rev Physiol* 50:465, 1988.

Leslie KK, Zuckerman DJ, Schruefer J, Burchell M, Smith J, Albertson BD: Oestrogen modulation with parturition in the human placenta, *Placenta* 15:79, 1994.

Lunenfeld B, Insler V: Follicular development and its control, *Gynecol Endocrinol* 7:285, 1993.

McLachlan RI et al: Circulating immunoreactive inhibin levels during the normal human menstrual cycle, *J Clin Endocrinol Metab* 65:954, 1987.

McNatty KP et al: The microenvironment of the human antral follicle: interrelationships among the steroid levels in antral fluid, the population of the granulosa cells, and the status of the oocyte in vivo and in vitro, *J Clin Endocrinol Metab* 49:851, 1979.

Miller FD, Chibar R, Mitchell BF: Synthesis of oxytocin in amnion, chorion and decidua: a potential paracrine role for oxytocin in the onset of human parturition, *Regulatory Peptides* 45:247, 1993.

Mitchell BF, Wong S: Changes in 17β,20α-hydroxysteroid dehydrogenase activity supporting an increase in the estrogen/progesterone ratio of human fetal membranes at parturition, *Am J Obstet Gynecol* 168:1377, 1993.

Rories C, Spelsberg TG: Ovarian steroid action on gene expression: mechanisms and models, *Annu Rev Physiol* 51:653, 1989.

Rossmanith WG: Contemporary insights into the control of the corpus luteum function, *Horm Metab Res* 15:192, 1993.

Steer PJ: The endocrinology of parturition in the human, *Ballieres Clin Endocrinol Metab* 4:333, 1990.

Turner RT, Riggs BL, Spelsberg TC: Skeletal effects of estrogen, *Endocr Rev* 15:275, 1994.

Ying S-Y: Inhibins, activins, and follistatins: gonadal proteins modulating the secretion of follicle-stimulating hormone, *Endocr Rev* 9:267, 1988.

Yong EL, Bairad DT, Yates R, Reichert LE Jr, Hillier SG: Hormonal regulation of the growth and steroidogenic function of human granulosa cells, *J Clin Endocrinol Metab* 74:842, 1992.

Books and Monographs

Adashi EY: The ovarian cycle. In Yen SSC, Jaffe RB, eds: *Reproductive endocrinology,* Philadelphia, 1991, WB Saunders.

Fisher DA: Endocrinology of fetal development. In Wilson JD, Foster DW, eds: *Williams textbook of endocrinology,* Philadelphia, 1992, WB Saunders.

Hiller SG: Cellular basis of follicular endocrine function. In Hiller SG, ed: *Ovarian endocrinology.* Oxford, 1991, Blackwell Scientific Publications.

Konigsberg D et al: Ovarian follicular maturation, and ovulation induction. In Degroot LJ et al, eds: *Endocrinology,* vol 3, New York, 1994, Grune & Stratton, Inc.

Marshall JL: Hormonal regulation of the menstrual cycle and mechanisms of anovulation. In Degroot LJ et al, eds: *Endocrinology,* vol 3, New York, 1994, Grune & Stratton, Inc.

Yen SSC: The human menstrual cycle: neuroendocrine regulation. In Yen SSC, Jaffe RB, eds: *Reproductive endocrinology,* Philadelphia, 1991, WB Saunders Co.

Multiple-Choice Review Questions

PART I
CELL PHYSIOLOGY

1. Which of the following statements about biological membranes is true?
   A. The phospholipid bilayer is quite permeable to water-soluble molecules with molecular weights between 200 and 500.
   B. Most lipid molecules are not free to move in the plane of the membrane.
   C. Flip-flop of lipid molecules from one monolayer to the other occurs infrequently.
   D. The carbohydrate chains of glycolipids and glycoproteins mostly face the cytosol.
   E. None of the above.

2. Which of the following statements about diffusion is true?
   A. The average time for a molecule to diffuse a particular distance is directly proportional to the first power of the distance.
   B. The rate at which substance X diffuses into a cell is proportional to the square of the area of the plasma membrane of the cell.
   C. The greater the solubility of a substance in nonpolar solvents, the smaller its rate of permeation across biological membranes.
   D. If protein A has molecular weight 100,000 and protein B has molecular weight 200,000, the diffusion coefficient of A will be twice that of B.
   E. None of the above.

3. A membrane separates chambers A and B. The membrane is permeable to water, but completely impermeable to NaCl. When chambers A and B are filled with water and 1 atm of hydrostatic pressure is applied to chamber A, water flows from A to B at 10 ml/min. Then chamber A is drained and refilled with a solution of NaCl with an osmotic pressure (as measured across a true semipermeable membrane) of 1 atm. Which of the following statements is true?
   A. No net water flow by osmosis will occur.
   B. A hydrostatic pressure of 1 atm applied to A would cause net water flow from A to B.
   C. With no hydrostatic pressure on side A, the net osmotic flow from B to A will be 10 ml/min.
   D. If the membrane were permeable to NaCl, the osmotic water flow from B to A would be 5 ml/min.
   E. None of the above.

4. Which of the following statements about active transport processes is true?
   A. They always transport the transported substance against a gradient of electrochemical potential.
   B. They require a direct link to metabolic energy.
   C. The energy for secondary active transport is provided by the gradient of electrochemical potential of another solute.
   D. Only primary active transport will be inhibited by metabolic poisons.
   E. None of the above.

5. Which of the following statements about the Na^+-K^+-ATPase is true?
   A. It pumps 2 Na^+ out of the cell and 3 K^+ into the cell for each ATP it hydrolyzes.
   B. It is electrogenic in some, but not all cells.
   C. In most cells it pumps Na^+ and K^+ against both chemical and electrical gradients.

D. It is driven by being phosphorylated by ATP.

E. None of the above.

6. The membrane that separates chamber A from chamber B is permeable only to anions. Chamber A contains 1 M KCl; chamber B contains 0.1 M KCl. In the steady state, which of the following statements is true?

A. The Nernst equation will apply to Cl^-, but not to K^+.

B. The potential difference across the membrane will be 60 mV, with side A negative with respect to B.

C. A significant amount of Cl^- (enough to diminish the concentration in A by at least 1 mmol, and increase it in B) will flow across the membrane to create the electrical potential difference.

D. K^+ will flow across the membrane in proportion to the flow of Cl^-.

E. None of the above.

7. The ionic distribution across the plasma membrane of a smooth muscle cell is as follows: the extracellular concentrations of K^+, Na^+, and Cl^- are 2.5, 120, and 120 mM, respectively. The cytoplasmic concentrations of K^+, Na^+, and Cl^- are 140, 10, and 12 mM, respectively. The resting membrane potential is -60 mV. Which of the following statements is true?

A. Na^+ tends to enter the cell because of both its concentration gradient and the membrane potential.

B. K^+ has a net tendency to enter the cell.

C. Cl^- is not in equilibrium across the membrane.

D. The Na^+-K^+-ATPase is not important in the steady state.

E. None of the above.

8. A membrane that separates chambers A and B is permeable to K^+, Cl^-, and water, but completely impermeable to X^-. Initially chamber A contains 0.1 M K^+ and 0.1 M X^- and chamber B contains 0.1 M KCl. Chamber A is pressure-jacketed so that a Gibbs-Donnan equilibrium can be obtained. Which of the following statements is true?

A. In attaining a Gibbs-Donnan equilibrium, Cl^-, but not K^+, will flow from side B to side A.

B. The amount of K^+ that flows from B to A will be larger, by an infinitesimal amount, than the amount of Cl^- that flows.

C. Both K^+ and Cl^- will be in equilibrium when the Gibbs-Donnan equilibrium has been attained.

D. When the Gibbs-Donnan equilibrium has been attained, the product of the concentrations of K^+ on side A and B will equal the product of the Cl^- concentrations on side A and side B.

E. None of the above.

9. Which of the following statements about resting membrane potentials is true?

A. The electrogenicity of the Na^+-K^+-ATPase makes no contribution to the resting membrane potential of certain types of cells.

B. The diffusion of K^+ and Na^+ down their electrochemical potential gradients contributes to the resting potential in only certain cell types.

C. Negative fixed charge in cells does not contribute via the Gibbs-Donnan equilibrium to the resting potential of most cells.

D. In skeletal muscle cells, the contribution of the electrogenic pump is small relative to the contributions of the ions diffusing down their electrochemical potential gradients.

E. None of the above.

10. The conductance of the plasma membrane of a resting cell to K^+ is 9 times that to Na^+. Cl^- is in equilibrium at the resting membrane potential. The contribution of the electrogenicity of the Na^+-K^+-ATPase is negligible. The extracellular concentrations of K^+ and Na^+ are 14 and 120 mM, respectively. The cytoplasmic concentrations of K^+ and Na^+ are 140 and 12 mM, respectively. Which of the following statements is true?

A. The resting membrane potential is about -48 mV.

B. We cannot estimate the resting potential without knowing the relative conductance to Cl^-.

C. Decreasing the extracellular $[K^+]$ will depolarize the cell.

D. Increasing the K^+ conductance will depolarize the cell.

E. None of the above.

11. The plasma membrane of a resting smooth muscle cell has a conductance to K^+ four times that to Na^+. The membrane Cl^- conductance is very small. The extracellular concentrations of K^+ and Na^+ are 13 and 110 mM, respectively. The cytoplasmic concentrations of K^+ and Na^+ are 130 and 11 mM, respectively. The resting membrane potential is -60 mV. Which of the following statements is true?

A. The resting membrane potential estimated

from the chord conductance equation is
−48 mV.
B. The electrogenic Na^+-K^+-ATPase might contribute the difference between the observed −60 mV and the resting potential calculated from the chord conductance equation.
C. Inhibiting the Na^+-K^+-ATPase by treating the cell with a large dose of ouabain would have no significant effect on the resting membrane potential.
D. If the plasma membrane were made equally conductive to Na^+ and K^+, the membrane potential would go to near −20 mV.
E. None of the above.

12. For a particular cell the electrogenicity of the Na^+-K^+-ATPase makes a negligible contribution to the resting membrane potential. Cl^- is in equilibrium at the resting potential. The extracellular concentrations of K^+, Na^+, and Cl^- are 2.5, 120, and 120 mM, respectively. The cytoplasmic concentration of K^+ and Na^+ are 140 and 9 mM, respectively. The resting membrane potential is −60 mV. Which of the following statements is true?
A. The intracellular $[Cl^-]$ is 3 to 4 mM.
B. K^+ and Na^+ are in equilibrium.
C. Increasing the conductance of the membrane to Cl^- would hyperpolarize the cell.
D. Increasing the extracellular $[K^+]$ would depolarize the cell.
E. None of the above.

13. Which of the following statements about the action potential in squid giant axons is correct?
A. The rising phase of the action potential is due to a rapid increase in the conductance of the plasma membrane to Na^+ and K^+.
B. The peak of the action potential reaches the equilibrium potential for Na^+.
C. The repolarizing phase of the action potential is due to a decrease in the sodium conductance and an increase in the potassium conductance.
D. The hyperpolarizing afterpotential is due to a prolonged decrease in the conductance to K^+.
E. None of the above.

14. Which of the following statements about the refractory periods of squid giant axon is true?
A. The absolute refractory period begins with the beginning of the action potential spike and ends when repolarization returns the cell to the resting membrane potential.

B. The relative refractory period begins near the peak of the action potential and ends when the hyperpolarizing afterpotential is over with.
C. The elevated K^+ conductance is principally responsible for refractoriness during the later part of the absolute refractory period.
D. Voltage-inactivation of Na^+ channels is the major cause of the absolute refractory period.
E. None of the above.

15. Which of the following statements about ion channels in squid giant axon is true?
A. K^+ channels show significant voltage inactivation.
B. Na^+ channels have an activation gate and an inactivation gate.
C. The activation gate and the inactivation gate of the Na^+ channel respond with similar kinetics to a depolarization of the cell membrane.
D. The K^+ channels are blocked by tetrodotoxin.
E. None of the above.

16. Which of the following statements about action potentials is true?
A. Smooth muscle cells have an appreciable number of fast Na^+ channels.
B. The plateau phase of the cardiac action potential is due to opening of channels that are conductive to Ca^{++}.
C. The initial spike of cardiac ventricular action potentials is due to Na^+ channels that open and inactivate slowly.
D. The Ca^{++} that enters a smooth muscle cell during the action potential plays no important role in excitation-contraction coupling.
E. None of the above.

17. Which of the following statements about the conduction of a subthreshold depolarization of a nerve or muscle cell is true?
A. A subthreshold depolarization is conducted by saltatory conduction.
B. It is conducted with decrement, which means that the depolarization gets smaller as the depolarization moves along the cell.
C. The length constant is the distance over which the depolarization decreases to 50% of its maximum size.
D. The length constant is inversely related to R_m/R_{in} (membrane resistance/internal resistance).
E. None of the above.

18. Which of the following statements about conduction in a myelinated axon is true?
    A. As a result of myelination, membrane resistance (R_m) is increased; this decreases the length constant.
    B. As a result of myelination, membrane capacitance (C_m) is increased; this contributes to increased conduction velocity.
    C. The velocity of conduction between nodes of Ranvier is very high compared to that in an unmyelinated axon.
    D. The action potential is regenerated all along the axon.
    E. None of the above.

19. Which of the following statements about neuromuscular transmission is true?
    A. A miniature endplate potential occurs following an action potential in the nerve terminal.
    B. Calcium influx into the nerve terminal initiates the chain of events that culminates in transmitter release.
    C. The acetylcholine receptor protein is a voltage-gated ion channel that opens when it binds acetylcholine.
    D. The desensitization of the acetylcholine receptor normally terminates the endplate potential.
    E. None of the above.

20. Which of the following statements about the acetylcholine receptor protein of the neuromuscular junction is correct?
    A. The acetylcholine receptor protein is a voltage-gated ion channel.
    B. The acetylcholine receptor is concentrated at the neuromuscular junction.
    C. The channel opened by acetylcholine is much more conductive to sodium than to potassium.
    D. Depolarization of the muscle cell promotes opening of the acetylcholine receptor protein's ion channel.
    E. None of the above.

21. Which of the following statements about synapses is true?
    A. Gap junctions contain ligand-gated ion channels.
    B. Synaptic delay is characteristic of chemical synapses.
    C. A neurotransmitter acts to increase or decrease the conductance of the postsynaptic membrane to one or more ions.

D. Chemical synapses require influx of Na^+ triggered by an action potential in the presynaptic nerve terminal.
    E. None of the above.

22. Which of the following statements about synapses on cat spinal motor neurons is true?
    A. An excitatory postsynaptic potential (EPSP) is due to opening of postsynaptic membrane channels that conduct primarily K^+.
    B. An inhibitory postsynaptic potential (IPSP) is due to opening of postsynaptic membrane channels that conduct Cl^-.
    C. Action potentials are usually generated in dendrites.
    D. Postsynaptic potentials that originate at synapses on the soma of the spinal motor neuron sum, but with significant decrement.
    E. None of the above.

23. Which of the following statements about signal transduction mechanisms is true?
    A. Many signal transduction pathways involve the binding of an intracellular agonist to a membrane receptor, which directly increases the intracellular level of a second messenger molecule.
    B. Heterotrimeric GTP-binding proteins (G proteins) are inhibited by agonist-bound membrane receptors.
    C. G protein–coupled membrane receptors have seven transmembane alpha-helices and belong to a protein family.
    D. Adenylyl cyclase is regulated positively, but not negatively, by G protein–mediated mechanisms.
    E. None of the above.

24. Which of the following statements about heterotrimeric GTP-binding proteins is true?
    A. The GDP-bound form of a G protein is the active form.
    B. Low-molecular-weight GTP-binding proteins are structurally very similar to heterotrimeric G proteins.
    C. In most cases the activated beta subunit interacts with an effector protein.
    D. Hydrolysis of GTP results in inactivation of the G protein.
    E. None of the above.

25. Which of the following statements about protein kinases is true?
    A. Protein tyrosine kinases do not share significant structural homologies with kinases that

phosphorylate serine or threonine residues on proteins.

B. Protein kinase C is frequently activated by signal transduction pathways involving hydrolysis of certain membrane phospholipids.

C. All known calcium-calmodulin–dependent protein kinases have multiple-substrate proteins.

D. Membrane receptors for serum growth factors are frequently serine or threonine protein kinases.

E. None of the above.

PART II
NERVOUS SYSTEM

26. The cell type responsible for the formation of myelin in the central nervous system is the
    A. astrocyte.
    B. ependymal cell.
    C. microglial cell.
    D. oligodendroglial cell.
    E. satellite cell.

27. The cell type that forms cerebrospinal fluid is the
    A. ependymal cell.
    B. neuron.
    C. oligodendroglial cell.
    D. satellite cell.
    E. schwann cell.

28. About how long would it take for an unmyelinated nociceptive primary afferent nerve fiber to conduct information about a damaging stimulus from the foot to the spinal cord in an adult person?
    A. 1 millisecond
    B. 10 milliseconds
    C. 100 milliseconds
    D. 1 second
    E. 10 seconds

29. Horseradish peroxidase is injected into a muscle in the foot in an experimental animal. Later, horseradish peroxidase is demonstrated immunohistochemically in the gray matter of the lumbosacral spinal cord. What type of cell would you expect to contain the horseradish peroxidase?
    A. Astrocytes
    B. Dorsal horn interneurons
    C. Motoneurons
    D. Oligodendroglia
    E. Spinothalamic tract cells

30. An automobile accident causes an injury of the sciatic nerve. As a consequence of the injury, you would expect
    A. death of all of the dorsal root ganglion cells whose axons were interrupted.
    B. release of antibodies to nerve growth factor from Schwann cells ensheathing damaged axons.
    C. regrowth of the axons distal to the injury at a rate of 400 mm/day.
    D. chromatolysis of motoneurons in the lumbosacral spinal cord.
    E. eventual complete restoration of sensory and motor function.

31. The cerebrospinal fluid (CSF)
    A. has a higher concentration of glucose than blood.
    B. contains more protein per unit volume than blood.
    C. is formed at a rate that is proportional to the CSF pressure.
    D. is absorbed at a rate that is proportional to the CSF pressure.
    E. is removed largely through dural sleeves of the spinal nerve roots.

32. The sensation of high-frequency vibration is signaled by
    A. Golgi tendon organs.
    B. Meissner's corpuscles.
    C. muscle spindles.
    D. nociceptors.
    E. pacinian corpuscles.

33. Raising the skin temperature to 52° C activates
    A. Meissner's corpuscles.
    B. Merkel cell endings.
    C. nociceptors.
    D. pacinian corpuscles.
    E. Ruffini endings.

34. The rate and magnitude of muscle stretch are encoded by
    A. Ruffini endings.
    B. pacinian corpuscles.
    C. muscle spindles.
    D. Meissner's corpuscles.
    E. Golgi tendon organs.

35. In a patient subjected to surgical replacement of the abdominal aorta, the arterial circulation of the spinal cord is compromised, resulting in damage to the white matter of the lateral and anterior funiculi,

but not of the posterior funiculi, at the level of T4. What functional deficit would be expected?

A. Ataxic gait
B. Cannot recognize numbers written on the toes
C. Failure to detect the vibrations of a tuning fork placed on the ankle
D. Inability to distinguish between warm and cold on the feet
E. Weakness of the arms

36. In the retina
A. rods are depolarized when light strikes the outer segment.
B. different rod photopigments discriminate between wavelengths.
C. rods are more sensitive to low intensities of light than are cones.
D. defective rods account for color blindness.
E. rods are most concentrated in the fovea.

37. A patient cannot see in the right upper quadrant of the visual field of either eye. The problem can best be explained by a lesion located in the
A. left middle frontal gyrus.
B. left optic tract.
C. left parietal lobe.
D. left temporal lobe.
E. optic chiasm.

38. In the cochlea
A. oscillations of the basilar membrane in response to high-frequency sound are greater near the apex of the cochlea than at the base.
B. the receptor potentials of the hair cells can be recorded as the cochlear microphonic potential.
C. cochlear nerve fibers discharge at the same frequency as the sound over the entire range of audible frequencies.
D. the intensity of sound is encoded by hair cells near the base of the cochlea.
E. olivocochlear efferents cause contractions of the basilar membrane.

39. When the head is rotated to the left,
A. neural activity from the ampulla of the left horizontal semicircular duct is increased.
B. the eyes deviate slowly to the left.
C. the discharge rate of sensory axons supplying hair cells in the utricular macula increases.
D. endolymph in the right horizontal semicircular duct shifts toward the utricle.
E. hair cells in the saccular macula are depolarized.

40. An α-motoneuron that innervates a postural muscle, such as the soleus muscle,
A. is excited monosynaptically by Golgi tendon organ afferents.
B. forms endplates on 3 to 6 skeletal muscle fibers.
C. contributes to the patellar reflex
D. belongs to a fast fatigable (FF) motor unit.
E. is inhibited disynaptically when the antagonist muscle is stretched.

41. A stroke affecting which of the following sites may cause conjugate deviation of the eyes toward the side of the lesion?
A. Basal ganglia
B. Cerebellum
C. Frontal eye fields
D. Horizontal gaze center
E. Motor cortex

42. A lesion of which of the following structures may result in incoordination, reduced postural tone, and pendular phasic stretch reflexes?
A. Midbrain locomotor system
B. Motor cortex
C. Premotor cortex
D. Cerebellum
E. Superior colliculus

43. Micturition in normal adults involves
A. inhibition of a pathway that descends from a center in the pons.
B. inhibition of the detrusor muscle by sacral parasympathetics.
C. sympathetic excitation of the internal sphincter muscle.
D. activation of mechanoreceptors in the bladder wall.
E. cholinergic transmission in the enteric plexuses.

44. The electroencephalogram
A. contains an alpha rhythm during quiet wakening.
B. has a 3 Hz spike and wave rhythm during REM sleep.
C. increases in frequency during deeper stages of non-REM sleep.
D. is characterized by delta waves during dreaming.
E. becomes isoelectric during jet lag.

45. Broca's aphasia
A. results from a lesion of the posterior superior temporal gyrus.

B. is characterized by neologisms.

C. is accompanied by poor comprehension of verbal or written commands.

D. never occurs in left-handed individuals.

E. involves a deficit in written, as well as verbal, expression.

PART III
MUSCLE

46. The power output of a muscle is
   A. maximal at a load of about 1.6 F_o (where F_o = maximum active force).
   B. zero in an isometric contraction.
   C. greatest at loads approaching zero.
   D. expressed in Newtons per square meter.
   E. greater in slow than in fast-twitch skeletal muscle fibers.

47. The determinant(s) of shortening velocity expressed as L_o/sec in a striated muscle is (arc)
   A. myosin isoform expressed in the cell.
   B. load.
   C. number of sarcomeres in the cell.
   D. cross-sectional area of the cell.
   E. both myosin isoform and load.

48. How many thin filament(s) are there for each end of the thick filament in striated muscle?
   A. One-half
   B. One
   C. Two
   D. Three
   E. Six

49. During "negative work" a muscle cell is
   A. contracting while lengthening.
   B. not hydrolyzing ATP during crossbridge cycling.
   C. bearing a force that exceeds the force generated in an isometric contraction.
   D. most subject to injury.
   E. All statements are true.

50. The I-band region of a skeletal muscle sarcomere DOES NOT contain
   A. myosin.
   B. cytoskeletal proteins.
   C. troponin.
   D. actin.
   E. tropomyosin.

51. Name the membrane system that is NOT involved in excitation-contraction coupling in striated muscle.

A. Microtubules
B. Sarcolemma
C. Sarcoplasmic reticulum
D. Motor endplate
E. Transverse (T-) tubules

52. The high excitability of the motor nerves in slow, red, oxidative, type I motor units is due to
   A. the cable properties of their small-diameter axons.
   B. the fact that these nerves only form neuromuscular junctions with a small number of muscle cells and require little synaptic input to release sufficient acetylcholine.
   C. the fact that comparatively few synaptic excitatory postsynaptic potentials are required to initiate an action potential.
   D. the presence of more excitatory synaptic input than is received by fast motor units.
   E. infrequent recruitment.

53. The force of a twitch is less than that in a tetanus because
   A. one action potential does not release sufficient Ca^{++} to saturate the troponin-binding sites.
   B. Ca^{++} reuptake into the sarcoplasmic reticulum is so rapid that there is insufficient time for all the troponin Ca^{++} binding sites to be occupied.
   C. crossbridge cycling is not fast enough to allow expression of the maximum possible force before Ca^{++} dissociates from troponin and is sequestered in the sarcoplasmic reticulum.
   D. not all the cells in the motor unit are recruited.
   E. the action potentials do not sum to give greater membrane depolarization.

54. Endurance (aerobic) exercise produces
   A. extensive hypertrophy of involved motor units.
   B. increased mitochondrial content of all slow type I motor units.
   C. conversion of some fast motor units to slow, oxidative motor units.
   D. formation of new slow, oxidative motor units.
   E. increased oxidative capacity in all motor units involved in the exercise.

55. Which cell type is not striated?
   A. Slow-twitch, oxidative skeletal motor units
   B. Fast-twitch, glycolytic skeletal motor units

   C. Cardiac muscle
   D. Muscles attached to the skeleton
   E. Muscle in the skeletal muscle vasculature

56. In a skeletal muscle cell the membrane system that lacks exposure to high Na^+ and low K^+ concentrations is
   A. the transverse (T-) tubular system.
   B. the sarcolemma.
   C. the motor endplate.
   D. the sarcoplasmic reticulum.
   E. none of the above

57. The maximum force of contraction in a skeletal muscle is generated by
   A. tetanization.
   B. activation of all recruitable motor units.
   C. increasing the amount of Ca^{++} released from the sarcoplasmic reticulum.
   D. tetanization and recruitment of all motor units.
   E. recruitment of only the fast, glycolytic motor units.

58. Smooth muscle has
   A. motor endplate membrane specializations.
   B. transverse (T-) tubules.
   C. motor nerves.
   D. gap junctions.
   E. sarcomeres.

59. In smooth muscle, crossbridge cycling rates and shortening velocities are not functions of
   A. load.
   B. myosin isoform expressed in a cell.
   C. myosin regulatory light chain phosphorylation.
   D. temperature.
   E. Ca^{++} binding to crossbridges.

60. Myoplasmic Ca^{++} transients in smooth muscle reflect
   A. Ca^{++} influx into the cell through potential-activated (gated) channels.
   B. Na^+/Ca^{++} exchange in the sarcolemma.
   C. inositol 1,4,5-trisphosphate (IP_3)–induced Ca^{++} efflux through channels in the sarcoplasmic reticulum.
   D. T-tubular depolarization–induced Ca^{++} release from the sarcoplasmic reticulum.
   E. Ca^{++} entry through sarcolemmal receptor-activated channels.

61. Compared with striated muscle, smooth muscle is characterized by

   A. a low force-generating ability.
   B. low crossbridge cycling rates.
   C. low economy of force maintenance.
   D. rapid fatigue.
   E. high work output.

62. Tone in smooth muscle is best described as
   A. trains of action potentials leading to summation of twitches into a large contraction.
   B. asynchronous contractions of a few motor units.
   C. cyclic contraction and relaxation.
   D. continuous, variable, submaximal contraction.
   E. an irreversible, maximal contraction (spasm).

63. Which structure is present in smooth muscle and lacking in skeletal muscle?
   A. Junctions between muscle cells
   B. Overlapping thick and thin filaments
   C. A cytoskeleton
   D. A sarcoplasmic reticulum
   E. Neuromuscular junctions (endplates)

64. Missing from the sliding filament/crossbridge mechanism in smooth muscle is
   A. force-length behavior.
   B. velocity-load relationships.
   C. presence of myosin-containing thick filaments.
   D. anatomical correlation between filament overlap and force.
   E. presence of thin filaments containing actin and tropomyosin.

65. The sarcoplasmic reticulum in smooth muscle
   A. is poorly developed or absent.
   B. forms an internal Ca^{++}-containing compartment surrounding the myofibrils.
   C. releases Ca^{++} when the T-tubules are depolarized.
   D. links the sarcolemma with the caveoli.
   E. is a continuous tubular network primarily located close to the plasma membrane.

66. The intracellular Ca^{++} concentration in smooth muscles is NOT modulated by
   A. activity of the Na^+-K^+ pumps in the sarcoplasmic reticulum.
   B. Na^+/Ca^{++} exchange with the extracellular fluid.
   C. voltage-gated Ca^{++} channels in the sarcolemma.
   D. receptor-mediated Ca^{++} mobilization.
   E. Ca^{++} channels in the sarcoplasmic reticulum.

67. Increases in cell [Ca^{++}] induce contraction in smooth muscle by
    A. Ca^{++} binding to troponin.
    B. activating myosin light-chain phosphatase.
    C. Ca^{++} binding to myosin light-chain kinase.
    D. Ca^{++} binding to calmodulin.
    E. activating Ca^{++} pumps.

68. Relaxation in smooth muscle can result from
    A. increased Ca^{++} conductance of channels in the sarcolemma.
    B. depolarization of the sarcolemma.
    C. receptor occupation leading to activation of phospholipase C and formation of inositol 1,4,5-trisphosphate (IP$_3$).
    D. increase in the activity of Ca^{++} pumps and exchangers.
    E. activation of myosin kinase.

PART IV
CARDIOVASCULAR SYSTEM

69. Conduction velocity of the cardiac impulse is slowest in
    A. atrial myocardial fibers.
    B. AV nodal fibers.
    C. Purkinje fibers.
    D. ventricular myocardial fibers.
    E. His bundle fibers.

70. The principal determinants of mean arterial pressure are
    A. cardiac output and peripheral resistance.
    B. arterial and venous capacitance.
    C. cardiac output and arterial capacitance.
    D. peripheral resistance and arterial capacitance.
    E. cardiac output and venous capacitance.

71. Injection of a drug that specifically increases ventricular contractility will decrease
    A. central venous pressure.
    B. mean arterial pressure.
    C. arterial pulse pressure.
    D. capillary blood flow.
    E. stroke volume.

72. During the steady, laminar flow of a newtonian fluid through a cylindrical tube, if the tube radius is tripled, flow will
    A. decrease by two thirds.
    B. remain unchanged.
    C. increase 3-fold.

D. increase 9-fold.
E. increase 81-fold.

73. When the velocity of flow in a blood vessel is 100 cm/sec and volume flow is 200 ml/sec, the cross-sectional area is
    A. 300 cm^2
    B. 100 cm^2
    C. 50 cm^2
    D. 2.0 cm^2
    E. 0.5 cm^2

74. The increase in arterial pulse pressure usually observed in an elderly hypertensive person is produced mainly by
    A. an increased stroke volume.
    B. an increased heart rate.
    C. a decreased cardiac output.
    D. an increased vagal activity.
    E. a decreased arterial compliance.

75. The main constituent of a blood clot is
    A. thrombin.
    B. fibrin.
    C. plasminogen.
    D. thrombinogen.
    E. thromboplastin.

76. Clot retraction is mediated by
    A. platelets.
    B. plasmin.
    C. urokinase.
    D. heparin.
    E. streptokinase.

77. The contribution of atrial contraction to ventricular filling is greatest
    A. when vagal activity is pronounced.
    B. at rapid heart rates.
    C. when atria and ventricles contract simultaneously.
    D. during atrial flutter.
    E. during third-degree AV block.

78. The circulatory variable that is maintained relatively constant by the baroreceptor reflex is
    A. heart rate.
    B. stroke volume.
    C. peripheral resistance.
    D. velocity of blood flow.
    E. mean arterial pressure.

79. Cardiac output (in liters per minute) divided by the heart rate (in beats per minute) equals
    A. cardiac index.

B. cardiac efficiency.
C. mean arterial pressure.
D. stroke volume.
E. blood velocity.

80. A reduction in arterial compliance results in
    A. a rise in systolic arterial pressure but a reduction in mean arterial pressure.
    B. a rise in systolic and in mean arterial pressure.
    C. a fall in systolic but a rise in diastolic arterial pressure.
    D. a rise in systolic and an even greater rise in diastolic arterial pressure.
    E. a rise in systolic but no significant change in mean arterial pressure.

81. The medullary vasomotor center is stimulated most effectively by
    A. decreased arterial blood oxygen tension.
    B. decreased arterial blood hydrogen ion concentration.
    C. increased arterial blood adenosine concentration.
    D. increased arterial blood carbon dioxide tension.
    E. increased arterial blood potassium ion concentration.

82. The arteriovenous shunts in the skin
    A. are insensitive to circulating catecholamines.
    B. possess a high degree of basal tone.
    C. dilate maximally when denervated.
    D. are less sensitive to sympathetic stimulation than are muscle arterioles.
    E. dilate in response to cooling of the hypothalamus.

83. The segment of the vascular bed responsible for local regulation of blood flow in most tissues is the
    A. distributing arteries.
    B. large veins.
    C. capillaries.
    D. venules.
    E. arterioles.

84. The most important function of the Starling mechanism in the heart is
    A. to provide an adequate cardiac output during sustained exercise.
    B. to ensure that the ventricles operate at an optimum length.
    C. to couple the efficiency of muscle contraction to the heart rate.

D. to match the output of one ventricle to that of the other.
E. to ensure that right and left atrial pressures are equal.

85. In response to an increase in carotid sinus pressure, the peripheral arterioles are dilated mainly by
    A. local dilator agents, such as lactic acid and CO_2.
    B. increased activity of sympathetic vasodilator fibers.
    C. increased activity of parasympathetic vasodilator fibers.
    D. decreased activity of sympathetic vasoconstrictor fibers.
    E. increased stimulation of alpha-adrenergic receptors.

86. The most important factor that regulates coronary blood flow is
    A. neural regulation by the autonomic nervous system.
    B. the intraventricular pressure during ventricular systole.
    C. autoregulation caused by local chemical factors.
    D. the aortic blood pressure.
    E. circulating epinephrine.

87. Lymph flow decreases in response to
    A. exercise.
    B. hemorrhage.
    C. protein leakage from capillaries.
    D. increased venous pressure.
    E. increased arterial pressure.

88. The principal factor that limits exercise performance is
    A. rate of respiration.
    B. depth of respiration.
    C. pumping capacity of the heart.
    D. oxygen consumption by the active muscles.
    E. oxygen saturation of arterial blood.

89. The Frank-Starling relation is essentially a restatement of
    A. the force-velocity relation of cardiac muscle.
    B. Poiseuille's Law of the heart.
    C. the length-tension diagram of cardiac muscle.
    D. Laplace's law of the heart.
    E. excitation-contraction coupling of cardiac muscle.

90. If total body oxygen consumption is 150 ml/min and the arterial and mixed venous O_2 levels are 19 and 4 ml O_2/100 ml blood, respectively, the cardiac output is
    A. 5000 ml/min.
    B. 1000 ml/min.
    C. 100 ml/min.
    D. 5 ml/min.
    E. 0.5 ml/min.

91. In an intact animal, stimulation of the left stellate ganglion
    A. reduces right atrial pressure but increases cardiac output.
    B. increases right atrial pressure and increases cardiac output.
    C. reduces right atrial pressure but increases left ventricular end-diastolic volume.
    D. prolongs the P-R interval.
    E. reduces right atrial pressure and decreases cardiac output.

92. Under physiological conditions capillary exchange is controlled mainly by
    A. capillary permeability.
    B. plasma oncotic pressure.
    C. vascular smooth muscle contraction.
    D. tissue hydrostatic pressure.
    E. lymphatic drainage.

93. Given that: P_{to} = tissue oncotic pressure, P_{po} = plasma oncotic pressure, P_{ph} = plasma hydrostatic pressure, and P_{th} = tissue hydrostatic pressure; then the driving force for capillary filtration equals
    A. $(P_{ph} - P_{po}) - (P_{th} - P_{to})$
    B. $(P_{th} - P_{po}) - (P_{ph} - P_{to})$
    C. $(P_{ph} - P_{to}) - (P_{th} - P_{po})$
    D. $(P_{po} - P_{ph}) - (P_{th} - P_{po})$
    E. $(P_{th} - P_{ph}) - (P_{to} - P_{po})$

PART V
RESPIRATORY SYSTEM

94. The important difference between the bronchi and bronchioles is that
    A. bronchioles have type 1 and type 2 squamous (thin, flat, pavement) epithelial cells, whereas the bronchi have columnar epithelium with cilia.
    B. the bronchiolar glands are situated directly beneath the lining epithelium (submucosa), whereas the glands of the bronchi are lar-

ger and external to the cartilage rings or plates.
    C. the smooth muscle of the bronchi is arranged longitudinally (along the axis of the airway), whereas the smooth muscle of the bronchioles is arranged circularly or spirally around the lumen of the airway.
    D. the bronchi are not directly attached to the connective tissue fibers of the lung tissue, whereas the bronchioles are directly embedded into the lung connective tissue structure.
    E. the pulmonary arteries and veins are always adjacent to the airways leading into and out of the ventilated lung units they subserve.

95. An eight-year-old, normal girl (30 kg body weight) has a resting breathing frequency of 20/min and a breath (tidal) volume of 80 milliliters. Her total alveolar ventilation is closest to
    A. 0.25 L/min.
    B. 0.6 L/min.
    C. 1.6 L/min.
    D. 2.4 L/min.
    E. unknown because the anatomical deadspace is not given.

96. The principal function of the lungs is to ensure that
    A. there is adequate distribution of inspired air and pulmonary blood flow to all lung units, so that gas exchange is accomplished with minimal energy expenditure.
    B. the hemoglobin in the erythrocytes leaving each lung unit is 100% saturated with oxygen.
    C. tissue cell oxygen consumption and carbon dioxide elimination are equal.
    D. each lung unit is perfused in proportion to its capillary blood volume and ventilated in proportion to its gas volume.
    E. diffusion into and out of blood is never rate-limiting for oxygen uptake or carbon dioxide release.

97. Normal inspiration is limited by
    A. sensory feedback to the respiratory centers in the medulla (brainstem) from mechanoreceptors in the chest wall and lung and various chemoreceptors sensitive to oxygen and carbon dioxide.
    B. the breathing frequency, about 12/min in resting adult humans, which limits each breath to 5 seconds' duration.
    C. the rise of abdominal pressure as the diaphragm contracts, especially in the supine

(dorsal recumbent) position, when the weight of abdominal contents pushes up on the diaphragm and tends to reduce end-expiratory lung volume.

D. the rise in pleural pressure as lung volume increases.

E. the mechanical limits of the chest wall (rib cage and diaphragm), which limits the maximal size of the thoracic cavity.

98. The normal resting oxygen consumption of adult humans is determined by the

A. fact that resting cardiac output is about 5 L/min, thereby limiting blood flow to some organs.

B. ability of tissue cells to extract oxygen from the capillary blood in the time available for exchange in the capillaries.

C. ability of pulmonary capillary blood to take up oxygen from alveolar gas in the time available (less than 1 second) for exchange.

D. fact that the P_{50} of adult human blood is normally 26 to 29 mm Hg, meaning that no more than 50% of the total blood oxygen concentration is available for cellular oxidative metabolism.

E. oxidative energy requirements of the body's cells, which are always met in the steady state.

99. The alveolar ventilation equation (the most important quantitative relationship in pulmonary physiology) states that

A. alveolar ventilation multiplied by the arterial P_{CO_2} is always a constant.

B. hyperventilation occurs when the arterial P_{CO_2} falls below 40 mm Hg.

C. oxygen uptake in the lung is normally diffusion-limited, whereas carbon dioxide elimination is not.

D. carbon dioxide elimination is ventilation-limited.

E. normally oxygen consumption cannot be altered by changing ventilation or by changing cardiac output.

100. Given arterial P_{CO_2} of 28 mm Hg and V_{CO_2} of 650 ml/min, the alveolar ventilation is about

A. 4 L/min.

B. 9 L/min.

C. 16 L/min.

D. 20 L/min.

E. It cannot be calculated because the inspired oxygen partial pressure is not given.

101. If an adult patient of average stature has a lung compliance of 0.33 L/cm H_2O and chest wall compliance of 0.15 cm H_2O, the total compliance of the respiratory apparatus is

A. about 0.2 L/cm H_2O.

B. about 0.4 L/cm H_2O.

C. high because the chest wall is abnormally distensible.

D. low because the lung is abnormally stiff.

E. normal.

102. A 17-year-old boy is admitted to the hospital with progressive failure of breathing as a result of a viral infection in his brain. An opening is made in his trachea in the lower part of his neck, and an endotracheal tube is inserted. The boy is placed on a ventilator that 15 times/min increases his lung volume by 0.5 L. If the boy's lung and chest wall compliances are each 0.2 L/cm H_2O over this range of lung volume change and if translung pressure is 5 cm H_2O at FRC,

A. pleural pressure at FRC will be about +5 cm H_2O.

B. pleural pressure at end-inspiration is about -2.5 cm H_2O.

C. pleural pressure cannot be calculated without knowing the value of FRC.

D. transdiaphragm-abdominal wall pressure at end-inspiration must be about +5 cm H_2O.

E. total pressure across the lungs and chest wall at end-inspiration is about +10 cm H_2O.

103. In clinical pulmonary edema resulting from lung injury by a toxic substance, plasma proteins enter the alveoli and inactivate the surfactant material. This fixes alveolar surface tension at a constant value of 50 dynes/cm. Which of the following descriptions of the pressure-volume curve of the lung best describes the new condition? (Refer to Figure 28-8).

A. The deflation limb of the curve for the liquid-filled lung will be displaced to the right (greater translung pressure at any given lung volume).

B. The deflation limb of the curve for the air-filled lung will be displaced to the right (greater translung pressure at any given lung volume).

C. The inflation limb of the curve for the air-filled lung will be displaced to the left (reduced translung pressure at any given lung volume).

D. The inflation limb of the curve for the air-filled lung will be normal.

E. The compliance of the respiratory system will be increased.

104. If pulmonary arterial pressure is 40 cm H_2O, alveolar pressure is 0, left atrial pressure is 20 cm H_2O, and cardiac output is 10 L/min, which of the following is the best estimate of pulmonary vascular resistance and what is the overall lung perfusion zone?
    A. 2.0 cm H_2O × min/L; zone 3.
    B. 0.5 cm H_2O × min/L; zone 3.
    C. 4.0 cm H_2O × min/L; zone 2.
    D. 0.5 cm H_2O × s/L; zone 2.
    E. 2.0 cm H_2O × s/L; zone 3.

105. A woman is brought to the intensive care unit at her community hospital. Chest roentgenograms show that her entire right lung (55% total lung mass) is consolidated (unventilated) as a result of a bacterial pneumonia. She has bluish (cyanotic) lips, fingernail beds, and skin. Subsequent measurements reveal her arterial oxygen tension to be 50 mm Hg (O_2 saturation = 85%) and her mixed venous oxygen tension to be 30 mm Hg (O_2 saturation = 57%), even though she is breathing 100% O_2. Her cardiac output is 7 L/min. Ignoring any correction resulting from dissolved O_2 in blood, the patient's venous admixture is approximately
    A. 55% of cardiac output.
    B. 15% of cardiac output.
    C. 6 L/min.
    D. 3 L/min.
    E. 1 L/min.

106. Figure 29-2 shows an idealized pulmonary pressure-flow curve for an adult man. Which of the following statements about the curve is most nearly correct?
    A. The flow resistance at rest is less than during exercise.
    B. The flow resistance at rest is greater than during exercise.
    C. The difference in flow resistance between rest and exercise is 0.5 mm Hg/L/min.
    D. The difference in flow resistance between rest and exercise is 2.0 mm Hg/L/min.
    E. Flow resistance cannot be calculated without knowing alveolar pressure.

107. Figure 29-5 shows a normal adult woman's V/Q distribution. The most important piece of information that can be gleaned from the figure is that
    A. venous admixture is normally about 1% of cardiac output.
    B. the range of possible V/Q ratios lies between 0 (wasted ventilation) and ∞ (venous admixture).

C. the mean V/Q is about 0.84.
    D. blood flow is greater than ventilation at all V/Q ratios greater than 1.
    E. ventilation is greater than blood flow at all V/Q ratios less than 1.

108. The normal regulation of local perfusion to lung ventilation units depends mainly upon
    A. alveolar CO_2 tension.
    B. the mixed venous blood (pulmonary arterial) hydrogen ion concentration (pH).
    C. alveolar O_2 tension.
    D. the mixed venous blood (pulmonary arterial) O_2 tension.
    E. the balance between sympathetic neural stimulation of vascular smooth muscle and parasympathetic stimulation of airway smooth muscle.

109. If alveolar oxygen tension equals 100 mm Hg and if the hemoglobin-oxygen equilibrium curve is shifted 10 mm Hg to the right because of acidosis (increased hydrogen ion concentration in blood), which of the following is the most likely to be the systemic arterial oxygen tension?
    A. 100 mm Hg
    B. 90 mm Hg
    C. 110 mm Hg
    D. 99%
    E. 95%

110. The blood hemoglobin concentration is 75 g/L in a 29-year-old mother with three children, all of whom sleep in an inadequately ventilated room heated by a small kerosene stove. Which of the following is closest to the steady-state HbO_2 and HbCO concentrations in her systemic arterial blood (Ca_{CO} and Ca_{O_2}) when alveolar P_{CO} and P_{O_2} are 0.5 and 100 mm Hg, respectively? (*Hint:* Calculate O_2 capacity; then use Figure 30-5 and related text to estimate the carbon monoxide saturation.)
    A. Ca_{CO} = 100 ml/L; Ca_{O_2} = 0 ml/L
    B. Ca_{CO} = 0 ml/L; Ca_{O_2} = 100 ml/L
    C. Ca_{CO} = 25 ml/L; Ca_{O_2} = 75 ml/L
    D. Ca_{CO} = 50 ml/L; Ca_{O_2} = 50 ml/L
    E. Ca_{CO} = 100 ml/L; Ca_{O_2} = 100 ml/L

111. In the absence of carbonic anhydrase, the rate of carbonic acid (H_2CO_3) dissociation into water and CO_2 in pulmonary capillary blood is too slow to reach a steady state with respect to alveolar P_{CO_2} in the time available for blood transit (<1 sec) through the pulmonary capillaries. Which of the

following blood gas relations is most nearly correct under this condition?

A. Arterial P_{CO_2} will fall below alveolar P_{CO_2}.
B. Arterial P_{CO_2} will be the same as alveolar P_{CO_2}, but the blood bicarbonate concentration will be lower than the alveolar gas bicarbonate concentration.
C. Arterial P_{CO_2} will rise above alveolar P_{CO_2}.
D. Alveolar ventilation will remain constant.
E. Alveolar ventilation will fall.

112. Normally, the venous blood leaving the myocardium in the coronary veins has an HbO_2 saturation of about 50% ($P_{O_2} = 29$ mm Hg; Bohr shift). If the systemic arterial blood P_{O_2} is 100 mm Hg and all other blood parameters are normal, the mean P_{O_2} in the heart muscle capillaries is closest to

A. 26 mm Hg.
B. 40 mm Hg.
C. 65 mm Hg.
D. 75 mm Hg.
E. 88 mm Hg.

113. In skeletal muscle during exercise the number of capillaries being perfused may be three times the resting number. One major consequence of having more capillaries with blood flow is that

A. tissue oxygen consumption is increased.
B. tissue oxygen consumption is decreased.
C. skeletal muscle vascular resistance is increased.
D. the diffusion path from blood to mitochondria is increased.
E. the diffusing capacity of the muscle capillary network is increased at least threefold.

114. If the arterial blood P_{CO_2} is 80 mm Hg, $[HCO_3^-]$ is 32 mEq/L, and P_{O_2} is 100 mm Hg, it is likely that

A. the carotid sinus baroreceptors are maximally stimulated.
B. the carotid body chemoreceptors are not stimulated or are depressed.
C. the medullary chemoreceptors are stimulated.
D. the medullary chemoreceptors are not stimulated or are depressed because the bicarbonate concentration in arterial blood is high.
E. A determination cannot be made without knowing the blood hydrogen ion concentration.

115. A middle-aged man with chronic obstructive pulmonary disease (emphysema) has an arterial P_{CO_2}

of 57 mm Hg, compared to the predicted normal value of 40 mm Hg at an alveolar ventilation of 4.3 L/min. If we assume the patient's carbon dioxide production is normal (200 ml/min), his alveolar ventilation is about

A. 3.0 L/min.
B. 8.0 L/min.
C. 3.8 L/min.
D. 6.1 L/min.
E. 1.8 L/min.

116. In a normal person during sleep the CO_2 ventilation tension-response curve (Figure 31-5) can be described as

A. having a steeper slope than in the same subject when awake.
B. being shifted leftward relative to the same person when awake.
C. having a flatter (less steep) slope than in the same subject when awake.
D. having a lower intercept P_{CO_2} (ventilation theoretically = 0) compared to the same subject when awake.
E. showing greater sensitivity than in the same subject when awake.

117. The partial pressures of oxygen and carbon dioxide interact in arterial blood and brain (Figure 31-7) in such a manner that

A. hypoxemia reduces the slope of the CO_2 ventilation dose-response curve.
B. hypercapnea decreases ventilation, when inspired oxygen concentration is normal.
C. hypoxemia has no effect on the slope of the CO_2 ventilation tension-response curve.
D. hypercapnea permits ventilation to increase more when inspired oxygen is reduced, compared to breathing air with no CO_2.
E. the threshhold P_{CO_2} of the CO_2 ventilation tension-response curve is decreased when inspired oxygen concentration increased.

118. If the diaphragm contracts more vigorously, it is likely that the slowly adapting lung stretch receptor activity will

A. increase inspiratory volume.
B. be greater than normal.
C. be less than normal.
D. be the same as normal.
E. cause the time for inspiration to lengthen.

PART VI
GASTROINTESTINAL SYSTEM

119. Which of the following statements about gastrointestinal smooth muscle is true?
    A. The electrogenic property of the Na^+-K^+-ATPase often makes a smaller contribution to the resting membrane potential than is the case in skeletal muscle.
    B. Cells of the circular layer have gap junctions, but are not well-coupled electrically.
    C. Resting membrane potentials in gastrointestinal smooth muscle are typically greater in magnitude than those in skeletal muscle.
    D. Contractions in gastrointestinal smooth muscle are stronger when action potentials are triggered near the crests of the slow waves.
    E. Tone is due to a baseline level of spontaneous action potentials in gastrointestinal smooth muscle.

120. Which of the following statements about the enteric nervous system is true?
    A. Excitatory motor neurons from myenteric ganglia release ACh, substance P, VIP, or NO onto smooth muscle cells of the circular and longitudinal layers.
    B. Interneurons in the enteric nervous system release substance P.
    C. Many stimulatory secretomotor neurons from the submucosal plexus release NO onto gland cells in the gastrointestinal tract.
    D. Vasodilator neurons may release norepinephrine onto mucosal blood vessels.
    E. Numerous sensory neurons, whose cell bodies are in the myenteric and submucosal ganglia, respond to mechanical and chemical stimuli.

121. Which of the following statements about the functions of the esophagus is true?
    A. The swallowing reflex is elicited when touch receptors on the posterior part of the hard palate are stimulated.
    B. The upper esophageal sphincter opens early on in the swallowing reflex and remains open for most of the time that the esophageal peristaltic wave requires to travel from the upper to the lower end of the esophagus.
    C. If food remains in the esophagus after a swallow, a new swallowing reflex occurs.
    D. In achalasia, the lower part of the body of the esophagus contracts spasmodically.
    E. The lower esophageal sphincter remains open while the peristaltic wave traverses the length of the esophagus.

122. Which of the following statements about gastric motility is true?
    A. Receptive relaxation in response to gastric filling is predominantly mediated by the vagus nerves.
    B. Strong gastric contractions begin in the middle of the fundus of the stomach and travel toward the antrum, gaining in strength as they travel.
    C. The frequency of slow waves in the stomach is about 8/min.
    D. Secretin and CCK enhance the force of contraction of the stomach.
    E. Contractions of gastric smooth muscle do not occur in the absence of action potentials.

123. Which of the following statements about gastric emptying is true?
    A. Fats tend to be emptied during the active contractile phase (phase III) of the migrating myoelectric complex (MMC).
    B. Acid in the duodenum elicits release of gastric inhibitory peptide, which slows the rate of gastric emptying.
    C. Indigestible objects 0.5 cm in size tend to be emptied into the duodenum during the active contraction phase (phase III) of the MMC.
    D. Fats and fat digestion products in the duodenum elicit the secretion of secretin, which slows the rate of gastric emptying.
    E. Gastrin released from duodenal G cells in response to peptides and amino acids decreases the force of gastric contractions.

124. Which of the following statements about the motility of the small intestine is true?
    A. Peristalsis is the most frequent type of contractile behavior in the small intestine of a fed individual.
    B. Because segmentation is under the control of the enteric nervous system, stimulating sympathetic nerves to the small intestine will have very little effect on segmentation.
    C. In a fasted individual the minute rhythm occurs in the small intestine.
    D. The MMC in the small intestine is characterized by 75 to 90 minutes of quiescence, punctuated by 3- to 6-minute periods of intense and propulsive contractions.

E. Long-range peristalsis occurs frequently in the small intestine.

125. Which of the following statements about colonic motility is true?
    A. Mass movements occur 1 to 3 times per day and sweep colonic contents toward the rectum.
    B. Haustral contractions are primarily controlled by the parasympathetic nervous system.
    C. The gastrocolic reflex is characterized by decreased colonic motility elicited by gastric distension and other stimuli arising in the stomach.
    D. The defecation reflex is evoked by distension of the sigmoid colon.
    E. The integrating center for the defecation reflex is in the enteric nervous system.

126. Which of the following statements about salivary secretion is true?
    A. The major control of salivary secretion is via the enteric nervous system.
    B. Secretory endpieces (acini) secrete a fluid that contains salivary amylase and has Na^+, K^+, and Cl^-, at concentrations similar to those in plasma, and HCO_3^- at levels much higher than in plasma.
    C. When salivary secretion is stimulated under physiological conditions, the concentration of bicarbonate in saliva rises.
    D. Stimulation of sympathetic nerves to salivary glands results in prolonged stimulation of salivation.
    E. Agonists that elevate cyclic AMP in salivary acinar cells stimulate acinar cells to secrete fluid and amylase; agonists that elevate intracellular Ca^{++} in acinar cells inhibit secretion of fluid and amylase.

127. Which of the following statements about gastric secretions is true?
    A. Oxyntic glands that contain parietal cells are located in the body and the pylorus of the stomach.
    B. The higher the flow rate of gastric juice, the higher its Cl^- concentration.
    C. Patients with ulcers in their stomachs secrete larger amounts of HCl than normal individuals.
    D. Glands in the pyloric glandular area contain G cells that secrete gastrin, cells that secrete mucus, and numerous chief cells and parietal cells.

E. The rate of HCl secretion during the cephalic phase is low, but the total amount of HCl secreted may be large.

128. Which of the following statements about gastric acid secretion is true?
    A. In an unstimulated parietal cell the amount of H^+-K^+-ATPase present is low.
    B. H^+ is secreted across the basolateral membrane of the parietal cell by the H^+-K^+-ATPase.
    C. HCO_3^- leaves the parietal cell at the basolateral membrane, and its downhill efflux powers the uphill entry of Cl^- into the parietal cell.
    D. Cl^- is secreted into the secretory canaliculus via a Cl^- active transport protein.
    E. H_2 receptor blockers inhibit HCl secretion by directly inhibiting the H^+-K^+-ATPase.

129. Which of the following statements about pancreatic secretion is true?
    A. The spontaneous secretion, produced by the acinar cells and the intralobular ducts, contains the enzyme component and bicarbonate levels that are higher than plasma bicarbonate.
    B. Secretin specifically stimulates the acinar cells and intralobular duct cells to produce a secretion that is high in bicarbonate concentration.
    C. During the gastric phase, CCK stimulates pancreatic acinar cells to secrete pancreatic enzymes; a juice that is low in volume, but high in enzyme content is produced.
    D. CCK is the principal physiological agonist of pancreatic acinar cell secretion; secretin does not influence the response of acinar cells to CCK.
    E. Pancreatic acinar cells are stimulated to secrete by agonists that elevate intracellular cyclic AMP, but not by agonists that elevate intracellular Ca^{++}.

130. Which of the following statements about bile is true?
    A. The more lecithin that is present in bile, the more cholesterol that can be held in the bile acid–lecithin–cholesterol mixed micelles.
    B. The bicarbonate-rich fluid component of bile is secreted by the epithelial cells of the bile ducts. CCK is the most important physiological agonist of this secretion.
    C. Conjugated bile acids have a higher critical

micelle concentration than unconjugated bile acids.

D. Bile acids returning to the liver in portal blood are a strong stimulus for secretion of bile acids by hepatocytes and also stimulate the de novo synthesis of bile acids.

E. The strongest physiological stimuli for emptying of the gallbladder are nerve impulses in branches of the vagus nerves that innervate the gallbladder.

131. Which of the following statements about digestion and absorption of carbohydrates is true?
  A. The only monosaccharides that are absorbed to an appreciable extent are glucose and galactose.
  B. α-Dextrinase (isomaltase) is the enzyme that is present in the cytosol of jejunal epithelial cells and that is responsible for cleaving the α-1,6-glycosidic linkage of branched starch molecules.
  C. Certain disaccharides are taken up by intestinal epithelial cells.
  D. A small fraction of the world's adult population is lactose intolerant.
  E. Sucrase and isomaltase (α-dextrinase) are synthesized as a single polypeptide chain.

132. Which of the following statements about digestion and absorption of proteins is true?
  A. Trypsinogen is activated by chymotrypsin that is secreted by the duodenal mucosa. Trypsin then activates the other proteases in pancreatic juice.
  B. Oligopeptidases present in pancreatic juice cleave peptides to produce smaller peptides and single amino acids.
  C. Neutral amino acids are transported across the brush border membrane by a single Na^+-powered secondary active transport protein with broad specificity for neutral amino acids.
  D. Dipeptides and tripeptides are taken up across the brush border membrane by a single type of H^+-powered secondary active transport protein.
  E. Small peptides are cleaved by cytosolic peptidases in the intestinal epithelial cells, so that single amino acids, and significant amounts of dipeptides and tripeptides, appear in portal blood.

133. Which of the following statements about intestinal salt and water transport is true?

A. The basolateral Na^+-K^+-ATPase is the only primary active transport protein present in intestinal epithelial cell plasma membranes.

B. Net absorption of bicarbonate occurs in the jejunum and the ileum, but bicarbonate is normally secreted into the colon.

C. The absorption of Na^+ along with sugars and amino acids in the jejunum provides the osmotic force for absorbing the major fraction of the water absorbed in the jejunum.

D. Tight junctions are leakiest in the ileum and tightest in the colon.

E. K^+ is passively absorbed in the small intestine, is usually absorbed in the net sense in the colon, but can be actively secreted in the distal colon.

134. Which of the following statements about intestinal handling of salts and water is true?
  A. The enteric nervous system plays a minor role in regulation of intestinal salt and water transport.
  B. Aldosterone enhances the absorption of Na^+ in the colon primarily by increasing the number of Na^+-K^+-ATPase molecules in the basolateral membrane.
  C. Cells in the crypts of Lieberkühn are stimulated to secrete Cl^- into the lumen by any agonist that elevates intracellular cyclic AMP in the crypt cells.
  D. The glucose that is present in oral rehydration solution promotes the absorption of Na^+, Cl^-, and water by the cells in the crypts of Lieberkühn.
  E. In the jejunum and ileum very little absorption of K^+ occurs via the paracellular pathway.

135. Which of the following statements about intestinal absorption of Ca^{++} is true?
  A. Ca^{++} is taken up across the luminal plasma membrane of small intestinal epithelial cells by an Na^+/Ca^{++} co-transport protein that uses the energy of the Na^+ gradient to take up Ca^{++} against an electrochemical gradient.
  B. The cytosolic Ca^{++} binding protein (calbindin) plays a key role in transporting Ca^{++} through the cytosol of the intestinal epithelial cell.
  C. A Ca^{++}-ATPase in the brush border membrane of the intestinal epithelial cell actively takes up Ca^{++} into the cell.
  D. Vitamin D acts to decrease uptake of calcium by mitochondria in intestinal epithelial cells.

E. Absorption of Ca^{++} is inhibited by parathyroid hormone.

136. Which of the following statements about intestinal absorption of iron is true?
    A. Iron is ingested as inorganic iron and as heme iron. A larger fraction of the ingested inorganic iron than of the heme iron is absorbed.
    B. Vitamin C inhibits iron absorption by forming a complex with ferrous (+2) iron.
    C. Iron bound to ferritin in intestinal epithelial cells is part of the absorbable pool of iron.
    D. Intestinal epithelial cells secrete a transferrin into the lumen of the intestine.
    E. The intestinal epithelial cells of iron-deficient animals contain more apoferritin than do the intestinal epithelial cells of iron-replete individuals.

137. Which of the following statements about intestinal absorption of water-soluble vitamins is true?
    A. Specific transport systems do not exist for intestinal absorption of most water-soluble vitamins.
    B. There is a significant hepatic store of vitamin B_{12}, so that even if ingestion of vitamin B_{12} totally ceases, a deficiency of vitamin B_{12} would not occur for several months.
    C. Ileal receptors do not recognize free vitamin B_{12}.
    D. Patients with pernicious anemia are deficient in secretion of intrinsic factor, but have normal rates of secretion of HCl and pepsinogens.
    E. The terminal jejunum is the principal site of vitamin B_{12} absorption.

138. Which of the following statements about intestinal digestion and absorption of lipids is true?
    A. Lipid absorption may occur when bile acid–lipid mixed micelles fuse with the brush border plasma membrane.
    B. The transport of bile acid–lipid mixed micelles across the intestinal unstirred layer is the rate-limiting step in lipid absorption.
    C. Triglycerides readily form mixed micelles with bile acids.
    D. Resynthesis of triglycerides occurs in the cytoplasm of intestinal epithelial cells.
    E. Emulsification of dietary lipids by bile acids serves to increase the rate at which brush border lipases can digest the lipids.

PART VII
RENAL SYSTEM

139. Through metabolism, an individual produces 900 mOsm/day of solute, which must be excreted by the kidneys. If this individual has a urine concentrating defect, and can only produce urine having a maximum osmolality of 300 mOsm/kg H_2O, what is the minimum volume of water that must be ingested to prevent a rise in body fluid osmolality? Assume that insensible water loss is 1.5 liters.
    A. 1.5 liters
    B. 3.0 liters
    C. 4.5 liters
    D. 6.0 liters
    E. 7.5 liters

For questions 140 and 141, consider the following graph, which shows the relationship between plasma [PAH] and PAH secretion:

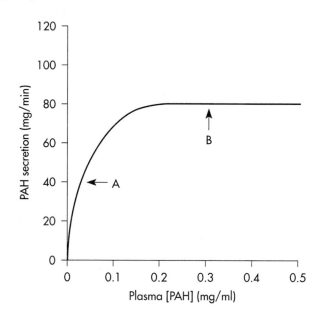

140. The amount of PAH filtered at the glomerulus is
    A. greater at point A than at point B.
    B. less at point A than at point B.
    C. the same at points A and B.

141. The clearance of PAH is
    A. greater at point A than at point B.
    B. less at point A than at point B.
    C. the same at points A and B.

142. Which of the following structures is the major barrier to the filtration of proteins across the glomerulus?

A. Fenestrated endothelium
B. Basement membrane
C. Visceral epithelial cells (podocytes)
D. Parietal epithelial cells
E. Mesangial cells

143. Hyponatremia (i.e., reduced plasma [Na^+]) occurs sometimes in individuals whose effective circulating volume is decreased. Which of the following factors contributes to the development of hyponatremia in this situation?
    A. Impaired ability of the kidneys to excrete solute-free water (C_{H_2O})
    B. Elevated levels of atrial natriuretic peptide (ANP)
    C. Increased excretion of Na^+ by the kidneys
    D. Decreased levels of antidiuretic hormone (ADH)
    E. Decreased levels of aldosterone

144. The volume of plasma from which all of a substance is removed (per unit time) and excreted into the urine is termed
    A. the renal plasma flow.
    B. the glomerular filtration rate.
    C. the renal clearance.
    D. the filtration fraction.
    E. the renal blood flow.

Assign the appropriate diagnosis from those listed below to the condition described in questions 145 through 148.
    A. *Metabolic acidosis with respiratory compensation*
    B. *Metabolic alkalosis with respiratory compensation*
    C. *Respiratory acidosis with renal compensation (chronic respiratory acidosis)*
    D. *Respiratory acidosis without renal compensation (acute respiratory acidosis)*
    E. *Metabolic acidosis and respiratory acidosis*

Use the following normal values:
  pH = 7.40
  [HCO_3^-] = 24 mEq/L
  P_{CO_2} = 40 mm Hg

145. An individual with an asthma attack: pH = 7.32; [HCO_3^-] = 25 mEq/L; P_{CO_2} = 50 mm Hg

146. An individual with diabetes mellitus who forgets to take insulin: pH = 7.29; [HCO_3^-] = 12 mEq/L; P_{CO_2} = 26 mm Hg

147. An individual with cardiopulmonary arrest: pH = 6.85; [HCO_3^-] = 10 mEq/L; P_{CO_2} = 60 mm Hg

148. An individual with a gastric ulcer who ingests large quantities of antacids: pH = 7.45; [HCO_3^-] = 30 mEq/L; P_{CO_2} = 45 mm Hg

149. A decrease in the effective circulating volume will result in which of the following:
    A. Increase in the GFR
    B. Increase in renin secretion
    C. Increase in ANP secretion
    D. Increase in solute-free water excretion
    E. Increase in urine flow rate

150. Which segment of the nephron contributes most to potassium excretion when dietary potassium is altered?
    A. Proximal convoluted tubule
    B. Descending limb of Henle's loop
    C. Proximal straight tubule
    D. Distal tubule
    E. Thick ascending limb of Henle's loop

151. According to the tubuloglomerular feedback theory, an increase in the flow of tubular fluid to the macula densa will result in
    A. a decrease in the glomerular filtration rate of the same nephron.
    B. an increase in renal blood flow.
    C. activation of the renal sympathetic nerves.
    D. an increase in proximal tubule solute and water reabsorption.
    E. a decrease in afferent arteriole resistance.

152. A drug that increases glomerular filtration rate may act by
    A. increasing the ultrafiltration coefficient, K_f.
    B. constricting the afferent arterioles.
    C. decreasing renal blood flow.
    D. increasing plasma protein concentration.
    E. increasing hydrostatic pressure in Bowman's space.

153. During antidiuresis, tubular fluid is hyperosmotic to plasma in the following part of the nephron:
    A. Thick ascending limb of Henle's loop
    B. Distal tubule
    C. Bend of Henle's loop
    D. Bowman's space.
    E. Proximal convoluted tubule

154. Potassium excretion is enhanced by
    A. an osmotic diuresis.
    B. acute metabolic acidosis.

C. hypoaldosteronism.

D. decreased tubular flow rate.

E. Ingestion of a low-potassium diet

155. During a 24-hour period, an individual excretes in the urine 60 mmol of NH_4^+, 40 mmol of titratable acid, and 10 mmol of HCO_3^-. If this individual is in acid-base balance, how much nonvolatile acid did he or she produce from metabolism?

A. 90 mmol/day

B. 100 mmol/day

C. 110 mmol/day

D. 120 mmol/day

E. 130 mmol/day

156. Intravenous infusion of 1 liter of which of the following solutions will lead to the largest increase in ECF volume?

A. Distilled water

B. Isotonic NaCl

C. Hypotonic NaCl

D. Hypertonic NaCl

E. Isotonic urea

157. Renal $PO_4^=$ excretion is enhanced by which of the following?

A. PTH

B. Decrease in the ECF volume

C. Metabolic alkalosis

D. Ingestion of a $PO_4^=$ deficient diet.

E. Deficiency of 1,25-dihydroxy vitamin D_3

158. PTH has which of the following effects on renal handling of Ca^{++}?

A. Increases the filtered load of Ca^{++}

B. Stimulates proximal tubule Ca^{++} reabsorption

C. Inhibits Ca^{++} reabsorption by the thick ascending limb

D. Stimulates Ca^{++} resorption by the distal tubule

E. Inhibits Ca^{++} reabsorption by the collecting duct

PART VIII
ENDOCRINE SYSTEM

159. Protein and catecholamine hormones share which of the following characteristics?

A. Synthesis requires the amino acid tryptophane.

B. At least one peptide bond is present.

C. Tight binding to a specific serum protein.

D. Initial interaction with a plasma membrane receptor.

E. Initial interaction with a nuclear receptor.

160. A decrease in the *sensitivity*, as opposed to the maximal responsiveness, of the body to the action of a hormone is caused by all of the following except

A. an increase in the rate of hormone degradation.

B. a decrease in hormone receptor affinity.

C. a decrease in hormone receptor number.

D. a decrease in the number of target cells.

E. an increase in the concentration of a competitive hormone antagonist.

161. Measurement of the respiratory quotient, $\dot{V}_{CO_2}/\dot{V}_{O_2}$, allows one to estimate

A. the proportions of fat and carbohydrate in the fuel mix.

B. the amount of glucose being disposed of by glycolysis.

C. the amount of glucose being formed by gluconeogenesis.

D. the efficiency of ventilation.

E. The amount of ATP being generated.

162. The metabolic adaptation to prolonged fasting includes each of the following except

A. a decrease in basal metabolic rate.

B. an increase in plasma ketoacids.

C. an increase in plasma glycerol.

D. a decrease in central nervous system glucose oxidation.

E. an increase in protein synthesis.

163. Insulin has which of the following actions on cells?

A. Increases free fatty acid uptake by muscle

B. Increases glucose output by the liver

C. Increases glucose uptake by the muscle

D. Increases glucose uptake by the brain

E. Increases free fatty acid release by adipose tissue

164. Glucagon secretion is stimulated by

A. a decrease in plasma ketoacids.

B. a decrease in plasma glucose.

C. an increase in plasma free fatty acids.

D. a decrease in physical exercise.

E. a high carbohydrate meal.

165. A deficiency of parathyroid hormone would result in

	Plasma calcium	Plasma phosphate	Urine phosphate
A.	Increased	Decreased	Increased
B.	Increased	Decreased	Decreased
C.	Decreased	Increased	Decreased
D.	Decreased	Increased	Increased
E.	Decreased	Decreased	Decreased

166. Each of the following would increase production of the most active form of vitamin D except
    A. increased renal 24-hydroxylase activity.
    B. increased renal 1-hydroxylase activity.
    C. increased hepatic 25-hydroxylase activity.
    D. increased parathyroid hormone levels.
    E. decreased phosphate levels.

167. Growth hormone secretion is inhibited by each of the following except
    A. somatomedin.
    B. somatostatin.
    C. glucose.
    D. amino acids.
    E. free fatty acids.

168. Antidiuretic hormone acts
    A. on the proximal renal tubule.
    B. via cyclic AMP.
    C. to increase free water clearance.
    D. to decrease blood pressure.
    E. in response to decreased plasma osmolality.

169. An essential amino acid is one that
    A. cannot be synthesized in the body.
    B. cannot be reabsorbed by the renal tubule.
    C. is required for the synthesis of certain proteins only.
    D. is poorly absorbed from the diet.
    E. can only be supplied by milk sources.

170. Thyroid hormone secretion involves each of the following processes except
    A. activation of a G-protein by TSH.
    B. proteolysis of thyroglobulin.
    C. exocytosis of thyroxine-containing secretory granules.
    D. oxidation of iodide.
    E. synthesis of thyroglobulin.

171. Triiodothyronine (T_3)
    A. decreases oxygen consumption.
    B. decreases sympathetic (adrenergic) nervous system tone.
    C. is secreted in greater amounts than is thyroxine (T_4).

    D. stimulates secretion of TSH.
    E. has a greater affinity for the thyroid hormone receptor than does thyroxine (T_4).

172. Loss of cortisol action would decrease each of the following except
    A. secretion of corticotropin-releasing hormone.
    B. liver glycogen content.
    C. urea production.
    D. neutrophils in blood.
    E. resorption of bone.

173. Aldosterone secretion is stimulated by increases in each of the following except
    A. potassium.
    B. sodium.
    C. renin.
    D. angiotensin-converting enzyme.
    E. ACTH.

174. Epinephrine decreases
    A. heart rate.
    B. cardiac contractility.
    C. basal metabolic rate.
    D. cutaneous blood flow.
    E. plasma glucose concentration.

175. Epinephrine secretion is stimulated by each of the following except
    A. hypotension.
    B. hypoglycemia.
    C. acetylcholine.
    D. hypothermia.
    E. diacylglycerol.

176. In a male, expression of the gene for antimüllerian hormone at the proper time is required for
    A. absence of a uterus.
    B. suppression of a cyclic pattern of gonadotropin secretion.
    C. development of the spermatic cord.
    D. development of a penis.
    E. absence of the female pattern of enclosing each developing germ cell within a follicle.

177. Secretion of gonadotropins is ordinarily inhibited by each of the following except
    A. estradiol.
    B. testosterone.
    C. calcium.
    D. melatonin.
    E. endorphins.

178. Loss of sertoli cell function would cause which of the following changes in plasma hormone concentrations?

	Testosterone	FSH	Inhibin
A.	Decreased	Increased	Increased
B.	No change	Increased	Decreased
C.	Increased	Decreased	No change
D.	Decreased	Increased	Increased
E.	Increased	Decreased	Decreased

179. Testosterone action on target cells below the neck leads to increases in each of the following except
    A. red blood cells.
    B. high-density lipoproteins.
    C. spermatogenesis.
    D. penile size.
    E. prostate size.

180. In humans, each of the following facilitates ovulation except
    A. pulsatile secretion of gonadotropin-releasing hormone.
    B. positive feedback effect of estradiol on LH secretion.
    C. positive feedback effect of progesterone on LH secretion.
    D. postcopulatory afferent impulses from the vagina to the hypothalamus.
    E. local release of prostaglandins and cytokines.

181. Estradiol
    A. increases basal body temperature.
    B. produces a thick, inelastic cervical mucus.
    C. inhibits bone resorption.
    D. decreases progesterone receptors.
    E. increases the glycogen content and tortuosity of endometrial glands.

182. Which of the following statements regarding hormone actions during pregnancy is correct?
    A. Human chorionic somatomammotropin (human placental lactogen) increases maternal responsiveness to insulin.
    B. Progesterone increases uterine contractility.
    C. Prolactin increases milk secretion.
    D. Human chorionic gonadotropin stimulates progesterone secretion by the corpus luteum.
    E. Estriol increases the firmness and tension of the pelvic ligaments to prevent rapid uterine expansion.

ANSWERS

Part I

1. C
2. E
3. C
4. C
5. D
6. A
7. A
8. C
9. D
10. A
11. B
12. D
13. C
14. D
15. B
16. B
17. B
18. C
19. B
20. B
21. B
22. B
23. C
24. D
25. B

Part II

26. D
27. A
28. D
29. C
30. D
31. D
32. E
33. C
34. C
35. D
36. C
37. C
38. B
39. A
40. E
41. C
42. D
43. D

44. A
45. E

Part III
46. B
47. E
48. C
49. E
50. A
51. A
52. C
53. C
54. E
55. E
56. D
57. D
58. D
59. E
60. C
61. B
62. D
63. A
64. D
65. E
66. A
67. D
68. D

Part IV
69. B
70. A
71. A
72. E
73. D
74. E
75. B
76. A
77. B
78. E
79. D
80. E
81. D
82. C
83. E
84. D
85. D
86. C
87. B

88. C
89. C
90. B
91. A
92. C
93. A

Part V
94. D
95. E
96. A
97. A
98. E
99. D
100. D
101. E
102. B
103. B
104. A
105. D
106. B
107. C
108. C
109. A
110. D
111. C
112. B
113. E
114. C
115. A
116. C
117. D
118. B

Part VI
119. D
120. E
121. E
122. A
123. C
124. D
125. A
126. C
127. B
128. C
129. A
130. A
131. E

132. D
133. C
134. C
135. B
136. D
137. C
138. B

Part VII

139. C
140. B
141. A
142. B
143. A
144. C
145. D
146. A
147. E
148. B
149. B
150. D
151. A
152. A
153. C
154. A
155. A
156. D
157. A
158. D

Part VIII

159. D
160. D
161. A
162. E
163. C
164. B
165. C
166. A
167. D
168. B
169. A
170. C
171. E
172. A
173. B
174. D
175. E
176. A
177. C
178. B
179. B
180. D
181. C
182. D

Index